COMPUTER ORGANIZATION AND ARCHITECTURE

Themes and Variations

Alan Clements

Teesside University

Australia • Brazil • Canada • Mexico • Singapore • United Kingdom • United States

Computer Organization and Architecture:
Themes and Variations, **First Edition**
Alan Clements

Publisher, Global Engineering:
Christopher M. Shortt

Acquisitions Editor: Swati Meherishi

Developmental Editor: Amy Hill, Collaborative
Concepts, LLC

Assistant Development Editor: Farah Naseem

Editorial Assistant: Tanya Altieri

Team Assistant: Carly Rizzo

Marketing Manager: Lauren Betsos

Media Editor: Chris Valentine

Content Project Manager: D. Jean Buttrom

Production Service: RPK Editorial Services, Inc.

Copyeditor: Shelly Gerger-Knechtl

Proofreader: Pat Daly

Indexer: Shelly Gerger-Knechtl

Compositor: MPS Limited

Senior Art Director: Michelle Kunkler

Cover and Internal Designer: Rokusek Design

Cover Image: © Dvpodt/Shutterstock

Rights Acquisition Director: Audrey Pettengill

Rights Acquisition Specialist, Text and Image:
Amber Hosea

Text and Image Permissions Researcher:
Kristiina Paul

Senior Manufacturing Planner: Doug Wilke

For product information and technology assistance, contact us
at **Cengage Customer & Sales Support, 1-800-354-9706
or support.cengage.com.**

For permission to use material from this text or product, submit
all requests online at **www.copyright.com**.

Library of Congress Control Number: 2012942868

ISBN-13: 978-1-111-98704-6
ISBN-10: 1-111-98704-1

Cengage
200 Pier 4 Boulevard
Boston, MA 02210
USA

Cengage is a leading provider of customized learning solutions
with employees residing in nearly 40 different countries and sales
in more than 125 countries around the world. Find your local
representative at: **www.cengage.com**.

To learn more about Cengage platforms and services, register or
access your online learning solution, or purchase materials for
your course, visit **www.cengage.com**.

Printed Number: 7   Print Year: 2024
Printed in Mexico

Dedication

Dedicated to the memory of my friend, colleague, and mentor
Tomás López Jiménez.

Contents

Preface

The twenty-first century is an age of scientific and technological wonders. Computers have proved to be everything people expected—and more. Bioengineering has unraveled the mysteries of the cell and enabled scientists to synthesize drugs that were inconceivable a decade ago. Nanotechnology provides a glimpse into a world where the computer revolution is combined with engineering at the atomic level to create microscopic autonomous machines that may, one day, be injected into the body to carry out internal repairs. Ubiquitous computing has given us cell phones, MP3 players, and digital cameras that keep us in touch with each other via the Internet. The computer is at the core of almost all modern technologies. This book explains how the computer works.

The discipline called *computing* has been taught in universities since the 1950s. In the beginning, computing was dominated by the large mainframe, and the subject consisted of a study of computers themselves, the operating systems that controlled the computers, languages and their compilers, databases, and business computing. Since then, computing has expanded exponentially and now embraces so many different areas that it's impossible for any university to cover computing in a comprehensive fashion. We have to concentrate on the essential elements of computing. At the heart of this discipline lies the machine itself: the computer. Of course, computing as a theoretical concept could exist quite happily without computers. Indeed, a considerable amount of work on the theoretical foundations of computer science was carried out in the 1930s and 1940s before the computer revolution took place. However, the way in which computing has progressed over the last 40 years is intimately tied up with the rise of the microprocessor. The Internet could not have taken off in the way it has if people didn't have access to very low-cost computers.

Since the computer itself has had such an effect on both the growth of computing and the path computing has taken, it's intuitively reasonable to expect that the computing curriculum should include a course on how computers actually work. University-level Computer Science and Computer Engineering CS programs invariably include a course on how computers work. Indeed, professional and course accreditation bodies specify computer architecture as a core requirement; for example, computer architecture is central to the joint IEEE Computer Society and ACM Computing Curriculum.

Courses dealing with the embodiment or realization of the computer are known by a variety of names. Some call them hardware courses, some call them computer architecture courses, and some call them computer organization courses (with all manner of combinations in between). Throughout this text, I will use the expression *computer architecture* to describe the discipline that studies the way in which computers are designed and how they operate. I will, of course, explain why this discipline has so many different names and point out that the computer can be viewed in different ways.

Like all areas of computer science, the field of computer architecture is advancing rapidly as developments take place in instruction set design, instruction level parallelism, cache memory technology, bus systems, speculative execution, multi-core computing, and so on. We examine all these topics in this book.

Computer architecture underpins computer science; for example, *computer performance* is of greater importance today than ever before, because even those who buy personal computers have to understand systems architecture in order to make the best choice.

Although most students will never design a new computer, today's students need a much broader overview of the computer than their predecessors. Students no longer have to be competent assembly language programmers, but they must understand how buses, interfaces, cache memories, and instruction set architectures determine the performance of a computer system.

Moreover, students with an understanding of computer architecture are better equipped to study other areas of computer science; for example, a knowledge of instruction set architectures gives students a valuable insight into the operation of compilers.

My motivation for writing this book springs from my experience in teaching an intermediate level course in computer architecture at the University of Teesside. I threw away the conventional curriculum that I'd inherited and taught what could be best described as *Great Ideas in Computer Architecture*. I used this course to teach topics that emphasized global concepts in computer science that helped my students with both their operating systems and C courses. This course was very successful, particularly in terms of student motivation.

Anyone writing a text on computer architecture must appreciate that this subject is taught in three different departments: electrical engineering (EE), electrical and computer engineering (ECE), and computer science (CS). These departments have their own cultures and each looks at the computer from their own viewpoint. EE and ECE departments focus on electronics and how the individual components of a computer operate. EE/ECE-oriented texts concentrate on gates, interfaces, signals, and computer organization. Many students in CS departments don't have the requisite background in electronics, so they can't follow texts that emphasize circuit design. Instead, computer science departments place more stress on the relationship between the low-level architecture of the processor and the higher-level abstractions in computer science.

Although it is near impossible to write a text optimized for use in both EE/ECE and CS departments, *Computer Organization and Architecture: Themes and Variations* is an effective compromise that provides sufficient detail at the logic and organizational levels for EE/ECE departments without including the degree of detail that would alienate CS readers.

Undergraduate computer architecture is taught at three levels: introductory, intermediate, and advanced. Some schools teach all three levels, some compress this sequence into two levels, and some provide only an introduction. This text is aimed at students taking first- and second-level courses in computer architecture and at professional engineers who would like an overview of current developments in microprocessor architecture. The only prerequisite is that the reader should be aware of the basic principles of a high-level language such as C and have a knowledge of basic algebra.

It is difficult to pitch a book at precisely the right level. Indeed, such an ideal level doesn't exist. Different students react in different ways to any specific text. If you make a book very focused and follow a narrow curriculum, you appeal only to students on a tiny handful of courses. *Computer Organization and Architecture: Themes and Variations* is well-suited to a wide range of courses, because it covers the basics and some of the more advanced topics in computer architecture.

Features of the Book

Why inflict yet another text on computer architecture on the world? Computer architecture is a fascinating topic. It's all about how you can take vast numbers of a single primitive element such as a NAND gate and make a computer. It's all about how common sense and technology meet. For example, the cache memory that makes processors so fast is conceptually no more complicated than the note on the back of an envelope. Equally, the way in which all processors operate uses a technique invented by Ford for car production: the *pipeline* or production line. I have tried to make the subject interesting and have covered a greater range of topics than absolutely necessary. For example, in this text we will look at memory devices that operate by moving an oxygen atom from one end of a crystal to the other.

The title of this text, *Computer Organization and Architecture*, emphasizes the structure of the complete computer system (CPU, memory, buses, and peripherals). The subtitle *Themes and Variations* indicates that there is a theme (i.e., the computer system) and also variations, for example, the different approaches to increasing the speed of a CPU or to organizing cache memory.

It is often easier to describe something in terms of what it isn't rather than what it is. This book is not concerned with the precise engineering details of microprocessor systems design, interfacing, and peripherals. It certainly isn't an assembly language primer. The central theme of this book is microcomputer *architecture* rather than microprocessor systems design. Computer architecture can be defined, for our present purposes, as the view of a computer seen by the machine language programmer. That is, a computer's architecture takes no account of its actual hardware or implementation and is concerned only with what it does. We will not consider some of the hardware and interfacing aspects of microprocessors, except where they impinge on its architecture (e.g., cache memory, memory management, and the bus).

The Target Architecture

Anyone writing an architecture text has to select a target architecture as a vehicle to teach the fundamentals of machine design and assembly language programming. Professors regularly debate with religious intensity the relative merits of illustrating a course with a real commercial processor or with a hypothetical generic processor. The generic machine is easy to understand and has a shallow learning curve. Students often find that absorbing the fine details of a real processor is time consuming and unrewarding. On the other hand, practical engineering is all about living with the limitations of the real world. Moreover, a real machine teaches students about the design decisions that engineers have to make in order to create a commercially viable product.

In the 1970s and 1980s DEC's PDP-11 minicomputer was widely adopted as a teaching vehicle. The PDP-11 gradually dropped out of the curriculum with the advent of 16-bit microprocessors such as the Motorola 68K. From the academic's point of view, the 68K (loosely based on the earlier PDP-11) was a dream machine, because its architecture is relatively regular and that made it easy for students to write programs in 68K assembly language. A casual observer might have expected the ubiquitous Intel IA32 family, which is found in most PCs, to have played a significant role in computer architecture education. After all, countless students get hands-on experience of Intel's processors. The 80x86 family has never really caught on in the academic world because its complex architecture grew in an *ad-hoc* fashion as each new member of the family was released and this presents students with an excessive burden. Some academics illustrate their course with a high-performance RISC processor, such as MIPS, which is both powerful and easy to understand. Such high-end RISC processors are found in workstations but are relatively unknown to many students (professors have observed that students often request PC-based technology due to their familiarity with the PC). However, RISC processors are used in both high-performance computers and most cell phones.

I have selected the ARM processor as a vehicle to introduce assembly language and computer organization. It is a processor that is powerful, elegant, yet easy to learn. Moreover, development tools for the ARM processor are widely available which means that students can write programs in ARM assembly language and run them in the lab or at home on their PCs.

A strong contender for the role of target architecture in a modern text is Intel's IA64 Itanium processor. This is a device of immense power and sophistication, yet its basic architecture is simpler than the 80x86 family. The rich and innovative features of the Itanium's architecture illustrate numerous concepts found in a computer architecture course–from the data stack to speculative execution, and from pipelining to instruction level parallelism (ILP). Consequently, I also introduce some features of this processor when we look at high-performance computing.

Computer Organization and Architecture: Themes and Variations isn't a conventional computer architecture text. I go beyond the conventional curriculum and cover material that is interesting, important, and relevant. One of my principal objectives is to provide students with an appropriate body of knowledge that they can absorb. All too often, students graduate from university with embarrassingly large gaps in their knowledge. I know of no other text that adopts my approach. For example, all computer architecture texts introduce floating-point arithmetic, yet very few discuss the codes for data compression required to store large volumes of textual and video information, and none describe the MP3 data compression that's at the heart of an entire industry. Similarly, computer architecture texts often lack coverage of architectural features intended to support multimedia applications. Some of the highlights of this text are described below.

History

Books on computer architecture usually have a section on the history of the computer. Such history chapters are often inaccurate and have received criticism from experts in the field. However, I feel that a history chapter is important, because a knowledge of computer history helps students appreciate how and why developments took place. By knowing where computers came from, students are better able to understand how they are likely to develop in the future. In this text, I provide a short overview of the history of computing and include further historical background in the supplementary web-based material that accompanies this book.

OS Support

The operating system is intimately bound up with computer architecture. *Computer Organization and Architecture: Themes and Variations* covers topics in architecture of interest to those who also study operating systems (e.g., memory management, context switching, protection mechanisms).

Multimedia Support

The most important driving force behind modern computer architectures is the growth of multimedia systems with their insatiable demand for high performance and high bandwidths. This text demonstrates how modern architectures have been optimized for multimedia applications. We look at the effect of multimedia applications on both the architectures of computers and the design of buses and computer peripherals, such as hard disks for use in audiovisual applications.

Input/Output Systems

Today's computers are not only much faster than their predecessors, but they also provide more sophisticated means of getting information into and out of the computer. I/O was of relatively little importance when the typical computer was interfaced only to a keyboard, modem, and printer. Computers are now routinely interfaced to peripherals, such as digital video cameras that demand massive data transfer rates. We will look at some of the modern, high-performance I/O systems, such as the USB and FireWire interfaces. We will also delve more deeply into input/output-related topics such as handshaking and buffering.

Computer Memory Systems

Memory is the *Cinderella* of the computer world. Without high-density, high-performance memory systems, neither low-cost desktop systems nor digital cameras with 32GB of storage would be possible. I have divided memory systems into two chapters: the first dealing with semiconductor memory and the second dealing with magnetic and optical memory. We will also take a look at some of the interesting emerging memory technologies, such as Ovonic memory and ferroelectric memory.

Approach

The books that I've most enjoyed are those where a little of the author's personality and view of the subject shines through. I hope that the same is true of this book. Computer architecture isn't something that can be expressed as a set of cold equations to be learned; it is a culture that has developed over the years. At conferences, you will meet academics who passionately argue the relative merits of this computer over that one. I would be a poor educator if I did not at least hint to students that computer architecture can be as much fad as fact.

It's also true that different authors emphasize different topics. For example, most authors stress the design of the processor and have relatively little to say about memory, buses, and peripherals–even though all of these elements are necessary to create a computer system. Perhaps some feel that one aspect of a computer is more intrinsically interesting than others. I provide more coverage of memory, buses, and interfaces than many texts, because I feel that these topics are as important as the processor itself. Similarly, I've included details of memory elements such as Ovonic devices, which store data by melting a bead of glass and then storing a 1 or 0 by selecting how fast it cools. This is such a remarkable example of engineering ingenuity that I felt I had to include it. I'd like students to share the enthusiasm I have for this subject.

I find that a significant shortcoming of many texts is the quality of diagrams and illustrations. All too often a figure has far too little annotation, and its meaning is entirely lost. I have drawn virtually all of the included diagrams myself, and I hope that they do a good job in illustrating the meaning of the text.

Here's an example of a diagram that illustrates the effect of a sequence of three instructions on a register. The purpose of the code is to take a byte from two registers and then concatenate them in a third register. The use of color makes it easy to see how the data is being processed.

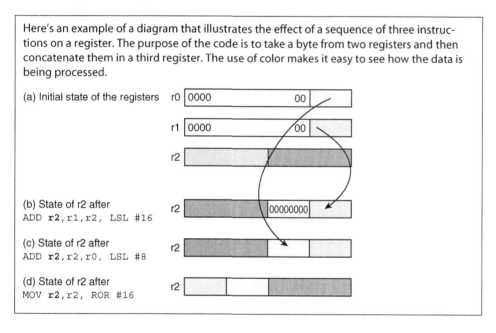

(a) Initial state of the registers   r0 | 0000   00 |

r1 | 0000   00 |

r2 |

(b) State of r2 after
ADD **r2**,r1,r2, LSL #16   r2 | 00000000 |

(c) State of r2 after
ADD **r2**,r2,r0, LSL #8   r2 |

(d) State of r2 after
MOV **r2**,r2, ROR #16   r2 |

Contents Overview

I have divided the book into five parts.

Part I: An Introduction to Computer Architecture covers the enabling material that allows us to discuss computer architecture. Chapter 1 takes an unusual approach to the introduction of a computer. I begin by presenting a problem, solving it, and then inventing a system to implement the solution–the computer. My aim is to demonstrate that the computer closely models the way in which we solve problems. Because so many topics in computer architecture are interrelated, I provide a brief overview of the computer system to allow me to mention topics, like cache memory, before I discuss them in detail.

We will look at the way in which information is encoded and represented in binary form; for example, I introduce the computer arithmetic of both signed and unsigned integers, demonstrate how floating-point numbers represent very large and very small quantities, and then briefly introduce more complex mathematical functions. An introduction to gates and logic design is also included in Chapter 2. This topic always presents the author of a text on computer architecture with some difficulty. Should logic be omitted because some students take a separate course on logic design? Or should logic be relegated to an appendix so that only those who require it need delve into it? I've decided to put it in this section, because it allows a natural progression from numbers to computers and helps students understand some of the material that follows. Even students who have already taken a course in logic should go over this chapter lightly. I end the section on logic design with a very simple proto-computer.

In an ideal world, Part I would include a detailed history of the way in which computers have developed from simple mechanical calculators to today's powerful processors, because all students should have an understanding of how the profession developed. The history of the computer is a fascinating story that begins with a desire to make tedious arithmetical calculations easier, then moves to the telegraph and transatlantic cable project that did so much to "shrink" the world. In the last century, modern computing grew out of the technology used to create the telephone system and was driven by the need for high-speed calculation by scientists, the business community, and the military. No other human artifact has developed as fast as the computer in terms of its increase in performance and its decrease in cost. It is said that if the automobile had developed as rapidly as the computer we'd be driving cars at many times the speed of light using a single drop of gasoline a year.

Unfortunately, I just can't do justice to the history of the computer here. Students have to cover a certain body of knowledge in a computer architecture course, and they have to develop a range of skills in order to become practitioners in their chosen discipline. Consequently, I have located the material on computer history on the book's companion website, the details of which can be found in the Supplements and Resources section below... All students are strongly encouraged to read this material. Computer scientists have a tendency to reinvent the wheel, so concepts developed one year sometimes appear in a different context the following year.

Part II: Instruction Set Architectures is the heart of this course and includes three chapters. I first introduce the concept of an instruction set and then examine some of the important issues in computer architecture, such as the data structures that are accessed and manipulated by computers. Chapter 3 introduces the ARM family of microprocessors that have a truly elegant architecture and include facilities allowing us to cover several interesting topics, such as ***predicated execution*** where an instruction is executed only if certain conditions are met. It would be nice to cover all of the variations in the architecture of the computer, but unless you are going to dedicate the rest of your life to that end, there is not time. In Chapter 4, we look at some of the variations in computer architecture. For example, memory indirect addressing modes that let you access complex data structures with remarkable ease, and special compressed-code processors that squeeze conventional code to a fraction of its normal size.

The final chapter of Part II looks at the way in which processors have been adapted to suit modern multimedia applications such as video encoding and decoding. In this chapter, we also take a brief look at the background of multimedia processing and explain why high-performance computing is so necessary.

Part III: Organization and Efficiency is concerned with computer organization; that is, how computers work and are organized internally. Chapter 6 begins with an introduction to performance, which is the way the speed of computers is measured and how computers are compared. Some might argue that performance should be introduced earlier because of its importance. However, because so many topics influence performance, I thought it is better placed here where the reader is able to understand more of the background.

Chapter 7 looks at how computers actually work. We begin by looking at microprogramming, which is a technique that makes it possible to design computers of arbitrary complexity and is still used to implement some of the more complex instructions in Intel's IA32 processors. Then we introduce pipelining, which is fundamental to all modern computer design. Pipelining mirrors the industrial production line where instructions are executed bit by bit as they flow through the computer. However, pipelining encounters problems when non-sequential instructions (branches) are encountered. The final part of Chapter 7 looks at how the problems caused by branch instructions can be limited.

Chapter 8 continues from where Chapter 7 left off. Here we look at how the performance of processors has been raised by executing instructions in parallel and even out-of-order. In Chapter 8, I introduce the very long instruction word computer (VLIW), which bundles several operations into a single instruction and executes them in parallel. As an example of modern high-performance computing, I introduce the remarkable IA64 Itanium computer that incorporates several powerful ways of accelerating the processor. You could even call the Itanium *a computer architecture course on a chip*.

Part IV: The Computer System turns our attention away from the central processing unit and devotes four chapters to the computer system, concentrating on memory, buses, and input/output. Chapter 9 focuses on two related topics: cache and virtual memory. Although the capacity of memory systems has expanded rapidly over the last few decades, their speed or access time has not improved at anything like the same rate as the CPUs. This situation has led to a bottleneck where memories can't provide data at the speed the CPUs can process it. Cache technology has been developed to make the best of a bad job by using a small amount of fast memory to do the work of a large amount of fast memory. I show how cache memories operate and describe their principal features. Chapter 9 also looks at virtual memory, which integrates main store with disk storage.

Chapters 10 and 11 cover memory technologies ranging from static semiconductor memory to disk and optical storage technologies.

Chapter 12 concludes Part IV by looking at the strategies used to get information into and out of the computer and then describing some of the modern high-speed interfaces used to support multimedia systems.

One of the key components of the modern computer is the bus, which distributes information between the various functional parts of the system. Indeed, the bus is one of the components most critical to the computer's performance. We look at the structure of computer buses, their functionality, and the way in which they permit competing devices to access the bus in multiprocessor systems. Finally, I describe one of the world's most popular high-performance buses: the PCI bus found in PCs.

Part V: The Near Future looks at the next direction in computer architecture, multiprocessing. Once upon a time, you could make a powerful computer by interconnecting an array of individual processors. Such computers had astronomical price tags. Today, it's possible to fabricate several processors on a single chip. Such multi-core processors are becoming increasingly popular, because they both increase processing power and reduce the amount of electrical energy needed. In Chapter 13, I provide a broad overview of multiprocessing covering the background, typical devices, interconnection topology, and the software of parallel processing.

Supplements and Resources

This book comes with a host of support material for both the instructor and the student. The supplements are housed on the book's companion website. To access the additional course materials, please visit www.cengagebrain.com. At the cengagebrain.com home page, search for the ISBN of your title (from the back cover of your book) using the search box at the top of the page. This will take you to the product page where these resources can be found.

Instructor Resources

An instructor's solutions manual (ISM) is available in both print and digital formats. The digital version is available to registered instructors at the publisher's website. This website also includes both a full set of PowerPoint slides containing all graphical images and tables in the text, and a set of LectureBuilder PowerPoint slides of all equations and example problems.

Student Resources

The student resources for this product include:

- a student workbook that provides a detailed introduction to writing programs in ARM processor assembly language and then running them on a simulator;
- additional study material, including a chapter on computer history and an overview of Karnaugh maps;
- useful links, including a link to download the student version of the Kiel simulator from ARM;
- code segments from the book; and
- handouts for all the lecture slides posted on the instructor website.

Acknowledgments

No one writes a textbook in a vacuum. All subjects have a history, background, and culture, and computer architecture is no exception. An author can go along with the flow or take a new direction. I could not have written this book without the contribution of all those texts that have preceded this, for example, the texts I used to study computer architecture myself when I was a student. Equally, I have to acknowledge the countless researchers that contributed to the body of knowledge making up computer architecture. The role of the writer is to take all of this information and create a path for students to follow. The writer has to decide what information is important and what can be omitted, what trends should be followed, and what trends can be relegated to the background. However, the writer is indebted to all those who have contributed to the body of knowledge.

Many people are involved with the production of a book as complex as this, but one stands out, the acquisitions editor. It was Swati Meherishi at Cengage Learning who began the long process of transforming a rough manuscript into the final polished article. An acquisitions editor has the ability to see beyond the manuscript and to understand how the book will fit into a complicated market. Swati had sufficient faith in me to help me endure hours of endless writing and rewriting. I'd like to thank Swati for putting up with me. The other key person in the creation of this book is the developmental editor, Amy Hill. Amy has spent a lot of time going through my manuscript to improve its structure and the way in which it all fits together. She has provided me with unfailingly good advice and gentle guidance. I appreciate the hard work and dedication of the entire team at Cengage Learning; without them this project would have remained a set of files on my hard disk. Also deserving of thanks are Rose Kernan and her team at RPK Editorial Services for their smooth management of all production related tasks, and Kristiina Paul for painstakingly researching and obtaining permissions for all third party material used in the book.

I must thank all the reviewers and editors that helped review and correct the manuscript. They made great suggestions about ways of improving it and helped point me in the right direction when I had misinterpreted source material.

I would like to acknowledge the technical reviewers and those who carefully read the text for errors and omissions. In particular, Loren Schwiebert did a great job of debugging Chapter 3. Sohum Sohoni of Oklahoma State University painstakingly provided a technical review of the entire manuscript. I must thank him for his hard work, especially for his excellent guidance on Chapter 8. My thanks also to Shuo Qin of the University of Southern California, who worked all the homework problems and checked the instructor's solution manual for errors.

The manuscript was reviewed at various stages of development by a number of instructors. I am thankful to all of them for their constructive comments. Below are the names of those who chose to be acknowledged:

Mokhtar Aboelaze
York University

Manoj Franklin
University of Maryland–College Park

Israel Koren
University of Massachusetts–Amherst

Mikko Lipasti
University of Wisconsin–Madison

Rabi Mahapatra
Texas A&M University

Xiannong Meng
Bucknell University

Prabhat Mishra
University of Florida

William Mongan
Drexel University

Vojin G. Oklobdzija
Erik Jonsson School of Engineering,
University of Texas-Dallas

Soner Onder
Michigan Technological University

Füsun Özgüner
Ohio State University

Richard J. Povinelli
Marquette University

Norman Ramsey
Tufts University

Bill Reid
Clemson University

William H. Robinson
Vanderbilt University

Carolyn J C Schauble
Colorado State University

Aviral Shrivastava
Arizona State

Sohum Sohoni
Oklahoma State University

Nozar Tabrizi
Kettering University

Dean Tullsen
University of California–San Diego

Charles Weems
Univ. of Mass.

Bilal Zafar
University of Southern California

Huiyang Zhou
North Carolina State University

Finally, I'd like to thank my wife Sue for her help in debugging the manuscript. Although Sue does not have a technical background, she has been very helpful in removing some of the ambiguities and clumsiness in my use of English.

Feedback from the readers, both critical and appreciative, is welcome. I would be particularly interested in suggestions for additional material that can be added to the companion website in order to help students with their studies. Please send your comments, concerns, and suggestions to globalengineering@cengage.com. You may also contact me directly at alanclements@ntlworld.com.

ALAN CLEMENTS

Paths through Computer Architecture

The simplest way to approach a book is to begin at page one and continue until you run out of pages. However, this is a large book and reading it from cover-to-cover may not be feasible. This text serves a wide range of courses and will be read by students from different backgrounds. Some readers may benefit from the following suggestions about how this book should be approached.

Although I could say that all the material in this book is important, it may not be equally important to each reader. Some readers may be familiar with the basics and can skip lightly over introductory material. Students may find some material going well beyond the *examinable portion* of their course. This doesn't mean that they should skip all such material. More advanced sections may be very relevant in a practical sense or they may be of importance later in the student's career. I ask my students, "If an employer interviews two almost identical students but one has gone beyond the minimal curriculum and has done some additional reading, which student is likely to get the job?"

In this document, we are going to make some suggestions about how this text can be approached. These include

- summary of the chapters in terms of the suitability of the material for various courses
- a pathway through the text for courses for both beginners and more advanced students
- the relationship between the book and the joint IEEE CS/ACM computing curriculum
- recurring themes within the book.

An expanded version of this document may be found on the book's companion website.

Summary of the Chapters

The following table lists each of the chapters and shows how its material fits into the scheme of things and whether that chapter or subsection contains material that can be regarded as elementary, advanced, of background interest.

Chapter	All	Important	Supplementary	Beginners	Advanced
1 The beginning				✓	
2 Arithmetic and digital logic				✓	
3 Architecture and organization	✓			✓	
4 Instruction set architectures	✓	✓	✓		✓
4.1 The stack and data storage	✓				
5 Computer architecture and multimedia		✓	✓		✓

Chapter	All	Important	Supplementary	Beginners	Advanced
6 Performance: Meaning and metrics	✓	✓		✓	
6.5 Benchmarks			✓		
7 Processor control	✓				
7.1 The generic processor				✓	
8 Beyond RISC: Superscalar, VLIW and the Itanium		✓	✓		✓
8.1 Instruction level parallelism	✓				
8.2 Binary translation			✓		✓
8.3 EPIC architecture		✓	✓		✓
9 Cache memory and virtual memory	✓			✓	
10 Main memory	✓				✓
11 Secondary storage	✓				
11.4 Secure memory and RAID	✓				
11.5 Solid state disk drives		✓	✓		✓
11.6 Magnetic tape			✓		
11.7 Optical		✓	✓		✓
12 Input/Output	✓				
12.1 Fundamental principles of I/O	✓			✓	
12.5 The bus	✓			✓	✓
12.6.1 Localized arbitration			✓		✓
12.6.2 Distributed arbitration			✓		✓
12.7.1 The PCI and PCI express bus			✓		✓
12.9 Serial interface buses		✓	✓		
13 Processor level parallelism	✓				
13.1 Power the final frontier		✓	✓		
13.3.1 Multiprocessor topologies	✓			✓	
13.5 Multithreading		✓	✓		✓
13.6 Multi-core processors	✓				

A Pathway for the Beginner and More Advanced Reader

I have been asked to suggest pathways through this text. The above material should be of help to readers and professors using this text. Here, I suggest two possible pathways. One is for students coming to this material for the first time, and one is for students who have taken lower-level courses on digital logic and similar topics. However, these are only suggested pathways. Each course will be different and each professor will have his or her individual priorities. The two lists below are not fine-grained in the sense that not all of each of the sections indicated should necessarily be taught in each course.

The Basic Course

Chapter 1	All									
Chapter 2	2.1	2.2	2.3	2.4	2.5	2.6	2.7	2.9	2.10	2.11
Chapter 3	All									
Chapter 4	4.1	4.2								
Chapter 6	6.1	6.2	6.3	6.4	6.6					
Chapter 7	7.2	7.3	7.4	7.5	7.6					
Chapter 9	All									
Chapter 10	10.1									
Chapter 11	11.1	11.3.2	11.7							
Chapter 12	12.1	12.2	12.5							

The Advanced Course

Chapter 3	All					
Chapter 4	4.1	4.2	4.4			
Chapter 5	All					
Chapter 6	6.1	6.2	6.3	6.4	6.6	
Chapter 7	All					
Chapter 8	8.1	8.3				
Chapter 9	All					
Chapter 10	10.1	10.2	10.3	10.4		
Chapter 11	11.1	11.2	11.3	11.4	11.5	11.7
Chapter 12	All					
Chapter 13	All					

Relationship to the Joint IEEE Computer Society/ ACM Computing Curricula

In 1991, the ACM and the IEEE Computer Society published a report on the computing curriculum as a guideline for those who construct courses in computer science. CC1991 provided a service to universities round the world by providing a framework for computer science courses. This work built upon ACM's *Curriculum '68*.

Nothing remains static in computing, and in 1998 the ACM and IEEE CS began the update of CC1991 to CC2001 to incorporate changes from the previous decade. I helped to update the architecture section of the 2001 report under the leadership of Dr. Willis King.

In 2007, work began on an interim revision to CC2001 to reflect current trends. The 2007 revision includes new material but not radically new areas – that will come in the next major report. I was largely responsible for creating the draft document for the architecture section of the updated curriculum and then it was edited by the committee. Below I have included the *learning objectives* of each of the components of the computer architecture section.

The parts of the 2007 updated curriculum are presented in black and the parts not covered in this text are in blue. In general, the parts that we do not cover are related to computer networks and communications, and computer peripherals. Note that the Computing Curricula do not intend all the topics of any given course area to be taught in a single course.

Digital Logic and Computer Arithmetic [Core]

Learning objectives:
- Design a simple circuit using fundamental building blocks.
- Appreciate the effect of AND, OR, NOT, and EOR operations on binary data.
- Understand how numbers, text, images, and sound can be represented in digital form and the limitations of such representations.
- Understand how errors due to rounding effects and their propagation affect the accuracy of chained calculations.
- Appreciate how data can be compressed to reduce storage requirements including the concepts of lossless and lossy compression.

Computer Architecture [Core]

Learning objectives:
- Describe the progression of computers from vacuum tubes to VLSI.
- Appreciate the concept of an instruction set architecture, ISA, and the nature of a machine-level instruction in terms of its functionality and use of resources (registers and memory).
- To understand the relationship between instruction set architecture, microarchitecture, and system architecture and their roles in the development of the computer.
- Be aware of the various classes of instruction: data movement, arithmetic, logical, and flow control.
- Appreciate the difference between register-to-memory ISAs and load/store ISAs.
- Appreciate how conditional operations are implemented at the machine level.
- Understand the way in which subroutines are called and returns made.
- Appreciate how a lack of resources in ISPs impacts on high-level languages and the design of compilers.
- Understand how, at the assembly language level, parameters are passed to subroutines and how local workspace is created and accessed.

Interfacing and Communication [Core]

- Appreciate the need for open- and closed-loop communications and the use of buffers to control dataflow.
- Explain how interrupts are used to implement I/O control and data transfers.
- Identify various types of buses in a computer system and understand how devices compete for a bus and are granted access to the bus.
- Be aware of the progress in bus technology and understand the features and performance of a range of modern buses (both serial and parallel).

Memory System Organization and Architecture [Core]

Learning objectives:
- Identify the memory technologies found in a computer and be aware of the way in which memory technology is changing.

- Appreciate the need for storage standards for complex data storage mechanisms such as DVD.
- Understand why a memory hierarchy is necessary to reduce the effective memory latency.
- Appreciate that most data on the memory bus is cache refill traffic.
- Describe the various ways of organizing cache memory and appreciate the cost-performance tradeoffs for each arrangement.
- Appreciate the need for cache coherency in multiprocessor systems.
- Describe the need for virtual memory; its relationship to the operating system, and the way in which physical addresses are translated into logical addresses.

Functional Organization [Core]

Learning objectives:
- Review of the use of register transfer language to describe internal operations in a computer.
- Understand how a CPU's control unit interprets a machine-level instruction – either directly or as a microprogram.
- Appreciate how processor performance can be improved by overlapping the execution of instruction by pipelining.
- Understand the difference between processor performance and system performance (i.e., the effects of memory systems, buses, and software on overall performance).
- Appreciate how superscalar architectures use multiple arithmetic units to execute more than one instruction per clock cycle.
- Understand how computer performance is measured by metrics such as MIPS or SPECmarks and the limitations of such metrics.
- Appreciate the relationship between power dissipation and computer performance and the need to minimize power consumption in mobile applications.

Multiprocessing and Alternative Architectures [Core]

Learning objectives:
- Discuss the concept of parallel processing and the relationship between parallelism and performance.
- Appreciate that multimedia values (e.g., 8-/16-bit audio and visual data) can be operated on in parallel in 64-bit registers to enhance performance.
- Understand how performance can be increased by incorporating multiple processors on a single chip.
- Appreciate the need to express algorithms in a form suitable for execution on parallel processors.
- Understand how special-purpose graphics processors, GPUs, can accelerate performance in graphics applications.
- Understand the organization of computer structures that can be electronically configured and reconfigured.

Performance Acceleration [Elective]

Learning objectives:
- Explain the concept of branch prediction and its use in enhancing the performance of pipelined machines.
- Understand how speculative execution can improve performance.
- Provide a detailed description of superscalar architectures and the need to ensure program correctness when executing instructions out-of-order.
- Explain speculative execution and identify the conditions that justify it.

- Discuss the performance advantages that multithreading can offer along with the factors that make it difficult to derive maximum benefits from this approach.
- Appreciate the nature of VLIW and EPIC architectures and the difference between them (and between superscalar processors).
- Understand how a processor re-orders memory loads and stores to increase performance.

Architecture for Networks and Distributed Systems [Elective]

Learning objectives:
- Explain the basic components of network systems and distinguish between LANs and WANs.
- Discuss the architectural issues involved in the design of a layered network protocol.
- Explain how architectures differ in network and distributed systems.
- Appreciate the special requirements of wireless computing.
- Understand the difference between the roles of the physical layer and data link layer and appreciate how imperfections in the physical layer are handled by the data link layer.
- Describe emerging technologies in the net-centric computing area and assess their current capabilities, limitations, and near-term potential.
- Understand how the network layer can detect and correct errors.

Devices [Elective]

Learning objectives:
- Understand how analog quantities, such as pressure, can be represented in digital form and how the use of a finite representation leads to quantization errors.
- Appreciate the need for multimedia standards and be able to explain in non-technical language what the standard calls for.
- Understand how multimedia signals usually have to be compressed to conserve bandwidths using lossless or lossy encoding.
- Discuss the design, construction, and operating principles of transducers such as Hall-effect devices and strain gauges.
- Appreciate how typical input devices operate.
- Understand the principles of operation and performance of various display devices.
- Study the operation of high-performance computer-based devices such as digital cameras.

New Directions in Computing [Elective]

Learning objectives:
- To appreciate the underlying physical basis of modern computing.
- Understand how the physical properties of matter impose limitations on computer technology.
- Appreciate how the quantum nature of matter can be exploited to permit massive parallelism.
- Appreciate how light can be used to perform certain types of computation.
- Understand how the properties of complex molecules can be exploited by organic computers.
- To get an insight into trends in memory design such as ovonic memory and ferromagnetic memories.

Recurring Themes

Some topics appear in several places or guises in computer architecture or any other subject. Often they are key elements of the discipline and sometimes they are topics that appear in other areas of computer science. Consequently, we have selected several of these recurring themes and have indicated where they occur in this text. This list is not, by any means, exhaustive.

Addressing Modes	3.2.2	3.7	4.5	8.3.4	12.1.1		
Amdahl's Law	6.4	13.2					
Arbitration	2.10.1	12.6.1	12.6.2	12.9.2			
Bus	1.6.2	2.11	12.5				
Cache Memory	9.1	13.4.2					
Conditional Branch	2.9.3	3.1.2	3.6.2	3.6.4	7.1	7.2	7.3.3
	7.4.1	7.4.2	7.4.3	7.4.4	7.5	7.6	8.3.7
Instruction Format	1.5	3.5.4,	3.7.7	4.3	4.6.1	8.3.1	8.3.3
Memory	1.4.5	1.6.1	6.1	8.3.8	10.1	13.4	
Power Consumption	13.1						
Predicated Execution	3.6.5	8.3.2	8.3.7				
Protocols	12.3	12.6.1	13.4.2				
Registers	3.2.1	3.3.1	8.3.2				
SIMD Processing	5.3	13.3	13.7.2				
Stack	3.8	3.10	3.11.3	4.1			
State Diagram	1.4.2	2.10.4	7.6	12.8	13.4.2		
Stored Program Computer	1.4	1.5	3.1				
Timing Diagrams	2.10.1	7.3.2	10.2.1	10.3	12.1	12.3	12.5.2
	12.6.1	12.9.2					
Virtual Memory	9.5						

About the Author

Alan Clements

Alan Clements was born in Lancashire, England, and studied Electronics at the University of Sussex. He was awarded a PhD at Loughborough University in equalizers for digital data transmission in 1976 at a time when microprocessors were just being introduced. By applying microprocessors to the problem of equalization, he became interested in computer design and joined the Department of Computer Science at the University of Teesside.

In the 1970s, literature on practical microcomputer design was comparatively rare and he wrote one of the first books in this area. Reviews were very positive and he went on to write two significant texts. *The Principles of Computer Hardware* was an undergraduate text covering the whole spectrum of computer hardware at an introductory level, with topics ranging from Boolean algebra to peripherals that measure rotational velocity. This text was written in a student-friendly style to encourage students to take an interest in computer architecture.

In the 1980s Alan wrote a definitive text on microprocessor systems design that bridged the gap between the academic and the practical by covering all stages in the design of a microcomputer and by providing actual circuits. Because of Alan's work in promoting microprocessor design, Motorola endowed Alan with a personal chair at Teesside in 1993.

Over the years, Alan became more and more interested in the problems of teaching computer architecture and became increasingly involved with computer science education. In 2001 he became chair of the Computer Society's international student competition, CSIDC, and in the same year received a National Teaching Fellowship in the UK, the UK's highest award for higher education. Alan has been actively involved with the Frontiers in Engineering Education conference and has been a guest editor of special editions on computer science education for two journals.

Alan has held several offices in the IEEE Computer Society including Editor in Chief of CS Press, Second Vice President of the CS, and Chair of the Educational Activities Board. He has also held visiting professorships in Heraklion, and Colorado State University.

Alan is active in curriculum design, has written papers on the future of computer architecture education, has worked with a consortium of US and European universities to exchange students, and worked on the CS/ACM 2001 computing curriculum project. His consultancies include work for the European Union, the UK government, Hitachi, and Sega.

In 2007 Alan was awarded the IEEE Computer Society's Taylor Booth award for education.

As well as teaching and writing, Alan is interested in photography and has had several public exhibitions of his work. He is also a private pilot and combines his love of flying with photography. His photographic work can be found at www.pbase.com/clements.

Alan retired from full-time teaching in 2010 to concentrate on the writing and photography parts of his life.

THE BEGINNING

I've called this book *Computer Systems Architecture: Themes and Variations* because we look at the basic principles governing all digital computer systems and discuss the variations between different types of computers. We cover the entire computer system rather than concentrate on just the CPU at the expense of memory and input/output mechanisms.

Readers will have come to this book via different paths. Some will be newcomers to the world of computer architecture and hardware, whereas others will have taken a previous course on digital design or may have done some reading on their own. In order to provide a common background for all readers, we begin by *inventing* a generic computer. A simple problem is provided, and then we ask what mechanism would be needed to solve the problem. I could have taken a more conventional approach by presenting a typical computer; however, this approach demonstrates that the computers we use today developed quite naturally from the need to solve problems. I'm no historian, but I feel that the structure of today's computer was all but inevitable because the basic principles of the computer as we know it today were described long before the computer age. It was 20th century technology that enabled the fabrication of practical computers.

Following this introduction, the book is divided into five broad sections.

Part I   *Foundations* introduces the concepts, history, and underlying technology of digital computers. Computer arithmetic and digital logic are covered in sufficient detail to allow students to appreciate how information is represented and manipulated within a computer and how basic functional units operate. The section on digital logic is not intended to enable the reader to construct complex digital circuits but to understand how a computer works at the level of primitive gates; indeed, we demonstrate how a few gates can be put together to build a simple computer.

Part II   *Instruction Set Architectures (ISAs)* looks at the programming model of a computer. In this section, we introduce the register model of a computer, its instruction types, and the addressing modes of a typical microprocessor family. We also describe its assembly language programming in order to give the reader an insight into how simple programs can be written and executed.

Part III   *Organization and Performance* describes how we measure the performance of computers. This topic has been placed later in this book than is traditional, because we have to cover the basic elements of a processor before examining how we measure performance. Part III is concerned with how computers are organized internally, and we discuss some of the ways in which the speed of computers has been enhanced by remarkably clever design techniques.

Part IV   *The System* covers the other parts of a computer that are required to convert the microprocessor chip into a complete system: for example, peripheral subsystems and the wide range of memory systems, storage devices, and buses available to the computer systems' designer. Today's technology puts more and more transistors on a single silicon chip, and a complete system on a chip is now possible. However, I stress the point that a computer architecture course should not overemphasize the role of the CPU at the expense of all the other elements of the system.

Part V   *Processor Level Parallelism* goes beyond the single-processor computer and introduces the notion of computers with multiple processors. In particular, we look at multicore processors that contain several individual processors on the same silicon chip. Multiprocessors have been adopted for exactly the same reasons that two horses were used to pull a chariot or four engines to drive an aircraft. We simply cannot breed a horse that is twice as strong as an existing horse.

Multiprocessor systems have been created because we appear to be approaching the limit of what we can do with one processor. Increasing the power of a processor further by raising its clock frequency produces a disproportionate increase in the heat generated. The final part of the book also introduces a range of topics associated with high-performance computing, such as multiprocessor topologies, special languages for multiprocessing, and graphics processing units (GPUs).

This book is entitled *Themes and Variations* because there are many ways of designing processors, memory systems, buses, and interfaces. We look at the principles governing the operation of a computer (the *themes*) and discuss some of the different approaches engineers have taken to the design of actual computers (the *variations*).

Three Uses of the Term *Architecture*

Although the term *architecture* appears over and over again in computer literature, it is used in three distinct ways.

Instruction Set Architecture (ISA) describes the programmer's abstract view of the computer and defines its assembly language and programming model. It is abstract because it does not take into account the implementation of the computer.

Microarchitecture describes the way in which an instruction set architecture is implemented. In other words, the microarchitecture is concerned with the internal design of a computer.

Systems Architecture is concerned with the complete system consisting of processor, memory, buses, and peripherals. Systems architecture is of interest to all who design computer systems and who have to arrange the components to create the most cost-effective solution to a problem.

The figure provides a map of the *territory* covered by our journey through computer systems architecture. This map introduces the parts of the computer system, from the *CPU* that carries out the information processing, to the *disk drives* (including pen drives and solid-state disks) that store huge quantities of data, and the *buses* (information highways) that transport data around the computer. Computer systems also include input and output devices such as a keyboard, mouse, display, printer (in a PC), a digital camera, or a GPS receiver (in a cell phone or navigation device). Courses in computer architecture often lack the time to cover these peripherals that range from the simple electromechanical mouse to the immensely complex GPS receiver.

The term *Central Processing Unit* (*CPU*) describes the part of a computer system responsible for reading instructions from memory and executing them. Today, it is largely synonymous with *microprocessor*. Many of the elements of the computer system in this diagram have one or more chapters devoted to them.

Students of computer architecture should be aware of the factors that affect computer design (shown in clouds in the figure). For example, *performance* is concerned with how fast a computer operates. Similarly, we are interested in *exception handling*, which is the mechanism that allows a computer to respond to external events, such as moving the mouse or striking a key. *Power consumption* is a key factor in computing today, because it must be minimized. High-performance computers have to reduce power consumption in order to prevent the processor from damage due to heat, and portable systems have to reduce power consumption in order to extend battery life.

Overview of Computer Systems Architecture

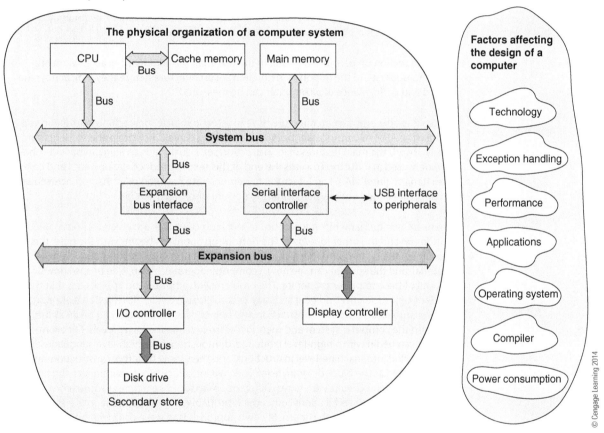

Why is This Book Organized Like This?

Choosing an effective way of organizing a text is not easy because there are so many conflicting factors. For example, I wanted to discuss computer history early in this book but could not simply because I would have been describing the history of the computer before the fundamental concepts needed by the reader had been introduced. A compromise is to begin with some of the milestones in the history of computing and then to save the rest of the story for later.

Different universities teach computer architecture and organization in different ways. However, all courses include a strong emphasis on the instruction set architecture (i.e., the computer's instruction set, register structure, and addressing modes). Consequently,

(continued)

I placed the section on the computer's instruction set architecture towards the beginning of this book.

I have covered two areas in more detail than is usual in other texts: memory systems and computer buses. This material is located towards the end of the book because not all courses will cover these topics in such detail. Colleges that do not may skip through this material quickly. However, to those who complain that I go into too much detail, I offer two defences. The first is that much of the growth in ubiquitous computing is due more to developments in memory and interface technology than to processor technology. The second is that I find some of the concepts inherent in memory/interface technology quite fascinating, and it is my job to communicate that to the student.

I put the section on performance later in the text than is traditional, because I want students to understand the notion of instruction set architecture before they look at how the speed and performance of a computer can be measured.

All teachers of computer architecture have to select a vehicle for teaching instruction sets. I chose the ARM microprocessor because its learning curve is not steep and students can easily absorb the main features of its instruction set. Equally, it is an elegant machine and is widely used in industry. Towards the end of this text, I introduce the powerful and complex Itanium (Intel's IA-64 architecture) to demonstrate how engineers have conceived an architecture aimed entirely at maximizing performance.

There are two fundamental approaches to teaching computer architecture: one is called *bottom-up* and the other *top-down*. The bottom-up approach begins with the gate, then the circuit, the system, and finally the computer. Each new layer builds on the previous material, and the student can see how a computer operates because he or she knows how all of the components operate. The problem with a bottom-up approach is that the student cannot see the end goal and may become lost in detail. The top-down approach fits well into today's world of abstraction and object-oriented design. You begin at the top with the computer system and then successively refine the lower levels. For example, we can begin with a high-level language, demonstrate how high-level languages are compiled into machine-level instructions, and then show how those instructions are implemented at the block diagram level. Then we introduce functional circuits and finally end up designing circuits with gates. Different professors take different approaches, and I would argue that the educational outcome is probably more dependent on the ability of the professor than the path chosen. My own approach is to use a combination of bottom-up and top-down designs.

Computer Systems Architecture

"Pinball games were constrained by physical limitations, ultimately by the physical laws that govern the motion of a small metal ball. The video world knows no such bounds. Objects fly, spin, accelerate, change shape and color, disappear and reappear. Their behavior, like the behavior of anything created by a computer program, is limited only by the programmer's imagination. The objects in a video game are representations of objects. And a representation of a ball, unlike a real one, never need obey the laws of gravity unless its programmer wants it to."

Sherry Turkle, 1984

"A programmer is demonstrating a supercomputer that combines artificial intelligence and speech-recognition to a panel of top brass at the Pentagon. A famous civil war battle is being simulated by the computer. At a critical point, a general interrupts and asks the computer, "Well, are you going to attack or retreat?" The computer answers, "Yes"."
"Yes what?" shouts the general.
"Sir, yes, sir", responds the computer.

Anon

"I think there is a world market for maybe five computers."

Thomas Watson, 1943

"We have a computer in Cambridge; there is one in Manchester and at the NPL. I suppose that there ought to be one in Scotland, but that's about all."

Professor Hartree, 1951

"There is no reason anyone would want a computer in their home."

Ken Olson, founder of Digital Equipment Corp., 1977

"Those who cannot remember the past are condemned to repeat it."

George Santayana, 1905

"The past is a foreign country; they do things differently there."

Harold Pinter

A computer is the most remarkable device ever invented, because there is seemingly no limit to what it can do or to the rate at which its performance grows month by month. A computer incorporated in a heart pacemaker can detect irregular heartbeats and shock the heart to correct the rhythm. A computer in an aircraft's navigation system can land a plane (like the Airbus 380) carrying over 800 people in fog so thick that the human pilot can't taxi clear of the runway after landing. In the movie industry, a computer can generate extras in crowd scenes indistinguishable from real people.

This text explains how the computer achieves all of these feats and how it is constantly improving and taking on new roles as the underlying technology advances. We define what we mean by a *computer system* and explain the meaning of the term *computer systems architecture*.

An important goal of this chapter is to explain what a computer actually does. Most people know what a computer can achieve, but few understand how it functions. We explain how the CPU and the other components of a computer work together to achieve all of the remarkable things modern computers do.

We provide brief overviews of important concepts (such as *data storage*) that you will encounter in later chapters. In particular, we emphasize that computer architecture is concerned with the computer system as a whole, as much as with the operation of the microprocessor that lies at the heart of the computer. Instead of starting with the description of a real computer, we introduce the computer by specifying a simple problem and then by asking what operations we need to carry out in order to solve this problem.

The material in the first part of this chapter sets the scene and explains what a computer does by asking what resources we need to solve a simple problem. Readers who have taken an introductory course may skip this section.

Where We're At

Texts on computer architecture and organization used to begin with pictures of the computer—big computers and little computers, old computers that you've seen in SF films like *2001*, and new computers. Today, we've passed the eighth decade of the computer revolution and the third decade of the personal computer or *ubiquitous* computer revolution. I'm not going to inflict yet another computer picture on the reader. We are entirely surrounded by computers and it's difficult to get away from them. There are now plenty of computers in an automobile, controlling everything from braking to engine ignition timing and winding down the window.

The computer revolution has grown so fast that science fiction films of only a few years ago are quite embarrassingly dated when it comes to the use of computers. I recently watched (again) one of the masterpieces of science fiction, Ridley Scott's *Blade Runner* made in 1982, where space travel was common and replicants that were better than the original humans could be grown in vats. Yet, in that movie, the data display terminals were chunky CRT-based devices with horrible low-resolution graphics. Similarly, *Blade Runner* had no web-style communications networks or evidence of the tiny cell phone with built-in video. These innovations caught science fiction writers unaware. We are in an age where the capability of computer technology has grown faster than the imagination of the writer (with honorable exceptions like the virtual universes of *The Matrix*).

The growth of computer technology has not been uniform across all of the various components or elements of a computer (data manipulation, data storage, data transmission, software, power consumption, etc.), and that is going to be a pervasive theme of this text. Indeed, disparities in the performance of computer components have reached remarkable levels. For example, in 1985 a microprocessor might be clocked at 10 MHz and use memory with an access time of 500 ns (i.e., 5 clock-cycle access time). Today, a processor may have a clock rate of 3 GHz and a memory access time of 50 ns (i.e., a 150 cycle clock-cycle access time). As we shall see, the change in relative speeds of processors and memory has forced designers to seek ways of mitigating the memory access time problem, such as multilevel cache memories (Chapter 9) and hyperthreading. (Chapter 13).

Some Successful Predictions

I shouldn't be too harsh on science fiction writers. Some of the predictions that have been successful are given here.

Robots—Karel Čapek introduced the word robot in his 1921 novel *Rossum's Universal Robots* (the word *robot* is derived from the Slavonic word for *work*).

Communication Satellites—A.C. Clark, an engineer turned science fiction writer, suggested the use of communication satellites in geosynchronous orbit in a technical journal in 1945.

Surveillance—George Orwell's dystopian novel *1984* envisaged a government that could follow people's every movement and render privacy impossible. That has already happened to a remarkable extent in some countries.

The Intelligent House—Ray Bradbury created a largely automated house in his 1950s *Martian Chronicles*.

Genetic Engineering—Aldous Huxley described the use of genetic engineering to modify humans in his 1932 novel *Brave New World*. This is an example of an unreasonable prediction, because he neither knew about DNA (which had not been discovered) nor about the computers needed to unravel DNA structures.

Personal Communications—While it's true that *Blade Runner* lacked personal communications technology, the communicator did exist in *Star Trek*.

The Replicator—In *Star Trek*, people use a replicator to make things, such as food. Such technology is now emerging. It's based on the inkjet printer and uses a device to spray a material to create a thin film in two dimensions. By spraying the same area over and over again, layers of a thin film can be built on top of each other to gradually construct a three-dimensional object. These replication devices are currently used to create prototypes from drawings. By spraying different materials (e.g., metals and insulators), it may become possible to create three-dimensional circuits.

First- and second-generation texts on computer architecture had very little to say about power consumption and dissipation. In 1980, you plugged your computer into a power connector and that was all there was to the subject, unless of course, you were one of the few who were designing satellite-based electronics. Systems were so bulky that few thought about portable computing, and processors did not have exorbitant power requirements. Today's world couldn't be more different. Miniaturization now allows the production of very sophisticated portable systems that have many orders-of-magnitude more speed and storage capability than yesterday's computers, together with high-resolution displays that fit into a pocket or brief case. This means that power is now a *first-class* design criterion. It is one of the limiting factors in the design of portable systems. The new holy grail of portability is the *working day* or the transatlantic flight. A portable system should be usable all day without needing a connection to the public electricity network—something that has been achieved with the latest generation of laptops and tablet-style computers (like iPad®).

Even when computers can be plugged in, power supply problems do not go away. If you put over a billion transistors on a single silicon chip and switch them at 4 GHz (four thousand million times a second), you generate a lot of heat in a tiny space, and you have to worry about removing that heat from the system. Indeed, in the final chapter in this book, we look at multicore processors that were developed largely because of the inability to clock processors faster than about 4 GHz without running into power dissipation problems.

A good example of the direction in which computing is heading was provided by Pawlowski, a Senior Research Fellow at Intel. In an article he wrote in 2007 on high-performance computing in healthcare, he pointed out that in 2004 medical imaging required about 1 GFlops (10^9 floating-point operations) to perform medical imaging. By 2010, it was estimated that 1 TFlops would be required, which is a thousand-fold increase over six years. This increase is required because medical imaging now uses higher-resolution images, better signal processing, and a greater number of images to allow rotation (change of viewing angle).

One of the most common applications of computing is the storage, processing, and display of moving images. Not very long ago, image processing was crude and limited. Today, high definition (high resolution) moving images are routinely stored and processed on even modest systems. This has required great leaps forward in processor architecture and organization, memory technology, data distribution via buses, instruction sets specially tailored to deal with the type of data used by images (and the algorithms used to compress images), storage mechanisms (such as flash memory), and hard disk and distribution systems (such as Blu-ray). All of these topics are now properly part of the modern curriculum in computer architecture.

1.1   What is Computer Systems Architecture?

The first concept we define is that of the *computer system*. The media have long since called a computer system a *microprocessor* or even a *chip*. In fact, a computer system consists of a CPU (*central processing unit*) that takes a program and executes it, plus memory that holds the program and data, and the subsystems required to turn a chip into a practical system. These subsystems facilitate communication between the CPU and external devices such as displays and printers as well as the Internet itself.

You will find the words *computer*, CPU, *processor*, *microprocessor*, and *microcomputer* in this and other texts. The part of a computer that actually executes a program is called a CPU or more simply a *processor*. A *microprocessor* is a CPU fabricated on a single chip of silicon. A computer that is constructed around a *microprocessor* can be called a *microcomputer*. These terms are frequently used interchangeably, although the use of *microcomputer* is declining, because there is no longer a need to distinguish between computers made from lots of individual integrated circuits and those made from single microprocessors (like the Intel Core i7).

Although the CPU is at the heart of the computer, the performance of a computer depends as much on its subsystems as on the CPU itself. There's no point in increasing the speed of a CPU if you can't move data around the computer efficiently. Computer scientists joke that improving the performance of a microprocessor just makes it *wait faster* for data from the memory or a disk drive. We will see later that the rate of improvement of the various components in a computer is a major problem for systems designers, because performance does not increase *uniformly* across all components. For example, the performance of processors themselves has continued to improve dramatically over several decades, whereas the performance (access time) of hard disks has remained largely static for three decades. This situation may now be coming to an end, as mechanical hard disks are being replaced by semiconductor-based hard disks called *solid state disks* (*SSDs*).

Figure 1.1 illustrates a simple generic computer such as a PC or workstation. As well as the CPU, there are elements that are found in almost every computer. Information (i.e., programs and data) is stored in a *memory*. A real computer uses different types of memory for different purposes. In Figure 1.1, we have *cache memory*, *main store*, and *secondary store*. Although Figure 1.1 shows cache memory external to the CPU, most processors include CPU on-chip.

All of these memory technologies hold information, but each operates in a different way. Some provide fast but expensive memory systems, whereas others are slow but cheap. Later, we cover these different memory systems because they are so important to the overall *performance* of the computer system. Without a suitable memory to hold programs, even the most

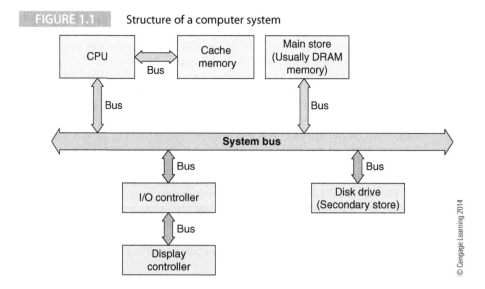

FIGURE 1.1   Structure of a computer system

© Cengage Learning 2014

powerful CPU would be useless; indeed, it was the development of low-cost flash memory that ensured the success of MP3 players, digital cameras, eBooks, and iPads.

It's probably true to say that memory systems have a broader spectrum of properties and are constructed with a wider range of technologies than any other component of a computer (excluding peripherals themselves). Cache memory is a special-purpose, high-speed memory that holds frequently used data. Main store holds the bulk of working data. Secondary store refers to the hard disks and CD-ROMs that archive large quantities of data so cheaply. Of course, when we use *data* here, we really mean *programs and data*. A large slice of this text is devoted to memory systems and their quite amazing technology.

The subsystems that make up a computer in Figure 1.1 are connected by the *buses* that move data from one part of a computer to another. Real buses such as the PCIexpress bus are complex subsystems. In Chapter 12, we describe the mechanisms or *protocols* that ensure data flows from one point to another without colliding with other data attempting to do the same thing.

Different Computers – Same Answer

If I ask my students, "Which is more accurate, an 8-bit computer or a 32-bit computer?" Some of them will say that the 32-bit computer is more accurate. From a computational point of view there is no difference between any two digital computers in terms of their computational ability. Of course, one computer may be faster or cheaper than another. But they are the same in terms of the problems that they can solve.

A computer is said to be *Turing complete* if it can simulate a Turing machine (an abstract computer model devised by Alan Turing). All today's computers are Turing complete, which means that all computers can solve the same set of problems, which in turn, means that a problem that can be solved on machine A can be solved on machine B.

Similarly, you can argue that if computer A can simulate computer B, then any program that can be executed by B can also be executed by the simulator running on A. A further corollary is that a problem that cannot be solved on a computer today cannot be solved by any future computer.

Several buses are shown in the system of Figure 1.1. For example, a hard disk may communicate with the computer via a different bus than that used to connect the CPU to its main memory. Some systems have bus *expansion interfaces* or *bridges* that allow the exchange of data between buses with different characteristics.

The subtitle of this book, *Themes and Variations*, is particularly appropriate when applied to buses. For example, the PCIe bus on a PC motherboard lets you plug new peripherals into your computer, and the USB and FireWire buses let you connect your computer to anything from a mouse to a Camcorder.

What is a Computer?

You can't define the term *computer* in isolation; you have to specify the *class* of computer. The computer that lies at the heart of systems ranging from the personal computer to a cell phone is a *von Neumann stored program digital computer* in honor of the mathematician Johann von Neumann, who was one of the first to quantify the structure of a computer. Other classes of computer are the *analog* computer, the *neural* computer, the *quantum* computer, and the *biochemical* computer. These process information in very different ways to the stored program machine and are beyond the scope of this text.

Figure 1.2 illustrates a programmable digital machine that receives an *input* that it processes to produce an *output*. In computer terms, *input* is the information you give to the computer, and *output* is the information it gives to you. For example, the input to a flight simulator is the movement of the joystick, and the output from the simulator is the video display of the world seen by the pilot.

FIGURE 1.2   The general-purpose computer

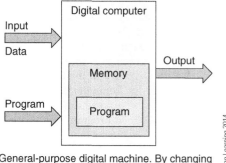

General-purpose digital machine. By changing the program, this machine can carry out any task capable of being performed by a computer.

© Cengage Learning 2014

The programmable machine of Figure 1.3 receives two types of input: the *data* it is to process and the *program* that defines exactly how the input data is to be processed. A program is nothing more than a sequence of operations that the computer carries out to accomplish a given task.

von Neumann

When von Neumann moved to the USA, he Anglicized his given name to *John*. He is also known as Janos, the Hungarian version of Johann. The term 'von Neumann computer' is now considered somewhat controversial, because some computer historians feel that credit for the structure of the modern computer should not be given solely to von Neumann.

The arrangement of Figure 1.2 is called a *general-purpose machine* because the *hardware* (i.e., the actual digital electronics) is directed by the program to perform the impressively wide range of activities that we associate with a computer. When we introduce computer history, we will find that the first digital computers were not general-purpose machines and were *hardwired* to solve specific tasks. Hardwired means that the computer's function (program) cannot be changed other than by physically rerouting the wires of the computer.

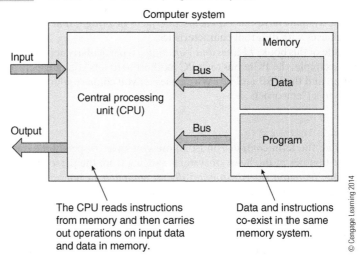

FIGURE 1.3   Structure of the stored program computer

The CPU reads instructions from memory and then carries out operations on input data and data in memory.

Data and instructions co-exist in the same memory system.

© Cengage Learning 2014

Figure 1.3 provides a more detailed view of the digital computer that can be divided into two parts: a *central processing unit* and a *memory system*. The CPU reads the program and carries out (or *executes*) the operations specified by the program. The memory system stores two types of information: the *program* and the *data* acted on or created by the program. It isn't necessary to store both the program and data in the *same* memory. However, most computers do store programs and data in a single memory system. As we have said, such computers are called *stored program machines*.

A computer is a black box[1] that moves *information* from one point to another and processes the information as it goes along. The word *information* refers to the data and the instructions held inside the computer. Figure 1.3 shows two *information-carrying* paths connecting the CPU to its memory. The lower path with the single arrowhead from the memory to the CPU indicates the route taken by the computer's *program*. The CPU uses these buses to read the sequence of commands that make up a program one by one from its memory. I must stress that this is a simple conceptual description, because a modern high-performance computer may copy several instructions from memory in one operation in order to improve performance.

The upper path in Figure 1.3 with arrows at *both* ends transfers data between the CPU and memory under the control of the program. Information flow is *bi-directional*; during a *write cycle*, data generated by the program flows from the CPU to the memory, where it is stored for later use. During a *read cycle*, the CPU requests data, which is transferred from the memory to the CPU.

Figure 1.4 illustrates program execution and is less complicated than it looks. The purpose of this diagram is to demonstrate how an instruction like Z=X+Y is read from the memory and passed to an *interpreter* unit that generates the control signals, causing this instruction to be carried out. In Chapter 7, we examine how the instruction is converted into the sequence of actions required to *interpret* it.

> **Background**
>
> The computer reads instructions from memory and interprets them (i.e., carries out or executes the actions they define). When an instruction is executed, it may read data from memory, operate on data, and store data back in memory.
>
> Registers are internal, single-element locations that hold data. The clock provides a stream of pulses and all internal operations take place when triggered by a clock pulse. The clock rate is a factor in determining the speed of the computer.

---

[1] I am using 'black box' rather ironically. Black box is a term used to describe a system that performs a function, but its internal operation can't be accessed and observed. The flight data recorder that stores an aircraft's operational parameters is also colloquially known as a black box.

FIGURE 1.4   Executing a program

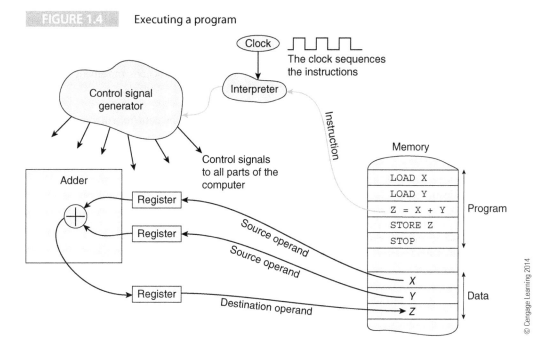

Imagine that a program reads two numbers (X and Y) from memory, adds them together, and then stores the result back in memory. The CPU must first *fetch* an instruction from memory. Once the CPU has analyzed or *decoded* the instruction, it has to read any data the instruction requires from memory. In this example, the first instruction, LOAD X, reads X from memory and stores it in a temporary holding place for data called a *register*. The second instruction reads Y from memory and stores that in another register. When the third instruction is fetched, the contents of the two registers are added, and the result is loaded in a third register. The fourth instruction transfers this sum to memory location Z.

It would be reasonably true to say that all a computer does is to read data from memory, perform operations on the data (addition, multiplication, etc.), and then put the results of any calculations back into the memory. The only other thing a computer can do is to test data (i.e., determine whether it is zero or not zero, negative or positive, etc.) and then choose between two alternative streams of instructions, depending on the result of the test.

An important theme of this text is the ways in which modern computers speed up the type of operations described in Figure 1.4. In Chapter 7, we will demonstrate how real computers begin processing a new instruction before they've finished the previous one. Computers like the AMD Phenom II, Xeon, or Core i7 can execute several instructions at the same time.

Few computers have two independent information paths between the CPU and memory as Figure 1.3 suggests. They have only one path along which information flows between the CPU and memory system, and data and instructions have to take turns using this path. We depict two paths in Figure 1.3 to emphasize that the memory system stores both the *instructions* that make up a program and the *data* used by the program. A computer that actually does store data and instructions in different memories (or has separate buses for data and instructions) is said to have a *Harvard architecture*.

Computer Instructions

Computer instructions involve a simple operation on data such as MOVE A, B that copies data from location *B* to *A* or an operation like ADD A, B, C that adds *B* and *C* to get *A*. Computers also implement the Boolean logical operations AND, OR, and NOT that we introduce in

(continued)

Chapter 2. One of the most important classes of computer operation is the *conditional operation* or *flow control instruction* that allows you to select one of two different courses of action (for example, if $x = 4$, then $y = 5$, else $y = 6$). Flow control instructions implement high-level language constructs like `IF...THEN...ELSE` or `REPEAT...UNTIL`.

In Figure 1.4, the operations that access memory are `LOAD` (copy data from memory to the CPU), `ADD`, and `STORE` (copy data from the CPU to memory).

Although a computer may be able to execute hundreds of different instructions, the following five basic instructions categorize all computer instructions. For example, the `ADD` instruction includes all data processing instructions such as subtract, multiply, and even *return the maximum of A and B*. Computer design is concerned with creating a set of instructions that perform computation efficiently. As we shall see, there is a trade-off between adding more and more clever instructions to perform all manner of complex calculations and the cost and complexity incurred by adding these instructions.

```
MOVE    A,B   move B to A
LOAD    A,B   put the value in memory location B in register A
STORE   A,B   put the value in register B in memory location A
ADD     A,B   add together A and B and put the result in A
TEST    A     test whether A is zero or not
BEQ     Z     if the last test was true, execute the code at location Z; otherwise continue
```

Computers perform operations on the entities: actual data itself such as the number 5, the contents of a register, or the contents of a memory location. It's as simple as that.

1.2   Architecture and Organization

This text is called *Computer Systems Architecture*. Having established what we mean by *computer* and by *system*, we must define the rather more elusive term *architecture*. *Computer architecture* implies *structure* and tells us something about the way in which the elements of a computer fit together. The problem of defining computer architecture arises because it is difficult to say exactly what we mean by structure, since different people view the computer from entirely different perspectives. For example, a secretary might view a computer as a clever word-processing machine, whereas a physicist might view it in terms of the behavior of electrons in crystals.

We are not interested in these two extreme viewpoints. Computer architecture is generally thought of as the *programmer's view of a computer*. What the programmer sees is an *abstract* view of a computer, and the actual hardware and implementation of the computer is hidden from him or her. For example, a programmer can instruct a computer to *add A and B* without knowing how this operation is actually going to be carried out. This abstract view of computer architecture is now generally called its ISA or *instruction set architecture*.

The Register

A register is a memory element that holds a single unit or word of data. A register is specified in terms of the number of bits it holds, which is typically, 8, 16, 32, or 64. Many of the computers we discuss in this text have either 32-bit or 64-bit-wide registers.

There is no fundamental difference between a register and a word in memory. The practical difference is that registers are located within the CPU and can be accessed more rapidly than external memory.

If you look at computer literature, you will encounter the term *organization* as frequently as *architecture*. Although architecture and organization are sometimes used interchangeably, they do have distinct meanings. A computer's *organization* represents the *implementation* of its *architecture*. A software engineer might say that computer organization is the *instantiation* of computer architecture (that is, it is the *abstract* made *real*). Today, the term *microarchitecture* is often used rather than organization. The difference between architecture and organization can be seen from an everyday example. A clock tells the time; its architecture is defined by the moving hands on a marked dial. Internally, the clock's organization may be mechanical with a flywheel or pendulum, electronic with a quartz crystal, or even externally controlled by radio waves.

The clock example demonstrates that a given architecture can be implemented with different organizations. For example, an ancient clock in a tower can be renovated by replacing its worn clockwork with an electric motor without the observer noticing that anything has changed.

> The code executed by a computer is expressed as strings of 1s and 0s and is known as *machine code*. Each type of computer will run only a specific machine code. The *human readable* version of machine code (e.g., ADD R0,Time) is called *assembly language*. Code that will run on entirely different types of computers and bears little relationship to the underlying computer architecture is called a *high-level language* (like C or Java). High-level languages have to be compiled into a computer's native machine code before they can be executed.

Similarly, a microprocessor with 32-bit *registers* may be organized internally as a 16-bit machine with 16-bit data highways that transfer information in packets of 16 bits at a time together with 16-bit functional units. If the programmer tells the computer to copy the 32-bit contents of register A into register B, he or she requests an operation on 32-bits. The machine executes the instruction as two 16-bit operations that are entirely invisible to the programmer. From this example, we can say that the machine has a 32-bit architecture but a 16-bit organization.

It would be wrong to divorce architecture and implementation entirely, because each has an influence on the other. A given architecture may be best suited to implementation using technology X and another architecture may be best suited to implementation with technology Y, even though you could implement each architecture using technology X or Y. In this text, we don't delve into the *lower levels of organization and implementation* (that is, the organization of the computer at the level of logic gates and circuit elements). That belongs to the electronic engineer. However, we do look at the *control unit* that is responsible for interpreting instructions, since it has had a strong impact on the development of computer architecture.

> An instruction set architecture includes *data types* (the number of bits per word and the interpretation of the bits), the registers used to hold temporary data, the type of instructions and their formats, and the addressing modes (ways of expressing the location of data in memory).

An ambiguity arises when we say that computer architecture describes the computer as seen by the programmer. What is a programmer? Someone who programs a computer in its *native assembly language* sees a very different machine from one who programs the machine in a high-level language like C or Java. Even the C programmer sees a radically different architecture from the Prolog or the LISP programmer.

> *Microcode* has nothing to do with microprocessors. Microcode defines the set of primitive operations that are carried out in order to interpret machine code. A typical machine instruction is ADD P,Q,R, whereas a microinstruction might be as simple as "move data from register *X* onto bus *Y*." Microinstructions are the province of the chip designer.

We cover microprocessor architectures at the register level (i.e., the assembly language programmer's view), introduce some of the aspects related to processor implementation (e.g., pipelining and cache memories), and also cover some of the aspects related to high-level languages (e.g., the stack and data structures). We concentrate on microprocessor architectures rather than mainframe architectures, because microprocessors are found in desktop and minicomputers, whereas the mainframe computer is increasingly employed in specialized applications. Today, it's probably wrong to talk about mainframe computers, as they belong to an era before the low-cost, small microcomputer. We use the term supercomputer to describe the immensely high-powered computer used by scientific and military applications that still requires its *own building* as the mainframe once did.

Convergence—Divergence—Which Way?

A theme of this text is the way in which computer architectures have been subject to the forces of both *convergence* and *divergence*. At times, computer architectures have diverged or grown apart. We saw that trend in the 1980s with the emergence of the RISC processor with its compact, high-performance structure that diverged from the traditional CISC structure of the Intel and Motorola architectures.

On the other hand, in recent years, RISC architectures (like the ARM) are converging with conventional CISC processors by introducing highly complex, special-purpose, and applications-oriented instructions. Computer architecture and organization are driven by the same economics as department stores—some specialize and sell a narrow range of products and some generalize and sell everything.

Similarly, CISC architectures are internally implemented by RISC-like microarchitectures that rely heavily on pipelining (the architectural feature probably most closely associated with RISC).

In this text, we use the term *architecture* to refer to the computer's abstract ISA (its instruction set) and the term *organization* to refer to the computer's implementation in terms of its actual hardware. However, organization will also be used to describe the entire computer, including its CPU, memory, buses, and input/output mechanisms. Finally, we will use the term *microarchitecture,* which has been popularized by Intel in the description of its Pentium and later processors to describe the realization of a CPU. Clearly, there is an overlap in the use of organization and microarchitecture.

The Dasgupta View

It's instructive to look at what others have said about computer architecture. In the late 1980s, Subrata Dasgupta was one of the first to discuss the relationship between architecture and organization in great detail. He talked of computer architecture as the *art, craft, and science of the design of computers* and coined the terms exo-architecture, endo-architecture, and micro-architecture to divide computer architecture into three levels.

Exo-architecture refers to the higher-level aspects of computer architecture such as data types and the instruction set. The prefix exo implies external and refers to the abstraction of the computer seen by the assembly level programmer. Endo-architecture refers to the internal organization of a computer and encompasses the performance of the computer's principal units, the way in which components are connected, and the way in which the flow of information is controlled. That is, the endo-architecture describes a processor at the level of its functional units including registers, adders, and control circuits.

Dasgupta's third component of computer architecture was the micro-architecture that is an expression of the actions necessary to execute a machine-level instruction (e.g., move data from one register to another). The operations carried out by the endo-architecture are realized by the micro-architecture. For example, an endo-architecture is concerned with what an adder does, whereas a micro-architecture is concerned with how the ALU does it. Thus, an exo-architecture is the programmer's abstract view of a computer that is implemented by an endo-architecture, which is itself realized by the execution of micro-programs that run on a micro-architecture.

We can summarize Dasgupta's view by saying that computers can be described in terms of hierarchical levels of abstraction. The exo-architecture is the highest level and represents the programmer's view of the computer. Endo-architecture represents the middle level that describes the organization of a computer in terms of its building blocks and their interconnection. Micro-architecture (the lowest level) describes how the building blocks are implemented by logic gates and is the province of the digital design engineer.

Chip Weams of UMASS feels that Dasgupta's layering is no longer relevant and makes the observation:

> "These three levels are non-standard, partly because they are poorly distinguished. The ISA/microarchitecture split is much easier to explain . . . The ISA is the contract between the architect and the programmer, which explains to the programmer how the machine appears to behave with respect to correctness of program execution. The microarchitecture is the internal implementation of this contract, but can result in different levels of performance. The Exo-architecture cannot fully describe the machine without referring to a subset of the Endo-architecture's elements (registers, memory, user/supervisor, traps, interrupts, branch hints, etc.) The common use of the term microarchitecture encompasses Dasgupta's use plus a large part of his endo-architecture."

1.2.1   Computer Systems and Technology

Figure 1.5 shows some of the factors that affect the design and performance of a computer. The box labeled *Technology* indicates the importance of the processes used to manufacture computer components (for example, the way chips are manufactured determines their speed

FIGURE 1.5   Factors affecting the computer designer

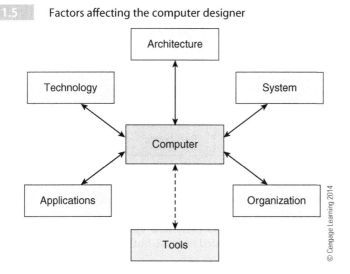

and power consumption). The speed of a computer plays a major role in the design of the rest of the system, because there's little point in having a fast processor and slow memory. Similarly, power consumption determines the range of applications of the computer (fixed location or mobile).

Since the 1970s, semiconductor technology has progressed according to Moore's law, which is the empirical observation that states that the number of devices per chip doubles every 18 months. This law has allowed chip manufacturers to begin the design of future processors even though the required manufacturing technology does not yet exist. The term "Moore's law" is widely referred to in articles on computing, although not everyone uses it with its original meaning (i.e., the number of components per chip doubles every 18 months). Colloquially, Moore's law is used to imply that the *performance* of processors doubles every 18 months or so.

The box labeled *Applications* in Figure 1.5 refers to the end use of the computer. Some computers are found in embedded control systems in automobiles, some in games machines, and others in the home and office. If different computers do different things, it is reasonable to assume that the intended application may have some influence on the design of a computer's architecture and its organization.

The box labeled *Tools* has been included to demonstrate that this is a factor in the design of computers although it is not part of the computer itself. Computers are used to design computers. Computer tools cover a range of software products, from packages that perform hardware design at the circuit level, to computer simulators, and to suites of programs used as *benchmarks* or test cases to compare the speeds of different computers.

State-of-the-art computers are constructed using state-of-the-art technology. Figure 1.6 illustrates some of the *technologies* of interest to the computer designer. *Device* technologies determine the speed of the computer and the capacity of its memory systems; these are the semiconductor technologies used to fabricate the processor and its main memory; magnetic technologies used for hard disks; optical technologies used for CD-ROMs, DVD, and Blu-ray drives; and also network links.

FIGURE 1.6   Computer technologies

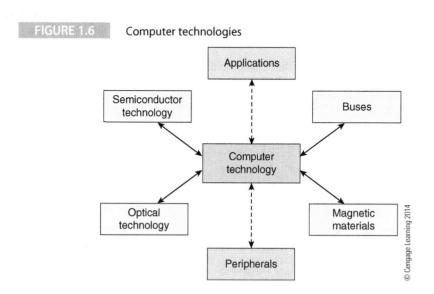

© Cengage Learning 2014

Figure 1.6 includes computer buses, because their structure, organization, and control have a profound effect on the performance of a computer. *Peripherals* (e.g., modems, keyboards, printers, and display systems) and *applications* (e.g., desktop publishing, graphics, and multimedia) are included because both of these affect the design of a computer system.

1.2.2   The Role of Computer Architecture in Computer Science

Some computer science students will graduate and find work in the computer industry, possibly working with applications such as databases, networks, web design, video post-production, or security. Although few graduates will ever design computer systems, you cannot separate the computer from computer science any more than you can isolate an airline pilot from aerodynamics and jet engine design. Even though a pilot may not need to know about aerodynamics or engine operation under most normal situations, he or she will find the knowledge most useful when things go wrong and the situation becomes distinctly abnormal.

A course on computer architecture provides an overview of how the computer works and what it does and teaches students the operation of a typical stored program computer. A good course should highlight the important issues facing today's designers and provide students with the tools they will need to carry out research. Moreover, the student who understands how a computer operates is better able to adapt to demanding situations. For example, an understanding of cache memory (Chapter 9) would enable a programmer to write faster programs by optimizing the way in which data is accessed.

The computer lies at the heart of computer science—without it, computer science would be a branch of theoretical mathematics. Students shouldn't regard the computer as just a black box that executes programs by magic. As well as the philosophical, there are concrete reasons why computer architecture is still relevant to all students of computer science and to practitioners.

An appreciation of computer architecture has important implications for the computer professional. Suppose a graduate enters the industry and is asked to select the most cost-effective computer for use throughout a large organization. Understanding how the individual elements of a computer contribute to its overall *performance* is vital. For example, is it better to spend $30 on doubling the size of the cache or $60 on increasing the clock speed by 200 MHz?

Computer architecture cannot be divorced entirely from software. The majority of processors are found not in PCs or workstations but in *embedded* applications such as MP3 players and cell phones. Those designing multiprocessors and real-time systems have to understand fundamental architectural concepts and the practical limitations of commercially available processors. Someone developing an automobile electronic ignition system may write his or her code in C, but they also might have to debug the system using a logic analyzer that displays the relationship between events taking place in the engine and the execution of the machine-level code.

Another reason for studying computer architecture is that it underpins many important ideas from other areas of the computer science curriculum. An appreciation of computer architecture helps students understand concepts in other areas of computer science via recurring themes. For example, studying how the computer provides architectural support for high-level languages reinforces aspects of the languages and compilers curriculum. In Chapter 4, we will look at how the underlying machine architecture supports *stack frames* and *parameter passing* in languages like C. Similarly, covering the design of computer buses introduces important topics such as *fairness versus priority* that are also found in courses on operating systems.

The Clock

We need to define the term *clock*. Most digital electronic circuits have a clock that generates a continuous stream of regularly spaced electrical pulses. It's called a clock because the pulses are used to time or sequence all events within the computer. For example, a processor might execute a new instruction each time a clock pulse arrives.

A clock is defined in terms of its *repetition rate* or *frequency*. Typical clock frequencies in computers range from 1 MHz to about 4.5 GHz. Clocks are also defined in terms of the

(continued)

width or duration of a clock pulse, which is the reciprocal of its frequency (that is, $f = 1/T$). For example a 1 MHz clock has a duration of 1 µs, and a 1 GHz clock has a duration of 1×10^{-9} s or 1 ns. A 5 GHz clock has a period of 200 ps (ps = picoseconds). Light travels approximately two inches in 200 ps.

Digital circuits whose events are triggered by a clock are called *synchronous*, because they are synchronized with a clock. Some events are asynchronous, because they can happen at any time. For example, if you move the mouse, it sends a signal to the computer. That is an asynchronous event. However, the computer may check the status of the mouse at each clock pulse, which is a synchronous event.

Solving New Problems from First Principles

I illustrate the importance of understanding underlying basic principles to my students by telling the following true story about a pilot who solved a new problem just by using all of the available information. A Cessna light aircraft was hopelessly lost over the Pacific Ocean with the pilot apparently doomed to the proverbial watery grave. The Cessna lacked any navigation equipment. Sunset was approaching, and it was highly probable that it would crash into the Pacific with little chance of the pilot surviving. The Cessna pilot had only a VHF (line-of-sight) radio with a 200 mile range and a magnetic compass. An Air New Zealand DC10 picked up his distress call, and the DC10 captain decided to hunt for the light aircraft without any obvious means of locating it.

At this stage, I ask my students, "What do you think the DC10 captain did?" Invariably, they are completely clueless; even pilots are not taught what to do in such a situation. The DC10 captain turned his aircraft to face the sun and asked the lost pilot to do the same. The DC10 reported the direction of the sun as 270° and the Cessna gave a bearing of 274°. This meant that the Cessna was to the south of the DC10, but it didn't tell the DC10 captain whether the Cessna was to his east or west. So, the DC10 asked the Cessna pilot to hold his hand out and say how many fingers the sun was above the horizon. The DC10 captain measured the sun two fingers above the horizon and the Cessna saw it four fingers above the horizon. If you think about it, the higher the sun in the sky, the closer it is to noon, so the local time for the DC10 was later than the Cessna's local time. This put the Cessna to the west of the DC10.

Now that the relative positions of the two aircraft had been established, the DC10 pilot flew towards the general direction of the Cessna and recorded the points at which the Cessna's radio began to fail. The figure illustrates how, by flying along a straight line, the DC10 could determine a chord in a circle whose radius is the 200 mile transmission range of the Cessna. The DC10 flies in a straight line, and the point at which the signal is first received and the point at which it is no longer received define a chord. The DC10 then reverses track, flies to the center of the chord, and turns at right angles to fly along the diameter. In this way, the Cessna was located and escorted to New Zealand.

I use this example because it is interesting, students with no knowledge of aviation can follow it, and it demonstrates sheer human ingenuity. The point I am attempting to make is that by understanding all aspects of their profession and by thinking about 'unusual circumstances', they are in a better position to solve real-world problems than those who lack in-depth understanding of the world they inhabit.

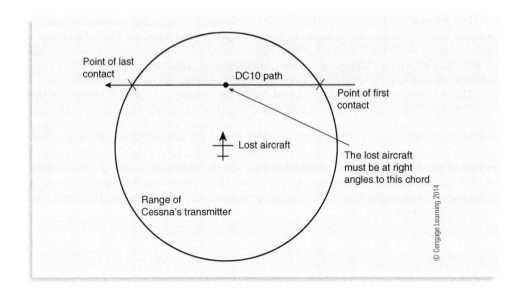

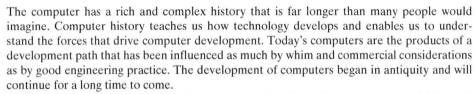

1.3   Development of Computers

The computer has a rich and complex history that is far longer than many people would imagine. Computer history teaches us how technology develops and enables us to understand the forces that drive computer development. Today's computers are the products of a development path that has been influenced as much by whim and commercial considerations as by good engineering practice. The development of computers began in antiquity and will continue for a long time to come.

Computer history is important because we need to know of past successes and failures if we are to make the best use of today's resources. For example, an important lesson from computer history is the role of backward compatibility in the development of the PC. Backward compatibility refers both to the need for new technology to be compatible with that of the past in order to allow users to easily upgrade computers and to the need for future computers to be compatible with the large existing codebase.

Moreover, some of the ideas that have emerged in the past only to be discarded may be more viable in the future as technology improves. It would be nice to begin with an extensive computer history. Since some students might find a large lump of history a little indigestible, we cover some of the significant milestones in the development of the computer in this chapter and add further details throughout the text. Here we deal only with the early history and background of the computer, because the details of modern developments are integrated into the rest of this book and the accompanying website.

1.3.1   Mechanical Computers

Humans are computational animals. Math was probably invented by cave dwellers not to solve Sudoku problems on cold, wet days but to measure fields, construct buildings, and send out tax demands. Romans used pebbles (or calculi in Latin) in little trays with depressions to represent numbers. Later, calculation was performed by sliding pebbles along a wire to aid addition or subtraction. Even I have visited stores in Central Asia where people used abacuses to rapidly perform calculations.[2]

---

[2]In 1960, Arthur C. Clarke wrote a science fiction story called "Into the Comet" where a spacecraft investigating a comet lost computer power and was unable to get home. A Japanese journalist on board remembered the abacus from his childhood and the crew used them to solve the equations necessary to get home.

The French mathematician Blaise Pascal developed a primitive mechanical adding machine in 1642 that essentially used clockwork to perform addition. By 1694, the German mathematician Gottfried Wilhelm Leibnitz had constructed a sophisticated mechanical calculator that performed addition, subtraction, multiplication, and division. These devices were not computers in today's sense of the word, because they could not be programmed.

The notion of *programming* developed during the industrial revolution because of the need for industrial control. In 1801, the Jacquard loom was constructed to weave pre-defined patterns in cloth automatically—something previously done by skilled workers. Punched wooden cards were used to control the weaving of patterns in textiles, where a hole or no hole at a certain point in the card determined whether the horizontal thread went in front of or behind the vertical threads. The punched card was the program, because a given pattern of holes specified a unique sequence of operations. Those operations generated patterns in cloth, but it was inevitable that it would occur to someone that the operations could be mathematical calculations.

Analog Control — The Governor

Automatic control predates the punched cards of the Jacquard loom. The very first machines of the industrial revolution were the steam engines that powered pumps in coal mines, looms in factories, and various transportation devices. The speed of a steam engine is controlled by the amount of steam entering its cylinders. You could control an engine's power by getting someone to open a valve when the speed dropped under a high load or to close a valve when the speed increased if the load fell. However, that was not easy, reliable, or safe.

James Watt introduced an automatic control mechanism to keep the speed of engines constant under varying loads. The analog speed governor is delightfully clever. A vertical spindle is rotated by the machine. The spindle has a horizontal arm from which two heavy metal balls hang on pivots. As the spindle rotates, the balls swing out—the faster the rotation the greater the movement. The arms from which the balls hang control the steam flow. If the speed drops, the balls move in and the steam is turned on. If the balls move out, the flow of steam is reduced. The speed now maintains an equilibrium point.

The principles behind Watt's governor are exactly those that control an aircraft's automatic landing system today.

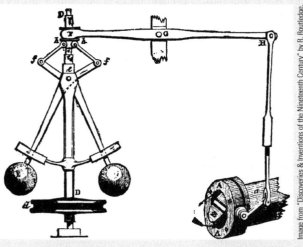

Image from "Discoveries & Inventions of the Nineteenth Century" by R. Routledge, 13th edition, published 1900.

In the early 19th century, there was a need for computers to construct the tables required for navigation. Computers did exist in the 19th century. In those days, a computer was a person employed to carry out arithmetic calculations such as the evaluation of trigonometric functions. Charles Babbage in England (1822) proposed a calculating machine called the *difference engine* to automate the evaluation of polynomials needed to calculate mathematical tables. Babbage never completed his difference engine; that was done by Per Georg Scheutz in Sweden (1855).

Babbage went on to design a machine called the *analytical engine*, which was only partially built (before his death in 1871) to test whether his concept was viable. The analytical engine was an entirely mechanical device that had an arithmetic processing unit to perform calculations, a means of inputting and outputting data, and a memory based on punched cards to store both data and programs.

Babbage's analytical engine is important because he conceived of the computer long before the age of electronics and some of the early pioneers of computing were inspired by Babbage's work (other pioneers were unaware of Babbage's work and therefore they re-invented the computer). Indeed, some claim that Babbage's analytical engine was more advanced and closer to the modern computer than the early electronic computers of the 1940s.

Augusta Ada King was an English writer and mathematician who worked with Babbage on his analytical engine. In 1842, she translated a memoir written by the Italian mathematician Luigi Menabrea on the analytical engine. As part of this work, she reproduced notes that included a means of using the analytical engine to compute Bernoulli numbers. Because of this work, Ada is generally regarded as the first person to write a computer program. It took another 100 years of technological progress before people really could write programs and then run them.

Principle of the Difference Engine

Babbage's difference engine employed *finite differences* to calculate polynomials of the form $a_0x + a_1x^1 + a_2x^2 + \ldots$. The table below demonstrates how you can use the method of finite differences to create a table of squares without having to use multiplication. The first column contains the natural integers 1, 2, 3, ... The second column contains the squares of these integers (i.e., 1, 4, 9, ...). Column 3 contains the *first difference* between successive pairs of numbers in column 2. For example, the first value is $4 - 1 = 3$, the second value is $9 - 4 = 5$, and so on. The final column is the *second difference* between successive pairs of first differences. As you can see, the second difference is always 2.

We can calculate the value of 8^2 using finite differences. We use the table in reverse by starting with the second difference and working back to the result. If the second difference is 2, the next first difference (after 7^2) is $13 + 2 = 15$. Therefore, the value of 8^2 is the value of 7^2 plus the first difference, (that is, $49 + 15 = 64$). We have generated 8^2 without using multiplication. This technique can be extended to evaluate other mathematical functions.

Number	Number Squared	First Difference	Second Difference
1	1		
2	4	3	
3	9	5	2
4	16	7	2
5	25	9	2
6	36	11	2
7	49	13	2

1.3.2   Electromechanical Computers

The development of the telegraph and the telephone later in the 19th century led to the design of automatic telephone switches and communication networks. Telephone switching mechanisms used electromechanical switches called *relays* that behave liked today's binary digital logic switching elements, and these could be used to construct electromechanical computers. The term *electromechanical* describes components that have moving parts but which are controlled electrically. For example, the relay uses a coil of wire that magnetizes a piece of iron which then activates a switch.

The introduction of the typewriter in 1867 and of the Hollerith census tabulator (punched card reader) in 1879 aided the development of the electromechanical computer. An electomechanical computer is the missing link between the mechanical era and the electronic era of vacuum tubes, transistors, and integrated circuits. Some regard Konrad Zuse as the inventor of the digital computer. His computers were constructed in Germany in the 1940s and used to design aircraft in World War II. However, much of his work was destroyed by Allied bombing. For a long time, Zuse's work was largely unknown in the computing world. Zuse's computer was the first programmable computer; other contemporary machines were not software programmable and were really automatic calculators. Zuse also developed the world's first programming language called Plankalkül.

In 1944, Howard Aiken at Harvard University developed the Mark I electromechanical computer to compute artillery trajectories. Mark I was an early programmable calculator but did not support conditional operations, so it was not a computer in today's terms.

1.3.3   Early Electronic Computers

High-speed automatic computation wasn't possible until the vacuum tube amplifier replaced the far slower relay. From 1937 to 1942, John V. Atanasoff constructed the first electronic computer (ABC) which was used to solve linear equations. Another early computer was Colossus, which was fully functioning by 1944 and was used at Bletchley Park to decode German Enigma messages in WW2. Colossus used vacuum tubes and was a true electronic digital computing machine, but it lacked a stored program and was dedicated to a single task.

In 1945, J. Mauchly and J. Eckert developed ENIAC, which was a vacuum-tube computer that could manipulate 10-digit decimal numbers, although it could not be programmed in the sense we use programming now. ENIAC carried out pre-determined operations that were hard wired into its circuits. ENIAC could be programmed only by rewiring, albeit by using patch cables and switches.

A more advanced machine, the EDVAC, was also developed by Eckert and Mauchly and featured a stored program. In England, researchers at Manchester University (1948) developed the Manchester Baby, the world's first operational *stored-program computer*. A stored-program computer or *von Neumann machine* forms the basis of today's computers and is characterized by a memory that holds both instructions and data. Baby was subsequently enhanced by Ferranti Ltd. to produce the EDSAC, which was Europe's first stored-program computer.

The development of the transistor at AT&T Bell Labs in 1948 created the semiconductor equivalent of a vacuum tube which was far smaller and consumed much less power. Once the

Colossus

One of the least-known early computers and a contender for the title *first computer* was Colossus, which was constructed in the UK in 1943. The reason for its obscurity was that all versions of it were destroyed by the British government and it was classified as an official secret until it was declassified in 1975.

Colossus was a special-purpose electronic digital computer using vacuum tubes designed to decode German cyphers during World War II. One of the key developers of Colossus was Terry Flowers, whose background had been in telephone switching networks.

Although Colossus had some of the characteristics of a general-purpose digital computer, it was programmed externally by switches and patches (wired links).

transistor had been created, it was inevitable that multiple transistors would be placed on a silicon chip to create an entire circuit.

By the mid-1960s, IBM had developed its System/360 architecture, which supported compatibility and interoperability across a product line that ranged from small business computers to large scientific machines. It was the IBM System/360 that led to the notion of *computer architecture* (i.e., the ISA).

1.3.4   Minicomputers and the PC Revolution

The mainframe computer was so large and expensive that it was *institutional*—bought only by large organizations and requiring a team of people to operate it. The development of integrated circuit technology paved the way for minicomputers [such as the PDP-8 produced by Digital Equipment Corporation (DEC)]. In the 1970s, departments within a university or small organizations could afford to buy and run a minicomputer.

The microprocessor was a natural step forward. Increased integration allowed all elements of a computer (apart from peripherals and memory) to be integrated on a single chip. By the 1970s, 8-bit microprocessors were available from Intel and Motorola. The first practical microcomputer kit, the Altair 8800, was marketed in 1975 by MITS, Inc. This was soon followed by the Apple I (1976) and Apple II (1977), which were the first commercially viable microcomputers marketed with useful software and peripherals.

Motorola developed the 68000 (a 32-bit microcomputer), upon which Apple based its innovative Macintosh computer in the early 1980s. Other 68000-based machines were developed by Atari (foreshadowing gaming technology) and Commodore (the Amiga was a forerunner of multimedia computers). For better or worse, IBM's adoption of the Intel architecture killed off Motorola's processors. There was a time when Motorola and Intel were neck-and-neck in the industry. Had IBM adopted the 68000 rather than the 8086, Motorola might have become what Intel is today. Motorola's 68000 (and its 8-bit processors) did not disappear. Freescale took over Motorola's microprocessors and marketed them under the name ColdFire. Motorola's processors live on but are largely found in embedded applications such as laser printers and automobiles.

IBM introduced the IBM personal computer (PC) in 1980 based on Intel technology and Microsoft operating software. Because of its *open architecture*, the PC was popular with third-party software and hardware developers. Intel expanded the 8080 microprocessor to include the 16-bit 80286 (1982) and 32-bit 80386 (1985), followed by the Pentium with a 64-bit data bus (1990). By 2000, Intel largely dominated the PC market by managing to keep ahead of its competitors with imaginative design, technological innovation, and aggressive marketing.

> **Open-Architecture**
>
> An open architecture can accept plug-in systems and devices from third-party manufacturers. Apple's computers were not open architecture, and Apple maintained control of all aspects of their computers at the expense of market penetration. You could argue that the openness of the IBM PC versus Apple was more legal than technical. IBM did not seek to use the law to protect the PC from clone makers, whereas Apple did

Interestingly, some manufacturers such as AMD and later the ill-fated Transmeta created logically equivalent versions of Intel's processors. Such chips execute the same machine code as Intel processors, but they do so by converting Intel's machine language into their own machine language before executing these *native* instructions. Some would say that even Intel's chips no longer execute Intel code, because instructions are internally translated into more primitive operations (although that is a simplification of the actual situation because not all instructions are translated). We look at the detailed internal operation of processors in Chapter 7.

1.3.5   Moore's Law and the March of Progress

The term Moore's law was coined by Carver Mead in 1975 following an observation by Gordon Moore that the number of components on an integrated circuit doubles every two years. This is, of course, an empirical observation, but the march of technology has indeed led to an

Out-of-Order Execution

Like instructions in a recipe, the instructions of a program are executed one-by-one in sequence and in the order in which they occur in a program. Consider the following sequence of algebraic expressions (the instructions).

1. $X = 2 * C$
2. $Y = C + 4$
3. $Z = X + Y$
4. $A = 4 * C$
5. $P = C - 3$

We can sometimes increase the speed of a computer by executing instructions out of their natural order. In this example, we can execute (4) and (5) at any point, but instruction (3) must be executed after (1) and (2) have been completed.

exponential growth in the number of transistors per chip over the last 40 years. This growth has been accompanied by a corresponding increase in the speed of integrated circuits. Increasing the number of transistors in an integrated circuit has led to a dramatic improvement in architectural sophistication and some very clever ways of accelerating performance.

In the 1990s, computer *instruction set architecture* was not advancing particularly significantly—the instruction set of the Pentium series is not that different to Intel's 80386. The success of the PC ensured that future processors had instruction sets that were backward compatible with older processors in order to maintain software compatibility across generations. However, the microarchitecture of Intel's processors was anything but static. Quite astonishing progress was made in the field of microarchitecture.

The 1980s saw a change in computer organization—called the *RISC revolution*—when designers attempted to create more streamlined processors. A key feature of RISC processors was the pipelining of instructions, where the processor is turned into an instruction execution factory with an automobile-style production line that can have four or more instructions in varying stages of execution at the same time. Although RISC processors were targeted at the high-end workstation market, they were never entirely successful, because the market for workstations was tiny compared to the PC market.

Intel threw itself with great enthusiasm into the development of ever more powerful processors and incorporated many aspects of the classic RISC processor into new versions of its own processors (the Pentium line of IA32 architecture chips).

Another change in computer organization was the introduction of *superscalar* processing and *out-of-order* execution. We will be looking at these topics later; at this point, all we need to say is that superscalar processing involves reading several instructions from memory and executing them in parallel. Out-of-order execution involves the execution of instructions in a different order to the order in which they appear in the program to speed up execution by not waiting for certain instructions to execute. Out-of-order execution allows an instruction later in the program to be executed if the current execution is waiting for resources that are currently in use. In terms of a recipe analogy, out-of-order execution is the same as making the desert while waiting for the main course to cook. Intel incorporated superscalar processing in its Pentium family.

Although the 1980s saw the great RISC versus CISC debate, computer scientists declared it a dead issue in the 1990s when the divergent RISC and CISC designs merged as RISC techniques were applied to CISC processors and CISC features added to RISC families.

1.3.6   The March of Memory Technology

High-speed processing is vital for today's computer applications, but without the fast, small, low-power, high-capacity memory needed to hold programs and data, the computer (as we know it) could not have emerged. In 1982, Albert Hoagland, who was one of the pioneers of digital data recording, said that Silicon Valley should be renamed *Iron Oxide Valley* in honor of the base-material covering the platter of disk drives.[3] I agree; it was the hard disk that made the PC revolution possible.

---

[3]A purist might argue that today's disks are coated with rather more esoteric materials than iron oxide and that the floppy disk popularized the IBM PC more than the hard disk in its early days. That is true, of course, but the spirit of Hoagland's comment is that data recording technology played just as important a role in the development of the computer as the well-known silicon chip.

In the 1930s John V. Atanasoff invented one of the first storage mechanisms, which was a rotating drum covered with capacitors that could hold a charge and store a 1 or 0. As the drum rotated, the capacitors passed under a row of contacts, and their values could be read. In the 1940s, the mercury ultrasonic delay line stored data as a series of ultrasonic sound pulses that travelled down a long narrow tube filled with mercury. When the pulses travelled from one end to the other, they were amplified and recirculated. This was true dynamic memory.

The first fast data storage device was invented by Frederick Williams at the University of Manchester in the UK. The cathode ray tube (used initially in radar displays and then in televisions) stored data on its face by illuminating a point with an electron beam to store a charge (this was an electronic version of Atanasoff's rotating drum). The first Williams tube could store only 1,024 bits but was later doubled to 2,048 bits.

In 1949, Forester developed the ferrite core memory in the US for the Whirlwind computer. The ferrite core is a tiny ring of a magnetic material that can be magnetized clockwise or counterclockwise. Ferrite core store became the mainstream memory for mainframes up to the 1970s. Indeed, ferrite core gave us the term *core store* that is still occasionally used to describe bulk memory. In the 1970s, semiconductor dynamic memory was developed as a replacement for core store and is now the standard means of data storage. Today, small DRAM modules (using several chips on a small circuit board) easily provide 8 GB of storage. We've come a long way from the 1,024 bits of the Williams storage tube (a factor of $2^{26} = 64$ million).

The hard disk used to archive programs and data has been around for a long time. IBM introduced a disk storage mechanism in 1956 in its RAMAC (*random access method of accounting and control*) that stored data by magnetizing the surface of a rotating disk. The RAMAC 305 disk stored about 5 MB of data and rotated at 1,200 rpm. Since then, the performance of disk drives has increased. The maximum capacity is now about 4 TB (2^{42} bytes), and a 1 TB portable external hard disk drive easily fits into a pocket (unlike the fridge-sized RAMAC). Because of the mechanical nature of a hard disk, its rotational speed is typically 7,200 rpm, which is only six times faster than the original RAMAC. Let me say this again: A modern disk drive has a million times the capacity of a RAMAC but is only six times faster. This little detail highlights one of the biggest bottlenecks in the development of the computer: The rate of technological advance is not constant across all the components of a computer. However, solid-state technology is now being used to replace hard disk drives with far faster entirely electronic drives (see Chapter 11).

Today, personal computers use optical storage (DVD or Blu-ray) to import programs or to archive data. Optical technology stores information along the track of a spiral grove on a transparent polycarbonate disk. Laser technology is used to read the information or to write it in the case of rewritable disks. CD technology was developed between 1958 (the invention of the laser) and 1978 (working optical storage). The DVD (introduced in 1997) is nothing more than an improved CD, and the Blu-ray disk (2006) is an improved DVD. In a few years, optical storage grew from a 600 MB disk to a 25 GB capacity. We look at storage technology in more detail later in this text.

1.3.7 Ubiquitous Computing

Today's world is characterized by the growth of *ubiquitous* or *pervasive* computing. These terms simply mean that computers are everywhere and in everything. For example, a modern automobile may have more than 50 different computers controlling everything from sophisticated satellite navigation systems to simple door interlocking mechanisms.

Ubiquitous computing is most visible in cell phones, MP3 players, digital cameras, and games consoles. An aspect of ubiquitous computing is the notion of *convergence,* where functionality is transported across the boundaries of various mobile devices. For example, the cell phone was devised as a communications system. However, since it had a computer, memory, keyboard, and display, it was easy to implement other functions such as an MP3 player and personal organizer. By incorporating more hardware (which is becoming ever cheaper), convergence can be extended to add GPS and digital camera functions to the cell phone.

Ubiquitous computing could also be called *power-aware* computing. The term power aware is synonymous with low-power computing. If systems are to be truly small and portable, they can't be connected to the power grid; they have to rely on batteries or other methods of generating their electrical power. Power constraints have led to a growth in the design of low-power circuitry and the means of reducing the power consumed by processors. Even the rise of multi-core processing is driven by power awareness, because you can increase computational power by using multiple cores more efficiently (electrically, speaking) than by raising the clock speed.

One of the products of power-aware computing is the eBook or electronic book that allows you to carry hundreds or even thousands of books in a device the size of a paperback. Although the eBook relies on low power CPU technology to do data processing and small, non-volatile high-speed flash memory to hold the books, it was the development of new power-aware display technology that made the eBook possible. Electronic ink (E-ink) uses millions of tiny microcapsules containing a clear fluid. Inside each microcapsule are tiny white and black particles with opposite electrical charges. By applying an electrostatic charge across the microcapsule, the white particles can be moved to the top, the black to the bottom, or vice versa. This allows the microcapsule to look white or black, respectively.

E-ink requires no power to operate it once the charge has been established (i.e., power is consumed only when *pages are changed*). Conventional LCD technology (e.g., that found in the iPad®) has the advantage of color displays and can be used in poor lighting conditions, but LCDs are power hungry because of the back lighting and are not easily visible in bright light.

1.3.8   Multimedia Computers

A feature of modern computers (both ubiquitous and traditional personal computers) is their ability to handle *multimedia*. Multimedia operations (processing and storing audio and video data) require large storage capacities and the ability to carry out vast numbers of simple repetitive operations on sound samples or pixels in real time. Improved computing power and storage technologies first made multimedia possible in desk top PCs and then in low-cost, light-weight personal devices (such as iPads).

In this text, we look at how modern processors exploit the nature of audio/video data (i.e., short wordlengths and repetitive operations) to enhance performance. In particular, we examine Intel's multimedia extensions (MMX) to its instruction set. We chose MMX to describe this topic because it was the first such addition to the Intel IA32 family. Later members of Intel's processor family have added further instruction extensions (such as SSE and SSE-2), and similar extensions have been added to AMD's processors.

As time passes, the degree of convergence is increasing with the introduction of media players where there is no longer any significant difference between a personal computer and a high-resolution video entertainment system.

1.4   The Stored Program Computer

This text looks at two high-performance computers in some detail: the ARM family and the Intel IA64 architecture. However, in this introduction, I don't want to treat the computer as a *fait acompli*. I want to show conceptually where it comes from. So, we are going to develop the structure of a very simple computer by posing a problem and then asking what we require to solve this problem. Although this problem is trivial, it illustrates the operations carried out by real programs and demonstrates the notion of instruction sequencing. The computer we will design to solve our problem will be a primitive version of real, stored-program computers.

1.4.1   The Problem

Consider the *string* of single-digit decimal numbers 23277366664792221 described by Figure 1.7. As you can see, there are *runs* of consecutive digits with the same value (for example, runs of 7s, 6s, and 2s). Our problem is simple: We want to find the length of the longest run of the same number. We will assume that the length of the string is greater than three in order to simplify the problem.

You can immediately see from Figure 1.7 that the answer is *four* because there are four 6s in a row in the string, but only two 7s and three 2s. We are going to design a *machine* that can take the string of Figure 1.7, read a digit at a time, and then tell you the *length of the longest run of consecutive digits of the same value*.

FIGURE 1.7   A string of digits

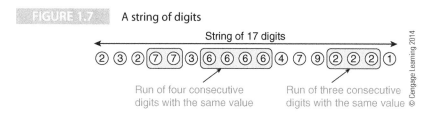

1.4.2   The Solution

If we start at the left-hand end of the string of digits and examine them one by one, at any point in the string, we are in one of two *states*: one state represents a *run* of digits with the same value and the other represents the *beginning* of a new run of one or more digits with the same value. In Figure 1.7, the digits in a run are enclosed in a shaded band to emphasize this point.

We can use the *state diagram* of Figure 1.8 to help us solve the problem. At any instant, a system may be in one of several possible states: For example, a person can be in the state *sleeping* or the state *awake*, or an aircraft may be in one of the three states: *climbing*, *descending*, or *level flight*. In the digital world, a change from one state to another occurs when an *event* (such as a clock pulse) takes place. State changes in Figure 1.8 take place when a new element in the string is read.

FIGURE 1.8   A state diagram for a run-length counter

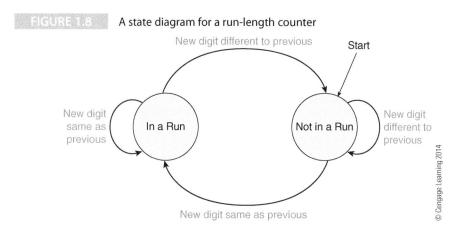

The two states in Figure 1.8 are labeled "*In a Run*" and "*Not in a Run.*" A *Run* state is defined as a condition in which two or more consecutive digits with the same value have been encountered. If we are in the "*Not in a Run*" state and the next new digit is different to the previous digit, we remain in the "*Not in a Run*" state. However, if the next digit is the same as the last one, we make a transition to the "*In a Run*" state, because we are at the second digit in a sequence. As long as each new digit is the same as the last one, each transition is back to this "*In a Run*" state. When the next digit is not the same as the last one, we make a transition to the "*Not in a Run*" state.

FIGURE 1.9   State changes when reading the string of Figure 1.7

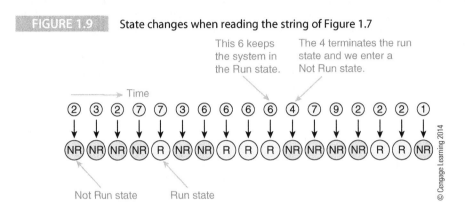

Figure 1.9 demonstrates the sequence of state changes as the digits of the string of Figure 1.7 are examined one by one. As you can see, a change of state occurs on the second digit of a run or on the digit that terminates a run.

Table 1.1 arranges the sequence of digits into a convenient form. The top row defines the *location* or *address* of each digit in the string S, and the next row provides the actual elements of S. For example, location 11 contains an element with the value of 4. The third row is the current run value; this begins with a question mark (?) because there is no current run value at the start of the string. As you can see, each element in the third row is the same as the previous element in the second row.

TABLE 1.1   Turning the String into a Table of Values

Position in String	1	2	3	4	5	6	7	8	9	10	11	12	13	14	15	16	17
Element Value	2	3	2	7	7	3	6	6	6	6	4	7	9	2	2	2	1
Current Run Value	?	2	3	2	7	7	3	6	6	6	6	4	7	9	2	2	2

Let's start at the left-hand end of the sequence of digits and look at the digits one by one. Each time we read a new digit we ask the question, "Have we just found a new run?" If we are at the start of a new run, the length of the current run is reset to one. If the digit is part of a run, the length of the current run is increased by one. We can now go through the digits in the string and record the current run length in Table 1.2, which varies from 1 to 4. Note that the *Current Run Value* in Table 1.2 is the value before the new element has been examined.

TABLE 1.2   The Current Run Length at Each Position Along the String of Digits

Position in String	1	2	3	4	5	6	7	8	9	10	11	12	13	14	15	16	17
Element Value	2	3	2	7	7	3	6	6	6	6	4	7	9	2	2	2	1
Current Run Value	?	2	3	2	7	7	3	6	6	6	6	4	7	9	2	2	2
Current Run Length	1	1	1	1	2	1	1	2	3	4	1	1	1	1	2	3	1

Whenever we start a new run, we ask whether the last run was *longer* than the longest run so far encountered. Table 1.3 shows how we do this by including a new row that keeps track of the value of the longest run so far. When we reach the 17th number at the right-hand end of the string, we can see that the maximum run encountered was four. Note that this problem assumes we know the length of the string and when to stop. If we didn't, we would need to know the last number, which would have to be unique.

TABLE 1.3   Expanding Table 1.2 to Include the Maximum Run Length

Position in String	1	2	3	4	5	6	7	8	9	10	11	12	13	14	15	16	17
Element Value	2	3	2	7	7	3	6	6	6	6	4	7	9	2	2	2	1
Current Run Value	?	2	3	2	7	7	3	6	6	6	6	4	7	9	2	2	2
Current Run Length	1	1	1	1	2	1	1	2	3	4	1	1	1	1	2	3	1
Maximum Run Length	1	1	1	1	2	2	2	2	3	4	4	4	4	4	4	4	4

© Cengage Learning 2014

1.4.3   Constructing an Algorithm

The next step is to provide an *algorithm* that tells us how to solve this problem in a clear and unambiguous fashion. As we step through the sequence of digits, we will need to keep track of some items of information. These items correspond, of course, to the rows of Table 1.3, and we give them the following *symbolic names* to make it easy for us to refer to them.

i	The current position in the string
New_Digit	The value of the current digit just read from the string of digits
Current_Run_Value	The value of the element in the current run
Current_Run_length	The length of the current run
Max_Run	The length of the longest run we've found so far

The following *pseudocode* expresses the actions we need to perform. The initialization operations that we perform once at the beginning are in black and the repetitive actions defined by the REPEAT . . . UNTIL are in blue. We've made a simplification to avoid complexity by omitting the index i and talking about the first or next digit instead of digit(i) or digit(i+1).

```
1.    Read the first digit in the string and call it New_Digit
2.    Set the Current_Run_Value to New_Digit
3.    Set the Current_Run_Length to 1
4.    Set the Max_Run to 1
5.    REPEAT
6.    Read the next digit in the sequence (i.e., read New_Digit)
7.    IF its value is the same as Current_Run_Value
8.            THEN Current_Run_Length = Current_Run_Length + 1
9.            ELSE {Current_Run_Length = 1
10.                Current_Run_Value = New_Digit}
11.   IF Current_Run_Length > Max_Run
12.           THEN Max_Run = Current_Run_Length
13.   UNTIL The last digit is read
```

This pseudocode employs two *constructs* found in many high-level computer languages: the REPEAT...UNTIL construct in lines 5 to 13 and the IF...THEN...ELSE construct. The REPEAT...UNTIL lets you carry out an action one or more times, and the IF...THEN... ELSE lets you select between one of two possible courses of action. The IF...THEN...ELSE construct is central to the operation of the digital computer, and you will encounter it many times and in many ways in this text. We will see how it is implemented at the hardware level and how it is expressed by several different computers. We will see how it has a negative effect on the performance of modern computers, and how some computers attempt to guess the outcome of this operation before it is actually computed. We also will see how some computers attempt to avoid it entirely by the use of predicated operations.

Algorithm, Code, and Pseudocode

An *algorithm* is a finite list of well-defined instructions that carries out an operation (e.g., calculating a function or performing a given task). Elementary CS texts say that a recipe is an algorithm for preparing a given dish.

A *program* is a set of computer instructions that implements an algorithm. That is, the program is an instance of an algorithm expressed in a certain way and intended for solution on a certain machine. A program may contain more information than is expressed by the algorithm, because the program has to set up an appropriate environment (just as a recipe doesn't tell you how to turn on the oven or the best place to buy butter).

Pseudocode falls between an algorithm and a program. Essentially, pseudocode is an algorithm expressed in an *ad-hoc* programming-like language. The purpose of pseudocode is to allow the writer to demonstrate an algorithm to the reader without demanding that the reader know a specific high-level language in detail. In this example, we use pseudocode operations like 'Read the first digit in the string.' The operation is entirely clear, and all of the details of how to actually read the data have been hidden away.

1.4.4   What Does a Computer Need to Solve a Problem?

Let's look at the operations performed while solving this problem. Since the actions on lines 1 to 13 are carried out *sequentially*, our computer must be able to perform operations in sequence. An observant reader might have noticed that we have chosen to solve the problem *sequentially* (a step-at-a-time). This approach mirrors both normal human behavior and the actual development of the computer. We could have taken a different approach by performing multiple actions in parallel. We have devised a general problem solving mechanism that can easily be adopted to solve other problems. We could have designed a mechanism that is capable of solving this problem only, but our goal is a machine that can be re-programmed to solve a range of problems. The first two lines of the algorithm are

```
1.    Read the first digit and call it New_Digit
2.    Set Current_Run_Value to New_Digit
```

In line 1, we read a digit from the string which must be stored somewhere in the computer's *memory*. The *symbolic name* New_Digit refers to the *location* of the number within the memory. The computer must be able to access this location whenever it needs to use the current value of the digit in the current run of digits.

The operation described by line 2 is an *assignment,* because it transfers a value to a variable. Similarly, lines 3 and 4 are also assignments that assign the value 1 to both Current_Run_Length and Max_Run. A computer must be able to read a value from memory, update it, and store it back in memory.

Line 5 consists of the word REPEAT and tells us that we are at the start of a group of operations that have to be executed one or more times. This group ends with the word UNTIL on line 13.

Lines 7, 8, and 10 demonstrate *conditional behavior*. That is, the type of operation carried out depends on the result of a test.

```
7.    IF its value is the same as Current_Run_Value
8.        THEN Current_Run_Length = Current_Run_Length + 1
9.        ELSE {Current_Run_Length = 1;
10.           Current_Run_Value = New_Digit}
```

In line 7, we *compare* the value of the digit read from the string with the digit in the current run (i.e., New_Digit is compared with Current_Run_Value). One of two operations is then carried out depending on the *outcome* of this comparison. One operation is specified by the text following the THEN in line 8 and the other by the text following the ELSE in lines 9 and 10. The curly braces { } enclosing lines 9 and 10 indicate that they are treated as a single entity in the ELSE path. Figure 1.10 illustrates this construct graphically.

Although this problem is relatively trivial, it contains all of the elements required to solve *any* problem. We have the assignments[4] that transfer information from one point to another; we have arithmetic operations such as addition and subtraction; and finally, we have the ability to select one of two courses of action depending on the result of a calculation (i.e., comparison).

Naming of Parts

Because some terms will pop up frequently when we are talking about both low-level and high-level languages, here are some brief definitions for reference and to help set the scene.

Constant—a value that does not change during the execution of a program; for example, if you say that the circumference of a circle can be expressed as $c = 2\pi r$, then both 2 and π are constants.

Variable—a value that can change during the execution of a program. In the previous example, both c and r are variables.

Symbolic name—we often refer to a variable or a constant by a name that makes it easier for us to remember and to understand. For example, we call the circumference of the circle C and we call its radius r. We give the irrational number 3.1415926 the symbolic name π. When a program is compiled into machine code, symbolic names are replaced by actual values.

Address—information in a computer is stored in memory locations and each location has a unique address. For example, the radius of a circle may be stored in address 1234. Rather than trying to remember actual address locations in memory, we give addresses symbolic names; in this case, the address may be called r.

Value and Location—When we refer to the expression $c = 2\pi r$, what is r? We (humans) see r as the symbolic name for the value of the radius, say 5. But, the computer sees r as the symbolic name for the address 1234, which has to be read to provide the actual value. However, if we write $r = r + 1$, do we mean $r = 5 + 1 = 6$ or do we mean $r = 1234 + 1 = 1235$? It is very important to distinguish between an address and its contents. This factor becomes significant when we introduce pointers.

Pointer—A pointer is a variable whose value is an address. If you modify a pointer, it points to a different value. This is not mysterious. In conventional arithmetic, we write x_i where i is really a pointer; we just call it an index. If you change the pointer (index), we can step through the elements of a table, array, or matrix and step x_1, x_2, x_3, and x_4.

---

[4]The assignment operator looks like the conventional equals operator = in arithmetic but indicates that the value on the right-hand side of the equation is transferred to the value on the left-hand side (that is, $y = x + 1$ states that the value of x has 1 added to it and result is transferred to y). Computer languages like Algol and Pascal use the symbol := rather than = to indicate assignment. Later we will use the arrow symbol ← to indicate an assignment.

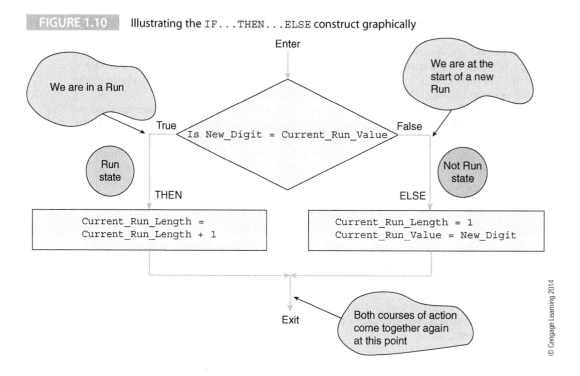

FIGURE 1.10   Illustrating the IF...THEN...ELSE construct graphically

© Cengage Learning 2014

There is little more to the computer than this. You could say that computer architecture is all about performing the type of operation we have described *rapidly* and *efficiently*.

Before we can continue with the design of a computer, we have to take a quick look at the *memory system* that stores both the program and the data used by the program.

1.4.5   The Memory

Human memory is a strange, not to mention *inexact*, thing. An *event* triggers the recall or *retrieval* of a data element that we call a *memory*. The event may be a question that someone asks you, or it may be something that *reminds* you of an episode that took place in the past. Often, we remember information only partially, or even incorrectly. Human memory seems to be keyed or accessed by matching an event against the stored information. That is, our memory is *associative* because we associate one memory with another. A computer's memory is very different and is best thought of as a *table* or *list* of stored items.

Figure 1.11 shows how the program to find the longest run of repeated values in a string of digits is stored in a hypothetical memory. I must stress that the program is conceptual rather than actual, because real computer instructions are rather more primitive than these. This figure is called a *memory map* and shows the location of information within the memory. It's a *snapshot* of the memory, because it represents the state of the memory at a particular instant. The memory map also includes the *variables* used by the program and the string of digits. Recall that the stored program computer stores instructions, variables, and constants all in the same memory.

Figure 1.11 demonstrates that each location in the memory contains either an *instruction* or a *data element*. Numbers 0 through 37 in the first column are *addresses* and express the position of data elements and instructions within the memory (addresses start from 0 rather than 1 because 0 is a valid identifier). The program is in locations 0 through 16, the variables in locations 17 through 20, and the data (the sting) in locations 21 through 37. You can regard the computer's memory as a table of data elements—the location of each element is its address. For example, memory location 3 contains the instruction "Set the Max_Run to 1," and location 20 contains the value of the element Max_Run. Locations 17 onward

FIGURE 1.11   Memory map of a program and its data

0	`i = 21`
1	`New_Digit = Memory(i)`
2	`Set Current_Run_Value to New_Digit`
3	`Set the Current_Run_Length to 1`
4	`Set the Max_Run to 1`
5	`REPEAT`
6	`i = i + 1`
7	`  New_Digit = Memory(i)`
8	`  IF New_Digit = Current_Run_Value`
9	`    THEN Current_Run_Length = Current_Run_Length + 1`
10	`    JUMP to 13`
11	`    ELSE Current_Run_Length = 1;`
12	`        Current_Run_Value = New_Digit`
13	`  IF Current_Run_Length > Max_Run`
14	`    THEN Max_Run = Current_Run_Length`
15	`UNTIL i = 37`
16	`Stop`
17	**New_Digit**
18	**Current_Run_Value**
19	**Current_Run_Length**
20	**Max_Run**
21	**2 (the first digit in the string)**
22	**3**
23	**2**
23	**7**
...	...
37	**1 (the last digit in the string)**

© Cengage Learning 2014

are in bold font to indicate that they contain the variables and the digits of the string we are operating on.

Please don't think that Figure 1.11 provides an actual program. We've had to hide some detail for the sake of simplicity. For example, we've introduced a jump instruction (in blue) in location 10 that tells the computer to ignore instructions in locations 11 and 12 and go directly to location 13. This action is necessary, because if we execute the THEN part of a loop, we have to bypass the ELSE part. However, we have demonstrated how you would access a digit by using the notation Memory(i) to indicate the *i*th location in memory. Similarly, the value of *i* is initialized to 21, and the loop stops when *i* gets to 37.

Figure 1.12 illustrates the organization of a *memory system*. The processor provides the memory with an address on the *address bus* and a control signal that selects either a *read* operation or a *write* operation (these are often called a read or a write *cycle*). In a *read cycle*, the memory responds by putting data onto the *data bus* that the CPU reads. In a *write cycle*, data on the data bus is stored in the memory. The point at which information enters or leaves memory (or any other functional part of a computer system) is called a *port*.

Although the memory in Figure 1.12 is simplified, it accurately describes a computer's memory where data and instructions are stored in consecutive locations. A real computer has

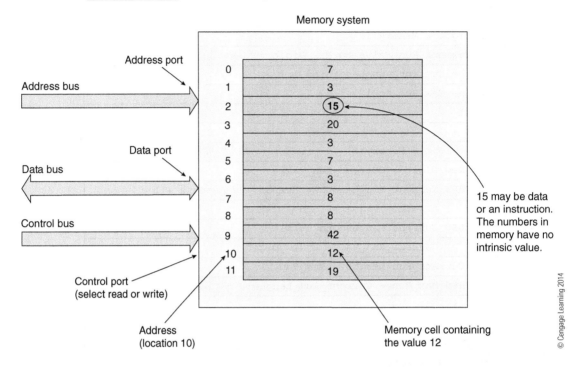

FIGURE 1.12   The memory system

Memory system

Address port

Address bus

Data port

Data bus

Control bus

Control port
(select read or write)

0	7
1	3
2	(15)
3	20
4	3
5	7
6	3
7	8
8	8
9	42
10	12
11	19

15 may be data
or an instruction.
The numbers in
memory have no
intrinsic value.

Address
(location 10)

Memory cell containing
the value 12

© Cengage Learning 2014

a hierarchy of memory systems (each may be implemented with a different technology). This hierarchy includes the very fast cache memory where frequently accessed data lives, main memory, and the much slower secondary storage where large quantities of data remain on magnetic disk or optical CDs or DVDs until the data is needed.

Because describing computer operations in words is often cumbersome, we introduce a notation called *register transfer language* (*RTL*) that makes it easy to define operations that take place within a computer.

RTL Notation

It is important to distinguish between a memory *location* and its *contents*. In RTL, we use square brackets [ ] to indicate the *contents* of a memory location. The expression

[15] = Max_Run

is interpreted as "memory location 15 contains the value of Max_Run."

The backward arrow symbol ← indicates a *data transfer*. For example, the expression

[15] ← [15] + 1

is interpreted as "the contents of memory location 15 are increased by 1 and the result put back in memory location 15." Consider the following three RTL expressions:

a. [20] = 5
b. [20] ← 6
c. [20] ← [6]

Expression (a) states that the contents of memory location 20 are equal to the number 5. Expression (b) states that the number 6 is put into (*copied* or *loaded* into) memory location 20. Expression (c) indicates that the contents of memory location 6 are copied into memory location 20. Note that the ← RTL symbol is equivalent to the conventional assignment symbol =, which is used in some high-level languages. RTL is not a computer language; it is a *notation* used to define computer operations.

1.5 The Stored Program Concept

We now introduce the stored program machine. What follows reasonably accurately describes the operation of many computers up to (about) the 1970s. Today's computers have departed from this simple model, because they perform internal operations in parallel (or even out-of-order) rather than sequentially. The following pseudocode expresses the fundamental action of a stored program machine.

```
Stored_program_machine

    Point to the first instruction in memory

    REPEAT
        Read the instruction at the memory location pointed at
        Point to the next instruction
        Decode the instruction read from memory
        Execute the instruction
    FOREVER
End
```

This pseudocode sequence tells us that a *memory reference* (i.e., a memory read) is required to fetch each instruction from memory. We can expand the action Execute the instruction to give

```
Execute the instruction
    IF the instruction requires data
        THEN fetch the data from memory
    END_IF
    Perform the operation defined by the instruction
    IF the instruction requires data to be stored in memory
        THEN store the data in memory
    END_IF
End
```

We can also express this sequence of actions in C as

```
InstructionPointer = 0;
do
{ instruction = memory[InstructionPointer];   /* read the instruction   */
  decode(instruction);                         /* decode the instruction */
  fetch(operands);                             /* fetch required data     */
  execute;                                     /* execute the instruction */
  store(results);                              /* store the result        */
} while (instruction != stop);
```

The execution of an instruction on this machine requires at least *two* memory accesses.[5] The first memory access is to read the instruction itself. The second access is either to read data from memory that is required by the current instruction or to put data in memory that previously has been created or modified in the CPU by an earlier instruction. Stored program machines are said to operate in a two-phase mode called a *fetch/execute* cycle.

Because it's necessary to access memory twice per instruction, the expression *von Neumann bottleneck* has been coined to indicate that one of the limiting factors of the stored program computer is the path between the CPU and the memory. Machines with different architectures have been devised to overcome this limitation, as we will discover in later chapters.

We now briefly look at the format of instructions that run on stored program machines. An intuitively reasonable instruction format for a stored program machine can be expressed in the form[6]

    Operation **Address1**,Address2,Address3

where Operation specifies the action the instruction is to carry out and Address1,Address2, and Address3 are the locations of the three operands in memory. In this generic instruction, the operands are the *addresses* of data and not the data itself.

Throughout this text, we use bold font to indicate the address that is the *destination* of data rather than its source. A typical three-operand instruction might be ADD **P**,Q,R, where *P*, *Q*, and *R* are the symbolic names of the addresses of three memory locations. There is no standard for the way in which the sequence of operands is written. Some computers use the notation ***destination*** source1, source2 and others use source1, source2, ***destination***.

The three-operand instruction format can be expressed in RTL notation as

    [Address1] ← [Address2] Operation [Address3]

The contents of the memory locations specified by Address2 and Address3 are operated on by the dyadic operation specified in the instruction (e.g., add, subtract, multiply, etc.). The result of the operation placed in the memory location is specified by Address1. Figure 1.13 illustrates this instruction, which requires *four* memory accesses (i.e., one to fetch the instruction, two to fetch the two source operands, and one to store the result). The instruction and operands *P*, *Q*, and *R* can lie *anywhere* within the memory.

FIGURE 1.13   Relationship between instruction and operands

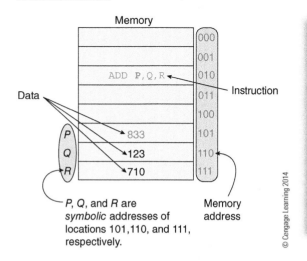

Recall how the instruction ADD **P**,Q,R and the three operands *P*, *Q*, and *R* are all in same memory. This is fundamental to the concept of the *stored program computer*.

Remember that *P*, *Q*, and *R* are the *symbolic addresses* of the three operands whose values are 833, 123, and 710, respectively, and whose binary memory addresses are 101, 110, and 111.

The computer instruction ADD **P**,Q,R is the symbolic form of the actual instruction that would be stored in memory as a binary sequence of 1s and 0s.

© Cengage Learning 2014

---

[5]This statement applies to computers where an instruction always accesses memory. Some computers access memory only to transfer data into or out of the CPU. All other instructions operate on data held within the CPU. This mechanism reduces the number of memory accesses, and we will return to it in later chapters.

[6]This instruction is hypothetical in the sense that it is not implemented by contemporary computers for technological reasons. We will soon see that real computers have simpler instructions.

Figure 1.14 illustrates the relationship between the instruction's four fields, the CPU, the memory, and the way in which the instruction is executed. There are only four instructions that the computer can execute and eight possible memory locations (only four locations are shown). The current instruction is ADD P,Q,R which adds the contents of Q to the contents of R and puts the result in P. This instruction has the binary encoding 10011010001 and the locations of P, Q, and R are 011, 010, and 001 (binary) or 3, 2, and 1 (decimal).

Figure 1.14 demonstrates how the op-code selects an operation (one of four), the source addresses select two memory locations, and the destination address selects the location into which the result is written. This diagram also shows the flow of information that takes place during the execution of the add instruction. We will later see how the op-code is converted into the sequence of actions that implement the intended instruction.

FIGURE 1.14   Interpreting the instruction ADD P,Q,R

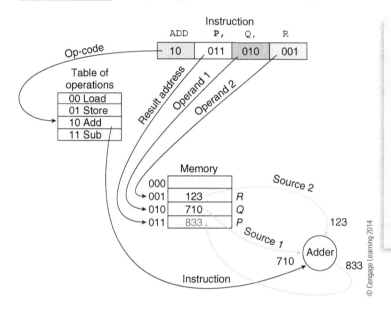

Visualizing the Three-Operand Instruction

The diagram in Figure 1.14 demonstrates how the four fields of the machine-level instruction are interpreted.

The op-code field ADD selects the instruction and controls the ALU.

The three address fields P, Q, and R each select a location in memory that takes part in the interpretation (execution) of the instruction.

© Cengage Learning 2014

Two Address Instructions

Some computers implement a two-address instruction format of the form

    Operation **Address1**, Address2

where Address2 is a source operand and Address1 is *both* a source and a destination operand. This format means that a source operand is accessed in memory and operated on, and the result is placed in the same location as the source operand. The RTL definition of ADD P,Q is

    [P] ← [P] + [Q]

A two-address instruction *destroys* one of the operands. That is, source operand P is replaced (overwritten) by the result. For most of this text, we adopt the convention that two-address instructions have the format

    Operation **destination**, source

When we look at practical computers, we will see that they do not generally allow you to use two memory addresses in the same instruction. Most computers (like the Pentium or the more modern Core i7 processors) specify one address in memory and a second address as a register. A register is a single-storage element in the computer with a name like r0, r1, r2, ..., r31 and is used to hold temporary data during calculations. We will have a lot more to say about registers when we look at the structure of computers.

One Address Instructions

The pressure to minimize instruction size prevented first-generation microprocessors from implementing a two-operand instruction. Typically, they employed a single address instruction format of the form

```
Operation address
```

Because only one address is provided and at least two addresses are required, the processor has to use a second operand that doesn't require an explicit address. That is, the second operand comes from a *register* once called the *accumulator* within the CPU. A *register* behaves like a memory location except that it is located within the CPU itself. The term *accumulator* is hardly used today, because most microprocessors now have several on-chip registers. Figure 1.15 demonstrates the flow of information during the execution of a single-operand instruction. The result remains in a register until another instruction transfers it to memory. Such a computer is hardly elegant, as the following sequence that implements $P = Q + R$ demonstrates.

```
LOAD   Q    ; Read Q into the accumulator
ADD    R    ; Add R to the accumulator
STORE  P    ; Store the accumulator in P
```

FIGURE 1.15   Single-operand instructions

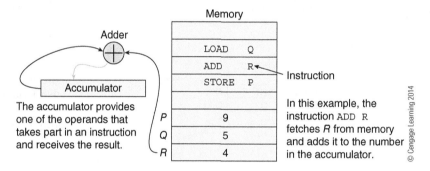

The accumulator provides one of the operands that takes part in an instruction and receives the result.

In this example, the instruction ADD R fetches R from memory and adds it to the number in the accumulator.

© Cengage Learning 2014

In this example, we have to load the accumulator with the first operand, add the second operand, and store the result in the third operand. Note how the accumulator acts as a bottleneck, because all of the data has to flow through it.

A major aspect of computer instruction sets is the *addressing mode,* which is the way we specify the location of an operand. For example, the location of an operand may be given *directly* (i.e., it's in location 1234) or *indirectly* (i.e., register 5 contains the address of the operand we need). We will look at addressing modes in detail in Chapters 3 and 4. However, when discussing computer architectures, we sometimes don't want to specify the actual addressing mode, so we use the generic term *effective address.* For example, we may say that register r1 is loaded with data from the effective address. We use *effective address* because it refers to all of the different ways of generating the address of an operand and lets us avoid specifying a specific address.

Computer Categories

As we have pointed out, computers have both main memories that store large quantities of data and instructions and temporary "scratchpad" registers on the CPU. First generation microprocessors had about six registers, whereas a typical modern microprocessor might have 16, 32, or 64 registers. We can categorize computers by the way in which their instructions act on data. If an instruction can read source operands from memory, perform an operation on the data, and store the result in memory, the computer is called *memory-to-memory.* Computers like the Intel IA32 (8086, Pentium, Core i7) that can perform operations on two operands where one is in memory and one is in a register (the result is written to either memory or a register) have *register-to-memory* architectures.

Some computers such as the ARM and MIPS perform operations only on the contents of registers and are called *register-to-register* computers. These computers have to use a LOAD instruction to put data in registers and a STORE instruction to transfer data from a register to memory. Because the LOAD and STORE operations are the only memory access instructions provided, such computers are frequently referred to as *load/store computers*. Later, we will see that register-to-memory computers fall into the class of computers known as CISC machines, and load/store computers fall into the class called RISC machines.

When we discuss issues in computer architecture, we will take examples from various microprocessors. However, because we would like students to understand how a program is constructed and executed, we are initially going to concentrate on one processor: the *ARM*.

1.6   Overview of the Computer System

To prepare for future chapters, we now take a brief look at the memory and bus systems needed to turn the CPU into a computer system. Computer scientists think of memory as a large array of elements accessed via an address. For example, if array M represents memory, then the ith element is represented by $M[i]$ or M_i. Memory is important because its size (i.e., storage capacity) determines the amount of data a program can store, and its speed (access time) determines how fast a program can process data. Over the past 50 years, the size of programs has increased, and the quantity of data programs use has increased even more rapidly. A good example is a flight simulator of the 1970s that was quite tiny and was concerned only with basic flight (straight and level, climbing, descending, and turning). A modern flight simulator still performs the basic operations, but it can generate complex views of both the cockpit and the outside world. Moreover, it may have a detailed map of most of the earth's surface. Such a program has grown from the region of 16 KB to several GB.

Computer technology has progressed, but some aspects of memory technology have lagged behind. For example, processor speed has outpaced the rate of increase in memory speed. Later chapters will describe how engineers attempt to hide the problem of slow memory. The term *memory wall* has been coined to hint that the memory performance ultimately limits processor performance.[7]

1.6.1   The Memory Hierarchy

Memory systems are fabricated with the widest range of technologies found anywhere in a computer. The fastest memory retrieves data in 10^{-9} s, and the slowest memory may take over 100 s, which is a range in performance of $1:10^{11}$. To put this in perspective, the ratio between my walking speed and the velocity of a missile is about $1:10^4$. Because of the growing disparity between processor and memory performance, designers have attempted to limit the effects of a relatively slow memory by fetching data from memory before it is needed in order to hide the *waiting time* (also called *latency*). Figure 1.16 provides the classic memory hierarchy diagram that shows the type of memory components in a computer, their speed (access time), and typical quantities in a PC. We have also included the ratios between speeds and memory sizes for each of the components. Registers hold the processor's working data, cache is a fast store for frequently used data, DRAM provides the bulk of working storage, and the hard disk archives programs and data. Note that the hard disk is forty million times bigger than the register storage and twenty million times slower. Before looking at memory, we'll provide a short introduction to cache memory, because it plays such a critical role in determining computer performance.

---

[7]W. A. Wulf, S. A. McKee, "Hitting the memory wall: Implications of the obvious," *Computer Architecture News*, Vol. 23, pp. 23–24, 1995.

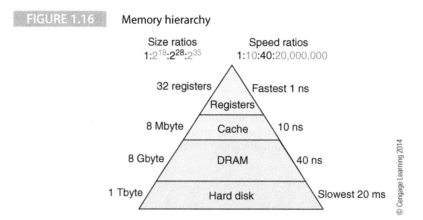

FIGURE 1.16   Memory hierarchy

Frequently used data is stored in very high-speed *cache memory* that has a much smaller access time than the main store. The overall performance of a computer might be as much influenced by its cache memory as by the architecture of the processor itself. Although cache memory is a complex topic because of the fine details of its implementation, its basic operation is very simple indeed. Cache memory contains a copy of frequently used data from the main store, just as a notebook or cellphone contains frequently used names and addresses that we can keep close to us. A cache memory system sits on the computer's address and data buses, eavesdropping on transactions between the CPU and memory.

Whenever the cache memory notices that an address from the CPU is the same as the address of a data element that it has saved, it shouts, "I've got that!", hands the data to the computer, and tells the memory not to bother with this access. In everyday life, if you want to phone a friend, you first look in your address book. If his or her name isn't there, you have to search the much larger telephone directory. Once you find your friend's number, you write it in your address book to save yourself time when you next need to phone them.

The analogy between cache and address book is nearly exact. The address book is effective because the number of people you phone regularly is a tiny subset of all those with phones, and you may find that 90% of the calls you make are to someone in the address book. Another good analogy is that when you buy an address book, there are no names in it, and you have to populate it with people's names when you first phone them. The address book illustrates another problem with cache memory design. Suppose you've filled the "S page" with all of your friends whose name begins with an "S." Now, if you meet someone else whose name begins with an "S," you have a problem. You look down the list of names beginning with "S" and say, "This guy hasn't sent me a birthday card in years; he's history." In cache terms, when the cache is filled with data elements, you must delete an old entry to make room for a new one.

There are other analogies between the human and cache worlds. You may carry a tiny address book containing information about your most important friends and have another one on the desk. When you want to call a friend, you first look up the name in your personal address book. If it isn't there, you try the desk copy. Multiple address books are analogous to levels of cache. A processor may access fast L1 (level 1) cache that's part of the CPU chip and expect information to be there 92% of the time. If the data isn't in the L1 cache, the larger and slower L2 (level 2) cache will be accessed. This might have a 98% probability of coming up with the data. If this access fails, the computer may even try yet another cache—the level 3 cache. If the data isn't there, the data will be fetched from the main memory.

There's yet another analogy between the cache and the address book. Suppose someone has three address books. When he makes a new friend, he writes the name in the three address books. If the friend moves, he may update one address book but forget to update the two others. When he phones this friend from the office, he may phone the old number because he's using the address book that didn't get updated. In Chapters 9 and 13, we will see that keeping the data in cache memory and disk *in step* is a major concern of computer designers.

We will also look at the main store that holds the programs as they are being executed. This is the immediate access memory which is composed of *volatile* semiconductor memory called dynamic random access memory (DRAM). The memory is volatile because data is lost when the power is removed. You can't use it for the long-term storage of programs. When we look at immediate access memory, we discuss some of the emerging memory technologies and concentrate on nonvolatile memory that retains data in the absence of power. This memory is of vital importance to portable applications such as MP3 players and digital cameras.

1.6.2   The Bus

The **bus** links together two or more functional parts of a computer and allows the exchange of data (for example, the bus between the CPU and its graphics card). Buses also link computers to external peripherals (for example, the USB bus that connects a printer to a computer). The bus is such a vital element in a computer system that we devote much of Chapter 12 to computer buses. Here we simply introduce the bus and highlight a few concepts.

Figure 1.17 illustrates the structure of a hypothetical system without a bus. Imagine that the blue circles are processing units that have to communicate with each other. In this example, some units communicate directly with only one other unit, whereas other units have to communicate with several devices. As you can see, the connections are complex and messy. Furthermore, if you were to modify the system by adding a new unit, you'd have to add new highways between the new device and every unit with which it has to communicate.

Figure 1.18 demonstrates the advantage of using a common bus to connect all units together. In this case, there is one single data highway, and each unit has an interface between itself and the common data highway.

FIGURE 1.17   An arbitrary interconnect structure—life without the bus

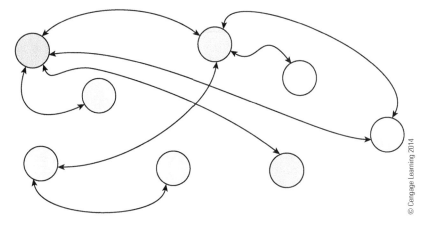

© Cengage Learning 2014

FIGURE 1.18   A common bus connecting all units

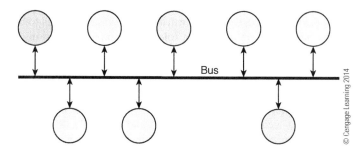

Bus

© Cengage Learning 2014

Bus Terminology

Width—The width of a bus is defined as the number of parallel data paths. A 64-bit bus can carry 64 bits (8 bytes) of information at a time. However, the same term also can be used to indicate the total number of wires (connections) that make up a bus. For example, a bus may have 50 information paths of which 32 carry data (the rest may be paths for control signals or even power lines).

Bandwidth—The bandwidth of a bus is a measure of the rate at which information can be transported across the bus. The bandwidth is expressed in either bytes per second or bits per second. Increasing the width of a bus while keeping the data rate constant increases the bandwidth.

Latency—Latency is the waiting period between a data transfer request and the actual data transmission. Typically, a bus' latency includes the time taken to arbitrate for the bus before transmission can take place.

The problem with the arrangement of Figure 1.18 is that only one device can communicate with another device at a time, because there's only one information path. If two devices want to use the bus simultaneously, they have to fight for control of the bus. We use the term *arbitration* to describe the process whereby two or more devices compete for a resource (in this case the bus). Some systems have a special unit called an *arbiter* that selects which of the competing devices is allowed to go ahead, while all the other contenders wait their turn.

Modern computers have more than one bus. There are buses within the chips themselves. There are buses between functional units (e.g., from the CPU to a memory), and there are buses between buses.

Figure 1.19 illustrates a system with multiple buses. The processor (dark blue) communicates directly with Bus *A*. A second bus, Bus *B*, is connected to Bus *A* via a bus interface unit. Why do we have two buses? First, multiple buses permit

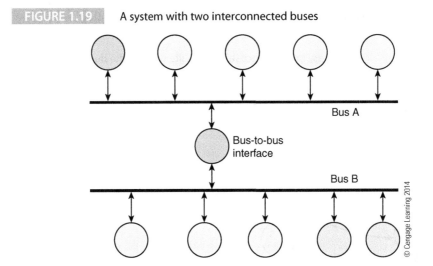

FIGURE 1.19   A system with two interconnected buses

Bus A

Bus-to-bus interface

Bus B

© Cengage Learning 2014

parallel operations. For example, two devices can communicate on Bus *A*, while a second pair of devices communicate with each other on Bus *B*. A more important reason is that the buses may have entirely different characteristics and operating speeds.

A computer may have a whole hierarchy of buses—each optimized for its intended application. The fastest buses in the computer are those that communicate with high-speed memory or video devices. The design and construction of these buses has to be optimized for pure speed. Some buses perform a different role (for example, by allowing you to connect lots of peripherals together). The USB and FireWire buses are typical examples of buses constructed from low-cost components that have been designed for functionality. These buses allow much longer path lengths than memory buses. The final part of this book will look at buses and their architectures in much greater detail.

Two recurring themes in this book (particularly when we discuss memory interfaces, buses, and input/output techniques) are standards and protocols. A standard is an agreed way of designing a system, specifying it, or categorizing any other aspect. Standards are vital in everyday life. For example, power sockets and plugs must conform to agreed standards in order to allow us to take electrical equipment from one place to another. Even things that we do not normally think about (such as the lighting and markings on airport runways and taxiways) are standardized so that a pilot can depart Beijing and arrive in Boston knowing that the interpretation of markings will be the same in both cities.[8] Computing standards specify the physical dimensions of plugs and sockets (e.g., USB), the voltage and current representing signals, and the sequence of signals during a data exchange.

Protocols are rather similar to standards but cover a narrower domain. A protocol governs the sequence of events that take place when two parties communicate. For example, when a computer sends data to a memory, the write protocol defines the sequence of signals—address, write command, and the data to be stored—together with the minimum and maximum duration of signals.

In the next chapter, we provide an overview of how numbers are represented in computers and the type of operations that can be applied to them. We also introduce basic circuit elements, gates, and flip-flops and show how they can be used to construct the basic circuits of a computer. Indeed, if we combine the introduction to the computer in this chapter with the simple circuits we introduce in the next chapter, the reader will have an insight into how the computer works. In spite of the remarkable complexity of today's digital computers, they are (in principle) very simple devices. The complexity lies in the fine details and in the ways in which we have attempted to speed up the operation of a computer (such as by predicting the outcome of tests before they are performed).

1.7 Modern Computing

Computer systems architecture is a continually developing field. Development itself is not evenly distributed across all of the aspects of computer architecture. For example, when we examine disk drives in Chapter 11, we will see that over the years disk *capacity* has grown tremendously, whereas disk *speed* or disk access time has remained almost constant. However, a new technology called the solid-state drive changed this situation, and disk speed started to grow too. Consequently, we focus on some areas more strongly than others.

Two areas of particular interest today are *performance* and *energy consumption*. A few decades ago the object was to get computing power—any computing power was better than none in the early days of the microprocessor. Today, the microprocessor is a mature device and ever more performance is demanded of it for the increasingly challenging roles it undertakes. This translates into areas such as high-definition and high-performance game playing. Consequently, this text is strongly concerned with performance: how we measure it and how we create it.

We are also living in a world of ubiquitous computing. Computers are everywhere—in supercomputer laboratories, universities, industry, the home office, almost all our appliances, televisions and display systems, music and entertainment systems, and cell phones. Such ubiquitous computing makes two demands. The first is connectivity. We have to be able to connect many of these devices together if they are to share data. For example, the digital camera, the MP3 player, the PC, and the TV are often linked by a wireless network. Another consequence of ubiquitous computing is the problem posed by power supplies (i.e., energy) for small devices like MP3 players. We want to fit them in our pockets, use them to watch video, and still expect them to run off a tiny battery for eight hours. Sometimes our expectations of computers rise faster than their performance. The need to reduce electrical power has forced designers to look for energy-efficient solutions to computing problems. In the final chapter,

---

[8]In the USA, road crossings sometimes have a counter that counts down to zero to indicate how many seconds you have left to cross the road before the traffic flows again. These are not common in Europe, but I once spied one in Amsterdam with 10 seconds remaining and thought I could cross the road if I shot forward like a bat out of hell. I got half way across the road and was almost squished by a tram. In Amsterdam, the counter reports the time remaining before the lights change to stop the traffic and not the time left for pedestrians to cross. Here's a case where different standards nearly curtailed my writing career.

we will see how limitations on power have led to the development of microprocessors with multiple on-chip processors called multicore processors.

Finally, some of the focus of research and development in computer systems architecture is driven by restrictions to what was the ever-expanding growth of computing. We have already used the term *wall* in the expression *memory wall,* which indicates that the growth in processor speed has far outstripped the growth in memory speed and results in processors having to wait for data to come from memory. At several points in this text, dealing with the consequences of the memory wall will be a dominant theme. Another wall is imposed by the limits to power dissipation (the inability to cool modern chips).

Summary

We have explained that this text is concerned with computer systems architecture; that is, the internal operation of a complete computer system from its central processing unit to its disk drives and peripherals.

In order to understand what is happening in computing, you have to appreciate the difference between computer *architecture* and computer *organization.* Computer architecture is concerned with the operation of the computer at the level of the assembly language programmer and represents an idealized, abstract view of the computer. Computer organization is concerned with how the architecture is implemented by real gates and circuits. Although architecture and organization are unrelated (in principle), they do influence each other.

We have introduced the stored program computer and demonstrated how it reads instructions from the memory and then executes them one-by-one, operating on data in the same memory as the instructions. We have solved a simple problem and demonstrated that the computer requires a relatively small variety of operations.

The stored-program computer is a simple and elegant device. It reads a binary pattern of 1s and 0s from memory devices and then carries out the operation defined by these 1s and 0s. The operations performed by a computer generally consist of an action such as add or subtract and the address or location of the data that is to take part in the operation. We have also seen that the computer needs to include some mechanism that provides conditional behavior; a means of using the result of an operation on data to determine which of two courses of action is to be followed next.

We also provided a very short history of the computer. The device that we call the computer was never invented. It grew out of a series of developments. Each new development took ideas developed in the past and used more modern technology to implement them. Early computers of the 1940s weren't computers by our standards, because they couldn't run different programs or were doomed to perform the same sequence of operations over and over again (that is, they didn't provide the necessary conditional behavior to allow a program to respond to its inputs).

By 1950, the computer was pretty much as it is today—at least in principle. What has changed is the way in which computers have developed in terms of their technology, internal organization, and peripherals.

Problems

1.1 What is the difference between a *dedicated* and *general-purpose* computer?

1.2 Is an aircraft's automatic pilot an example of a dedicated or a general-purpose computer?

1.3 We said that the pattern of 1s and 0s used to represent an instruction in a computer has no intrinsic meaning. Why is this so and what is the implication of this statement?

1.4 Why is the performance of a computer so dependent on a *range* of technologies such as semiconductor, magnetic, optical, chemical, and so on?

1.5 Modify the algorithm used in this chapter to locate the longest run of non-consecutive characters in the string.

1.6 I was once criticized for saying that Charles Babbage was the inventor of the computer. My critic argued that Babbage's proposed computer was entirely mechanical (wheels, gears, and mechanical linkages) and that a real computer has to be electrical. Was my critic correct?

1.7 What is the effect of the following sequence of RTL instructions? Describe each one individually and state the overall effect of these operations. Note that the notation [*x*] means the contents of memory location *x*.

  a. $[5] \leftarrow 2$
  b. $[6] \leftarrow 12$
  c. $[7] \leftarrow [5] + [6]$
  d. $[6] \leftarrow [7] + 4$
  e. $[5] \leftarrow [[5] + 4]$

1.8 What are the differences between RTL, machine language, assembly language, high-level language, and pseudocode?

1.9 What is a *stored-program machine*?

1.10 I would maintain that *conditional behavior* is the key element that makes a computer a computer. Conditional behavior is implemented at the machine level by operations such as BEQ XYZ (branch to instruction *XYZ*) and at high-level language level by operations such as If x = y then do THIS else do THAT. Why are such conditional operations so necessary to computing?

1.11 What are the relative advantages of one-address, two-address, and three-address computer architectures?

1.12 What is the difference between a computer's *architecture* and its *organization*?

1.13 Can you think of other systems besides computers that may be said to have both an *architecture* and an *organization*?

1.14 What is the difference between an exo- and an endo-architecture?

1.15 Over the years, has more computer progress been made in computer architecture or computer organization?

1.16 What is the semantic gap and what is its importance in computer architecture? You will need to use the Internet or library to answer this question.

1.17 What is the difference between human memory and computer memory?

1.18 What is the von Neumann bottleneck?

1.19 Suppose Intel did not develop the first microprocessor. Was the microprocessor inevitable?

1.20 Identify as many enabling technologies as you can that were required before the computer could be constructed.

1.21 Suppose Babbage had succeeded in creating a general-purpose mechanical computer that could operate at, say, one operation per second. What effect,

if any, do you think it might have had on Babbage's Victorian society?

1.22 Use the method of finite differences to calculate the value of 15^2.

1.23 Extend the method of finite differences in table below to calculate the value of 8^3 and 9^3.

Number	Number squared	First difference	Second difference
1	1		
2	4	3	
3	9	5	2
4	16	7	2
5	25	9	2
6	36	11	2
7	49	13	2

1.24 Suppose you decided to try and make computers more 'human' and introduce the 'random element.' How would you do that?

1.25 Computers always follow blind logic. Executing the same program always gives the same results. That's what the computer books say. But is it true? My computer can appear to behave differently on different occasions. Why do you think that this might be so?

1.26 The value of X is 7. Some computer languages (or notations) interpret $X + 1$ as 8 and others interpret it as Y. Why?

1.27 Carry out the necessary research and write an essay on the history of the development of computer memory systems (e.g., CRT memory, delay-line stores, ferrite core stores, etc.)

1.28 Of all the early computers, which do you think should be called the first computer if you are judging the world by today's standards?

1.29 In what applications have computers been most successful? And in what applications have they been least successful or even useless?

1.30 Why is the bus so important to a computer?

1.31 It is common to hear the argument that the development of the CPU (microprocessor) in terms of its size, power, and speed has driven the computer revolution. What other aspects of the computer system have driven the computer revolution?

1.32 Where do you think the bottlenecks (limitations or barriers) to the growth of the computer (in terms of its computational power and its abilities/application) lie?

1.33 Is Moore's law a law?

1.34 Why do you think that Moore's law 'exists'? What drives it or makes it possible?

Computer Arithmetic and Digital Logic

"I often say that when you can measure what you are speaking about and express it in numbers you know something about it; but when you cannot express it in numbers your knowledge is a meager and unsatisfactory kind: it may be the beginning of knowledge but you have scarcely, in your thoughts, advanced to the stage of science, whatever the matter may be."

Lord Kelvin

"You cannot control what you cannot measure."

Tom deMarco, 1982

"The Devil is in the details."

English proverb

"You can have data without operations but you can't have operations without data."

Anon

"Round numbers are always false."

Samuel Johnson

We've introduced the computer and explained that it operates on data stored in memory. Now we look more closely at the data that it processes and the logic elements used to construct its internal circuits. Later, in Part II, we look at how the central processor itself executes instructions.

We begin with the representation of information and digital circuits, because these topics have important implications for much of what follows. The way in which we represent information—be it numerical data, textual data, or multimedia (such as speech and video)—has profound implications for the architecture of a computer. The representation of numbers determines the type of operations computers execute, the way in which data is transferred from place to place, and the characteristics required of storage elements. Similarly, the properties of digital circuits affect the way in which we design computers.[1] In other words, we don't design digital computers using two-state binary logic because we want to—we do it because it is cost-effective.

Because both digital arithmetic and logic circuit design are complete subjects in their own right, here we can do little more than provide an overview and appreciation. Chapter 2

---

[1] If logic elements and storage devices with the three states −1, 0, and +1 were available at the same cost as today's two-state elements and at the same density (elements per chip), it is *possible* that the organization of computers would be somewhat different from those of today.

begins with the representation of numbers and computer arithmetic. Computers represent numbers and all other information in binary form simply because it is easy to design and construct complex binary circuits in large volumes economically. We look at binary integers and topics such as the representation of negative numbers, fractional values, and irrational numbers like π. We also look at the basic arithmetic operations of addition, subtraction, multiplication, and division.

A section is devoted to the *floating-point arithmetic* that lets us handle large and small numbers like 1.3453×10^{23} in scientific calculations. Floating-point arithmetic is important, because its implementation determines the speed of a computer when it is carrying out complex graphical transformations and image processing. Moreover, the nature of floating-point arithmetic has important consequences for the *accuracy* of calculations. After describing floating-point numbers, we briefly consider some of the implications for the accuracy of chained calculations of the form $(p - q)(x + y)$, as well as the calculation of transcendental and trigonometric functions.

The second section of Chapter 2 provides an introduction to computer logic and the circuits used to construct digital systems. Because digital logic is often the subject of an entire course, here we cover only topics of general importance in computer architecture: a basic knowledge of Boolean algebra is needed to understand logical operations at the machine level. An appreciation of the way in which logic circuits operate is required to explain how a computer functions at the level of information flow within the machine. Moreover, a knowledge of the properties of gates helps us to appreciate how functionality is implemented at the circuit level. Finally, an understanding of synchronous circuits is necessary to help us explain how one of the major techniques used to accelerate computers called pipelining is implemented. By the end of this section, the reader should have some idea of how simple logic elements can be combined to create a simple digital computer.

Anything we can describe can be converted into binary form and processed by a computer. Computer data can represent anything from the very text of this book, to the numbers in an arithmetic computation, to the music from a CD-ROM, or to a movie. One computer might help a Hollywood artist to create a virtual crowd scene, whereas another might calculate someone's mortgage payment; however, both computers carry out the same fundamental operations on identical types of data elements: the bits.

In this chapter, we examine how information is represented by strings of 1s and 0s. In particular, we look at the representation of numbers and the way in which numbers are manipulated.

We use the term *computer* without any fear of ambiguity. Four or more decades ago you had to prefix *computer* by either *analog* or *digital* to distinguish between the two entirely different types of computer then available—the analog computer and the digital computer. We begin by explaining why computers are digital, why computer designers have universally adopted the binary system, and how information is represented inside computers.

Computers were originally designed as calculating machines to make it easier for people to perform tedious arithmetic operations. Because we represent numbers in decimal form, we explain why numbers are converted from their everyday decimal representation into a form that can be stored and processed in the computer. We also show how computers represent and handle negative as well as positive numbers. This chapter describes how addition and subtraction are performed, and we look at the much more complex operations of multiplication and division.

We should stress that there is nothing magical about binary numbers and binary arithmetic. Digital computers and binary arithmetic have arisen simply because the corresponding technology is cost-effective to produce. Computer arithmetic is the same as everyday arithmetic; if you have seven apples and you eat one, six remain. Whether you use decimal arithmetic, count on your fingers, slide beads along an abacus, or use a computer, the result is the same.

Very large or very small numbers such as 1992347119845, 0.0000000000000000 0000342, 1.234×10^9 or -1.3428×10^{-12} are handled by means of a system called *floating-point arithmetic*. We show how such numbers are stored and processed by digital computers.

However, because a floating-point number may be an *approximation* to its actual value, we discuss the nature of errors in arithmetic operations and how they are propagated during compound calculations. We also include an introduction to the way in which computers generate mathematical functions such as a square root, sine or cosine.

The second part of this chapter introduces the logic elements that are used to construct a digital computer. Although we do not have the space to do this topic the justice it deserves, we are able to demonstrate how remarkably simple digital elements called gates can be arranged to create a digital computer.

2.1   What is Data?

Data is information such as numbers, text, computer programs, music, images, symbols, moving images, DNA codes, etc. Indeed, information is anything that can be stored and manipulated by the computer. In this section, we introduce the bit and demonstrate how it can be used to represent information.

2.1.1   The Bit and Byte

The smallest quantity of information that can be stored and manipulated inside a computer is the *bit*, which is a contraction of the words *BInary digiT*. A bit can take the value of 0 or 1 and is indivisible, because it can't be subdivided into smaller units of information.

Digital computers store information in their memories in the form of groups or strings of bits called *words*; for example, the string 01011110 represents an 8-bit word. By convention, we write a binary string with the least-significant bit at the right-hand end.

It would be a little easier to explain how computers work if computers operated in decimal arithmetic just as we do. We could simply employ the arithmetic used in our everyday lives. To build such a decimal computer would require electronic circuits capable of storing and manipulating the ten decimal digits 0, 1, 2, 3, 4, 5, 6, 7, 8, and 9. We can't yet build low-cost circuits that can reliably distinguish between ten different voltage levels. We can build low-cost electronic circuits capable of distinguishing between two voltage levels that we call 0 and 1.

Data

There isn't space to go into the philosophical details of the three related concepts: *data*, *information*, and *knowledge*. The term *data* means a measurement or a set of measurements that represent values or variables.

For the purpose of this text, we use the term *data* to indicate the binary information stored in a computer. This data may represent numbers, instructions, images, or any other object that can be expressed digitally.

Information is often used synonymously with data. However, the term *information* also implies that the data has some form of intrinsic value. For example, in information theory, the value (or entropy) of information is related to its probability. Information theory was developed by Claude Shannon and is of importance because it tells us how fast we can transmit information over a noisy channel.

Knowledge is the structure the human brain applies to information and is at a much higher level of abstraction than information and data.

Computers don't usually operate on a single bit at a time; they operate on groups of bits.[2] A group of eight bits is known as a *byte*. Today's microprocessors are byte-oriented with word lengths that are integer multiples of eight bits (i.e., their data elements and addresses are 8, 16, 32, 64, or 128 bits). A word is described as being 2, 4, or 8 bytes long, because its bits can be formed into 2, 4, or 8 groups of eight bits, respectively.

In general, the more bits a computer can process at once, the faster it is. As computer technology has grown both more powerful and cheaper, the groups of bits processed by a computer have become larger. The first microprocessors in the 1970s could deal with only four bits at a time. By the early 1990s, 64-bit microcomputers were beginning to enter the personal computer market. Some graphics cards operate on units of data that are 128 or 256 bits wide.[3]

Some computer manufacturers employ the term *word* to mean a 16-bit value (as opposed to a byte, which is an 8-bit value) and *longword* to mean a 32-bit value. Other manufacturers use the term *word* to refer to a 32-bit value and *halfword* to refer to a 16-bit value. Throughout this, text we will generally use *word* to mean the basic unit of information operated on by a computer.

2.1.2   Bit Patterns

We have said that a string of bits can represent any data element. It is quite natural to ask how many bits it takes to represent such a data element. If we want to represent the hour of the day electronically as one of 24 different values (i.e., 0 to 23), how many bits do we need, and how do we assign bit patterns to these numbers?

Figure 2.1 demonstrates how we can generate a sequence of binary values using one bit, two bits, three bits, and four bits. Starting at the left-hand side of Figure 2.1 with a single bit, we can move along one of two paths—up represents the bit in a 0 state and down represents the bit in a 1 state. That is, a single bit provides two possible states: 0 and 1. Adding a second bit provides four paths from the starting point to states 00, 01, 10, and 11. Adding a third bit generates eight paths to states 000, 001, 010, 011, 100, 101, 110, and 111. Finally, adding a fourth bit provides sixteen states from 0000 to 1111.

Each time a bit is appended to the number, the total number of paths doubles. Four bits give 16 paths, 5 bits 32 paths, and so on. An n-bit word provides 2^n different paths or bit patterns. An 8-bit byte has $2^8 = 256$ possible values, and a 16-bit word has $2^{16} = 65,536$ different values. To represent any quantity that can have up to n values as a binary number, we need to select the minimum number of bits m to be $n \leq 2^m$. For example, if we were expressing integer percentages (i.e., $n = 0, \ldots, 100$), m would be 7 because $100 \leq 2^7$. We have not indicated how these 2^m patterns of m bits are best arranged. If you look at Figure 2.1, you will see that the bits are arranged in the same *positional* way we count (Section 2.2 will discuss this in detail). There are several different ways of representing numerical values, as we shall soon see.

Representing Information

An n-bit word can be arranged into 2^n unique bit patterns, as Figure 2.1 demonstrates for $n = 1, 2, 3,$ and 4. What then do the n bits of a binary word represent? The simple answer is *nothing*, because no intrinsic meaning is associated with a pattern of 1's and 0's. How a particular bit pattern is interpreted depends only on the meaning given to it by the programmer. Before we introduce binary arithmetic, we briefly look at binary code in general before considering numeric codes. The following are some of the entities that a word may represent.

---

[2] We need to qualify this statement. If a computer has a 32-bit word, then adding two 32-bit words requires that 32 pairs of bits be added together. However, since addition involves a carry-in from the two digits on the right, you really do have to add bits together—a bit at a time. In practice, the programmer sees this as a single operation on two words. The fine details of the addition are hidden from the computer user. The same is true for most other operations. However, some computers allow you to operate on a specific bit of a word; for example, you can set it to 0 or to 1, flip its values from 0 to 1 or 1 to 0, or test whether it is a 1 or 0 and use the outcome of the test later in the program.

[3] A 128-bit unit of data in a graphics card doesn't represent a single data element but a set of individual pixels that are handled and processed in parallel. We discuss this concept in detail when we describe multimedia applications.

FIGURE 2.1   The binary tree

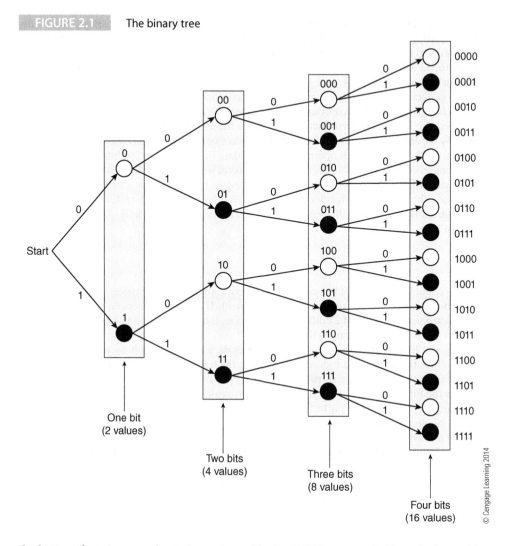

One bit
(2 values)

Two bits
(4 values)

Three bits
(8 values)

Four bits
(16 values)

© Cengage Learning 2014

An Instruction An operation to be performed by the CPU is represented by a single word in computers with 32-bit or longer words (8-bit and 16-bit computers employ a sequence of words to represent a single instruction).[4] The relationship between the bit pattern of an instruction and what it does is determined by the designer of the computer. For example, the sequence of bits that means 'add *A* to *B*' on one computer might have an entirely different meaning on another computer.

A Numeric Quantity A word or a sequence of words may represent a numerical quantity. Numbers can be represented in many formats (for example, BCD integer, unsigned binary integer, signed binary integer, binary floating point, complex integer, complex floating point, double precision integer, etc). The byte 10001001 may represent the value −119 in one system, 137 in another system, and 89 in yet another system. Programmers have to perform operations on data according to their type; there is nothing to stop you from multiplying the binary representation of 'John' by the number 8.[5]

---

[4] Some 16-bit computers have instructions varying between 16 bits and 80 bits, depending on whether or not the instruction includes memory addresses.

[5] In a low-level language, you can perform any operation on any data—no matter how inappropriate the operation. High-level languages employ a mechanism called *typing* to ensure that an inappropriate operation cannot be applied to data. For example, you cannot multiply a character code by a number.

TABLE 2.1   The ASCII Code

		0 000	1 001	2 010	3 011	4 100	5 101	6 110	7 111
0	0000	NULL	DCL	SP	0	@	P	`	p
1	0001	SOH	DC1	!	1	A	Q	a	q
2	0010	STX	DC2	"	2	B	R	b	r
3	0011	ETX	DC3	#	3	C	S	c	s
4	0100	EOT	DC4	$	4	D	T	d	t
5	0101	ENQ	NAK	%	5	E	U	e	u
6	0110	ACK	SYN	&	6	F	V	f	v
7	0111	BEL	ETB	'	7	G	W	g	w
8	1000	BS	CAN	(	8	H	X	h	x
9	1001	HT	EM	)	9	I	Y	i	y
A	1010	LF	SUB	*	:	J	Z	j	z
B	1011	VT	ESC	+	;	K	[	k	}
C	1100	FF	FS	,	<	L	\	l	|
D	1101	CR	GS	–	=	M	]	m	}
E	1110	SO	RS	.	>	N	^	n	~
F	1111	SI	US	/	?	O	_	o	DEL

© Cengage Learning 2014

A Character   A character is a member of a set of elements called an *alphabet*. The alphanumeric characters of the Latin or Roman alphabet (A to Z, a to z, 0 to 9) and the symbols *, −, +, !, ?, etc. have been assigned binary patterns so that they can be stored and manipulated within the computer. One particular code widely used throughout the computer industry is the ISO 7-bit character code or the ASCII code (American Standard Code for Information Interchange), which represents a character by 7 bits, allowing $2^7 = 128$ different characters. 96 characters are printing characters. The remaining 32 characters are non-printing characters that carry out special functions such as carriage return, backspace, line feed, etc. Table 2.1 defines the relationship between the bits of the ASCII code and the character it represents. Because computers are byte-oriented, it is usual to convert 7-bit ASCII values to eight bits by appending a zero to the most significant bit—we shall follow this approach.

To convert an ASCII character into its 7-bit binary code, you read the upper-order three bits of the code from the column in which the character falls and the lower-order four bits of code from the row. Table 2.1 numbers the rows and columns in both binary and *hexadecimal* forms (we'll introduce hexadecimal numbers shortly). For example, the ASCII representation of the letter "Z" is given by $5A_{16}$ or 1011010_2.

The ASCII codes for the decimal digits 0, 1, 2, 3, 4, 5, 6, 7, 8, and 9 are 30_{16}, 31_{16}, 32_{16}, 33_{16}, 34_{16}, 35_{16}, 36_{16}, 37_{16}, 38_{16}, and 39_{16}, respectively. For example, the symbol for the number 4 is represented by the ASCII code 00110100_2, whereas the binary value for 4 is represented by 00000100_2. When you hit the key "4" on a keyboard, the computer receives the input 00110100 and not 00000100. Whenever you read input from a keyboard or send output to a display, you have to convert between the codes for the numbers and the values of the numbers. In high-level language, this translation takes place automatically.

The two left-hand columns of Table 2.1, representing ASCII codes 0000000 to 0011111, do not contain letters, numbers, or symbols. These columns are non-printing codes that are used either to control printers and display devices or to control data transmission links. Data link control characters such as ACK (acknowledge) and SYN (synchronous idle) are associated with communications systems that mix the text being transmitted with the special codes required to regulate the flow of the information. Such systems are not as popular as they once were, and data links are controlled by other mechanisms.

> ## ASCII Control Characters
>
> Not that long ago, typewriters and printers were electromechanical devices and several of the ASCII control characters were used to move the paper or carriage (back space, line feed, form feed, and carriage return). There is even a bell code to ring a bell and draw the operator's attention. Some special characters are still used. For example, the escape redefines the meaning of following characters.
>
> The data link control characters such as STX (start of text) were required by specific data communications protocols that are used relatively little today.

The 7-bit ASCII code provides 128 characters and has been extended to the 8-bit ISO 8859-1 Latin code to include accented characters such as Å, ö, and é. Because this code is not suited to many of the world's languages, a 16-bit code called Unicode was designed to represent the characters of languages such as Chinese and Japanese. The first 256 characters of Unicode map onto the ASCII character set, making ASCII to Unicode conversion very easy. Java has adopted Unicode as the standard means of character representation.

Sometimes it is necessary to extend or to enhance a character code (for example, to include the symbols for characters in a different language). Existing character codes can be extended by either employing more bits per character or by means of escape sequences. The 7-bit ISO/ASCII character set can be extended to eight bits to provide two 128-character sets. If the most-significant bit is set to zero, the remaining seven bits indicate one of the 128 normal ISO/ASCII characters. If, however, the most-significant bit is set to one, the remaining seven bits can indicate any one of 128 new characters (e.g., non-Latin alphabets or even graphics symbols).

Another way of extending a character set is by using an escape sequence that employs a special character to indicate that the following character (or sequence of characters), which is to be interpreted in a different way from the normal or *default* character set. For example, the ISO/ASCII escape character, ESC, can be used to indicate that the character following the ESC has a new meaning.

Images, Sound, and Vision Digital computers process large quantities of data that represent sound, still images, and video. They may say that a picture is worth a thousand words, but that is not true in the digital world. If we assume that the average word is five ASCII characters and a space, a thousand words is 6,000 characters or approximately 6 Kbytes. A digital SLR camera with a full-frame sensor takes raw (unencoded) images that are about 30 Mbytes *each*. In this case, an image is worth 5,000 times more than a thousand words.

The fundamental unit of an image is the *pixel* (picture element), which may be 8 bits (monochrome) or 24 bits (three primary colors). A high-resolution image may be larger than 4 K by 3 K pixels. The situation with moving images is even worse, because video is transmitted as a sequence of still images that are updated, say, 60 times a second (times/s).[6] In practice, the situation is not quite as bad as these figures would indicate, because both still and moving images are compressed to reduce their size—often by a factor of ten or more. Still images are compressed using JPEG algorithms. Moving images are compressed using MPEG algorithms. We will return to this topic when we introduce special-purpose instruction sets for dealing with multimedia. Images require large quantities of storage,

---

[6] The rate at which moving images are transmitted is a little complicated. A television image or *frame* is often *interlaced* and transmitted as two consecutive *fields* of odd and even lines. Interlacing reduces the effect of flicker, because 60 half images a second looks better than 30 full images. Movies sometimes use 24 full frames/second. European television uses 25 frames of 50 interlaced fields/second in contrast with the 30 frames of 60 interlaced fields in the United States (because of the power line frequency of 50 Hz in Europe and 60 Hz in the U.S.A.).

high-bandwidth transmission paths, and very large amounts of processing power if they are to be operated on in real time. It would be reasonable to state that progress in high-performance computer technology over the last decade has been driven by the needs of multimedia. As an example of the scale of problems involving multimedia, consider the design of the in-flight entertainment on the A380 Airbus that can carry about 800 passengers. In theory, all 800 people could be using the system to select one of more than 50 movies—all in real time.

The storage and processing of sound was once challenging for the computer designer, but this is no longer a significant problem in comparison with visual images, because the bandwidth required by audio is well within the capacity of all modern processing, storage, and transmission devices. The Nyquist theorem states that sound can be reconstituted if the original waveform is sampled at least twice as fast as the highest frequency signal in the audio stream. Sixteen-bit sampling is sufficient for all but the highest quality audio. Consequently, sampling sound at 32 K times/s would require about $2^{15} \times 2$ bytes/s = 2^{16} bytes/s, which presents no problem today. Sound can be compressed using *psychoacoustic encoding* (see Chapter 5) with an algorithm such as MP3 to occupy less than a tenth of its initial bandwidth.

There is one significant difference between the encoding of multimedia signals and most other forms of data. Data normally has to be encoded accurately with no loss of information in either its storage or processing. No one expects random errors to occur in text files or in your credit card bill. Such encoding is called *lossless*, and it indicates that, no matter how many times you encode and decode it, it will always remain the same (a good example of lossless encoding is the zip-encoding used to compress text files by substituting short tokens for frequently occurring words or groups of letters). The encoding of images and audio (MPEG, JPEG, and MP3) uses *lossy encoding*, which means that encoding causes an irreversible loss of quality. Such lossy encoding exploits the principle that humans are not unduly worried about a loss of detail if they cannot readily perceive it.

2.2   Numbers

The numbers used to count things (i.e., 1, 2, 3, 4, . . .) are called *natural* numbers, because they don't depend on mathematics—there are three stars in Orion's belt whether or not there is anyone on the Earth to count them. We use the decimal or *denary* system to count things, because this base has the ten symbols 0 to 9 (corresponding to the number of fingers we have).

Not all numbers are natural. We have invented *negative* numbers to handle my bank balance. We have created *real* numbers to describe fractions such as 123.456 and 13/14. Real numbers themselves are divided into *rational* and *irrational* numbers. A rational number can be expressed as a fraction (e.g., 7/12), whereas an irrational number can't be expressed as one integer divided by another; for example, π or $\sqrt{2}$.

The modern number system, which includes a symbol to represent zero, was introduced in Europe from the Hindu-Arabic world in about 1400.[7] This system represents decimal numbers by means of *positional notation*, where the value or *weight* of a digit depends on its location within a number. The value of the 6 in 1261 is ten times as great as the value of the 6 in 126. The decimal number 1261 is equal to $1 \times 1000 + 2 \times 100 + 6 \times 10 + 1 \times 1$. In positional notation, the value by which a digit is multiplied when it moves one place left in a number is called the *base* or *radix*.

---

[7] Just as today's computer scientists battle over the supremacy of Windows and Linux, there was a period in European history when Roman numerals and Hindu-Arabic numerals coexisted. If you wanted an accountant to balance your books, you had to go to someone skilled in either the use of traditional Roman numerals or the new symbols recently arrived from the east. Hindu-Arabic numbers were popularized by Leonardo de Pisa in 1202, although Roman numerals lingered for several centuries. Hindu-Arabic numbers were not in common use until about 1550.

2.2.1 Positional Notation

In positional notation, the n-digit integer N is written as a sequence of digits in the form

$$a_{n-1}, a_{n-2}, \ldots, a_i, \ldots, a_1 \, a_0$$

The a_i's are *digits* that can take one of b values (where b is the base). For example, in base 10, we can write $N = 278 = a_2 \, a_1 \, a_0$, where $a_2 = 2$, $a_1 = 7$, and $a_0 = 8$.

Positional notation can be extended to express real values by using a *radix point* (e.g., decimal point in base ten arithmetic or binary point in binary arithmetic) to separate the integer and fractional part. A real value in decimal arithmetic is written in the form 1234.567. If we use n digits in front of the radix point and m digits to the right of the radix point, we can write $a_{n-1}, a_{n-2}, \ldots, a_i, \ldots, a_1, a_0 \, . \, a_{-1}, a_{-2}, \ldots, a_{-m}$.

The value of this number expressed in positional notation in the base b is defined as

$$N = a_{n-1}b^{n-1}, \ldots, + a_1 b^1 + a_0 b^0 + a_{-1}b^{-1} + a_{-2}b^{-2}, \ldots, + a_{-m}b^{-m}$$

$$= \sum_{i=-m}^{i=n-1} a_i b^i$$

A positional number is equal to the sum of its digits, each of which is multiplied by a weight according to its location within the number. For example, the decimal number 1982 is equal to $1 \times 10^3 + 9 \times 10^2 + 8 \times 10^1 + 2 \times 10^0$. The value of b^0 is 1 for any base b.

Extending this notation to base two, the binary number 10110.11 is defined as $1 \times 2^4 + 0 \times 2^3 + 1 \times 2^2 + 1 \times 2^1 + 0 \times 2^0 + 1 \times 2^{-1} + 1 \times 2^{-2}$ or in decimal as $16 + 4 + 2 + 0.5 + 0.25 = 22.75$.

To distinguish between decimal, binary, and hexadecimal bases, we employ the subscripts 10, 2, and 16, respectively, to indicate the base of a number (e.g., 1234_{10}, 10101011_2, and $12A3_{16}$). However, if the number base is obvious, we will often omit the subscript.

Computer scientists are interested in four bases: decimal, binary, octal, and hexadecimal (the term hexadecimal is often abbreviated to hex). The octal base is little used today, so we will not discuss it further. Table 2.2 shows the digits used by each of these bases. Because the hexadecimal base has 16 digits, letters A through F indicate values between ten and fifteen. For example, $12CF_{16}$ consists of the digits one, two, twelve, and fifteen. The hexadecimal base is used by people to express long strings of binary digits. For example, the 8-bit binary number 10001001 is equivalent to the hexadecimal number 89, which is easier to remember.

TABLE 2.2   The Digits Employed by Four Number Bases

System	Base	Set of Digits
Decimal	$b = 10$	$a = \{0, 1, 2, 3, 4, 5, 6, 7, 8, 9\}$
Binary	$b = 2$	$a = \{0, 1\}$
Octal	$b = 8$	$a = \{0, 1, 2, 3, 4, 5, 6, 7\}$
Hexadecimal	$b = 16$	$a = \{0, 1, 2, 3, 4, 5, 6, 7, 8, 9, A, B, C, D, E, F,\}$

Warning!

Decimal positional notation cannot record all fractions exactly; for example, 1/3 is 0.33333333333 ... 33. The same is true of binary positional notation. Note that some fractions that can be represented in decimal cannot be represented in binary; for example, 0.1_{10} cannot be converted exactly into binary form.

There are many ways of representing numbers—consider BCD, which is the *binary coded decimal* that was once widely used by calculators and computers for financial calculation. BCD uses the 4-bit codes 0000 to 1001 represent decimal values 0 through 9. Strings of decimal digits are represented by BCD values (for example 1948 as 0001 1001 0100 1000). BCD is inefficient, because binary codes 1010 to 1111 are unused and decimal arithmetic is more cumbersome than binary arithmetic. Because BCD is no longer a significant factor in computer design, we will omit it from this text.[8]

---

[8] Curiously, decimal arithmetic is making a comeback. High-speed decimal arithmetic logic is used in IBM's POWER6 processors to handle transactions in financial databases.

2.3   Binary Arithmetic

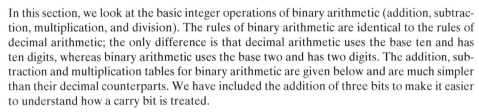

In this section, we look at the basic integer operations of binary arithmetic (addition, subtraction, multiplication, and division). The rules of binary arithmetic are identical to the rules of decimal arithmetic; the only difference is that decimal arithmetic uses the base ten and has ten digits, whereas binary arithmetic uses the base two and has two digits. The addition, subtraction and multiplication tables for binary arithmetic are given below and are much simpler than their decimal counterparts. We have included the addition of three bits to make it easier to understand how a carry bit is treated.

Addition	Subtraction	Multiplication	Addition (three bits)
$0 + 0 = 0$	$0 - 0 = 0$	$0 \times 0 = 0$	$0 + 0 + 0 = 0$
$0 + 1 = 1$	$0 - 1 = 1$ (borrow 1)	$0 \times 1 = 0$	$0 + 0 + 1 = 1$
$1 + 0 = 1$	$1 - 0 = 1$	$1 \times 0 = 0$	$0 + 1 + 0 = 1$
$1 + 1 = 0$ (carry 1)	$1 - 1 = 0$	$1 \times 1 = 1$	$0 + 1 + 1 = 0$ (carry 1)
			$1 + 0 + 0 = 1$
			$1 + 0 + 1 = 0$ (carry 1)
			$1 + 1 + 0 = 0$ (carry 1)
			$1 + 1 + 1 = 1$ (carry 1)

These tables define what happens when you add, subtract, or multiply two *single-bit* binary values. Adding or subtracting two bits may generate a carry or borrow, just as in decimal arithmetic (e.g., $4 + 8 = 2$, carry 1).

Real computers use 8-, 16-, 32-, and 64-bit numbers, and arithmetic operations must be applied to all the bits of a word. When you add two binary words, you add pairs of bits, a column at a time, starting with the least-significant bit. Any carry-out has to be added to the next column on the left. Consider the following four examples of 8-bit binary addition. Note that we have included carries (when they are 1) in blue.

Example 1	Example 2	Example 3	Example 4
00101010	10011111	00110011	01110011
+01001101	+00000001	+11001100	+01110011
1	11111		111 11
01110111	10100000	11111111	11100110

When subtracting two binary numbers, you have to remember that $0 - 1$ results in the difference 1 and a borrow from the column on the left. Consider the following examples of binary subtraction (borrows in blue). Note that in Example 5, we have reversed the subtraction (smaller from larger) as we would do in conventional arithmetic. Computers do not operate in this way as we shall soon see.

Example 1	Example 2	Example 3	Example 4	Example 5
01101001	10011111	10111011	10110000	01100011
−01001001	−01000001	−10000100	−01100011	−10110000
	1	1	1 111	1 1111
00100000	01011110	00110111	01001101	−01001101

Decimal multiplication is difficult—we have to learn multiplication tables from $1 \times 1 = 1$ to $9 \times 9 = 81$. Binary multiplication requires a simple multiplication table that multiplies two bits to get a single-bit product.

$0 \times 0 = 0$
$0 \times 1 = 0$
$1 \times 0 = 0$
$1 \times 1 = 1$

The following example demonstrates the multiplication of 01101001_2 (the multiplier) by 01001001_2 (the multiplicand). The product of two n-bit words is a $2n$-bit value. You start with the least-significant bit of the multiplier and test whether it is a 0 or a 1. If it is a zero, you write down n zeros; if it is a 1, you write down the multiplier (this value is called a partial product). You then test the next bit of the multiplicand to the left and carry out the same operation—in this case, you write zero or the multiplier one place to the left (i.e., the partial product is shifted left). The process is continued until you have examined each bit of the multiplicand in turn. Finally, you add together the n partial products to generate the product of the multiplier and the multiplicand.

> Note that a computer does not perform arithmetic in this way by creating columns of numbers and adding them up. If this algorithm were mechanized, partial products would be added to a running total.

Multiplicand	Multiplier	Step	Partial products
0100100**1**	01101001	1	0 1 1 0 1 0 0 1
010010**0**1	01101001	2	0 0 0 0 0 0 0 0
01001**0**01	01101001	3	0 0 0 0 0 0 0 0
0100**1**001	01101001	4	0 1 1 0 1 0 0 1
010**0**1001	01101001	5	0 0 0 0 0 0 0 0
01**0**01001	01101001	6	0 0 0 0 0 0 0 0
0**1**001001	01101001	7	0 1 1 0 1 0 0 1
01001001	01101001	8	0 0 0 0 0 0 0 0
		Result	0 0 1 1 1 0 1 1 1 1 1 0 0 0 1

Computers don't perform multiplication the way we do. Later in this chapter we introduce some of the ways in which computers mechanize multiplication.

Range, Precision, Accuracy, and Errors

Before we continue on to the representation of negative numbers, we need to introduce four vital concepts in computer arithmetic. When we process text, we expect the computer to *get it right*. We all expect computers to process text accurately and we would be surprised if a computer suddenly started spontaneously spelling words incorrectly. The same consideration is not true of numeric data. Numerical errors can be introduced into calculations for two reasons. The first cause of error is a property of numbers themselves and the second is an inability to carry out arithmetic operations exactly. We now define three important terms used in computer arithmetic that have important implications for both hardware and software architectures: range, precision and accuracy.

Range—The variation between the largest and smallest values that can be represented by a number is a measure of its range; for example, a natural binary number in n bits has a range from 0 to $2^n - 1$. A two's complement signed number in n bits can represent numbers in the range -2^{n-1} to $+2^{n-1} - 1$. When we talk about floating-point real numbers that use scientific notation (e.g., 9.6124×10^{-2}), we take *range* to mean how large we can represent numbers to how small we can represent them (e.g., 0.2345×10^{25} or 0.12379×10^{-14}). Range is particularly important in scientific applications when we represent astronomically large values, such as the size of the galaxy or a banker's bonus, to microscopically small numbers such as the mass of an electron.

Precision—The precision of a number is a measure of how well we can represent it. For example, π cannot be exactly represented by a binary or a decimal real number—no matter how many bits we take. If we use five decimal digits to represent π, we say that its precision is 1 in 10^5. If we take 20 digits, we represent to one part in 10^{20}.

Accuracy—The difference between the representation of a number and its actual value is a measure of its accuracy. For example, if we measure the temperature of a liquid as 51.32

and its actual temperature is 51.34 , the accuracy is 0.02. It is tempting to confuse accuracy and precision. They are not the same. For example, the temperature of the liquid may be measured as 51.320001 which is a precision of eight significant figures, but if its actual temperature is 51.34, the accuracy is only to three significant figures.

Errors—You could say that an error is just a measure of accuracy (that is, error = true value – actual value). This is true. However, what matters to us as computer designers, programmers, and users is how errors arise, how they are controlled, and how their effects are minimalized.

A good example of the problems of precision and accuracy in binary arithmetic arises with binary fractions. Any decimal integer can be exactly represented in binary form given sufficient bits for the representation (i.e., a sufficiently large range). In positional notation the bits of a binary fraction are $0.1_2 = 0.5, 0.01_2 = 0.25, 0.001_2 = 0.125, 0.0001_2 = 0.0625_{10}, \ldots$. It is a property of binary numbers that not all decimal fractions can be represented exactly in binary form however many bits are used. For example, $0.1_{10} = 0.000110011001100110011\ldots._2$. No matter how many bits you use, you cannot represent 0.1_{10} in binary form. In 32 bits you can achieve a precision of 1 in 2^{32}.

Forgetting that decimal fractions cannot be represented accurately can have serious consequences. Probably the most documented failure of decimal/binary arithmetic is the Patriot missile failure. A Patriot missile is an antimissile device that is fired at an enemy missile. It is intended to detonate and release about 1,000 pellets in front of its target at a distance of 5 to 10 m. Any further away and the chance of sufficient pellets being able to destroy the target is very low.

A Patriot missile's ground-based tracking radar detects and tracks the incoming target. The Patriot's software uses 24-bit precision arithmetic and the system clock is updated every 0.1 second. The tracking accuracy is related to the absolute error in the accumulated time; that is, as the software runs using a 24-bit clock, the error in the estimated position of the target gradually increases. This should not be a problem, because a Patriot battery was expected to be active for a relatively short period.

In 1991 during the first Iraq war, a Patriot battery at Dhahran had been operating for over 100 hours. During that time, the accumulated error in the clock had reached 0.3433 s, which corresponds to an error in the estimation of the target position (a SCUD missile flying at Mach 5) of about 667 m. In this case the incoming SCUD was not intercepted and 28 U.S. soldiers were killed. Of course, it is not possible to state that the missile would have been intercepted had the error not occurred. But it is possible to say that the error guaranteed that the system would fail.

Although we have highlighted a dramatic example of the consequences of a failure to appreciate the effects of finite precision in binary arithmetic, the effects of range, precision, and accuracy in binary arithmetic must always be taken into account.

2.4 Signed Integers

Although negative numbers can be represented in many different ways, computer designers have adopted three techniques: sign and magnitude representation, two's complement representation, and biased representation. Each of these mechanisms has its own advantages and disadvantages.

2.4.1   Sign and Magnitude Representation

An n-bit word has 2^n possible different values from 0 to $2^n - 1$. For example, an eight-bit word can represent the numbers $0, 1, \ldots, 254, 255$. One way of representing a negative number is to take the most-significant bit and reserve it to indicate the sign of the number. The usual convention is to use 0 to represent positive numbers and 1 to represent negative numbers.

We can express the value of a sign and magnitude number as $(-1)^S \times M$, where S is the number's sign bit and M is its magnitude. If $S = 0$, $(-1)^0 = +1$ and the number is positive. If $S = 1$, $(-1)^1 = -1$ and the number is negative. For example, in eight bits, we can interpret the two numbers 00001101 and 10001101 as

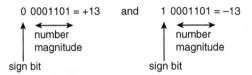

The range of a sign and magnitude number in n-bits is given by $-(2^{n-1} - 1)$ to $+(2^{n-1} - 1)$. An 8-bit number can represent from -127 (11111111) to $+127$ (01111111). An objection to this system is that it has two values for zero:

$$00000000 = +0 \text{ and } 10000000 = -0$$

The sign and magnitude representation of numbers is not used in integer arithmetic because it requires separate adders and subtractors. Other ways of representing negative numbers remove the need for separate adders and subtractors as we shall soon see. Sign and magnitude numbers are used in floating-point arithmetic.

2.4.2   Two's Complement Arithmetic

Microprocessors use the two's complement system to represent signed integers because it allows us to perform *subtraction* by the *addition* of a complement. You subtract 5 from 7 by *adding* the two's complement of 5 to 7.

Complementary Arithmetic

Because several interrelated concepts are involved in this section, the reader may have difficulty understanding where we are going. This box provides an overview. A number and its *complement* add up to a constant; for example in nines complement arithmetic a digit and its complement add up to nine; the complement of 2 is 7 because $2 + 7 = 9$. In n-bit binary arithmetic, if P is a number then its two's complement is Q and $P + Q = 2^n$.

In binary arithmetic, the two's complement of a number is formed by inverting the bits and adding 1. For example, the two's complement of 01100101 is $10011010 + 1 = 10011011$. We are interested in complementary arithmetic because subtracting a number is the same as adding a complement. So, to subtract 01100101 from a binary number, we just add its complement 100111011.

The two's complement of an n-bit binary value, N, is defined as $2^n - N$. If $N = 5 = 00000101$ (8-bit arithmetic), the two's complement of N is given by $2^8 - 00000101 = 100000000 - 00000101 = 11111011$. Note that 11111011 represents -00000101 (-5) or $+123$, depending only on whether we interpret the bit pattern 11111011 as a two's complement integer or as an unsigned integer.

The following example demonstrates two's complement arithmetic in 8-bit arithmetic. We begin by writing down the two's complement representations of +5, −5, +7 and −7.

$+5 = 00000101$ $-5 = 11111011$ $+7 = 00000111$ $-7 = 11111001$

We can now add the binary value for 7 to the two's complement of 5.

$$
\begin{array}{ll}
00000111 & 7 \\
+\,11111011 & -\,5 \\
\hline
100000010 & 2
\end{array}
$$

The result is 100000010 in *nine* bits. If we forget about the leftmost bit (called the *carry bit*), the result is $00000010_2 = +2$, which is the number we want. Now consider the addition of −7 to +5.

$$
\begin{array}{ll}
00000101 & 5 \\
+\,11111001 & -\,7 \\
\hline
11111110 & -\,2
\end{array}
$$

The result is 11111110 (the carry bit is 0). The expected answer is –2; that is, $2^8 - 2 = 100000000 - 00000010 = 11111110$. Once again, this is the result we are looking for.

Two's complement arithmetic is not magic. Consider the calculation $Z = X - Y$ in n-bit arithmetic which we do by *adding* the two's complement of Y to X. The two's complement of Y is defined as $2^n - Y$. We get $Z = X + (2^n - Y) = 2^n + (X - Y)$.

In other words, we get the desired result, $X - Y$, together with an *unwanted* carry-out digit (i.e., 2^n) in the leftmost position that is discarded.

Complementing a number twice results in the original number; for example, $-5 = 2^8 - 00000101 = 11111011$. Complementing again, we get: $-(-5) = 100000000 - 11111011 = 00000101 = 5$. That is, we have $-x = 2^n - x$ and $-(-x) = 2^n - (2^n - x) = x$. Consider the addition of all the combinations of positive and negative values for a pair of numbers.

Let $X = 9 = 00001001$ and $Y = 6 = 00000110$.

$-X = 100000000 - 00001001 = 11110111$
$-Y = 100000000 - 00000110 = 11111010$

$$
\begin{array}{llll}
1. \;\; +X & +9 & 00001001 \\
 +Y & +6 & +00000110 \\
\hline
& & 00001111 = 15
\end{array}
\qquad
\begin{array}{llll}
2. \;\; +X & +9 & 00001001 \\
 -Y & -6 & +11111010 \\
\hline
& & 100000011 = +3
\end{array}
$$

$$
\begin{array}{llll}
3. \;\; -X & -9 & 11110111 \\
 +Y & +6 & +00000110 \\
\hline
& & 11111101 = -3
\end{array}
\qquad
\begin{array}{llll}
4. \;\; -X & -9 & 11110111 \\
 -Y & -6 & +11111010 \\
\hline
& & 111110001 = -15
\end{array}
$$

All four examples give the result we'd expect when the result is interpreted as a two's complement number. Example 3 calculates $-9 + 6$ by adding the two's complement of 9 to 6 to get −3 expressed in two's complement form. The two's complement representation of −3 is given by $100000000 - 00000011 = 11111101$.

Example 4 evaluates $-X + -Y$ to get −15 but with the addition of a 2^n term. The two's complement representation of −15 is given by $100000000 - 00001111 = 11110001$. When both numbers are negative we have $(2^n - X) + (2^n - Y) = 2^n + (2^n - X - Y)$. The first part of this expression is the redundant 2^n, and the second part is the two's complement representation of $-X - Y$. The two's complement system works for all possible combinations of positive and negative numbers.

Calculating Two's Complement Values

The two's complement system would not be so attractive if it weren't for the ease with which two's complements can be formed. Consider the two's complement of N, which is defined as

$2^n - N$. Let's rearrange the expression $2^n - N$ by subtracting 1 from the 2^n and adding it to the result.

$$2^n - 1 - N + 1 = \underset{n \text{ places}}{\underleftrightarrow{111 \dots 1}} - N + 1$$

For example, in eight bits ($n = 8$) we have

$$2^8 - N = 100000000 - N = 100000000 - 1 - N + 1 \text{ (after rearranging)}$$
$$= 11111111 - N + 1$$

The expression $11111111 - N$ is easy to evaluate. Consider the ith bit of N; that is, n_i. If $n_i = 0$, then $1 - 0$ is 1. Similarly, if $n_i = 1$, then $1 - 1 = 0$. Clearly, $1 - n_i = \bar{n}_i$. It's therefore easy to evaluate the two's complement of N. All you have to do is invert the bits and add 1 to the result. For example, in five bits we have

$$7 = 0\,0\,1\,1\,1$$
$$-7 = \bar{0}\,\bar{0}\,\bar{1}\,\bar{1}\,\bar{1} + 1 = 11000 + 1 = 11001$$

Evaluating two's complement numbers in this fashion is attractive because it's easy to perform with hardware.

Properties of Two's Complement Numbers

1. The two's complement system is a true complement system in that $+X + (-X) = 0$.
2. There is one unique zero $00, \dots, 0$.
3. The most-significant bit of a two's complement number is a sign bit. The number is positive if the most-significant bit is 0 and negative if it is 1.
4. The range of an n-bit two's complement number is from -2^{n-1} to $+2^{n-1} - 1$. For $n = 8$, the range is -128 to $+127$. The total number of different numbers is $2^n = 256$ (128 negative, zero, and 127 positive).
5. The hardware for two's complement addition and subtraction is the same because subtraction is performed by adding the complement.

Arithmetic Overflow

We've just stated that the range of two's complement numbers in n bits is from -2^{n-1} to $+2^{n-1} - 1$. Consider what happens if we violate this rule by carrying out an operation whose result falls outside the range of values that can be represented by two's complement numbers. In a five-bit representation, the range of valid signed numbers is -16 to $+15$. Consider the following examples.

Case 1		Case 2	
$5 =$	00101	$12 =$	01100
$+7 =$	00111	$+13 =$	01101
12	$01100 = 12_{10}$	25	$11001 = -7_{10}$

(as a two's complement number)

In Case 1, we get the expected answer of $+12_{10}$, but in Case 2, we get a *negative* result because the sign bit is '1'. If the answer were regarded as an unsigned binary number, it would be $+25$, which of course is the correct answer. However, once the two's complement system has been chosen to represent signed numbers, all answers must be interpreted in this light.

Similarly, if we add together two negative numbers whose total is less than -16, we also go out of range. For example, if we add $-9 = 10111_2$ and $-12 = 10100_2$, we get

$-9 =$	10111
$-12 =$	$+10100$
-21	101011 gives a positive result $01011_2 = +11_{10}$

Both examples demonstrate *arithmetic overflow* that occurs during a two's complement addition if the result of adding two positive numbers yields a negative result, or if adding two negative numbers yields a positive result. If the sign bits of A and B are the same but the sign bit of the result is different, arithmetic overflow has occurred. If a_{n-1} is the sign bit of A, b_{n-1} is the sign bit of B, and s_{n-1} is the sign bit of the sum of A and B, then overflow is defined by the following logical expression (we introduce logic expressions in the next section).

$$V = a_{n-1} \cdot b_{n-1} \cdot \overline{s_{n-1}} + \overline{a_{n-1}} \cdot \overline{b_{n-1}} \cdot s_{n-1}$$

In practice, real systems detect overflow from the carry bits into and out of the most-significant bit of an adder; that is, $V = C_{in} \neq C_{out}$. Arithmetic overflow is a consequence of two's complement arithmetic and shouldn't be confused with carry-out, which is the carry bit generated by the addition of the two most-significant bits of the numbers.

2.5   Introduction to Multiplication and Division

As well as addition and subtraction, computers have to perform multiplication and division. Both these operations are far more complex than addition/subtraction and take longer to perform (or require more complex hardware). Here, we cover only the fundamentals of multiplication and division.

2.5.1   Shifting Operations

Before we look at how multiplication is performed in binary arithmetic, we are going to describe the *arithmetic shift*[9] that treats the bit pattern to be shifted as a two's complement value. In a shift operation, all the bits of a word are shifted one place left or right (e.g., the bit string 00101100 becomes 01011000 if shifted one place left and 00010110 if shifted one place right). Some computers can shift a string of bits more than one place at a time.

If the bit pattern represents a positive two's complement integer, shifting it left multiplies it by 2. Consider the string 00100111 that represents the decimal value 39. Shifting all the bits of this string one place left, gives 01001110, which represents the decimal value 78. Figure 2.2a describes the arithmetic shift left. A zero enters into the vacated least-significant bit position and the bit shifted out of the most-significant bit position is recorded in the computer's *carry flag* which is represented by C in Figure 2.2. The carry flag is a one-bit memory used by the computer to store the state of the carry bit.

FIGURE 2.2   Arithmetic shift operations

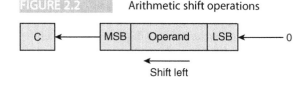

Shift left

(a) Arithmetic shift left:

A zero enters the least-significant bit position and the most-significant bit is copied into the carry flag. For example: 11000101 becomes 10001010 after shifting one place left.

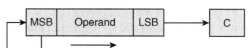

Shift right

(b) Arithmetic shift right:

A copy of the most-significant bit enters the most-significant bit position. All other bits are shifted one place right. The least-significant bit is copied into the carry flag.

For example; 00100101 becomes 00010010 after shifting one place right, and 11100101 becomes 11110010 after shifting one place right.

© Cengage Learning 2014

[9] We will see later that computers implement different types of shift: logical, arithmetic, circular (rotate), and extended.

Suppose that 11100011 is shifted one place left to become 11000110. The original number, 11100011, when interpreted as a two's complement value represents −29. After shifting one place left, the result 11000110 represents the decimal value −58.

Now consider the effect of shifting a number right to divide it by 2. If we take 00001100 (i.e., 12_{10}), we get 00000110 (i.e., 6_{10}) after shifting it one place right. Note that if we shift 00001101 (i.e., 13_{10}) one place right we get 00000110 which is also 6. The least-significant bit has been lost.

Shifting a negative number right requires some care. If we were to apply a simple right shift to the negative number 11100010, the result would be 01110001 which is clearly incorrect because it is positive when interpreted as a two's complement value. In order to maintain the sign of a two's complement number, the sign-bit is replicated when it is shifted right as Figure 2.2(b) demonstrates. Consider 11100010 (i.e., −30). Shifting this one place right while propagating the sign-bit gives 11110001, which is equal to −15.

Why does replicating the sign-bit work when dividing a two's complement number by 2 by shifting it right? A positive number is defined as $0xxxx, \ldots, xx$, where the x's are 1s or 0s. If we divide the number by 2, we get $00pppp, \ldots, pp$. Suppose we convert the number to its two's complement form, we get $1yyyy, \ldots, yy + 1$ (where each y is the complement of the corresponding x). Now, suppose we convert $00pppp, \ldots, pp$ to negative format. We now get $11qqqq, \ldots, qq + 1$. As you can see, the sign bit is propagated in both positive and negative cases.

2.5.2   Unsigned Binary Multiplication

Consider the product of a multiplier and a multiplicand using the human *pencil and paper algorithm*. We stress *human* because computers don't implement this algorithm the way we do— they add up totals as they proceed. The multiplier bits are examined, one at a time, starting with the least-significant bit. If the current multiplier bit is one the multiplicand is written down, if it is zero then *n* zeros are written down instead. Then the next bit of the multiplier is examined, but this time we write the multiplicand (or zero) one place to the left of the last digits we wrote down. Each of these groups of *n* digits is called a *partial product*. When all partial products have been formed, they are added up to give the result of the multiplication.

10×13 Multiplier   = 1101_2
               Multiplicand = 1010_2

```
        1010
        1101
```

1010	Step 1	first multiplier bit	= **1,**	write down multiplicand
0000	Step 2	second multiplier bit	= **0,**	write down zeros shifted left
1010	Step 3	third multiplier bit	= **1,**	write down multiplicand shifted left
1010	Step 4	fourth multiplier bit	= **1,**	write down multiplicand shifted left
10000010	Step 5	add together four partial products		

The result, $10000010_2 = 130_{10}$, is eight bits long. The multiplication of two *n*-bit numbers yields a 2*n*-bit product. As we have said, digital computers don't implement the algorithm in the above way, as this would require the storing of *n* partial products, followed by the simultaneous addition of *n* words. A better technique is to add up the partial products as they are formed. A possible algorithm for the multiplication of two *n*-bit unsigned binary numbers is given in Figure 2.3. Consider the example of 1101×1010 using the algorithm of Figure 2.3. The mechanization of the product of 1101×1010 is presented in Table 2.3.

2.5.3   High-speed Multiplication

Multiplication by shifting and adding is relatively slow. Practical computers accelerate multiplication in various ways; for example, special-purpose logic arrays can be constructed to generate the product of two numbers directly without shifting operands.

FIGURE 2.3   An algorithm for multiplication

Step a.   Set a counter to n.

Step b.   Clear the $2n$-bit partial product register.

Step c.   Examine the rightmost bit of the multiplier (initially the least-significant bit). This bit is underlined in Table 2.3. If it is one, add the multiplicand to the n most-significant bits of the partial product.

Step d.   Shift the partial product one place to the right.

Step e.   Shift the multiplier one place to the right (the rightmost bit is, of course, lost).

Step f.   Decrement the counter and repeat from step c until the count is 0 after n cycles. The product is in the partial product register.

© Cengage Learning 2014

TABLE 2.3   Mechanizing Unsigned Multiplication from Figure 2.3

Cycle	Step	Counter	Multiplier	Partial Product
		Multiplier = 1101_2	Multiplicand = 1010_2	
	a and b	4	1101	00000000
1	c	4	110<u>1</u>	10100000
1	d and e	4	0110	01010000
1	f	3	0110	01010000
2	c	3	011<u>0</u>	01010000
2	d and e	3	0011	00101000
2	f	2	0011	00101000
3	c	2	001<u>1</u>	11001000
3	d and e	2	0001	01100100
3	f	1	0001	01100100
4	c	1	000<u>1</u>	10000010
4	d and e	1	0000	10000010

© Cengage Learning 2014

Multiplying Negative Numbers

The multiplication algorithm we've just discussed is valid only for unsigned values. Consider the product of a positive number X with a negative number $-Y$, where the two's complement representation of $-Y$ is $2^n - Y$.

If we use two's complement arithmetic, the product $X(-Y)$ is defined as $X(2^n - Y) = 2^nX - XY$. The *desired* result, $-XY$, is negative and should be represented in $2n$-bit two's complement form by $2^{2n} - XY$. Note that the most-significant bit is 2^{2n}, rather than 2^n, because the multiplication of two n-bit numbers automatically yields a double length $2n$-bit product. In order to get the correct two's complement result we have to add a correction factor of $2^{2n} - 2^nX = 2^n(2^n - X)$ to the result to give $2^nX - XY + 2^n(2^n - X) = 2^{2n} - XY$ (correct result for the product $-X \cdot Y$ in two's complement form). This correction factor is the two's complement of X scaled by 2^n. Generating and adding such a correction factor slows the multiplication process. Let's do an example.

In 4-bit arithmetic, say we wish to multiply X by $-Y$ where $X = 3$ and $Y = -2$. The binary value of 3 is 0011, and the value of 2 is 0010. The two's complement of 2 is $1101 + 1 = 1110$. So, let's multiply 0110 by 1110. We get 00101010, which is $+42$ in eight bits. Our algebraic calculation demonstrated that a simple multiplication would yield a result that was too large by -2^nX. The value of $-X$ is 1101 (two's complement) and multiplying it by 2^4 gives

(continued)

11010000. To get the correct answer, we add the product to the correction factor to get 00101010 + 11010000 = 11111010. This is, of course, the two's complement of 6 which is the correct answer.

Booth's Algorithm

The classic approach to the multiplication of signed numbers in two's complement form is provided by *Booth's algorithm*, which works for two positive numbers, one negative and one positive, or both negative. Booth's algorithm is similar to conventional unsigned multiplication but with the following differences. Two bits of the multiplier are examined together to determine which of the following three courses of action is to take place next.

1. If the current multiplier bit is 1 and the next lower order multiplier bit is 0, subtract the multiplicand from the partial product.
2. If the current multiplier bit is 0 and the next lower order multiplier bit is 1, add the multiplicand to the partial product.
3. If the current multiplier bit is the same as the next lower order multiplier bit, do nothing.

   Note 1. When adding in the multiplicand to the partial product, discard any carry bit generated by the addition.

   Note 2. When the partial product is shifted, an arithmetic shift is used and the sign bit propagated.

   Note 3. Initially, when the current bit of the multiplier is its least-significant bit, the next lower-order bit of the multiplier is assumed to be zero.

Table 2.4 demonstrates the multiplication of +15 by −13 (01111 × 10011) in 5-bit arithmetic. The result is 1100111101_2, which corresponds to −195.

TABLE 2.4	Demonstration of Booth's Algorithm

Step	Multiplier Bits	Partial Product
		0000000000
Subtract multiplicand	10011<u>0</u>	1000100000
Shift partial product right		1100010000
Do nothing	1001<u>1</u>	1100010000
Shift partial product right		1110001000
Add multiplicand	100<u>1</u>1	10101101000
Shift partial product right		0010110100
Do nothing	10<u>0</u>11	0010110100
Shift partial product right		0001011010
Subtract multiplicand	1<u>0</u>011	1001111010
Shift partial product right		1100111101

Some programmers avoid multiplication by using clever tricks that involve the relatively fast operations of shifting and adding. Consider examples of multiplication by 10 and by 9.

$10P = 2 \times (2 \times 2 \times P + P)$; that is, shift P left twice, add P, shift the result left.
$9P = 2 \times 2 \times 2 \times P + P$; that is, shift P left three times and add P to the result.

Multiplication can be carried out by means of a *look-up table* in which all the possible results of the product of two numbers are stored in read-only-memory. That is, you simply use the values of X and Y to look up the contents of location X,Y to read the result $X \cdot Y$. For example, an 8-bit by 8-bit multiplier requires a table with 16 inputs (i.e., two 8-bit words) and an array of 2^{16} locations each holding a 16-bit product. Multiplying 00001010 by 00111100 involves reading location 0000101000111100 to get 0000001001011000.

The disadvantage of this technique is the exponential increase in the size of the ROM as the number of bits in the multiplier and multiplicand increases. An n-bit ROM-based multiplier requires a capacity of $2n \times 2^{2n}$ bits; for example an 8-bit multiplier must store 16×2^{16} = 1,048,576 bits.

You can reduce the size of the multiplier look-up table by a simple technique. Suppose we require the product of two 16-bit numbers A and B. We can express 16-bit A as two 8-bit numbers side by side, A_u, A_l, where A_u represents the eight most-significant bits of A and A_l the eight least-significant bits. If $A = 1111000010101010$, $A_u = 11110000$ and $A_l = 10101010$. We can express A as $A_u \times 256 + A_l$ and B as $B_u \times 256 + B_l$. Now, consider the product $A \times B$.

$$A \times B = (A_u \times 256 + A_l)(B_u \times 256 + B_l)$$
$$= 65536A_uB_u + 256A_uB_l + 256A_lB_u + A_lB_l$$

This expression requires the evaluation of four 8-bit products (A_uB_u, A_uB_l, A_lB_u, A_lB_l), the shifting of the products by 16 or 8 positions (i.e., multiplication by 65,536 or 256), and the addition of four partial products. We have performed 16-bit multiplication with 8-bit multipliers and four additions. You should appreciate that there are very many ways of using hardware to speed up multiplication.

2.5.4 Division

Division is performed by repeatedly subtracting the divisor from the dividend until the result is either zero or less than the divisor. The number of times the divisor is subtracted is called the quotient, and the number left after the final subtraction is the remainder. That is,

dividend/divisor = quotient + remainder/divisor

Before we consider binary division let's examine decimal division using the traditional pencil and paper technique. The following example illustrates the division of 575 by 25.

$$\overset{\text{quotient}}{\text{divisor }\lceil\text{dividend}}\qquad 25\,\lceil\overline{575}$$

The first step is to compare the two digits of the divisor with the most-significant two digits of the dividend and ask how many times the divisor goes into these two digits. The answer is 2 (i.e., $2 \times 25 = 50$), and 2×25 is subtracted from 57. The number 2 is entered as the most-significant digit of the quotient to produce the situation here.

$$\begin{array}{r} 2 \\ 25\lceil\overline{575} \\ \underline{50} \\ 7 \end{array}$$

The next digit of the dividend is brought down and the divisor is compared with 75. As 75 is an exact multiple of 25, a three can be entered in the next position of the quotient to give

$$\begin{array}{r} 23 \\ 25\lceil\overline{575} \\ \underline{50} \\ 75 \\ \underline{75} \\ 00 \end{array}$$

As we have examined the least-significant bit of the dividend and the divisor was an exact multiple of 75, the division is complete, and the quotient is 23 with a zero remainder. A difficulty associated with division lies in estimating how many times the divisor goes into the partial dividend (i.e., 57 was divided by 25 to produce 2 remainder 7). Consider, the same example using unsigned binary arithmetic.

$25 = 11001_2$ $575 = 1000111111_2$

```
      11001⟌1000111111
            11001
```

The five bits of the divisor do not go into the first five bits of the dividend, so a zero is entered into the quotient and the divisor is compared with the first six bits of the dividend.

```
            01
      11001⟌1000111111
            11001
            001010
```

The divisor goes into the first six bits of the dividend once, to leave a partial dividend 001010(1111). The next bit of the dividend is brought down to give

```
            010
      11001⟌1000111111
            11001
            010101
            11001
```

The partial dividend is less than the divisor, and a zero is entered into the next bit of the quotient. The process continues as follows.

```
            10111
      11001⟌1000111111
            11001
            101011
            11001
            100101
            11001
            11001
            11001
            00000
```

In this case the partial quotient is zero, so that the final result is 10111, remainder 0.

Restoring Division

The traditional pencil and paper algorithm we've just discussed can be implemented in digital form with little modification. The only real change is to the way in which the divisor is compared with the partial dividend. People do the comparison mentally whereas computers must perform a subtraction and test the sign of the result. If the subtraction yields a positive result, a one is entered into the quotient, but if the result is negative a zero is entered into the quotient and the divisor added back to the partial dividend to restore it to its previous value.

A suitable algorithm for restoring division is as follows.

1. Align the divisor with the most-significant bit of the dividend.
2. Subtract the divisor from the partial dividend.
3. If the resulting partial dividend is negative, place a zero in the quotient and add back the divisor to restore the partial dividend.
4. If the resulting partial dividend is positive, place a one in the quotient.
5. Perform a test to determine end of division. If the divisor is aligned so that its least-significant bit corresponds to the least-significant bit of the partial dividend, stop. The final partial product is the remainder. Otherwise, continue with step 6.
6. Shift the divisor one place right. Repeat from step 2.

The flowchart corresponding to this algorithm is given in Figure 2.4. Let's use this algorithm to divide 01100111_2 by 1001_2, which corresponds to 103 divided by 9 and should yield a quotient 11 and a remainder 4. Table 2.5 illustrates the division process, step by step.

FIGURE 2.4   The flowchart for restoring division

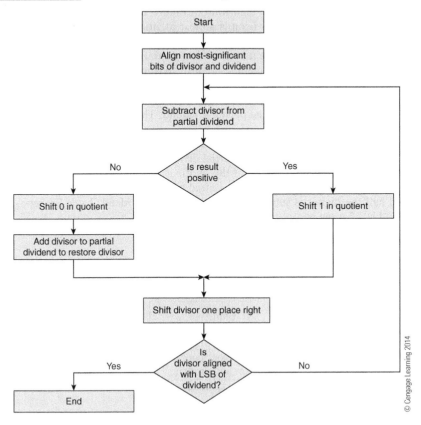

TABLE 2.5   Example of Restoring Division for 1001|0100111

Step	Description	Partial Dividend	Divisor	Quotient
		01100111	00001001	00000000
1	Align	01100111	01001000	00000000
2	Subtract divisor from partial dividend	00011111	01001000	00000000
4	Result positive—shift in 1 quotient	00011111	01001000	00000001
5	Test for end			
6	Shift divisor one place right	00011111	00100100	00000001
2	Subtract divisor from partial dividend	−00000101	00100100	00000001
3	Restore divisor, shift in 0 in quotient	00011111	00100100	00000010
5	Test for end			
6	Shift divisor one place right	00011111	00010010	00000010
2	Subtract divisor from partial dividend	00001101	00010010	00000010
4	Result positive—shift in 1 in quotient	00001101	00010010	00000101
5	Test for end			
6	Shift divisor one place right	00001101	00001001	00000101
2	Subtract divisor from partial dividend	00000100	00001001	00000101
4	Result positive—shift in 1 in quotient	00000100	00001001	00001011
5	Test for end			

Non-Restoring Division

We can modify the restoring division algorithm of Figure 2.4 to reduce the time taken to perform a division. *Non-restoring* division is almost identical to the restoring algorithm. The only difference is that the restoring operation is eliminated. In restoring division, one half the divisor is subtracted in the next cycle after a partial dividend has been restored by adding back the divisor. Each cycle includes a shift-divisor-right operation that is equivalent to dividing the divisor by two. The restore divisor operation in the current cycle followed by the subtract half the divisor in the following cycle is equivalent to a single operation of add half the divisor to the partial dividend. That is, $D - D/2 = +D/2$ where D is the divisor.

Figure 2.5 gives the flowchart for non-restoring division. After the divisor has been subtracted from the partial dividend, the new partial dividend is tested. If it is negative, zero is shifted into the least-significant position of the quotient and half the divisor is added back to the partial dividend. If it is positive, one is shifted into the least-significant position of the quotient and half the divisor is subtracted from the partial dividend. Table 2.6 repeats the example of Table 2.5 using non-restoring division.

FIGURE 2.5   Flowchart for non-restoring division

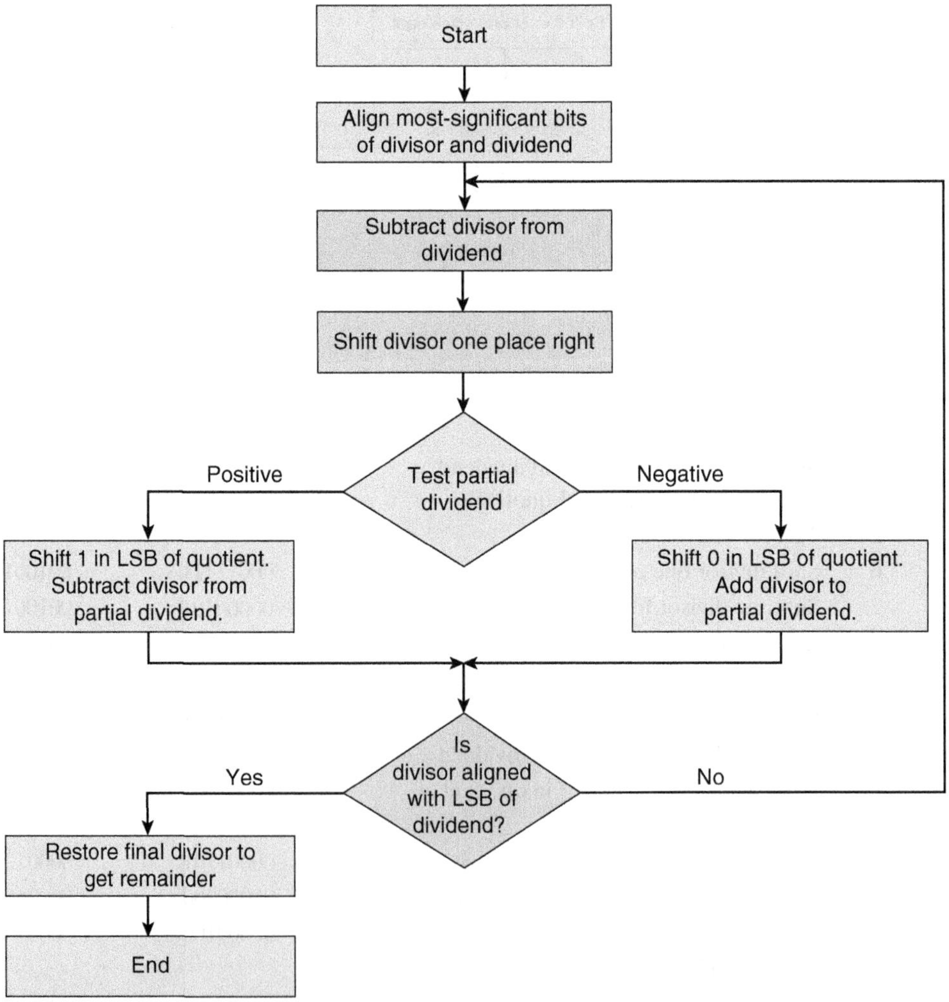

TABLE 2.6   An Example of Non-Restoring Division for 1001|01100111

Step	Description	Partial Dividend	Divisor	Quotient
		01100111	00001001	00000000
1	Align divisor	01100111	01001000	00000000
2	Subtract divisor from partial dividend	00011111	01001000	00000000
3	Shift divisor right	00011111	00100100	00000000
4	Test partial dividend—enter 1 in quotient and subtract divisor from partial dividend	−00000101	00100100	00000001
6	Test for end of process	−00000101	00100100	00000001
3	Shift divisor right	−00000101	00010010	00000001
5	Test partial dividend—enter 0 in quotient and add divisor to partial dividend	00001101	00010010	00000010
6	Test for end of process	00001101	00010010	00000010
3	Shift divisor right	00001101	00001001	00000010
4	Test partial dividend — enter 1 in quotient and subtract divisor from partial dividend	00000100	00001001	00000101
6	Test for end of process	00000100	00001001	00000101
3	Shift divisor right	00000100	0000100.1	00000101
4	Test partial dividend—enter 1 in quotient and subtract divisor from partial dividend	−00000000.1	0000100.1	00001011
6	Test for end of process	−00000000.1	0000100.1	00001011
7	Restore last divisor	00000100	0000100.1	00001011

2.6   Floating-Point Numbers

Having looked at integers, the next step is to discuss *floating-point* arithmetic, the arithmetic of *real* numbers. Real numbers are the set of all rational and irrational numbers. Floating-point arithmetic lets you handle the very large and very small numbers found in scientific applications (as opposed to financial or commercial calculations). Unlike integer arithmetic, the results of floating-point calculations are not, generally, exact. The result of a floating-point calculation on one chip may be different to the result of a floating-point calculation on another. We will explain why floating-point arithmetic does not yield exact answers and discuss some of the pitfalls the programmer must be aware of.

A computer with an *n*-bit wordlength is capable of handling unsigned integers in the range 0 to $2^n - 1$ in a *single word*. Larger integers can be created by chaining words together. For example, a 32-bit computer can work with 64-bit numbers by concatenating two 32-bit words (one as the upper half of the 64-bit value and the other as the lower half). Scientists and engineers frequently handle numbers that have an immense range of values (e.g., from the mass of an electron to the mass of a star). Such numbers are represented and processed by means of *floating-point* arithmetic, so called because the binary point is not located at a *fixed* position in the number. A floating-point value is stored as *two* components: a number and the *location* of the binary point within the number. We sometimes use the term *radix point* to avoid having to worry about whether we mean decimal point or binary point.

Floating-point is also called *scientific notation*, because scientists use it to represent large and small numbers. Examples of decimal floating-point numbers are 1.2345×10^{20}, 0.4599×10^{-50}, -8.5×10^3. In decimal arithmetic, scientific numbers are written in the form mantissa $\times 10^{\text{exponent}}$, where the *mantissa* describes the number and the *exponent* scales it by the appropriate power of ten.

A binary floating-point number is represented by mantissa $\times 2^{\text{exponent}}$. For example, 101010.111110_2 can be represented by 1.01010111110×2^5, where the significand is 1.01010111110 and the exponent 5 (the exponent is 00000101 in 8-bit binary arithmetic). Today, the term *mantissa* has been replaced by *significand* to indicate the number of significant bits in a floating-point number. Because a floating-point number is defined as the *product* of two values, a floating-point value is not unique; for example $10.110 \times 2^4 = 1.011 \times 2^5$.

Over the years, many systems have been used to represent the significand and exponent of floating-point numbers in computers. In the 1970s, a standard means of representing floating-point numbers rapidly replaced most system-specific formats. The *IEEE 754 standard format for floating-point numbers* provides three representations: single precision for 32-bit numbers, double precision for 64-bit numbers, and quad precision for 128-bit numbers.

Normalization of Floating-Point Numbers

An IEEE-754 floating-point significand is always *normalized* (unless it is equal to zero) and is in the range $1.000 , \dots , 0 \times 2^e$ to $1.111 , \dots , 1 \times 2^e$, where e is the exponent. A normalized number always begins with a leading 1. Normalization allows the highest available precision by using all significant bits. If a floating-point calculation were to yield the result $0.110 , \dots , \times 2^e$, the result would be normalized to give $1.10 , \dots , \times 2^{e-1}$. Similarly, the result $10.1 , \dots , \times 2^e$ would be normalized to $1.01 , \dots , \times 2^{e+1}$.

Normalizing a significand takes full advantage of the available precision. For example, the unnormalized 8-bit significand 0.0000101 has only four significant bits, whereas the normalized 8-bit significand 1.0100011 has eight significant bits.

Biased Exponents

The significand of an IEEE format floating-point number is represented in *sign and magnitude* form; that is, a single sign-bit indicates whether the number is positive or negative. The exponent is represented in a *biased* form, by adding a constant to the true exponent. Suppose an 8-bit exponent is used and all exponents are biased by 127. If a number's exponent is 0, it is stored as $0 + 127 = 127$. If the exponent is -2, it is stored as $-2 + 127 = 125$. A real number such as 1010.1111 is normalized to get $+1.0101111 \times 2^3$. The true exponent is $+3$, which is stored as a biased exponent of $3 + 127$; that is 130_{10} or 10000010 in binary form.

The advantage of the biased representation of exponents is that the most negative exponent is represented by zero. The floating-point value of zero is represented by $0.0 , \dots , 0 \times 2^{\text{most negative exponent}}$ (see Figure 2.6). By choosing the biased exponent system we arrange that zero is represented by a zero significand and a zero exponent as Figure 2.6 demonstrates.

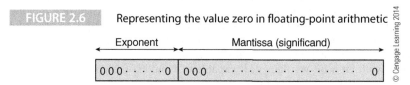

FIGURE 2.6   Representing the value zero in floating-point arithmetic

© Cengage Learning 2014

2.6.1   IEEE Floating-Point Numbers

A 32-bit single precision IEEE 754 floating-point number is represented by the bit sequence

S EEEEEEE 1.MMMMMMMMMMMMMMMMMMMMMMM

where *S* is the *sign bit* that tells you whether the number is positive or negative, *E* the eight-bit *biased exponent* that tells you where to put the binary point, and *M* the 23-bit *fractional significand*. If you count the bits in this number, you will find that there are 33 not 32 bits. The leading 1 in front of the significand is omitted when the number is stored in memory. Only the fractional part of the significand represented by the *M*s is stored in memory (the reason for this will soon become clear). Figure 2.7 illustrates the structure of a 32-bit floating-point number.

FIGURE 2.7   Structure of a 32-bit IEEE floating-point number

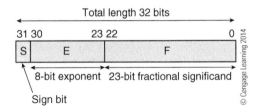

The S-bit is a *sign bit* that determines the sign of the number. If $S = 0$, the number is positive, and if $S = 1$, it is negative. The exponent is the power of two that scales the significand in a floating-point number and is *biased* by 127. For example, if the floating-point number is $+1.11001\ldots0 \times 2^{12}$, the exponent is stored as $12 + 127 = 139_{10} = 10001011_2$.

The significand of an IEEE floating-point number is always *normalized* in the range $1.0000\,,\ldots,00$ and $1.1111\,,\ldots,11$, unless the floating-point number is zero, in which case it is represented by $0.000\,,\ldots,00$. Because the significand is *always* normalized and *always* begins with a leading 1, it is not necessary to include the leading 1 when the number is stored in memory. A non-zero IEEE format floating-point number X is defined as

$$X = -1^S \times 2^{E-B} \times 1 \cdot F$$

where,
S = sign bit, 0 = positive significand, 1 = negative significand
E = exponent biased by B
F = fractional significand (note that the significand is $1 \cdot F$ and has an implicit leading one)

We have already pointed out that the special value zero is represented by $S = 0$, $E = 0$, and $M = 0$ (that is, the representation of 0 is all zeros).

An IEEE floating-point value is said to be *lexicographically ordered* because the order of two numbers is the same, independently of whether they are treated as floating-point values or as signed integer values. This characteristic means that numbers can be compared using simple logic circuits that do not depend on the complexities of floating-point representation.

Consider the following example of how a 32-bit IEEE floating-point number X is *unpacked* into a sign bit, a biased exponent, and a significand. $X = 11000001100110011000000000000000$. Unpacking the three fields gives

$S = 1$, $E = 10000011$, and $F = .00110011000000000000000$

To get the actual significand, we insert a leading one in front of the stored *fractional* significand when we unpacked it to get 1.0011001100000000. The number is therefore equal to $-1.0011001100000000 \times 2^{10000011-01111111} = -1.0011001100000000 \times 2^4 = -10011.0011$.

IEEE Floating-Point Format
ANSI/IEEE 754-1985 specifies basic and extended floating-point formats and a limited set of arithmetic operations (add, subtract, multiply, divide, square root, remainder, and compare).

An important concept in the IEEE standard is the NaN, or *Not a Number*. A NaN is a special entity provided by the IEEE standard that permits the representation of formats outside the IEEE standard. The use and definition of NaNs is system dependent and you can use the bits of a NaN to convey any information you want.

The IEEE standard defines three floating-point formats: *single, double* and *quad* (see Table 2.7). In 32-bit single IEEE format, the maximum exponent $E_{\max}$ is $+127$ and the minimum exponent $E_{\min}$ is -126 rather than $+128$ to -127 as we might expect. The special value $E_{\min} - 1$ (i.e., -127) is used to encode zero and $E_{\max} + 1$ is used to encode plus or minus infinity or a NaN.

When a number is unpacked, the number of bits in its exponent and significand is increased to fill the available space and the format is said to be *extended*. By extending the format, the

TABLE 2.7   IEEE Floating-Point Formats

	Single Precision	Double Precision (Single Extended)	Quad Precision (Double Extended)
Field width in bits			
S = sign	1	1	1
E = exponent	8	11	15
L = leading bit	1 (not stored)	1 (not stored)	1 (stored)
F = fraction	23	52	111
Total width	32	64	128
Exponent			
Maximum E	255	2047	32,767
Minimum E	0	0	0
Bias	127	1023	16,383
E_{max}	127	1023	16,383
E_{min}	−126	−1022	−16,382

S = sign bit (0 for a negative number, 1 for a positive number)
L = leading bit (always 1 in a normalized, non-zero significand)
F = fractional part of the significand
The range of exponents is from the minimum $E + 1$ to the maximum $E − 1$
The number is represented by $−1^S \times 2^{E − \text{exponent}} \times \text{L.F}$
Zero is represented by the minimum exponent, $L = 0$, and $F = 0$, for all three formats.
The maximum exponent, $E_{max} + 1$ represents signed infinity for all three formats.

range and precision of the floating-point number are increased. For example, when a 32-bit floating-point number is unpacked, the 23-bit fractional significand is increased to 24 bits by including the leading 1 and then the significand is extended to 32 bits (either as a single 32-bit word or as two 16-bit words). All calculations are then performed using the 32-bit extended precision significand. After a sequence of floating operations has been carried out in the extended format, the floating-point number is repacked and stored in memory in its basic form.

The minimum exponent is numerically smaller than the largest exponent in order to ensure that the reciprocal of the smallest number will not generate an overflow condition. Of course, the converse of this is that the reciprocal of the largest number will underflow. An underflow is less serious than an overflow.

Figure 2.8 illustrates the IEEE single-precision format. The special cases of the exponent $E = 0$ and $E = 255$ are used to encode zero, denormalized small values, plus or minus infinity, and NaNs.

FIGURE 2.8   IEEE floating-point number space for a single-precision number

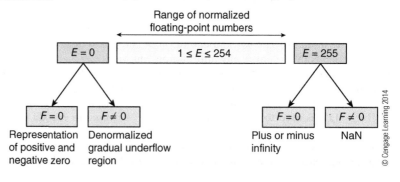

Example of Decimal to Binary Floating-Point Conversion

Let's convert 4100.125_{10} into a 32-bit single-precision IEEE floating-point binary number.

Convert 4100.125 into a fixed-point binary by first converting the integer part to get $4100_{10} = 1000000000100_2$ and then the fractional part to get $0.125_{10} = 0.001_2$. Therefore, $4100.125_{10} = 1000000000100.001_2$.

The next step is to normalize 1000000000100.001_2 by moving the binary point to the left until the significand is in the form 1.xxx. The exponent is incremented for each place the binary point is moved. This gives $1.000000000100001 \times 2^{12}$. We can now put together the final value.

- The sign bit, S, is 0 because the number is positive.
- The stored exponent is the true exponent plus 127; that is, $12 + 127 = 139_{10} = 10001011_2$.
- The stored significand is 00000000010000100000000 (the leading 1 is stripped and the significand expanded to 23 bits).
- The final number is therefore, 01000101100000000010000100000000, or in hexadecimal format, 45802100_{16}.

Binary Floating-Point to Decimal Conversion

Let's carry out the reverse operation. Consider the hexadecimal value $C46C0000_{16}$. In binary form, this number is 11000100011011000000000000000000. The first step is to *unpack* the number into *sign bit*, *biased exponent*, and *fractional significand*.

- $S = 1$
- $E = 10001000$
- $F = 11011000000000000000000$

As the sign bit is 1, the number is negative. We subtract 127 from the biased exponent 10010000_2 to get the actual exponent $1001000_2 - 01111111_2 = 00000111_2 = 7_{10}$. The fractional significand is $.11011000000000000000000_2$. Reinserting the leading one gives $1.11011000000000000000000_2$. The number is $-1.11011000000000000000000_2 \times 2^7$, or -11101100_2 (i.e., -236.0_{10}).

Now, consider $C4962800_{16} = 11000100100101100010100000000000_2$. We can divide this into S, E, and F parts as

$S = 1, E = 10001001 = 137_{10} = 137 - 127 = 10_{10}$ (true), $M = 1.F = 1.001011000101_2$.

The binary number is $-10010110001.01_2 = -1201.25$

Characteristics of IEEE Floating-Point Numbers

Let's look as some of the less obvious characteristics of floating-point numbers. Consider the difference between floating point numbers F_1 and F_2 where the binary patterns for these numbers differ only in the least-significant bit position; that is,

$$d = F_2 - F_1 = e_2 \, x \, 1.f_2 - e_1 \times 1.f_1 = 2^{e2-b} \times 1.f_2 - 2^{e2-b} \times 1.f_1 = 2^{e-b} \times 0.000 , \ldots , 1$$

The difference is $2^{-p} \times 2^{e-b}$, where p is the number of bits in the significand and b is the bias in the exponent. When the exponent is large, the difference between consecutive floating-point values is large, and when b is small, the difference is smaller.

Another feature of floating-point numbers concerns their behavior near to zero. Figure 2.9 demonstrates a floating point system with a two-bit exponent with a 2-bit stored significand. The value zero is represented by 00 000. The next positive *normalized* value is represented by 00 100 (i.e., $2^{-b} \times 1.00$ where b is the bias).

FIGURE 2.9   Floating-point numbers near zero

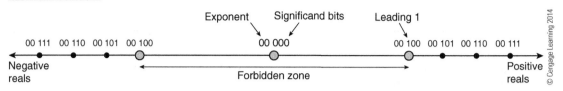

Figure 2.9 demonstrates that there is a *forbidden zone* around zero where floating-point values can't be represented because they are not normalized. This region where the exponent is zero and the leading bit is also zero, is still used to represent valid floating-point numbers. However, such numbers are unnormalized and have a lower precision than normalized numbers, thus providing gradual underflow.

Other Features of IEEE Floating-Point Numbers

- The IEEE standard specifies that the default rounding technique should be towards the nearest value. If the number is equally distant from two nearest representable values, the value whose least-significant bit is zero is taken; that is, the rounding is towards the even value. The standard requires that three other rounding modes must be available (round toward zero, and round towards positive or negative infinity).

- The four comparisons specified by the IEEE format are equal to, less than, greater than, and *unordered*. The latter case, unordered, arises when one of the operands is a NaN.

- The IEEE format specifies five *exceptions*. An exception or interrupt is a request for attention that forces the computer to take a different course of action. The exceptions defined by the IEEE format are

  Invalid Operation—This exception arises when the programmer attempts something illegal such as an operation on a NaN or an addition or subtraction with infinity, or an attempt to calculate the square root of a negative number.

  Division by Zero—An exception is raised whenever an attempt is made to divide a number by zero (because the result is infinity).

  Overflow—Overflow exception occurs when the result is larger than the largest value that can be stored. Overflow can be dealt with by terminating the calculation or by using saturation arithmetic (holding the value at its maximum). Overflow in floating-point arithmetic is not the same as overflow in signed integer arithmetic.

  Underflow—Underflow exception occurs when the destination operand is smaller than the smallest value that can be stored; that is, the result is smaller than 2^{Emin}. Underflow can be handled by setting the smallest number to zero or by representing the number as an unnormalized value less than 2^{Emin}.

  Inexact—An inexact exception occurs when a round-off error is committed by an operation.

- NaNs *propagate* when used in expressions; that is, if part of an expression is a NaN, the result is a NaN.

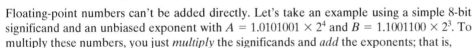

2.7   Floating-Point Arithmetic

Floating-point numbers can't be added directly. Let's take an example using a simple 8-bit significand and an unbiased exponent with $A = 1.0101001 \times 2^4$ and $B = 1.1001100 \times 2^3$. To multiply these numbers, you just *multiply* the significands and *add* the exponents; that is,

$$A \cdot B = 1.0101001 \times 2^4 \times 1.1001100 \times 2^3$$
$$= 1.0101001 \times 1.1001100 \times 2^{3+4}$$
$$= 1.000011010101100 \times 2^8.$$

Now let's look at addition. If these two floating-point numbers were to be added *by hand*, we would automatically align the binary points of A and B as follows.

$$\begin{array}{r} 10101.001 \\ + \ 1100.1100 \\ \hline 100001.1110 \end{array}$$

However, as these numbers are held in a normalized floating-point format, the computer has the following problem of adding

$$\begin{array}{r} 1.0101001 \times 2^4 \\ + 1.1001100 \times 2^3 \end{array}$$

The computer has to carry out the following steps to equalize exponents.

Step 1.   Identify the number with the smaller exponent.
Step 2.   Make the smaller exponent equal to the larger exponent by dividing the significand of the smaller number by the same factor by which its exponent was increased.
Step 3.   Add (or subtract) the significands.
Step 4.   If necessary, normalize the result (post normalization).

In this example, $A = 1.0101001 \times 2^4$ and $B = 1.1001100 \times 2^3$. The exponent of B is smaller than that of A. We increase B's exponent by 1 to make it equal to A's exponent. Since adding 1 to B's exponent doubles the size of the number, we have to halve the number by shifting B's significand one place right to divide B's significand by 2. These two operations preserve the size of B and create a new value of B equal to 0.110011×2^4. We can now add A to the denormalized B.

$$\begin{array}{rl} A = & 1.0101001 \times 2^4 \\ B = & +0.1100110 \times 2^4 \\ \hline & 10.0001111 \times 2^4 \end{array}$$

The result of the addition is not normalized because its integer part is 10_2. Consequently, the result needs post-normalizing. We shift the significand one place right and add 1 to the exponent to get 1.00001111×2^5. The answer is expressed to a precision of *eight* significant figures after the binary point, whereas A and B are each expressed to a precision of seven significant figures.

The procedure for the addition of floating-point numbers is given in Figure 2.10 as a flowchart. A few points to note about this flowchart are given here.

1. Because the exponent sometimes shares part of a word with the significand, it is necessary to separate (unpack) them before the process of addition can begin.
2. If the two exponents differ by more than $p + 1$, where p is the number of significant bits in the significand, the smaller number is too small to affect the larger and the result is effectively equal to the larger number. For example, $1.1010 \times 2^{60} + 1.01 \times 2^{-12}$ is effectively 1.1010×2^{60} because the exponents differ by 72.
3. During post-normalization the exponent is checked to see if it is less than its minimum value or greater than its maximum value to test for *exponent underflow* and *overflow*,

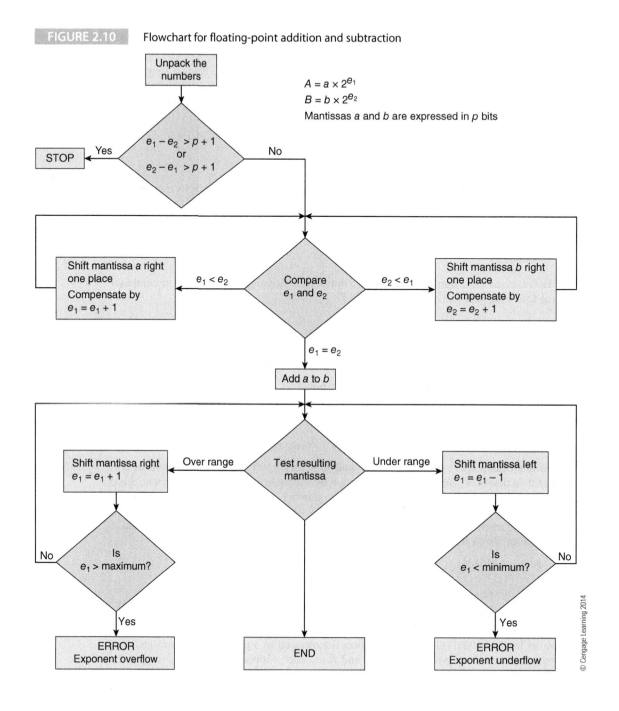

FIGURE 2.10   Flowchart for floating-point addition and subtraction

$A = a \times 2^{e_1}$

$B = b \times 2^{e_2}$

Mantissas a and b are expressed in p bits

respectively. Exponent underflow would generally lead to the number being made equal to zero, whereas exponent overflow would result in an error condition and might require the intervention of the operating system.

Rounding and Truncation Errors

Because floating-point arithmetic can lead to an increase in the number of bits in the significand, you need a means of keeping the number of bits in the significand constant. The simplest technique is called *truncation* and involves dropping unwanted bits. For example, truncating 0.1101101 to four significant bits gives 0.1101. Truncating a number creates an *induced error* (i.e., an error has been induced in the calculation by an operation on the number) that is *biased*, because the number after truncation is always smaller than the number that was truncated.

A better technique for reducing the number of bits is *rounding*. If the value of the lost digits is greater than half the least-significant bit of the retained digits, 1 is added to the least-significant bit of the remaining digits. Consider rounding to four significant bits the following numbers

$0.1101\underline{011} \rightarrow 0.1101$ The three bits removed are 011, so do nothing
$0.1101\underline{101} \rightarrow 0.1101 + 1 = 0.1110$ The three bits removed are 101, so add 1

Rounding is preferred to truncation because it is more accurate and gives rise to an unbiased error. Truncation always undervalues the result, leading to a systematic error, whereas rounding sometimes reduces the result and sometimes increases it. The major disadvantage of rounding is that it requires a further arithmetic operation to be performed on the result.

Figure 2.11 describes rounding mechanisms. The simplest rounding mechanism is truncation or rounding towards zero. In rounding to nearest, the closest floating-point representation to the actual number is used. In rounding to positive or negative infinity, the nearest valid floating-point number in the direction positive or negative infinity respectively is chosen. When the number to be rounded is midway between two points on the floating-point continuum, IEEE rounding specifies the point whose least-significant digit is zero (i.e., round to even).

FIGURE 2.11   Rounding mechanisms

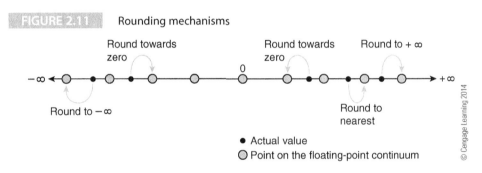

© Cengage Learning 2014

Floating-point implementations use three special-purpose bits to aid the rounding process. An m-bit significand may be represented by $1.m_1, m_2, \ldots, m_m \ G \ R \ S$, where G is a guard bit used to temporarily increase the precision of the floating-point value, R is a rounding bit to aid rounding, and S is a sticky bit. The sticky bit is the logical OR of all bits to the right of the R-bit and is called this because once it is set (to indicate one or more sticky bit 1s to the right) it remains set. These three bits are used by rounding algorithms to determine the value of the rounding bit.

2.8 Floating-Point Arithmetic and the Programmer

Integer operations are exact and repeatable; for example, the integer product $x \cdot y$ yields the same result on every computer. The precise behavior of a floating-point unit in one computer can differ from that in another machine, although the introduction of the IEEE floating-point standard has improved the situation considerably.

The details of floating-point arithmetic (i.e., floating-point accuracy and the way in which expressions are evaluated) cannot always be hidden from the user, because he or she must be aware of the implications. In general, some of the considerations concerning floating-point users affect only highly-specialized numeric applications; few programmers will ever calculate precise mid-course corrections for a spacecraft on the way to Mars. On the other hand, there are circumstances when the very nature of a problem magnifies the effect of floating-point errors. For example, calculating your position using GPS when the satellites are almost in-line leads to a situation where tiny errors in the source data can lead to large errors in the calculated result.

Consider the expression $z = x^2 - y^2$ that calculates the difference between two squares, where x, y, and z, are real values. We can evaluate this expression either as $x^2 - y^2$ or as $(x + y) \cdot (x - y)$. Integer arithmetic would get the same result using either expression, but floating-point arithmetic may give different values.

Why should it matter how the calculation is performed? Consider first $x^2 - y^2$. If the values of x and y are greatly different and $x \gg 1$ and $y \ll 1$, then the difference between x^2 and y^2 will be correspondingly greater. Consequently, the subtraction of $x^2 - y^2$ may generate an error because the value of x^2 will be very large and the value of y^2 very small. In order to perform the subtraction, the exponent of y^2 will have to be increased to the same as that of x^2. The significand of y^2 will have to be shifted right and some of its precision will be lost.

Now consider the product $(x + y)(x - y)$. In this case, the error in the evaluation of $(x - y)$ will be much less and the final result more accurate. Let's demonstrate this operation in eight-digit decimal arithmetic where

$x = 5.9995998 \times 10^2$ and $y = 1.0002010 \times 10^{-1}$.

$x^2 = (5.9995998 \times 10^2)^2$

$\quad = 35.99519776016004 \times 10^4$

$\quad = 3.5995198 \times 10^5$ (after rounding to 8 digits)

$y^2 = (1.0002010 \times 10^{-1})^2$

$\quad = 1.000402040401 \times 10^{-2}$

$\quad = 1.0004020 \times 10^{-2}$ (after rounding to 8 digits)

The addition is performed after scaling the smaller number so that it has the same exponent as the larger number. In order to increase the exponent of y^2 by 7, we have to shift its significand right seven places. The subtraction can now be performed.

$$\begin{array}{r} 3.5995198 \times 10^5 \\ -0.00000010004020 \times 10^5 \\ \hline 3.59951969995980 \end{array}$$

After rounding the result to eight digits, we get 3.5995197×10^5. Now let's perform the calculation in the form $(x + y)(x - y)$. We begin by performing the appropriate scaling to generate $(x + y)$ and $(x - y)$ as

$$\begin{array}{r} 5.9995998 \times 10^2 \\ +0.0010002010 \times 10^2 \\ \hline 6.0006000010 \end{array} \qquad \begin{array}{r} 5.9995998 \times 10^2 \\ -0.0010002010 \times 10^2 \\ \hline 5.9985995990 \end{array}$$

When rounded to eight digits these numbers become 6.0006000×10^2 and 5.9985996×10^2. The product of these numbers is $3.599519675976 \times 10^5$ or 3.5885197×10^5 rounded to eight digits. As you can see, the two results are different simply because we have carried out the same calculation in two different ways.

The effect of roundoff errors on calculations has important implications for compiler technology. For example, the familiar associative laws of algebra no longer hold and $(a + b) + c$ is not the same as $a + (b + c)$. Different compilers can yield different results with the same calculation on the same machine using the same data, if expressions are reordered.

Similar problems occur when dealing with mixed precision (i.e., single- and double-precision) arithmetic. Suppose you evaluate $x \cdot y + z$, where x and y are single-precision values and z is double precision. The operations take place in single-precision arithmetic after z has been converted to single-precision format. However, the operations may take place in double-precision arithmetic after converting x and y to double precision and then converting the final answer to single-precision form. The effect of roundoff errors may make these two results different.

The IEEE standard requires that the result of addition, subtraction, multiplication, and division be computed exactly and then rounded to the nearest floating-point number using a round-to-even method.

2.8.1   Error Propagation in Floating-Point Arithmetic

Continuing with the theme of errors in floating-point calculations, we now examine what happens when you perform chains of calculations with floating-point values. Errors are introduced into floating-point arithmetic either as a *generated error* caused by rounding or as a *propagated error* introduced when an error is propagated through a chain of calculations. Let's look at the propagated error. If there is an error e_y in Y, what is the error in Z if Z is a function of Y? Consider first the addition operation $Z = X + Y$.

Suppose that X' is the value of X used by the computer and is defined as $X + R_x$, where X is the true value of X and R_x is the error introduced by rounding. Similarly, assume that $Y' = Y + R_y$. The *relative errors* in X and Y are defined as R_x/X and R_y/Y, respectively.

When the programmer performs the operation $X + Y$, the computer generates $X + Y + R_x + R_y$. This expression indicates that the roundoff errors accumulate during addition. The relative error in the sum is given by $(R_x + R_y)/(X + Y)$. If X and Y are approximately the same, the relative error is R_x/X.

The product operation $X \cdot Y$ yields $(X + R_x) \cdot (Y + R_y) = X \cdot Y + X \cdot R_y + Y \cdot R_x + R_x \cdot R_y$. We can assume that the errors R_x and R_y are small, and therefore, the term $R_x \cdot R_y$ may be neglected. The roundoff error approximates to $X \cdot R_y + Y \cdot R_x$. The relative error in the product is given by $(X \cdot R_y + Y \cdot R_x)/(X \cdot Y)$, which is $R_y/Y + R_x/X$. If X and Y are approximately the same, the relative error is $2 \cdot R_x/X$. Note that multiplication yields a greater relative error than addition.

The propagated error for a polynomial $f(x) = a_0 + a_1x + a_2x^2 + a_3x^3 + $ can be approximated by $f^{(1)}(x)R_x$, where $f^{(1)}(x)$ is the first differential of the polynomial $f(x)$.

By means of the *Taylor series* for a polynomial, we can write an expression for $f(x)$ about x_0 as $f(x) = f(x_0) + f^{(1)}(x_0)(x - x_0) + f^{(2)}(x_0)(x - x_0)^2/2 + \ldots$. If we let $x - x_0$ be R_x and assume that all powers of $R_x > 1$ are negligible, we can write $f(x) = f(x_0) + f^{(1)}(x_0)R_x$. Consequently, the propagated error is given by $f(x) - f(x_0) = f^{(1)}(x_0)R_x$. Consider the error in calculating the function $2x^3 + 4x^2 + 3x + 2$ at $x = 2$. The derivative of this polynomial is $6x^2 + 8x + 3$, and the estimated error in the calculation when $x = 2$ is $(24 + 16 + 3)R_x = 43R_x$.

Significance errors arise when two nearly equal numbers are subtracted. Consider $z = x - y$, where $x = 1.234567$ and $y = 1.234521$. Both operands have seven significant figures, but the result is $x - y = 0.000046$, which has only two significant figures. Suppose that z is then used in a further calculation, such as $p = q/z$. We are going to use arithmetic accurate to seven places with an operand accurate to only two places. The programmer has to take great care when using a computer to perform floating-point operations.

2.8.2   Generating Mathematical Functions

Some financial and most scientific and engineering calculations require more complex operations than simple addition, subtraction, multiplication, and division. These functions range from the square root, to trigonometric operations such as $\sin(x)$ and $\cos(x)$, and to hyperbolic functions like $\cosh(x)$. Such functions are not necessarily generated directly by means of dedicated hardware. In this section, we provide an insight into how higher-level functions can be generated using basic arithmetic operations. Some floating-point units do provide hardware support for the generation of a subset of scientific functions via the basic mathematical operations.

Although this is not a text on mathematics, we should describe how scientific functions are generated—partially for the sake of completeness and partially because of the implications for the design of computers, arithmetic units, and instruction sets. Any continuous function such as $\sin(x)$ or $\sqrt{x}$ can be expressed in terms of a polynomial expansion about a reference point x_0 as

$$f(x) = f(x_0) + f^{(1)}(x_0)(x - x_0) + f^{(2)}(x_0)(x - x_0)^2/2! + f^{(3)}(x_0)(x - x_0)^3/3! + \ldots$$

The notation $f^{(n)}(x_0)$ is defined as the nth order derivative of $f(x)$ at x_0. This expression for $f(x)$ is the Taylor series and requires the summation of an infinite number of terms in order to calculate $f(x)$. In practice, the number of terms required to evaluate $f(x)$ to an acceptable level of precision can be fairly small. The Taylor series for $f(x) = \sin(x)$, is given by $x - x^3/3! + x^5/5! - x^7/7! + \ldots$.

You can obtain a value for $\sin(x)$ summing the first n terms in the appropriate series. The value of n depends on how rapidly successive terms converge towards zero. Sometimes the series converges rapidly and only four or five terms are required. Sometimes the convergence is slow and the series cannot be used because the time required to perform the calculation would be prohibitive.

Arithmetic units that provide scientific functions sometimes store the coefficients of the series required to generate a particular function in a *lookup table* in read-only memory to avoid having to calculate them each time a function is generated. For example, an arithmetic unit designed to calculate $\sin(x)$ using the Taylor series would store the coefficients 1/3!, 1/5!, and so on in the look-up table.

In practice, the coefficients provided by the Taylor series do not provide the best accuracy for a function of x over a range of values of x (e.g., for x from 0 to $\pi/2$ in the calculation of the sine of x). A real arithmetic unit would use coefficients that provide a better accuracy for $\sin(x)$ over a given range of values for x. The accuracy of some series deteriorates rapidly as the variable reaches its limits. For example, a series that generates $f(x)$ for $0 \leq x < 1$ might provide a poor approximation when x is close to 1. Some series, such as the *Chebyshev series*, are better behaved than the Taylor series, and the error is distributed more evenly over the range of x. Unlike the Taylor series, the Chebyshev series does not provide a poor approximation at the limits of the variable. Moreover, the Chebyshev series converges faster than the Taylor series, and fewer terms are required to generate an approximation.

The Chebyshev series for a function of x is defined as the sum of a series of Chebyshev polynomials:

$$f(x) = \tfrac{1}{2}C_0 + C_1T_1(x) + C_2T_2(x) + C_3T_3(x) \ldots \quad \text{for } -1 \leq x \leq 1.$$

The values of C_0, C_1, C_2, etc. are the Chebyshev coefficients for the function $f(x)$ and $T_1(x)$, $T_2(x)$, etc. are *Chebyshev polynomials*. The value of $T_n(x)$ is defined as $\cos(n \cdot \mathrm{acos}(x))$, and the first few Chebyshev polynomials are

$$T_0(x) = 1$$
$$T_1(x) = x$$
$$T_2(x) = 2x^2 - 1$$
$$T_3(x) = 4x^3 - 3x$$
$$T_4(x) = 8x^4 - 8x^2 + 1$$
$$T_5(x) = 16x^5 - 20x_3 + 5x$$

We can express a function of $f(x)$ by the series

$$\tfrac{1}{2}C_0 + C_1x + C_2(2x^2 - 1) + C_3(4x^3 - 3x) + C_4(8x^4 - 8x^2 + 1)$$
$$+ C_5(16x^5 - 20x^3 + 5x) + \ldots$$

If we take only six terms of the Chebyshev polynomial, the terms of this equation can be regrouped to give

$$f(x) = a_0 + a_1x + a_2x^2 + a_3x^3 + a_4x^4 + a_5x^5$$

where

$$a_0 = \tfrac{1}{2}C_0 - C_2 + C_4$$
$$a_1 = C_1 - 3C_3 + 5C_5$$
$$a_2 = 2C_2 - 8C_4$$
$$a_3 = 4C_3 - 20C_5$$
$$a_4 = 8C_4$$
$$a_5 = 16C_5.$$

For any required function such as $\sin(x)$, the values of the coefficients $a_0, a_1, \ldots$ can be stored in a read-only memory in the computer's floating-point unit.

The purpose of this section has been to demonstrate where complex mathematical functions *come from* and how they can be generated by summing a series of terms until the required accuracy is achieved.

Using Functions to Generate New Functions

It is not necessary to employ either series or iterative techniques to generate scientific functions. Many functions can be derived directly from other functions. For example, the value of e^{-x} is given by $e^{-x} = -\sinh(x) + \cosh(x)$. The two hyperbolic functions $\sinh(x)$ and $\cosh(x)$ can be generated by a floating-point unit (using the appropriate series). Similarly, we can calculate natural logarithms from $\log_e(x) = 2\tanh^{-1}(x-1)/(x+1)$.

We have just demonstrated that a computer can derive the type of mathematical functions used in calculations ranging from the financial to scientific indirectly in a number of ways. What matters to the programmer is that the way in which calculations are performed can affect the final answer. What matters to the designer is that there are opportunities for creating arithmetic units that can accelerate the calculation of complex numerical functions.

Iterative Techniques

Another means of generating mathematical functions involves *iterative techniques*. An iterative technique uses an approximate value of a function, y_i, to generate a better value, y_{i+1}. This better value is then used to obtain an even better value, and so on. The Newton-Raphson formula states that

$$y_{i+1} = y_i - f(y_i)/f^{(1)}(y_i)$$

where y_i is the current estimate, y_{i+1} is the next estimate, f is a function and f' is the first derivative or differential of the function at y_i. The figure below illustrates this equation.

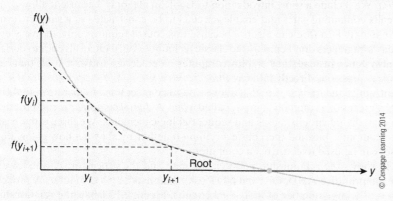

Consider the Newton-Raphson function for the square root of x:

$$y_{i+1} = y_i - \tfrac{1}{2}(y_i - x/y_i)$$

Here y_i represents the *i*th estimate of the square root of x, and y_{i+1} is the next and better estimate. Suppose we wish to generate the square root of 2 (i.e., $x = \sqrt{2}$). We will use the initial guess that $y_0 = 1$. At the first iteration

$$y_1 = y_0 - \tfrac{1}{2}(y_0 - x/y_0) = 1 - \tfrac{1}{2}(1 - 2/1) = 1 - \tfrac{1}{2}(-1) = 1.5$$
$$y_2 = y_1 - \tfrac{1}{2}(y_1 - x/y_1) = 1.5 - \tfrac{1}{2}\,(1.5 - 2/1.5) = 1.5 - \tfrac{1}{2}\,(0.1666667) = 1.4166667$$
$$y_3 = y_2 - \tfrac{1}{2}(y_2 - x/y_2) = 1.4166667 - \tfrac{1}{2}(1.4166667 - 2/1.4166667) = 1.41421569$$

(continued)

After only three iterations, the result is very close to the true value of $\sqrt{2} = 1.41421356$. Similarly, we can calculate the reciprocal of a number x iteratively with the expression $y_{i+1} = y_i(2 - x \cdot y_i)$. Calculating a reciprocal is important because it can be used to perform division, since $A/B = A \cdot 1/B$.

2.9 Computer Logic

We now turn our attention to *gates*, the very stuff of which computers are made. We introduce gates and circuits and demonstrate what you can do with gates. When we later describe how computers operate, you will be able to visualize the processes that are taking place within the computer at the level of the bit and byte.

Computers are constructed from two basic circuit elements—*gates* and *flip-flops*, known as *combinational* and *sequential* logic elements, respectively. A combinational logic element is a circuit whose output depends only on its current inputs, whereas the output from a sequential element depends on its past history as well as its current inputs. A sequential element can *remember* its previous inputs and is therefore also a memory element. Sequential elements themselves can be made from simple combinational logic elements.

The practical significance of gates is that they are very cheap to manufacture with integrated circuit technology allowing you to put millions of gates on a single silicon chip. Consequently, we can easily build digital systems composed of millions of gates. We take a bottom-up approach to digital systems and demonstrate what a single gate can do before showing how you can use several gates to create a functional unit. Later, we will examine how functional units are put together to create a computer.

The behavior of digital circuits can be described in terms of a formal notation called *Boolean algebra*. We include a short introduction to Boolean algebra[10] because it allows you to analyze circuits containing gates and enables circuits to be constructed in a simpler form. We don't delve deeply into Boolean algebra because this course requires only an overview. Moreover, today's engineers don't employ the classic techniques of Boolean algebra to design complex digital systems. Instead, they employ computer-aided design software tools that can turn algorithmic expressions directly into circuits.

We also introduce the *tri-state* gate that allows you to connect lots of separate digital circuits together by means of a common highway called a *bus*. A digital computer is composed of nothing more than digital circuits, buses and sequential logic elements. We discuss the basic features of the computer bus and then put everything together by showing how gates, flip-flops, and buses can be used to construct a computer.

Figure 2.12 shows the relationship between the gate and the processor. At the lowest level, we have the primitive AND, OR, and NOT gates that make up a computer. A processor is composed of a large number of individual circuits. Figure 2.12 labels the relationship between the gate and the circuit as *wiring*; that is, the gates are wired together to make circuits and this relationship is permanent.[11]

At the highest level in Figure 2.12 is the processor that *interprets* the machine-level language. Machine language is the native language of the computer that is represented in human readable form by assembly language. The connection between the processor and the

Levels of hardware

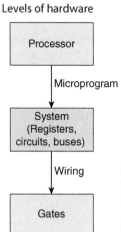

© Cengage Learning 2014

---

[10] It is common to use the term *Boolean algebra* when talking about computer logic and digital circuits. However, there are an infinite number of Boolean algebras. A Boolean algebra is a set together with operations and variables that are subject to a group of axioms (laws). The Boolean algebra we are interested in has a set with two members, 0 and 1.

[11] Not all systems have a fixed relationship between gates and circuits because the interconnections can be changed as the computer runs. Such a system is said to have programmable or *reconfigurable* hardware. Reconfigurable hardware employs the same technology used to store data in flash memory. In flash memory, 1s and 0s are stored, whereas in reconfigurable hardware, connections or non-connections are stored.

Boolean Algebra

Boolean algebra provides a means of manipulating logical expressions. Boolean algebra can be used to take a logical expression and *simplify* it; that is, convert it into a different form with fewer terms. Why should we want to do this? Because terms are implemented by gates, and the simpler an expression is, the less logic we need to implement the Boolean algebra in silicon. Less logic means fewer parts, faster testing, and greater reliability. Moreover, we can use Boolean algebra to create a circuit with particular types of gate (some gates are easier to fabricate or are faster than others).

Few designers today would directly apply Boolean simplification techniques to their circuits. Modern computer-based design tools take the math out of Boolean algebra and let you specify a circuit at its functional level. The design tool then does all the hard work of circuit design and minimization. Indeed, when using modern programmable logic, it is possible to feed equations into a computer and configure a programmable logic device.

Boolean algebra is an algebra with variables, the constants 0 and 1, and the logical operations AND, OR, and inversion. The conventional math operations of multiplication and division do not exist in Boolean algebra. The following table defines the nine *postulates* of Boolean algebra used to derive all other Boolean operations. Note that the $+$ operator is the logical OR and the $\cdot$ (dot) is the logical AND. These symbols have been chosen because OR and AND operators share some of the properties of the corresponding arithmetic operators as we shall soon see.

1	$X + 0 = X$	ORing a variable with 0 has no effect on the variable
2	$X + 1 = 1$	ORing a variable with 1 gives 1
3	$X + X = X$	ORing a variable with itself has no effect on the variable
4	$X + \bar{X} = 1$	ORing a variable and its complement is 1
5	$X \cdot 1 = X$	ANDing a variable with 1 has no effect on the variable
6	$X \cdot 0 = 0$	ANDing a variable with 0 gives 0
7	$X \cdot X = X$	ANDing a variable with itself has no effect on the variable
8	$X \cdot \bar{X} = 0$	ANDing a variable with a complement gives 0
9	$\bar{\bar{X}} = X$	The complement of a complement cancels out

Boolean variables follow the normal commutative, associative, and distributive laws of conventional algebra. For example, we can write

$X + Y + Z = Z + Y + X$	(commutative law)
$X \cdot Y = Y \cdot X$	(commutative law)
$X + (Y + Z) = (X + Y) + Z$	(associative law)
$X \cdot (Y \cdot Z) = (X \cdot Y) \cdot Z$	(associative law)
$X + Y \cdot Z = (X + Y) \cdot (X + Z)$	(distributive law)
$X \cdot (Y + Z) = X \cdot Y + X \cdot Z$	(distributive law)
$\overline{X + Y} = \bar{X} \cdot \bar{Y}$	(de Morgan's law)
$\overline{X \cdot Y} = \bar{X} + \bar{Y}$	(de Morgan's law)
$X + \bar{X} \cdot Y = X + Y$	

circuits is the *microprogram*,[12] which is an interpreter that converts individual machine-level instructions into a sequence of actions that can be carried out by the circuits composed of individual gates.

---

[12] In a later chapter, we will see that it is possible to design a computer where each machine-level instruction is translated into one or more primitives called micro-operations. A micro-operation performs actions such as moving data from one register to another. Microprogramming is not generally employed to implement today's computers, although some microprocessors are partially micro-programmed, with some instructions being executed directly and some interpreted as microinstructions.

2.9.1   Digital Systems and Gates

Digital computers are *binary* systems that use the two symbols 0 and 1 (sometimes called *false* and *true*, or low and high, or off and on). The physical representation of these symbols can be made as unlike each other as possible to give the maximum discrimination between the two digital values. Computer logic uses a two-level signal to represent 0s and 1s.

It is not widely appreciated outside the engineering community that there is more than one type of silicon chip. Several technologies are used to manufacture digital devices, and we speak of a gate as belonging to a specific *logic family*. Each of these families has its own characteristics: some are faster than others, some are cheaper to manufacture, some use less power, and some are less sensitive to radiation (making them suitable for use in space vehicles). As time passes, some families become less popular with designers and new families are introduced to reflect advances in technology. In the 1970s and 1980s, the standard logic family, TTL, operated with a 5 V power supply. By the 1990s, newer families with much lower operating voltages became available. Today, logic elements can operate from voltages as low as 1 V.[13]

A typical logic family operating from a 3 V power supply represents the logical 0 state by an output in the range from 0 to 0.1 V and the logical 1 state by an output in the range 2.9 to 3.0 V. An input in the range from 0 to 0.9 V is recognized as a logical 0, and a voltage in the range from 2.1 to 3.0 V as a logical 1. Although we talk about two logic values, these are represented by two distinct ranges of voltage.

Logic Values

Two fundamental conventions are used to describe logic systems: one is called *positive* logic and the other *negative* logic. Throughout this text, unless explicitly stated, we employ positive logic in which the logical 1 state is the electrically high state of a gate. This high state can also be called the true state, in contrast with the low state that is the false state.

Each logic state has an inverse or complement that is the opposite of its current state. The complement of a true or one state is a false or zero state, and vice versa. By convention we use an overbar to indicate a complement; for example X and $\overline{X}$.

A signal can have a constant value or a variable value. If it is a constant, it always remains in that state. If it is a variable, it may be switched between the states 0 and 1. A Boolean constant is frequently called a literal.

The signal level (i.e., electrically high or low) that causes a variable to carry out the function suggested by its name is arbitrary. If a high level causes the action, the variable is called active-high. If a low level causes the action, the variable is called active-low. Thus, if an active-high signal is labeled STOP, a high level will initiate the action. If the signal is active-low $\overline{STOP}$, a low level will trigger the action.

The term *asserted* indicates that a signal is placed in the level that causes its activity to take place; for example, if we say that START is asserted, we mean that it is placed in a high state to cause the action determined by START. Similarly, if we say that $\overline{LOAD}$ is asserted, we mean that it is placed in a low state to trigger the action.

2.9.2   Gates

All digital computers can be constructed from only three fundamental types of gates, called AND, OR, and NOT gates, together with flip-flops. Because flip-flops themselves can be

---

[13] K. Ragini and B. K. Madhavi, "Ultra low-power digital-logic circuits in sub-threshold for biological applications," *Journal of Theoretical and Applied Information Technology*, 2005, pp. 584–590.

constructed from gates, it follows that all computers can be constructed from gates alone. Moreover, because one gate, the NAND gate, can be used to synthesize AND, OR, and NOT gates, we can say that any computer can be constructed from nothing more than a large number of NAND gates. We begin our discussion of gates by introducing the three fundamental gates and then describe some gates that are derived from the fundamental gates.

Fundamental Gates

Figure 2.13 shows a black box with two *input* terminals, A and B, and a single *output* terminal C. This device takes the two logic values at its input terminals and produces an output that depends only on the states of the inputs and the nature of the logic element. At any time the inputs and outputs can be in only one of two states: 0 or 1.

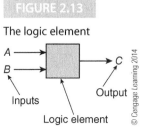

FIGURE 2.13

The logic element

The AND Gate

The behavior of a gate is described by its *truth table* that defines its output for each of the possible inputs. Table 2.8a provides the truth table for the two-input AND gate. If one input is A and the other B, output C is true (i.e., 1) if and only if inputs A and B are both 1. The AND function is represented in equations by the *dot*, so that the operation A AND B is written $A \cdot B$. Figure 2.14 gives the circuit symbol for a two-input AND gate and a three-input AND gate. An AND gate can have any number of inputs. Table 2.8b gives the truth table for an AND gate with *three* inputs A, B, and C and an output $D = A \cdot B \cdot C$. In this case, D is 1 only when inputs A, B, and C are each 1 simultaneously.

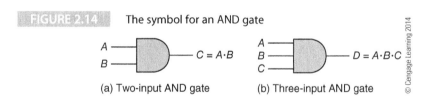

FIGURE 2.14   The symbol for an AND gate

(a) Two-input AND gate   (b) Three-input AND gate

TABLE 2.8   Truth Table for the AND Gate

B	A	C = A·B
0	0	0
0	1	0
1	0	0
1	1	1

(a) Two-input AND gate

C	B	A	D = A·B·C
0	0	0	0
0	0	1	0
0	1	0	0
0	1	1	0
1	0	0	0
1	0	1	0
1	1	0	0
1	1	1	1

(b) Three-input AND gate

The OR Gate

Another fundamental gate is the OR. Table 2.9 gives the truth tables for two-input and three-input OR gates. The output of an OR gate is 1 if either one or more of its inputs are 1. In fact, the only way to make the output of an OR gate go to a logical 0 is to set all its inputs to 0. The OR function is represented by the $+$ symbol, so that the operation A OR B is written $A + B$. Figure 2.15 gives the circuit symbol for the two-input OR and three-input OR gates.

TABLE 2.9   Truth Table for the OR Gate

B	A	$C = A + B$
0	0	0
0	1	1
1	0	1
1	1	1

(a) Two-input OR gate

C	B	A	$D = A + B + C$
0	0	0	0
0	0	1	1
0	1	0	1
0	1	1	1
1	0	0	1
1	0	1	1
1	1	0	1
1	1	1	1

(b) Three-input OR gate

© Cengage Learning 2014

FIGURE 2.15   The symbol for an OR gate

A —⟩ $C = A + B$ A, B, C —⟩ $D = A + B + C$

(a) Two-input OR gate   (b) Three-input OR gate

© Cengage Learning 2014

The · and + operators used for AND and OR functions, respectively, behave rather like multiplication and addition in conventional algebra. The equation $P = A{\cdot}B + C{\cdot}D$ states that P is true if either A and B are true or if C and D are true. As in the case of conventional algebra, the AND operator has a higher priority than the OR operator and $P = A{\cdot}B + C{\cdot}D$ is the same as $P = (A{\cdot}B) + (C{\cdot}D)$.

The Inverter

The third basic gate is the simplest of all gates, the *inverter* or *complementer* that has a *single* input and output. Its output is the *logical complement* of its input; if the input is A, the output is written NOT A. Figure 2.16 gives the circuit symbol and truth table for an inverter. If the input is 1, the output is 0 and vice versa. The small circle in Figure 2.16 indicates inversion—whenever you see such a circle in a logic diagram, the signal passing through it is inverted.

As an aid to understanding the operation of basic gates, Table 2.10 provides the truth tables for AND and OR gates for both constant and variable inputs.

FIGURE 2.16   The symbol and truth table for an inverter

A —▷∘— $\overline{A}$

(a) Symbol for inverter

A	$\overline{A}$
0	0
0	1

(b) Truth table of inverter

© Cengage Learning 2014

TABLE 2.10   Truth Table for AND and OR Gates with Both Constant and Variable Inputs

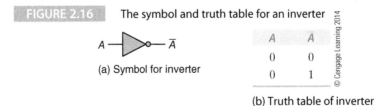

AND		OR	
Constant	Variable	Constant	Variable
$0{\cdot}0 = 0$	$A{\cdot}0 = 0$	$0 + 0 = 0$	$A + 0 = A$
$0{\cdot}1 = 0$	$A{\cdot}1 = A$	$0 + 1 = 1$	$A + 1 = 1$
$1{\cdot}0 = 0$	$A{\cdot}\overline{A} = 0$	$1 + 0 = 1$	$A + \overline{A} = 1$
$1{\cdot}1 = 1$	$A{\cdot}A = A$	$1 + 1 = 1$	$A + A = A$

© Cengage Learning 2014

Having introduced the three basic gates, we can point out that any combinational logic circuit can be constructed from AND, OR, and NOT gates. A circuit can be represented in one of two forms. Figure 2.17 shows the *sum of products* circuit where the input variables or their complements are used to form product terms by AND gates. These product terms are ORed together in a single OR gate. Figure 2.18 illustrates a *product of sums* circuit that uses OR gates to generate the sum terms and a single AND gate to form the product of the sums.

Derived Gates—the NOR (Not OR), NAND (Not AND), and Exclusive OR

Figure 2.19 shows the circuit symbols for three gates derived from a combination of AND, OR, and NOT gates;[14] the NOR gate, the NAND gate, and the Exclusive OR gate. The truth

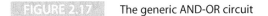

FIGURE 2.17   The generic AND-OR circuit

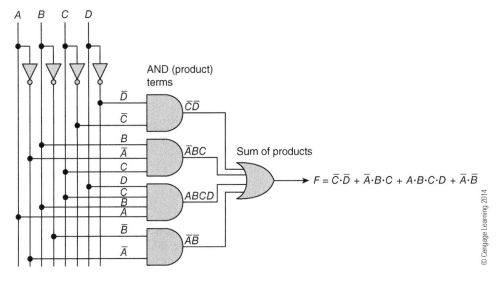

© Cengage Learning 2014

FIGURE 2.18   The generic OR-AND circuit

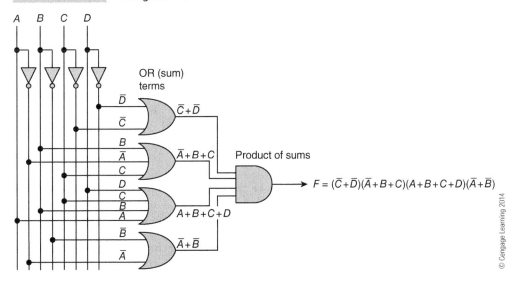

© Cengage Learning 2014

---

[14] From a mathematical point of view, the AND, OR, and NOT gates are fundamental and NAND gates, etc. can be derived from them. However, from the point of view of the circuit designer, the NAND gate is fundamental (easiest to fabricate) and other gates are derived from that.

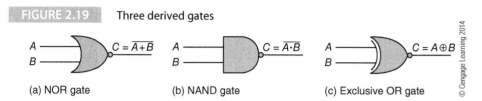

FIGURE 2.19   Three derived gates

(a) NOR gate   (b) NAND gate   (c) Exclusive OR gate

© Cengage Learning 2014

tables for these gates are given in Table 2.11. The term *NOR* gate is a contraction of the words *Not* and *OR*, because a NOR gate is equivalent to an OR gate followed by an inverter. The symbol for the NOR gate in Figure 2.19a is the same as the OR gate of Figure 2.15, except that the NOR gate has a *bubble* at its output to show that the output is inverted. If you look at the truth table for the NOR gate (Table 2.11a), you will see that its output is true if and only if both inputs are false.

The NAND gate, Figure 2.19b, is equivalent to an AND gate followed by an inverter (i.e., **NOT AND**). Table 2.11b shows that the output of a NAND gate is false if both inputs are true and true otherwise. Any digital circuit can be constructed from NOR gates only (or NAND gates only).

TABLE 2.11   Truth Table for the NOR Gate, NAND Gate, and Exclusive OR Gates

A	C	$C = \overline{A + B}$	. .	A	B	$C = \overline{A \cdot B}$	. .	A	B	$C = A \oplus B$
0	0	1		0	0	1		0	0	0
0	1	0		0	1	1		0	1	1
1	0	0		1	0	1		1	0	1
1	1	0		1	1	0		1	1	0

(a) The NOR gate   (b) The NAND gate   (c) The XOR gate

© Cengage Learning 2014

The Exclusive OR function of Figure 2.19c is often written XOR or EOR and uses the symbol $\oplus$ (e.g., $C = A \oplus B$). As Table 2.11c demonstrates, the difference between an *exclusive* OR and a conventional *inclusive* OR gate arises when *both* inputs are true (that is, $1 + 1 = 1$), whereas $1 \oplus 1 = 0$. An XOR gate can be constructed from two inverters, two AND gates and an OR gate, as Figure 2.20 demonstrates because $A \oplus B = \overline{A} \cdot B + A \cdot \overline{B}$.

FIGURE 2.20   Constructing an XOR circuit from AND, OR, and NOT gates

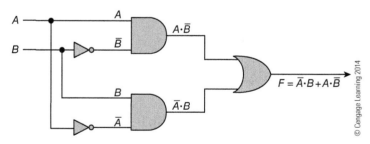

© Cengage Learning 2014

Table 2.11(c) demonstrates that the XOR function is a *difference* detector—the output is true if the inputs are different (i.e., 0 and 1 or 1 and 0). This circuit is also a primitive two-bit adder because $0 + 0 = 0, 0 + 1 = 1, 1 + 0 = 1$, and $1 + 1 = 0$ (if we neglect the carry-bit).

Because of the way in which the XOR function is implemented, it is easy to imagine that the exclusive OR is a function of two variables only. Indeed, many texts imply that an XOR function has only two inputs. You can have exclusive OR functions of any number of variables. For example $F = A \oplus B \oplus C$. XOR gates can be used as a building block in binary adders and as programmable inverters (if one input of an XOR gate is C and the other X, then the output is X if $C = 0$ (pass through) or $\overline{X}$ if $C = 1$ (invert).

2.9.3   Basic Circuits

Several gates can be interconnected to create a *circuit*. All you have to remember is that the *output* of a gate can be connected to the *input* of one or more other gates, but you cannot connect the outputs of two or more gates together.[15] Because the output of a gate *drives* anything

Transistors and Gates

Computers are constructed from functional units. Functional units are constructed from gates. Gates are constructed from transistors. Therefore, computers are constructed from transistors. Lots and lots of 'em—a microprocessor chip can have over a billion transistors on a single chip of silicon.

This is not an electronics text. However, it is helpful to show how real gates are implemented. The basic element of the integrated circuit is the *transistor*. A transistor is an electronic switch with three terminals called *gate, source* and *drain*. Current either flows between two terminals (source and drain) or it does not flow between two terminals depending on the voltage at a third terminal (the gate). Transistors fall into two categories: the *bipolar* transistor (the type first made at Bell Labs in 1947) which is found in circuits like amplifiers and the *field effect* transistor that is used in digital circuits. Field effect transistors can be *p-type* or *n-type* depending on their physical structure. Modern digital logic is constructed using CMOS technology (complementary metal oxide semiconductor) which uses pairs of transistors—one *n*-type and one *p*-type—in series. At any instant either the *n*-type is switched off or the p-type is switched off and there is virtually no electrical conduction through the pair. This feature ensures that CMOS technology uses very little power indeed.

The diagram below illustrates a single transistor and shows the circuits of an inverter, NAND gate, and NOR gate (AND and OR gates are made from NAND and NOR gate plus inverters). A transistor with an *inward pointing arrow* is an *n*-type device [e.g., T2 in figure (a)] and a transistor with an *outward pointing arrow* is a *p*-type device [e.g., T1 in figure (a)]. An *n*-type transistor is turned off by a low voltage, and a *p*-type transistor is turned off by a high voltage at its gate.

If you look at the circuit of the inverter (a), there are two transistors, T1 and T2, connected in series between the ground, 0 V, and positive, V_{dd}, power rails. If the input is at a low voltage, transistor T2 is turned off and T1 is turned on, which connects the output to V_{dd}. If the input voltage is high, transistor T2 is on and T1 is turned off, which connects the output to ground. Consequently, the output voltage is the inverse of the input voltage.

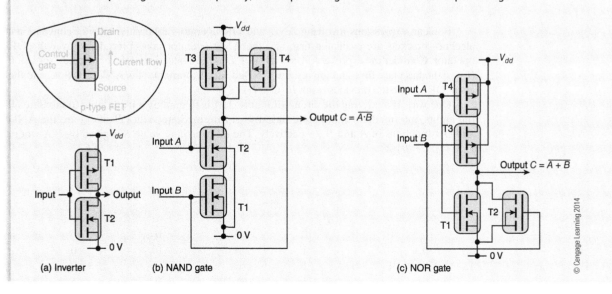

(a) Inverter   (b) NAND gate   (c) NOR gate

© Cengage Learning 2014

---

[15] The output of a gate is determined by its inputs. If you connected outputs A and B from gates A and B together, you could have the situation in which A was low and B high. Since they are wired together both outputs could not physically be in the same state—in reality one of the gates could be burnt out. However, there are special gates, called tri-state gates, whose outputs can be connected together because only one gate at a time attempts to drive its output to a 0 or a 1.

(continued)

Diagram (b) demonstrates a NAND gate. Remember that the output is the negated AND of the inputs; that is, the output is low only if both inputs are high. As you can see, transistors T1 and T2 are in series and both of them are turned on by a high voltage on input *A* and a high voltage on input *B*. When they are both turned on, the output goes low (i.e., inputs 1,1 give an output 0).

Diagram (c) is a NOR gate where the output is low if either input is high or the output is high if both inputs are low. In this case, transistors T1 and T2 are connected in parallel. If both inputs are low, neither conducts and the output is high.

connected to that output to the logic level defined by the gate's inputs, connecting two outputs together leads to an *undefined* logic state if one output is 1 and the other 0.

Figure 2.21 describes a circuit with four gates, labeled G1, G2, G3, and G4. Lines that cross each other in a diagram *without a dot* at their intersection are not connected together—lines that meet at a dot are connected. This circuit has three inputs A, B, and X and an output C. It also has three intermediate logical values; the outputs of gates G1, G2, and G3 are labeled P, Q, and R, respectively. We can treat a gate as a *processor* that operates on its inputs according to its logical function; for example, the inputs to AND gate G3 are P and X, and its output is $P \cdot X$. Because $P = A + B$, the output of G3 is $(A + B) \cdot X$. Similarly, the output of gate G4 is $R + Q$, which is $(A + B) \cdot X + A \cdot B$.

FIGURE 2.21   A circuit with four gates

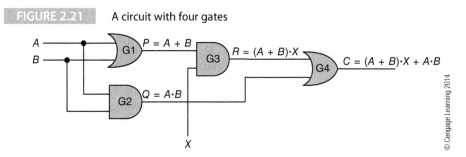

© Cengage Learning 2014

Boolean expressions involving AND and OR operators follow the rules of conventional algebra—brackets are evaluated first, and the AND operator takes precedence over the OR operator. Consider the expression $A \cdot C + B \cdot (C + D)$. The evaluation of the bracketed expression $(C + D)$ takes place first, and the result is ANDed with B. Variable A is ANDed with C, and the result ORed with the previous result $B \cdot (C + D)$.

One way of analyzing the circuit of Figure 2.21 is to employ a truth table to tabulate all the inputs, intermediate values, and outputs— . Intermediate values P and Q are the AND and OR functions of A and B, respectively. The intermediate value R is given by $P \cdot X$ where $P \cdot X = (A + B) \cdot X$.

TABLE 2.12   Truth Table for Figure 2.21

Inputs			Intermediate Values			Output
X	A	B	$P = A + B$	$Q = A \cdot B$	$R = (A + B) \cdot X$	$C = Q + R$
0	0	0	0	0	0	0
0	0	1	1	0	0	0
0	1	0	1	0	0	0
0	1	1	1	1	0	1
1	0	0	0	0	0	0
1	0	1	1	0	1	1
1	1	0	1	0	1	1
1	1	1	1	1	1	1

© Cengage Learning 2014

The output of the circuit is $C = (A + B) \cdot X + A \cdot B$, as Figure 2.21 demonstrates. But what does this *mean*? If you look at Table 2.12, you can find a pattern. When the input X is 0, the output C is identical to $A \cdot B$ (i.e., A AND B). When input $X = 1$, the output is identical to $A + B$ (i.e., A OR B). We have just constructed a *programmable logic element* that performs an AND operation or an OR operation depending on the state of one of its inputs, X. Suppose X is connected to a signal that comes from an *instruction* in a computer, the operation carried out would be a logical AND or an OR depending on the bit pattern of the instruction. In a real computer, the circuit would be replicated for each bit of a word and all control inputs would be connected together. We now introduce some important circuits.

The Half Adder and Full Adder

If you take a look at Table 2.11c, you can see that the XOR gate calculates the *sum* of two bits as we have already pointed out. Table 2.13 gives the truth table of a *half adder* that adds bit A to bit B to produce a sum S and a carry C, and Figure 2.22 shows the possible structure of a two-bit adder. The carry bit is generated by ANDing the two inputs.

A single two-bit half adder is of limited practical use, because computers have to add together two m-bit words; that is, bit a_i must be added to bit b_i together with a carry-in from stage $i - 1$ of the adder to produce sum s_i and carry-out c_i to stage $i + 1$. We therefore need a so-called *full adder* that adds two bits together with a carry-in to produce a sum and a carry-out.

Inversion Bubbles

By convention, inverters are often omitted from circuit diagrams and bubble notation is used. A small bubble is placed at a gate's input to indicate inversion. In the circuit below, the two AND gates form the product of NOT A AND B, and A AND NOT B.

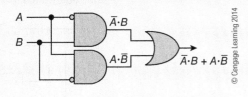

© Cengage Learning 2014

FIGURE 2.22   The two-bit adder (the half adder)

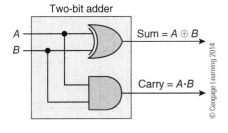

© Cengage Learning 2014

TABLE 2.13

Truth Table of a Half Adder

A	B	Sum	C
0	0	0	0
0	1	1	0
1	0	1	0
1	1	0	1

© Cengage Learning 2014

If one XOR gate adds two bits to produce a sum, we can take the output of an XOR gate and feed it into a second XOR gate together with a carry-in to produce the sum of three bits. But how do we generate a carry-out? If you add three bits, you will get a carry if either two or three of the inputs are 1. Figure 2.23 demonstrates how two half adders are used to create a full adder that adds together three bits to produce a sum and a carry-out. The carry-out bit is generated by ORing the carry bits generated by the two half adders.

FIGURE 2.23   The full adder

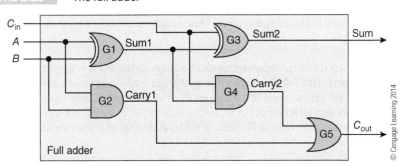

© Cengage Learning 2014

Is the circuit of Figure 2.23 really that of a full adder? One way of determining whether this is a full adder circuit is to write down its truth table and compare the outputs with those of a full adder—Table 2.14. The left-most three columns in Table 2.14 cover all eight possible values of the three bits to be added (C_{in}, A, and B), and the next two columns display the carry-out, C_{out}, and sum of a full adder that implements the binary addition C_{in} plus A plus B. The right-hand five columns in Table 2.14 provide the intermediate values (Carry1, Sum1, Carry2) generated by the circuit of Figure 2.23 and the two outputs (C_{out} and Sum). As you can see, this circuit does indeed implement the function of a full adder.

One Bit of an ALU

The purpose of this chapter is to demonstrate how gates are used to construct the CPU and other digital systems. Although we have covered only the very basics, at this stage we can demonstrate the basic concept of an ALU, arithmetic and logical unit. The diagram below describes one bit of a very primitive ALU that can perform five operations on two bits A and B (XOR, AND, OR NOT A and NOT B). The actual function performed is determined by the three-bit control signal F_2, F_1, F_0.

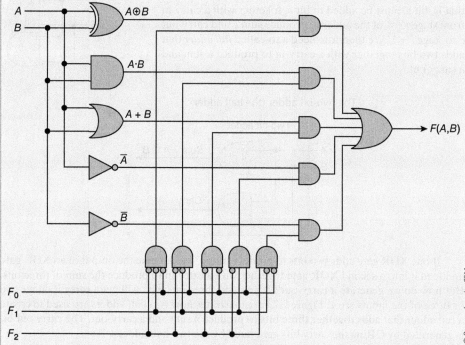

In order to implement a real ALU, this circuit would have to be replicated 32 times for a 32-bit word. Note that this slice through an ALU cannot perform arithmetic as we have not included carry-in and carry-out paths. That would require only a minor extension to the circuit.

The way in which the circuit operates should be obvious. The five functions are generated by the five gates on the left. On the right, five AND gates are used to gate the selected function to an OR gate to provide the output. The five AND gates along the bottom decode the function select input F_2, F_1, F_0 into one-of-five lines, which are used to gate the required function to the output. The little bubbles at the inputs of the five AND gates act as inverters.

TABLE 2.14			Truth Table for the Circuit of Figure 2.24						
Inputs			Full Adder		Full Adder Circuit				
C_{in}	A	B	C_{out}	Sum	Carry1	Sum1	Sum2	Carry2	C_{out}
0	0	0	0	0	0	0	0	0	0
0	0	1	0	1	0	1	1	0	0
0	1	0	0	1	0	1	1	0	0
0	1	1	1	0	1	0	0	0	1
1	0	0	0	1	0	0	1	0	0
1	0	1	1	0	0	1	0	1	1
1	1	0	1	0	0	1	0	1	1
1	1	1	1	1	1	0	1	0	1

© Cengage Learning 2014

We can represent a full adder in a different way. Figure 2.24 shows that a full adder can be realized by a three-input EOR circuit that takes the exclusive OR of three bits (i.e., the output is true if and only if an odd number of inputs are true). The carry function is generated by three AND gates and an OR gate. We leave it to the reader to demonstrate that this is indeed a full adder.

FIGURE 2.24   Alternative full adder circuit

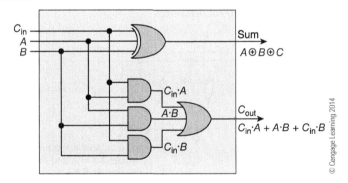

© Cengage Learning 2014

We need m full adder circuits to add two m-bit words *in parallel*,[16] as Figure 2.25 demonstrates. Each of the m full adders adds bit a_i to bit b_i, together with a carry-in from the stage on its right, to produce a carry-out to the stage on its left. You might think that the least-significant stage that adds a_0 to b_0 doesn't require a full adder, because it doesn't have a carry-in (the carry-in in Figure 2.25 is shown as 0). In practice, a parallel adder does have a carry-in to the first stage, C_{in}, for three reasons. The first is that if B is set to 0 and C_{in} set to 1, the parallel adder performs the operation A plus 1. The second is that if the carry-in to the least-significant stage C_{in} is set to the carry-out of a previous addition, the circuit can be used to perform *chained arithmetic*. The third reason is that C_{in} can be used to facilitate *subtraction*, as we shall soon see.

Suppose we make one of the inputs of an XOR gate a *control* input, X, and feed the binary value a_i to the other input. If X is 1, the output is 0 if a_i is 0 and 1 if a_i is 1. When X is 0, the

---

[16] This circuit is traditionally called a parallel adder because all the bits of the two words to be added are presented to it at the same time in contrast with the serial added where two bits of a word are added and then the next two bits are added in exactly the same way humans perform addition. However, the circuit is not truly parallel because bit a_i cannot be added to bit b_i until the carry-in bit c_{i-1} has been calculated by the previous stage. This is a *ripple through* adder because addition is not complete until any carry bit has rippled through the circuit which, in the worst case, corresponds to $111, \ldots, 11 + 000, \ldots, 01 = (1) 000, \ldots, 00$. Real adders use high-speed *carry look-ahead* circuits to generate carry bits more rapidly and speed addition.

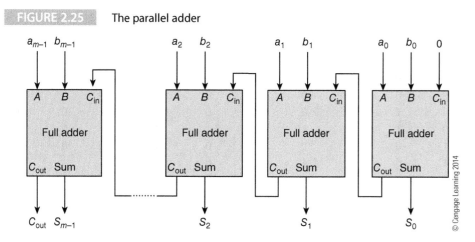

FIGURE 2.25   The parallel adder

XOR gate *copies* the other input to its output. When X is 1, the output is 1 if a_i is 0, and 0 if a_i is 1. When X is 1, the XOR inverts the input. Figure 2.26 shows how m XOR gates implement an m-bit programmable inverter.

Figure 2.27 demonstrates how we can place a *programmable inverter* (i.e., an XOR gate) in series with one of the inputs of each parallel adder to create an adder/subtractor. The control input is connected to the carry-in bit of the least-significant stage. When the control

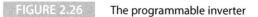

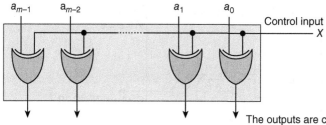

FIGURE 2.26   The programmable inverter

The outputs are copies of the respective inputs if $X = 0$, and the complements of the inputs if $X = 1$

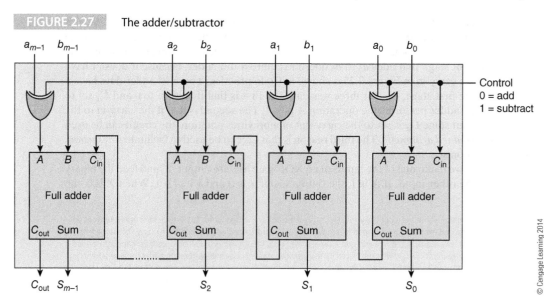

FIGURE 2.27   The adder/subtractor

input is 0, the bits of A are passed unchanged to the adder and the circuit calculates A plus B. When the control input is set to 1, the programmable inverter negates the bits of A and adds 1 to the carry-in of the first stage. Because the two's complement of a number is formed by inverting its bits and adding 1, this circuit performs *addition* when the control bit is 0, and *subtraction* when it is 1. In a computer, the control bit might be derived from the instruction being executed (e.g., ADD or SUB).

The Decoder

Let's look at some of the interesting things you can do with a few gates. The circuit of Figure 2.28 has three inputs A, B, and C, and eight outputs Y_0 to Y_7. The three inverters generate the complements of the inputs A, B, and C. Each of the eight AND gates is connected to three of the six lines $\overline{A}$, A, $\overline{B}$, B, $\overline{C}$, C (each of the three variables must appear in either its true or complemented form).

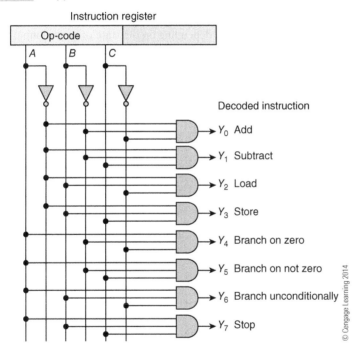

FIGURE 2.28   Application of a decoder

© Cengage Learning 2014

The output of gate Y_0 is $\overline{A} \cdot \overline{B} \cdot \overline{C}$ and is 1 if all inputs to the AND gates are 1 (i.e., $\overline{A} = 1$, $\overline{B} = 1$, and $\overline{C} = 1$). Therefore, Y_0 is 1 when $A = 0$, $B = 0$, and $C = 0$. If you examine the other AND gates, you will see that each gate is enabled by one of the eight possible combinations of A, B, and C.

This circuit is called a three-line to eight-line *decoder*, because it converts a three-bit binary value of A, B, and C into one of 2^3 outputs. Table 2.15 gives a truth table for this circuit, which is also called a *decoder* because it can take, for example, the op-code field of an instruction register and decode it into individual instructions as Figure 2.28 demonstrates. In this example, we have converted a 3-bit op-code into eight *actions*.

The Multiplexer

Figure 2.29 describes a two-line to one-line *multiplexer* constructed from NAND gates. We've used a NAND to synthesize an inverter because, if the inputs of a NAND gate are connected together, it behaves exactly like an inverter. Figure 2.29 also provides a truth table for this

TABLE 2.15	The Decoder									
Inputs			Outputs							
A	B	C	Y_0	Y_1	Y_2	Y_3	Y_4	Y_5	Y_6	Y_7
0	0	0	1	0	0	0	0	0	0	0
0	0	1	0	1	0	0	0	0	0	0
0	1	0	0	0	1	0	0	0	0	0
0	1	1	0	0	0	1	0	0	0	0
1	0	0	0	0	0	0	1	0	0	0
1	0	1	0	0	0	0	0	1	0	0
1	1	0	0	0	0	0	0	0	1	0
1	1	1	0	0	0	0	0	0	0	1

© Cengage Learning 2014

circuit that has three inputs and an output. The truth table shows that when input $C = 0$ the output is given by A, and when $C = 1$, the output is given by B. This circuit is an *electronic switch* that selects input A or input B, depending on the state of control input C. Figure 2.30 demonstrates the same circuit constructed of AND, OR, and NOT gates.

FIGURE 2.29	The two-input multiplexer and its truth table

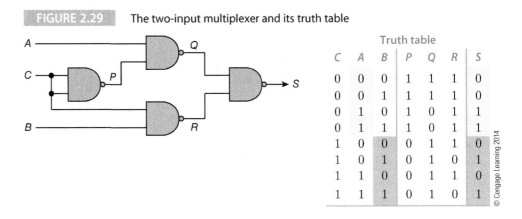

Truth table

C	A	B	P	Q	R	S
0	0	0	1	1	1	0
0	0	1	1	1	1	0
0	1	0	1	0	1	1
0	1	1	1	0	1	1
1	0	0	0	1	1	0
1	0	1	0	1	0	1
1	1	0	0	1	1	0
1	1	1	0	1	0	1

© Cengage Learning 2014

FIGURE 2.30	Alternative representation of the two-input multiplexer

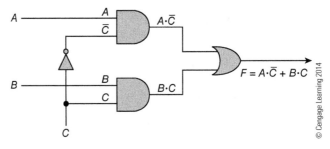

© Cengage Learning 2014

The Voting Circuit

Figure 2.31 illustrates a three-input *majority logic* (or voting) circuit whose output corresponds to the state of the majority of the inputs. This circuit uses three two-input AND gates labeled G1, G2, and G3 and a three-input OR gate labeled G4 to generate an output F.

Table 2.16 tabulates the output F in Figure 2.31 for all the eight possible combinations of the three inputs A, B, and C. We have also included columns for the outputs of the three AND gates.

FIGURE 2.31   The majority logic circuit

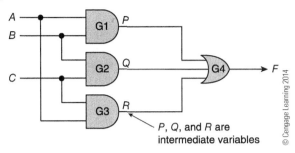

P, Q, and R are intermediate variables

© Cengage Learning 2014

TABLE 2.16   Truth Table for the Majority Logic Circuit of Figure 2.31

| Inputs | | | Intermediate Values | | | Output |
A	B	C	$P = A \cdot B$	$Q = B \cdot C$	$R = A \cdot C$	$F = P + Q + R$
0	0	0	0	0	0	0
0	0	1	0	0	0	0
0	1	0	0	0	0	0
0	1	1	0	1	0	1
1	0	0	0	0	0	0
1	0	1	0	0	1	1
1	1	0	1	0	0	1
1	1	1	1	1	1	1

© Cengage Learning 2014

The three intermediate signals P, Q, and R are $P = A \cdot B$, $Q = B \cdot C$, and $R = A \cdot C$. We can write down the output function, F, as the logical OR of P, Q, and R; that is, $F = P + Q + R$. Substituting the expressions for P, Q, and R gives $F = A \cdot B + B \cdot C + A \cdot C$. By visually inspecting Table 2.16, we can see that the output is true if two or more of the inputs A, B, and C are true. This circuit implements a *majority logic* function whose output takes the same value as the majority of inputs.

Figure 2.32 shows another simple circuit using four four-input AND gates and an OR gate. This circuit detects the 3-of-4 condition in which exactly three of four variables are asserted high. Because there are four possible combinations of 3-out-of-4 (ABC, ABD, ACD, and

FIGURE 2.32   The 3-of-4 detector

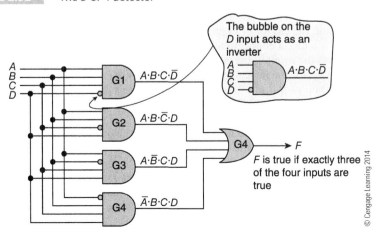

The bubble on the D input acts as an inverter

$A \cdot B \cdot C \cdot \bar{D}$

$A \cdot B \cdot C \cdot \bar{D}$

$A \cdot B \cdot \bar{C} \cdot D$

$A \cdot \bar{B} \cdot C \cdot D$

$\bar{A} \cdot B \cdot C \cdot D$

F is true if exactly three of the four inputs are true

© Cengage Learning 2014

BCD), we require four *four*-input gates. A four-input gate is required because we have to detect exactly three of the four variables true; that is, we need to detect $\overline{A}\cdot B\cdot C\cdot D$, $A\cdot\overline{B}\cdot C\cdot D$, $A\cdot B\cdot\overline{C}\cdot D$, and $A\cdot B\cdot C\cdot\overline{D}$.

Because each AND gate requires an inverter, we require an inverter in the path between an input variable and one of the AND gate's inputs. Figure 2.32 shows an alternative means of representing an inverter. An inverting bubble is shown at the appropriate inverting input of each AND gate. This inverting bubble can be applied at the input or output of any logic device.

The Prioritizer

Figure 2.33 describes the *prioritizer*, which is a circuit that deals with *competing* requests for attention and is found in multiprocessor systems where several processors can compete for access to memory. Suppose that the prioritizer's five inputs x_0 to x_4 are connected to the outputs of five devices that can make a request for attention. In Figure 2.33, input x_4 has the highest priority. That is, device i can put a logical 1 on input x_i to request attention at priority level i. If several inputs are set to 1 at the same time, the prioritizer sets only one of its outputs to 1, and all the other outputs remain at 0. For example, if the input is $x_4, x_3, x_2, x_1, x_0 = 00110$, the output $y_4, y_3, y_2, y_1, y_0 = 00100$, because the highest level of input is x_2.

FIGURE 2.33   The priority circuit

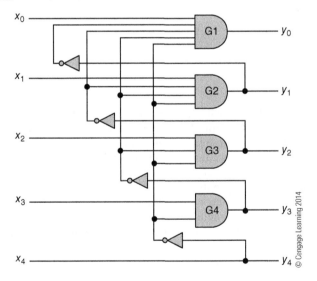

© Cengage Learning 2014

If you examine the circuit of Figure 2.33, you can see that output y_4 is equal to input x_4 because there is a direct connection. If x_4 is 0, then y_4 is 0; and if x_4 is 1, then y_4 is 1. The value of y_4 is fed to the *input* of the AND gates G3, G2, and G1 in the lower priority stages via an inverter. If x_4 is 1, the logical level at the inputs of the AND gates is 0, which disables them and forces their outputs to 0. If x_4 is 0, the value fed back to the AND gates is 1 and therefore they are not disabled by x_4. Similarly, when x_3 is 1, gates G2 and G1 are disabled, and so on.

We have described just a handful of digital circuits here out of the millions of digital circuits in existence. However, the circuits we have described are quite representative of the type of combinational digital circuit used to build computers. We introduce the other type of circuit used to construct computers, the flip-flop, in Section 2.10 and then provide a short introduction to more formal methods of constructing sequential systems. Before that, we provide a short section on Boolean algebra for readers with no prior knowledge of this topic.

How a Computer Implements Decisions

The purpose of this panel is to demonstrate how remarkably little and relatively simple logic is required to implement the action that makes a computer a computer—the ability to take decisions. We will be looking at this topic in more detail when we discuss the structure and programming model of the CPU in later chapters. However, we have introduced the concept of conditional behavior in the first chapter.

Suppose we wish to implement an operation like IF $(a = b)$ THEN do this ELSE do that. Two actions are necessary. The first is a data operation involving the comparison of a and b, and the second is a control action that selects between one of two courses of action (the THEN code and the ELSE code).

The following diagram illustrates how these concepts can be implemented in logic. In this example, we have used NAND logic to demonstrate that you can build a circuit with just one gate. The left-hand box is the comparator, which is just a simple exclusive OR (XOR) that gives a 1 output if the inputs are different and a 0 output if they are the same. Of course, this compares only two single bits. To compare two m-bit words, you'd need m XORs in parallel and would then test all XOR outputs for 0 (inputs the same) or any input 1 (inputs different).

The output of the comparators is used to compare a multiplexer that selects between two values: one is the address of the code to execute the THEN construct and the other is the code to execute the ELSE construct. Note that practical computers implement an IF-THEN-ELSE construct by jumping to a non-sequential block of code if the THEN part is true, or continuing from the current position if the ELSE part is true. In this simple example, the address multiplexer chooses between two single-bit addresses. In order to handle an n-bit address, you would need n multiplexers in parallel.

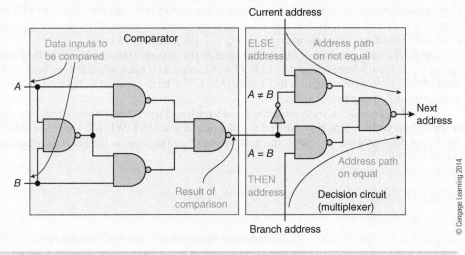

2.10   Sequential Circuits

All the circuits we've looked at have one thing in common: Their outputs are determined only by the inputs and the configuration of the gates. These circuits are called *combinational* circuits. We now look at a circuit whose output is determined by its inputs, the configuration of its gates, and its *previous state*. Such a device is called a *sequential circuit* and has the property of *memory*, because its current state is determined by its previous state. The fundamental sequential circuit building block is known as a *bistable* because its output can exist in one of two stable states. By convention, a bistable circuit that responds to the state of its inputs at any time is called a *latch*, whereas a bistable element that responds to its inputs only at certain times is called a *flip-flop*. The three basic types of bistable we describe here are the RS, the D, and the JK. After introducing these basic sequential elements we describe elements that are constructed from flip-flops or latches: the *register* and the *counter*.

2.10.1   Latches

Figure 2.34 provides the circuit and symbol of a simple latch (output P is labeled $\overline{Q}$ by convention). The output of NOR gate G1, P, is connected to the input of NOR gate G2. The output of NOR gate $G2$ is Q and is connected to the input of NOR gate G1. This circuit employs *feedback*, because the input is defined in terms of the output; that is, the value of P determines Q, and the value of Q determines P.

FIGURE 2.34   Feedback in a logic circuit

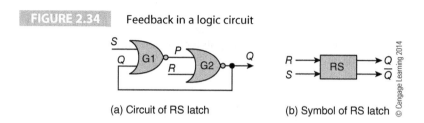

(a) Circuit of RS latch   (b) Symbol of RS latch

© Cengage Learning 2014

Assume that the S input to gate G1 and the R input to gate G2 are currently 0. If we also assume that output $Q = 1$, the output of gate G1 is 0 (because its inputs are 0, 1). The inputs to gate G2 are both 0 and its output, Q, is 1, which is what we assumed. Now suppose that $Q = 0$ initially. The output of gate G1 is 1, and the inputs to G2 are 0 and 1. The output of G2 is therefore 0, which is also what we assumed. In other words, this circuit *maintains* its current state with $Q = 0$ or $Q = 1$. It's a *memory element*.

Inputs R and S are *control inputs* that change the current state of the latch. If Q is 1, setting input R to 1 forces output Q to 0. Q will remain at 0 even when R goes back to 0. Similarly, if $Q = 0$, setting input S to 1 forces Q to 1. The latch is called an RS latch because its output can be reset to 0 or set to 1.

Table 2.17 provides a truth table for the RS latch of Figure 2.34. Whenever $R = 1$, the output Q is reset to 0. Similarly, when $S = 1$ the output is set to 1. When R and S are both 0, the output does not change; that is, the new output Q^+ is the same as the old (i.e., previous) output Q.

TABLE 2.17   Truth Table for the Circuit in Figure 2.34

Inputs			Output
R	S	Q	Q⁺
0	0	0	0
0	0	1	1
0	1	0	1
0	1	1	1
1	0	0	0
1	0	1	0
1	1	0	?
1	1	1	?

© Cengage Learning 2014

The output of the circuit is currently Q, and the new inputs to be applied to the input terminals are R, S. When these inputs are applied to the circuit, its output is given by Q^+. For example, if the current output Q is 1, and the new values of R and S are 1, 0, then the new output, Q^+, will be 0. This value of Q^+ becomes the next value of Q when new inputs R and S are applied to the circuit.

If both R and S are simultaneously 1, the output is undefined (hence the question marks in truth Table 2.17), because the output can't be set and reset at the same time. As the RS latch is implemented by two NOR gates, the output X must go low when $A = B = 1$ and *both* outputs Q and $\overline{Q}$ are 0 simultaneously. In practice, the user of an RS latch should avoid the condition $R = S = 1$.

Table 2.18 shows an alternative truth table for an RS latch. When $R = S = 0$, the new output Q^+ is simply the old output Q. In other words, the output doesn't change state and remains in its previous state as long as R and S are both 0. The inputs $R = S = 1$ result in the output $Q^+ = X$, where X indicates an indeterminate or undefined condition.

TABLE 2.18   An Alternative Truth Table for the RS Latch

Inputs		Output	Description
R	S	Q^+	
0	0	Q	No change
0	1	1	Set output to 1
1	0	0	Reset output to 0
1	1	X	Forbidden[17]

© Cengage Learning 2014

An RS latch is constructed from two cross-coupled NAND gates just as easily as from two NOR gates. Figure 2.35 illustrates a two-NAND gate latch whose truth table is given in Table 2.19. The significant difference between the NOR gate latch of Figure 2.34 and the NAND gate latch of Figure 2.35 is that the inputs to the NAND gate latch are active-low. The *no change* input to the NAND gate latch is $R, S = 1, 1$; the output is cleared by forcing $R = 0$ and set by forcing $S = 0$; the forbidden input state is $R, S = 0, 0$.

FIGURE 2.35   RS latch constructed from two cross-coupled NAND gates

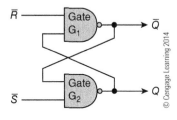

© Cengage Learning 2014

TABLE 2.19   Truth Table for an RS Latch Constructed from NAND Gates

Inputs		Output	Comment
R	S	Q^+	
0	0	X	Forbidden
0	1	1	Reset output to 0
1	0	0	Set output to 1
1	1	Q	No change

© Cengage Learning 2014

---

[17] Here the term *forbidden* indicates that this input should not be allowed to happen because the state of the flip-flop will be either undefined or, possibly, random.

Clocked RS Flip-flops

The RS latch of Figure 2.35 responds to its inputs according to its truth table. Sometimes we want the RS latch to ignore its inputs until a specific time. The circuit of Figure 2.36 demonstrates how we can turn the RS latch into a clocked RS flip-flop.

A normal RS latch lies in the inner box in Figure 2.36. Its inputs, R' and S', are derived from the external inputs R and S by ANDing them with a clock input C. As long as $C = 0$, the inputs to the RS latch, R' and S', are clamped at 0 and the output of the RS latch remains constant.

FIGURE 2.36   The clocked RS flip-flop

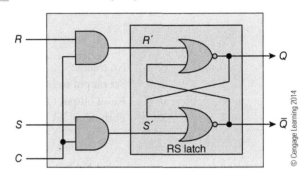

© Cengage Learning 2014

Whenever $C = 1$, the external R and S inputs to the circuit are transferred to the latch so that $R' = R$ and $S' = S$, and the RS latch responds accordingly. The clock input may be thought of as an inhibitor, restraining the latch from acting until the right time.

The Arbiter—An Application of RS Latches

You've seen television quiz shows in which each member of a team has a buzzer that they press to show they are ready to answer a question. The first person to push the button sounds the buzzer and locks out all the other contestants. The same circuit is widely used in computing to decide which of two competing devices may gain access to a resource such as a bus. You might find this circuit in a system with several processors that can communicate with each other via a common bus.

The following figure illustrates the circuit of an arbiter with *request* inputs R_0 and R_1 and *grant* outputs G_0 and G_1. A third input, Reset, is used to initialize the arbiter and to clear it after a round of arbitration. When one or more requesters assert their respective request inputs, the arbiter grants access to the first device to request attention.

Suppose that the Reset line has been asserted active-high. Both RS latches will be reset with the two grant outputs G_0 and G_1 inactive-low. Note that the $\overline{Q}$ output from each flip-flop is fed back to the AND gate in the input circuit of the *other* flip-flop. The effect of this cross coupling is that each flip-flop enables the other flip-flop's S input if its own Q output is low.

If, say, input R_0 is asserted high while request R_1 is low, flip-flop FF_0 is set and the G_0 output goes high to claim the resource for R_0. Because of the cross-coupling between the $\overline{Q}$ output of FF_0 and the AND gate in the R_1 input path, asserting R_1 now has no effect on the circuit because R_0 is locking it out.

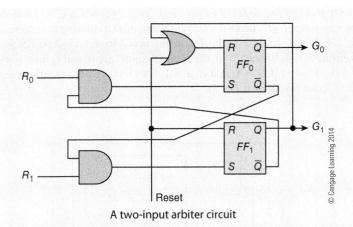

A two-input arbiter circuit

A fundamental problem of all arbiters is dealing with the almost simultaneous assertion of request inputs. If both R_0 and R_1 are asserted at the same time (when the circuit is in a reset condition) both S inputs will go high and both latches will be set. The OR gate removes the problem by detecting when FF_1 is set and using it to reset FF_0. Consequently, non-simultaneous requests are granted and the winning request locks out the other, and simultaneous requests lead to a failure of the circuit that is corrected by the output of FF_1 correcting the output of FF_0. Note that the resolution of simultaneous requests is resolved in favor of R_1.

Practical flip-flops have three types of clock input. The arrangement in Figure 2.36 implements a *level-sensitive clock* in which the flip-flop is clocked whenever the clock is asserted. An *edge-clocked* flip-flop is clocked instantaneously on the rising (or falling edge) of the clock. A *master-sla*ve flip-flop captures data on one edge of the clock but does not change its output until the following clock edge. We will soon see that many sequential circuits require clocked devices to ensure reliable operation.

D Flip-flop

Figure 2.37 illustrates the D flip-flow that has a D (data) input and a C (clock) input. Setting the C input to 1 is called *clocking* the flip-flop. D flip-flops can be level sensitive, edge triggered, or master-slave.

FIGURE 2.37   The D flip-flop

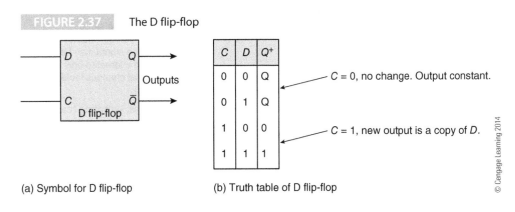

(a) Symbol for D flip-flop   (b) Truth table of D flip-flop

When the D flip-flop is clocked, the value at its D input is captured and transferred to its Q output, which remains constant until the next time it is clocked. The Q^+ output column in the truth table of Figure 2.37 indicates the *new* output when the flip-flop is clocked. When

$C = 0$ the output is the *remembered* or *previous* output Q. When $C = 1$ and the flip-flop is clocked, the new output, Q^+, takes the current value of D, replacing the old value of Q.

Figure 2.38 demonstrates a D flip-flop constructed from a clocked RS flip-flop and an inverter to ensure $S = \overline{R}$. When $C = 0$, the R and S inputs are 0, and Q does not change state. When $C = 1$, $R, S = 0, 1$ if $D = 1$; and $R, S = 1, 0$ if $D = 0$. Consequently, when $C = 1$, Q is set to 1 if $D = 1$ and is set to 0 if $D = 0$.

FIGURE 2.38   Circuit of a D flip-flop

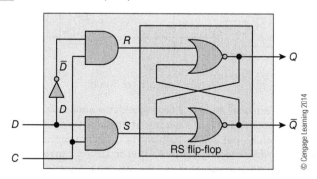

Before demonstrating applications of D flip-flops, we introduce the *timing diagram* that explains the behavior of sequential circuits. A timing diagram shows how a *cause* creates an *effect*. Figure 2.39 shows how we represent a signal as two parallel lines at 0 and 1 levels. These parallel lines don't imply that a signal is both 0 and 1. They imply that this signal may be 0 or 1 (we are not concerned with which level the signal is in). What we are concerned with is the point at which a signal changes its state.

FIGURE 2.39   Timing diagram conventions

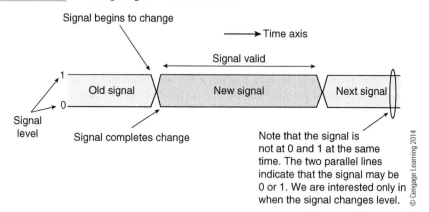

Figure 2.40 demonstrates a fundamental application of D flip-flops: sampling (capturing) a time-varying signal. Three processing units A, B, and C each take an input and operate on it to generate an output after a certain delay. New inputs, labeled i, are applied to processes A and B at the time t_0. A process can be anything from a binary adder to a memory device.

The timing diagram in Figure 2.40 shows that process A has a shorter delay than process B. The output from process A is valid at t_1, and the data from the output of process B is valid at t_2. All valid signals are shown in blue. The outputs of processes A and B are applied to process C, whose output is not valid until time t_3.

If a D flip-flop is placed at the output of process C and is clocked four units of time after $t = 0$, the desired data will be latched into the flip-flop and held constant until the next clock pulse. We will assume that the D flip-flops are *rising edge-triggered* and latch the input on the

FIGURE 2.40   Capturing the output of a system

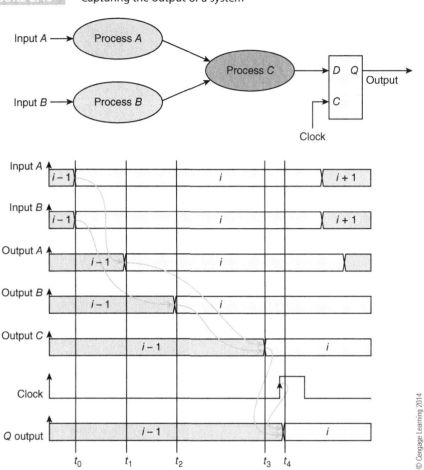

rising edge of the clock. Between clock pulses, the outputs of the flip-flops are processed by the logic elements and the new data values are presented to the inputs of flip-flops.

After a suitable time delay (longer than the time taken for the slowest process to be completed), the flip-flop is clocked at time t_4. The output of process C is held constant until the next time the flip-flop is clocked. Such a clocked system is called *synchronous*, because processes are started simultaneously on each new clock pulse. An asynchronous system is one in which the end of one process signals (i.e., triggers) the start of the next.

The D-flip-flop captures the data after it has settled down after processing by units A, B, and C. Clearly, the time span $t_4 - t_0$ must be long enough to account for the worst-case delay a signal can incur while flowing through the processes.

Figure 2.41 extends Figure 2.40 to demonstrate *pipelining*, in which flip-flops separate processes A and B in a digital system by acting as a *barrier* between them. The flip-flops in Figure 2.41 are edge-triggered and all are clocked at the same time.

Assume that all the flip-flops have just been clocked at time t_i. At point A in Figure 2.41, flip-flop 1 has captured the input to process A and holds it constant until the next time flip-flop 1 is clocked. Flip-flop 1 is a barrier between the input process that generates a current input and process A that operates on the input. At point B, the output from process A (corresponding to the new input) becomes valid.

On the next clock pulse at time t_{i+1}, a new input is clocked into flip-flop 1. At the same time, the output from process A (due to input i) is clocked into flip-flop 2 at point C. While

FIGURE 2.41   Pipelining using flip-flops

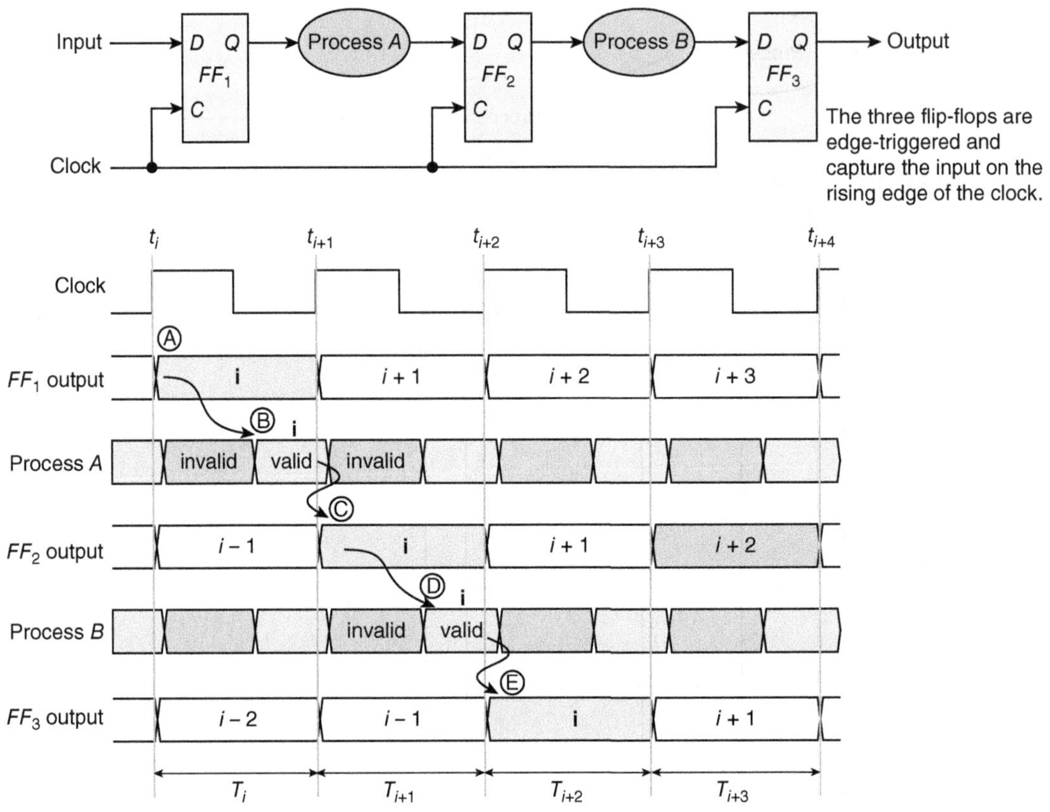

The three flip-flops are edge-triggered and capture the input on the rising edge of the clock.

process A is operating on input $i + 1$ in time slot T_{i+1}, process B is operating on the data i held in flip-flop 2. At point D, the output from process B becomes valid.

On the third clock pulse at time t_{i+2}, the output from process B is latched into flip-flop 3, which is the output flip-flop. At this point, E, which is the data due to input i, is valid at the output. The effect of the flip-flops is threefold. First, they hold data constant between stages (i.e., processes). Second, they delay the output of the data (in this case, it takes two clock cycles for the current input to appear at the output). Third, and most importantly, this arrangement allows operations to take place *in parallel*. While process A is operating on data i, process B is operating on data $i - 1$. This mechanism is called *pipelining*, because you can view the system in Figure 2.32 as a pipeline through which data flows while being processed at various stages along the pipe. We return to the theme of pipelining later when we discuss ways in which the speed of a computer can be increased.

The JK Flip-Flop

The third flip-flop is the JK, whose symbol and truth table is given in Figure 2.42. This is the most versatile of all flip-flops. Figure 2.43 demonstrates how a JK flip-flop can be constructed from an RS latch and two three-input AND gates. As long as the JK flip-flop is not *clocked*, its Q output remains constant and is independent of the values of J and K.

When the flip-flop is clocked with values of $J = 0$ and $K = 0$, the Q output remains in its previous state. When the flip-flop is clocked with values of $J = 1$ and $K = 0$, the Q output goes to 1. When the flip-flop is clocked with values of $J = 0$ and $K = 1$, the Q output goes to 0. So far, the JK flip-flop behaves exactly like a *clocked* RS flip-flop.

FIGURE 2.42   The JK flip-flop

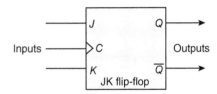

J	K	Q	Q⁺
0	0	0	Q
0	0	1	Q
0	1	0	Q
0	1	1	Q
1	0	0	Q
1	0	1	0
1	1	0	1
1	1	1	0

$J = K = 1$ toggles output. © Cengage Learning 2014

FIGURE 2.43   Construction of a basic JK flip-flop

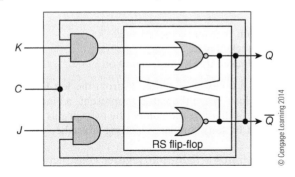

© Cengage Learning 2014

When the JK flip-flop is clocked and both inputs J and K are 1, the Q output *toggles* (that is, it changes state so that $Q+ = \overline{Q}$). In plain English, the output of the JK flip-flop can remain in its previous state, be forced to a 1 or a 0, or toggled.

Most JKs are *master-slave* flip-flops, because they consist of two flip-flops in series. When a master-slave flip-flop is clocked by, say, a low-to-high transition at its clock input, the data at its inputs is captured by the master flop-flop. The output of this master flip-flop is set according to the data at its input.

The output of the master-slave flip-flop does not change because it comes from the slave part of the flip-flop, which is not affected by the low-to-high transition at its clock input. When the clock makes a high-to-low transition, data from the master flip-flop is captured by the slave flip-flop and only now does the output of the device change state. The master-slave mechanism ensures that the flip-flop's output state cannot affect its input in circuits where the output of a flip-flop is used to generate the flip flop's next input; for example, in a counter where the current count i is fed back to generate the next count $i + 1$.

2.10.2   Registers

RS, D, and JK flip-flops are the building blocks of sequential circuits such as registers and counters. The register in Figure 2.44 is an m-bit storage element that uses m flip-flops to

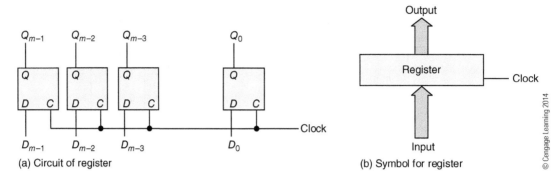

FIGURE 2.44   The register

(a) Circuit of register

(b) Symbol for register

store an m-bit word. Indeed, a memory is nothing more than a large array of registers. In Figure 2.44, the clock inputs of the flip-flops are connected together so that all the flip-flops are clocked at the same instant. When the register is clocked, the word at its D inputs is transferred to its Q output and held constant until the next clock pulse.

Shift Register

By modifying the structure of a register we can build a *shift register* whose bits are moved one place right every time the register is clocked. For example, the binary pattern 01110101 becomes

00111010 after the shift register is clocked once,
00011101 after it is clocked twice,
00001110 after it is clocked three times, and so on.

After the first shift, a 0 has been shifted in from the left-hand end, and the 1 at the right-hand end has been lost. When the pattern represents a binary number, shifting it one place right has the effect of dividing the number by two (just as shifting a decimal number one place right divides it by ten). Similarly, shifting a number one place left multiplies it by 2.

Figure 2.45 demonstrates a right-shift register constructed from D flip-flops. The Q output of each flip-flop is connected to the D input of the flip-flop on its right. Clock inputs are connected together and each flip-flop is clocked simultaneously. When the ith stage is clocked, output Q_i takes on the value from the stage on its left (i.e., $Q_i \leftarrow Q_{i+1}$). Data presented at the input of the left-hand flip-flop, D_{in}, is shifted into the $(m-1)$th stage at each clock pulse. The flip-flops must either be *edge-triggered* or *master-slave* flip-flops to ensure that data does not ripple through all stages as soon as the register is clocked.

FIGURE 2.45   The shift register

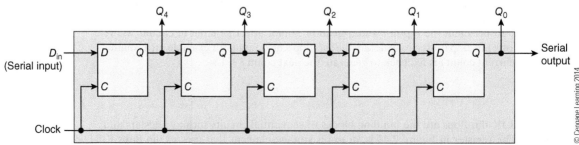

FIGURE 2.46   Timing diagram of a shift register

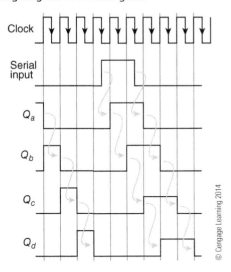

Figure 2.46 illustrates a shift register whose initial value $Q_a Q_b Q_c Q_d = 0100$. As the register is clocked, you can see the data being shifted through the register on each successive clock pulse. Note that there is also a serial input to the shift register that is captured and moved through the register as it is clocked.

Suppose you were to take a shift register and connect its most-significant bit output to its input and preset it to $1000, \ldots, 00$. On the first clock, the sequence $1000, \ldots, 00$ would become $0100, \ldots, 00$. The 1 is shifted right, and the 0 that falls off the right-hand end is returned to the left-hand end. On the next clock, the sequence would become $0010, \ldots, 00$. Eventually, the 1 would reach the least-significant bit position, and the count would be $0000, \ldots, 01$. Now, on the next clock, the 1 would be fed back to the right-hand stage of the shift register to give $1000, \ldots, 00$, which is the pattern we started with. That is, we can use the shift register as a *sequencer*[18] without any additional logic (except the logic necessary to ensure that the initial value is $1000, \ldots, 00$).

Shift registers perform several roles in a computer. By shifting left, they *multiply* a binary integer by two, and by shifting right, they *divide* an unsigned integer by 2 (a left shift is implemented by connecting the output of a flip-flop to the input of its left-hand neighbor). If you wish to access a bit within a number, you can shift the number left or right until the bit is shifted out into the carry bit. Then you can execute a conditional branch on the state of the carry bit. Shift registers can also be used to convert a parallel word into a sequence of bits. If you load an m-bit word into a shift register and clock it m times, the output of the shift register consists of m pulses. This feature is used to transmit data serially—a bit at a time—over a serial (i.e., single line) data link.

Left-Shift Register

Let's look at a left-shift register using JK flip-flops (Figure 2.47). The input of the ith stage, J_i, is connected to the output of the $(i - 1)$th stage so that, at each clock pulse, $Q_i \leftarrow J_{i-1}$. The same is done with the K input, so $\overline{Q}_i \leftarrow K_{i-1}$. Using the same data as the previous example, get 01110101 becomes

11101010 after one shift left and
11010100 after two shifts left

---

[18] A sequencer is a device that steps through a predefined sequence of values. For example, a 3-bit sequencer may generate the sequence 0,1,4,7,6,2,0,1,4,7,6,2,....

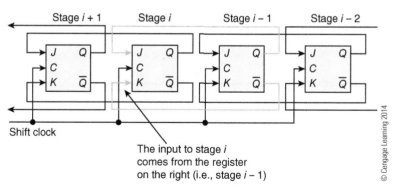

FIGURE 2.47   The left-shift register

Computer instruction sets include logical, arithmetic, and circular shifts. These operations all shift bits left or right and the only difference between them concerns what happens to the bit shifted in. In an arithmetic shift, the sign of a two's complement number is preserved when it is shifted right. In a circular shift, the bit shifted out of one end becomes the bit shifted in at the other end. Table 2.20 describes what happens when the 8-bit value 11010111 undergoes three types of shift. We look at shifts in more detail when we introduce instruction sets.

TABLE 2.20   The Effect of Logical, Arithmetic, and Circular Shifts

Shift Type	Shift Left	Shift Right
Original bit pattern before shift	11010111	11010111
Logical shift	10101110	01101011
Arithmetic shift	10101110	11101011
Circular shift	10101111	11101011

Sometimes it is necessary to load data into a shift register. In this case, it has to behave as both a normal register (parallel load of all D inputs) and a shift register. Figure 2.48 shows such a right-shift register with a parallel load capacity. A two-input multiplexer, composed of two AND gates an OR gate and an inverter, switches a flip-flop's D input between the output of the previous stage to the left (*shift* mode) and the load input (*load* mode). The control inputs of all

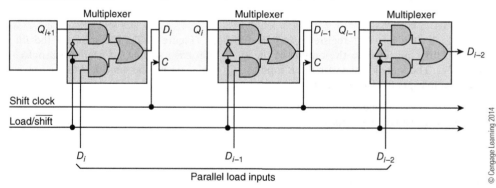

FIGURE 2.48   Shift register with a parallel load capability

multiplexers are connected together to provide the mode control, labeled load/$\overline{shift}$. If load/$\overline{shift}$ = 0 the operation performed is a shift and if load/$\overline{shift}$ = 1 the operation performed is a load.

2.10.3   Asynchronous Counters

A counter does what its name suggests: it *counts*. Simple counters count up or down through the natural binary sequence, whereas more complex counters may step through an arbitrary sequence. When a sequence terminates, the counter starts again at the beginning. A counter with n flip-flips cannot count through a sequence longer than 2^n.

Figure 2.49 gives the circuit diagram of a 3-bit *counter* using JK flip-flops. The input to the first flip-flop is a clock (i.e., a continuous stream of pulses). The Q output of each flip-flop is connected to the clock input of its neighbor. The J and K inputs of all flip-flops are permanently connected to a logical 1 level, forcing each flip-flop to change state when it is clocked. We will assume that each JK flip-flop is clocked on the *falling* or *negative* edge of the clock pulse (the small circle at each clock input indicates that the flip-flop is triggered by the *negative* or falling edge of a clock).

The binary counter

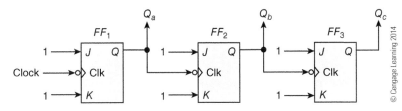

The timing diagram of Figure 2.50 analyzes the behavior of the counter. The uppermost trace is the clock at the input to the first flip-flop, FF_1. Since both J and K inputs are permanently connected to a logical 1 level, this flip-flop toggles on each falling edge of the clock pulse. Figure 2.50 marks the negative edge of each clock pulse with a downward arrow. The output of FF_1, Q_a, changes state on each falling edge of the clock.

Waveform Q_a is a copy of FF_1's clock input but at *half* the frequency (i.e., two pulses of the clock are required for each pulse of the Q_a output). The Q_a output from FF_1 is fed to FF_2's

Timing diagram for the counter of Figure 2.49

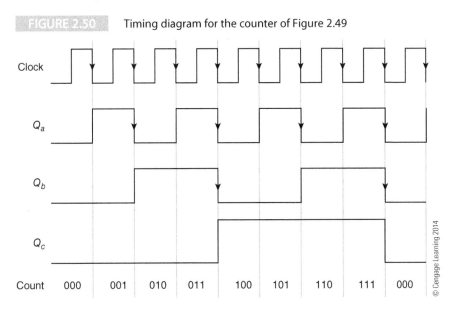

clock input that produces another waveform at FF_2's Q_b output. In this case, Q_b is half the frequency of Q_a. Output Q_b is fed into the clock input of the third flip-flop to produce a third waveform Q_c. Figure 2.50 demonstrates that each flip-flop divides the incoming clock by two.

If we write down the three binary outputs from flip-flops FF_3, FF_2, and FF_1 at each clock pulse and the sequence starts at 000, the successive values are 001, 010, 011, 100, 101, 110, and 111—then the sequence repeats. The outputs of the flip-flops constitute the *natural binary sequence*.

The counter described by Figure 2.49 is *an asynchronous, binary, ripple-through, count-up counter*. It is *asynchronous* because the flip-flops change state when they are clocked, in contrast to a synchronous counter where all flip-flops change state at the same time because a master clock triggers all changes of state.[19] The counter is binary because it counts sequentially through the binary sequence 0 to 2^{n-1}. You can design counters that count up to an arbitrary value and then return to zero, for example digital watches that count up to 12 before resetting.

The counter is described as *ripple-through* because a change of state always begins at the least-significant bit end and ripples through the flip-flops. For example, if the current count in a 5-bit counter is 01111, the counter will become 10000 on the next clock. However, the counter will very rapidly go through the sequence 01111, 01110, 01100, 01000, and 10000 as the 1-to-0 transition of the first stage propagates through the chain of flip-flops. Figure 2.51 shows the ripple-through effect in the 3-bit counter of Figure 2.49 by including the delay between a flip-flop being clocked and its output changing state. The counts in blue (110 and 100) are the spurious counts that appear momentarily on a change of 111 to 000. Finally, this is an up counter, because it counts from zero to 2^{n-1}. Note that if you were to design the same circuit with the $\overline{Q}$ of one stage connected to the clock input of the next stage, the counter would count down from 2^{n-1} to zero.

FIGURE 2.51   The ripple-through effect

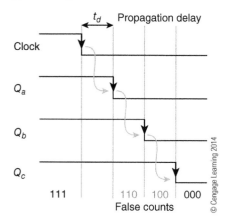

Using a Counter to Create a Sequencer
We can combine the counter with the multiplexer (i.e., three-line to eight-line decoder) to create a sequence generator that produces a sequence of eight pulses T_0 through T_7—one after another. Figure 2.52 provides the structure of such a sequencer. As the counter in Figure 2.52 cycles through states 000 to 111, the three-bit output from the flip-flops Q_c, Q_b, and Q_a is decoded into one of eight values T_0 through T_7 by the decoder. Figure 2.52 also provides a timing diagram for the eight outputs of the decoder. If we regard a cycle as equivalent to eight clock pulses, the start of a cycle is signified by T_0 being pulsed. Then T_1 is pulsed, and so on.

---

[19] Of course, even in a synchronous system, there will be small variations in the times at which outputs change, because the clock signal will arrive at different flip-flops at different times due to delays in the clock signal caused by the wiring. However, these considerations are beyond the scope of this text.

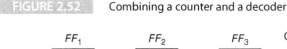

Combining a counter and a decoder

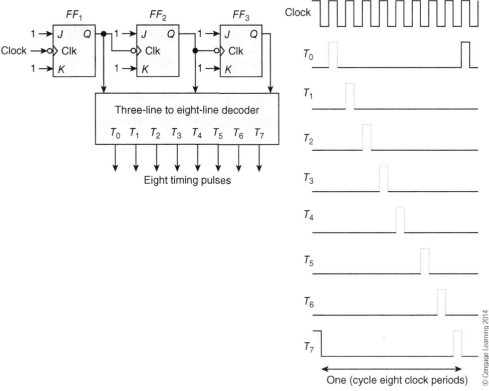

Such a circuit might be used by a computer to step through the primitive *microinstructions* required to execute an instruction. For example, timing pulse T_0 may be used to copy the program counter to the memory address register; T_1 used to read the contents of the main store; T_2 used to increment the contents of the program counter; T_3 used to copy the instruction from memory into the memory buffer register; and so on. We will discuss this topic in greater detail at the end of this chapter.

The circuit of Figure 2.52 would not be used as a sequencer in a real system because the simple *ripple-through counter*, which is composed of the three flip-flops, goes through intermediate states when carries are generated that propagate through the registers. These carries would generate very short spurious pulses from the timing generator (i.e., glitches). A practical system would use a synchronous counter.

2.10.4   Sequential Circuits

Any system with internal memory and external inputs (such as the flip-flop) can be said to be in a *state* that is a function of its internal and external inputs. A state diagram shows some (or all) of the possible states of a given system. A labeled circle represents each of the states and the states are linked by unidirectional lines showing the paths by which one state becomes another state.

State Diagrams

The state diagram is an important tool that we saw in the first chapter. We will later use it when discussing computer organization. Recall that it defines all of the states a system can be in and the transitions between states.

Figure 2.53 gives the state diagram of a JK flip-flop that has just two states, S_0 and S_1. S_0 represents the state $Q = 0$, and S_1 represents the state $Q = 1$. The transitions between states S_0 and S_1 are determined by the values of the JK inputs at the time the flip-flop is clocked. In Figure 2.53, we have labeled the flip-flop's input

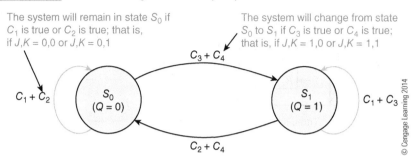

FIGURE 2.53   The state diagram of a JK flip-flop

The system will remain in state S_0 if C_1 is true or C_2 is true; that is, if $J,K = 0,0$ or $J,K = 0,1$

The system will change from state S_0 to S_1 if C_3 is true or C_4 is true; that is, if $J,K = 1,0$ or $J,K = 1,1$

$C_3 + C_4$

$C_1 + C_2$

S_0 ($Q = 0$)

S_1 ($Q = 1$)

$C_1 + C_3$

$C_2 + C_4$

© Cengage Learning 2014

TABLE 2.21

Relationship Between JK Inputs and Conditions C_1 to C_4

J	K	Condition
0	0	C_1
0	1	C_2
1	0	C_3
1	1	C_4

© Cengage Learning 2014

TABLE 2.22

Excitation Table of a JK Flip-Flop

Inputs		Transition
J	K	$Q \rightarrow Q^+$
0	d	$0 \rightarrow 0$
1	d	$0 \rightarrow 1$
d	1	$1 \rightarrow 0$
d	0	$1 \rightarrow 1$

© Cengage Learning 2014

states C_1 through C_4. Table 2.21 defines the four possible input conditions C_1, C_2, C_3, and C_4 in terms of J and K.

Figure 2.53 shows that conditions C_3 or C_4 cause a transition from state S_0 to state S_1. Similarly, conditions C_2 or C_4 cause a transition from state S_1 to state S_0. Condition C_4 causes a change of state from S_0 to S_1 and also from S_1 to S_0. This is, of course, the condition $J = K = 1$, which causes the JK flip-flop to toggle its output. Some conditions cause a state to change to *itself* (that is, there is no overall change). Thus, conditions C_1 or C_2, when applied to the system in state S_0, have the effect of leaving the state unchanged.

Synchronous machines are composed of flip-flops that are all clocked at the same time (i.e., synchronously). The outputs of all flip-flops in a synchronous machine become valid at the same time, and the ripple-through effect associated with asynchronous counters is absent. Synchronous machines can be easily designed to count through any arbitrary sequence just as well as the natural sequence 0, 1, 2, 3, ...

We design a synchronous machine by means of a state diagram and the excitation table for the appropriate flip-flop (either RS or JK). An *excitation table* is a version of a flip-flop's truth table arranged to display the input states required to force a given output transition. Table 2.22 illustrates the excitation table of a JK flip-flop. For example, Table 2.22 tells us that the J and K inputs should be 1 and "d" to force the Q output of a JK flip-flop to make the transition from 0 to 1 the next time it is clocked. The symbol "d" indicates a *"don't care"* condition.

If $J = 1$ and $K = 0$, the flip-flop is set when it's clocked, and Q^+ becomes 1. If $J = 1$ and $K = 1$, the flip-flop is toggled when it's clocked and the output $Q = 0$ is toggled to $Q = 1$. The state of the K input doesn't matter when we wish to set Q^+ to 1 when $Q = 0$ and $J = 1$. It should now be clear why all the transitions in the JK's excitation table have a *"don't care"* input—a given state can be reached from more than one starting point.

The next step in designing a state machine is to construct a truth table for the system to determine the JK inputs required to force a transition to the required next state for each of the possible states in the table. It is much easier to explain this step by example rather than by algorithm.

Suppose we require a counter that follows the seven-state sequence 0, 2, 4, 6, 7, 5, and 3 and then repeats. As there are seven states, three flip-flops are required to define each state uniquely. Table 2.23 provides a truth table for this example. The leftmost column gives the sequence the counter follows at each new clock pulse, and the column headed by *Output* gives the three-bit output corresponding to each of the seven states.

The column labeled *Next State* gives the state following the current state. We have to determine the values of the J and K inputs required to force the desired next state when the flip-flop is clocked. Consider the first state $Q_a, Q_b, Q_c = 0,0,0$ and the next state $Q_a, Q_b, Q_c = 0, 1, 0$. The transitions are

$$Q_c \qquad 0 \rightarrow 0$$
$$Q_b \qquad 0 \rightarrow 1$$
$$Q_a \qquad 0 \rightarrow 0$$

TABLE 2.23   Truth Table for a Synchronous Counter

	Output			Next State			J, K Inputs Required to Force Transition					
Count	Q_c	Q_b	Q_a	Q_c	Q_b	Q_a	J_c	K_c	J_b	K_b	J_a	K_a
0	0	0	0	0	1	0	0	d	1	d	0	d
2	0	1	0	1	0	0	1	d	d	1	0	d
4	1	0	0	1	1	0	d	0	1	d	0	d
6	1	1	0	1	1	1	d	0	d	0	1	d
7	1	1	1	1	0	1	d	0	d	1	d	0
5	1	0	1	0	1	1	d	1	1	d	d	0
3	0	1	1	0	0	0	0	d	d	1	d	1
0	0	0	0	0	1	0	0	d	1	d	0	d

© Cengage Learning 2014

TABLE 2.24   Truth Tables for the J, K Inputs for the System of Table 2.23

Q_c	Q_b	Q_a	J_c	Q_c	Q_b	Q_a	K_c	Q_c	Q_b	Q_a	J_b	Q_c	Q_b	Q_a	K_b	Q_c	Q_b	Q_a	J_a	Q_c	Q_b	Q_a	K_a
0	0	0	0	0	0	0	d	0	0	0	1	0	0	0	d	0	0	0	0	0	0	0	d
0	0	1	x	0	0	1	x	0	0	1	x	0	0	1	x	0	0	1	x	0	0	1	x
0	1	0	1	0	1	0	d	0	1	0	d	0	1	0	1	0	1	0	0	0	1	0	d
0	1	1	0	0	1	1	d	0	1	1	d	0	1	1	1	0	1	1	d	0	1	1	1
1	0	0	d	1	0	0	0	1	0	0	1	1	0	0	d	1	0	0	0	1	0	0	d
1	0	1	d	1	0	1	1	1	0	1	1	1	0	1	d	1	0	1	d	1	0	1	0
1	1	0	d	1	1	0	0	1	1	0	d	1	1	0	0	1	1	0	1	1	1	0	d
1	1	1	d	1	1	1	0	1	1	1	d	1	1	1	1	1	1	1	d	1	1	1	0

© Cengage Learning 2014

Table 2.22 tells us that the transition $0 \rightarrow 0$ occurs when $J, K = 0$, d, and the transition $0 \rightarrow 1$ when $J, K = 1$, d. We can use this to write down the J, K inputs required by the three flip-flops to make the first transition. Table 2.23 gives all the values of J, K for this counter.

We now have six Boolean equations to solve for the six J, K inputs in terms of the flip-flop's outputs. Table 2.24 gives six truth tables for all of the J and K inputs. Note that the sequence of the variables $Q_a, Q_b,$ and Q_c is the natural order. Moreover, since state $Q_a, Q_b, Q_c = 0, 0, 1$ does not occur in the state machine's output sequence, we can put x in the J and K columns for this state. In Table 2.24 both d and x represent "*don't care*" states.

The final step is to solve these equations. Figure 2.54 gives the corresponding Karnaugh maps from which we can write down the equations for the JK flip-flop's inputs. The Karnaugh map provides a graphical means of solving Boolean equations for up to five variables and is described on the we site companying this book.

$$J_c = Q_b \cdot \overline{Q_a} \qquad\qquad K_c = \overline{Q_b} \cdot Q_a$$
$$J_b = 1 \qquad\qquad\qquad K_b = \overline{Q_c} + Q_a$$
$$J_a = Q_c \cdot Q_b \qquad\qquad K_a = \overline{Q_c}$$

Figure 2.55 provides a circuit diagram for this synchronous counter.

FIGURE 2.54   Karnaugh maps for Table 2.23

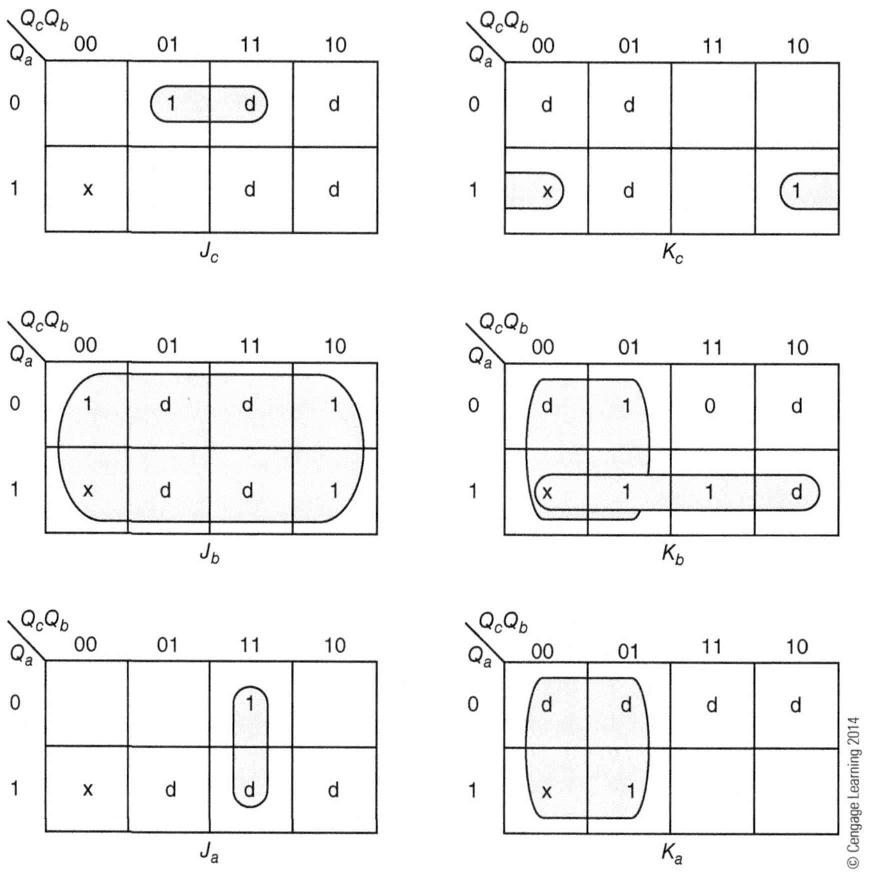

FIGURE 2.55   Circuit of the synchronous counter of Table 2.23

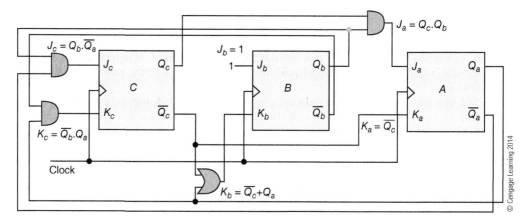

2.11   Buses and Tristate Gates

The final section in this chapter brings together some of the circuits we've just covered and hints at how a computer operates by moving data between registers and by processing data. Now that we've built a register out of D flip-flops, we can construct a more complex system with several registers. By the end of this section, you should have an inkling of how computers execute instructions. First we need to introduce a new type of gate — a gate with a *tristate* output.

In a computer, we want to put the output from one of several registers on a bus in order to transmit the data to another register or functional unit (e.g., ALU). The tristate output lets us do the seemingly impossible and connect several outputs together. Figure 2.56 gives the symbol for two tristate gates. Figure 2.56a is a tristate gate with an *active-high* control input E (E stands for *enable*). When E is asserted high, the output of the gate Y is equal to its input X. Such a gate is acting as a buffer and transferring a signal without modification. When E is inactive-low, the gate's output Y is *internally disconnected*, and the gate does not drive the output. That is, you cannot say what the output Y is, because it is disconnected from the gate.

FIGURE 2.56   The tristate gate (tristate buffer)

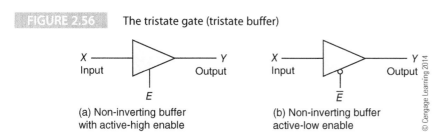

(a) Non-inverting buffer with active-high enable

(b) Non-inverting buffer active-low enable

© Cengage Learning 2014

If output Y is connected to a bus, the signal level at Y when the gate is disabled is that of the bus. Consequently, this output state is called *floating*, because the output floats up and down with the traffic on the bus. There are other circuit elements that can also drive a bus; here we are interested only in tristate devices.

Figure 2.56b illustrates another tristate gate, except that this time the control input is active-low and the gate is enabled when its control input $\overline{E}$ is low.

You can connect two or more circuits with tristate outputs to the same line as long as only one tristate output is enabled at any instant. Figure 2.57 shows a system with four registers connected to a common bus. The registers in Figure 2.57 have been simplified — only one bit is shown to avoid clutter. Each register is edge-clocked or is a master-slave flip-flop. The Q output of each register is connected to the bus by means of a *buffer* with a tristate output. This

FIGURE 2.57   Registers and buses

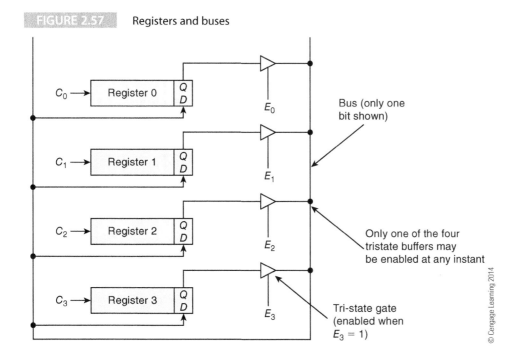

© Cengage Learning 2014

gate passes the signal at its input terminal to its output terminal *unchanged* when the enable input is 1 state. When the enable input is 0, the tristate gate *disconnects* its output pin from the rest of the circuit. In Figure 2.57 only *one* of the enable signals E_1, E_2, E_3, and E_4 may be set to a logical 1 at any instant.

If two buffers are enabled simultaneously and they both attempt to drive the bus in different directions (one to high and one to low), the state of the bus will be undefined, and the buffers may be physically damaged. If no buffer is enabled, the state of the bus will also be undefined. However, in this case, a resistor is used to passively pull the bus up into an inactive-high level. In other words, the bus has a default level.

The bus in Figure 2.57 is connected to the D input of each of the four registers. When one of the registers is clocked, the data on the bus is copied into that register. For example, if you set E_2 to 1 and then clock C_3, the contents of register R2 are placed on the bus and copied into register R3. In terms of register transfer language, the operation is represented by

    [R3] ← [R2]

Registers, Buses, and Functional Units

We can now put things together and show how very simple operations can be implemented by a collection of logic elements, registers, and buses. The system we construct will be able to take a simple 4-bit binary code IR_3,IR_2,IR_1,IR_0 and cause the action it represents to be carried out.

Figure 2.58 demonstrates how we can take the four registers and a bus to create a simple functional unit that executes MOVE instructions. A MOVE instruction is one of the simplest computer instructions that copies data from one location to another; in high level language terms it is equivalent to the assignment $Y = X$. The arrangement of Figure 2.58 employs two two-line to four-line decoders to select the source and destination registers used by an

FIGURE 2.58   Controlling the bus

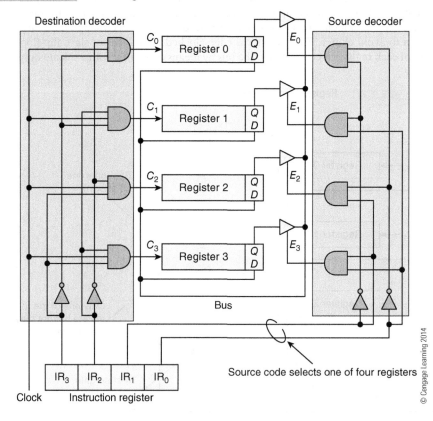

© Cengage Learning 2014

instruction. This structure can execute a machine level operation such as MOVE **Ry**, Rx that is defined as

[Ry] ← [Ri]

The instruction register in Figure 2.58 uses two bits IR_1, IR_0 to select the *source* of the data. The two-line to four-line decoder on the right-hand side of the diagram decodes bits IR_1, IR_0 into one of four signals: E_0, E_1, E_2, or E_3. The register enabled by the source code puts its data on the bus, which is fed to the D inputs of all four registers. The 2-bit destination code IR_3, IR_2 from the instruction register is fed to the decoder on the left-hand side of Figure 2.58 to generate one of the four clock signals C_0, C_1, C_2, or C_3. All of the AND gates in this decoder are enabled by a common clock signal, so the data transfer does not take place until this line is asserted.

The circuit of Figure 2.59 can only move data from one register to another. Figure 2.59 extends the system further to include multiple buses and an ALU (arithmetic and logic unit).

FIGURE 2.59   The registers, buses, and ALU

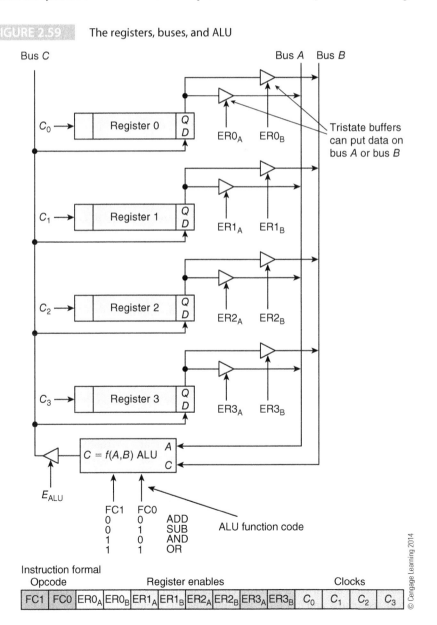

© Cengage Learning 2014

An operation is carried out by enabling two source registers and putting their contents on bus A and bus B. These buses are connected to the input terminals of the arithmetic and logical unit that produces an output depending on the function the ALU is programmed to perform. We have already seen how functions can be selected (e.g., AND/OR, adder/subtractor, programmable NOT). The output of the ALU is put onto bus C and the data clocked into the appropriate register. The operation performed by the circuit of Figure 2.60 is controlled entirely by the clock signals, the tristate gate enable signals, and the ALU function select signals.

By selecting appropriate bus driver enables, register clocks, and ALU function codes, we can program Figure 2.59 to carry out *machine-level* or *assembly level* instructions. Table 2.25 describes several machine-level instructions in terms of the signals required to synthesize them. Consider the operation ADD **R2**,R1 that adds the contents of register R1 to register R2. Buffer control signals $ER1_A$ and $ER2_B$ are set to 1 to put the contents of register 1 on bus A and register 2 on bus B. The ALU's function code is set to FC1, FC0 = 0, 0 to select addition and buffer control E_{ALU} is set to 1 to put the output from the ALU on bus C. Finally, register R2 is clocked by setting C_2 to 1 to capture the result.

TABLE 2.25   Microinstructions

Operation	Bus Driver Enables									Register Clocks				ALU Control	
	ERO_A	ERO_B	$ER1_A$	$ER1_B$	$ER2_A$	$ER2_B$	$ER3_A$	$ER3_B$	E_{ALU}	C_0	C_1	C_2	C_3	FC1	FC2
ADD **R2**,R1	0	0	1	0	0	1	0	0	1	0	0	1	0	0	0
SUB **R2**,R0	1	0	0	0	0	1	0	0	1	0	0	1	0	0	1
MOVE **R1**,R0	1	1	0	0	0	0	0	0	1	0	1	0	0	1	1

That's it. The computer really is as simple as we've described. Of course, the fine details of an actual computer are a little more complex, but the basic principles are the same.

Summary

Computers use a two-state binary system to represent information, because it is the most cost-effective way of doing things. We have looked at how numbers are represented as both unsigned and signed integers. Modern computers invariably use two's complement notation to represent signed integers, because adding two complement values automatically performs subtraction if one (or both) of the numbers is negative.

We have also looked at the representation of floating-point numbers used in scientific calculations (such as rendering an image in graphics). Because floating-point numbers are not exact representations of real numbers, we also looked at the issue of errors in floating-point numbers and their consequences. We briefly discussed the way in which computers generate complex arithmetic functions — such as cosines or tangents — partially to demonstrate that there are many ways of performing such calculations and that there is a tradeoff between cost, speed, and accuracy.

We have demonstrated that digital circuits can be constructed from three basic gates: the AND, OR, and NOT gates. We have shown that the NAND gate alone (or the NOR gate) can be used to create any digital circuit. The purpose of this section has been to give readers an insight into the underlying logic of computers. However, modern computers are no longer designed at the gate level—software is available that allows systems to be described in high-level terms (e.g., in C-like languages). Such software is able to directly create logic circuits from the high-level specification.

Digital systems are divided into two classes: combinational and sequential. A combinational circuit uses gates, and its output is a function of its inputs, for example, an adder circuit.

A sequential circuit uses bi-stable elements such as D flip-flops, and its output is a function of its inputs and its previous state; that is, a sequential circuit has memory. We have looked at sequential logic (both elements such as flip-flops and circuits such as registers) because these are of importance to those studying computer architecture and organization. Operations that are performed by shift registers in hardware are found as machine-level instructions in microprocessors.

An important logic element that we introduced at the end of this chapter is the tristate bus driver. A device with a tristate output can be connected to a bus and either enabled or disabled according to the state of its control input. When enabled, a gate with a tristate output drives the bus into a 1 or a 0 state. When disabled, the tristate output is disconnected from the bus. Most memory elements have tristate outputs to enable the one containing the data being accessed to drive the bus. Tristate logic (and similar devices) permit the design of systems with buses that steer data between register, functional units, memory devices, and I/O systems. Finally, we demonstrated how combinational and sequential logic elements plus tristate bus drivers could be used to perform the type of functions we associate with a computer (e.g., data transfer).

Problems

2.1 Why is binary arithmetic employed by digital computers?

2.2 We said that binary values have no intrinsic information (that is true of all other number representations). The *Voyager* I spacecraft, containing samples of human music and other messages, was the first human artifact to leave the solar system to travel to the stars. How is it possible to communicate with aliens in binary form if there is no intrinsic meaning to the data?

2.3 How much more inaccurate is binary integer arithmetic than decimal integer arithmetic? Can the accuracy of binary computers be improved to make them as accurate as decimal computers?

2.4 Why are computers byte-oriented?

2.5 Calculations are to be performed to a precision of 0.001%. How many bits does this require?

2.6 What are the decimal equivalents of the following values (assume positional notation and unsigned integer formats)?
a. 11001100_2 b. 11001100_3
c. 11001100_4 d. 11001100_{-2}

2.7 Why do we have octal and hexadecimal arithmetic?

2.8 Convert the following decimal numbers into (a) binary and (b) hexadecimal forms.
a. 25   b. 250
c. 2500   d. 25555

2.9 Convert the following unsigned binary numbers into decimal form.
a. 11   b. 1001
c. 10001 d. 10011001

2.10 Convert the following hexadecimal numbers into decimal form.
a. AB   b. A0B
c. 10A01 d. FFAAFF

2.11 Convert the following hexadecimal numbers into binary format.
a. AC   b. DF0B
c. 10B11 d. FDEAF1

2.12 Convert the following fractional decimal numbers into 16-bit unsigned binary form. Use eight bits of precision.
a. 0.2   b. 0.046875
c. 0.1111 d. 0.1234

2.13 Perform the following calculations in the stated bases.
a. 00110111_2 b. 00111111_2
   $+01011011_2$ $+01001001_2$

c. 00120121_{16} d. $00ABCD1F_{16}$
   $+0A015031_{16}$ $+ 0F00800F_{16}$

2.14 What is arithmetic overflow? When does it occur and how can it be detected?

2.15 The n-bit two's complement integer N is written a_{n-1}, $a_{n-2}, \ldots a_1, a_0$. Prove that (in two's complement notation) the representation of a signed binary number in $n + 1$ bits may be derived from its representation in n bits by repeating the leftmost bit. For example, if $n = -12 = 10100$ in five bits or $n = -12 = 110100$ in six bits.

2.16 Convert 1234.125 into 32-bit IEEE floating-point format.

2.17 What is the decimal equivalent of the 32-bit IEEE floating-point value CC4C0000?

2.18 What is the difference between overflow in the context of two's complement numbers and overflow in the context of floating-point numbers?

2.19 In the *negabinary system* an i-bit binary integer, N, is expressed using positional notation as:

$$N = a_0 \times -1^0 \times 2^0 + a_1 \times -1^1 \times 2^1 + \cdots + a_{i-1} \times -1^{i-1} \times 2^{i-1}$$

This is the same as conventional natural 8421 binary weighted numbers, except that alternate positions have the additional weighting $+1$ and -1. For example, $1101 = (-1 \times 1 \times 8) + (+1 \times 1 \times 4) + (-1 \times 0 \times 2) + (+1 \times 1 \times 1) = -8 + 4 + 1 = -3$. The following 4-bit numbers are represented in negabinary form. Convert them into their decimal equivalents.

a. 0000   b. 0101
c. 1010   d. 1111

2.20 Perform the following additions on 4-bit negabinary numbers. The result is a 6-bit negabinary value.

a. 0000   b. 1010
   +0001   +0101

c. 1101   d. 1111
   1011   1111

2.21 Arithmetic overflow occurs during a two's complement addition if the result of adding two positive numbers yields a negative result or if adding two negative numbers yields a positive result. If the sign bits of A and B are the same but the sign bit of the result is different, arithmetic overflow has occurred. If a_{n-1} is the sign bit of A, b_{n-1} is the sign bit of B, and s_{n-1} is the sign bit of the sum of A and B, then overflow is defined by

$$V = a_{n-1} \cdot b_{n-1} \cdot \overline{s_{n-1}} + \overline{a_{n-1}} \cdot \overline{b_{n-1}} \cdot s_{n-1}$$

In practice, real systems detect overflow from $C_{\text{in}} \neq C_{\text{out}}$ to the last stage. That is, we detect overflow from

$$V = \overline{c_n} \cdot c_{n-1} + c_n \cdot \overline{c_{n-1}}.$$

Demonstrate that this expression is correct.

2.22 What is the difference between a *truncation* error and a *rounding* error?

2.23 Positive and negative numbers can be represented in many ways in a computer. List some of the ways of representing signed numbers. Can you think of any other ways of representing signed values?

2.24 What is a NaN and what is its significance in floating-point arithmetic?

2.25 Write down the largest base-5 positive integer in n digits and the largest base 7 number in m digits. It is necessary to represent n-digit base-5 numbers in base 7. What is the minimum number m of digits needed to represent all possible n-digit base-5 numbers? *Hint*: The largest m-digit base-7 number should be greater than, or equal to, the largest n-digit base-5 number.

2.26 What is the largest three-digit number in base 13?

2.27 You are evaluating $x^2 - y^2$ where $x = 12.1234$ and $y = 12.1111$. If arithmetic operations are carried out to six significant figures (in decimal), does it matter whether you evaluate this expression as $x^2 - y^2$ or $(x + y)(x - y)$?

2.28 You are evaluating the function $x^4 + 4x^2 + 10x + 2$ at $x = 2$. What is the estimated error if the error in x is Rx?

2.29 What does the following ASCII-encoded message mean? Each character is given in hexadecimal form.

$$43, 6F, 6D, 70, 75, 74, 65, 72, 2E$$

2.30 Floating-point arithmetic is seldom used to perform financial calculations. Why?

2.31 Modern computers use unsigned integer arithmetic, fixed-point arithmetic, two's complement arithmetic, and floating-point arithmetic.

a. By examining a binary number, can you tell which number system it represents?
b. Why are there so many ways of representing numeric values?
c. Do we need them all?

2.32 For each for the following numbers, state the base in use; that is, what is p, q, r, s, t, u?

a. $100001_p = 33_{10}$
b. $25_q = 13_{10}$
c. $25_r = 23_{10}$
d. $25_s = 37_{10}$
e. $1010_t = 68_{10}$
f. $1001_u = 126_{10}$

2.33 A digital logic element represents the high state with an *output* of between 2.8 and 2.95 V. The same logic element will see an input high state as a voltage in the range of 2.1 to 3.0 V. What is the reason for this difference? What are the practical implications?

2.34 Draw a truth table to represent the intermediate values and output of the circuit in Figure P2.34.

FIGURE P2.34

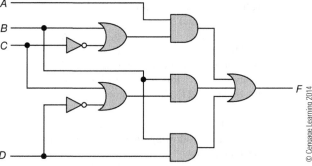

2.35 In a digital system, what is the meaning of *negative logic*?

2.36 The signal $\overline{\text{RESET}}$ is asserted. What does this statement mean?

2.37 A digital system has four one-bit inputs D, C, B, A and an output F. The input represents a 4-bit number in the range of 0 through 15, where A denotes the least-significant bit. The output F is true if the binary input is divisible by 3, 4, or 7. Construct a truth table to represent this system and construct a logic circuit to implement it.

2.38   Consider the circuit in Figure P2.38 that takes a 4-bit binary input and encodes it. Construct a truth table for the 16 inputs 0000 to 1111 and examine the output code. What is the characteristic feature of this code?

FIGURE P2.38

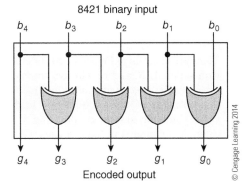

8421 binary input

Encoded output

2.39   A logic circuit has two 2-bit natural binary inputs A and B. A is given by A_1, A_0, where A_1 is the most-significant bit. Similarly, B is given by B_1, B_0, where B_1 is the most-significant bit. The circuit has three outputs, X, Y, and Z. This circuit compares A with B and determines whether A is greater or less than B or is the same as B. The relationship between inputs A and B and outputs X, Y, and Z is as given in Table P2.39. Design a circuit to implement this function.

TABLE P2.39

Condition	X	Y	Z
$A > B$	1	0	0
$A < B$	0	1	0
$A = B$	0	0	1

2.40   Draw a truth table for the circuit in Figure P2.40 and explain what it does.

FIGURE P2.40

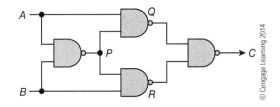

2.41   The circuit in Figure P2.41 receives three pairs of inputs a_0, b_0, a_1, b_1, ..., a_3, b_3 and produces a 4-bit output c_0, c_1, ..., c_3. Analyze the circuit and explain what it does in plain English.

FIGURE P2.41

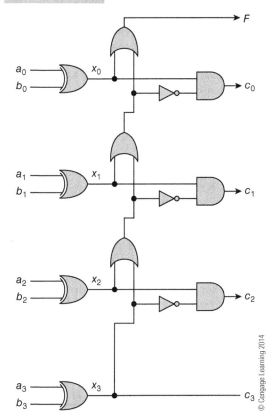

2.42   Demonstrate that all logic circuits can be constructed from NOR gates by building an inverter, an AND gate, and an OR gate from one or more NOR gates.

2.43   Design a logic circuit to implement the logical function $X = A \cdot B \cdot C + \overline{A} \cdot B \cdot C$

2.44   Simplify the following Boolean expressions.

a. $F = \overline{A} \cdot B \cdot C \cdot \overline{D} + A \cdot B \cdot C \cdot \overline{D} + A \cdot \overline{B} \cdot C \cdot D + A \cdot B \cdot C \cdot D$

b. $F = \overline{A} \cdot B \cdot C + A \cdot B \cdot C + A \cdot \overline{C} \cdot D + A \cdot B \cdot C \cdot D$

c. $F = A \cdot \overline{C} \cdot D + A \cdot B \cdot C \cdot D + A \cdot B \cdot C$

2.45   It is possible to have n-input AND, OR, NAND, and NOR gates, where $n > 2$. Can you have an n-input XOR gate for $n > 2$? Explain your answer with a truth table.

2.46   Draw the truth table of a full subtractor that directly subtracts bit A from B together with a borrow-in to produce a difference D and a borrow-out.

2.47   Design a D flip-flop using only an RS latch and simple gates. Include a circuit diagram, timing diagram, and truth table in your answer.

2.48   Without the tristate gate, it would be almost impossible to design a modern computer. Why is this so?

2.49 The circuit in Figure P2.49 consists of four JK flip-flops. Inputs *J* and *K* are not shown, because it is assumed that they are both permanently connected to a logical 1. These JK flip-flops are positive edge-triggered (i.e., they change state on the rising edge of the clock). Note that these flip-flops have a CLR (clear) input that sets *Q* to 1 when CLR = 1. What does this circuit do?

FIGURE P2.49

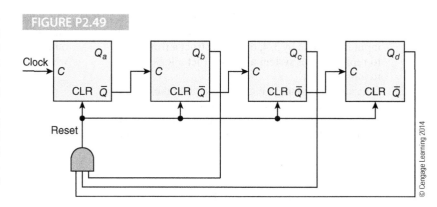

2.50 Consider the circuit in Figure P2.50. Assume that all gates are implemented in silicon by NAND gates, NOR gates and inverters. Each NAND gate, NOR gate, or inverter has an internal delay of 0.4 ns. A logic transition at an input may cause a change at the output (depending on other inputs and the circuit). The time for a transition at the input to appear as a corresponding transition at an output depends on the circuit path and the nature of the gates. What is the longest circuit path through this circuit? What is the worst case delay that a signal experiences going through the circuit?

2.51 The state diagram in Figure P2.51 describes a system with states *A*, *B*, and *C*. The system is initialized in State *A*. The state transition notation *x/yz* indicates that an input *x* causes a transition in the direction shown and the system outputs the value *yz*. For example, an input 1 in State *A* causes a transition to State *B*, and the system outputs 01.

(a) How many flip-flops would be required to construct this system?

(b) If the system receives the input 0001001100101011110010, what would the output be?

FIGURE P2.50

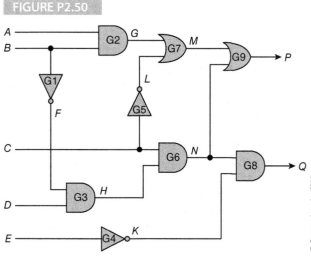

FIGURE P2.51

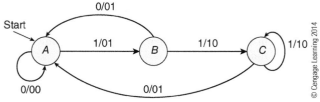

2.52 The structure in Figure P2.52 contains registers, ALUs, multiplexers, tristate gates, and buses and is essentially a more elaborate form of the register, ALU structure we introduced in this chapter. As a computing structure, what are the advantages of this over the simpler system we described earlier?

FIGURE P2.52

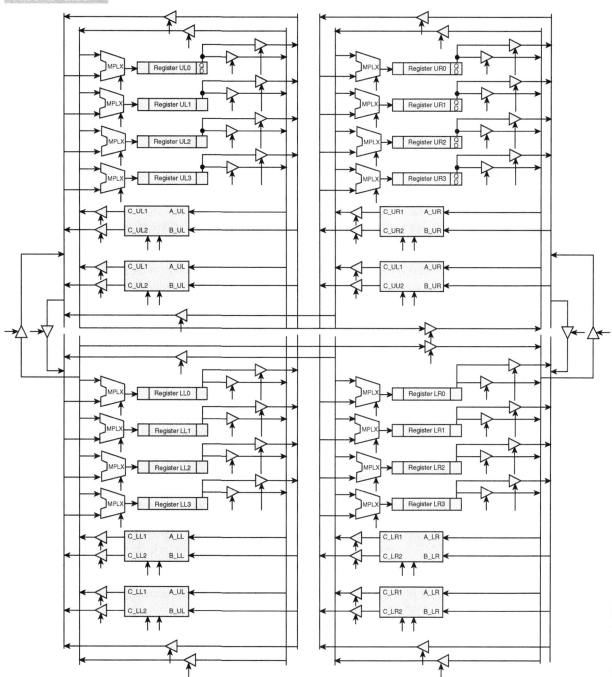

II

INSTRUCTION SET ARCHITECTURES

In Part I, we introduced the computer and covered the background material needed to let us look at the central processing unit. Part II comprises Chapters 3, 4, and 5 and is concerned with the *programmer's view of the computer*, rather than its internal organization. We examine the computer's operation at the *machine level*, which is the level of its *native* instructions. In Part II, we are interested in *what* a computer does rather than *how* it works.

Part II begins with an introduction to the architecture of the *stored program computer* and its *instruction set*. We show how information flows between the various parts of a computer as instructions are read from memory and executed. To help us explain what a computer does, we can use either a real off-the-shelf computer or a hypothetical teaching vehicle. Real machines reflect life's little complexities, irregularities, and inconsistencies and may present readers with a steep learning curve. Idealized teaching computers are nice and regular and have no nasty complexities or flaws—but they are artificial and fail to demonstrate the trade-offs that computer designers are forced to make. We have decided to use a real computer in this text called ARM.

ARM processors, from ARM Holdings, Ltd., are widely used in *embedded* systems such as cell phones and MP3 players. ARM provides an excellent vehicle for teaching computer architecture, because it is simple to understand and yet incorporates very powerful features. Indeed, I would maintain that the ARM processors have an elegant architecture. ARM is an ideal teaching vehicle, because ARM processors are doing so well in the marketplace, and because of the sheer range of their applications.

There are variations in the architecture of processors. Sometimes the variation is slight (e.g., the architectural difference between members of the same family such as the Core i5 and the Core i7 is relatively small). Sometimes the variation between processors is greater (for example, the difference between two RISC processors such as MIPS and SPARC). Sometimes the difference between processors is quite considerable, such as the difference between an Intel Core i7 and the ARM, because one processor has a classical CISC architecture and the other has a RISC architecture.[1]

It would be nice to cover the differences between various processors as we go along. For example, when ARM's register set is introduced, we could talk about variations between register models such as Intel's IA32 processors or Freescale's 68K family (the 68K has been renamed ColdFire). Such an approach is not practical, and the reader would soon get lost in the fine detail. In Chapter 3, we treat the ARM as a basic theme, and in Chapter 4, we look at some of the *variations* between processors.

Although we could cover computer architecture just by defining an instruction set, that would be as sensible as learning to fly an aircraft by studying only the laws of

---

[1] We have not yet introduced the terms CISC and RISC. For our current purposes we can say that RISC architectures have regular instruction sets with fixed-size instructions, and the only operations permitted on memory locations are load a register and store a register. CISC architectures have variable-length instructions, irregular instruction sets, and permit operations on memory such as add the contents of location P to register Q and store the result in memory location P. Examples of RISC processors are ARM, MIPS, SPARC, and examples of CISC architectures are the Pentium and its derivatives, and the Freescale 68K family.

aerodynamics and then climbing into an aircraft for the first time and setting off solo. Courses on computer architecture frequently have a lab component where a machine's assembly language can be studied and investigated. We have integrated understanding ARM's ISA with writing programs for the ARM that run on a PC. To learn more about what computers do, we concentrate on writing practical programs in *assembly language*. The intent of this book is not to create expert assembly language programmers but to demonstrate how a microprocessor's underlying architecture supports high-level languages and how architectural resources are used. By understanding the structure of assembly language instructions, we are in a better position to appreciate some of the trade-offs that engineers and architects face when designing a particular ISA.

Chapter 4 sets two objectives. The first is to look at the *stack*, which is so important in the design of functions, the management of local variables, and parameter passing. We demonstrate how a microprocessor's instruction set can support the stack, pointers, parameter passing and the allocation of space to local variables in a programming environment. However, because ARM's architecture does not implement all of the stack handling mechanisms incorporated in some CISC processors, we also take a look at another processor and its approach to stack handling. In Chapter 4, we look at machine-level instructions in greater depth and demonstrate some of the interesting operations provided by the ARM and other processors.

The final chapter in Part II, Chapter 5, is devoted to architectural support for *multimedia*, which is primarily concerned with the processing and manipulation of sound and images. We begin by briefly looking at several examples from the area of multimedia, such as graphics processing and image compression, in order to give readers a feel for the type of operations that modern processors have to perform. Multimedia applications drive today's computer development in previously unimagined ways. For example, a video clip contains more information than might have passed through the CPU of an IBM System/360 processor during its lifetime (i.e., mid-1960s to 1970s). We look at practical aspects of multimedia and describe how special-purpose instruction sets, such as Intel's *short vector SIMD* instructions, efficiently process audio and video signals.

CHAPTER 3

Architecture and Organization

"Everything should be made as simple as possible, but not simpler."
Albert Einstein

"Computers in the future may weigh no more than 1.5 tons."
Popular Mechanics, 1949

"Anyone can build a fast CPU. The trick is to build a fast system."
Seymour Cray

A computer's *instruction set architecture* (ISA) describes the assembly language programmer's view of a computer and stresses its functionality rather than its internal organization or implementation. The ISA tells us *what* the machine does, whereas its organization tells us *how* it works. In later chapters, we cover processor organization and explain how a given ISA may be implemented. The objectives of this chapter are as follows.

- Examine the stored program machine and show how an instruction is executed.
- Introduce instruction formats for *memory-to-memory*, *register-to-memory*, and *register-to-register* operations.
- Demonstrate how a processor implements *conditional behavior* by selecting one of two alternative actions, depending on the result of a test.
- Describe a set of computer instructions and show how computers access data (*addressing modes*).
- Introduce ARM's development system and show how ARM programs are written.
- Demonstrate how the ARM uses *conditional execution* to implement efficient code.

Because realistic ISAs can have a steep learning curve, we begin with a generic computer architecture that has the essential features of real machines but lacks their complexity. Moreover, the computer structure we use is based on a *register, bus,* and *ALU* model that is well-suited to describing instruction execution. Later, we will introduce another way of demonstrating computer structure that is well-suited to explaining pipelining—one of the techniques incorporated in all modern processors and described in Chapter 7.

3.1 Introduction to the Stored Program Machine

A microprocessor like the ARM has a *stored program architecture* that locates programs and data in the *same* memory and operates in a *fetch-execute mode*, whereby instructions are read from memory, decoded, and executed sequentially. Such a computer has registers, an arithmetic and logical unit (ALU), memory, and buses that link these components.

You might wonder how a program is loaded into memory in the first place. It is either always there stored in the type of read-only memory we cover in Chapter 10, or it is loaded from disk by the operating system.

Registers are storage elements located within the CPU that behave just like storage elements in memory. Some CPUs, such as those found in embedded applications, have no more than 20 registers; other computers might have well over 100 registers. Registers have *names* such as r0, r1 , . . . , r15 (ARM names); AX, BX, CX, DX, SP, BP, SI (Intel names); or D0, D1 , . . . , D7 (Freescale names) rather than *addresses*. A computer instruction uses relatively few bits in its op-code to reference a register; the register-select field of an instruction might be 3 to 5 bits, depending on how many programmer-visible registers there are. Because main memory has much greater storage capacity than a register set (e.g., 4 GB), memory is referenced by an address that might be 32 or 64 bits long. A computer with 32 general-purpose registers requires only 5 bits in the instruction to specify a register, whereas it takes 32 bits to uniquely access one of 4G locations in main memory.[2]

Registers in a CPU perform several functions. Some registers are *scratchpad* registers and hold data elements or the *addresses* of data elements (i.e., they contain *pointers*). Some registers perform special functions such as a loop counter that counts the number of times round a loop, and some record the current status of the processor. The most important register in a CPU is its *program counter* (PC) that contains the address of the *next instruction* to be executed; that is, the program counter keeps track of the execution of a program. The PC is sometimes called the *instruction pointer*, which better indicates its function.

Computer instructions have many formats. For the sake of simplicity, we will assume that our generic computer provides the following three instruction formats:[3]

```
LDR register_destination, memory_source
STR register_source, memory_destination [4]
Operation register_destination, register_source1, register_source2
```

For example, LDR **r1**,1234, STR r3,**2000**, ADD **r1**,r2,r3, and SUB **r3**,r3,r1 are all valid instructions.

The LDR[5] instruction copies data from memory into a register, and the STR performs the reverse operation by transferring the contents of a register to a memory location. For example, LDR **r1**,1234 reads the contents of memory location 1234 into register r1, and STR r2,**5000** writes the contents of register r2 to memory location 5000.

---

[2] This assumes that the computer can access individual *bytes* in memory. If the computer accesses only 32-bit *words*, then only 30 bits are required to specify the word.

[3] Recall that, throughout this text, the destination operand is depicted in a bold font because there is no standard for the way in which assembly language instructions are written. Some processors write an instruction in the form opera-tion source, **destination** and others write operation **destination**, source. By using a bold font to indicate the destination operand, you don't have to worry which processor's convention is being used.

[4] It would be consistent to write this operation as STR **memorydestination** , registersource with the destination operand on the left. However, as the ARM always puts the register on the left in its load and store operations, we have adopted that format to avoid readers having to change notation when moving from the simple processor to the ARM.

[5] It would make sense to use the mnemonics LOAD and STORE rather than LDR and STR. However, I have used LDR and STR for consistency with ARM's instruction set.

The third instruction type has *three* operands, each of which references a register. The op-code[6] part of the instruction, denoted by operation, defines the activity to be performed (e.g., ADD, SUB, AND) by the CPU. The three *operand fields* following the op-code specify the registers taking part in the operation. The *source* operands specify where the data comes from, and the *destination* operand specifies where the result is to be written. This three-operand *register-to-register* instruction format is typical of RISC processors such as the ARM, MIPS, and PowerPC, but it differs from the two-operand format of CISC microprocessors such as Intel's Pentium family.

Note that when ADD **r1**,r2,r3 adds the contents of register r2 to register r3 then writes the sum to register r1, the contents of registers r2 and r3 remain unchanged.

Let's examine how an instruction is retrieved from memory and executed by our simple computer. Figure 3.1 provides a block diagram of the stored program machine introduced in Chapter 1. The central processing unit consists of an arithmetic and logical unit (ALU), registers, and buses. The CPU also includes a control unit which is a hardware subsystem responsible for fetching and decoding an instruction then using the decoded information to control data flow between registers and functional units such as the ALU, which is an input/output interface. Chapter 7 looks at how instructions are decoded and executed in greater detail.

FIGURE 3.1   Fundamental structure of a computer

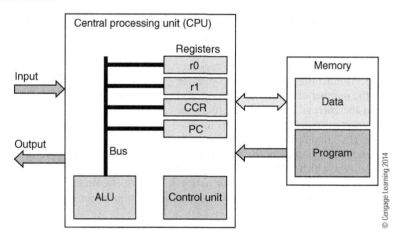

Figure 3.2 illustrates the structure of a hypothetical stored program machine.[7] The most lightly shaded parts of the diagram can be ignored for the time being. The registers in this diagram are defined as follows.

MAR   The memory address register stores the *address* of the location in main memory that is currently being accessed by a read or write operation.

MBR   The memory buffer register stores data that has just been read from main memory or data to be immediately written to main memory.

PC   The program counter contains the address of the next instruction to be executed. Thus, the PC *points* to the location in memory that holds the next instruction.

IR   The instruction register stores the instruction most recently read from main memory. This is the instruction currently being executed.

---

[6] The terms op-code and instruction are occasionally used interchangeably. In this book, instruction denotes the most primitive machine-level action that can be specified in a program. An instruction uses the op-code to indicate the specific action (e.g., ADD, MOVE, STR) and also includes operands.

[7] The machine is hypothetical in two important senses. It has a simplified instruction set, and it executes an instruction to completion before beginning the next one. Later, we will demonstrate how instruction execution can be overlapped and the next instruction can begin execution before the current one has been completed. This uses a mechanism known as pipelining, which is fundamental to all computers.

FIGURE 3.2   Partial structure of a hypothetical stored program machine

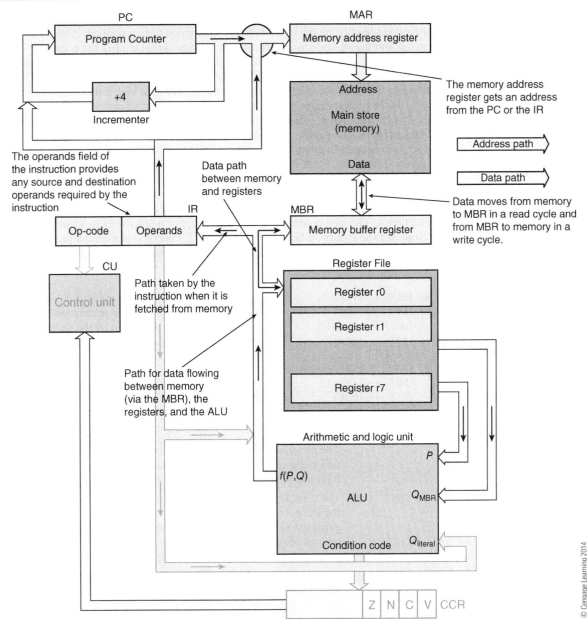

r0-r7   The register file is a set of eight general-purpose registers r0, r1, r2, . . . , r7 that store temporary (working) data (for example, the intermediate results of calculations). A computer requires at least one general-purpose register. Our simple computer has *eight* general-purpose registers.

As well as registers, the computer in Figure 3.2 has *buses* that transfer information between registers and between registers and the arithmetic and logic unit (ALU). The ALU operates on data to calculate the function of one operand (*monadic* or *unary* operation) or two operands (*dyadic* or *binary* operation). Examples of monadic operations are negation, incrementing, and clearing (setting to zero); examples of dyadic operations are addition and logical operations such as AND and XOR. We will later encounter instructions with three operands such as $a + b \cdot c$ that forms the product of b and c and adds it to a.

The control unit (CU) in Figure 3.2 *interprets* the instruction in the IR; that is, it causes the instruction specified by the op-code of the instruction to be executed. It uses a stream of clock pulses and the op-code to generate all of the signals that control the computer's buses, memory, registers, and functional units. Recall from Chapter 2 that the control unit uses tristate gates and register clocks to move data around the computer.

In Figure 3.2, dark blue indicates *address buses*, and light blue indicates *data buses*. An *address* is the binary representation of the location of a word in memory. The processor in Figure 3.2 is a register-to-register architecture, which requires three register addresses when it executes an instruction such as ADD **r1**,r2,r3. When this computer transfers data between a register and memory, it requires two operands: the register and the memory location. A load register instruction,[8] such as LDR **r1**,1234, is defined in RTL as [r1] ← [1234]. We introduced RTL in Chapter 1; recall that the ← symbol indicates the direction of data transfer and [] indicates the contents of a register or memory location.

The following RTL notation shows how the processor in Figure 3.2 reads LDR **r1**,1234 then executes it. The first column (e.g., FETCH) supplies a *label field* to identify the line of code. Note that the rightmost field that begins with a semicolon is a comment field designed to aid understanding. In this text, we use the convention that the semicolon separates code from comment (an ARM convention).

```
FETCH   [MAR] ← [PC]              ;copy PC to memory address register
        [PC]  ← [PC] + 4          ;increment PC to point to next instruction
        [MBR] ← [[MAR]]           ;read the instruction pointed at by the address in MAR
        [IR]  ← [MBR]             ;copy the instruction now in the MBR to the IR

LDR     [MAR] ← [IR(address)]     ;copy the operand address from the IR to the MAR
        [MBR] ← [[MAR]]           ;read the operand value from memory into the MBR
        [r1]  ← [MBR]             ;move the operand to register r1
```

Let's examine each of these low-level operations in greater detail.

[MAR] ← [PC]   The *program counter*, which contains the address of the next instruction to be executed, is copied to the memory address register, where it is used to access memory to read the instruction to be executed.

[PC] ← [PC] + 4   The constant 4 is added to the contents of the program counter so that the PC points to the *next instruction* to be executed. Instructions are executed sequentially unless *a flow control instruction* modifies the natural sequence. The increment is 4 rather than 1 because ARM's instructions are each four bytes long.

[MBR] ← [[MAR]]   The contents of the memory location pointed to by the contents of the MAR (memory address) are copied to the memory buffer register (MBR). Since the MAR contains the address of the next instruction to be executed, the MBR will now hold the op-code of the current instruction. The notation [[MAR]] denotes *the contents of the memory location whose address is contained in MAR*.

[IR] ← [MBR]   The contents of the memory buffer register are copied to the instruction register. The IR now contains the *instruction* to be executed. These instruction bits are used by the control unit to generate all the signals that implement the instruction.

---

[8] Load/store computers like the ARM do not support direct loading of data from a specific memory location such as 1234. Instead, a memory location is specified indirectly via a pointer in a register (for example LDR **r1**, [r2] , where r2 contains the actual memory address). CISC processors like the IA32 do permit direct memory accesses. We use direct memory addressing here because it is so easy to understand.

Register Visibility

There are three types of register. General-purpose registers hold temporary variables created during a calculation. The ARM processor has 16 general-purpose registers called r0, r1, …, r15. Registers r14 and r15 are general purpose in the sense that the programmer accesses them using the same instructions that access r0 to r13. However, r14 and r15 have special roles in the ARM processor architecture.

Special-purpose registers perform specific functions. For example, the PC points at the next instruction to be executed. Other special-purpose registers are the status register, stack pointer, and the CPU identification register, which stores the manufacturer's name and model.

Some registers are invisible to the programmer, are not part of the processor's architecture, and are not directly accessible by programmers. For example, the instruction register, IR, and memory address register, MAR, are programmer-invisible registers. These are required to implement the computer and are not part of its ISA.

Use of Multiplexers to Implement Data Flow

The following diagram demonstrates how part of the conceptual circuit of Figure 3.2 would be implemented using the multiplexers we described in Chapter 2. Multiplexers are used to control the flow of data to the PC and the MAR. The multiplexers are controlled by signals from the control unit.

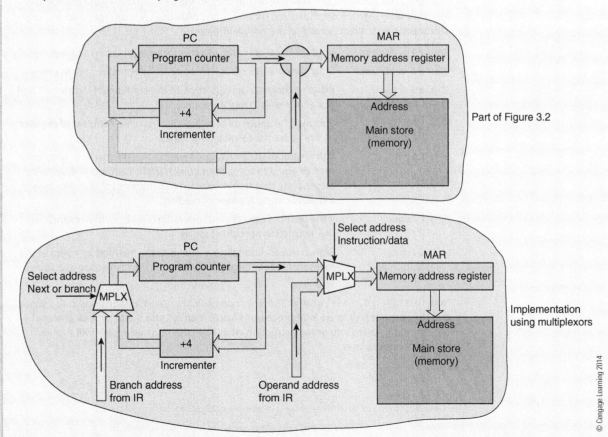

© Cengage Learning 2014

Why is the Program Counter Incremented by 4?

Computer memory is *byte addressed* where bytes are numbered 0, 1, 2, ... However, 32-bit microprocessors use 32-bit instructions and 32-bit data words. Thus, the PC must be incremented by 4 after each fetch, since 4 bytes × 8 bits/byte = 32 bits = 1 instruction word.

These four RTL actions constitute the *fetch phase* of the processing cycle, whereby an instruction is read from memory and the program counter incremented in readiness for fetching the following instruction. The second group of three actions, labeled LDR, comprises the *execution phase* of the processing cycle. Each instruction begins with the same fetch phase. However, each instruction's execute phase is determined by the pattern of bits in the IR for that instruction. Figure 3.3 illustrates the sequence of actions that takes place during fetch and execute of the LDR **r1**, 1234 instruction.

3.1.1   Extending the Processor: Dealing with Constants

We've demonstrated how a computer accesses data from memory using an operation like LDR **r1**, 1234 where 1234 refers to the contents of memory location 1234. Suppose we want to load the *number* 1234 *itself* into register r1. Such a number is called a *literal* operand.[9]

A *literal* is a value that is used directly in an operation, as opposed to the *contents of a memory location*, and is indicated by prefixing a number with a hash symbol (#).[10] The assembly language instruction LDR **r1**, 200 loads register r1 with the *contents* of memory

---

Some Instructions

Before continuing, we need to introduce a few simple instructions that can be used in examples. The following are ARM-like instructions in the sense that they use the ARM assembly language format. However, recall that the ARM does not support load and store instructions with direct (absolute) memory addresses.

LDR **r0**, address	Load the contents of the memory location at *address* into register r0.
STR r0, **address**	Store the contents of register r0 at the specified *address* in memory.
ADD **r0**, r1, r2	Add the contents of register r1 to the contents of register r2 and store the result in register r0.
SUB **r0**, r1, r2	Subtract the contents of register r2 from the contents of register r1 and store the result in register r0.
BPL target	If the result of the previous operation was positive (i.e., greater than or equal to zero), then branch to the instruction at address *target*. Note that target is stored as a *numeric value*—here we use a *symbolic name* for ease of reading.
BEQ target	If the result of the previous operation was zero, then branch to the instruction at address *target*.
B target	Branch unconditionally (i.e., *jump*) to the instruction stored at the memory address target. This executes the instruction at address target.

Note that LDR **r0**, address and STR r0, **address** are pseudoinstructions in the sense that they are not part of an ARM processor's instruction set and the assembler automatically translates them into other instructions that have the same effect. We will explain why this happens later.

---

[9] The terms *immediate* value and *literal* value are used interchangeably in computer science literature.

[10] The # symbol indicates a literal in ARM and Freescale assembly languages but is not universal.

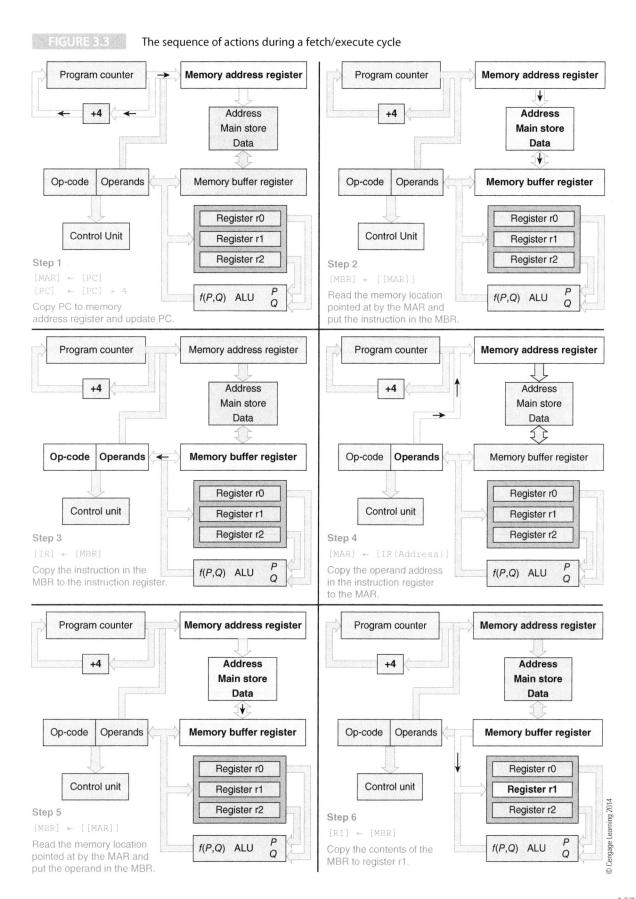

FIGURE 3.3   The sequence of actions during a fetch/execute cycle

Step 1
[MAR] ← [PC]
[PC] ← [PC] + 4

Copy PC to memory
address register and update PC.

Step 2
[MBR] ← [[MAR]]

Read the memory location
pointed at by the MAR and
put the instruction in the MBR.

Step 3
[IR] ← [MBR]

Copy the instruction in the
MBR to the instruction register.

Step 4
[MAR] ← [IR(Address)]

Copy the operand address
in the instruction register
to the MAR.

Step 5
[MBR] ← [[MAR]]

Read the memory location
pointed at by the MAR and
put the operand in the MBR.

Step 6
[R1] ← [MBR]

Copy the contents of the
MBR to register r1.

© Cengage Learning 2014

137

location 200 and is represented in RTL by [r1] ← [200]. The assembly language instruction LDR **r1**,#200 loads register r1 with the *constant* 200 and is represented in RTL by [r1] ← 200. Another example of the use of a literal operand is ADD **r0**,r1,#25, which means that the value 25 is added to the contents of r1 and the sum deposited in register r0.

Figure 3.4 illustrates the new data paths required to implement literal operands. A path from the instruction register, IR, routes a literal operand to the register file, MBR, and ALU. For example, when ADD **r0**,r1,#25 is executed, the operand to be added to register r1 is routed from the operand field of the IR, rather than from the memory system via the MBR.

FIGURE 3.4   Information paths for literal operands

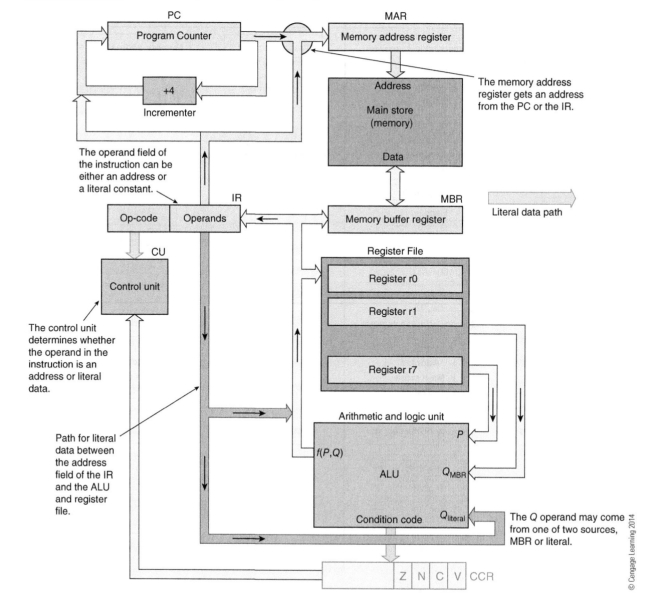

3.1.2   Extending the Processor: Flow Control

Flow control refers to any action that modifies the strict instruction-by-instruction sequence of a program. In other words, it refers to a computer's ability to execute instructions out of their normal sequence. In general, flow control refers to branches and jumps[11] to a specific point in a program, subroutine/procedure calls and returns, interrupts, and operating system calls. It is important to stress that *flow control* is the key element in a computer's ability to make decisions and choose between multiple courses of action.

An example of flow control is *conditional behavior* that allows a processor to select one of two possible courses of action. We encountered this concept in Chapters 1 and 2, and now we demonstrate how it appears at the assembly language level. A computer implements conditional behavior by testing the outcome of an operation and then following one of two paths through the program; that is, the test determines which of two different addresses is to be loaded into the program counter. In turn, that leads to one of two possible instructions being fetched from memory and loaded into the instruction register.

Figure 3.5 shows the information paths of the computer that are required to implement conditional behavior. A *conditional instruction* like BEQ results in either continuing program execution normally with the next instruction in sequence pointed to by PC + 4 or loading the program counter with a new value and executing a *branch* to another region of code. The following fragment of code demonstrates a conditional branch.

```
         SUBS    r5,r5,#1        ;Subtract 1 from r5
         BEQ     onZero          ;IF zero then go to the line labeled 'onZero'
notZero  ADD     r1,r2,r3        ;ELSE continue from here
         .
         .
         .
onZero   SUB     r1,r2,r3        ;Here's where we end up if we take the branch
```

In this example, the first instruction SUBS r5,r5,#1 subtracts 1 from the contents of register r5.[12] After completing this operation, the number remaining in r5 may or may not be zero. We are now going to test for zero.

The next instruction BEQ onZero forces a branch or jump to the line labeled 'onZero' if the outcome of the last operation was zero. So, if r5 contains zero, a jump is made to the line containing SUB r1,r2,r3 and execution continues from there. Otherwise, the next instruction in sequence after the BEQ (i.e., the ADD) is executed. This code implements: if zero then r1 = r2 - r3 else r1 = r2 + r3.

Let's look at Figure 3.5 and follow the information paths that implement conditional behavior. Essentially, the new information paths link the ALU with the program counter and allow the ALU to determine which of two possible instructions get executed next. The condition on which a branch is taken (or not taken) is the outcome of an operation performed by the ALU. For example, a branch can be made if the result of the last operation was zero or negative or the carry bit was set. Information from the ALU in Figure 3.5 is fed to the *condition code register* (*CCR*), which stores the various conditions that can be tested (e.g., zero, negative, positive). That is, when the ALU performs an operation, it updates the zero, carry, negative, and overflow bits in the CCR.

The CCR is connected to the control unit. The control unit is responsible for decoding instructions and generating the necessary control signals to execute instructions. When a conditional branch like BEQ is executed, the control unit selects the desired condition from the CCR for testing (in this case the Z-bit). If the tested condition is false, the CPU continues execution with the next instruction (e.g., pointed to by PC + 4) in sequence as normal. If the

---

[11] The terms branch and jump are generally synonymous, although some use a branch to indicate one of two possible paths and a jump to indicate an unconditional *goto*.

[12] Eagle-eyed readers will note that the mnemonic for subtract is SUBS rather than SUB. ARM processors do not update the condition code register automatically. The programmer has to append an S to the instruction in order to ensure updating.

FIGURE 3.5   Implementing conditional behavior at the machine level

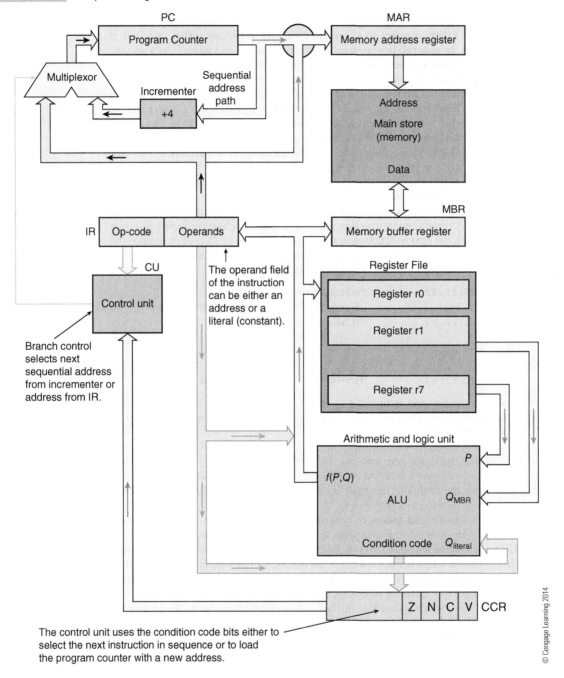

tested condition is true, a jump is made to the address specified by the branch field of the conditional instruction, called the *branch target address* (BTA).

A typical conditional instruction is BPL Error (branch on positive), which forces a jump to a region of code labeled 'Error' if the result of the previous operation was positive. In Figure 3.5, an *address bus* between the *address field* of the instruction register and the *program counter* permits the loading of a *non-sequential* address into the PC.

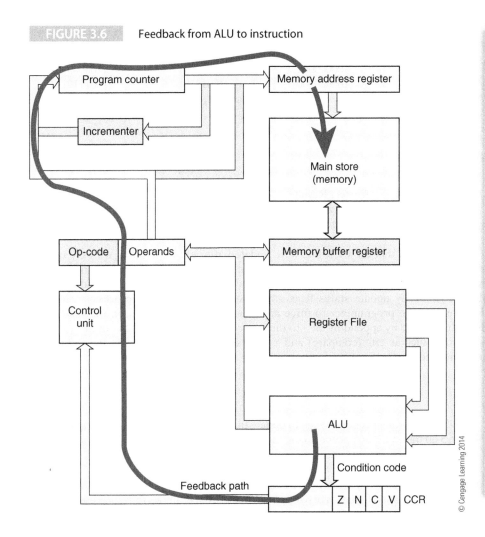

FIGURE 3.6   Feedback from ALU to instruction

Let's start with the program counter that fetches a conditional branch instruction from an address in memory. The branch instruction reads the contents of the condition code register whose bits are determined by the result of the previous instruction.

When executed, the conditional branch instruction performs one of the following two actions:

- If the bit in the CCR being tested is false, the processor fetches the next instruction at [PC] + 4.
- If the bit in the CCR being tested is true, then the program counter is loaded with a new address from the operand field of the instruction register, and a jump is made to this new address.

© Cengage Learning 2014

Figure 3.6 demonstrates how the outcome of an ALU operation is fed back to the PC to select an instruction that implements transfer of control.

Status Information

Let's look more closely at how conditional branches are implemented. When the computer performs an operation, it stores *status* or *condition* information in the CCR[13] of Figure 3.5. The processor records whether the result is zero (Z), negative in two's complement terms (N), generated a carry (C), or arithmetic overflow (V). Consider the following examples of 8-bit addition on the bits of the CCR. We also provide the same examples in decimal with the effect on the condition code bits in plain English.

---

[13] Computer manufacturers use different terms to describe the status register. I use CCR (condition code register) to describe the outcome of a computer operation. ARM calls this *the current processor status register*, CPSR, and Intel calls it the status register.

Example 1	Example 2	Example 3	Example 4
00110011	11111111	01011100	11011100
+01000010	+00000001	+01000001	+11000001
01110101	100000000	10011101	110011101
$Z = 0, N = 0$	$Z = 1, N = 0$	$Z = 0, N = 1$	$Z = 0, N = 1$
$C = 0, V = 0$	$C = 1, V = 0$	$C = 0, V = 1$	$C = 1, V = 0$
51	−1	92	−36
+66	+1	+65	−63
117	0	−99	−99
Correct result	Correct Result	Incorrect result	Correct Result
	Result zero	Result negative	Result negative
	Carry generated	Overflow generated	Carry generated

When we said that the CCR bits were updated after each operation, we were being *a little economical with the truth*. The actual situation is much more complex, and it varies markedly from processor to processor. CISC processors, like the Intel IA32 and Freescale 68K, automatically update status flags after each operation. RISC processors, like the ARM, require the programmer to force an update of the status flags. In the case of the ARM, this is done by appending an S to the instruction (for example SUBS or ADDS). Some instructions, such as CMP (compare) and TST (test), automatically update the status flags and do not require the S suffix.

To Update the CCR or to Not Update the CCR?

The RISC programmer chooses when to update the CCR bits either explicitly by appending an S on the end of an opcode (in the case of ARM) or implicitly by performing a comparison with a CMP instruction.

CISC processors generally update the CCR after each operation automatically. However, there are significant differences between CISC families. For example, some processors update the CCR following a load instruction and some don't. The 68K has both data and address registers. If you perform an operation on a data register, the CCR is updated. If you perform an operation on an address register, the CCR is not updated. The rationale is that operations on address registers yield pointers, and you don't want the act of updating a pointer to modify the CCR.

The advantage of automatic CCR updating after each instruction is that you don't need to provide an update/don't update control bit in the instruction word (instruction bits are very precious). The disadvantage is that you may wish to retain the status across several instructions rather than destroying the status after each and every operation.

Example of a Branch Instruction

We next demonstrate how to use the conditional branch operation BEQ address to implement a high-level language construct. First, recall that the op-code field of the BEQ instruction in the IR selects a bit in the CCR to be tested (e.g., Z, N, or C). If the tested bit is 1, then PC is loaded with a new value (i.e., the branch target address). Otherwise, PC is not modified. The assembly language construct

```
BEQ address                    ; branch to address if zero flag is set
```

is expressed in RTL as IF [Z] = 1 THEN [PC] ← <address> where [Z] indicates the Z-bit of the CCR. Consider the high-level language code fragment

```
X = P - Q
IF X ≥ 0    THEN X = P + 5
            ELSE X = P + 20
```

We can translate this into ARM code using the subset of ARM instructions defined earlier in the panel. In the following code, lines in blue typeface are *assembler directives* that provide an *environment* for the assembly language code. Assembler directives are not executed themselves. The DCD directive reserves memory locations for the three operands. For example, the line P DCD 12 equates the symbolic name P to the current memory location and stores the value 12 in that location. That is, we preload memory with the number 12 and give it a symbolic name P. This assembler directive is equivalent to the C language statement int P = 12. We look at assembler directives in more detail when discussing assembly language.

```
        LDR   r0,P          ;Load r0 with the contents of memory location P[14]
        LDR   r1,Q          ;Load r1 with the contents of memory location Q
        SUBS  r2,r0,r1      ;Subtract the contents of Q from P to get
                               X = P − Q[15]
        BPL   THEN          ;IF X ≥ 0 then execute the 'THEN' part
        ADD   r0,r0,#20     ;ELSE Add 20 to the contents of r0 to get P + 20
        B     EXIT          ;Skip past 'THEN' part to 'EXIT'
THEN    ADD   r0,r0,#5      ;Add 5 to r0 to get P + 5
EXIT    STR   r0, X         ;Store r0 in memory location X
        STOP
P       DCD   12           ;These three lines reserve memory space for
Q       DCD   9            ;the three operands P, Q, X. The memory
X       DCD                ;locations are 36, 40, and 44, respectively.
```

This sequence of assembly-language instructions can be expressed in RTL notation.

```
        LDR   r0,P          ;[r0] ← [P]
        LDR   r1,Q          ;[r1] ← [Q]
        SUBS  r2,r0,r1      ;[r2] ← [r0] − [r1]
        BPL   THEN          ;IF [r2] ≥ 0 [PC] ← THEN
ELSE    ADD   r0,r0,#20     ;[r0] ← [r0] + 20
        B     EXIT          ;[PC] ← EXIT
THEN    ADD   r0,r0,#5      ;[r0] ← [r0] + 5
EXIT    STR   r0,X          ;[X] ← [r0]
```

Figure 3.7 illustrates how this code could be executed on the hypothetical computer discussed previously. We look at two cases:

Case 1: $P = 12$, $Q = 9$, and the branch is *taken* (control is transferred to the branch target address);

Case 2: $P = 12$, $Q = 14$, and the branch is not taken (control is transferred to PC + 4).

Figure 3.7 provides two memory maps for this code fragment. The only difference between the two maps is the value of Q in memory location 40. Below each memory map is the instruction sequence executed by the corresponding code. The value of the PC is shown

---

[14] Remember that LDR r0, P is a pseudoinstruction and not actually implemented by ARM processors. We use it because it aids understanding (all CISC processors do implement this form of instruction; that is, load register with the contents of a specified memory location).

[15] ARM processors do not update the condition flag values after an operation unless you append an S after the mnemonic to explicitly perform an update.

FIGURE 3.7   Illustration of conditional execution

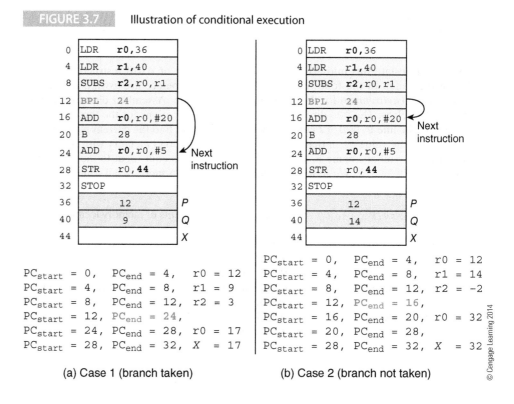

(a) Case 1 (branch taken)   (b) Case 2 (branch not taken)

at the beginning and end of each instruction, together with register contents at the end of the instruction.

Let's look at another example of the use of conditional branching in the mechanization of a loop that calculates $1 + 2 + 3 + \ldots + 20$. In this case, a counter is incremented from 1 to 20. On the final pass, the count becomes 21. The operation[16] CMP r0,#21 compares the counter value in r0 with the literal 21 by subtraction. The next operation BNE Next makes a branch back to the instruction labeled Next unless the previous result was zero. On the 20th iteration, the result becomes zero, and the branch is not taken and the loop exited.

```
        LDR     r0,#1        ;Put 1 in register r0 (the counter)
        LDR     r1,#0        ;Put 0 in register r1 (the sum)
Next    ADD     r1,r1,r0     ;REPEAT: Add the current count to the sum
        ADD     r0,r0,#1     ;  Add 1 to the counter
        CMP     r0,#21       ;  Have we added all 20 numbers?
        BNE     Next         ;UNTIL we have made 20 iterations
        STOP                 ;If we have THEN stop
```

The next step is to look at the ARM processor in depth. However, before we do that we need to look at the concept of an instruction set architecture (ISA) in a little more detail.

---

[16] Remember that the op-code is CMP not CMPS. Since the purpose of a compare is to set up the condition codes prior to a branch, it is not necessary to append an S to the mnemonic.

The Branch—Conceptually (yet again. . .)

Before we leave conditional branching, we are going to demonstrate once more how a branching operation might be implemented at the logic level to help you appreciate how computer operations can be implemented with simple logic devices. The figure below provides a hypothetical branching circuit. Bold lines indicate 16 or more program counter bits; that is, there are 16 or more multiplexers operating in tandem—one for each bit of the address. Although simplified, this diagram does illustrate the principles involved (the nature of conditional branching, the role of the program counter and condition code register, and the relationship between hardware and software). The program counter receives its next input either from the incrementer (next sequential address) or from the branch address field of the current instruction (the branch target address). The source of the program counter's input is controlled by a multiplexer.

The program counter input multiplexer is controlled by the op-code field of the instruction register. If the current instruction is a conditional branch, then a bit in the op-code field enables the two-input AND gate that controls the program counter multiplexer. When the output of the AND gate is 0, the input to the program counter is the next sequential address; otherwise, the input to the program counter is the branch address.

The second field in the instruction is a *select condition field* that specifies a branch on zero (Z flag), branch on negative (N), branch on carry set (C), or branch on overflow (V). A two-to-four decoder transforms the 2-bit select condition field into one of four control signals. Each of these control signals is fed to a four-input condition multiplexer that selects the appropriate bit from the condition code register and routes it to the AND gate that controls the multiplexer.

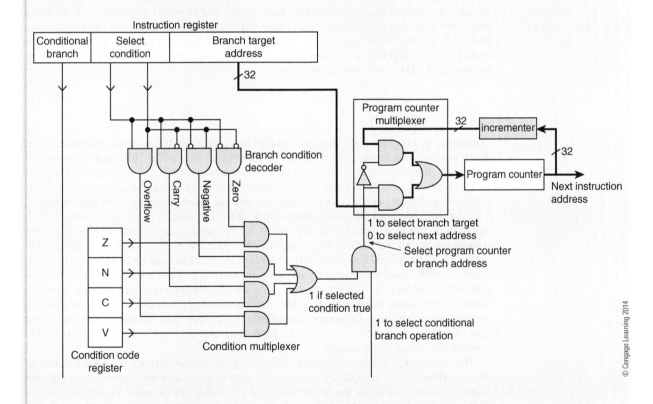

3.2   The Components of an ISA

Before we look at a real microprocessor in detail, we have to say a little more about the three elements of an ISA: its register set, its addressing modes, and its instruction formats. Together, these define the assembly language level programmer's view of the processor. There are, in fact, two assembly language programmers: the *human* and the *compiler*. The human programmer writes assembly language either as a student in class studying computer architecture or as a professional who is writing code to program a computer or is debugging a system at the assembly language level. Most low-level or machine code is produced automatically by a compiler that generates code from a high-level language.

The distinction between human and compiler is important in the design of computer instruction sets. It is probable that a human can write short fragments of code efficiently using an assembly language. Humans have ingenuity and can be inventive and creative. Humans can exploit a machine's instruction set and produce remarkably compact or efficient code. It is difficult to produce very large programs in this way. Moreover, assembly language written by one person using ingenious design features is often impenetrable and unreadable by another. Humans can write high-level language code more efficiently and reliably than low-level code. Consequently, in today's world, most programmers write in a high-level language and leave the production of machine code to the compiler.

Compilers create machine code automatically from a high-level language. Their automatic methods of code translation are often poor at exploiting the features of a particular machine. Although a computer architect may construct an ingenious machine-level instruction, a compiler might never use it. Compilers just cannot always exploit all of a processor's features. An architecture that is good for humans may not be good for compilers. In the 1970s and 1980s when microprocessors were being developed, manufacturers created interesting and exotic machine-level instructions that were later found to be virtually useless when the output from compilers was examined. This observation was one of the driving forces behind the so-called RISC revolution of the 1980s.

3.2.1   Registers

Conceptually, a register is the most useless part of a computer. It's not even necessary, and it doesn't do any computing. Indeed, one of the microprocessors of the 1970s lacked an on-chip register set; it just used a group of memory locations as a register set. A single pointer register on the chip contained the address of the registers in memory. Registers are, however, necessary for the *efficient* performance of computers and the practical design of instruction sets.

You could design a computer instruction with the format ADD P = Q + R where P, Q, and R are locations in memory. Assuming a 16-bit operation code (the ADD part) and a 32-bit address space, this instruction would require $16 + 32 + 32 + 32 = 112$ bits, as depicted in Figure 3.8a. Typical real computer instructions are 16 or 32 bits wide and a 112-bit instruction is not generally feasible. Moreover, accessing memory is problematic, because you have to transmit the 32-bit address from the CPU to a specific memory chip and perform a certain amount of electronic housekeeping (called address decoding). It is also important to appreciate that the access time of memory components is relatively long compared with the access time of on-chip registers. In general, the more data that is directly accessible within CPU registers, the faster the processor.

Real computers implement on-chip storage in registers that perform the same function as memory elements; the only difference being the ease of access and the response time. On-chip registers require only a few instruction bits to specify them. For example, a computer with an 8-bit op-code and eight on-chip registers (using three bits to access one of r0, r1, r2 , . . . , r7) can implement ADD P = Q + R with $8 + 3 + 3 + 3 = 17$ bits. A more realistic example is given in Figure 3.8b of a computer with three 5-bit operand address fields allowing 32 registers to be addressed. The 32-bit word allows the remaining 17 bits to be devoted to the instruction op-code plus additional control fields.

FIGURE 3.8   Demonstration of operand address widths

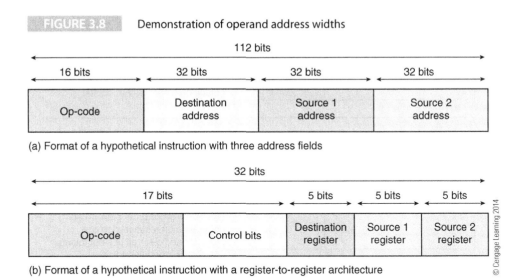

(a) Format of a hypothetical instruction with three address fields

(b) Format of a hypothetical instruction with a register-to-register architecture

Since a computer has only a few on-chip registers, it is necessary to transfer data between memory and registers with instructions that load data into registers and then store the contents of registers in memory (we've already used these LDR and STR operations). Computers that apply all data-processing operations to the contents of registers are called *load* and *store* computers to indicate that the only memory operations are those that transfer data to and from registers. These computers invariably fall into the RISC group of computers.

The size of a register (its bit width) is often the same as the maximum size of the data operation performed by the CPU. For example, a 16-bit computer may have 16-bit-wide registers that can add together two 16-bit values in a single operation.

Some computers permit operations on a *subset* of a register. For example, if a 32-bit register contains four bytes ABCD, a 16-bit operation on that register may act on bytes C and D while leaving bytes A and B unchanged.

Some registers treat their contents as two's complement values, and the result of an operation on a subsection of the register is sign-extended to the entire word. For example, if register r1 contains 12345678_{16}, adding 00002122_{16} in a 16-bit signed operation results in $0000779A_{16}$. However, if we add 00003122_{16} to 12345678_{16}, we get $FFFF879A_{16}$ because the 16-bit negative result is sign-extended to a 32-bit negative value. Let's verify this. The original 12345678_{16} is converted to 16 bits for addition; that is 5678_{16}. To this is added the 16-bit value 3122_{16} to give $5678_{16} + 3122_{16} = 879A_{16}$. Since this is 1000011110011010_2 in binary and the leading bit is set, it is sign-extended to $11111111111111111000011110011010_2 = FFFF879A_{16}$ in 32 bits. Figure 3.9 illustrates some of the possible effects of subregister operations.

General-Purpose Versus Special-Purpose Registers

The IA32 architecture used by the 8086 had a rather *ad-hoc* bunch of 16-bit registers called AX, BX, CX, and DX that could be divided into pairs (e.g., AH, AL) and used as byte registers. It also had four index (pointer) registers and four segment registers (used to break through the 64K-page limit imposed by 16-bit pointer registers). The point I am making is that the 8086's registers were highly specialized and the programmer had to remember what each register did and which instructions used that register. For example, register C was a special-purpose counter register. The 8086's register architecture was designed to ensure a measure of compatibility with Intel's earlier 8-bit processor—the 8080. The compatibility was weak, because 8080 code could not run on an 8086. However, it was anticipated that it would be easy to automatically map 8080 code onto the 8086 architecture because of the degree of similarity.

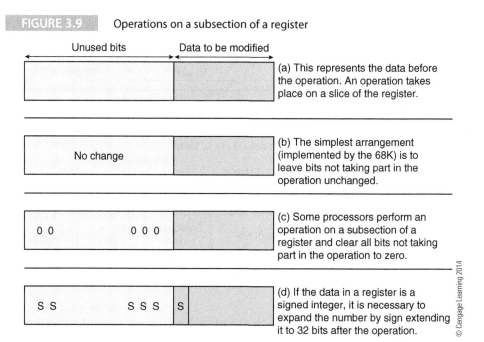

FIGURE 3.9   Operations on a subsection of a register

Unused bits   Data to be modified

(a) This represents the data before the operation. An operation takes place on a slice of the register.

No change

(b) The simplest arrangement (implemented by the 68K) is to leave bits not taking part in the operation unchanged.

0 0   0 0 0

(c) Some processors perform an operation on a subsection of a register and clear all bits not taking part in the operation to zero.

S S   S S S   S

(d) If the data in a register is a signed integer, it is necessary to expand the number by sign extending it to 32 bits after the operation.

© Cengage Learning 2014

Providing special-purpose registers means that you don't always need to allocate bits to specify them in an instruction word. For example, if a register is defined as a counter, an instruction that increments the counter doesn't need a register address, because the counter is bound to the instruction. Using dedicated registers leads to more compact code, a feature that was of vital importance in the early days of the microprocessors when memory was millions of times more expensive than today. Yes, *millions* of times!

Intel's then competitor, Motorola,[17] developed the 68K processor shortly after the 8086. This was a 32-bit machine with 32-bit registers. Motorola took a different approach to register architecture than Intel and used general-purpose registers, eight data registers D0 to D7, and eight address (pointer) registers A0 to A7. The registers were entirely interchangeable in the sense that whatever you did with D0 you could do with D1, etc. Of course, taking this approach means that any register-based instruction requires a 3-bit register select field to distinguish between {D0, D1, D2, . . . , D7}. It is also interesting to note that Motorola went to the trouble of distinguishing between data and address registers—something not taken up by RISC processors like the ARM. This was a very controversial decision, and at the time, there was a lot of debate. Some didn't like being told that A0 can hold only an address, and they objected that separate data and address registers were inefficient if you had, say, 12 data elements and four addresses. On the other hand, it allows instructions to treat address and data values appropriately. For example, arithmetic operations on address registers always sign-extend the result to 32 bits, because an address is a single entity whereas a data value may be partitioned into fields. Both arguments have their merit, but the general-purpose approach won the day.

The ARM has 16 general-purpose registers r0 to r15. Registers r0 to r13 are interchangeable and behave alike. Registers r14 and r15 have additional functions (r14 is the link register that holds subroutine return addresses and r15 is the program counter). Although you are free to use r13 in any way you desire, good programming practice requires that r13 be reserved for use as a stack pointer.

---

[17] Motorola renamed its Semiconductor Products Sector in 2004, and we will now refer to Freescale processors rather than Motorola processors.

3.2.2   Addressing Modes—an Overview

Instructions operate on data, and that data has to be moved to wherever it is to be processed. The ways in which the data is specified are collectively called *addressing modes*. In principle, there are three fundamental addressing modes, although there are a great many variations. These fundamental addressing modes are

- Literal
- Direct
- Indirect

We have already met these addressing modes. The simplest form of addressing is *literal addressing* where the operand is part of the instruction itself. Consider an operation $P = Q + 5$. Here, the 5 is a literal operand because it forms part of the instruction; that is, it isn't stored in a memory location or a register because it's part of the instruction itself. A corollary of literal addressing is that it is constant and cannot be changed during the execution of a program. The operation $P = Q + 5$ always adds 5 to Q. The value of Q may change whereas the constant 5 never changes.

This addressing mode is also called *immediate addressing* because the operand is immediately available (it doesn't have to be read from a register or memory). Computers use literal addressing to set up constants that do not change during the execution of a program (e.g., loop counts or limits).

As we have already seen, the ARM processor specifies a literal operand by means of the prefix #. For example, ADD **r1**,r2,#5 performs the operation $[r1] \leftarrow [r2] + 5$.

The second addressing mode is *direct addressing* (also called absolute addressing). This addressing mode provides the address of an operand as part of the instruction. For example, ADD **P**,Q,R indicates that the contents of memory location Q are added to the contents of location R and the result put in location P.

You could argue that direct addressing is the fundamental addressing mode of the stored program (or von Neumann) computer where an instruction is read during its fetch phase and then executed in its execute phase when the operands specified in the instruction are read from memory. Direct addressing is indeed widely used by CISC computers of the Intel IA32 (e.g., Pentium) or 68K varieties. For example, the IA32 instruction mov **ax**, [2468h] copies the contents of memory location 2468_{16} into register ax. Similarly, the Freescale 68K instruction ADD 1234,**D2** adds the contents of memory location 1234 into register D2. Note how the Intel and Freescale assembly instruction formats differ in several ways—from the register naming, to the order of the operands, to the way in which a direct address is specified. Unfortunately, assembler formats were never standardized.

Load and store computers like ARM do not implement direct addressing. All memory operands must either be specified as a literal or indirectly via a pointer in a register.

The third addressing mode has many names—we will call it *indirect addressing*, or more strictly, *register indirect addressing*. In register indirect addressing, the instruction provides the address of the register containing the address of the operand. As you can see, obtaining an operand requires three accesses: reading the instruction, reading the register containing the operand address, and finally, reading the actual operand.

The register containing the address of the operand is referred to as a pointer register. Load and store computers like the ARM use these addressing modes to access operands in memory. For example, you can write in ARM code LDR **r1**, [r2] to indicate that the contents of the memory location whose address is in register r2 is moved into register r1. In ARM, Intel IA32, and Freescale 68K assemblers we can write

```
LDR     r1, [r2]      ;Copy the contents of memory pointed at by register r2 into
                        register r1
MOV     ax, [bx]      ;Copy the contents of memory pointed at by register bx into
                        register ax
MOVE    (A5),D2       ;Copy the contents of memory pointed at by address register
                        A5 into data register D2
```

Register indirect addressing is useful for accessing tables and arrays because you can operate on the pointer register to access an element in the array.

There are many variants of register indirect addressing. The most common form is register indirect addressing with displacement, where the address of an operand is specified as the contents of a register plus a constant or offset. Typical forms of this addressing mode are

```
LDR   r2,[r3,#8]      ;Copy the contents of memory pointed at by register r3 plus 8
                       into register r2
MOV   ax,[12,bx]      ;Copy the contents of memory pointed at by register bx plus 12
                       into register ax
MOVE  (16,A5),D2      ;Copy the contents of memory pointed at by address register
                       A5 plus 16 into data register D2
```

This addressing mode includes a literal constant that is set up at the time the program is written. Register indirect addressing with displacement (or offset) allows you to use a pointer to the base of a data area and then an offset to indicate a given element in the array. Figure 3.10 describes the sequence of addressing modes.

FIGURE 3.10   Progressive sequence of addressing modes

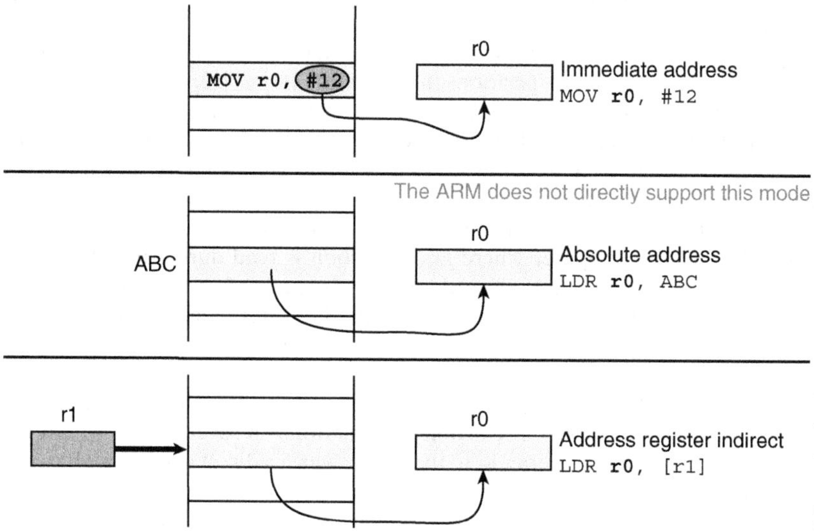

© Cengage Learning 2014

Program Counter Relative Addressing

Register indirect addressing allows you to specify the location of an operand with respect to a register. For example, LDR r0, [r1, #16] specifies that the operand is 16 bytes on from r1. Now, suppose that we use r15, the PC, to generate an address and write LDR r0, [PC,#16]. In this case the operand is 16 bytes on from the PC or 8 + 16 = 24 bytes from the current instruction. (The ARM's PC is always eight bytes on from the current instruction because of a mechanism called *pipelining* where it automatically fetches the next instruction before the current one has been executed).

The ability to use the PC in program counter relative addressing is an important feature of ARM's architecture. It allows you to generate the address of an operand with respect to the program that accesses it. If you relocate the program and its data elsewhere in memory; the relative offset does not change. We will find that PC relative addressing is a fundamental element of the ARM's ability to use pseudoinstructions to manipulate 32-bit constants.

Memory and Register Addressing

We have pointed out that there are no fundamental differences between registers and memory locations. The practical difference lies in the relative speed of access and the number of bits used to specify a register or memory location. In practice, this means that computer instructions in the era of 8-bit computers were not able to implement long memory addresses. With few exceptions, that meant that an instruction could provide only one memory address.

A corollary of this is that memory-to-memory operations, where a source and destination operand is in memory, cannot be supported. Consequently, microprocessors generally provide three instruction modes:

Memory-to-register — The source operand is in memory and the destination operand is in a register

Register-to-memory — The source operand is in a register and the destination operand is in memory

Register-to-register — Both operands are in registers.

The Freescale 68K did actually support a memory-to-memory operation (but only for MOVE) instructions in which both source and destination addresses could be in memory. This instruction was 10 bytes (80 bits) long because it required a 16-bit instruction word followed by two 32-bit addresses.

3.2.3   Instruction Formats

At the heart of an instruction set, architecture is the computer instruction itself. An instruction specifies the next operation to be performed—although we will later discover that some processors like the Itanium (IA64) series have instructions that specify multiple operations. For the purpose of this introduction, we will consider an instruction to be a word that specifies what action is to be performed together with any additional information (*i.e.*, literals, registers, or addresses).

RISC computers like the ARM and MIPS have instructions that suffer from a terrible tyranny; the need to fit an instruction into a word. If a computer has a 32-bit word and 32-bit registers, instructions will be 32 bits wide. Consequently, the instruction set designer is constrained by the instruction wordlength. CISC machines solve the problem of a fixed wordlength by allowing an instruction to extend over several consecutive words. For example, a 68K processor uses five consecutive 16-bit words to hold an op-code (16-bit), a source (32-bit), and a destination (32-bit) address. We look at multilength instructions in the next chapter. Here we point out that multiple length instructions pose immense challenges for the computer designer. We will consider only fixed wordlength computers in this chapter.

Let's look at an example in which a designer decides that all instructions in a 32-bit processor will have three fields and be of the form `Operation,literal_value,memory_location`. Now suppose the designer has determined that a literal operand in the range 0 to 900 is required and that the size of the memory is to be 250,000 locations. How many unique instructions can be implemented using this arrangement?

The literal operand requires 10 bits to encode because $2^{10} > 900 > 2^9$, and the memory operand requires 18 bits because $2^{18} > 250,000 > 2^{17}$. These two fields take up $10 + 18 = 28$ bits, which leaves $32 - 28 = 4$ bits with which to specify an actual operation. This computer can, therefore, have an instruction set that contains only $2^4 = 16$ operations. This example demonstrates that there is a trade-off between the number of instructions (op-codes), the range of literals, and the size of the memory that can be directly addressed. Increase the range of one and you have to reduce the range of the other.

The history of computer design is littered with attempts to break the cruel tyranny of the equation that states *op-code bits plus operand bits = computer wordlength*. However, even this example can demonstrate that a little ingenuity can go a long way. Recall that we said the literal field ranges from 0 to 900. Using a 10-bit field to accommodate the literal provides a range from 0 to 1,023 (0 to $2^{10} - 1$). Consequently, the 123 literal values from 901 to 1,023 can be assigned to new instructions. Of course, these new instructions have to perform operations that don't require a literal operand. If ever there were a domain in which the expression '*there's no such thing as a free dinner*' were true, it's in the realm of instruction set design.

3.2.4   Op-codes and Instructions

Instructions can be grouped or classified in several ways. The most important single factor in the design of computer architectures is the number of operand addresses per instruction. For example, an instruction set that implemented ADD **r1**,r2,r3 would be a *three-address* machine, whereas a computer that implemented ADD **r1**,r2 would be a *two-address machine*. Here we introduce three-, two-, one-, and zero-address machines. Instructions also can be grouped according to the nature of their operation. For example, we use *data movement* that copies data from one location to another, *data processing* that operates on data, and *flow control* that modifies the order in which instructions are executed.

Consider the following examples of instructions with zero to three operands. In these examples, operands *P*, *Q*, and *R* may be memory locations or registers.

Operands	Instruction	Effect
Three	ADD **P**,Q,R	Add Q to R and put the result in P
Two	ADD **P**,Q	Add Q to P and put the result in P
One	ADD P	Add P to an accumulator and put the result in the accumulator
Zero	ADD	Pop the top two items off the stack, add them and push the result

A *three-address* instruction can be written operation **destination,** source1, source2, where operation defines the nature of the instruction, source1 is the location of the first operand, source2 is the location of the second operand, and **destination** is the location of the result. Microprocessors don't implement three *memory* address instructions for the reasons we have already stated. A typical RISC processor allows you to specify three *register* addresses in an instruction by providing three 5-bit operand address fields, as Figure 3.11 demonstrates.[18] We'll use the ADD instruction to add together the four values in registers r2, r3, r4, and r5. This code is typical of RISC processors like the ARM.

```
ADD   r1,r2,r3              ;r1 = r2 + r3
ADD   r1,r1,r4              ;r1 = r1 + r4
ADD   r1,r1,r5              ;r1 = r1 + r5 = r2 + r3 + r4 + r5
```

FIGURE 3.11   The three address instruction

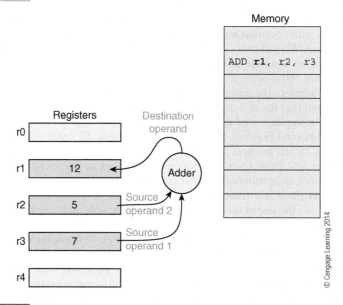

---

[18] Recall that ARM has 16 general-purpose registers (actually 13, since r15 and r14 have special functions, and r13 is reserved for use as a stack pointer). ARM has a richer instruction set than some RISC processors and has a reduced register set to provide more bits for the op-code.

Two Address Machines

A CISC machine (like the Pentium or 68K) has a *two-address* instruction format. Clearly, you can't execute $P = Q + R$ with two operands. You can execute $Q \leftarrow P + Q$. One operand appears *twice*: first as a source and then as a destination. The operation ADD P,**Q** performs the operation $[Q] \leftarrow [P] + [Q]$. The price of a two-operand instruction format is the destruction from the overwriting of one of the source operands. Most computer instructions can't directly access two memory locations. Typically, the operands are either two registers or one register and a memory location. For example, the 68K ADD instruction can be written as

Instruction	RTL definition	Mode
ADD D0,**D1**	[D1] ← [D1] + [D0]	Register-to-register
ADD P,**D2**	[D2] ← [D2] + [P]	Memory-to-register
ADD D7,**P**	[P] ← [P] + [D7]	Register-to-memory

One Address Machines

A one address machine specifies just one operand in the instruction. The second operand is a fixed register called an *accumulator* that doesn't have to be specified. For example, the operation one-address instruction ADD P means $[A] \leftarrow [A] + [P]$. The notation $[A]$ indicates the contents of the accumulator. The simple operation $R = P + Q$ can be implemented by the following fragment of 8-bit code from a first-generation 6800 8-bit processor.

```
LDA   P     ;load accumulator with P
ADD   Q     ;add Q to accumulator
STA   R     ;store accumulator in R
```

Eight-bit machines have one-address architectures. As you can imagine, 8-bit code is verbose, because you have to load data into the accumulator, process it, and then store it to avoid it being overwritten by the next data processing instruction. One-address machines are still used widely in embedded controllers in ultra-low-cost, low-performance systems, such as toys.

Zero Address Machines

A zero address machine uses instructions that do not have an address at all. A zero address machine operates on data that is at the top of a stack and zero address machines are normally referred to as *stack machines*. Of course, a pure zero address machine is not practical. Load and store instructions are needed to read data from memory and to put data in memory.

Operations with one operand (monadic operations such as negate, clear, increment, decrement) act on the element at the top of the stack whereas operations with two operands (dyadic operations such as add, multiply, logical OR) pull the top two elements off the stack, perform the operation and then push the result. The operation used to add an item to the stack is normally called PUSH, and the operation used to remove an element from the stack is called POP. The code used to evaluate the expression $Z = (A + B) \cdot (C - D)$ might be written as

PUSH A	Push A on stack
PUSH B	Push B on stack
ADD	Pull the top two items off the stack, add them, and push $A + B$ on the stack
PUSH C	Push C on the stack
PUSH D	Push D on the stack
SUB	Subtract top two items and push $C - D$ on the stack
MUL	Multiply top two items on the stack; that is $(C - D)$ and $(A + B)$ and push the result
POP Z	Pull the top item off the stack (the result)

Stack machines can also handle Boolean logic. Consider the operation if $(A < B)$ or $(C = D)$. This expression must yield a Boolean result—true or false. This can be expressed as

```
PUSH A      Push A on stack
PUSH B      Push B on stack
LT          Pull A and B and perform comparison between them; push true or false
            depending on the result
PUSH C      Push C
PUSH D      Push D
EQ          Pull C and D and test for equality; push true or false
OR          Pull top two items off stack (Boolean values); perform OR and push the
            result
```

The Boolean value on the stack can be used with a branch on true or a branch on false command, as in the case of any other computer.

As well as the conventional data processing operations, zero address machines also include data manipulation operations to facilitate stack programming. For example, there are instructions that duplicate the top item on the stack or exchange (swap) the top two items on the stack.

In 1980, Charles Moore invented a stack-based programming language called Forth for control applications (e.g., telescopes, robots, etc). The high-level language Java is compiled into a low-level stack based language called *bytecode* that is then interpreted on real machines. Some ARM processors incorporate Jazelle DBX (Direct Bytecode eXecution), which enables them to directly execute bytecode. Executing bytecode in hardware (rather than interpreting it) increases the ARM's performance when running Java applications.

One-and-a-Half Address Machines

The Intel IA32 family and the Freescale family are frequently called *one-and-a-half address* machines because their instructions specify two operands, where one operand is an address in memory and the other is a register. The register address is ironically called a *half address*, because there are only a handful of registers compared with gigabytes of memory space. The following 68K code demonstrates the evaluation of the expression $(A + B)(C - D)$.

```
MOVE A, D0     ;Load A from memory into register D0
ADD  B, D0     ;Add B from memory into register D0
MOVE C, D1     ;Load C from memory into register D1
SUB  D, D1     ;Subtract D from memory from register D1
MULU D0, D1    ;Multiply register D1 by D0
MOVE D1, X     ;Store register D1 in memory location X
```

Compare this with the following generic code of a one address accumulator-based machine:

```
LDA A     ;Load A from memory into the accumulator
ADD B     ;Add B from memory into the accumulator
STA P     ;Store the accumulator in memory location P
LDA C     ;Load C from memory into the accumulator
SUB D     ;Subtract D from memory from the accumulator
MUL P     ;Multiply the accumulator by P from memory
STA X     ;Store the accumulator in memory location X
```

Now that we've looked at the basic structure of processors and the elements of an ISA, we can turn our attention to a real microprocessor—the ARM. The architecture of ARM processors has developed over the years, and there are now several classes of ARM processors—each class intended for a particular segment of the market, such as embedded microcontrollers or high-performance computers. In this chapter, we discuss A-class devices, such as the A8, A9, and A15.

3.3 ARM Instruction Set Architecture

The 1980s were the age of the RISC processor that was driven by academics who had analyzed the performance of existing microprocessors and found them wanting. When researchers analyzed the code executed by microprocessors, they made some interesting observations. For example, they discovered that existing processors in the 1980s had instructions that were hardly ever executed. Such instructions wasted valuable real estate on the silicon chip. RISC (*reduced instruction set*) processors were suggested as a means of increasing the efficiency of micro-processors. Computer designers went back to first principles.

For a time, the RISC versus CISC argument raged, but it eventually died out. The Intel archi-tecture was here to stay because of its massive user base. Moreover, Intel put a lot of effort into expanding the IA32 architecture to better fit its role, while improving its microarchitecture to enhance its computational efficiency. Although RISC processors didn't sweep away CISC proces-sors, companies like Acorn Computers in the UK were able to design powerful RISC processors with a tiny fraction of the resources available to giants like Intel and Freescale.

We next examine ARM's ISA and introduce the programming of the ARM in its native assembly language. We use the ARM as a vehicle for teaching basic principles, because it is a simple and elegant machine that lacks the steep learning curve of many other processors.

ARM is a 32-bit machine with a register-to-register architecture and load/store instructions that move data between memory and registers. All operands are 32 bits wide, except for several multi-plication instructions that generate a 64-bit product stored in two 32-bit registers. First-generation ARMs supported 32-bit words and unsigned bytes, whereas later versions also support 8-bit signed bytes and 16-bit signed and unsigned half-words.[19]

ARM

The ARM architecture is the intellectual property of ARM Hold-ings, based in Cambridge, England. The company was founded in 1990 as Advanced RISC Machines by Acorn Computers, Apple Computers, and VLSI Technology.

The acronym ARM originally came from Acorn RISC Machine. Acorn Computers Ltd. built the BBC Microcomputer which was based on the 8-bit 6502 and which was used extensively in British high schools. Engineers at Acorn were inspired by the Berkeley RISC project to design their own RISC processor, the Acorn RISC Machine (ARM).

The second generation ARM designed by Acorn was a 32-bit microprocessor built with only 20,000 transistors—half the number used by Freescale in their 68K. In 1990, the ARM project led to a new company called Advanced RISC Machines Ltd.

Unlike Intel, AMD, and Freescale, ARM does not build chips but licenses its core processors for use in (*systems on chips* (SoCs) and microcontrollers. By 2008, over 10 billion ARM cores had been built, and it was estimated that five billion cores would be shipped in 2011.

ARM technology is licensed by semiconductor companies, including Intel, Freescale, Texas Instruments, Philips, Fujitsu, and Sony.

ARM has been a remarkable success story in a world where many other microprocessors became popular for a few years and then decayed into obscurity. The 68K was once widely thought of as more elegant and powerful than Intel's 8086 (indeed, the 68K was a true 32-bit machine at a time when the 8086 was a 16-bit machine). The 68K was adopted by Apple's Mac, the Atari, and Amiga computers, which were all major players in the home computer market. How could the humble 8086 possibly compete? IBM selected the 8086 for its new personal computer and the rest is history.

Not only has ARM survived, it has prospered and successfully targeted the world of mobile devices such as netbooks, tablets, and cell phones. ARM processors have incorpo-rated some interesting architectural features that have given it a competitive advantage over its rivals. Before we describe ARM's instructions, we discuss its register set, because all data processing instructions operate on the contents of registers.

---

[19] In ARM terminology, 8 bits is a *byte*, 16 bits is a *half-word*, and 32 bits is a *word*.

3.3.1   ARM's Register Set

Figure 3.12 illustrates ARM's 16 *programmer-visible* registers (r0 to r15) and its *status register*. There are 14 general-purpose registers r0 to r13. Register r14 stores a subroutine return address, and r15 contains the program counter. Because r15 is accessible by the programmer, you can perform computed branches (e.g., the type of operation used in a case statement in high-level languages). The ARM is unusual for a RISC processor in having only 16 general-purpose registers. Sixteen registers require a 4-bit address that saves three bits per instruction over RISC processors with 32-register architectures (5-bit address). This approach allows the ARM to have a richer instruction set than some RISC processors. Register r13 is *reserved* for use by the programmer as the stack pointer. Unlike special-purpose registers r14 and r15, whose additional functions are implemented by the ARM's hardware, r13 is a stack pointer only if the programmer uses it as a stack pointer.

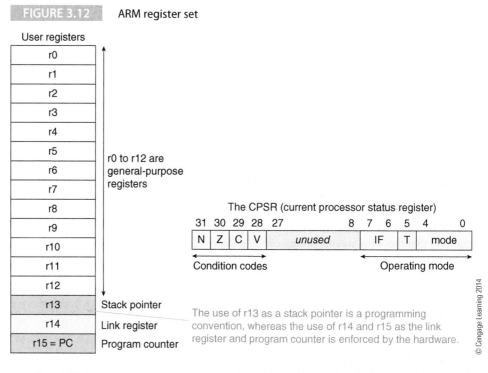

FIGURE 3.12   ARM register set

The ARM's *current program status register* (CPSR) contains Z (zero), N (negative), C (carry), and V (overflow) flag bits and is similar to the condition code or status register in other processors. The least-significant eight bits of the CPSR contain system information, such as ARM's status and interrupt handling mechanisms, which are covered later in this book.

As we have just said, by implementing only 16 registers, ARM processors can afford a rich instruction set. For example, some instructions implement a *four operand* format. Consider ADD **r1**,r2,r3,LSL r4 and MLA **r1**,r2,r3,r4. We will look at these instructions later.

Figure 3.12 is simplified in one important way. Some of the registers have multiple *versions* or instances. We do not have to worry about this feature here. However, when the ARM encounters an exception, it switches register banks automatically to avoid forcing the operating system to save registers that were in use prior to the interrupt. We will return to this point later.

3.3.2   ARM's Instruction Set

There has been remarkably little change in instruction set architecture between today's high-performance machines and first-generation microprocessors. The instructions implemented by the ARM are broadly the same as those of the Pentium or 68K and can be grouped into several basic types, although individual writers categorize instruction sets in different ways. We will choose rather broad groupings of data movement, arithmetic, logical, shift, and program control.

Table 3.1 describes the instruction set of the ARM's register-to-register architecture. At first sight, this instruction set hardly seems exciting. Indeed, the ARM appears to have a rather minimal instruction set with some glaring omissions; that is, shift operations. We will soon discover that Table 3.1 doesn't tell the whole story.

TABLE 3.1   ARM Data Processing, Data Transfer, and Compare Instructions

Instruction	ARM Mnemonic	Definition
Addition	ADD **r0**,r1,r2	$[r0] \leftarrow [r1] + [r2]$
Subtraction	SUB **r0**,r1,r2	$[r0] \leftarrow [r1] - [r2]$
AND	AND **r0**,r1,r2	$[r0] \leftarrow [r1] \cdot [r2]$
OR	ORR **r0**,r1,r2	$[r0] \leftarrow [r1] + [r2]$
Exclusive OR	EOR **r0**,r1,r2	$[r0] \leftarrow [r1] \oplus [r2]$
Multiply	MUL **r0**,r1,r2	$[r0] \leftarrow [r1] \times [r2]$
Register-to-register move	MOV **r0**,r1	$[r0] \leftarrow [r1]$
Compare	CMP r1,r2	$[r1] - [r2]$
Branch on zero to label	BEQ label	$[PC] \leftarrow$ label (jump to label)

© Cengage Learning 2014

Updating ARM Condition Codes

As we have already pointed out, unlike most CISC architectures, the ARM does not automatically update its status flags after an arithmetic or logical operation. ARM provides an *update-on-demand* mode, whereby condition codes are updated only if the current mnemonic has an S suffix. For example, ADD **r1**,r2,r3 performs an addition without updating status flags, whereas ADDS **r1**,r2,r3 updates status flags. This facility allows the programmer to perform a test and then to carry out other instructions without changing the status flags.

```
SUBS   r1,r1,#1      ;subtract 1 from r1 and set status bits
ADD    r2,r2,#4      ;increment the counter (don't update status)
MUL    r5,r3,r4      ;form the product of r3 and r4 (don't update status)
BEQ    Error         ;if r1 is zero then deal with problem
```

In this example, two instructions (ADD and MUL) are performed between the subtraction and the conditional branch that acts on the result of the subtraction. Because an S is not suffixed to the ADD and MUL, neither of these operations update the processor status bits.

3.4   ARM Assembly Language

All processors have an *assembly language* that was crafted by the chip's designers, although third-party suppliers may well create different versions of the processor's assembly language. You have already seen fragments of ARM assembly language, and now we introduce some of the features that enable you to write programs that will run in an ARM environment. ARM instructions are written in the form

```
Label  Op-code operand1, operand2, operand3   ;comment
```

Consider the following example of a loop.

```
Test_5 ADD   r0,r1,r2            ;calculate TotalTime = Time + NewTime
       SUBS  r7,#1              ;Decrement loop counter
       BEQ   Test_5            ;IF zero THEN goto Test_5
```

The Label field is a user-defined label that can be used by other instructions to refer to that line (for example, by a conditional branch). Note that it doesn't matter whether there are one or more spaces after the commas in argument lists; you can write operand1,operand2 or operand1, operand2. Any text following a semicolon is regarded as a comment field and is ignored by the assembler. The comment field is used to improve the readability of a program. Here we provide only a very basic introduction to the ARM assembler. Commercial assemblers are more sophisticated and provide a range of facilities to help you write assembly-language programs. For example, *macros* allow you to repeat sequences of machine level instructions, and conditional assembly includes or omits parts of a program at assembly time.

Let's look at a simple fragment of ARM code. Suppose we wish to generate the sum of the cubes of numbers from 1 to 10. We can use the multiply and accumulate instruction as given here.

```
      MOV   r0,#0             ;clear total in r0
      MOV   r1,#10            ;FOR i = 1 to 10 (count down)
Next  MUL   r2,r1,r1          ;    square number
      MLA   r0,r2,r1,r0       ;    cube number and add to total
      SUBS  r1,r1,#1          ;    decrement counter (set condition flags)
      BNE   Next              ;END FOR (branch back on count not zero)
```

This fragment of assembly language is *syntactically* correct and implements the appropriate algorithm. However, it is not yet a program that we can run. For example, we have to specify where the code goes in memory and define what resources it requires. An assembly program is composed of two types of statements: *executable instructions* that are executed by the computer and *assembler directives* that tell the assembler something about the environment. For example, the assembly directive END is not executed—it simply tells the assembler that the end of the program has been reached. Assembly directives tell the assembler where the code is to be located in memory, allocate storage space to variables, and set up initial data that the program might need during its execution.

3.4.1   Structure of an ARM Program

We begin with a program that can be executed on an ARM computer or a PC with an ARM cross-development system. The following fragment of code demonstrates the structure of the simple program we described above that forms the cubes of the first ten integers. The text in blue represents assembly directives rather than executable ARM code.

```
      AREA ARMtest, CODE, READONLY
      ENTRY
      MOV   r0,#0             ;clear total in r0
      MOV   r1,#10            ;FOR i = 1 to 10
Next  MUL   r2,r1,r1          ;    square number
      MLA   r0,r2,r1,r0       ;    cube number and add to total
      SUBS  r1,r1,#1          ;    decrement loop count
      BNE   Next              ;END FOR
      END
```

The AREA assembler directive defines a section of code. In this example, the section is called ARMtest and has the attributes CODE and READONLY.[20] The assembler directive ENTRY tells the assembler where to find the instruction that is to be executed first. The END

---

[20] A region of memory that can be written to is designated READWRITE.

FIGURE 3.13   Assembling an assembly language program using Kiel's ARM IDE

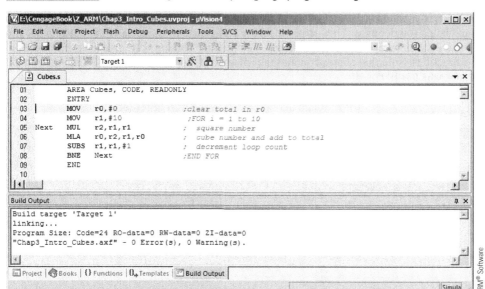

assembler directive is mandatory and tells the assembler that the end of the program has been reached. Figure 3.13 shows how this source code program looks in the editor window of the ARM integrated development system from Keil. Details of the development system are given on the website accompanying this text. Note that the figures from the ARM development system in this chapter use color. Consequently, we have provided copies of the figures on this book's web site.

After the assembly language program has been written, it can be assembled and executed. We can also look at the *disassembled code* (see Figure 3.14). This window gives the source code and the disassembled code. Disassembly means that the code generated by the assembler is converted back into source form to show you both code and instructions, The disassembled code shows the memory address of the instruction on the left, the code generated by the instruction, and then the disassembled instruction itself. For example, at address x00000008, we have the code E0020191 that corresponds to the instruction MUL r2,r1,r1. This example also helps demonstrate that instructions/code are stored in memory not as text but in binary form; in this case, the instruction is 0xE0020191 or 11110000000000000100000000 110010001$_2$.

FIGURE 3.14   The disassembly window with the hexadecimal code generated by the program

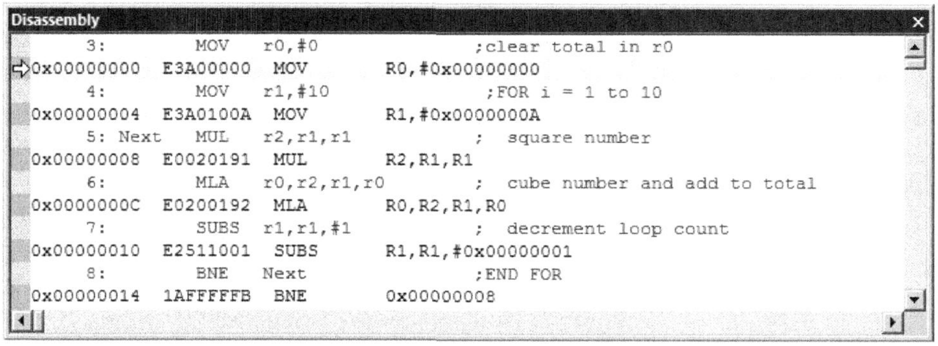

```
Disassembly                                                                    ×
     3:        MOV    r0,#0              ;clear total in r0
⇨0x00000000 E3A00000  MOV      R0,#0x00000000
     4:        MOV    r1,#10             ;FOR i = 1 to 10
 0x00000004 E3A0100A  MOV      R1,#0x0000000A
     5: Next   MUL    r2,r1,r1            ;  square number
 0x00000008 E0020191  MUL      R2,R1,R1
     6:        MLA    r0,r2,r1,r0         ;  cube number and add to total
 0x0000000C E0200192  MLA      R0,R2,R1,R0
     7:        SUBS   r1,r1,#1            ;  decrement loop count
 0x00000010 E2511001  SUBS     R1,R1,#0x00000001
     8:        BNE    Next                ;END FOR
 0x00000014 1AFFFFFB  BNE      0x00000008
```
ARM® Software

Figure 3.15 shows the state of the program during its execution. I have removed all unnecessary windows from the display to leave the register window and the code window. The register window shows the contents of each register during the execution of the program and helps you debug it. The arrow on Figure 3.15 points to the icon of the button you click on to execute instructions. Figure 3.16 shows the same windows after we have executed four instructions. Note that r1 now contains the value $0x0A$, the PC contains $0x0C$, and r2 contains $0x64$, which is 100 or 10^2.

FIGURE 3.15   The program during execution and its register set

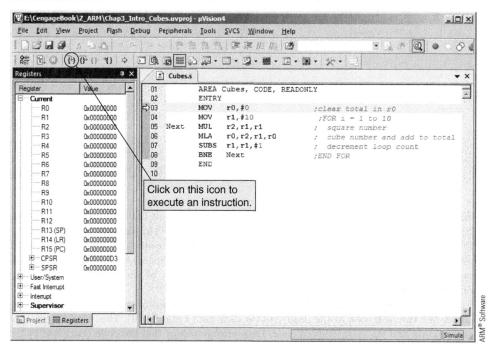

FIGURE 3.16   The program after the execution of two instructions

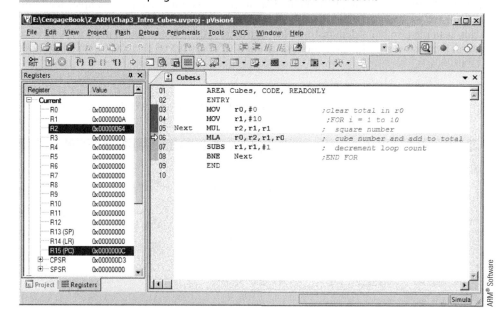

3.4.2   The Assembler – Practical Considerations

Before looking at the ARM's architecture in greater depth, we need to examine some of the conventions used in writing assembly language programs and to show how memory space can be reserved for constants and variables. We have already seen that literals used as operands are prefixed by a # symbol. Numbers are regarded as decimal unless prefixed by 0x, which indicates hexadecimal. For example, MOV r0,#0x2C. ASCII characters are indicated by using single quotes. For example,

```
CMP     r0,#'A'                  ;was it a letter 'A'?
```

Two important assembler directives are EQU that binds a name to a value and DCD that allows you to preload memory with data before a program runs. The EQU directive is very easy to understand. It binds (i.e., attaches) a numeric value to a name. Suppose you write:

```
Tuesday EQU 2
```

This assembly directive binds the string Tuesday to the value 2. If you write ADD r1,r2,#Tuesday it is treated by the compiler as ADD r1,r2,#2. The reason for equating a name to a value is to enhance the readability of a program. If you equate Tuesday to 2 whenever you write Tuesday, the assembler simply replaces it by 2.

Assembly language programs (like high-level language programs) use constants, variables, and data structures. These are located somewhere in memory. Computer users never have to worry about this, as the operating system and system software automatically take care of the allocation of memory space to data. The high-level programmer usually has to allocate memory space to data by means of declarations. For example, consider the C code given here.

```
int    x;
int    y;
char   key;
int    p = 4;
```

In this example, four data elements are defined and storage space allocated. Three are integer variables and one is a character variable. Note that storage is allocated for the integer value *p* and the memory is pre-loaded with the initial value 4.

The ARM assembler directive DCD reserves memory space for constants and variables. Consider the following example.

```
Value1 EQU   12              ;associate name Value1 with 12
Value2 EQU   45

Table  DCD   Value1          ;store the word 12 in memory
       DCD   Value2          ;store the word 45 in memory
```

The mnemonic DCD reserves a 32-bit word of storage in memory and loads whatever value the expression to the right of DCD yields into that location. In this case, we've bound 'Value1' to the number 12, and therefore, the binary value 00000000000000000000000000001100 will be stored at this location. The memory location used will be the next location in sequence (i.e., storage directives store data in memory sequentially).

The location counter is advanced by four bytes so that the next DCD or instruction will be placed in the next word in memory. The term "*location counter*" refers to the pointer to the next location in memory when a program is being assembled and is similar, in concept, to the program counter.

You don't have to use 32-bit values. The assembler directives DCB and DCW store a byte and a 16-bit halfword in memory, respectively. For example,

```
Q1     DCB   25              ;store the byte 25 in memory
Q2     DCB   42              ;store the byte 42 in memory
Tx2    DCW   12342           ;store the 16-bit value 12342 in memory
```

Although you could use DCD to store text strings in memory, it would be rather clumsy. The ARM assembler provides a simpler mechanism that uses the = symbol followed by a string in double quotes. The string can be followed by other byte values, each of which is separated by a comma. For example,

```
Mess1 =     "This is message 1", 0
Mess2 =     "This is message 2", &0C, &0A
      ALIGN
```

The assembler also lets you write

```
Mess1 DCB   "This is message 1", 0
```

In this case, the ASCII string "This is message 1" is stored in memory followed by the value 0. This string is followed by the second string and then the bytes $0C_{16}$ and $0A_{16}$. Note the use of the assembler directive ALIGN in this fragment of code. Since the ARM aligns all instructions and words on 32-bit word boundaries, the ALIGN directive tells the assembler to align whatever follows on the next word boundary. In other words, if you store three 8-bit characters in memory, the ALIGN command would skip a byte to force the next address to a 32-bit boundary. The following fragment of code provides a demonstration of storage allocation and the use of the ALIGN directive.

```
      AREA Directives, CODE, READONLY
      ENTRY
      MOV     r6,#XX          ;load r6 with 5 (i.e., XX)
      LDR     r7,P1           ;load r7 with the contents of location P1[21]
      ADD     r5,r6,r7        ;just a dummy instruction
      MOV     r0, #0x18       ;angel_SWIreason_ReportException
      LDR     r1, =0x20026    ;ADP_Stopped_ApplicationExit
      SVC     #0x123456       ;ARM semihosting (formerly SWI)
XX    EQU     5               ;equate XX to 5
P1    DCD     0x12345678      ;store hex 32-bit value 1345678
P3    DCB     25              ;store the byte 25 in memory
YY    DCB     'A'             ;store byte whose ASCII character is A in memory
Tx2   DCW     12342           ;store the 16-bit value 12342 in memory
      ALIGN                   ;ensure code is on a 32-bit word boundary
Strg1 =       "Hello"
Strg2 =       "X2", &0C, &0A
Z3    DCW     0xABCD
      END
```

This code includes what some professors call *magic code*, which is represented by the three lines in blue. We use the term magic code because it is system dependent; that is, it is a function of the environment or operating system, and the code has to be learned because there is often a different set of instructions for each commercial system. These instructions move data into register r0 and r1 and then execute SVC #0x123456 (we explain the format of the LDR r1, =0x20026 later). The effect of this code is to halt the processor. The values loaded into r0 and r1 are parameters, and the SVC instruction is a call to the operating system. Incidentally, the SVC instruction used to be called the SWI (software interrupt) instruction. If you don't like magic code, you can terminate a program with an infinite loop by writing

```
Stop  B     Stop            ;infinite loop!
```

---

[21] This instruction looks as if it loads r7 with the contents of memory location P1. Recall that ARM processors don't support such an addressing mode. This instruction is a pseudoinstruction that will be assembled into suitable ARM code. We discuss pseudoinstructions later in this section.

Figure 3.17 shows the simulator output corresponding to this code, and Figure 3.18 provides a memory map demonstrating the allocation of memory. If you look at the disassembly window of Figure 3.17, you will see that the code we wrote ends at memory location 0x00000010 (the SVC instruction). After that, we have the data we loaded into memory with the assembler directive. Note that the disassembler has attempted to disassemble the numbers we loaded in memory, leading to the gibberish code on the right. Finally, note that memory location 0x00000029 contains the byte 0x00 that has been inserted by the assembler to ensure that the next data value, the half word 16-bit value 0xABCD, lies on a 16-bit boundary.

Figure 3.19 shows another session of ARM debugging. In this case, a different IDE has been used—several development systems are available. Here, there is a code window, a register widow, and a memory window that lets you examine memory as code is executed.

FIGURE 3.17   Allocating data to memory

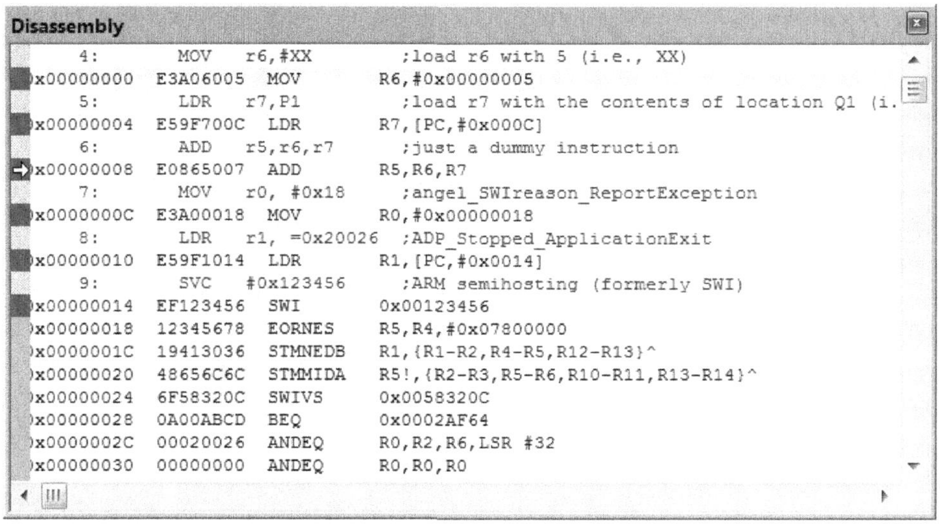

```
Disassembly                                                                     ☒
      4:        MOV    r6,#XX          ;load r6 with 5 (i.e., XX)
■x00000000 E3A06005  MOV     R6,#0x00000005
      5:        LDR    r7,P1           ;load r7 with the contents of location Q1 (i.
■x00000004 E59F700C  LDR     R7,[PC,#0x000C]
      6:        ADD    r5,r6,r7        ;just a dummy instruction
➡x00000008 E0865007  ADD     R5,R6,R7
      7:        MOV    r0, #0x18       ;angel_SWIreason_ReportException
■x0000000C E3A00018  MOV     R0,#0x00000018
      8:        LDR    r1, =0x20026    ;ADP_Stopped_ApplicationExit
■x00000010 E59F1014  LDR     R1,[PC,#0x0014]
      9:        SVC    #0x123456       ;ARM semihosting (formerly SWI)
■x00000014 EF123456  SWI     0x00123456
 x00000018 12345678  EORNES  R5,R4,#0x07800000
 x0000001C 19413036  STMNEDB R1,{R1-R2,R4-R5,R12-R13}^
 x00000020 48656C6C  STMMIDA R5!,{R2-R3,R5-R6,R10-R11,R13-R14}^
 x00000024 6F58320C  SWIVS   0x0058320C
 x00000028 0A00ABCD  BEQ     0x0002AF64
 x0000002C 00020026  ANDEQ   R0,R2,R6,LSR #32
 x00000030 00000000  ANDEQ   R0,R0,R0
◄ ▐█▌                                                                          ►
```

FIGURE 3.18   Allocating data to memory—the memory map

000000000018	12	Word 0x12345678
000000000019	34	
00000000001A	56	
00000000001B	78	
00000000001C	19	Byte 25
00000000001D	41	Byte 'A'
00000000001E	30	Half Word 12342
00000000001F	36	
000000000020	H	String "Hello"
000000000021	e	
000000000022	l	
000000000023	l	
000000000024	O	
000000000025	x	String "X2"
000000000026	2	
000000000027	0C	Byte 0x0C
000000000028	0A	Byte 0x0A
000000000029	00	Forced alignment
00000000002A	AB	Half Word 0xABAC
00000000002B	CD	

FIGURE 3.19   Snapshot of the ARM debugger window

3.4.3   Pseudoinstructions

We now introduce the pseudoinstruction[22] which is an instruction that is available to the programmer but which is not part of the processor's ISA. A pseudoinstruction is a form of shorthand that allows a programmer to express an action simply and then let the assembler generate the appropriate code.

CISCs like the Pentium can load addresses and 16- or 32-bit values into registers. Because of ARM's regular 32-bit instruction format, it's impossible to directly implement such instructions. For example, you cannot write MOV **r0**, #0x1234567 to load register r0 with the 32-bit hexadecimal value 01234567. Fortunately, ARM's assembler provides a shortcut that simplifies loading constants.

One of the ARM's most useful *instructions* is ADR that loads a destination register with an address. The format is ADR **r**destination, label, where label is the label of the effective address in the program. Although you will see ADR in programs, this is not a *real* instruction, because there's no instruction that can load a 32-bit constant into one of the ARM's registers. ADR is a *pseudooperation* or *virtual instruction* that you use and then let the assembler figure out how to generate the actual machine code to do the same thing. A pseudoinstruction is really a form of shorthand that the assembler fleshes out and relieves the programmer of some *housekeeping*. In general, the ADR makes use of the ARM's ADD or SUB instruction together with *PC relative addressing* to generate the required address.

---

[22] This is a slightly more complex topic that uses concepts we haven't yet fully introduced; the reader can skip this and return to it later. The main point is that the ARM has two pseudo-instructions that load a 32-bit value into a register by letting the assembler generate the actual code needed to do this.

The following code fragment demonstrates the use of the ADR pseudoinstruction.

```
        ADR    r1,MyArray                ;set up r1 to point to MyArray
         .
        LDR    r3,[r1]                    ;read an element using the pointer
         .            .
MyArray DCD    0x12345678
```

In this example, the operation ADR r1,MyArray loads register r1 with the 32-bit address of MyArray using the appropriate code generated by the assembler. The programmer does not have to know how the assembler generates suitable code to implement the ADR.

Another useful pseudoinstruction is LDR rd, = value. In this case, the compiler generates the code that allows register rd to be loaded with the stated value; for example,

```
    LDR r0, =  0x12345678
```

loads r0 with 12345678_{16}. The assembler uses a MOV or MVN if it can, or it uses an LDR r0,[pc,#offset] instruction to access the appropriate constant 12345678_{16} that is stored in a *literal pool* or *constant pool* somewhere in memory.[23]

Let's look at how pseudoinstructions are treated by the ARM development system. Consider the following code fragment. This is just dummy code intended to illustrate a point; it doesn't have any purpose.

```
        AREA ConstPool, CODE, READONLY
        ENTRY
        LDR    r0,=0x12345678           ;load r0 with a 32-bit constant
        ADR    r1,Table                 ;load r1 with the address of Table
        ADR    r2,Table1                ;load r2 with the address of Table1
        LDR    r3, = 0xAAAAAAAA         ;load r3 with a 32-bit constant
        LDR    r4,P3                    ;what does this do?

Table   DCD    0xABCDDCBA               ;dummy data
Table1  DCD    0xFFFFFFFF
P3      DCD    0x22222222
```

Figure 3.20 illustrates the use of the LDR ri,= and ADR pseudoinstructions by providing a snapshot of the source code when the ARM debugger is running. This snapshot is from the disassembly window of the debugger. The first memory location of our program is at 0x00000000. The instruction LDR r0,=0x12345678 has been assembled as LDR r0,[PC,#0x0018]. Now, since the ARM's PC is eight bytes in advance of the current instruction (because of the way in which the ARM is pipelined), the address of the operand loaded into r0 is $8_{16} + 18_{16} = 20_{16}$. If you look at location 0x00000020, you will find it contains the 32-bit constant 0x12345678. The assembler has placed a constant in memory in the literal pool and generated a PC-relative instruction to access it.

The next instruction is ADR r1,Table, which is a pseudoinstruction that is intended to load register r1 with the address of the line labeled by Table. If you look at Figure 3.20, you will see that ADR r1,Table has been assembled as ADD r1,PC,#0x00000008. The address of this instruction is 0x00000004, and therefore, the value of the PC must be eight bytes greater; that is, 0x0000000C. If we add 8 to this, we get the value 0x00000014. From Figure 3.20, we can see that memory location 0x00000014 contains 0xABCDDCBA, which is the contents of the memory location Table.

Finally, we included the instruction LDR r4,P3, which is not an ARM instruction, because you can't use direct addressing with the ARM. So, what happened and why wasn't

---

[23] We have not yet encountered program counter relative addressing. All you need know here is that the address of an operand is specified with respect to the current PC; for example, the effective address [PC,#12] indicates that the operand is 12 bytes onward from the location currently pointed at by the PC.

FIGURE 3.20   Code using pseudoinstructions

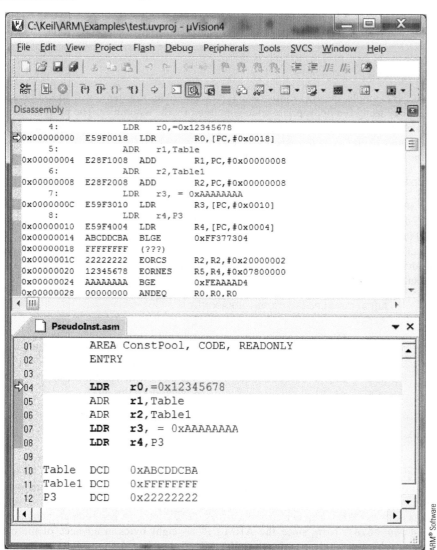

this *illegal* instruction rejected by the assembler? If you examine Figure 3.20, you will see that LDR **r4**, P3 has been treated as a pseudoinstruction and assembled as the legal instruction LDR **r4**, [PC, #0x0004]. This instruction loads r4 with the contents of the memory location given by PC + 4; that is 0x10 (the instruction address) + 0x08 (the PC is always eight higher) + 4 (the offset in the instruction) = 0x0000001C. If we look at location 0x0000001C, we see that it contains 0x22222222, which is the value of P3. In other words, although ARM lacks direct addressing, we can write LDR **r0**, address, and the assembler will generate the appropriate code. If you are confused by the statement that the address generated by LDR **r4**, [PC, #0x0004] is [PC] + 4 + 8 and wonder where the 8 came from, you need to remember two things. First, the PC is automatically incremented to point to the next instruction (that's 4 taken care of). Second, ARM processors are pipelined (discussed in Chapter 7), which means they fetch the next instruction before they've finished the current one. Pipelining accounts for the second 4, and therefore, the actual PC is 4 + 4 = 8 bytes on from the current instruction.

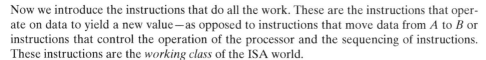

3.5   ARM Data-processing Instructions

Now we introduce the instructions that do all the work. These are the instructions that operate on data to yield a new value—as opposed to instructions that move data from *A* to *B* or instructions that control the operation of the processor and the sequencing of instructions. These instructions are the *working class* of the ISA world.

3.5.1   Arithmetic Instructions

Let's begin with ARM's arithmetic instructions that perform operations on data representing *numeric quantities*.

- Addition   ADD
- Subtraction   SUB
- Negation   NEG
- Comparison   CMP
- Multiplication   MUL
- Shifting[24]   LSL, LSR, ASL, ASR, ROL, ROR

Addition and Subtraction

As well as the 'plain vanilla' of *adding two words together*, most microprocessors implement an *add with carry-in* instruction that adds two operands together with the carry bit from the condition code register. This instruction facilitates *chain arithmetic* on word lengths greater than the intrinsic word length of the computer. We will demonstrate chain arithmetic first, using simple one-digit decimal addition. Consider the three examples below.

Case 1	Case 2	Case 3
Single digit addition	Two-digit addition with no carry Result incorrect	Two-digit addition with carry Result correct
4	56	56
+3	+27	+27
7	73	83

In Case 1, we add two single digits together and get the expected result. In Case 2, we add the two least-significant digits 6 + 7 and get the result 3, setting the carry bit.Then we add the two most-significant digits 5 + 2 to get 7. The result (73) is incorrect, because the *carry out* resulting from 6 + 7 in the 1s column is not added to the ten's column. Case 3 solves this problem by including the carry out from the one's column when we add the digits in the ten's column.

Figure 3.21 demonstrates how the *add with carry* instruction implements chain arithmetic. In Figure 3.21a, a conventional ADD **r2**,r0,r1 adds the contents of source registers r0 and r1, then it stores the result in destination register r2. The carry-out generated by this operation is stored in the carry bit.

Now, suppose that we are using an ARM with its 32-bit architecture and have to deal with 64-bit integers and perform 64-bit addition. Figure 3.21b demonstrates the application of carry propagation to 64-bit arithmetic using 32-bit registers. Two 64-bit numbers have been assigned to register pairs r1, r0 and r3, r2.

We add r0 + r2 storing the result in r4 using ADDS **r4**,r0,r2. The carry out is stored in the carry bit of the CCR (note that we append the S to ADD to force the carry bit to be updated). When we add the two most-significant pairs of numbers with ADC **r5**,r1,r3, the carry bit is added to the sum of r1 and r3. The mnemonic *ADC* means *add with carry*. We can extend this principle to perform extended-precision arithmetic with integers of arbitrary

---

[24] Some would group shifting with arithmetic operations because shifts can be used to multiply or divide by 2. Equally, some include shifts with logical operations because they operate on groups of bits.

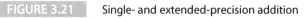

FIGURE 3.21   Single- and extended-precision addition

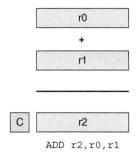

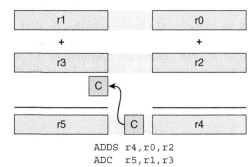

ADD r2,r0,r1

(a) Single-precision addition. When r0 is added to r1, the result is loaded into r2, and the carry bit is loaded into the carry flag.

ADDS r4,r0,r2
ADC r5,r1,r3

(b) Double-precision extended addition. When r0 is added to r2, any carry out is stored in the the carry bit. When r1 is added to r3, the carry bit is added to their sum. In other words, the carry out generated by ADDS r4,r0,r2 becomes the carry in used by ADC   r5,r1,r3.

© Cengage Learning 2014

length. If you want to implement addition with 200-bit integers, you can. The ARM also provides an SBC or *subtract with carry* instruction to support extended-precision subtraction.

The ARM provides an unusual *reverse subtract* instruction RSB. The subtract SUB **r1**,r2,r3 is defined as [r1] ← [r2] - [r3], whereas the reverse subtract operation RSB **r1**,r2,r3 is defined as [r1] ← [r3] - [r2]. This variation may seem pointless because you can perform [r3] - [r2] by writing SUB **r1**,r3,r2. However, reverse subtraction is useful because the ARM treats its two operands *unequally*; later we will see that the ARM can scale the second operand. Moreover, reverse subtraction is also useful with literal operands. For example, SUB **r1**,r2,#10, calculates [r2] - 10, whereas reverse subtraction, RSB **r1**,r2,#10, performs 10 - [r2].

Negation

Negation subtracts a number from zero. For example, the negative of r0 is 0 - [r0]. ARM does not have a negation instruction as such. However, you can use reverse subtraction because RSB **r1**,r1,#0 is equivalent to NEG r1.

Another interesting and unusual operation provided by ARM is *move with complement*. For example MVN **r0**,r1 copies the *logical complement* of the contents of register r1 into register r0. This operation implements *logical complement* (bit reversal), not *arithmetic complement* (sign reversal).

Comparison

The comparisons necessary to implement conditional behavior (i.e., loops and IF . . . THEN constructs) can be *implicit* or *explicit*. An implicit comparison takes place when you write SUBS **r1**,r1,#1 because a 1 is subtracted from r1 and then the Z-bit of the CCR is set if the result is zero.

An explicit comparison takes place when you compare P and Q by executing CMP Q,P, which evaluates $Q - P$ but does not store the result. For example, CMP r0,r1 evaluates [r0] - [r1] and then updates the status bits accordingly. A comparison modifies the contents of the *condition code register* (CCR), which later can be tested to determine whether execution continues in sequence or a branch is taken.

Consider the following example.

```
            CMP   r1,r2          ;is r1 = r2?
            BEQ   DoThis         ;if equal then goto DoThis
            ADD   r1,r1,#1       ;else add 1 to r1
            B     Next           ;don't forget to jump past the then part
            .
DoThis      SUB   r1,r1,#1       ;subtract 1 from r1
Next    . . .                    ;both forks end up here
```

Multiplication

A multiplication operation takes two m-bit operands and forms their $2m$-bit product. Doubling the bit length of the result creates a problem. Computers either automatically truncate the result so that a 32-bit by 32-bit multiplication yields a 32-bit result or it is necessary to construct an instruction with four operands: two for the source operands and two for the destination operands (the upper and lower halves of the result). Moreover, although addition and subtraction provide the correct result with two's complement values, multiplication and division do not. That is, you cannot use the same multiplication operation for both signed and unsigned values.

> ### ARM Multiplication
>
> ARM includes several forms of multiplication instruction in addition to the 32-bit MUL.
>
> | | |
> |---|---|
> | UMULL | Unsigned long multiply (Rm x Rd yields 64-bit product in two registers) |
> | UMLAL | Unsigned long multiply-accumulate |
> | SMULL | Signed long multiply |
> | SMLAL | Signed long multiply-accumulate |

Some microprocessors provide both signed and unsigned multiplication operations and 16-bit and 32-bit full multiplication with a 32- or 64-bit product. Further details on ARM's multiplication options are given in the panel. Note also that ARM's multiplication operation does not fall into the same category as other operations, such as ADD or AND, because you can't multiply by a constant and you can't shift the second operand (see later).

The ARM's multiply instruction, MUL **Rd**,Rm,Rs, calculates the product of two 32-bit signed integers stored in 32-bit registers Rm and Rs, then deposits the result in 32-bit register Rd, which stores the 32 lower-order bits of the 64-bit product. You should, of course, ensure that the result does not go out of range. For example, the ARM assembly code that multiplies 121 by 96 is

```
MOV   r0,#121      ;load r0 with 121
MOV   r1,#96       ;load r1 with 96
MUL   r2,r0,r1     ;r2 = r0 x r1
```

Unfortunately, you can't use the *same* register to specify both the destination Rd and the operand Rm, because ARM's implementation uses Rd as a temporary register during multiplication. This is a feature of the ARM processor.

ARM has a *multiply and accumulate* (MLA) instruction that performs a multiplication and adds the product to a running total. The MLA instruction has a four-operand form: MLA **Rd**,Rm,Rs,Rn, whose RTL definition is [Rd] ← [Rm] x [Rs] + [Rn]. A 32-bit by 32-bit multiplication is truncated to the lower-order 32 bits.

ARM's *multiply and accumulate* supports the calculation of an inner product by performing one multiplication and addition per instruction. The inner product is used in multimedia applications. For example, if vector **a** consists of n components $a_1, a_2, \ldots, a_n$ and vector **b** consists of the n components $b_1, b_2, \ldots, b_n$, the *inner product* of **a** and **b** is the scalar value $s = a \cdot b = a_1 \cdot b_1 + a_2 \cdot b_2 + \ldots + a_n \cdot b_n$.

The following code fragment shows how the multiply and accumulate instruction is used to form the inner product between two n-component vectors Vector1 and Vector2. Although we have not yet covered the ARM's addressing modes, the following example includes the instruction LDR **r0**,[r5],#4 that loads register r0 with an element from the array pointed at by register r5 and then updates r5 to point at the next element.

```
        MOV    r4,#n          ;r4 is the loop counter
        MOV    r3,#0          ;clear the inner product
        ADR    r5,Vector1     ;r5 points to vector 1
        ADR    r6,Vector2     ;r6 points to vector 2

Loop    LDR    r0,[r5], #4    ;REPEAT read a component of A and update the pointer
        LDR    r1,[r6], #4    ;   get the second element in the pair
        MLA    r3,r0,r1,r3    ;   add new product term to the total (r3 = r3 + r0·r1)
        SUBS   r4,r4,#1       ;   decrement the loop counter (and remember to set CCR)
        BNE    Loop           ;UNTIL all done
```

Figure 3.22 gives a snapshot of this program after it has been executed. We have set up an environment and have used two four-element vectors with the values (1, 2, 3, 4) and (2, 3, 4, 5) to give an inner product of $1 \cdot 2 + 2 \cdot 3 + 3 \cdot 4 + 4 \cdot 5 = 40$ or 28_{16}. As you can see from Figure 3.22, the value in register r3 is indeed 28. This diagram also provides a memory window showing the two vectors in memory.

FIGURE 3.22   Snapshot of the state of the simulator after the execution of the inner product program

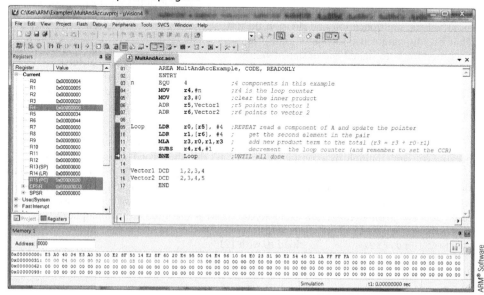

Division

Division operations suffer from similar problems to multiplication; a $2m$-bit value divided by an m-bit value leaves a quotient and a remainder. Similarly, two separate division instructions are required for signed and unsigned division. The ARM does not implement a division operation (in its basic models), and the programmer has to write suitable division routines in software.

3.5.2   Bitwise Logical Operations

Logical operations are known as *bitwise* operations because they are applied to the individual bits of a register. Although there are sixteen possible logical operations on two Boolean variables, microprocessors generally support only AND, OR, NOT and EOR operations. The following examples illustrate logical operations on r1 = 11001010_2 and r0 = 00001111_2. We've used 8-bit arithmetic for simplicity.

Logical instruction[25]	Operation	Final value in r2
AND **r2**,r1,r0	11001010.00001111	00001010
OR **r2**,r1,r0	11001010+00001111	11001111
NOT **r2**, r1	$\overline{11001010}$	00110101
EOR **r2**,r1,r0	11001010⊕00001111	11000101

Although ARM lacks an explicit NOT instruction, you can perform a NOT by using an EOR with the second operand equal to $FFFFFFFF_{16}$ (32 ones in a register) because the value

---

[25] The ARM does not have a NOT instruction of this form.

of $x \oplus 1$ is $\bar{x}$. A NOT operation can also be implemented with the *move negated instruction* MVN, which copies the *logical complement* of a value into a register.

A typical application of logical operations might be to merge groups of bits, which is an operation that is commonly used to pack more than one variable into a register or memory location. Suppose that register r0 contains the 8 bits bbbbbbxx, register r1 contains the bits bbbyyybb, and register r2 contains the bits zzzbbbbb, where x, y, and z represent the bits of desired fields and the b's are unwanted bits. We wish to pack these bits to get the final value zzzyyyxx. We can achieve this using the following code.[26]

```
AND  r0,r0,#2_00000011     ;Mask r0 to two bits xx
AND  r1,r1,#2_00011100     ;Mask r1 to three bits yyy
AND  r2,r2,#2_11100000     ;Mask r2 to three bits zzz
OR   r0,r0,r1              ;Merge r1 and r0 to get 000yyyxx
OR   r0,r0,r2              ;Merge r2 and r0 to get zzzyyyxx
```

Using Logical Operations on Bits

Suppose we have an 8-bit binary string *abcdefgh* and we wish to clear bits *b* and *d*, set bits *a*, *e*, and *f*, and toggle (invert) bit *h*. We can do this with a sequence of AND, OR, and EOR operations.

```
AND  r0,r0,#2_10101111     ;Clear bits b and d to get a0c0efgh
OR   r0,r0,#2_10001100     ;Set bits a, e, f to get 10c011gh
EOR  r0,r0,#2_00000001     ;Toggle h to get 10c011gh̄
```

ARM provides a *bit clear* instruction, BIC, that ANDs its first operand with the *complement* of its second operand. For example, BIC r0,r1,r2 implements [r0] ← [r1]·[$\overline{r2}$]. BIC copies bits from the first operand to the destination when the corresponding bit of the second operand is 0 and clears bits when the corresponding bit of the second operand is 1. If r1 = 10101010 and r2 = 00001111, then BIC r0,r1,r2 yields

$$10101010 \cdot \overline{00001111} = 10101010 \cdot 11110000 = 10100000$$

The BIC instruction can be used to clear the low-order byte of a register. BIC r0,r1,#0xFF copies the 32 bits in register r1 to register r0 and then clears bits 0 to 7 of the data in r0.

3.5.3   Shift Operations

Shift operations move the bits of a word one or more places right or left. However, when you shift a string of bits, the bits at one end drop off and new bits enter at the other end. Consider the following examples.

Source string	Direction	Number of shifts	Destination string
0110011111010111	Left	1	1100111110101110
0110011111010111	Left	2	1001111101011100
0110011111010111	Left	3	0011111010111000
0110011111010111**1**	Right	1	0011001111101011
011001111101011**11**	Right	2	0001100111110101
0110011111010**111**	Right	3	0000110011111010

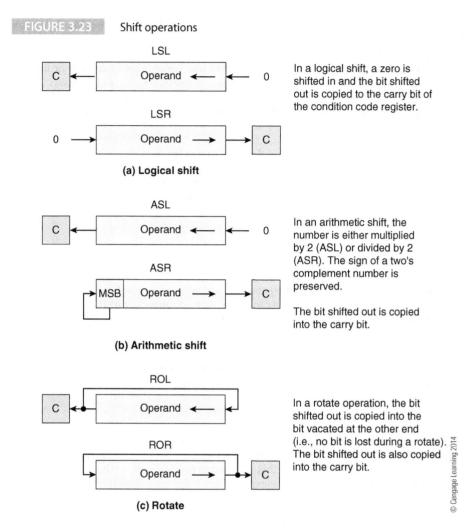

FIGURE 3.23   Shift operations

In a logical shift, a zero is shifted in and the bit shifted out is copied to the carry bit of the condition code register.

(a) Logical shift

In an arithmetic shift, the number is either multiplied by 2 (ASL) or divided by 2 (ASR). The sign of a two's complement number is preserved.

The bit shifted out is copied into the carry bit.

(b) Arithmetic shift

In a rotate operation, the bit shifted out is copied into the bit vacated at the other end (i.e., no bit is lost during a rotate). The bit shifted out is also copied into the carry bit.

(c) Rotate

© Cengage Learning 2014

The bits in the destination string in blue are the bits shifted in and the bits in the source string in bold are the bits lost (dropped) after the shift. The type of shift we have illustrated here is called a logical shift. Figure 3.23 illustrates the three basic types of shift operation: *logical*, *arithmetic*, and *rotate*.

All microprocessors provide a logical shift operation. Some support a single-bit left or right shift, and others support multi-bit shifting. If the number of bits to be shifted is coded as a *constant* in the instruction, the shift is called a *static* shift, because it can't be changed at run-time. If the number of bits to be shifted is specified by the *contents of a register*, it's a *dynamic* shift, since the shift amount can be modified while a program is running.

A typical application of logical shifting is to extract a bit pattern from within a word. Suppose we have an 8-bit string bxxxxbbb, where the x's represent the bits to be extracted and the b's denote "don't-care" values. We can extract and right-justify the required field as follows (note that this code is for illustration and is not ARM code).

```
LSR    r0,r0,#3,                ;Shift r0 three places right to get 000bxxxx
AND    r0,r0,# 2_00001111       ;Mask out unwanted bits to get 0000xxxx
```

ARM is unique in the sense that it lacks explicit shift instructions but includes them as part of existing instructions. ARM processors allow you to shift the second operand in register-to-register instructions. For example, ADD **r0**,r1,r2 LSL #4 performs four logical shifts left on the contents of register r2 and then adds the result to register r1. ARM's shifting mechanism does not add to the cycle time of an instruction and therefore is free. We look at ARM shifts after describing variations in shift operations in general.

Arithmetic Shift

An arithmetic shift treats the operand as a signed two's complement value, which is either divided by two (shift right 1 bit) or multiplied by two (shift left 1 bit), as Figure 3.23b demonstrates. The purpose of arithmetic shifts is to preserve the sign of two's complement numbers when they take part in shifting operations that represent multiplication or division by powers of 2.

An *arithmetic shift left* is effectively the same as a *logical shift left*. An *arithmetic shift right* preserves the sign bit that is replicated after each right shift. For example, if the 8-bit value 11001110 is shifted left arithmetically, it becomes 10011100 (LSB padded with zero), whereas if it is shifted right, it becomes **1**1100111 (sign bit replicated). An *m*-bit arithmetic shift left has the effect of multiplying an integer by a factor of 2^m and an *m*-bit shift right has the effect of dividing it by 2^m.

Although, in principle, there is no difference between an arithmetic shift left and a logical shift left, there are subtle differences between processors. For example, a processor might update all status bits after an *arithmetic* shift left, but it does not update the arithmetic overflow flag after a *logical* shift left.

Rotate

The rotate operation treats the contents of a register as a ring where the LSB and MSB are adjacent, as in Figure 3.23c. When a shift takes place, the bit that is shifted *out* at one end is shifted *in* at the other end. In contrast to logical or arithmetic shifts, a rotate operation does not discard bits. For example, if the 8-bit value 11001110 is rotated left, it becomes 10011101 (the bit in bold font has been shifted out of the msb position into the lsb position). In Figure 3.23c, the bit shifted out is also copied into the carry register. Microprocessors generally take this approach, because it allows you to perform a conditional operation on the state of the bit shifted out.

Some microprocessors implement a variation on the rotate called an *extended rotate* or a *shift through carry*. This operation is the same as a rotate, except that the carry bit is included in the rotation, as Figure 3.24 demonstrates. The bit shifted out is copied into the carry bit, and the old value of the carry bit becomes the new bit shifted in. This operation is used in chained arithmetic (it's the analog of the *add with carry* and *subtract with borrow* operations).

© Cengage Learning 2014

FIGURE 3.24   The rotate through carry

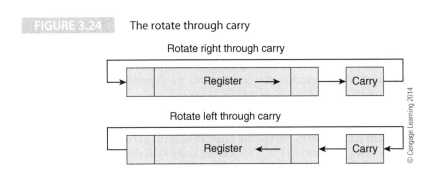

Implementing a Shift Operation on the ARM

ARM's implementation of shift operations is unique in the sense that there is no explicit shift operation. Instead, shifting is combined with other data processing operations because the second operand can be shifted before it is used. Figure 3.25 illustrates ARM's shifting mechanism where a barrel shifter is introduced into the second operand data path. A barrel shifter moves data left or right by combinational logic and does not use shift registers. Bits are simply copies from the input terminals to the output terminals. Consequently, a shift operation

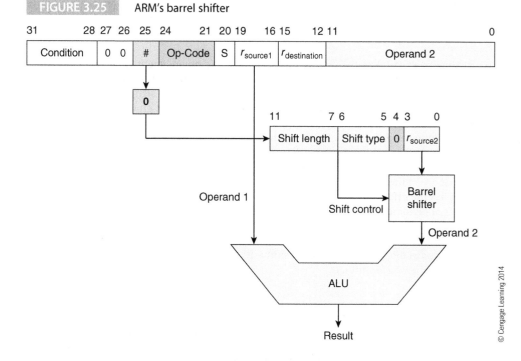

FIGURE 3.25   ARM's barrel shifter

© Cengage Learning 2014

is implemented without incurring an additional cycle time. Consider an example of the ADD operation that uses a shifted second operand. We can write

```
ADD r0,r1,r2, LSL #1
```

In this case, a logical shift left is applied to the contents of r2 before they are added to the contents of r1. Consequently, this operation is equivalent to

```
[r0] ← [r1] + [r2] x 2
```

If you wish to apply a shift operation to a register without any other data processing, you can use the move instruction and write

```
MOV r3,r3, LSL #1
```

Because you can perform *dynamic shifts*, you may also write MOV **r4**,r3, LSL r1, which moves the contents of r3 left by the value in r1 before putting the result in r4. Suppose a number in r0 is of the form 0.00000010101111 . . . and you want to normalize it to 0.101. . . . If register r1 contains the exponent, we can execute MOV **r0**,r0,LSL r1 to perform the normalization operation in a single cycle.

ARM supports both *static* and *dynamic* shifts apart from RRX which allows only one single shift. A static shift determines the number of places shifted when the code is written, whereas a dynamic shift allows you to change the number of shifts at runtime when the code is executed. You can write MOV **r3**,r3, LSL r2 to logically shift r3 the number of places left specified by the contents of r2. The value in r2 is treated as a modulo 32 number, because you can't shift more than 32 positions.

ARM implements only the following five shifts (the programmer must synthesize the rest).

LSL	logical shift left
LSR	logical shift right
ASR	arithmetic shift right
ROR	rotate right
RRX	rotate right through carry (one shift)

Although there is no rotate left operation, you can readily implement it by means of a rotate right operation. The following example demonstrates the rotation, both left and right, of a 4-bit value. After four rotations, the number is unchanged. As you can see, there is symmetry between left and right rotations. For a 32-bit value, an n-bit shift left is identical to a $32-n$ bit shift right.

Rotate right		Rotate left	
1101	start	1101	start
1110	rotate right 1	1011	rotate left 1
0111	rotate right 2	0111	rotate left 2
1011	rotate right 3	1110	rotate left 3
1101	rotate right 4	1101	rotate left 4

There is also no *rotate left through carry* operation. You can synthesize a rotate left through carry by means of ADCS `r0`,r0,r0. This may seem strange, because it adds the contents of r0 to the contents of r0 together with the carry bit and puts the result in r0 (that is, it generates $2 \times [r0] + C$). Now consider a rotate left through carry. Shifting left is equivalent to multiplying by 2. Moving the carry bit into the least-significant position is equivalent to adding the carry bit to get $2 \times [r0] + C$. Note that appending S to the instruction forces the CCR to be updated, which ensures that any carry out is loaded into the C-bit. Consequently, ADCS `r0`,r0,r0 and RXL r0 are equivalent.

This example demonstrates an important truth about assembly language. You can perform interesting operations by seemingly obscure code sequences. This feature makes assembly language popular with some hackers but anathema to many computer scientists. It makes assembly language difficult to understand and, therefore, has implications for programmer productivity, code reliability, and the ability to debug existing code.

Operand Scaling and C

This ability to scale operands by shifting is useful when dealing with pointers in C. For example, suppose `*P_int` is a pointer to a 32-bit integer x, and we want to increment the pointer to access an item that is offset in memory from the pointer. Consider

```
*P_int = *P_int + 4 * offset;
```

If pointer `*P_int` is in register r0 and the desired offset is in r1, we can compute the offset to x by

```
ADD r0,r0,r1, LSL #2
```

The offset is added to r0 conventionally, but it is scaled by 4 by using a shift operation.

3.5.4   Instruction Encoding—An Insight Into the ARM's Architecture

The constraints on instruction encoding faced by the designer give an insight into how instruction sets are constructed. Let's take a brief look at how the ARM's shift operations are encoded. Although this chapter emphasizes ARM's instruction set architecture rather than its internal organization, the way in which instructions are encoded is important because it determines how the underlying microarchitecture is designed.

Figure 3.26 illustrates the binary encoding of an ARM data processing instruction. This follows the general pattern of other RISC architectures with an op-code, two register operands, and a third multi-purpose operand. Register operands r_{source} and $r_{destination}$ define a first source operand and a destination register. The second source operand is encoded by bits 0 to 11.

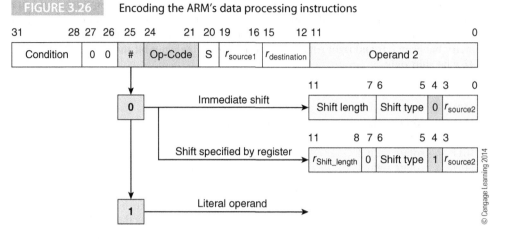

FIGURE 3.26   Encoding the ARM's data processing instructions

Depending on the state of bit 25, bits 0 to 11 provide either a third operand register or a literal. If bit 25 of the op-code is a 0, operand 2 provides both a second operand register and a shift operation. Bit 4 determines whether the number of bits to be shifted is a literal or is specified by the contents of a fourth register. If bit 4 of the instruction is 0 (specifying a literal shift), bits 7 to 11 define the shift amount (0 to 31 bits), while bits 5 and 6 specify the type of shift.

3.6   ARM's Flow Control Instructions

A computer executes instructions in strict sequence.[27] Flow control modifies this default sequential execution. We have already introduced the *unconditional branch* that forces a jump to a nonsequential point in the program and the *conditional branch* that forces a branch on the outcome of a test. Here we introduce the subroutine *call* and *return* instructions that also modify the flow of control by jumping to a block of instructions, executing them, and then returning to the point after the subroutine call.

3.6.1   Unconditional Branch

ARM's unconditional branch instruction has the form B target, where target denotes the *branch target address* (BTA), which is the address of the next instruction to be executed. The following fragment of code demonstrates how the unconditional branch is used.

```
..    do this        ;Some code
..    then that      ;Some other code
      B    Next       ;Now skip past the next instructions
..                    ;...the code being skipped past
..                    ;...the code being skipped past
Next  ..              ;Target address for the branch, denoted by label Next
```

In a high-level language, the unconditional branch is called a *goto*, and its use is considered poor programming style. However, this is difficult to avoid in assembly (low-level) programming, because high-level constructs such as *if...then...else* are not supported in the native instruction set.

The unconditional branch is not an exciting instruction. Invariably, it is used to return to a common point after one particular pathway through a program has been executed. The unconditional branch is really a '*get me back*' instruction.

---

[27] This statement is not strictly true. Modern microprocessors execute instructions in parallel and even out-of-sequence in order to improve performance. However, they give the appearance of sequential execution, because any out-of-order execution they perform must not modify a program's semantics (meaning).

3.6.2   Conditional Branch

ARM's conditional branches are similar to those of other RISC and CISC processors. They consist of a mnemonic B$_{cc}$ and a target address, where the subscript defines one of 16 conditions that must be satisfied for the branch to be taken and the target address is the location of the place in the code where execution continues if the branch is taken. A typical conditional example of conditional behavior in a high-level language is given by the following construct.

```
IF (X == Y)
    THEN Y = Y + 1;
    ELSE Y = Y + 2
```

In this example, a test is performed (i.e., the comparison of X and Y) and one of two courses of action is carried out, depending on the outcome of the test. We can translate this into ARM assembly language as

```
        CMP  r1,r2      ;assume r1 contains y and r2 contains x: compare them
        BNE  plus2      ;if not equal then branch to the else part
        ADD  r1,r1,#1   ;if equal fall through to here and add one to y
        B    leave      ;now skip past the else part
plus2   ADD  r1,r1,#2   ;ELSE part add 2 to y
leave   ...             ;continue from here
```

The *conditional branch* instruction tests flag bits in the processor's condition code register, then takes the branch if the tested condition is true. Since the condition code register always includes a zero bit (Z), negative bit (N), carry bit (C), and overflow bit (V), there are eight possible conditional branches based on the state of a single bit (four that branch on true and four that branch on false). Table 3.2 defines all the ARM's conditional branches. Note that there is a branch always and a branch never instruction.

Branch instructions can refer to *signed* or *unsigned* data. Consider the four-bit values $x = 0011$ and $y = 1001$. Suppose we want to branch if y is greater than x. If we are using unsigned arithmetic, $x = 3$ and y is 9, so $y > x$. However, if we regard these as signed values, then $x = 3$

TABLE 3.2   ARM's Conditional Execution and Branch Control Mnemonics

Encoding	Mnemonic	Branch on Flag Status	Execute on condition
0000	EQ	Z set	Equal (i.e., zero)
0001	NE	Z clear	Not equal (i.e., not zero)
0010	CS	C set	Unsigned higher or same
0011	CC	C clear	Unsigned lower
0100	MI	N set	Negative
0101	PL	N clear	Positive or zero
0110	VS	V set	Overflow
0111	VC	V clear	No overflow
1000	HI	C set and Z clear	Unsigned higher
1001	LS	C clear or Z set	Unsigned lower or same
1010	GE	N set and V set, or N clear and V clear	Greater or equal
1011	LT	N set and V clear, or N clear and V set	Less than
1100	GT	Z clear, and either N set and V set, or N clear and V clear	Greater than
1101	LE	Z set, or N set and V clear, or N clear and V set	Less than or equal
1110	AL		Always (default)
1111	NV		Never (reserved)

and $y = -7$, so $y < x$. Clearly, we have to select an unsigned comparison for unsigned arithmetic and a signed comparison for signed arithmetic.

Some microprocessors have synonyms for certain conditional branch operations; that is, the same branch condition has *two* mnemonics. For example, branch on carry set (BCS) can often be written BHS (branch on higher or same) because carry set implements the $\geq$ operation in unsigned arithmetic. Similarly, BCC can be written BLO (branch on lower) because the carry clear implements an unsigned comparison that is lower.

3.6.3   Compare and Test Instructions

ARM has four instructions in its test-and-compare group (CMP, CPN, TST, TEQ). These instructions *explicitly* update the condition code flags, so it's not necessary to append an S to any of them. We have already met the compare instruction.

- The *test equivalent* instruction (TEQ) determines whether two operands are equal. If they are equal the Z-bit is set high, otherwise the Z-bit is set low. For example, TEQ r1,r2 performs the RTL operation [r1] - [r2], then sets the Z-bit high if the values of r1 and r2 are identical. TEQ is similar to the conventional compare (CMP), except that TEQ does not affect overflow flag status during the test and modifies only the Z-bit. In contrast, CMP updates the overflow flag.
- The *test* instruction (TST) also compares two operands by ANDing them together and updating the flag bits accordingly. We can use TST to test individual bits of a word. For example, because bit 5 of a lower-case ASCII character is set to 1, we can determine if an ASCII character in r0 is lower-case, as

```
TST r0, #2_00100000    ;AND r0 with 00100000 to test the state of bit 5
BEQ     LowerCase      ;If bit 5 is 1 then jump to code dealing with lowercase
```

In this example, the character in r0 is ANDed with the binary literal 00100000_2 and we can branch on *zero* result (branch *not taken*) or non-zero result (branch *taken*).

- ARM's *compare* instruction subtracts the second source operand from the first, then updates the condition code. For example, CMP r1,r2 evaluates [r1] - [r2], then sets the N, Z, C, and V bits of the CPSR.
- The *compare negated* instruction, CMN, complements the second operand before performing the comparison. For example, CMN r1,r2 evaluates [r1] - (-[r2]) and sets the CPSR. Note that the value of [r1] - (-[r2]) is the same as [r1] + [r2].

3.6.4   Branching and Loop Constructs

Nothing illustrates the concept of flow control better than the classic loop constructs that are at the core of structured programming. The following demonstrate the structure of the FOR, WHILE, and UNTIL loops.

The FOR Loop

```
        MOV   r0,#10          ;set up the loop counter
Loop    code ...              ;body of the loop

        SUBS  r0,r0,#1        ;decrement loop counter and set status flags
        BNE   Loop            ;continue until count zero–branch on not zero
        Post  loop ...        ;fall through on zero count
```

The WHILE Loop

```
Loop        CMP   r0,#0       ;perform test at start of loop
            BEQ   WhileExit   ;exit on test true
            code ...          ;body of the loop

            B     Loop        ;Repeat WHILE true
WhileExit   Post  loop ...    ;Exit
```

The UNTIL loop

```
Loop        code ...                    ;body of the loop

            CMP    r0,#0                ;perform test at start end of loop
            BNE    Loop                 ;Repeat UNTIL true
            Post   loop ...             ;Exit
```

Combination Loop

This loop combines the features of all three loops. The FOR part specifies a maximum count that limits the number of excursions round the loop. The WHILE part tests an initial condition in r1 and allows an immediate exit if the condition is not true. The UNTIL part allows an exit if r2 is true at the end of the loop.

```
            MOV    r0,#10               ;set up the loop counter
LoopStart   CMP    r1,#0                ;start with the WHILE test
            BEQ    ComboExit            ;if true then exit loop

            Code   ...                  ;body of the loop

            CMP    r2,#0                ;test UNTIL condition
            BEQ    ComboExit            ;if true then exit loop
            SUBS   r0,r0,#1             ;decrement loop counter and set status flags
            BNE    LoopStart            ;continue until count zero–branch on not zero
ComboExit   Post   loop ...            ;Exit
```

In the next section, we look at what I regard as one of the most interesting features of the ARM family, their ability to perform conditional execution; that is, to execute or not execute an instruction based on the current processor status.

3.6.5   Conditional Execution

One of ARM's most unusual features is that each instruction is *conditionally executed*. We can associate an instruction with a logical condition (one of the 16 in Table 3.2). If the stated condition is true when the instruction is ready to be executed, the instruction is executed. Otherwise it is bypassed (*annulled* or *squashed*). Up to now all ARM instructions have been executed because we have associated them with the default case *execute always*. For example, we could write ADDAL r0,r1,r2 to indicate that we want ADD to be executed always. No one would write this instruction because the AL condition is the default.

ARM's instruction encoding in Figure 3.26 uses bits 28 to 31 to select the condition that must be satisfied if the instruction is to be executed. As well as the default case *always*, another special case is *branch never* which is reserved by ARM for future expansion. The assembly language programmer indicates the conditional execution mode by appending the appropriate condition to a mnemonic. For example, the mnemonic

```
ADDEQ   r1,r2,r3
```

specifies that the addition is performed only if the Z-bit in the condition code register is set because a previous result was zero. The RTL form of this operation is

```
IF Z = 1 THEN [r1] ← [r2] + [r3]
```

There is, of course, nothing to stop you combining conditional execution and shifting because the branch and shift fields of an instruction are independent. You can write

```
ADDCC   r1,r2,r3 LSL r4
```

which is interpreted as IF C = 0 THEN $[r1] \leftarrow [r2] + [r3] \times 2^{[r4]}$

ARM's conditional execution mode makes it easy to implement conditional operations in a high-level language. Consider the following fragment of C code.

```
if (P == Q) X = P - Y ;
```

If we assume that r1 contains P, r2 contains Q, r3 contains X, and r4 contains Y, then we can write

```
CMP     r1,r2           ;compare P == Q
SUBEQ   r3,r1,r4        ;if (P == Q) then r3 = r1 - r4
```

Notice how this operation is implemented without using a branch by squashing instructions we don't wish to execute rather than branching round them. In this case, the subtraction is squashed if the comparison is false. Now consider a more complicated example of a C construct with a compound predicate:

```
if ((a == b) && (c == d)) e++;
```

```
CMP     r0,r1           ;compare a == b
CMPEQ   r2,r3           ;if a == b then test c == d
ADDEQ   r4,r4,#1        ;if a == b AND c == d THEN increment e
```

The first line, CMP r0,r1, compares a and b. The next line, CMPEQ r2,r3, executes a conditional comparison only if the result of the first line was true (i.e., a == b). The third line, ADDEQ r4,r4,#1, is executed only if the previous line was true (i.e., c == d) to implement the e++. Without conditional execution, we might write

```
        CMP     r0,r1           ;compare a == b
        BNE     Exit            ;exit if a =! b
        CMP     r2,r3           ;compare c == d
        BNE     Exit            ;exit if c =! d
        ADD     r4,r4,#1        ;else increment e
Exit
```

As you can see, this conventional approach to compound logical conditions requires five instructions. You can also handle some testing with multiple conditions. Consider:

```
if (a == b)  e = e + 4;
if (a < b)   e = e + 7;
if (a > b)   e = e + 12;
```

Using the same register assignments as before, we can use conditional execution to implement this as

```
CMP     r0,r1           ;compare a == b
ADDEQ   r4,r4,#4        ;if a == b then e = e + 4
ADDLE   r4,r4,#7        ;if a < b then e = e + 7
ADDGT   r4,r4,#12       ;if a > b then e = e + 12
```

Once again, using conventional non-conditional execution, we would have to write something like the following to implement this algorithm.

```
        CMP     r0,r1           ;compare a == b
        BNE     Test1           ;not equal try next test
        ADD     r4,r4,#4        ;a == b so e = e+4
        B       ExitAll         ;now leave
Test1   BLT     Test2           ;if a < b then
        ADD     r4,r4,#12       ;if we are here a > b so e = e + 12
        B       ExitAll         ;now leave
Test2   ADD     r4,r4,#7        ;if we are here a < b so e = e + 7
ExitAll
```

This is rather less elegant than the previous version.

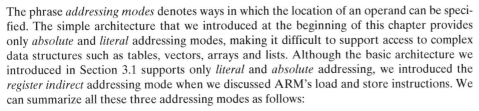

3.7   ARM Addressing Modes

The phrase *addressing modes* denotes ways in which the location of an operand can be specified. The simple architecture that we introduced at the beginning of this chapter provides only *absolute* and *literal* addressing modes, making it difficult to support access to complex data structures such as tables, vectors, arrays and lists. Although the basic architecture we introduced in Section 3.1 supports only *literal* and *absolute* addressing, we introduced the *register indirect* addressing mode when we discussed ARM's load and store instructions. We can summarize all these three addressing modes as follows:

Mnemonic	RTL form	Description
ADD **r0**,r1,#Q	[r0] ← [r1] + Q	*Literal*: Add the integer Q to contents of register r1
LDR **r0**,Mem	[r0] ← [Mem]	*Absolute*: Load contents of memory location Mem into register r0. This addressing mode is not supported by ARM but is supported by all CISC processors
LDR **r0**,[r1]	[r0] ← [[r1]]	*Register Indirect*: Load r0 with the contents of the memory location pointed at by r1

We now describe each of ARM's seven addressing modes (illustrated in Figure 3.27), in greater detail. Once more, we should stress that the ARM lacks a simple memory direct (i.e., absolute) addressing mode and does not have an LDR **r0**, address instruction that implements direct addressing to load the contents of a memory location denoted by address into a register. CISC processors like the Pentium and 68K support memory direct addressing (see panel).

Direct (Absolute) Addressing—A Reminder

In direct addressing, the address field of the instruction provides the actual memory address of the operand in the current instruction. Direct addressing permits a programmer to specify variables (i.e., symbolic user-selected names), that change dynamically during the course of a program's execution. Some call this addressing mode *absolute addressing*, because the instruction provides the actual address of the operand.

The high-level language statement *P* = *Q* + *R* can be translated into a *one-and-a-half-address* generic microprocessor assembly language, as follows:

Mnemonic	RTL Description	Description in Words
MOV **r0**,Q	[r0] ← [Q]	Load contents of memory location Q into register r0
ADD **r0**,R	[r0] ← [r0] + [R]	Add contents of memory location R to register r0
MOV **P**,r0	[P] ← [r0]	Store the result in memory location P

Here, P, Q, and R are symbolic names representing the locations of the variables in main memory. Each instruction employs direct addressing to access its operand.

Absolute addressing suffers from significant limitations. It doesn't support *relocatable code*, which can be moved anywhere in memory without computing new operand addresses. Nor does it support *re-entrant code*, which can be interrupted by the operating system, then reused by the interrupting program without overwriting data used by a previous instance of the code. Relocatable code is important in *multiprocessing*, where several programs stored in memory can be active in a processor at the same time. Similarly, re-entrant code is useful in embedded processing where frequent interrupts can complicate the status of running processes.

FIGURE 3.27   Summary of addressing modes that can be used by ARM

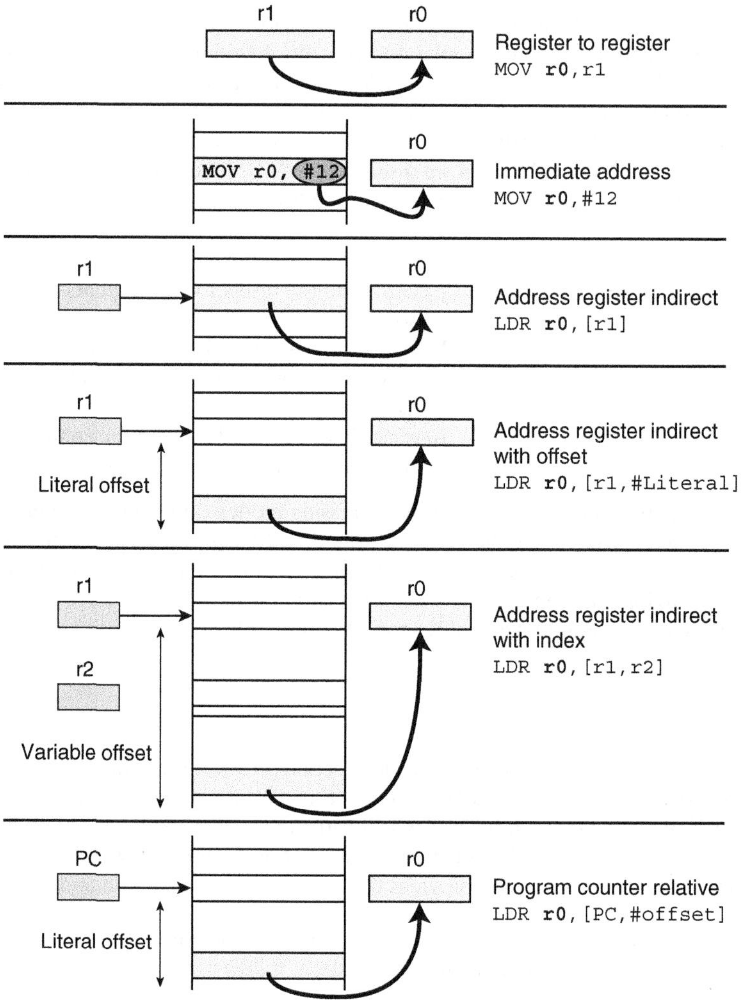

© Cengage Learning 2014

3.7.1   Literal Addressing

We've already covered this topic, so we will provide an example of its use before we demonstrate how the ARM implements it. Literal addressing is used by high-level language (HLL) constructs that specify a constant rather than a variable, such as:

```
IF I > 25 THEN J = K + 12,
```

where the constants 12 and 25 can be specified by literal addressing. We can express this as

```
                        ;assume I is in r0, J in r1, and K in r2
      CMP    r0,#25      ;Compare I with the value 25
      BLE    Exit        ;IF I ≤ 25 THEN exit
      ADD    r1,r2,#12   ;        ELSE add 12 to K
Exit                     ;...
```

We can simplify the code by using conditional execution as follows.

```
      CMP    r0,#25      ;Compare I with the value 25
      ADDGE  r1,r2,#12   ;IF I ≤ 25 THEN exit
      ADDGT
```

ARM's Way

ARM implements literals in a way that is profoundly different to other processors like the IA32, 68K, or MIPS. ARM provides a 12-bit literal field in instructions but doesn't provide a 12-bit unsigned literal in the range 0 to 4095 or a signed literal in the range from -2048 to $+2047$, as you might expect. Instead, the ARM provides an 8-bit literal that can be *scaled by a power of 2*. You could even say that ARM provides a type of floating-point literal. Figure 3.28 illustrates ARM's literal operand encoding structure. When bit 25 of an op-code is 0, the ARM performs a shift operation. When bit 25 is 1, the operand 2 field encodes a 12-bit literal, which is subdivided into two parts: an 8-bit literal and a 4-bit alignment.

FIGURE 3.28   Diagram of ARM's literal operand encoding

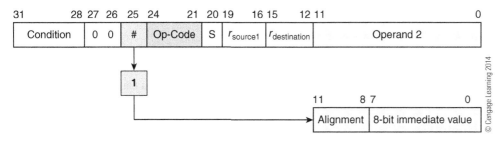

© Cengage Learning 2014

The four most-significant bits of the literal field specify the literal's alignment within a 32-bit word. If the 8-bit immediate value is N and the 4-bit alignment is n in the range 0 to 12, then the value of the literal is given by $N \times 2^{2n}$. Thus, ARM provides an 8-bit literal that is scaled by a power of 2. Of course, this is analogous to the way that floating-point numbers are represented and stored.

Let's look at an example of the way in which ARM literals are scaled. In each of the four cases in Figure 3.29, the 8-bit literal is 11010110. This diagram demonstrates how the alignment field moves the literal within a 32-bit frame. Remember that the number in the alignment field has to be *doubled* to get the number of places by which the literal is rotated *right* (yes, it's the number of right shifts that are recorded).

Suppose you wish to clear all but the most-significant byte (bits 24 to 31) in a register. You can AND the register with the literal $FF000000_{16}$ to clear the specified bits, as demonstrated:

```
AND r0,r0,#0xFF000000
```

Although you can't specify a 32-bit literal directly, the ARM's scaling mechanism represents the constant $FF000000_{16}$ by the 8-bit value FF_{16} shifted left 24-bits (i.e., shift right by 8 bits).

FIGURE 3.29   ARM's literal operand encoding

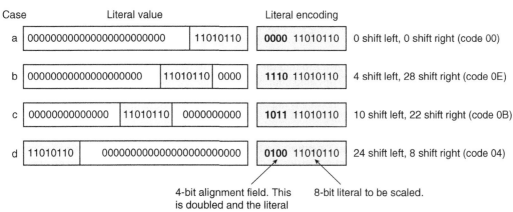

© Cengage Learning 2014

The *move negative instruction*, MVN **r0**,r1,#literal, can specify an unshifted constant in the range 0xFFFFFF00 to 0xFFFFFFFF. The programmer is not expected to worry about how shifted constants are generated. That's the job of the assembler. The scaling of literals is not a pseudooperation. ARM users can, at first, be confused by the idea that a literal can be expressed in 8 bits, and these 8 bits then can be fitted anywhere into a 32-bit slot.

FIGURE 3.30   Example of ARM literal encoding

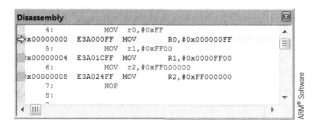

Figure 3.30 provides a final demonstration of literal operand encoding using the three examples below. We have presented the disassembly window so that you can see the code generated by the assembler. For example, MOV **r1,** #0x0000FF00 generates the code E3A01CFF$_{16}$.

```
MOV   r0,#0xFF
MOV   r1,#0x0000FF00
MOV   r2,#0xFF000000
```

Figure 3.28 tells us that the rightmost 12 bits of an instruction define the literal. If we take MOV **r1,**#0xFF00, the literal is encoded as CFF$_{16}$, which is 1100 11111111$_2$, and the shift value is C$_{16}$ or 12$_{10}$. The actual shift is twice this value or 24 and is implemented as a *rotate right*. Twenty-four rotate right operations are equivalent to eight rotate left operations, which are required to shift the FF$_{16}$ to FF00$_{16}$. Finally, if we look at 0xFF000000, we see that a six-position hexadecimal shift left has been applied, which corresponds to 24 binary shifts left. However, a rotate right is used and is equivalent to a right rotation through eight positions. The stored scaling constant is half this; that is, 4. If you look at the instruction encoding in Figure 3.30 on line 6, you will see that the constant is encoded as 4FF.

3.7.2   Register Indirect Addressing

We have already encountered this addressing mode where the location of an operand is held in a register. It is called *register indirect addressing*, because the instruction specified the register where a pointer to the actual operand can be found. In ARM literature, this addressing mode is called *indexed addressing*. Some people call this *base addressing*. Register indirect addressing mode requires *three* read operations to access an operand:

- Read the instruction to find the pointer register
- Read the pointer register to find the operand address
- Read memory at the operand address to find the operand

Register indirect addressing is important because the contents of the register containing the pointer to the actual operand can be modified at runtime, and therefore, the address is a variable, allowing access to data structures such as arrays, lists, matrices, vectors, and tables.

Figure 3.31 illustrates memory indirect addressing using the ARM *load* instruction LDR **r1,** [r0]. The pointer register r0 contains the value *n* and thus *points to* or *references* location *n* in memory. LDR **r1,** [r0] loads the contents of the memory location pointed at by register r0 into register r1.

Suppose we next execute ADD **r0,**r0,#4 to increment the contents of the pointer register r0 by 4. Since the ARM is byte addressed and the addresses of consecutive words differ by

FIGURE 3.31   Register indirect addressing

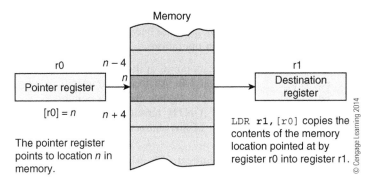

The pointer register points to location *n* in memory.

LDR **r1**, [r0] copies the contents of the memory location pointed at by register r0 into register r1.

© Cengage Learning 2014

FIGURE 3.32   Effect of incrementing the pointer register

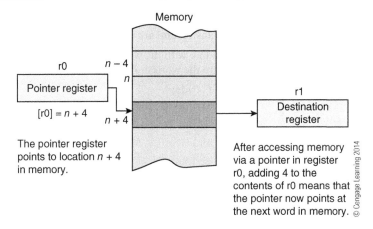

The pointer register points to location *n* + 4 in memory.

After accessing memory via a pointer in register r0, adding 4 to the contents of r0 means that the pointer now points at the next word in memory.

© Cengage Learning 2014

4, r0 now points to the *next* 32-bit word in memory, as shown in Figure 3.32. The two operations LDR **r1**, [r0] and ADD **r0**, r0, #4 are defined in RTL as

```
[r1] ← [[r0]]          ;Read the memory location pointed at by r0
                       ;[[r0]] indicates the contents of the memory pointed at by r0
[r0] ← [r0] + 4        ;Point to the next location by incrementing the pointer
```

Consider the following example where a table of seven entries represents the days of the week. D_1 represents Monday, D_2 represents Tuesday, etc. If D_i is day i then D_{i+1} represents the next day. In order to move from one day to the next, all we need do is increment index i. This is why we need *variable* addresses.

```
       ADR  r0,week                    ;r0 points to array week
       ADD  r0,r0,r1, LSL #2           ;r0 now points at the day whose value is in r1
       LDR  r2, [r0]                   ;read the data for this day into r2
Week   DCD                             ;data for day 1
       DCD                             ;data for day 2
         .
       DCD                             ;data for day 7
```

In this example, the index of days is assumed to be 0 to 6. We have to multiply the day index by 4, because the array is an array of words (4-byte values) and consecutive day addresses are separated by 4.

Character strings provide a good example of the use of pointer-based addressing. Suppose we want to find the location of a certain character within a string. We can write the following code. This isn't ARM code, because we haven't yet introduced byte operations. I have used subscripts to indicate the size of operands because I wish to demonstrate the difference between data operations and pointer operations.

```
        LDR₃₂   r0,#String          ;r0 points at the string
Loop    LDR₈    r1,[r0]             ;REPEAT Read character
        ADD₃₂   r0,r0,#1            ;    Update character pointer
        CMP₈    r1,#Terminator      ;UNTIL terminator found (Terminator is the end-
                                         of-line character)

        BNE     Loop
```

Some computers combine register indirect addressing with *pointer updating* so that the pointer is automatically moved to the next item in memory after the pointer has been used. For example, the 68K family uses the notation (A0)+ to indicate that A0 is a pointer to the operand and that the pointer is to be incremented after the operand has been accessed (a so-called post-incrementing or post-indexing mode). We will soon see that ARM too implements auto-updating pointers.

ARM implements load and store operations, LDR and STR, which provide register-to-memory and memory-to-register data transfer, and are written as

Load   LDR   r0,[r1]   ;load r0 with the word pointed at by r1
Store   STR   r2,[r3]   ;store the word in r2 in the location pointed at by r3

Consider the following fragment of C code:

```
for (i = 0; i < 21; i++)
    {
    j[i] = j[i] + 10;
    }
```

The values 0, 21, and 10 in this program are constants specified via immediate addressing during assembly. We can translate the previous high-level code into ARM assembly language.

```
        MOV   r0,#0         ;Set counter i in r0 to initial value zero
        ADR   r8,j          ;Index register r8 points to array j (pseudoinstruction)
Loop    LDR   r1,[r8]       ;REPEAT Get j[i]
        ADD   r1,r1,#10     ;        Add 10 to j[i]
        STR   r1,[r8]       ;        Save j[i]
        ADD   r0,r0,#1      ;        Increment loop counter i
        CMP   r0,#21        ;        Compare loop counter with terminal value + 1
        BNE   Loop          ;UNTIL i = 21
```

Note that we have counted up from 0. Had we loaded r0 with 21, we could have used a SUBS r0,r0,#1 to decrement the counter, followed by a BNE Loop to save an instruction.

Addressing Modes: A Note on Terminology

Because computer science is a relatively recent discipline and one that is often industry-led, some of its vocabulary has not yet settled. Nowhere is that more true than in the province of addressing modes. Below are some examples of the terminology.

Name	Alternative name	ARM example	68K example
Literal	Immediate	MOV r0,#4	MOVE #4,D0
Direct	Absolute		MOVE MemAdr,D0
Register indirect	Indexed, base	LDA r0,[r1]	MOVE (A1),D0
Register indirect with offset	Pre-indexed, base with displacement	LDR r0,[r1,#4]	MOVE (4,A0),D0

Register indirect post-incrementing	Post-indexing, autoindexing	LDR r0,[r1],#4	MOVE (A1)+,D0
Register indirect pre-incrementing	Pre-indexed, autoindexing	LDR r0,[r1,#4]!	MOVE -(A0),D0
Register indirect	Double register indirect Register indexed	LDR r0,[r1,r2]	MOVE (A0,D2),D0
Register indirect with scaling	Double register indirect with scaling Register indexed with scaling	LDR r0,[r1,r2,LSL #2]	

3.7.3 Register Indirect Addressing with an Offset

ARM supports a memory-addressing mode where the *effective address* of an operand is computed by adding the *contents of a register* to a *literal offset* coded into the load/store instruction. This addressing mode is often called *base plus displacement addressing*. ARM's literal offset is 12 bits. It is a true 12-bit literal and not the 8-bit scaled value. Figure 3.33 illustrates this concept with the instruction LDR **r0**, [r1,#4]. Here, the effective address *is* given by the sum of the contents of the pointer register r1 plus the offset 4; that is, the operand is four bytes on from the address specified by the pointer.

FIGURE 3.33 Register indirect addressing with an offset

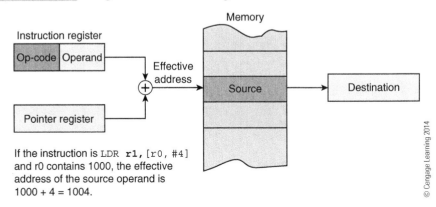

If the instruction is LDR **r1**, [r0, #4] and r0 contains 1000, the effective address of the source operand is 1000 + 4 = 1004.

© Cengage Learning 2014

This addressing mode allows you to synthesize absolute addressing, because if you put zero into register r1, the effect of LDR **r0**, [r1,#8] is the same as LDR **r0**, [#8]. However, since the offset is an 8-bit scaled constant, such absolute addressing is not a terribly useful facility.

Let's look at a simple but typical example of offset addressing. The following fragment of code demonstrates the use of offsets to implement array access. Because the offset is a constant, it cannot be changed at runtime.

```
Sun     EQU  0                  ;offsets for days of the week
Mon     EQU  4
Tue     EQU  8
 .
Sat     EQU  24

        ADR  r0, week           ;r0 points to array week
        LDR  r2,[r0,#Tue]       ;read the data for Tuesday into r2
```

```
Week    DCD                              ;data for day 1 (Sunday)
        DCD                              ;data for day 2 (Monday)
        DCD                              ;data for day 3 (Tuesday)
        DCD                              ;data for day 4 (Wednesday)
        DCD                              ;data for day 5 (Thursday)
        DCD                              ;data for day 6 (Friday)
        DCD                              ;data for day 7 (Saturday)
```

Figure 3.34 shows the code generated by this program and the disassembled code. The values in the registers are those at the end of the code execution. You can see that Tuesday and Wednesday have been added together.

ARM allows you to specify the offset as a second register so that you can use a dynamic offset that can be modified at runtime (See Figure 3.35).

```
LDR  r2,[r0,r1]                ;[r2] ← [[r0] + [r1]]  load r2 with the memory
                                location pointed at by r0 plus r1
LDR  r2,[r0,r1,LSL #2]         ;[r2] ← [[r0] + 4 x [r1]]  Scale r1 by 4
```

In the second example, register r1 is scaled by 4. This allows you to use a scaled offset when dealing with arrays. For example, if r0 points to array X and r1 contains index i, then the address of element i is $X + 4i$. We can access this element using LDR r2,[r0,r1,LSL #2].

FIGURE 3.34   Example of register indirect addressing using offsets

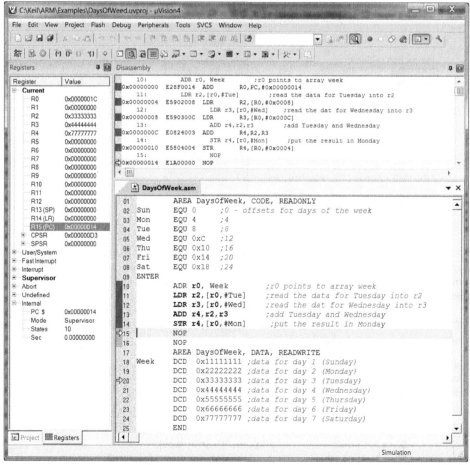

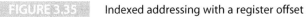

FIGURE 3.35   Indexed addressing with a register offset

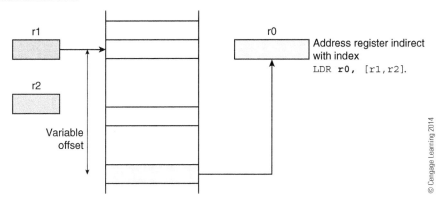

We have demonstrated that the register-indirect addressing mode can be extended by adding a literal offset or a register offset to the base register, as shown in Figure 3.35. In ARM terminology, this *base register plus offset* addressing mode is called *pre-indexing*, because the offset is added to the pointer *before* the operand is accessed. The ARM instruction LDR **r0**, [r1, #8] specifies the *preindexed* addressing mode, and the effective address of the operand is given by [r1] + 8. Here, *pre-indexing* denotes addition of the offset #8 to the base register r1 *before* it is applied to memory during the read phase of the *load* operation.

The pre-indexing addressing mode can be used to access element i in array X. For example,

```
ADR   r0,X               ;register r0 points at array X
LDR   r1,[r0,i]          ;get element i
```

Let's turn this into executable code and look at it in the simulator. Figure 3.36 provides a snapshot of the simulator output after the two lines of the code have been executed. We have provided a dummy array X complete with some data. Note that the value of the index *i* is 2 but has been equated to 2*4, because each element in the array is four bytes, so that the byte offset is 8. As you can see, after the code has been executed, register r0 contains 0x00008088 (the address of the first word in the array), and r1 contains 0xabababab, which is the third element in the array (remember that the first element is at offset 0 so that the element with index 2 is the third element).

When 12 Really is 12

Recall that ARM processors specify a 12-bit constant as a literal operand (for example, ADD **r0**, r1, #123). This constant is an 8-bit value scaled by a 4-bit shift which allows you to have constants like 0xAB, 0xAB0, 0xAB000, etc.

However, when ARM processors specify a literal offset as part of an indexed address, it is a true 12-bit value. Thus, literals used by ARM processors can be 8 bits scaled or a true 12 bits.

In summary, pre-indexed register indirect addressing uses the effective address [r0, #d] or [r0, r1] and the offset (#d or r1) is added to the base register to create an effective address before the access is made. The base register remains unmodified.

FIGURE 3.36   Example of the use of pre-indexed register indirect addressing

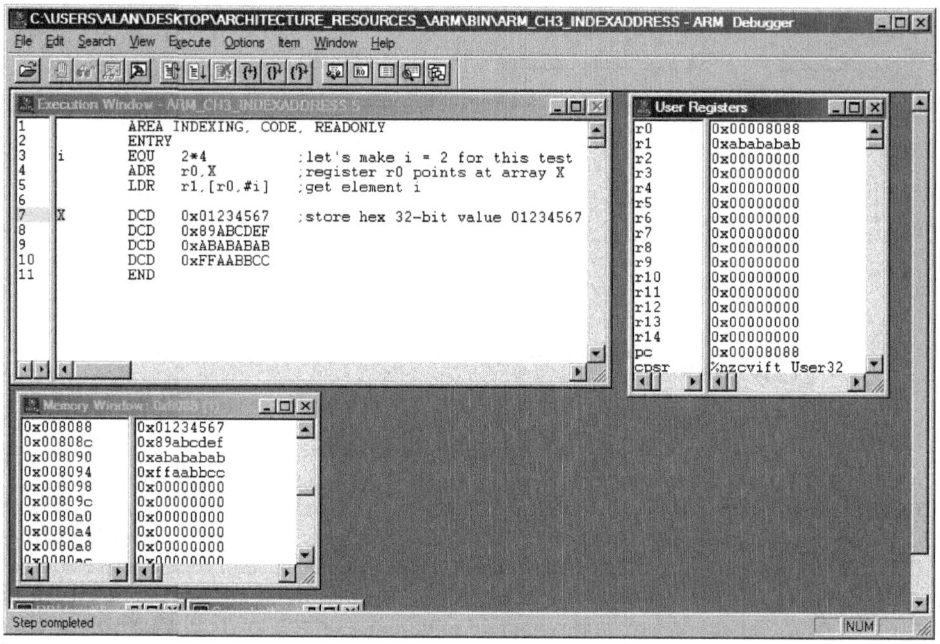

3.7.4   ARM's Autoindexing Pre-indexed Addressing Mode

We now look at ARM's ability to automatically change the contents of the index or base register during register indirect addressing in order to facilitate the reading of sequential data in structures likes vectors, tables, and arrays. Various terms are used to describe these addressing modes. For example, *autoincrementing* and *autodecrementing* are used to emphasize that the pointer is changed automatically, and *pre-indexing* and *post-indexing* stress when the increment is made. In this section, we will use ARM's terminology.

Because elements in an array, table, or similar data structure are frequently accessed sequentially, ISA designers implemented autoindexing addressing modes in which the pointer register is automatically adjusted to point at the next element *before* or *after* it is used. ARM implements two autoindexing modes by adding the offset to the base (i.e., pointer register). The difference between these two modes is the point at which the base register is incremented, either *before* or *after* that memory access has been made.

68K Autoindexing

The 68K has a simple autoindexing scheme. A base/index register is used that is automatically updated on each use. Only two variations are permitted: pre-decrementing and post-incrementing. For example,

MOVE -(A0),D0   ;Decrement pointer A0 and then load D0 with the memory location pointed at by A0

MOVE (A0)+,D0   ;Load D0 with the memory location pointed at by A0, then increment pointer A0

ARM's *autoindexing pre-indexed* addressing mode is indicated by appending the suffix ! to the effective address. Consider the ARM instruction:

```
LDR  r0,[r1,#8]!    ;load r0 with the word pointed at by register r1 plus 8
                    ;then update the pointer by adding 8 to r1
```

The RTL definition of this instruction is given by

[r0] ← [[r1] + 8]   Access the memory 8 bytes beyond the base register r1
[r1] ← [r1] + 8   Update the pointer (base register) by adding the offset

This *autoindexing mode* does not incur additional execution time, because it is performed in parallel with memory access. Consider the following example of the addition of two arrays.

```
Len      EQU   8                    ;let's make the arrays 8 words long
         ADR   r0,A - 4             ;register r0 points at array A
         ADR   r1,B - 4             ;register r1 points at array B
         ADR   r2,C - 4             ;register r2 points at array C
         MOV   r5,#Len              ;use register r5 as a loop counter
Loop     LDR   r3,[r0,#4]!          ;get element of A
         LDR   r4,[r1,#4]!          ;get element of B
         ADD   r3,r3,r4             ;add two elements
         STR   r3, [r2,#4]!         ;store the sum in C
         SUBS  r5,r5,#1             ;test for end of loop
         BNE   Loop                 ;repeat until all done
```

Each time the operation `LDR r3, [r0, #4]!` is executed, the operand is fetched from the address pointed at by r0 plus 4, and then the value of r0 is increased by 4 to point at the next element ready for the next excursion round the loop. We have to set the pointers to 4 bytes before the start of each array initially, because the pointer is incremented before its use. Fortunately, the assembler allows us to write `ADR r0, A-4` (that is, the assembler generates the address of label A and subtracts 4 from it before using it to assemble the instruction). Figure 3.37 demonstrates the register map and memory map of the system after executing the code.

FIGURE 3.37   Register, code, and memory data after executing code using autoindexing

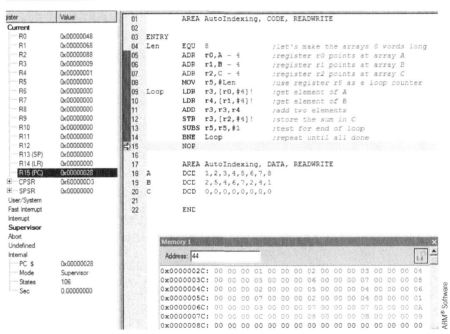

3.7.5   ARM's Autoindexing Post-Indexing Mode

ARM also provides an autoindexing *post-indexing* addressing mode that first accesses the operand at the location pointed to by the base register, then increments the base register. Consider the code fragment:

```
LDR    r0, [r1], #8        ;load r0 with the word pointed at by r1
                           ;now do the post-indexing by adding 8 to r1
```

Post-indexing is denoted by placing the offset *outside* the square brackets (e.g., [r1],#8). The RTL definition of this instruction is

[r0] ← [[r1]]   Access the memory address in base register r1
[r1] ← [r1] + 8   Update the pointer (base register) by adding the offset

Figure 3.38 illustrates ARM's variations on indexed addressing. In each case, the base register is r1, the offset is 12, and the destination register is r0.

FIGURE 3.38   Register indirect addressing with offset

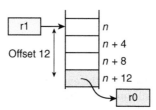

(a) LDR **r0**, [r1,#12]
Offset added to base register to generate effective address. Operand accessed at effective address. Base register remains unchanged.

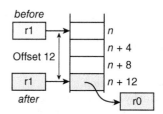

(b) LDR **r0**, [r1,#12]!
Offset added to base register to generate effective address. Operand accessed at effective address. Base register updated after access.

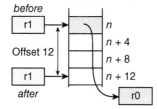

(c) LDR **r0**, [r1], #12
Effective address specified by base register. Operand accessed at effective address. Offset added to base register after the access.

© Cengage Learning 2014

3.7.6 Program Counter Relative (PC-Relative) Addressing

Any ARM register can be used to implement register indirect addressing. However, register r15 is not just *any* register; it's the program counter. If you use r15 as a pointer register to access an operand, the resulting addressing mode is called *program counter relative address-ing* and the location of the operand is specified with respect to the current code location. This means that you can move the code and its associated data to a different location in memory, without having to recalculate operand addresses.

Suppose we execute the instruction LDR **r0**, [r15,#100]. The operand is specified as 100 bytes (25 words) from the contents of register r15. Thus, the operand is 100 bytes from the current position.[28] Recall that the assembler makes use of PC-relative addressing to imple-ment pseudoinstructions; for example, it can generate the 32-bit address of a near-by address by using the current PC value and adding in an offset.

---

[28] Note that this is not the same as 100 bytes from the instruction with the PC-relative addressing mode because the PC is incremented after it has been used in a fetch cycle. Furthermore, the state of the PC is also affected by the ARM's pipelining mechanism that overlaps operations. ARM's PC is 8 bytes on from the current instruction.

3.7.7 ARM's Load and Store Encoding

Figure 3.39 illustrates the format of the ARM's load and store instructions. Memory access operations have a conditional execution field, bits 31 through 28 of the op-code, and can be conditionally executed like other ARM instructions. This facility makes it possible to write code like

```
                    ;if (a == b) then x = p else x = q
CMP     r1,r2       ;if (a == b)
LDREQ   r3,[r4]     ;then x = p
LDRNE   r3,[r5]     ;else x = q
```

Op-code bit 20 selects the direction of data transfer; that is, whether the instruction is a load or store. Bit 25 (the # bit) determines whether the offset is the contents of a register with optional shifting or a 12-bit constant. Bit 22 selects the operand size, and determines whether the ARM is executing a 32-bit word transfer or an 8-bit byte transfer. Whenever a byte is loaded into a 32-bit register, bits 8 through 31 are set to zero (i.e., the byte is not sign-extended). Later versions of the ARM family have extended the ISA to permit sign-extension.

In Figure 3.39, the base register r_{base} is a memory pointer, and the U-bit defines whether the effective address should be calculated by *adding* or *subtracting* the offset. The P- and W-bits determine how indexing is implemented. The W-bit determines whether the base register is to be updated at the end of the current instruction. Using W = 1, the base register is updated. The P-bit controls whether the offset is added to the base register *before* or *after* the

FIGURE 3.39 Format of ARM's load and store instructions

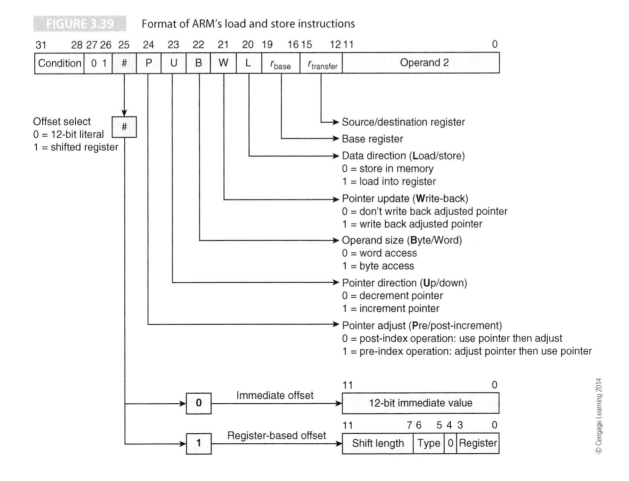

© Cengage Learning 2014

effective address is calculated. Because the P-bit specifies addition or subtraction of an offset, ARM is able to use addressing modes, such as

```
        LDR    r0,[r1, +r2]              ;effective address is [r1] + [r2]
and     LDR    r0,[r1, -r2]              ;effective address is [r1] - [r2]
```

The ability to specify whether the second component of an effective address should be added to or subtracted from the base register is unusual, but this can be useful in some applications. Table 3.3 summarizes the ARM's register-based addressing modes.

TABLE 3.3   Summary of ARM's Indexed Addressing Modes

Addressing Mode	Syntax	Effective Address	Final Value in r1
Pre-indexed, base unchanged	LDR r0,[r1,#d]	[r1] + d	[r1]
Pre-indexed, updated base	LDR r0,[r1,#d]!	[r1] + d	[r1] + d
Post-indexed, updated base	LDR r0,[r1],#d	[r1]	[r1] + d

© Cengage Learning 2014

The formal (Backus-Naur format, called BNF) grammatical definition of ARM's load/store instructions is

```
LDR|STR{cond}{B} Rd,[Rn, offset]{!}
```

or

```
LDR|STR{cond}{B} Rd,[Rn],offset
```

where the curly braces { } denote an optional field and | denotes an alternative. The optional {B} field indicates a byte operand. I have used this notation because it is found in ARM's literature and provides a useful means of defining the format of expressions. Consider the instructions:

```
LDR    r0,[r1, r2,LSL #4]!     ;the effective address is r1 + r2 x 2⁴
STRB   r9, [r5,-r7,ASR #2]     ;the effective address is r5 - r7 x 2⁻²
```

As a further example, consider the binary string 01010111001000100100000100000110, which represents an ARM instruction. We can partition this instruction into fields described by Figure 3.24, to get the decoding shown in Table 3.4, which yields STRPL r4, [r2, -r6,LSL #2]!. If we assemble and execute this instruction using the ARM simulator, we find that its stored op-code is 57224106_{16}.

3.8   Subroutine Call and Return

A subroutine (also called a procedure, a function, or, in Java-speak, a *method*) is a body of code that can be called and executed, and then a return is made to the instruction following the calling point. Strictly speaking, a *subroutine* is a block of code that is called and executed, whereas a *function* is called and returns a parameter. Moreover, when a subroutine is called, parameters may be passed between the calling body and subroutine.

As well as discussing subroutines in this section, we will deal with two issues. One is how parameters are passed to and from subroutines, and the other is how a return is made from a subroutine to the calling point. Chapter 4 deals with subroutines and functions in much greater detail and discusses the use of local workspace and parameter passing.

Figure 3.40 illustrates a subroutine call made by a typical CISC[29] processor. The instruction BSR Proc_A *calls* the subroutine Proc_A. The processor saves the address of the next

[29] The ARM does not have a RTS instruction. It uses a different means of returning from a subroutine call.

Field Name	Value	Action	Interpretation
TABLE 3.4		Decoding the ARM Instruction STRPL r4, [r2,-r6,LSL #2]!	
Condition	0101	PL	Execute on positive
Op-code	01		Defines load/store instruction
#	1	Operand 2 format	Operand is a shifted register
P	1	Pre/post adjust	Adjust pointer before using
U	0	Pointer direction	Decrement pointer
B	0	Byte/word	This is a word access
W	1	Pointer writeback	Update pointer after use
L	0	Load/store	Store data in memory
r_{base}	0010	Base register	r2 is the base (pointer) register
$r_{transfer}$	0100	Source/destination	r4 is the source in this store instruction
Shift length	00010	Shift length	Shift the register 2 places
Shift type	00	Logical shift left	Shift the offset in r6 left
Op-code	0		
Shift register	0110	Specifies register to be shifted	r6 is shifted left twice

© Cengage Learning 2014

instruction to be executed in the calling code in a safe place and loads the program counter with the target address, Proc_A, of the first instruction in the subroutine. Loading the PC with a non-sequential value forces a branch to the subroutine. At the end of the subroutine, a *return from subroutine instruction,* RTS, causes the processor to return to the point immediately following the subroutine call.

FIGURE 3.40   The subroutine call and return (this mechanism is not used by ARM processors)

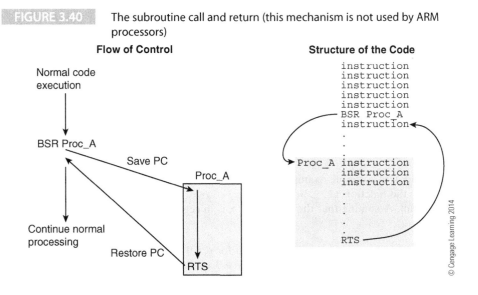

© Cengage Learning 2014

CISC processors use a stack mechanism to provide hardware support for subroutine calls and returns. RISC processors do not generally provide full hardware support for stack-based subroutine calls and leave subroutine handling to the programmer. We cover these topics in more detail in this section.

3.8.1   ARM Support for Subroutines

Like other RISC processors, ARM processors do not provide a fully automatic subroutine call/return mechanism like CISC processors, such as the Intel IA32 or the Freescale 68K. ARM's *branch with link* instruction, BL, automatically saves the return address in register r14. The format of the branch instruction (Figure 3.41) has an 8-bit op-code with a 24-bit signed program counter relative offset. Since the destination of a branch is a word address, branch addresses are aligned on 32-bit word boundaries. Consequently, the 24-bit offset is first shifted left twice to convert the word-offset address to a byte address. Then the 26-bit byte address is sign-extended to 32 bits, and this 32-bit byte offset is added to the contents of the program counter. Because the branch address offset is 26 bits (i.e., 24 + 2), conditional branches span the address range PC ± 32 Mbytes. In other words, you can branch forward or backward up to 32 Mbytes from the current location.

FIGURE 3.41   Encoding ARM's branch and branch-with-link instructions

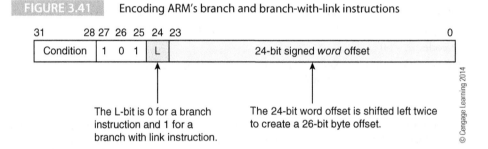

The branch with link instruction behaves like the corresponding branch instruction but also copies the return address (i.e., address of the next instruction to be executed following a return) into the link register r14. If you execute:

```
BL      Sub_A           ;branch to "Sub_A" with link
                        ;save return address in r14
```

the ARM executes a branch to the target address specified by the label Sub_A. It also copies the program counter held in register r15 into the link register r14 to preserve the return address. At the end of the subroutine you return by transferring the return address in r14 to the program counter. You don't need a special return instruction; you can simply write:

```
MOV     pc,lr           ;we can also write this MOV r15,r14
```

Let's look at a simple example of the use of a subroutine. Suppose that you wanted to evaluate the function if x > 0 then x = 16x + 1 else x = 32x several times in a program. Assuming that the parameter x is in register r0, we can write the following subroutine.

```
Func1   CMP     r0,#0           ;test for x > 0
        MOVGT   r0,r0, LSL #4    ;if x > 0 x = 16x
        ADDGT   r0,r0,#1         ;if x > 0 then x = 16x + 1
        MOVLE   r0,r0, LSL #5    ;ELSE if x ≤ 0 THEN x = 32x
        MOV     pc,lr           ;return by restoring saved PC
```

We've made use of conditional execution here. The only thing needed to turn a block of code into a subroutine is an entry point (the label 'Func1') and a return point that restores the address saved by the BL in the link register. Consider the following.

```
LDR    r0, [r4]           ;get P
BL     Func1              ;P = (if P > 0 then 16P + 1 else 32P) First function call
STR    r0, [r4]           ;save P
.
. some code
.
LDR    r0, [r5,#20]       ;get Q
BL     Func1              ;Q = (if Q > 0 then 16Q + 1 else 32Q) Second function call
STR    r0, [r5,#20]       ;save P
```

Figure 3.42 demonstrates the execution of this code in the ARM simulator. We have used dummy data for the two calls; first with P = 3 and then with Q = −1 (FFFFFFFF$_{16}$). At the end of execution, the two memory locations for P and Q contain the expected values of 49 (31$_{16}$) and −32 (FFFFFFE0$_{16}$). These two values are stored after the data at addresses 0x4C and 0x50, respectively. Note that we used indexed addressing with displacement to store the results in memory e.g., STR r4, [r0,#8].

FIGURE 3.42   Demonstrating a subroutine call

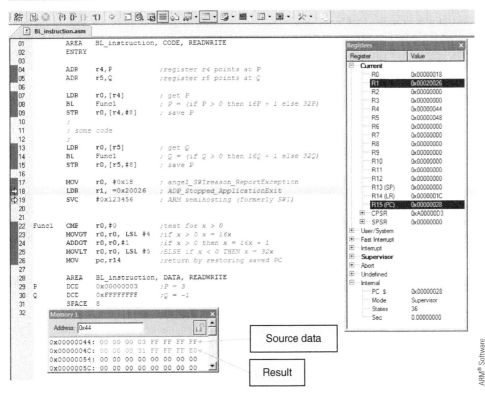

ARM® Software

3.8.2   Conditional Subroutine Calls

Because the branch with link instruction can be conditionally executed, ARM provides a full set of conditional subroutine calls. For example,

```
CMP    r9,r4              ;if r9 < r4
BLLT   ABC                ;then call subroutine ABC
```

The mnemonic BLLT is made up of B (branch unconditionally), L (branch with link), and LT (execute on condition less than).

3.9   Intermission: Examples of ARM Code

We now provide several fragments of ARM assembly language to highlight the ARM's instruction set. Some examples are taken from ARM's literature and demonstrate interesting aspects of a particular ISA. Computer users should be aware of the properties of the underlying hardware because it is the hardware that determines the performance of a computer system, whether it be an ISA or a cache memory access mechanism.

3.9.1   Extracting the Absolute Value

Suppose we wish to obtain the absolute value of a signed integer; that is, if $x < 0$ then $x = -x$. This fragment of code uses the TEQ instruction and a reverse subtract operation.

```
TEQ    r0,#0          ;compare r0 with zero
RSBMI  r0,R0,#0       ;if negative then 0 − r0 (note use of reverse subtract)
```

3.9.2   Byte Manipulation and Concatenation

Sometimes, it is necessary to rearrange the order of the bytes in a word. Figure 3.43 demonstrates how individual bytes in a word can be manipulated. Suppose we wish to take the low-order bytes from registers r0 and r1 and put them in the higher-order 16 bits of r2 without overwriting or modifying the lower-order 16 bits of r2. If r0 contains 00000078_{16}, r1 contains $000000EF_{16}$, and r2 contains 11223344_{16}, the final result in r2 will be $78EF3344_{16}$.

Assume that, initially, the high-order bits of r0 and r1 are all zeros. Figure 3.43 shows how we can move the bytes using just three instructions.

```
ADD r2,r1,r2, LSL #16.
ADD r2,r2,r0, LSL #8
MOV r2,r2,    ROR #16
```

The key to these operations is the ARM's ability to both *process* data and to *shift* data in a single instruction. The first operation is ADD r2,r1,r2, LSL #16 which adds r1 to r2 after r2 has been shifted left 16 times. The 16-bit left shift moves the lower-order 16-bits of r2 into the upper-order 16-bits and clears the lower-order 16 bits. We have achieved two things. We've preserved the old lower-order half of r2, and we've cleared the new lower-order 16 bits ready for inserting the bytes from r0 and r1.

The addition part of ADD r2,r1,r2, LSL #16 adds in r1 to create the situation shown in stage (b) of Figure 3.43. The next instruction, ADD r2,r2,r0, LSL #8, inserts the

FIGURE 3.43   Example of byte manipulation

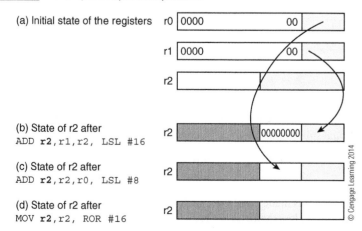

(a) Initial state of the registers

(b) State of r2 after
ADD r2,r1,r2, LSL #16

(c) State of r2 after
ADD r2,r2,r0, LSL #8

(d) State of r2 after
MOV r2,r2, ROR #16

low-order byte of r0 into bits 8 to 15 of r2 because r0 is first shifted left by 8 bits. Since zeros are shifted into r0, this operation doesn't affect bits 0 to 7 of r2 (see Figure 3.44). At this stage we have inserted the bytes from r0 and r1 into r2.

If we wish to swap the upper- and lower-order 16 bits of r2, we can execute MOV **r2**,r2,ROR #16. This operation copies r2 to itself after performing a 16-bit rotation. Since the rotate operation is non-destructive, we lose no information, and the bits shifted out of the high-order 16 bits are shifted into the low-order 16 bits of r2. The final result in Figure 3.43d has the bytes from r0 and r1 concatenated in the upper-order 16 bits of r2 and the lower-order 16 bits of r2 remain unchanged. Figure 3.44 shows the output of the simulator when this fragment of code is executed.

FIGURE 3.44   Manipulating bytes

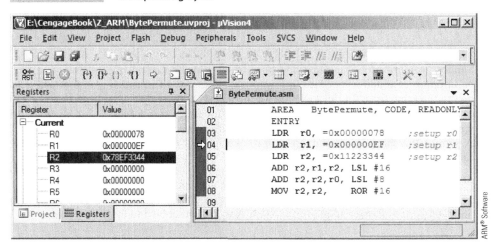

3.9.3   Byte Reversal

This is another example of data manipulation that you might use to store *big-endian* data in *little-endian* format (*Endianism* is discussed in Section 3.11.1 and concerns the arrangement of bytes in memory—the big endian format stores a word with the most-significant bytes at the lowest memory address, and the little endian format stores a word with the least-significant byte at the lowest address). Suppose the data we wish to re-order, 0xABCDEFGH, is in r0, and r1 is a working register. The following code (taken from ARM literature) implements this operation, which generates the new sequence 0xGHEFCDAB (i.e., the bytes have been reversed but not the nibbles in the bytes). The comment fields for each of these operations show what's happening to the data.

```
EOR    r1,r0,r0, ROR #16        ;A⊕E, B⊕F, C⊕G, D⊕H, E⊕A, F⊕B, G⊕C,
                                  H⊕D
BIC    r1,r1,      #0x00FF0000   ;A⊕E, B⊕F, 0, 0, E⊕A, F⊕B, G⊕C, H⊕D
MOV    r0,r0,ROR #8             ;G,H,A,B,C,D,E,F
EOR    r0,r0,r1, LSR #8         ;r1 after LSR #8 is 0,0, A⊕E, B⊕F, 0, 0,
                                  E⊕A, F⊕B
                                 ;G,H,A ⊕A⊕E, B⊕B⊕F, C,D,E⊕
                                  E⊕A,F⊕F⊕B
                                 ;G,H,E,F,C,D,A,B
```

Note that the expression A ⊕ A ⊕ E is equivalent to E, because A ⊕ A is 0, and 0 ⊕ E = E.

3.9.4   Multiplication by $2^n - 1$ or $2^n + 1$

The ARM's ability to shift an operand before using it in an addition or subtraction provides a convenient way to multiply by $2^n - 1$ or $2^n + 1$. Consider the following fragment of code that exploits both this feature and conditional execution.

```
;IF x > y THEN p = (2^n + 1)q
;          ELSE IF (x = y)  p = 2^n·q
;                    ELSE p = (2^n - 1)·q

        CMP     r2,r3                   ;Compare x and y
        ADDGT   r4,r1,r1, LSL #n        ;IF > calculate p = q · (2^n + 1)
        MOVEQ   r1,r1,    LSL #n        ;IF = calculate p = q · 2^n
        RSBLT   r4,r1,r1, LSL #n        ;IF < calculate p = q · (2^n - 1)
```

3.9.5   The Use of Multiple Conditions

Suppose you are processing text and have to examine commands that might sometimes be in upper-case and sometimes in lower-case. One way of proceeding is to convert all text to the same form. In this example we'll convert to lower-case text. Bit 5 of an ASCII character is zero for upper-case letters, and one for lower-case letters. It is easy to detect upper-case letters, because they are contiguous, beginning with 'A' and ending with 'Z'. Assuming the character to convert is in r0 and the remaining bits of r0 are all clear, we can write

```
        CMP     r0,#'A'                 ;Are we in the range of capitals?
        RSBGES  r1,r0,#'Z'              ;Check less than Z if greater than A. Update flags
        ORRGE   r0,r0,#0x0020           ;If A to Z then set bit 5 to force lower-case
```

The first instruction checks whether the character is 'A' or greater. If it is, the second line checks that the character is less than 'Z'. Note that this test is performed only if the character in r0 is greater than 'A' and that we are using reverse subtraction because we wish to test whether 'Z' − char is positive. The mnemonic is *"if greater than or equal to then reverse subtract and update the status bits on the result."* Finally, if we are in range, the conditional OR instruction is executed and an upper- to lower-case conversion is performed.

3.9.6   With Just One Instruction…

I came across a rather unusual application of an ARM instruction that both demonstrates the power of the ARM instruction set and the ability to write code whose immediate meaning is rather less than clear. Consider the operation BIC r0,r0,r0, ASR #31.

What does this do? The BIC forms the logical AND between the first operand and the logical complement of the second operand. In this case *both* the first operand and the second operand are specified as register r0. The ASR #31 shifts the second operand right 31 places using an arithmetic shift. The effect of an arithmetic shift right is to propagate the sign-bit; in this case, after 31 shifts the second operand will consist only of 32 copies of the sign bit. If the number was positive, the second operand will be 0. If the number was negative, the second operand will be 111, … , 11.

Since BIC complements the second operand, if r0 initially contains a positive number the AND will be carried out between the value in r0 and the compliment of 000, … , 00, which is r0. If r0 contains a negative number, the AND will be between r0 and 0000, … , 00, which is zero. That is, this operation implements:

```
        If (x < 0)   x = 0;
```

3.9.7 Implementing Multiple Selection

Consider the switch statement in a high level language. For example,

```
switch (i) {
    case 0:  do action;   break;
    case 1:  do action 1; break;
    .

    .
    case n:  do action n; break;
    default: exception
}
```

We can exploit the ARM's program counter relative addressing mode to implement this function. Register r0 contains the selector *i* (i.e., the case number).

```
        ADR    r1, Case            ;load r1 with the address of the jump table
        CMP    r0,#maxCase         ;better see if the switch variable is in range
        ADDLE  pc,r1,r0, LSL #2    ;if OK then jump to the appropriate case
                                   ;default exception handler here

        .

        .
Case  B    case0                  ;from the case table jump to the actual code
      B    case1
      B    casen
```

The key to this is the instruction ADDLE **pc**,r1,r0, LSL #2 which is a computer goto statement (i.e., it modifies the value of the PC and therefore changes the next instruction). This instruction is conditionally executed so that if the switch variable in r0 (tested by the previous statement) is out of range, the jump statement is not executed and the code for the default case is executed. If the case variable is in range, it has to be multiplied by 4 because each jump address in the case table is 4 bytes. Fortunately, we don't have to multiply the case number in r0 by 4 because we can shift r0, the second operand, two places left. The instruction loads the program counter with the value in r1 (i.e., the address of the case table) plus the case offset, which results in a jump to that location. At this location in the case table, we place a branch instruction whose target is the actual code that implements that particular case.

3.9.8 Simple Bit-Level Logical Operations

Suppose we have a 4-bit code, p, q, r, s, (xxxxxxxxxxxxxxxxxxxxxxxxxxxxpqrs$_2$) in the least-significant bits of a register and we wish to implement the algorithm

$$\text{if } ((p == 1) \ \&\& \ (r == 1)) \ s = 1;$$

If a word containing bits p, q, r, and s is in r0 and we use r1 as a working register, we can write

```
    ANDS   r1,r0,#0x8         ;clear all bits in r1 and copy p from r0
    ANDNES r1,r0,#0x2         ;if p = 1 clear all bits in r1 except the r bit
    ORRNE  r0,r0,#1           ;if r = 1 then s = 1
```

Note how we have used the ARM's compound or nested condition testing because the third line is executed only if both of the previous conditions are true. Note also that the execution of the second instruction, ANDNES **r1**,r0,#0x2, is dependent on the outcome of the first instruction and that the S bit is set to make the third instruction dependent on the outcome of this instruction.

3.9.9 Hexadecimal Character Conversion

We sometimes have to carry out hexadecimal character conversion to translate between a 4-bit binary value and the appropriate ASCII character. That is, we need to convert a value in the range 0000_2 (0_{16}) to 1111_2 (F_{16}) into the corresponding codes for "0" to "F". The ASCII codes for the digits "0" to "9" are 30_{16} to 39_{16}, respectively, and the ASCII codes for "A" to "F" are \$41 to \$46, respectively. The following algorithm converts the numbers in the range 0 to 9 to ASCII by adding 30_{16} and then deals with values in the range 10 to 15 by adding an additional 7.

```
character = hexValue + $30
if (character > $39) character = character + 7

        ADD     r0,r0,#0x30     ;add 0x30 to convert 0 to 9 to ASCII
        CMP     r0,#0x39        ;check for A to F hex values
        ADDGE   r0,r0,#7        ;if A to F then add 7 to get the ASCII
```

3.9.10   Character Output in Hexadecimal

When writing assembly language programs, you often wish to output the contents of a register on the console in hexadecimal form. In order to do this, you have to use the algorithm we described above repetitively. The following subroutine prints the contents of register r1 on the console in hexadecimal form using an operating system call to perform the printing. The Keil simulator does not support console I/O.

```
        MOV     r2,#8               ; REPEAT (8 times with r2 as loop counter)
NxtDig  MOV     r0,r1, LSR #28      ;   get 4 bits
        ADD     r0,r0,#0x30         ;   convert this nibble to a character
        CMP     r0,#0x39
        ADDGE   r0,r0,#7
        SVC     0                   ;   call O/S to print character
        MOV     r1,r1, LSL #4       ;   move the bits one nibble left
        SUBS    r2,r2, #1           ;   decrement the loop counter
        BNE     NxtDig              ; Until all 8 nibbles printed
```

This uses an operating system call SVC 0 to perform the printing operation – this instruction provides a mechanism by which the ARM simulator can interact with the user. The SVC takes a numeric parameter (in this case 0) that can be used by the software interrupt handler. The following example also uses this technique.

3.9.11   To Print a Banner

The following subroutine prints a banner (an ASCII-encoded string) that is terminated by the null byte 00_{16}. In this case the dummy string is followed by a line feed and carriage return to move the cursor to the next line. This code uses the software interrupt, SVC, to call an operating system function to do the printing. Note the use of the two consecutive tests for the terminator. This is a feature of the ARM development system and not of the ARM architecture itself. When the character is picked up, SVCNE WriteC, displays it if it was not a zero. The next instruction, BNE Bnner1, tests the same condition to decide whether to branch. We haven't discussed the LDRB instruction yet. This loads a byte into a register. Note that the post increment is one byte rather than 4.

```
Banner  ADR     r1,String       ;r1 points at the string to be printed (note use of ADR)
Bnner1  LDRB    r0,[r1], #1     ;pick up a character and increment the pointer
        CMP     r0,#0           ;was the character a zero (the terminator)?
        SVCNE   WriteC          ;if it wasn't then print it (using O/S SWI function)
        BNE     Bnner1          ;if it wasn't then jump back after printing
        MOV     pc,r14          ;if it was a zero then return

String =        "This is a test",&0A,&0D,0
WriteC EQU      0               ;the SVC code for print a character is 0
```

Accessing ARM Control Registers

ARM's control register contains status and system information bits, and both the user and operating system need to access them. Two instructions access the CPSR status register: MRS **rd**, CPSR copies the CPSR to register rd, and the inverse operation MSR **CPSR**, rs copies the contents of register rs, or a literal, to the CPSR.

If ARM is operating in its user mode, MSR **CPSR**, rs cannot be used to copy a value to CPSR that would change the status bits in CPSR. In the user mode, only the condition code flags Z, N, C, and V can be changed. You can also use the forms MRS **rd**, SPSR and MSR **SPSR**, rs to access the SPSR (saved processor status registers).

Suppose you wish to clear the carry bit C in the CPSR, for example, on return from a subroutine to indicate whether or not an error occurred. The following ARM code implements this.

```
ExitOK MRS   r0,CPSR           ; read the CPSR into r0
       BIC   r0,r0, #0x2000    ; clear bit 29, the carry bit
       MSR   CPSR,r0           ; put it back in the CPSR
       MOV   r15,r14           ; return (indicating no error) by restoring the
                                 saved PC
```

3.10   Subroutines and the Stack

We now demonstrate how the stack supports subroutine call and return mechanisms and introduce ARM's stack-handling instructions. The stack is a data structure, a *last in first out queue*, LIFO, in which items enter at one end and leave in the reverse order. Stacks in microprocessors are implemented by using a *stack pointer* to point to the top of the stack in memory. As items are added to the stack (pushed), the stack pointer is moved up, and as items are removed from the stack (pulled or popped) the stack pointer is moved down.

A Stack and Its Ups and Downs

A confusing aspect of stacks is the meaning of *up*, *down* and *growth*. A computer stack is analogous to the stack in everyday life. If you have a pile of magazines and put one on the top of the stack, it grows. The same is true of computer memory; pushing a new item to the stack causes it to grow.

I regard the top of a sheet of paper as *up* and the bottom as *down*. When I draw a computer stack, I show it growing up towards the top of the paper. However, because we generally number lines from the top of the paper down, we speak of the stack growing even though the stack grows upward towards *lower* addresses. For example, if the top of the stack is at address 0x001234 and we push a return address on the stack, the stack grows *up* to address 0x001230.

Some writers reverse the convention and put high addresses at the top of the page and the stack grows down toward low addresses at the bottom of the page! So, whether a stack grows up or down depends on how you write it on the page.

(continued)

Now, it gets worse. Stacks can be arranged to grow either toward high addresses or toward low *addresses*. This depends on the way in which the stack is implemented by the computer. When someone says that the stack moves up, are they speaking of the direction on the paper or whether the stack address is increasing numerically? It's difficult to tell. Consequently, you have to be aware of these issues when reading about stacks and when implementing them.

Figure 3.45 demonstrates four ways of constructing a stack. The two design decisions you have to make when implementing a stack are whether the stack grows *up toward low* memory as items are pushed (Figures 3.45a and b) or whether the stack grows *down toward high* memory as items are pushed (Figures 3.45c and d). Note that the term TOS means *top of stack* and indicates the next item on the stack. This diagram shows the stack being used to store a return address after a subroutine call.

The other decision is whether the stack pointer points at the item currently at the top of the stack (Figures 3.45a and c) or whether it points to the first free location above the top of the stack (Figures 3.45b and d). The actual arrangement of any stack is not of importance; all that matters is that it behaves consistently. The practical issues in stack design are whether the addressing mode used to push data on the stack is autoincrementing or autodecrementing and whether the stack adjustment takes place before or after the data is pushed.

FIGURE 3.45   Possible stack structures

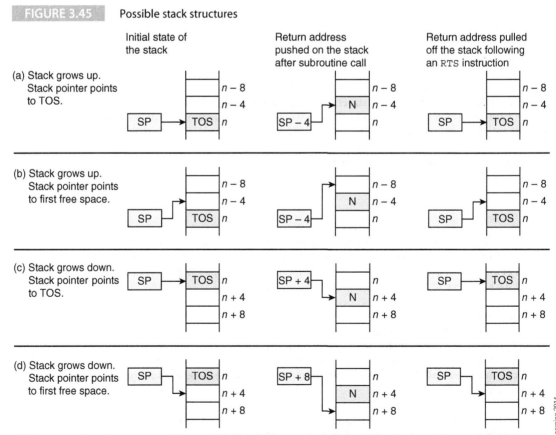

TOS = top of stack (the element at the top of the stack). By convention, up refers to the top of the page and lower addresses. This can be confusing. These diagrams assume byte-addressed memory and that each element on the stack is 32 bits (4 bytes).

Figure 3.46 illustrates a stack where the stack pointer points at the top item on the stack and, when an item is added to the stack (pushed), the stack pointer is decremented. When an item is moved from the stack, the item is removed at the location indicated by the stack pointer and the stack pointer incremented. By the way, such an operation is said to be *atomic* in the sense that it cannot be interrupted (i.e., the computer can't operate on the stack pointer, do something else and then access the stack). This is necessary to ensure the integrity of the stack. We return to this concept in Chapter 13. We can define the actions push and pop with relation to the stack pointer, SP, as

```
PUSH:    [SP]   ← [SP] – 4        ;move stack pointer up one word
         [[SP]] ← data            ;push data onto the stack

POP:     data   ← [[SP]]          ;pull data off the stack
         [SP]   ← [SP] + 4        ;move stack pointer up one word
```

Note that the stack pointer is decremented and incremented by four because we are following the convention that the memory is byte addressed and stack items are one word (four bytes) long.

FIGURE 3.46   Using the stack to save a return address

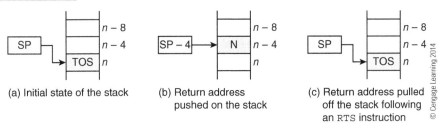

(a) Initial state of the stack

(b) Return address pushed on the stack

(c) Return address pulled off the stack following an RTS instruction

© Cengage Learning 2014

3.10.1   Subroutine Call and Return

A subroutine call can be implemented by pushing the return address on the stack and then jumping to the branch target address. Typically, this operation is implemented by JSR target or BSR target by CISC processors. Because the ARM does not implement this operation, you could use the following to synthesize this instruction.[30]

```
                     ;assume that the stack grows towards low addresses and the
                     ;SP points at the next item on the stack.
SUB   r13,r13,#4     ;pre-decrement the stack pointer
STR   r15, [r13];    ;push the return address on the stack
B     Target         ;jump to the target address
...                  ;return here
```

Once the code or body of the subroutine has been executed, a *return from subroutine* instruction, RTS, is executed, and the program counter restored to the point it was at after the BSR Proc_A instruction had been fetched. The effect of RTS instruction is

```
RTS: [PC]   ← [[SP]]        ;Copy the return address on the stack to the PC
     [SP]   ← [SP] + 4      ;Adjust the stack pointer
```

In Figure 3.46, the stack moves up by 4 because each address occupies four bytes. Because the ARM does not support a stack-based subroutine return mechanism, you would have to write:

```
LDR   r12,[r13],#+4    ; get saved PC and post-increment the stack pointer
SUB   r15,r12,#4       ;fix PC and load into r15 to return
```

---

[30] We will see later that the ARM has block move instructions that can efficiently transfer one or more registers to or from a stack.

This operation reads the return address at the top of the stack and increments the stack pointer to pull the saved PC off the stack and restore it. However, the saved PC has to be adjusted because it is pointing four bytes after the actual return (because of the ARM's internal pipelining). The PC is then loaded into r15 to force a return from subroutine. Figure 3.47 demonstrates the output of the ARM simulator after running this code. We have included the code window, the disassembly window, the register windows, and the memory window after the code has been executed.

If you look at the memory window, you will see that the value left on the stack at address 0x70 is 0x00000014 which is the saved PC. Since the saved PC is 4 on from the return address, the actual return address is 0x00000010. Now, if you look at the disassembled code, you will see that the instruction at 0x00000010 is MOV r0, #0xFF which is the correct return point.

In Practice

Although the above method of calling a subroutine would work, there is a better mechanism that uses ARM's block move instruction and which conforms to ARM's programming standard; that is;

```
STMIA sp! {r6,lr}      ;push r6 and link on the stack
  . . .                ;subroutine code here
LDMDB sp! {r6,pc}      ;pull saved r6 and get PC
                       ;we push the return address
                       ;and put it in the PC to return
```

The stack will be covered in greater detail in Chapter 4.

FIGURE 3.47   Demonstrating the ARM subroutine call and return

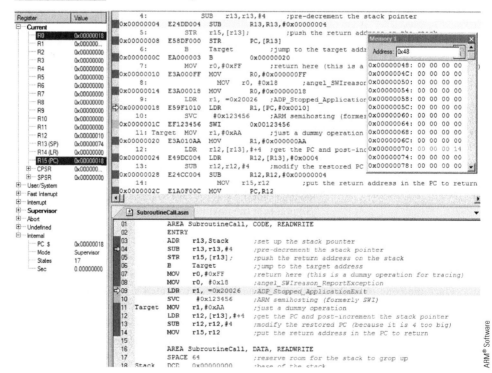

ARM® Software

3.10.2   Nested Subroutines

An important feature of the stack is its ability to support *nested subroutines* in which one subroutine calls another. Subroutines are nested when one subroutine fits entirely within another (that is, if you call a subroutine, you always return to the next instruction immediately after the calling point). It is considered poor programming practice to call a subroutine from one point and then return from that subroutine to an entirely different point.

FIGURE 3.48   An example of nested subroutines

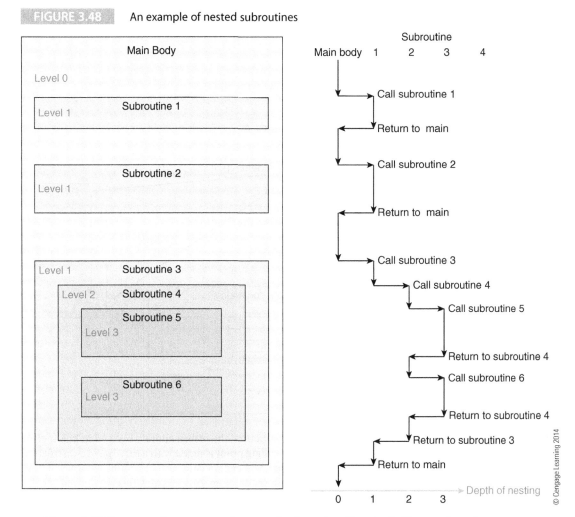

Figure 3.48 illustrates the concept of a nested subroutine. The shaded boxes show the way in which one subroutine is nested in another and lines to the right of the figure demonstrate the flow of control between the various subroutines in this example. This arrangement assumes a CISC processor with a stack-based subroutine call and return-mechanism.

Figure 3.49 demonstrates how a subroutine can be nested by using the stack to handle return addresses. The main program calls subroutine A and pushes the return address on the stack. During the execution of subroutine A, a second and nested subroutine B is called. The return address in subroutine A is pushed on the stack and a branch made to subroutine B. Figure 3.49 shows the state of the stack during the call and return sequence. As you can see, return addresses are pulled off the stack in the reverse order in which they were pushed.

We have not looked at the idea of parameter passing in any detail here—we are saving that for the next chapter. However, it is worth pointing out that the stack is used not only to manage subroutine return addresses but also parameters passed to and from a subroutine as well as local variables used by the subroutine.

3.10.3   Leaf Routines

ARM literature often refers to *leaf routines*. A leaf routine doesn't call another routine (that is, it's at the end of the tree). When programming a CISC processor with its general-purpose stack mechanism and subroutine call and return instructions, you don't have to worry about return addresses. RISC processors that have no direct stack support for subroutine calls and returns force the programmer to be aware of this subtlety.

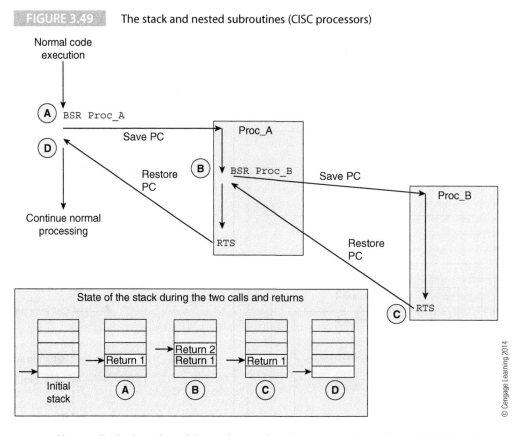

FIGURE 3.49   The stack and nested subroutines (CISC processors)

If you call a leaf routine with a BL instruction, the return address is saved in link register r14 rather than the stack. A return to the calling point is made with a MOV **pc**, lr instruction. However, if the routine is not a leaf routine, you cannot call another routine without first saving the link register. The following code fragment demonstrates how this is achieved.

```
        BL      XYZ                 ;call a simple leaf routine
        .
        .
        BL      XYZ1                ;call a routine that calls a nested routine
        .
        .
XYZ     . . .                       ;code (this is the leaf routine)
        .
        MOV     pc,lr               ;copy link register into PC and return
XYZ1    STMFD   sp!, {r0-r4,lr}     ;save working registers and link register
        .
        BL      XYZ                 ;call XZY – this overwrites the old link register
        .
        LDMFD sp!,{r0-r4,pc}        ;restore registers and force a return
```

Subroutine XYZ is a leaf subroutine that does not call a nested subroutine, and therefore, we don't have to worry about the link register, r14, and we can return by executing MOV **pc**, lr. However, subroutine XYZ1 contains a call to a nested subroutine, and we have to save the link register in order to return from XYZ1. The simplest way of saving the link register is to push it on the stack. In this case, we use a *store multiple registers* instruction and also save registers r0 to r4. When we return from XYZ1, we restore the registers and load the saved r14 (the return address in the link register) into the program counter. The STMFD and LDMFD are ARM instructions that push and pull, respectively, a block of registers on/off the stack. We examine these instructions later in this chapter.

3.11   Data Size and Arrangement

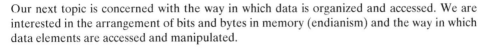

Our next topic is concerned with the way in which data is organized and accessed. We are interested in the arrangement of bits and bytes in memory (endianism) and the way in which data elements are accessed and manipulated.

3.11.1   Data Organization and Endianism

You would think that the way in which numbers are stored in memory is a trivial matter (that is, you just store them one after another). For example, a decimal computer could store the number 1984 as the sequence 1,9,8,4 or as 4,8,9,1. However, the numbering of bits and bytes can lead to incompatibilities between microprocessor families that store data in different ways. Let's begin by looking at byte numbering. Figure 3.50 shows how bytes in memory are numbered from 0 to $2^n - 1$.

Word numbering is universal, and all computers number the first word in memory *word* 0 and the last word $2^n - 1$. However, *bit* numbering can vary between processors. Figure 3.51a shows right-to-left numbering, similar to how we write numbers with the least-significant digit on the right. Most microprocessors (e.g., ARM, Intel, and Freescale) number the bits of a word from the least-significant bit (lsb), which is bit 0, to the most-significant bit (e.g., msb), which is bit $m - 1$, in the same way. Some microprocessors, such as the PowerPC, reverse this scheme, as illustrated in Figure 3.51b.

FIGURE 3.50   Numbering the bytes of a memory array

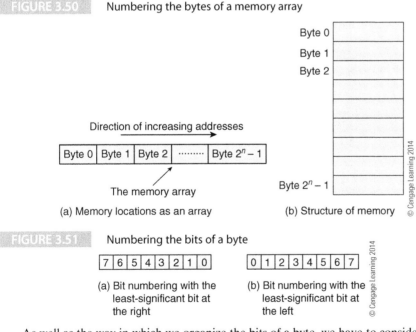

Direction of increasing addresses

| Byte 0 | Byte 1 | Byte 2 | ········· | Byte $2^n - 1$ |

The memory array

Byte 0

Byte 1

Byte 2

Byte $2^n - 1$

(a) Memory locations as an array

(b) Structure of memory

© Cengage Learning 2014

FIGURE 3.51   Numbering the bits of a byte

| 7 | 6 | 5 | 4 | 3 | 2 | 1 | 0 |

| 0 | 1 | 2 | 3 | 4 | 5 | 6 | 7 |

(a) Bit numbering with the least-significant bit at the right

(b) Bit numbering with the least-significant bit at the left

© Cengage Learning 2014

As well as the way in which we organize the bits of a byte, we have to consider the way in which we organize the individual bytes of a word. Figure 3.52 demonstrates that we can number the bytes of a word in two ways. We can either put the most-significant byte at the *highest byte address* of the word or we can put the most-significant byte at the *lowest address* in a word. The ordering is called *big endian*[31] if the most-significant element goes in at the lowest address and *little endian* if it goes in at the highest address.

---

[31] In Jonathan Swift's *Gulliver's Travels*, war breaks out between those opening an egg at its big versus little end, causing many deaths. See also the classic paper, "On Holy Wars and a Plea for Peace" [Cohen80].

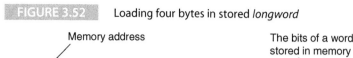

FIGURE 3.52   Loading four bytes in stored *longword*

Memory address

The bits of a word stored in memory

n	$n+1$	$n+2$	$n+3$
bits 24 – 31	bits 16 – 23	bits 8 – 15	bits 0 – 7

n	$n+1$	$n+2$	$n+3$
bits 0 – 7	bits 8 – 15	bits 16 – 23	bits 24 – 31

(a) The most-significant byte is stored at the lowest address (big endian)

(b) The least-significant byte is stored at the lowest address (little endian)

© Cengage Learning 2014

Intel's microprocessors store the most-significant byte at the highest address (a little endian arrangement), and Freescale's microprocessor's store the most-significant byte at the lowest address (a big endian arrangement). Although you could easily regard this issue as almost trivial and a matter of fashion, it does have important implications for both the programmer and the hardware designer. First, it can lead to programming errors and second it complicates interfacing Intel-based systems to Freescale-based systems. If you interface big endian and little endian processors (e.g., when they share a block of memory in a multiprocessor network) you have to ensure that bytes are transferred in the correct order either by software or by special-purpose interface hardware that rearranges the bytes as they are moved.

Figure 3.53 illustrates big endian and little endian *byte* organization in ARM memory. ARM is unusual in the sense that it can be configured either as a big endian or a little endian processor. When using the ARM simulator we have to select a mode – throughout this chapter, we have used big endian ordering. In Figure 3.53, we have generated some dummy ARM code complete with constants. The code and disassembly are the same for both modes. However, if you look at the code and data in memory as in Figures 3.53a and 3.53b, you will see that the ordering of bytes is different.

FIGURE 3.53   Big endian and little endian byte ordering

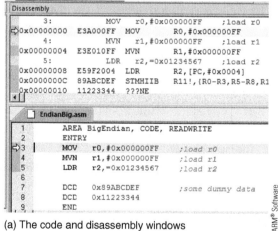

(a) The code and disassembly windows

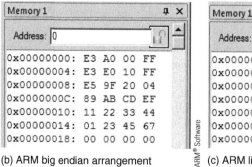

(b) ARM big endian arrangement

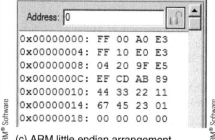

(c) ARM little endian arrangement

ARM® Software

3.11.2   Data Organization and the ARM

Let's look at the way in which ARM stores, manipulates, and accesses data in a little greater detail. As we know, ARM's memory is byte addressable and consecutive 32-bit word addresses differ by 4. Words *must* always be aligned on four-byte word boundaries. Whenever ARM fetches an *instruction* from memory, the lowest-order two bits of the address are always zero; that is, the instruction address is in the form xxx , … , xxx00$_2$, which preserves alignment. Sixteen-bit values (*halfwords*) are aligned on halfword boundaries. Figure 3.54 illustrates the ARM's big- and little-endian word and halfword alignment. The ARM endian configuration is set in hardware by hard-wiring the appropriate pin to a logical one or zero level.

FIGURE 3.54   ARM (a) little-endian and (b) big-endian modes

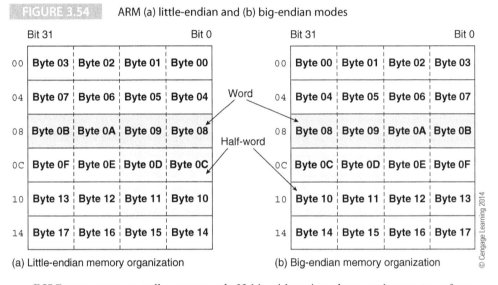

(a) Little-endian memory organization   (b) Big-endian memory organization

© Cengage Learning 2014

   RISC processors generally support only 32-bit arithmetic and you can't operate on fewer bits in a register. The only 8-bit or 16-bit support comes in the form of load and store operations and the ability to sign-extend integers that we introduce in the next section (see the panel on 16-bit operations).

   The ability of some processors to set condition codes on 8- or 16-bit operations is very nice. If you are using a 32-bit processor to do 16-bit arithmetic operations, you have to do your own checking for overflow and carry-out. For example, if r0 is used as a destination operand during a 16-bit operation on an unsigned value, you can check for a carry-out into bit 17 by executing, say, MOVS r_{temp},r0,LSR #16. This operation shifts r0 right by 16 places, so if bits 16 to 31 were all 0s, the result deposited in the temporary register would be zero, and the Z-bit would be set. If carry-out had occurred, the value in the temporary register would be non-zero and the Z-bit not set.

   Testing a signed 16-bit value for overflow is more complicated. Recall that arithmetic overflow occurs when the result of an arithmetic operation goes outside the range of values that can be represented. This happens when two positive values are added and the sign bit is 1, or when two negative values are added and the sign bit is 0. If we have a 16-bit value in a 32-bit register and we do 32-bit arithmetic operations, the ARM detects overflow from the state of bit 31. If we're interested in 16-bit arithmetic we have to detect overflow from the state of bit 15. Consider the following four examples of 16-bit addition using 32-bit arithmetic.

Case 1

Addition of two positive numbers
(no overflow)

```
 00000000000000000010100000000000
 00000000000000000100110000000000
+00000000000000000101010000000000
```

Case 2

Addition of two negative numbers
(no overflow)

```
 11111111111111111011100000000000
 11111111111111111101110000000000
+11111111111111111001010000000000
```

Case 3

Addition of two positive numbers (overflow)

```
 00000000000000000111100000000000
 00000000000000000101110000000000
+00000000000000001101010000000000
```

Case 4

Addition of two negative numbers (overflow)

```
 11111111111111111011100000000000
 11111111111111111001110000000000
+11111111111111110101010000000000
```

We can detect arithmetic overflow with the following code which sets the Z-bit if the number is in range.

```
MOVS    r1,r0, ASR #15
CMNNE   r1, #1
```

The MOVS r1,r0,ASR #15 operation shifts the result in r1 right fifteen places. The next operation CMNNE r1, #1 is executed if the previous result generated a Z-bit = 0. This operation performs a comparison between r1 and –1; that is, it tests r1 for the value 111 , … , 11 (all 1s). Consider the effect of this code on the four cases above.

Case 1

```
MOVS    r1,r0,ASR #15       ;r1 = 00000000000000000101010000000000, Z-bit is 1
CMNNE   r1, #1              ;Because Z-bit is 1, this instruction is not executed,
                            ;Z-bit stays at 1 to signal no overflow
```

Case 2

```
MOVS    r1,r0,ASR #15       ;r1 = 11111111111111111001010000000000, Z-bit is 0
CMNNE   r1,r1, #1           ;Z-bit 0, the instruction is executed
                            ;Z-bit is set to 1 to signal no overflow
```

Case 3

```
MOVS    r1,r0,ASR #15       ;r1 = 00000000000000001101010000000000, Z-bit is 0
CMNNE   r1,r1, #1           ;Z-bit 0, the instruction is executed
                            ;Z-bit stays at 0 to signal overflow
```

Case 4

```
MOVS    r1,r0,ASR #15       ;r1 = 11111111111111110101010000000000, Z-bit is 0
CMNNE   r1,r1, #1           ;Z-bit 0, the instruction is executed
                            ;Z-bit stays 0 to signal overflow
```

Life gets even more complex when you consider the issue of *alignment* (i.e., a word is aligned with its address at an integer multiple of its size) and *endism*. In principle, these issues are part of the fine detail of systems design, although in practice engineers can spend an awful amount of time dealing with such problems. ARM's own literature devotes quite a few pages to these issues; for example, suppose you have to load a 16-bit half word into a register and you don't know the alignment of the data when you write a program (i.e., you don't know whether the pointer will be odd or even when the program runs). The following fragment of code fetches the unsigned half word by performing two byte accesses and then concatenating the data. Register r2 points at the first byte and register r0 receives the half word (register r1 is used as a scratchpad).

```
LDRB    r0, [r2,#0]         ;get the byte pointed at by r2
LDRB    r1, [r2,#1]         ;get the second byte
ORR     r0,r0,r1, LSL #8    ;concatenate the two bytes
```

Note how the logical OR instruction takes the byte in r1 at the *higher address*, shifts it by 8 bits left, and then concatenates it in r0 with the byte from the lower address. Since the byte at the lower address forms the lower bits of the register, this is a *little-endian* move. The equivalent big-endian code is

```
LDRB   r0,[r2,#0]        ;get the byte pointed at by r2
LDRB   r1,[r2,#1]        ;get the second byte
ORR    r0,r1,r0, LSL #8  ;concatenate the two bytes
```

We have had to shift the byte from the lower address left before performing the concatenation. Loading a 32-bit word from an arbitrary alignment is more complex as the following code from ARM demonstrates. We assume that the address of the word is in r0; that the operating mode is little endian, and the result will be stored in r1. If the word was aligned on a word boundary, we could simply write LDR   r1,[r0]. However, we have to fetch *two* words from memory and then assemble the word from its components according to the actual alignment.

```
BIC    r2,r0,#3          ;get the address and mask to the alignment
                         bits 1 and 0.
LDMIA  r2,{r1,r3}        ;load r1 and r2 with the 64-bits containing the word.
AND    r2,r0,#3          ;now get the alignment (offset of the word) in bytes
MOVS   r2,r2,LSL #3      ;now get the alignment (offset) in bits
MOVNE  r1,r1,LSL r2      ;read the data from the lower 32 bits
RSBNE  r2,r2,#32         ;remaining offset 32 - offset
ORRNE  r1,r1,r3, LSL r2  ;get the top 32 bits and combine with lower bits
```

Figure 3.55 illustrates a misaligned little endian word that is loaded in memory at byte 5. The left-hand view shows word-aligned memory and the right hand view shows byte-aligned memory. The last two bits of each byte address are in bold to illustrate the effect of the above code that uses these bits to extract the word's alignment. This code first strips off any byte offset and obtains the word at that address and the word at the next word-aligned address. This guarantees that we have extracted the word from memory. The next step is to use the offset to shift the two words to get the required bytes in the correct places. Finally, we put the two words together by means of an OR operation. Note that the last three instructions are all conditional – if the offset was zero we have an aligned word and do not need to do anything more. We will walk through the code and demonstrate how it reads this byte from

FIGURE 3.55   Example of a misaligned word load

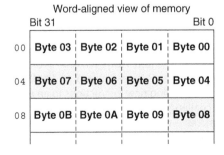

memory. Figure 3.56 demonstrates the execution of this code in the simulator. The panel extends the discussion of ARM's data arrangement by describing operations that access 16-bit values.

```
                              ;Initially r0 contains 5
    BIC    r2,r0,#3           ;r2 contains 000101 . NOT 111100 = 000100
    LDMIA  r2,{r1,r3}         ;load r1 and r3 with the 64-bits containing
                               the word.
                              ;r1 = b7 b6 b5 b4 , r3 = bB bA b9 b8
    AND    r2,r0,#3           ;r2 = 01
    MOVS   r2,r2,LSL #3       ;r2 = 0100
    MOVNE  r1,r1,LSR r2       ;r1 = 00 b7 b6 b5 (after shifting right 8 places)
    RSBNE  r2,r2,#32          ;r2 = 28
    ORRNE  r1,r1,r3, LSL r2   ;r3 shifted 28 places left is b8 00 00 00
                              ;r1 is 00 b7 b6 b5 OR b8 00 00 00 = b8 b7 b6 b5
```

FIGURE 3.56   Demonstrating an unaligned read operation using an ARM simulator

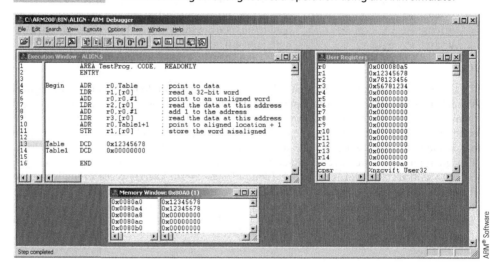

ARM® Software

ARM's Halfword Data Transfers

Like first-generation RISC processors, ARM is word-oriented. However, ARM's role is frequently in embedded systems where much of the data is byte-oriented, and its ISA was modified to support 16- and 8-bit data transfers. Adding new instructions to an existing instruction set presents the designer with a challenge. Indeed, ARM's literature is most candid when it states "... *half word data transfers ... are somewhat shoe-horned into the instruction space ...*" The following figure shows the structure of a half-word/signed byte instruction, which is similar to that of other memory reference instructions. You can appreciate how messy this instruction format is from the way in which the literal format has split the 8-bit value into two chunks in bits 0 to 3 and bits 8 to 11 of the opcode. These instructions lack the full functionality of ARM's existing data instructions, because the immediate offset is limited to eight bits and you can't use a scaled register offset.

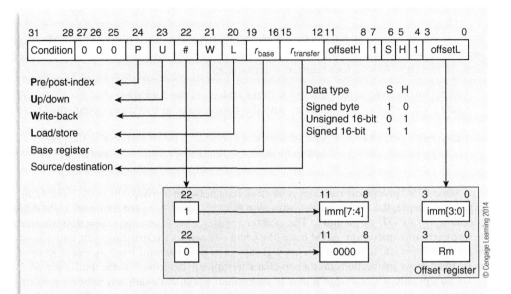

Load and store operations are handled differently. When a 16-or 8-bit value is loaded into a register, it is either zero- or sign-extended. Because memory is byte-addressed, data is stored in 1, 2, or 4 byte-wide locations, and therefore zero- or sign-extending is meaningless. The signed byte register load instruction loads a byte into a register and sign-extends the 8-bit value to 32 bits. Half-word versions allow you to load a 16-bit value into a register and to either sign-extend it or to zero-extend it. These instructions are modified versions of the existing LDR and STR instructions. Their assembly language forms are

LDRH	Load halfword (16 bits) into register
LDRSH	Load signed halfword (16 bits) into register
LDRSB	Load signed byte into register
STRH	Store halfword in memory
STRB	Store byte in memory

Like all ARM instructions, these instructions can be conditionally executed. For example, the mnemonic LDREQSH **r0**, [r1], #2 means *"if the Z-bit is set then load register r0 with the signed half-word at the address pointed at by r1, sign-extend this operand value to 32 bits, and then post-increment r1 by 2"*. ARM's literature demonstrates how you might use the signed 16-bit data load to transform an array of 16-bit signed values into an array of 32-bit signed values.

```
        ADR     r0,Source16      ;load address of source array
        ADR     r1,Dest32        ;load address of destination array
        ADR     r2,SourceEnd     ;2 bytes beyond end of source array
Loop    LDRSH   r3,[r0], #2      ;REPEAT read signed 16-bit half-word
        STR     r3,[r1], #4      ;    and store it at destination
        CMP     r0,r2            ;UNTIL all transferred
        BLT     Loop             ;
```

There's nothing surprising about this fragment of code. Instead of using a loop counter, this code stops when the last address is two bytes beyond the end of the first array. If the source array is 256 half-words at address 1000_{16}, the source array takes 512 bytes and extends to 1400_{16}, and the terminating value will be 1402_{16}. The additional 2 is required because of the post-indexing addressing mode.

3.11.3   Block Move Instructions

We now introduce one of ARM's least RISC-like instructions: its ability to copy a block of registers to or from memory with a single instruction. The following conventional ARM code demonstrates how to load four registers from memory.

```
ADR   r0,DataToGo    ;load r0 with the address of the data area
LDR   r1,[r0],#4     ;load r1 with the word pointed at by r0 and update the
                        pointer
LDR   r2,[r0],#4     ;load r2 with word pointed at by r0 and update the pointer
LDR   r3,[r0],#4     ;and so forth for the remaining registers r3 and r5...
LDR   r5,[r0],#4
```

Some CISC processors can copy a block of data between a group of registers and memory. For example, the 68K uses the instruction MOVEM {D0-D7},-(A7) to store eight data registers, D0 to D7, in memory. The pointer register, A7, is automatically decremented before each store operation. ARM has a *block move-to-memory* instruction, STM, and a *block move-from-memory*, LDM, that can copy groups of registers to and from memory. Both of these block move instructions take a two-character suffix to describe *how* the data is accessed.

Conceptually, a block move is easy to understand, because it's simply a '*copy the contents of these registers to memory*' or vice versa. In practice, it's complex, because ARM provides a full set of options that determine how the move takes place (for example, whether the registers are moved from high-to-low or low-to-high addresses or whether the memory pointer is updated before or after a transfer). Let's start by moving the contents of registers r1, r2, r3, and r5 into sequential memory locations with

```
STMIA r0!,{r1-r3, r5}   ;note the syntax of this and all block moves with the
                           register list in curly braces
```

This instruction copies registers r1 to r3 and r5 into sequential memory locations using r0 as a pointer with auto-indexing (indicated by the ! suffix). The suffix IA indicates that index register r0 is *incremented after* each transfer with the data transfer in order of increasing addresses. Although ARM's block mode instructions have several variations, *ARM always stores the lowest numbered register at the lowest address*, followed by the next lowest numbered register at the next higher address, and so on (e.g., r1 then r2, r3, and r5 in the preceding example). Figure 3.57 gives the ARM simulator output after we've loaded this instruction and executed it.

If you examine Figure 3.57, you can see we have tested the STMIA instruction by providing it with an environment: dummy values for the registers to be stored, a pointer register to the destination of the registers to be stored, and an initial value for the pointer. We used a SPACE directive to save or reserve 20 bytes of storage after the program for the data to grow upward. There are two blocks of data in the memory window: one at 0x38 and one at 0x4C. The second block at 0x4C is the constant pool set up by the LDR r1,=0x11111111 etc., pseudo instructions. This data is loaded before the program runs. The block of data at 0x38 contains the four registers stored in memory. If you look at r0 in the list of registers, you will see 0x48, which is the next free address in memory that confirms r0 is incremented after each memory store operation.

The next example implements a block transfer between a region of memory and multiple registers using the load multiple registers instruction. Note also the difference in assembler syntax between the LDR, STR pair and LDMIA, STMIA pair. The single load/store instructions put the register first and then the effective address, whereas the multiple load/store instructions put the pointer register first and then the list of registers to be stored. Moreover, the multiple load/store registers do not put the pointer register in square brackets.

```
LDMIA r0,{r3,r4,r5,r9}   ;load r3 from location pointed at by r0
                         ;load r4 from location pointed at by r0 + 4
                         ;load r5 from location pointed at by r0 + 8
                         ;load r9 from location pointed at by r0 + 12
```

FIGURE 3.57   Using the simulator to demonstrate a `STMIA` instruction

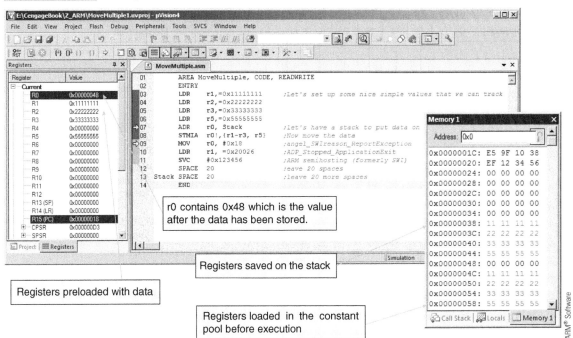

Note that this is *not* a stack operation, because the pointer register r0 is not updated after the data transfer. When using block moved in conjunction with a stack pointer, the stack pointer must be updated by appending a ! to the base register (i.e., stack pointer). Had we written STMIA r13!,**{r3,r4,r5,r9}**, the value of the system stack pointer r13 would be incremented by four as each register was moved. Figure 3.58 illustrates the encoding of the ARM processor's block move instructions.

Block Moves and Stack Operations

As we've said, ARM's block move instruction is versatile because it supports the four possible stack modes shown in Figure 3.45. The differences among these modes are the *direction* in which the stack grows (up or *ascending* and down or *descending*) and whether the stack pointer points to the item currently at the top of the stack or the next free item on the stack. CISC processors with hardware stack support generally provide only one fixed stack mode. The ARM's literature uses four terms to describe stacks:

1. FD   *full descending*   Figure 3.59a
2. FA   *full ascending*   Figure 3.59b
3. ED   *empty descending*   Figure 3.59c
4. EA   *empty ascending*   Figure 3.59d

It is important to note that ARM uses the terms ascending and descending to describe the growth of the stack toward higher or lowers addresses, respectively, and *not* whether it grows up or down on the page. A stack is described as *full* if the stack pointer points to the top element of the stack. If the stack pointer points to the next free element above the top of the stack, then the stack is called *empty*.

Table 3.5 associates ARM's block-mode transfer suffixes with each case in Figure 3.59. We assume, as usual, that low addresses are at the top of the page. For example, stack type 1 grows up towards low addresses, and the stack pointer points at the item at top-of-stack.

FIGURE 3.58   Encoding ARM's block move instructions

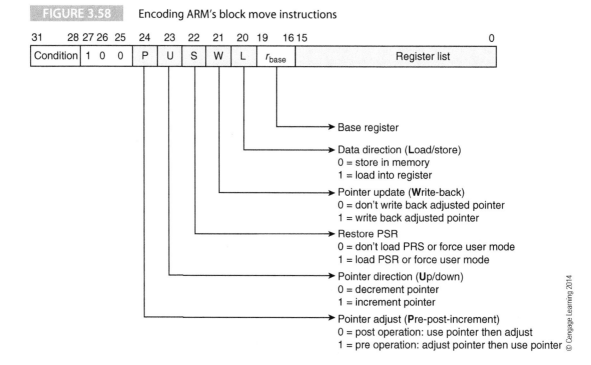

ARM has *two* ways of describing stacks, which can be a little confusing at first. A stack operation can be described either by *what* it does or *how* it does it. For example, if you decide to implement one of the most popular stacks that points at the top item on the stack and which grows towards lower addresses, it is a *full descending stack*, FD (the type used in this text). Consequently, we can write STMFD **sp!**, {r0,r1} when pushing r0 and r1 on the stack, and we can write LDMFD sp!, **{r0,r1}** when popping r0 and r1 off the stack. A full descending stack is *implemented* by first decrementing the pointer and then storing data at that address

TABLE 3.5   Stack Types and the ARM Block Move Instruction Suffixes

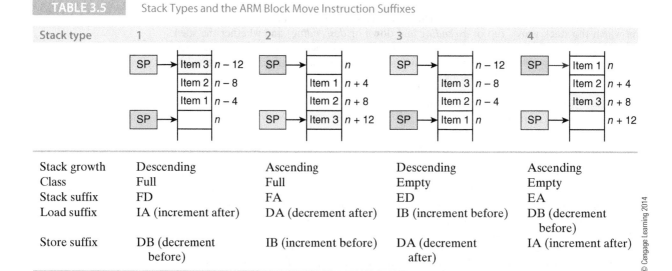

Stack type	1	2	3	4
Stack growth	Descending	Ascending	Descending	Ascending
Class	Full	Full	Empty	Empty
Stack suffix	FD	FA	ED	EA
Load suffix	IA (increment after)	DA (decrement after)	IB (increment before)	DB (decrement before)
Store suffix	DB (decrement before)	IB (increment before)	DA (decrement after)	IA (increment after)

FIGURE 3.59   ARM's four stack modes

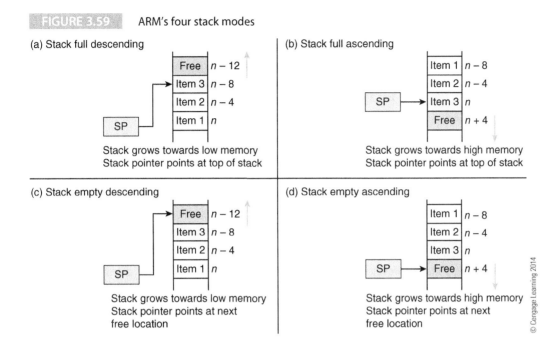

(a) Stack full descending

Free	$n-12$
Item 3	$n-8$
Item 2	$n-4$
Item 1	n

Stack grows towards low memory
Stack pointer points at top of stack

(b) Stack full ascending

Item 1	$n-8$
Item 2	$n-4$
Item 3	n
Free	$n+4$

Stack grows towards high memory
Stack pointer points at top of stack

(c) Stack empty descending

Free	$n-12$
Item 3	$n-8$
Item 2	$n-4$
Item 1	n

Stack grows towards low memory
Stack pointer points at next
free location

(d) Stack empty ascending

Item 1	$n-8$
Item 2	$n-4$
Item 3	n
Free	$n+4$

Stack grows towards high memory
Stack pointer points at next
free location

© Cengage Learning 2014

(push data) or by reading data at the stack address and then incrementing the pointer (pull data). We therefore can write STMDB **sp!**,{r0,r1}or LDMIA sp!, **{r0,r1}** instead of LDMFD/STMFD.

Applications of Block Move Instructions

One of the most important applications of the ARM's block move instructions is in saving registers on entering a subroutine and restoring registers before returning from a subroutine. Consider the following ARM code:

```
        BL      test                ;call subroutine test, save return address in r14
        .
test    STMFD   r13!,{r0-r4,r10}    ;subroutine test, save six working registers

        .
                body of code

        .
        LDMFD   r13!,{r0-r4,r10}    ;subroutine completes, restore the registers
        MOV     pc,r14              ;copy the return address in r14 to the PC
```

We can reduce the size of this code because the instruction MOV **pc**,r14 is redundant. Why? Because if you are using a block move to restore registers from the stack, you can also include the program counter. We can now write:

```
test    STMFD r13!,{r0-r4,r10, r14}   ;save the working registers and return address
                                            in r14

        :
        LDMFD  r13!,{r0-r4,r10,r15}   ;restore working registers and put r14 in the PC
```

At the beginning of the subroutine, we push the link register r14 containing the return address onto the stack, and then at the end we pull the saved registers, including the value of the return address, which is placed in the PC, to effect the return. Nice.

The block move provides a convenient means of copying data between memory regions. In the next example, we copy 256 words from **table 1** to **table 2**. The block move instruction allows us to move eight registers at once, as the following code illustrates.

```
         ADR     r0,table1       ; r0 points to source (note the pseudo-op ADR)
         ADR     r1,table2       ; r1 points to the destination
         MOV     r2,#32          ; 32 blocks of 8 = 256 words to move
Loop     LDRFD   r0!,{r3-r10}    ; REPEAT Load 8 registers in r3 to r10
         STRFD   r1!,{r3-r10}    ;        store the registers at their destination
         SUBS    r2,r2,#1        ;        decrement loop counter
         BNE     Loop            ; UNTIL all 32 blocks of 8 registers moved
```

3.12   Consolidation—Putting Things Together

We now provide an extended example of an ARM program in assembly language. When developing assembly language programs, you often need a simple means of getting data into and out of the program for testing purposes. This statement is especially true if you are running the code in a development environment such as a PC. The ARM assembler solves this problem with the help of the SWI (software interrupt) instruction. Because SWI has been renamed SVC, we will use SVC from now on.

We deal with interrupts in a later chapter—here we provide a short overview in order to help you understand the ARM's software interrupt instruction that is used to call an operating facility. Executing an SVC saves the return address in r14_svc, saves the CPSR in SPSR_svc, enters the supervisor mode, disables interrupt requests, and forces a jump to memory location 08_{16}. The interrupt handler must ensure that the program counter and condition codes are restored correctly at the end of exception handling. Registers r13 and r14 are banked so that when an exception occurs the new registers r13 and r14 are switched in. ARM literature refers to the new r14 in the SVC mode as r14_svc. Similarly, the saved processor status register is stored in SPSR_svc.

The SVC instruction takes a 24-bit immediate operand. This operand is not used by the instruction itself, but the exception handler may access it to determine, say, the type of software interrupt. That is, when a software interrupt instruction is encountered, the SVC handler code may read the binary pattern of the SVC that caused the exception to determine the nature of the software interrupt.

ARM's assembler (in conjunction with the ARM development system) defines some software interrupts that can be used by the programmer to develop software. For example, the code 0 is used to display the ASCII-encoded character in register r0 in the console. All you have to write is the following code.

```
         SVC     0            ;print the character in r0
```

In practice, you are more likely to write the rather more clear code:

```
WriteC   EQU     0            ;code for "print a character on console" function
         .
         .
         SVC     WriteC       ;call the operating system to print the character in r0
```

We should stress that these SVC functions are a feature of a specific development system and are not part of the ARM's instruction set architecture. These functions do not work on the Kiel development system, which is intended for use with embedded systems (i.e., not having a PC style keyboard and display).

Four-Function Calculator Program

Suppose we wish to write a skeleton program to perform the action of a simple four-function calculator. The user enters an expression of the form 123 + 4567 = and the program prints the result. Note that we will assume all values are positive and that all inputs and the result are within the ARM's 32-bit range.

Since the program has to deal with a variable number of digits, we can read in decimal digits and stop accumulating them when we reach either an operator + | - | * | / | or an equals. The first level pseudocode for this problem is

```
Get first number and terminator
Save number as operand 1 and save terminator as operator
Get second number and terminator
Switch (operator)
{  Case of +: do addition
   Case of -: do subtraction
   Case of *: do multiplication
   Case of /: do division }
Output the result
{  While valid digit
       divide result by 10
       stack remainder
    endWhile }
Print the stacked digits
```

The only complex part of this code is the output mechanism. If the result is 1234, dividing by 10 yields a remainder 4. We push that on the stack. Dividing by 10 again yields 3, which we also push on the stack. Eventually, the stack contains 1, 2, 3, and 4. We can now print the numbers in the correct order 1, 2, 3, 4. The following is the source code developed from the above algorithm.

```
        AREA ARMtest, CODE, READONLY

WriteC  EQU    &0                       ;OS code to write a character to console
ReadC   EQU    &4                       ;OS code to read a character from the
                                         console
Exit    EQU    $11                      ;OS code to exit

        ENTRY

calc    MOV    r13,#0xA000              ;initialize the stack pointer
        BL     NewLn
        BL     input                    ;get first number and terminator
        MOV    r2,r0                    ;save terminator (i.e., operator)
        MOV    r3,r1                    ;save first number
        BL     NewLn
        BL     input                    ;get second number and terminator
        MOV    r4,r0                    ;save terminator
        BL     NewLn
        BL     math                     ;do the calculation
        CMP    r4,#'h'
        BLEQ   outHex
        BLNE   outDec                   ;display the number
        BL     NewLn
        BL     getCh
        CMP    r0,#'y'
        BL     NewLn
        BEQ    calc
        SVC    Exit                     ;end

input                                   ;read string of digits and accumulate
                                         total in r1
                                        ;return with non-valid digit terminator
                                         in r0
```

```
          MOV    r0,#0                        ;clear input register
          MOV    r1,#0                        ;clear accumulated total
next      STR    r14,[sp,#-4]!                ;save link register on the stack
          BL     getCh                        ;get a character in r0
          LDR    r14,[sp],#4
          CMP    r0,#'0'                       ;test for digit in the range 0 to 9
          MOVLT  PC,r14                        ;exit on less than '9'
          CMP    r0,#'9'                       ;is the digit above '9'?
          MOVGT  pc,r14                        ;if it is, then exit
          SUB    r0,r0,#0x30                   ;else convert ASCII char to digit
          MOV    r4,r1                         ;need to fix MUL limitation
          MOV    r5,#10                        ;MUL can't use a literal
          MUL    r1,r4,r5                      ;multiply previous total by 10
          ADD    r1,r1,r0                      ;and add in new digit
          B      next                          ;continue

getCh     SVC    ReadC                         ;char input
          MOV    pc,r14                        ;return

putCh     SVC    WriteC                        ;char print
          MOV    pc,r14                        ;return

math      CMP    r2,#'+'                       ;Here we check the operator
          ADDEQ  r1,r1,r3
          CMP    r2,#'-'
          SUBEQ  r1,r3,r1
          CMP    r2,#'*'
          MOVEQ  r4,r1                         ;fix MUL
          MULEQ  r1,r4,r3
          MOV    pc,r14

outHex                                        ;print the result in r1 in hex format
          STMFD  r13!,{r0,r1,r8,r14}
          MOV    r8,#8
outNxt    MOV    r1,r1,ROR #28                 ;get ms nibble in ls position
          AND    r0,r1,#0xF                    ;get nibble to print in r0
          ADD    r0,r0,#0x30                   ;convert hex to ASCII
          CMP    r0,#0x39
          ADDGT  r0,r0,#7
          STR    r14,[sp,#-4]!                 ;save link register on the stack
          BL     putCh                         ;print it
          LDR    r14,[sp],#4                   ;restore link register
          subs   r8,r8,#1
          bne    outNxt
          LDMFD  r13!,{r0,r1,r8,pc}

outDec                                        ;print the result in r1 in decimal form
          STMFD  r13!,{r0,r1,r2,r8,r14}        ;save working registers
          MOV    r8,#0
          MOV    r4,#0                         ;number of digits
outNxt    MOV    r8,r8, LSL #4
          ADD    r4,r4,#1                      ;count the digits
          BL     div10
          ADD    r8,r8,r2                      ;insert remainder (least significant digit)
          CMP    r1,#0                         ;if quotient zero then all done
          BNE    outNxt                        ;else deal with next digit
outNx1    AND    r0,r8,#0xF
```

```
            ADD     r0,r0,#0x30
            MOVS    r8,r8,LSR #4
            BL      putCh
            SUBS    r4,r4,#1              ;decrement counter
            BNE     outNx1               ;repeat until all printed
outEx       LDMFD   r13!,{r0,r1,r2,r8,pc}  ;restore registers and return

div10                                    ;divide r1 by 10
                                         ;return with quotient in r1, remainder in r2
            SUB     r2,r1, #10           ;
            SUB     r1,r1,r1, LSR #2
            ADD     r1,r1,r1, LSR #4
            ADD     r1,r1,r1, LSR #8
            ADD     r1,r1,r1, LSR #16
            MOV     r1,r1,     LSR #3
            ADD     r3,r1,r1, ASL #2
            SUBS    r2,r2,r3, ASL #1
            ADDPL   r1,r1,#1
            ADDMI   r2,r2,#10
            MOV     pc,r14

NewLn                                    ;newline
            STMFD   r13!,{r0,r14}        ;stack registers
            MOV     r0,#0x0D             ;carriage return
            SVC     WriteC               ;char print
            MOV     r0,#0x0A             ;line feed
            SVC     WriteC               ;char print
            LDMFD   r13!,{r0,pc}         ;restore and return

            END
```

Summary

In this chapter, we have introduced the basic stored program concept and demonstrated the flow of information during the execution of machine-level instructions. We began with a simple hypothetical computer that was a very much cut-down version of the ARM that we employ as the main teaching machine in this text. We expanded the basic computer to demonstrate how both literal operands and branch instructions were handled. In Chapter 7, we demonstrate how such a machine can be implemented at the hardware level, and in Chapter 8, we show how modern computers increase their speed by overlapping the execution of instructions by pipelining them (that is, we can be fetching the next instruction while executing the current one).

Most of this chapter has been concerned with the ARM's architecture. The ARM is a machine that is delightfully easy to understand and yet is elegant enough to introduce many interesting aspects of instruction set design. The ARM incorporates most of the arithmetic and logical operations of conventional processors, but it also includes ingenious features such as the ability to shift one of its operands and predicated execution. Unlike other classic RISC processors, the ARM implements a very sophisticated stack mechanism with its block move instructions.

We have spent some time covering the ARM's assembly language. Although we could just have listed the ARM's instructions and defined its addressing modes, we demonstrated how you would actually employ some of these instructions, because all too often their use is not obvious.

Problems

3.1 Why is the program counter a *pointer* and not a *counter*?

3.2 Explain the function of the following registers in a CPU.
a. PC
b. MAR
c. MBR
d. IR

3.3 For each of the following 6-bit operations, calculate the values of the C, Z, V, and N flags.

a. $\begin{array}{r} 001011 \\ +001101 \\ \hline \end{array}$ b. $\begin{array}{r} 111111 \\ +000001 \\ \hline \end{array}$ c. $\begin{array}{r} 000000 \\ -111111 \\ \hline \end{array}$

d. $\begin{array}{r} 101101 \\ +011011 \\ \hline \end{array}$ e. $\begin{array}{r} 000000 \\ -000001 \\ \hline \end{array}$ f. $\begin{array}{r} 111110 \\ +111111 \\ \hline \end{array}$

3.4 The classic processors flags are C, N, V, and Z. Conditional branches may be made on these flags being true or false (branch on zero or not zero). Branches may be on combinations of flags (branch on greater than or equal to zero). Can you think of any other conditional branch conditions that could be included?

3.5 The ARM's r13 and r14 registers are *overlapped* (banked or windowed), and a separate register pair exists for each of the exception modes. What do we mean when we say these registers are *overlapped*? Explain why the ARM's designers have taken this approach.

3.6 ARM's literature often describes its assembly language instruction syntax in BNF notation. Suppose an instruction is described in BNF as having the syntax:

```
This|That{B}{S}{,P|Q}
```

Give examples of possible legal instructions using this format.

3.7 The ARM puts the program counter in register r15, making it visible to the programmer. Someone writing about the ARM stated that this *exposed the ARM's pipeline*. What did he mean, and why? *Note*: You may have to look at the section on pipelining to answer this.

3.8 What are the relative advantages and disadvantages of general-purpose registers compared to separate address and data registers?

3.9 What is a *misaligned operand*? Why are misaligned operands such a problem in programming?

3.10 Why does the ARM provide a reverse subtract instruction RSB r0,r1,r2 that implements [r0] = [r2] − [r1] when the normal subtraction instruction SUB r0,r2,r1 will do exactly the same job?

3.11 If r1 = 11110000111000101010000011111101 and r2 = 00000000111111110000111100001111, what is the value of r3 after executing BIC r3,r1,r2?

3.12 If r1 = 0FFF$_{16}$ and r2 = 4, what is the value of r3 after each of the following instructions has been executed (assume that each instruction uses the same data)?
a. MOV r3,r1, LSL r2
b. MOV r3,r1, LSR r2
c. MVN r3,r1, LSL r2
d. MVN r3,r1, LSR r2

3.13 If r1 = 00FF$_{16}$ and r2 = 4, what is the value of r0 after each of the following instructions has been executed (assume that each instruction uses the same data)?
a. ADD r0,r1,r1, LSL #2
b. ADD r0,r1,r1, LSL #4
c. ADD r0,r1,r1, ROR #4

3.14 What is the effect of the instruction MOV r0,r0, ASR #31?

3.15 Write an ARM assembly language routine to count the number of 1s in a 32-bit word in r0 and return the result in r1.

3.16 A word consists of the bytes b4, b3, b2, and b1. Write a function to re-order (transpose) these bytes in the order b1, b3, b2, and b4.

3.17 ARM instructions have a 12-bit literal. Instead of permitting a word in the range 0 to $2^{12} - 1$, the ARM uses an 8-bit format for the integer and a 4-bit alignment field that allows the integer to be shifted in steps of 2. What are the advantages and disadvantages of this mechanism in comparison with a straight 12-bit integer?

3.18 Write one or more ARM instructions that will clear bits 20 to 25 inclusive in register r0. All the other bits of r0 should remain unchanged.

3.19 This is a classic problem of assembly language programming. Write a sequence of ARM instructions that swap the contents of registers r0 and r1 without using any additional registers or memory storage (that is, you can't move r1 to a temporary location).

3.20 What is the difference between the TEQ and CMP and CMN comparison instructions?

3.21 What are the advantages and disadvantages of the use of the ARM's BL (branch and load) subroutine call mechanism in comparison with the conventional CISC BSR (branch to subroutine) mechanism?

3.22 Write ARM code to implement the following C operation.

```
int s = 0;
    for ( i = 0; i < 10; i++) {
         s = s + i*i;)
```

3.23 What is the effect of the following addressing mode?

```
STR r0,[r2,r3,ROR #3]!
```

3.24 What is the meaning of each of the P, U, B, W, and L bits in the encoding of an ARM memory reference instruction?

3.25 What is the binary encoding of the following instructions?
    a. STRB **r1**,[r2]
    b. LDR **r3**,[r4,r5]!
    c. LDR **r3**,[r4],r5
    d. LDR **r3**,[r4,#-6]!

3.26 What is the effect of LDR **r0**,[r5,r6, LSL r2]?

3.27 You go for a job as a computer architecture designer. At your interview, you tell the panel that you have some real neat ideas for some cool processors. For example, you have an instruction that can set an individual bit in memory. The format of your instruction is

```
BITSET r0,r1,[(r2),r3,r4 LSR r5, #n]
```

    This instruction sets a bit in memory located r0 plus r1 bits from the word address whose address is given by the contents of memory location r2, plus the contents of register r3, plus the contents of register r4, and shifted left by register r5 plus the constant *n*.
    You don't get the job. Why?

3.28 What effective address is generated by the instruction LDR **r0**,[r2,-r3, LSL #1]?

3.29 Some machines have a find-first-one instruction that counts the location of the first bit set to 1 within the word. Write an ARM sequence of instructions that takes the word in r0 and puts the locations of the first bit set to 1 in r1. Count from the left, so that if bit 31 is set, the value returned should be 0. If bit 0 is set, the value returned is 31. If no bit is set, the value returned should be 32.

3.30 What is the meaning of *sign-extension* in the context of copying data from one location to another?

3.31 Why is sign-extension an important issue when we use a LDR instruction to load a register from memory but of no importance when we use an STR to store a register in memory?

3.32 Assume that r2 contains the initial value 00001000_{16}. Explain the effect of each of the following six instructions, and give the value in r2 after each instruction executes.
    a. STR   r1,**[r2]**
    b. STR   r1,**[r2, #8]**
    c. STR   r1,**[r2, #8]**!
    d. STR   r1,**[r2] #8**
    e. STR   r1,**[r2, r0, LSL #8]**
    f. STMFD r2!,**{r1,r2}**

3.33 Most RISC processors do not include a block move instruction. What are the advantages and disadvantages of the ARM's LDM and STM instructions?

3.34 What is the effect of executing STMIB **r13!**, {r0-r2,r4}? Draw a picture of the state of the stack pointed at by r13 before and after this operation.

3.35 The two pairs of instructions LDMIA, STMDB and (LDMFD, STMFD) do exactly the same things. Why do these two pairs have different or *alternative* mnemonics? Why does the first pair have different suffixes IA and DB? Why does the second pair have the same suffix FD?

3.36 Without using the ARM's multiplication instruction, write one or more instructions (using ADD, SUB, and shifting) to multiply by the following integers.
    a. 33
    b. 1025
    c. 4095

3.37 A word consists of the bytes b4, b3, b2, and b1. Write a function to invert the bits of b3 and clear the bits of b2, leaving all other bits unchanged.

3.38 Write suitable ARM code to implement

```
if x = y call PQR else call ZXY
```

3.39 Write an ARM assembly language program that scans a string terminated by the null byte 0x00 and copies the string from a source location pointed at by r0 to a destination pointed at by r1.

3.40 Write a program to copy text from the location pointed at by r0 to the location pointed at by r1. The copied version must be in reverse order. Assume the string you are copying is non-null (it contains at least one character) and that it is terminated.

3.41 Write a program to reverse a character string with an odd number of characters pointed at by r0. Assume the string you are copying is non-null (it contains at least one character) and that it is terminated by 0x0D (the number of characters included the terminator). You must not use an additional memory buffer (i.e., you can use only registers and the string storage itself).

3.42 Repeat Problem 3.39, but when you transfer the string, remove any occurrences of the word "the." For example, "and the man said" would become "and man said."

3.43 Repeat Problem 3.39, but when you transfer the string, reverse the sequence of any words beginning with "t." For example, "and the man said" would become "and eht man said."

3.44 What does the following code do?

```
TEQ    r0,#0
RSBMI  r0,r0,#0
```

3.45 What is the meaning of the following mnemonics (and what do they do)?

   a. LDRSH

   b. RSBLES

   c. CMPS

   d. MRS   **r0**,CPSR

3.46 What is the meaning of *sign-extension* in the context of copying data from one location to another?

3.47 What is wrong with the following instruction?

   MLA **r0**,r0,r1,r2

3.48 What, in the contest of assembly language, is a *pseudo-operation*?

3.49 Explain the meaning of the following two ARM pseudo-operations. What do they do and why have they been implemented?

   a. ADR **r0**,table

   b. LDR = **r0**, 0x1234FEDC

3.50 Suppose you execute LDR, **r0**=0x12345678 on an ARM machine followed by STR r0, **[r1]** where r1 = 0x1000. Now, suppose you do a byte read to the same address with LDRB **r2**,[r1]. What would the value be in r2 if (a) the ARM was big-endian configured and (b) it was little-endian configured?

3.51 Write an ARM assembly language program to determine whether a string of characters with an odd length is a palindrome (for example, *mom*) under the following constraints.

   a. The string of ASIC-encoded characters is stored in memory.

   b. At the start of the program, register r1 contains the address of the first character in the string, and r2 contains the address of the last character. On exit from the program, register r0 contains a 0 if the string is not a palindrome, and 1 if it is.

3.52 How would you deal with the previous problem if the palindrome could have an odd or even length?

3.53 A singly linked list (Figure P3.53), whose first element is pointed at by r0, consists of elements whose head is a 32-bit address pointing to the next element in the list and a variable-length tail. The tail may be of any length greater than four bytes. The last element in the list points to the null address 0. Write a program to search the list for an element whose tail begins with the word in data register r1. On success, set r4 to 0xFF, and r0 should contain the address of the desired record. On failure, load r1 with 0xFFFFFFFF.

3.54 Explain what this fragment of code does instruction by instruction and what purpose it achieves (assuming that register r0 is the register of interest). Note that the data in r0 must not be 0 on entry.

```
       MOV     r1,#0
loop   MOVS    r0,r0,LSL #1
       ADDCC   r1,r1,#1
       BCC     loop
```

3.55 Consider the following Java constructs. Express each in ARM assembly language. Assume all variables are single-bit Booleans and are in registers r0 = A, r1 = B, r2 = C, and r3 = D. *Note*: The Java operators &, ¦, ! are AND, OR, and NOT, respectively. The operators && and ¦¦ are AND and OR operators that support short-circuit evaluation (that is, if the expression yields false AND or true OR further evaluation is halted).

   a. A = (B & C) ¦ (!D);

   b. A = (B && C) ¦¦ (!D);

3.56 The following is a loop expressed in ARM code. The code is wrong. Why?

```
       MOV r0,#10      ;Loop counter – ten
                        times round the loop
Next   ADD r1,r1,r0    ;add loop counter to
                        total
       SUB r0,r0,#1    ;decrement loop counter
       BEQ Next        ;continue until all done
```

3.57 We need to swap the following registers. Do this using *block moves*.

Before	After
r1	r3
r2	r4
r3	r5
r4	r6
r5	r7
r6	r1
r7	r2

3.58 Why is the assembly language structure (format) of ARM's block move instructions inconsistent with ARM's normal assembler conventions? If you were redesigning ARM's assembly language, how might you express these operations (remember, changing the assembly language does not change the

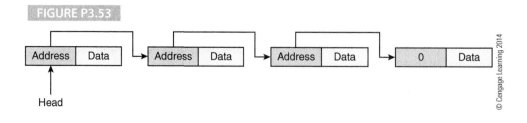

FIGURE P3.53

Head

© Cengage Learning 2014

architecture—it affects only programmer productivity and accuracy).

3.59 Write a function (subroutine) that inputs a data value in register r0 and returns value in r0. The function returns $y = a + bx + cx^2$, where a, b, and c are parameters built into the function (i.e., they are not passed to it). The subroutine also performs clipping. If the output is greater than a value d, it is constrained to d (clipped). The input in r0 is a positive binary value in the range 0 to 0xFF. Apart from r0, no other registers may be modified by this subroutine.

3.60 A computer has three eight-element vectors in memory, **Va**, **Vb**, and **Vc**. Each element of a vector is a 32-bit word. Write the code to calculate all elements of **Vc** if the ith element is given by

$$\textbf{Vc}_i = \tfrac{1}{2}\,(\textbf{Va}_i + \textbf{Vb}_i)$$

3.61 Register r15 is the program counter. You can use it with certain instructions such as a MOV (e.g., MOV pc,r14). However, r15 cannot be used in conjunction with most data processing instructions. Why?

Instruction Set Architectures–Breadth and Depth

"It's tough being at the top."
Anon

"Plus ca change, plus c'est la meme chose."
French proverb

"Don't reinvent the wheel, just realign it."
Anthony J. D'Angelo

"Bill Gates is a very rich man today. . . and do you want to know why? The answer is one word: versions."
Dave Barry

"I just invent, then wait until man comes around to needing what I've invented."
R. Buckminster Fuller

The previous chapter represents one of the principal *themes* of this book—the architecture of a microprocessor together with an introduction to its assembly language. In this chapter, we introduce some *variations*, and we also look at the ARM processor architecture in greater depth. You can think of this chapter as both broadening and deepening our understanding of ISAs and their applications. The purpose of this chapter is not to dazzle readers with a bewildering array of different processor architectures as if we were teaching the conjugation of irregular Spanish verbs. We are demonstrating the rich variety of instruction set architectures in both the ARM processor and traditional CISC processors. There are good practical reasons for doing this. In later life, students will be exposed to different types of processors: some may design engine controllers in automobiles, some may write device drivers for applications in workstations, and so on. It is necessary that students have some understanding of the variations possible in ISA design.

There is another reason for introducing some of the architectural features we cover here. In computing, the wheel is continually being re-invented and techniques that have fallen out of favor sometimes appear in later generations of processor. Although we could have expanded this chapter endlessly, we have to be *realistic* and cover topics that are both interesting and illustrative of ways of carrying out machine-level operations. In this chapter, we look at the following topics.

The Stack—The stack is probably the most fundamentally important data structure in computer science. We've already introduced the stack and explained how it can manage subroutine return addresses. Here, we go into greater detail and demonstrate how the stack is used to store a subroutine's local data (stack frames) and how parameters are passed to and from subroutines via the stack. We also demonstrate how a language like C uses the stack. This section is important, because real-world computers must provide effective support for high-level languages if they are to execute code efficiently. This section is an extension of the previous chapter and looks at the ARM architecture in greater depth.

Exceptions and Protected Modes—Computers invariably implement a means of interrupting the normal flow of execution when an external device requires attention or whenever certain types of error (exception) occur and the intervention of the operating system is required. Chapter 12 looks at the hardware details of interrupt handling. Here, we describe how the exception handing is incorporated in the ARM processor's instruction set architecture. We also introduce the notion of the *supervisor* mode and demonstrate how one CISC family implemented a supervisor or protected mode state for running operating system code.

MIPS—It is always instructive to see how others do things. MIPS is a venerable RISC processor that has a history approximately as long as the ARM processor. We briefly look at MIPS because it is remarkably similar to the ARM processor but has significant differences and tradeoffs. For example, the ARM processor has fewer general-purpose registers but supports conditional execution.

Data Manipulation, Packing, and Bit Fields—Because data movement is the most common computer operation, we look at some interesting examples of data manipulation, such as packing. The highpoint of the CISC architecture was reached when *bit field instructions* were introduced. (At least in my opinion!) I have to admit that I was quite fascinated when I first encountered the bit field instruction that had a massively complex instruction format and could perform operations on arbitrary strings of bits anywhere in memory. Bit field instructions treat the entire memory space as a contiguous sequence of bits, allowing you to take a slice (the bit-field) of between 1 and 32 bits from anywhere in memory and to perform certain operations on these bits. The bit field has applications in areas from data structures to graphics. Enthusiasm for bit field instructions declined at the height of the RISC revolution when it appeared to be more cost-effective to perform the operations in software using primitive machine operations than to provide hardware support for them.

In this section, we also take a look at the notion of bounds checking, which is the ability to determine whether an array element (or other value) is in or out of bounds.

Memory Indirect Addressing—We have already introduced register indirect addressing in which a register points to an operand in memory. In *memory indirect addressing*, the address of the operand is in a memory location. Indeed, one of the ISAs we examine supports combined register indirect and memory indirect addressing, which means that a *register points to a pointer in memory*. This addressing mode makes it remarkably easy to handle complex multidimensional data structures.

Compressed Code—An unusual feature of an ARM processor is its ability to operate in a 16-bit mode where it runs a compressed version of its instruction set. This mode is called the ARM processor's *Thumb state*, and it is used to reduce the cost of ARM processor-based embedded systems. We also look at other code compression techniques.

Because this chapter provides variations on the theme of instruction set architectures, it is not necessary to read all the material in order to understand later chapters. Some students may wish to skip some of the material here. However, the purpose of this chapter is to demonstrate both variations in architecture and some of the ways in which the underlying architecture supports high-level languages (e.g., parameter passing in C). If students have just a basic knowledge of an ISA, they will be missing much of the bigger picture and be less able to appreciate some of the limitations an ISA imposes on, for example, the compiler writer.

Although this text uses the ARM processor as an example of a computer's ISA, here we look at examples of other processors. In particular, we discuss features of Motorola's 68K family that has a particularly rich ISA—especially in terms of instructions such as bit-field operations, complex addressing modes, and the support for array bounds checking.

68K Registers

In this section, we refer to the 68K family, which is a classic CISC architecture introduced by Motorola in 1979 and later marketed by Freescale Semiconductor. The 68K has a general-purpose register set with sixteen 32-bit registers. Its architecture was influenced by DEC's PDP-11, which was the dominant minicomputer architecture of the time.

The 68K is unusual in that eight general-purpose registers A0 to A7 are used as *pointer registers* and eight registers D0 to D7 are used as *data registers*. These two sets of registers are essentially the same, but address registers treat their contents as a 32-bit pointer, whereas data registers treat their contents as arbitrary data (for example, you can't apply logical and shift operations to address registers). Moreover, operations on address registers do not automatically set the condition code flags—unlike operations on data registers. Address registers cannot take part in byte operations, as an address is considered indivisible.

Historical Background

Developments in computer architecture have always been influenced by factors such as architectural and technological innovation, the need to maintain backward compatibility with previous members of a family, the changing requirements of users, and fashions in design. In the 1970s and early 1980s, progress in *commodity* microprocessor architectures was driven by Intel and Motorola. By the mid-1980s, the RISC architectures developed at IBM, Stanford, and Berkeley seemed poised to kill off conventional complex instruction set architectures of the 68K and 80x86 families. The casual observer could have been forgiven for thinking that the conventional CISC, such as the Intel IA32 family, was nearing the end of its life.

All we need say here is that the so-called RISC revolution abandoned the complex instruction formats of the 68K and IA32 processors, threw away convoluted and infrequently used instructions and addressing modes, employed large register sets, and permitted only two memory-based operations: load and store. A key feature of RISC machines is the *overlapping* or *pipelining* of instruction execution. As soon as one instruction is read into the computer, the next instruction is fetched from memory while the current instruction is being decoded. Pipelining thrives on simple, regular instruction formats and doesn't go well with complex, variable-length CISC instruction formats. However, we should point out that Intel has done a remarkable job in taking its IA32 architecture and applying pipelining techniques to the underlying CISC ISA. Motorola also applied RISC techniques to its 68K line.

In the 1980s, the arguments in favor of RISC processors appeared to be overwhelming. However, pure RISC machines like MIPS and SPARC didn't sweep all other architectures away because the power of history was too strong. Too much effort had been invested in ISAs like the Intel IA32 family for people to throw everything away and start again, particularly when an operating system plus one or two software packages cost more than the average desktop PC today. Although Apple abandoned the 68K family in favor of the PowerPC RISC, Intel continued to develop its 80x86 family because of the enormous market provided by the IBM PC and its clones. Today, the IA32 architecture still dominates the market for PCs, and Apple has abandoned their PowerPC and followed the IA32 path.

CISC manufacturers took notice of the development of the RISC. They incorporated the best features of RISC technology into their CISC designs. For example, Motorola used pipelining in its 68040 and 68060 and ColdFire families and dropped some of the more complex instructions from its 68K architecture. In order to ensure backward compatibility, the 68060 detected 68020 opcodes and called operating system utilities to interpret them using less exotic machine-level instructions.

Intel did everything it could to make their IA32-based Pentium architectures as RISC-like as possible. Some of the companies that cloned IA32 architectures sidestepped the CISC-RISC debate by constructing processors with RISC cores and RISC instruction sets, but executed IA32 instructions by translating them into native RISC-code inside the chip.

By the late 1990s, the CISC architecture was back in fashion, experiencing a *rebirth* as Intel introduced its MMX instruction set and then its *streaming enhancements* to provide facilities required to exploit the new multimedia applications. Even pure RISC processors, such as the SPARC, introduced instruction set extensions to enhance the handling of sound and images. In fact, some RISC processors, such as the PowerPC, had instruction sets that exceeded the complexity of earlier complex instruction set processors. Today, there's no clear *meaningful* distinction between CISC and RISC processors. A better distinction is between load/store and memory-to-register processors. In this text, we have chosen a load/store processor—the ARM processor—as a teaching vehicle, because it is rather easy to learn its assembly language and it is widely used in a range of applications, such as in Apple's iPad.

We begin this chapter with a look at the *bit field* data type and some of the operations provided by processors to support bit fields. We start with the bit field simply because it represents a distinct change from the data elements we have encountered so far and because they are interesting. Moreover, although bit field support declined after the 68020, some contemporary processors are now adding bit field support.

4.1   The Stack and Data Storage

We now look at a topic of great importance in computing: the stack. In this section, we both increase our understanding of the ISA of ARM processors and take a quick look at how a CISC processor provides support for stack operations. Let's begin by looking at some background issues concerning data storage, procedures, and parameter passing. High-level language programmers make use of variables to represent data elements that we can think of as variables or as *abstract data cells*. The data cell is abstract because it may hold any type of data element defined by the programmer (e.g., a byte, an array, or a record). As far as the programmer is concerned, the abstract data cell has all of the properties of a real memory cell: It can be read from or written to (i.e., data may be assigned to it). A variable is assigned a name by the programmer. The process of associating the name of a variable with its storage location is called *binding* (binding does much more than simply connecting a name to a variable).

In addition to its name, a variable has a *scope* associated with it. The scope of a variable defines the range of its visibility or *accessibility* within a program. For example, a variable declared within a procedure might be *visible* within that procedure but *invisible* outside the procedure. That is, the variable can be accessed inside the procedure, but any attempt to access it outside the procedure would result in an error. Figure 4.1 illustrates the scope of variables in a block-structured, high-level language. This allows you to define variables that are visible in only the current or lower-level procedures (or modules). Block structured languages include Algol 60 (the granddaddy of programming languages), Pascal (a popular teaching language of the 1980s), C, Ada (a language commissioned by the U.S. Department of Defense for use in safety-critical applications and now used in applications such as air traffic control, railway transport, and banking), and Java.

FIGURE 4.1   The concept of scope

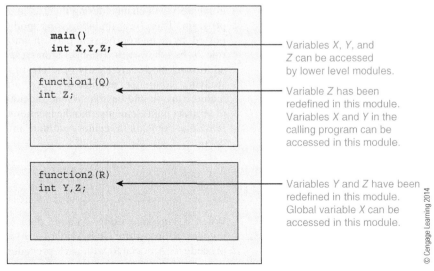

A variable has a *lifetime*. Variables are declared by assigning names to them and by reserving storage for them. From the assembly language programmer's point of view, variables exist from the moment they are loaded into memory to the moment when the program stops running. However, in many high-level languages, a variable is bound to a storage area for a specific period. Once the variable's lifetime has expired, the storage area is released and may be bound to another variable. Later, we look at the way in which temporary storage is allocated to variables in procedures.

The process of allocating storage to variables at runtime is called *dynamic allocation* and is an important feature of languages like Java. Some languages such as COBOL and FORTRAN employ *static allocation* and make all *bindings* at compile time. Variables are assigned to storage values when the program is compiled. Each variable is assigned a location when the program is translated from high-level language into machine code, and each variable keeps its location in memory until the program is terminated.

A corollary of static allocation is that it does not permit *recursion*. A procedure is *recursive* if it can call itself. Recursion frequently offers an elegant means of solving a problem; for example, the following fragment of C code evaluates factorial N, where $N! = N \times (N-1)!$. Recursion requires dynamic storage, since a new copy of the procedure's local variables must be made each time it is called.

```
int Factorial(int n)
{
  if(n == 1)
    return 1;
  else
    return n * Factorial(n-1);
}
```

High-level languages like Java employ a *stack* to hold variables as they are created and then released. Data structures can be allocated statically or dynamically. Languages like LISP let you create *dynamic* data structures at runtime whose size can be modified as the program runs.

What is the relevance of all this to computer architecture? If the allocation of storage is made at *runtime* during the execution of a program, the addresses of variables are clearly dynamic. This, in turn, means that the location of data elements in memory will change during the execution of a program. Therefore, the addressing modes supported by a microprocessor play an important role in handling such dynamic addresses. These addressing modes are clearly of interest to those studying computer architecture. An effective architecture should be able to map the addresses of abstract data elements onto the locations of the real data elements in memory without an undue delay.

Procedure, Subroutine, Function

The terms procedure, subroutine, and function are used in texts and papers on both high-level and low-level programming—sometimes interchangeably. Strictly speaking, these terms are not interchangeable because there are differences between them, although these differences tend to vary according to the domain (e.g., C programming or FORTRAN programming).

- A *subroutine* is a piece of code called from a main program that is executed and then a return is made to the calling point. Low-level languages invariably provide a subroutine call and return mechanism. For example, BSR and RTS in 68K code and BL and mov pc,lr in ARM code.
- A *procedure* is an extension to a subroutine that can use input and output parameters. The term procedure is not used in C but is associated with Pascal.
- A *function* is often defined as a subroutine that returns a value. The C language uses the term function exclusively (i.e., never subroutine or procedure).

4.1.1   Storage and the Stack

When a language using dynamic data storage invokes a procedure, it is said to *activate* the procedure. Associated with each procedure and

each invocation of a procedure is an *activation record* containing all of the information necessary to execute the procedure. You can regard an activation record as a procedure's view
of the world. Languages that support recursion use dynamic storage, because the amount of
storage required changes as the program is executed. Storage must be allocated at *runtime*.
Figure 4.2 illustrates the concept of an activation record.

Temporary storage is needed to evaluate expressions such as $X = (A + B) \cdot (C - D)$,
because the intermediate result $A + B$ must be stored somewhere, while $C - D$ is being
calculated. The activation record described by Figure 4.2 is also known as a *frame*. After an
activation record has been used, executing a return from procedure *deallocates* or frees the
storage taken up by the record. We now look at how frames are created and managed at the
machine level and demonstrate how two pointer registers are used to implement efficiently
activation record creation and deallocation.

FIGURE 4.2   The activation record

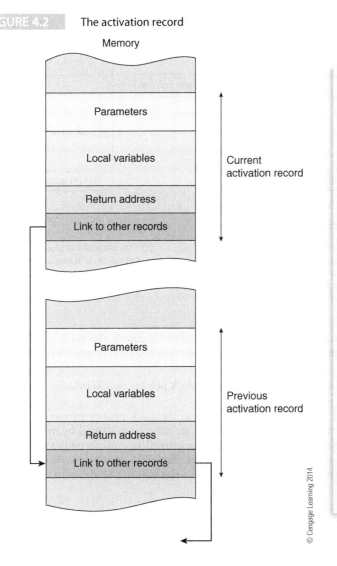

Stack Pointer and Frame Pointer

There are two pointers associated
with stack frames that must not
be confused, the stack pointer (SP)
and frame pointer (FP). In general,
all CISCs maintain a hardware SP
that is automatically adjusted when
BSRs or RTSs are executed. RISC
processors like the ARM do not
have an explicit SP, although by
convention r13 is used as the ARM's
programmer-maintained stack
pointer.

The stack pointer always points
to the top of the stack. The frame
pointer points to the base of the
current stack frame. The stack
pointer may change during the execution of the current procedure, but
the frame pointer will not change.
Data in the stack frame may be
accessed with respect to either the
stack pointer or the stack frame. By
convention, r11 is used as a frame
pointer in ARM environments, and
A6 is used in 68K environments.

© Cengage Learning 2014

The Stack Frame and Local Variables

Procedures often require *local workspace* for their temporary variables. The term *local* means that the workspace is private to the procedure and is never accessed by the calling program or by other subroutines. If a procedure is to be made re-entrant[1] or used recursively, its local variables must be bound up not only with the procedure itself but with the occasion of its use. In other words, each time the procedure is called a new workspace must be assigned to it. If a procedure is allocated a fixed region of workspace and is interrupted and called by the interrupt routine, any data in fixed locations will be overwritten by the procedure's re-use.

The stack provides a mechanism for implementing the dynamic allocation of workspace. Two concepts associated with dynamic storage techniques are the *stack frame* (SF) and the *frame pointer* (FP). The stack frame is a region of temporary storage at the top of the current stack. Figure 4.3 demonstrates how a *d*-byte stack frame is created by moving the stack pointer up by *d* locations at the start of a subroutine. Throughout this section, we assume that the stack pointer grows up towards low addresses and is always pointing at the item currently at the top of the stack. Some stacks point to the next free (empty) element above the stack.

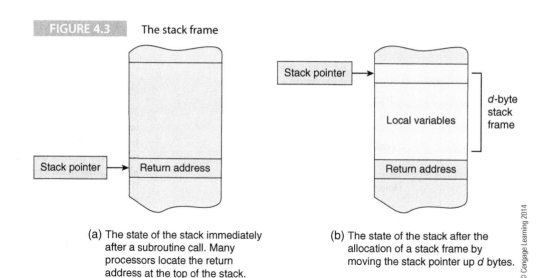

FIGURE 4.3   The stack frame

(a) The state of the stack immediately after a subroutine call. Many processors locate the return address at the top of the stack.

(b) The state of the stack after the allocation of a stack frame by moving the stack pointer up *d* bytes.

© Cengage Learning 2014

Because the stack grows towards the low end of memory, the stack pointer is decremented to create a stack frame. For example, reserving 100 bytes of memory is achieved by

```
SUB  r13,r13,#100          ;move the stack pointer up 100 bytes
```

Remember that the programmer uses r13 as the stack pointer *by convention*. Before a return from subroutine is made, the stack frame must be collapsed by moving the stack pointer down again with ADD **r13**,r13,#100. In general, operations on the stack frame are *balanced*; that is, if you put something on the stack frame you have to remove it. Consider the following simple example of an procedure. Note—this may not be the most efficient code—can you see why?

---

[1] A re-entrant subroutine can be interrupted and used by the interrupting program without any of its state information being destroyed by its reuse.

```
Proc  SUB  r13,r13,#16      ;move the stack pointer up 16 bytes
      Code                  ;some code
      STR  r1,[r13,#8]      ;store something in the frame 8 bytes below
                             top-of-stack
      Code                  ;some more code
      ADD  r13,r13,#16      ;adios stack frame
      MOV  pc,r14           ;time to go home... restore the PC to return
```

The temporary variables on a stack frame can be accessed using the stack pointer. In Figure 4.4a, variable XYZ is 12 bytes below the stack pointer, and we can access XYZ via the effective address [r13,#12]. Because the stack pointer is free to move as other information is added to (or removed from) the stack, it is better to construct a stack frame with a fixed pointer that is independent of the stack pointer.

FIGURE 4.4   Accessing variables in the stack frame

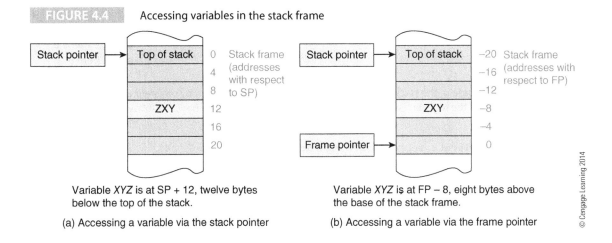

Variable *XYZ* is at SP + 12, twelve bytes below the top of the stack.

(a) Accessing a variable via the stack pointer

Variable *XYZ* is at FP – 8, eight bytes above the base of the stack frame.

(b) Accessing a variable via the frame pointer

© Cengage Learning 2014

Figure 4.4b illustrates a stack frame with a *frame pointer* (FP). In this case, the frame pointer points to the bottom of the stack frame and is independent of the stack pointer (that is, if data is pushed on the stack, the stack pointer changes but the frame pointer remains where it is). The variable can now be accessed via the frame pointer at [r11,#-8] if we assume that r11 is the frame pointer.

Mechanizing Stack Frames with Link and Unlink Instructions

The 68K supports stack frames via its LINK and UNLK instructions that create or collapse a stack frame in one operation. A LINK instruction pushes the frame pointer on top of the stack to preserve its old value, then the current stack pointer is loaded into the frame pointer. By convention, A6 is used as a frame pointer. The stack pointer is then moved up *d* bytes to create a stack frame. The frame pointer is now pointing to its previous value at the bottom of the stack, and the stack pointer is pointing at the top of the stack frame. For example, LINK FP,#-12 creates a 12-byte stack frame, and UNLK FP collapses it.

The following figure shows a stack frame during subroutine *A* in part (a) and after a second subroutine *B* has been invoked in part (b). The frame pointer now contains the value of the stack pointer immediately before the creation of stack frame *B*. As you can see, nested subroutines build successive stack frames on the stack.

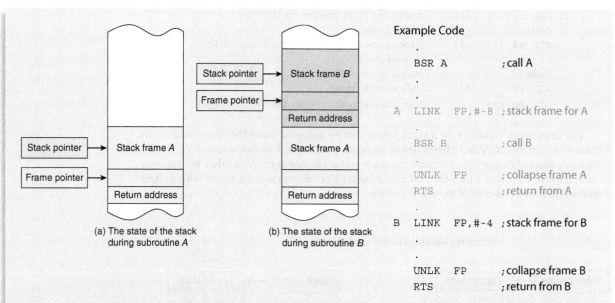

Example Code

```
        BSR   A            ; call A
        .
        .
        .
A   LINK    FP,#-8   ; stack frame for A
        .
        BSR   B            ; call B
        .
        UNLK    FP         ; collapse frame A
        RTS                ; return from A
        .
B   LINK    FP,#-4   ; stack frame for B
        .
        .
        UNLK    FP         ; collapse frame B
        RTS                ; return from B
```

(a) The state of the stack during subroutine *A*

(b) The state of the stack during subroutine *B*

Because FP points to the base of the stack frame, all local variables can be accessed by register indirect addressing with displacement, where FP is the register used to access them. At the end of subroutine *B*, the following sequence is executed:

```
UNLK    FP    De-allocate subroutine B's stack-frame
RTS           Return to calling point
```

The UNLK instruction collapses the stack frame. The stack pointer is first loaded with the contents of the frame pointer that contains the value of the stack pointer just before stack frame *B* was created. By doing this, stack frame *B* collapses. The next step is to pop the top item off the stack and place it in FP to return both the stack and the contents of FP to the points they were in before LINK was executed.

Following the UNLK, a return to subroutine *A* is made. The execution of subroutine *A* continues from the point at which it left off. The LINK and UNLK instructions have been included to support recursive procedures.

You might wonder where the terms LINK and UNLK come from. Each time a LINK is executed, the current contents of the frame pointer are pushed on the stack, and the new frame pointer points at the old frame pointer on the stack. This arrangement constitutes the data structure known as a *linked list*.

The ARM processor lacks a link instruction that creates a stack frame or an unlink instruction that collapses it when you leave. You have to do things the hard way. To create a stack frame, you could push the old link pointer on the stack and then move up the stack pointer by *d* bytes by executing

```
SUB    sp,sp,#4       ;move the stack pointer up by a 32-bit word
STR    fp, [sp]       ;push the frame pointer on the stack
MOV    fp,sp          ;move the stack pointer to the frame pointer to point at the base
SUB    sp,sp,#8       ;move stack pointer up 8 bytes (we have made d equal to 8)
```

In this code, fp represents the frame pointer. The frame pointer points at the base of the frame and can be used to access local variables in the frame. By convention, register

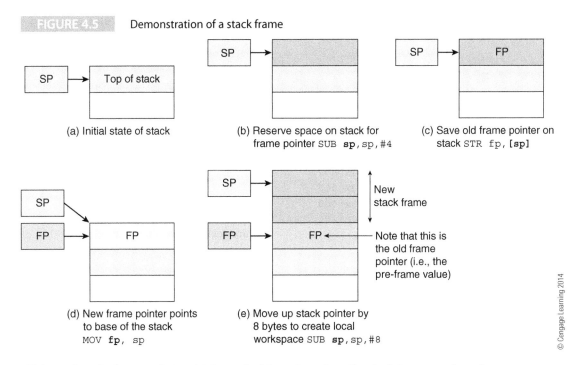

FIGURE 4.5   Demonstration of a stack frame

(a) Initial state of stack

(b) Reserve space on stack for frame pointer SUB **sp**,sp,#4

(c) Save old frame pointer on stack STR fp, [**sp**]

(d) New frame pointer points to base of the stack MOV **fp**, sp

(e) Move up stack pointer by 8 bytes to create local workspace SUB **sp**,sp,#8

© Cengage Learning 2014

r11 is used as the frame pointer. At the end of the subroutine, the stack frame can be collapsed by:

```
MOV    sp,fp         ;restore the stack pointer
LDR    fp,[sp]       ;restore old frame pointer from the stack
ADD    sp,sp,#4      ;move stack pointer down 4 bytes to restore stack
```

Figure 4.5 demonstrates the way in which the stack frame grows instruction by instruction. Note that the old frame pointer appears *twice*: once as the old/previous stack frame on the stack and once as the current stack frame pointing to the base of the stack frame. In practice, we would use the pre-decrementing multiple store instruction, STMFD, to push both the link register (containing the return address) and the frame pointer on that stack with

```
STMFD  sp!,  {lp,fp}   ;restore old link register from the stack
SUB    sp,sp,#4        ;move stack pointer down 4 bytes
```

Example of an ARM Processor Stack Frame

The following code demonstrates how you might set up a stack frame on an ARM processor. We push a register on the stack, call a subroutine, save the frame pointer and link register, create a one-word frame, access the parameter, and then return to the calling point.

```
          AREA TestProg, CODE, READONLY
          ENTRY
Begin
Main      ADR    sp,Stack        ;set up r13 as the stack pointer
          MOV    r0,#124         ;set up a dummy parameter
          MOV    fp,#123         ;dummy frame pointer
          STR    r0,[sp,#-4]!    ;push the parameter
          BL     Sub             ;call the subroutine
          LDR    r1,[sp]         ;retrieve the data
Loop      B      Loop            ;wait here (endless loop)
```

```
Sub       STMFD    sp!,{fp,lr}    ;push frame-pointer and link-register
          MOV      fp,sp          ;frame pointer at the bottom of the frame
          SUB      sp,sp,#4       ;create the stack frame (one word)
          LDR      r2,[fp,#8]     ;get the pushed parameter
          ADD      r2,r2,#120     ;do a dummy operation on the parameter
          STR      r2,[fp,#-4]    ;store it in the stack frame
          ADD      sp,sp,#4       ;clean up the stack frame
          LDMFD    sp!, {fp,pc}   ;restore frame pointer and return

          DCD      0x0000         ;clear memory
          DCD      0x0000
          DCD      0x0000
          DCD      0x0000
Stack     DCD      0x0000         ;start of the stack (stack grows towards lower
                                   addresses)

          END
```

Figure 4.6 demonstrates the behavior of the stack during the code's execution. Figure 4.6a depicts the stack's initial state. In Figure 4.6b, the parameter has been pushed on the stack. In Figure 4.6c, the frame pointer and link register have been stacked by STMFD sp!,{fp,lr}. Recall that STMFD provides a stack that grows towards low addresses, and the stack pointer points to the current item at the top of the stack.

In Figure 4.6d, a four-byte word has been created at the top of the stack. Finally, Figure 4.6e demonstrates how the pushed parameter is accessed and moved to the new stack frame using register indirect addressing with the frame pointer.

FIGURE 4.6   The behavior of the stack during the execution of the code

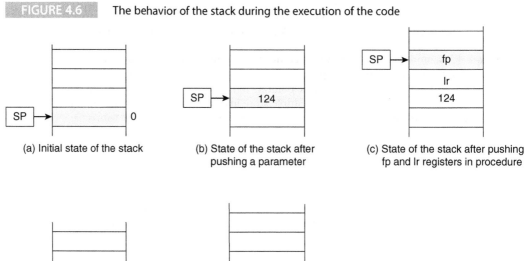

(a) Initial state of the stack

(b) State of the stack after pushing a parameter

(c) State of the stack after pushing fp and lr registers in procedure

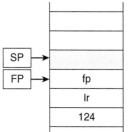

(d) State of stack after creating 4-byte space on the stack

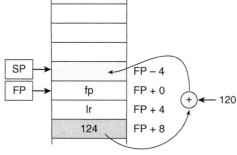

(e) State of stack after the sequence
```
LDR  r2,[fp,#8]   ;get parameter
ADD  r2,r2,#120   ;add 120
STR  r2,[fp,#-4]  ;store sum in stack frame
```

Figure 4.7a provides a snapshot of the output of an ARM processor development system that shows the contents of the registers and the state of the stack after the code has been loaded into the simulator. In Figure 4.7b, we have executed code up to the subroutine call. You can see that the stack pointer (r13) points at 0x08CC and that this location contains 0x7C (the value in r0 pushed on the stack). In Figure 4.7c, we have executed up to the ADD instruction. You can see that the stack pointer is at 0x80C0 and that the link register and old frame pointer have been pushed on the stack. In Figure 4.7d, the subroutine has been completed, and we have returned to the calling program. Figure 4.7e shows the state at the completion of the program.

4.1.2   Passing Parameters via the Stack

Having introduced the basics of procedures and memory allocation, we now look at some details and explain how parameters are passed via the stack to a procedure. In particular, we look at how the underlying machine architecture supports procedures.

You can pass a parameter to a procedure in two ways: by *value* or by *reference*. In the former, a copy of the actual parameter is transferred, whereas in the latter, the address of the parameter is passed between the program and then procedure/function. This distinction is important, because it affects the way in which parameters are handled. When passed by value, the procedure receives a *copy* of the parameter. If the parameter is modified by the procedure, the new value does not affect the previous value of the parameter elsewhere in the program. In other words, passing a parameter by value causes the parameter to be cloned and the clone to be used by the procedure. The clone never returns from the procedure.

When a parameter is passed by reference, the procedure receives a *pointer* to the parameter. In this case, there is only one copy of the parameter, and the procedure is able to access this unique value because it knows the address of the parameter. If the procedure modifies the parameter, it is modified globally and not just within the procedure.

FIGURE 4.7   Snapshots of the state of the registers and memory during the execution of the code

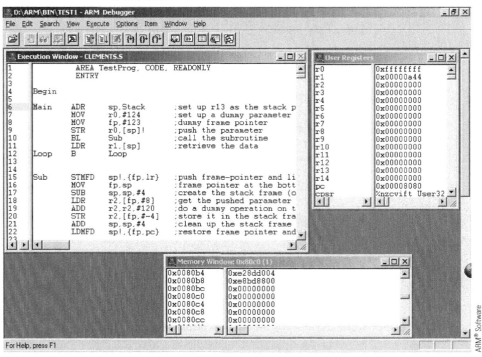

(a) Snapshot of the initial register and memory state on loading the code

FIGURE 4.7   Snapshots of the state of the registers and memory during the execution of the code (*Continued*)

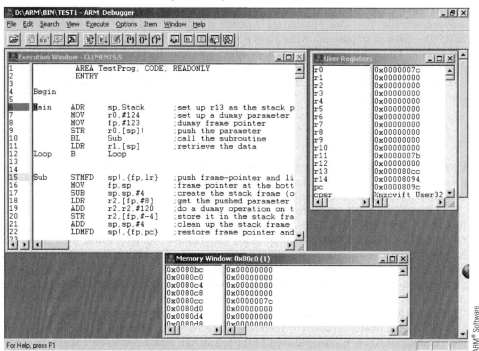

(b) Snapshot of the register and memory state on entering the subroutine

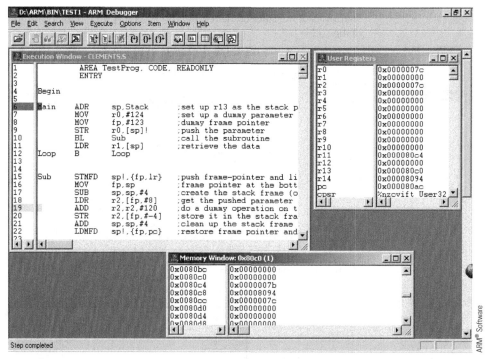

(c) Snapshot of the register and memory state at the end of the program

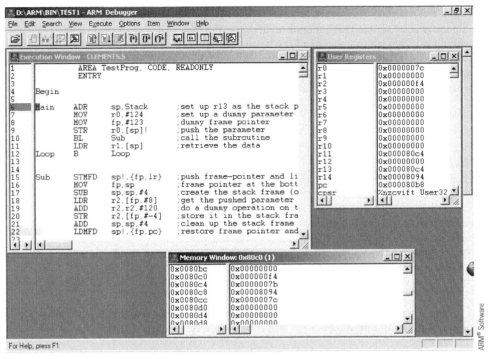

(d) Snapshot of the register and memory state before exiting the subroutine

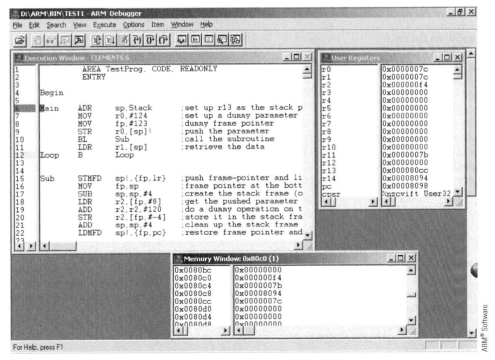

(e) Snapshot of the register and memory state at the end of the program

Pointers and C

A pointer is a variable that contains an address. You can write programs in some languages without ever having to understand the nature and use of pointers—this certainly is not true of C. Indeed, C is famous for being pointer-oriented. In C, you have to indicate explicitly that a variable is a pointer by prefixing it with an asterisk. Consider the following C declarations where x is an integer variable and y is a pointer to x.

```
int x;
int *y;
```

The operator `*` indicates that y is a pointer and `int` indicates that the pointer is pointing at an integer. C requires you to declare pointers to be the same type as the data they access for a very good reason—the compiler needs to know the size of each object pointed at.

When you put an asterisk in front of a pointer, you are said to be *dereferencing* the pointer (i.e., accessing the data being pointed at). For example, the expression `p = *q` means "assign the value pointed at by pointer q to variable p."

Having created a pointer, you have to initialize it. In order to bind pointer y to value x, we perform the operation

```
y = &x;
```

The `&` operator yields the address of a variable. You can combine the declaration of a pointer with its initialization as

```
int x = 12;       /* declare x as an integer variable = 12 */
int *P_x = &x;    /* declare P_x as a pointer to integer x */
```

Consider the following example of an input *polling loop* written in C. A polling loop is a feature of an input/output (I/O) mechanism. You read the status of a device in a loop and exit the loop when the device is ready to take part in a data transfer. We look at I/O in Chapter 12.

```
void main(void)
/* REPEAT
      read input device status byte
   UNTIL the device is ready
   Read data from device
*/
 {
  int x;
  int *P_port;                              /* create a pointer to the port */
  P_port = (int*) 0x4000;                   /* set the pointer to the port */
  do { } while ((*P_port & 0x0001) == 0);   /* wait for port to become ready */
  x = *(P_port + 1);                        /* read data */
 }
```

This code reads the contents of the *memory-mapped* input device. Memory-mapped I/O devices appear to the programmer exactly like any other memory location (see Chapter 12). We declare a variable `*P_port` that points at the memory-mapped input/output port at location 4000_{16}. We need to cast the variable to an integer pointer by writing the statement

```
P_port = (int*) 0x4000.
```

We have to cast the variable to an integer, because the C compiler needs to know the type of object the pointer is pointing at. In this example, address 4000_{16} is a status port that tells us whether the I/O device is ready, and address 4002_{16} is the location through which the data is passed. The actual polling loop is expressed in the form

```
do { } while ((*P_port & 0x0001) == 0);
```

The do...while loop performs the null action { }, while the ready bit of the port is 0. Once the port's ready bit is set to 1, the polling loop is exited, and the character is read from the device by means of the statement x = *(P_port + 1). Note that the pointer offset is 1 because an integer variable in C is two bytes, so (P_port + 1) points two bytes beyond P_port.

Functions and Parameters

Let's examine how parameters are passed to a function when we compile the high-level function swap(int a, int b) that is *intended* to exchange two values.

```
void swap (int a, int b)      /* this function swaps the values of a and b */
   { int temp;
     temp = a;                /* copy a to temp, b to a, and temp to b */
     a = b;
     b = temp;
   }
void main (void)
   { int x = 2, y = 3;
     swap (x, y);             /* swap a and b */
   }
```

We said that this code was intended to swap values; in fact, it fails to do this. In order to determine why the program won't work, we are going manually cross-compile this code and then trace its behavior. The following output is from a non-optimizing compiler; that is, it does not generate efficient code. For example, if you load a literal into memory, read it, and put it somewhere else in memory, it will do just that. An optimizing compiler will not re-read data from memory that it already has in a register.

The structure of the program is a C function, swap, followed by a main function. In main, a stack frame is first set up to hold the two variables to be swapped. Then, these are read and pushed onto the stack by value before the function swap is called.

In function swap, a second stack frame is set up to hold the temporary variable while the two values are being swapped. Then, the swap takes place using both parameters on the stack and the temporary variable in the current stack frame. Finally, the stack frame in swap is removed, and a return back to main made. On returning to main, its stack frame also is collapsed and the program halted.

```
        AREA SwapVal, CODE, READONLY
Stop    EQU         0x11                    ;code for program termination and exit
        ENTRY
        MOV         sp,#0x1000              ;set up stack pointer
        MOV         fp,#0xFFFFFFFF          ;set up dummy fp for tracing
        B           main                    ;jump to the function main
;       void swap (int a, int b)
;       Parameter a is at [fp]+4
;       Parameter b is at [fp]+8
;       Variable temp is at [fp]-4
swap    SUB         sp,sp,#4                ;Create stack frame: decrement sp
        STR         fp,[sp]                 ;push the frame pointer on the stack
```

```
            MOV        fp,sp                  ;frame pointer points at the base
            SUB        sp,sp,#4               ;move sp up 4 bytes for temp
;           {
;           int temp;
;           temp = a;
            LDR        r0,[fp,#4]             ;get parameter a from the stack
            STR        r0,[fp,#-4]            ;copy a to temp on the stack frame
;           a = b;
            LDR        r0,[fp,#8]             ;get parameter b from the stack
            STR        r0,[fp,#4]             ;copy b to a
;           b = temp;
            LDR        r0,[fp,#-4]            ;get temp from the stack frame
            STR        r0,[fp,#8]             ;copy temp to b
;           }
;                                             ;Collapse stack frame created for swap
            MOV        sp,fp                  ;restore the stack pointer
            LDR        fp,[fp]                ;restore old frame pointer from stack
            ADD        sp,sp,#4               ;move stack pointer down 4 bytes
            MOV        pc,lr                  ;return by loading link register into PC
;           void main (void)
;           Variable x is at [fp]+4
;           Variable y is at [fp]+8
main                                          ;Create stack frame in main for x, y
            SUB        sp,sp,#4               ;move the stack pointer up
            STR        fp,[sp]                ;push the frame pointer on the stack
            MOV        fp,sp                  ;the frame pointer points at the base
            SUB        sp,sp,#8               ;move sp up 8 bytes for 2 integers
;           {
;           int x = 2, y = 3;
            MOV        r0,#2                  ;x = 2
            STR        r0,[fp,#-4]            ;put x in stack frame
            MOV        r0,#3                  ;y = 3
            STR        r0,[fp,#-8]            ;put y in stack frame
;           swap (x, y);
            LDR        r0,[fp,#-8]            ;get y from stack frame
            STR        r0,[sp,#-4]!           ;push y on stack
            LDR        r0,[fp,#-4]            ;get x from stack frame
            STR        r0, [sp,#-4]!          ;push x on stack
            BL         swap                   ;call swap, save return address in link
                                              ; register
;           }
            MOV        sp,fp                  ;restore the stack pointer
            LDR        fp,[fp]                ;restore old frame pointer from stack
            ADD        sp,sp,#4               ;move stack pointer down 4 bytes
            SWI        Stop                   ;call O/S to terminate the program
            END
```

Let's go through this code again, observing the state of the stack. The main function initializes the parameters, puts them in a stack frame, and pushes their values on the system stack prior to calling function swap. Figures 4.8a to d show the state of the stack at four stages during the execution of this program. Function swap duly swaps the parameters, exactly as it is designed to do. Unfortunately, the parameters are swapped only *within the function*,

FIGURE 4.8   Passing values to a subroutine by value

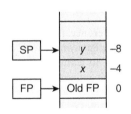

(a) State of the stack in main after
creating stack frame with:
```
SUB  sp,sp,#4
STR  fp,[sp]
MOV  fp,sp
SUB  sp,sp,#8
```

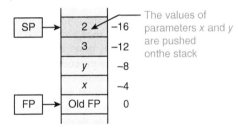

The values of
parameters x and y
are pushed
onthe stack

(b) The stack in main after putting two
parameters in the stack frame with:
```
MOV  r0,#2
STR  r0,[fp,#-4]
MOV  r0,#3
STR  r0,[sp,#-8]
```
Then pushing two parameters on the stack
```
LDR  r0,[fp,#-8]
STR  r0,[sp,#-4]!
LDR  r0,[fp,#-4]
STR  r0,[sp,#-4]!
```

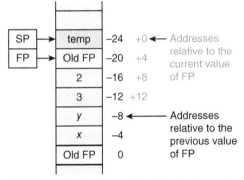

← Addresses
relative to the
current value
of FP

← Addresses
relative to the
previous value
of FP

(c) The stack after the creation of a stack frame in
swap. The new stack frame is four bytes deep
and holds the variable temp. The frame is
created by:
```
SUB  sp,sp,#4
STR  fp,[sp]
MOV  fp,sp
SUB  sp,sp,#4
```

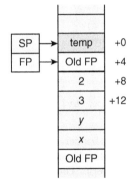

(d) The stack after executing the body of swap.
Note that all data is referenced to FP.
```
LDR  r0,[fp,#4]
STR  r0,[fp,#-4]
LDR  r0,[fp,#8]
STR  r0,[fp,#4]
LDR  r0,[fp,#-4]
STR  r0,[fp,#8]
```

© Cengage Learning 2014

because they are passed by value and are thus copies of the variables in main. The parameters
are swapped only in the function swap's stack frame, which is collapsed when we exit swap.
The variables in main remain unchanged. Instead, we must swap the variables in the calling
function.

By stepping through the code and following Figure 4.8, we see that the main function
creates a stack frame containing two locations for variables x and y; then it pushes copies of
the variables on the stack before calling function swap. The offsets on the right-hand side of
the memory maps in Figure 4.8 are given with respect to the frame pointer.

In Figure 4.8c, the memory map illustrates the state of the system after a stack frame has
been created in the function swap. Note that variables are now accessed with respect to the
new frame pointer. This is denoted by two sets of offsets, which appear on the right-hand side
of Figure 4.8c: One set of offsets gives the locations of the variables with respect to the value of

fp in main and the other offsets state locations with respect to the value of fp in swap. Figure 4.8d shows the four instructions in swap that carry out the data exchange. Inspection of the code reveals that the copies of x and y in the function's stack frame are indeed swapped, but nothing happens to the values of x and y in the calling function main.

We next look at how parameters are passed to functions by reference (i.e., address) and how we can write procedures that change parameters in the calling environment. C uses a *call-by-value* mechanism in which the value of a variable is passed to a function *in one direction only*—the actual parameter is not modified in the calling environment. If you pass parameters to a function and then modify them inside the function, their values outside the function do not change, as the previous example demonstrated when copies of the values we wanted to swap were passed to a function that swapped them but did not return the values. This can be remedied by passing the *address* of the parameters, so parameters can be accessed in the calling environment.

Pass-by-Reference

The function swap from the preceding example can readily be modified to exchange two parameters by calling swap(&a, &b) to pass the *addresses* of parameters a and b to the called function swap, as shown in the following HLL code.

```
void swap (int *a, int *b)   /* swap two parameters in calling program */
    {  int temp;
       temp = *a;
       *a = *b;
       *b = temp;
    }
void main (void)
    {  int x = 2, y = 3;
       swap(&x, &y);   /* call swap and pass addresses of parameters */
    }
```

In the function's header, we specify int *a and int *b to indicate values that are *pointers to* variables a and b. The statement temp = *a assigns the value pointed at by pointer a to integer variable temp. The statement *b = *a assigns the value pointed at by pointer a to the *location* pointed at by pointer b. Let's analyze the ARM processor code for this procedure, where blue typeface highlights the parts of this program that differ from the previous example.

```
            AREA SwapVal, CODE, READONLY
Stop    EQU         0x11                    ;code for program termination and exit
        ENTRY
        MOV         sp,#0x1000              ;set up stack pointer
        MOV         fp,#0xFFFFFFFF          ;set up dummy fp for tracing
        B           main                    ;jump to main function
;           void swap (int *a, int *b)
;           Parameter a is at [fp]+4
;           Parameter b is at [fp]+8
;           Variable temp is at [fp]-4
swap    SUB         sp,sp,#4                ;create stack frame: decrement sp
        STR         fp, [sp]                ;push the frame pointer on the stack
        MOV         fp,sp                   ;the frame pointer points at the base
        SUB         sp,sp,#4                ;move sp up 4 bytes for temp
;           {
;           int temp;
;           temp = *a;
        LDR         r1,[fp,#4]              ;get address of parameter a
```

```
          LDR          r2,[r1]              ;get value of parameter a
          STR          r2,[fp,#-4]          ;store parameter a in temp in stack frame
;         *a = *b;
          LDR          r0,[fp,#8]           ;get address of parameter b
          LDR          r3,[r0]              ;get value of parameter b
          STR          r3,[r1]              ;store parameter b in parameter a
;         b = temp;
          LDR          r3,[fp,#-4]          ;get temp
          STR          r3,[r0]              ;store temp in b
;         }
          MOV          sp,fp                ;Collapse stack frame: restore sp
          LDR          fp,[fp]              ;restore old frame pointer from stack
          ADD          sp,sp,#4             ;move stack pointer down 4 bytes
          MOV          pc,lr                ;return by loading link register contents into PC
;         void main (void)
;         Variable x is at [fp]-4
;         Variable y is at [fp]-8
main      SUB          sp,sp,#4             ;Create stack frame: move sp up
          STR          fp,[sp]              ;push the frame pointer on the stack
          MOV          fp,sp                ;the frame pointer points at the base
          SUB          sp,sp,#8             ;move sp up 8 bytes for two integers
;         {
;         int x = 2, y = 3;
          MOV          r0,#2                ;x = 2
          STR          r0, [fp,#-4]         ;put x in stack frame
          MOV          r0,#3                ;y = 3
          STR          r0, [fp,#-8]         ;put y in stack frame
;         swap (&x, &y)                     ;call swap, pass parameters by reference
          SUB          r0,fp,#8             ;get address of y in stack frame
          STR          r0, [sp,#-4]!        ;push address of y on stack
          SUB          r0,fp,#4             ;get address of x in stack frame
          STR          r0, [sp,#-4]!        ;push address of x on stack
          BL           swap                 ;call swap – save return address in lr
;         }
          MOV          sp,fp                ;collapse frame: restore sp
          LDR          fp,[fp]              ;restore old frame pointer from stack
          ADD          sp,sp,#4             ;move stack pointer down 4 bytes
          SWI          Stop
          END
```

In the function main, the addresses of the parameters are pushed on the stack by means of the following instructions.

```
          SUB          r0,fp,#8             ;get address of y in the stack frame
          STR          r0, [sp,#-4]!        ;push the address of y on the stack
          SUB          r0,fp,#4             ;get address of x in the stack frame
          STR          r0, [sp,#-4]!        ;push the address of x on the stack
```

In the function swap, the address of parameter a (i.e., x) is popped off the stack by means of

```
          LDR          r1,[fp,#4]           ;get the address of parameter a
```

FIGURE 4.9   Passing values to a subroutine by reference

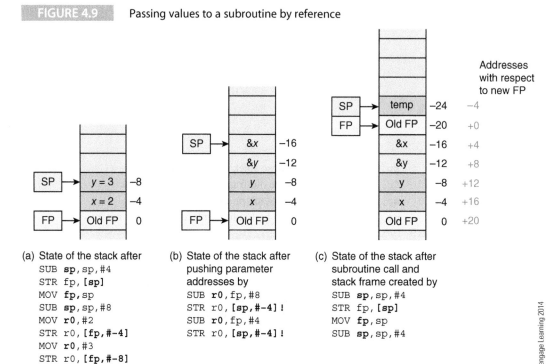

(a) State of the stack after
```
SUB  sp,sp,#4
STR  fp,[sp]
MOV  fp,sp
SUB  sp,sp,#8
MOV  r0,#2
STR  r0,[fp,#-4]
MOV  r0,#3
STR  r0,[fp,#-8]
in function main
```

(b) State of the stack after
pushing parameter
addresses by
```
SUB  r0,fp,#8
STR  r0,[sp,#-4]!
SUB  r0,fp,#4
STR  r0,[sp,#-4]!
```

(c) State of the stack after
subroutine call and
stack frame created by
```
SUB  sp,sp,#4
STR  fp,[sp]
MOV  fp,sp
SUB  sp,sp,#4
```

© Cengage Learning 2014

The operation `temp = *a` is implemented by

```
LDR     r2,[r1]            ;get the value of parameter a
STR     r2, [fp,#-4]       ;store parameter a in temp in the stack
                            frame
```

Figure 4.9 illustrates the state of the stack during execution of the preceding code.

Using Recursion

The following example, based on material from ARM Holding's literature, puts together several things we've learned in this and the previous chapter. We call a function to convert the contents of a register into a string of binary digits and then print them. We call a function recursively and use the *procedure call standard* as a convention to assign registers across function calls. This program borrows two routines from the ARM processor's literature [ARMdui0021A]: a subroutine `utoa` that converts an unsigned integer into a string of decimal digits and a subroutine `div10` that divides a number by 10.

ARM Procedure Call Standard

ARM literature includes the *ARM Procedure Call Standard* (APCS) that defines how programmers should use registers and pass information between procedures (in the same way that MIPS defines aliases for registers). The APCS is followed by the ARM C compiler, and it is important to the assembly language programmer if you wish to interface to the ARM's C routines or to read/debug code generated by the C compiler. For example, the APCS indicates which registers are general-purpose registers and how the stack is structured. The following table defines the APCS used by those programming ARM processors.

Register	APCS Name	APCS Role
r0	a1	Argument 1/integer result/scratch pad
r1	a2	Argument 2/scratch pad
r2	a3	Argument 3/scratch pad
r3	a4	Argument 4/scratch pad
r4	v1	Register variable 1
r5	v2	Register variable 2
r6	v3	Register variable 3
r7	v4	Register variable 4
r8	v5	Register variable 5
r9	sb/v6	Static base/register variable 6
r10	sl/v7	Static base/register variable 7
r11	fp	Frame pointer
r12	ip	Intra-procedure-call scratch register
r13	sp	Base of current stack frame
r14	lr	Link address/scratch register
r15	pc	Program counter

© Cengage Learning 2014

The program sets up a dummy value in location Convert, calls utoa, and prints the result using the ARM processor's software interrupt. Because we have used the APC standard, registers are renamed so that a1 to a4 are general-purpose argument or working registers, and v1 to v5 are variable registers whose values are preserved across function calls. As usual, sp, lr, and pc are the stack pointer, link register, and program counter, respectively, in resisters r13, r14, and r15.

On entry to utoa, the unsigned integer to convert is in argument register a2, and a pointer to the buffer that will hold the character representation of the string is in register a1. On exit from this routine, register a1 points at the next location immediately after the string.

When we call div10 to perform a division by 10, the values in a1 and a2 are *saved across the call* by putting them in registers v1 and v2 (both v1 and v2 are saved on the stack on entry to the recursive subroutine utoa).

Subroutine utoa operates by taking a value x and then calling subroutine div10 to divide it by 10 to obtain a quotient. If we take 10 times the quotient and subtract it from the original number, we get remainder $r = p - 10 \cdot \text{quotient}(p/10)$. To avoid using the multiply instruction to calculate 10 times the quotient, we write

```
SUB  v2,v2,a1,  LSL #3 subtracts 8 times a1 from v2 and the next operation
SUB  v2,v2,a1,  LSL #1 subtracts 2 times a1 from v2 to get v2 = v2 − 10 · a1
```

Having obtained the least-significant digit in the range from 0 to 9, we convert it to '0' to '9' by adding the constant 30_{16}. After converting the first digit, utoa is called recursively until the quotient is zero, at which point the process is complete.

```
        AREA DecimalConversion,  CODE,    READONLY
        ENTRY
ToDec   ADR    r0,Convert            ;point to data to convert
        LDR    a2,[r0]               ;load argument register a2 with the number to
                                      convert
        ADR    a1,String             ;load argument register a1 with the buffer address
        BL     utoa                  ;call conversion routine
```

```
               ADR     r1,String              ;point to the result string
               MOV     r2,#10                 ;print the result (ten digits maximum for
                                               0xFFFFFFFF)
   PrtLoop     LDR     r0,[r1], #1            ;get a character and advance the pointer
               SWI     0                      ;print the character
               SUBS    r2,r2,#1               ;decrement the loop counter
               BNE     PrtLoop                ;repeat until 10 digits printed
               SWI     17                     ;exit (call O/S function 0x11)
   utoa        STMFD   sp!,{v1,v2,lr}         ;convert register to decimal string - save
                                               registers
               MOV     v1,a1                  ;save parameter a1 because div10 will overwrite
                                               them
               MOV     v2,a2                  ;save parameter a2
               MOV     a1,a2                  ;div10 expects a parameter in a1
               BL      div10                  ;call div10 to do a1 = a1/10
               SUB     v2,v2,a1, LSL #3       ;subtract 10 x a1 from v2 (a2 = a2 − 10a1)
               SUB     v2,v2,a1, LSL #1       ;note we multiply by 10 by doing
                                               8p + 2p = 10p
               CMP     a1,#0                  ;is the quotient zero yet?
               MOVNE   a2,a1                  ;if not zero save it in a2
               MOV     a1,v1                  ;save the pointer in a1
               BLNE    utoa                   ;if not zero then call this routine recursively
               ADD     v2,v2,#'0'             ;convert final digit to ASCII by adding 0x30
               STRB    v2,[a1],#1             ;store this digit at the end of the buffer
               LDMFD   sp!,{v1,v2,pc}         ;restore registers and return from recursive
                                               function
   div10       SUB     a2,a1,    #10          ;subroutine to divide a1 by 10
               SUB     a1,a1,a1, LSR #2       ;return with quotient in a1, remainder in a2
               ADD     a1,a1,a1, LSR #4       ;magic division! Multiply by 1/10 = 0.1
               ADD     a1,a1,a1, LSR #8
               ADD     a1,a1,a1, LSR #16
               MOV     a1,a1,    LSR #3
               ADD     a3,a1,a1, ASL #2
               SUBS    a2,a2,a3, ASL #1
               ADDPL   a1,a1,    #1
               ADDMI   a2,a2,    #10
               MOV     pc,r14                 ;return with quotient in a1
   Convert     DCD     0x12345678             ;dummy data
   String      DCD     0x0                    ;location of result
               END
```

The divide-by-10 routine requires some comment. Some members of the ARM processor family lack a divide instruction; it's necessary to do division the hard way by shifting and adding (the Cortex M3/M4 processors do have division instructions). However, the following routine divides a number by ten in a most ingenious way by exploiting the processor's ability to add or subtract a shifted operand. Consider the following sequence of four instructions that is the key to this code.

```
   SUB     a1,a1,a1, LSR #2
   ADD     a1,a1,a1, LSR #4
   ADD     a1,a1,a1, LSR #8
   ADD     a1,a1,a1, LSR #16
```

If we assume that the number in register a1 is 1, these instructions generate successively:

```
SUB a1,a1,a1,LSR #2     ;1 − 1 × 2⁻² = 0.11
ADD a1,a1,a1,LSR #4     ;0.11 + 0.0011 = 0.110011
ADD a1,a1,a1,LSR #8     ;0.110011 + 0.00000000110011 = 0.11001100110011
ADD a1,a1,a1,LSR #16    ;0.11001100110011 + 0.0000000000000000 +
                        0.11001100110011
                        ;= 0.110011001100110011001100110011
```

Let's look at this more carefully. If you convert the decimal value 0.1_{10} into binary, you get the recurring binary sequence $0.110011001100110011001100110011_2$. This is the same number we generated by the sequence of instructions with shifted operands. Consequently, these operations perform a division-by-10 by multiplying a binary value by the sequence that generates 0.1_{10}.

The next section takes a brief look at privileged modes and exceptions, which is a topic closely related to operating systems and input/output techniques.

4.2   Privileged Modes and Exceptions

The next topic, interrupts and exceptions, belongs both to the world of hardware and interfacing (Chapter 12) and to the world of architecture and ISA. Interrupts and exceptions are events that force the computer to stop *normal* processing and to invoke a program called an *exception handler* (usually part of the operating system) to deal with the exception. Exceptions are raised in response to internal hardware or software errors or by peripherals seeking attention. Programmers can use software interrupt instructions to call operating system functions such as input or output operations. We deal with interrupts in greater detail in Chapter 12.

Let's look at exceptions. At any instant, an ARM processor is operating in one of the modes described in Table 4.1. The five low-order bits of the CPSR define the current mode. The normal operating mode is the *user* mode. A switch between modes takes place whenever an interrupt or exception occurs. Each of these modes has its own saved program status register, SPSR, which is used to hold the current CPSR when the exception occurs. When an exception switches in new registers r13 and r14, *the new register set* (or *bank*) is indicated by the name given in Table 4.1.

TABLE 4.1	The ARM Processor's Operating Modes and the Name of Its Register Sets		
Operating Mode	**CPSR[4:0]**	**Use**	**Register Bank**
User	**10000**	**Normal user model**	**user**
FIQ	10001	Fast interrupt processing	_fiq
IRQ	10010	Interrupt processing	_irq
SVC	10011	Software interrupt processing	_svc
Abort	10111	Processing memory faults	_abt
Undef	11011	Undefined instruction processing	_und
System	11111	Operating system	user

© Cengage Learning 2014

Exceptions—An Overview

As this is such an important topic, here we provide an overview and a reminder of the concepts involved. Exceptions are like subroutines that are *jammed* into code at runtime. Exceptions generally use similar call-and-return mechanisms to subroutines; the major difference being that the call address is supplied by the processor's hardware. Typically, a processor decodes the exception type and reads a pointer in memory that indicates the start of the exception handling routine. Also, some processors save the current status

(continued)

word (as well as the return address), because an exception should not alter the processor status.

As well as hardware interrupts, there are page-fault interrupts due to memory access errors, user-supplied operating system calls, illegal instruction exceptions (e.g., an illegal op-code), and divide-by-zero exceptions. Exceptions are invariably handled by operating system software.

Some processors change their operating mode when an exception occurs. This mode can be a privileged mode in which certain operations are forbidden in order to protect the integrity of the operating system.

Figure 4.10 shows the registers where those in dark blue are banked and are associated with specific operating modes. As you can see from Figure 4.10, registers r13 and r14 are replicated in each of the operating modes. For example, if a supervisor exception occurs, the new registers r13 and r14 are called r13_SVC and r14_SVC, respectively.

Of course, when writing ARM processor code, registers r13_SVC and r14_SVC are still written as r13 and r14. It's important to remember that when you switch to the supervisor mode, the *user-mode values* of r13 and r14 are no longer available. The supervisor mode has its own private registers: r13 and r14. This mechanism means that the programmer does not have to save r13 and r14 each time an exception occurs (unless it is a nested exception).

Exceptions are generated by both internal and external events. An external event is an interrupt request (IRQ) and includes fast interrupt requests (FIQs), the reset, and page faults. Internal exceptions include software interrupts and undefined instructions.

FIGURE 4.10   The ARM processor's banked register set

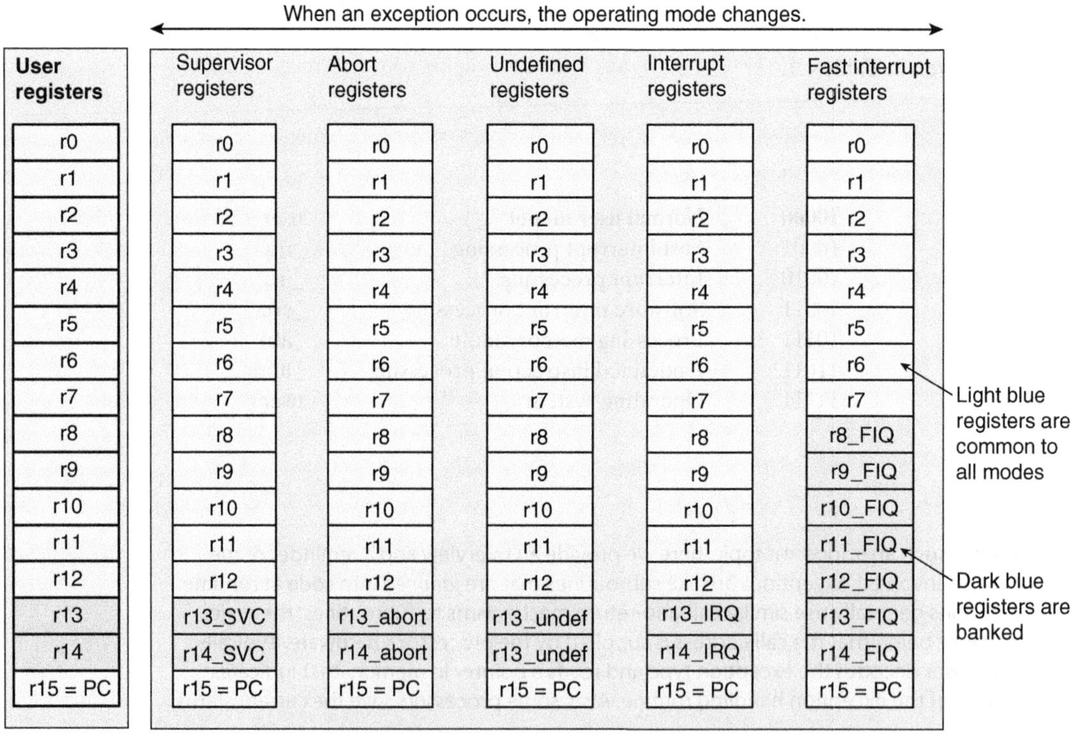

When an exception occurs, the operating mode changes.

User registers	Supervisor registers	Abort registers	Undefined registers	Interrupt registers	Fast interrupt registers	
r0	r0	r0	r0	r0	r0	
r1	r1	r1	r1	r1	r1	
r2	r2	r2	r2	r2	r2	
r3	r3	r3	r3	r3	r3	
r4	r4	r4	r4	r4	r4	
r5	r5	r5	r5	r5	r5	
r6	r6	r6	r6	r6	r6	
r7	r7	r7	r7	r7	r7	Light blue registers are common to all modes
r8	r8	r8	r8	r8	r8_FIQ	
r9	r9	r9	r9	r9	r9_FIQ	
r10	r10	r10	r10	r10	r10_FIQ	
r11	r11	r11	r11	r11	r11_FIQ	
r12	r12	r12	r12	r12	r12_FIQ	Dark blue registers are banked
r13	r13_SVC	r13_abort	r13_undef	r13_IRQ	r13_FIQ	
r14	r14_SVC	r14_abort	r13_undef	r14_IRQ	r14_FIQ	
r15 = PC	r15 = PC	r15 = PC	r15 = PC	r15 = PC	r15 = PC	

When an exception occurs, the ARM processor completes the current instruction (unless the instruction execution itself was the cause of the exception) and then enters an exception-processing mode. The sequence of events that then takes place is as follows.

1. The operating mode is changed to the mode corresponding to the exception. For example, an interrupt request would select the IRQ mode.
2. The address of the instruction following the point at which the exception occurred is copied into register r14. That is, the exception is treated as a type of subroutine call, and the return address is preserved in the link register.
3. The current value of the current status processor status register (CPSR) is saved in the SPSR of the new mode. For example, if the exception is an interrupt request, CPSR gets saved in SPSR_irq. It is necessary to save the current processor status, because an exception must not be allowed to modify the processor status.
4. Interrupt requests are disabled by setting bit 7 of the CPSR. If the current exception is a fast interrupt request, further FIQ exceptions are disabled by setting bit 6 of the CPSR.
5. Each location in the exception table contains an instruction that is executed first in the exception handling routine. This instruction is normally a branch operation (for example, B myHandler). This would load the program counter with the address of the corresponding current exception handler.

Table 4.2 defines the memory locations accessed by the ARM processor's exceptions. Each memory location contains the first instruction of the appropriate exception handlers; this implies, of course, that this table should be in read-only memory.

TABLE 4.2   Exception Vectors

Exception	Mode	Vector Address
Reset	SVC	0x00000000
Undefined instruction	UND	0x00000004
Software interrupt (SWI)	SVC	0x00000008
Prefetch abort (instruction fetch memory fault)	Abort	0x0000000C
Data abort (data access memory fault)	Abort	0x00000010
IRQ (normal interrupt)	IRQ	0x00000018
FIQ (fast interrupt)	FIQ	0x0000001C

© Cengage Learning 2014

If you examine Table 4.2, you will note that the vectors are *non-contiguous*, because the entry at memory location 14_{16} is missing. When the ARM processor was first introduced, the location was used by the misaligned word exception. Later versions of the ARM processor can handle misaligned addresses without calling an exception handler.

After the exception has been dealt with by a suitable handler, it is necessary to return to the point at which the exception was called (of course, if the exception was fatal, a return is no longer possible).

In order to return from an exception, the information that defines the pre-exception mode must be restored (that is, the program counter and CPSR). Unfortunately, returning from an exception is not as trivial a matter as it might seem. If you restore the PC first, you are still in the exception-handling mode. On the other hand, if you restore the processor status first, you are no longer within the exception-handling routine, and there is no way in which you can restore the CPSR.

You can't use a *normal* sequence of operations to return from an exception because it involves a change of operating mode. Two exception return mechanisms are provided: one for the case in which the return address has been stored in the banked r14 and the other for the case in which the return address has been pushed on the stack. Moreover, the return mechanism depends on the type of exception being handled.

If you are returning from an exception where the return address is in the link register, you can execute instructions described in Table 4.3, where MOVS and SUBS are special versions of the normal instructions used when the destination register is the PC. You have to

TABLE 4.3	ARM Return from Exception Operations
Exception Type	**Instruction to Return to User Mode**
SWI, undefined instruction	MOVS **pc**,r14
IRQ, FIQ	SUBS **pc**,r14,#4
Data abort to repeat the faulted instruction	SUBS **pc**,r14,#8

modify the value of the PC when returning from an IRQ, FIQ, or a data abort. In the former case, the PC has to be wound back by four. In the latter case, the PC has to be wound back by eight in order to repeat the instruction that was faulted.

If the exception handler has copied the return address on the stack, you have to use a slightly different mechanism. Under normal circumstances, you would return from a subroutine with a stacked pc by means of an instruction such as

```
LDMFD r13!, {r0-r4, pc}
```

where r0-r4 is the list of registers to be restored. If you wish to unstack the saved registers and restore the CPSR at the same time, you have to use the special version of this instruction

```
LDMFD r13!, {r0-r4, pc}^      ;restore r0 to r4, return and restore CPSR
```

The ^ symbol after the register list indicates that the CPSR is to be restored at the same time the program counter is restored. The program counter was not modified at the point at which it was restored. You have to modify the PC *before* you stack it!

68K User and Supervisor Modes

The 68K family adopts an interesting approach to exception handing. Following a hard reset at power up, the 68K begins operation in its supervisor mode. This mode is indicated by an S-bit in the processor status word. Moreover, it is indicated by the signal level at one of its pins. External hardware (memory and peripherals) can detect whether the 68K is in its supervisor or user mode. In this way, the regions of memory can be made accessible only to the operating system.

The operating system can make a transition to the user mode by clearing the S-bit. Once in the user state, any exception forces a transition back to the supervisor state. In the user mode, a programmer cannot set the S-bit to invoke the supervisor mode, as that would cause an exception and force a return to the supervisor state. You may think that this is what the user intended when trying to set the S-bit. However, when an exception causes an operating system intervention, the operating system takes control of the computer.

An important feature of the 68K is that there are two stack pointers, A7: one for the *user* and one for the *supervisor* mode. If an error occurs in the user mode and the system crashes, any resulting exception will invoke the supervisor mode, and its stack pointer will still be intact making a recovery possible.

4.3   MIPS: Another RISC

MIPS is a classic RISC architecture developed by John Hennessy at Stanford University in 1980 to exploit the best aspects of RISC philosophy in an efficient 32-bit processor. Hennessey left Stanford in 1984 to set up the MIPS Corporation. Like Intel's IA32 architecture, MIPS has gone through several generations and is available in 64-bit versions. MIPS is important because it has been widely used to support the teaching of computer architecture.

We introduce it here, because MIPS makes an interesting contrast with the ARM processor. MIPS is found in a wide range of embedded and mobile applications and in some games systems (for example, PlayStation®).

MIPS has a conventional 32-bit load-and-store ISA and 32 general-purpose registers. Register r0 is unusual, because it holds a zero and cannot be changed. This is an important feature of MIPS, because it allows the programmer easy access to zero and an ability to *suppress* a register in an instruction.

MIPS Instruction Format

Figure 4.11 illustrates three MIPS instruction formats: R-type specifies register-to-register operations, I-type provides a 16-bit literal operand, and J-type is used for direct jump instructions. There is also a C-type for co-processor operations that we do not discuss here.

FIGURE 4.11   MIPS instruction formats

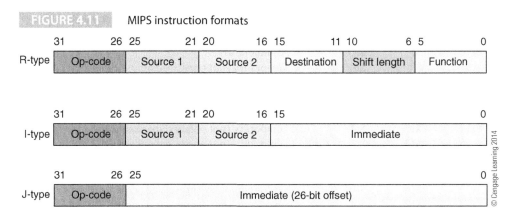

The most common instruction format is the R-type, which provides a register-to-register data processing operation very much like the ARM processor's corresponding instruction. A significant difference between MIPS and ARM processors is that MIPS can specify one of 32 registers, whereas the ARM processor provides only 16 registers. A typical R-type instruction is add r1,r2,r3. By convention, MIPS assembly language instructions are written in lower case. MIPS lacks two important ARM processor mechanisms: conditional execution and the ability to shift the second operand.

The I-type instruction format concatenates three fields from the R-type instruction to create a 16-bit literal or immediate field that can be used to provide a constant in operations like *add immediate* or an offset in register indirect (indexed) addressing modes. The 16-bit literal may be signed or unsigned, permitting a range of −32,768 to +32,767 or 0 to 65,535. Unlike the ARM processor, the literal cannot be scaled. A typical I-type operation is addi r1,r2,4. MIPS appends an i to the opcode to indicate a literal, whereas the ARM processor uses the # symbol to prefix a literal. These differences refer to the assembler grammar and not the ISA of the processors.

Because MIPS uses 16-bit literals, depositing a 32-bit word into a register is easily done by loading two consecutive literals. A *load upper immediate* instruction, lui, deposits a 16-bit literal into the upper-order 16-bits of a register and clears the lower-order 16 bits to zero. For example, lui $1,0x1234 loads register r1 with 0x12340000. A logical OR with a 16-bit immediate operand can now be used to access the lower-order 16 bits. For example, ori $1,0xABCD will set r1 to 0x1234ABCD. Note that $0 through $31 are MIPS names for registers r0 through r31.

The J-type instruction format uses unconditional jumps and provides a 26-bit literal that is used to construct a branch offset. Because MIPS is word (32-bit) oriented, the branch offset is shifted left twice before using it to provide a 28-bit byte range of 256 Mbytes.

The MIPS register set is conventional, and apart from r0 being fixed at 0, it has no special-function registers like the ARM processor. MIPS assembly language initially looks odd to those who have first learned another assembly language, because its registers are written $0, $1, \ldots$ rather than r0, r1, $\ldots$ Table 4.4 describes the MIPS register set and gives the alternate registers names used by programmers (entirely analogous to the ARM processor's procedure call standard).

TABLE 4.4		MIPS Register Naming Conventions	
Name	MIPS Name	Assembly Name	Use
r0	$0	$zero	Constant 0
r1	$1	$at	Reserved for assembler
r2	$2	$v0	Expression evaluation and results of a function
r3	$3	$v1	Expression evaluation and results of a function
r4	$4	$a0	Argument 1
r5	$5	$a1	Argument 2
r6	$6	$a2	Argument 3
r7	$7	$a3	Argument 4
r8	$8	$t0	Temporary (not preserved across call)
r9	$9	$t1	Temporary (not preserved across call)
r10	$10	$t2	Temporary (not preserved across call)
r11	$11	$t3	Temporary (not preserved across call)
r12	$12	$t4	Temporary (not preserved across call)
r13	$13	$t5	Temporary (not preserved across call)
r14	$14	$t6	Temporary (not preserved across call)
r15	$15	$t7	Temporary (not preserved across call)
r16	$16	$s0	Saved temporary (preserved across call)
r17	$17	$s1	Saved temporary (preserved across call)
r18	$18	$s2	Saved temporary (preserved across call)
r19	$19	$s3	Saved temporary (preserved across call)
r20	$20	$s4	Saved temporary (preserved across call)
r21	$21	$s5	Saved temporary (preserved across call)
r22	$22	$s6	Saved temporary (preserved across call)
r23	$23	$s7	Saved temporary (preserved across call)
r24	$24	$t8	Temporary (not preserved across call)
r25	$25	$t9	Temporary (not preserved across call)
r26	$26	$k0	Reserved for OS kernel
r27	$27	$k1	Reserved for OS kernel
r28	$28	$gp	Pointer to global area
r29	$29	$sp	Stack pointer
r30	$30	$fp	Frame pointer
r31	$31	$ra	Return address (used by function call)

MIPS load and store instructions are `lw` (load word) and `sw` (store word). Addressing modes are minimal, and MIPS provides only a register indirect with offset addressing mode. For example, `lw $1,16($2)` implements $[\$1] \leftarrow [16+[\$2]]$, which is entirely analogous to the ARM processor's `LDR r1,[r2,#16]`. MIPS lacks the complex addressing modes of CISCs and the ARM processor's block move instructions. However, direct memory addressing is possible if you use register r0 (because that forces a 16-bit absolute address), and program counter-relative addressing is supported.

Conditional Branches

MIPS handles conditional branches in a markedly different way from the ARM processor. Recall that an ARM processor branch depends on the state of processor condition code bits set or cleared by a previous instruction. MIPS provides *explicit* compare and branch instructions. For example, `beq r1,r2,label` compares the contents of register r1 with r2 and branches to `label` on equality. MIPS lacks the set of 16 conditional branches provided by CISC processors (and the ARM processor) and implements only

```
beq  $1,$2,label    ;Branch on equal
bne  $1,$2,label    ;Branch on not equal
blez $1,$2,label    ;Branch on less than or equal to zero
bgtz $1,$2,label    ;Branch on greater than zero
```

An interesting MIPS instruction is the *set on condition*. For example. The *set on less than* instruction `slt $1,$2,$3` performs the test [$2] < [$3] and then sets $1 to 1 if the test is true and to 0 if the test is false. This instruction turns a Boolean condition into a value in a register that can later be used by a conditional branch or as a data value in an operation. We will meet this concept again when we introduce multimedia operations in Chapter 5. A typical example of the use of `slt` is

```
slt  $1,S2,$3       ;if $2 < $3 THEN $1 = 1 ELSE $1 = 0
bne  $1,$0,Target   ;branch on $1 not zero (that is, branch on $2 < $3)
```

There is also an `sltu` operation that performs the same operation on unsigned numbers, and there are `slti` and `sltui` versions that have immediate operands.

4.3.1   MIPS Data Processing Instructions

MIPS data processing operations are generally very similar to the ARM processor's data processing instructions. One small difference is that MIPS provides explicit shift operations that provide either a fixed length shift with a literal shift field or a dynamic shift with a register shift field. For example,

```
sll  $1,$2,4        ;Shift $2 left logically 4 places and put the result in $1
sllv $1,$2,$3       ;Shift $2 left logically the number of places in $3 and put the
                     result in $1
```

Note that a different instruction is required for static and dynamic shifts. This is a feature of the assembler rather than the ISA.

Table 4.5 describes MIPS's data processing operations. The Boolean operations and, or, not, and xor are conventional, although nor is also implemented, which is rather unusual. Add and subtract are conventional, although the unsigned versions are unusual (they differ only in that the overflow flag is not set on arithmetic overflow). The MIPS add instruction generates an *exception* or *software interrupt* on arithmetic overflow (i.e., an operating system call). For this reason, many prefer to use the `addu` instruction. There are no explicit extended addition operations that add in the carry bit from a previous operation. In order to perform extended arithmetic, you have to isolate the carry bit from a lower-order addition and add it in to the next higher-order addition. For example,

```
addu  $1,$3,$5      ;Add low order words of $2,$1 = $4,$3 + $6,$5
sltu  $2,$1,$5      ;Get the carry in bit
addu  $2,$2,$4      ;Add in first most-significant word
addu  $2,$2,$6      ;Add in second most-significant word
```

The key to this sequence of operations is the `sltu`, meaning *set less than unsigned*, which compares $1 < $5 and sets $1 to 1 if true. This test is comparing the sum of the two lower-order words with one of them. If the sum is less than one of them, a carry must have been generated, and a 1 is loaded into $2 (the higher-order word of the sum) before adding the two higher-order words. If this seems confusing, consider a decimal example in two digits. Suppose we are adding 3 + 4. The sum is 9 and 9 > 4 (no carry). Now, if we add 8 + 4, we get 2 and 2 < 4 (carry generated).

TABLE 4.5	MIPS Data Processing Operations	
Mnemonic	**Operation**	
sll	Shift left logical	
srl	Shift right logical	
sra	Shift right arithmetic	
mult	Multiply	
multu	Multiply unsigned	
div	Divide	
divu	Divide unsigned	
add	Add	
addu	Add unsigned	This performs an addition except that the overflow flag is not set on overflow
sub	Subtract	
subu	Subtract unsigned	This performs a subtraction except that the overflow flag is not set on overflow
and	Logical AND	
or	Logical OR	
xor	Exclusive OR	
nor	Logical NOR	

© Cengage Learning 2014

Data processing instructions that can take a literal as the second operand are addi, addiu, andi, ori, slti, sltui, and xori.

Both signed and unsigned multiplication are provided that generate 32-bit by 32-bit products yielding a 64-bit result. The MIPS approach to multiplication (and division) is unusual. Other processors generally specify the source and destination registers as part of the instruction. A full 32-bit multiplication would require four registers: two for the source and two for the upper and lower words for the 64-bit product. MIPS takes a different approach and provides two dedicated registers *hi* and *lo* that receive the two halves of the result. Of course, this approach requires dedicated instructions to access these dedicated registers:

mfhi	Move from hi	mfhi $1	Copies high-order word to $1
mflo	Move from lo	mflo $1	Copies low-order word to $1
mthi	Move to hi	mthi $1	Copies high-order word from $1 to hi
mtlo	Move to lo	mtlo $1	Copies low-order word from $1 to lo

Pseudoinstructions

Like ARM, MIPS assemblers support pseudoinstructions that are really renamed operations. For example, the pseudoinstruction li $1,0x1234 is rendered as the actual MIPS instruction addi $1,$0,0x1234. This works because $0 is always zero, so adding 0x1234 to it and putting the result in $1 is equivalent to moving 0x1234 to $1. Similarly, move $1,$2 is a pseudoinstruction rendered by add $1,$0,$2.

Flow Control

MIPS provides a *jump and link* operation jal $1,Target, where Target is a 16-bit signed branch offset. A branch is executed to Target and the return address stored in $1. Unlike the ARM, there is no dedicated link register, and it's impossible to access the program counter directly. Consequently, you have to provide a return from procedure instruction (*jump register*) jr $1, which loads $1 into the PC to execute a return.

MIPS Example

As a simple example of MIPS code, consider the following where we perform a simple vector operation $Y = s \cdot X$, where X and Y are eight-component vectors and s is a scalar with the value 8.

```
int i;
int X[8], Y[8];
for (i = 0; i < 8; i = i++){
  Y[i] =  X[i] * 8;}
        lui     $1,upperX       ;put upper 16 bits of address of X in $1
        ori     $1,$1,lowerX    ;now put in the lower 16 bits of X
        addi    $3,$0,#8        ;$3 is the loop counter
Loop:   lw      $4,0($1)        ;Repeat: get xi
        sll     $4,$4,3         ;  shift xi left three times to multiply by 8
        sw      $4,32($1)       ;  Y[i] = X[i] * 8 (Y is 8 × 4 = 32 bytes from X)
        addi    $1,$1,4         ;  increment pointer to next element
        subi    $3,$3,1         ;  decrement loop counter
        bne     $3,$0,Loop      ;Until all done
```

This MIPS code is not too dissimilar to the following ARM processor code.

```
        adr     r1,X            ;Load r1 with address of X
        mov     r2,#8           ;set up loop counter
Loop    ldr     r1,[r0],#4      ;Repeat: get xi
        mov     r1,r1, LSL #3   ;  shift xi left three times to multiply by 8
        str     r1,[r0,#28]     ;  Y[i] = X[i] * 8 (note offset 28 not 32)
        subs    r2,r2,#1        ;  decrement loop counter
        bne     Loop            ;Until all done
```

We have been able to compress the ARM processor code by using autoindexing.

Other Loads and Stores

MIPS provides several load and store instructions that let you load and store 8-bit byte values into a register; that is

lb	load	byte	lb	$1,12($2)	[$1] ← [12 + [$2]]	sign extend to 32 bits
lbu	load	byte unsigned	lbu	$1,12($2)	[$1] ← [12 + [$2]]	zero extend to 32 bits
lh	load	halfword	lh	$1,12($2)	[$1] ← [12 + [$2]]	sign extend to 32 bits
lhu	load	halfword unsigned	lhu	$1,12($2)	[$1] ← [12 + [$2]]	zero extend to 32 bits
sb	store	byte	sb	$4,64($6)	[64 + [$6]] ← [$4]	store a byte in memory
sh	store	halfword	sh	$4,64($6)	[64 + [$6]] ← [$4]	store a halfword in memory

MIPS and the ARM Processor

It's tempting to ask, "Which is the better: MIPS or the ARM processor?" There's no easy answer for several reasons. Both MIPS and the ARM processor are available in several versions with different architectures and different clock rates. MIPS has far more internal registers that can reduce memory traffic considerably, particularly for arithmetic operations. The ARM processor supports conditional execution, which makes it easier to write compact code. Array handling is easier for programmers using ARM processors because of its wealth of addressing and autoindexing modes. On the other hand, MIPS does provide a true 16-bit constant in literal operations. However, in 2011, the ARM processor was the better selling chip, and some games processors were replacing MIPS with ARM processors.

4.4   Data Processing and Data Movement

In this section, we are going to look at some of the aspects of data movement ranging from packing and shifting data elements, to processing groups of bits, to checking that data elements are within the correct bounds. We look at processors other than the ARM. The point of this section is not to teach new ISAs but to demonstrate the variety in the approach to computer design.

The most frequent computer operation is *data movement*, which is the copying of data from one point to another. As we have seen, computers have load and store instructions and register-to-register data transfers. Sometimes, you have to do more than copy data from one place to another. You may wish to modify the order of the bytes in a 32-bit word as they are moved, or you may need to move data from consecutive memory locations to consecutive *odd* or *even* locations (required by memory-mapped peripherals). Consequently, we look at some of the instructions provided for the movement and packing of data.

Consider the problem of *endianism*. One system may represent four bytes {A, B, C, and D} as ABCD, whereas another system may represent that same data as DCBA. Intel's IA32 processors provide a BSWAP reg32 instruction that performs a big-endian to little-endian transformation that converts the string of 32 bits [31...24, 23...16, 15...8, 7...0] to the form [7...0, 15...8, 23...16, 31...24]; that is, the byte sequence ABCD becomes DCBA.

You could implement an instruction that permutes the four bytes of a 32-bit value in any order. Suppose that their natural order is written 4321. The hypothetical instruction PERM 1234,**R0** performs an endian transformation, because the bytes are written back in reverse order. Similarly, PERM 1324,**R0** swaps the outer bytes and PERM 2233,**R0** would swap the innermost two bytes and also copy them to the outermost bytes. Some of the multimedia extensions that we discuss in Chapter 5 do indeed support this type of byte manipulation or *shuffling*.

Figure 4.12 describes an IA32 data movement instruction xlat (translation) that takes no parameters because it employs two specific registers: the 8-bit register al and 16-bit base register bx. As you can see from Figure 4.12, base register bx points to a region of memory and register al contains an 8-bit offset. When xlat is executed, the contents of al are added to bx to give an effective address (al is being used as an offset). The 8-bit operand at this effective address is then loaded into al. In other words, the offset is used to look up the data element at that offset in a table and then that value replaces the offset. The following code fragment demonstrates how xlat is used. The third instruction makes reference to ds, which is the register that points to the data segment in the IA32's segmented memory system.

```
mov  al,4          ;load index into al
lea  bx,table      ;set up base address of table
xlat               ;get byte at ds + bx + al in al
```

xlat operates only with tables of up to 256 byte values. It can be used, for example, to convert one code into another. If bx contains the address of the code conversion table and al contains the code to be looked up, simply executing xlat performs the code conversion.

FIGURE 4.12   Effect of the xlat instruction

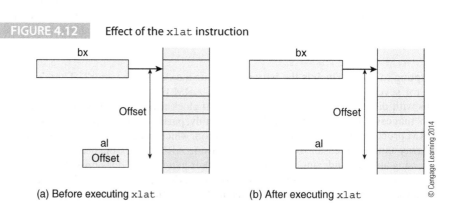

(a) Before executing xlat   (b) After executing xlat

© Cengage Learning 2014

Data Movement—An Overview

The data move or copy instruction is the most important of all computer operations because it is the most frequently executed type of instruction. Move instructions are also classified by the *size* of the data they move (e.g., 8, 16, or 32 bits) and the source and destination of the operands. Some processors specify the size of the data transfer explicitly with a special mnemonic (e.g., STW = store word, STB = store byte). Others use an *extension* or *suffix* to specify the size of the data transfer (e.g., MOVE.B = move byte, MOVE.W = move word). Sometimes the size of the data transfer is specified implicitly by the source or destination (e.g., LDA = load 8-bit accumulator, LDX = load 16-bit X register).

The figure below illustrates some variations in the move instruction beginning with the basic transfer of data between a register and memory in Figures (a) and (b). The source and destination of operands taking part in move instructions may be an internal register or a location in memory. All processors permit register-to-memory, memory-to-register, and register-to-register moves. Few microprocessors permit direct memory-to-memory moves.

Some processors implement instructions that exchange the contents of two registers. For example, EXG X,S might copy the X register to the stack pointer and vice versa. Some processors implement an instruction that can be applied to two fields of a single register. The contents of one field are swapped with the contents of the other field. For example, SWAP X exchanges the two halves of a register. Figure (d) illustrates the effect of an instruction that exchanges the contents of a pair of registers, and Figure (e) describes an instruction that swaps over the two halves of a register. Of course, you could devise an instruction that allows you to arbitrarily shuffle the bytes of a register, as (f) demonstrates.

Not all computer literature employs a consistent terminology: the terms swap, exchange, and transfer are interchangeable to some extent. We can summarize move operations in the following way.

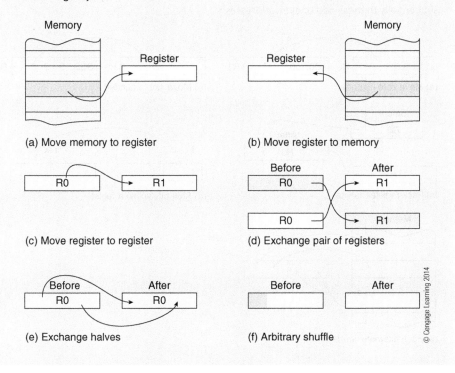

(a) Move memory to register

(b) Move register to memory

(c) Move register to register

(d) Exchange pair of registers

(e) Exchange halves

(f) Arbitrary shuffle

(continued)

Type	Data Transfer
Data transfer	Register → register
Data transfer	Register → memory
Data transfer	Memory → register
Data transfer	Memory → memory
Data exchange	$Register_A → Register_B$; $Register_B → Register_A$
Data swap	$Register_{Field\_A} → Register_{Field\_B}$; $Register_{Field\_B} → Register_{Field\_A}$

Some move instructions act on data when it is transferred. When a two's complement value is widened and an m-bit value is represented in n bits where $n > m$, the sign-bit is copied to the new bits. For example, the 8-bit value 10001100 would be represented in 16 bits by 1111111110001100. Some computers such as the IA-32 have a special move and sign extend, MOVSX, instruction that copies a source operand to a register and sign-extends an 8-bit value to 16 or 32 bits or a 16-bit value to 32 bits.

The following Figure (a) demonstrates a move instruction where the destination is wider than the source and the destination is padded with leading zeros, whereas in (b), the source data is moved and the destination padded with the sign bit.

Figure (c) demonstrates a packing instruction that takes the least-significant word in two registers and packs them into a third register. Figure (d) describes a general data movement operation in which a field of a register is shifted within the word. As this form of data movement is very important, we shall soon revisit it when we look at shift operations.

Figure (e) illustrates a rather curious form of data movement operation. Four bytes in consecutive word locations are moved into the four bytes of a register (or vice versa). This seemingly strange operation is used by microprocessors where an input or output device is assigned consecutive odd or even byte addresses and it is necessary to transfer bytes sequentially to these odd or even addresses.

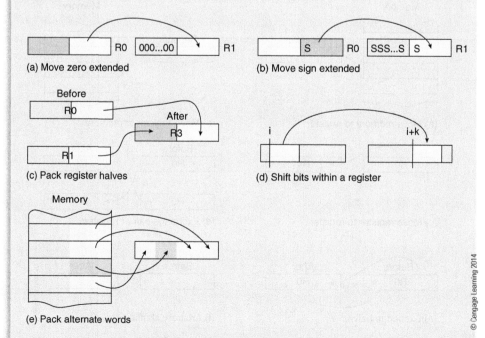

(a) Move zero extended

(b) Move sign extended

(c) Pack register halves

(d) Shift bits within a register

(e) Pack alternate words

This instruction demonstrates the strengths and weaknesses of the classic CISC philosophy. A single instruction xlat performs an operation normally requiring two operations (i.e., add the index to the base register and perform a register-indirect move). xlat is a compact instruction because it doesn't require operands (the use of the bx and al registers is implicit). On the other hand, xlat demonstrates the weakness of the CISC philosophy. It is used only in one specific application and is inflexible (the operand size is fixed and it can be used only with the al and bx registers).

4.4.1 Indivisible Exchange Instructions

We now look at a class of instructions that are required by operating systems in a very specific application: the synchronization of processes. Some data move instructions provided by both CISC and RISC processors look, at first glance, rather strange. For example, IA32 processors provide a *compare and exchange* instruction cmpxchg that uses three operands (one implicit and two explicit). Its format is cmpxchg reg,reg or cmpxchg mem,reg, and the operands may be 8-, 16-, or 32-bit values. This instruction compares the al, ax, or eax accumulator with the first operand, sets the zero flag if they are equal, and then copies the second operand into the first. If the accumulator and first operand are not equal, cmpxchg copies the first operand into the accumulator. We can describe the effect of cmpxchg **bx**,cx as

```
IF [ax] = [bx] THEN [z] ← 1, [bx] = [cx]
              ELSE [z] ← 0, [ax] = [bx]
```

The most important feature of the cmpxchg instruction is its *indivisibility*; that is, it is always executed to completion and cannot be interrupted. Although you could synthesize cmpxchg in terms of primitive instructions, these instructions can be interrupted before the sequence has been completed. The cmpxchg instruction is used in multitasking or multiprocessing environments where two processes may request the same resource. Without an indivisible or *atomic* instruction, each process may read the resource's status, find it free, and then claim it. If this happens, the resource becomes double-booked. Indivisible instructions such as cmpxchg first perform a test (in this case the comparison between the accumulator and the first operand) and then immediately perform one of two possible actions as a result of that test. No other process or processor may intervene between the start of the test and the completion of the following action.

> ### The Semaphore
>
> A semaphore is a flag that is used to provide a signal to processes and is necessary when two processes competing for a resource each ask if it's free almost simultaneously. Suppose process A and process B ask if resource Q is free (the resource may be a disk drive). Say the resource is free. Process A finds that it's free, and process B also finds that it's free. Then process A and process B each claim the resource and the system may lock up or crash.
>
> A semaphore solves this problem. When process A asks if it is free, the semaphore is locked and cannot be accessed again until after process A has cleared it to claim it. Semaphores are important in databases to avoid two processes from accessing the same item simultaneously.

The 68K family has an indivisible *test and set* instruction with the format TAS <ea>, where <ea> is the address of an operand in memory. This instruction tests the byte at the specified address and sets the negative and zero flag bits of the condition code register accordingly. The overflow and carry flags are cleared. The most-significant bit of the operand, bit 7, is set to 1. The RTL definition of TAS ea is

```
IF [ea]    = 0 THEN Z ← 1
IF [ea(7)] = 1 THEN N ← 1
[ea(7)] ← 1
```

These operations are indivisible, and the CPU executes a *read-modify-write* cycle in which the operand is read, modified, and written back to memory in one operation. The initial test determines the status of a *semaphore* flag. If the resource associated with the flag is free, the flag bit is set and the resource claimed before any other device, processor, or process can claim the resource.

The ARM processor has an indivisible swap, SWP, instruction that exchanges a word between a register and memory. For example,

```
ADR    r1,flag         ;r1 points to the flag (semaphore)
MOV    r0,#1           ;set r0 to 1
SWP    r0,r0,[r1]      ;perform the swap
CMP    r0,#0           ;test the result (check if memory locked)
```

When an ARM processor executes a swap instruction it also asserts a hardware signal called LOCK to indicate that the data transaction cannot be interrupted (the same is true of the 68K's TAS instruction where the address strobe remains asserted for both the read and write cycles instead of being negated between consecutive read and write cycles). We will return to this topic in Chapter 13 when we introduce multiprocessors.

4.4.2 Double-Precision Shifting

As we have seen, shift operations move all the bits of a register one or more places left or right; consequently, the maximum number of bits you can perform a shift over is equal to the length of a register. Sometimes you have to perform a shift over a larger number of bits (for example, when performing extended-precision arithmetic). Some processors do provide an extended shift in which the carry bit is included in this shift, allowing you to implement a multiple-precision shift in which the bit shifted out of one register is shifted into the carry and then into the second register taking part in the shift. The IA32 provides two double-precision instructions shld and shrd (shift-left double and shift-right double) that take a pair of operands and shift both simultaneously. The left-shift forms of the instruction are

```
shld operand1,operand2,immediate      ;"immediate" defines number of shifts
shld operand1,operand2,cl             ;register cl allows dynamic shifts
```

Operand 2 must be a 16- or 32-bit register. Operand 1 can be a register or a memory location. The number of places shifted may be expressed as a literal or as a dynamic value in the cl register. The shld operation first makes a temporary internal copy of operand 2, and then shifts the bits of operand 1 by the appropriate number of bits. The temporary copy of operand 2 is also shifted left, and the bits that fall out are shifted into operand 1; that is, operands 2 and 1 are treated as a single entity for the purpose of the shift. Because the temporary copy of operand 2 took part in the shift, the value of operand 2 itself is not affected by this operation. Thus, a shld **ax**,P,8 instruction shifts the contents of register ax left by eight places and copies the most-significant 8 bits of P into the least-significant 8 bits of ax. Figure 4.13 illustrates the effect of a shld **ax**,P,8 instruction.

FIGURE 4.13   Using the shld instruction

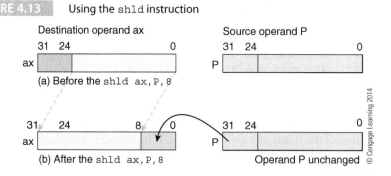

(a) Before the shld ax,P,8

(b) After the shld ax,P,8

Operand P unchanged

© Cengage Learning 2014

This double-length shift instruction can be used to pack data from several sources into a single register. Suppose we wish to pack register bx with 5 bits from memory location P, 7 bits from location Q, and 4 bits from location R. These bits are packed in the order PQR, where P is the most-significant 5 bits. We can use the code:

```
mov    ax,P        ;read the high-order bits from P into the accumulator
shld   bx,ax,5     ;copy the high-order 5 bits from ax into bx
mov    ax,Q        ;read the middle bits from Q into the accumulator
shld   bx,ax,7     ;copy the middle-order 7 bits from ax into bx
mov    ax,R        ;read the low-order bits from R into the accumulator
shld   bx,ax,4     ;copy the low-order 4 bits from ax into bx
```

Figure 4.14 illustrates the effect of these instructions. As you can see, we have to load the ax register with data three times in the above fragment of code.

FIGURE 4.14   Using `shld` to pack data

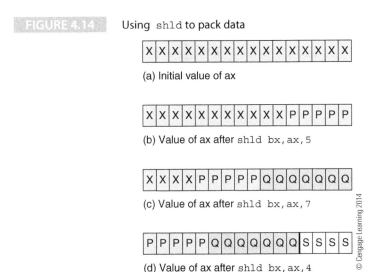

(a) Initial value of ax

(b) Value of ax after `shld bx,ax,5`

(c) Value of ax after `shld bx,ax,7`

(d) Value of ax after `shld bx,ax,4`

© Cengage Learning 2014

4.4.3   Pack and Unpack Instructions

Packing and unpacking data implies moving multiple data elements into a single register or memory location (packing) or moving one data element into multiple registers or memory locations (unpacking). Let's look at an example from the 68K ISA that implements PACK and UNPK instructions. Both of these instructions act on the lower-order 16-bits of a 32-bit register. Figure 4.15 illustrates the action of PACK D0,**D1**,#literal. As you can see, the PACK instruction takes the four 4-bit values in register D0 (in this case 3432_{16}) and converts them to the two 4-bit values (in this case, 42_{16}). This instruction is designed to facilitate the conversion between unpacked ASCII characters and packed BCD data. In this example, the two ASCII characters 4 and 2 from the keyboard, corresponding to codes 34_{16} and 32_{16}, respectively, are converted into the BCD equivalent 42_{10}. Note that the conversion process allows a 4-bit literal to be added to each of the source 4-bit words. In this case, the constants are all zero.

Figure 4.16 describes the inverse of the PACK instruction, UNPK, that takes two hexadecimal nibbles in the low-order byte of a word and converts them into two 8-bit values. In this case, the two nibbles are moved into consecutive bytes, and a constant is added to the result. If you are converting BCD values to ASCII character codes, you execute the instruction UNPK D0,**D1**,#$3030, because a BCD digit is converted to its corresponding ASCII code by adding 30_{16}.

These PACK and UNPK instructions are ingenious and save a tiny amount of effort in converting between BCD and ASCII values. However, these instructions are of no significant

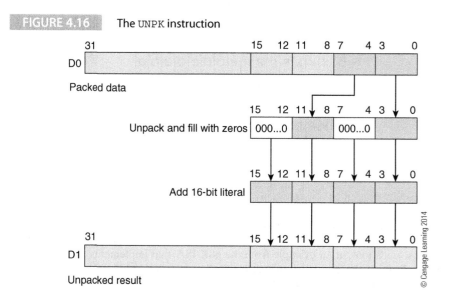

FIGURE 4.15 The `PACK` instruction

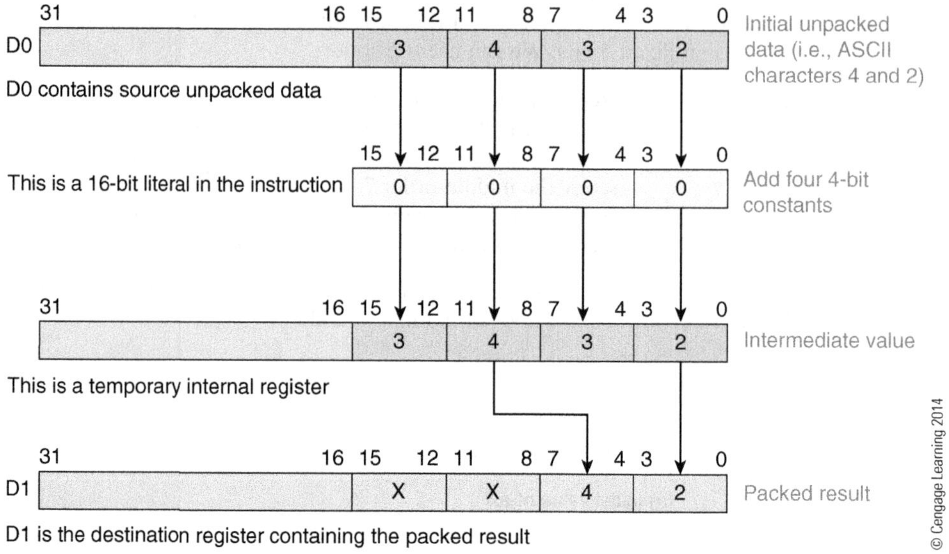

31
16 15 12 11 8 7 4 3 0
D0
3
4
3
2
Initial unpacked data (i.e., ASCII characters 4 and 2)
D0 contains source unpacked data
This is a 16-bit literal in the instruction
0
0
0
0
Add four 4-bit constants
Intermediate value
This is a temporary internal register
D1
X
X
4
2
Packed result
D1 is the destination register containing the packed result
© Cengage Learning 2014

FIGURE 4.16 The `UNPK` instruction

31
15 12 11 8 7 4 3 0
D0
Packed data
Unpack and fill with zeros
000...0
000...0
Add 16-bit literal
D1
Unpacked result
© Cengage Learning 2014

value, and it's quite difficult to see why silicon "real estate" was ever devoted to their implementation. It does however, tell us something about the *mindset* of computer designers in the late 1970s.

4.4.4 Bounds Testing

We're now going to look at an instruction that performs a very useful function by checking whether a value is within a predetermined range or not. When working with data structures such as arrays and tables, you need to know whether the element you are accessing falls within the array. An array access error occurs when the index (location) of an element is incorrectly computed at run time. Array errors are sometimes deliberately used by those writing malware to inject malicious code in a program. A problem can arise if the value of the array element is computed incorrectly and a data value is accessed outside the range of the array. Some high-level languages test that the subscript of an array being accessed is within its correct bounds (the C language does not provide such testing). The 68020 implements a *bounds checking* operation, CHK2, which determines whether an array subscript is within its correct range. If an out-of-range condition is detected, the operating system is invoked to deal with the situation; that is, a trap or exception is called.

Typically, an array subscript is compared against its upper and lower limits using two tests and two conditional branches to determine whether the address is within range. We can do the same with the single 68020 instruction CHK2 as follows.

```
          LEA     Array,A0      ;Address register A0 contains the base address of
                                 the array
          ADDA    D0,A0         ;The element index in D0 is added to the base and
   *                             A0 now points to the required element
          CHK2.L  Bounds,A0     ;Do a bounds check on pointer A0
          MOVE    (A0),D1       ;Read the required element

Bounds    DC.L    Lower         ;Store the lower bound in memory
          DC.L    Upper         ;followed by the upper bound
```

In this case, we require only one instruction to perform both the *upper* and *lower* bounds check. CHK2.L Bounds,A0 compares the value in A0 first with the lower bound at the address given by Bounds and then compares the value in A0 with the address given by Bounds+4. Here the bounds are 32-bit, four-byte values. If the value in A0 is within range, nothing happens. If it is outside the range defined by the bounds, an exception is generated and the operating system must deal with the recovery. The 68020 also provides a CMP2 instruction that has the same format as the CHK2 instruction but which sets the carry flag to signal an out-of-range error.

An interesting aspect of the CMP2 and CHK2 instructions is that they can test values in both address and data registers and that they work with 8-bit, 16-bit, or 32-bit bounds. Moreover, they can test for both *signed* and *unsigned* bounds. If you are testing a region of memory, the bounds might be, say, $801000 to $801FFF. If you are testing an array subscript that runs from -128 to $+127$, the bounds might be $80 to $7F. The processor automatically determines whether the bounds are signed or unsigned. For example, the pairs $20, $30 or $A0, $AF would be interpreted as unsigned, whereas the pairs $80, $20 or $B2, $C4 would be regarded as signed.

Figure 4.17 illustrates the relationship between the bounds specified in the CHK2 instruction and the range of valid values. In each case, we are testing register A0 against a pair of bounds. In the first two examples, the range is unsigned. Having looked at some of the facilities offered by complex instructions, we now examine how some processors provide a means of accessing data structures via memory indirect addressing modes.

Although the bounds check instructions are an excellent idea, they too were dropped from the 68K architecture because they were not cost-effective. In today's world where memory accesses are expensive in terms of clock cycles, the 68K's boundary checking instructions, as implemented, just do not make sense. The panel here introduces the *memory wall*, which

Just Another Brick in the Wall

Although we have not covered computer organization, microarchitecture, and memory technologies yet, we have to make an important point. The good news is that main store DRAM memory is getting faster year by year. The bad news is that it is not getting faster as fast at the processor. Consequently, the performance of processors is increasing at a faster rate than memory.

The increasing gap between processor and memory performance is called the *memory wall* and indicates a potential barrier to the progress of computer systems design. The term is attributed to Wulf and McKee, who wrote the classic paper "Hitting the Memory Wall: Implications of the Obvious" in 1994. Cache memory technology has somewhat mitigated the problem—but only to a limited extent.

From the point of view of computer architecture and ISAs, the implication of the memory wall is dramatically simple. Avoid memory access like the plague.

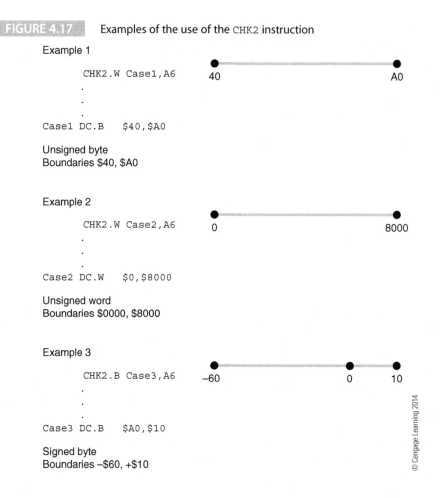

FIGURE 4.17   Examples of the use of the CHK2 instruction

Example 1

```
        CHK2.W Case1,A6
        .
        .
        .
Case1 DC.B    $40,$A0
```

Unsigned byte
Boundaries $40, $A0

Example 2

```
        CHK2.W Case2,A6
        .
        .
        .
Case2 DC.W    $0,$8000
```

Unsigned word
Boundaries $0000, $8000

Example 3

```
        CHK2.B Case3,A6
        .
        .
        .
Case3 DC.B    $A0,$10
```

Signed byte
Boundaries –$60, +$10

© Cengage Learning 2014

indicates that memory accesses are going to become increasingly more expensive in terms of access time. Consequently, an instruction such as CHK2 that requires the reading of a pair of bounds is just not cost-effective in today's world.

4.4.5   Bit Field Data

We begin with the *bit field*, which is a data structure that is just an arbitrary string of bits of any length, although real bit fields are restricted to a maximum width determined by the processor's register width. You can use bit fields to represent information that doesn't fit into a convenient 8-, 16-, 32-, or 64-bit package like characters, integers, and floating-point values. For example, a certain 19-bit bit field might represent a packed data value consisting of three individual fields of three bits, seven bits, and nine bits. Equally, it may represent a line of pixels in an image. Or it may describe the state of sectors (free/taken) in a disk directory. There is no limit to the use of bit fields.

You could say that 8-bit microprocessors imposed the tyranny of the byte on data structures, restricting data to units of eight bits severely limits both storage and data processing operations to those that act on multiples of a byte. There's no fundamental reason why we cannot consider memory as a long string of bits. However, bit fields are not widely implemented because of the additional complexity bit field operations impose on the underlying hardware. Since memory is physically byte-oriented with 8-, 16-, 32-, or 64-bit buses, an access to a bit field that spans several words may require multiple consecutive memory accesses, which degrades performance.

Because a bit field is nothing more than a string of consecutive bits, we can define a bit field in terms of two parameters: its *width* or length w and its location in memory q. The value of q is, of course, expressed in bits. For example, we could define a 56-bit bit field x as beginning 92,345 bits away from the first bit in memory and extending from bit 92,346 to bit 92,401. An alternative way of specifying bit fields involves a compromise between bits and bytes—it uses a byte address to specify a location in memory and then a bit offset from this location to specify the bit field's position with respect to the designated byte.

Figure 4.18 illustrates a bit field specified by a byte address plus an offset; the bit field starts 11 bits from bit 0 in byte i in memory and is 10 bits wide. We have numbered the bytes in memory from right to left and used a little endian arrangement for both bytes and bits.

© Cengage Learning 2014

FIGURE 4.18   The bit field

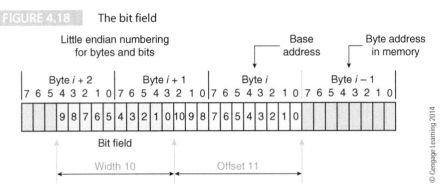

The structure in Figure 4.18 is *little endian consistent*. The bytes are numbered from the least-significant byte (on the right) as are the bits of a byte. The offset of the bit field from byte i begins at bit 0 of that byte. The offset bits are also numbered right-to-left, as are the bits of the bit field. We will soon see that not all processors follow this convention.

In theory, the size of a bit field can vary from 0 to the total number of bits in the memory system. Because bit fields are processed in registers within the computer, the maximum bit field size is the number of bits that can be held in a register or processed by the CPU (for example, 32). Practical bit field implementation runs into the limitations imposed by byte boundaries; that is, the bit field is the point at which computer architecture and computer organization collide.

Few microprocessors provide instructions that directly allow you to manipulate bit fields. Here we look at the 68020 that provides support for the bit fields described by Figure 4.19. Because the 68K family uses *little endian* bit numbering and *big endian* byte numbering, bit fields present a dilemma. Do the bits follow the 68K's *bit* numbering or *byte* numbering conventions? Figure 4.19 demonstrates that the bit field location is defined with respect to the most-significant bit of byte i (byte i is called the *base byte* and is the effective address of the bit field specified in instructions), and the bits of the bit field are numbered in the reverse sense to

> ### Bit Field Numbering
>
> The bits of a byte are numbered from right to left, but the bits of a bit field are numbered from left to right.
>
> In this example, the byte address of the bit field is i. The offset of the bit field is 11, because its first bit is 11 bits from the mostsignificant bit of byte i.
>
> The 10-bit bit field itself straddles bytes $i + 1$ and $i + 2$.
>
> Note that this numbering does not apply to the 68020 microprocessor.

the bits of a byte; that is, the bit field follows the big endian numbering convention. The least-significant bit of a bit field begins at bit 7 of the base byte. Let's repeat this. The bits of a bit field are numbered in reverse order with respect to the bits of a byte.

Let's look at an example of a typical 68020 instruction, where the *bit field insert operation*, BFINS, copies the bit field in data register Dn to memory. The mnemonic form is BFINS Dn, **<ea>{offset:width}**. The bit field is stored *offset* bits from the byte specified by the effective address <ea>. Consider the way in which BFINS D0, **1234{11:10}** is interpreted. The least-significant 10 bits in register D0 are copied into the main store, starting at 11 bits (i.e., the offset) from bit 7 of the base byte address 1234 (Figure 4.19

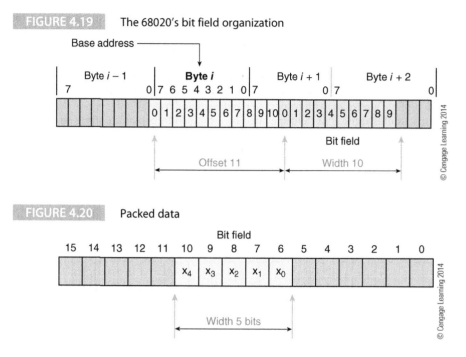

FIGURE 4.19   The 68020's bit field organization

FIGURE 4.20   Packed data

has the same offset:width values, which should help visualize this operation). The 68020 allows you to specify bit widths dynamically by using a data register. For example, you can write BFINS D0,**1234{D3:D4}**.

Let's look at how a bit field operation could be used in practice. Figure 4.20 demonstrates a 5-bit data element x that is packed within a 16-bit word in memory. Suppose we wish to extract this bit field. Without bit field operations, we would typically load the data into a register, shift the data right to put it in the least-significant bit position, and then clear the remaining bits to zero.

```
MOVE   PQRS,D0                    ;Get the 16 bits of packed data at location
                                   PQRS into register D0
LSR    #6,D0                      ;Shift D0 six places right to right-justify the bit
                                   field into D0-D5
AND    #%0000000000011111,D0      ;Clear all other bits of register D0. The %
                                   indicates a binary value
```

The 68020's *bit field extract* instruction, BFEXTU, performs this operation in one instruction:

```
BFEXTU   PQRS{5:5},D0        Get the packed data
```

Note that the bit field offset is 5, because the position of the bit field is measured from the most-significant bit of the base byte (i.e., bit 15 of the word). The first bit of the bit field is bit x_4, which is five bits to the right of bit 15. Yes, it makes my head ache too.

Typical bit field operations allow you to read a bit field from memory, to insert a bit field in memory, to clear/set/toggle all of the bits of a bit field, and to test a bit field. Figure 4.21 demonstrates how the 4-bit bit field in bits 6 to 3 of memory location 1000 can be moved to bits 4 to 1 of memory location in two instructions by

```
BFEXTU $1000,{1:4},D0        ;Read the source bit field into D0
BFINS  D0,$1003,{3:4}        ;Store the bit field in memory
```

Recall that the offsets 1 and 4 are specified with respect to bit 7 of the byte at the base address. BFINS is a bit field insert instruction that loads a bit field into memory. The following list describes the 68K's bit field instructions.

FIGURE 4.21   Using bit field instruction to move bits

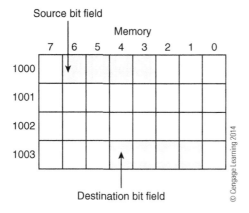

BFEXTU BFEXTU <ea>{offset:width},**Dn** The *extract a bit field unsigned* instruction
copies a bit field from memory and deposits it in a data register. This is the bit field
equivalent of the conventional MOVE or LOAD instruction. The bit field is loaded
into the low-order bits of the data register, and the high-order bits are set to zero.

BFEXTS BFEXTS <ea>{offset:width},**Dn** The *extract a bit field signed* instruction also
deposits a bit field in a register. When the bit field is moved into the data register,
it is sign-extended to 32 bits.

BFINS BFINS   Dn, **<ea>{offset:width}** The *insert bit field* instruction copies a bit
field from a data register to memory and is the bit field equivalent of a STORE.

BFTST BFTST   <ea>{offset:width} The *bit field test instruction* tests the specified bit
field and sets the CCR accordingly. The *N*-bit is set if the most-significant bit of the
bit field is one, and the *Z* bit is set if all the bits of the bit field are zero.

BFCLR BFCLR   <ea>{offset:width} The *bit field clear* instruction tests the bit field
exactly like BFTST and then clears all the bits of the bit field.

BFSET BFSET   <ea>{offset:width} The *bit field set* instruction tests the bit field
exactly like BFTST and then sets the bits of the bit field to one.

BFCHG BFCHG   <ea>{offset:width} The *bit field test and change* instruction behaves
exactly like a BFTST instruction, except that the bits of the bit field are all inverted
after the test.

BFFFO BFFFO   <ea>{offset:width},**Dn** The *find first one in bit field* instruction performs a calculation on the bits of a bit field. The bit field at the specified address is
read and scanned, and the *location* of the first 1 bit in the bit field is then loaded into
the specified data register. Note that the location of the first bit set to one is defined as
the offset of the bit field plus the location of the bit within the bit field. If no 1 is found
(i.e., the bit field was all zeros), the value returned is the offset plus the field width.

Figure 4.22 demonstrates the power of the BFFFO instruction using a 21-bit bit field that
begins in byte 1001. Suppose we wish to locate the position of the first bit set to a 1 within this
bit field. If $1000 is the base byte, the operation BFFFO $1000{10:21},**D0** scans the bit field
and determines the location of the first 1 (i.e., bit 15 of the bit field) and loads the value 25 into
register D1. The value is 25, because it's the location of the first 1 in the field plus the offset 10.

Bit fields are important in areas such as computer graphics, because the bit field can be
used to specify part of a line in a display. A bit field can be used to describe a *free sector list* in
a disk's directory. If you imagine a bit field representing a sector list, each bit in the bit field
is associated with a sector on the disk. If a bit is set, the corresponding sector is in use (i.e.,
belongs to a file). If the bit is clear, the corresponding sector is available for incorporation in a
file. Similarly, bit field operations can be used to count the number of 1s in a string or to locate
the most-significant bit in floating point operations. Motorola dropped bit field instructions
in their later derivatives of the 68K ColdFire family and used traps (operating system calls)

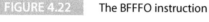

FIGURE 4.22   The BFFFO instruction

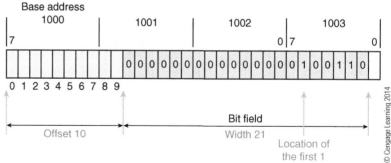

to emulate bit field operations. In the next section, we take a brief look at how computers can mechanize a simple operation called the looping construct.

ARM Bit Fields

Some versions of ARM processors support limited bit field operations. Later in this chapter, we introduce the ARM's *thumb mode* that allows it to behave as if it had a 16-bit instruction set architecture with a two-address format. The second-generation Thumb architecture provides a *bit field insert* operation of the form

```
BFI r0,r1, #bitpos, #bitwidth
```

that inserts a range of bits from one register into another (reducing an action that would require three instructions into a single operation).

In addition to the bit field insert, the ARM's Thumb-2 state provides a *bit field clear* and both signed and unsigned *bit field extracts*. These operations apply only to data in registers unlike the 68020 that can directly operate on memory.

4.4.6 Mechanizing the Loop

Our next topic is the *loop*, one of the most common programming constructs. Although all processors provide conditional branches to support loops, some have ways of making the common looping construct more efficient. First, let's look at the basic loop that can be expressed in pseudocode as

```
Preset loop variable
REPEAT
       Perform some action
       Decrement loop variable
UNTIL loop variable = 0
```

We can readily implement this in ARM processor code by

```
      MOV   r0,#10        ;Set up a loop counter ready to count down
Next                      ;   Body of loop
      SUBS  r0,r0,#1      ;   Decrement the loop counter and set the status bit
      BEQ   Next          ;REPEAT: until zero
```

The key actions in a loop construct are *decrement a counter* and then *branch back on nonzero*. Intel's IA32 architecture implements an explicit loop instruction that performs a decrement and branch using the cx register as a loop counter[2] and, therefore, requires no explicit reference to a register. The instruction is LOOP, and it just takes a target address as an operand. The IA32 code is

```
        MOV  cx, count          ;set up count value
Next    Do something            ;body of the loop
        LOOP Next               ;decrement counter and branch on non-zero
```

The 68K provides a sophisticated *decrement and branch instruction*, DBRA, which behaves like the IA32's loop instruction but allows the programmer to specify one of eight loop counters and one of *two* exit points. The loop can be terminated when the loop count has been exhausted or when a specific condition has been detected. We can express the action of DB$_{condition}$ **Di**,target as

```
IF condition TRUE THEN EXIT                    ;exit on condition true
                ELSE [Di] ← [Di] - 1           ;
                IF [Di] = -1 THEN EXIT          ;or loop count exhausted
                            ELSE [PC] ← target
```

This instruction terminates the loop on a count of -1 rather than 0. The following fragment of code demonstrates how a DBCS (decrement and branch on carry set) might be used to add together ten numbers but terminate if integer overflow occurs.

```
        MOVE  #10,D0            ;Set up loop counter ready to count down
        CLR   D1               ;Clear the total in register D1
        LEA   Table,A0         ;Point to the list of numbers
Next    ADD   (A0)+,D1         ;REPEAT: Add in the next number
        DBCS  D0,Next          ;UNTIL all added OR overflow
```

Without the DBCS instruction, the body of the loop would require four instructions. Note that ADD (A0)+,**D1** adds the contents of memory pointed at by register A0 to register D1 and then increments the pointer in A0. The LEA (load effective address) instruction loads a pointer into an address register; in this case, it loads address register A0 with the address 'Table.' Finally, the 68K has an explicit, clear instruction that clears a data register by setting its contents to zero; that is, CLR D0 results in [D0] ← 0.

4.5   Memory Indirect Addressing

We now introduce *memory indirect addressing*, which is a means of implementing complex data structures. Register indirect addressing uses a pointer to access the required operand. In *memory indirect addressing*, a register provides a pointer to *a pointer in memory*. The actual operand is accessed by reading this *second* pointer and accessing the element at the address given by the pointer. *Four* memory/register accesses are required: read the instruction, read the register containing the pointer to memory, read the memory containing the pointer to the operand, and access the operand.

Figure 4.23 illustrates *memory indirect addressing*, where a pointer register contains the 32-bit value 1234_{16}. The contents of the target address specified by this pointer are 122488_{16} and are used as a second pointer to access the actual operand. If the initial pointer register is R1, the destination register is R2, and the instruction is a move, we can express this operation in RTL as

```
[R2] ← [[[R1]]]
```

---

[2] The original Intel X86 architecture had four general-purpose 16-bit registers called *AX, BX, CX,* and *DX*. Unlike the 68K, the X86's registers also performed specific functions. For example, the *CX* register can also be used as a counter. The advantage of dedicating a register to counting functions is that you don't have to specify it in counting instructions, which makes the op-code shorter because you don't need a counter-select field.

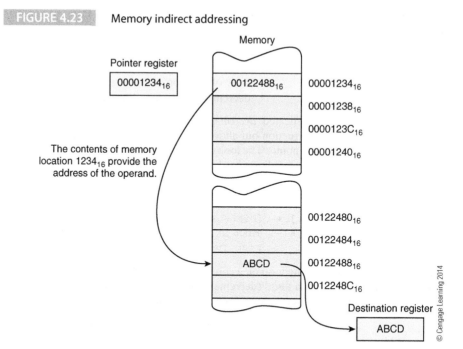

FIGURE 4.23   Memory indirect addressing

This expression defines the *three* memory references required to get the operand: read the contents of R1, read the contents of the memory pointed at by the pointer in R1, and read the contents of memory using the pointer accessed in memory. Perhaps a better way of presenting the RTL is to break up the sequence into more manageable bits.

```
Pointer1  ←   [R1]          ;This is the pointer in the register
Pointer2  ←   [Pointer1]    ;This is the pointer in memory
Operand   ←   [Pointer2]    ;This is the final operand
[R2]      ←   Operand       ;Store the operand in the destination register
```

The only processors to support this general memory indirect addressing mode, the 68020, 68030, and 68040, use the assembly language format:

```
MOVE  [(A1)],D2        [D2]  ←  [[[A1]]]
```

Address register A1 is the pointer register, and data register D2 is the destination operand register. This addressing mode is the degenerate form of a more general 68020 memory indirect addressing mode. Before we look at the fine details of this addressing mode, we need to ask the question, "Why do we need it?"

Register indirect addressing with its single pointer is useful for dealing with arrays or tables of simple data values. Consider Figure 4.24, where a data structure consists of a sequence of consecutive 16-bytes values. The pointer register contains 1234_{16}, corresponding to the first item in the structure. To access item 2, you have to add 16 to the value in the pointer register. Processors use a *register indirect with offset mode* to add a constant to a pointer. For example, the ARM processor lets you write LDR **r1**, [r0,#16], and the 68K lets you write MOVE (16,A0),**D1**. Some computers go one step further and permit *double indexing*, where the effective address is given by the sum of two registers. For example, LDR **r1**, [r2,r3] copies the data pointed at by the sum of the contents of r2 and r3 into register r1. Such double indexing allows you to use one register to point to the head of the data structure and the other to hold the required item's offset from the head of the structure.

Unfortunately, not all data structures are as well ordered as that in Figure 4.24, where the size of each of the data items is the same. Figure 4.25 illustrates the situation where each of the four items has a different size. We can't step through this data structure item-by-item just by adding a fixed constant to the pointer register.

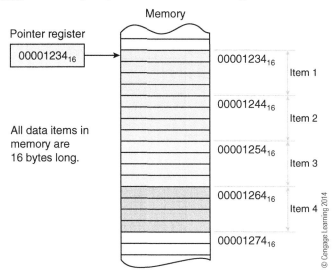

FIGURE 4.24   Accessing a *regular* data structure with register indirect addressing

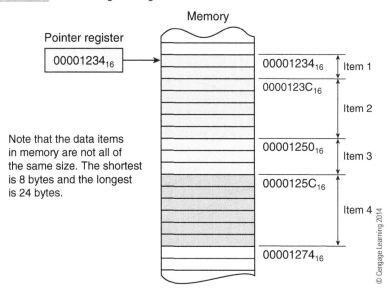

FIGURE 4.25   Accessing an irregular data structure

We can take an alternative approach to accessing data in structures with records of varying length. Figure 4.26 uses a pointer register that points to a *table of pointers*. Each of these pointers points to the actual record in memory. You can step through the data items simply by incrementing the base pointer by four, because the base pointer steps through the table of pointers.

If memory indirect addressing is used and register A0 is the base pointer, we can access the first element of a data item by MOVE [(A0)],**D0**. Executing ADD #4,**A0** results in A0 being incremented by four bytes and pointing at the pointer to the next element.

Figure 4.27 shows how the 68020 accesses *both* the pointer table and target data structure using ([Offset$_{inner}$,A0],Offset$_{outer}$), which has *three* parameters. Let's walk through this. First, the base pointer register A0 and inner displacement are added together to locate the pointer in the table of pointers that points at the data structure. The pointer is read from memory and the outer displacement is added to it to access the desired target operand. The effective address is ([Offset$_{inner}$,A0],Offset$_{outer}$]), which is expressed in RTL as

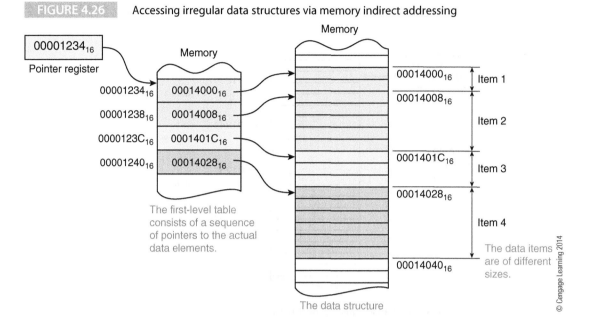

FIGURE 4.26   Accessing irregular data structures via memory indirect addressing

FIGURE 4.27   Accessing an operand within the target data structure

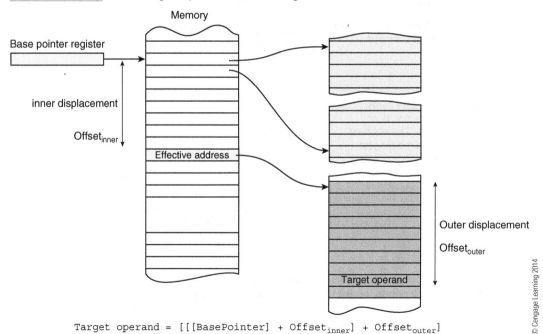

Target operand = [[[BasePointer] + Offset$_{inner}$] + Offset$_{outer}$]

[[[A0] + Offset$_{inner}$] + Offset$_{outer}$],

where A0 is the base pointer, Offset$_{inner}$ is the constant offset into the pointer table, and Offset$_{outer}$ is the constant offset into the target structure. In practice, the situation is slightly more complex; the contents of an index register (either an address or a data register) can be added to either inner or outer displacement but not both. The index register can be scaled by 1, 2, 4, or 8.

In order to demonstrate the power of memory indirect addressing, we will provide two further examples. One is a jump table that can be used to implement the case construct, and the other offers a means of accessing complex data structures.

Using Memory Indirect Addressing to Implement a switch Construct

A common construct in many high-level languages is the switch that allows you to invoke one of *n* functions depending on the value of a variable. Suppose you were constructing a CPU simulator. You might have an inner interpreter that looks something like the following to select one of four cases.

```
Switch (operation)
{
  case LOAD:  { LOAD  code; break:}
  case STORE: { STORE code; break:}
  case ADD  : { ADD   code; break:}
  case BEQ  : { BEQ   code; break:}
}
```

Figure 4.28 illustrates a possible data structure for this construct where a table in memory holds pointers to the functions. The required function is executed by loading the appropriate pointer into the program counter. Let's implement a switch construct using a conventional CISC architecture such as the 68K. We can use memory indirect addressing to call the required subroutine by executing

```
JSR ([A0,D0*4])     ;Call the subroutine specified by D0 (Table base in A0)
```

That is, [PC] ← [[A0] + 4 x [D0]].

FIGURE 4.28   The jump table

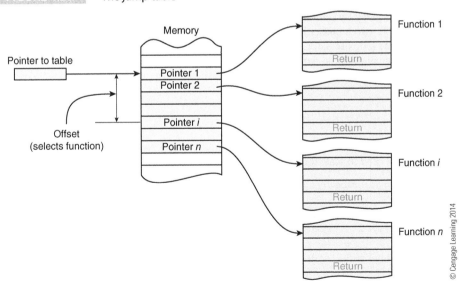

© Cengage Learning 2014

In this example, the contents of register A0 are added to the contents of register D0 multiplied by 4 to create the effective address of the required subroutine. The 32-bit address at this location is read and loaded into the program counter to call the subroutine. Figure 4.29 illustrates the situation in which A0 points at the table of pointers and register D0 contains the index 6.

Memory indirect addressing provides an elegant solution with two parameters and a single instruction, JSR ([A0,D0*4]). The scaling factor 4 that forms part of the instruction turns the case number into a four-byte (32-bit) address offset. Without memory indirect addressing, we would have written something like: This code requires *four* instructions and overwrites base register A0.

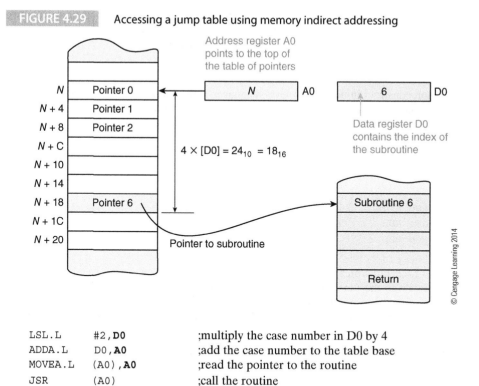

FIGURE 4.29   Accessing a jump table using memory indirect addressing

```
LSL.L    #2,D0          ;multiply the case number in D0 by 4
ADDA.L   D0,A0          ;add the case number to the table base
MOVEA.L  (A0),A0        ;read the pointer to the routine
JSR      (A0)           ;call the routine
```

Before we complete this topic, we demonstrate the versatility of the 68020's addressing mode in all its glory by demonstrating the optional parameters. Figures 4.30 and 4.31 illustrate the 68020's two memory indirect addressing modes. These addressing modes differ only in the use of the index register. It was not possible to make the addressing mode fully flexible and supporting all possible variations, as that would have required *six* parameters to specify an address. Instead, we are provided with two basic options: preindexing and postindexing.

SWITCH Revisited—The ARM Version

Here's an example of the switch construct using ARM code. Note how the built-in shift operation is used to scale the case number.

```
switch (Number)
{
    Case 0:   code 1;   break;
    Case 2:   code 2;   break;

    Case m-1: code m-1; break;
}

                                          ; Assume case number in r0 initially
            ADR  r1,CaseTab               ; load r1 with pointer to table of cases
            LDR  r15,[r1,r0,LSR #2]        ; jump to selected case
            .

            .
    CaseTab DCD  case 0                    ; code for case 0
            DCD  case 1                    ; code for case 1
```

FIGURE 4.30   Preindexed memory indirect addressing

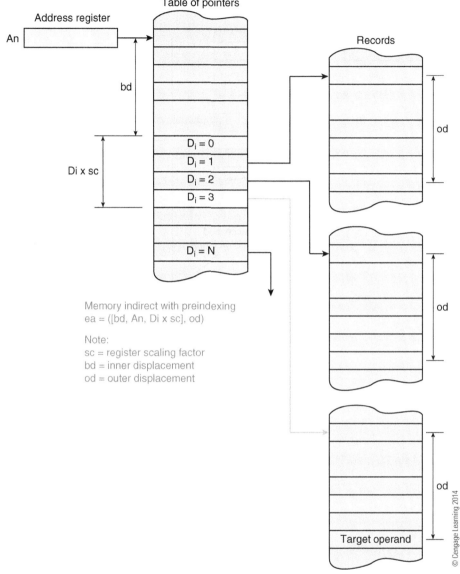

Figure 4.30 illustrates memory indirect *preindexing* addressing where the index register selects one of the pointers in the table of pointers. Because the index register can be changed at run time, we can change the pointer and, therefore, select any record.

Figure 4.31 illustrates memory indirect addressing with *postindexing*, where the index register is added to the pointer to the record that lets us select a particular element within a record dynamically. Because the 68020's memory indirect-addressing modes permit only one index register, the programmer is forced to decide whether the pointer in the first-level table is to be made a runtime variable or whether the index into a target record in the second-level table is to be a runtime variable.

In Figures 4.30 and 4.31, sc is a scaling factor that can be used to scale the contents of the index register. For example, the effective address ([20,A3],D4*4,64) takes the contents of A3 and accesses the 32-bit value at [A3] + 20, where the 32-bit value at this address is a pointer. Then, the contents of D4 multiplied by 4 are added to the pointer plus 64. This final value is the location of the desired operand.

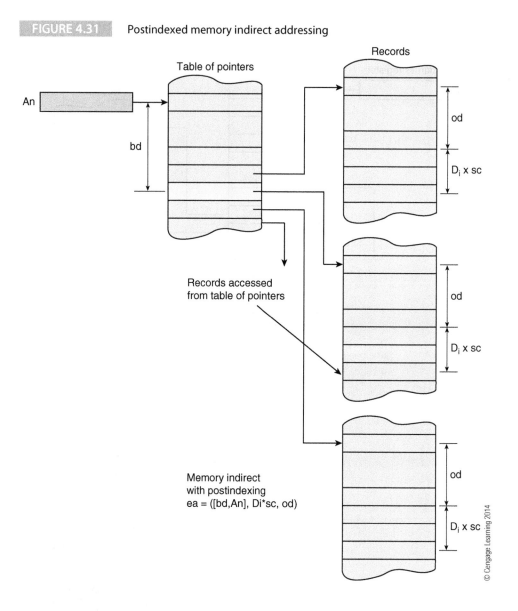

FIGURE 4.31   Postindexed memory indirect addressing

Table of pointers

Records

An

bd

Records accessed
from table of pointers

Memory indirect
with postindexing
ea = ([bd,An], Di*sc, od)

od

D_i x sc

od

D_i x sc

od

D_i x sc

© Cengage Learning 2014

Using Memory Indirect Addressing to Access Records

Suppose we have a set of records indexed by day and that each record contains up to six 32-bit items. Assume also that the table of pointers to the records contains 64 bytes of data that describe the table (Figure 4.32). We have constructed a region of memory with the 64 bytes of data used by the table of pointers followed by the pointers themselves. Each pointer points to the appropriate day's six results.

Register A0 in Figure 4.32 points to the base of the memory devoted to the data structure, which may include other items as well as the pointers to the records. The base displacement, bd, is the offset to the start of the list of days with respect to the start of the region of data—that is, the first day's entry is at address [A0] + bd.

The index of the day to be selected is in data register D0. Since each entry in the table of pointers is a four-byte value, we have to scale the contents of D0 by 4. The effective address of the pointer to the selected record is therefore [A0] + bd + 4*[D0]. The processor reads this pointer, which points to the start of the day's record. Suppose we want to know the value of item 5. The outer displacement provides us with a facility to do this. When the processor reads the pointer from memory, it adds the outer displacement to it to calculate the effective

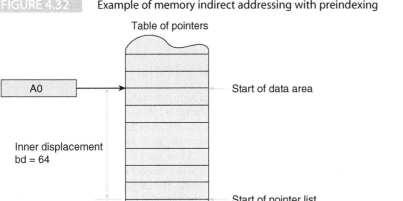

FIGURE 4.32   Example of memory indirect addressing with preindexing

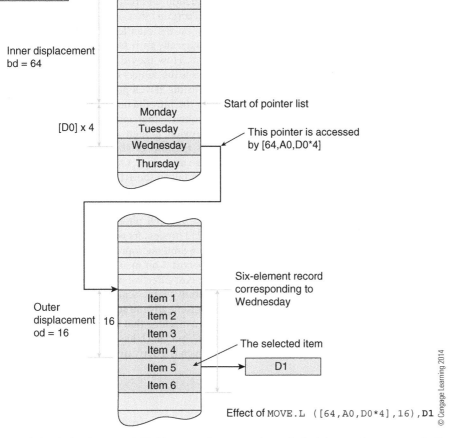

Effect of MOVE.L ([64,A0,D0*4],16),D1

address of the desired operand. If this example were to be coded without using memory indirect addressing, its assembly form might look like

```
LSL.L    #2,D0                ;Multiply the student index by 4
LEA      (64,A0,D0.L),A1      ;Calculate address of pointer to record
MOVEA.L  (A1),A1              ;Read the actual pointer
ADDA.L   #4,A1                ;Calculate address of CS result
MOVE.B   (A1),D1              ;Read the result
```

The same calculation can be carried out using memory indirect addressing with preindexing:

```
MOVE.B   ([64,A0,D0.L*4],4),  D1
```

As we have said, such complicated memory indirect-addressing modes are no longer supported—even though they can provide remarkably compact and elegant machine-level code. Indeed, some argue that such complicated addressing mechanisms cannot easily be exploited by compilers and are ultimately of little value. Still, if you combine the 68020's memory indirect addressing with its bit field instructions, you could write

```
BFFFO    ([128,A3,D4*4],64){D2,D3},D7
```

which has eight parameters. Such an instruction is immensely powerful (we leave it as an exercise to the student to synthesize it from basic machine level operations). However, it is complicated to implement, relatively slow, reduces the rate of instruction execution, and is very difficult for compilers to exploit fully.

4.6   Compressed Code, RISC, Thumb, and MIPS16

We now introduce a rather unusual topic in instruction set architectures, the notion of *compressed code*. Essentially, the ARM processor and certain other processors are able to *change gear* and take on the mantle of a different architecture in order to magically turn from a 32-bit machine into a 16-bit machine. How they do this and why they should want to do this is the subject of this section.

An interesting development in computer architecture has been the introduction of the compressed RISC. It is now axiomatic that modern CISC/RISC 32/64-bit processors provide a high throughput in workstation environments because of their wide buses, high clock rates, and large memories. One of the last bastions of the CISC world to fall to the RISC has been the embedded processor. The embedded processor is found in applications ranging from the laser printer to the MP3 player, digital camera, the cell phone, and the toy. For a long time, embedded processors were modern versions of second generation 8-bit microprocessors. An 8-bit processor used very low-cost 8-bit memory, peripherals, and buses in order to be cost-effective in a competitive market.

Compression

You can compress the size of text files by replacing commonly occurring strings with short symbols. Analogous techniques can be used to compress the instructions of a processor to create a dense form of code. The price paid for compressed code is a more limited instruction set. The benefit of compressed code is the ability to use 8-bit wide buses and peripherals. Embedded applications such as cell phones, personal organizers, and palmtops cannot afford the luxury of wide buses where cost is at a premium. Here we look at the compressed ARM instruction set called the Thumb and the compressed MIPS called the RISC16.

RISC manufacturers wanted to enter the lucrative large-scale embedded processor market, but their 32-bit machines were simply not cost-effective in such applications. A compromise was to create the *compressed RISC*, which is a machine that has many features of a RISC architecture, is compatible with a traditional RISC, but which has a much shorter wordlength. One of the first such machines was the Thumb, which is a derivative of the ARM architecture.

4.6.1   Thumb ISA

We cover the Thumb state because its architecture demonstrates a high level of ingenuity. Thumb takes the ARM processor's 32-bit instruction set and forces it into a 16-bit mold while remaining within the spirit of the ARM processor's instruction set architecture. The ARM processor's Thumb state gives the designer the best of both 16-bit and 32-bit worlds: The processor can execute both compressed 16-bit Thumb code and normal 32-bit code. This sleight of hand is achieved by putting the required ARM processor code in small 32-bit wide memories and then locating everything else in low-cost, 16-bit wide memories. Thumb code is 26% smaller than ARM processor code if optimized for performance and 32% smaller if optimized for size. When optimized for performance, Thumb code can achieve 98% of the performance of native ARM processor code.

Figure 4.33 describes Thumb's register set much as we described the ARM processor's register set in Chapter 3. In Thumb state, the programmer has unrestricted access to registers r0 to r7, the stack pointer (r13), the link register (r14), and the program counter (r15). Registers r8 to r12 exist but can be accessed only by special instructions. Most Thumb instructions employ a two-address format similar to conventional CISC ISAs. Thumb instructions are all 16 bits wide and use some ARM architecture concepts. For example, several instructions provide conditional execution.[3]

---

[3] Philippe Robin, ARM Ltd., Embedded Linux Conference 2007.

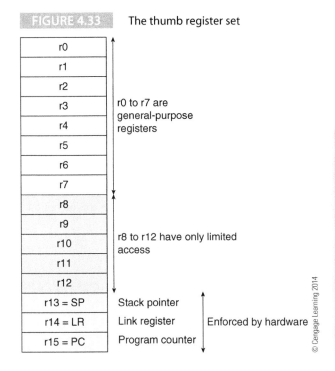

FIGURE 4.33   The thumb register set

r0 to r7 are general-purpose registers

r8 to r12 have only limited access

r13 = SP   Stack pointer
r14 = LR   Link register   Enforced by hardware
r15 = PC   Program counter

© Cengage Learning 2014

Thumb Registers

Registers r0 to r7 can be accessed by general-purpose Thumb instructions.

Registers r8 to r12 cannot be accessed normally other than by special-purpose ARM instructions.

Registers r13 to r15 are special-purpose system registers. The stack pointer is a traditional CISC-style stack pointer that is automatically incremented or decremented as data is pulled off or pushed on the stack by means of the Thumb instructions POP and PUSH. The Thumb state stack is a full descending stack.

Versions of the ARM processor family supporting the Thumb state provide a *T-bit* in bit 5 of the CPSR. When the T-bit is set to 1, the processor interprets the code as 16-bit Thumb instructions; otherwise, the code is executed normally. Following a reset, the ARM processor enters its default native state.

Thumb state is entered by executing the BX instruction (branch and exchange) that sets the T-bit in the CPSR and executes a jump to the specified location. The same instruction is used to switch from Thumb state back to ARM processor state. Its format is BX Rm, where register Rm contains the target address of the Thumb code to be executed.

When BX is executed, the least-significant bit of Rm is tested. If it is set to a 1, the processor switches to its Thumb state and begins executing code at the address in Rm aligned to a half-word (16-bit) boundary. If the least-significant bit of Rm is 0, a jump is made to the address in Rm aligned to a word (32-bit boundary), and the ARM processor continues execution in its normal default state.

Design Decisions

Anyone can move from a small home to a larger one, but moving from a large house to a small home is always traumatic because you have to decide what to throw away. Designers of the ARM architecture faced the same problem when designing the Thumb state. Just what should be thrown away? You can get rid of clutter and luxuries, but you can't remove essentials.

Removing registers would cut down on the number of bits in an op-code but would change the architecture substantially and ensure Thumb-state and ARM processor-state incompatibility. The compromise is to retain the original register set and redefine the way in which it is accessed. The eight registers r0 to r7 of the ARM processor architecture are mapped directly into registers r0 to r7 of the Thumb state. Registers r14 and r15 (link register and program counter) remain the same, except they can't be explicitly accessed, and new instructions are required to access them. Register r13 can be used as a stack pointer in the ARM processor architecture (by convention). In the Thumb state, r13 is defined as a hardware stack pointer and it now has auto-decrementing and incrementing modes.

Registers r8 to r12 (shaded gray in Figure 4.33) lead a twilight existence. Most instructions can't access them—only the most frequently used instructions can access these registers. This strategy allows the instruction set designer to use a 3-bit register selection field most of the time, while allowing the programmer to access extra registers in special, but common, cases.

As you would expect, the Thumb state has abandoned the luxury of conditional execution to save 4 bits per instruction. Many of the data processing instructions in the Thumb state use a two-address format (like the CISC processors) to avoid encoding a third operand. Similarly, the luxury of a shifted second operand has been dropped and a new set of explicit shift instructions added. Finally, the greatest saving has been made by drastically reducing the size of immediate operands.

Figure 4.34 shows the encoding of the Thumb's data processing instructions. You can see how immediate values have been reduced to 3-bit, 7-bit, and 8-bit values. The eight instruction formats in Figure 4.33 are given next. Recall that the BNF notation ADD|SUB indicates that ADD and SUB are alternatives. Any element separated by a vertical line represents an option.

1.	ADD	**Rd**,Rn,Rm	; (ADD|SUB)
2.	ADD	**Rd**,Rn,#imm3	; (ADD|SUB)
3.	ADD	**Rd**|**Rn**,#imm8	; (ADD|SUB|MOV|CMP)
4.	LSL	**Rd**|**Rn**,#imm8	; (LSL|LSR|ASR)
5.	MVN	**Rd**|**Rn**,Rn|Rs	; (MVN|CMP|CMN|TST|ADC|SBC|NEG|MUL|LSL|
			; LSR|ASR|ROR|AND|EOE|ORR|BIC)
6.	ADD	**Rd**|**Rn**,Rm	; (ADD|CMP|MOV) high registers
7.	ADD	**Rd**,SP|PC,#imm8	; (ADD)
8.	ADD	**SP**,SP,#imm7	; (ADD|SUB)

In each case, we provide a sample instruction format and give the instructions that can take this format on the right. The ADD and SUB instructions have the widest variety of formats

FIGURE 4.34   Encoding the Thumb-state data processing instructions

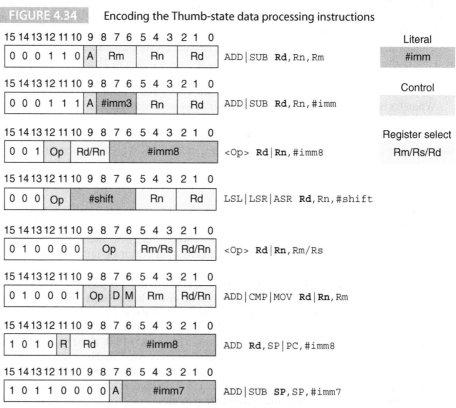

and only instruction format 5 (two-operand, register-to-register) supports the general class of data processing instructions.

The normal way of updating the ARM processor's status bits after an instruction is to append an S to the instruction (and set the S-bit in the op-code). In the Thumb state, this philosophy is dropped, and the conventional technique used by 8-bit processors (and the IA32/68K architectures) is adopted; that is, data processing instructions operating on registers r0 to r7 always update the condition code bits.

Instruction format 6 is used to access the high-order registers (r8 to r12), and data processing operations on these registers do not affect the flag bits, except of course, for the CMP instruction.

The ADD and SUB instructions provide the widest variety of formats. Instruction format 5 (two-operand, register-to-register) is the only instruction supporting general data processing.

Figure 4.35 describes the encoding of the Thumb state's branch instructions. A *conditional* branch has an 8-bit offset, whereas an *unconditional* branch can afford an 11-bit offset. This branch encoding allows short-range branching for conditionals within small loops and if-then-else constructs.

FIGURE 4.35   Encoding the Thumb-mode branch instructions

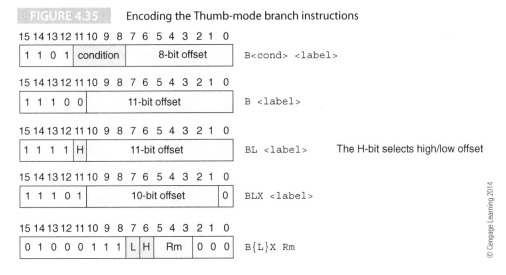

The subroutine call instruction, *branch with link* (BL), poses a special problem. Any substantial piece of code can be expected to require long-distance subroutine calls, and therefore, a short literal is unlikely to provide the necessary range of target addresses. The solution adopted is to employ a *branch with link instruction* with an 11-bit offset and then to *repeat* the instruction to get a second 11-bit offset that can be concatenated to create a 22-bit offset. The philosophy is to allow this instruction pair to be interrupted without harmful side effects.

When the first instruction is executed with the H-bit in the op-code clear, the link register is used as a temporary register to hold the partial branch target address that is given by the PC plus the high-order target shifted 12 places left. The shift is by 12 bits because, in Thumb state, all instructions are 16-bit aligned on a half-word boundary. The following algorithm describes this action:

1. $H = 0$ $lr = pc + \text{sign-extended offset} \times 2^{12}$
2. $H = 1$ $pc = lr + \text{offset} \times 2^1; lr = pc + 3$

When the second instruction of the pair is executed, the low-order part of the target address is added to the partial sum in the link register, and the result is loaded in the program counter to implement the branch. The return address is loaded into the link register.

When writing ARM/Thumb programs, you have to tell the assembler what state you are using. You indicate the type of code to the assembler by means of the directives

CODE32 (ARM code) and CODE16 (Thumb code). The default directive is CODE32. You might write:

```
        ADD     r1,r2,r3    ; Dummy instruction to show we are in ARM state
        ADR     r0, This + 1 ; Generate address of Thumb section
                            ; Adding 1 to the address forces bit 0 in r0 high
        BX      r0          ; Off we go – branch and change to Thumb state
        CODE16              ; Assemble Thumb instructions
This                        ; Arrive here in Thumb state
        ADD     r1,r2       ; Dummy instruction to show we are in Thumb
                              state
        ADR     r0, That    ; Generate address in ARM section (even address)
        BX      r0          ; Off we go again – back to ARM code
                            ;
        CODE32              ; Assemble ARM code
That                        ; Arrive here executing ARM code
```

Thumb's load and store operations, described in Figure 4.36, follow a similar pattern to the corresponding ARM processor instructions except that the displacement specified by the immediate offset is relatively small (either 5 or 8 bits). Byte, half-word, and word transfers are supported. The offset is scaled to suit the size of the data transfer. For example, if the 5-bit offset is 12 and the effective address is [r0, #12] where r0 contains 1000, then a byte will be accessed at location 1012, a half-word at location 1024, and a word at location 1048, because the offset is automatically multiplied by the size of the operand.

FIGURE 4.36   Encoding the Thumb-state data transfer instructions

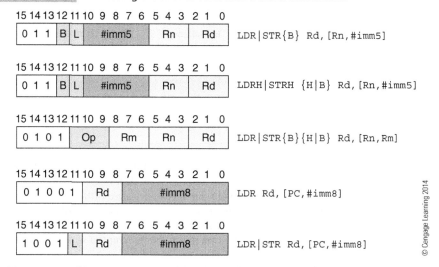

You can provide program counter relative addressing with an 8-bit signed offset with LDR Rd, [PC,#imm8]. This special format is required because the Thumb state can't directly access the program counter in r15. This addressing mode is clearly intended to load local constants rather than to store data (in any case, much of the Thumb code will be in ROM). Consequently, there is no STR form of this instruction.

The more general LDR Rd, [SP,#imm8] and STR Rd, [SP,#imm8] forms of this instruction permit data accesses with respect to the stack pointer.

The Thumb instruction set also includes multiple memory move instructions, although the range of variations is not as great as in the ARM processor architecture (this is probably a blessing). Figure 4.37 describes the two basic forms of the block register move instructions.

The 16-bit instruction format allows you to move only registers r0 to r7; you can't move any of the higher-order registers.

The STMIA Rn!, {registerList} instruction lets you copy the block of registers specified by registerList to the memory location pointed at by register Rn.

FIGURE 4.37   Encoding the Thumb-mode multiple register transfer instructions

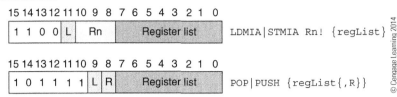

The only mode permitted is *increment after,* which indicates that a register is stored at the location pointed at by Rn and then the register is incremented by 4 after the register has been moved. The lowest numbered register is stored first at the lowest memory address (i.e., the initial starting address in the pointer register).

LDMIA **Rn!**, {registerList} copies data from memory to registers. The lowest numbered memory address is first loaded into the lowest numbered register, the pointer incremented by 4, and the next load carried out. The STMIA and LDMIA instructions are inverse operations in the sense that an STMIA **Rn!**, {registerList} can be followed immediately by an LDMIA **Rn!**, {registerList} without a change in the state of the system.

The other block move is the PUSH and POP pair of instructions that are true inverses of each other in the sense that a PUSH followed by a POP leaves the state of the system unchanged. These instructions do not require a register to be specified nor do they require the "!" suffix, because by definition, they access the stack pointed at by r13, the stack pointer.

> ## PC Relative Addressing
>
> Register-indirect addressing uses a register to provide the address of an operand. Because the contents of the register can be changed, the effective address becomes a variable allowing dynamic data structures to be accessed at run time.
>
> If the pointer register is the program counter itself, the target address is specified with respect to the current instruction. This addressing mode is used universally for branching to permit relative branches, which means code can be relocated without recalculating target addresses.
>
> By using PC relative addressing to access data operands, code can be made fully reloadable (since the location of data is specified with respect to the current location) and located in read-only memory.

The syntax for the register list is registerList{,R}, where the {,R} field is optional and R may be sp or pc. For example, you can write PUSH {r0-r4,lr} and PULL {r0-r4,pc}. The R-field in the instruction provides an ingenious means of adding the program counter or link register to the block of registers being transferred.

We have covered the Thumb mode for several reasons. First, it presents an interesting approach to ISA design and is in keeping with the theme and variations subtitle of this book. Second, it has helped elevate ARM Holdings' position in the world of embedded computing to an industry user. Finally, it demonstrates tradeoffs between code density and performance.

4.6.2   MIPS16

The MIPS16 is analogous to Thumb; it too was developed to provide a 16-bit processor while keeping compatibility with its big brothers: the MIPS-I and MIPS-III architectures. The secret of MIPS16 is the way in which MIPS-III 32-bit instructions are mapped onto the MIPS16 16-bit instruction set. Figure 4.38 demonstrates how this is achieved[4] for the I-format MIPS instruction. We do not cover the MIPS 32-bit architecture here, as we are interested only in the way in which a 32-bit ISA is mapped onto a 16-bit ISA.

Compressing MIPS code is achieved by treating the MIPS instructions set like salami and slicing bits off. The already slim MIPS instruction set is further reduced by dropping one of the op-code bits. Second, the number of registers is reduced from 32 to 8, saving two register specifier bits per register. Finally, the size of the immediate value in the I-format instruction is reduced from 16 bits to 5.

---

[4]K. Kissell, "MISP16: High-density MIPS for the Embedded Market," MIPS Technologies, Inc., Technical report, 1997.

© Cengage Learning 2014

FIGURE 4.38   MIPS16 compressed op-code format

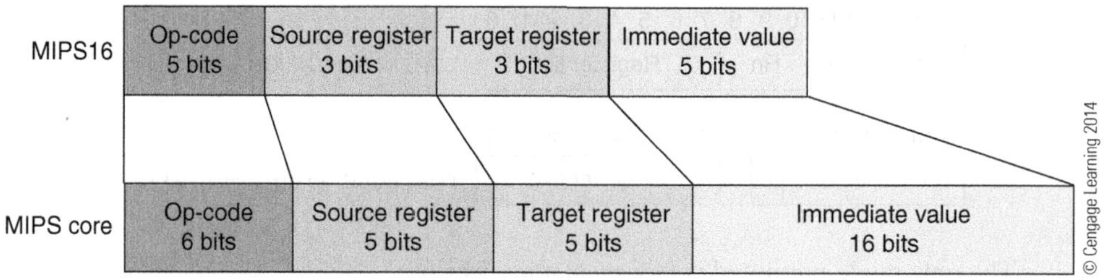

MIPS16	Op-code 5 bits	Source register 3 bits	Target register 3 bits	Immediate value 5 bits

MIPS core	Op-code 6 bits	Source register 5 bits	Target register 5 bits	Immediate value 16 bits

MIPS16 employs the classic two-address mode instruction in which the source and destination registers are the same; that is, one of the two source operands is overwritten by the result.

The severe compression required to fit a 32-bit instruction set into a 16-bit word requires new instructions to cope with the problems caused by such a small register set and the tiny 5-bit literal field. The MIPS16 has an extended instruction that does not execute an operation but simply provides an 11-bit literal that can be concatenated with the 5-bit literal of the following instruction. This mechanism is, of course, a marginally more elegant version of the CISC's multiple length instruction.

Like Thumb, the MIPS16 implements a hardware stack pointer and allows loads and stores relative to the stack pointer—another feature more associated with CISCs than RISCs. When a load or store is performed with respect to the stack pointer, the offset is eight bits because the redundant register field can be concatenated with the literal.

Figure 4.39 shows how the MIPS16 registers are mapped onto the MIPS core register set. Although the MIPS16 has only eight visible registers, the other $32 - 8 = 24$ MIPS registers can be accessed via special move instructions that copy data between the MIPS core and MIPS16 register sets.

The MIPS16 supports branches on any register being equal or not equal to zero with `BEQZ rx,immediate` and `BNEZ rx,immediate` instructions. The branch instruction takes the 8-bit signed literal that forms part of the instruction, shifts it left one bit, and adds it to the contents of the program counter to create a relative address. A branch takes place if the contents of the specified register are zero (`BEQZ`) or not zero (`BNEZ`).

Figure 4.39 shows a new MIPS16 register called the T register, which is not part of the core MIPS. This register is needed to support conditional execution in conjunction with the `BTEQZ immediate` and `BTNEZ immediate` instructions. These instructions operate exactly like the corresponding `BEQZ` and `BNEZ`, except that the register tested is the T register. The T register is set or cleared by the MIP16's *set on less than* instruction. You can be forgiven for wondering why the T register has been implemented. Suppose you wish to compare two registers. You can use the `SLT R1,R2` instruction to perform the comparison and the `BTEQZ` or `BTNEZ` to implement the branch.

4.7   Variable-Length Instructions

The final section in this chapter looks at a feature of 8-bit microprocessors that was carried over into the CISC generation of processors like the IA32 and 68K architectures—the *variable-length instruction*. Although RISC families (e.g., PowerPC, SPARC, MIPS, and ARM) have a fixed-length instruction, 8-bit microprocessors and CISC processors of the 68K and IA-32 families have variable-length instructions. Each instruction may occupy a different number of bytes and the program counter is not incremented by a constant value during each fetch cycle. If instructions are not all the same length, it becomes very difficult for the processor to look ahead at the instruction stream. For example, the processor can't read the tenth instruction in line, because it doesn't know where its boundaries lie. Moreover, variable-length instructions make it very difficult to design the decode stage, because you have to look at the entire instruction rather than a single field.

FIGURE 4.39   Mapping MIPS16 registers on to the MIPS core

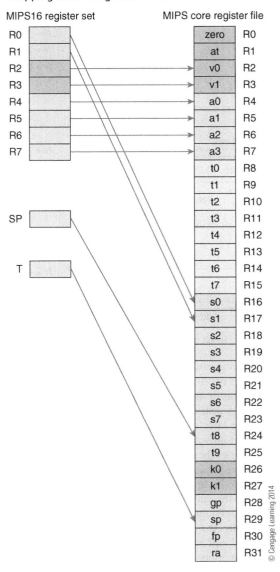

Since it's not possible to design a realistic 8-bit instruction that includes both an operation code and an operand address, a simple solution is to implement instructions that are multiples of 8 bits. However, if all instructions were 32-bit, the code density would be lower. When 8-bit microprocessors first appeared, the prime consideration was the massively high cost of main store (more expensive than memory today by about six orders of magnitude). A solution was to make some instructions that required no, or very short operands, one-byte. Instructions were two bytes if they required an 8-bit operand. An instruction that required a 16-bit memory reference was three bytes, and so on. Such variable-length instructions first appeared in mainframes.

Figure 4.40 demonstrates a technique of providing effective instruction lengths of 8, 16, or 24 bits; this is a hypothetical example based on an 8-bit microprocessor. The instruction is divided into discrete bytes and these bytes are stored in consecutive 8-bit memory locations. For example, a 24-bit instruction is stored in memory as three consecutive bytes. Of course, this 24-bit instruction has to be read from memory by executing three read cycles during the instruction fetch phase.

FIGURE 4.40   Variable-length instructions

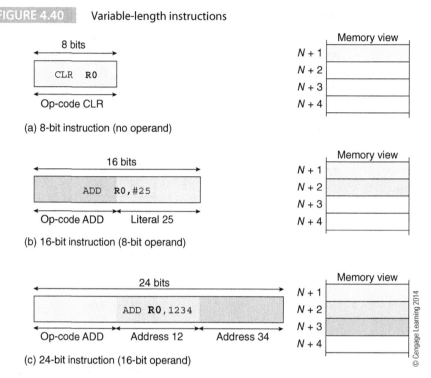

(a) 8-bit instruction (no operand)

(b) 16-bit instruction (8-bit operand)

(c) 24-bit instruction (16-bit operand)

A typical 8-bit microprocessor represents each operation by a single byte (some 8-bit microprocessors implement two-byte op-codes). First-generation 8-bit microprocessors didn't assign all $2^8 = 256$ possible op-codes to valid instructions, leaving gaps in the instruction set. These so-called *holes* in instruction sets where a bit pattern didn't correspond to a valid op-code were available for later expansion of the instruction set. Microprocessor designers employed some of the previously unassigned op-codes as a pointer to a new group of up to 256 op-codes. That is, when one of these op-codes is encountered in the instruction set, the following byte is read to determine the actual operation (creating a 16-bit instruction). The Z80 that was derived from Intel's 8080 is the best example of this approach to instruction set design.

When the CPU reads the current instruction during its fetch cycle, it first examines the instruction's class or type and then fetches zero, one or two further bytes before beginning the execute phase. Conceptually, the computer still has a fetch phase and an execute phase, but the fetch phase is extended and now reads a variable-length instruction from the main store. This arrangement has been adopted by all 8-bit microprocessors. You could regard the *fetch extended operand phase* as part of the processor's execute phase, since the CPU cannot begin to fetch an extended operand until it has decoded the current instruction.

Variable-length instructions have two important advantages. The first is that a nominally eight-bit machine can employ instructions of an arbitrary complexity; that is, you can construct a machine to implement any instruction set, no matter how large or complex. Similarly, a variable-length address field means that you can design a processor to access programs and data structures of any size. The second advantage of variable-length instructions is that frequently used instructions can sometimes be constrained to one byte and the size of the resulting program minimized. For example, suppose a system often executes the operation *add the small integers* 1, 2, 3, *or* 4 to the contents of a register. This instruction could be encoded as the 8-bit string xxxxxxyy, where the x's represent the op-code *add a small literal operand to the register* and the two bits yy represent the actual operand to be added. The 68K instruction set uses the term *quick* to describe an instruction with a small literal operand. For example, ADDQ #3,D0 adds 3 to the contents of register D0.

Practical 8-bit microprocessors don't employ infinitely extensible instructions and operand addresses. Op-codes are allocated one byte to provide up to 256 instructions and addresses are limited to 16 bits to provide a 64K byte address space. Later, 8-bit microprocessors like the Zilog Z80 used 16-bit op-codes to expand their instructions beyond the 256 limit provided by a pure 8-bit instruction set.

Consider a second-generation 8-bit microprocessor with an 8-bit op-code and a 16-bit address. The total number of bytes occupied by a program, S, is therefore $S = l + 2m + 3n$, where l is the number of 8-bit instructions, m is the number of 16-bit instructions, and n is the number of 24-bit instructions. The average number of bytes per instruction is therefore $S/(l + m + n)$.

Just as there are advantages in employing variable-length instructions, there are disadvantages. Throughput is reduced because the external program and data memory might have to be accessed as many as five times per instruction (three times to read the three bytes of a 24-bit instruction and twice to access the specified operand if it is a 16-bit value).

A less obvious disadvantage of multi-length instructions is the additional complexity required to decode instructions into one, two, or three-byte formats. Figure 4.41 demonstrates four of the ways in which op-codes in a processor with variable size instructions might be decoded. In Figure 4.41a, two bits of the op-code are reserved to indicate the instruction length in terms of the number of bytes. This mechanism is both fast and easy to implement. However, it wastes *op-code space*. For example, if few instructions are one byte long, the quarter of the instruction space prefixed by 00 is allocated to instructions with no extension.

Figure 4.41b demonstrates a better approach based on Huffman encoding. The most common operand size is given the prefix 0 so that only one bit in the instruction is allocated to instruction size. The next most common operand size is given the prefix 01, which requires two bits to specify instruction size, and so on. This arrangement is efficient in systems where at least half of the instructions have the most popular size, at least a quarter of the instructions have the next

FIGURE 4.41   Decoding the instruction size for variable-length instructions

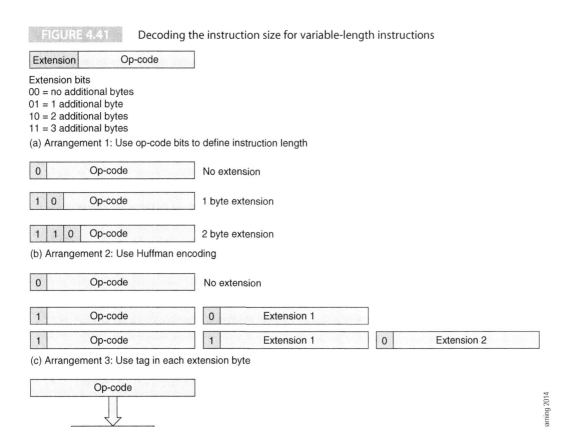

Extension bits
00 = no additional bytes
01 = 1 additional byte
10 = 2 additional bytes
11 = 3 additional bytes
(a) Arrangement 1: Use op-code bits to define instruction length

(b) Arrangement 2: Use Huffman encoding

(c) Arrangement 3: Use tag in each extension byte

(d) Arrangement 4: Use Look-up table

© Cengage Learning 2014

Huffman Codes

A Huffman code is a variable-length code where the code words have differing lengths. Huffman encoding is efficient when some words are used more frequently than others, because you can assign short codes to frequently used words. A good example of a Huffman code is the Morse code that assigns frequently used letters (in English) short symbols. For example, the letter 'e' is a single dot, whereas the letter x is −·−.

As well as using Huffman encoding to reduce the average size of machine code, Huffman encoding is used in JPEG image encoding to compress images.

most popular size, and so on. However, by using a variable number of bits to indicate instruction size, it follows that the number of bits left to define the op-code must also be variable. Consequently, Huffman encoding creates problems in the design of the op-codes themselves.

The decoding mechanisms of Figures 4.40a and 4.40b are fixed when the processor is first manufactured. Figure 4.41c demonstrates a scheme that is infinitely extensible. A tag bit in the op-code is tested during the fetch cycle. If that bit is 0, no extension is required. If the bit is a 1, an extension byte is fetched from memory. If the corresponding tag bit in the extension is 0, fetching terminates. If the tag bit in the extension is 1, another extension byte is fetched, and so on.

This mechanism provides for an infinite number of extension bytes, although it may be slow, because you cannot determine the total number of extension bytes until you read them from memory in sequence.

Figure 4.41d demonstrates another way of obtaining the instruction size, where a look-up table in ROM is used to decode the instruction. This arrangement means that any op-code bit pattern can be associated with any instruction size, thus removing many of the problems of instruction encoding. However, the decoding ROM takes up valuable real estate on the chip and slows instruction execution because it requires an internal memory access before an instruction can be fully decoded.

Decoding Variable-Length Instructions

We can demonstrate variable length instruction decoding using a size field as follows. Assume that each word fetched is a byte. In the code that follows, MAR (memory address register) is a pointer to the address being accessed in memory, and MBR (memory buffer register) is the register that holds data read from memory or to be written into it. These concepts are discussed in Chapter 7.

```
MAR  = PC;               /* PC to MAR                        */
PC   = PC + 1;           /* increment PC                     */
MBR  = memory[MAR];      /* get next instruction             */
IR   = MBR;              /* copy MBR to IR                   */
size = (IR&0xC0)>>6;     /* extract size field from op-code  */
switch (size)
{ case 0: {break;}       /* no extension so exit here        */
  case 1: {MAR = PC;     /* one-byte extension so do fetch   */
           PC  = PC + 1;
           MBR = memory[MAR];
           Ext1 = MBR; break;}
  case 2: {MAR = PC;     /* two byte extension do 2 fetches  */
           PC  = PC + 1;
           MBR = memory[MAR];
           Ext1 = MBR;      /* first extension               */
           MAR = PC;        /* get next byte in second fetch */
           PC  = PC + 1;
           MBR = memory[MAR];
           Ext2 = MBR; break;)
  case 3: {       };     /* and so on...                     */
}
```

We can implement the Huffman decoding scheme of Figure 4.41b by testing the operand extension bit in the instruction. If it is 0, we have finished. If it is 1, we perform another fetch cycle and then test the next extension bit, and so on.

```
MAR = PC;                   /* fetch cycle: PC to MAR        */
PC  = PC + 1;               /* increment PC                  */
MBR = memory[MAR];          /* get next instruction          */
IR  = MBR;                  /* copy MBR to IR                */
if (IR&0x80==0);            /* test first size bit of op-code */

                            /* if 0 then no extension so exit */
else
  {                         /* require at least 1 byte more   */
    MAR = PC;               /* second fetch cycle: PC to MAR  */
    PC  = PC + 1;           /* increment PC                   */
    MBR = memory[MAR];      /* get next instruction           */
    Ext1 = MBR;            /* save first extension byte      */
    if (IR&0x40==0);        /* test for 10 or 11 prefixes     */
                            /* if prefix 10 exit with 1 byte  */
    else
      {                     /* require at least 1 byte more   */
      MAR = PC;             /* PC to MAR                      */
      PC  = PC + 1;         /* increment PC                   */
      MBR = memory[MAR];    /* get next instruction           */
      Ext2 = MBR;          /* save second extension byte     */
      if (IR&0x20==0)       /* test for 100 or 110 prefixes   */
         ;                  /* if prefix 100 exit with 2 bytes */
      else                  /* continue in this way testing   */
        {and so on}
      };
```

We can implement the tag-bit scheme of Figure 4.41c in much the same way as the Huffman decoder. All we need do is to read a byte and continue accumulating extensions until the tag bit is 0.

```
MAR = PC;                   /* fetch cycle: PC to MAR        */
PC  = PC + 1;               /* increment PC                  */
MBR = memory[MAR];          /* get next instruction          */
IR  = MBR;                  /* copy MBR to IR                */
if (IR&0x80==0)             /* test tag bit of op-code       */
  ;                         /* if 0 then no extension so exit */
  else
    {                       /* require at least 1 byte more   */
      MAR = PC;             /* second fetch cycle: PC to MAR  */
      PC  = PC + 1;         /* increment PC                   */
      MBR = memory[MAR];    /* get next instruction           */
      Ex1 = MBR;           /* save first extension byte      */
      if (Ex1&0x80==0)      /* test tag bit in first extension */
         ;                  /* if tag 0 no more extension     */
      else
        {                   /* require at least 1 byte more   */
        MAR = PC;           /* PC to MAR                      */
```

```
PC  = PC + 1;        /* increment PC                   */
MBR = memory[MAR];   /* get next instruction           */
Ex2 = MBR;           /* save second extension byte      */
if (Ex2&0x80==0)     /* test tag bit in second extension */
   ;                 /* if tag 0 no more extension       */
else                 /* continue in this way testing     */
   {and so on}
};
}
```

Sixteen-bit architectures of the 1970s and 1980s had to use multiple-length instructions because 16-bits were not sufficient to express an op-code and an operand. Only the advent of the 32-bit processor made it possible to incorporate both an op-code and operand within a single instruction. Even so, the 32-bit architecture imposed a register-to-register architecture because it is impossible to provide a two-operand format even with 32-bits.

Summary

The world of instruction set architectures is an interesting one. On one hand, you can write any program using just a handful of low-level instructions (in the limit, you can write all programs using one single machine-level instruction). On the other hand, a real-world designer has to weigh up the factors of performance, code density, backward compatibility, and competing processors. Even the advertising executive sometimes has to be taken into account, because in the early days of the microprocessor, instructions were publicized that looked good in a brochure but which added little to performance. To be fair, we have moved on from the early days of the microprocessor in terms of design objectives.

The previous chapter provided an introduction to an elegant ISA from ARM. This chapter has examined several aspects of instruction architectures: bit fields, data movements and packing, and complex memory indirect addressing modes.

Part of this chapter has been devoted to the stack and the way in which it is used to provide the programmer with local storage for functions and with a means of passing parameters between a function and the program that called it. In particular, we have looked at the relationship between the C language used by systems programmers and the underlying machine. We have used a cross-compiler to examine the way in which the underlying architecture supports the high-level language.

The final part of this chapter has considered how semiconductor manufacturers have employed code compression techniques to make the code of processors like those from ARM or MIPS more compact and therefore more cost-effective in applications such as cell phones and PDAs.

Problems

4.1 What is a stack frame and why is it so important?

4.2 What is the difference between a *stack pointer* and a *frame pointer*?

4.3 In the context of ARM processors, what is the difference between a *link register* and a *frame pointer*?

4.4 The following code gives an example of a C program that calls a function.

```
void adder(int a, int *b)
{
*b = a + *b;
}
void main (void)
{
int x = 3, y = 4;
adder(x, &y);
}
```

This 68000 program creates the following output from a compiler. The panel describes some of the instructions that are not obvious.

```
*1 void  adder(int a, int *b)
          Parameter a is at 8(FP)
          Parameter b is at 10(FP)
adder

          LINK        FP,#0
*2        {
*3        *b = a + *b;
          MOVEA.L   10(FP),A4
          MOVE      (A4),D1
          ADD       8(FP),D1
          MOVE      D1,(A4)

*4        }
          UNLK      FP
          RTS
*5
```

```
*6        void main (void)
          Variable x is at -2(FP)
          Variable y is at -4(FP)
main
          LINK        FP,#-4
*7        {
*8          int x = 3, y = 4;
          MOVE      #3,-2(FP)
          MOVE      #4,-4(FP)
*9          adder(x, &y);
          PEA       -4(FP)
          MOVE      #3,-(A7)
          JSR       adder
*10
*11       }
          UNLK      FP
          RTS
```

Draw the state of the stack immediately after function adder is called in function main (i.e., the return address is on the top of the stack but no code in adder has yet been executed). Carefully label all items on the stack and give their location with respect to the current value of the frame pointer.

Draw a sequence of diagrams (i.e., memory maps) showing what happens to the stack as function adder is executed. Explain the action of each instruction that modifies the state of the stack. Show how parameters are passed to and from the function and how stack-based values are accessed.

4.5 What makes a complex instruction *complex*, and a simple instruction *simple*? Is the notion of complex

68000 Instructions

68K instructions operate on bytes, words or 32-bit longwords and are indicated by suffixes .B, .W, .L, respectively. No suffix indicates the default .W (two bytes).

MOVEA.L 10(FP),A4 copies the contents of the memory location 10 bytes on from the contents of the frame pointer in A6 to address register A4.

PEA pushes a long word (32-bit) value onto the system stack pointed at by A7. This stack pointer points to the element at the top of the stack and is pre-decremented by 4 before a new longword (32 bits) is pushed. The 68K stack corresponds to ARM's FD (full descending stack).

The effect of PEA -4(SP) is to calculate the effective address defined by [SP] – 4 and push that on the stack. SP and A7 are synonymous.

LINK FP,#0 creates a stack frame of zero bytes! This is done to create a link between this function and other functions. The LINK FP,#0 pushes the old frame pointer on the stack and loads the new frame pointer with the address of the stacked frame pointer. Note that the frame pointer and stack pointer are pointing at the same location in this case.

The UNLK FP instruction collapses the stack frame by loading the stack pointer with the frame pointer (pointing to the base of the frame) and then pulling the old frame pointer off the stack and restoring the frame pointer.

and simple in the realm of instruction set architecture meaningful?

4.6 We said that any computer program can be constructed using a single instruction that performs a subtraction and branches on negative. Discuss the accuracy of this statement.

4.7 SBN a,b,c is defined as: [a] = [a] – [b]; if [a] ≤ 0, then branch to c. Using only a SBN **destination**, source, target, (subtract source from destination and branch to target on negative) instruction, implement the following primitives

    a. MOVE **X**, Y   ;copy the contents of memory location Y to location X

    b. ADD **X**, Y   ;add the contents of memory location Y to location X

    c. IF (X ≥ 0)Y ← 2Y ;if the contents of location X are greater than zero, then Y← 2Y.

4.8 Eight-bit microprocessors and classic CISC computers like the Intel IA32 and 68K families have variable-length instructions. What are the advantages of computer architectures with instructions of different lengths? What are the disadvantages?

4.9 Investigate the variations in multiply and divide instructions supported by three different microprocessor families. Why is there more variation in the ways in which multiplication and division are performed than in, for example, addition?

4.10 Data movement requires that data be moved from a source to a destination. What are the practical issues that a designer must be aware of when implementing a data move instruction?

4.11 Some data movement instructions re-order the sequence of bits that comprise the data elements they are moving. Why?

4.12 Suppose you have a basic instruction set with typical primitive arithmetic and logical operations (*add*, *sub*, *and*, *or*, *not*, *xor*, *lsl*, and *lsr*). How would you move a 32-bit word from memory location a to location b and reorder the bytes from *PQRS* to *SQRP* using only this instruction set?

4.13 Two instructions provided by the 68K CISC processors are LEA (load effective address) and PEA (push an effective address on the stack). What do these instructions do, and how are they used in practice?

4.14 Why do some processors provide support for double-precision shifting operations? What other double-precision operations are supported by processors?

4.15 What is *bounds testing*, and how does the 68020 implement it? If you were a processor designer, would you implement bounds testing? Can you think of other ways of incorporating bounds testing in an architecture?

4.16 Investigate several microprocessors, and describe how they support the mechanization of control loops.

4.17 What instructions are missing from microprocessor instruction sets? This question asks you to think about the type of operations computers perform, and to think about what primitives could usefully be incorporated to accelerate the calculation.

4.18 What is memory indirect addressing, and how is it used?

4.19 Why do so few processors implement memory indirect addressing?

4.20 The 68020 supports both preindexed and postindexed memory indirect addressing modes. What, in this context, are *preindexing* and *postindexing*? If only one of these addressing modes could be implemented, which do you think is the better alternative? Explain your reasoning with examples.

4.21 The 68020's postindexed memory addressing mode can be expressed by the effective address ([64,A0],D3*2,24)

    a. Both graphically and in words, demonstrate how this effective address is calculated.

    b. What are the inner and outer displacements in this effective address?

    c. What is index register scaling, and why is it implemented?

4.22 What is *compressed code*, in the sense of *Thumb*?

4.23 Do compressed architectures represent an advance or a regression in computer development? Use examples to justify your answer.

4.24 Both ARM and MIPS processors implement reduced or compressed instruction sets. Can you suggest other ways of compressing instruction sets?

4.25 We said that a problem with variable length instructions is that you cannot determine instruction boundaries further down the instruction stream, because you have to decode each instruction individually to determine its length. Modern high-performance computers look ahead in the code in order to deal with branch addresses or to perform out-of-order execution, and so on. Can you think of a mechanism that would allow multiple length instructions but still allow you to determine instruction boundaries without fully decoding each instruction in the stream?

4.26 The 68K family has separate general-purpose address and data registers. What are the differences between these two types of registers? Since most other computers with general-purpose register sets do not distinguish in hardware between *data* and *addresses*, was Motorola's idea a good one or a bad one? What are the strengths and weaknesses of enforcing the differences between addresses and data in hardware?

4.27 You have a CISC architecture that employs bit field instructions to perform some very clever graphical operations on image data. These bit field instructions account for 5% of the code executed. However, the bit field instructions increase the cost of the silicon die by 20% and prevent the use of advanced pipelining and microarchitectural techniques. If the bit field instructions are dropped, the chip can operate at twice the rate. Typically, a bit field instruction can be replaced by seven conventional machine-level instructions. Would you advise keeping or getting rid of bit field instructions?

4.28 Why does *endianism* raise its ugly head so starkly when we look at the notion of bit fields? Illustrate your answer with reference to the 68020's bit field instructions.

4.29 Consider a computer with a bit field instruction of the form BFFFO R1, (R2), {R3:R4} that locates the first bit set to one in the specified bit field. After this instruction has been executed, register R1 contains the position of the first bit set to one, or the length of the bit field, if the bit field is all zeros. Register R2 contains the address of a word in memory, register R3 contains the offset of the bit field from the word pointed at by R2, and R4 contains the width of the bit field from 1 to 32 (empty bit fields are not supported).

For a real or hypothetical machine, write suitable machine code to implement this operation using basic (non-bit-field) operations. Assume that dynamic shifting is permitted. State any assumptions that you need to make about the problem (e.g., bit/byte numbering).

4.30 Suppose that a BFFFO operation is applied to the data structure in Figure P4.30. What value would be loaded into the destination register assuming that the base byte address is 800?

4.31 Suppose that multiple-precision shifts did not exist and that you could use only logical shifts. How would you implement a double-precision shift by using only single-precision shifts?

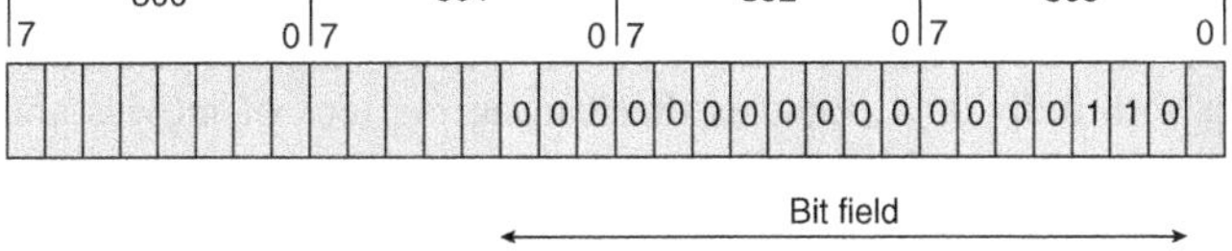

4.32 The 68020's CHK2 and CMP2 instructions are rather nice because they let you compare a register against two bounds in order to perform array subscript checking at run time. Unfortunately, they are expensive, because they require two words to be read from memory (the upper and lower bounds) and that is a time-consuming operation. Can you think of a way of implementing such an instruction without incurring such massive penalties?

4.33 A 68020 assembly language instruction that uses a bit field operation with a complex effective address is:

```
BFFFO ([8,A0,D1*4],4){D2,D3},D4
```

Write a suitable sequence of ARM processor operations that would perform the same action. Assume that the target address is a 32-bit word.

4.34 If you have an *m*-bit register, loading an *m*-bit constant presents a bit of a problem. Discuss how ISA designers have gone about solving this problem in both CISC and RISC domains.

4.35 Why are RISC programs that perform similar functions to CISC programs generally larger?

4.36 How do the instruction encodings of ARM processors and MIPS processors differ?

4.37 You are designing a new processor that is similar to members of the ARM family, but are not happy with the ARM processor's approach to literal constants. Recall that the instruction has a 12-bit field that specifies a 4-bit scale factor and an 8-bit literal. You have a new idea, the *clut* or *constant lookup table*. Your knowledge of computer architecture tells you that the range of literals used by most programs is very small, but you also know that large constants and pointers are needed. The *clut* is a table with 2^{12} entries that is set to 0 to 4,095 sequentially at startup so that you can specify a literal in the range 0 to 4,095. However, by means of a special *clut set instruction* cluts the programmer can reload the table with any 32-bit values they wish, up to 4,096 in total. If you are using less than 4096 literals you can access them with a 12-bit literal field in the instruction which becomes a pointer into the clut table. Is the clut instruction a great idea or is it a bit clumsy?

FIGURE P4.30

Computer Architecture and Multimedia

"A picture is worth a thousand words."
Anon

"Many hands make light work."
Traditional

"The world today doesn't make sense, so why should I paint pictures that do?"
Pablo Picasso

"Time is my greatest enemy."
Evita Peron

So far, we've looked at the instruction sets of microprocessors by first introducing the ARM in Chapter 3 and then focusing on selected aspects of ISAs in Chapter 4. Now we complete our introduction to ISAs by looking at a development in microprocessors that has had a profound effect on computer technology from instruction set design to the development of hard disks. This development is *multimedia* and it has had a profound effect on the way in which computer technology is continuing to develop. We begin this chapter by setting the historical background to these developments, showing why sophisticated mathematical processing is necessary in audio-visual applications, and then we look at enhancements to ISAs designed to cope with these applications.

Over a decade ago, a colleague left the university at which I was teaching to work for a large multinational involved in consumer electronics. When I bumped into him again, he asked me, "Do you know what the driving force behind much of modern digital electronics is?" I said, "No", and he replied that it was the demand for in-room movies in hotel rooms. A large hotel with hundreds of rooms may have hundreds of guests requesting movies that they wish to watch in real time. Just think about the sheer quantity of data that has to be accessed, manipulated and transmitted. I have flown on the Airbus A380, which is capable of accommodating 800 passengers—all with individual consoles that can access tens of movies and play, rewind, and fast-forward through them in real-time.

In the 1970s, the microprocessor was little more than a programmable control element that helped to automate washing machines or to build simple desk-top calculators. In the days of the first-generation 8-bit microprocessor, instruction sets were limited, and few microprocessors could even implement 8-bit integer multiplications. One of the first popular applications of microprocessors was in word processing, because it's easy to handle 8-bit ASCII-encoded mono-spaced characters like courier.

As time passed, new microprocessor applications appeared, such as the spreadsheet that helped turn the microprocessor into an indispensable office tool. The growth of desktop publishing allowing you to use a wide range of fonts and escape the manual typewriter with its monospaced courier font was a step further along the road to widespread personal computing.

We have passed through the age of desktop publishing and arrived at the age of *multimedia*. Multimedia is a blanket term for computer applications that process sound and images in real time and particularly for applications that integrate sound and vision. Few multimedia applications are more spectacular than the retrieval, processing, and display of movies.

Multimedia applications have an insatiable demand for storage capacity, processing power, low latency, and bandwidth. In an era of simple ASCII word-processing, 8-bit chips, and 8 KB memories, observers asked why anyone should ever want drives larger than 10 MB. Recall that a single ASCII character occupies a byte, five characters make up the average word in English, and there are about 500 words/page or about 3K bytes/page. Consequently, a 10 MB drive would store about 3,000 pages. When personal computers advanced and Microsoft's DOS became available, the same pundits asked why we'd ever want more than 100 MB. Microsoft's Windows was hiding round the corner waiting to utter its first words: *feed me!* Today, we have drives with capacities over 3 TB and, yes, the same people are still asking whether we really need more capacity. A single high-resolution digital photograph in uncompressed format can occupy over 100 MB; a clip of video from a camcorder can eat up 300 MB in a short time. There's never been a period in the history of computing when that didn't demand more storage, higher performance, greater bandwidth, and lower latency. Of course, new technologies such as the cloud are changing the bandwidth/capacity relationship by allowing personal devices to store data remotely in the *cloud*.

The first part of this chapter looks at some of the areas of computing that require high performance and storage. We introduce concepts from computer graphics, digital signal processing, and image compression to illustrate the nature of the operations carried out by today's computers. The purpose of this section is not to teach students how to implement graphics operations or to design MPEG decoders, it's to demonstrate why some application have such an insatiable demand for computing power and what led computer designers to develop multimedia extensions to their ISAs.

Multimedia applications have common characteristics; they involve simple operations on sets of primitive data elements in parallel. These are called *short vector* operations. The main part of this chapter examines how semiconductor manufacturers have incorporated short vector operations in the ISAs of their processors. We begin with Intel's *multimedia extensions* and describe some of the more modern versions, now called *streaming extensions*, which are implemented by Intel and other processor families.

We return to multimedia at the end of this text when we look at multicore processors and the highly parallel processing systems used in graphics display cards.

5.1 Applications of High-Performance Computing

When you design an instruction set architecture, you have to choose what data-processing operations should be included in the instruction set. Clearly, everyday integer arithmetic operations such as addition, subtraction, multiplication, division, and bit shifting are *apparently* necessary. Not all of these operations are necessary. For example, multiplication and division can be implemented by adding/subtracting and shifting. Early microprocessors didn't provide multiplication and division. You had to write your own subroutines for multiplication and division. Even today, not all microprocessors support division. Our next step is to ask where arithmetic operations appear in computing.

Digital computers were first constructed to perform numerical calculations for scientific purposes (mathematical tables), military purposes (field gun firing tables), and commercial applications. Although some computers in the 1960s and 1970s were intended for scientific calculations, most mainframe computers were used for commercial data processing that did not require the high-speed mathematics used by scientists in modeling systems such as the atmosphere.

Division by Multiplication

As an example of the use of other instructions to implement division, consider the following iterative process that performs division using multiplication, addition, and shifting.

Suppose we divide a dividend N by a divisor D to obtain a quotient Q, so that $Q = N/D$. First, scale D so that $\frac{1}{2} \leq D < 1$ by shifting D left or right and recording the number of shifts. Define Z in terms of D as $Z = 1 - D$ where $0 < Z \leq \frac{1}{2}$. We can write $Q = N/D = KN/KD$. Suppose that $K = 1 + Z$, then

$$Q = \frac{N}{D} = \frac{N(1 + Z)}{D(1 + Z)} = \frac{N(1 + Z)}{(1 - Z)(1 + Z)} = \frac{N(1 + Z)}{1 - Z^2}$$

If we now repeat the process with $K = (1 + Z^2)$, we get

$$Q = \frac{N(1 + Z)}{1 - Z^2} \cdot \frac{1 + Z^2}{1 + Z^2} = \frac{N(1 + Z)(1 + Z^2)}{1 - Z^4}$$

This process may be repeated n times with the result that

$$Q = \frac{N}{D} = \frac{N(1 + Z)(1 + Z^2)(1 + Z^4)\ldots(1 + Z^{2^{n-1}})}{1 - Z^{2^{n-1}}}$$

Since $Z < 1$, the value of $Z^{2^{n-1}}$ approaches zero as n is increased and Q is given by

$$Q = N(1 + Z)(1 + Z^2)(1 + Z^4)\ldots(1 + Z^{2^{n-1}})$$

For 8-bit precision, n need be only 3, and if $n = 5$, the quotient yields a precision of 32 bits. As the divisor was scaled to lie between $\frac{1}{2}$ and unity, the corresponding quotient, Q, calculated from the above formula, must be scaled by the same factor to produce the desired result.

Let's take an example. Suppose we wish to evaluate $Q = 7/9$ using this technique. If we scale in decimal, we get $Q = N/D = 0.7/0.9$ and $Z = 1 - D = 1 - 0.9 = 0.1$. We can now write $Q = 0.7(1 + 0.1)(1 + 0.1^2)(1 + 0.1^4)(1 + 0.1^6)\ldots = 0.7 \times 1.1 \times 1.01 \times 1.0001 \times 1.000001 = 0.7777785$. The actual value of $.7/.9$ is 0.77777777777777. We have achieved five decimal places of precision with just three iterations.

The personal computer revolution of the early 1980s was not about scientific applications or data processing. Personal computers allowed people to perform word processing, to store their information, and to play computer games at home. These applications did not *initially* require intensive mathematics. For example, word processing using a fixed-pitch typeface such as Courier is easy to implement as each character is separated from its neighbor by, say, 0.1 inch. If you insert four characters in a string of text, all other characters are moved along by 0.4 inch. As word processing developed and non-monospaced fonts appeared (e.g., Times and Arial), adding or deleting a character required very many more arithmetic operations to repaginate text whenever it is modified. It's even more complicated when you take into account *kerning* (changing the spacing of characters when certain characters appear next to each other such as V and M). Computer games of that era had even more demanding mathematical requirements.

Let's look at how computer graphics led the way in requiring high-performance. Remember we are not attempting to provide a course in computer graphics or any other aspect of modern computer applications. In this section, we demonstrate where the performance bottlenecks lie in high-performance computing and then demonstrate how conventional instruction sets have been developed to cope with them.

Computer Graphics

We begin by demonstrating why graphics is such a demanding application. Three-dimensional objects can be modeled as polygons and translated (moved), scaled, and rotated about each of the three coordinate axes. Figure 5.1 demonstrates the rotation of a point about each coordinate axis and provides the corresponding *rotation matrix* that is applied to each point of an object to be rotated. If a point is rotated about all three axes, all three transformations must be carried out.

FIGURE 5.1   Three-dimensional rotations

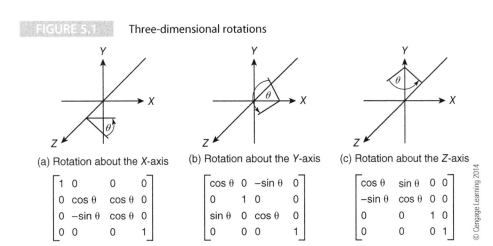

(a) Rotation about the *X*-axis   (b) Rotation about the *Y*-axis   (c) Rotation about the *Z*-axis

$$\begin{bmatrix} 1 & 0 & 0 & 0 \\ 0 & \cos\theta & \cos\theta & 0 \\ 0 & -\sin\theta & \cos\theta & 0 \\ 0 & 0 & 0 & 1 \end{bmatrix} \qquad \begin{bmatrix} \cos\theta & 0 & -\sin\theta & 0 \\ 0 & 1 & 0 & 0 \\ \sin\theta & 0 & \cos\theta & 0 \\ 0 & 0 & 0 & 1 \end{bmatrix} \qquad \begin{bmatrix} \cos\theta & \sin\theta & 0 & 0 \\ -\sin\theta & \cos\theta & 0 & 0 \\ 0 & 0 & 1 & 0 \\ 0 & 0 & 0 & 1 \end{bmatrix}$$

© Cengage Learning 2014

Computer images are approximations to reality, where objects are modeled as polygons; the more polygons, the more accurate the image. Modern computers use tens of thousands of polygons to represent a single image. Modifying a single frame may require a matrix transformation to be applied to each of the vertices of all these polygons, which represents a very large amount of computation indeed. These calculations involve little more than addition, subtraction, and multiplication.

First-generation computer graphics were crude and provided a wholly unrealistic view of the world. Second-generation graphics like those found on high-performance personal computers of the mid 1990s were much better and were capable of generating excellent moving images. Even these systems could not create *realistic* images. Modeling realistic objects requires two important elements—*lighting* and *texture*. When we view a scene, we see it because it is *illuminated*. A light source such as the sun generates light rays that travel to all the points in the scene and then to our eyes. The amount of light reflected by each point in

Multiplying Two Matrices

```
for (int i=0; i < a; i++)
    for (int j=0; j < b; j++)
        P[i][j] = 0;
        for (int k=0; k < c; k++)
            P[i][j]+= Q[i][k] * R[k][j];
```

the scene depends on the *angle* between the light source and the observer, the *texture* of the object, and the nature of the light source itself. If we are to create a realistic image, it must mimic the characteristics of the original.

A realistic system has to take account of *indirect* as well as *direct* illumination. A light ray might hit a reflecting object and then hit an element of the scene before reaching our eyes. Consequently, we view an object from both *direct illumination* due to the light source and from *indirect illumination* from scattered light. Figure 5.2 shows two *circles* that have been filled in and then illuminated from a point source of light to create the impression of *spheres* with a surface texture. The only difference between these two spheres is the parameters used to define the way in which the circles have been filled and the location of the light source.

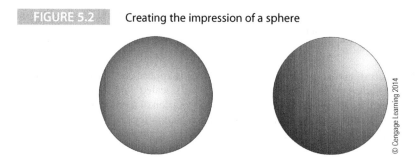

FIGURE 5.2   Creating the impression of a sphere

Computers deal with illumination by *ray-tracing*. Figure 5.3 shows a light source, a viewport, an object, and an observer. Light from the source moves outwards in every direction and some of it strikes the object. A ray of light hitting the object is reflected according to the laws of optics and passes through the viewport to the observer's eye. The intensity of the light at the point it passes through the viewport is mapped onto the corresponding pixel on the display device. To calculate the color and intensity of each pixel, you have to trace each of the light rays from the light source.

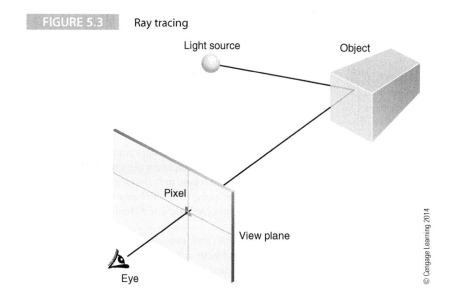

FIGURE 5.3   Ray tracing

The only rays that matter are those that end up in the observer's eye. Practical ray-tracing algorithms perform the inverse operation and start with the eye and the viewport. A ray that ends up at the eye is followed back through the viewport to the object that is visible to the eye. The light leaving the object is traced back to the light source.

The object's color depends on the light reflected from it and the color of the light source. Calculating the color of the ray leaving an object is difficult, because the world is not filled with mirror-like surfaces that obey the simple laws of reflection. Real-life objects have surfaces with varying degrees of *roughness*. When light hits a rough surface, it is reflected or *scattered* in many directions to produce a diffuse cone of light. Some objects are *transparent* and permit the transmission of light through them. When light passes from one medium to another (e.g., air-to-glass) it is *refracted* or bent. By taking the properties of real objects into account when ray-tracing, it is possible to generate realistic images, although it requires a very large amount of calculation.

Drawing images, manipulating them, and dealing with texture, reflection, and refraction requires vast numbers of arithmetic operations. These are largely additions and multiplications and are well suited to the short vector operations we describe later. However, we should also point out that the sheer volume of arithmetic processing required in digital image processing has led to the development of special-purpose graphical image processors that are now incorporated in graphics cards. In the next section, we look at the nature of some of the operations required in image processing.

In recent years, computer graphics has made tremendous advances and has had to cope with the demands of today's high definition displays that require even more real-time processing. In Chapter 13, when we look at multiprocessors, we will see that the demand for CPU power has increased to the extent that some graphics processing has now been moved from the CPU to the *graphics processing unit* (GPU) in the video display card.

5.1.1   Operations On Images

As well as processing images made up of graphics objects that are manipulated individually, we have to process complex images composed of pixels; that is the type of images created by digital still cameras and later processed by applications like Photoshop®. In order to demonstrate the nature of digital image processing, we look at two examples of operations provided by all image processing packages: noise filtering and contrast enhancement.

Noise Filtering

First, let's look at *noise*. Images from video sensors in cameras suffer from the effects of *noise* or random signals. Each sensor element generates a signal due to the light falling on it (the wanted signal) and a random signal due to the motion of charged electrons in the sensor (unwanted noise). The effect of noise is significant at low light levels when the signal-to-noise ratio is low. Noise is called *snow* because it consists of random speckles. One way of reducing the effect of noise is to take the *average value* of an area of $m \times m$ pixels centered on the current pixel. If the brightness of the current pixel is above or below this average value, its value can be moved to bring it closer to the average value. A pixel that's much brighter or darker than its near neighbors no longer stands out after its brightness has been adjusted towards the mean. This operation is applied pixel-by-pixel to the entire image by adding together the value of a pixel and its eight immediate neighbors (after scaling to determine the level of noise reduction) and then normalizing the result. The total number of operations per image is very large—one for each pixel. As in the case of image transformations, the operations are simple and repetitive.

> ### Noise Filtering Variations
>
> In practice, there are several algorithms for the removal of noise from images, some of which are more effective than others. Instead of taking a linear average of local pixel intensities, a Gaussian noise filter gives the neighboring pixels a Gaussian weighting. Some filters detect and remove peak noise because noise is sometimes impulsive—noise values are greater than surrounding pixel values.

Contrast Enhancement

Images taken by digital cameras sometimes suffer from low contrast and the image appears uniformly gray. This is particularly true of images taken from aircraft where light scattered by dust in the air degrades contrast. A simple way of enhancing contrast is to scale each pixel's brightness to force an image to display the full range of brightness levels. If a and b are the lowest and highest values of pixels in the enhanced image and c and d are the lowest and highest values of the pixels in the actual image, we can scale the *old* pixel P_{old} to get a *new* value by

$$P_{\text{new}} = (P_{\text{old}} - c)\left(\frac{b - a}{d - c}\right) + a$$

Scaling is adversely affected by a single pixel whose value is much higher or lower than the others. A better way of performing contrast enhancement is to create a *histogram* of pixel values by examining all of the pixels of an image and plotting the number of occurrences of a given value against the value. Figure 5.4 provides a histogram of a low-contrast image with 256 gray scales. The vertical axis gives the normalized relative frequency of each pixel intensity.

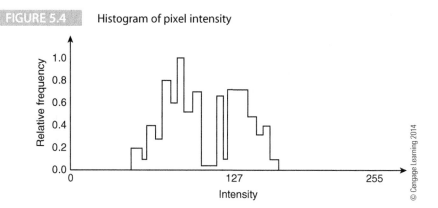

FIGURE 5.4   Histogram of pixel intensity

© Cengage Learning 2014

As you can see from Figure 5.4, the range of pixel values is not from 0 to 255, but from about 50 to 150. That is, the lightest parts of the image are not white and the darkest parts are not black. You can rescale the histogram so that pixels now range in value from 0 to 255. Rescaling to widen the range of pixel values increases the image's contrast.

Edge Enhancement

Sometimes we want things to *stand out* (for example, the leaves on a tree). We can do this by emphasizing their edges in a process called *edge enhancement*; this is often called *sharpening* by image processing packages. The image is examined around each pixel, and the value of the pixel is enhanced if the pixel forms part of an edge. Edge enhancement employs a process called *convolution* to accentuate high frequencies. In image processing, the term *high frequency* indicates a rapid change of intensity, which you would expect at the edge of an image. One way of performing edge enhancement is to take a matrix of $m \times m$ pixels centered on a given pixel and multiply the matrix by a matrix that enhances contrast. A typical edge-enhancing matrix is

$$
\begin{matrix}
-1 & -1 & -1 \\
-1 & 9 & -1 \\
-1 & -1 & -1
\end{matrix}
$$

Figure 5.5a shows a low-contrast image of an aircraft and Figure 5.5b the same image after edge enhancement. In Figure 5.6a, we have taken a low-contrast aerial scene, enhanced the contrast, and performed edge enhancement to obtain the much improved image in Figure 5.6b.

FIGURE 5.5   Histogram, contrast and edge enhancement—before

Alan Clements

Before

FIGURE 5.6   Histogram, contrast and edge enhancement—after

After

Once more, we can see that a multimedia application requires vast numbers of simple, regular, and repetitive operations involving multiplication, addition, and subtraction. Division is sometimes necessary, but the need for division can be much reduced by means of scaling and other techniques—indeed, division can be performed indirectly using multiplication and an iterative algorithm as described on page 300.

Lossy Compression

We now turn to one of the most important applications of processing power. *Image compression* is used to reduce the size of audio and video image files. Some data compression mechanisms are *lossless* because you take data, compress it, and uncompress it, so the final version is an *exact* copy of the original (e.g., ZIP files). The limit to the amount by which you can compress data is determined by the level of *redundancy* (repetition) in the data. Indeed, truly *random* data can't be compressed at all, because there is no redundant information to remove.

Data can be highly compressed if you accept a loss of information. If an image is digitized using lossy compression, losing some information has little effect on the *perceived* quality of the picture. Lossy compression relies on the nature of human perception. Sometimes we can detect tiny degradations in information quality, whereas sometimes the loss of large amounts of information may go unnoticed. For example, the human eye is more sensitive to *luminance* (the level of illumination or brightness) than to *chrominance*. Consequently, the color detail in a TV picture can be much more highly degraded than the brightness—without the image quality becoming unacceptable. Three of the most important data compression technologies are JPEG, MPEG, and MP3, which we describe next.

JPEG

Although there are many lossy compression algorithms, three of the most widely known systems are JPEG (used to compress still images), MPEG (used to compress moving images), and MP3 (used to compress sound).

JPEG is an acronym for *Joint Photographic Expert Group*, which is a body set up to standardize compression algorithms. JPEG compresses a still image by converting the image information into a different *domain*, filtering the information (the lossy part) in that domain, and then using a lossless compression technique to reduce the size of the data still further. Figure 5.7 illustrates the sequence of operations in JPEG compression. The source image is digitized and converted into an array of pixels. Each 8-bit source pixel is stored as a level of brightness in the range of 0 to 255. Color images are divided into *layers* with each layer containing pixel information for one of the primary colors.

JPEG compression begins by dividing the image into blocks of 8×8 pixels (the array size 8×8 is a compromise based on speed, accuracy, and computational complexity). Each block of 64 pixels is then transformed from information in the *spatial* domain to information in the *frequency* domain.

FIGURE 5.7   JPEG compression

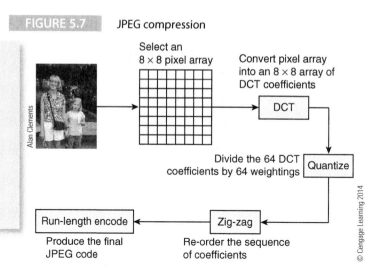

The box labeled DCT (discrete cosine transform) converts pixels in the spatial domain into points in the frequency domain; that is, a set of points is converted into a set of frequencies. Redundancy is much easier to remove from data in the frequency domain.

The source image exists in the *spatial* domain and is simply the pixels that make up the image mapped out in space on a screen or on paper. The *frequency* domain consists of waveforms. Transformations between the time domain and frequency domain have been common in sound engineering for a long time. For example, the Fourier transform allows us to take any arbitrary waveform and decompose it into an infinite series of sine and cosine waves of differing amplitudes. Converting time-varying waveforms into a sequence of sine waves allows us to carry out certain types of signal processing, such as removing *scratches* on vinyl disks.[1]

A similar process of domain conversion can be applied to the 8 × 8 array of pixels in the spatial domain. You can define 64 *discrete cosine transform* (*DCT*) basis functions and use them to construct any 8 × 8 pixel image. Figure 5.8 illustrates these 64 base functions.

FIGURE 5.8   The 8 × 8 DCT basis functions

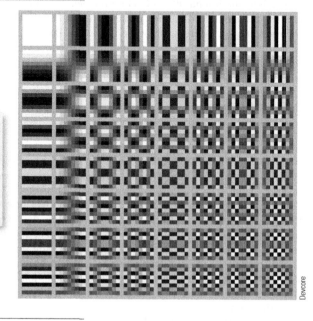

An array of 8 × 8 pixels can be decomposed into the sum of these 64 DCT basis functions, each of which is multiplied by a weighting factor.

---

[1] Until the advent of digital media, music was often recorded on black plastic disks along the length of a spiral track. Data was stored as the sound vibrations themselves (i.e., it was an analog medium) in the form of indentations in the wall of the track. Today's students encounter vinyl disks only when their parents or grandparents get divorced and fight over the custody of 1970's Beatles records.

Any 8-pixel by 8-pixel *image* can be expressed as the weighted sum of all the basis functions in Figure 5.8. That is, an 8×8 matrix of intensity pixels can be replaced by an 8×8 matrix of the coefficients of each of the basis functions. The first basis function in Figure 5.8 represents the average intensity of the block of 8×8 pixels and is known as the DC level of the block.

Once a discrete cosine transformation has been applied to the block of pixels, we have an 8×8 matrix of coefficients. This transformation has no effect on the size of the data set; all we've done is to exchange spatial information for frequency information.

However, there is a very big difference indeed between the spatial and frequency domain matrices. The 8×8 matrix that contains the DCT coefficients corresponding to a 64-pixel array has an important property. The DCT coefficients are grouped so that the upper left of the matrix contains the low-frequency components that dominate most images, and the lower right contains the high-frequency components that are normally of a much lower magnitude. If the brightness of pixels in a block is fairly uniform (as you might get in part of the sky or a leaf), then most of the coefficients in the bottom right-hand corner, corresponding to the basis functions that are most rapidly changing, will be very small or zero. In other words, for certain types of image, many of the coefficients in the transformed matrix are zero or nearly zero, and these coefficients are grouped in one region of the matrix.

The next step is to *quantize* the DCT coefficients by dividing each of the 64 DCT coefficients by one of 64 constants. It is this step that makes the JPEG compression lossy and contributes so much to the compression. The quantization matrix in Table 5.1 gives the 64 constants used in JPEG encoding. The elements in the bottom right (the high-frequency region of the DCT array) are larger. Dividing by a large number reduces the weight of the corresponding DCT coefficient. Indeed, for most of the time, the quantized DCT coefficients in the lower right of the block will be zero. Removing the high-frequency components from the picture has a much less disturbing effect to the human viewer than removing the low-frequency components. Remember that the (high-frequency) coefficients are in the bottom right-hand side of the matrix and that these are often close to zero; dividing them by large numbers further reduces their significance.

TABLE 5.1		The JPEG Quantization Matrix					
16	11	10	16	24	40	51	61
12	12	14	19	26	58	60	55
14	13	16	24	40	57	69	56
14	17	22	29	51	87	80	62
18	22	37	56	68	109	103	77
24	35	55	64	81	104	113	92
49	64	78	87	103	121	120	101
72	92	95	98	112	100	103	99

© Cengage Learning 2014

The first coefficient is called the DC level, which indicates the *average intensity* of the current 8×8 block of pixels. Instead of transmitting this average value, the value recorded is the difference between the current and previous block.

Once the DCT coefficients have been quantized, the 8×8 matrix is converted into a 64×1 vector (that is, a string of 64 coefficients). The string of coefficients is formed by taking a zig-zag path through the 8×8 array, as Figure 5.9 demonstrates. This step concentrates the energy (i.e., the larger coefficients) at one end of the vector. For example, the vector might be $0.2, 0.7, 0.4, 0.9, 0.3, 0.5, 0.1, 0.05, 0, 0, 0, \ldots, 0$.

Because there is a high probability that the quantized and reordered DCT coefficient vector will include a few nonzero coefficients and a lot of zeros, *run-length encoding* is used to compress the 64-component vector. The vector is encoded as a sequence of (*skip, value*) pairs, where *skip* is the number of zeros to skip over and *value* is the next nonzero element. The final value in the sequence is 0, 0. The last stage in JPEG encoding is to take the run-length

FIGURE 5.9   The JPEG DCT coefficient reordering zig-zag

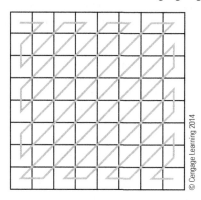

© Cengage Learning 2014

encoded values and then create the final code by means of Huffman encoding. Huffman encoding is a lossless encoding mechanism that assigns short codes to frequently occurring values in a message and long codes to infrequently occurring values.

MPEG

The digital video coding standards MPEG-1 to MPEG-4 (Moving Pictures Experts Group) are one of the triumphs of modern digital engineering and can compress moving images by a factor of over 100, sufficient to transmit them or to store them on DVDs, hard drives, and even pen drives. The original MPEG-1 standard was completed in 1991 and was intended for video storage on CD at bit rates of up to 1.5 Mbps. In 1994, MPEG-2 was launched to encode NTSC and PAL encoded television signals. This standard is also known as the ISO standard 13818 and the International Telecommunications Union standard H.262.

Two types of compression can be performed on moving images: *spatial* compression used in JPEG and *temporal* compression, which exploits the fact that successive images in a movie or TV picture are often almost the same. Like JPEG, MPEG performs a discrete cosine coding to compress the image by removing redundancy. However, it has a *dynamic* aspect, because it examines consecutive frames to remove the redundancy between them. For example, if a video scene is static with no movement, successive frames will be identical. MPEG coding does not transmit the same frame over and over again. The video data is transmitted as a series of frames like any other moving image. The frames are divided into I-frames, P-frames, and B-frames. The I-frame is a stand-alone frame that is compressed much like a JPEG image. The P-frame is a *predictively encoded* frame that describes the difference between the new frame and the previous frame. The B-frame is another form of predictively encoded frame that has the option of using forward or backward interpolation. B-frames have a good coding efficiency and are more effective in scenes where a moving object reveals hidden areas. A typical frame encoding sequence might be I, B, B, P, B, B, I, B, B, P, ...

In order to perform MPEG coding and decoding, it is necessary to perform a JPEG-style compression on individual frames in real-time and to store successive frames in order to perform the predictive (differential) encoding. This requires a lot of calculation in a short time. Until relatively recently, you had to buy dedicated video processing hardware to handle such a workload. Today, high-performance processors can code and decode MPEG data unaided. Indeed, processors in devices like the iPad can cheerfully handle MPEG images.

MP3

MP3 is the audio equivalent of JPEG and is used to compress audio. The algorithm has become so ubiquitous that a device may be called an MP3 player whether it uses MP3 encoding or not. There are now several alternatives to MP3 encoding such as Windows media (WMA) and advanced audio coding (AAC).

In 1987, the German Fraunhofer Institut began to develop a *perceptual coding* scheme for use in Digital Audio Broadcasting (DAB). The compression technique they developed became standardized as *ISO-MPEG Audio Layer-3*, which is now abbreviated to MP3. This compression algorithm can reduce the bit rate required to transmit or store digital audio by a factor of ten or more. A low-quality telephone conversation can be compressed by a factor of almost 100:1 at a bit rate of 8 kbits/s, whereas CD quality sound is compressed by a factor of 16:1 at a bit rate of 96 kHz.

Just as JPEG works by removing high-frequency information from images, MP3 works by removing sounds that contribute little to the whole—as far as the human observer is concerned. MP3 uses *psychoacoustic encoding*, because it depends entirely on the way in which humans perceive sound.

Everyone has heard the expression *"It was so quiet that you could hear a pin drop."* Scientists at the Fraunhofer Institut exploited this maxim in MP3 decoding by using its corollary. You can't hear a pin drop in a noisy environment, therefore, the noise of a pin dropping can be removed from the audio signal without any apparent degradation.

A loud sound *masks* a quiet sound because you don't notice low intensity sounds in the presence of loud ones. Suppose you have a signal at 1.2 kHz at a level of 0 dB and you add a signal at 1.1 kHz at a level of -20 dB (i.e., the added signal has a power level ten times lower than the 1.1 kHz signal). Adding this low-level signal has little effect on the overall sound. However, if you added a 4 kHz signal at a level of -20 dB, you *would* notice this signal, because a large frequency difference between the signals causes the brain to detect the much higher frequency, even though it is a the low-level signal. MP3 applies a similar technique to the encoding of stereo signals. Instead of having entirely independent left and right channels, MP3 uses a middle channel that contains the sum of the channels $(L + R)$ and a difference signal $(L - R)$ that can be used to recreate the stereo effect.

An MP3 encoder takes the digitized audio and divides the sound into 32 frequency bands. A modified discrete cosine transform performs the band-splitting operation. The signal energy in each of the bands is then compressed by using a psychoacoustic model (built into the coder). The encoder then assembles the output of each of the 32 filter banks into a frame. Because of the masking effect, the information content of some of the banks can be reduced. Encoding bits are not allocated to each of the bands equally. A process called *bit allocation* decides which of the sub-bands should receive the bits used to encode the signal.

The data itself is encoded using Huffman encoding (short codes for frequently appearing symbols). Thus, MP3 uses lossy psychoacoustic encoding to remove energy that does not contribute to the overall sound, followed by lossless Huffman encoding that removes redundancy in the data. The next section introduces *digital signal processing*, which is used to operate on time-varying data in applications that encompass image processing to fetal heart monitoring. Digital signal processing is at the heart of all modern control systems from automobile engine control to aircraft automatic landing systems.

Digital Signal Processing

Digital signal processing (DSP) is referred to in the joint ACM/IEEE Computing Curriculum 2005 as *"The field of computing that deals with digital filters, time and frequency transforms, and other digital methods of handling analog signals."* DSP is not a topic normally covered in computer architecture texts, and DSP is all but invisible to most students taking CS and IT degrees. Yet DSP is probably the most important of all computing applications in today's world.

DSP and the CAT

One of the first practical applications of digital signal processing was in processing a series of X-rays, each taken from a different angle, to create a 3D model. The machine that carried this out was called a CAT scanner (computed axial tomography), and in 1979, its two principal contributors, Hounsfield and Cormack, were awarded the Nobel prize for medicine.

DSP covers the branch of computer science (computer engineering) that deals with the processing and interpretation of signals from electrocardiograms (EKGs), MRI scanners, seismometers, ultrasound scanners, oil rigs (when exploring), and so on. These are applications that are vital to the world's economy and to our healthcare systems.

Almost any system that takes analog signals, converts them into digital form, and processes them can be regarded as a branch of DSP. As well as the previous examples, we can add applications ranging from an automobile's engine control to the Airbus's fly-by-wire computers. Here, we can only introduce DSP and discuss aspects of its impact on computer architecture, for example, by promoting multiply-and-add operations or by ensuring that loops are efficient. One of the reasons for the public invisibility of DSP is that it requires an understanding of a graduate-level branch of math using discrete Fourier transforms. The Fourier transform allows us to take a series of time-varying signals, convert them into signals in the *frequency domain* (remember JPEG and the DCT), process them in the frequency domain, and then convert them into signals in the time domain. Such signals can represent speech or vision but also can be the time-varying value of the Dow-Jones index, the average temperature of the earth, the rainfall fifty miles upstream of a dam, or nerve impulses picked up by a heart pacemaker.

DSP Applications

Digital signal processing is one of the most ubiquitous of all modern digital technologies. DSP is carried out by conventional processors, dedicated special-purpose processors with ISAs optimized for DSP applications, or FPGs (field programmable gate arrays). Typical DSP applications are

- Cellular phones
- Digital pagers
- Audio systems, (MP3, AAC decoding, equalization, sound effects).
- HDTV (High definition television)
- Automobile control (active suspension, engine control, anti-skid)
- Disk drives (digital filtering for pulse detection)
- Active noise cancellation

DSP has its immediate origins in the 1960s when the *adaptive equalizer* was designed to maximize the bandwidth of transmission paths across the global analog-based telecommunications network of the 1960s and 1970s. In the 1960s, high-speed wide-bandwidth optical transmission lines didn't exist, and public-switched telephone data links had a bandwidth of only 3,000 Hz.

An equalizer in a communications network performs roughly the same function as an equalizer in a Hi-Fi system; it tailors the bandwidth of the channel to suit circumstances. In the switched telephone networks of the 1960s and 1970s, equalizers had to automatically adapt to the characteristic of the line as different routes were set up by dialing. That is, the equalizer had to be able to self-optimize to random communications paths in real time. Although the theory of equalization is not trivial, the hardware is remarkably so. Figure 5.10 describes the basic element of digital signal processing using conventional DSP terminology and symbols.

FIGURE 5.10   The fundamental element of a DSP block

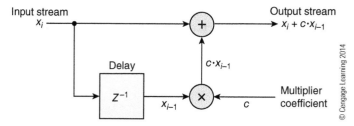

© Cengage Learning 2014

A DSP system uses a sequence of *sampled values* of a signal. For example, the signal level of an analog signal such as music is picked up by a microphone and sampled at over 40,000 times a second. These samples can be multiplied by constants and variables and delayed in *unit delay* blocks. A delay is equal to one clock period and the mathematical operator that denotes a delay is written z^{-1}. If you were to write an expression in *Z-transform* notation $y_i = 0.8x_i + 0.2x_{i-1}z^{-1} + 0.1x_{i-2}z^{-2}$, it would indicate that the ith value of the output, y_i, was given by 0.8 multiplied by the ith value of the input, x_i, plus 0.2 of the input before that, x_{i-1}, plus 0.1 of the input before that, x_{i-2}.

The basic DSP building block of Figure 5.10 can be used to construct digital filters that transmit or stop a range of frequencies. For example, you could use a digital filter with an EKG to separate out the heartbeat of a fetus from that of the mother. There are two classic forms of digital filter; Figure 5.11 illustrates the finite impulse response filter (FIR) and Figure 5.12 illustrates the infinite impulse response filter, IIR.

FIGURE 5.11   Structure of the finite impulse response (FIR) filter

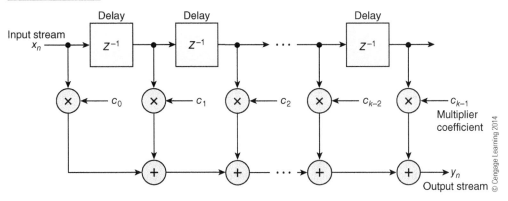

FIGURE 5.12   Structure of the infinite impulse response (IIR) filter

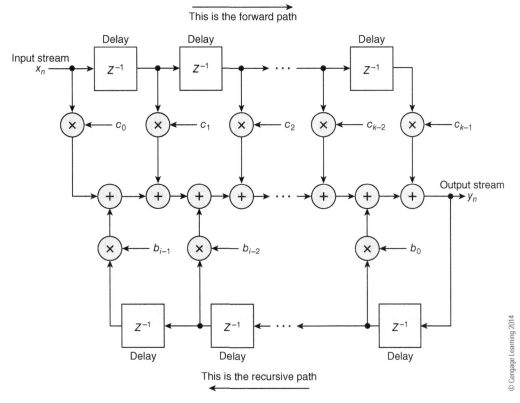

The fundamental architectural difference between the FIR and IIR is that the IIR is *recursive*; that is, the output is defined in terms of itself. For example, we can write the general expression

$$Y_i = c_0 x_i + c_1 x_{i-1} z^{-1} + c_2 x_{i-2} z^{-2} + \dots + b_0 y_{i-1} z^{-1} + b_1 y_{i-2} z^{-2} +$$

The practical difference between FIR and IIR networks is that the IIR filter can be more effective for a given number of stages, but it is harder to design (i.e., to choose the correct multiplier coefficients) and can be unstable (i.e., oscillate under certain circumstances).

Of course, the regular structure of digital signal processors makes it easy to implement them with dedicated hardware. Equally, they are easy to implement with general-purpose microprocessors, because the required operations are simply additions, multiplications, and delays. A delay is synthesized by writing a value into a register or memory and then reading it back.

DSP Architectures

Recall that digital filters are synthesized by evaluating

$$y_i = \sum x_i \cdot a_{i-j}$$

The key action in this calculation is the repeated multiplication of pairs of numbers and their summation, which suggests that multiply-accumulate operations (MACs) similar to the ARM's MLA would feature prominently in any dedicated DSP architecture. Similarly, the variable subscripts imply that the ability to handle vectors is vital and that the vectors should be able to model long sequences of sample values (the data streams).

In the 1980s, semiconductor manufacturers introduced the first generation of dedicated DSP processors; for example, TI introduced the TSM32010 programmable integer DSP chip operating at 5 MIPS, and Motorola introduced the 56000 family in the 1980s. Table 5.2 gives some of the features of these two first-generation DSPs.

TABLE 5.2	Features of Two Early DSP Chips	
Feature	TI TSM32010 Family	Motorola 56000 Family
Data	16-bit (accumulator 32-bit)	24-bit (accumulator 56-bit)
Architecture	Harvard	Harvard
Arithmetic	Fixed-point	Fixed-point
Instruction set	Multiply-accumulate	Multiply-accumulate

© Cengage Learning 2014

Dedicated digital signal processors are designed for high-volume low-cost applications, and that implies that they are highly optimized (if they were not, they would not have a significant advantage over conventional processors). The principal features of DSP architectures are

- Specialized instruction set – support for multiply and accumulate
- Harvard architecture – separate instruction and data paths
- Multiple buses
- Multiple memory banks
- Special-purpose memory access (e.g., buffers) to synthesize delays
- Support for auto-incrementing addressing and circular modes
- Lack of multiuser OS support and memory management facilities
- 16-bit or 24-bit fixed point arithmetic
- No cache memory

At this stage, we are going to introduce a microprocessor, the SHARC family, which has been specifically designed to implement digital signal processing systems. Such DSP processors are highly optimized for their intended application and provide special instructions, addressing modes and even separate data/instruction memories (i.e., the Harvard architecture). We introduce the DSP chip because some of its attributes are finding their way into more conventional ISAs (for example, multiply and accumulate operations).

The SHARC Family of Digital Signal Processors

There is no shortage of dedicated digital signal processors with which to illustrate this text. We have selected SHARC from Analog Devices because this is a DSP processor that has stood the test of time and is now in its fourth generation following its introduction in 1994. SHARC has several architectural features that are now common to many of today's high performance processors.

SHARC has a *Harvard architecture* that it fully exploits. For example, by decoupling the data flow from the instruction flow, it becomes possible to implement 48-bit instructions and 32-bit integers (data). SHARC is a VLIW (*very long instruction word*) processor that encodes multiple operations in a single 48-bit instruction, allowing it to apply its parallel processing capabilities on multiple data elements. SHARC provides both integer and floating-point arithmetic. Floating-point uses extended 40-bit arithmetic and 40-bit registers to maintain precision during chained calculations. The same registers are used for 32-bit integers.

We don't go into the detail of VLIW processors and parallelism here, because those topics are covered in Chapters 8 and 13, respectively. In this section, we are interested in how computer applications have had an impact on instruction set architectures. SHARC is a load-and-store computer that specifies a load register operation with `Ri = DM(<address>)` and a store with `DM(<address>) = Ri`. The notation `<address>` indicates any legal expression that generates a valid address, while DM indicates *data memory*, and PM indicates *program memory*. SHARC lets you perform a load from data memory and program memory in parallel. For example, `r0 = DM(I0,M0) r1 = PM(I4,M4)` loads data registers r0 and r1 from data and program memory in the same cycle (using post-indexed addressing).

> ## Harvard Architecture
>
> The term Harvard Architecture is a reference to the Harvard Mark 1 computer or Automatic Sequence Controlled Calculator built by Howard Aiken and installed at Harvard University in 1944. This was an electromechanical computer.
>
> The Harvard Mark 1 was controlled by a program on 24-channel paper tape. Program and data were stored separately, in contrast to the von Neumann machine that stores programs and data in the same memory. Today, the expression Harvard machine is applied to computers that store programs and data separately.

SHARC's architecture includes two processing elements, *PEx* and *PEy*, plus two data address generators, *DAG1* and *DAG2*. Recall that we said that digital signal processing invariably requires buffer-based *cyclical addressing* where data address generators serve the special addressing needs of DSP applications. Each of the two data address generators keeps track of up to eight pointers (and their associated base address registers and buffer-length registers). Figure 5.13 describes the SHARC's two DAG register sets. The I registers are post-modified and updated by the M registers. For example, the notation `R4 = DM(I1,M2)` indicates that register R4 is loaded from memory by the location pointed at by I1 and then register M2 is used to update the pointer.

Let's look at some common integer operations. The standard arithmetic and logical operations are what you would expect and just highlight some of the more interesting instructions.

`Rn = ABS Rx`	Determines the absolute value of the integer value in *Rx* to be returned in *Rn*.
`Rn = MIN(Rx,Ry)`	Returns the smaller of two integer values in *Rx* and *Ry*. The result is returned in *Rn*.
`Rn = MAX(Rx,Ry)`	Returns the larger of two integer values in *Rx* and *Ry*. The result is returned in *Rn*.
`Rn = CLIP Rx,BY Ry`	Returns the integer value in *Rx* if the absolute value of the operand in *Rx* is less than the absolute value of the operand in *Ry*. Otherwise, $|Ry|$ is returned if *Rx* is positive and $-|Ry|$ if *Rx* is negative.

Because the loop is one of the most important constructs in DSP, SHARC implements a dedicated loop operation that makes use of a special-purpose counter register. Its assembly language form is

```
LCNTR = n, DO loop UNTIL LCE:
Loop:
```

FIGURE 5.13   The SHARC's two data address generator register sets

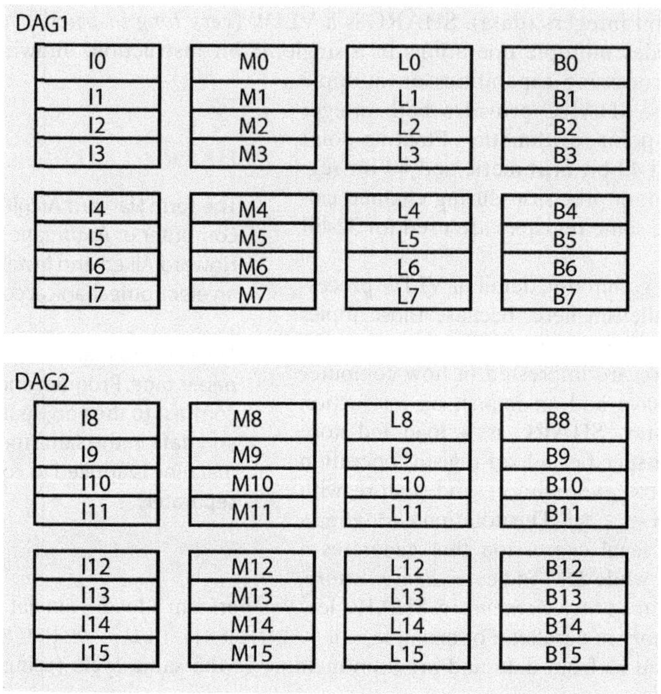

DAG1

I0	M0	L0	B0
I1	M1	L1	B1
I2	M2	L2	B2
I3	M3	L3	B3

I4	M4	L4	B4
I5	M5	L5	B5
I6	M6	L6	B6
I7	M7	L7	B7

DAG2

I8	M8	L8	B8
I9	M9	L9	B9
I10	M10	L10	B10
I11	M11	L11	B11

I12	M12	L12	B12
I13	M13	L13	B13
I14	M14	L14	B14
I15	M15	L15	B15

© Cengage Learning 2014

The text in blue represents a loop constant and a user-supplied label, respectively. The text in black is part of the instruction itself. LCNTR indicates the dedicated loop counter register, and LCE indicates *loop counter expired*. As well as a constant loop terminator, you can specify a dynamic terminator in a register using, for example, LCNTR = R10, DO loop UNTIL LCE. Note also that the value following DO can be a label, 24-bit address, or a 24-bit signed offset from the PC (i.e., a relative address). Loops can be nested. There is more than one loop counter, and each time a loop is set up, the LCNTR is pushed on the stack where it becomes the CURLCNTR (current loop counter).

The basic C code for a simple FIR digital filter (neglecting the circular buffering) can be expressed as

```
for (i = 0, i < m, i++)
    s = s + c[i] * x[i];
```

This can be translated into SHARC code (neglecting loop setting up operations) as

```
        LCNTR = N, DO FIRloop UNTIL LCE:
        R1 = DM(I0,M0), R2 = PM(I8,M8)
FIRloop: R3 = R1*R2, R4 = R4 + R3
```

Having looked at some of the operations performed in signal processing, graphics, and image processing, the remainder of this chapter is concerned with the extensions to CISC and RISC architectures that were implemented to deal with both the type of data elements used in multimedia processing and the type of operations themselves.

5.2   Multimedia Influences—Reinventing the CISC

By the mid-1990s, some observers thought that the trend towards simplicity as embodied by RISC processors such as MIPS, SPARC, ARM, and the PowerPC would spell the end for the traditional CISC instruction sets of the IA32 and 68K families. Indeed, newer members of the

68K family gradually dropped some of this architecture's complex architectural features, leaving the 68060 with a simpler version of the 68020's instruction set. Motorola decided that it was more cost-effective to streamline the 68060's internal organization by using RISC techniques and emulating some of the complex instructions in software than to retain the full 68020 architecture. In fact, with the introduction of special-purpose instructions, an opposite trend has emerged with a tendency toward greater complexity.

We now explain how processors have extended their architectures to incorporate facilities to enhance the rate at which audio and video data can be processed. Intel was one of the first companies to emphasize audio-visual data processing when it called the enhancements to its IA32 architecture *multimedia extensions*. We look at the type of operations required by multimedia applications and demonstrate how architectures have been adapted to facilitate the processing of audio and video data. In general, the architectures of existing processor families have been adapted to facilitate multimedia processing.

SIMD Operations

Single instruction multiple data (SIMD) indicates that a single operation is applied to several data elements in parallel.

For example, a 64-bit register might hold four 16-bit elements that can be added to another four elements in a register in one operation. The figure below demonstrates how four 16-bit additions take place in parallel.

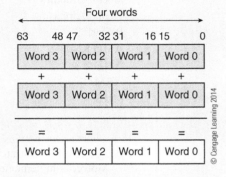

Architectural Progress

Today's 32- and 64-bit microprocessors are the result of years of progress. They have become faster, have more registers, and have wider data highways, but their instruction set architectures have remained remarkably constant. In particular, architectures developed in the 1990s were well-suited to scientific and business applications, but their performance is disappointing when dealing with applications such as audio, video, and telephony. The problem with conventional processor ISAs is the lack of dedicated facilities for working with audio or video data structures. For example, the audio compact disc uses 16-bit integer data, which is a format that doesn't make effective use of modern 32- or 64-bit architectures. Similarly, digital video often uses 8-bit red-green-blue triplets that can be processed with 8-bit operations.

Up to the mid-1990s, processors offered little support at the instruction level for the intrinsic operations in typical multimedia applications. This was a particularly grievous omission, because multimedia audio and video processing applications often have inner loops that require countless millions of relatively simple primitives to be executed. In 1996, Intel announced a *multimedia extension technology*, called MMX, that added innovative architectural features to the IA32 architecture. This move was surprising in that Intel was increasing instruction set complexity when the trend was in the other direction. Yet, at the same time, it made good sense to incorporate facilities that would give the IA32 a strongly competitive edge. Since then, each new generation of a microprocessor family has added special-purpose *multimedia friendly* instructions to its core of existing instructions.

MMX technology evolved from Intel's previous i860 architecture, which was an early, general-purpose processor designed to provide support for graphics *rendering*. As we have seen, when the real world is modeled graphically, a complex object such as a human face is composed of tens of thousands or even millions of polygons. Rendering is the final stage of the modeling process where the surface of an object is made to look realistic by adding texture and lighting effects to these polygons. The i860 processor provided instructions that operated on multiple adjacent data operands in parallel (for example, four adjacent pixels of an image). MMX belongs to what is commonly called *short-vector SIMD* technology. *Single instruction multiple data* (SIMD) refers to a class of architectures that perform the same operation in parallel on multiple data elements (we return to this in Chapter 13). Similarly, the

TABLE 5.3	First-Generation Short-Vector Processors
Processor	**Vector Extension Name**
Sun UltraSPARC	VIS (Visual Instruction Set)
Hewlett-Packard PA-RISC	MAX (Multimedia Acceleration eXtensions)
Intel Pentium	MMX (MultiMedia eXtensions)
Intel Pentium	SSE (Streaming SIMD extensions)
Intel Core i7	SSE4 (Streaming SIMD extensions)
Intel Sandy Bridge processor	AVX (Advanced Vector Extensions)
Silicon Graphics	MDMX (MIPS Digital Media eXtension)
Digital Alpha	MVI (Motion Video Instructions)
PowerPC	AltiVec
AMD K6-2	3Dnow!
AMD	XOP, FMA4, CVT16
ARM	NEON
MIPS	DSP ASE

© Cengage Learning 2014

term *short vector* implies that the multiple data consists of few components (typically eight bytes). Table 5.3 illustrates some of the processors designed for multimedia applications.

Motorola implemented one of the most complete implementations (at that time) of short-vector SIMD technology in its AltiVec architecture incorporated in the PowerPC G4 processor. Although Motorola dropped out of the mainstream microprocessor business and sold off its semiconductor arm to FreeScale, IBM (the originator of the Power architecture) still supports AltiVec. AltiVec was implemented in a separate unit allowing concurrent operation of existing integer and floating-point units. AltiVec's 162 instructions provided wide-field shift, pack and unpack instructions, merge operations that interleave data, and a permute operation capable of arbitrarily selecting data elements from two source vectors and ordering them in the destination register. AltiVec provided 32 new 128-bit registers.

Sun's UltraSPARC short vector extensions (VIS) were targeted at video processing. VIS integer operations use the SPARC's floating-point registers. Video information at the pixel level is stored as four 8-bit or 16-bit integer values representing the colors red, green, blue and the alpha information for the pixel. These values define the amounts of red, green, and blue in the pixel together with its transparency. The alpha value in the range $0 \leq \alpha \leq 1$ represents *transparency*. The types of operations provided by the VIS extension to SPARC are

- Pixel expand and packing
- SIMD logical operations
- SIMD add, multiply, and compare
- Alignment and edge handling
- Merge
- Pixel distance

The *pixel expand and pack* instructions are inverse operations that convert data between 8-bit and 16-bit pixel representations and vice versa. The add/subtract instructions provide for two or four 16-bit add/subtracts and one or two 32-bit add/subtracts. The multiply instruction can multiply four 16-bit values by four 8-bit values.

The `align` instruction allows the processor to access pixels in the middle of a 64-bit word. If an image starts or ends on pixels that are not aligned on 64-bit boundaries, the `edge` instruction masks off all the unused pixels.

As we have just seen, a key element of modern video processing is the *compression* used to reduce the number of bits required to encode images and the bandwidth needed to transmit them. Because human perception is less sensitive to color change than to changes in intensity, you can store color information at a lower resolution than intensity information. Consequently, algorithms for video compression separate the intensity information from the color information. The merge instruction is used to convert pixel information from packed to *planar* form, and then the pixel add and multiplication instructions are used to process the intensity and color information separately.

Recall that the compression of moving images relies on the fact that the information content of successive frames often remains almost the same. You calculate the offset between the two blocks (regions of an image) to get a motion vector. The error between the current block and the similar block in the previous frame is encoded and transmitted along with a motion vector for the block.

The UltraSPARC VIS architecture provides hardware support for *motion estimation* by performing comparisons between successive regions of the image to obtain a motion value that minimizes the estimation error. The error is calculated by summing the differences for each pixel in the region between a reference frame and a newer frame. UltraSPARC can operate on eight pixels at a time by summing the absolute difference between eight pairs of pixels. This operation requires eight subtractions, eight absolute values, eight additions, a load of eight pixels, an align of eight pixels, and one final addition. Table 5.4 illustrates some of the architectural features of short vector processors.

TABLE 5.4   Features of Short-Vector Processors

Architecture	MIPS MDMX	Intel MMX	SUN VIS	HP MAX-2
Extension feature				
Operands	3 to 4	2	3	3
Vector registers	32 FP	8 FP	32 FP	32 Integer
Integer storage sizes				
Eight 8-bit	✓	✓	✓	
Four 16-bit	✓	✓	✓	✓
Two 32-bit		✓		
Integer computation sizes				
Eight 8-bit	8- or 24-bit	✓	✓	
Four 16-bit	16- or 48-bit	✓	✓	✓
Two 32-bit		✓		
Arithmetic operations				
Vector/Scalar	✓			
Accumulation	✓			
Saturation	✓	✓	✓	✓
Vector floating point	✓			
Multiply	✓	✓	✓	
Multiply and add	✓			✓
Distance			✓	

© Cengage Learning 2014

The once radical technology MMX is now mainstream, and each iteration of the Intel IA32 family has added more instruction set extensions. *Multimedia extensions* became *streaming extensions* and then *advanced vector extensions*. Intel's rival AMD provided its own extensions called *3DNow!* But, in 2010, AMD announced that it would not support 3DNow! in future processors (presumably because of Intel's success in the market place).

It may seem that adding new instructions to a processor flies in the face of backward compatibility. This would be true if all new software used new streaming instructions and that software would not run on older machines. In practice, the problem is solved via a programming applications interface. An application requests a service of the operating system that uses streaming extensions. The application itself does not have to run code containing streaming instructions. If the processor does not implement streaming extensions, the requested operation is executed using legacy code.

We now look at Intel's MMX technology. Although MMX has been superseded by richer and more versatile streaming extensions to the IA32 core architecture, it provides an excellent vehicle for introducing the nature of SIMD processing because of its small core of new operations.

5.3   Introduction to SIMD Processing

Intel's multimedia extensions provide the IA32 architecture with a SIMD facility that lets you apply a single instruction to multiple data elements. For example, you can multiply four pairs of 8-bit integers in parallel or simultaneously add *eight pairs of bytes*. Figure 5.14 illustrates the new data types that are all 64-bit values in registers MM0 to MM7.

When introduced in 1997, the IA32 was cursed by the need to maintain backward compatibility. A prime requirement of MMX technology was that it was to be compatible with existing operating systems, and the IA32 architecture could not be extended to include new registers, new condition codes, or new exception handing features. In short, MMX technology was designed to be added to the IA32 without changing anything. Such an approach soon proved to be untenable, and the early architectural restrictions were removed as multimedia technology continued to advance with each new generation of Intel processors.

To avoid adding *new* registers, MMX had to adopt a *registerless* structure such as a stack or to make existing registers serve a dual function. Intel, like the UltraSPARC, uses floating-point registers to provide MM0 to MM7 that share the same silicon as the floating-point register set. Therefore, you can't perform MMX operations and floating-point operations at the same time. The floating-point registers are 80 bits wide, although the MMX instructions use only the first 64 bits of these registers.

An MMX register can be partitioned into eight 8-bit values, four 16-bit values, or two 32-bit values, as Figure 5.14 demonstrates. Table 5.5 summarizes the MMX instruction set where instructions have typical CISC register-to-register two-address formats. MMX mnemonics are prefixed by P to indicate a *packed operation*. For example, the PADD instruction performs a packed add. The suffix b, w, or d indicates a byte, word, or doubleword operation, respectively. For example, a PADDb **MM0**, MM1 instruction simultaneously adds eight pairs of bytes in register MM1 to the eight bytes in register MM0 and deposits the eight sums in MM0. Because these instructions can take options, we will write the base mnemonic in uppercase and use lowercase to indicate an option. For example,

```
        PADDb MM0,MM1
or      PSUBw MM3,MM0
```

When PADDb adds the eight bytes in MM1 to the corresponding bytes in MM0 and deposits the result in MM0, eight carry-out bits are generated, and *these are discarded and not recorded*.

FIGURE 5.14   MMX data types

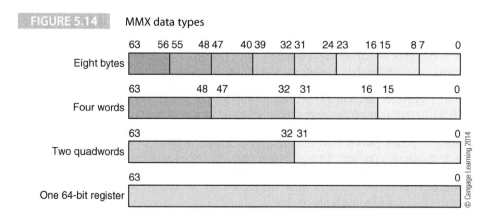

TABLE 5.5	Summary of MMX Instructions	
Mnemonic	**Options**	**Description**
PADD (b,w,d)	Wraparound, saturate	Add packed eight bytes, four 16-bit words, or two 32-bit words.
PSUB (b,w,d)	Wraparound, saturate	Subtract packed eight bytes, four 16-bit words, or two 32-bit words.
PCMPEQ (b,w,d) PCMPGT (b,w,d)	Equal, greater than	Compares packed bytes/words/double-words. Result is a mask of all 1s if true or all 0s if false in the destination register.
PMULLW PMULHW		Multiplies four packed 16-bit values in parallel. You can select either the high- or low-order 16 bits of the four 32-bit results.
PMADDWD		Multiplies four packed, signed, 16-bit words and adds together adjacent pairs of 32-bit results in parallel. The result is a 32-bit doubleword.
PSRA (w,d) PSLL (w,d,q) PSRL (w,d,q)	Shift count in a register or an immediate	Shifts arithmetic right, logical left, logical right packed four words, two double-words, or the entire 64-bit quadword.
PUNPCKL (bw,wd,dq) PUNPCKH (bw,wd,dq)		Merges packed eight bytes, four 16-bit words, or two 32-bit doublewords with interleaving.
PACKSS (wb,dw)	Always saturate	Packs doublewords to words or bytes in parallel.
PAND		PAND performs a bitwise logical AND.
PANDN		PANDN performs a bitwise logical NAND.
POR		POR performs a bitwise logical OR.
PXOR		PXOR performs a bitwise exclusive OR.
MOV (d,q)		Move 32 or 64 bits between memory and a MMX register.
EMMS		Clear floating-point register tag bits.

© Cengage Learning 2014

Before you can process data in an MMX register, you have to load the data. The MMX architecture provides two new move instructions, MOVd and MOVq, to copy 32-bit or 64-bit values between MMX registers and memory. The MOVd version transfers a 32-bit double-word, and the MOVq version transfers a 64-bit quadword. You can also transfer data between an MMX register and another IA32 register if the operand size is 32-bits.

The last MMX mode instruction in Table 5.4 is different, because it's not a data-processing instruction. EMMS empties the floating-point register tag bits and is used to accelerate the transition between MMX and floating-point modes. We will return to this instruction later. We next look at the nature of packed MMX operations in greater detail.

Packed Operations

All MMX instructions (apart from the special EMMS instruction used to facilitate a change of MMX to floating-point mode) have a source and a destination operand. For example, PADDsb **MM0**,MM1 performs the parallel addition of eight pairs of bytes using *saturating arithmetic*. Saturating arithmetic clamps numbers at the lower or upper limit when values go out of range (for example, in eight bits 250 + 20 = 255, which we discuss further in the next section). The source in register MM1 is added to the destination in MM0 and the result deposited in MM0. Note the structure of the mnemonic: P (packed) + ADD (operation) + s (arithmetic mode) + b (data size). Another suffix used by MMX instructions is u, which indicates an *unsigned* operation. Although two's complement and unsigned addition/subtraction follow the same rules, in saturating arithmetic, the upper and lower limits are different (00_{16} and FF_{16} unsigned and 80_{16} and $7F_{16}$ signed), which means you have to indicate a signed or unsigned operation. Figure 5.15 illustrates the effect of PADDb **MM1**,MM0 where eight pairs of bytes in two MMX registers are added together simultaneously.

FIGURE 5.15   MMX addition: executing PADDb **MM1**,MM0

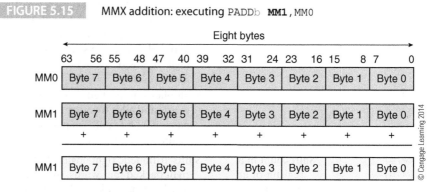

Since the eight additions in a packed MMX addition take place *simultaneously*, an 8-bit vector is required to hold the eight carry bits generated by the additions. As we've already pointed out, no such vector exists, and the programmer has no access to carries generated in MMX operations. Consequently, additions and subtractions using MMX instructions do not modify the state of the processor's carry bit.

Examples of MMX Multiplication from the Web

These two examples have been provided to demonstrate how MMX code looks. In the first example, a constant is added to each component of a 24-bit vector (that is, $x_i = x_i + c$ for $I = 0$ to 23). This uses MMX and conventional Intel IA32 code, but the meaning is clear.

```
        movq  mm1,c          ; load constant into mm1 (8 copies)
        mov   cx,3           ; set up loop counter for three trips 8 x 3 = 24
        mov   esi,0          ; set pointer to 0 (use as index into vector)
Next:   movq  mm0,x[esi]     ; load 8 bytes into mm0 using indexed addressing
        paddb mm0,mm1        ; now do 8 bytes of the vector addition
        movq  x[esi],mm0     ; store 8 bytes of result in x
        add   esi,8          ; increment index by 8
        loop  L1             ; Intel's loop construct using register cx
        emmes                ; Say goodbye to MMX and release the registers
                               to the FP unit
```

The second example, by Andreas Jonsson, is a demonstration of the use of SIMD code to blend two 32-bit video aRGB channels. The blending factor is an integer in the range from 0 to 255, and each channel is blended using output = $(a*fa + b*(255 - fa))/255$. Note that division by 255 is not attempted in this code. The division is by 256 using shifting and then adding 1 if the factor is above 127.[*]

```
; DWORD LerpARGB(DWORD a, DWORD b, DWORD f);
; load the pixels and expand to 4 words
      movd        mm1, [esp+4]     ; mm1 = 0 0 0 0 aA aR aG aB
      movd        mm2, [esp+8]     ; mm2 = 0 0 0 0 bA bR bG bB
      pxor        mm5, mm5         ; mm5 = 0 0 0 0 0 0 0 0
      punpcklbw   mm1, mm5         ; mm1 = 0 aA 0 aR 0 aG 0 aB
      punpcklbw   mm2, mm5         ; mm2 = 0 bA 0 bR 0 bG 0 bB
; load the factor and increase range to [0-256]
      movd        mm3, [esp+12]    ; mm3 = 0 0 0 0 faA faR faG faB
      punpcklbw   mm3, mm5         ; mm3 = 0 faA 0 faR 0 faG 0 faB
      movq        mm6, mm3         ; mm6 = faA faR faG faB [0 - 255]
      psrlw       mm6, 7           ; mm6 = faA faR faG faB [0 - 1]
      paddw       mm3, mm6         ; mm3 = faA faR faG faB [0 - 256]
; fb = 256 - fa
      pcmpeqw     mm4, mm4         ; mm4 = 0xFFFF 0xFFFF 0xFFFF 0xFFFF
      psrlw       mm4, 15          ; mm4 =   1    1    1    1
      psllw       mm4, 8           ; mm4 = 256 256 256 256
      psubw       mm4, mm3         ; mm4 = fbA fbR fbG fbB
; res = (a*fa + b*fb)/256
      pmullw      mm1, mm3         ; mm1 = aA aR aG aB
      pmullw      mm2, mm4         ; mm2 = bA bR bG bB
      paddw       mm1, mm2         ; mm1 = rA rR rG rB
      psrlw       mm1, 8           ; mm1 = 0 rA 0 rR 0 rG 0 rB
; pack into eax
      packuswb    mm1, mm1         ; mm1 = 0 0 0 0 rA rR rG rB
      movd        eax, mm1         ; eax = rA rR rG rB
      ret
```

Saturating Arithmetic

A particularly interesting aspect of MMX instructions is their treatment of overflow and underflow in integer arithmetic. Conventional integer arithmetic uses a *wraparound* mode where (for example) in eight bits the addition of 1 to 11111111 results in 00000000 and a carry out. The term wraparound indicates that the maximum value wraps around to the minimum value.

MMX instructions can operate in a conventional wraparound mode, but this does not model well the physical reality of multimedia operations. Suppose we use an 8-bit value to represent the intensity of a blue pixel. If 11111111 represents the brightest blue and 00000000 represents no blue, the *physical reality* is 11111111 + 1 = 11111111 (that is, you can't get bluer than blue). Similarly, 00000000 − 1 = 00000000, because you can't get less blue than no blue. In both cases, the maximum or minimum value is *constrained* during arithmetic operations. This philosophy of *clamping* a value at its maximum or minimum corresponds well to physical reality. If you are representing part of a scene that is fully white, you can't make it even whiter. Operations that clamp a value at its limits and don't permit wraparound are called *saturating* operations. Some writers used the term *saturation arithmetic* rather than saturating arithmetic.

Let's look at Table 5.5 in more detail. All MMX instructions (except for multiplication) take place in a single cycle. Some MMX operations may operate in a wraparound or a saturated mode. If the operation is both signed and saturated, the upper and lower values are $7F_{16}$ and 80_{16} (in byte arithmetic). If the operation is unsigned and saturated, the upper and lower values are 00_{16} and FF_{16} (in byte arithmetic). Consider the following unsigned byte addition operation, PADDusb MM0, MM1, where the initial vales in MM0 and MM1 are

77012345F0AAF01F and 11FFEE00002387CE

In PADDusb, the u indicates unsigned, s indicates saturating arithmetic, and b indicates byte operands. If we split the hexadecimal digits into eight groups of two, we can perform the additions and record any sum that exceeds FF_{16} as FF_{16}.

```
 77 01 23 45 F0 AA F0 88
+11 FF FF 00 7A 23 87 CE
 88 FF FF 45 FF CD FF FF
```

The sum of four pairs of numbers exceeds FF_{16}, and the result is clamped at FF_{16}. Now suppose we perform the same operation using *signed saturating arithmetic*. In this case, we get

```
 77 01 23 45 F0 AA F0 88
+11 FF FF 00 7A 23 87 CE
 7F 00 22 45 6A CD 80 80
```

Values that have saturated at 80_{16} (most negative) and $7F_{16}$ (most positive) are presented in bold face blue. Later, we demonstrate how saturating arithmetic can be used to perform clipping without using branch instructions.

Here's an example of *branchless* arithmetic that uses saturating arithmetic to perform an operation that would normally require branching. Suppose that P, Q, and R are each eight byte vectors, and we wish to calculate the absolute value $P = |Q - R|$. Conventionally, we can write

```
IF (Q > R)
  THEN P = Q - R
  ELSE P = R - Q
```

This algorithm can be implemented efficiently using saturating arithmetic. If we perform both subtractions $Q - R$ and $R - Q$, the operation that results in a *positive* value gives the correct result. However, the operation that yields a negative value results in *zero* due to the saturation arithmetic limiting the value to zero. Consequently, one of the two values $Q - P$ and $P - Q$ is zero. Since one of these operations yields a positive result and the other yields a zero result, ORing the two values together yields the correct result. If MM0 contains P and MM1 contains Q, we can write

```
MOVQ     MM2,MM0     ;make a copy of P in MM2
PSUBusb  MM0,MM1     ;compute the difference P – Q; result 0 if P < Q
PSUBusb  MM1,MM2     ;compute the difference Q – P; result 0 if Q < P
POR      MM0,MM1     ;OR differences together (one of which is 0)
```

Branches Considered Harmful

Up to now, we have considered branches as perfectly normal instructions. In Chapter 7, we will see that the branch instruction can be detrimental to the performance of computers.

Modern processors employ pipelining, which is a technique that overlaps the execution of instructions. For example, while the current instruction is being executed, the previous one is being decoded, and the one before that is being fetched. If the current instruction is a branch that is taken, all of those partially executed instructions behind it are no longer needed, and the work done on them so far is thrown away.

The need to reject the partially executed instructions is called a *control hazard* and, consequently, programming mechanisms that avoid branches are a good thing.

This fragment of code demonstrates that we can perform eight absolute difference operations on eight pairs of bytes in four cycles without having to use branch operations. This improves the processor's efficiency by removing a potential *control hazard* associated with a branch. We return to this point in Chapter 7 when we look at processor organization.

Packed Shifting

The MMX's logical and shift instructions operate on the individual subdivisions of the 64-bit MMX register. Figure 5.16 illustrates the effect of a parallel shift left on four words in a register. Each of the four 16-bit words in register MM0 is shifted by 4 bits (register MM1 contains the shift count) using PSLLw **MM0**,MM1. The vacated bits of each word are filled with zeros. As in the case of addition, carry-outs are not recorded by parallel shift operations. Note that logical bitwise operators applied to the whole MMX register as subdivisions are meaningless, because the result is the same whether the register is divided into bytes or any other unit.

FIGURE 5.16   The packed shift operation

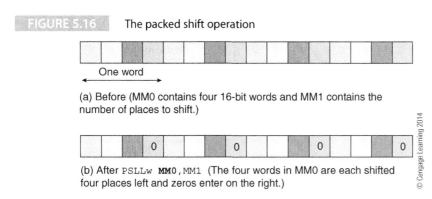

One word

(a) Before (MM0 contains four 16-bit words and MM1 contains the number of places to shift.)

(b) After PSLLw **MM0**,MM1 (The four words in MM0 are each shifted four places left and zeros enter on the right.)

© Cengage Learning 2014

Packed Multiplication

The MMX architecture provides two types of multiplication: a conventional multiplication instruction that forms the products of pairs of numbers in parallel and a powerful *multiply-and-add* operation that forms the sum of products. We have already seen that the ARM has a multiply-and-add instruction that multiplies two numbers and adds the result to a third operand. Multiply-and-add is widely used in vector arithmetic to calculate the inner product of two vectors.

The PMULhw and PMULlw instructions each multiply four signed, packed, 16-bit words to generate four 32-bit products. Because four double-length 32-bit products will not fit in a 64-bit MMX register, you must use PMULhw to select the *upper-order* 16-bits of each product or PMULlw to select the *lower-order* 16-bits of each product. Figure 5.17 demonstrates the effect

FIGURE 5.17   Packed multiplication

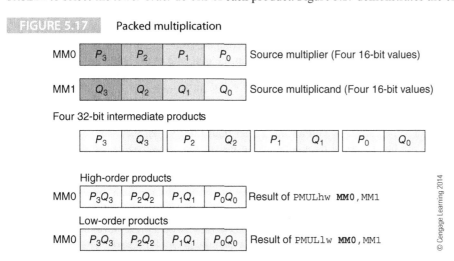

© Cengage Learning 2014

of these two instructions on the contents of two MMX registers. Note how the MMX instruction set provides a higher level of generality for add-and-subtract instructions (8- to 32-bit additions), whereas the multiplication instructions are constrained to 16-bit signed arithmetic.

The packed multiply-and-add instruction, PMADDwd, also multiplies four pairs of signed 16-bit words. However, the two pairs of adjacent 32-bit products are added together to create two 32-bit sums. If the source words are P_3, P_2, P_1, P_0 and Q_3, Q_2, Q_1, Q_0, the products are $P_3 \cdot Q_3$, $P_2 \cdot Q_2$, $P_1 \cdot Q_1$, $P_0 \cdot Q_0$. The adjacent pairs of products are then summed to get $(P_3 \cdot Q_3 + P_2 \cdot Q_2)$, $(P_1 \cdot Q_1 + P_0 \cdot Q_0)$. Figure 5.18 describes this operation.

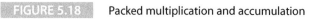

FIGURE 5.18   Packed multiplication and accumulation

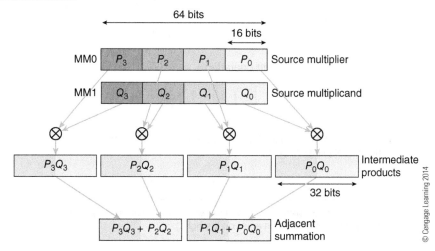

Parallel Comparison

MMX instructions allow you to compare two values, although both the way in which values are compared and how you use the results of the comparison are anything but conventional. Because the MMX architecture doesn't modify the IA32's architectural state, the result of a test cannot be permitted to set one of the processor's flags. Consequently, any test or comparison has to operate by setting or clearing the bits of a register. Moreover, since the MMX is operating on words in parallel, the nature of a vector comparison is different to a scalar comparison because there are multiple results.

The MMX architecture provides two comparison instructions: one performs a comparison for equality and the other provides a *greater than* test. These are two of the most widely used tests (it isn't feasible to provide all of the 16 standard Boolean tests implemented by most processor architectures). Both of these tests can operate on byte, word, and doubleword values.

Intel has adopted an interesting approach to testing that adopts the same philosophy as MIPS where the *set less than instruction*, slt, compares the contents of two registers and sets the contents of a third to 1 or 0, depending on the outcome of the tests. Instead of setting a flag bit in a status register, the outcome of an MMX test either fills the destination operand with all 0s or with all 1s. The condition *true* yields all 1s, and the condition *false* yields all 0s. For example, the test of equality with the values 5 and 9 would yield 00000000 (in a byte comparison), whereas a test with the values 7 and 7 would yield 11111111. In other words, the outcome of a comparison is a data value rather than a flag value.

The data bits returned by a comparison instruction allow you to use the result of the comparison as a *mask* in a logical operation, and hence, avoid the need to execute an explicit conditional branch. Suppose we have a region of an image that we wish to *posterize*; that is, we set the region to one color if the intensity of the color is greater than (or less than) a given threshold. Posterization creates regions of an image with a constant color.

```
PCMPGTb  MM0,MM1       ;compare 8 pixels in MM1 with a preset level in MM0
PAND     MM1,MM0       ;set regions less than the preset level to 0
```

The PACKssdw instruction

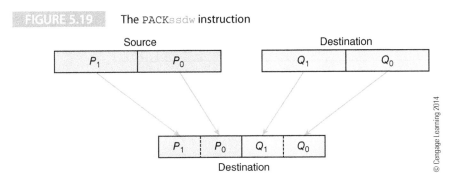

The first operation, PCMPGTb, compares eight bytes in register MM0 with eight bytes in MM1 by evaluating the difference [MM0] − [MM1]. Each of the eight comparisons yields FF_{16} if $[MM0_{byte\_i}] > [MM1_{byte\_i}]$ or 00_{16} if $[MM0_{byte\_i}] \leq [MM1_{byte\_i}]$. Consider the following example.

Source pixels in MM1	8F C2 30 34 40 F1 A3	
Preset level in MM0	50 50 50 50 50 50 50	
Comparison mask MM1	FF FF 00 00 00 FF FF	after PCMPPGTb **MM0**,MM1
Final result in MM1	8F C2 00 00 00 F1 A3	after PAND **MM1**,MM0

Packing and Unpacking

Packing and unpacking instructions are provided to perform conversions between data sizes; that is, 32 bits to 16 bits, or 16 bits to 8 bits. As its name suggests, the *pack* instruction packs data into an MMX register by converting words into bytes (or doublewords into words). When data is packed, it loses precision and may have to be truncated using either signed or unsigned saturating arithmetic. The *pack-with-signed saturating arithmetic* instruction, PACKss, packs and saturates the signed data elements from the source and the destination operands and writes the signed results to the destination operand (the signed, saturated mode is the *only* option with this instruction).

The PACKssdw instruction packs two signed 32-bit words from the source operand and two signed words from the destination operand into four signed 16-bit values in the destination register, as Figure 5.19 demonstrates. If the signed value of a word is larger or smaller than the valid range of a signed 16-bit integer, the value is saturated and clamped at $7FFF_{16}$, if it is positive, or 8000_{16} if it is negative.

Figure 5.20 demonstrates the PACKsswb version of this instruction that packs two signed words from the source operand and two signed words from the destination operand into four signed bytes in the destination register. In this case, the saturated values are clamped at $7F_{16}$ (positive) or 80_{16} (negative).

The inverse operation to packing is *unpacking*, where data is expanded. However, the PUNPCK instruction is not a simple inverse of the parallel pack instruction. The parallel unpack

The PACKsswb instruction

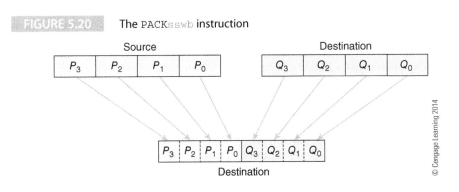

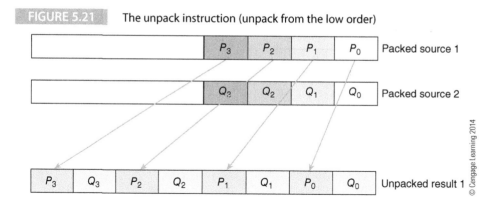

FIGURE 5.21   The unpack instruction (unpack from the low order)

instruction should really be thought of as a parallel merge instruction that operates with three data sizes: byte-to-word (bw), word-to-doubleword (wd), and doubleword-to-quadword (dq). When a word is unpacked, it occupies twice the number of bits. Each MMX instruction allows you to unpack from either the upper half of a word or the lower half of a word. Consequently, there are six variations on the unpack instruction

Figure 5.21 demonstrates the effect of the PUNPCKLbw instruction that unpacks bytes to words. Here, two MMX registers containing four bytes (in the low-order half of the register) are unpacked, and the eight bytes are placed in the destination register.

You can perform some processing with the unpack instruction (for example) by setting one of the registers to zero. Consider the following example, where MM1 initially contains $0123456789ABCDEF_{16}$. This operation results in $000089AB000CDEF_{16}$ being loaded into MM1.

```
PXOR        MM0,MM0     ;clear register MM0
PUNPCKLwd   MM1,MM0     ;unpack/merge MM1 into MM0
```

Coexisting with Floating-Point

As we've said, MMX technology uses the IA32's existing floating-point registers. This action is called *register aliasing*, because the same registers are known by two names (floating-point and MMX). Figure 5.22 illustrates the relationship between the floating-point and MMX registers. The floating-point registers are not explicitly accessible. They form part of an eight-deep stack, and floating-point operations act on the top of this stack. The MMX registers are explicitly addressable by the MMX instructions.

Whenever you write to an MMX register, bits 0 to 63 are affected, and bits 64 to 80 (the exponent field of floating-point numbers) are automatically set to 1s. An all-1s exponent is defined as a NaN (not a number) when interpreted as a floating-point value and prevents the contents of an MMX register from being treated as a valid floating-point value.

The floating-point registers are arranged as a stack and accessed via a stack mechanism. Consequently, floating-point registers can't be accessed at random. It is considered good practice to keep the stack clean. After pushing floating-point values on the stack and performing arithmetic, you should clean up the stack at the end of the sequence of calculations. Each floating-point register has an associated tag field that indicates when the register is in use. When a new floating-point value is pushed, the corresponding tag is validated. When a register is popped off the stack, the tag is cleared. These tags can be used by the operating system when saving the machine context. If a tag is set, that register has to be saved; if a tag is clear, that register is not in use and need not be saved.

Although incorporating the MMX registers adds a new layer of complexity, Intel implements a simple scheme. All tags are set the first time an MMX register is accessed, which ensures that all MMX registers are saved on a context switch from MMX to floating-point

FIGURE 5.22   Mapping MMX registers onto floating-point registers

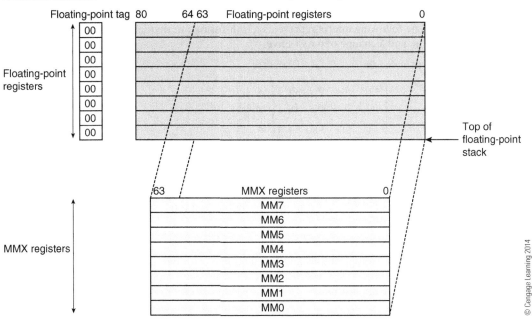

use. The EMMS instruction clears all tag bits prior to ending an MMX sequence and beginning floating-point operations. Failing to use the EMMS instruction does not lead to incorrect operation, but it may lead to inefficient operation.

Not all Intel processors supported MMX technology. Sometimes it's necessary to determine which processor the code was being run on. The IA32 instruction CPUID returns the CPU *identity* in the edx register. It's a sign of the times that Intel's documentation on the CPUID instruction is longer than the documentation describing the entire instruction sets of first generation microprocessors. If MMX technology is implemented, a flag bit is set. The following fragment of C code demonstrates how you can test for an MMX enabled processor. Note that the behavior of CPUID is controlled by a parameter that is accessed in the eax register.[2]

```
bool isMMXSupported()
{
   int fSupported;
   __asm
   {
   mov    eax,1              ;The CPUID needs a parameter in eax
   cpuid                     ;ask CPU for identity information
   and    edx,0x800000       ;mask status information to MMX flag
   mov    fSupported,edx     ;copy flag to variable
   }
   if (fSupported != 0)      ;now return the state of the flag
   return true;
   else
   return false;
}
```

[2] [Intel485] Intel Processor Identification and the CPUID instruction, Intel, Application note AP-485

5.3.1   Applications of SIMD Technology

We have already mentioned how homogeneous transformations can be used to manipulate images. The following matrix represents a transformation that involves rotation and scaling, translation, and a change of perspective.

$$
\begin{bmatrix} x' \\ y' \\ z' \\ w' \end{bmatrix} =
\begin{bmatrix}
a_0 & a_1 & a_2 & a_3 \\
b_0 & b_1 & b_2 & b_3 \\
c_0 & c_1 & c_2 & c_3 \\
d_0 & d_1 & d_2 & d_3
\end{bmatrix}
\begin{bmatrix} x \\ y \\ z \\ w \end{bmatrix}
$$

Rotate and scale ↗   Translate ↗   Perspective ↗

© Cengage Learning 2014

Consider, for example, the new value of x' that is given by $x' = a_0x + a_1y + a_2z + a_3w$. The calculation of x' requires four multiplications and three additions. Using MMX instructions, we can write

```
PMADDwd MM0,MM1   ;perform a₀x + a₁y and a₂z + a₃w
```

to perform four multiplications and two additions with one instruction. We are now going to demonstrate how MMX instructions can be used to implement multimedia operations *elegantly* in three applications.

Chroma Keying

Everyone is familiar with the effect of *chroma keying*—even if they haven't heard of the term. Chroma keying lets you merge two images. For example, we see it in action every day when the weather forecaster stands in front of a synoptic chart that we know is not actually there. Chroma keying lets you cut out a complex image (such as a person) and then superimpose it on another image. Figure 5.23 demonstrates the effect of chroma keying. Figure 5.23a shows the image of a woman and Figure 5.23b shows the image of a flower. In Figure 5.23c the image of the woman has been superimposed on that of the flower. It's as if the woman were standing in front of the flower. These images are reproduced here in monochrome. The woman actually is sitting in front of a blue background.

Although chroma keying looks as if it requires some applied magic, it's achieved by a remarkably simple trick. The woman in Figure 5.23 is placed against a dark blue background. If this image is scanned and converted into pixels, a pixel will either be blue (the background) or not blue (part of the desired image to be superimposed). If any part of the woman's image is the same color as the blue background, part of the second image will bleed through.

To form the composite image, we just read a pixel from image 1 and the corresponding pixel from image 2 and then determine the final pixel for the composite image 3 as

```
if pixelₐ = blue      then pixel_c = pixel_b
                      else pixel_c = pixelₐ
```

FIGURE 5.23   **Chroma keying.** Reprinted with permission of Intel Corporation. © 1996 IEEE. Reprinted, with permission, from Alex Peleg, U.Weister, MMX Technology Extension to Intel Architecture. P49. IEEE Micro August 1996.

(a) First image against a blue background

(b) Second image that is to appear in the background

(c) Composite image
The first image appears in front of the second image

For the entire image, we can write

```
for (i = 0; i < lastPixel; i++) {
    if (imageOne[i] == blue) compositeImage[i] = imageTwo[i];
    else                     compositeImage[i] = imageOne[i];
}
```

Using SIMD extensions, we can process groups of eight pixels in parallel. Furthermore, we can use the comparison operation to provide a bit mask without having to resort to conditional branches. Consider the following fragment of code.

```
                        ;initially MM1 contains the blue mask (eight blue pixels)
MOVEq    MM3,image1     ;read 8 pixels from the woman's image into MM3
MOVEq    MM4,image2     ;read 8 pixels from the image of the flower into MM4
PCMPEQb  MM1,MM3        ;compare the woman's image with blue pixels to create a mask
PAND     MM4,MM1        ;retain the image of the flower where image 1 is blue
PANDN    MM1,MM3        ;retain the image of the woman where image 1 is not blue
POR      MM4,MM1        ;combine the two images
```

SIMD Instructions Used in the SMID Extension Program

```
MOVEq    MM3,image1     ;Load an MMX register from memory. The q denotes quad 64 bits
MOVEq    MM4,image2
PCMPEQb  MM1,MM3        ;A parallel compare for equality over the 8 bytes in two MMX
                         registers.
                        ;When two bytes are equal, 0xFF is loaded into the destination
                        ;register. If they are unequal, the value 0x00 is loaded.
PAND     MM4,MM1        ;AND the source operand with the inverted destination operand.
PANDN    MM1,MM3        ;This instruction performs an AND and inverts the result.
POR      MM4,MM1        ;An OR is performed between the two 64-bit registers
```

These instructions process eight pixels in just six machine cycles. Figure 5.24 demonstrates the effect of PCMPEQb MM1,MM3, where MM1 initially contains the blue mask that's used to mask out the woman in image 1. That is, each pixel in MM1 is set to the same value as the blue background.

After the parallel compare instruction PCMPEQb MM1,MM3 has been executed, register MM1 contains a bit mask. The bits in each byte are all ones if the pixel from image 1 was part of the blue background, or the bits are all zeros if the pixel was part of the woman. We can use this mask to select either the image of the woman or the flower when we create a composite image.

FIGURE 5.24   Using a blue bit mask to create an image mask. Reprinted with permission of Intel Corporation. © 1996 IEEE. Reprinted, with permission, from Alex Peleg, U.Weister, MMX Technology Extension to Intel Architecture. P49. IEEE Micro August 1996.

The effect of executing PCMPEQb MM1, MM3

MM1	blue	blue	blue	blue	blue	blue	blue	blue	The blue mask (each pixel set to blue)

MM3	x_7	x_6	x_5	x_4	x_3	x_2	x_1	x_0	Image 1 (the woman) Pixels in blue represent blue

MM1	00	00	FF	00	00	FF	FF	FF	The mask register after the compression (FF = blue pixel in image 1)

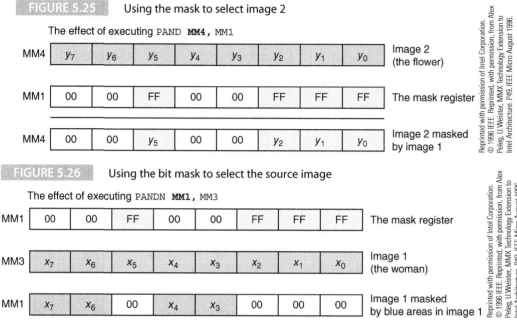

FIGURE 5.25   Using the mask to select image 2

FIGURE 5.26   Using the bit mask to select the source image

Figure 5.25 demonstrates how we use the mask register with image 2 to mask out the bits of image 2 (the flower), where the corresponding bits of image 1 are *not* blue. The PAND **MM4**, MM1 instruction retains bits in register MM4 when the corresponding mask bits are 1. At the end of this operation, the pixels in MM4 correspond to the flower where the background in image 1 is blue.

Figure 5.26 uses the negated AND operation PANDN **MM1**, MM3 to copy pixels of image 1 for which the corresponding mask bits are 0. At the end of this operation, register MM1 contains pixels from the image of the woman with 0s where the background was blue.

Having now created the two partial images with 0s where pixels are masked out, we can merge these images with a logical OR operation POR **MM4**, MM1, as Figure 5.27 demonstrates.

Fade In and Out

Another video processing application that can make good use of SIMD extensions is image combining or image merging. Most people are familiar with the *dissolve shot* on video where one image gradually fades out while another image fades in. If we combine image A (the old image) with image B (the new image), we get

$$\text{output} = \text{fade} \times \text{image A} + (1 - \text{fade}) \times \text{image B}$$

In this equation, the variable *fade* represents the amount of the old image in the composite image and is in the range from 0 to 1. The speed at which the value of *fade* is ramped down from 1 to 0 determines how long the dissolve takes. We will re-label image A as A and image B as B and rearrange the equation to get

$$\text{output} = \text{fade} \times (A - B) + B$$

FIGURE 5.27   Combining the two masked images

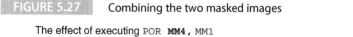

The effect of executing POR **MM4**, MM1

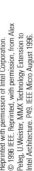

Peleg, et al.[3] demonstrate the implementation of a dissolve algorithm that uses unpacking, parallel multiplication, and re-packing.

Alex Peleg, Sam Wilkie, Uri Weiser, "Intel MMX for Multimedia PCs," Communications of the ACM, January 1997, Vol 40, No 1, pp. 25–38. DOI: 10.1145/242857.242865 © 1997 Association for Computing Machinery, Inc. Reprinted by permission.

PXOR	**MM7**, MM7	;set MM7 to 0 (we need a dummy 0 later)
MOVq	**MM3**, fade	;load the fade value (replicated in four words)
MOVd	**MM0**, imageA	;load pixels from source image A
MOVd	**MM1**, imageB	;load pixels from source image B
PUNPCKlbw	**MM0**, MM7	;unpack 8-bit pixels, convert to words in MM0
PUNPCKlbw	**MM1**, MM7	;unpack 8-bit pixels, convert to words in MM1
PSUBw	**MM0**, MM1	;subtract image B from image A
PMULhw	**MM0**, MM3	;multiply this difference by the fade value
PADDwq	**MM0**, MM1	;add result to image B to get fade $\times (A - B) + B$
PACKUSwb	**MM0**, MM7	;repack 16-bit results to byte form

Two MOVd instructions take the pixels from images A and B and load them into registers MM0 and MM1 for processing. The two PUNPACKlbw instructions are unpack operations that expand the byte-values to word values. This operation is necessary, because the later multiplication operation acts only on word values. Note that PUNPACKlbw requires two source operands (in this case, source operand 2 is the dummy register MM7 that contains zero).

The three operations PSUBw, PMULhw, and PADDwq do all of the processing work by calculating the new image element as output = fade $\times$ image A + $(1 - $ fade$) \times$ image B.

The final operation, PACKUSwb, packs the word length values into bytes. Figure 5.28 illustrates this sequence of operations. Note that this processing applies to only one set of four pixels of one color. We would have to repeat these operations once for each of the three color values (and the alpha value if that is used).

FIGURE 5.28   **Merging images.** Alex Peleg, Sam Wilkie, Uri Weiser, "Intel MMX for Multimedia PCs," Communications of the ACM, January 1997, Vol 40, No 1, pp. 25–38, DOI: 10.1145/242857.242865 © 1997 Association for Computing Machinery, Inc. Reprinted by permission.

Source A	A_3	A_2	A_1	A_0	MM0 after MOVd **MM0**, imageA,
Source B	B_3	B_2	B_1	B_0	MM1 after MOVd **MM0**, imageB,
Difference A – B	$(A_3 - B_3)$	$(A_2 - B_2)$	$(A_1 - B_1)$	$(A_0 - B_0)$	PSUBw **MM0**, mm1 (calculate difference between images)
Fade value	fade	fade	fade	fade	mm3 contains replicated fade value
Fade × difference	fade × $(A_3 - B_3)$	fade × $(A_2 - B_2)$	fade × $(A_1 - B_1)$	fade × $(A_0 - B_0)$	PMULhw **MM0**, mm3 (fade times difference)
Source B	B_3	B_2	B_1	B_0	MM1
Combined image	$Output_3$	$Output_2$	$Output_1$	$Output_0$	PADDw **MM0**, MM1
Repacked result			$Output_3$ $Output_2$ $Output_1$ $Output_0$		PACKuswb **MM0**, MM1

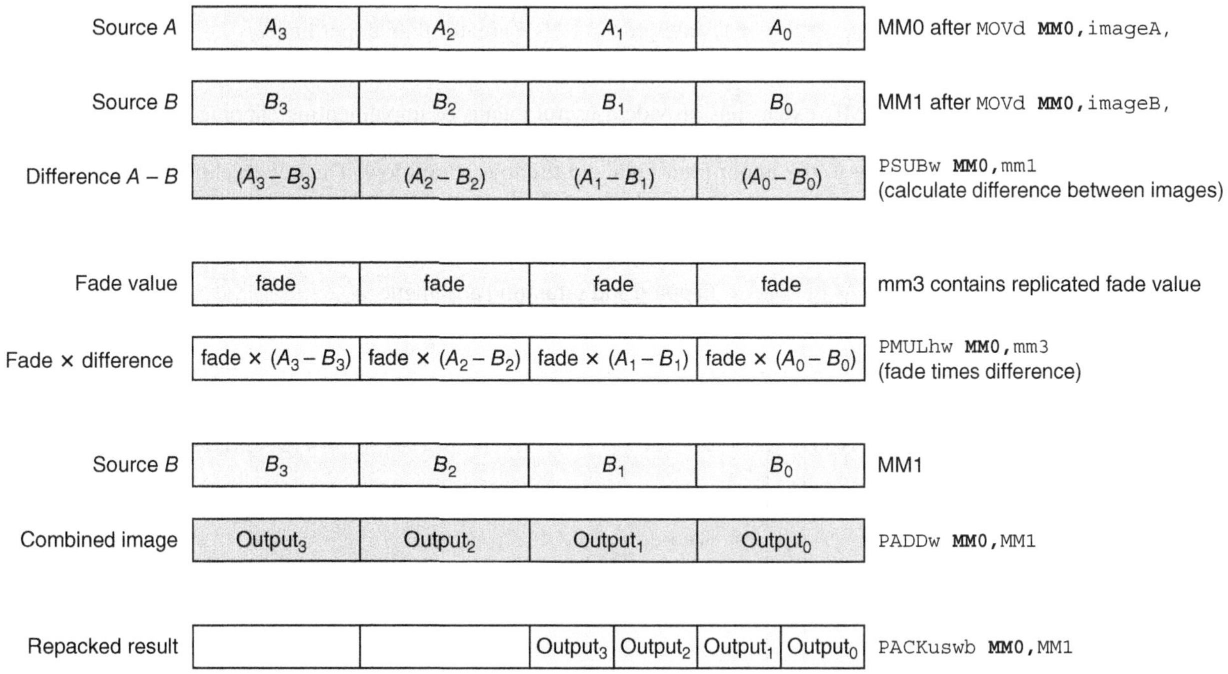

[3] Alex Peleg, Sam Wilkie, and Uri Weiser, "Intel MMX for Multimedia PCs," Communications of the ACM, January 1997, Vol 40, No 1, pp. 25–38.

Clipping

An important operation in graphics is *clipping*, which constrains a variable to lie within a fixed range. For example, if the upper and lower values for a variable X are 4 and 20, then the value of X must always fall within this range. If $X = 15$ and we add 7, then we get $X = 20$ because 20 is the upper limit of X and all higher values are clamped to 20. This operation should remind you of *saturating arithmetic*.

A constrained value is used in graphics when one image obscures another. Therefore, it's necessary to know when to clip or hide parts of the obscured background image behind the foreground. If a value x is clipped to be greater than x_{low} and higher than x_{high}, we can write

```
if x < x_low then x = x_low
          else if x > x_high then x = x_high
```

In conventional assembly language terms, we might have to write something like the following generic CISC code.

```
        CMP  x,xLow        ;Is x less than the lower limit?
        BLT  FixLow        ;If so, then fix it
        CMP  x,xHigh       ;Is x greater than the upper limit?
        BGT  FixHigh       ;If too high then fix it
        BRA  Exit          ;If we reach here then it's in range
FixLow  MOV  x,x_low       ;Constrain to the lower limit and exit
        BRA  Exit
FixHigh MOV  x,xHigh       ;Constrain to the higher limit and exit
Exit
```

We can do a better job by exploiting ARM's conditional execution. Consider the following code.

```
        CMP    r0,#xLow     ;Is x less then lower limit?
        MOVLT  r0,#xLow     ;If so then fix at lower limit
        CMP    r0,#xHigh    ;Is x greater than upper limit?
        MOVGT  r0,#xHigh    ;If so then fix at upper limit
```

SIMD extensions provide a useful means of implementing clipping without using conditional branch operations. The following fragment of code demonstrates unsigned word clipping to the upper limit high and the lower limit low. The key to clipping is the application of *saturating* arithmetic. Figure 5.29 illustrates the relationship between the clipping and

FIGURE 5.29   Clipping and saturating arithmetic

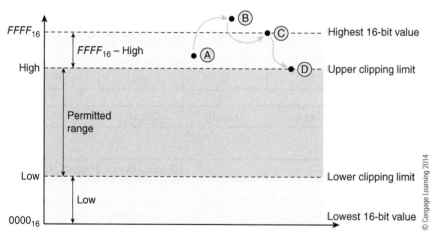

saturation limits and demonstrates an example of clipping. In the following code, the option **u** indicates unsigned, **s** indicates saturated and **w** indicates a word (16-bit) operation.

```
PADDusw   MM0,0xFFFF-high           ;clip to high
PSUBusw   MM0,0xFFFF-high+low       ;clip to low
PADDw     MM0,low                   ;adjust result
```

Let's walk through this code. The first instruction, PADDusw **MM0**,0xffff-high, adds a constant to the source value in MM0 to be clipped. This constant is the maximum value in 16-bits minus the upper clipping limit. If the number to be clipped is above the upper limit, it is pushed to the upper limit and held there by the saturation mechanism. The second operation moves the number down, performing the same operation at the lower limit. The final operation adds back the lower limit that was previously subtracted. This sequence of operations performs

$$x = x + 0xFFFF - \text{high} - 0xFFFF + \text{high} - \text{low} + \text{low} = x$$

That is, if the value of x is in range, it is unaffected by this chain of operations. If x is out of range, it is clipped due to the saturation mechanism. Consider the following example using 4-bit unsigned integer arithmetic with a range from 0 to 15. We will use the clipping limits high = 12 and low = 5. Let's see what happens with the test cases in Table 5.6 (Figure 5.29 also demonstrates clipping by showing the stages A, B, C, and D that take place when a value above the upper limit is clipped).

TABLE 5.6   Using Saturating Arithmetic to Implement Clipping

Operation	Case 1	Case 2	Case 3
Initial Value	7	14	3
PADDusw **MM0**,15-12	10	$17 \rightarrow 15$	6
PSUBusw **MM0**,15-12+5	2	7	$-2 \rightarrow 0$
PADDw **MM0**,5	7	12	5

Note: All arithmetic is saturated 4-bit with a range 0 to 15 and the operation $17 \rightarrow 15$ indicates a true result of 17 clamped to the saturation limit 15.

© Cengage Learning 2014

This example demonstrates that the test case of 7 (in range) yields the result 5. However, the test value 14 is above the maximum range, and that yields the clipped result 12. Similarly, the value 3 is below the lower limit and the clipped result is 5.

5.4   Streaming Extensions and the Development of SIMD Technology

Since the introduction of the first generation of SIMD extensions, development has been continuous, and the handful of MMX extensions has been replaced by generation after generation of new extensions. However, what matters is that the basic principles are still the same: the ability to manipulate multiple elements in a single register and to perform operations that are of importance in multimedia or in other domains such as cryptography.

A short time is a long time in the microprocessor world. Intel had hardly launched its MMX architecture before it moved on and introduced its *Streaming SIMD Extensions* (*SSE*) in 1999. These Streaming SIMD Extensions first appeared in the Pentium III and have two components: a group of integer instructions known as *New-media Instructions* that operate on the MMX register set and a set of new *floating-point* SIMD operations that operate on an entirely new set of registers.

To a considerable extent, the Pentium III's new media instructions are MMX extensions that didn't make it onto silicon the first time around. Some of these instructions are extended

Efficiency of SIMD Computation

How effective are SIMD extensions in enhancing computer architectures? The examples and fragments of code we have introduced are impressive. Intel's published benchmarks claim performance increases of 4.6:1 for an application involving video processing when a processor with MMX is compared with a processor without MMX at the same clock rate.

A paper by Talla, et al.* raises a note of caution when evaluating SIMD extensions by arguing that the benefits of SIMD calculation can be swamped by other computational overheads. Talla's paper introduces the notion of SIMD efficiency, which is defined as the ratio of execution cycles strictly necessary for the computation to the number of execution cycles actually executed. For example, suppose you were evaluating the efficiency of a matrix multiplication algorithm on a certain machine with parallel processing. Since the matrix multiplication of two $N \times N$ matrices has a computational complexity of $O(N^3)$, we can assume that an 8×8 matrix multiplication will take 512 cycles—assuming a pipelined multiplier adder.

If the multiplier and adder use SIMD technology and four operations can be performed per cycle, SIMD efficiency is $512/128 = 4$. Talla states that a real machine takes 2,500 cycles, corresponding to a SIMD efficiency of $512/2,500$ or 20%, and asks why SIMD efficiency should be so low?

Talla provides a fragment of code to demonstrate the effect of overheads in multimedia processing with SIMD extensions. There are 29 lines of code, and each line has an annotation describing its function (load/address overhead, address overhead, initialization overhead, load overhead, true computation, SIMD reduction, SIMD conversion overhead, store overhead, branch, and branch overhead). Of these 29 lines, only four of them are labeled *true computation*.

The overheads identified by Talla are the calculation of addresses required to access the data structure being processed, loads and stores (moving data to and from memory), and branch overheads.

Another paper by Ma, et al.** also looks at Intel's streaming SIMD extensions in 3D geometry processing and suggests speedups in the range from 3.0 to 3.8, which are very respectable. Ma suggests that factors such as the structure of data in memory and careful prefetching is necessary to obtain optimum gains for SIMD extensions.

Although first generation SIMD extensions like MMX did suffer from limitations (not least because of the need to share register sets), over the years all major processor designers have incorporated SIMD extensions, which implies that their impact is beneficial. However, like all architectural enhancements and innovations, the user has to be aware of their limitations. Because streaming extensions can have considerable overheads, their effectiveness may depend very much on the nature and the scale of the application.

---

*Deepu Talla, Lizy Kurian John, and Doug Burger, "Bottlenecks in multimedia processing with SIMD style extensions and architectural enhancements," *IEEE Transactions on Computers*, Vol 52, No. 8, August 2003, pp. 1015–1031.

**Wan-chun Ma and Chia-lin Yang, "Using Intel Streaming SIMD Extensions for 3D Geometry Processing," *Proceedings of the 3rd IEEE Pacific-Rim Conference on Multimedia*, 2002.

MMX instructions, and some provide floating-point facilities. One commentator writing on the introduction of SSE stated it was really Intel's attempt at *getting MMX right the second time round*. You have to appreciate that Intel was locked into a battle with those making IA32 compatible processors (such as AMD) and that delaying innovation until it was perfect was not an option. This is probably less true today, since Intel appears to have emerged as the undisputed winner in the PC-architecture wars even though AMD is still a major player.

MMX's operations (the parallel add, subtract, multiply, and multiply accumulate) helped accelerate some multimedia application. However, MMX provided little support for the operations required for video processing and video encoding/decoding. The *new-media* instructions of the SSE architecture addressed this shortcoming.

TABLE 5.7		Some Example of SSE Integer Operations
Operation	Mnemonic	Action
Extract word	PEXTRw	Read one of four 16-bit words from an MMX register and copy it to the lower half of a 32-bit register.
Insert word	PINSRw	Load a word from the lower-order word of a 32 bit register and load it into one of the four words in an MMX register.
Mask byte to integer	PMOVMSKB	This instruction takes the sign-bit of each of the eight bytes in an MMX register and copies them to the low-order byte of a register. This 8-bit word is, therefore, a mask composed of 8 sign bits.
Packed shuffle word	PSHUFL	This is a three-operand instruction with source and destination operands plus an 8-bit immediate value. The shuffle instruction takes two source and destination registers containing four words and then shuffles words according to the encoding specified by the immediate operand. The new sequence of words is loaded into the destination register. That is, you can cut words out of the two registers and paste them into the destination.

© Cengage Learning 2014

New-media instructions include the usual suspects—data transfer and packing and data processing. Some of the new data transfer/packing instructions are given in Table 5.7. SSE data processing instructions are rather more exotic than the original MMX instructions. For example, the *packed sum of absolute differences* instruction, PSADbw, takes two source operands and calculates the absolute difference between each of the eight pairs of bytes in the source and operand. These eight absolute differences are summed and loaded into the low-order, 16-bit word of the destination operand. The effect of PSADbw **MM0**,MM1 is therefore

$$\sum_0^7 |a_i - b_i|$$

where a_i and b_i are bytes from registers MM0 and MM1. If the source and destination values are

```
01   03   10   45   F1   34   FA   D2
11   CA   0D   FF   00   00   98   A1
```

we get the eight differences

```
-10  -C7   03  -46  +F1  +34  +62  +31
```

The absolute values of these numbers are 10 C7 03 46 F1 34 62 31, and their sum is $02D8_{16}$. The sum of absolute differences instruction is important in motion estimation between successive frames in video. The PSADbw instruction improves MMX operation by a factor of two.[4]

Another powerful new media instruction calculates the average of two data values. The PAVGb MM_d, MM_s (or PAVGb MM_d, m64 where the source is memory) instruction operates on bytes and the PAVGw MM_d, MM_s form operates on words. The *packed average* instruction adds the unsigned data elements of the source operand to the unsigned data elements of the destination register, together with the carry-in. The results of the parallel additions are each independently right-shifted by one bit position. The high-order bits of each element are filled with the carry bits of the corresponding sum.

Some operations require you to determine the maximum or the minimum of a pair of values (for example, when clipping or limiting signals). The new media instructions provide four operations that are variations on this theme. The *packed signed word maximum* instruction, PMAXsw MM_d, MM_s, compares four pairs of words and returns the maximum value of each of the pairs. The PMAXub variant performs the same operation with eight bytes using unsigned arithmetic. The corresponding PMINsw and PMINub instructions return the minimum values of four words or eight bytes, respectively. It has been reported that the PMIN instruction can speed speech recognition processing by a factor of 19%, because calculating the minimum value of pairs of elements is one of the most common operations in some speech recognition algorithms.

5.4.1   Floating-point Software Extensions

The SSE architectural extensions provide floating-point operations and use eight *new* 128-bit registers, XMM0 to XMM5. These registers can't be constrained to the IA32's existing architecture like the MMX registers. The new SIMD floating-point registers require a change to the IA32's fundamental architecture, and the operating system has to be aware of this change. That is, these registers cannot be hidden from the operating system and exception processing. The new *architecturally visible* processor state required by the streaming extensions was one of the most significant extensions of the IA32 family's state architecture since the 80386. We are not counting addition of floating-point to the 486 (which was implemented externally by a coprocessor in 80386 systems) or the addition of atomic operations capable of supporting multiprocessing.

The new SSE state has its own dedicated interrupt vector (for handling numerical exceptions) and a new control and status register, MXCSR, that defines the operating mode and shows SSE status flags. The old MMX state is, of course, still buried in the floating-point state.

The eight XMM registers are each 128 bits wide and hold four 32-bit single precision floating-point values. SIMD floating-point operations act either on all four floating-point values (packed mode) or just the least-significant pair of floating-point numbers (scalar mode). Because the XMM floating point registers are separate from the MMX registers, SIMD floating-point and MMX technology (or conventional floating-point) operations can be carried out concurrently without having to worry about saving registers.

The floating-point SIMD operations are analogous to their integer counterparts (for example, SSE provides floating-point addition, subtraction, division, multiplication and comparison).

Two floating-point operations of interest are RCP and RSQRT that calculate a reciprocal and reciprocal square root, respectively. These instructions obtain the reciprocal (or reciprocal square root) by means of a look-up table and are very fast. However, the result from the look-up table is accurate only to about 12 bits, which is sufficient in some audio and visual applications. If these instructions don't provide the required level of accuracy, you can use a single cycle of *Newton-Raphson iteration* to calculate a reciprocal (or reciprocal square root) to about 22 bits. Recall that we discussed iterative techniques in Chapter 2 when discussing floating-point arithmetic.

---

[4] S.K. Raman, V. Pentkovski, and J. Keshava, "Implementing Streaming SIMD Extensions and the Pentium III Processor," *IEEE Micro*, July-August 2000, pp. 47–57.

AMD 3Dnow! Technology

Although Intel developed the IA32 architecture over several decades, the success of the PC made it inevitable that other companies would attempt to win a share of the market. You can patent certain aspects of a processor, such as the way in which it implements multiprocessing, but you can't patent an instruction set. Consequently, anyone can make a chip that reads IA32 architecture op-codes and executes them to yield the same results as a native Intel IA32 processor.

In the 1990s, several companies made functionally equivalent versions of the IA32 architecture—often aimed at the low-cost market because Intel, with its near monopoly, didn't have to worry too much about price cutting. By the late 1990s, AMD was Intel's only serious competitor (apart from the short-lived Transmeta that burst onto the scene in 2000 with a low-power IA32 chip aimed at the laptop market). AMD launched its K6-2 microprocessor with its 3DNow! technology, managing to beat Intel to the marketplace with SIMD floating-point extensions.

AMD's 3DNow! technology incorporates the MMX architecture together with AMD's own SIMD floating-point extensions. Like Intel, AMD uses the IA32 architecture's floating-point registers to provide MMX registers during MMX operations. However, AMD continued to use this model with its floating-point extensions. That is, the same floating-point registers are used to hold two 3DNow! technology 32-bit packed IEEE single-precision floating-point values.

AMD's floating-point format supports only one rounding mode (round to nearest even), and all conversions between integer and floating-point use truncation. Like the MMX technology, AMD's floating-point operations do not generate exceptions. Moreover, floating-point operations that would normally lead to underflow and overflow conditions result in saturated values. All inputs and outputs smaller than the minimum normalized value that can be represented are flushed to zero.

The 3DNow! floating-point instruction set is described below. As you can see, there are few surprises. In 2000, 3DNow! was extended, and some DSP-specific operations added. Ultimately, 3DNow! was doomed, because both its floating-point and integer registers used existing IA32 floating-point registers. AMD announced that it was no longer supporting 3DNow! in 2010.

Mnemonic	Instruction Description
PFADD	Packed floating-point addition
PFSUB	Packed floating-point subtraction
PFSUBR	Packed floating-point reverse subtraction
PFACC	Packed floating-point accumulation
PFCMPGE	Packed floating-point comparison, greater than or equal
PFCMPGT	Packed floating-point comparison, greater than
PFCMPEQ	Packed floating-point comparison, equal
PFMIN	Packed floating-point minimum
PFMAX	Packed floating-point maximum
PIF2D	Packed floating-point double, 32-bit integer to floating-point conversion
PF2ID	Packed floating-point to 32-bit integer double conversion
PFRCP	Packed floating-point reciprocal approximation
PFRSQRT	Packed floating-point reciprocal square root approximation
PFMUL	Packed floating-point multiplication
PFRCPIT1	Packed floating-point reciprocal iteration first step
PFRSQIT1	Packed floating-point reciprocal square root iteration second step
PFRCPIT2	Packed floating-point iteration second step
PMULHRW	Packed floating-point 16-bit integer multiply with rounding
PAVGUSP	Packed floating-point packed 8-bit unsigned integer average

5.4.2   Intel's Third Layer of Multimedia Extensions

Intel's Pentium 4 introduced yet another new batch of *SIMD Streaming Extensions* called SSE2 that extended both the MMX and the SSE architectures by adding 144 new instructions and increasing the number of data types. The new data types are a 128-bit packed double-precision floating-point, a 64-bit quadword integer, and four 128-bit integer data types. The packed floating-point type allows two IEEE 64-bit double-precision, floating-point values to be packed into one double quadword. The 64-bit quadword integer supports both signed and unsigned values, and the 128-bit integer allows two quadwords, four doublewords, eight words, or 16 bytes integers to be packed into a double quadword. Table 5.8 provides a brief overview of the history of Intel's extensions to its IA32 architecture.

The double-precision, floating-point instructions provide data movement, arithmetic, comparison, conversion, logical, and shuffle instructions. Floating-point SSE2 instructions can move packed double-precision floating-points and perform arithmetic operations on them, as well as converting between double- and single-precision floating-point formats.

Some of the new integer instructions are PADDQ (packed quadword add), PMULUDQ (unsigned doubleword multiply), PSHUFD (shuffle packed doublewords in an XMM register), and MOVQ2DQ (move integer data from MMX to XMM registers). All existing 64-bit MMX and SSE integer instructions are extended to operate on 128-bit operands in the XMM registers.

The Pentium's new multimedia-friendly SSE2 architecture is rather messy, having picked up a lot of baggage along the way that is no longer optimum. You can't convert MMX codes into SSE2 code just by replacing register name MM0 with XMM0, etc. You can't use MOVQ to access the 128-bit registers; you have to use a MOVAPD instruction instead. Similarly, you have to use new versions of shuffle and shifting instructions.

TABLE 5.8   History of the Development of Intel's Streaming Extensions

IA32	Intel Pentium III Processor Intel Pentium II Processor Intel Pentium Processor	**SSSE3**	Quad-Core Intel Xeon 73XX Intel Core 2 Quad 6XXX Intel Core 2 Duo 7XXX Intel Core 2 Solo 2XXX Intel Pentium dual-core
SSE2	Intel Xeon processors Intel Pentium 4 processors	**SSE4.1**	Intel Xeon 74XX series Quad-Core Intel Xeon 54XX Dual-Core Intel Xeon 52XX Intel Core 2 Quad 9XXX Intel Core 2 Duo E7200

SSE3	Dual-Core Intel Xeon 70XX
	Dual-Core Intel Xeon 2.8
	Intel Xeon processors with
	  SSE3
	Intel Core Duo
	Intel Core Solo
	Intel Pentium D
	Intel Pentium 4 processors
	  with SSE3
		SSE4.2	Intel Core i7 Processors
			Intel Core i5 Processors
			Intel Core i3 Processors
			Intel Xeon 55XX series

© Cengage Learning 2014

5.4.3   Intel's SSE3 and SSE4 Instructions

Intel's SSE3 extensions were introduced in 2004 in the Pentium 4E. They use 128-bit registers and add no new data types, but they do add 14 new instructions aimed at applications such as floating-point to integer conversion, complex arithmetic, video encoding, and thread synchronization. In 2006, they were extended to the SSSE3 instruction set in the *Core Duo* with the addition of a further 16 op-codes. The term Core Duo indicates that two CPUs are collocated on the same chip to increase processor power. We discuss this in more detail in Chapter 13.

Complex arithmetic is important in digital signal processing applications, particularly in the realm of Fourier transforms. A complex value z is represented by $a + bi$, where i is the square root of -1. The product of two values $a_1 + b_1i$ and $a_2 + b_2i$ is given by $a_1a_2 - b_1b_2 + (a_1b_2 + a_2b_1)\,i$. SSE 3 provides support for complex arithmetic. ADDSUBPS **OperandA** OperandB, where Operand A contains coefficients $a_3, a_2, a_1,$ and a_0 and Operand B contains $b_3, b_2, b_1,$ and b_0, generates $a_3 + b_3, a_2 - b_2, a_1 + b_1, a_0 - b_0$.

Figure 5.30 illustrates the PSHUFB **mm,** m64 instruction that takes a string of eight bytes in the destination register and re-orders them according to the contents of the corresponding byte in the source register. A source byte states where each byte of the destination is to come from in the source. This arrangement allows you to both *permute* bytes and to *repeat* them. For example, the sequence 0x12345678 could be rearranged as 0x12785612 using permutation and repetition. However, if a byte in the source has its most-significant bit set (0xFF and 0x80 in Figure 5.30) then the corresponding byte in the destination is cleared.

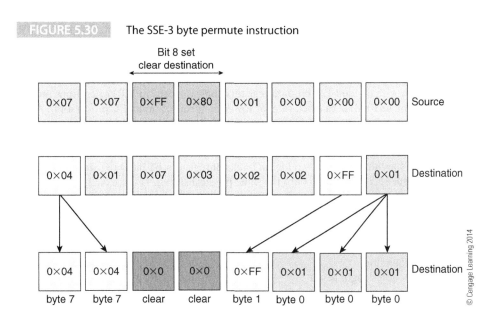

FIGURE 5.30   The SSE-3 byte permute instruction

In 2007, Intel released the next set of streaming extensions, SSE4, with the Core2 Duo. These were intended to support codecs (coding and decoding audio/video data streams) and cryptography. They extended the streaming instructions by providing more support for unaligned data objects. Fifty-four instructions known as SSE4.1, and a further seven instructions known as subset SSE4.2 were added to the Core i7 processor.

These extensions extend the goals originally set by MMX extensions—the ability to perform parallel operations on data representing audio and video data. A good example is the MPSADBW instruction that computes eight offset sums of absolute differences (i.e. $|x_0 - y_0| + |x_1 - y_1| + |x_2 - y_2| + |x_3 - y_3|, |x_0 - y_1| + |x_1 - y_2| + |x_2 - y_3| + |x_3 - y_4|, \ldots$). This operation is central to the design of HDTV codecs and can be used to evaluate an 8×8 block of pixel differences in only seven cycles.

SSE4 also provides string processing operations that can be used in applications ranging from text processing to scanning data for viruses. For example, the new instructions allow you to perform multiple compare and search operations in a single instruction. A string-processing instruction compares two strings to produce a bit mask, where a bit of the mask is 1 if the corresponding bytes are equal. For example, the string "MyNameIsNotAllan" matched with "MiNameIzNotEllan" would produce 1011111011101111. The matching process can be modified by a control integer in the instruction to perform other operations, such as matching a source string with any characters in the second string wherever they may occur. For example, matching "MyNameIsNotAllan" with "aeiou" would give 0001011001010010. You can also match ranges of characters. For example, if the second string is "azAZ", a match will take place if a character is $a-z$ or $A-Z$. You can use this to test data for illegal characters—in this case, numbers and punctuation.

Another useful instruction is POPCNT, which counts the number of bits set to 1 in the source operand and returns the number of bits in the destination operand. The source operand may be 16, 32, or 64 bits. The 64-bit format is POPCNT **r64_d**,r64_s.

Intel's ISA extensions have come a long way from the days of the original multimedia extensions and have provided a wide range of functionality to ISAs to better help them cope with multimedia processing in applications ranging from MPEG video decoding to cryptography. It does raise an interesting point – if ISAs were redesigned to incorporate extended instruction sets (i.e., if the IA32 ISA was designed today using available technology), what would it look like?

5.4.4   ARM Family Multimedia Instructions

Members of the ARM family have also implemented instruction extensions. The Cortex-A ARM processor has an *advanced SIMD architecture extension*, called NEON. NEON technology operates with 64- and 128-bit data registers and can perform MP3 operations on processors operating at 10 MHz (a clock rate far lower than conventional PC processors). ARM family SIMD extensions, also called DSP extensions, allow low-cost, low-speed processors to have cutting-edge architectures that permit the mass manufacture of very low-cost consumer entertainment and automotive systems.

The NEON architecture has two *logical* banks of registers: 64-bit D registers and 128-bit Q registers (Figure 5.31) that map onto the same physical storage (Figure 5.32).

The ARM processor ISA maintains its three-register format with the new instructions but introduces new suffixes to specify data types. Some of ARM's instruction modifiers are

Q	Saturating arithmetic
R	Rounding
D	Double-length result
H	Half-length result
.I16	16-bit operand

A typical operation is VADD.I16 **D0**,D1,D2, which adds four pairs of 16-bit values in registers D1 and D2 and puts the four 16-bit results in D0. Consider the mixed 16-bit/32-bit multiplication operation VMUL.I32.S16 **Q0**,D2,D3 that forms the product of four pairs of

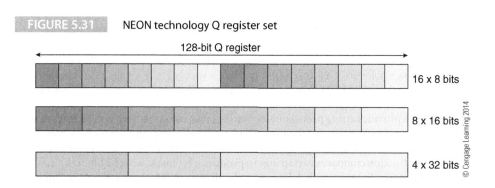

FIGURE 5.31   NEON technology Q register set

128-bit Q register

16 x 8 bits

8 x 16 bits

4 x 32 bits

© Cengage Learning 2014

FIGURE 5.32   Mapping D and Q registers on to each other

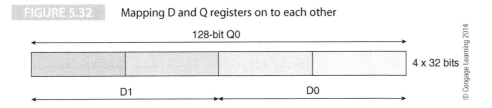

16-bit integers and returns the four 32-bit products in the 128-bit register Q0 (which is, of course, registers D1 and D0). Figure 5.33 demonstrates this operation.

NEON can use both D and Q registers in the same instruction, because elements can be *promoted* or *demoted* (ARM terminology). Long operations promote data types to double-length (for example, in multiplication). Narrow operations have the reverse effect, where the result is half the length of the source. A wide operation promotes the elements of the second operand (for example, a 16-bit operand plus a 32-bit operand is promoted to 32 bits).

Figure 5.34 demonstrates the inverse operation VSHR.I16.I32 **D0,Q1,#5**. Here, four 32-bit values in 128-bit register Q1 are shifted five places right and the four 32-bit results are truncated to 16 bits and loaded in 64-bit register D0.

NEON can load multiple values, as Figure 5.35 demonstrates. For example, VLD3.16 **{D0,D1,D2}**,[R0]! indicates that 64-bit registers D0, D1, and D2 should each be loaded with four 16-bit elements. This instruction allows you to move array elements that have multiple components per element.

FIGURE 5.33   16x16 bit multiplication and promotion to 32-bit products

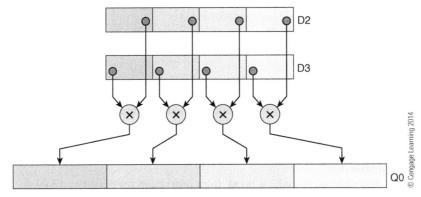

FIGURE 5.34   32-bit shifting and demotion to 16-bit results

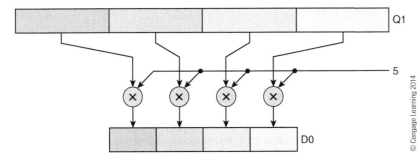

FIGURE 5.35   Loading and storing data structures

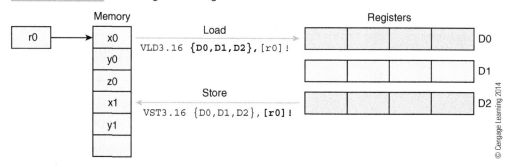

An example of the use of ARM's SIMD instructions is given next. This is taken from ARM's literature.

```
int a[256], b[256], c[2560;
foo () {
 int i:
 for (i=0; i<256;i++){
     a[i]=b[i]+c[i];
 }
}
```

The inner loop of this C-code is as follows.

```
loop VLD1.32    {d0,d1}, [r0]!
     SUBS       r3,r3,#1
     VLD1.I32   {d2,d3}, [r1]!
     VADD.I32   q0,q0,q1
     VST1.32    {d0,d1}, [r2]!
     BNE        loop
```

We've now completed the introduction to instruction set extensions for multimedia applications. The purpose of this chapter has not been to teach graphics and video processing, nor has its purpose been to study multimedia programming in depth. The purpose of this chapter is to demonstrate the nature of the applications that require huge amounts of computation and the way in which computer manufacturers have responded to the demand for multimedia support—not by designing entirely new processors but by extending the ISAs of existing processors.

In the next chapter, we take a break from instruction set design and look at how the performance of computers can be measured so that commercial computers can be compared and the most cost-effective bought or so that the effectiveness of changes in ISAs or computer organization can be measured. After looking at performance, we turn to computer organization in Chapter 7 when we ask how computers are designed.

Summary

The first part of this chapter introduced the idea of multimedia, and we looked at the notion of lossy compression. Lossy compression is at the heart of multimedia applications, because without it current technology would simply not be able to handle the massive volume of information represented by uncompressed images. Lossy compression relies on human perception to remove information that is of little significance to the human observer.

The second part of this chapter looked at the way in which modern processors have adapted to today's applications by implementing instruction sets optimized for multimedia processing. These instruction-set enhancements increase a processor's throughput for applications such as video and audio processing and image compression. The key to these operations is the SIMD or single-instruction, multiple-data mode in which two or more data elements are operated on in parallel using the same operation. For example, eight bytes can be added in parallel to eight bytes to create an eight-byte result. We have covered some of the interesting aspects of these instructions—not least of which is their ability to implement branchless computing by setting a mask following a conditional operation and then using the mask to determine which of two or more operations gets carried out.

Problems

5.1   What is multimedia?

5.2   A computer lacks a division instruction and performs iteratively using the technique described on page 300. Suppose we wish to evaluate 327/940. How many cycles of iteration would it take to get an accuracy of ten decimal places?

5.3   What is the difference between lossless and lossy compression?

5.4   Is the popular ZIP compression technique lossy or lossless?

5.5   What are the characteristics of data streams (i.e., the type of data) that are used in lossy compression techniques?

5.6   A $10'' \times 8''$ color picture is to be compressed using a lossy compression algorithm with a compression ratio of 20:1. If the image has a resolution of 72 dpi and each dot is made up of a trio of color pixels (each pixel has 4,096 levels), what is the size of the image in bits?

5.7   Why does converting 64 pixels to 64 DCT functions help to compress a JPEG image?

5.8   Why are SIMD operations so common in multimedia applications?

5.9   You are building a hotel with 100 rooms. Each room is to have a video-on-demand television that displays movies in HDTV resolution of 1920×1080 pixels at 30 frames/s. If each pixel is 24 bits and you can compress video data by a factor of 100, determine the following.

   a. What is the maximum bit rate that can be demanded if you assume that, in the worst case, 70% of the rooms will be watching television?

   b. What is the required disk storage capacity if the hotel provides 250 movies on demand, each lasting on average 90 minutes?

   c. When a viewer selects a new movie, the server is interrupted and an appropriate routine is executed. The switching routine is composed of approximately 15,000 instructions that take approximately one cycle/instruction on a processor running at 4 GHz. If a user can cope with a latency of 80 ms (i.e., a delay in the data stream of less than 80 ms is not visually disturbing), how many simultaneous switch-overs can the system handle at any instant?

5.10 Why did Intel *initially* take the decision to implement its multimedia extensions without modifying the processor's state architecture (i.e., by not implementing new registers, condition codes, or changes in exception processing)?

5.11 What is *saturating arithmetic* and what are its advantages and disadvantages in typical multimedia applications?

5.12 What is the effect of PCMPEQB **MM0**,MM1?

5.13 What is the effect of PCMPGTW **MM0**,MM1?

5.14 ARM processors perform predicated operations; for example ADDEQ performs an addition only if the Z-bit is set. Multimedia instructions that operate with multiple independent words don't set the condition code bits. For example, Intel's comparisons are used to set subwords to all 0s or all 1s that can later be used as masks in logical operations. Suppose you were proposing to add something like the MMX extensions to an ARM-like instruction set with a three or four register instruction format, how do you think you could take advantage of predicated execution?

5.15 If MMX register MM0 contains $0012ABFF34807F6A_{16}$ and MM1 contains $F20361111888890A_{16}$, what is the effect of executing each of the following instructions?

a. PADDusb   **MM0**,MM1
b. PADDub   MM0,MM1
c. PADDsb   **MM0**,MM1
d. PSUBsb   **MM0**,MM1
e. PSUBub   **MM0**,MM1
f. PADDusw   **MM0**,MM1

5.16 What is the effect of each of the following instructions? Assume that MM0 contains $0012ABFF34807F6A_{16}$ and MM1 contains $F20361111888890A_{16}$ at the start of each operation.

a. PAND   **MM0**,MM1
b. PACKuswb   **MM0**,MM1
c. PCMGTb   **MM0**,MM1
d. PCMGTw   **MM0**,MM1
e. PSRAw   **MM0**,5
f. PSRAb   **MM0**,5

5.17 If MM0 contains 0x0001 0002 0003 0004 and MM1 contains 0x0005 0006 0007 0008, what is the effect of PMADDWD **MM0**,MM1? .

5.18. The MMX architecture does not include conditional branch instructions. How then are conditional operations implemented by MMX?

5.19 Investigate the special-purpose multimedia facilities provided by some of today's computer manufacturers.

5.20 What is clipping and how can the MMX architecture be used to facilitate clipping operations?

5.21 The Intel Pentium has a CPUID (processor identification) instruction. Investigate this instruction and suggest ways in which it may be used.

5.22 Consider the following loop that adds a constant to a vector (we discussed this earlier). There's quite a lot of overhead associated with the solitary SIMD instruction. Suppose you were designing a new ISA that implemented operations like paddb. How would you make the code more efficient?

```
        movq    mm1,c        ;load constant into mm1
                               (8 copies)
        mov     cx, 3        ;set up loop counter for
                               three trips 8 × 3 = 24
        mov     esi, 0       ;set pointer to 0 (use as
                               index into vector)
Next:   movq    mm0,x[esi]   ;Repeat: load 8 bytes
                               into mm0 using indexed
                               addressing
        paddb   mm0, mm1     ;   now do 8 bytes of
                               the vector addition
        movq    x[esi],mm0   ;   store 8 bytes of
                               result in x
        add     esi,8        ;   increment index by 8
        loop    Next         ;Until all done
```

5.23 In Chapter 4, we learned that MIPS provides signed and unsigned addition and that some processors provide signed and unsigned multiplication. In this chapter, we encountered saturating arithmetic. It does seem rather sensible to *tag* data or instructions with the type of operand. What would be the advantages, disadvantages, and implications of such a step?

5.24 What is the effect of the PACKssdw **MM0**,MM1 instruction if initially MM0 contains 0xE000001200000611 and MM1 contains 0x00102222FFFFFFFF?

5.25 You decide to add a new architectural feature to a processor by creating some new instructions; that is, you are extending its ISA. What consequences could these additions have for the existing ISA?

5.26 Consider the waveform in Figure P.5.26. If it were applied to a simple DSP with the transfer function $y_i = 0.7x_i + 0.3x_{i-1}$, what would the output look like? Assume the data is 0.0, 0.0, 0.20, 0.5, 0.8, 1.0, 0.85, 0.55, 0.3, 0.15, 0.05, 0.1, 0.25, 0.50, and 0.24

FIGURE P5.26

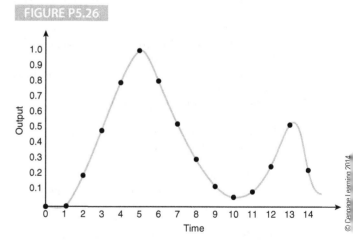

5.27 What lessons did Intel's MMX and AMD's 3DNow! extensions teach us about ISAs?

5.28 You are asked to design a new processor with a 64-bit word. Taking advantage of advances in technology, you decide that you can allocate an extra 5 bits to each word. That is, data words in registers and memory will occupy 69 bits. The additional 5 bits will be used to describe the data. How might you allocate these additional bits (i.e., what functions would you assign to them) in the light of what you have read in the previous chapters?

5.29 Explain what the following fragment of code achieves. Note that the data is *signed* and that the packed shift right arithmetic instruction operates on word (16-bit) operands.

```
MOVQ    MM0,MM1
PSRAW   MM0,15
PXOR    MM0,MM1
```

5.30 Consider the following block of operations that might be found inside a loop. Explain what the instructions do and what operation is being performed on the data.

```
MOVQ    MM1,A       ;move 8 pixels of
                     image A
MOVQ    MM2,B       ;move 8 pixels of
                     image B
MOVQ    MM3,MM1
PSUBSB  MM1,MM2
PSUBSB  MM2,MM3
POR     MM1,MM2
```

III

ORGANIZATION AND EFFICIENCY

Part II introduced the computer and its instruction set architecture. In Part III, our theme is the *performance* of computers, both in terms of how we measure it and how we achieve it. Chapter 6, *Performance—Meanings and Metrics*, introduces the notion of computer performance and describes how we measure it; a task that is rather harder than you might first imagine. Here we learn about the *metrics* of computer performance using benchmarks such as MIPS and SPEC. The variations covered in this chapter are the differing ways in which we measure and interpret performance.

Chapter 7 concentrates on the *organization* and internal structure of processors. We begin with *microprogramming* and demonstrate how instructions can be interpreted and executed. Here we tie together instruction set architectures with the gates and buses we learned about in Part I. Microprogramming provides an elegant explanation of how the arbitrary strings of bits representing instructions are translated into the very actions that execute op-codes. Although the heyday of microprogramming has passed, it is still used in some processors to interpret complex op-codes.

One of the techniques that has done so much to improve the performance of computers is *pipelining*, which is the overlapping of the execution of instructions. A pipelined processor starts the next instruction as soon as the current instruction has begun its journey through the pipeline. Because of the importance of pipelining, we devote a considerable section to pipelining and its effect on performance. In particular, we show how pipelining is subject to limitations, such as the effect of conditional branches that force you to abandon work in progress when you jump to another part of the program. We introduce *branch prediction*, which is a means of guessing whether a branch is going to be taken or not in order to speed up processing.

Chapter 8 is the final section in Part III and looks at ways of making computers even faster using *superscalar* techniques whereby the processor uses multiple functional units to execute several instructions at the same time. Executing instructions in parallel poses new problems for the computer designer, because we have to ensure that the meaning (i.e., semantics) of a program is preserved—even though its instructions may be executed out of order.

The last part of Chapter 8 introduces the architecture of Intel's IA-64 *EPIC* processor, which is a superscalar processor with a wealth of interesting features. *Explicitly Parallel Instruction Computing* (EIPC) indicates that the compiler is responsible for dealing with the effects of out-of-order computation rather than the processor's hardware.

6

Performance— Meaning and Metrics

"To aim is not enough; you must hit."
German proverb

". . . computer hardware progress is so fast. No other technology since civilization began has seen six orders of magnitude in performance-price gain in 30 years."
Fred Brooks, Jr

"There is nothing so useless as doing efficiently that which should not be done at all."
Peter F. Drucker

"Well, you're obviously being totally naive of course," said the girl, "When you've been in marketing as long as I have, you'll know that before any new product can be developed it has to be properly researched. We've got to find out what people want . . ."
"And the wheel," said the Captain, "What about this wheel thingy? It sounds a terribly interesting project."
"Ah," said the marketing girl, "Well, we're having a little difficulty there."
"Difficulty?" exclaimed Ford. "Difficulty? What do you mean, difficulty? It's the single simplest machine in the entire Universe!"
"The marketing girl soured him with a look. "Alright, Mr. Wiseguy," she said, "If you're so clever, you tell us what colour it should be."
Hitchhiker's Guide to the Galaxy, Douglas Adams

"Vogons. They are one of the most unpleasant races in the galaxy. Not actually evil, but bad-tempered, bureaucratic, officious, and callous. They wouldn't even lift a finger to save their own grandmothers from the ravenous Bug-Blatter Beast of Traal without orders signed in triplicate, sent in, sent back, lost, found, queried, subjected to public inquiry, lost again."
Hitchhiker's Guide to the Galaxy, Douglas Adams

"Rimmer: Was there any damage?
Holly: I don't know. The damage report machine has been damaged."
Red Dwarf

"Fair benchmarking' can be less of an oxymoron—if the people using benchmark results know what tasks the benchmarks really perform and what they measure."

Reinhold P. Weicker

"Computers should be made of glass—they're so full of bottlenecks."

Alan Clements March 2001

"The great thing in the world is not so much where we stand, as in what direction we are moving."

Oliver Wendell Holmes

A little old man goes into a computer store and a sales assistant approaches him.

"Can I help you, sir?"
"Yes. I'm looking for a new computer."
"You've come to the right place. Is there anything you're looking for in particular?"
"Yes, I want one with a SPEC 2006 rating of better than 350 and a total power dissipation of 150 watts. Oh yes, and I'd like a GPU with a rating of 1 TFLOPS"
"Would that be the one with the silver case or the black case, sir?"

In this chapter, we introduce one of the most interesting and yet problematic topics in computer architecture: performance. It's interesting because few people seem to agree on how to measure it or how to interpret the results. It's problematic because, as the little story above illustrates, its use can be incongruous in everyday situations and also because performance specifies only one aspect of a computer—an aspect that may be of the least importance in many applications.

A popular theme in folk culture is asking for something and not getting what you really wanted. When King Midas was granted a wish, he requested that everything he touched be turned to gold. Sadly, this included his food and his family. Today, like King Midas, we want performance from our computers. But what exactly is it, and what price do we pay for it?

A dictionary definition of *performance* is *how successful someone or something is.* Such a definition is, of course, almost entirely useless, because it simply replaces the word *performance* with the word *successful* and leaves us none the wiser.

You might think that I am being deliberately obscure, because everyone knows that in the computer world *performance* indicates *speed.* One computer performs better than another if it is faster. Unfortunately, things are not as simple as this definition might suggest. I have a PC that can execute a program more rapidly than, say, a little iPad. Suppose we compare the performance of my PC with the iPad by running a utility such as a calendar. If we measure the time from switch-on to reading the result, the iPad is far faster, because the PC has an intolerably long boot-up period whereas the iPad springs into life the

Performance—Who Cares?

Several broad groups of people are interested in computer performance. First, corporate users who buy massive numbers of computers, banks, insurance companies, and government departments need to obtain the best value for money and require objective criteria by which contending systems can be compared.

Second, knowledgeable enthusiasts want the best computer for game playing or multimedia applications. They need to know how various computers will handle the programs they run.

Third, chip manufacturers need an objective means of assessing innovations they are considering incorporating. For example, which is the most important objective: an increase in cache size or to improve branch prediction algorithms? Without an understanding of the way in which end users will evaluate their processors, the chip manufacturer is in the dark.

Fourth, software and compiler writers need to understand the relationship between hardware and software in order to create the optimum machine level code.

Finally, academics teach computer design and carry out research. They have to have a means of quantitatively evaluating new designs.

moment it is switched on. Any advantage the PC has in terms of *raw processing speed* is lost in the delay or *latency* it experiences during initialization.

Although this example is trite, it demonstrates that defining a computer's performance is far from simple. In this section, we explain what we mean by *performance* and show how we can compare one computer with another. We also introduce the concept of the *benchmark*, which is a program or a suite of programs used to compare one computer with another by measuring how long each computer takes to execute the benchmark. An important objective of this chapter is to explain why computer performance is such a difficult and illusive notion.

Early undergraduate texts on microprocessors and computer architecture and organization had surprisingly little to say on the subject of performance. It wasn't until Patterson and Hennessey wrote *Computer Organization and Design* and made computer performance a central theme that this topic began to achieve the attention it deserved. There are several reasons why this situation arose.

- Microprocessors grew out of the available technology. The first microprocessor, Intel's 4004, was developed as part of a calculator project.
- Other manufacturers like Motorola rushed to build competing processors. Simply getting into production was the main driving force.
- First-generation microprocessor chips were largely used as control elements to implement logic algorithms (i.e., they were not used as general-purpose computers because memory, storage devices, and peripherals had not then been developed).
- In some ways, there was not a great demand for computer power, because many of the applications were in humble devices like washing machine control units.
- The body of knowledge associated with performance was largely concerned with mainframe computers that could be said to inhabit a different universe.

Even today, performance is not a simple concept. A large organization that has to buy hundreds of thousands of computers for their banking empire is going to be most interested in getting the best deal for their money and it needs objective criteria by which competing bids can be compared. Small office and home office computer users are in a very different situation; they do not have the means to evaluate computers and interpret the results. Moreover, the meaning of performance is often bound up with the intended application. For example, suppose one computer can perform an action in 0.01s and another can do it in 0.001s. Although there is a factor of ten difference in the performance, there may be no *perceived* difference in the behavior of the computers. For many computer users, other factors such as the weight of the computer, its endurance when operating from a battery and its Internet connectivity are more important than sheer performance. Many students today use a notebook computer rather than a desktop computer because they can take it to lectures.

Performance is important to those who run cutting edge applications, such as computer games and video applications. Because of this, popular computer magazines intended for the consumer have defined their own standards of performance that are different to the more formal criteria found in the commercial and academic worlds.

In this chapter, we introduce Moore's law, an empirical rule that states that progress in semiconductor technology will take place year after year at a constant exponential rate. Then we look

A Gloomy Comment

Almost nowhere in computer science is there more scope for controversy than in performance measurement. In 1995 the following statement about performance measurement was made by Gustafson and Snell, computer scientists at the Ames DoD Laboratory, Iowa. Although the statement is dated and undeniably harsh, it expresses the views of senior members of the academic community.

"It doesn't work anymore. Most algorithms do more data motion than arithmetic, and most current computers are limited by their ability to move data, not to do arithmetic. While there has been much hand-wringing over misreporting of performance results, there has not been a constructive proposal of what should be done instead. Scientists and engineers express surprise and frustration at the increasing rift between nominal speed (as determined by nominal MIPS or Mflop/s) and actual speed for their applications. Use of memory bandwidth figures in Mbytes/s is too simplistic because each memory regime (registers, primary cache, secondary cache, main memory, disk, etc.) has its own size and speed; parallel memories compound the problem."

at the structure of a computer system and point out that overall performance is determined by its subsystems and not by any component alone. Next we introduce *Amdahl's law* that might be considered the evil twin of Moore's law because Amdahl's law appears to put a limit on the performance that can be achieved by means of parallelism.

The next section examines some of the ways in which we can measure performance, concentrating on the CPU itself. We demonstrate how commonly used metrics can be misleading and show that clock speed, in particular, is a poor predictor of computer performance. Then, we look at a range of performance measures and provide examples of some of the contemporary ways in which personal computers are compared. Finally, we introduce the SPEC metric that is widely used by industry, researchers, and academics to compare the performance of different computers.

6.1   Progress and Computer Technology

In a static world, there would be no need to worry about performance because there would be little change from day to day. In the real world, technology is developing at a phenomenal rate with improvements in the performance of existing systems and the emergence of new systems; for example, hard drives are continuing to improve and solid state drives (that use an entirely different storage technology) are beginning to appear. Before considering performance itself, we take a brief look at the march of progress in computer technology.

Moore's Law

Moore's law is probably the most quoted and, in my view, most misunderstood law of computing. Moore's law is not a law; it's an observation. Gordon Moore was director of Fairchild Semiconductor's Research and Development Laboratories in the mid-1960s. In 1965, he wrote what was to become a classic paper in *Electronics* (Volume 38, Number 8, April 19, 1965) called "Cramming more components onto integrated circuits." At that time, the integrated circuit was relatively new and the number of devices per chip was miniscule. See Figure 6.1.

Because semiconductor devices are fabricated by photolithographic processes (i.e., the circuit is projected onto a chip, the chip is etched and then heated in gases containing so-called impurity ions that locally change the properties of the surface), there is no direct relationship between the cost of manufacture and the number of devices on a chip. Moore observed that

FIGURE 6.1   The basics of Moore's law

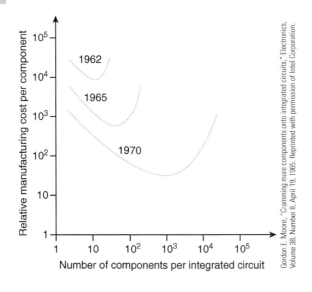

Gordon E. Moore, "Cramming more components onto integrated circuits," *Electronics*, Volume 38, Number 8, April 19, 1965. Reprinted with permission of Intel Corporation.

Semiconductor Size

The figure below illustrates the structure of a MOSFET transistor on silicon. Its essential dimensions are its width and length. Reducing these dimensions both increases the speed of the transistor and enables more to be placed on a single chip.

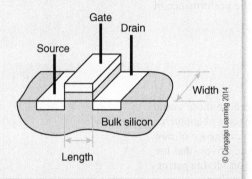

a consequence of this manufacturing process is that the cost per device is inversely proportional to the number of devices. The cost of chip manufacture rises rapidly as the *size* of the individual chips increases. Moore made two observations. The first is that increasing the number of devices (transistors) per chip forces the cost per device down until the size of the chip must be increased to accommodate extra devices, at which point the cost per device increases. For any technology (manufacturing process) there is a minimum device cost. The second observation is that the march of technology is continually increasing the yield and maximum size of chips, which means that the maximum number of devices per integrated circuit is increasing year by year and the minimum cost per device is decreasing. Moore suggested that the maximum number of devices required to achieve a minimum cost would double every year, although he changed the prediction to every two years in 1975.

The term *Moore's law* was not used by Gordon Moore himself, but was coined by Carver Meade in 1970. There's no precise formulation of Moore's law, which suggests an exponential increase in the number of components on a chip. Over time, Moore's law has also come to imply a doubling in the performance of digital systems every 18 months.

Moore's law is based on an observation of the progress of semiconductor technology over four decades. The trends that Gordon Moore observed in the 1960s have continued largely unbroken until 2010. This is something that is probably unique in human endeavor. If a man in 1960 was sauntering along the street and he increased his speed in accordance with Moore's law, he would be moving at 33,554,432 miles/hour in 2010, which is nearly 1,000 miles/s.

Moore's law represents the triumph of a cyclic and multifaceted development process. It is multifaceted because developments in all aspects of semiconductor technology are taking place (materials, purification, photolithography, X-ray and ion-beam imaging), and it is cyclic because new advances make it possible to create more advances. For example, the very first chips were designed and laid out by human engineers. Today, design, testing, layout, and verification are performed using computer aided design.

Moore's law is not a *law* because Moore's observation has no basis in physics or the natural laws. Perhaps it should be renamed *Moore's luck,* because it represents the longest winning streak in history. Consequently, it's natural to ask how long Moore's law can continue. Clearly, it can't go on infinitely because conventional semiconductor technology will soon reach the limits set by the atomic structure of matter and the limits set by quantum mechanics where currents can no longer be considered to flow smoothly as we reach currents made up of a handful of electrons. Already we have seen the maximum clock rates of processors begin to slow down as heat dissipation is rising faster than clock rate and it is no longer possible to cool chips adequately.

Semiconductor Progress

At the heart of the computer revolution is the microprocessor, which is a complex circuit fabricated on a single silicon chip. Figure 6.2 displays the number of transistors per chip as a function of time. In barely five decades, the chip density has gone from in the region of two thousand to two billion devices.

Figure 6.3 describes the progress in Intel's mainstream microprocessors that power most of the world's personal computers. I met an engineer from Intel at a conference who told me that Intel was so confident of the continuation of Moore's law in the medium future that they could begin the design of the next generation of microprocessors even though, at the time, the required fabrication technology did not yet exist. Progress was so reliable that engineers could assume that by the time the next generation of processors had been designed, suitable technologies would be in place.

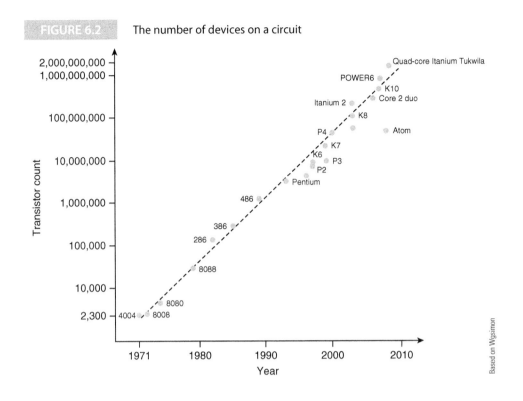

FIGURE 6.2   The number of devices on a circuit

Based on Wgsimon

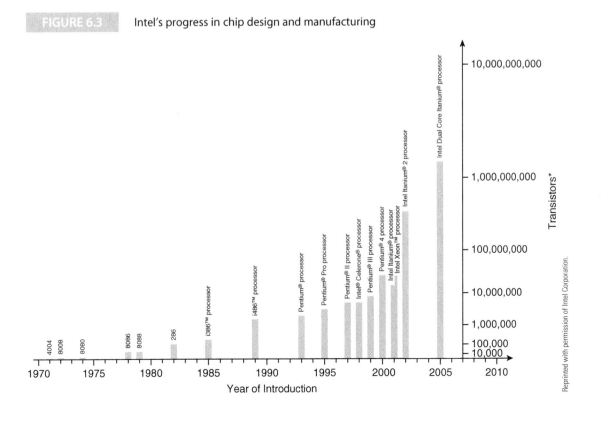

FIGURE 6.3   Intel's progress in chip design and manufacturing

Reprinted with permission of Intel Corporation.

Figure 6.4 provides a three-dimensional illustration of the growth of semiconductor technology by plotting both transistor density and the number of transistors/chip against time.

Memory Progress

We look at semiconductor memory in detail in Chapter 10. Here, we simply point out that, like the computer, the density of semiconductor memory has grown exponentially. Of course, you'd expect that, because microprocessor and memory technology go hand-in-hand. Figure 6.5 demonstrates the growth in DRAM memory capacity over 50 years. DRAM is the mainstream memory component used to provide the bulk of main memory storage in PCs and is able to store one bit of data per transistor.

In order for computer systems to reap the benefits of technological progress, it is necessary that progress take place across all the components of a computer; that is, there's little point in making processors faster and faster if the data they need cannot be read from memory or moved

FIGURE 6.4   **3D illustration of microprocessor progress.** "De-Mystifying Software Performance Optimization," Intel® Software Network, http://software.intel.com/en-us/articles/de-mystifying-software-performance-optimization/. Reprinted with permission of Intel Corporation.

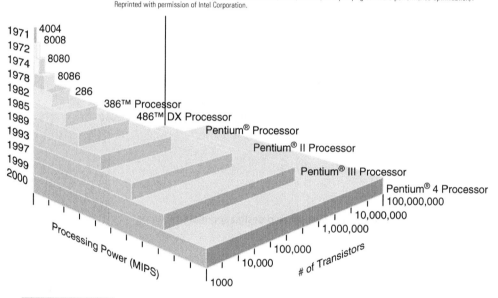

FIGURE 6.5   The growth of memory capacity

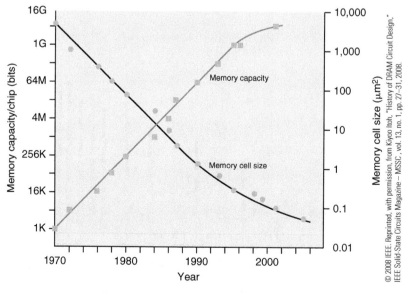

© 2008 IEEE. Reprinted, with permission, from Kiyoo Itoh, "History of DRAM Circuit Design," IEEE Solid-State Circuits Magazine – MSSC , vol. 13, no. 1, pp. 27–31, 2008.

from place to place via buses sufficiently rapidly. We will see in later chapters that progress is not uniform and that bottlenecks are developing because some technologies (such as hard disk transfer times) are lagging behind processor technologies. We will also discover that processor manufacturers improve CPU organization in order to help overcome deficiencies elsewhere; a good example of this is the introduction of multithreaded processors that allow the CPU to switch to another stream of instructions if the current stream is stalled waiting for data from memory.

As well as semiconductor memory, personal computers and workstations currently rely on magnetic recording technologies and hard disk drives (covered in Chapter 10). These devices too have seen a dramatic increase in storage density as Figure 6.6 demonstrates. However, because of their mechanical nature that uses one or more spinning platters, the access time and data transfer rates of hard disks have hardly changed since their introduction. Figure 6.7 provides an interesting demonstration of the progress of disk technology in terms of the annual year-on-year incremental improvement as a function of price.

FIGURE 6.6   Hard disk trends – areal density growth by year

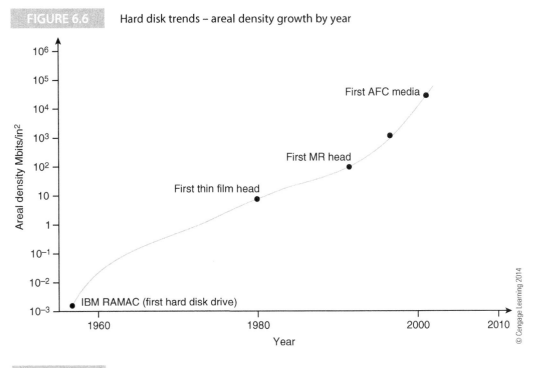

FIGURE 6.7   Disk improvement as a function of price

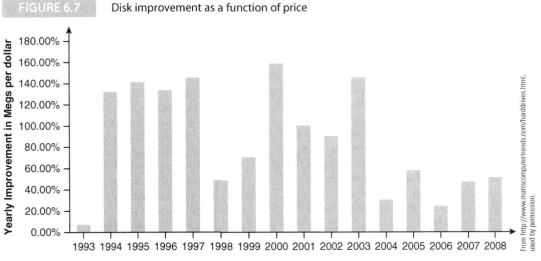

6.2   The Performance of a Computer

Before looking at how performance is measured, we need to emphasize that the performance of a computer is not dependent only on its central processing unit, but on other components of the computer system itself. Although this statement is largely self-evident, we have to make it because so much of the glamor of computers has been focused on the CPU. We've all seen the sticker *Intel Inside* on laptops that stresses CPU performance. No one has ever seen a sticker *Part No. 143-44-NQ4G inside*. In this section, we briefly look at some of the elements of a computer that determine its performance.

Time and Perception

The time taken to execute a job or batch of jobs can be measured precisely and used as a metric of performance. However, we should keep in mind that time is also subjective. Suppose computer A takes five minutes to load the data it needs, ten minutes to execute it, and four minutes to close down. Computer B takes two minutes to load the data, eleven minutes to process it and seven minutes to close down. Which is the faster computer?

Since computer A takes 5 + 10 + 4 = 19 minutes and B takes 2 + 11 + 7 = 20 minutes, clearly A is the faster computer. However, B might seem the faster computer because I can go for a coffee while it's closing down. In other words, we may be more interested in the perceived time than the actual time. The latency in switching programs (for example, waiting for a text box to pop up) may be far more annoying than the total time to completion.

Who is interested in computer performance? Designers, engineers and academics create the computer; the marketing team promotes and sells the computer; shareholders receive profits from its sales; and finally, there's the user of the computer. As you can imagine, performance means different things to different people. Here we are most interested in the ways in which the performance of a computer can be described primarily as an aid to those intending to adopt that computer.

The metrics that define a computer's performance are used by the designer to improve the computer's architecture and organization by locating bottlenecks and eliminating them or minimizing their effects. The folks in marketing and sales that hawk the computer don't want fancy benchmarks. All they want is a single number that is higher than their competitors'. The shareholders haven't even heard of benchmarks, but they do know a dividend when they see it. The end user, well he or she just wants value for money.

Figure 6.8 shows the structure of a typical PC from the point of view of the systems designer. Real computers range in complexity from a single-chip controller in a point-and-shoot camera to complex parallel systems with multiple processors and distributed resources. The performance of a computer is dependent on factors like cache memory (fast memory that holds frequently used data), main memory that holds programs and data, buses that allow sub-modules to communicate with each other, and secondary storage such as hard disks and CD ROM drives. We look at all these systems in later chapters. All you need appreciate here is that every one of these systems contributes to the overall performance of the computer and you cannot maximize the performance of a computer by improving only one of its components.

The designer's view of the computer is characterized by a mass of parameters and specifications that have to be refined and optimized to suit any specific computer application. Only by generating appropriate metrics can anyone optimize the system being designed. In principle, you can analyze the performance of a computer mathematically by modeling all its features. However, computers are generally too complex to model sufficiently accurately.

Figure 6.8 demonstrates that improved computer performance can come from many sources. Indeed, this figure only hints at some of these sources. Below the level of the CPU, the device physicist and semiconductor engineer create intrinsically faster devices; for example, the maximum clock rate can be increased by reducing the propagation delay of signals through the gates in

Balance

It is intuitively reasonable that we concentrate on the processor in this text. As we've pointed out, the processor is not the only factor in determining performance. In particular, there must be a *balance* between the processor and the memory system. Data has to be supplied at the rate the processor needs it otherwise the processor's potential cannot be achieved. The data path between the memory and the cache and between the cache and the CPU must be sufficient to keep the processor supplied with instructions and to handle load and store instructions.

FIGURE 6.8   Performance, the computer, and the designer

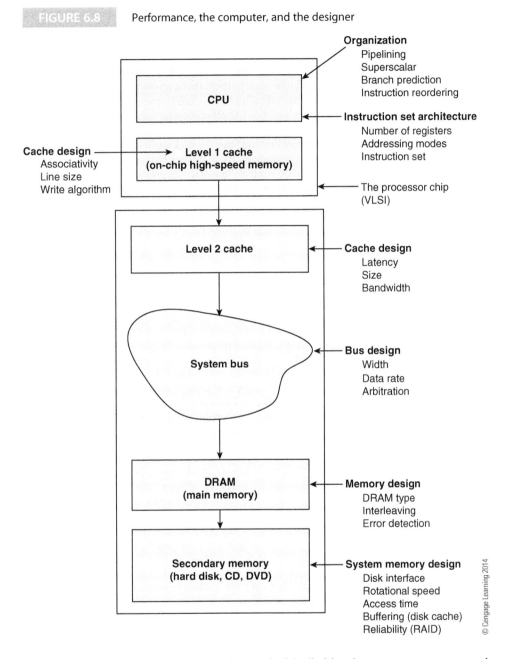

© Cengage Learning 2014

a chip. Today, *heat dissipation* is one of the principle limiting factors on computer speed. The heat density (watts per cubic centimeter) in a modern high-speed processor chip now approaches the same level as the heat density of the core of a nuclear reactor! This statement is true but not quite as awesome as it seems because chips are relatively small and the total heat dissipated by a powerful CPU is of the order of 100 W.

Above the level of the CPU, the software engineer develops compiler technology to better exploit the underlying architecture of the processor. Indeed, in today's world the computer designer and the compiler writer have to work hand-in-hand.

The issue of performance can sometimes become positively surreal; for example, I've seen published benchmarks for motherboards where the speed of the batch of boards being tested varies over a range of one or two percent. Such tiny variations become meaningless in the face of factors such as cache memory that have a much greater effect on performance. Clearly, there is no point in spending a lot of time choosing between board A and board B on the basis of performance if there is little difference between them. Similarly, we can debate the relative merits of SCSI and IDE bus adapters as interfaces between a computer system and a hard disk drive. However, the speed of these bus interfaces may be so much greater than the speed of the hard disk drives connected to them that even moderate variations in the characteristics of the interface have little effect on the overall system performance.

It's easy to state what the designer should do to optimize performance. In the real world the designer doesn't normally have carte blanche to create an entirely new system. The computer is invariably part of a manufacturer's product line and has to be backwards compatible with the code of previous products in the same family. Moreover, the computer has to work with existing software over which the designer may have no control. Similarly, the processor has to operate with other components such as memory and system buses that you cannot optimize to suit your CPU.

Finally, there's another design constraint that is not normally mentioned in polite society. The processor will be interfaced to a bus or memory that may be covered by a patent. If you make the wrong design decision you could find Mr. Intel's lawyer knocking at your door clutching a patent.

6.3   Computer Metrics

We're going introduce some of the ways of comparing the performance of various computers. However, we have to state here that computer metrics can be notoriously unreliable. We begin by introducing some of the terminology of performance and then demonstrate why clock frequency alone is not a reliable indicator of performance. One of my favorite quotations on numbers is from William Thomson (1824–1907), the British scientist whose work on transmission lines underpins all modern electronics. Thomson, who later became Lord Kelvin, said:

> "When you can measure what you are speaking about, and express it in numbers, you know something about it. But when you cannot measure it, when you cannot express it in numbers, your knowledge is of a meager and unsatisfactory kind. It may be the beginning of knowledge but you have scarcely in your thoughts advanced to the state of science."

A Comment on Kelvin

Chip Weems pointed out that there is another side to Lord Kelvin's famous comment: "Given some pretty numbers, we often stop thinking (as demonstrated by the many physics students who compute a terminal velocity for a falling object that exceeds the speed of light, and never notice the discrepancy). This is especially true in architecture research, where nearly all results ignore sensitivity analysis and are usually no more significant than the background noise in the performance measurement of a real system."

Lord Kelvin's words are most apt when they are applied to performance. We need numbers to measure performance. The big problems are, "What numbers?" and "How do we interpret these numbers?"

Before we can select metrics for computer performance, we require criteria by which we can judge the metrics. David Lilja[1] provides a list of criteria by which we can judge the metrics of computer performance: linearity, reliability, repeatability, ease of measurement, consistency, and independence.

The *linearity* criterion suggests that a metric should be linear; that is, increasing the performance of a computer by a fraction x should be reflected by an increase of fraction x in the metric. If computer A has a metric of, say 200, and is twice as fast as computer B, then computer B's metric should be 100. Not all metrics are linear. For example, the Richter earthquake scale, the level of sound intensity (decibels), and camera aperture (F stops) are all logarithmic. Since computer performance is increasing year by year at an exponential rate, I would argue that performance should also be expressed in a logarithmic fashion.

---

[1] Lilja, David J., *Measuring Computer Performance*, Cambridge University Press, Cambridge, 2000

Performance metrics should be *reliable* and correctly indicate whether one computer is faster than another. You could also call this property *monotonicity*; that is, an increase in the value of a metric should indicate an increase in the speed of the computer and never vice versa. This isn't true of all metrics. Sometimes, a computer may have a metric implying a higher-level performance than another computer when its performance is worse. This situation arises when there is a poor relationship between what the metric actually measures and the way in which the computer operates. We will soon see that clock-rate and MIPS (instruction execution rate) are two notoriously unreliable metrics.

A good metric should be *repeatable* and always yield the same result under the same conditions. Not all computer systems are deterministic (deterministic means responding in the same way to the same data) because the number of parameters that affect a computer's performance is very large indeed and you can't always achieve the same results for each test run of a program. If this seems hard to believe, consider the following. Suppose you are carrying out a test that requires the reading of data from a disk. In the first run, the data may be about to fall under the read head at the time it is required, and therefore, the data will be immediately ready. In the next run of the same test, the data may have just passed under the read head and the system will have to wait for a complete rotation to access the data. In a third run, the data might be cached in RAM, and the system will entirely bypass the disk's hardware. Consequently, we can have three runs of a test yielding three different metrics using the same data.

Lilja's *ease of measurement* criterion is self-explanatory. If it is difficult to measure a performance criterion, few users are likely to make that measurement. Moreover, if a metric is difficult to measure, an independent tester will have great difficulty in confirming it.

A metric is *consistent* if it is precisely defined and can be applied across different systems. Perhaps this metric should be called *universality* or *generality* to avoid confusion with *repeatability*. Consistency can be very difficult to achieve if the metric measures a feature of a specific processor and that feature is not constant across all platforms. Using clock rate as a metric demonstrates a lack of consistency, because the relationship between clock rate and performance is not consistent across different platforms. The relationship between clock rate and the performance of a PowerPC processor is not the same as the corresponding relationship between clock rate and a Core i7 processor. An example of a consistent metric that was commonly used to indicate the performance of graphics cards in PCs in the late 1990s was the maximum number of frames per second at which the *Quake* game could refresh the display.

Finally, Lilja states that a good metric should be *independent of commercial influences*. If computer manufacturers defined performance metrics, they might be tempted to select a criterion that shows their processor in a better light than their competitors' processors. In particular, the manufacturer might select a metric that emphasizes a specific feature of their processor that's lacking in competitors' devices, even though this feature may have little or no overall effect on system performance in practice. The next step is to define some terms associated with performance.

6.3.1   Terminology

Computer performance has its own characteristic terminology, and we need to define a few of these terms before we continue.

Efficiency

A computer is, of course, always executing instructions unless it is in a halt state or suspended state. However, a computer may not always be executing *useful* (i.e., application level) instructions because, for example, it may be repeatedly going round a polling loop waiting for data from a peripheral (when there are other tasks waiting to be executed). The efficiency of a computer is an indication of the fraction of time that it is doing useful work. The definition of efficiency is

$$\text{Efficiency} = \frac{\text{total time executing useful work}}{\text{total time}} = \frac{\text{optimal time}}{\text{actual time}}$$

For example, if a computer takes 20 s to perform a computational task and 5 s is taken waiting for a disk that has been idle to spin up to speed, the efficiency is 20 s/(20 s + 5 s) = 20/25 = 80%.

It's not easy to determine the efficiency of a computer system because it's difficult to measure the time spent executing useful work. The notion of efficiency can be problematic when evaluating the performance of a computer, because it's important to measure things that contribute directly to throughput. For example, if a computer goes round a polling loop faster, it doesn't improve the system's performance. All you are doing is to make the computer *wait faster*.

Throughput

The *throughput* of a computer is a measure of the amount of work it performs per unit time. For example, a bus's throughput is measured in megabits/s, whereas a computer's throughput is measured in instructions per second. The upper limit to a system's throughput can normally be determined from basic system parameters. For example, if a computer has a 500 MHz clock and it can execute up to two instructions in parallel per clock cycle and each instruction takes 1, 2, or 4 clock cycles, then the upper limit on throughput occurs when all instructions are being executed in parallel in one cycle; that is 10^9 instructions/s. Here we are assuming that the best a computer can do is one instruction per cycle. We will see in Chapter 8 that *superscalar processors* employ instruction level parallelism (ILP) to execute more than one instruction per clock cycle. Note that the definition of throughput includes the term *amount of work* because instruction execution is meaningful only if the instructions are performing useful calculations; a computer executing an endless stream of NOPs (no operations) may be operating at its peak rate but is achieving nothing other than to wait. We will soon see that *instructions per second* is a very poor indicator of the actual performance of a computer.

Latency

Latency is the delay between activating a process (for example, a memory write or a disk read, or a bus transaction) and the start of the operation; that is, latency is the *waiting time*. Latency is an important consideration in the design of rotating disk memory systems where, for example, you have to wait on average half a revolution for data to come under the read-write head. In some computer applications, the effects of latency may be negligible in comparison with processing time. In some systems, the effects of latency may have an important effect on system performance. Note that some define latency at the time to *finish* a process.

Relative Performance

We are interested in how one computer performs with respect to another. The relative performance of computers A and B is the inverse of their execution times; that is

$$\text{Performance}_{A\_to\_B} = \frac{\text{performance}_{ComputerA}}{\text{performance}_{ComputerB}} = \frac{\text{execution time}_{ComputerB}}{\text{execution time}_{ComputerA}}$$

If system A executes a program in 105 s and system B executes the same program in 125 s, we can calculate the value of n as 125/105 = 1.190. You can say that machine A is 19% faster than B.

The objective of the computer designer is to create a system with the greatest possible throughput.[2] When you are trying to improve a system, you are often most interested in how much better the new system is in comparison with the old system. The *old* system may be a previous machine, the same machine without the improvement, or even a competitor's machine, and is called the reference machine or baseline machine. The speedup ratio is a measure of *relative performance* and is defined as

$$\text{Speedup ratio} = \frac{\text{execution time on reference machine}}{\text{execution time}}$$

If a reference machine takes 100 seconds to run a program and the test machine takes 50 seconds, the speedup ratio is 100/50 = 2.

---

[2] No it isn't—the computer designer's goal is to maximize corporate profit, which may not necessarily lead to the computer with the highest performance. There is little point in designing a fast computer if no one buys it. For example, if a certain instruction set architecture is very fast on Linux-based operating systems but has a poor performance under Windows, it would probably not be a commercial success.

Time and Rate

Benchmarks can be expressed as the *time* required to execute a task or as the *rate* at which tasks are executed. For example, one benchmark may yield a time of 20 s, whereas another benchmark may yield a rate of 12 tasks/s. Indeed, the computer game *Quake* has become a popular benchmark for PCs with the figure of merit being the rate at which frames are displayed by the processor (although the *Quake* frame rate is probably a reasonable indication of how your computer performs relative to other computers running *Quake*, it is not a good general benchmark). They say people feel more comfortable with metrics that increase numerically with performance (i.e., rate) rather than those that reduce with performance (time).

Time and *rate* benchmarks don't behave in the same way with respect to averaging. Suppose we benchmark a computer and get execution times for tasks *A* and *B*, respectively, of 2 and 4 seconds. We can also say that the rates at which tasks *A* and *B* execute are 0.5/s and 0.25/s, respectively.

The average *execution time* is $\frac{1}{2}(2 + 4) = 3$ s. The average rate of execution for the tasks is $\frac{1}{2}(0.5 + 0.25) = 0.375$. The average execution time is 3 s, corresponding to an average rate of $1/3$ s $= 0.333$/s, which is not the same as the 0.375/s that we calculated by averaging the rates. The reason for this difference between the two averages is easy to find; times and rates are *reciprocals* and you can't apply arithmetic averaging to a reciprocal. Suppose we measure two *times x* and *y*. The average time is $\frac{1}{2}(x + y)$. These times correspond to *rates* $1/x$ and $1/y$ and the average value of these rates is therefore $\frac{1}{2}(1/x + 1/y)$. The time corresponding to this average rate is $1/(\frac{1}{2}(1/x + 1/y))$ which is $2xy/(x + y)$ and is not the same as $\frac{1}{2}(x + y)$ unless $x = y$. Jacob and Mudge[3] provide an interesting and extensive discussion of the meaning of times and rates in a University of Michigan technical report. We will return to the concept of averaging later. We now discuss some of the criteria used to express the performance of computers: clock rates, MIPs, and benchmarks.

6.3.2   Clock Rate

The most obvious indicator of a computer's performance is its *clock rate*, the speed at which fundamental operations are carried out within the computer. For example, you may buy a 3.0 GHz processor in the hope that the 3.0 GHz processor is faster than the 2.8 GHz version. Using the clock rate as a metric to compare processors is better than nothing—but only just. It is probably the worst metric by which to judge computers.

Figure 6.9 illustrates the CPU clock. At each clock cycle, the processor carries out an internal operation. At first sight, it's tempting to think that the processor's performance is proportional to its clock rate, and therefore clock rate is a precise metric.

FIGURE 6.9   The CPU's clock

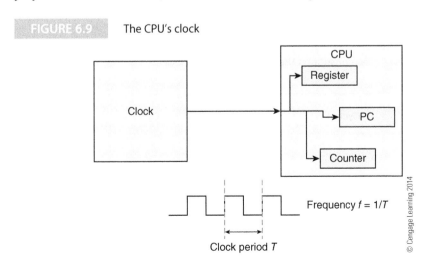

© Cengage Learning 2014

[3] Jacob Bruce and Mudge Trevor, *Notes on Calculating Computer Performance*, University of Michigan Tech Report CSE-TR-231-95, March 1995.

There are so many flaws in this argument that it's difficult to know where to begin. We can start by pointing out that there is no *single* clock in most computers. There are several clocks in nearly every computer system. Some systems may have entirely independent clocks (i.e., there's a separate clock generator for each functional part such as the CPU, the bus, and the memory). Moreover, the CPU itself may have several functional units, each driven by its own clock. Some systems have a single master clock that generates pulses at the highest rate required by any circuit and all other clocks run at a sub-multiple of this frequency. Another point worth mentioning is that some processors have variable clock rates. For example, *mobile* processors designed for use in laptops can reduce the clock rate to conserve power. This is called *clock throttling* or *dynamic frequency scaling* and is now used in other processors today. Similarly, some processors switch to a lower clock speed if the core temperature rises and the chip is in danger of overheating.

Some digital circuits employ *clock doubling* or even clock *quadrupling*. Figure 6.10 demonstrates how a clock running at a rate $2F$ can be derived from a clock running at rate F. If you delay a clock by 90° and then perform an XOR operation between the original and delayed clocks, you get a new clock at twice the frequency of the original clock.

Increasing clock speed doesn't necessarily lead to a *linear* increase in performance because some actions such as accessing memory require a minimum amount of time. Consider the situation depicted by Figure 6.11 where three cycles are required to perform an operation. Assume that the first two clock cycles in Figure 6.11 are *scalable* and we can increase the clock speed and safely assume that whatever is done in a clock cycle is still done if the cycle time is reduced. However, the third cycle in Figure 6.11 includes an operation that requires a

FIGURE 6.10   Clock doubling

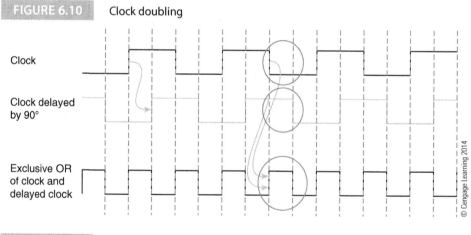

Clock

Clock delayed by 90°

Exclusive OR of clock and delayed clock

FIGURE 6.11   Demonstrating the non-scalability of clocks

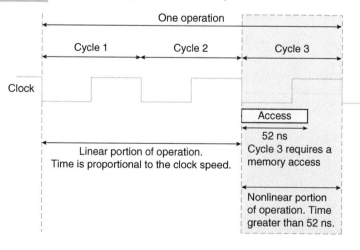

In cycle 3 the system has to carry out a task that takes 52 ns. You can speed up the clock rate and reduce the cycle time up to the point that the cycle takes 52 ns. The only way that you can further reduce clock cycle time is to increase the number of cycles occupied by the 52 ns operation.

minimum time of 52 ns for its completion. If the clock cycle is less than 52 ns, cycle three must be stretched either by slowing the clock or by replacing cycle three by two or more cycles.

In practice, it's difficult to alter the clock rate dynamically from clock pulse to clock pulse. Most systems implement a mechanism that extends an operation by an integer number of cycles called *wait states*. In the example of Figure 6.12, the cycle time is 30 ns and the 52 ns memory access requires two cycles. The operation now takes 4×30 ns to complete.

The non-scalable clock property of systems described by Figure 6.11 demonstrates why some computers become unstable if you try and *overclock* them (a mechanism beloved of hackers who clock processors at a rate higher than that specified by the manufacturer).

Table 6.1 relates the total time required to perform the operation to the clock cycle time. If the clock cycle time is 100 ns, cycles 1 and 2 each require 100 ns. Similarly, cycle 3 requires one clock cycle of 100 ns, which exceeds the minimum by 48 ns. In this case, three 100 ns clocks are required for the operation and the total time is 300 ns.

FIGURE 6.12   The effect of introducing wait states

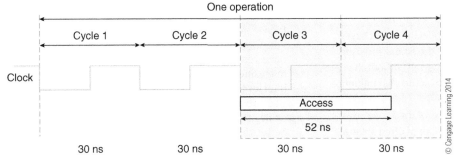

If you now reduce the clock cycle time, the total time required is proportional to the clock cycle time, until you reach a clock cycle time of 50 ns. Once the clock cycle time drops below 52 ns, an extra clock cycle must be inserted during phase three because the minimum access times of 52 ns can't be accommodated. In this case, when phase three is extended to 100 ns by the inclusion of a 50 ns wait cycle, the total time required is now 200 ns.

Table 6.1 provides the total time required as well as the *linear time* that assumes that the system scales linearly with clock speed. The table also gives the ratio of linear time to total

TABLE 6.1   Total Time Required to Execute the Operation of Figure 6.11 as a Function of Clock Cycle Time

Clock	Cycles 1 & 2	Clocks/cycle 3	Cycle 3	Total Time	Linear Time	Ratio: LT/TT
100	200	1	100	300	300	1.000
80	160	1	80	240	240	1.000
60	120	1	60	180	180	1.000
50	100	2	100	200	150	0.750
40	80	2	80	160	120	0.750
30	60	2	60	120	90	0.750
20	40	3	60	100	60	0.600
10	20	6	60	80	30	0.325
8	16	7	56	72	24	0.333
6	12	9	54	66	18	0.273
4	8	13	52	60	12	0.200
0	0	∞	52	52	0	0.000

Is Megahertz Enough?

The following is taken from an IDC White Paper written by Shane Rau and sponsored by AMD (www.idc.com).

PC buyers usually rely on the clock speed (megahertz) of a PC's microprocessor to determine their purchasing decision. Because the industry lacks a simple, universally accepted way to judge performance, users have become conditioned to substituting clock speed to gauge how fast their applications will run. This practice has grown common over many years.

- The popularization of the PC among general consumers has increased the available pool of buyers unfamiliar with factors in PC performance.
- The growth of the direct model of PC purchases has made it more likely that the actual end user will buy a PC for himself or herself without the help of a third party familiar with factors that influence PC performance.
- The increasing sophistication of the PC exposes the buyer to a growing number of often arcane technical specifications, from which clock speed promises a convenient escape.

time that expresses the degradation of the system due to its nonscalability. As the cycle time falls, the effect of the delay becomes more and more pronounced and eventually dominates the system's performance.

Figure 6.13 illustrates the relationship between the clock period and the total time required by the system of Figure 6.11 to execute an operation. As you can see, the total time becomes almost independent of clock speed when the clock period drops below about 10 ns.

Clock rate is a poor indicator of performance because there's no simple relationship between clock rate and system performance across *different* platforms. Although system *A* may exhibit a moderately linear relationship between clock rate and performance over a given range of clock speeds, system *B*'s performance may have a very different relationship between its clock rate and performance. In other words, clock rate might help us to compare two type *A* systems or two type *B* systems, but it does not allow you to compare a type *A* system with a type *B* system. I once read, to my horror, a review in a popular computer magazine that stated the G4 found in an Apple Mac was inferior to a Pentium processor, because the G4's clock rate was substantially less than the Pentium's clock rate. Such an argument is nonsense. The Pentium may or may not be better than the G4. The reviewer is wrong to presume that performance is directly related to clock speed across different platforms and that a processor with a high clock speed is faster than one with a lower clock speed (see the sidebar).

FIGURE 6.13   Relationship between clock period and throughput for the system of Figure 6.11

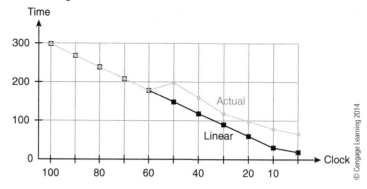

© Cengage Learning 2014

Some systems require more clock cycles per operation than others. For example, one CISC-based processor required about 32 clock cycles to execute a single instruction, whereas a RISC processor might require 1.5 clock cycles. A superscalar processor might perform two instructions per clock cycle. If all three processors have the same clock rate, the CISC is 16 times slower than the RISC system and up to 32 times slower than the superscalar processor. This is assuming, of course, that the instructions running on the three processors are approximately the same in terms of the work they do.

An interesting comment on clock rates was provided by the behavior of Intel and AMD. Both companies market processors for the PC market. AMD's processors are not

IA32 clones *internally*; they interpret IA32 instructions using a RISC-like core architecture. Consequently, it is unfair to compare the clock rate of Intel and AMD processors directly. However, competitive pressures locked Intel and AMD into a clock-rate battle. AMD was first to break the 1 GHz barrier and Intel was first to reach 2 GHz. In the fall of 2001, AMD introduced its *Athlon XP 1800+* processor that was specified in terms of an *effective clock-speed rating*. AMD quoted the *equivalent* clock speed of their XP 1800+ as 1.8 GHz even though this device had an actual clock speed of 1.533 GHz. AMD's argument is that 1.8 GHz is a better representation of the processor's capability than 1.533 GHz.

Although I am sympathetic to AMD's plight (i.e., the basic clock rate does not indicate the processor's actual performance), implying a higher notional clock rate than the physical clock rate is a dangerous path to take, because it could lead to manufacturers making wildly exaggerated claims.

The Clock and the Consumer

It may well be that the clock rate as a prime metric will live on, or it may become increasingly irrelevant as the internal architectures and organization of processors that execute IA32 code continue to diverge and clock rates continue to stagnate. However, clock rates may still feature in advertising for practical reasons. Computers are not only bought by corporations and academics but are now a commodity item. It's easier to promote computers with a single easy-to-understand metric like *clock rate* than it is to educate the public on the meaning of more sophisticated metrics such as SPEC. You are not going to hear someone in a store say to the sales assistant, "Forget the clock rate. What's the effect of the compiler on load latencies?" Since clock rates have stagnated in recent years, pure clock rate has ceased to be a premier advertising point in manufacturers' literature. However, microprocessor families are still usually available in three or more versions depending on clock rate. For example, in 2010, Intel's Core i7 was available in versions at 2.66 GHz, 2.80 GHz, 3.06 GHz, and 3.20 GHz at a price ratio of 1:1.138:1.251:2.439. A high-performance *extreme edition* was available with a clock rate of 3.33 GHz at a cost ratio of 4.184 (the extreme edition also includes more cores and additional on-chip cache). These figures demonstrate that there is a considerable premium for the member of a family with the highest clock rate.

In some ways, the power barrier has rendered clock speed arguments rather moot. Since about 2008, clock speed has ceased to increase dramatically because the limits of power dissipation have been reached and power dissipation is proportional to the *square* of the clock frequency. As we shall see in Chapter 13, manufacturers have directed their efforts towards multi-core processors rather than faster processors. However, clock speeds will rise again if power consumption falls because of the introduction of new semiconductor materials or because of circuit innovations such as asynchronous clocking where circuits are not driven by a master clock and one event triggers another. Asynchronous circuits can operate at up to 70% lower power levels than their clocked counterparts, but design, testing, and verification is problematic. Moreover, asynchronous circuits have several technical advantages over clocked circuits. For example, they reduce the problem of metastability that can plague clocked circuits. Although ARM did create a largely experimental asynchronous version of its processor, called Amulet, it never entered commercial production.

6.3.3   MIPS

A slightly better metric than clock rate is *millions of instruction per second (MIPS)*. This metric removes the discrepancy between systems with different numbers of clocks per operation by measuring instructions per second rather than clocks per second. For a given computer,

$$\text{MIPS} = \frac{n}{t_{\text{execute}} \times 10^6}$$

where n is the number of instructions executed and t_{execute} is the time taken to execute them.

The MIPS rating is a poor metric that fails for the same reason as the clock rate. A MIPS rating tells you only how fast a computer executes instructions, but doesn't tell you what is actually achieved by the instructions being executed. Consider the following hypothetical example of computation on two computers A and B, where computer A has a load/store architecture without a multiplier and computer B has a memory-to-register architecture. Computer A is more verbose than B (Table 6.2). Both the computers evaluate the expression $z = 4(x + y)$.

TABLE 6.2   The Expression $z = 4(x + y)$ Executed on Two Hypothetical Computers

Computer A (LOAD/STORE)			Computer B (Memory-Register)		
LDR	**r1**,(r0)	;load x	LDR	**r1**,(r0)	;load x
LDR	**r2**,(4,r0)	;load y	ADD	**r1**,(4,r0)	;x+y
ADD	**r2**,r1,r2	;x+y	MUL	**r1**,#4	;4(x+y)
ADD	**r2**,r2,r2	;2(x+y)	STR	r1,(**8,r0**)	;store z
ADD	**r2**,r2,r2	;4(x+y)			
STR	r2,(**8,r0**)	;store z			

© Cengage Learning 2014

Suppose computers A and B have the same MIPS and execute code equally rapidly. If you relied solely on MIPS as a metric of performance, you'd conclude that the computers offer the same performance. As you can see, computer B is faster than computer A because only four instructions are required to do the work (this argument is based on the assumption that all instructions take the same time). In practice, computer A might be faster than B because memory-to-register architectures are slower than register-to-register architectures.

A CISC processor with instructions for multimedia-oriented operations such as Intel's streaming extensions might achieve a low MIPS rating while executing complex instructions that would take tens of primitive instructions on a machine without multimedia extensions.

The MIPS metric is also sensitive to the way in which a compiler generates code. The duration of a single instruction is cycles $\times t_{cycle}$, where *cycles* is the number of machine cycles required to execute the instruction and t_{cycle} is the cycle time (usually the clock period). The total execution time for a program is given by

$$t_{execution} = t_{cycle} \times \sum n_i \times c_i$$

where n_i is the number of times instruction, i occurs in the program, and c_i is the number of cycles required by instruction i. If we plug this formula into the equation for MIPS, we get

$$\text{MIPS} = \frac{n}{t_{cycle} \times \sum n_i \times c_i \times 10^6}$$

This expression tells us that the MIPS value is affected by the instruction mix (i.e., the nature of n_i) and the length of each instruction executed (i.e., the c_i term). If an instruction mix consists of a large number of instructions with one cycle, the MIPS value may be high, even though the actual code is executed more slowly than an instruction mix that contains a lot of multiple-cycle instructions.

Consider the following example. A program is compiled to run on a computer and the compiler generates two million one-cycle instructions and one million two-cycle instructions. If we assume that the cycle time is 10 ns, the time taken is given by

$$2 \times 10^6 \times 1 \times 10\,\text{ns} + 1 \times 10^6 \times 2 \times 10\,\text{ns} = 4 \times 10^6 \times 10\,\text{ns} = 4 \times 10^{-2}\,\text{s}$$

Now suppose that a different compiler generates code for the same problem but with 1.5 million one-cycle instructions and 1.2 million two-cycle instructions. In this case, the time required to execute the code is

$$1.5 \times 10^6 \times 1 \times 10\,\text{ns} + 1.2 \times 10^6 \times 2 \times 10\,\text{ns} = 3.9 \times 10^6 \times 10\,\text{ns}$$

$$= 3.9 \times 10^{-2}\,\text{s}$$

As you can see, the second compiler generated faster code. Now let's evaluate the MIPS for each case. In the first case, the MIPS is given by

$$\frac{n}{t_{cycle} \times \sum n_i \times c_i \times 10^6}$$

That is

$$MIPS = 3 \times 10^6/(10 \text{ ns} \times (2 \times 10^6 \times 1 + 1 \times 10^6 \times 2) \times 10^6)$$

$$= 0.75 \times 10^2 = 75 \text{ MIPS}$$

In the second case, the MIPS is given by

$$MIPS = 2.7 \times 10^6/(10 \text{ ns} \times (1.5 \times 10^6 \times 1 + 1.2 \times 10^6 \times 2) \times 10^6)$$

$$= 0.69 \times 10^{-2} = 69 \text{ MIPS}$$

The faster computer has a lower MIPS even though it is clearly superior to the slower computer. This failure of the MIPS metric is inevitable because instruction throughput takes no account of how much work each instruction actually performs.

Instruction Cycles and MIPS

Recall that the time required to execute code is given by $CPU_{execution\ time}$ = instructions $\times$ clock cycles/instruction $\times$ clock period. Each of the three terms is dependent on a particular set of design criteria, and each of these criteria can be independently optimized. Consider the number of instructions per program. The size of a program depends primarily on the instruction set architecture. We have already seen that a CISC processor can execute complex instructions, such as operations on bit fields that require many separate instructions on a RISC processor. Another factor determining the number of instructions to be executed is the nature of the compiler that generates the instructions. Some compilers can generate more efficient code than others.

The number of clock cycles per instruction is a function of the processor's internal organization. We look at this aspect of computer organization in the next chapter. However, there is a relationship between instruction complexity and the number of clock cycles required to execute an instruction. Consider the design of a shifter that moves bits one or more places left or right. A computer engineer can design a shifting circuit to shift an m-bit word left by p bits in p clock cycles by using a shift register or in one clock cycle by using a *barrel shifter*. A shift register uses a chain of flip-flops to move data one bit per clock pulse, whereas a barrel shifter uses logic gates to shift a bit pattern by selecting a path between the input and output terminals of the shifter. A barrel shifter gains speed at the expense of complexity because it requires a large array of gates to route bits between the input and output. In this case, there is a simple trade-off between time and circuit complexity.

Some operations can't be accelerated just by using more hardware (for example, instructions with complex addressing modes or bit field instructions that act on an arbitrary sequence of bits in memory that may cross several byte boundaries). These operations are inherently serial in nature and can't be fully *parallelized*.

The average number of clock cycles per instruction is determined by the instruction mix (i.e., the relative number of 1-cycle, 2-cycle, 3-cycle instructions etc.) and by the organization or logic design of the processor. The average number of clock cycles per instruction (CPI) is

$$CPI = \sum_{i=1}^{N} F_i \cdot C_i$$

F_i is the fraction of instructions taking C_i clock cycles to execute. In a *non-scalar* processor, the value of i ranges from 1 (one instruction per cycle) to N. The value of N is the length of the longest instruction in terms of clock cycles. In Chapter 8, we introduce the *superscalar* processor with multiple execution units that can achieve CPIs less than unity.

The clock period is governed by three factors: the device physics, the ability to dissipate heat, and the chip's logic design. The rate at which signals propagate through semiconductor devices can be improved only by shrinking the device or by changing the semiconductor's electronic properties. When a signal changes level in a chip, it is necessary to charge inter-electrode capacitances of the transistors on the chip.[4] This process requires energy and raising the switching rate increases the power consumption. Since all energy eventually ends up as heat, increasing the clock rate causes the chip to become hotter. This heat can be dissipated only via conduction to the outside. Consequently, the maximum clock rate is often limited by the ability of a system to dissipate heat.

The maximum clock rate is also determined by the chip's internal logic design. If each event inside the chip has to be completed within a single clock cycle, the maximum clock rate will be determined by the longest propagation path through logic devices within the chip.

Consider the following example. A benchmark runs on a hypothetical RISC processor to give the results in Table 6.3. We can obtain the cycles per instruction for each class of operation by multiplying the relative frequency of the instructions by the cycles per instruction to get the values in Table 6.4.

TABLE 6.3   Relative Instruction Frequency and Cycle Count for a Computer

Machine Operation	Relative Frequency	Cycles per Instruction
Arithmetic/logical instruction	53%	1
Register load operation	20%	4
Register store operation	7%	2
Unconditional branch instruction	12%	1
Conditional branch instruction	8%	2

© Cengage Learning 2014

TABLE 6.4   Calculating the Average CPI for the System in Table 6.3

Machine Operation	Frequency	Cycles	CPI
Arithmetic/logical instruction	53%	1	0.53
Register load operation	20%	4	0.80
Register store operation	7%	2	0.14
Unconditional branch instruction	12%	1	0.12
Conditional branch instruction	8%	2	0.16
Average cycles per instruction			1.75

© Cengage Learning 2014

We can also relate the relative instruction frequencies to the percentage of time an instruction class takes by multiplying the instruction class frequency by the number of cycles and dividing the result by the average cycles per instruction (i.e., 1.75). In Table 6.5, the register load instruction takes up 20% of the code but is responsible for 45.71% of the processor cycle time; this is telling us that loading registers from memory is expensive.

Figure 6.14 sums up the relationship between clock rate, MIPS, instruction count, and computer systems design. The concentric layers represent the components of a computer from its technology (i.e., fabrication in silicon) to the programs that run on it. The three ovals in blue represent the factors that affect the performance of the system, such as clock rate. This figure demonstrates (approximately) the relationship between the design layers and the performance factors. For example, the manufacturing technology determines the maximum clock rate, whereas the compiler has no effect on the clock rate.

---

[4] Computers are made up of large numbers of transistors (electronic switches) fabricated on a silicon chip. These transistors have three electrodes: terminals separated by an insulator. When a pulse is applied to an electrode, that electrode is charged up and electrons stored on its surface, or the charge is dissipated. It is this movement of charge that contributes to one of the sources of chip heating.

TABLE 6.5   Calculating the Average Time Spent Executing Each Instruction Class

Machine Operation	Frequency	Average Cycles	Average Time
Arithmetic/logical instruction	53%	$1 \times .53 = 0.53$	30.29%
Register load operation	20%	$4 \times .2 = 0.80$	45.71%
Register store operation	7%	$2 \times .07 = 0.14$	8.00%
Unconditional branch instruction	12%	$1 \times 0.12 = 0.12$	6.86%
Conditional branch instruction	8%	$2 \times .08 = 0.16$	9.14%

© Cengage Learning 2014

FIGURE 6.14   Factors affecting computer performance

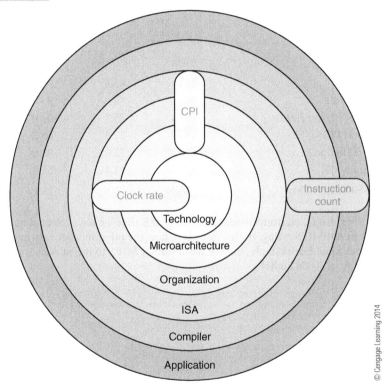

© Cengage Learning 2014

6.3.4   MFLOPS

MFLOPS indicates *millions of floating-point operations per second* and is similar to MIPS, except that MFLOPS counts only floating-point operations. In principal, the same objections to MIPS apply to the MFLOPS metric and therefore you might expect MFLOPS to be as poor an indicator of performance as MIPS. In fact, MFLOPS is a better metric than MIPS because MFLOPS measures the *work* done rather than instruction throughput. MIPS counts *all* instructions executed by a computer, many of which perform no useful work in solving a problem (e.g., data movement operations). MFLOPS considers only *floating-point operations* that are at the heart of the algorithm being implemented.

No matter how fast a computer is, its MFLOPS metric will be zero if the program contains no floating-point operations. MFLOPS is most useful in comparing computers used in scientific applications when numerical calculations dominate the computation. Giladi[5] points

[5] Giladi, Ran., "Evaluating the MFLOPS Measure," *IEEE Micro*, August 1996, pp. 69–75.

out that even in scientific calculation, the nature of the program and its data has a profound influence on the MFLOPS rating. For example, when comparing two computers, one system might yield a better rating for calculations involving sparse matrices and the other might yield better results for calculations involving full matrices. MFLOPS is also useful in comparing special-purpose processors such as DSP (digital signal processing) chips that are used in embedded applications to process audio and visual signals in real time.

The MFLOPS metric isn't easy to calculate because all computers don't implement floating-point arithmetic in the same way. One computer might use dedicated hardware to calculate a trigonometric function such as a sine, whereas another computer might evaluate $\sin(x)$ by directly evaluating the appropriate series for $\sin(x)$.

Table 6.6 takes seven computers from the list of machines that Giladi presents for the years 1988–1995. Giladi compares the computers on the basis of various metrics. We have selected the fastest machine by MIPS rating from each of the vendors appearing in Giladi's list. Table 6.6 demonstrates that the same relative order of performance holds for all the published metrics.

TABLE 6. 6   Metrics for Seven Machines. © 1996 IEEE. Reprinted, with permission, from Ran Giladi, "Evaluating the Mflops Measure," IEEE Micro, vol. 16 no. 4, pp. 69–75, 1996.

Vendor	MIPS	MFLOPS	SPECfp92	Clock
HP 755/125	213	45.4	195.7	125
DEC 10000/610	202	42	193.6	200
Sun SPARCstation 20/61	167.4	35.3	102.8	60
Sanar Sunar WS 10/51	135.5	27.3	65.2	50
SGI R4400 Indigo2 Ex	120	22.6	93.6	75
Intel Xpress/MX	112	11.4	58.7	66.7
IBM RS5000/220/M20	36.4	6.5	29.1	33

Table 6.7 is also taken from Giladi's list, except this time we have selected machines with an almost constant MIPS rating. In this case, note how the other metrics vary markedly from machine to machine. For example, the HP has a similar MIPS to the SPARCstation 5/70 but its Mflops rating is 300% higher.

TABLE 6. 7   Comparing Machines of Nearly Equal MIPS Ratings. © 1996 IEEE. Reprinted, with permission, from Ran Giladi, "Evaluating the Mflops Measure," IEEE Micro, vol. 16 no. 4, pp. 69–75, 1996.

Vendor	MIPS	MFLOPS	SPECfp92	Clock
HP T500	100	35	170.2	90
DEC 3000/300L	100.1	12.2	63.6	100
Sun SPARCstation 5/70	100.3	13.1	47.3	70
Intel Xpress/MX	112	11.4	58.7	66.7

In Table 6.8, we have selected computers from Giladi's table to demonstrate conflicting results between benchmarks; for example, the Sun SPARCstation 20/61 has the lowest clock rate and the highest MIPS and MFLOPS. If you compare the DEC 3000/300L with the HP 750, the former wins on the MIPS rating and the latter on the MFLOPS rating.

TABLE 6. 8   Contradictory Benchmarks. © 1996 IEEE. Reprinted, with permission, from Ran Giladi, "Evaluating the Mflops Measure," IEEE Micro, vol. 16 no. 4, pp. 69–75, 1996.

Vendor	MIPS	MFLOPS	SPECfp92	Clock
DEC 3000/300L	100.1	12.2	63.6	100
HP 750	76	22	92	66
Sun SPARCstation 5/70	100.3	13.1	47.3	70
Sun SPARCstation 20/61	129.4	29.6	84.8	50

What does all this tell us? It tells us that MIPS, MFLOPS, and clock speed do not provide uniformly reliable metrics. Sometimes, as we have seen, the ordering of the various metrics is constant across different computers and sometimes they even contradict each other. Before we look at a more universal benchmark, we take a break and discuss a law that is at the heart of computer performance.

6.4 Amdahl's Law

The most famous *law* governing computer performance is *Amdahl's law*. It's also an *infamous* law because it appears to place a limit on the maximum performance increase that can be achieved by optimizing a computer's subsystems. Clearly, improving part of a system can increase the system's overall performance. For example, you can reduce the time of a journey that involves a flight followed by a car ride by increasing the car's speed without changing the length of time spent in the aircraft.

Amdahl's law describes the performance increase you get when a program is run in a system where some of the operations can take place in parallel. Amdahl's law tells you what performance increase you get for greater parallelism. Amdahl's law is applicable to any system where you are interested in the effect of local improvements on the system globally. You could say that Amdahl's law highlights the effects of bottlenecks in a system.

Figure 6.15 illustrates the effect of parallelizing part of an activity. The diagram demonstrates that the serial (irreducible) part of the process remains the same while the parallel (reducible or improvable) part of the system is reduced. Ultimately, system performance is dominated by the serial part of the system and the motto of the computer designer has become *make the common case fast*.

FIGURE 6.15   Illustration of Amdahl's law

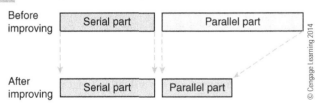

Suppose a computer executes a program on a single processor in time t_s seconds. If we have p processors and the program is divided into p equal chunks, the same program will run in t_s/p seconds. In reality, it is difficult to divide a program up into equal parts that can be executed in parallel and part of the program may not be susceptible to parallelization. Assume that a fraction of the program f_p can run on the p processors and a fraction f_s can run only on one processor. The execution time on the parallel computer system T_p will be:

$$T_p = t_s \cdot f_s + t_p \cdot f_p$$

Since $f_s + f_p = 1$ and $t_p = t_s/p$, we can write $T_p = t_s(f_s + (1 - f_s)/p)$. The speedup ratio for this system is the ratio of the speed without parallelization to the speed with parallelization; that is, $S = t_s/T_p$.

$$S = \frac{1}{f_s + \dfrac{1 - f_s}{p}} = \frac{p}{pf_s + 1 - f_s}$$

Amdahl's law tells us that the speedup varies from 1 ($f_s = 1$ with no parallelism) to p ($f_s = 0$ with perfect parallelism). More importantly, it teaches the *law of diminishing returns*. If we cannot reduce f_s, there is little point in increasing the value of p.

Consider the two cases in Table 6.9 in which $f_s = 0.2$ and $f_s = 0.1$, respectively, for various degrees of parallelism. As you can see from Table 6.9, the speedup ratio falls off rapidly once the serial part of the process begins to dominate the equation. In the limit, an infinite number of processors cannot make the speedup ratio greater than the reciprocal of the fraction of time devoted to serial processing.

TABLE 6.9   Effect of Amdahl's Law

Processors	Speedup Ratio $f_s = 0.2$	Speedup Ratio $f_s = 0.1$
1	1	1
2	1.667	1.818
3	2.143	2.500
4	2.500	3.077
5	2.778	3.571
10	3.571	5.263
100	4.808	9.174
∞	5.000	10.00

© Cengage Learning 2014

Amdahl's law illustrates the effect of a bottleneck on system performance and shows that there is a limit beyond which attempts at further improvement are futile, unless the bottleneck can be removed. A popular formulation of Amdahl's law is

$$S = \frac{1}{1 - \text{fraction}_{\text{enhanced}} + \dfrac{\text{fraction}_{\text{enhanced}}}{\text{speedup}_{\text{enhanced}}}}$$

Examples of the Use of Amdahl's Law

A computer used in scientific computation has a floating-point unit and an integer unit. The normal instruction mix in terms of time is 70% floating-point, and 30% integer and other instructions. The engineers say that they can increase the speed of the floating point unit by a factor of 3 but that would increase the cost of the chip by 25%. The sales people report that users will pay an additional 25% if the overall performance is doubled. Is it worth making the modification?

If we plug the figures into the equation with $\text{fraction}_{\text{enhanced}} = 0.7$ and $\text{speedup}_{\text{enhanced}} = 3$, we get

$$S = \frac{1}{1 - 0.7 + \dfrac{0.7}{3}} = 1.875$$

The speedup is 1.875, which falls below the value required by the sales people to break even.

Let's look at a second example of Amdahl's law where we have an operation that consists of six sequential operations or processes as described in Figure 6.16. Five of the processes are accelerated using the factors given in Figure 6.16. This diagram presents the change in the total time after each process has been accelerated. As you can see, by the time the bottom line has been reached, the six processes are now of roughly similar durations. We have reached the state at which the common case has been eliminated and there is nothing left to accelerate.

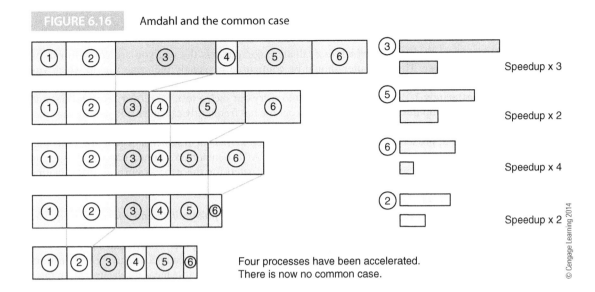

FIGURE 6.16   Amdahl and the common case

Speedup x 3

Speedup x 2

Speedup x 4

Speedup x 2

Four processes have been accelerated.
There is now no common case.

© Cengage Learning 2014

Amdahl Revisited

Although Amdahl's law appears to be an iron rule of parallelism, not everyone is so convinced. John Gustafson of the Sandia National Laboratories in the United States suggested that Amdahl's law is flawed by pointing out that numerical calculations at Sandia have achieved performance increases due to parallelism greater than Amdahl's law would suggest.* Gustafson argues that Amdahl's law is predicated on the assumption that the fraction of time spent executing code in parallel is independent of the number of processors, and that this assumption is not always true in practice. Gustafson's assertion is that *the problem scales with the number of processors*. If a problem is increased in size, the fraction of time spent performing actions serially may not scale linearly.

In 1996, Yuan Shi** published a paper that established a mathematical equivalence between Amdahl's and Gustafson's laws and stated that they were different formulations of the same basic law. Shi's conclusion is

> *"…the use of the serial percentage concept in parallel performance evaluation is misleading. It has caused nearly three decades of confusion in the parallel processing community. This confusion disappears when processing times are used in the formulations. Therefore, we suggest that time-based formulations would be the most appropriate for parallel performance evaluation."*

Perhaps the best way of looking at Amdahl's law is that it defines the lower bound for the speedup of a system using parallelism in the event that gains due to either a reformulation of the problem or gains due to scaling the problem cannot be realized.

---

*Gustafson, J.L., *Reevaluating Amdahl's Law*, CACM, 31(5), 1988. pp. 532–533.

**Yuan Shi, *Reevaluating Amdahl's Law and Gustafson's Law*, Temple University, October 1996.

6.5   Benchmarks

The ideal program with which to evaluate a computer is the one you are going to run on that computer. The computer should be configured in exactly the same way in which you are going to use it (i.e., same memory size, cache, I/O configuration, and operating system). Unfortunately, it's normally impractical to carry out such a test. Moreover, workstations and personal computers don't run a single program, but frequently execute a mix of programs that changes from moment to moment.

One approach to benchmarking is based on *kernels* or fragments of real programs that require intense computation, such as the LINPACK benchmark. Another approach is to run *synthetic* benchmarks that are programs constructed for the express purpose of evaluating computer performance and which purport to be similar to the type of code that users might actually execute.

Benchmarks can be divided into two categories: *fine-grained* and *coarse-grained*. The granularity of a benchmark is a function of the object being measured; for example, a benchmark that measures the performance of a complete computer system can be considered coarse-grained, whereas a benchmark that measures the performance of, say, branch instructions, can be considered fine-grained.[6] Fine-grained benchmarks measure the execution rates of specific instructions rather than entire applications or kernels. Such benchmarks are sometimes called *microbenchmarks*.

Krishnaswamy and Scherson describe a formal approach to benchmarking by treating a computer as a block box that is accessible only through an interface consisting of instructions. They specify a performance vector that consists of instruction classes; $\mathbf{x} = [x_1\, x_2 \ldots x_n]^T$, where x could be, $[x_{\text{ALU}}\ x_{\text{FP}}\ X_{\text{mem}}\ X_{\text{cond}}]^T$, for example. A second vector, called a *workload vector*, consisting of the number of instructions executed by each of the instruction classes is defined as $\mathbf{a} = [a_1\, a_2 \ldots a_n]^T$. The workload vector is a property of the benchmark being used. The total time required to execute a benchmark is given by $t = \mathbf{a}^T\mathbf{x}$. Krishnaswamy and Scherson claim that this fine-grained geometric benchmarking helps analyze the way in which instruction types influence the behavior of the computer. However, such a complex tool is suited more for academic research than practical use. Moreover, the end user is as interested in the contribution of memory to performance as the CPU.

Benchmarks must be interpreted carefully. You have to know what is being measured and how the measurements have been carried out. A set of benchmarks may demonstrate that one computer is far better than another. In practice, these benchmarks may be entirely misleading if your workload does not match the benchmarks. Suppose the benchmarks rely very heavily on pure CPU processing power and the target application involves a large database and much disk-based I/O. The benchmarks may have little relevance to the application.

For many computer users, benchmarks are rather irrelevant since even today's entry-level computer is largely untaxed by the workload of the average user. Word-processing, spreadsheets, web surfing, and modest levels of image processing can be performed with little noticeable delay. Indeed, for many users the bottleneck is the Internet connection which is beyond their control. It's the gaming and graphics community that most require very high levels of computational performance.

LINPACK and LAPACK

LINPACK is a collection of Fortran subroutines that solve linear equations and linear least-squares problems (and perform other calculations on matrices). LINPACK was designed for supercomputers in use in the 1970s and early 1980s and has now been replaced by LAPACK, which is designed to run efficiently on shared-memory, vector supercomputers.

---

[6] Krishnaswamy, Umesh and Scherson, Isaac, D., "A Framework for Computer Performance Evaluation Using Benchmark Sets," *IEEE Transactions on Computers*, Vol. 49, No. 12, December 2000, pp. 1325–1338.

The LINPACK benchmark was not originally intended as a standard means of evaluating computers. It was included as an appendix in the 1979 LINPACK Users' Guide.[7] This appendix used one of the LINPACK programs to determine the time to solve a problem involving a matrix of size 100. Solving a system of n linear equations requires $2n^3/3 + 2n^2$ floating-point additions and subtractions.

Other benchmarks have been developed (such as *Dhrystone* and *Whetstone*) that measure a computer's integer and floating-point performance, respectively. These are now largely obsolete, and the popular SPEC benchmarks are probably the most widely used.

Kaivalya Dixit[8] has suggested that successive waves of benchmarks have had an historic effect on computer performance. Dixit maintains that the Linpack benchmarks were the inspiration behind load/store units and compiler *tricks*; Dhrystone benchmarks influenced the design of string instructions; and SPEC benchmarks improved cache design and branch predictors. Dixit made three interesting comments on benchmarks:

1. The biggest problem with benchmarks is that they represent yesterday's workloads. It takes time to select, test, and agree the use of a particular benchmark. Such a process takes a long time in an age where computer hardware and applications are changing daily.
2. Benchmarks help to move bottlenecks, not remove them.
3. *Benchmarketing* is taking over from benchmarking.

We now look at some of the benchmarks used to measure computer performance, beginning with benchmarks used to categorize PC performance. Then we look at some traditional benchmarks before introducing the SPEC benchmark that is supported by the computer industry and which changes with time in order to keep up the progress in performance.

Benchmarks may also lead to a form of positive feedback in the processor development cycle. If the speed of an operation is increased to improve a computer's benchmark, that instruction will be used more often in the future. In turn, that will lead to increasing pressure to improve the instruction's efficiency even further.

Oracle Applications Standard Benchmark

As well as general-purpose benchmarks, there are benchmarks that are highly focused on specific applications; for example, the database. Oracle has created a benchmark that simulates real applications on a large system. The workload represents

> *"the normal daily activity of a mid-market sized company, with a day-time batch load of roughly 30%. The database used is a synthetic database of substantial size. … The Oracle Applications Standard Benchmark is a realistic workload mix that accurately represents a common customer scenario, both with a high volume of OLTP (online transaction processing) users and a substantial batch component."*

The Oracle benchmark measures total system performance because it simulates a complete system including its ramp up, steady state and ramp-down operating phases over a substantial period (e.g., one hour). The benchmark unit is a *user count* that measures the number of users that the system can sustain while keeping response times within defined limits.

---

[7] Dongarra J.J, Bunch JR., Moler C., and Stewart G.W., *LINPACK Users' Guide*, SIAM, Philadelphia, 1976.

[8] Dixit, Kaivalya, M., "Performance SPECulations—Benchmarks, Friend or Foe," *Seventh International Symposium on High Performance Computer Architecture*, Monterrey, Mexico, January 20–24, 2001.

The Case for Scalability Benchmarks

The problem with benchmarks is that they tell you how good a system is; they don't tell you anything about its *incremental performance*. What happens as the workload is increased? Some systems demonstrate a severe drop in performance when the load grows (we cover some of these aspects of systems behavior when we introduce memory systems).

A paper by Weyuker and Avritzer in IBM Systems Journal (Vol. 41, No. 1, 2002) suggests a metric for predicting performance under a growing workload. Their metric is called *Performance Nonscalability Likelihood* (PNL) where *scalability* refers to a system's ability to handle larger workloads. The advantage of such a metric is that it helps locate and eliminate performance bottlenecks in a system.

One possible way of generating a scalability benchmark would be to increase the number of tasks and then measure the reduction in performance per task as a function of the number of tasks. For example, performance may be *superlinear* at low numbers of tasks (efficiency grows as parallelism is exploited) and then fall off as the number of tasks is increased. As some point, the performance will drop significantly when the overhead in running multiple tasks grows faster than the number of tasks (this is called *threshing* when it occurs due to page faults in an operating system).

PC Benchmarks

From the 1990s onward, a large number of *ad-hoc* benchmarks appeared in popular personal computing journals. These benchmarks were intended to simulate the type of work people would carry out at home or in small offices and were often weighted towards computationally intensive programming such as rendering images or video compression.

PC benchmarks are not often created by academic groups or international bodies like the SPEC benchmarks we look at in the next section. PC benchmarks are constructed by magazines that review PCs or by those who have similarly oriented websites. A good example is Mark Prieur's website http://www.behardware. com. PC benchmarks are frequently aimed at those who use PCs in demanding applications such as state-of-the-art computer games or in multimedia applications. Figure 6.17 presents the results of tests on a large number of processors, both Intel and AMD, when rendering an image using 3D Studio Max 2010.

The interesting feature of these figures is that each successive generation of microprocessors is faster than the previous generation, and that within a given generation the speed increases with clock frequency. Moreover, the slowest members of a new generation of processors may be slower than the fastest members of a previous generation—but you should remember that the fastest members of any generation often have a very high price premium.

Figure 6.18 provides another example of processor benchmarks from PC Games Hardware. In this case, the parameter being measured is the frame rate of a popular computer game. The frame rate gives an indication of how long it takes to move from one video epoch to another and is a popular metric in PC-based journals and websites. Figure 6.19 provides benchmarks for a range of high-performance processors.

Comparison of High-Performance Processors

We now provide another example of processor performance from HEXUS. Table 6.10 illustrates basic data for high-performance IA32 architectures in late 2010. Because we have not covered all the terminology used in Table 6.10, the panel provides brief definitions of some of these terms. Later chapters go into more detail about multi-core processors, multithreading and cache memory.

Figure 6.20 from the HEXUS review gives the power consumption for the processors when operating in idle and full power modes. Power consumption is now so important that processor reviewers include it because it has implications for system design. Some of the highest performance PCs require water cooling systems to remove heat from processors operating at the limits of performance, especially if *overclocking* is being used. Overclocking is the term used to describe operating a processor at a higher clock rate than that guaranteed by the manufacturer.

Figure 6.21 provides two of the graphs presented by HEXUS: rate information (frames per second) for video encoding, and time information for the wPrime benchmark. What these figures demonstrate is the spread of performance between economy chips (e.g., Core i5), mainstream processors (Core i7 920), and cutting-edge processors optimized for the highest

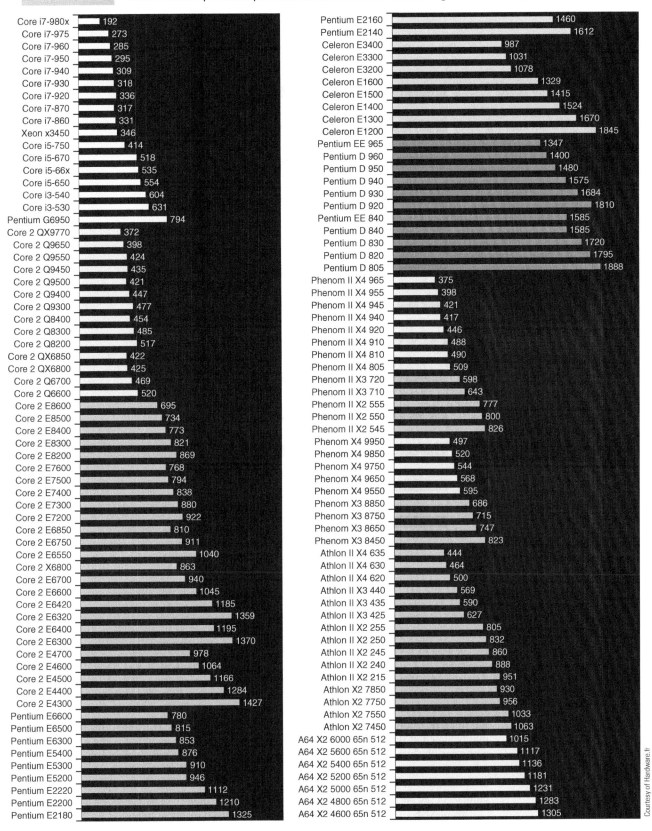

FIGURE 6.17   Intel and AMD processor performance (3D Studio Max, rendering time in seconds)

Processor	Time
Core i7-980x	192
Core i7-975	273
Core i7-960	285
Core i7-950	295
Core i7-940	309
Core i7-930	318
Core i7-920	336
Core i7-870	317
Core i7-860	331
Xeon x3450	346
Core i5-750	414
Core i5-670	518
Core i5-66x	535
Core i5-650	554
Core i3-540	604
Core i3-530	631
Pentium G6950	794
Core 2 QX9770	372
Core 2 Q9650	398
Core 2 Q9550	424
Core 2 Q9450	435
Core 2 Q9500	421
Core 2 Q9400	447
Core 2 Q9300	477
Core 2 Q8400	454
Core 2 Q8300	485
Core 2 Q8200	517
Core 2 QX6850	422
Core 2 QX6800	425
Core 2 Q6700	469
Core 2 Q6600	520
Core 2 E8600	695
Core 2 E8500	734
Core 2 E8400	773
Core 2 E8300	821
Core 2 E8200	869
Core 2 E7600	768
Core 2 E7500	794
Core 2 E7400	838
Core 2 E7300	880
Core 2 E7200	922
Core 2 E6850	810
Core 2 E6750	911
Core 2 E6550	1040
Core 2 X6800	863
Core 2 E6700	940
Core 2 E6600	1045
Core 2 E6420	1185
Core 2 E6320	1359
Core 2 E6400	1195
Core 2 E6300	1370
Core 2 E4700	978
Core 2 E4600	1064
Core 2 E4500	1166
Core 2 E4400	1284
Core 2 E4300	1427
Pentium E6600	780
Pentium E6500	815
Pentium E6300	853
Pentium E5400	876
Pentium E5300	910
Pentium E5200	946
Pentium E2220	1112
Pentium E2200	1210
Pentium E2180	1325

Processor	Time
Pentium E2160	1460
Pentium E2140	1612
Celeron E3400	987
Celeron E3300	1031
Celeron E3200	1078
Celeron E1600	1329
Celeron E1500	1415
Celeron E1400	1524
Celeron E1300	1670
Celeron E1200	1845
Pentium EE 965	1347
Pentium D 960	1400
Pentium D 950	1480
Pentium D 940	1575
Pentium D 930	1684
Pentium D 920	1810
Pentium EE 840	1585
Pentium D 840	1585
Pentium D 830	1720
Pentium D 820	1795
Pentium D 805	1888
Phenom II X4 965	375
Phenom II X4 955	398
Phenom II X4 945	421
Phenom II X4 940	417
Phenom II X4 920	446
Phenom II X4 910	488
Phenom II X4 810	490
Phenom II X4 805	509
Phenom II X3 720	598
Phenom II X3 710	643
Phenom II X2 555	777
Phenom II X2 550	800
Phenom II X2 545	826
Phenom X4 9950	497
Phenom X4 9850	520
Phenom X4 9750	544
Phenom X4 9650	568
Phenom X4 9550	595
Phenom X3 8850	686
Phenom X3 8750	715
Phenom X3 8650	747
Phenom X3 8450	823
Athlon II X4 635	444
Athlon II X4 630	464
Athlon II X4 620	500
Athlon II X3 440	569
Athlon II X3 435	590
Athlon II X3 425	627
Athlon II X2 255	805
Athlon II X2 250	832
Athlon II X2 245	860
Athlon II X2 240	888
Athlon II X2 215	951
Athlon X2 7850	930
Athlon X2 7750	956
Athlon X2 7550	1033
Athlon X2 7450	1063
A64 X2 6000 65n 512	1015
A64 X2 5600 65n 512	1117
A64 X2 5400 65n 512	1136
A64 X2 5200 65n 512	1181
A64 X2 5000 65n 512	1231
A64 X2 4800 65n 512	1283
A64 X2 4600 65n 512	1305

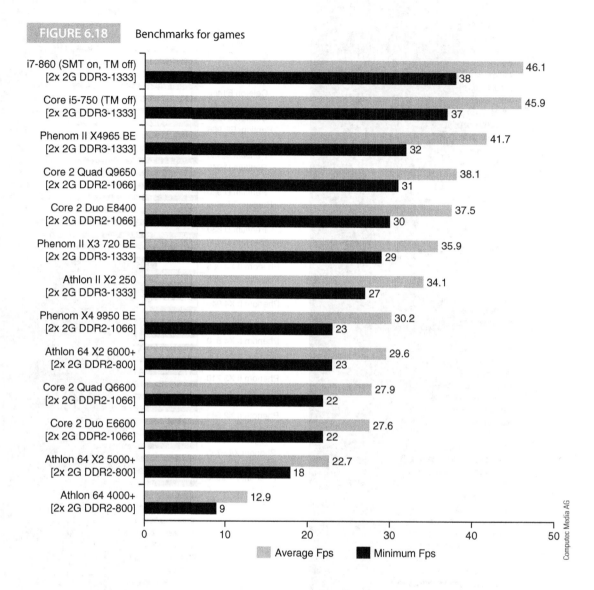

FIGURE 6.18   Benchmarks for games

possible performance (Core i7 980X EE). These processors have relative performance values of 0.81:1.00:1.80 (using the video encoding rates) and relative price ratios of 0.79:1.00:4.18 (using data at the time of the review). The ratios are normalized to the *mainstream* i7 920 processor. The performance to price ratio for these processors is 1.03:1.00:0.43, which demonstrates that you pay a lot for the cutting edge.

PCMARK7 A Commercial Benchmark for PCs

Futuremark, a company set up in Finland in 1997, has developed benchmarking software primarily aimed at the high-performance PC market. One target audience is the games player who wants the highest performance possible to cope with the ever-increasing demands of modern computer games. Futuremark has written performance measurement tools to suite both enthusiast and professional sectors of the PC market, and its business model provides free editions of the basic performance tools with an upgrade path to more comprehensive tools.

PCMARK7 from Futuremark was released in 2011 and is intended to be used in a very specific environment, PCs running Windows 7 with the DirectX 9 (and later) compatible graphics card. Benchmarks are targeted on typical PC activities; for example, the basic

FIGURE 6.19   Benchmarks reported by a commercial benchmarking tool

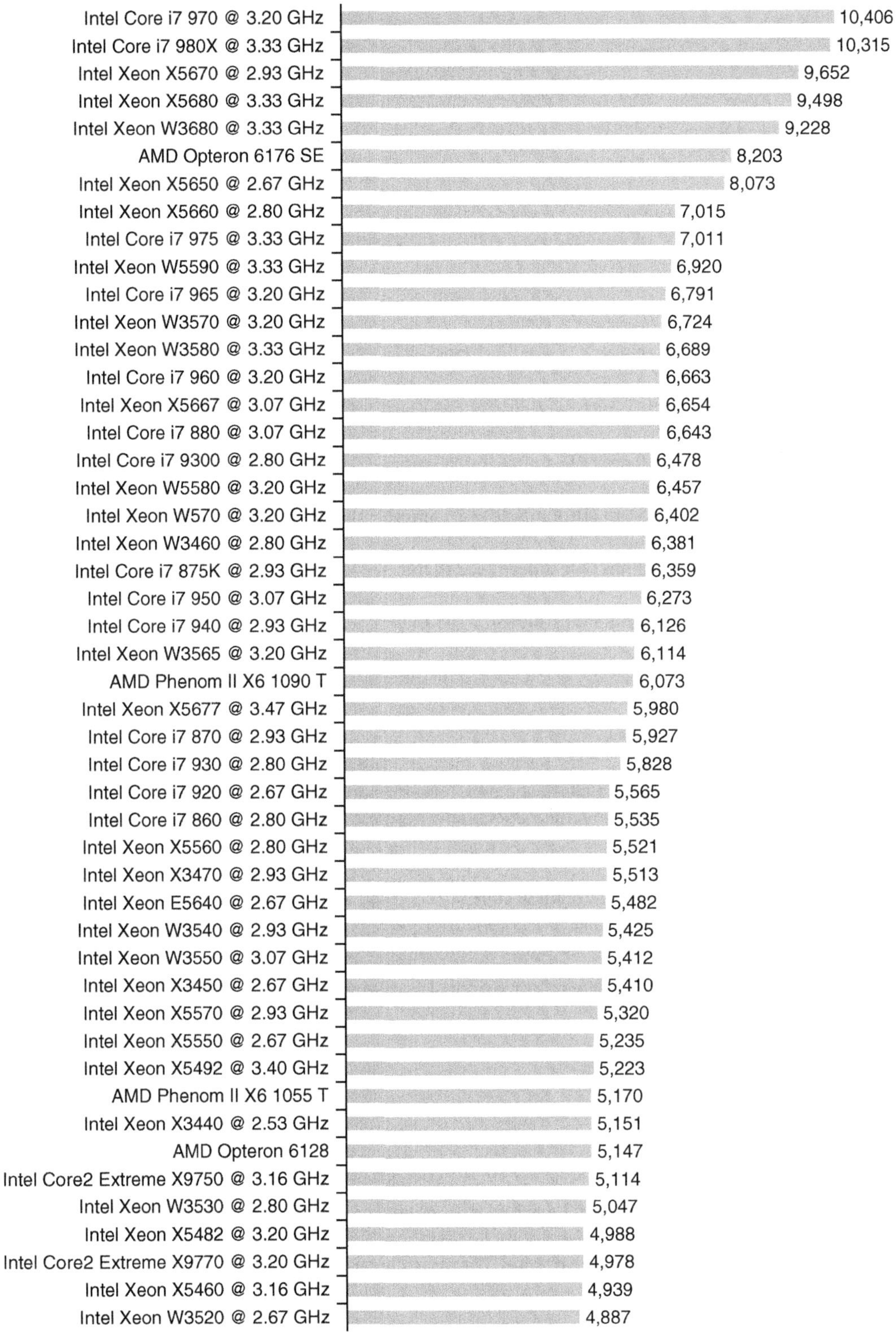

Intel Core i7 970 @ 3.20 GHz	10,406
Intel Core i7 980X @ 3.33 GHz	10,315
Intel Xeon X5670 @ 2.93 GHz	9,652
Intel Xeon X5680 @ 3.33 GHz	9,498
Intel Xeon W3680 @ 3.33 GHz	9,228
AMD Opteron 6176 SE	8,203
Intel Xeon X5650 @ 2.67 GHz	8,073
Intel Xeon X5660 @ 2.80 GHz	7,015
Intel Core i7 975 @ 3.33 GHz	7,011
Intel Xeon W5590 @ 3.33 GHz	6,920
Intel Core i7 965 @ 3.20 GHz	6,791
Intel Xeon W3570 @ 3.20 GHz	6,724
Intel Xeon W3580 @ 3.33 GHz	6,689
Intel Core i7 960 @ 3.20 GHz	6,663
Intel Xeon X5667 @ 3.07 GHz	6,654
Intel Core i7 880 @ 3.07 GHz	6,643
Intel Core i7 9300 @ 2.80 GHz	6,478
Intel Xeon W5580 @ 3.20 GHz	6,457
Intel Xeon W570 @ 3.20 GHz	6,402
Intel Xeon W3460 @ 2.80 GHz	6,381
Intel Core i7 875K @ 2.93 GHz	6,359
Intel Core i7 950 @ 3.07 GHz	6,273
Intel Core i7 940 @ 2.93 GHz	6,126
Intel Xeon W3565 @ 3.20 GHz	6,114
AMD Phenom II X6 1090 T	6,073
Intel Xeon X5677 @ 3.47 GHz	5,980
Intel Core i7 870 @ 2.93 GHz	5,927
Intel Core i7 930 @ 2.80 GHz	5,828
Intel Core i7 920 @ 2.67 GHz	5,565
Intel Core i7 860 @ 2.80 GHz	5,535
Intel Xeon X5560 @ 2.80 GHz	5,521
Intel Xeon X3470 @ 2.93 GHz	5,513
Intel Xeon E5640 @ 2.67 GHz	5,482
Intel Xeon W3540 @ 2.93 GHz	5,425
Intel Xeon W3550 @ 3.07 GHz	5,412
Intel Xeon X3450 @ 2.67 GHz	5,410
Intel Xeon X5570 @ 2.93 GHz	5,320
Intel Xeon X5550 @ 2.67 GHz	5,235
Intel Xeon X5492 @ 3.40 GHz	5,223
AMD Phenom II X6 1055 T	5,170
Intel Xeon X3440 @ 2.53 GHz	5,151
AMD Opteron 6128	5,147
Intel Core2 Extreme X9750 @ 3.16 GHz	5,114
Intel Xeon W3530 @ 2.80 GHz	5,047
Intel Xeon X5482 @ 3.20 GHz	4,988
Intel Core2 Extreme X9770 @ 3.20 GHz	4,978
Intel Xeon X5460 @ 3.16 GHz	4,939
Intel Xeon W3520 @ 2.67 GHz	4,887

Benchmarks reported by PassMark's benchmarking tool. From http://www.cpubenchmark.net/high_end_cpus.html

TABLE 6.10   Details of Eleven Processors in a Review by HEXUS

Model Number	Cores / Threads	GHz Clock	Turbo Boost (max)	Process	Die Size	Cache	Memory Support	TDP
Phenom II X4 965 BE	4/4	3.40	N/A	45 nm (Deneb)	258 mm²	2 MB 6 MB L3	L2 DDR3-1,333+	95 W
Phenom II X6 1055T	6/6	2.80	3.30	45 nm (Thuban)	346 mm²	3 MB 6 MB L3	L2 DDR3-1,600+	125 W 95 W
Phenom II X6 1090T	6/6	3.20	3.60	45 nm (Thuban)	346 mm²	3 MB 6 MB L3	L2 DDR3-1,600+	125 W
Core i5 661 (IGP)	2/4	3.33	3.60	32 nm (Clarkdale)	81 mm²	512 KB 4 MB L3	L2 DDR3-1,333	87 W
Core i5 750	4/4	2.67	3.20	45 nm (Lynnfield)	296 mm²	1 MB 8 MB L3	L2 DDR3-1,333	95 W
Core i7 860	4/8	2.80	3.46	45 nm (Lynnfield)	296 mm²	1 MB 8 MB L3	L2 DDR3-1,333	95 W
Core i7 870	4/8	2.93	3.60	45 nm (Lynnfield)	296 mm²	1 MB 8 MB L3	L2 DDR3-1,333	95 W
Core i7 920	4/8	2.67	2.93	45 nm (Bloomfield)	263 mm²	1 MB 8 MB L3	L2 DDR3-1066	130 W
Core i7 975 EE	4/8	3.33	3.60	45 nm (Bloomfield)	263 mm²	1 MB 8 MB L3	L2 DDR3-1066	130 W
Core i7 970	6/12	3.20	3.46	32 nm (Westmere)	248 mm²	1.5 MB 12 MB L3	L2 DDR3-1066	130 W
Core i7 980X EE	6/12	3.33	3.60	32 nm (Westmere)	248 mm²	1.5 MB 12 MB L3	L2 DDR3-1066	130 W

Courtesy of HEXUS (Link: http://hexus.net/qazgm)

Processor Terminology

Table 6.10 gives the details of some high-performance processors. As we have not yet covered the background, this panel briefly describes the significance of the parameters.

Cores—This defines the number of CPUs per chip. Each CPU is independent of the others.

Thread—A thread is a stream of code that is being executed. A processor can execute multiple threads in parallel on its cores, or it can switch between threads; for example, if one thread is waiting for external memory, another thread can continue.

Turbo boost—The ability to increase clock frequency for short periods without damaging the chip physically.

Process—The minimum size of features (lines) on the silicon chip in nanometers. Visible light is 400-700 nm which means features are one tenth the size of a light wave. Process size indicates just how close to the atomic level modern devices are. The size of an atom is 0.1 to 0.5 nm.

Die size—The physical dimension of the chip. Larger chips are harder to fabricate.

Cache—Cache memory is local high-speed memory and vital to a processor's performance. There are three levels of cache memory; L1, L2, L3. Level L1 is searched first, then L2, then L3 (and finally main store). The more cache the better.

Memory support—Defines the bandwidth of memory that can be used.

TDP—The power consumption of the chip. The lower the power, the better.

FIGURE 6.20   Power consumption by processor in both idle and load modes

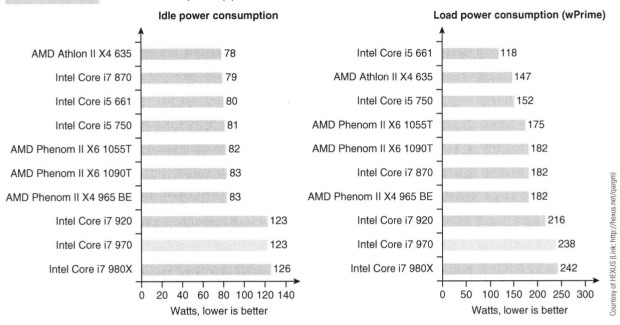

Idle power consumption

Processor	Watts
AMD Athlon II X4 635	78
Intel Core i7 870	79
Intel Core i5 661	80
Intel Core i5 750	81
AMD Phenom II X6 1055T	82
AMD Phenom II X6 1090T	83
AMD Phenom II X4 965 BE	83
Intel Core i7 920	123
Intel Core i7 970	123
Intel Core i7 980X	126

Watts, lower is better

Load power consumption (wPrime)

Processor	Watts
Intel Core i5 661	118
AMD Athlon II X4 635	147
Intel Core i5 750	152
AMD Phenom II X6 1055T	175
AMD Phenom II X6 1090T	182
Intel Core i7 870	182
AMD Phenom II X4 965 BE	182
Intel Core i7 920	216
Intel Core i7 970	238
Intel Core i7 980X	242

Watts, lower is better

Courtesy of HEXUS (Link: http://hexus.net/qazgm)

FIGURE 6.21   Performance by processor. The left-hand parameters are rates and the right-hand parameters are times

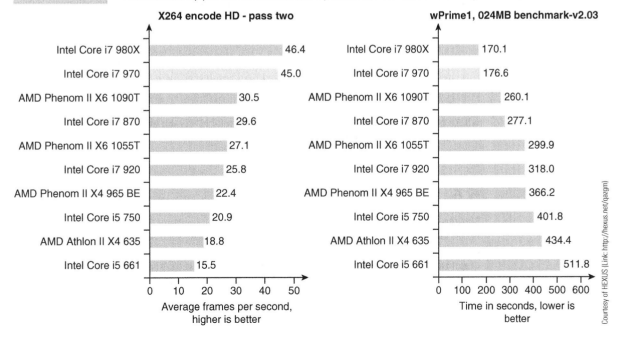

X264 encode HD - pass two

Processor	Average frames per second
Intel Core i7 980X	46.4
Intel Core i7 970	45.0
AMD Phenom II X6 1090T	30.5
Intel Core i7 870	29.6
AMD Phenom II X6 1055T	27.1
Intel Core i7 920	25.8
AMD Phenom II X4 965 BE	22.4
Intel Core i5 750	20.9
AMD Athlon II X4 635	18.8
Intel Core i5 661	15.5

Average frames per second, higher is better

wPrime1, 024MB benchmark-v2.03

Processor	Time in seconds
Intel Core i7 980X	170.1
Intel Core i7 970	176.6
AMD Phenom II X6 1090T	260.1
Intel Core i7 870	277.1
AMD Phenom II X6 1055T	299.9
Intel Core i7 920	318.0
AMD Phenom II X4 965 BE	366.2
Intel Core i5 750	401.8
AMD Athlon II X4 635	434.4
Intel Core i5 661	511.8

Time in seconds, lower is better

Courtesy of HEXUS (Link: http://hexus.net/qazgm)

PCMark test examines storage, video playback and transcoding, image manipulation, web browsing and decrypting.

PCMARK7 provides a separate *lightweight test* aimed at benchmarking low-performance systems such as entry level desktops, tablets and notebooks where the end user is unlikely to run a lot of tests concurrently or to use computationally intensive programs. Other benchmarks performed by PCMARK7 are as follows.

- An *entertainment test* that focuses on applications such as recording, viewing, streaming and transcoding movies,
- A *creativity test* that focuses on activities such as image processing and video editing,
- A *productivity test* that focuses on storage, web browsing, and text editing,
- A *computation test* that is designed to isolate computational performance by concentrating on video operations such as downscaling,
- A *storage test* that isolates the performance of the PC's storage system. This test includes individual tests that import pictures, use Windows media center, and start applications,
- A *secondary storage* test that is identical to the storage test but allows you to test a specific drive; for example, for the purpose of comparing a conventional drive with a solid state drive, SSD.

A particular feature of PCMARK7 is that advanced graphics functions relying on GPUs (graphics processor units) in graphics cards are also tested. This is important because a GPU may have far more computational power than the processor on the motherboard (we look at GPUs in Chapter 13). A typical benchmark is *GPU Cloth* that renders twelve waving flags that are simulated on the GPU. The computation is done in vertex and geometry shaders. Each flag is modeled as a grid of vertices. Each vertex is connected to its eight neighbors with springs. Animated wind and gravity affects the behavior of each flag.

PCMARK7 delivers a single score for each test based on the SPEC methodology (described next); that is, a PCMARK7 score is obtained by taking the *geometric mean* of individual benchmarks and normalizing the result with respect to a base machine. In this case the base machine is a high-end PC with a solid state disk. We discuss the geometric mean, which is one way of averaging results, in the next section.

As well as a performance benchmark, PCMARK7 provides a measure of the spread of results due to the variations in a multitasking operating system and the non-deterministic nature of some hardware (e.g., disk drives). According to Futuremark's PCMARK7 white paper:

$$w_n = \text{mean}_i(w_{i,n})$$

where $w_{i,n}$ is the workload result, n is the workload index, i is the workload run which is repeated three times, and the residual workload result n is given by

$$r_n = \text{mean}_i(\text{abs}(w_{i,n} - w_n))/w_n$$

where mean( ) is the arithmetic mean function and abs( ) is the absolute function.

The residual score is given by $r = \text{mean}(r_n)$.

6.6   SPEC

Various organizations exist round the world to offer the consumer *unbiased* advice, whether it be about the quality of wines or the safety of automobiles. Such an organization has arisen to create benchmarks. The *Standard Performance Evaluation Corporation* (SPEC) is a nonprofit organization with the status of a charity that's been formed to *establish, maintain, and endorse a standardized set of relevant benchmarks that can be applied to the newest generation of high-performance computers*. It is important to note that the SPEC benchmarks are supplied as source code and are compiled. The user is responsible for compiling the benchmarks before they are run. SPEC allows two types of compilation. One type of compilation is restrictive in terms of compiler flags and switches and that provides a *base* result. The other type of compilation is more liberal and allows switches to be optimized for each individual program. This provides a *peak* metric.

The SPEC consortium develops suites of benchmarks intended to measure computer performance and new benchmarks are created by SPEC as technology develops. SPEC members develop test programs. Of course, some members might be tempted to create benchmarks that exploit special features of their own products. However, scrutiny by other members of the consortium helps to reduce any product bias in the benchmarks created by SPEC. That doesn't keep them from changing the benchmark to hide deficiencies that they all share; for example, benchmarks that showed poor branch predictor performance in CPU95 were dropped in CPU2000. Branch predictors were generally weak at the time, and it was in nobody's interest to continue demonstrating that.

Lilja points out that SPEC benchmarks can have an effect on compiler writers. Since compiler writers are likely to use SPEC suites to test their compilers, it is probable that some compiler writers will tweak or *tune* their compilers to the characteristics of the SPEC programs.

Over the years SPEC suites have changed to remove some of their inefficiencies and errors; for example, in 2001 the SPEC CPU95 benchmark was declared obsolete or *retired* and replaced by CPU2000. In turn, CPU2000 was replaced by CPU2006. The benchmark programs that form a SPEC suite can vary over a wide range of applications. For example, from Lisp interpreters to compilers to data-compression programs in older SPECs, and from quantum mechanics to fluid dynamics in more recent SPECs. The complexity of applications grows with each new set of SPEC benchmarks to reflect the increasing demand on computers, as does the size of the data sets on which they operate.

The first SPEC benchmarks appeared in 1988 with SPEC89, which is a suite of ten programs. The next major change was in 1992, when SPEC CINT92 (6 integer programs) and SPEC CFP92 (14 floating-point programs) were published. In 1995, SPEC CINT95 and PSEC CFP95 (8 integer, 10 floating-point programs, respectively) were introduced. The fourth major revision to the benchmarks was SPEC CPU2000 with SPEC CINT2000 and SPEC CFP2000 (12 integer and 14 floating-point programs). SPEC 2000 also introduced benchmarks in C, C++, Fortran 77, and Fortran 90. SPEC 2006 uses 12 integer programs and 17 floating-point programs (we look at SPEC 2006 in more detail later in this section).

The rapid growth in computer technology in the late 1990s caused SPEC to expand its range of benchmarks by including the new market created by Java. Programs in Java are normally compiled to the so-called *bytecode* and then executed on a bytecode interpreter known as the Java virtual machine, JVM. In 1998 SPEC

The Eight SPEC Figures

There are eight SPEC figures quoted for each machine tested. There are integer and floating-point measurements, time and rate measurements, and base and peak measurements.

Integer and floating-point measurements refer to suites of integer and floating-point benchmarks. Time and rate measurements refer to the time to run benchmarks or the rate at which multiple benchmarks are run.

Base and peak results refer to measurements taken under conservative or optimum conditions, respectively. The base test requires that all members of a program suite are tested under identical conditions using the same compiler flags (and in the same order). Peak measurements allow the tester to optimize compiler flags for each benchmark.

Abuse of Benchmarks

Daniel Citron[*] wrote one of the classic papers on benchmarks in 2003 when he discussed MisSPECulation (his term). Essentially, Citron argues that many academics abuse SPEC benchmarks and do not employ them in the way intended by those who created the benchmarks.

The SPEC benchmarks consist of sets of both integer and floating-point benchmarks and data sets upon which the benchmarks operate. Running all the benchmarks with full data can take a long time. This is particularly true in the case of machines simulated in software that run one or two orders of magnitude slower than the real machine would (if constructed in silicon).

Consequently, some researchers do not use the entire SPEC benchmarks. Citron points out that (in the sample he measured) out of 173 published papers, the researchers had used all benchmarks in only 23 instances. Moreover, he states that the majority of papers using truncated suites of benchmarks did not mention the fact. Such an approach can invalidate the research results because the benchmarks are no longer the *balanced* set provided by SPEC.

---

[*] Daniel Citron, MisSPECulation: partial and misleading use of SPEC CPU2000 in computer architecture conferences, *ISCA'03 Proceedings of the 30th Annual International Symposium on Computer Architecture*, ACM, New York, NY, USA ©2003.

released SPECjvm98, a benchmark suite that measures computer system performance for Java virtual machine client platforms.

By 2001, SPEC was writing very specific benchmarks for CPU-intensive applications; for example the *SPEC Application Performance Characterization project group* (SPECapc) released the first standardized benchmark for evaluating performance of systems running 3D Studio MAX R3.1 in February 2001. This benchmark contains four applications that test the underlying system running Studio MAX such as "An architectural visualization containing more than a million polygons, with multiple objects and light sources, glass walls for refraction and opacity tests, and multiple textures."

The advantage of a benchmark such as SPEC is that it is widely available and it is relatively easy to obtain. Moreover, SPEC provides real results on real data rather than simply relying on metrics such as clock rate and MIPS or MFLOPS that, at best, do not tell the whole story and, at worst, are positively misleading.

SPEC Methodology

We now describe how SPEC results are arrived at. This is a rather controversial area and several academic papers have been written to explain why the SPEC methodology is unreasonable. In the next section we look at some of the background to the averaging of benchmarks (the source of the controversy).

The SPEC methodology is to measure the time required to execute each program in the test suite. For example, the times might be $T_{p1}, T_{p2}, T_{p3}, \ldots$, where the subscripts $p1$, $p2$, etc. refer to the components of the suite.

The SPEC suite is also executed on a so-called *standard basis machine* to give the values $B_{p1}, B_{p2}, B_{p3}, \ldots$. The measured times are divided by the reference times to give the values $Tp_1/B_{p1}, T_{p2}/B_{p2}, T_{p3}/B_{p3}, \ldots$. This step normalizes the execution times of the components of the SPEC suite. Finally, the *normalized* times are averaged to generate a final SPEC metric for the machine being tested. As we have pointed out, the way in which a set of results is averaged is controversial. Here, we will simply state that the individual SPEC benchmarks are averaged by taking the *geometric mean*, rather than the arithmetic mean. The geometric mean is calculated by multiplying together n values and then taking the nth root; for example, the arithmetic mean of the values 4, 5, and 6 is $(4 + 5 + 6)/3 = 3$ and the geometric mean is $\sqrt[3]{4 \times 5 \times 6} = 3.915$.

SPEC uses a historical Sun system, the *Ultra Enterprise 2*, that was introduced in 1997 as a reference machine. This reference machine uses a 296 MHz UltraSPARC II processor and is the same reference machine used by the CPU2000 benchmark. Although the CPU2000 and CPU2006 reference machines are nominally the same, the more recent CPU2006 reference machine has better cache memory. It seems to me intuitively wrong that such an old machine has been chosen for normalization because of the considerable advances in computer organization in recent years.

The advantage of normalizing a test machine with respect to a standard machine is that the effect of large differences in the times taken by individual benchmarks is reduced. Let's look at a hypothetical example. Suppose that a reference machine takes 10, 100, and 5 seconds to perform tasks A, B, and C, respectively. Now, imagine that three machines M1, M2, and M3 are tested and give the results specified in Table 6.11. The right-hand column gives the total execution times for all four machines. As you can see, test machine M2 has the best case execution time, M3 the second best, and M1 lags considerably behind because of the time taken to execute task C.

Table 6.12 gives the same results normalized to the reference machine. Now, if you examine Table 6.12, you will see that the total normalized times make M3 the worst machine,

TABLE 6.11   Test Results for Three Hypothetical Machines

	Task A	Task B	Task C	Total
Reference machine	10	100	5	115
Machine M1	10	200	5	215
Machine M2	20	100	5	125
Machine M3	20	100	20	140

TABLE 6.12   Test Results for Three Hypothetical Machines after Normalization

	Task A	Task B	Task C	Total
Machine M1	1	2	1	4
Machine M2	2	1	1	4
Machine M3	2	1	4	7

© Cengage Learning 2014

and M1 and M2 have the same performance. Why has this happened? It's because normalization has reduced execution times to ratios with respect to the reference machine, rather than absolute times which may be dominated by a test program that has a large data set.

The SPEC benchmarks are sometimes called the *best of a bad lot* or the *best metrics in the absence of anything that is more successful*. A particular criticism of SPEC is that it is CPU intensive and doesn't test the computer system. In particular, the effect of memory systems is not fully taken into account. Moreover, the SPEC tests are not necessarily representative of the type of workload found in a multitasking environment—although the situation appears to have improved with the introduction of SPEC CPU2006.

The SPEC organization has been criticized for periodically changing its benchmarks. You could say that this prevents all systems being compared against an agreed baseline. Equally, you could say that the nature of computers is changing and the way in which they are used is changing, and therefore it is essential to update the basis for comparison.

> ### Balancing Data Streams
>
> It is intuitively reasonable that we concentrate on the processor in this text. However, we have pointed out that the processor is not the only factor in determining performance. In particular, the processor needs a stream of instructions and data to keep it supplied; otherwise the processor's potential cannot be achieved. The data path between the memory and the cache, and between the cache and the CPU, must be sufficient to keep the processor supplied with instructions and to handle load and store instructions. When this state is achieved, the workload is said to be *balanced*.

The SPEC benchmarks are sometimes said to be given in *unitless* values because the actual measurements for a given machine are divided by the same measurements for the reference or baseline machine. For example, if the test machine gives the values 1s, 10s, and 10s and the reference machine gives 2s, 10s, and 20s, the normalized unitless results are ½, 1, and ½. However, we will see that these values do have the properties of either *time* or *rate* with respect to averaging and that it is necessary to employ the appropriate averaging technique.

The SPEC benchmarks have proved very successful in some areas. Figure 6.22[9] gives the number of papers on computer architecture submitted to various conferences on themes

FIGURE 6.22   The use of SPEC benchmarks in published papers. © 2003 IEEE. Reprinted, with permission, from Hennessy, J.; Citron, D.; Patterson, D.; Sohi, G. "The Use and Abuse of SPEC: An ISCA Panel," IEEE Micro, vol. 23, no. 4, pp. 73–77, July/Aug. 2003.

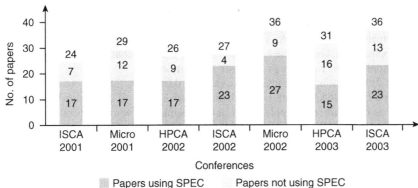

[9] "The Use and Abuse of SPEC: An ISCA Panel", *IEEE Micro*, July–August 2003, Vol. 23, No. 4.

SPEC CPU2006 and the Academic Community

One of the key users of SPEC benchmarks is the computer architecture research community. A paper by Sarah Bird et al. (Performance Characterization of SPEC CPU Benchmarks on Intel's Core Microarchitecture base processor) provides an insight into how SPEC benchmarks are used. This paper looks at CPU2006 benchmarks as applied to Intel's Core processors (in particular the *Woodcrest* realization). Both CPU2006 integer and floating point benchmarks are examined in terms of their interactions with the processor architecture; for example, the benchmarks are broken down in terms of instruction mix. It is instructive to see the percentages of branches, loads and stores in the 12 integer and 17 floating-point benchmarks.

In the integer benchmarks, the percentage of branches ranges from 7.5% to 27.3%; the percentage of loads ranges from 14.4% to 35%, and the percentage of stores ranges from 4.6% to 17.7%. For the floating-point benchmarks, the corresponding ranges are 0.2 to 16.4% (branches), 23.3% to 46.5% (loads), and 3.0% to 14.5% (stores). These figures tell you what aspects of the computer architecture are being tested (or *stressed*) by the benchmarks.

A significant difference between CPU2006 and CPU2000 is that CPU2006 benchmarks have larger input datasets and run longer than the earlier benchmarks; for example, CPU2000 programs run for 56 to 170 s, whereas CPU2006 programs run from 563 to 1590 s in an Intel Core 2 Duo.[10] The new benchmarks provide a better test of the capabilities of modern processors.

related to computer architecture between 2001 and 2003. These papers are divided into two categories; those that use the SPEC benchmark and those that don't. Clearly, SPEC is employed by a large majority of researchers.

As you might expect, academics use benchmarks to test the computers they are designing in order to determine whether the new computer is faster or not than other computers. Since it is not economically feasible to construct each new computer physically, researchers simulate a new architecture in software on an existing computer and then measure the speed of the simulated computer. Of course the simulated computer runs very much slower than its real equivalent would, which means that it is often impossible to use some real-world benchmarks because of the large amount of computation required.

The SPEC CPU2006 Benchmarks

Table 6.13 describes the SPEC CPU2006 integer benchmark suites. These are written in either C or C++ , two languages that are widely used in academia, the graphics world, and high-performance computing. As you can see, the benchmarks cover both commercial and scientific applications. One of the key features of CPU2006 is that the benchmarks represent real-world applications rather than synthetic applications.

The 17 floating-point benchmarks are described by Table 6.14. These are all written in C, C++, and FORTRAN which is one of the oldest high-level computer languages and is still widely used in numerical applications in the engineering world. Figure 6.23 provides an example of the SPEC CPU2006 results for an Intel Core i7 processor in summary form, and Table 6.15 provides the results of the actual runs.

The final part of this chapter introduces a topic that has become rather controversial and well-discussed in academic literature: benchmark averaging. Just how should benchmarks be averaged?

---

[10] T.K. Prakash, and L. Peng, ISAST Transactions on Computers and Software Engineering, No. 1, Vol 2, 2008.

TABLE 6.13   The SPEC 2006 Integer Reference Benchmarks

Benchmark	Language	Application	Brief Description
400.perlbench	C	Programming Language	Derived from Perl V5.8.7. The workload includes SpamAssassin, MHonArc (an email indexer), and specdiff (SPEC's tool that checks benchmark outputs).
401.bzip2	C	Compression	Julian Seward's bzip2 version 1.0.3, modified to do most work in memory, rather than doing I/O.
403.gcc	C	C Compiler	Based on gcc Version 3.2, generates code for Opteron.
429.mcf	C	Combinatorial Optimization	Vehicle scheduling. Uses a network simplex algorithm (which is also used in commercial products) to schedule public transport.
445.gobmk	C	Artificial Intelligence: Go	Plays the game of Go, a simply described but deeply complex game.
456.hmmer	C	Search Gene Sequence	Protein sequence analysis using profile hidden Markov models (profile HMMs).
458.sjeng	C	Artificial Intelligence: Chess	A highly-ranked chess program that also plays several chess variants.
462.libquantum	C	Physics/Quantum Computing	Simulates a quantum computer, running Shor's polynomial-time factorization algorithm.
464.h264ref	C	Video Compression	A reference implementation of H.264/AVC, encodes a videostream using two parameter sets. The H.264/AVC standard is expected to replace MPEG2.
471.omnetpp	C++	Discrete Event Simulation	Uses the OMNet++ discrete event simulator to model a large Ethernet campus network.
473.astar	C++	Path-Finding Algorithms	Pathfinding library for 2D maps, including the well known A* algorithm.
483.xalancbmk	C++	XML Processing	A modified version of Xalan-C++, which transforms XML documents to other document types.

TABLE 6.14   The SPEC 2006 Floating-Point Reference Benchmarks

Benchmark	Language	Application Area	Brief Description
410.bwaves	Fortran	Fluid Dynamics	Computes 3D transonic transient laminar viscous flow.
416.gamess	Fortran	Quantum Chemistry	Gamess implements a wide range of quantum chemical computations. For the SPEC workload, self-consistent field calculations are performed using the Restricted Hartree Fock method, Restricted open-shell Hartree-Fock, and Multi-Configuration Self-Consistent Field.
433.milc	C	Physics/Quantum Chromodynamics	A gauge field generating program for lattice gauge theory programs with dynamical quarks.
434.zeusmp	Fortran	Physics/CFD	ZEUS-MP is a computational fluid dynamics code developed at the Laboratory for Computational Astrophysics (NCSA, University of Illinois at Urbana-Champaign) for the simulation of astrophysical phenomena.
435.gromacs	C, Fortran	Biochemistry/ Molecular Dynamics	Molecular dynamics, i.e., simulate Newtonian equations of motion for hundreds to millions of particles. The test case simulates protein Lysozyme in a solution.
436.cactus ADM	C, Fortran	Physics/General Relativity	Solves the Einstein evolution equations using a staggered-leapfrog numerical method.

(Continued)

TABLE 6.14	Continued		
Benchmark	Language	Application Area	Brief Description
437.leslie3d	Fortran	Fluid Dynamics	Computational Fluid Dynamics (CFD) using Large-Eddy Simulations with Linear-Eddy Model in 3D. Uses the MacCormack Predictor-Corrector time integration scheme.
444.namd	C++	Biology / Molecular Dynamics	Simulates large biomolecular systems. The test case has 92,224 atoms of apolipoprotein A-I.
447.dealII	C++	Finite Element Analysis	Deal.II is a C++ program library targeted at adaptive finite elements and error estimation. The test case solves a Helmholtz-type equation with non-constant coefficients.
450.soplex	C++	Linear Programming, Optimization	Solves a linear program using a simplex algorithm and sparse linear algebra. Test cases include railroad planning and military airlift models.
453.povray	C++	Image Ray-Tracing	Image rendering. The test case is a 1280 × 1024 anti-aliased image of a landscape with some abstract objects with textures using a Perlin noise function.
454.calculix	C, Fortran	Structural Mechanics	Finite element code for linear and nonlinear 3D structural applications. Uses the SPOOLES solver library.
459.Gems FDTD	Fortran	Computational Electromagnetics	Solves the Maxwell equations in 3D using the finite-difference time-domain (FDTD) method.
465.tonto	Fortran	Quantum Chemistry	An open source quantum chemistry package, using an object-oriented design in Fortran 95. The test case places a constraint on a molecular Hartree-Fock wavefunction calculation to better match experimental X-ray diffraction data.
470.lbm	C	Fluid Dynamics	Implements the "Lattice-Boltzmann Method" to simulate incompressible fluids in 3D.
481.wrf	C, Fortran	Weather	Weather modeling from scales of meters to thousands of kilometers. The test case is from a 30km area over 2 days.
482.sphinx3	C	Speech recognition	A widely-known speech recognition system from Carnegie Mellon University.

TABLE 6.15	SPEC 2006 Sample Published Benchmarks (actual run times)

Benchmark*	Base						Peak					
	Seconds	Ratio	Seconds	Ratio	Seconds	Ratio	Seconds	Ratio	Seconds	Ratio	Seconds	Ratio
400.perlbench	562	17.4	**540**	**18.1**	539	18.1	455	21.5	475	20.6	**455**	**21.5**
401.bzip2	**733**	**13.2**	732	13.2	764	12.6	729	13.2	765	12.6	**730**	**13.2**
403.gcc	621	13.0	**622**	**12.9**	623	12.9	456	17.7	**473**	**17.0**	474	17.0
429.mcf	395	23.1	396	23.0	**396**	**23.0**	316	28.8	333	27.4	**329**	**27.7**
445.gobmk	595	17.6	**595**	**17.6**	594	17.7	562	18.7	**575**	**18.2**	587	17.9
456.hmmer	**277**	**33.7**	277	33.7	278	33.6	267	34.9	**256**	**36.4**	256	36.5
458.sjeng	**679**	**17.8**	678	17.8	679	17.8	627	19.3	**628**	**19.3**	656	18.5
462.libquantum	151	137	151	137	**151**	**137**	151	137	151	137	**151**	**137**
464.h264ref	799	27.7	**800**	**27.7**	801	27.6	**777**	**28.5**	759	29.2	781	28.3
471.omnetpp	505	12.4	**506**	**12.4**	526	11.9	418	14.9	399	15.7	**417**	**15.0**
473.astar	554	12.7	**554**	**12.7**	577	12.2	**569**	**12.3**	569	12.3	571	12.3
483.xalancbmk	**305**	**22.6**	295	23.4	306	22.5	**305**	**22.6**	295	23.4	306	22.5

*Results appear in the order in which they were run. Bold underlined text indicates a median measurement.

FIGURE 6.23   SPEC 2006 sample published benchmark (overall results)

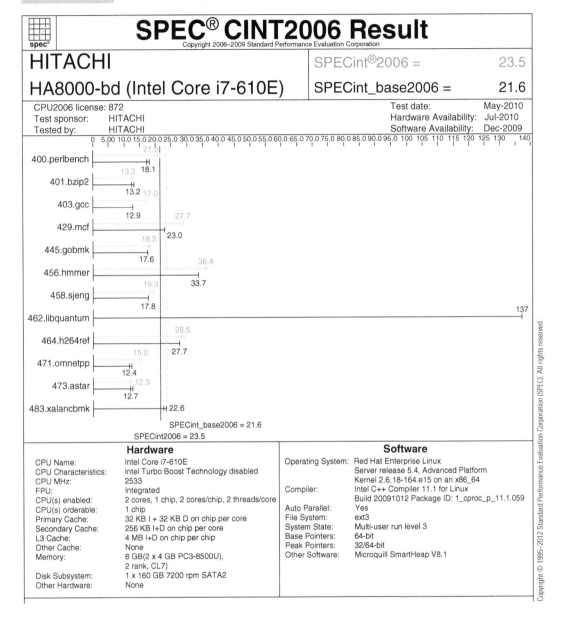

SPEC and Power

Because power is now a first-class consideration in computer design, it's becoming a component of benchmarking. SPEC has created a standard to measure a server's power and performance across multiple loads, SPECpower_ssj2008. The metric reports the server's performance in ssj_ops (*server side Java operations per second*) divided by the power consumed in watts. This benchmark is primarily intended for servers in commercial applications where power consumption is of importance; that is, the cost of buying the power at a time when energy costs are rising, the cost of cooling and ventilation to remove the power, and the additional cost due to increased failure rate in hot environments.

Figure 6.24 from Dell provides an example of the normalized performance per watt for four commercial servers. Figure 6.25 from SPEC gives the output of a particular test. As you

FIGURE 6.24   SPECpower_ssj2008 performance per watt

SPEC—The Payoff

SPEC benchmarks are not simply abstract figures that languish in databases yearning for daylight. No, life is sometimes breathed into them by the corporate communications division. Consider the following from HP that compares its servers with those of its competitors.

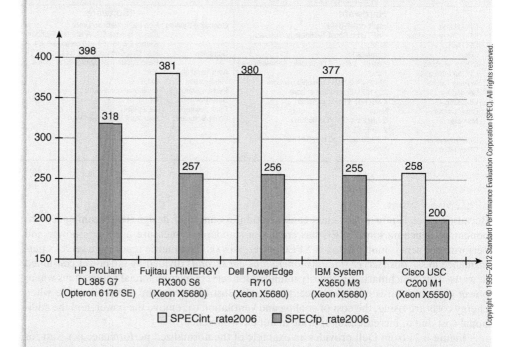

SPECpower_ssj2008 performance

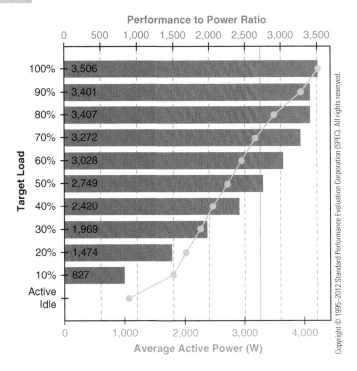

can see it consists of a graph of the power consumption against the total load on the server and a bar chart giving the performance to power ratio for each load level.

6.7   Averaging Metrics

We stated that the SPEC benchmark tests the performance of a computer by running several test programs and then *averaging* the normalized results of the tests. We did not say what we meant by *averaging*, although we have pointed out the problem of averaging times and rates.

By its very nature, a computer will normally be used in multiple roles unless it is part of an embedded system. Because of the relationship between the instruction set architecture, the compiler and the program, it follows that the behavior of a machine will depend on the program it is running. Clearly, a machine with fast floating-point hardware may beat another machine when carrying out numerical integration, but it may be slower when, say, performing data compression or list searching.

If we assemble programs to test the features of computers that are thought of as currently desirable, we have to *combine* the results from several tests into a *single figure of merit*, a process called, in everyday English, averaging.

There is more than one way of averaging a series of numbers. The simplest average is the *arithmetic mean* and is calculated by adding together a series of values and dividing the sum by the number of elements in the series; that is

$$A_{\text{arithmetic}} = \frac{1}{n}\sum x_i$$

We use the arithmetic average in daily life. If you watch three movies of 90 minutes duration, 96 minutes and 75 minutes, it's reasonable to say that that the average length of the movies was 87 minutes. That's intuitive. Now suppose we travelled in an aircraft at 500 mph followed by a car at 60 mph and a train at 100 mph. If we work out the average speed in the same way, we get 220 mph, which intuitively doesn't seem correct.

Consider the effect of averaging benchmarks 1.0, 1.2, 1.2, 0.8, 1.4. Their sum is 5.6 and the arithmetic average is 1.12. Now consider the series 1.2, 1.6, 5.2, 0.8, 0.7. In this case, the sum is 9.5 and the arithmetic average is 1.9. The results are presented below for ease of comparison.

1.0, 1.2, 1.2, 0.8, 1.4	Sum = 5.6	Average = 1.12
1.2, 1.6, **5.2**, 0.8, 0.7	Sum = 9.5	Average = 1.90

If you look at these two series, you will find that they are similar with the exception of a single large value of 5.2 in the second series. This large value dramatically affects the mean and begs the question, "*Should we permit a single out-of-line result to have such an effect on the average?*"

There are many ways of averaging data values and each technique has its particular advantages and disadvantages. The averaging technique sometimes used to extract a single benchmark from a series of computers is called the *geometric* mean.

Geometric Mean

The geometric mean of two numbers is the square root of their product; that is the geometric average of p and q is $\sqrt{(p \cdot q)}$. Suppose a test value yields the results 100 and 150, the geometric average is $\sqrt{(100 \times 150)} = 122.47$, and the arithmetic average is 125. If we have n test values, the geometric mean is the nth root of the product of the n benchmarks.

The geometric mean of n values is the length of the diagonal of an n-dimensional cube that has the same volume as an n-dimensional rectangle whose sides are the lengths of the n values. Because the geometric mean *multiplies* benchmarks, you get the same result by doubling benchmark x and halving benchmark y as you get by doubling benchmark y and halving benchmark x (because $\frac{1}{2}x \times 2y = 2x \times \frac{1}{2}y = xy$). Jacob and Mudge point out that the geometric mean is not an intuitive way of computing the mean of a series of performance metrics.

SPEC uses the geometric mean to arrive at a single figure using the normalized execution times of the components the benchmark suite. Consider the examples in Table 6.16 where systems 1, 2, and 3 take part in three tests and the results are all times.

TABLE 6.16				Tests on Two Machines Compared to Two Reference Machines	
	Test 1	Test 2	Test 3	Arithmetic Average	Geometric Mean
System 1	10	10	10	10	10
System 2	1	12	17	10	5.89
System 3	10	5	20	11.67	10

© Cengage Learning 2014

We have chosen these figures to demonstrate the difference between arithmetic and geometric mean. The three benchmarks for system 1 and system 2 have identical arithmetic averages. However, the low result for test 1 in system 2 has almost halved the geometric mean. Now consider system 3. Here we have the same geometric mean as system 1 but a higher arithmetic mean.

Now suppose we have two machines yielding benchmarks of 100, 100, 2 and 100, 200, 1, respectively. Because the individual benchmarks are multiplied together to calculate a geometric mean, these two sets of figures have the same geometric mean. Do we really regard the improvement of 100 to 200 in the same light as 1 to 2?

Let's take this comparison of arithmetic and geometric means a little further and see what happens when we have one result out of line in a series of results. Suppose that six tests for a base machine and systems 1 and 2 yield the test results in Table 6.17.

We have taken five identical results and added a sixth result—in one case the result is ten times better than the others and in the other case the sixth result is ten times worse than the average. We have also included the *Base* system that does not have the sixth anomalous result. You can immediately see that the arithmetic mean is more strongly affected by the high anomaly than the geometric mean, whereas the geometric mean is more affected by the low anomaly. This is to be expected, of course, because of the way in which the means are formed.

System	Test Results	Arithmetic Mean	Geometric Mean
Base	10 10 10 10 10 10	10	10
1	10 10 10 10 10 100	25	14.68
2	10 10 10 10 10 1	8.5	6.81

TABLE 6.17   Arithmetic and Geometric Means

© Cengage Learning 2014

One of the claimed advantages of the geometric mean is that *the mean of the ratios is the ratio of the means*. This tells us that it doesn't matter whether we take the geometric mean of unnormalized data and then normalize it, or whether we normalize the data and then take the geometric mean. Consider the example of Table 6.18.

TABLE 6.18   Geometric Means

	t_1	t_2	t_3	Geometric Mean
Evaluation machine	2	2	4	2.52
Reference machine	4	6	9	6
Normalized results	2	3	2.26	2.38

© Cengage Learning 2014

Table 6.18 tells us the geometric mean of the evaluation machine's times, the geometric mean of the reference machine's times, and the geometric mean of the normalized times. If we take the geometric mean of the unnormalized times from the reference machine (i.e., 6) and divide it by the geometric mean of the times from the evaluation machine, we get the expected value of $6/2.52 = 2.38$.

Harmonic Mean

We have already demonstrated that the average of *times* is not the same as the average of *rates*. If you wish to calculate the average of a series of tests that yield *rates*, you have to use the *harmonic* mean. The harmonic mean of n values is calculated by summing the reciprocals of the n terms and taking $1/n$ of the reciprocal of the result. For two variables the harmonic mean is $2/(1/x + 1/y)$. For n variables the harmonic mean is

$$\frac{n}{\left[\sum \frac{1}{x_i}\right]}$$

Consider the three examples of three tests in Table 6.19. How can we interpret these results? If the figures correspond to *times*, the lowest average time is best and the arithmetic mean is appropriate. Here, cases 1 and 2 give the same result. If the figures correspond to *rates*, the highest rate is best and the harmonic mean is appropriate; now case 2 is best.

TABLE 6.19   Illustration of the Difficulty of Comparing Means

	Test 1	Test 2	Test 3	Arithmetic	Geometric	Harmonic
Case 1	10	10	10	10	10	10
Case 2	1	12	17	10	5.89	12.392
Case 3	10	5	20	11.67	10	8.571

© Cengage Learning 2014

The geometric mean adopted by SPEC is highly controversial, with some members of the computing community strongly opposed to it. Others provide broad support and point out that the geometric mean produces a constant rank order among the machines, and often it is rank order that matters more to people than absolute benchmark values.

Weighted Means

It is intuitively reasonable to use a set of benchmarks to measure the performance of your system if the benchmarks are used in the same ratios by your application; that is, the benchmark suite should look the same as your applications. In a *weighted mean*, the various components of a benchmark are multiplied by a factor that indicates their contribution to the whole. Let's look at an example. Suppose you are a NASA engineer who is evaluating workstations to replace existing computers. Table 6.20 gives the time your existing workstations spends on various tasks.

TABLE 6.20	Time Taken by a Workstation to Carry Out Various Activities		
Operation		Symbol	Time
Orbital dynamic calculations		C_{OD}	50%
Performing aerodynamic calculations		C_{AD}	25%
Image processing		C_{IP}	15%
Microsoft flight simulator		C_{MS}	10%

© Cengage Learning 2014

Suppose that the times taken to execute these programs on a machine being evaluated are given as

t_{COD}	30 minutes
t_{CAD}	25 minutes
t_{CIP}	70 minutes
t_{CMS}	65 minutes

A weighted average for this system can be calculated from the sum of the products of the weights (i.e., percentage use) defined in Table 6.19 and the times; that is

$$t_{weighed} = 0.50 \times 30 + 0.25 \times 25 + 0.15 \times 70 + 0.10 \times 65 = 34.35 \text{ min}$$

Summary

Once we were happy with a computer if it worked. Today, we are interested in performance because some want to design faster machines and because some wish to buy machines with the highest performance for the lowest cost. We have looked at various ways of comparing the performance of machines. A processor's clock rate is the crudest of all metrics and different processors can't be compared on the basis of clock speed at all, simply because clock speeds tell us nothing about how the processor behaves internally. The MIPS metric goes one step further by telling us how fast instructions get executed but it tells us nothing about how much work the instructions actually do.

We have looked at the specialist PC benchmarks designed for the *power user* or *enthusiast*. These benchmarks are rarely referred to by academics, but they provide practical guidance to large numbers of computer users and are highly focused on real applications.

Professional program suites like SPEC are popular in the academic and commercial worlds, although their derivation is based on many assumptions. It is true that, over the years, the SPEC benchmarks have become increasingly realistic. However, the use of a relatively slow machine to normalize SPEC results does seem strange to me because many of the new architectural features will not exist on the standard machine. On the other hand, the use of an older machine as a standard for normalization will highlight progress in architecture.

We have mentioned that power is becoming a key criterion in computer design, partly because of the growth of battery-based mobile computing and partially because of the need to cool computing equipment. We can expect to see power become increasingly important as a performance metric and companies advertising computers on the basis of their processing power per watt.

Some readers might feel a little disappointed at the end of this section on performance, because we have not been able to define a simple single measure of performance. Of course not! Each user has his or own way of defining performance. What matters is that the current performance measurement techniques contribute to progress in computer design.

Problems

6.1 What is *performance* in the context of computer systems and why is it so difficult to define?

6.2 A system consists of a CPU, cache memory, main store, and hard disk drive. Where are time and effort best spent improving the system's performance? What factors affect your answer?

6.3 Should metrics for computer performance be linear or non-linear? For example, if a linear metric has a value X, the metric $2X$ would imply twice the performance, whereas if the metric were logarithmic, the metric $2X$ would imply a ten-fold increase in performance.

6.4 A data transmission system transmits data in the form of a master frame containing 16 sub-frames. Each sub-frame includes a 1024-bit data word and a 12-bit error-correcting code. The master frame itself contains a 32-bit error correcting code. What is the efficiency of this system?

6.5 The time taken by machines A, B, and C to execute a given task is

A 16 m, 9 s
B 14 m, 12 s
C 12 m, 47 s

What is the performance of each of these machines relative to machine A?

6.6 Why is clock rate a poor metric of computer performance? What are the relative strengths and weaknesses of clock speed as a performance metric?

6.7 The timing diagram in Figure P6.7 illustrates a system in which operations occur as three consecutive clock cycles. Actions taking place in clock cycle 1 are scalable; that is, if the clock cycle time changes, the actions can be speeded up or slowed down correspondingly. In cycle 2, the action *process 1* requires 25 ns and in clock cycle 3 the action *process 2* requires 32 ns. If the clock cycle is less than the time required for process 1 or process 2, then one or more wait cycles have to be inserted for the process to complete.

What is the time to complete an operation if the clock cycle time is

a. 50 ns   b. 40 ns   c. 30 ns
d. 20 ns   e. 10 ns

6.8 What are the relative strengths and weaknesses of the MIPS as a metric of computer performance?

6.9 Can you think of a better metric than MIPS or clock speeds that gives a good impression of the power of a processor (without having to use benchmarks).

6.10 How is it possible for one computer with a low MIPS rating to have a better performance in practice than a computer with a high MIPS rating?

6.11 Overclocking a computer means operating it at a higher clock rate than that specified by its manufacturer; for example, a 2 GHz chip might be clocked at 2.1 GHz to squeeze more performance out of it.

Does overclocking disprove the famous aphorism "There's no such thing as a free lunch," or is there a hidden cost? If so, what is the *cost* of overclocking?

6.12 The following figures define the typical operating parameters of a processor.

Operation	Frequency	Cycles
Arithmetic/logical instructions	45%	1
Register load operations	20%	3
Register store operations	10%	2
All branch instructions	25%	2

If the clock rate could be reduced by 15%, it would require only 2 cycles to perform a register load. Would that be a good idea?

6.13 A computer has the following parameters.

Operation	Frequency	Cycles
Arithmetic/logical instructions	65%	1
Register load operations	10%	5
Register store operations	5%	2
Conditional branch instructions	20%	8

If the average performance of the computer (in terms of its CPI) is to be increased by 20% while executing the same instruction mix, what target must be achieved for the cycles per conditional branch instruction?

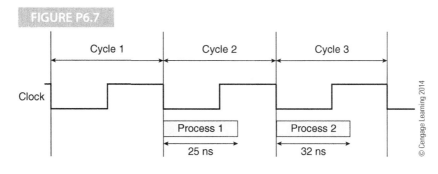

FIGURE P6.7

© Cengage Learning 2014

6.14 A program is run on a computer with the following parameters.

Clock cycle time	10 ns
Instructions with 1 cycle	70%
Instructions with 2 cycles	20%
Instructions with 3 cycles	10%

What is the MIPS rating of this computer?

6.15 For the following data, what is the average number of cycles per instruction?

Operation	Frequency	Cycles
Arithmetic/logical instructions	45%	1
Register load operations	18%	5
Register store operations	10%	2
Unconditional branch instructions	7%	1
Conditional branch instructions	20%	6

© Cengage Learning 2014

6.16 In a particular system, a CPU is used for 78% of the time and a disk drive for 22% of the time. A designer has two options:
a. improve the disc performance by 40% and the CPU performance by 20%
b. improve the disc performance by 10% and the CPU performance by 80%

Which is the better option, and why?

6.17 For the following systems that have both serial and parallel activities, calculate the speedup ratio.
a. 10 processors $f_s = 0.1$
b. 100 processors $f_s = 0.1$
c. 5 processors $f_s = 0.4$
d. 100 processors $f_s = 0.01$

6.18 A system has a single core processor that costs $150. Suppose that adding more cores to the chip costs $10 per additional processor. (*Note*: For this system, the value of f_s is 0.10).

If it is considered worthwhile adding cores until the incremental speedup ratio increases by less than 5% over the original (unmodified) performance, what is the optimum number of processors? What percentage increase in cost is required to achieve this performance?

6.19 A computer employed in arithmetic processing uses a software division routine. A program runs for two minutes on this machine with division taking 60% of the total time. If we wish to add a dedicated division unit in order to increase the performance of the computer by a factor of two, how much faster do we have to make the hardware division unit than the existing division mechanism?

6.20 A system containing several operational units may have multiple enhancements. In Figure P6.20, a system consists of a process S followed by processes P_1 and P_2. Both processes P_1 and P_2 can be enhanced. If process P_1 is enhanced by factor f_1 and process P_2

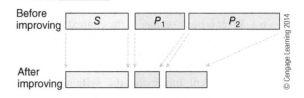

FIGURE P6.20

Before improving: S | P_1 | P_2

After improving:

© Cengage Learning 2014

is enhanced by factor f_2, what is the speedup ratio of the system? Note that enhancement is defined as the old speed of the process divided by its new speed.

6.21 A program is executed in 200 ms during which 250 million instructions are executed. What is the average MIPS for this program?

6.22 A coprocessor is added to a computer to speed the execution time of string-processing instructions by a factor of 3.5. What fraction of the execution time must use these string-processing instructions in order to achieve an average speedup of 1.5?

6.23 Consider the current high-performance desktop computer and the laptop (or notebook). Suppose you wish to increase the performance of both machines. Do you think that the same elements of the system (in both desktop and notebook forms) need improving equally, or do you think that the task of improving each system is different? Is so, why?

6.24 Amdahl or Gustafson? Consider the following example. A physical process involving a flat area 100 units by 100 units is being simulated. Processing of the units can take place in parallel. However, there is a border region 10 units wide that goes round the square where parallel processing cannot take place because of relationships with the external world. In other words, parallel processing can take place in the inner square but not in the 10-unit wide strip/border round it.
a. If parallel processing is used, what speedup is possible?
b. Suppose that the system is scaled to 200 by 200 units. The border region remains at 10 units. What speedup can parallel processing now achieve?

6.25 A computer spends 25% of its time accessing a hard disk. It spends 20% of the time doing floating point. The hard disk is replaced by two disks operating in parallel and the floating point unit is replaced by one four times faster. The speed up is given by

$$S = \frac{1}{f_s + \dfrac{1 - f_s}{p}} = \frac{p}{pf_s + 1 - f_s}$$

So, the speedup for the disk is $S_{disk} = 2/(2 \times 0.75 + 1 - 0.75) = 2/1.75 = 1.429$.

The speedup for the floating-point unit is $S_{\text{floating-point}} = 4/(4 \times 0.80 + 1 - 0.80) = 4/1 = 4/3.4 = 1.176$. The total speedup ratio is the product of the individual speedups which is $1.429 \times 1.176 = 1.681$. Is this answer correct?

6.26 Someone decided to use the following C code as part of a benchmark to determine the performance of a computer including its memory. It has two potential faults. What are they?

```
for (i = 0; i < 100; i++){
    p = q * s + 12345
    x = 0.0;
    for (j = 0; j < 60000; j++){
        x = x + A[j] * B[j];
    }
}
```

6.27 You are redesigning a system. You can replace the existing single processor by two P processors or by four Q processors. However, the P processors are able to run 80% of the code in parallel, whereas the Q processors are able to run 50% of the code in parallel. Which is the better option?

6.28 An operation can be speeded up by applying two different optimizations, O_1 and O_2. These optimizations operate on different parts of the process and there is no overlap. If O_1 speeds up fraction f_1 of the program by S_1 and O_2 speeds up fraction f_2 of the program by S_2, what is the overall speedup?

6.29 You manufacture a computer that executes a program in 50 minutes whereas your competitor's takes 45 minutes. How are you going to sell (advertise) your processor?

6.30 What are the relative advantages and disadvantages of arithmetic, geometric, and harmonic means as methods of averaging benchmarks?

6.31 For two benchmarks, x and y, show that their arithmetic mean is always higher than, or the same as, the geometric mean.

6.32 The SPEC benchmarks present results with respect to a standard machine by normalizing the benchmarks. That is, a set of benchmarks is run on a reference machine and the times obtained for each of the benchmarks. When a test machine is benchmarked, its times are divided by the results on the reference machine. What are the advantages and disadvantages of giving benchmarks with respect to a reference machine?

6.33 Two computers and a reference machine produce the following results.

Machine	Benchmark 1	Benchmark 2	Benchmark 3
Reference	150 s	65 s	95 s
A	120 s	40 s	65 s
B	70 s	35 s	80 s

Present the results in a normalized form and provide benchmarks for machines A and B.

6.34 In 2013, a woman with a small business at home is going to buy a desktop computer to handle her correspondence and diary/calendar, to allow her to email colleagues and to deal with her tax. Being a sensible person, she decides to get the best computer she can afford and Google's 'computer performance'. Should she use clock speed, MIPS, or SPEC as a metric?

Processor Control

"This is the place."
Brigham Young

"A place for everything, and everything in its place."
Samuel Smiles

"Out of clutter, find simplicity. From discord, find harmony. In the middle of difficulty lies opportunity."
Albert Einstein

"The secret of all victory lies in the organization of the non-obvious."
Marcus Aurelius

"Efficiency is doing things right; effectiveness is doing the right things."
Peter F. Drucker

"Prediction is very difficult, especially about the future."
Niels Bohr

"Got any ideas?"
Captain Sullenburger, 22 seconds before ditching in the Hudson

"Actually not."
First officer Skile's response

How do computers work, and why are they so fast? This is the place where we provide some of the answers to these two questions. The theme of this chapter is the internal operation of microprocessors and some of the ways in which designers have improved their performance. In particular, we explain how the performance of a computer can be significantly improved by overlapping the execution of instructions by means of *pipelining*, which is a mechanism once associated with RISC processors but now employed by all contemporary processors. Unfortunately, pipelining is highly sensitive to certain types of instruction or to sequences of machine-level instructions that degrade performance. The final part of this chapter examines why the efficiency of pipelining is so dependent on the nature of the code being executed, and describes some of the techniques employed to overcome the inherent limitations of pipelining.

In this chapter, we cannot cover all the techniques that have been used to improve performance so we restrict ourselves to pipelining that sets the maximum performance limit at one instruction per cycle. In the next chapter, we introduce the *superscalar* processor that increases performance beyond *one instruction per cycle* by employing multiple execution units to execute instructions in parallel. In Chapter 13, we look at the multicore processor that improves performance simply by replicating the entire CPU itself in order to create a multiprocessor on a chip.

This chapter covers a lot of material because we have to introduce the organization of a computer both at a conceptual level that demonstrates how binary-encoded instructions can be read and interpreted and at a more practical level that illustrates some of the techniques employed to implement computers. Moreover, there is more than one way of describing computer organization: We can look at the processor as a general-purpose digital machine controlled by an internal sequencer (this approach more closely models early CISC machines), or we can present the computer as a set of digital building blocks (memory, registers, arithmetic blocks, logic blocks, and multiplexers) through which an instruction flows during its execution. This latter model is in keeping with RISC organization and uses the instruction bits directly to control the operation of the functional units. Following the *theme and variations* philosophy of this text, we introduce both approaches to computer implementations.

We begin by demonstrating how registers, functional units (ALUs and adders), tristate gates, and buses can be used to execute instructions and process data. We show how *microprogramming* is used to convert *ad-hoc* instruction codes into the very actions that execute them. Compilers for high-level languages transform programs into streams of low-level or machine-level instructions. Microprogramming converts machine-level instructions into sequences of micro-operations. The micro-operation is a primitive hardware operation such as *latch data in a register* or *copy data from bus A to bus B*. Microprograms that interpret machine-level instructions are built into the hardware of the microprocessor and are not user accessible, although it is possible to design processors where the user can change the microprogram and hence the instruction set.

Computer Hierarchy

A computer can be regarded as a hierarchical system with multiple levels of abstraction. We can view the digital computer at the levels of abstraction described below. In Chapter 3, we were interested at the low-level language layer. In this chapter, we are interested in the microarchitecture layer.

Application Level—At this level, the computer appears to be a device that performs a function. For example, a computer running a GPS mapping package appears just like a satnav device.

High-Level Language Level—At this level, the computer appears to be *machine independent* and executes a high-level language. All computers executing the same language are identical (in principal) and differ only in terms of performance.

Low-Level Language—At this level, the computer is *architecture dependent* and the machine code executed will run only on one particular class of computer (e.g., Intel IA32 code on a Core i7).

Microarchitecture—This level represents the physical organization of the computer in terms of registers, functional units, and buses. The microarchitecture may be unique to a particular instance of a microprocessor (i.e., two microprocessors share the same low-level language but different microarchitectures). Normally, this level is not accessible by the end user. However, modern programmable logic does allow users to modify the microarchitecture of processors constructed from programmable logic. This chapter is concerned with the microarchitecture level.

(continued)

Gate Level—Below the microarchitecture lie the individual gates that determine the ultimate speed of the processor.

Device Physics Level—The device physics is the lowest level and is determined by the electronic properties of the material used to fabricate the gates.

The following diagram illustrates these levels and those concerned with them. It is an inverted pyramid representing the relative numbers of those involved with each level. For example, there are a large number of computer users but relatively few people designing transistors (i.e., IC fabrication). Note also that the microprogram element may be absent if the low-level language is directly interpreted (as it is in many RISC processors).

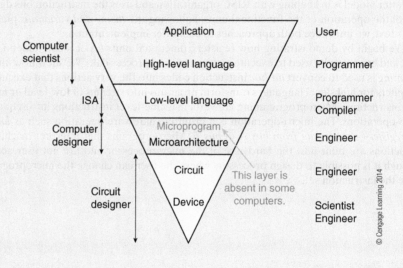

Microprogramming is rarely used to implement *fully* today's CISC processors, although processors like the Pentium still use microprogramming to execute some of their most complex instructions. We describe microprogramming because it so neatly explains how computers can execute any arbitrarily complex operation. After describing the general-purpose microprogrammed computer, we introduce the *single-cycle RISC* organization that executes a machine-level instruction in one clock cycle.

The structure of this machine makes it easy to understand how instructions are executed on pure RISC processors with 32-bit (or longer) instruction words that have a regular encoding. (See the panel). The structure of this processor will exemplify an important feature of modern processor operation: the so-called *stored-program von Neumann machine* no longer accurately models processor architectures. Today's processors have *Harvard architectures* because they have separate instruction and data stores that can be accessed in parallel. Although contemporary processors are nominally stored-program machines with common instruction and data memory, the use of high-speed on-chip instruction and data *cache memories* means that the processor can overlap instruction and data access. We look at cache memory in Chapter 9.

Having demonstrated a simple *flow-through*, single-cycle machine, we look at the multi-cycle machine that increases performance by pipelining where, at any one time, a processor may be executing four or more instructions, each in a different state of completion. Pipelining is entirely analogous to the production line in automobile manufacturing. Although pipelining is associated with RISC processors, it was widely used long before the advent of RISC.

Moreover, Intel employed pipelining most aggressively to increase the performance of its Pentium processors (although more recent Intel processors have shorter pipelines than some of the Pentium processors). We also provide a short discussion of the ARM's internal pipelining.

The remainder of this chapter examines how computer designers overcome the effects of branches in pipelined processors. When a branch instruction is encountered, a jump is made to a different place in the program (assuming the branch is taken). Consequently, instructions following the branch that have been brought into the pipeline and have started execution have to be rejected. In the next chapter we look at *superscalar processors* that are able to execute more than one instruction per clock cycle by using parallel processing units.

Regular Instruction Sets

A regular instruction format indicates that there is a strong correlation between the bit pattern of the instruction and the operation to be performed. Typically, there are few different instruction classes (e.g., load, store, register-to-register, and branch) and all instructions within a class have the same format. Moreover, the instruction format always uses the same bits to encode the source and destination registers. Such a regular instruction format makes it easy to derive the control signals that implement an operation directly from the op-code. CISC processors have variable-length instructions (e.g., from 16 to 80 bits), and a myriad of different encodings — so many that writing assemblers for processors like the 68K was not a trivial task. Because of the CISC's complicated instruction encoding, it requires complex logic or a look-up table to convert the op-code into the signals required to implement the instruction.

7.1   The Generic Digital Processor

We are going to return to *first principles* and begin with the structure of a generic general-purpose digital machine that can be used to execute any instruction set; that is, there is no relationship between the structure of the machine and the instruction encoding. Figure 7.1 describes a very simple computer with three buses and two general-purpose registers, R0 and R1. This structure is only a modest enhancement to the type of bus and register arrangement we described in Chapter 2 and can be extended to any number of registers and buses. These are the only two registers that are *user visible* in machine-level instructions; all other registers are invisible to the programmer. The CPU has a program counter (PC) that contains the address of the next instruction to be executed, a memory address register (MAR) that holds the address of the data element being read from memory or written into memory, a memory buffer register (MBR) that contains the actual data to be written into memory or read from memory, and an instruction register (IR). The instruction register contains both the op-code of the instruction read from memory and the address of the operand required by the instruction.

We have chosen this particular structure because it can implement both register-to-register and memory-to-register/register-to-memory instructions.

The CPU of Figure 7.1 has three buses A, B, and C. All registers receive data from the ALU via bus A. All data has to pass through the ALU to bus A. All registers (except the MAR) can drive data onto bus B via their appropriate tristate driver, but only the MBR and general-purpose registers can put data onto bus C.

The ALU uses three control inputs F_2, F_1, F_0 to define the eight ALU functions in Table 7.1. For example, if $F_2, F_1, F_0 = 0, 1, 1$, the A output of the ALU is $C + 1$. We have provided a very small set of ALU operations to keep the system simple, adding an extra function bit would give us 16 different operations, allowing us to add multiplication and logical operations. Some ALU functions are symmetric; for example, code 000 copies the data on bus B to bus A, whereas code 001 copies the data on bus C to bus A. There is a lack of symmetry with respect to subtraction; function code 111 subtracts the data on bus B from the data on bus C. There is no code for the inverse operation *subtract bus C from B*.

The ALU has a *condition code register*, CCR, whose V-bit is set if arithmetic overflow occurs, whose N-bit is set if the result is negative, whose C-bit is set if a carry is generated, and whose Z-bit is set if a zero result is recorded. In this example, only the Z-bit is used.

FIGURE 7.1   Structure of a primitive computer in terms of buses and functional units

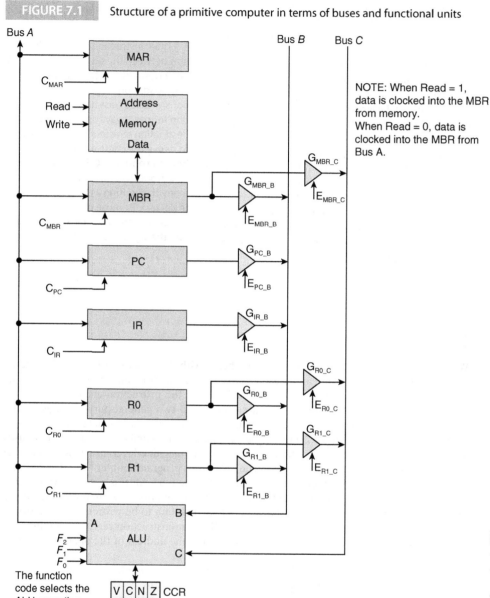

NOTE: When Read = 1, data is clocked into the MBR from memory.
When Read = 0, data is clocked into the MBR from Bus A.

TABLE 7.1   ALU Function Codes for the Computer of Figure 7.1 (A, B, and C refer to the buses)

F_2	F_1	F_0	Operation	
0	0	0	Copy bus B to Bus A	$A = B$
0	0	1	Copy bus C to Bus A	$A = C$
0	1	0	Copy bus $B + 1$ to bus A	$A = B + 1$
0	1	1	Copy bus $C + 1$ to bus A	$A = C + 1$
1	0	0	Copy bus $B - 1$ to bus A	$A = B - 1$
1	0	1	Copy bus $C - 1$ to bus A	$A = C - 1$
1	1	0	Copy bus B + bus C to bus A	$A = B + C$
1	1	1	Copy bus C − bus B to bus A	$A = C - B$

© Cengage Learning 2014

© Cengage Learning 2014

Table 7.2 defines all the micro-operations that can be applied to this structure. These are divided into memory operations, clocks (latching registers), tristate control signals, and the ALU function. All machine-level operations can be implemented in terms of sequences of these micro-operations.

TABLE 7.2	Microinstruction Set for the CPU of Figure 7.1	
Operation Class	Operation (Mnemonic)	Operation (Name)
Memory	Read	Read from MBR memory
Memory	Write	Write from memory to MBR
Clock	C_{MAR}	Clock MAR
Clock	C_{MBR}	Clock MBR
Clock	C_{PC}	Clock PC
Clock	C_{IR}	Clock IR
Clock	C_{R0}	Clock register R0
Clock	C_{R1}	Clock register R1
Enable	E_{MBR_B}	Enable MBR onto B bus
Enable	E_{PC_B}	Enable PC onto B bus
Enable	E_{IR_B}	Enable IR onto B bus
Enable	E_{R0_B}	Enable R0 onto B bus
Enable	E_{R1_B}	Enable R1 onto B bus
Enable	E_{MBR_C}	Enable MBR onto C bus
Enable	E_{R0_C}	Enable R0 onto C bus
Enable	E_{R1_C}	Enable R1 onto C bus
ALU function	F_2,F_1,F_0	Set ALU function

© Cengage Learning 2014

Table 7.3 defines a very basic machine-level instruction set for this computer. We are going to create a register-to-memory ISA. The instruction set includes: data transfer, arithmetic, and program flow-control operations. Note the lack of symmetry; we can add R0 to R1 and put the result in R1, but not add R1 to R0 and put the result in R0. Similarly, we can subtract R0 from R1 but not vice versa. Such restrictions are typical of real computers because the number of bits in an op-code is limited. We have also restricted addressing to absolute addresses (i.e., indexing or pointer-based addressing is not supported). Similarly, operations with literal operands have not been implemented.

The next step is to demonstrate how the machine-level instructions of Table 7.3 are interpreted. In what follows, we assume that at the start of each instruction execution, the op-code is in the instruction register and that the op-code contains two fields: the instruction to be executed and the address of the operand in memory accessed by the instruction or the branch address. The number of bits in an operand field must be less than the number of bits in the MBR (because the IR holds both an op-code and operand).

TABLE 7.3	Machine-Level Instructions for the CPU of Figure 7.1	
Op-Code	Name	Operation (Defined in RTL)
0 0 0	LOAD **R0**,M	$[R0] \leftarrow [M]$
0 0 1	LOAD **R1**,M	$[R1] \leftarrow [M]$
0 1 0	STORE **M**,R0	$[M] \leftarrow [R0]$
0 1 1	STORE **M**,R1	$[M] \leftarrow [R1]$
1 0 0	ADD **R1**,R0	$[R1] \leftarrow [R1] + [R0]$
1 0 1	SUB **R1**,R0	$[R1] \leftarrow [R1] - [R0]$
1 1 0	BRA T	$[PC] \leftarrow T$
1 1 1	BEQ T	IF $[Z] = 1$ THEN $[PC] \leftarrow T$

© Cengage Learning 2014

7.1.1   The Microprogram

Let's begin by implementing the machine-level operation LOAD R0,M defined in RTL as $[R0] \leftarrow [M]$. This instruction requires that the contents of memory location M be accessed and copied to register R0. Because the address M is in the instruction register, gate G_{IR_B} must be enabled to put the address on bus B (there is no alternative bus) and route it to the memory address register, MAR, that holds the address of the element being accessed in memory.

Because the MAR is connected only to the A bus, the value of the address M has to be moved from the B bus via the ALU to route $A \leftarrow B$ using the ALU bypass function $F_2,F_1,F_0 = 0,0,0$. Once address M is on the A bus, it can be clocked into the MAR to point to the next location to be accessed. The memory control signal can then be set to read and the memory buffer register, MBR, clocked to capture the data. A similar sequence of actions now has to be carried out to copy the data in the MBR to R0 (enable $G_{MBR\_B}, F_2,F_1,F_0 = 0,0,0$, clock R0). We can express these actions as follows, where each line represents a clock cycle.

$E_{IR\_B} = 1, F_2,F_1,F_0 = 0,0,0, C_{MAR}$;Copy IR to MAR

$Read = 1, C_{MBR}$;Read memory and copy data to MBR

$E_{MBR\_B} = 1, F_2,F_1,F_0 = 0,0,0, C_{R0}$;Copy MBR to R0

These operations constitute a *microprogram* that is stored in a read-only memory called a *control store*. A microprogrammed control unit reads the microinstructions from the control store and uses them to interpret machine level instructions. The effect of first microinstruction is illustrated in the box.

TABLE 7.4   Micro-Operations that Interpret the Instruction Set of Table 7.3

Op-code	Operation	Micro-operations	Phase	Explanation
	Fetch cycle	$E_{PC\_B} = 1, F_2,F_1,F_0 = 0,0,0, C_{MAR}$	T_0	Copy PC to MAR
		$E_{PC\_B} = 1, F_2,F_1,F_0 = 0,1,0, C_{PC}$	T_1	Increment PC
		$Read = 1, C_{MBR}$	T_2	Read memory to MBR
		$E_{MBR\_B} = 1, F_2,F_1,F_0 = 0,0,0, C_{IR}$	T_3	Copy op-code in MBR to IR
000	LOAD R0,M	$E_{IR\_B} = 1, F_2,F_1,F_0 = 0,0,0, C_{MAR}$	T_0	Copy IR to MAR
		$Read = 1, C_{MBR}$	T_1	Read operand from memory
		$E_{MBR\_B} = 1, F_2,F_1,F_0 = 0,0,0, C_{R0}$	T_2	Copy MBR to R0
001	LOAD R1,M	$E_{IR\_B} = 1, F_2,F_1,F_0 = 0,0,0, C_{MAR}$	T_0	Copy IR to MAR
		$Read = 1, C_{MBR}$	T_1	Read operand from memory
		$E_{MBR\_B} = 1, F_2,F_1,F_0 = 0,0,0, C_{R1}$	T_2	Copy MBR to R1
010	STORE M,R0	$E_{R0\_B} = 1, F_2,F_1,F_0 = 0,0,0, C_{MBR}$	T_0	Copy R0 to MBR
		$E_{IR\_B} = 1, F_2,F_1,F_0 = 0,0,0, C_{MAR}$	T_1	Copy IR to MAR
		$Write = 1$	T_2	Write R0 (in MBR) to memory
011	STORE M,R1	$E_{R1\_B} = 1, F_2,F_1,F_0 = 0,0,0, C_{MBR}$	T_0	Copy R1 to MBR
		$E_{IR\_B} = 1, F_2,F_1,F_0 = 0,0,0, C_{MAR}$	T_1	Copy IR to MAR
		$Write = 1$	T_2	Write R1 (in MBR) to memory
100	ADD R1,R0	$E_{R0\_B} = 1, E_{R1\_C} = 1, F_2,F_1,F_0 = 1,1,0, C_{R1}$	T_0 T_1	Send R0, R1 to ALU, add and latch result in R1
101	SUB R1,R0	$E_{R0\_B} = 1, E_{R1\_C} = 1, F_2,F_1,F_0 = 1,1,1, C_{R1}$	T_0 T_1	Send R0, R1 to ALU, subtract and latch result in R1
110	BRA T	$E_{IR\_B} = 1, F_2,F_1,F_0 = 0,0,0, C_{PC}$	T_0	Copy address in IR to PC
111	BEQ T	$E_{IR\_B} = 1, F_2,F_1,F_0 = 0,0,0,$ IF Z = 1 THEN C_{PC}	T_0	Put address in IR on bus and latch into PC if Z-bit = 1

Microprogramming the Detail

This figure is a copy of Figure 7.1 except that the paths that implement the first microoperation of the load instruction are highlighted. This micro-operation copies the instruction address in the instruction register into the memory address register, where it is used to access the actual instruction from memory. In order to do this, the address must be gated from the IR onto Bus *B*. Then the ALU function is set to pass through (input P to output R), and finally the memory address register is clocked to capture the address on Bus *A*.

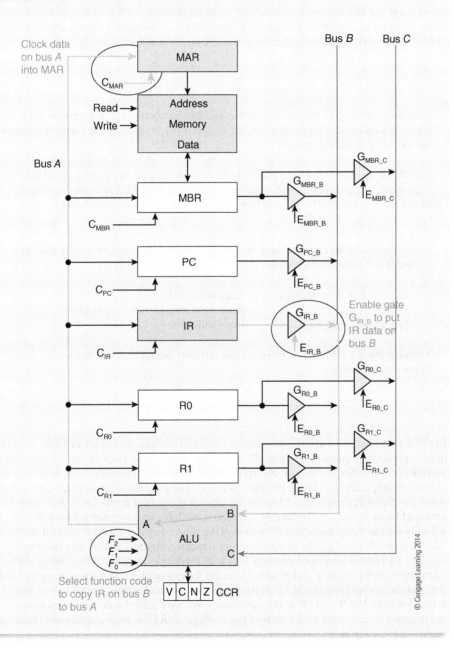

Table 7.4 lists the micro-operations required to interpret each of the instructions in Table 7.2 plus the micro-operations corresponding to a fetch cycle. As you can see, each micro-program is executed in one to four cycles numbered T_0 to T_3. This table reveals some interesting features of the computer structure of Figure 7.1. Some operations are longer and more complex than others. Is this because these operations are intrinsically more complex, because we are not writing the best sequence of micro-operations, or because the structure of the system in Figure 7.1 is not optimum for this instruction set?

Microprogramming was popular for a period in the 1980s when AMD produced a set of *bit-slice* components that allowed anyone to create an architecture of their own choosing (no matter how complex). However, the growth of the microprocessor as a commodity component and the use of low-cost standard operating systems killed off the ad-hoc computer. Computer manufacturers have expanded instruction sets by adding short vector multimedia extensions, ensuring that current microprocessor families will have a long life.

Example of Microprogramming

Microprogramming permits an infinitely extensible instruction set. Suppose we require a machine level instruction to carry out [R0] ← [[R0] + [R1]] that reads the memory location pointed at by R0 plus R1 into R0.

How do we do this? We need to add R0 to R1 and use the sum as a pointer. As R0 is going to be modified by this operation, it makes sense to add R1 to R0. We move R0 to bus B, R1 to bus C, set the ALU to add and then latch the ALU output into R0; that is, $E_{R0\_B} = 1$, $E_{R1\_C} = 1$, $F_2, F_1, F_0 = 1,1,0$, C_{R0}.

Having generated an address, we need to copy it to the MAR, perform a read, and copy the result in the MBR to R0; that is,

$E_{R0\_B} = 1$, $F_2, F_1, F_0 = 0,0,0$, C_{MAR} ;Copy R0 to MAR
Read = 1, C_{MBR} ;Read memory and copy data to MBR
$E_{MBR\_B} = 1$, $F_2, F_1, F_0 = 0,0,0$, C_{R0} ;Copy MBR to R0

Although this is a primitive architecture, we can carry out arbitrarily complex operations by microprogramming.

Modifying the Processor Organization

Let's take another look at the computer of Figure 7.1, its instruction set in Table 7.2, and the micro-operation sequences of Table 7.3 and demonstrate how both the processor structure and the corresponding microcode can be modified to achieve objectives such as greater speed or greater simplicity (i.e., lower production costs). The purpose of this exercise is to demonstrate the flexibility inherent in the design of such a bus-based processor. In Figure 7.1, we arranged buses B and C to be largely symmetric, we have an asymmetric instruction set (i.e., R1 is always the destination, and R0 is always the source register for addition and subtraction). We can go further and create an even simpler structure, Figure 7.2, with only two buses.

The instruction execution phase begins with [MAR] ← [IR]. Including this as part of the instruction fetch sequence makes sense, because we can implement it for nothing by latching the operand address into the MAR at the same time it is latched into the IR (all we have to do is to clock both MAR and IR at the same time). We have added a dedicated incrementer for the program counter and have removed bus C. Register R1 has been permanently hard-wired to the ALU's C input.

The memory output is connected to the IR via multiplexer M_MBR so that an instruction can be captured directly from memory. A temporary register, T, has been added to support more complex operations. This register is *architecturally invisible* and is not part of the processor's ISA; that is, it can be used to implement operations that require temporary storage but the programmer cannot explicitly access this register.

The ALU's pass-through function has been dropped to save one ALU control signal and Table 7.5 gives the new ALU function set. Three multiplexers have been added: one to allow bus A to receive data from the ALU or from bus B, one to allow the PC to receive data from bus A or from an incrementer, and one to allow the MBR to receive from the memory or from bus B.

In Figure 7.2, a fetch cycle is defined below. Note that we can shorten tristate enables from E_{PC_B} to E_{PC} because there is only one destination for all tristate gates. Moreover, the program counter is incremented by 1. In a real machine it would be incremented by the size of an instruction in bytes (typically 4).

TABLE 7.5

Simplified ALU Control Functions

F_2	F_1	Operation
0	0	$A = B + 1$
0	1	$A = B - 1$
1	0	$A = B + C$
1	1	$A = B - C$

© Cengage Learning 2014

RTL	Micro-operations	
[MAR] ← [PC]	$E_{PC} = 1, M\_ALU=1, C_{MAR}$	;PC to bus B to bus A to MAR
[PC] ← [PC] + 1	$M\_PC = 1, C_{PC}$	;PC to incrementer to PC via multiplexer
[IR] ← [M[MAR]]	Read=1, M_MBR = 0, C_{IR}, C_{MBR}	;Read data at address in MAR latch in IR and MBR

We have clocked the instruction into both the instruction register and memory buffer register. Since the PC incrementer is permanently wired to the PC, we were able to generate the next address by M_PC = 1, C_{PC}. If you look at the CPU structure, you will notice that incrementing the PC and reading memory can take place in parallel as these two operations require no common resources. Therefore, we can reduce the fetch phase to:

$E_{PC} = 1, M\_ALU= 1, C_{MAR}$;PC to bus B to bus A to MAR
Read=1, M_MBR = 0, $C_{IR}, C_{MBR}, M\_PC = 1, C_{PC}$;Read instruction, latch in IR, MBR and increment PC

Table 7.6 gives the table of 17 micro-operations for the computer of Figure 7.2. All actions are completed in either two or three cycles.

The last line of Table 7.6 contains the value Z in the clock PC column, C_{PC}. During a conditional branch, the next address (i.e., branch target address from the instruction register) is

TABLE 7.6   Micro-Operations Required to Interpret the Instruction Set of Table 7.3

Operation		Phase	Gate Enables and Multiplexer Controls								Clock Registers						ALU	Memory
			E_{MBR}	E_{PC}	E_{IR}	E_{R0}	E_{R1}	M_PC	M_MBR	M_ALU	C_{MAR}	C_{MBR}	C_{PC}	C_{IR}	C_{R0}	C_{R1}	$F_1 F_0$	R/W
Fetch		T_0	0	1	0	0	0	0	0	1	1	0	0	0	0	0	x x	1
(read instruction)		T_1	0	0	0	0	0	1	0	0	0	1	1	1	0	0	x x	1
LOAD	**R0**,M	T_0	0	0	1	0	0	0	0	1	1	0	0	0	0	0	x x	1
		T_1	0	0	0	0	0	0	0	0	0	1	0	0	0	0	x x	1
		T_2	1	0	0	0	0	0	0	1	0	0	0	0	1	0	x x	1
LOAD	**R1**,M	T_0	0	0	1	0	0	0	0	1	1	0	0	0	0	0	x x	1
		T_1	0	0	0	0	0	0	0	0	0	0	0	0	0	0	x x	1
		T_2	1	0	0	0	0	0	0	1	0	0	0	0	0	1	x x	1
STORE	R0,**M**	T_0	0	0	1	0	0	0	0	1	1	0	0	0	0	0	x x	1
		T_1	0	0	0	1	0	0	0	1	0	1	0	0	0	0	x x	1
		T_2	0	0	0	0	0	0	0	0	0	0	0	0	0	0	x x	0
STORE	R1,**M**	T_0	0	0	1	0	0	0	0	1	1	0	0	0	0	0	x x	1
		T_1	0	0	0	0	1	0	0	1	0	1	0	0	0	0	x x	1
		T_2	0	0	0	0	0	0	0	0	0	0	0	0	0	0	x x	0
ADD	**R1**,R0	T_0	0	0	0	1	0	0	0	0	0	0	0	0	0	1	1 0	1
SUB	**R1**,R0	T_0	0	0	0	1	0	0	0	0	0	0	0	0	0	1	1 1	1
BRA	T	T_0	0	0	1	0	0	0	0	1	0	0	1	0	0	0	x x	1
BEQ	T	T_0	0	0	1	0	0	0	0	1	0	0	Z	0	0	0	x x	1

© Cengage Learning 2014

FIGURE 7.2   Alternative CPU structure

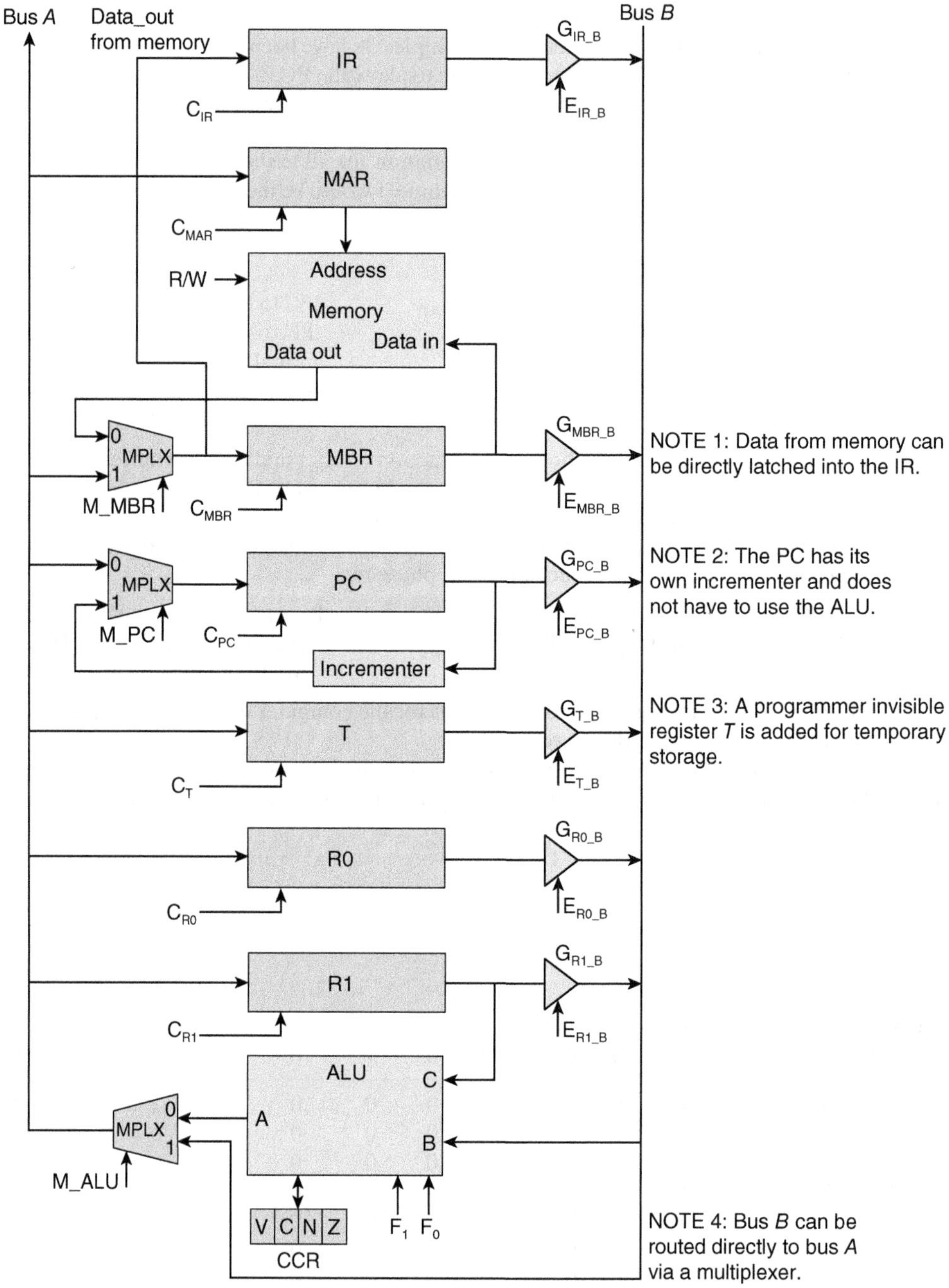

NOTE 1: Data from memory can be directly latched into the IR.

NOTE 2: The PC has its own incrementer and does not have to use the ALU.

NOTE 3: A programmer invisible register T is added for temporary storage.

NOTE 4: Bus B can be routed directly to bus A via a multiplexer.

© Cengage Learning 2014

placed on the bus to the program counter. If the Z-bit is 1, the PC is clocked and it receives a new address.

Figure 7.3 provides yet another variation on the structure of Figures 7.1 and 7.2. This is the deluxe version with four buses, two at the register inputs and two at the register outputs. We have had to increase the number of multiplexers to allow registers to receive data from bus A or bus B. The new structure does not add new instructions. Any machine level instruction can be synthesized with the arrangements of Figures 7.1 and 7.2. What we have done is to increase the level of performance (assuming all other factors are equal) by allowing parallel

FIGURE 7.3   A four bus CPU structure

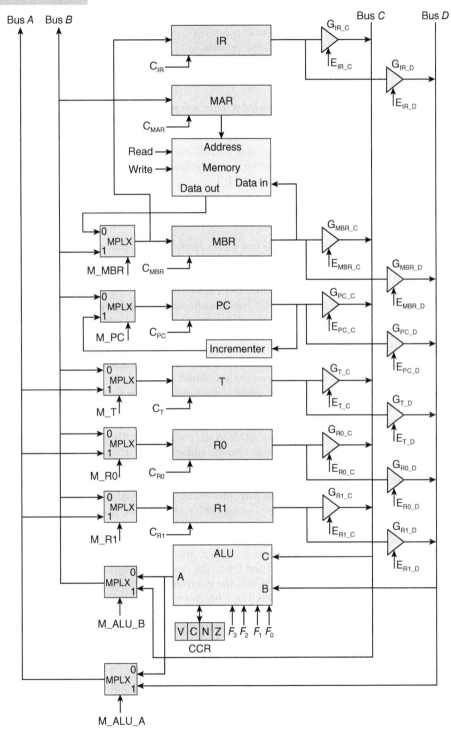

© Cengage Learning 2014

operations. For example, you could swap registers R0 and R1 by routing R0 to bus C and R1 to bus D, copying bus C to bus B and bus D to bus A, putting bus A onto R0 and bus B onto R1 and then clocking both registers. The purpose of this diagram is to demonstrate that a CPU can be extended by adding the appropriate resources.

Suppose you wished to implement the operations

$$[R0] \leftarrow [R0] + [R1]$$
$$[T] \leftarrow [R1]$$

The simple architecture of Figure 7.1 would require several microoperations in order to carry this out. The structure of Figure 7.3 could do it in one cycle because we can put R0 on bus C and R1 on bus D. The ALU is set to add and the sum on output A(B,C) is passed through multiplexer M_ALU_B to bus B where it is latched into register R0 via multiplexer M_R0. Because R1 is already on bus D, it can be routed to bus A via multiplexer M_ALU_A and then clocked into register T via multiplexer M_T.

7.1.2   Generating the Microoperations

The next step is to ask how we might run a microprogram to generate the control signals required to interpret machine-level operations; that is, how do we turn an op-code into a sequence of 17-bit values for each possible op-code and fetch cycle.

A versatile and regular way of generating control signals is to use a microprogrammed control unit, a technique invented in 1951 at Manchester University in the UK by Maurice Wilkes. Figure 7.4 provides the structure of a microprogrammed control unit. This should give you a sense of déjà vu, because it's *a computer within a computer*, albeit a specialized computer with a wide instruction word. Because it lacks an ALU and register set, you could equally think of a microprogrammed control unit as a very sophisticated *sequencer*. A microprogrammed control unit runs a program whose input is the machine-level op-code to be executed and whose output is the bus enables, multiplexer controls, clocks, and signals that control the processor.

Let's start at the beginning. The bit pattern (op-code) of the next machine-level instruction to be executed is fed to a *mapping ROM* in the control unit. This mapping ROM is just a look-up table that translates an op-code into the address of the first microinstruction of the microcode that interprets the machine-level operation. The use of a mapping ROM frees the designer from having to worry about instruction encoding, but at a price. The mapping ROM adds a decoding delay in the instruction path. Moreover, you can add new instructions to your computer at any time, simply by writing the appropriate microcode, putting it into the microprogram ROM (control store), and then using an unallocated op-code in the mapping ROM to point to the new microcode.

Suppose that the microprogram counter now contains the address of the first line of the microprogram that executes an instruction fetch. The corresponding microcode is looked up in the microprogram ROM and loaded into the microinstruction register. In Figure 7.4, the microinstruction register has three fields. The *CPU control field* provides the bits that control the registers, memory, ALU, and tri-state gates of the computer. The *next microinstruction address* field of the microinstruction register provides the address of the next microinstruction to be executed if a branch is to be made. The *condition code select* field selects which condition code bit is to be tested to determine whether a branch is to be made (the conditions include branch always and branch never).

If the condition selected is *branch never* and the microprogram counter generates the next sequential address. If, for example, the *condition select field* selects the Z-bit, it is sent to the microprogram counter. If the Z-bit is true, the contents of the *next microinstruction address* field is routed to the microprogram counter and loaded to force a branch in the microprogram. If the Z-bit is false, the microprogram counter is not reloaded with a new value and execution of the microprogram continues normally.

The microprogram controlled unit of Figure 7.4 communicates with the CPU via three ports: the op-code from the machine-level instruction register, the condition code from the

FIGURE 7.4 The microprogrammed control unit

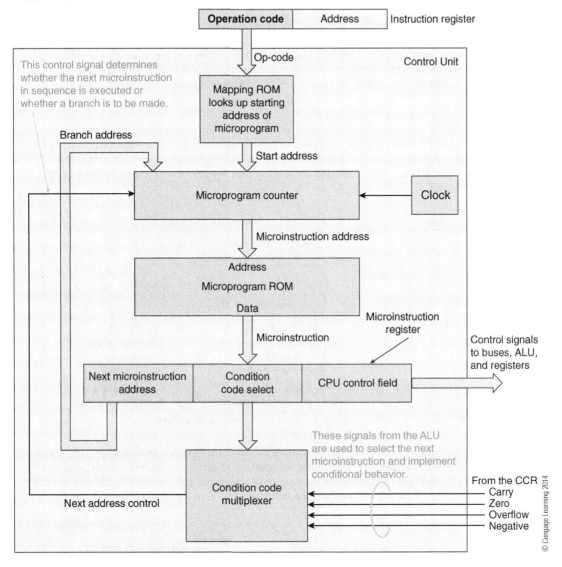

CCR, and the microprogram control signals that go to all parts of the CPU to control the flow of data.

Microprogramming was popular in the 1980s, when main store was expensive and it was cost-effective to execute dense code by interpreting complex instructions with microprograms. As the cost of main store has plummeted and its speed has increased, interpreting machine-level instructions in microcode is no longer such an attractive idea. Microprogramming is still used by processors such as the Pentium to implement some more complex instructions.

In the next section, we look at the structure of a rather different computer, one whose architecture provides a register-to-register (load and store) instruction set. We will see that, although we have stated that architecture and organization are orthogonal to each other, there is sometimes a natural affinity between architecture and organization. We are going to describe the architecture and organization of a RISC-like machine (i.e., such as ARM or MIPS), beginning with the description of a single-cycle implementation and later introducing pipelining and multicycle processor organization.

Vertical and Horizontal Microprogramming

The microinstruction field may be very long because it includes all the tristate bus drivers, register clocks, ALU function inputs, memory controls, and so on. It is not unusual for a microinstruction to be over 100 bits wide. There are two approaches to dealing with microinstructions. One is called *horizontal* encoding and the other *vertical* encoding. In a system with horizontal encoding, the bits of the microinstruction directly control the CPU. For example, if there are eight registers that can put data on a bus, there will be eight enable signals in the microinstruction, one for each register.

In vertical encoding, a secondary decoder is required between the microinstruction register and the actual clock/enable/control signals. For example, if there are eight registers that can put data on a bus, the microinstruction register uses three bits to select one of eight registers and a decoder converts the 3-bit register selection code into one-of-eight select signals; that is, additional logic must be used to convert the 3-bit register select code into one-of-eight enable signals, one for each register.

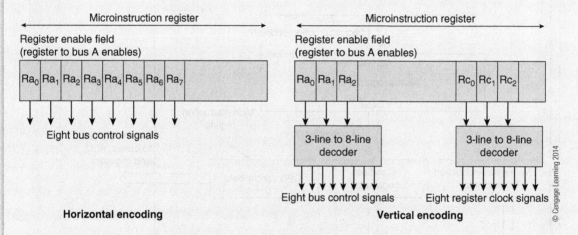

Horizontal encoding   **Vertical encoding**

© Cengage Learning 2014

A vertically encoded instruction register is slower than a horizontally encoded instruction register because additional delays are required for the decoding. However, in a horizontally encoded microinstruction, operations can be carried out in parallel (e.g., you can select multiple registers as the destination of an operand). This is not possible with vertical encoding where, say, a register select field defines one specific register. A real system might use a mixture of vertical and horizontal encoding.

Nanoprogramming

The microprogram takes up real-estate on the silicon chip that could be used by other circuits (e.g., cache memory). Nanoprogramming provides a means of compressing microprograms by compressing the microcode. Suppose a microinstruction is 100 bits long. That means that there are 2^{100} possible microinstructions, an astronomically large number. In a real computer the total number of different microinstructions might be tiny. Nanoprogramming takes the actual microinstructions used in the control store and puts them in a small memory. The code in the control store is then replaced by pointers to the actual microinstructions. This approach reduces the number of bits in the control store at the cost of an extra stage in the microinstruction translation process.

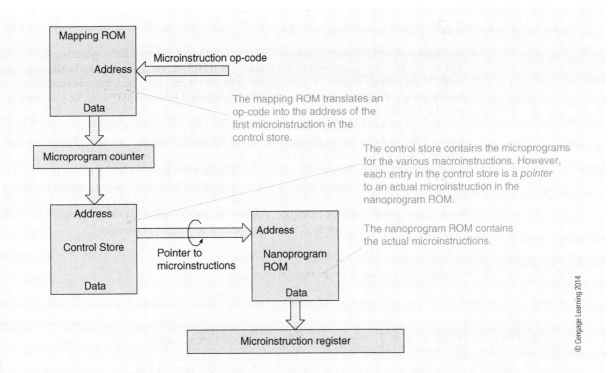

The mapping ROM translates an op-code into the address of the first microinstruction in the control store.

The control store contains the microprograms for the various macroinstructions. However, each entry in the control store is a *pointer* to an actual microinstruction in the nanoprogram ROM.

The nanoprogram ROM contains the actual microinstructions.

© Cengage Learning 2014

Consider a hypothetical example. Suppose a computer has an 8-bit op-code and 200 unique machine-level instructions, each requiring four 150-bit-wide microinstructions. It is observed that there are only 120 unique microinstructions. Let's calculate two values: (a) the number of bits of control store required for conventional microprogramming and (b) the number of bits required if nanoprogramming is used.

Solution (a)

Two hundred machine-level instructions using four microinstructions require $200 \times 4 = 800$ microinstructions.

The mapping ROM uses an 8-bit op-code to select one of 800 microinstructions which requires a 256 word $\times$ 10-bit ROM (i.e., 2,560 bits). The mapping ROM is ten bits wide because $2^{10} = 1,024$ which is the next power of 2 greater than 800.

The control store requires 800 words $\times$ 150 bits $= 120,000$ bits. The total number of bits is the control store plus the mapping ROM $= 120,000 + 2,560 = 122,560$ bits.

Solution (b)

There are 120 unique microinstructions each of which is 150 bits wide in the nanoprogram ROM; this requires a storage capacity of $120 \times 150 = 18,000$ bits.

There are 120 unique microinstructions that require a 7-bit address to select one of them ($2^7 = 128$). The control store contains 800 microinstructions, each requiring a 7-bit pointer which takes $800 \times 7 = 5,600$ bits.

The required storage using nanoprogramming is 2,560 bits (mapping ROM) + 5,600 bits (control store) + 18,000 bits (nanoprogram store) $= 26,160$ bits that reduces the storage by about 80%.

This example demonstrates that nanoprogramming is able to reduce total control store requirements from 122,560 bits to 26,160 bits at the cost of the time taken to access the nanoprogram store.

7.2 RISC Organization

We begin with a gentle introduction to the organization of a simple RISC processor to provide a firm basis for later discussion of the features of RISC architectures that make them so amenable to pipelining. Figure 7.5 presents the instruction encoding of a hypothetical processor that is not too dissimilar to commercial machines like ARM and MIPS. We've just made it a little bit simpler.

The six most-significant bits of the op-code, bits 31-26, define the instruction, one bit for each instruction; that is, instructions are of the form 100000..., 010000, 001000..., etc. Of course, this is wasteful of bits, but it does simplify instruction encoding. The instructions are: Load (L), Store (S), Branch (B), Jump (J), Register-to-register (R2R), and Reserved (X). Bit 25 is the *literal bit* and is labeled #. This indicates whether the second source register or the literal field is to be used in the instruction. Bits 24 to 21 provide a parameter that can be used by any instruction (for example, to define the type of register-to-register operation (add, subtract, AND, OR, etc), or to define the condition in a branch operation).

FIGURE 7.5   The instruction format of a hypothetical computer

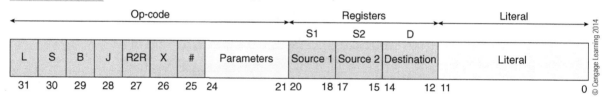

The three register select fields S1, S2, and D in bits 20 to 12 choose the registers that are going to take part in the current operation. We have used 3-bit register fields permitting eight registers (i.e., $2^3 = 8$). Finally, the 12-bit field in bits 11 to 0 provides a literal that can be used by instructions. The problem set at the end of this section will ask you to discuss this arrangement and to suggest alternatives.

Some of the typical operations that can be supported by this instruction set are

Assembly Form	RTL Definition	
LDR **D**,(S1,S2)	[D] ← [M([S1] + [S2])]	Register indirect with double indexing
LDR **D**,(S1,#L)	[D] ← [M([S1] + L)]	Register indirect with literal offset
STR **(S1,#L)**,S2	[M([S1] + L)] ← [S2]	Register indirect with literal offset
ADD **D**,S1,S2	[D] ← [S1] + [S2]	Register to register
ADD **D**,S1,#L	[D] ← [S1] + L	Register to register with literal operand
BEQ L	[PC] ← [PC] + 4 + L	This is a relative conditional branch
JMP (S1,#L)	[PC] ← [S1] + L	This is an unconditional jump to a computed address

As you can see, these instructions cover many of the instruction types provided by the ARM (although we have not included conditional execution—another exercise for the student).

Figure 7.6 illustrates one of the most important components of modern processors, the *multiport memory* that contains the general-purpose register file. The bus-based computer of Figure 7.1 can implement any arbitrary number of registers simply by wiring them to a bus via suitable tristate gates. However, what is needed in high-performance computers is an array of fast registers that can be accessed in parallel. Dedicated high-speed register files have been designed that can be incorporated into VLSI chips or bought as stand-alone components. These devices are also called *multiport memories* because they have several access ports that can be used in parallel. For example, the system of Figure 7.6 has three address ports, which

FIGURE 7.6   The multiport register file

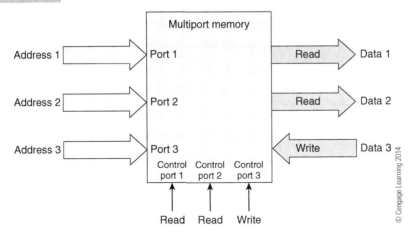

means that three registers can be addressed at the same time. You can read two source registers simultaneously and write to the destination register. With such a memory, you can implement an operation like ADD D, S1, S2 that adds the contents of register S1 to register S2 and puts the result in register D in one clock cycle.

We are now going to construct a processor by beginning with the basic structure and adding features until we have a complete processor. Let's call this a *flow-through* processor because you can imagine an instruction flowing through it from the program counter on the left to the data memory on the right (this metaphor will help when we introduce pipelining). Figure 7.7 describes a simple processor that uses a multiport memory as a register file. We have assumed a three-register-address structure typical of RISC processors like MIPS and the ARM. This is an incomplete system as it cannot yet implement literal operands or flow control (unconditional and conditional branches, subroutine calls and returns). It can implement register-to-register, load, and store operations.

Assume that, initially, the program counter (PC) contains the address of the next instruction to be executed. This address is fed to the instruction memory and back to the PC via an incrementer where 4 is added to point to the next instruction (we assume a 32-bit instruction and byte-addressed memory).

The instruction is fed to the instruction memory that provides the 32-bit op-code corresponding to the current instruction. In Figure 7.7 the three register address fields are sent to the register file to read source registers S1 and S2. The contents of these registers are fed

FIGURE 7.7   Core structure of a single cycle *flow-through* processor

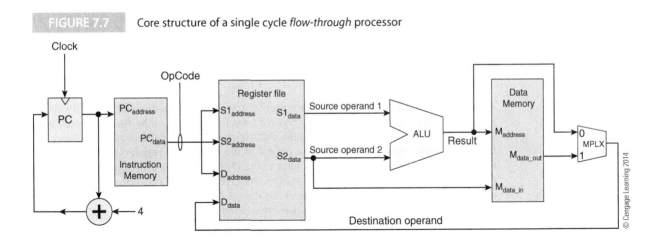

to the ALU where they are used to generate a value in a register-to-register operation (e.g., ADD, AND, SUB, OR), or they can be added together to create an effective address for use in load and store instructions.

If the operation is register-to-register, the result from the ALU is fed back to the register file via the multiplexer to be written in the destination port of the register file. If the operation is a load, the output of the ALU provides the effective address of the operand which is used to access the operand from the data memory, which is fed, via the multiplexer, to the register file. If the operation is a store, the S2 register output of the register file is fed to the data memory's data input. Note that the store operation can't use the S2 source register as an address pointer in this implementation.

Let's now look at a fuller implementation of a load/store processor using the op-code format of Figure 7.5. Figure 7.8 includes two major additions over the structure of Figure 7.7. The first is the ability to deal with a literal in the op-code. Because this literal is only 12 bits wide, it has to be converted to 32 bits before being used by the ALU, which requires a sign-extender block in the path between the literal field in the op-code and the ALU. It's also necessary to place a multiplexer at the input to the ALU to select between the S2 operand from the register file and the literal. The second addition to the basic CPU is the four-way input multiplexer to the PC that selects between one of three paths in order to support branch and jump instructions. We are now going to examine the operation of the processor by looking at how various classes of instruction are executed.

FIGURE 7.8   More detailed structure of a flow-through processor

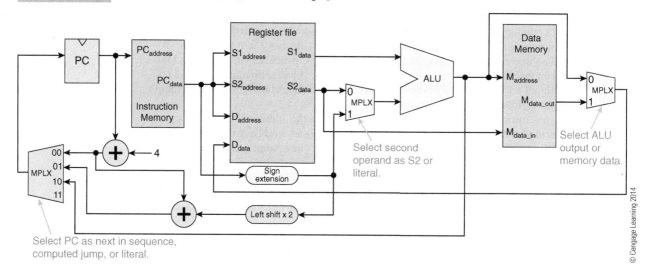

7.2.1   The Register-to-register Data Path

The simplest instructions to execute are register-to-register operations as Figure 7.9 shows. Here, and in successive figures, we have grayed out information paths and functional units not taking part in the operation. For register-to register instructions of the form Operation D,S1,S2, all we need do is to feed the output of the two source registers from the register file to the ALU and to route the result back to the D (destination) write input of the register file. A multiplexer is required at the input of the ALU register file D port because information can be written into the register file from both the ALU and the data memory during a load operation. Note that the PC control block is not needed for this and load/store operations, and the PC input is [PC] + 4 to point to the next instruction.

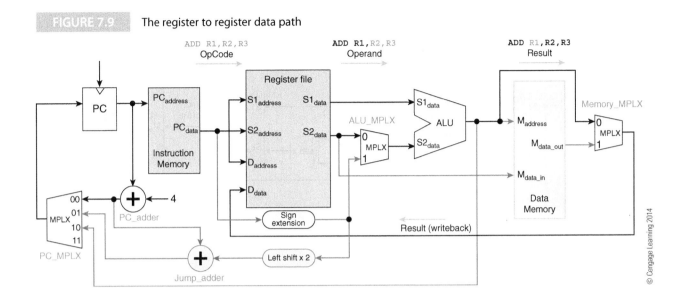

FIGURE 7.9   The register to register data path

Load and Store Operations

Figures 7.10 and 7.11 describe the data flow in load and store operations, respectively. In Figure 7.10, the instruction LDR **D**, (S1,#L) is the address of the operand and is computed as the contents of source register S1 plus a literal L from the instruction; that is, [S1] + L. This effective address from the ALU is used to look up the actual operand in memory, which is routed through the memory multiplexer to the destination port of the register file where it is stored at address D.

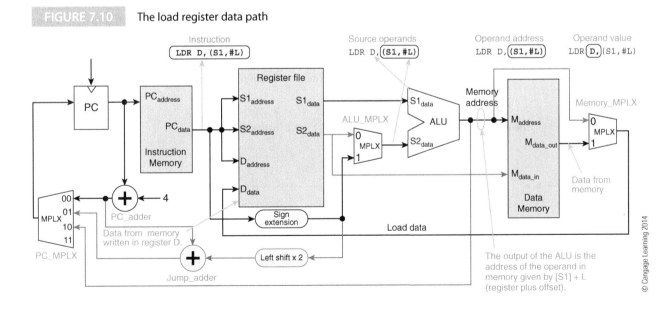

FIGURE 7.10   The load register data path

The arrangement in Figure 7.11 shows the information flow when executing STR S2, (S1,#L). This diagram is similar to that of Figure 7.10; the difference is the direction of data transfer. The memory address [S1] + L is calculated exactly as in a load, but source operand S2 from the register file is fed to the data input port of the memory.

FIGURE 7.11   The store register data path

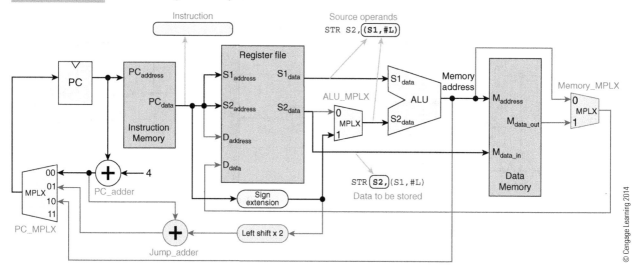

Jump and Branch Operations

Recall that the terms *branch* and *jump* both describe a means of changing the instruction flow in a computer from the normal sequential flow. Generally, the term *branch* implies a jump to a location relative to the current value of the program counter, whereas a *jump* represents a jump to an absolute location. Moreover, the branch is invariably conditional; that is, it takes place if a defined condition exists (e.g., branch on zero, or branch on carry). We first look at the computed jump, JMP (S1,#L), and then the relative conditional branch BEQ L. Figure 7.12 demonstrates the information flow during a jump operation.

The unconditional jump operation executes the next instruction at (S1,#L), which means that we have to get the contents of register S1, add offset L from the instruction, and then jam the result in the program counter. The 12-bit literal offset in the instruction is first sign-extended to 32 bits. The contents of source operand register S1 are read from the register file and added to the offset by the ALU. The result is sent to the PC multiplexer and then to the program counter. Consequently, the next instruction is fetched from memory location [S1] + L.

FIGURE 7.12   The execution of a computed jump JMP (S1,#L) instruction

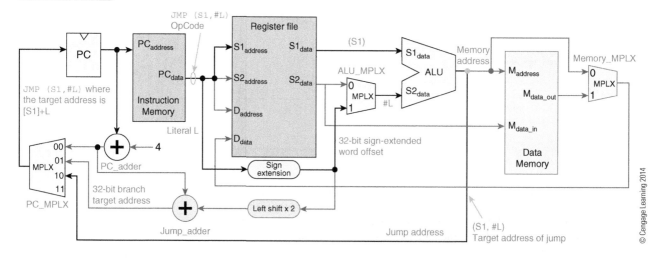

FIGURE 7.13   The execution of a conditional branch instruction.

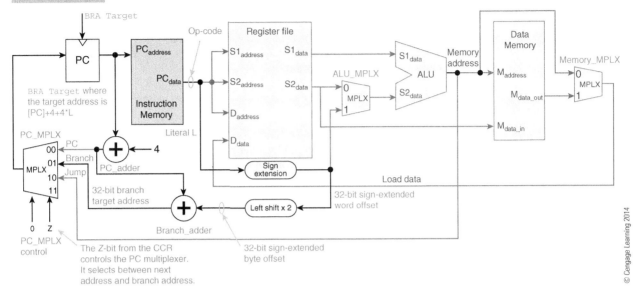

Figure 7.13 demonstrates the operation of the conditional branch instruction, BEQ L. In this case we use program counter relative addressing; that is, the branch target address is the contents of the program counter plus the offset in the instruction.

In Figure 7.13, you can see that the register file, ALU, and memory data paths are not used in implementing a conditional branch. All you need to do is to take the program counter and add the offset in the instruction. However, we have to sign-extend the offset and convert it from a *word* offset to a *byte* offset. The offset must be multiplied by 4 by shifting it two places left because the computer is a 32-bit machine and addresses are separated by 4. This provides a branching range of $-2,043$ to $+2,052$. Note that the PC is incremented by 4 before the offset is added.

All we have to do to implement a conditional branch (i.e., branch on zero) is to use the Z-bit from the computer's condition code register to determine whether the next address is the branch target address or the next address in sequence. In Figure 7.13, one of the bits of the PC multiplexer is wired to the Z-bit and that selects whether the branch is taken.

7.2.2 Controlling the Single-cycle Flow-through Computer

The next step is to examine how the computer we have constructed goes about translating machine-level op-codes into the actions that carry them out. What signals do we need to control the system of Figures 7.8 to 7.13? First, we have to point out that we are looking at the basic principles only and have not included interrupt and exception handling or instruction and data memory control (how instructions and data get into memory). Moreover, in a real computer, both the instruction and data memory are invariably not the computer's actual main memory system but its on-chip cache (we discuss caches in Chapter 9). Here, we are interested only in an overview of the way in which an op-code is interpreted.

Table 7.7 describes the control signals used by the structure of Figure 7.8. As you can see, remarkably few control signals are required. There are four ALU control signals allowing 16 ALU operations, four multiplexer control signals, and a memory write.

TABLE 7.7		Control signals required by the flow-through (single cycle) computer		
Multiplexers			ALU	Memory
ALU	1 (operand S2/L)		4 (Function)	1 (Write register file)
PC	2 (continue, branch, jump)			1(Write data memory)
Memory	1 (ALU/memory data)			

© Cengage Learning 2014

Four multiplexer control signals are necessary because there are three multiplexers; two have one control signal and the PC multiplexer requires two control signals because it has to select one of three paths. The instruction memory needs no control signals because it is fed an address from the PC and it delivers an instruction. We will see in Chapter 9 that instructions are mostly provided by a high-speed instruction cache that performs the same function as the instruction memory in Figure 7.14.

The register file has two source operand read ports: S1 and S2. No control signals are needed for these, because the register file simply supplies the data being specified at the S1 and S2 address inputs (if the register file is not in use, the S1 and S2 outputs are ignored). However, a write control signal is required to write the data into the destination port D during a load operation or a register-to-register operation. Finally, the data memory requires a single write signal when data is written into it during a store operation.

A single clock is needed to latch the new value into the PC at the end of a cycle and to latch data in the register file and data memory.

In Table 7.8, we give the signals required to implement each of the processor's instructions. The box to the right reminds us of the instructions and their definitions. Remember that the 4-bit ALU field can be used to specify one of 16 operations in a register-to-register operation, and it can be used to select one of 16 conditions in a conditional branch operation.

Instruction Set Reminder

LDR D,(S1,S2)	[D] ← [M([S1]+[S2])]
LDR D,(S1,#L)	[D] ← [M([S1]+L)]
STR (S1,#L),S2	[M([S1]+L)] ← [S2]
ADD D,S1,S2	[D] ← [S1]+[S2]
ADD D,S1,#L	[L] ← [S1]+L
BEQ L	[PC] ← [PC]+4+L
JMP (S1,#L)	[PC] ← [S1]+L

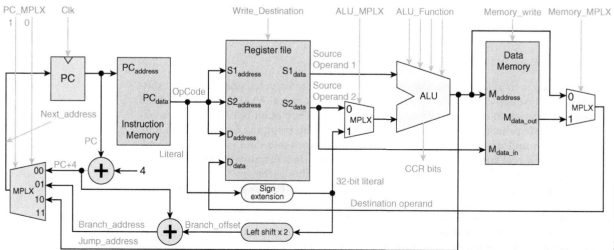

FIGURE 7.14   Adding control signals to the computer

© Cengage Learning 2014

TABLE 7.8	Implementing the Instruction Set

LOAD   `LDR D,(S1,S2) or LDR D,(S1,#L)`

PC_MPLX	= 00	(next instruction)
ALU_MPLX	= #	(select S2 or literal)
Write_Destination	= 1	(write operand into register file)
ALU	= ADD	(to calculate effective operand address of S1 + S2 or S1 + L)
Memory_MPLX	= 1	(read data memory)

STORE   `STR S1,(S2,#L)`

PC_MPLX	= 00	(next instruction)
ALU_MPLX	= 1	(select literal)
Write_Destination	= 0	(read operand from memory)
ALU	= ADD	(to calculate effective operand address of S1 + L)
Memory_Write	= 1	(write to data memory)

REGISTER-TO-REGISTER   `Op D,S1,S2`

PC_MPLX	= 00	(next instruction)
ALU_MPLX	= #	(select S2 or literal)
ALU	= function	(perform the operation defined by the operation field of the instruction)
Memory_MPLX	= 0	(select ALU output for register file)
Write_Destination	= 1	(write operand into register file)

BRANCH   `BRA Target, Bcc Target`

PC_MPLX	= 01	(select branch address)
Branch_Control	= function	(use function bits to select branch condition)

JUMP   `JMP (S1,#L)`

PC_MPLX	= 10	(select computed jump address)
ALU_MPLX	= #	(select S2 or literal)
ALU	= ADD	(to calculate effective jump address of S1 + S2 or S1 + L)

© Cengage Learning 2014

Figure 7.15 demonstrates the additional logic necessary to implement multiple conditional branches. The Z, N, C, and V bits from the ALU are fed to a combinational logic block that uses the branch control from the instruction to select one of 16 branches. The outcome is used to control an AND gate whose output is 1 if both a branch operation has been selected and the chosen condition is true. Otherwise, the output is zero, which forces a next address. We force the PC multiplexer to use the next PC address rather than the branch address whenever a branch is not taken. The PC_ALU multiplexer is coded as follows.

Bit 1	Bit 0	
0	0	Next instruction (default, no branch/jump)
0	1	Branch
1	0	Jump
1	1	Not used

FIGURE 7.15   The branch control block

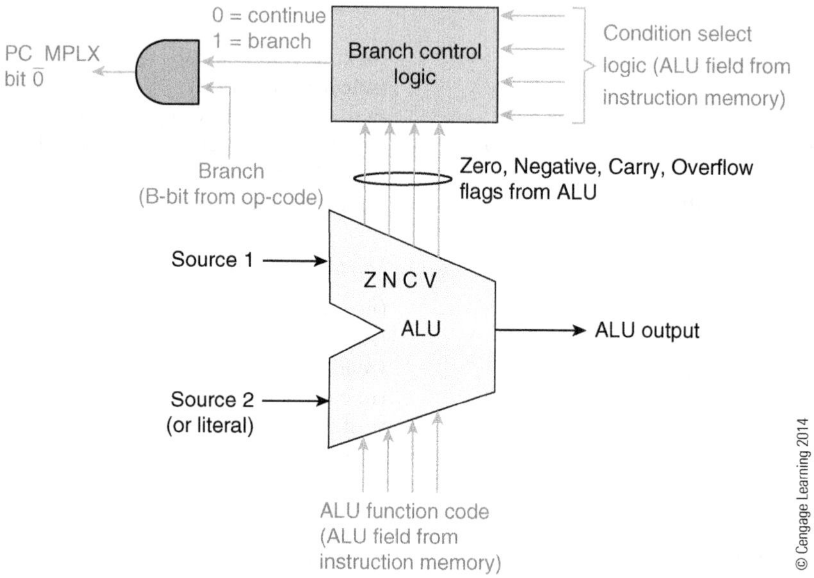

PC MPLX
bit $\overline{0}$

0 = continue
1 = branch
Branch control logic

Condition select logic (ALU field from instruction memory)

Branch
(B-bit from op-code)

Zero, Negative, Carry, Overflow flags from ALU

Source 1 →

Z N C V

ALU

→ ALU output

Source 2
(or literal) →

ALU function code
(ALU field from instruction memory)

© Cengage Learning 2014

Execution Time

We will now briefly consider how long it takes to execute an instruction. In a flow-through single-cycle computer, it does not matter how long it takes to execute the average instruction. What matters is how long it takes to execute the *longest* (slowest) instruction, because that determines the cycle time. Let's say this again, the longest (most complex) instruction determines the cycle time which means that less complex instructions have more time allocated to them than is actually necessary. Complex instructions kill performance.

The longest instruction is the load because the op-code has to be read from memory, the register file accessed to calculate the effective address of the operands, the execution unit used to add the literal offset to the register, the operand fetched from the data store, and, finally, the operand must be stored in the register file. Figure 7.16 demonstrates the information flow

FIGURE 7.16   Longest path length during the execution of a load instruction

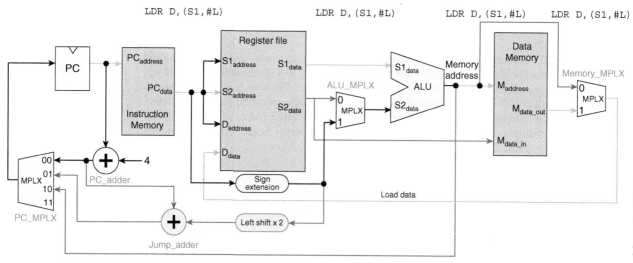

© Cengage Learning 2014

during a load operation (information paths are shown in dark blue). The sequence of actions and their times during a read cycle are

1. PC setup time (from clock to PC output stable), t_{PC}
2. Read instruction (from PC stable to op-code stable), t_{Imem}
3. Read operand (from op-code stable to operand stable), t_{RF}
4. Time to add S1 and S2 in the ALU, t_{ALU}
5. From address from ALU stable to memory operand stable, t_{Dmem}
6. Memory MPLX time, t_{MPLX}
7. Data setup time of the destination operand in the register file, t_{RF_s}

The cycle time is, therefore, the sum of these.

$$T_{cycle} = t_{PC} + t_{Imem} + t_{RF} + t_{ALU} + t_{Dmem} + t_{MPLX} + t_{RF\_s}.$$

7.3   Introduction to Pipelining

In this section, we introduce the idea of *pipelining* in which the execution of instructions is overlapped to improve efficiency. Although pipelining is the hallmark of the RISC processor, it must be stressed that pipelining is used by many of today's digital systems and all modern microprocessors. Indeed, Intel's Pentium family and its successors owe much of their high performance to extensive pipelining. We discuss the nature of pipelining generally before looking at the structure of a simple pipelined processor. It is important to note that we are leaving the single clock cycle processor structure we have just described where an instruction flows through it and are introducing a structure that slices an instruction into parts that are executed separately.

Figure 7.16 illustrates the *machine cycle* of a microprocessor that executes an instruction of the form ADD R0,R1,R2 (i.e., [R0] ← [R1] + [R2], where R0, R1, and R2 are registers). A clock cycle is the smallest event that can happen in a computer and a machine cycle is the time required to execute an instruction. This instruction is executed in five phases:

Instruction Fetch—Read the instruction ADD R0,R1,R2 from the system memory and increment the program counter.
Instruction Decode—Decode the instruction read from memory during the previous phase. The nature of the instruction decode phase is dependent on the complexity of the instruction set. A regularly encoded instruction can be rapidly decoded with two levels of gating, whereas a complex instruction format might require ROM-based look-up tables to implement the decoding.
Operand Fetch—The operands specified by the instruction are read from registers R1 and R2 in the register file and latched into flip-flops.
Execute—The operation specified by the instruction is carried out.
Operand Store—The result of the execution phase is written into the operand destination. This may be an on-chip register or a location in external memory. In this case the result is stored in register R0.

Each of the above five phases may take a specific time (although the time taken is normally an integer multiple of the system's master clock period). Some instructions may require less than five phases. For example, CMP R1,R2 compares R1 and R2 by subtracting R2 from R1 to set the condition codes and does not need an operand store phase (although, of course, the condition code flags do have to be updated).

Suppose a processor steps through the five phases of Figure 7.17 sequentially. At any instant only 20% of the processor is active. The other 80% making up the remaining four stages is idle; for example, when an instruction is in the execution phase, the operand store logic is waiting for the result with bated breath while it does nothing. A better approach to the design of a processor is to overlap or to *pipeline* the various phases of instruction execution.

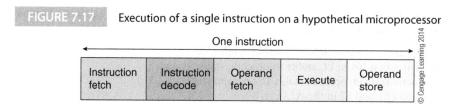

FIGURE 7.17   Execution of a single instruction on a hypothetical microprocessor

Let's look at early RISC processors, the Berkeley RISC I and RISC II that were instrumental in changing the way in which ISAs are constructed and computers are designed at the microarchitecture level. Figure 7.18 illustrates the RISC I's pipeline.

Instruction execution has been divided into two 500 ns slots, and the instruction is fetched in the first slot. In the second 500 ns slot the operands are fetched, execution takes place, and the result is written back. Pipelining is implemented by beginning the next instruction as soon as the current fetch phase has been completed. In this way, the speed of the processor is effectively doubled without changing the technology or speeding up the clock. The increase in speed is obtained by using the processor's functional parts more efficiently rather than raising the clock rate.

Figure 7.19 demonstrates the more sophisticated pipeline of the RISC II. The instruction fetch phase is reduced to 330 ns and its internal operations are divided into two stages: an operand fetch phase and execute phase followed by an operand store phase. In this case, three instructions can be overlapped.

The optimal length of a pipeline is technology-dependent[1] and a function of both the architecture and of the code being executed. As we shall see, the length of the pipeline has important implications for the way in which certain classes of instruction are handled (e.g., branches and other instructions that control program flow). The ratio of the instruction-fetch cycle time to the sum of the operand fetch plus the execute and operand store time plays an important part in determining the optimum length of the pipeline. Although the pipeline of RISC style processors is generally short, Intel's Pentium Pro had a 10-stage pipeline and the Pentium 4 had a 20-stage pipeline (various versions of the Pentium family had pipelines up to 31 stages).

FIGURE 7.18   The RISC I pipeline

FIGURE 7.19   The RISC II pipeline

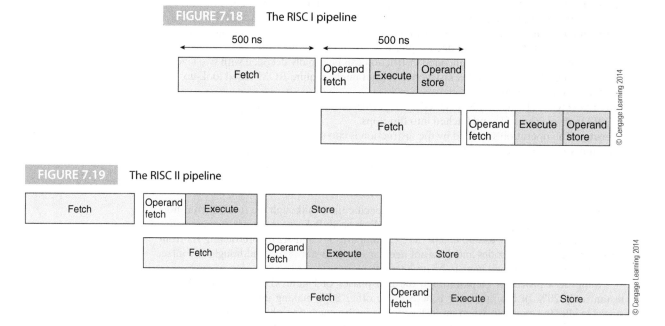

---

[1] V. M. Milutinovic (editor), *High Level Language Computer Architecture*, Computer Science Press, 1989, ISBN 0-88175-132-4.

A History of RISC Processors

The IBM/360 can be regarded as the grandfather of all modern processors, because it introduced the notion of computer architecture and the ISA and it ultimately led to Intel, Motorola, Zilog, and other microprocessor families in the 1970s.

In retrospect, the 360 and its grandchildren were what we would now call CISC processors. They had complex instruction sets at a time when main store was very expensive and you wanted instructions to do as much work as possible. In 1974, a new architecture emerged from IBM's Watson Research Center, a 32-bit RISC called the 801 project. This architecture was never sold commercially (in fact, it was never even built), but it did lead to IBM's line of RISC processors in the mid-1980s.

The RISC architecture caught the attention of the academic world. By1985, microprocessors were too complex (at the design level) for professors and students alike, and the RISC architecture once again put the processor in the classroom because students could understand the RISC microarchitecture and even design their own.

RISC architecture soon became associated with two giants of the computer world, John Hennessey at Stanford and David Patterson at Berkeley. Hennessey designed the MIPS and Patterson designed the Berkeley RISC I and II (research and teaching machines) that were to provide the foundation for Sun Microsystem's SPARC architecture. SPARC was unusual in that it used a windowed register system that allowed the programmer to access a new set of registers each time a subroutine was called (but only up to a depth of 8).

MIPS sold well because of its adoption by Nintendo in games consoles and was also enthusiastically adopted by many universities as a teaching vehicle. ARM (originally the Acorn RISC machine) was designed in the UK in 1983 and rapidly became the most popular RISC processor in embedded applications. MIPS, SPARC, and ARM illustrate a simple fact about the design of RISC architectures: All three were designed without the cost and workforce of organizations like Intel and Motorola.

Today, the acrimonious debate of RISC versus CISC that took place in the 1990s has vanished. RISC microarchitectural techniques are incorporated in all processors today. The Intel IA32 family still reigns supreme in the PC world because of Intel's ability to continually reinvent its microarchitecture. Simple 8-bit CISC microprocessors are used in ultra-low-cost applications, and more powerful 32-bit RISC in high-end applications such as smart phones.

Figure 7.20 illustrates a five-stage pipeline that takes five clock cycles to execute an instruction completely. By the time an instruction has been executed, the following four instructions are in various stages of execution. At each successive clock pulse, a new instruction completes its execution.

FIGURE 7.20   The five-stage pipeline

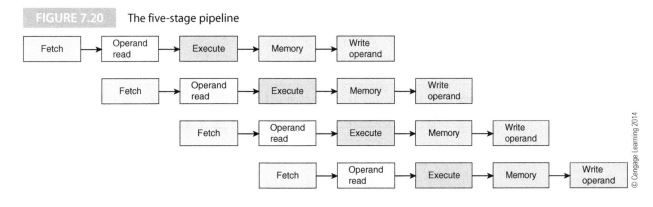

© Cengage Learning 2014

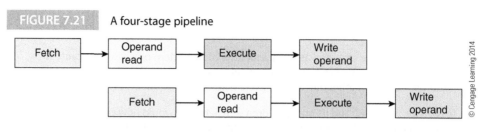

FIGURE 7.21   A four-stage pipeline

We will omit the instruction decode phase in future examples. It's not necessary in typical RISC architectures as decoding can be combined with the operand fetch phase. We will use the simple four-stage pipeline of Figure 7.21 to illustrate some of the problems that arise in pipelined systems. In the remainder of this chapter, we will use examples of short pipelines because they are easier to understand.

Real pipelines vary widely in terms of length from three stages to the thirty stages of one of Intel's Pentiums. Figure 7.23 describes the ARM11 pipeline which has eight stages. Like many real microprocessors, the ARM11 pipeline has parallel stages because a single linear pipeline would be too long to accommodate all operations. As you can see from Figure 7.22, there are three parallel pipelines following the common first four stages. One pipeline deals with conventional register-to-register operations, one implements the multiply and accumulate instruction, and one deals with load and store operations. Like most microprocessors, different versions of ARM have different pipeline structures. For example, the ARM7 has a three-stage pipeline and the ARM10 has a six-stage pipeline.

Figure 7.23 describes the SPARC T1 pipeline. This has six stages. The second stage is labeled *thread select* and chooses which of up to four threads is to run next. We look at multithreading in more detail in Chapter 13. All we need say here is that it provides a means of increasing processor performance by replicating registers and the PC so that several threads may in the processor at any instant. Only one thread at a time is active and being executed. All other threads are dormant and exist only as state information in the registers. However, if a thread is switched, all the registers of the current thread are switched out and the registers of the new thread switched in. Multithreading improves performance by switching threads when a resource is unavailable (for example, because the processor is waiting for a load from memory to complete). Multithreading hides the effects of latency.

FIGURE 7.22   The ARM pipeline

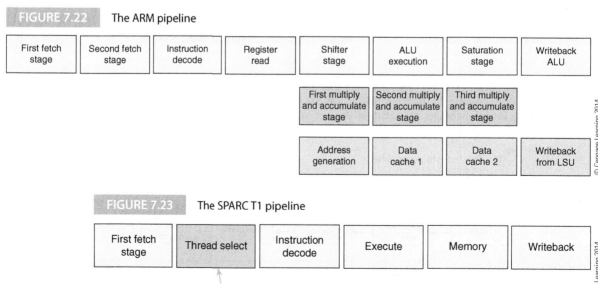

FIGURE 7.23   The SPARC T1 pipeline

7.3.1 Speedup Ratio

The performance of a pipeline is expressed in terms of its *speedup ratio*; that is, the ratio of the speed of the system with a pipeline to the speed of the same system without a pipeline. If a system divides an operation into n pipelined stages, it takes n cycles to perform the first operation. Thereafter, a new operation is completed on each clock cycle. Once the pipeline is full a new operation is completed every clock cycle. In practice, this cannot be achieved for several reasons. The two most important reasons are the effect of branches on the operation of a pipeline and the effect of data dependency (e.g., when an instruction requires an operand generated by a previous instruction and that operand has not yet been computed and stored). We look at these problems in detail later in this chapter. The Holy Grail of the pipeline designer is a one cycle per instruction execution rate.

If i operations are performed with this pipeline, the time taken is $i + (n - 1)$ cycles. Without pipelining, the system would require $n \cdot i$ cycles. The speedup ratio is therefore

$$S = \frac{n \cdot i}{i + (n - 1)} = \frac{n}{1 + \dfrac{n - 1}{i}} \text{cycles}$$

In the limit, when $i = 1$ the value of S is 1, and when $i = \infty$, the speedup is n. However, before we go into an analysis of pipelines, we are going to return to circuits and flip-flops to demonstrate how pipelining is implemented in principle.

7.3.2 Implementing Pipelining

Having looked at the notion of pipelining, we now demonstrate how it can be incorporated in a processor. We are going to construct a processor step by step, beginning with the program counter. The structure of the computer will be similar to that of the flow-through computer, but with the addition of pipeline stages. Essentially, the only significant difference between a flowthough and a pipelined processor is the addition of flip-flops or registers between stages.

Figure 7.24 illustrates a pipelined stage. Stage i begins with a D flip-flop. In order for the system to work correctly, this flip-flop must be a master-slave device to isolate its input and output. We will assume that the Q output of the D flip-flop changes on the rising edge of the clock.

FIGURE 7.24 The pipelined stage

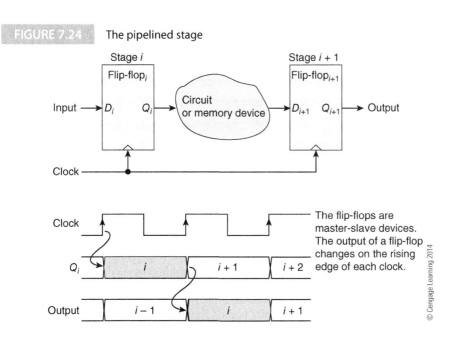

© Cengage Learning 2014

The output from stage i is fed to a circuit (i.e., logical network, process or a subsystem) that operates on the data captured in flip-flop i. This circuit may be an adder, a multiplier, or even a memory that takes in an address and looks up the data at that location. What is important, however, is the requirement that the output of the circuit be valid *before* the next clock pulse. The output of the circuit is captured by flip-flop $i + 1$ on the next clock pulse. After a period T, the flip-flops of the pipeline are clocked and the output from the circuit is captured by the stage $i + 1$ flip-flops. Flip-flop $i + 1$ captures the output of the circuit and holds it constant while the next stage is using it.

The timing diagram below the pipeline stage in Figure 7.24 demonstrates its operation. The outputs of flip-flops change on the rising edge of the clock and a signal is held constant between successive clock pulses.

When describing the operation of a pipelined computer, the program counter is a convenient place to start. Figure 7.25 shows the program counter and program memory stage in the hypothetical processor we are developing. At each clock pulse, the contents of the program counter are incremented by 4 because the processor has 32-bit instructions but is byte addressed. While the output of the PC is constant, the corresponding instruction is read from the instruction memory and clocked into the instruction flip-flop on the next clock pulse. The timing diagram shows that in clock cycle i the program counter is pointing at the current instruction, and in the next clock cycle, the binary pattern corresponding to this instruction is latched and available for decoding.

In Figure 7.25, the program counter and the instruction register are the two latched stages. We assume that the memory is not clocked. The memory receives an address and delivers the data stored at that location (i.e., it translates PC into op-code).

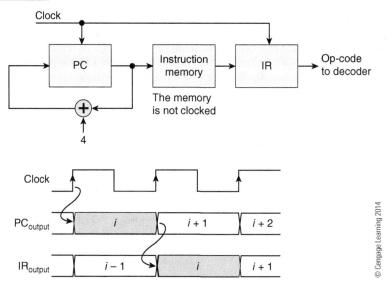

FIGURE 7.25   Reading the instruction in a pipelined processor.

As we have seen, a processor contains a *register file* which is the set of general-purpose registers used to hold temporary variables and is, essentially, a small memory. The register file is a critical component in a processor because it has to do more work than other parts of the CPU. In each cycle the register file has to perform three actions. It must supply *two* operands for use in the current operation and it must store the result generated by a previous operation. As we have already seen, such a memory has *six* ports; the two source addresses and the source outputs, the destination address and the destination data input; this is also call a memory with two read ports and two write ports).

From PC to Operands

Let's look at the first few stages of the pipeline from PC to source operand latches. We will assume that the current operation is register-to-register. In Figure 7.26, we have extended the computer by adding the register file and its output flip-flops. The notation OA_1 means *operand address* 1 and OV_1 means *operand value* 1. Assume that the initial value in the program counter is *i*. Figure 7.26 shows that in the third cycle the program counter contains instruction $i + 2$ and the source operand flip-flops contain the actual operands to be used by instruction *i*. Because of the effects of pipelining, at this point, the instruction register is pointing at the operands required by the *next* instruction, and the program counter is reading the instruction *after* that.

FIGURE 7.26   Reading the values of the operands

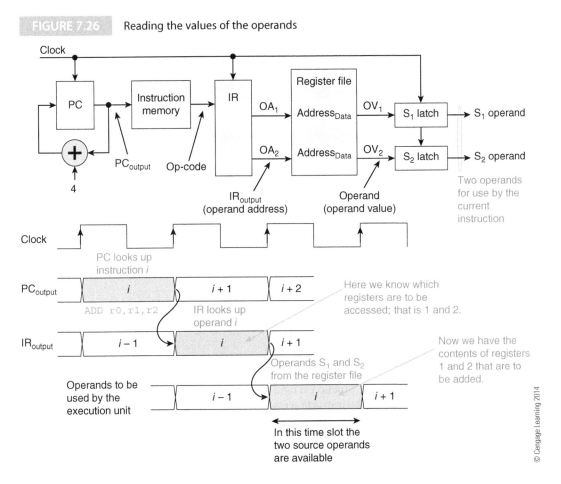

Figure 7.27 extends the pipelining one stage further to include the ALU that operates on the two source operands and the result flip-flop that stores the result (i.e., output of the ALU) generated by the instruction. This result must, of course, be written back into the register file during a register-to-register operation.

To store the result (destination operand) in the register file, you might be tempted to think that all we have to do is to send the result to the register file and clock it in using the destination address in the instruction. Unfortunately, we've forgotten something important. During the period that the output of the ALU is valid, the destination operand in the instruction register is not the same destination operand that generated this result. Because of the effects of pipelining, the destination address currently in the instruction register corresponds to two instructions later than the one whose result has just been calculated; that is, the destination operand is out of phase with the destination address because of the different delays in the operand *address* path and the operand *value* path.

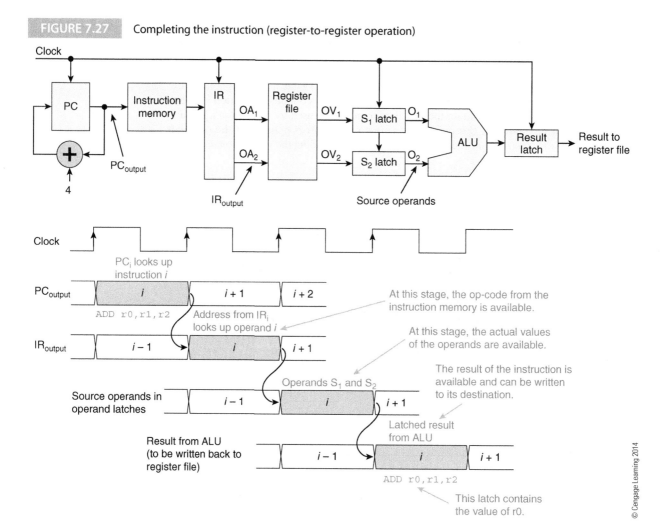

FIGURE 7.27   Completing the instruction (register-to-register operation)

Figure 7.28 shows the situation at time T_3 when instruction $i + 3$ is in the program counter and the result generated by instruction i is in the result register. At this time operand address $i + 2$ is in the instruction register. But we need operand address i. The only way to deal with this situation is to include a delay in the path between the destination address from the instruction register and the register file.

Figure 7.29 extends the circuit further to include the flip-flops that provide the delays necessary to ensure that the destination operand address arrives at the register file at the same time as the destination operand. We have added two flip-flops (i.e., registers) in the path between the destination address file of the instruction register and the destination operand address of the register file.

There's a further timing problem that we have to deal with. We don't want the op-code to get to the ALU before the data gets there, so we have to delay the op-code to ensure that instruction i reaches the ALU at the same time its operands reach the ALU. In Figure 7.29, a single delay is inserted in the op-code path between the IR and the ALU. Figure 7.30 provides a timing diagram for Figure 7.29. As you can see, the ALU op-code is present while the operands are present at the ALU and the destination address is available at the register file at the same time as the result.

Implementing Branch and Literal Operations

The next embellishments to the pipelined architecture are described by Figure 7.31, where we've added two new features. The first is the path between the literal field in the instruction

FIGURE 7.28   The state of the pipeline after four clock pulses

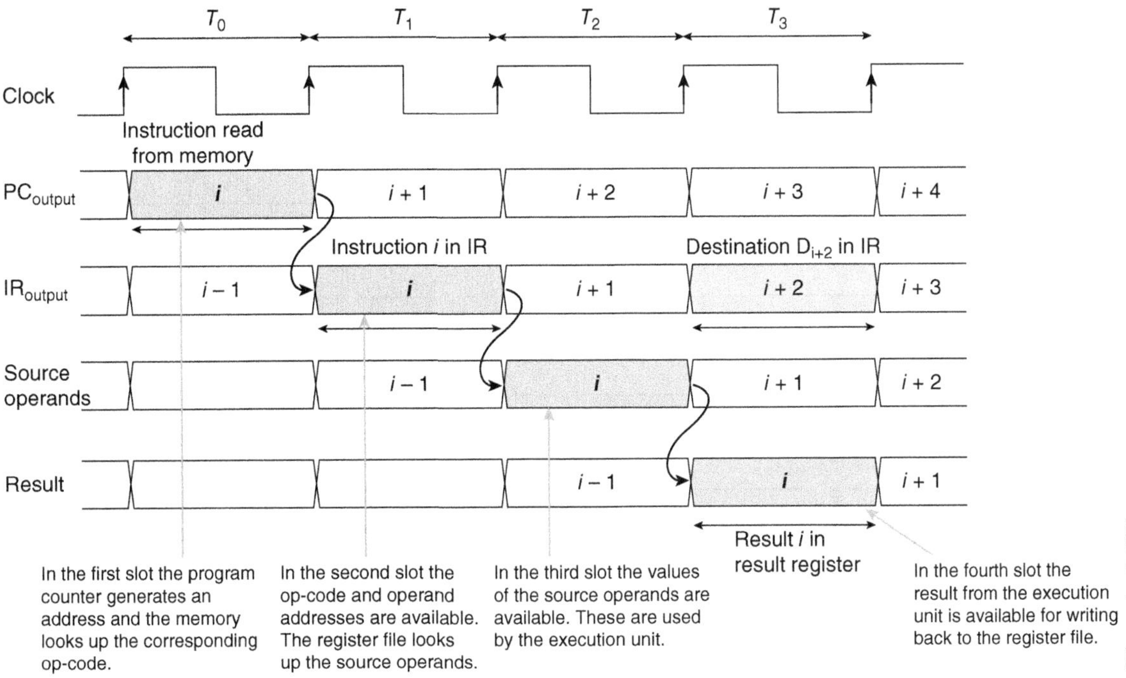

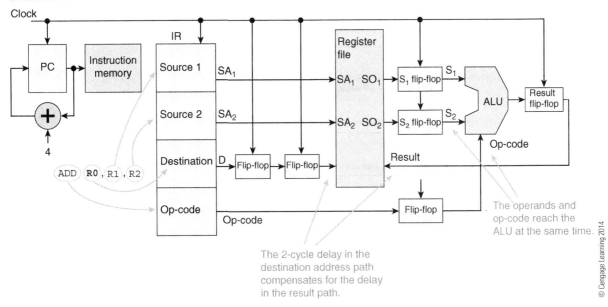

Clock

Instruction read from memory

PC$_{output}$: $i-1$ | i | $i+1$ | $i+2$ | $i+3$ | $i+4$

Instruction i in IR   *Destination D_{i+2} in IR*

IR$_{output}$: $i-1$ | i | $i+1$ | $i+2$ | $i+3$

Source operands: $i-1$ | i | $i+1$ | $i+2$

Result: $i-1$ | i | $i+1$

Result i in result register

In the first slot the program counter generates an address and the memory looks up the corresponding op-code.

In the second slot the op-code and operand addresses are available. The register file looks up the source operands.

In the third slot the values of the source operands are available. These are used by the execution unit.

In the fourth slot the result from the execution unit is available for writing back to the register file.

© Cengage Learning 2014

FIGURE 7.29   Compensating for delays in the pipeline

Clock

PC | Instruction memory | IR

Source 1 — SA$_1$
Source 2 — SA$_2$
Destination — D — Flip-flop — Flip-flop
Op-code — Op-code

+4

ADD R0, R1, R2

Register file: SA$_1$ SO$_1$ — S$_1$ flip-flop — S$_1$
SA$_2$ SO$_2$ — S$_2$ flip-flop — S$_2$
Result
Flip-flop

ALU — Result flip-flop

Op-code

The operands and op-code reach the ALU at the same time.

The 2-cycle delay in the destination address path compensates for the delay in the result path.

© Cengage Learning 2014

register and the ALU needed to implement literal operations such as ADD **r1**,r2,#123 or LDR **r1**, [r2,#20]. This data path requires a single delay element because the path for data operands from the register file to the ALU incurs one cycle delay.

We have also added a path between the PC and the ALU to allow the computation of relative branch addresses and a path between the output of the ALU and the PC to allow conditional branches.

FIGURE 7.30   Timing diagram of Figure 7.29

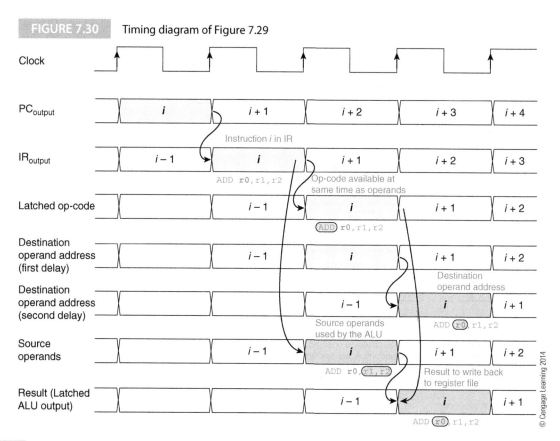

FIGURE 7.31   Extending the pipelined architecture by adding a literal path and a branch facility

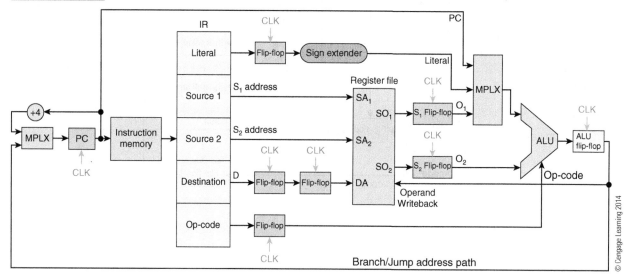

We have not included a data memory in order to simplify the diagram. The data memory gets its operand address from the ALU, because addresses are calculated. Data from memory can be stored in the register file, or data from the register file stored in the memory as Figure 7.32 demonstrates.

Figure 7.33 provides a timing diagram for the store operation. As you can see, the second flip-flop in the data path between the register file and data memory is necessary to ensure

FIGURE 7.32   Extending the pipelined architecture by adding data memory

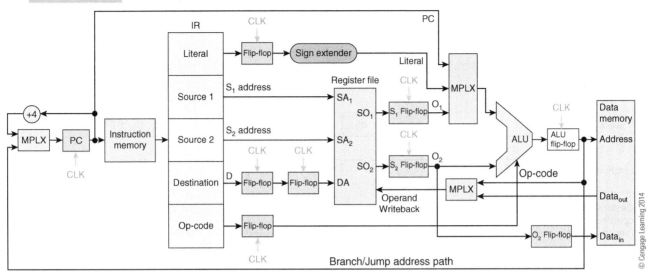

FIGURE 7.33   Timing of a store memory access

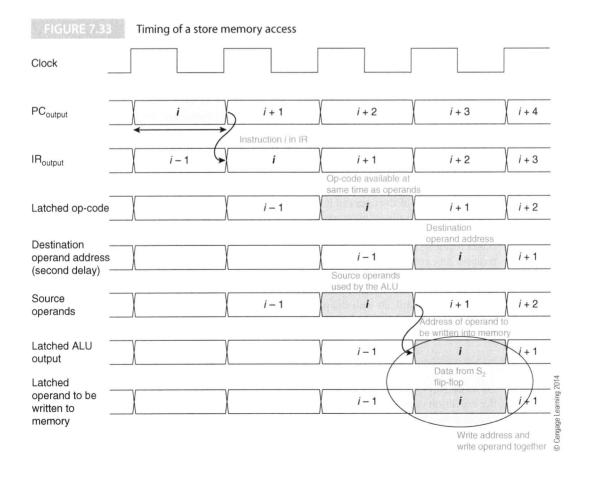

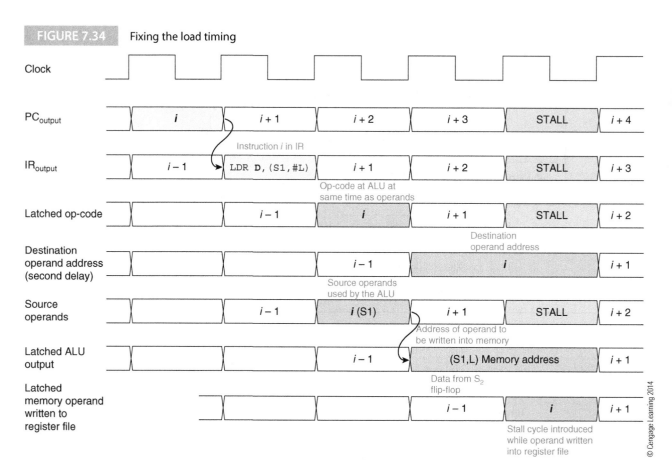

FIGURE 7.34   Fixing the load timing

© Cengage Learning 2014

that the data to be written to the memory is available at the same time as the operand address which has been delayed because of the time taken to compute it in the ALU.

Figure 7.34 gives the timing diagram for a load operation. Here we run into a problem. The operation cannot be completed in four clock cycles because of the extra delay incurred in reading the operand in the data path.

The problem is solved by introducing a *stall cycle* in the pipeline where all stages are frozen to provide time for the data memory to provide data. When the ALU has calculated the effective address of the operand in memory, the address is used to access the data memory to obtain the operand from memory. During the next clock period, that address is latched into the register file. However, during this clock period, the clock is not applied to all the clocked stages resulting in a missing a clock cycle (i.e., a stall). In other words, because a memory load requires an extra operation that cannot be arranged in parallel with existing operations, the processor has to be halted while the memory is accessing the data. This represents one of the most significant limitations to processors today.

We have now dealt with the problems of delays in a pipelined processor. Unfortunately, we've only just begun to deal with the problems. In the next section we look at the interrelationship between the pipeline and the code being executed.

7.3.3   Hazards

As we know, pipelining dramatically increases the performance of a processor by overlapping the execution of instructions. Unfortunately, it isn't possible to implement pipelining without encountering problems commonly called *hazards* that reduce the pipeline's throughput. In this section we explain why all stages of a pipelined processor cannot be kept continually active and why *bubbles* or *stalls* have to be introduced into the pipeline.

A *data hazard* refers to a situation in which the processing of one instruction depends on the data created by a previous instruction that is still in the pipeline. The *control hazard* arises when a branch is taken and all the partially executed instructions in the pipeline have to be thrown away. We look at control hazards first. Later in this chapter we examine how the effects of the control hazard, the greatest source of inefficiency in pipelined processors, can be reduced.

Another hazard is the *structural hazard* that occurs when two activities require the same resource simultaneously. Memory can be a structural hazard if two instructions attempt to access it at the same time and it cannot grant simultaneous access. We do not discuss this form of hazard further in this chapter as it is more important in architectures that execute multiple instructions simultaneously (*superscalar processors*), which are introduced in Chapter 8.

Most of this section is concerned with the control hazard caused by branch operations because overcoming it is critical to the efficient operation of a pipeline.

A pipeline is an ordered structure that thrives on regularity. Consider Figure 7.35 in which a sequence of instructions is being executed in a pipelined processor. In this section we will use short pipelines to illustrate hazards simply to make the diagrams easier to read. The same principles apply to pipelines of any length.

When the processor encounters a *branch instruction* that is taken, it is forced to reload its program counter with a new value; that is, the *target* address of the branch. Reloading the program counter with a non-sequential address means that all the useful work performed by the pipeline must now be rejected, since the instructions immediately following the branch are not going to be executed. Note that calculating the target address is not a trivial task, because most microprocessors use *program counter relative addressing*. A typical branch, say BEQ XYZ, doesn't use an absolute (actual) target address but a relative target address expressed as the number of bytes from the address in the program counter. Consequently, a target address must be calculated by adding the offset in the branch instruction to the program counter. Clearly, this takes time and the new target address may not be available until late in the instruction cycle.

Branches are not the only instructions that cause problems in a pipelined system. Subroutine calls, returns, traps, and exceptions all modify the sequential execution of instructions and may introduce bubbles if the branch resolution is delayed.

Cost of Branches

Although we look at the effect of branches in more detail later in this chapter, we should provide some idea of the effect of branches on the efficiency of a pipeline here.

Suppose that 20% of instructions are branches, 80% of branches are taken, and the additional penalty incurred by a taken branch is 4 cycles. The additional number of cycles per instruction is 20% × 80% × 4 = 0.64; that is, the number of cycles per instruction without branches is 1.00 and with branches it is 1.64.

FIGURE 7.35   Pipeline bubbles caused by branch instructions

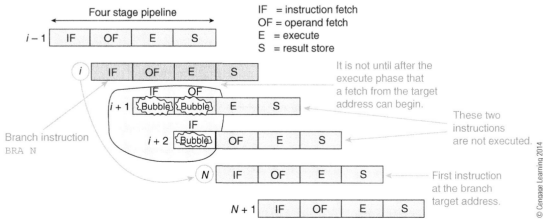

When data in a pipeline is rejected or the pipeline is held up by the introduction of idle states, we say that a *bubble* has been introduced. Another term for a bubble is a *pipeline stall*. The longer the pipeline, the more instructions that must be rejected once the branch is encountered.

Delayed Branch

As program flow control instructions are so frequent (typically 5 to 30% of the instructions in a program), any realistic processor using pipelining must do something to overcome the problem of bubbles caused by this class of instructions. The Berkeley RISC machines reduced the effect of bubbles by refusing to throw away the instruction immediately following a branch. That is, the instruction following a branch is *always* executed. Consider the effect of the following sequence of instructions executed on a machine implementing a delayed branch.

```
ADD    R3,R2,R1    ;[R3] ← [R2] + [R1]
B      N           ;[PC] ← [N]                  ;Goto address N
ADD    R5,R4,R6    ;[R5] ← [R4] + [R6]          ;Executed in the delay slot
ADD    R7,R8,R9    ;Not executed because branch taken
```

The processor calculates R3 = R2 + R1 and then encounters the branch. Because the instruction immediately following the branch, ADD R5,R4,R6, is also executed. RISC literature generally refers to this mechanism as the *delayed jump* or the *branch-and-execute* technique.[2]

Unfortunately, it's not always possible to arrange a program in such a way as to include a useful instruction immediately after a branch. Whenever this happens, the compiler must introduce a *no operation* (NOP) instruction after the branch and accept the inevitability of a bubble.

Figure 7.36 demonstrates how a Berkeley RISC II processor with its three-stage pipeline implements the delayed branch. The branch described in Figure 7.36 is a *computed branch* whose target address is calculated during the execute phase of the instruction cycle.

The delayed jump was implemented by MIPS and SPARC, but not all RISC processors incorporate it (e.g., ARM). In general, the use of a delayed jump mechanism is not popular today. In the next chapter, we will encounter OOO (out-of-order) execution where the processor reorders the sequence in which instructions are executed. OOO execution complicates the design of delay slots and many feel that it is not worth the effort of including delay slots in the microarchitecture.

FIGURE 7.36   The RISC II delayed jump mechanism

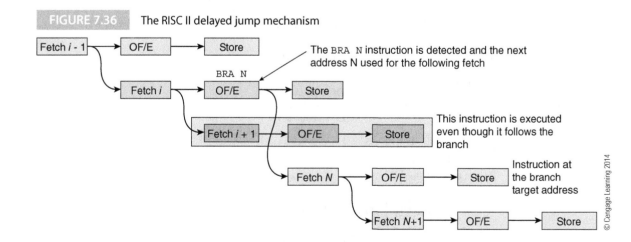

[2] *Reduced Instruction Set Architectures for VLSI*, The MIT Press, 1985, ISBN 0-262-11103-9.

You could also argue that the delay slot is included to mitigate a problem at the microarchitecture level and yet it is visible to the programmer at the ISA level (because the programmer has to be aware of the nature of the branch delay slot and either has to fill it with a valid instruction or use a NOP). The use of a branch delay slot also increases the complexity of exception handling schemes. We next look at *data hazards* caused by the interdependency of streams of instructions before returning to the effect of the branch penalty.

Data Hazards

Data dependency arises when the outcome of the *current* operation is dependent on the result of a *previous* instruction that has not yet been executed to completion. Data hazards arise because of the need to preserve the order of the execution of instructions. Consider the following fragment of code.

```
ADD     R2,R1,R0        [R2] ← [R1] + [R0]
SUB     R4,R5,R6        [R4] ← [R5] − [R6]
AND     R9,R5,R6        [R9] ← [R5] · [R6]
```

In this case, there's no data hazard and the order in which the instructions are executed does not matter. However, had the second instruction been SUB R4,R5,R2, a hazard would have arisen because R2 is a source operand in this instruction and a destination operand in the previous instruction. Data hazards are classified into three types according to the sequence of actions that causes the associated hazard; that is,

RAW	Read after write	(also called *true data dependency*)
WAW	Write after write	(also called *output dependency*)
WAR	Write after read	(also called *anti data dependency*)

The most important form of hazard is RAW,[3] where a read operation takes place after a write. Consider the following example of a RAW hazard where a programmer wishes to carry out the calculation:

```
X = (A + B) AND (A + B - C)
```

Assuming that A, B, C, X and two temporary values, T1 and T2, are in registers,[4] we can write

```
ADD     T1,A,B      ;[T1] ← [A] + [B]        ;T1 will not be available for 4 cycles
SUB     T2,T1,C     ;[T2] ← [T1] − [C]       ;this uses T1 which has not yet been
                                              generated
AND     X,T1,T2     ;[X] ← [T1] · [T2]       ;this uses T2 which has not yet been
                                              generated
```

In the code, we've written the operands that result in a RAW hazard in blue; the second instruction uses T1 before the first instruction has committed it to memory. In this example there are two RAW hazards, because in the second and third instructions, T2 is used before it has been stored. Figure 7.37 demonstrates the execution of this code using a four-stage pipeline. Ideally, this code should be executed in 6 clock cycles (4 for the first instruction and 1 cycle for each of the following two instructions). However, Figure 7.37 shows that ten cycles are required.

---

[3] The RAW hazard arises because it degrades the performance of non-superscalar pipelined machines. WAW and WAR hazards do not degrade the performance of a pipeline unless out-of-order execution is implemented (see Chapter 8).

[4] I have used variable names rather than registers to make the code easier to follow. Some assemblers allow you to rename registers as if they were variables to enhance code readability.

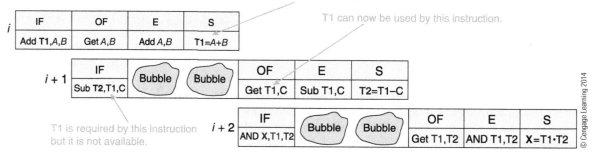

FIGURE 7.37   The effect of data dependency during a RAW

Instruction $i + 1$, SUB **T2**,T1,C, in Figure 7.37 begins execution during the operand fetch phase of the previous instruction. However, instruction $i + 1$ cannot continue on to its operand fetch phase, because the very operand it requires does not get written back to the register file, and will not be written back for another two clock cycles; that is, we have a read-after-write hazard and can't do the read operand for instruction $i + 1$ until we've done the write operand for instruction i.

Consequently, bubbles must be introduced in the pipeline while instruction $i + 1$ waits for its data. In a similar fashion, the following logical AND operation also introduces a RAW hazard as it too requires the result of a previous operation which is in the pipeline.

Example of RAW Hazards

Because this is such an important concept, we provided a further example. Consider the following code and indicate the RAW hazards if the pipeline has four stages (IF, OF,OE,OS).

1. ADD **r1**,r2,r3
2. ADD **r4**,r2,r3
3. ADD **r5**,r2,r4
4. ADD **r6**,r3,r4
5. ADD **r7**,r2,r1
6. ADD **r8**,r8,r3
7. ADD **r9**,r7,r1

Clock cycle	1	2	3	4	5	6	7	8	9	10	11	12	13
1. ADD **r1**,r2,r3	IF	OF	OE	OS									
2. ADD r4,r2,r3		IF	OF	OE	OS								
3. ADD **r5**,r2,r4			IF	S	S	OF	OE	OS					
4. ADD **r6**,r3,r4				IF	OF	OE	OS						
5. ADD r7,r2,r1					IF	OF	OE	OS					
6. ADD **r8**,r8,r3						IF	OF	OE	OS				
7. ADD **r9**,r7,r1							IF	S	OF	OE	OS		

Instruction 3 has to add two stalls while waiting for r4 from instruction 2. However, instruction 4 that also needs r4 does not have to wait because r4 has already been stored. Instruction 7 needs r7 generated by instruction 5. However, because of the intervening instruction, only one stall is introduced.

© Cengage Learning 2014

The *write after read* hazard differs from the RAW hazard in more than the obvious way; there is no syntactic dependency between the variables in the instructions of a WAR hazard. Consider the following sequence.

```
ADD R1,R2,R3
SUB R2,R4,R5
```

In this case, the first instruction generates a new value for R1 and writes it to the register file having read source operands R2 and R3 (the read part of the hazard occurs when operand R2 is read). The following subtract instruction writes the result to register R2 (the write part of the hazard); the same register is used by the previous instruction. Figure 7.38 demonstrates such a write after read operation. This does not create a pipeline stall or bubble because the read does not hold up the following write operation. Consequently, in normal circumstances a WAR hazard does not arise. A hazard would arise, of course, if the second instruction was executed before the first instruction (this is a factor to be considered in the superscalar processors we look at in the next chapter).

FIGURE 7.38   The WAR hazard

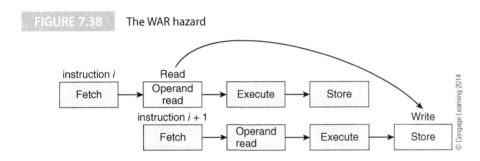

The third hazard, the *write after write* (WAW) is also unlikely to occur because it would require an operation sequence such as

```
ADD R1,R2,R3
SUB R1,R4,R5
```

Such a sequence is unlikely because destination operand R1 is generated by the first instruction and overwritten by the second instruction. We will see in Chapter 8 that WAW hazards can occur in superscalar processors that perform out-of-order execution.

Let's return to the most commonly occurring form of data hazard, the RAW *true dependency*. When an instruction is executed, the destination operand is available as soon as it leaves the ALU. However, it is not available for use by another instruction until after it has been written back to the register file. Figure 7.39 demonstrates how *internal forwarding* reduced the effects of data dependency by passing an operand directly to the next instruction. The example provided uses a four-stage pipeline where the following sequence of operations is executed.

1. ADD	R3,R1,R2	;[R3] ← [R1] + [R2]		
2. ADD	R6,R4,R5	;[R6] ← [R4] + [R5]		
3. SUB	R9,R1,R2	;[R9] ← [R1] − [R2]		
4. ADD	R7,R3,R4	;[R7] ← [R3] + [R4]	;RAW hazard because R7 is required by	
			the next instruction	
5. ADD	R8,R1,R7	;[R8] ← [R1] + [R7]		

In this example, instruction 4 uses an operand generated by instruction 1 (i.e., the contents of register R3). However, because of the intervening instructions 2 and 3, the destination operand generated by instruction 1 has time to be written into the register file before it is read as a source operand by instruction 4.

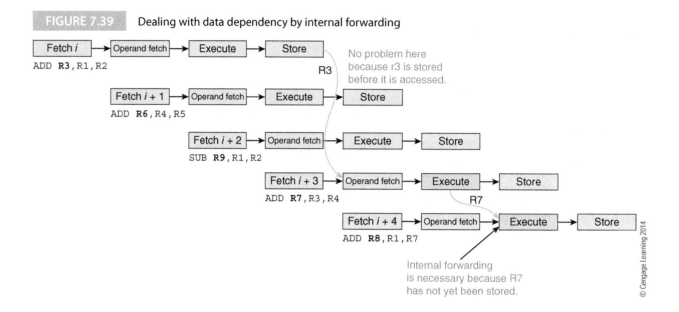

FIGURE 7.39   Dealing with data dependency by internal forwarding

Instruction 4 generates a destination operand R7 that is required as a source operand by the next instruction. If the processor were to read the source operand requested by instruction 5 from the register file, it would see the old value of R7. By means of *internal forwarding* the processor transfers R7 from instruction 4's execution unit directly to the execution unit of instruction 5 (see Figure 7.40).

Figure 7.40 demonstrates how a RAW hazard can be detected between two consecutive instructions. Comparators compare the previous destination address with the two current source addresses. If either (or both) of them matches, the appropriate operand is copied directly from the execution unit rather than the register file. Figure 7.41 illustrates how internal forwarding can be implemented. The result is fed back directly from the output of the ALU to the two source operand latches via two multiplexers. These multiplexers allow the inputs to the source operand latches to come from the register file or from the ALU. In practice, internal forwarding is more complex because provisions have to be made for interrupts and exceptions.

FIGURE 7.40   Detecting data dependency

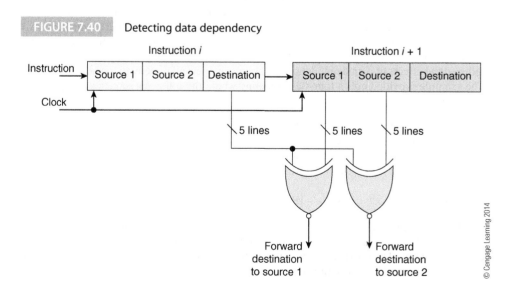

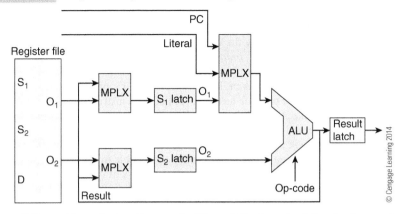

FIGURE 7.41   Implementing internal forwarding

Figure 7.42 provides a timing diagram for Figure 7.41 when the following instructions are executed.

```
ADD r1,r2,r3
ADD r4,r1,r5
```

In this example, the second instruction requires the operand r1 generated by the first instruction. Figure 7.42 demonstrates how in time slot $i + 1$ operand r5 is obtained from the source latch and operand r1 is obtained from the ALU output.

FIGURE 7.42   Implementing internal forwarding: timing diagram and data flow

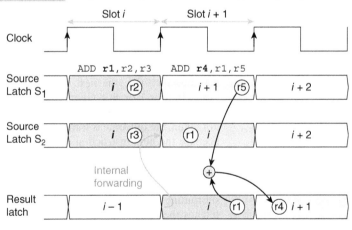

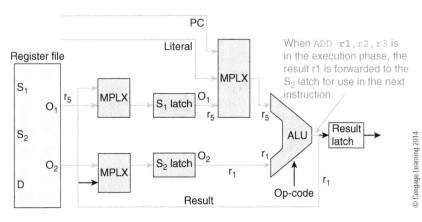

7.4   Branches and the Branch Penalty

We're now going to look at the branch in greater detail, calculate how much it costs in terms of lost cycles, stalls, and discuss how the effects of a branch can be reduced by predicting its outcome and fetching data from the *branch target address*—before we even know whether it will or will not be taken.

A branch instruction, when taken, loads the processor's program counter with a new non-sequential value, and the pipeline has to be refilled with instructions following the branch target address. The cost (i.e., the additional number of clock cycles) of executing an operation that causes a non-sequential flow of control is known as the *branch penalty*.

In this section, we look at instructions that modify the flow of control and ask ourselves how we can go about reducing or even eliminating the bubble in the RISC's pipeline caused when a branch is taken; that is, we are concerned with ways of reducing the branch penalty. Some of the techniques involve limiting the *damage* done by a branch and some techniques attempt to predict the outcome of a branch before it has been executed.

Before we look at the branch penalty, it's worthwhile looking at the branch itself. Several types of instruction modify the flow of control; for example, the unconditional branch, the conditional branch, the subroutine call, and the subroutine return. Internally generated traps and exceptions and externally generated interrupts also modify the flow of control. Subroutine call and returns are not normally regarded as branch operations from the computer architect's point of view, but they have similar characteristics from the computer designer's point of view; that is, they also incur a branch penalty.

The unconditional branch (e.g., BRA target) is always taken and forces execution to continue at the target address. An unconditional branch is equivalent to the high-level language goto and its outcome is known at *compile-time*.

The outcome of a conditional branch is determined by the state of one or more flag bits in the processor's condition code register (or some equivalent mechanism) and is therefore not known until *runtime*. The conditional branch may be taken (i.e., the next instruction is at the target address) or not taken (the next instruction in sequence is executed). When a branch is not taken, the outcome is sometimes called *in line* because the next instruction immediately following the branch is executed.

A subroutine call is a type of unconditional branch that saves the return address. Similarly, a subroutine return is an unconditional branch that fetches the target address from a register or the stack. Some computers support conditional subroutine calls and returns.

Grohoski reports that about one-third of all branches are unconditional, one-third are conditional *loop-closing* branches, and one-third are other conditional branches. A loop-closing branch terminates a loop construct and is taken for the first $n-1$ cycles of an n-cycle loop.

DeRosa and Levy[5] discuss the relative merits of so-called one-instruction and two-instruction branches. In a two-instruction branch, an explicit instruction sets up the branch (for example, a CMP or a TEST instruction). These instructions update the values of the processor's condition code flag bits that determine whether a conditional branch is to be taken.

Compile Time and Run Time

Compile time refers to the state of a program when it is first translated into machine code. At compile time certain details are already known; for example, the value of constants that are defined by operations such as ADD **r1**, r2, #5.

Run time refers to the state of a program as it is being executed. The state of variables is known only at run time because their values are not created until the program runs. In the previous example, r1 is not known at compile time because we don't know the value of r2.

The outcome of unconditional branches is known at compile time. However, the outcome of conditional branches is not known until the condition on which the branch depends has been evaluated.

[5] J. DeRosa and H. M. Levy, "An evaluation of branch architectures," *ISCA '87 Proceedings of the Annul International Symposium on Computer Architecture ACM*, 1987, pp. 10–16.

Sima, et al.[6] use the term *result state* to describe the situation defined by the processor's condition codes. Many mainstream architectures adopt this result state approach to branching (e.g., VAX, 68K family, Pentium family, SPARC, PowerPC). Cragon[7] calls instructions that require a separate instruction to set condition codes *CC branches*, and instructions that perform a test and then branch on the outcome '*TB branches*' (i.e., test and branch). A typical branch sequence on a generic processor is

```
CMP   R0,R1        ;if R0 = R1
BEQ   Same         ;THEN ...
```

The *result state approach* is well-suited to strictly in-order sequential systems (i.e., test data, set condition, test condition and branch on condition). Systems with multiple execution units cannot support a simple 'result state branch' mechanism because the result state may be dependent on *out-of-order execution*. We will look at out-of-order instructions later—here all we need point out is that some processors with multiple instruction execution units are able to execute instructions in an order that is different to the order in which they appear in memory. Such difficulties must be resolved by means of hardware that ensures that a branch is made on the appropriate condition.

A single branch instruction (or *direct check* operation) performs both the comparison and the branch itself in one operation; for example, the HP Precision Architecture, HP-PA, has an *add and branch instruction* that performs an addition and then branches on the result. Digital's Alpha architecture supports a single branch instruction that explicitly tests the contents of a register; for example,

```
BEQ R4,Loop
```

tests the contents of register R4 and then branches to Loop if R4 contains zero. MIPS also provides a single branch instruction BEQ r1,r2,target that branches to target if registers r1 and r2 are equal.

The HP-PA also provides a *skip-next* instruction that can be used to *squash* (i.e., ignore or nullify) the following instruction. If the condition tested is true, the following instruction is not executed. The following example of HP-PA code is taken from Corporaal[8] where the construct

```
If (a == b)
     c = c - 1;
else d = d + 10;
```

is translated into

```
sub,<>    r1,r2,r0       //compare a and b, nullify next instruction if not equal
addi,tr   -1,r3,r3       //c := c − 1, always nullify next instruction
addi      10,r4,r4       //d := d + 10
```

This fragment of code is executed in only three cycles. We have already seen that the ARM processor provides conditional execution for all instructions; that is, any instruction can be squashed if it does not satisfy the condition defined by the most-significant 4 bits of its opcode. For example, ADDEQ **r1**,r2,r3 is executed only if the state of the Z-bit represents zero.

7.4.1   Branch Direction

At first sight, you might be tempted to think that a conditional branch has a 50:50 chance of being taken. This is not so. When a branch is used to mechanize a loop, the branch might be

---

[6] D. Sima, T. Fountain, and P. Kacsuk, *Advanced Computer Architectures: A Design Space Approach,* Addison-Wesley, 199.

[7] H. G. Cragon, *Memory Systems and Pipelined Processors,* Jones and Bartlett, 1996.

[8] H. Corporaal, *Microprocessor Architectures from VLIW to TTA,* Wiley, 1998.

taken thousands of times before the loop is finally exited. Corporaal summarizes the results of several papers dealing with the outcome of conditional branches. The lower bound on the frequency of taken branches is 57% and the upper bound 99%. A typical value is about 80%; that is, 80% of all branches are taken. A corollary of this statement is that more emphasis should be placed on the processing of branches that are taken rather than those that are not (unless, of course, not taking a branch incurs an excessive penalty).

7.4.2   The Effect of a Branch on the Pipeline

The branch reduces the efficiency of a pipelined architecture by introducing bubbles or pipeline stalls. We now analyze the effect of pipeline bubbles and discuss some of the attempts that have been made to reduce them. Table 7.9 summarizes the effect of a branch instruction for an architecture with a four-stage pipeline (fetch instruction, fetch operand, execute, store operand). The first column gives the clock cycle number, and the other columns show which instruction each of the four pipeline stages is currently handling.

At cycle 0, instruction $i - 5$ is in the final stage of its execution (i.e., the operand store phase). At the same time, the previous instruction $i - 4$ is being executed and instruction $i - 2$ has just entered the first stage of the pipeline, the instruction fetch phase.

Assume that instruction i is a *branch* instruction that forces a jump to the instruction at location N. This branch instruction is first fetched into the pipeline in cycle 2 and reaches the execute phase in cycle 4. If we make the assumption that a branch instruction can first be acted upon at the end of its execution phase, the branch is taken in cycle 5.

In cycle 5, the instruction at branch target address N is read into the processor. However, Table 7.9 demonstrates that two instructions $i + 1$ and $i + 2$, following the branch, are in the pipeline and have to be aborted or *squashed* when the branch is taken. It is not until cycle 8 that the pipeline is completely filled again. The period during which the pipeline is unfilled (following a branch) is called the pipeline's *start-up latency*, and the effect of a branch that creates a bubble is called an instruction *misfetch penalty*. The branch instruction in Table 7.9 causes a delay of two cycles between the start of instruction i and the start of the next instruction to execute (i.e., the instruction at target address N).

In practice, the actual situation depends on several factors; for example, we have assumed that a condition has to be evaluated and a fetch instruction from the branch target address cannot begin until the end of the execute phase. If the branch is unconditional with an absolute address, the processor can begin fetching the instruction at the target address as early as the following instruction decode phase. Table 7.10 demonstrates the effect of an early branch resolution. In this case, only one cycle is lost. Clearly, the detection of branch instructions can be used to reduce branch penalties by taking action at the earliest possible time. Processors once detected branches only *after* the instruction had been brought into the pipeline and decoded. Schemes have been proposed that give advance warning of a branch instruction as

TABLE 7.9	The Effect of a Branch Instruction on a RISC's Pipeline		
Cycle	Fetch Instruction	Fetch Operand	Execute	Store
0	$i - 2$	$i - 3$	$i - 4$	$i - 5$
1	$i - 1$	$i - 2$	$i - 3$	$i - 4$
2	i	$i - 1$	$i - 2$	$i - 3$
3	$i + 1$	i	$i - 1$	$i - 2$
4	$i + 2$	$i + 1$	i	$i - 1$
5	N			i
6	$N + 1$	N		
7	$N + 2$	$N + 1$	N	
8	$N + 3$	$N + 2$	$N + 1$	N
9	$N + 4$	$N + 3$	$N + 2$	$N + 1$

TABLE 7.10		The Effect of Early Branch Resolution		
Cycle	Fetch Instruction	Fetch Operand	Execute	Store
2	*i* {branch instruction}	*i* − 1	*i* − 2	*i* − 3
3	*i* + 1 {bubble}	*i* {branch resolved}	*i* − 1	*i* − 2
4	*N*	*i* + 1 {bubble}	*i*	*i* − 1
5	*N* + 1	*N*	*i* + 1 {bubble}	*i*
6	*N* + 2	*N* + 1	*N*	*i* + 1 {bubble}
7	*N* + 3	*N* + 2	*N* + 1	*N*
8	*N* + 4	*N* + 3	*N* + 2	*N* + 1

© Cengage Learning 2014

Alternative view of the Pipeline

Table 7.9 presents one way of looking at the pipeline. Computer engineers use two views, the only difference being the axes. The diagram below presents an alternative view of Table 7.9.

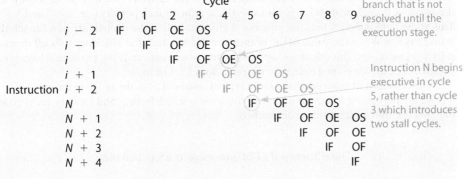

© Cengage Learning 2014

it enters the pipeline. The Itanium processor that we describe in Chapter 8 uses *hints* in the instruction to help the processor prepare for the branch. We will return to this topic when discussing actual processors.

7.4.3   The Cost of Branches

If we are going to reduce the effect of branch instructions on the performance of pipelined processors, we need a *metric* or *figure of merit* that describes the performance of the system. Because we cannot know how many branches a given program contains, or whether each branch is to be taken, we have to construct a *probabilistic model* for the system. We will make the following assumptions.

1. Each non-branch instruction is executed in one cycle.
2. The probability that a given instruction is a branch is p_b.
3. The probability that a branch instruction will be taken is p_t.
4. If a branch is taken, the *additional* penalty is b cycles.
5. If a branch is not taken, there is no penalty and only 1 cycle is required.

Since the probability that an instruction is a branch plus the probability that an instruction is not a branch must add up to 1, we can immediately state that the probability that an instruction is not a branch instruction is $1 - p_b$.

The average number of cycles required by an instruction during the execution of a program is the sum of the cycles taken for non-branch instructions, plus the cycles taken by branch instructions that are taken, plus the cycles taken by branch instructions that are not taken.[9] That is

$$T_{ave} = (1 - p_b)\cdot1 + p_b\cdot p_t\cdot(1 + b) + p_b(1 - p_t)\cdot1 = 1 + p_b p_t b$$

This expression, $1 + p_b p_t b$, tells us that the number of branch instructions, the probability that a branch is taken, and the penalty per branch instruction, all contribute to the cost of a branch. If we replace $p_b p_t$ by p_e (the effective probability of a branch), the average number of cycles per instruction is given by $1 + p_e b$. The efficiency of a RISC processor, E, can be defined as

$$E = \frac{average\_cycles\_per\_instruction\_without\_branch\_instructions}{average\_cycles\_per\_instruction\_with\_branch\_instructions} \times 100\%$$

That is

$$E = \frac{1}{1 + p_e b} \times 100\%$$

Figure 7.43 illustrates the relationship between the efficiency of a RISC processor and the effective branch probability p_e for two values of the branch penalty (i.e., $b = 2$ and $b = 6$). This graph demonstrates that the pipeline is efficient only if the probability of a branch that is taken is low. When the value of the branch penalty b is high, the efficiency falls off dramatically as p_e increases. This effect is exactly what you would expect. If the pipeline is long (i.e., b is high), even an occasional branch in the code does a lot of harm.

Observations of real code have given typical values of p_e in the range in the range 0.06 to 0.2. We are now going to examine some of the ways in which both p_e and b can be reduced and the effect of the branch penalty diminished.

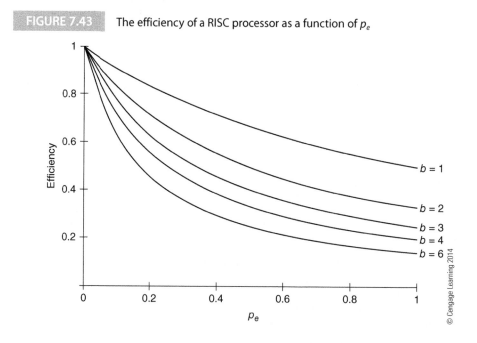

FIGURE 7.43   The efficiency of a RISC processor as a function of p_e

[9] D. J. Lilja, "Reducing the Branch Penalty in Pipelined Processors," *IEEE Computer*, July 1988, pp. 47–54.

Branches Demonstrated Harmful

In this section we calculate the cost of branches in terms of the number of additional cycles (stalls) required by branches and then demonstrate how correctly predicting the outcome of a branch can reduce its cost. To set the scene, here is an interesting demonstration of the effect of branches described by Igor Ostrovsky, a software developer at Microsoft, who ran a fragment of code on a computer that was designed to generate a repeatable pattern of branches (for example, always taken, always not taken, taken, not taken alternating, and so on). His timings, given in the table below are instructive. There is a factor of six between different branch patterns. These results demonstrate why the effects of branches are so significant and why (as we shall soon see) branch prediction is an important part of all modern processor design. The code used by Igor was

```
for (int i = 0; i < max; i++) if (<condition>) sum++;
```

Condition	Branch Pattern	Time (ms)
(i & 0x80000000) == 0	T repeated	322
(i & 0xffffffff) == 0	F repeated	276
(i & 1) == 0	TF alternating	760
(i & 3) == 0	TFFFTFFF…	513
(i & 2) == 0	TTFFTTFF…	1675
(i & 4) == 0	TTTTFFFFTTTTFFFF…	1275
(i & 8) == 0	8T 8F 8T 8F …	752
(i & 16) == 0	16T 16F 16T 16F …	490

Courtesy of Igor Ostrovsky.

The table demonstrates that the *best* branch construct was always taken and the worst, the pattern TTFFTT which would indicate that the branch predictor was most sensitive to mispredicting this sequence. The diagram below illustrates the number of mispredictions made by each of the loops.

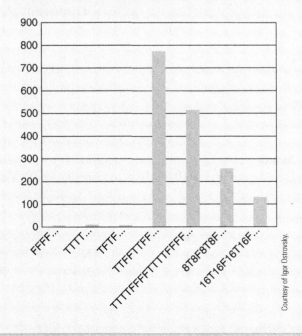

Courtesy of Igor Ostrovsky.

7.4.4 The Delayed Branch

The simplest way of dealing with branches is to do nothing; that is, the pipeline is *frozen* as soon as a branch is detected and *unfrozen* when the branch is resolved and instruction fetching continues at the target address. Table 7.11 shows the effect of freezing a pipeline on a branch that is not taken and on a branch that is taken. In Table 7.11(a), we have assumed that the branch is detected during the *read operand* phase in cycle −3. The pipeline becomes unfrozen when the branch progresses to its *execute* phase in cycle −2. In Table 7.11(b) the branch is taken and the pipeline not unfrozen until cycle −1. This approach to branching is inefficient.

Let's calculate the cost of branches for the arrangement of Table 7.11. A branch that is not taken incurs a penalty of 1 extra cycle and a branch that is taken incurs a penalty of 3 extra cycles. The average number of cycles per instruction is, therefore,

$$T_{ave} = 1 \cdot (1 - P_b) + 2 \cdot P_b(1 - P_t) + 4 \cdot P_b P_t = 1 + P_b + 2P_b P_t$$

If we take values for P_b of 0.2 and for P_t of 0.8 (20% of instructions are branches and 80% of branches are taken), we get

$$T_{ave} = 1 + 0.2 + 2 \times 0.2 \times 0.8 = 1.52.$$

This equation represents a performance degradation of 52% due to branches.

As we have already seen, some processors implement a *delayed branch*, which is a technique adopted by RISC II, MIPS, Am29000, and the IBM 801 processors. Since the instruction immediately following a branch has been almost executed by the time the branch is taken, it seems intuitively reasonable to let it be completely executed.[10] That is, you place an instruction after the branch, and this instruction is always executed in parallel with the branch.

TABLE 7.11   Freezing the Pipeline for Branches that are not Taken and Branches that are Taken

(a) Branch Not Taken

Cycle	Fetch Instruction	Read Operand	Execute	Store Operand
−5	$i − 1$	$i − 2$	$i − 3$	$i − 4$
−4	i = **Branch**	$i − i$	$i − 2$	$i − 3$
−3	$i + 1$	i (freeze)	$i − 1$	$i − 2$
−2	•	•	i (unfreeze)	$i − 1$
−1	$i + 2$	$i + 1$		i
0	$i + 3$	$i + 2$	$i + 1$	
+1	$i + 4$	$i + 3$	$i + 2$	$I + 1$

(b) Branch taken

Cycle	Fetch Instruction	Read Operand	Execute	Store Operand
−5	$i − 1$	$i − 2$	$i − 3$	$i − 4$
−4	i = **Branch**	$i − 1$	$i − 2$	$i − 3$
−3	$i + 1$	i (freeze)	$i − 1$	$i − 2$
−2	•	•	i	$i − 1$
−1	•	•	•	i (unfreeze)
0	N			
+1	$N + 1$	N		
+2	$N + 2$	$N + 1$	N	
+3	$N + 3$	$N + 2$	$N + 1$	N

---

[10] I have assumed that for a taken branch the pipeline becomes unfrozen at the end of the *store result* stage. It would be possible to devise a scheme that unfroze the pipeline at the end of the execute phase to save a cycle.

TABLE 7.12   The Effect of a Delayed Branch on a RISC's Pipeline

Cycle	Fetch Instruction	Read Operand	Execute	Store Result
-5	$i-1$	$i-2$	$i-3$	$i-4$
-4	**i {branch taken}**	$i-1$	$i-2$	$i-3$
-3	$i+1$ {always executed}	i	$i-1$	$i-2$
-2		$i+1$	i	$i-1$
-1			$i+1$	i
0	N			$i+1$
$+1$	$N+1$	N		
$+2$	$N+2$	$N+1$	N	
$+3$	$N+3$	$N+2$	$N+1$	N
$+4$	$N+4$	$N+3$	$N+2$	$N+1$

© Cengage Learning 2014

Table 7.12 demonstrates the effect of the delayed branch on a four-stage pipeline. Instruction $i+1$ is always executed even though it follows the branch. As you can see, the branch penalty is reduced by one cycle. It is, of course, possible to extend this principle and execute two more instructions following the branch.

Consider the average number of cycles per instruction for the 4-stage pipeline with delayed branching of Table 7.12. When a branch is not taken, the branch penalty is *zero* because instruction $i+1$ is executed and the pipeline is not held up assuming the pipeline is not frozen. When a branch is taken, only two cycles are lost because instruction $i+1$ is always executed.

$$T_{ave} = 1 \cdot (1 - P_b) + 1 \cdot P_b(1 - P_t) + 3 \cdot P_b P_t = 1 + 2 \cdot P_b P_t$$

If we take values for P_b of 0.2 and P_t of 0.8, we get $T_{ave} = 1 + 2 \times 0.2 \times 0.8 = 1.32$, which represents an improvement over simply freezing the pipeline.

From the programmer's point of view, it seems strange to locate an instruction that is to be executed *before* a branch *after* the branch. Consider the fragment of code expressed as hypothetical RISC code:

```
R = P + Q
C = B - A
goto NEXT
```

The corresponding RISC code is

```
ADD   R,P,Q   ;[R]  ← [P] + [Q]
BRA   NEXT    ;[PC] ← NEXT
SUB   C,B,A   ;[C]  ← [B] - [A] ;always executed in the delay slot
```

As you can see, the RISC code places the subtract instruction immediately after the branch. The subtraction is automatically executed because it has almost finished execution by the time the branch is taken.

Sima, et al.[11] go as far as to say,...*delayed branching reverses the execution sequence of the branch instruction and the instruction placed in the delay slot*. In other words, delayed branching requires an *architectural redefinition of the execution sequence* of instructions compared with traditional von Neumann architectures.

The delayed branch mechanism is effective if the compiler can find an *isolated* instruction to place after the branch. I use the term *isolated* (i.e., branch independent) to emphasize that you cannot put just any instruction after the branch. For example, you cannot use an instruction whose execution determines the outcome of the branch. The instruction must be one that's executed whether or not the branch is taken. If no instruction is available, the compiler

---

[11] D. Sima, T. Fountain, P. Kacsuk, and *Advanced Computer Architectures: A Design Space Approach*, Addison-Wesley, 1997.

must insert a NOP (no operation) to keep the pipeline operating. A suitable instruction that can be placed after the branch can be found in approximately 60% of cases. As we said earlier, the delayed branch is less popular today because of its side effects on superscalar processors and exception handling mechanisms.

The HP Precision Architecture has a one instruction delay following a branch. HP-PA branch instructions include a one-bit *nullification field* that determines how the instruction following the branch is to be treated. When set, the nullification bit allows a branch instruction to conditionally ignore the instruction in the delay slot depending on the branch outcome. Consider the branch at the end of a loop construct. This branch is normally taken and is only not taken when the loop is exited. By making the execution of the instruction in the delay slot dependent on whether the branch is taken or not, we can put a loop instruction in the delay slot. When a branch is taken to the start of the loop, the loop instruction is executed. However, when the loop is exited at the end of the loop, this instruction that belongs *in* the loop is not executed. Figure 7.44 demonstrates the effect of an HP-PA branch. Figure 7.44a shows the code sequence and Figure 7.44b illustrates how the delay slot is executed when the branch is taken back to the loop entry point. In Figure 7.44c the branch is not taken and the delay slot is nullified.

FIGURE 7.44   The effect of the HP-PA's delay slot nullification instruction

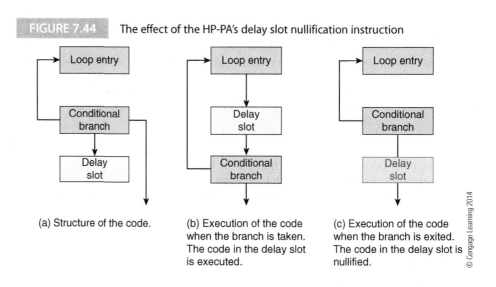

(a) Structure of the code.

(b) Execution of the code when the branch is taken. The code in the delay slot is executed.

(c) Execution of the code when the branch is exited. The code in the delay slot is nullified.

© Cengage Learning 2014

The effect of the delayed branch has been studied by DeRosa and Levy[12]. They report that compiler optimization allows between 40 and 60 percent of delay slots to be filled following a *conditional* branch and 90 percent of slots to be filled following an *unconditional* branch. DeRosa and Levy use the term *basic block size* to indicate the average number of instructions between consecutive (taken) branch instructions. The average length of a block is given by $1/(P_b \cdot P_t)$, where P_b is the probability of an instruction being a branch and P_t the probability of a branch being taken. They state that delayed branches will have a relatively small impact on computers with a large block size and that RISC architectures tend to have larger block sizes than CISC architectures. DeRosa and Levy also point out that the delayed branch mechanism incurs a penalty in terms of wasted memory space, because about 35 percent of delay slots must be filled with useless NOP instructions.

Let's re-evaluate the formula for the average number of cycles taken by an instruction when delayed branching is used and not all delay slots can be filled by useful instructions. Assume that fraction f_d of branches are followed by useful instructions in the delay slot and that a branch that is not taken requires one cycle.

[12] J. DeRosa and H. M. Levy, "An evaluation of branch architectures", *ISCA '87 Proceedings of the 14th Annual International Symposium on Computer Architecture, ACM*, 1987, pp. 10–16.

$$T_{\text{ave}} = (1 - p_b) \cdot 1 + p_b(1 - p_t) \cdot 1 + p_b \cdot p_t \cdot (1 + b - 1) f_d + p_b \cdot p_t \cdot (1 + b)(1 - f_d)$$

Note that in $1 + b - 1$, the -1 term is due to the cycles saved because of the delay slot.

$$T_{\text{ave}} = 1 - p_b + p_b - p_b p_t + p_b \cdot p_t \cdot f_d b + p_b \cdot p_t - p_b \cdot p_t \cdot f_d + p_b \cdot p_t \cdot b - p_b \cdot p_t \cdot b \cdot f_d$$

$$= 1 + p_b \cdot p_t \cdot b(1 - f_d)$$

The branch penalty term, b, has to be modified if delayed branching is employed, because a filled delay slot incurs no penalty (in practice, the penalty depends on the length of the pipeline and the point at which the branch is resolved). When a branch is resolved in a real machine depends on the actual architecture; for example, how the branch target address is generated and how the condition codes on which the branch depends are accessed.

If we assume $b = 1$, $p_b \cdot p_t = 0.2$, and $f_d = 0.8$, we have $T_{\text{ave}} = 1 + 1 \times 0.2 \times (1 - 0.8) = 1.04$ with delayed branching and $T_{\text{ave}} = 1 + 1 \times 0.2 = 1.2$ without delayed branching.

7.5   Branch Prediction

Cycles are lost because the pipeline contains partially executed instructions that aren't executed when a branch is taken (this process is called *flushing* the pipeline). If we knew that a branch instruction was going to be taken *before* it was executed, we could start filling the pipeline with instructions from the *branch target address*. For example, if the instruction BRA N is encountered, the processor can start fetching instructions at locations $N, N + 1, N + 2$, etc., as soon as the branch instruction is fetched from memory and decoded. In this way, the pipeline is always filled with useful instructions.

Prediction mechanisms works well with unconditional branches like BRA N. Conditional branches pose a problem. Consider the *branch to N on zero-bit set* instruction BEQ N. Should the processor make the assumption that the branch will not be taken and fetch instructions in sequence, or should it make the assumption that the branch will be taken and fetch instructions at the branch target address N? As we have already said, conditional branches are required to implement various types of high-level language construct. Consider the following fragment of high-level language code.

```
IF (J < K) I = I + L;
   (FOR T = 1; T <= I; T++) {
       .
       .
   }
```

The first conditional operation causes J to be compared with K. Only the nature of the problem will tell us whether J is often less than K, or whether J is often larger than K, or whether J is less than K for half the time and greater than K for half the time.

The second conditional operation in this fragment of code is provided by the FOR construct that tests a counter at the end of the FOR loop and then decides whether to jump back to the body of the construct or to terminate to loop. In this case, you could probably bet that the loop is more likely to be repeated than exited. Some loops are executed thousands of times before they are exited. Therefore, it might be a shrewd move to look at the *type* of conditional branch and then either fill the pipeline from the branch target if you think that the branch will be taken, or fill the pipeline from the instruction after the branch if you think that it will not be taken.

Various schemes have been devised to reduce the cost of branches by *predicting* the outcome of the branch. Before we look at these schemes, we will show how their effectiveness can be calculated. If we attempt to predict the behavior of a system with two possible outcomes (branch taken or branch not taken), there are four possibilities.

1. Predict the branch is taken and the branch is taken—successful outcome.
2. Predict the branch is taken and the branch is not taken—unsuccessful outcome.
3. Predict the branch is not taken and the branch is not taken—successful outcome.
4. Predict the branch is not taken and the branch is taken—unsuccessful outcome.

Let's apply a cost to each of these four possible outcomes, Table 7.13. Note that we can speak of *cost* or *penalty* when referring to branches. The cost is the total number of cycles required and the penalty is the additional number of cycles over the no-penalty situation; that is, penalty = cost − 1.

TABLE 7.13   The Branch Penalty

Result	Prediction	Branch Cost	Branch Penalty
Branch taken	Branch taken	a	$a - 1$
Branch taken	Branch not taken	b	$b - 1$
Branch not taken	Branch taken	c	$c - 1$
Branch not taken	Branch not taken	d	$d - 1$

© Cengage Learning 2014

Let's calculate the average penalty for a particular system. To do this we need more information. The first thing we need to know is the probability that an instruction will be a branch (as opposed to any other category of instruction). Assume that the probability that an instruction is a branch is p_b. The value of p_b can be measured from either static or a dynamic instruction counts. The next thing we need to know is the probability that the branch instruction will be taken, p_t. Finally, we need to know the accuracy of the prediction. Let p_c be the probability that a branch prediction is correct. Figure 7.45 illustrates all the possible outcomes of an instruction.

The branch cost (i.e., the number of cycles required) for a branch instruction is

$$C_{ave} = a(p_{branch\_predicted\_taken\_and\_taken}) + b(p_{branch\_predicted\_not\_taken\_but\_taken})$$
$$+ c(p_{branch\_predicted\_taken\_but\_not\_taken}) + d(p_{branch\_predicted\_not\_taken\_and\_not\_taken})$$

The cost of an instruction is given by $(1 - p_b) + p_b C_{ave}$.

FIGURE 7.45   The branch prediction tree

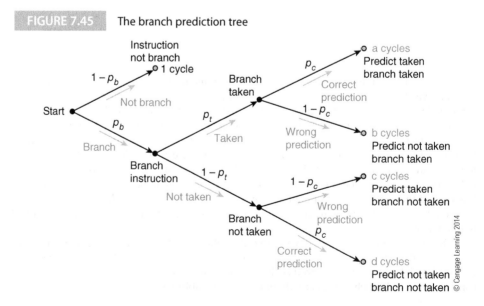

© Cengage Learning 2014

By using the principle that if one event or another must take place, their probabilities add up to unity, we can write

$(1 - p_b)$ = probability that an instruction is not a branch.

$(1 - p_t)$ = probability that a branch will not be taken.

$(1 - p_c)$ = probability that a prediction is incorrect.

The average number of cycles per branch is given by following each possible path through the trellis from the start and multiplying the probabilities and cost together, and then summing all paths; that is,

$$C_{ave} = a \cdot (p_t p_c) + b p_t \cdot (1 - p_c) + c \cdot (1 - p_t) \cdot (1 - p_c) + d \cdot (1 - p_t) \cdot p_c$$

This expression is not intuitively helpful. We can make two assumptions to help us to simplify this expression. The first is that $a = d = N$ (i.e., if the prediction is correct the number of cycles is N). The other simplification is that $b = c = B$ (i.e., if the prediction is wrong the number of cycles is B). The average number of cycles per instruction is therefore, $1 - p_b + p_b \cdot C_{ave}$; that is,

$$(1 - p_b) + p_b \cdot [N \cdot p_t p_c + B \cdot p_t \cdot (1 - p_c) + B \cdot (1 - p_t) \cdot (1 - p_c) + N \cdot (1 - p_t) \cdot p_c]$$
$$= (1 - p_b) + p_b \cdot [N \cdot p_c + B \cdot (1 - p_c)].$$

If we make the further assumption that there is no penalty for a correct prediction (i.e., $N = 1$), we get:

$$(1 - p_b) + p_b \cdot [1 \cdot p_c + B \cdot (1 - p_c)].$$

Static and Dynamic Branch Prediction

Two approaches to implementing branch prediction are *static branch prediction* and *dynamic branch prediction*. *Static branch prediction* assumes that a branch is always taken or never taken. Since observations of real code have demonstrated that branches have a greater than 50% chance of being taken, the simplest static branch prediction mechanism is to fetch the next instruction from the branch target address as soon as the branch instruction is detected.

A better method of statically predicting the outcome of a branch is to observe how it behaves in actual code, because some branch instructions are taken more or less frequently than other branch instructions. That is, you examine lots of code, determine how branches are used by compilers and make a branch prediction on the basis of these observations. Lilja claims that using the op-code to predict the outcome of a branch results in a 75% accuracy.

Static branch prediction can be extended by assigning a bit of the branch's op-code as a branch prediction flag. The prediction bit is set or cleared by the compiler, depending on whether the compiler estimates that the branch is most likely to be taken. You can provide this prediction *manually* in some assembly languages by appending, for example, the suffix h to an instruction to hint that the branch will be taken. This technique has a prediction accuracy in the range of 74 to 94%.[13] The Itanium that we discuss in the next chapter uses *hints* that can be appended to a branch instruction; for example, the hint . sptk (static, taken) indicates that the compiler has predicted that the branch will be taken, whereas the hint. dptk (dynamic, taken) indicates that the compiler predicts branch taken with less certainty.

Dynamic branch prediction techniques operate at runtime using the past behavior of the program to predict its future behavior. Suppose the processor maintains a table of branch instructions that contains information about the likely behavior of each branch. Each time a branch is executed, its outcome (i.e., taken or not taken) is used to update the corresponding entry in the table. The processor uses the table to determine whether to take the next instruction

---

[13] D. R. Ditzel and H.R. McLellan, "Branch Folding in the CRISP Microprocessor: Reducing the Branch Delay to Zero," *Proc 14th Ann. Symp. Computer Architecture*, 1987, pp. 2–9.

from the branch target address or from the next address in sequence. Lee and Smith[14] report that a single-bit branch predictor provides an accuracy of over 80 percent and a five-bit predictor provides an accuracy up to 98 percent. We now look at dynamic branch prediction in greater detail.

7.6   Dynamic Branch Prediction

There are many ways of implementing a run-time branch prediction mechanism. Predicting the future involves using information about the past to make an *educated guess*. We can create a flag and set it to the last outcome of a branch. If a branch was *not taken*, the flag would be set to N (i.e., Not) and the next prediction would be not taken. In other words, the prediction mechanism has a one-bit memory and the decision is to do what we did last time. Although this strategy may work well in loops, it's a poor predictor when there are occasional changes of branch direction and it gets the prediction 100% wrong when faced with the sequence TNTNTNT, ... However, if the string of branch decisions is TTTNNNNTTTTTNNN, a one-bit predictor, assuming an initial T (Taken) state would predict TTTTNNNN4TTTTTNN. The states in blue are the mispredictions.

If you use two bits of branch history, you can have a more sophisticated predictor. Indeed, as you add more bits to the predictor, it is able to make better predictions.

Let's begin with the simple and elegant 2-bit dynamic branch prediction algorithm that was incorporated in Intel's Pentium processor.[15] A two-bit register records four possible states of the predictor (see Figure 7.46) and uses the *recent history* of a branch to predict its future behavior. This branch predictor operates on a *per branch* basis; that is, a separate state counter is required for each branch in a program.

Consider the two-bit dynamic branch predictor of Figure 7.46 which attempts to predict the trend of a branch. This mechanism can accept one incorrectly predicted branch without changing its next prediction. However, if it gets a prediction wrong twice in succession, it reverses its prediction.

The state machine in Figure 7.46 represents a binary counter that counts up or down from 00 to 11. The two states 11 and 10 in Figure 7.46 represent branch *taken* states, and two states 01 and 00 represent branch *not taken* states. This is a saturating counter because the state 00 remains at 00 if it is decremented and the state 11 remains at 11 if it is incremented. This predictor introduces the notion of a *strongly* taken (or a strongly not taken) state. If a given branch instruction is taken (or not taken) two or more times in succession, the state machine assumes the *strongly* taken 11 state (or the strongly not taken 00 state).

Now, if a branch occurs that breaks the sequence, the state machine moves from a strong state to a *weak* state but it does not change its prediction. Suppose that the machine is in a strongly taken state, 11. It will predict that the next branch will also be taken. If, however, the next branch is not taken, it moves to the weakly taken state 10. In this state, it will still predict the branch will be taken in spite of the last misprediction.

In a weak state, the previous branch direction is assumed to be aberrant and untypical. On the next branch, one of two things may happen. Either the previous trend will be reinforced, causing a move back to the strongly taken state, or a second incorrectly predicted branch will put the counter in its weakly not taken state, 01, and will now change the direction of its next branch prediction. If a third branch is in the same *not taken* direction, the state machine will move all the way to its strongly *predict not taken* state, 00. The state machine in Figure 7.46 is also called a *saturating counter* because when a count goes up from 10 to 11 (or down from 01 to 00), further counts do not change the state (i.e., the counter is saturated). Recall that we first encountered the notion of saturation when dealing with the MMX instruction set.

[14] J.K.F. Lee and A.J. Smith, "Branch Prediction Strategies and Branch Target Buffer Design," *Computer*, Jan. 1984, pp. 6–22.

[15] M. Bekerman and A. Mendelson, "A performance analysis of Pentium processor systems", *IEEE Micro*, Vol. 15, No. 5, October 1995, pp. 72–83.

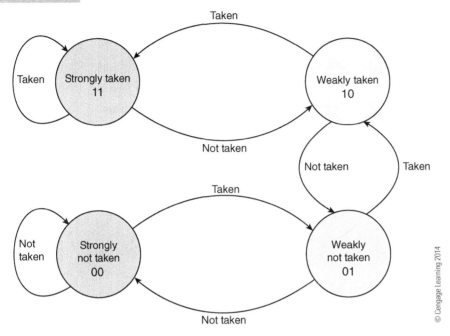

FIGURE 7.46   Saturating counter branch predictor state diagram

Consider the following example. Suppose a particular branch records the following sequence of outcomes (T = branch taken, N = branch not taken).

T T N T T T T N N N N N T N N N T T T T N T T T T

In this example, 16 branches are taken and 10 branches are not taken. If we assume that a branch that is taken costs 4 cycles and that a branch that is not taken costs 1 cycle, the total cost of this sequence is $16 \times 4 + 10 \times 1 = 74$ cycles.

Let's use the saturating counter to predict whether or not to take a branch. We will assume that the initial state of the branch predictor is ST (strongly taken). The outcome in the sequence below is recorded as *C* for a correct prediction and *W* for a wrong prediction.

Actual	T	T	T	N	T	T	T	T	N	N	N	N	N	T	N	N	N	T	T	T	T	N	T	T	T	T
State	ST	ST	ST	WT	ST	ST	ST	ST	WT	WN	SN	SN	SN	WN	SN	SN	SN	WN	WT	ST	ST	WT	ST	ST	ST	ST	ST
Prediction	T	T	T	T	T	T	T	T	T	N	N	N	N	N	N	N	N	N	T	T	T	T	T	T	T	T
Outcome	C	C	W	C	C	C	C	W	W	C	C	C	W	C	C	C	W	W	C	C	W	C	C	C	C

In this case the dynamic predictor makes 19 correct predictions and 7 wrong predictions and the total number of cycles is $19 \times 1 + 7 \times 4 = 47$ cycles.

The branch prediction state machine of Figure 7.46 is not the only possible model. Other state machines can be used to predict whether the next branch is likely to be taken on the basis of its past history. Figure 7.47 illustrates another possibility, a *two-bit predictor with hysteresis*, that makes a transition from a weakly predicted state to the opposite strongly predicted state on the first incorrectly predicted transition. Below are two sequences and their treatment by the predictors of Figures 7.46 and 7.47. Note how the hysteresis counter is able to deal with a glitch in the sequence by not changing state.

Saturating counter (Figure 7.46)

Actual	T	T	T	N	N	T	N	T	N	N
State	ST	ST	ST	WT	WN	WT	WN	WT	WN	
Prediction	T	T	T	T	N	T	N	T	N	
Outcome	C	C	W	W	W	W	W	W	C	

Counter with hysteresis (Figure 7.47)

Actual	T	T	T	N	N	T	N	T	N	N
State	ST	ST	ST	WT	SN	WN	SN	WN	SN	
Prediction	T	T	T	T	N	N	N	N	N	
Outcome	C	C	W	W	W	C	W	C	C	

FIGURE 7.47   Alternative 2-bit branch prediction state machine

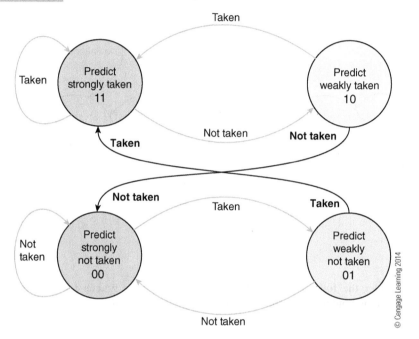

© Cengage Learning 2014

7.6.1   Branch Target Buffer

An important concept associated with branch penalty reduction is a special-purpose memory called a *branch target buffer* (BTB) or *branch target cache* that stores information about the branch instructions that are currently *active* in a program. We deal with caches in Chapter 9; all we need say here is that a cache is a fast memory holding frequently-used data that can be accessed far more rapidly than main store. The information cached can include: the address of the branch, the predicted outcome of the branch, its past history, the branch target address, and a copy of the instruction at the target address.

Figure 7.48 illustrates a simple form of BTB in which the address of the next instruction is stored together with a bit that indicates whether this branch is likely to be taken.[16]

Table 7.14 illustrates the operation of the branch target buffer as described by Kavi. In Table 7.14(a) instruction *i* fetched in cycle 0 is a *conditional branch*. Assume that this instruction has been cached in the BTB and that its target address is available immediately. The next instruction can be loaded from the target address provided by the BTB and we do not need to execute a new instruction fetch cycle. Consequently, there is no penalty for the branch.

Table 7.14(b) demonstrates what happens when the branch prediction stored in the BTB is wrong. Instructions are fetched from the target address cached in the BTB as soon as the branch is encountered. However, when the original conditional branch instruction reaches the

---

[16] K.M. Kavi. "Branch folding for conditional branches", *IEEE CS Technical Committee on Computer Architecture (TCCA) Newsletter*, Dec. 1997, pp. 4–7.

FIGURE 7.48   The simple branch target buffer

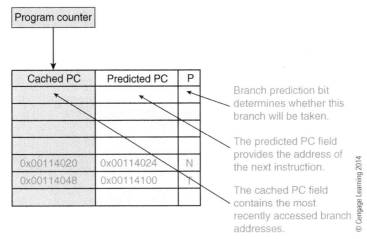

Program counter

Cached PC	Predicted PC	P
0x00114020	0x00114024	N
0x00114048	0x00114100	T

Branch prediction bit determines whether this branch will be taken.

The predicted PC field provides the address of the next instruction.

The cached PC field contains the most recently accessed branch addresses.

© Cengage Learning 2014

TABLE 7.14   Using the BTB to Reduce Pipeline Bubbles

(a) Branch correctly predicted

Cycle	Instruction	Fetch Instruction	Fetch Operand	Execute Instruction	Store Operand
0	Conditional branch	i	$i-1$	$i-2$	$i-3$
1	Predicted instruction	N	i	$i-1$	$i-2$
2	Predicted instruction + 1	$N+1$	N	i	$i-1$
3	Predicted instruction + 2	$N+2$	$N+1$	N	i
4	Predicted instruction + 3	$N+3$	$N+2$	$N+1$	N

(b) Branch mispredicted (M is the address where execution begins following misprediction)

Cycle	Instruction	Fetch Instruction	Fetch Operand	Execute Instruction	Store Operand
0	Conditional branch	i	$i-1$	$i-2$	$i-3$
1	Predicted instruction	N {bubble}	i	$i-1$	$i-2$
2	Predicted instruction + 1	$N+1$ {bubble}	N {bubble}	i	$i-1$
3	Mispredicted instruction	M	$N+1$ {bubble}	N {bubble}	i
4	Mispredicted instruction + 1	$M+1$	M	$N+1$ {bubble}	N {bubble}

© Cengage Learning 2014

execution unit, the misprediction is detected and the program counter has to be reloaded with the correct target address (labeled M in Table 7.14(b)). This misprediction costs two cycles.

Kavi proposed an ingenious modification to the conventional BTB that, essentially, *hedges the bet*. As well as storing the branch target address alongside each cached branch address, the instruction at the mispredicted address is also stored (Figure 7.49). If a branch is correctly predicted, instructions are fetched from the target address as in Table 7.14(a). If a branch is mispredicted, we don't detect it until the branch instruction is in its execute phase. Because we've cached the instruction at the mispredicted address, we have immediate access to this instruction and we can begin to execute it without having to issue a new fetch cycle.

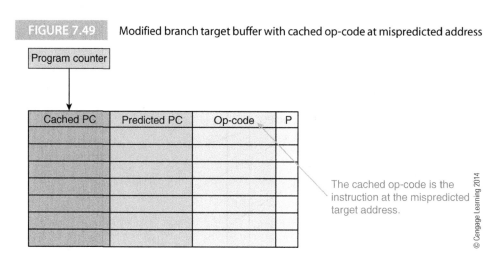

FIGURE 7.49   Modified branch target buffer with cached op-code at mispredicted address

Program counter

Cached PC	Predicted PC	Op-code	P

The cached op-code is the instruction at the mispredicted target address.

© Cengage Learning 2014

TABLE 7.15   The Effect of Caching the Op-Code at the Mispredicted Address

Cycle	Instruction	Fetch instruction	Fetch operand	Execute instruction	Store operand
0	Conditional branch	i	$i - 1$	$i - 2$	$i - 3$
1	Predicted instruction	N {bubble}	i	$i - 1$	$i - 2$
2	Mispredicted instruction	M	N {bubble}	i	$i - 1$
3	Mispredicted instruction + 1	$M + 1$	M	N {bubble}	i
4	Mispredicted instruction + 2	$M + 2$	$M + 1$	M	N {bubble}
5	Mispredicted instruction + 3	$M + 3$	$M + 2$	$M + 1$	M

© Cengage Learning 2014

Table 7.15 demonstrates what happens when we are using the modified branch target buffer after a misprediction. The mispredicted instruction is immediately available after the branch has been resolved (that's in Cycle 2 in Table 7.15) because it's been cached. The program counter is automatically adjusted to access the *second* instruction in the mispredicted instruction path because we've taken the first instruction from the buffer.

Some branch target buffers cache a single instruction at the branch target address, whereas others cache several consecutive instructions. Equally, some branch target buffers store only the target address of the branch.

The AMD Am29000 RISC processor employs a BTB arranged as a two-way set-associative cache that can hold 128 instructions (i.e., 512 bytes).[17] Each entry in the cache contains the first *four* instructions at the branch target address. Whenever a *non-sequential* fetch occurs, the address for the fetch operation is transmitted to the BTB at the same time the address is transmitted to the memory via the Am29000's memory management unit. If the target instruction for the fetch is currently in the BTB, it is fetched for decoding in the next cycle.

If, however, the target instruction is not currently in the BTB cache, the instruction must be fetched from the external memory and the BTB updated. The Am29000 implements a random replacement algorithm based on the state of the processor's clock. If a second branch instruction occurs in the four instructions that are cached, the branch is executed before the BTB is filled and the BTB contains less than four instructions.

---

[17] *Am29000 User's Manual*, Advanced Micro Devices, Sunnyvale, CA.

Because the entry for a cached branch contains less than four instructions, the outcome of subsequent accesses to this block depends on whether the branch is taken or not. If a cached branch is taken, it doesn't matter that the entry in the cache contains less than four instructions. If the branch is not taken, the *missing* instructions must be fetched from the external memory.

The Am29000 also has an instruction *fetch-ahead* mechanism. When a non-sequential instruction is encountered, the BTB is accessed and a hit indicates that the four instructions at the branch target address are already on-chip. The Am29000 calculates an address four instructions beyond the branch target address (i.e., the address of the next instruction immediately following those that are cached) and begins to fetch instructions into its *instruction prefetch buffer*. This fetch-ahead mechanism keeps the bus between the processor and external memory busy.

AMD's literature states that without the BTB a branch that is taken results in a one-cycle branch execution time plus a five-cycle penalty because of the need to refill the instruction pipeline. If the average instruction mix contains 20% branches that are taken, the average cycle time for an Am29000 would be 0.2×6 cycles $+ 0.8 \times 1$ cycle $= 2.0$ cycles; that is, the effect of branches halves the processor's overall performance. The BTB has an average hit-rate of 60% and its overall performance is therefore, $0.8 \times 1 + 0.2 \times (0.4 \times 6 + 0.6 \times 1) = 0.2 \times 3 + 0.8 = 1.4$ cycles.

Recall that the average number of cycles required to execute an instruction is $T_{ave} = 1 + p_b p_t b$, where p_b is the probability that an instruction is a branch, p_t is the probability that a branch will be taken, and b is the branch penalty. This formula can be extended to take account of the BTB to give

$$T_{ave} = 1 + p_b p_t ((1 - p_m)b_1 + b_2 p_m),$$

where b_1 is the branch penalty when the non-sequential instruction is in the BTB, b_2 is the branch penalty when the non-sequential instruction is not in the BTB, and p_m is the probability of a miss when the BTB is accessed.

Calder and Grunwald[18] suggest that the effectiveness of the BTB can be increased by storing only branches that are taken in the BTB. If a branch is not cached in the BTB and is not taken, it is not entered in the BTB. A branch that is taken is always entered in the BTB. The philosophy behind this approach is that a branch that is not taken does not increase the branch penalty as much as a taken branch, because the next instruction has already been fetched into the pipeline. By excluding branches that are not taken, the BTB will have more room for the branches that are taken.

Branch target buffers and branch predictors are different entities. However, both schemes can be combined to improve efficiency. Entries in the BTB can be extended to include branch histories as well as the instruction(s) at the target address.

7.6.2   Two-Level Branch Prediction

The pressure to reduce the branch penalty led to a burst of research activity in the early 90s into ways of accurately predicting the outcome of conditional branches. Predicting the outcome of a given branch isn't as simple as predicting the outcome of a horse race—nor is it as scientific as reading tarot cards or tea leaves. Predicting the outcome of a branch is tricky. Consider,

```
IF x < 50 then y = y + 4;
```

You cannot predict the outcome of this branch without knowing the value of x. If you encounter this operation a second time, predicting the same outcome as last time may be a good strategy, because the value 50 might be a limit that is exceeded only occasionally. Life gets

[18] B. Calder and D. Grunwald, "Fast & accurate instruction fetch and branch predication", *IEEE Proceeding of the 21st Annual Intl. Symposium on Computer Architecture*. 1994, pp. 2.11,

even more complex when branches *interact* with each other; that is, when the outcome of one branch affects the outcome of a succeeding branch. Consider,

```
IF x < 50 then y = y + 4;
.

.
IF y = 7 then z = 2;
```

The second branch tests y and makes a decision. However, the first branch tests x and makes a decision that modifies y. Consequently, the first branch may *affect* the outcome of the second branch; that is, there is a degree of correlation between branches.

We need a means of observing the interaction between multiple branches. Figure 7.50 demonstrates the notion of the *branch pattern history*, BPH, used by some adaptive predictors. When a branch is encountered, the outcome of the branch is recorded in a shift register. In this example, there are three branches in the program, A, B, and C, and the shift register is seven bits deep; that is, only the most recent seven branch outcomes are recorded. Suppose that the

Global Behavior

The following diagram demonstrates what capturing global behavior means. Each of the boxes contains a branch that appears in the program. Note that some branches appear several times because they can be reached by different paths through the program. Below each branch is the decision box that represents taken or not taken.

We have drawn two lines through this network, one in solid blue and one in broken blue. Each of these represents a particular path through the code and is labeled by the path in terms of the branch history (i.e., 111 or 010). As you can imagine, a knowledge of the path is likely to help make future predictions.

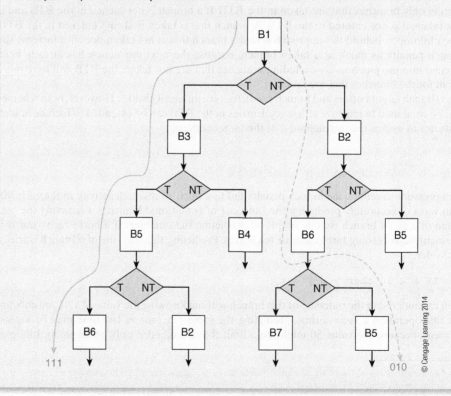

FIGURE 7.50   Recording branch sequences—the branch pattern history

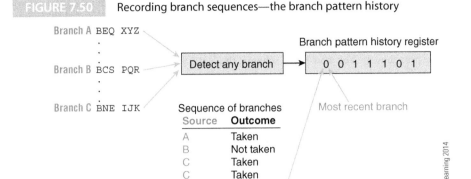

branches are executed in the sequence *ABCCBAC* and the outcomes of the corresponding branches are T,N,T,T,T,N,N, respectively. This sequence puts 0011101 in the branch pattern history register which can be used as a key in a prediction mechanism (note that the TNTTTNN and 0011101 sequences are written in reverse order).

Figure 7.51 demonstrates the way in which the branch pattern history can be used to make a prediction by indexing into a table of 2-bit state machines (we use only three bits of BPH to keep the diagram simple). Assume that the outcomes of the last three branches are: taken, not taken, taken. This sequence provides the vector 101_2 and entry 5 in the table is accessed. Suppose that the state machine in that entry is currently in the *weakly taken* state. The next branch will be predicted as taken. If, now this branch is indeed taken, the state counter is changed to *strongly taken* and the new value in the shift register is updated to 110. If the branch were not taken, the state predictor would be changed to *weakly not taken* and the new value in the shift register would become 010. A more practical arrangement might record the last 12 branches. This mechanism has been demonstrated to be effective. However, it is rather conceptually difficult to see how it actually works. The branch pattern history buffer contains a sequence of bits that correspond to a particular path through the code. You can think of this path as a *signature* that represents that path. That signature indexes into a table of saturating state machines that learn to follow that signature.

FIGURE 7.51   Using the pattern history to index into a table of individual state machines

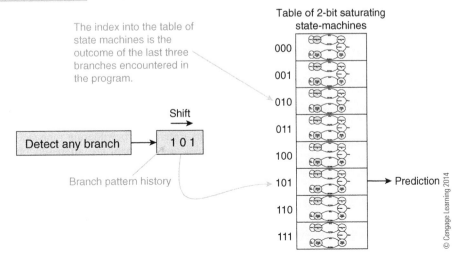

The branch pattern history makes a prediction based only on the previous sequence of branches. It doesn't take into account *where* the branches are in memory. Yeh and Patt[19] proposed extensions to simple branch prediction tables that use *branch-correlation* or *two-level branch prediction*. This prediction mechanism takes account of both the *history* and the *location* of the branches. In Yeh and Patt's scheme, the last *k* branches encountered provide the first-level branch execution history, and the last *j* occurrences of a specific branch provide the second-level branch execution history.

Figure 7.52 illustrates the concept of a two-level predictor. The first level of prediction is obtained from the low-order bits of the program counter. This information is related to the *location* of the current branch in the program (we've used 3 bits for simplicity). Note that we do not use the two least-significant PC bits as these are always 00 in a byte-addressed 32-bit machine.

FIGURE 7.52   Two-level branch prediction

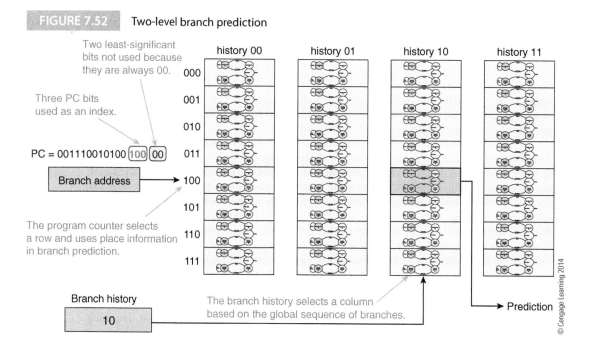

The second level of information is related to the recent history of all branches (here we've used just two bits). The address bits and the history bits select the appropriate state machine to provide the next prediction. Real systems don't simply use the low-order bits of the program counter as an index into the tables, they use a more complex hashing mechanism based on the low-order bits of the PC.

Research papers written on two-level branch predictors describe a bewildering range of possible prediction mechanisms, all based on the same theme. One variation is *global* versus *local* (i.e., individual). In Figure 7.52 the branch history is *global* because there is only one register that

Branch Predictor Naming

Yeh and Patt proposed a three letter naming scheme for two level branch predictors. The letters are:

- G, P, S for global, per branch, per set history.
- A, S for pattern history type (adaptive or static)
- g, p, s for pattern history table organization (global, per branch, per set).

Possible two-level predictors are GAg, GAs, GAp, PAg, PAs, PAp, SAg, SAs, and SAp. These are not the only possible two-level structures.

---

[19] Tse-Yu Yeh and Y. N. Pratt, "Alternative implementations of two-level adaptive branch predictions", *19th Annual International Symposium of Computer Architecture*, Portland, OR, December 1992, pp. 124–134.

accommodates all branches. We can also construct a *per branch* history table that stores the sequence of outcomes for each branch. In this case, we would use the branch as an index into the table of state machines.

Branch prediction literature employs a specific three-character notation to describe two-level predictors. For example, the notation GAg uses *G* to indicate a single *global* branch history register, *A* to indicate *adaptive*, and *g* to indicate a single pattern history table. Figure 7.53 describes a GAp predictor. In this case, there is still a single global branch history register but there is a *per branch* pattern history table, hence the *p* notation. In this notation, *p* indicates "per branch or per instruction" and *G* indicates global.

If we have a table of state machines just indexed by the address of the branch, the predictor is represented by PAg because there is a per *address* index into the table but a global pattern history table. Figure 7.53 describes the PAg and PAp predictors.

FIGURE 7.53   The PAg and PAp branch outcome predictors

(a) The PAg predictor with a global PHT

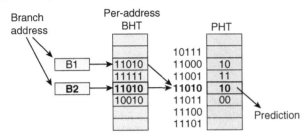

(b) The PAp predictor with a per branch PHT

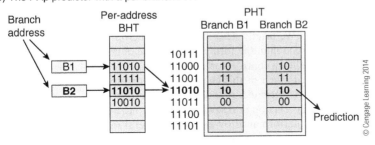

© Cengage Learning 2014

Combining Instruction Addresses and Branch History

Two simple ways of combining both branch address and branch history information have been proposed that provide a vector into the PHT (pattern history table), without the cost and complexity of a two-level branch predictor. The *gselect* predictor takes the least-significant *p* bits of the *word* branch address and replaces them with the *p*-bit global branch pattern history value; for example, if the current branch address is 000111101000 and the 6-bit branch history is 011110, the index into the PHT is 0001**011110**. Combining an *address* with a *sequence* of bits that determine whether the last *k* branches were taken or not taken does seem strange, but it is effective.

The *gshare* predictor operates in a similar fashion except that the least-significant *p* bits of the branch address and the *p* bits of the global history register are applied to *p* XOR gates to generate a composite vector. If we use the previous example with the current branch address of 000111101000 and the branch history is 011110, the index into the PHT is 0001**100100**. Note that the two least-significant bits of the branch address are grayed out as these are always zero and don't take part in the branch prediction process.

A particularly interesting predictor is the *tournament predictor* that includes a predictor predictor; yes, it has three predictors and one of them predicts which of the other two should be used next. Consider a system with two different predictors, P1 and P2, one basing its output on the program's local branch history and one on the global branch history. As each branch is encountered, both predictors guess the next outcome. The third predictor, P3, looks at the two outcomes and selects one. The selection algorithm is simple. At start up, P3 initially selects P1 or P2 as the correct output, say, P1. If the prediction made by P2 differs to that made by P1, P3 remains with P1's prediction. However, if P1 gets the next prediction wrong and P2 gets it right, P3 switches to accepting P2's prediction.

Studies on the performance of branch prediction techniques make interesting reading. Amit,[20] et al. summarize the results of work carried out on branch prediction. The performance increase obtained by using a 2-bit predictor rather than a 1-bit predictor is surprisingly small. Moreover, correlation-based schemes are superior to simple counter-based schemes only for sufficiently large branch target buffers, BTBs. Indeed, they stress that branch prediction strategy is secondary to the BTB hit ratio; that is, resources are better spent increasing the size of the branch target buffer rather than trying to reliably predict the outcome of branches. However, the branch prediction accuracy generally ranges from 95 to 100% accuracy.

Summary

Up to this chapter, we've been concerned with the instruction architecture and have paid little attention to how processors operate or are organized internally. Now we've looked at what happens inside the silicon when a program is executed. We began by describing the arrangement of registers, buses, and functional units in a processor that are used to implement a digital computer.

The control unit is a key element of a computer that is responsible for interpreting machine instructions. We briefly described how a microprogrammed control unit interprets a machine-level instruction in the form of a sequence of microoperations that are stored in a read-only memory called a control store.

We have devoted much of this chapter to the organizational feature that has done so much to improve the performance of today's computer, *pipelining*. Pipelined computers divide the execution of an instruction into several stages (up to 31 in the case of processors like the Pentium) and boost performance by executing several instructions at once. We started by describing a flowthrough computer and demonstrating how an instruction can be executed in a single cycle by following the path from the program counter to the memory where an operand is written back. Then we looked at the structure of a *pipelined computer* where functional units are separated by registers that hold data while each functional unit is processing its data.

Unfortunately, pipelining runs into a difficulty when non-sequential branch, subroutine calls, or subroutine returns are encountered. Once a jump to a non-sequential location is executed, all the partially completed instructions in the pipeline following the current instruction have to be discarded. The effect of branch instructions can severely degrade the performance of a RISC processor.

This chapter has looked at some of the ways in which the so-called branch-penalty can be overcome. Some systems always execute the instruction immediately following a branch in order to maximize the use of the pipeline—this is called a delayed branch. Some processors attempt to guess the outcome of a branch and start reading instructions from the branch target address. In the next chapter we take another step into computer organization and look at how superscalar processors increase throughput by means of multiple processing units.

---

[20] [Amit95] Amit Mital and Barry Fagin, "The Performance of Counter- and Correlation-Based Schemes for Branch Target Buffers," *Transactions on Computers,* December 1995 (Vol. 44, No. 12), pp. 1383–1393.

Problems

7.1 For the microprogrammed architecture of Figure P 7.1, give the sequence of actions required to implement the instruction ADD **D0**, **D1** which is defined in RTL as

[D1] ← [D1] + [D0].

You should describe the actions that occur in plain English (e.g., "Put data from this register on that bus") and as a sequence of events (e.g., Read = 1, E_{MSR}). The following table defines the effect of the ALU's function code. Note that all data has to pass through the ALU (the copy function) to get from bus B or bus C to bus A.

F_2	F_1	F_0	Operation	
0	0	0	Copy P to bus A	$A = P$
0	0	1	Copy Q to bus A	$A = Q$
0	1	0	Copy $P + 1$ to bus A	$A = P + 1$
0	1	1	Cop $Q + 1$ to bus A	$A = Q + 1$
1	0	0	Copy $P - 1$ to bus A	$A = P - 1$
1	0	1	Copy $Q - 1$ to bus A	$A = Q - 1$
1	1	0	Copy bus $P + Q$ to bus A	$A = P + Q$
1	1	1	Copy bus $P - Q$ to bus A	$A = P - Q$

© Cengage Learning 2014

FIGURE P7.1   Architecture of a hypothetical computer

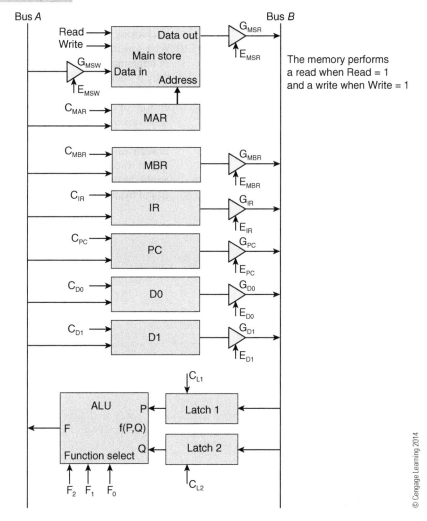

The memory performs a read when Read = 1 and a write when Write = 1

© Cengage Learning 2014

7.2 For the architecture of Figure P7.1 write the sequence of signals and control actions necessary to implement the fetch cycle.

7.3 Why is the structure of Figure P7.1 so inefficient?

7.4 Why is the ALU instruction set of Figure P7.1 so inefficient?

7.5 For the architecture of Figure P7.1, write the sequence of signals and control actions necessary to execute the instruction ADD M, **D0** that adds the contents of memory location M to data register D0 and deposits the results in D0. Assume that the address M is in the instruction register IR.

7.6 This question asks you to implement *register indirect addressing*. For the architecture of Figure P7.1, write the sequence of signals and control actions necessary to execute the instruction ADD (D1), **D0** that adds the contents of the memory location pointed at by the contents of register D1 to register D0, and deposits the results in D0. This instruction is defined in RTL form as [D0] ← [[D1]] + [D0].

7.7 This question asks you to implement *memory indirect addressing*. For the architecture of Figure P7.1, write the sequence of signals and control actions necessary to execute the instruction ADD [M], **D0** that adds the contents of the memory location pointed at by the contents memory location M to register D0, and deposits the results in D0. This instruction is defined in RTL form as [D0] ← [[M]] + [D0].

7.8 This question asks you to implement *memory indirect addressing with index*. For the architecture of Figure P7.1, write the sequence of signals and control actions necessary to execute the instruction ADD [M,D1], **D0** that adds the contents of the memory location pointed at by the contents memory location M plus the contents of register D1 to register D0, and deposits the results in D0. This instruction is defined in RTL form as [D0] ← [[M]+[D1]] + [D0].

7.9 For the microprogrammed architecture of Figure P7.1, define the sequence of actions (i.e., micro-operations) necessary to implement the instruction TXP1 (D0)+,D1 that is defined as:

[D1] ← 2*[M[D0]] + 1
[D0] ← [D0] + 1

Explain the actions in plain English and as a sequence of enables, ALU controls, memory controls and clocks. This is quite a complex instruction because it requires a register-indirect access to memory to get the operand and it requires multiplication by two (there is no ALU multiplication instruction). You will probably have to use a temporary register to solve this problem and you will find that it requires several cycles to implement this instruction. A *cycle* is a sequence of operations that terminates in clocking data into a register.

7.10 Why was microprogramming such a popular means of implementing control units in the 1980s?

7.11 Why is microprogramming so unpopular today?

7.12 Figure P7.12 from the text demonstrates the execution of a conditional branch instruction in a flow-through computer. The grayed out sections of the computer are not required by a conditional branch instruction. Can you think of any way in which these unused elements of the computer could be used during the execution of a conditional branch?

FIGURE P7.12   Architecture of a hypothetical computer

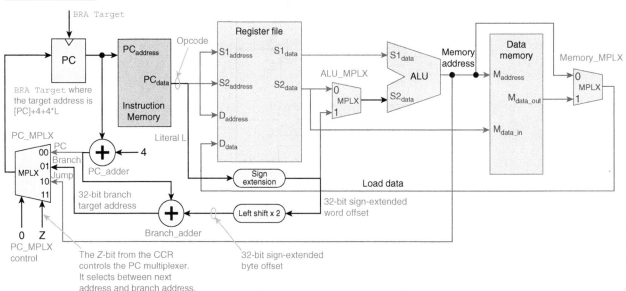

7.13 What modifications would have to be made to the architecture of the computer in Figure P7.12 to implement predicated execution like the ARM?

7.14 What modifications would have to be added to the computer of Figure P7.12 to add a conditional move instruction with the format MOVZ **r1**,r2,r3 that performs [r1] ← [r2] if [r3] == 0.

7.15 What modifications would have to be made to the architecture of the computer in Figure P7.12 to implement operand shifting (as part of a normal instruction) like the ARM?

7.16 Derive an expression for the speedup ratio (i.e., the ratio of the execution time without pipelining to the execution time with pipelining) of a pipelined processor in terms of the number of stages in the pipeline m and the number of instructions to be executed N.

7.17 In what ways is the formula for the speedup of the pipeline derived in the previous flawed?

7.18 A processor executes an instruction in the following six stages. The time required by each stage in picoseconds (1,000 ps = 1 ns) is given for each stage.

IF	instruction fetch	300 ps
ID	Instruction decode	150 ps
OF	Operand fetch	250 ps
OE	Execute	350 ps
M	Memory access	700 ps
OS	Operand store (writeback)	200 ps

a. What is the time to execute an instruction if the processor is not pipelined?

b. What is the time taken to fully execute an instruction assuming that this structure is pipelined in six stages and that there is an additional 20 ps per stage due to the pipeline latches?

c. Once the pipeline is full, what is the average instruction rate?

d. Suppose that 25% of instructions are branch instructions that are taken and cause a 3-cycle penalty, what is the effective instruction execute time?

7.19 Both RISC and CISC processors have registers. Answer the following questions about registers.

a. Is it true that a larger number of registers in any architecture is always better than a smaller number?

b. What limits the number of registers that can be implemented by any ISA?

c. What are the relative advantages and disadvantages of dedicated registers like the IA32 architecture compared to general purpose registers like ARM and MIPS?

d. If you have an m-bit register select field in an instruction, you can't have more than 2^m registers. There are, in fact, ways round this restriction.

Suggest ways of increasing the number of registers beyond 2^m while keeping an m-bit register select field.

7.20 Someone once said, "RISC is to hardware what UNIX is to software." What do you think this statement means and is it true?

7.21 What are the characteristics of a RISC processor that distinguish it from a CISC processor? Does it matter whether this question is asked in 2015 or 1990?

7.22 What, in the context of pipelined processors, is a *bubble* and why is it detrimental to the performance of a pipelined processor?

7.23 To say that the RISC philosophy was all about reducing the size of instruction sets would be wrong and entirely miss the point. What enduring trends or insights did the so-called RISC revolution bring to computer architecture including both RISC and CISC design?

7.24 There are RAW, WAR, and WAW data hazards. What about RAR (read-after-read)? Can a RAR operation cause problems in a pipelined machine?

7.25 Consider the instruction sequence in a five-stage pipeline IF, OF, E, M, OS:

1. ADD **r0**,r1,r2
2. ADD **r3**,r0,r5
3. STR r6, **[r7]**
4. LDR **r8**,[r7]

Instructions 1 and 2 will create a RAW hazard. What about instructions 3 and 4? Will they also create a hazard?

7.26 A RISC processor has a three-address instruction format and typical arithmetic instructions (i.e., ADD, SUB, MUL, DIV etc). Write a suitable sequence of instructions to evaluate the following expression in the minimum time:

$$X = \frac{(A + B)(A + B + C)E + H}{G + A + B + D + F(A + B - C)}$$

Assume that all variables are in registers and that the RISC does not include a hardware mechanism for the elimination of data dependency. Each instance of data dependency causes one bubble in the pipeline and wastes one clock cycle.

7.27 Figure P7.27 gives a partial skeleton diagram of a pipelined processor. What is the purpose of the flip-flops (registers) in the information paths?

7.28 Explain why branch operations reduce the efficiency of a pipelined architecture. Describe how branch prediction improves the performance of a RISC processor and minimizes the effect of branches?

7.29 Assume that a RISC processor uses branch prediction to improve its performance. The following table gives the number of cycles taken for predicted and actual branch outcomes. These figures include both the cycles taken by the branch itself and the branch penalty associated with branch instructions.

FIGURE P7.27   Structure of a Pipelined Processor

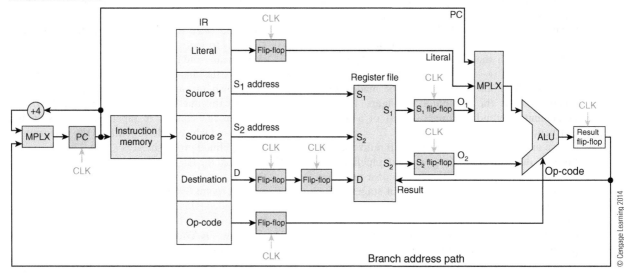

© Cengage Learning 2014

	Actual	
Prediction	Not taken	Taken
Not taken	1	4
Taken	2	1

If p_b is the probability that a particular instruction is a branch, p_t is the probability that a branch is taken, and p_w is the probability of a wrong prediction, derive an expression for the average number of cycles per instruction, T_{ave}. All non-branch instructions take one cycle to execute.

7.30 IDT application note AN33 gives an expression for the average number of cycles per instruction in a RISC system as:

$$C_{ave} = P_b(1 + b) + P_m(1 + m) + (1 - P_b - P_m)$$

where:

p_b = probability that an instruction is a branch
b = branch penalty
p_m = probability that an instruction is a memory reference
m = memory reference penalty

Explain the validity of this expression. How do you think that it might be improved?

7.31 RISC processors rely (to some extent) on on-chip registers for their performance increase. A cache memory can provide a similar level of performance increase without restricting the programmer to a fixed set of registers. Discuss the validity of this statement.

7.32 RISC processors best illustrate the difference between architecture and implementation. To what extent is this statement true (or not true)?

7.33 A RISC processor executes the following code. There are no data dependencies.

```
ADD  r0,r1,r2
ADD  r3,r4,r5
ADD  r6,r7,r8
ADD  r9,r10,r11
ADD  r12,r13,r14
ADD  r15,r16,r17
```

a. Assuming a four-stage pipeline (fetch, operand fetch, execute, and write) what registers are being read during the sixth clock cycle and what register is being written?

b. Assuming a five-stage pipeline (fetch, operand fetch, execute, write, and store) what registers are being read during the sixth clock cycle and what register is being written?

7.34 A RISC processor executes the following code. There are data dependencies but no internal forwarding. A source operand cannot be used until it has been written.

```
ADD  r0,r1,r2
ADD  r3,r0,r4
ADD  r5,r3,r6
ADD  r7,r0,r8
ADD  r9,r0,r3
ADD  r0,r1,r3
```

a. Assuming a four-stage pipeline fetch, operand fetch, execute, and result write, what registers are being read during the tenth clock cycle and what register is being written?

b. How long will it take to execute the entire sequence?

7.35 A RISC processor has an eight-stage pipeline: F D O E1 E2 MR MW WB (fetch, decode, register read operands, execute 1, execute 2, memory read, memory write, and result writeback to register). Simple logical and arithmetic operations are complete by the end of E1. Multiplication is complete by the end of E2. How many cycles are required to execute the following code assuming that internal forwarding is not used?

```
MUL  r0,r1,r2
ADD  r3,r1,r4
ADD  r5,r1,r6
ADD  r6,r5,r7
LDR  r1,[r2]
```

7.36 Repeat the previous problem assuming that internal forwarding is implemented?

7.37 Consider the same structure as problem 7.35 but with the following code fragment. Assume that internal forwarding is possible and an operand can be used as soon as it is generated. Show the execution of this code.

```
LDR  r0,[r2]
ADD  r3,r0,r1
MUL  r3,r3,r4
ADD  r6,r5,r7
STR  r3,[r2]
ADD  r6,r5,r7
```

7.38 The following table gives a sequence of instructions that are performed on a four-stage pipelined computer. Detect all hazards. For example if instruction m uses operand r2 generated by instruction m-1, then write m-1,r2 in the RAW column in line m.

Number	Instruction	RAW	WAR	WAW
1	Add r1,r2,r3			
2	Add r4,r1,r3			
3	Add r5,r1,r2			
4	Add r1,r2,r3			
5	Add r5,r2,r3			
6	Add r1,r6,r6			
7	Add r8,r1,r5			

© Cengage Learning 2014

7.39 Consider the following code:

```
LDR  r1,[r6]      ;Load r1 from memory.
                   r6 is a pointer
ADD  r1,r1,#1     ;Increment r1 by 1
LDR  r2,[r6,#4]   ;Load r2 from memory
ADD  r2,r2,#1     ;Increment r2 by 1
ADD  r3,r1,r2     ;Add r1 and r2 with
                   total in r3
ADD  r8,r8,#4     ;Increment r8 by 4
STR  r2,[r6,#8]   ;Store r2 in memory
SUB  r2,r2,#64    ;Subtract 64 from r2
```

The processor has a five-stage pipeline F O E M S; that is, instruction fetch, operand fetch, operand execute, memory, and operand writeback to register file.

a. How many cycles does this code take to execute assuming internal forwarding is not used?

b. How many cycles does this code take to execute assuming internal forwarding is used?

c. How many cycles does the code take to execute assuming that it is reordered (no internal forwarding)?

d. How many cycles does the code take to execute assuming reordering and internal forwarding?

7.40 Why do conditional branches have a greater effect on a pipelined processor than unconditional branches?

7.41 Describe the various types of *change of flow-of-control* operations that modify the normal sequence in which a processor executes instructions. How frequently do these operations occur in typical programs?

7.42 Consider the following code:

```
     MOV  r0,#Vector   ;point to Vector
     MOV  r2,#10       ;loop count
Loop LDR  r1,[r0]      ;Repeat: get element
     SUBS r2,r2,#1     ;decrement loop count
                        and set Z flag
     MUL  r1,r1,#5
     STR  r1,[r0]      ;save result
     ADD  r0,r0,#4     ;point to next
     BNE  Loop         ;until all done
                        (branch on Z flag)
```

Suppose this ARM-like code is executed on a four-stage pipeline with internal forwarding. The load instruction has one cycle penalty and the multiply instruction introduces two stall cycles into the execute phase. Assume the taken branch has no penalty.

a. How many instructions are executed by this code?

b. Draw a timing diagram for the first iteration showing stalls. Assume internal forwarding.

c. How many cycles does it take to execute this code?

7.43 Branch instructions may be taken or not taken. What is the relative frequency of taken to not taken, and why is this so?

7.44 What is branchless computing?

7.45 What is a delayed branch and how does it contribute to minimizing the effect of pipeline bubbles? Why are delayed branch mechanisms less popular then they were?

7.46 How does branch prediction reduce the branch penalty?

When calculating the cost of a branch, we derived two expressions. One was $1 - b.p_e$ and the other was $(1 - p_b) + p_b \cdot [1 \cdot p_c + B \cdot (1 - p_c)]$. Demonstrate that these are the same result.

7.47 A pipelined computer has a four-stage pipeline: fetch/decode, operand fetch, execute, writeback. All operations except load and branch do not introduce stalls. A load introduces one stall cycle. A non-taken branch introduces not stalls and a taken branch introduces two stall cycles. Consider the following loop.

```
for (j=1023; j > 0; j--) {x[j]=x[j]+2;}
```

a. Express this code in an ARM-like assembly language (assume that you cannot use autoindexed addressing and that the only addressing mode is of the register indirect of the form [r0]).
b. Show a single trip round the loop and indicate how many clock cycles are required.
c. How many cycles will it take to execute this code in total.
d. How can you modify the code to reduce the number of cycles?

7.48 Suppose that you design an architecture with the following characteristics:
- Cost of a non-branch instruction   1 cycle
- Fraction of instructions that are branches 20%
- Fraction of branches that are taken   85%
- Fraction of delay slots that can be filled 50%
- Cost of an unfilled delay slot   1 cycle

For this architecture, determine the following.

a. Calculate the average number of cycles per instruction.
b. Calculate the improvement (as a percentage) if the fraction of delay slots that are filled can be increased to 95%.

7.49 A pipelined processor has the following characteristics:
- Loads   18%
- Load stall (load penalty)   1 cycle
- Branches   22%
- Probability a branch is taken   80%
- Branch penalty on taken   3 cycles
- RAW dependencies   20% of all instructions except branches
- RAW penalty   1 cycle

Estimate the average cycles per instruction for this processor.

7.50 What is the difference between *static* and *dynamic* branch prediction?

7.51 A processor has a branch-target buffer. If a branch is in the buffer and it is correctly predicted, there is no branch penalty. The prediction rate is 85 correct. If it is incorrectly predicted, the penalty is 4 cycles. If the branch is not in the buffer, and not taken, the penalty is two cycles. Seventy percent of branches are taken. If the branch is not in the buffer and is taken the penalty is three cycles. The probability that a branch is in the buffer is 90%. What is the average branch penalty?

7.52 How can the compiler improve the efficiency of some processors with branch prediction mechanisms?

7.53 Consider the following two streams of branch outcomes (T = taken and N = not taken). In each case, what is the simplest form of branch prediction mechanism that would be effective in reducing the branch penalty?

a. T, T, T, T, T, N, T, T, T, T, T, T, T, T, N, T, T, T, T, T, N, T, T, T, T, T, T, T, N, T, T, T, T, T

b. T, T, T, T, T, N, N, N, N, N, N, N, N, N, T, T, T, T, T, T, T, T, T, T, T, N, N, N, N, N, N, N, N

7.54 A processor uses a 2-bit saturation-counter dynamic branch predictor with the states strongly taken, weakly taken, weakly not taken, and strongly not taken. The symbol T indicates a branch that is taken and an N indicates a branch that is not taken. Suppose that the following *predicted* sequence of branches is recorded:

TTTNTX

What is the value of X?

7.55 The following sequence of branch outcomes is applied to a saturating counter branch predictor TTTNTTNNNNTNNNTTTTTNTTTNNTTTTNT. If the branch penalty is two cycles for an incorrectly predicted branch, how many additional cycles does the system incur for the above sequence of 30 branches? Assume that the predictor is initially in the strongly predicted taken state.

7.56 The state diagram in Figure P7.56 represents one of the many possible 2-bit state machines that can be used to perform prediction. Explain, in plain English, what it does.

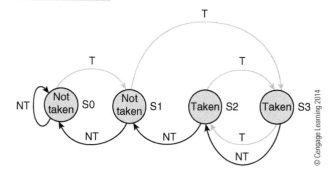

7.57 What is a branch target buffer, and how does it contribute to a reduction of the branch penalty?

7.58 Consider the 4-bit saturating counter as a branch predictor with 16 states from 1111 to 0000? Describe in words the circumstances where such a counter might be effective.

7.59 Draw the state diagram of a branch predictor using a 3-bit saturating counter? Under what circumstances do you think such a predictor might prove effective?

7.60 Given the branch sequence TTTTNTTNNTTTTTNNN NNNNNNTNTTTTTTTTTTTT and assuming that the 3-bit saturating predictor starts in its saturated T state, what will the predicted sequence be?

7.61 The following code is executed by an ARM processor:

```
        MOV    r0,#4
B1      MOV    r2,#5
        SUB    r2,r2,r0
B2      SUBS   r2,r2,#1
        BNE    B2          ;Branch 1
        SUBS   r0,r0,#1
        BNE    B1          ;Branch 2
```

Assume that a 1-bit branch predictor is used for both branch 1 and branch 2 and that both predictors are initially set to N. Complete the following table by running through this code.

	Branch 1			Branch 2	
Cycle	Branch prediction	Branch outcome	Cycle	Branch prediction	Branch outcome
1	N	N	1	N	T
2			2		
3			3		
4			4		
5					
6					
7					
8					
9					
10					

© Cengage Learning 2014

Repeat the same exercise with the same initial conditions but assume a 2-bit saturating counter branch predictor.

7.62 A processor executes all non-branch instructions in one cycle. This processor implements branch prediction, which incurs an additional penalty of 2 cycles if the prediction is correct and 4 cycles if the prediction is incorrect.

   a. If conditional branch instructions occupy 15% of the instruction stream, and the probability of an incorrect branch prediction is 20%, what is the average number of cycles per instruction?

   b. If the same processor is to run no less than 28% slower than a machine with a zero branch penalty when up to 20% of the instructions are conditional branches, what level of accuracy must the branch prediction achieve on average?

7.63 A computer has a branch target buffer, BTB. Derive an expression for the average branch penalty if the following apply.

   • A branch not in the BTB that is not taken incurs a penalty of 0 cycles.
   • A branch not in the BTB that is taken incurs a penalty of 6 cycles.
   • A branch in the BTB that is not taken incurs a penalty of 4 cycles.
   • A branch in the BTB that is taken incurs a penalty of 0 cycles.
   • The probability that a branch instruction is cached in the BTB is 80%.
   • The probability that an instruction not in the BTB is taken is 20%.
   • The probability that an instruction in the BTB is taken is 90%.

7.64 A RISC processor implements a subroutine call using a link register (i.e., the return address is saved in the link register). The cost of a call is 2 cycles and the return costs 1 cycle. If a subroutine is called from another subroutine (i.e., the subroutine is nested), the contents of the link register must be saved and later restored. The cost of saving the link register is 6 cycles, and the cost of restoring the link register is 8 cycles. Assume that a certain instruction mix contains 20% subroutine calls and returns (i.e., 10% calls, 10% returns). The probability of a single subroutine call and return without nesting is 60%. The probability that a subroutine call will be followed by a single nested call is 40%. Assume that the probability of further nesting is vanishingly small. What is the overall cost of subroutine calls? The average call of all other instructions is 1.5 cycles. What is the average number of cycles per instruction?

7.65 Why is the literal in the op-code sign-extended before use (in most computer architectures)?

7.66 Why is the address offset shifted two places left in branch/jump operations in 32-bit RISC-like processors?

7.67 Assume a five-stage pipeline (instruction fetch, operand fetch, execute, memory, write-back). For the following code show any stalls and indicate where operand forwarding would be needed.

```
ADD    R9,R9,R8
MUL    R1,R2,R3
LDR    R5,(4,R1)
SUB    R5,R5,R1
ADD    R7,R8,R9
MUL    R7,R1,R5
```

CHAPTER

8

Beyond RISC: Superscalar, VLIW, and Itanium

"Every film should have a beginning, middle and end—but not necessarily in that order."

Jean-Luc Godard

"Sentence first—verdict afterwards."

The Queen, *Alice in Wonderland*

"It's life, captain, but not life as we know it."

Spock

We now look at how computer performance has been pushed beyond the limit of *one instruction per cycle* by exploiting pipelining and introduce *superscalar* processing that uses multiple ALUs to enable instructions to be executed in parallel. I have heard superscalar computers referred to as *post-RISC processors*. In addition to superscalar processors, we describe the *very long instruction word* (VLIW) processor that encodes multiple operations in a single instruction. At the end of this chapter, we provide an introduction to Intel's IA64 architecture, *Itanium*, that combines several powerful architectural features to enhance its performance. In Chapter 13, we introduce two further techniques that have been introduced to enhance performance: *hyperthreading* and *multi-core* processing.

Earlier, we introduced performance and examined how pipelined processors have struggled to achieve the goal of *one instruction per cycle*. In this final chapter of Part III, we look at how the *superscalar* processor has exceeded the RISC ideal of one-cycle-per-instruction and is designed to execute multiple instructions per cycle. By using multiple execution units, a superscalar processor can complete several instructions in parallel during a single clock cycle. Consequently, superscalar processors achieve a better than one cycle per instruction throughput using *instruction level parallelism, ILP*. The performance of superscalar processors is limited by the need to schedule instructions for parallel execution without data or control conflicts (see the sidebar).

The second part of this chapter takes a look at Intel's Itanium family (i.e., the IA64 architecture) that provides superscalar performance by forcing the compiler or programmer to schedule instructions for parallel execution rather than the processor's internal hardware. The IA64 is a *very long instruction word* (VLIW) processor that puts three operations in each instruction word. The VLIW is

Parallel Execution

Superscalar processors rely on arranging instructions so that they can be executed simultaneously. Consider:

```
ADD r1,r2,r3
ADD r5,r4,r1    ;cannot be executed until r1 created
ADD r6,r5,r7    ;cannot be executed until r5 created
```

The above code cannot be executed in parallel because of data dependency. However, the following fragment can be executed in parallel.

```
ADD r1,r2,r3
ADD r4,r5,r6    ;no waiting for r5 and r6
ADD r7,r8,r9    ;no waiting for r8 and r9
```

simpler than the superscalar processor because the job of finding instructions to be executed in parallel is delegated to the compiler.

Architecture, Organization, Performance, and Challenges

The instruction sets of today's processors are remarkably similar to ISAs dating back to the 1980s. On the other hand, their organization has changed dramatically. This change in organization, the way we implement processors, has considerably increased computer performance. Having said that, we must point out that the underlying technology (faster transistors and higher clock speeds) has also provided a considerable increase in computer performance over the last three decades.

Although a purist could maintain that organization and architecture are independent of each other, they are not. A particular architecture may have a preferred organization. For example, pipelining is easier to implement in RISC than in CISC architectures, although Intel has managed to successfully marry CISC architectures with deep pipelining.

Superscalar organization extends the performance gains of pipelining by implementing multiple, parallel execution units. Pipelining and superscalar technologies are *orthogonal* in the sense that they are independent of each other.

Superscalar organization creates new challenges for the systems designer, because of structural hazards—conflicts that occur when different instructions request the same resources. For example, if two additions are computed sequentially, the state of the zero flag can be tested separately after each addition. However, if the additions are performed in parallel, then what meaning is associated with testing the status flag?

Overview of Chapter 8

We begin with an overview of pipelining and then introduce the traditional superscalar processor in Section 8.1. Following this we look at the VLIW processor and the EPIC architecture before introducing the Itanium processor in Section 8.3. I have chosen the Itanium for several reasons. First, it is the marriage of two very different companies, Intel and Hewlett Packard. Second, the Itanium embodies some very interesting architectural features. However, the computer world has been rather slow to adopt Itanium, and Intel's Pentium derivatives continue to gain strength; not least because of the rate at which performance has increased over the years. Finally, the cost of developing an advanced processor like the Itanium has to be amortized over relatively few devices compared with the development cost of the Pentium derivatives which is amortized over countless millions of chips.

8.1 Superscalar Architecture

Before introducing superscalar processors, let's look at how pipelining can be extended. Figure 8.1a illustrates *pipelining* where instruction execution is overlapped. Figure 8.1b illustrates *superpipelining* where some stages are subdivided. For example, the fetch stage has been divided into I_{fetch1} and I_{fetch2}, whereas the decode stage remains a single operation.

Figure 8.1 doesn't really tell you what is happening because it omits *time*. Figure 8.2 changes the timescale to demonstrate the effect of superpipelining by allowing some operations to take place in half a clock cycle. As you can see, the overall throughput is increased. Some stages such as instruction execution take the same time in both pipelined and superpipelined systems, whereas instruction decoding is faster in the superpipelined processor.

Figure 8.2 allocates the same time to two-phase operations as the conventional pipeline does to single-phase operations in Figure 8.1. For example, the time for $I_{decode1}$ plus $I_{decode2}$ is

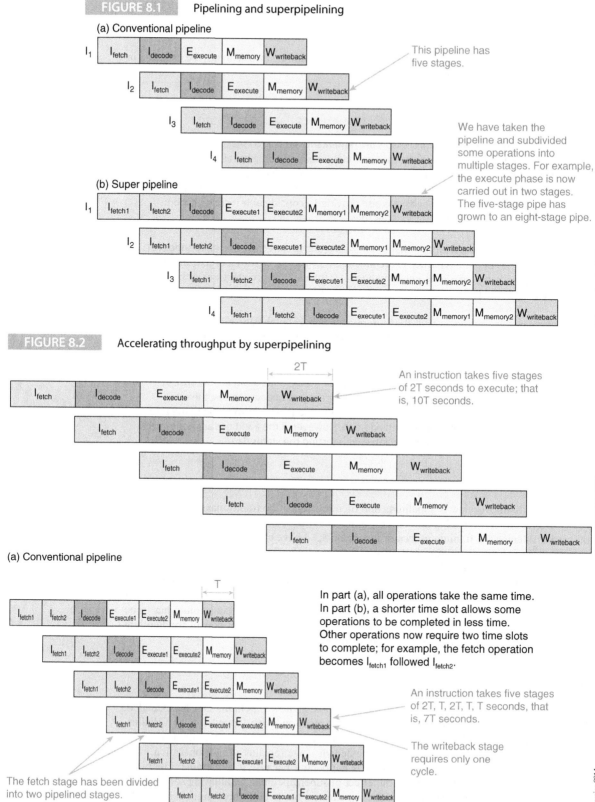

FIGURE 8.1 Pipelining and superpipelining

(a) Conventional pipeline

I_1: I_{fetch} | I_{decode} | $E_{execute}$ | M_{memory} | $W_{writeback}$

I_2: I_{fetch} | I_{decode} | $E_{execute}$ | M_{memory} | $W_{writeback}$

I_3: I_{fetch} | I_{decode} | $E_{execute}$ | M_{memory} | $W_{writeback}$

I_4: I_{fetch} | I_{decode} | $E_{execute}$ | M_{memory} | $W_{writeback}$

This pipeline has five stages.

We have taken the pipeline and subdivided some operations into multiple stages. For example, the execute phase is now carried out in two stages. The five-stage pipe has grown to an eight-stage pipe.

(b) Super pipeline

I_1: I_{fetch1} | I_{fetch2} | I_{decode} | $E_{execute1}$ | $E_{execute2}$ | $M_{memory1}$ | $M_{memory2}$ | $W_{writeback}$

I_2: I_{fetch1} | I_{fetch2} | I_{decode} | $E_{execute1}$ | $E_{execute2}$ | $M_{memory1}$ | $M_{memory2}$ | $W_{writeback}$

I_3: I_{fetch1} | I_{fetch2} | I_{decode} | $E_{execute1}$ | $E_{execute2}$ | $M_{memory1}$ | $M_{memory2}$ | $W_{writeback}$

I_4: I_{fetch1} | I_{fetch2} | I_{decode} | $E_{execute1}$ | $E_{execute2}$ | $M_{memory1}$ | $M_{memory2}$ | $W_{writeback}$

© Cengage Learning 2014

FIGURE 8.2 Accelerating throughput by superpipelining

2T

I_{fetch} | I_{decode} | $E_{execute}$ | M_{memory} | $W_{writeback}$

I_{fetch} | I_{decode} | $E_{execute}$ | M_{memory} | $W_{writeback}$

I_{fetch} | I_{decode} | $E_{execute}$ | M_{memory} | $W_{writeback}$

I_{fetch} | I_{decode} | $E_{execute}$ | M_{memory} | $W_{writeback}$

I_{fetch} | I_{decode} | $E_{execute}$ | M_{memory} | $W_{writeback}$

An instruction takes five stages of 2T seconds to execute; that is, 10T seconds.

(a) Conventional pipeline

© Cengage Learning 2014

T

I_{fetch1} | I_{fetch2} | I_{decode} | $E_{execute1}$ | $E_{execute2}$ | M_{memory} | $W_{writeback}$

I_{fetch1} | I_{fetch2} | I_{decode} | $E_{execute1}$ | $E_{execute2}$ | M_{memory} | $W_{writeback}$

I_{fetch1} | I_{fetch2} | I_{decode} | $E_{execute1}$ | $E_{execute2}$ | M_{memory} | $W_{writeback}$

I_{fetch1} | I_{fetch2} | I_{decode} | $E_{execute1}$ | $E_{execute2}$ | M_{memory} | $W_{writeback}$

I_{fetch1} | I_{fetch2} | I_{decode} | $E_{execute1}$ | $E_{execute2}$ | M_{memory} | $W_{writeback}$

I_{fetch1} | I_{fetch2} | I_{decode} | $E_{execute1}$ | $E_{execute2}$ | M_{memory} | $W_{writeback}$

I_{fetch1} | I_{fetch2} | I_{decode} | $E_{execute1}$ | $E_{execute2}$ | M_{memory} | $W_{writeback}$

In part (a), all operations take the same time. In part (b), a shorter time slot allows some operations to be completed in less time. Other operations now require two time slots to complete; for example, the fetch operation becomes I_{fetch1} followed I_{fetch2}.

An instruction takes five stages of 2T, T, 2T, T, T seconds, that is, 7T seconds.

The writeback stage requires only one cycle.

The fetch stage has been divided into two pipelined stages.

(b) Superpipeline

© Cengage Learning 2014

Gmicro/500–Twin Pipeline Processor

The Gmicro/500 was a processor that included both RISC and CISC attributes. It has a five-stage pipeline as shown (in fact, there were two parallel but not identical pipes that permitted superscalar operation). This diagram demonstrates that, without any hardware help, a branch would cause a three-cycle bubble.

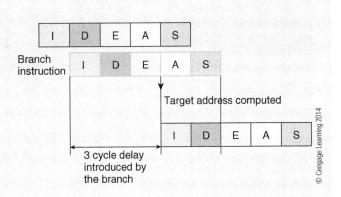

The Gmicro/500 can generate the address of an instruction in three ways. An address may be the next sequential address, an address from the branch target buffer, or an address from a first-in-first-out queue of return addresses. The branch target buffer is arranged as a two-way set-associative cache of 32 entries. Each entry consists of a branch instruction address and a branch target address.

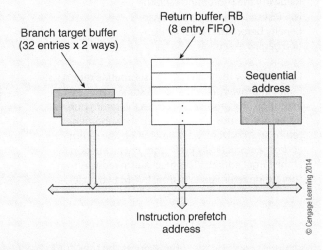

Consider the effect of the Gmicro/500's branch target buffer on the execution of an unconditional branch instruction or a subroutine call. The instruction prefetch address of the unconditional branch accesses the BTB to obtain the appropriate target address (note that the Gmicro500's BTB supplies the target address rather than the instruction at that address). This address is available at the end of the branch instruction's I phase. The branch target address automatically prefetches the corresponding target instruction in the next phase. The processor can then execute this instruction without delay; that is, there is a zero-cycle branch penalty if the branch is taken.

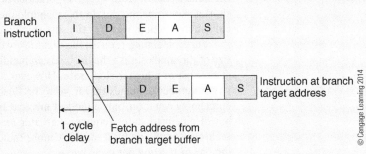

Another source of an instruction prefetch address is the eight-entry return buffer, RB, used to implement a fast return from subroutine mechanism. Consider the timing of a return from subroutine instruction, RTS, without the use of the return buffer. The processor pulls the subroutine return address from the stack in external memory during the memory-access stage A of the pipeline. This address is used to fetch the target instruction (i.e., the first instruction following the subroutine call). The RTS instruction incurs a five-cycle penalty before the target instruction begins execution.

Consider how the Gmicro/500's subroutine return buffer improves performance. In stage D of the RTS execution, the processor accesses the return buffer. If a hit occurs, the processor reads the subroutine return address from the return buffer. The return address automatically prefetches the instruction at the target address. This mechanism reduces the subroutine return penalty from five to two cycles.

(continued)

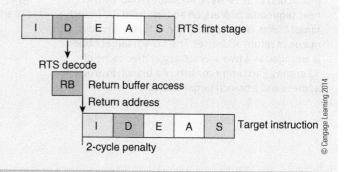

the same as the time for I_{decode}. Operations not requiring as much time can be fitted into the shorter superpipeline slot. In this example the pipelined processor takes 10T seconds to execute an instruction, whereas the superpipelined version takes 7T, where T is the cycle time.

Superpipelining reduces the *granularity* of the processes in the pipeline by reducing the size of the smallest task that a stage can handle. Average instruction throughput is increased at the cost of a higher clock rate. However, the branch penalty is increased because of the larger number of stages that have to be flushed on a mispredicted branch. At best, superepipelining can offer only a modest increase in performance. Moreover, increasing the number of pipelined states requires a greater *interstage delay,* because the extra registers have their own setup and hold times. A more radical approach to pipelining is required to break the one instruction per cycle barrier.

The only way that we can increase throughput beyond the one cycle per instruction barrier is to employ multiple execution units and multiple pipelines. Figure 8.3 illustrates a processor with *two* parallel five-stage pipelines that can double the throughput without increasing the clock rate; the panel gives details of an early two-pipeline processor called the GMicro/500. Both pipelines are not always used; sometimes it's impossible to find two instructions to execute in parallel.

An arrangement with *m* parallel pipelines is called an *m-way superscalar processor.* However, you cannot expect an *m*-fold increase in performance. In practice, it's even harder to approach the *m* instructions per cycle ideal for an *m*-way superscalar processor than it is to approach the one instruction per cycle goal for a pipelined processor. The Pentium P5, introduced in 1993, had two integer pipelines but its second pipeline could be used for only about 30% of the time. The original Pentium introduced in 1993 with a speed of 60 or 66 MHz

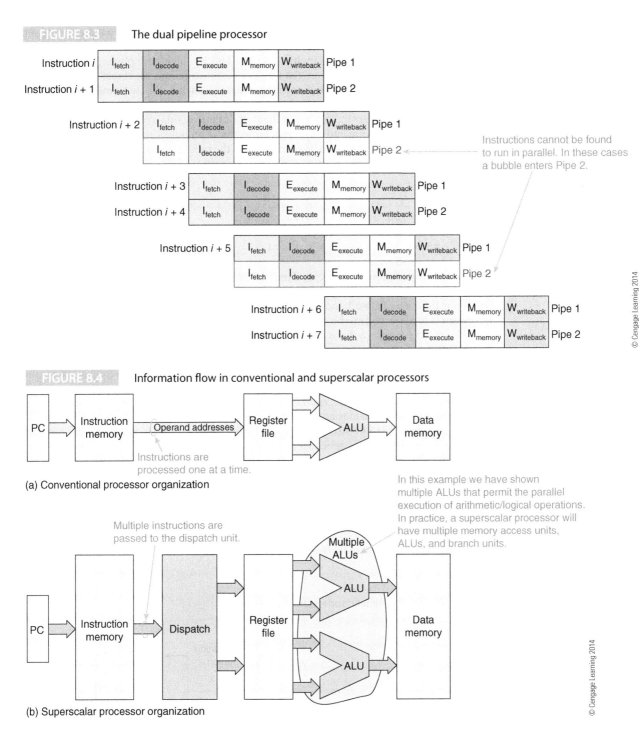

FIGURE 8.3 The dual pipeline processor

Instructions cannot be found to run in parallel. In these cases a bubble enters Pipe 2.

FIGURE 8.4 Information flow in conventional and superscalar processors

Instructions are processed one at a time.

(a) Conventional processor organization

In this example we have shown multiple ALUs that permit the parallel execution of arithmetic/logical operations. In practice, a superscalar processor will have multiple memory access units, ALUs, and branch units.

Multiple instructions are passed to the dispatch unit.

(b) Superscalar processor organization

© Cengage Learning 2014

had two five-stage integer pipelines U and V and a six-stage floating-point unit. The U and V pipelines were not identical, because the U pipe included a shifter.

Superscalar processors are not normally implemented by taking a RISC and just increasing the number of pipelines. They are implemented by adding a pool of resources such as multiple processing units to allow parallel instruction overlap. Figure 8.4a illustrates the information flow in a conventional pipelined processor, and Figure 8.4b shows the information flow in a generic superscalar processor with two ALUs.

The superscalar processor of Figure 8.4b takes multiple instructions from the instruction source and then *dispatches* them in parallel to a multiported register file, where the operands are looked up and transmitted to multiple ALUs. This is a highly simplistic picture because it entirely omits a key consideration: the way in which instructions are *dispatched* in parallel. In a real system, instructions come from several sources, such as the cache memory and local instruction buffers, but they all have to be executed in the correct *semantic order*. The expression *correct semantic order* indicates that the order of instruction execution must not change the meaning of the program and end result. However, in an out-of-order processor the actual order of instruction execution can be changed from the program order as long as the meaning of the program remains unaltered. Consider the example:

Case 1	Case 2
ADD **r1**,r2,r3	ADD **r1**,r2,r3
ADD **r4**,r1,r3	ADD **r5**,r6,r7
ADD **r5**,r6,r7	ADD **r4**,r1,r3

Suppose the instruction sequence in Case 1 is the original program sequence. In Case 2, we have re-ordered this sequence. However, the semantic meaning has not been changed; that is, executing the code of Case 1 and Case 2 will yield the same result. The object of a superscalar processor is to allow code to be executed in parallel without changing the semantic order.

Figure 8.5 provides a more realistic structure of a superscalar processor that consists of five elements.

- **Instruction Fetch**—The instruction fetch stage obtains the instructions from memory (the cache or any other special-purpose mechanism used to hold instructions) to be executed.
- **Instruction Decode**—The *instruction decode and register rename* stage decodes the instructions and renames any resources required by the instructions. *Register renaming* reduces the effect of bottlenecks caused by the limited number of user-visible registers by using temporary registers; that is, the name of a register is changed at runtime to avoid conflicts between instructions using the same register.
- **Instruction Issue**—The *instruction issue* stage forwards instructions to the execution units and ensures that the maximum number of instructions possible is executed in parallel. The two terms *issue* and *dispatch* describe the forwarding of instructions for execution. *Dispatch* indicates that an instruction is transmitted to functional units for execution, and *issue* indicates that an instruction is transmitted to an execution unit even though that instruction may not yet be ready for immediate execution.

FIGURE 8.5   The generic superscalar processor

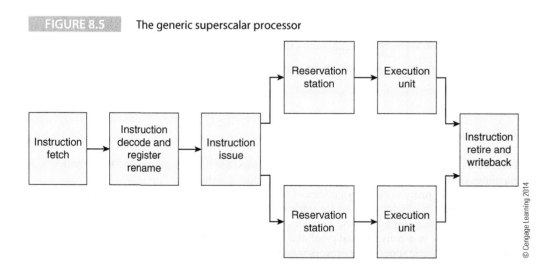

- **Reservation Stations**—The *reservation stations and execution units* are the instruction buffers and ALUs (or other functional devices) that interpret the stream of op-codes. The reservation stations dispatch instructions to the associated processing units only when all the required resources are available.
- **Instruction Retire**—The *instruction retire and writeback* stage is responsible for writing back the result to registers and ensuring that instructions are completed in the correct order. This stage also communicates with the reservation stages when instruction completion frees resources.

The ultimate performance of a superscalar processor is determined by the number of execution units. These execution units may be ALUs (integer and floating-point), multipliers, load-and-store controllers, and branch controllers. The actual limit on performance will be determined by the nature of the code (data dependencies) and the ability of the processor to optimally schedule instruction execution.

In-Order and Out-of-Order Execution

When instructions are executed strictly sequentially in program order, they are said to be processed *in-order*. If instructions are not executed in program sequence with one instruction being executed *before* the previous instruction, they are said to be processed *out-of-order*. Parts of a superscalar pipeline may operate *in-order*, and other parts operate *out-of-order*. Note that out-of-order execution is not a characteristic of superscalar processing but a technique that can be exploited to enhance superscalar processing; that is, out-of-order execution is not a fundamental requirement of superscalar processing; it is a necessary evil.

In Figure 8.5 the *instruction retire stage* is responsible for ensuring that instructions are *completed* in-order whether or not they were *executed* out-of-order. Consider the following example that demonstrates the concept of in-order and out-of-order execution:

Instruction	Case 1	Case 2	Case 3
1	ADD **R1**,R2,R3	ADD **R1**,R2,R3	ADD **R1**,R2,R3
2	ADD **R4**,R1,R7	ADD **R6**,R3,R8	ADD R8,R6,R2
3	ADD **R5**,R1,R2	ADD **R4**,R1,R7	ADD **R4**,R1,R7
4	ADD **R6**,R3,R8	ADD **R5**,R1,R2	ADD **R5**,R1,R2
5	ADD **R8**,R6,R2	ADD **R8**,R6,R2	ADD R6,R3,R8

Case 1 demonstrates a sequence of instructions that could generate hazards (data dependencies) when executed in a pipelined computer because of the two read before write hazards, that is, R1 in instruction 2 and R6 in instruction 5. Case 2 demonstrates the effect of introducing out-of-order execution by reordering the instruction stream to remove these dependencies. In this case the semantics of the code is unaltered. In Case 3, out-of-order execution has also been applied to remove the dependencies. However, the semantics of the code is changed because R6 is used in instruction 2 but is not created until instruction 5. This demonstrates that out-of-order execution is possible, but only with care.

Let's look at an example that illustrates superscalar execution by demonstrating the effect of both in-order and out-of-order execution. Consider the following code fragment. The comments in the rightmost column indicate the restrictions placed on the execution of the corresponding instruction. For example, the load requires two cycles, and there is only one multiplier (a resource dependency) that prevents parallel multiplications.

I1	x = b*	LDR X,[B]	a load takes two cycles
I2	b = b + 1	ADD **B**,B,#4	(note increment a pointer by 1 adds 4 bytes)
I3	y = c.g	MUL Y,C,G	only one multiplication at a time
I4	d = c.x	MUL D,C,X	multiply (only one at a time)
I5	f = d - q	SUB F,D,Q	RAW data dependency with I4 which generates D
I6	e = p + g	ADD **E**,P,G	
I7	x = a.f	MUL X,A,F	multiply (only one at a time)
I8	c = c + 1	ADD **C**,C,#1	
I9	e = e + 1	ADD **E**,E,#1	

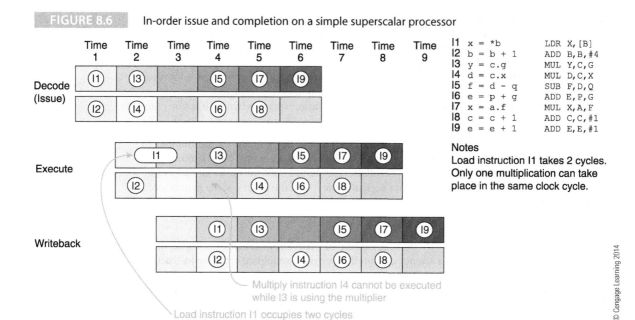

FIGURE 8.6   In-order issue and completion on a simple superscalar processor

I1	x = *b	LDR X, [B]
I2	b = b + 1	ADD B, B, #4
I3	y = c.g	MUL Y, C, G
I4	d = c.x	MUL D, C, X
I5	f = d - q	SUB F, D, Q
I6	e = p + g	ADD E, P, G
I7	x = a.f	MUL X, A, F
I8	c = c + 1	ADD C, C, #1
I9	e = e + 1	ADD E, E, #1

Notes
Load instruction I1 takes 2 cycles.
Only one multiplication can take place in the same clock cycle.

Multiply instruction I4 cannot be executed while I3 is using the multiplier

Load instruction I1 occupies two cycles

Figure 8.6 illustrates a superscalar model where instructions are issued and executed in-order in a simple two-way superscalar machine. This machine has only three stages: decode/issue, execute, and write-back (store operand). A non-superscalar RISC processor with a three-stage pipeline would take four cycles to complete the first instruction (the load), and 12 cycles to execute all the instructions. (Twelve cycles are required because of the three-cycle latency, eight successive cycles, and the one-cycle stall caused by I1.)

The restrictions on the execution of this code (apart from data dependencies) are that the load instruction I1 takes two cycles and only one multiplier unit is available, so two multiplications cannot be performed in parallel. There are two execution units allowing two operations to take place simultaneously. A superscalar machine must preserve the semantics of the code, and its meaning (outcome) must not be changed by modifying instruction order.

Figure 8.6 demonstrates that the superscalar architecture achieves a reduction of only three cycles over a single three-stage pipeline. Initially, instructions I1 and I2 are issued in parallel. These flow to the two execution units. However, because I1 has a two-cycle latency, I2 can be executed but cannot be retired (written back) in the next cycle. It must wait until I1 has been completed. If I2 were permitted to write its results before I1 had finished and an interrupt were taken, the architectural state in registers and memory would be in error. Instructions I1 and I2 are retired in time slot 4. The hardware dependencies between multiplication instructions I3 and I4 ensure that only one pipe can be used in time slots t4 and t5; that is, superscalar processing cannot be exploited. Moreover, the RAW hazard between instructions I4 and I5 ensures that I4 must be executed before I5. Although we have implemented superscalar processing, we have achieved a speedup of 12/9 = 1.3, rather than the desired 2 that could have been achieved if both execution units had been fully used.

Let's now modify the processor to permit a degree of *out-of-order processing*. Figure 8.7 illustrates in-order issue and out-of-order completion. In this case, I2 is completed before I1, and the number of cycles reduced by one. Instructions I1 to I9 are executed in eight cycles with out-of-order processing. An instruction can *complete* out of order but is never *retired* out of order, since retiring involves updating the state of the machine.

Figure 8.8 further extends this example by permitting both *out-of-order issue* and *out-of-order completion*. In this case, we have used a three-way superscalar model with different code from the previous example and have included two loads and a multiplication. In Figure 8.8, we have given the issued code below the original code. Note that the writeback (retire) stage ensures that instructions are retired in order.

FIGURE 8.7   In-order issue and out-of-order completion (execution)

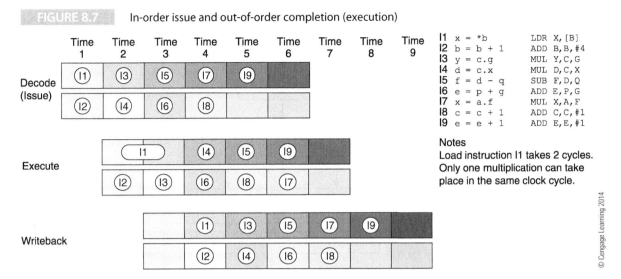

I1 x = *b   LDR X,[B]
I2 b = b + 1   ADD B,B,#4
I3 y = c.g   MUL Y,C,G
I4 d = c.x   MUL D,C,X
I5 f = d - q   SUB F,D,Q
I6 e = p + g   ADD E,P,G
I7 x = a.f   MUL X,A,F
I8 c = c + 1   ADD C,C,#1
I9 e = e + 1   ADD E,E,#1

Notes
Load instruction I1 takes 2 cycles.
Only one multiplication can take
place in the same clock cycle.

Figure 8.9 illustrates information flow in a conceptual out-of-order superscalar processor. The front end of the processor resembles a conventional RISC processor and ensures a continuous flow of instructions. The middle part consists of the *reservation stations* and the pipelines of the execution units, where some pipelines have one stage such as integer operations and others have multiple stages such as floating-point operations. This middle stage operates in an out-of-order fashion. Finally, the end stage of the superscalar processor consists of the reorder buffer and the retire (i.e., writeback) stage, where instructions are completed in order.

Comment from Bob Colwell

"Out-of-order, OOO, completion is one unholy can of worms. OOO retirement means that the visible architectural state will not always be the same. Branch mispredictions, cache effects, bus contention, paging effects, and many other things will cause the same sequence of code to exhibit different states when interrupted." Constructing a processor with OOO is not an easy task.

FIGURE 8.8   Out-of-order issue and out-of-order completion

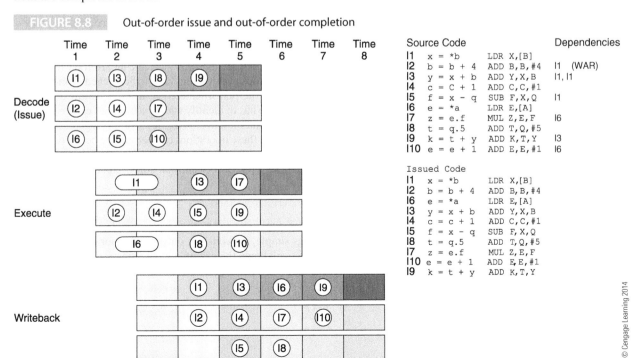

Source Code			Dependencies
I1	x = *b	LDR X,[B]	
I2	b = b + 4	ADD B,B,#4	I1 (WAR)
I3	y = x + b	ADD Y,X,B	I1,I1
I4	c = C + 1	ADD C,C,#1	
I5	f = x - q	SUB F,X,Q	I1
I6	e = *a	LDR E,[A]	
I7	z = e.f	MUL Z,E,F	I6
I8	t = q.5	ADD T,Q,#5	
I9	k = t + y	ADD K,T,Y	I3
I10	e = e + 1	ADD E,E,#1	I6

Issued Code
I1 x = *b   LDR X,[B]
I2 b = b + 4   ADD B,B,#4
I6 e = *a   LDR E,[A]
I3 y = x + b   ADD Y,X,B
I4 c = c + 1   ADD C,C,#1
I5 f = x - q   SUB F,X,Q
I8 t = q.5   ADD T,Q,#5
I7 z = e.f   MUL Z,E,F
I10 e = e + 1   ADD E,E,#1
I9 k = t + y   ADD K,T,Y

© Cengage Learning 2014

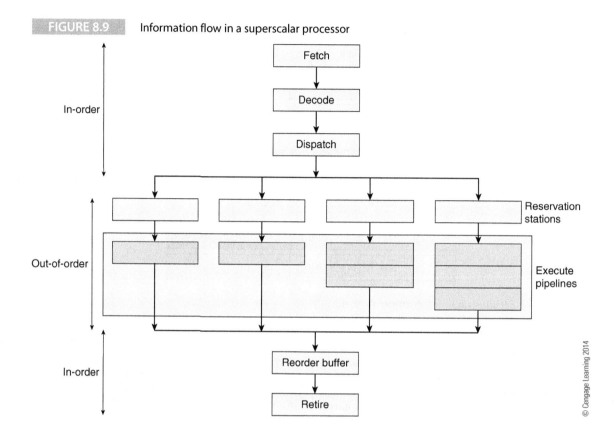

FIGURE 8.9   Information flow in a superscalar processor

To understand the strengths and weaknesses of superscalar processors, let's take another look at *instruction level parallelism*. Figure 8.9 illustrates the information flow in a generic superscalar processor. The front-end consisting of the fetch/decode/issue stages is like that of a conventional RISC/CISC processor and is concerned with ensuring a continuous flow of instructions. The middle section consists of the *reservation stations* and the pipelines of the execution units (integer pipelines have one stage and floating-point pipelines have multiple stages).

The strength of a superscalar processor is that it is able to execute an arbitrary number of instructions in parallel. The weakness is that those instructions have to be fetched from memory, which requires a corresponding increase in the performance (i.e., transfer bandwidth) of the *upstream* stages, and the processor hardware has to identify which instructions can be executed in parallel without changing a program's semantics. Moreover, it has to manage resources by optimally allocating them to instructions. All these operations can perhaps be better performed by the compiler. In the next section, we look more closely at the nature of instruction level parallelism and demonstrate how renaming registers at run time can improve performance.

8.1.1   Instruction Level Parallelism (ILP)

We now take a fragment of code and examine how the use of instruction level parallelism can improve throughput. To keep things simple, we consider only a processor that can execute two instructions at the same time. We will begin the example by demonstrating the execution of the code on a non-superscalar processor. Consider the following fragment of code that performs the vector addition $\mathbf{C} = \mathbf{A} + \mathbf{B}$.

```
for (i = 0; i < n; i++) {
  c[i] = a[i] + b[i];
}
```

We can translate this algorithm into generic load/store architecture code as follows.

```
Start ldr   r1,a      ;put base address of array A in pointer r1
      ldr   r2,b      ;put base address of array B in pointer r2
      ldr   r3,c      ;put base address of array C in pointer r3
      ldr   r4,#n     ;load the loop count for n cycles in r4
Loop  ldr   r5,[r1]   ;REPEAT get a[i]
      ldr   r6,[r2]   ;        get b[i]
      stall           ;        wait for r6 to load from array B
      add   r5,r5,r6  ;        c[i] = a[i] + b[i]
      str   r5,[r3]   ;        store c[i]
      add   r1,r1,#4  ;        increment pointer to a[i]
      add   r2,r2,#4  ;        increment pointer to b[i]
      add   r3,r3,#4  ;        increment pointer to c[i]
      sub.s r4,r4,#1  ;        decrement loop counter
      bne   Loop      ;until loop completed
```

This code employs three array pointers that are explicitly updated at the end of each iteration. Instructions take one cycle except for the load r6 that takes two cycles when a stall cycle is introduced forcing the next operation to wait for the data to be retrieved. A superscalar processor with two processing units could execute the main loop in the following way. In order to simplify the problem, we have assumed that the two processing units can carry out all operations. In practice, this may not be the case and it will be necessary to schedule the code to make optimum use of the available resources.

```
         Processing Unit 1          Processing Unit 2

Loop  ldr     r5,[r1]                              ;REPEAT get a[i]
      ldr     r6,[r2]                              ;get b[i]
      stall
      add     r5,r5,r6   add    r1,r1,#4  ;calculate c[i]; update a ptr
      str     r5,[r3]    add    r2,r2,#4  ;store c[i]; update b ptr
      sub.s   r4,r4,#1   add    r3,r3,#4  ;dec loop ctr; update c ptr
      bne Loop                            ;until loop completed
```

We have assumed that two loads cannot be executed in parallel and that a load has a two-cycle latency. We have appended a .s suffix to the subtraction mnemonic because it is necessary to define which operation will update the condition code register.

This level of instruction parallelism has converted a loop from 10 cycles into a loop that executes in seven cycles. By reordering the code we can further reduce the number of cycles to five per iteration as follows. This code assumes that a register can be used as a pointer and updated in one cycle.

```
         Processing Unit 1          Processing Unit 2

Loop  ldr     r5,[r1]    add    r1,r1,#4
      ldr     r6,[r2]    add    r2,r2,#4
      stall              sub.s  r4,r4,#1
      add     r5,r5,r6   add    r3,r3,#4
      str     r5,[r3]    bne    Loop
```

By reordering the code, the number of cycles has been reduced to five per iteration. Clearly, the generation of the code has a significant effect on the level of the parallelism that can be exploited.

We can exploit superscalar processing further by *loop unrolling*. In the following example, two iterations are executed per loop. To make the code more readable we use the notation that, for example, register r5 contains a[i] and register r5a contains a[i+1]. In this example we perform two iterations in eight cycles, corresponding to one iteration in four cycles. A combination of superscalar processing, instruction reordering and loop unrolling has allowed us to go from ten cycles per iteration to four cycles per iteration.

	Processing Unit 1		Processing Unit 2	
Loop	ldr	**r5**, [r1]	add	r1,r1,#4
	ldr	**r6**, [r2]	add	r2,r2,#4
	ldr	**r5a**,[r1a]	add	r1a,r1a,#4
	ldr	**r6a**,[r2a]	add	r2a,r2a,#4
	add	**r5**,r5,r6	sub.s	r4,r4,#1
	str	r5,**[r3]**	add	r5a,r5a,r6a
	str	r5a,**[r3a]**	add	r3a,r3a,#4
	add	**r3**,r3,#4	bne	Loop

Data Dependencies and Register Renaming

We have seen that the performance of pipelined systems is degraded by *data dependency*. Data dependency is a crucial factor in instruction level parallelism, as all too frequently an instruction cannot be executed because an operand is not currently available. One approach to reducing data dependency involves a mechanism called *register renaming* (i.e., the use of *virtual registers*), which dates back to the late 1960s when it was used by IBM.

Some data dependencies can be removed by the allocation of extra registers at run time. As its name suggests, *register renaming* is a means of creating apparently new registers by reassigning the names of physical registers. Consider the following example that calculates:

```
a = b + c;
p = q + r;
1      ldr    r1, [r2]      ;get b from memory
2      add    r3,r1,r4      ;add c to b from register r4 to get a
3      ldr    r1, [r5]      ;get q from memory
4      add    r6,r1,r7      ;add r to q from register r7 to get p
```

Consider the *history* of r1. It is loaded with an operand in line 1 and that operand is used in line 2. The instruction in line 3 *reuses* register r1 in the calculation of p because its contents are no longer needed after the value of a has been calculated in line 2. However, pipelined and superscalar processors run into a problem because the first use and the subsequent reuse of register r1 may overlap.

Register renaming solves the problem by using a pool of registers that is larger than the visible register set. In the next example the second use of register r1 is detected and r1 mapped onto register rr1 causing data dependency to disappear. Register renaming is an internal operation that is invisible to the programmer.

```
1      ldr    r1, [r2]      ;get b from memory
2      add    r3,r1,r4      ;add c from register
3      ldr    rr1, [r5)     ;get q from memory      note register renaming
4      add    r6,rr1,r7     ;add r from register      note register renaming
```

Consider the following example on a generic machine (taken from Asanovic/Devadas). This code loads data and evaluates a simple expression.

```
1      ldr    r1, [r0]      ;get x
2      ldr    r2, [r3]      ;get y in r2 (first use of r2)
3      mlt    r4,r1,r2      ;z = x·y
4      sub    r5,r1,#4      ;q = x − 4
5      div    r2,r1,r5      ;z = x/(x − 4) (reuse of r2)
6      add    r6,r4,r2      ;s = x·y + x/(x − 4)
```

Figure 8.10 illustrates graphically this sequence of instructions being executed. We have used a notation that gives the program line number of each instruction, the operation being

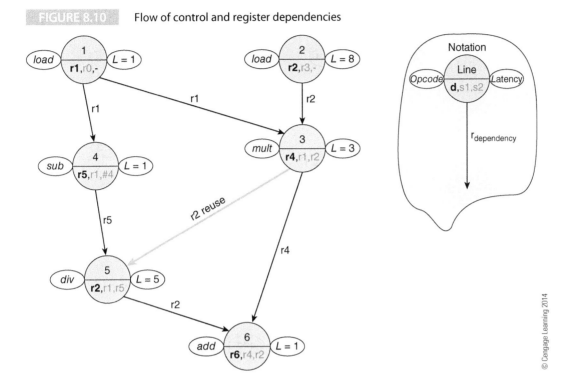

FIGURE 8.10   Flow of control and register dependencies

© Cengage Learning 2014

performed, the latency of the instruction, the destination operand, and the source operands. Lines drawn between instructions indicate dependencies. For example, instruction 3 uses registers r1 and r2 from instructions 1 and 2, respectively.

Figure 8.10 demonstrates the effect of the *false data dependency* and the reuse of register r2 in instruction 5 by the heavy blue line. As you can see, the execution process along the path 1, 4, 5, 6 is held up by the path 2, 3, 6. The term *false* data dependency indicates that the dependency is not inherent but is caused by a write-after-write (WAW) or write-after-read (WAR) operation.

False data dependencies are of the WAW and WAR types, where an operand in a register can be created before a previous operation has finished using that register. Such false dependencies can be removed by recognizing that a register is in use and using a new register in place of the specified operand. The same philosophy that is used to implement virtual memory can be applied to register management.

By renaming register r2 as register r7 in instructions 5 and 6 (Figure 8.11), the delay imposed on instruction 5 is removed. The arrangement of Figure 8.11 permits a significant improvement by means of out-of-order execution and register renaming. As you can see, the execution of instructions 1, 4, and 5 can take place in parallel with the execution of instructions 2 and 3.

Register renaming can be implemented in several ways. One technique is called *explicit register renaming* because all write operations to a register create a new register. For example, ADD r1,r2,r3 would automatically result in ADD **rr1**,r2,r3, where rr1 is the new register. Explicit register renaming can be implemented by having a physical register file larger than the logical register file (i.e., the user-visible registers). A mapping table translates logical register addresses to physical register addresses (i.e., it maps r1 to rr1). A physical register is released back to the pool when it is not being used by any instructions in the instruction window.

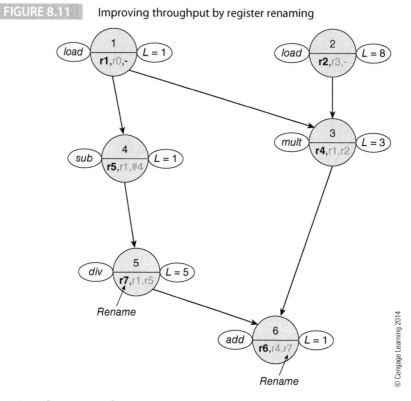

FIGURE 8.11   Improving throughput by register renaming

© Cengage Learning 2014

8.1.2   Superscalar Instruction Issue

We now briefly discuss how instructions in a superscalar processor are issued; that is, forwarded to the execution units. Sima, Fountain, and Kacsuk[1] in their definitive work employ the term *design space* to indicate the range of solutions that are available to the designer. For example, the instruction issue design space includes all possible solutions to the way in which instructions can be issued. Figure 8.12 illustrates some of the factors that have to be considered when designing a superscalar processor.

In the absence of *control and data dependencies* of the types we discussed in Chapter 7, instruction issue wouldn't be a problem. All you'd have to do is to throw instructions at ALUs and load/store controllers as fast as they could be processed. In the real world, control dependencies caused by branches and procedure calls, and data dependencies, have to be overcome in order to exploit the potential gains of instruction level parallelism. Moreover, the number of *data dependencies* rises as the issue rate increases. Increasing the number of instructions that you issue increases the chance that an instruction will need another instruction's destination operand. For example, when two three-operand instructions are executed in parallel, we have

op r_i, r_j, r_k
op r_p, r_q, r_r

and there are *four* possible conflicts between destination operand r_i and source operands r_q and r_r, and between destination operand r_p and source operands r_j and r_k. Suppose we execute the following three instructions in parallel

op r_i, r_j, r_k
op r_p, r_q, r_r
op r_s, r_t, r_u

In this case, each source operand has to be checked against two destination operands and there are 12 potential hazards.

[1]D. Sima, T.J. Fountain, and Kacsuk, *Advanced Computer Architectures*, Addison Wesley, Harlow, 1997.

FIGURE 8.12   Design space and issue policies

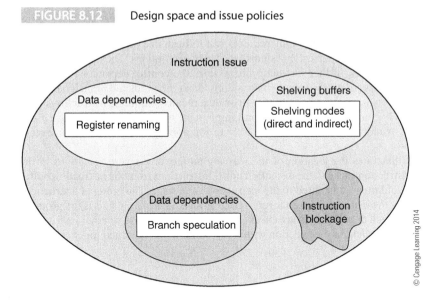

© Cengage Learning 2014

Data Dependencies

True data dependencies are an intrinsic property of programs and cannot be eliminated by either hardware or software techniques. A true data dependency is a RAW (read-after-write hazard); for example, in the code below, the ADD instruction cannot be executed until after the load has taken place.

```
LDR  r3,[r10]       ;get data into r3
ADD  r5,r3,r4       ;now add the value in r3 to r4
```

There are two *false* data dependencies that can be removed by hardware or software techniques. An *output dependency* occurs when two instructions write to the same location. This creates a write-after-write (WAW) hazard if the second instruction performs a write after the first instruction, for example,

```
ADD  r3,r1,r2       ;add r1 and r2
SUB  r5,r3,r4       ;subtract r4 from r3 and put the result in r5
ADD  r3,r7,r8       ;now reuse r3 in the next calculation
```

Register r3 is the destination for an operand in the first line and this operand is used in the second line. Register r3 is re-used as a destination operand in the third instruction. If the write to r3 caused by ADD r3,r7,r8 occurs before the read in SUB r5,r3,r4, an error will occur.

The second false dependency is the *antidependency* which is a write-after-read hazard (WAR). This dependency occurs when an instruction uses a location as a source operand and a following instruction reuses the location as a destination, for example,

```
ADD  r3,r1,r2       ;add r1 and r2
ADD  r1 r5,r4       ;reuse r1 to hold the sum of r53 and r4
```

When the first instruction is executed, the value of r1 is read. The next instruction re-uses r1 as a destination register. If the write to r1 in the second instruction takes place before the read in the first instruction, an error will occur.

Control Dependencies

Superscalar processors are even more susceptible to the effects of control dependencies than scalar pipelines. Taking a branch not only requires you to flush one pipeline, it requires you to flush multiple pipelines. Superscalar systems have only two options. They can detect control dependency and block further instruction issue until the control dependency has been resolved, or they can *speculatively* execute instructions along the predicted branch path.

Figure 8.12 refers to *shelving buffers* that provide a means of holding instructions until all their resources become available. If you don't implement shelving, you have to use *direct issue* whereby instructions are forwarded directly to the appropriate functional or execution units.

Figure 8.13 illustrates the location of the *shelving buffers* or *reservation stations* in the chain from the instruction cache to the execution units. Instructions from the instruction buffer are first decoded and then transmitted to the instruction issue stage. The group of instructions being processed by the issue mechanism is referred to as an *issue window*. This is the group of instructions over which dependencies are checked in order to determine whether instructions can be executed in parallel. Any instruction within this window can be issued for execution.

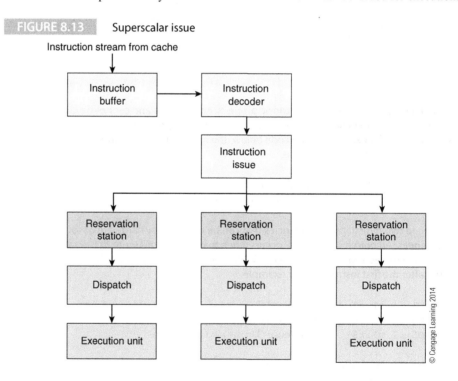

FIGURE 8.13   Superscalar issue

In shelved or indirect issue, instructions are forwarded to buffers called reservation stations that sit in front of execution units (e.g., ALU, integer logic unit, load or store unit, etc.). The reservation station postpones dependency checking until a later stage in the execution of an instruction. Within the reservation station, instructions are checked to see whether they have all the necessary resources needed to allow them to be dispatched to the functional units. Actually implementing a reservation station is no easy task because it requires the coordination of instruction, registers, and functional units. Several techniques have been devised to implement reservation stations based on Tomasulo's algorithm or scoreboarding. These techniques are beyond the scope of this text.

Figure 8.12 highlights *instruction blockage* that refers to the ways and means of dealing with the situations in which instruction issue is stalled or blocked. This stage handles the

Data Speculation: The Final Frontier?

Computer designers try everything they can to speed up calculation and reduce unnecessary latencies in computation. We've seen how control speculation is used to predict the outcome of a branch and to fetch instructions from the target address. A branch predictor has only one outcome to predict: *taken* or *not taken* (of course, if the branch is predicted as taken the target address must also be predicted).

Data speculation involves predicting the value of a variable before that value has been calculated (or read) and using the speculated value until the actual value is obtained. If the speculated and actual values are the same, all is well. If they differ, operations using the speculated value have to be repeated with the actual value. Using speculation with 16-bit integers gives you 65,536 possible values to choose from. At first sight data speculation is not feasible; however, consider the code fragment

```
LDR    r1, [r9]      ;read r1
ADD    r1,r1,#4      ;increment r1 by 4
STR    [r9],r1       ;save new value
ADD    r5,r5,r1      ;update r5
```

With data speculation you can execute ADD r5,r5,r1 *before* the value of r1 has been determined by the previous three instructions. If you examine the three instructions that generate r1 closely, you will see why data speculation may be possible.

Several RISC processors use register r0 to hold zero in order to provide a simple way of loading the constant zero or to suppress a register (e.g., ADD r1,r2,r0 becomes MOVE r1,r2). If zero is used so frequently, you could argue that the simplest data speculation would be to always assume that the value of any variable is zero and to use zero in speculative calculations.

If we call using the value 0 a *zero-order predictor*, the first-order predictor would use a constant; that is, it would assume that a value does not change each time it is used. The use of a constant value in data speculation can yield a correct prediction in a surprisingly high number of cases.

In many instances, a variable increases by a fixed amount or *stride* (for example, 0, 1, 2, 3, 4… or 4, 8, 12, 16, 20, …). We could define a second-order predictor as a mechanism that speculates that the next value of a variable will be the previous value plus a constant. More complex stride predictors may take account of *repetitive* behavior such as 1, 2, 3, 4, 1, 2, 3, 4, 1, 2, …, where a sequence is repeated (for example, the value of a counter within a nested loop).

It is, of course, impossible to deal with arbitrary sequences of a variable unless they are repetitive (for example, 20, 15, 47, 95, 20, 15, 47, 95, 20, 15, 47, 95, . . .).

Sazeides and Smith have reported that *last value* (i.e., constant) prediction provides an accuracy of from 23% to 61% with an average of about 40%. They give the accuracy of stride predictors as 38% to 80% with an average of about 56%. They also report that context-based predictors provide prediction accuracies of 56% to over 90% with an average of about 78%. The context-based predictor is derived from the finite context methods used in text compression and predicts the next value based on a finite number of preceding values.

effects of (true) data dependency and ensures the optimum issuing of instructions. The issue blocking stage also deals with in-order and out-of-order execution.

Issue blocking is also concerned with two types of issue window. A fixed issue window takes a group of instructions at a time and executes them before the window is refilled. A sliding window provides a fixed length buffer where new instructions enter the execution window as previous instructions are executed and exit that window.

Examples of Superscalar Processors

We conclude this section by looking at two major processors that led the way in implementing superscalar technology: the DEC Alpha machine and the Intel Pentium.

The Alpha

The Digital Equipment Corporation, or Digital as it became known as, created the PDP series of minicomputers that dominated academic computing in the 1970s. From the PDP sprang the VAX, which became one of the most popular minicomputers of the 1980s.

Although slow to adapt to the PC-dominated world of the 1990s, Digital designed one of the world's fastest processors, the superscalar Alpha in 1992. The first-generation 21064 Alpha was a 64-bit load/store superscalar RISC with 32 general-purpose integer and 32 floating-point registers that dispatched two instructions per clock cycle at a clock rate of 200 MHz. Its instruction set architecture is similar to other mainstream RISCs such as MIPs, with some interesting features such as a conditional branch that tests a designated register rather than a condition code register. It has a conditional move instruction; for example, CMOVEQ R1,R2,**R3** performs the *move on zero* operation if [R1] = 0 THEN [R3] ← [R2]. Like ARM, this instruction gives the Alpha a *branchless* programming facility. For example, we can extract the maximum of a pair of values in registers R1 and R2 and put the maximum value in R1 by

```
CMPLT    R1,R2,R3      Compare R1 < R2 and put 1 in R3 if true
CMOVLBS  R3,R2,R1      Conditional move on low bit of R3 set
```

Sad History of DEC and the Alpha

The Alpha was too late to save Digital, which was forced to sell off some of its divisions in the early 1990s. Digital became involved in patent battles with Intel, and, in 1998, sold its semiconductor production facilities to Intel to settle the litigation. As a condition of the sale, the Federal Trade Commission forced Digital to make its Alpha technology available to Intel's rivals, Advanced Micro Devices and IBM.

In 1995, Digital launched its second generation 21164 Alpha with a 600 MHz clock and a four-way superscalar organization that could dispatch four instructions per clock to achieve a maximum throughput of 2.4 billion instructions per second.

In 1998, Digital ceased to exist when Compaq paid $9 billion to acquire Digital, which was then completely absorbed into Compaq. The same year saw the third generation Alpha, the 21264, that could issue up to six instructions per cycle (in bursts). In 2002, Compaq announced that it was pulling the plug on its Alpha processors. Processor development would continue into 2003 and systems development into 2004. The Alpha was consigned to the dustbin of history. Compaq switched its focus from the Alpha and MIPs to Intel's IA64 Itanium. The merger of Hewlett Packard and Compaq in May 2002 may have been one of the factors that influenced the future of the Alpha. Some web commentators implied that Compaq's dropping the Alpha was a veritable act of treachery; others said that it represented good business sense in a harsh economic world. Digital's woes were, perhaps, due to its attempting to back too many horses at once (VAX and MIPS). Compaq found itself in the same position in 2002 when it was simultaneously promoting the Alpha, MIPS and Xeon processors. The Xeon is an Intel chip built round the Pentium core specifically designed for use in high-end workstations. The Xeon has an enhanced cache system and multiprocessor communications interface, facilitating its use in clusters of four or eight processors.

People grieve over the passing of the Alpha for several reasons. Some grieve because it was one of the fastest processors of the 1990s, and some because Digital was the *Rolls Royce* of the computer world. The Alpha is not the only innovative processor to have had an untimely end. AMD's 29000 RISC series and Motorola's 88000 RISC are two other victims.

Figure 8.14 illustrates the structure of the Alpha 21264's pipeline. The Alpha did not employ any spectacularly new architectural feature, but owed its performance to both good architectural and good circuit design. For example, when the 21264 fetches a block of four instructions from the instruction cache, the cache also provides a prediction of the location of the next block of four instructions. These predictions are updated dynamically as the code executes.[2] The 21264 employs a sophisticated branch prediction mechanism that uses either *local* or *global* information as part of its prediction strategy. This branch prediction scheme dynamically adapts to the property of the program by selecting either a local or a global prediction strategy.

The local history predictor is a 1024-entry table indexed by the instruction address from the program counter. Each entry in the local table contains a 10-bit branch history sequence. The global predictor consists of a 4096-entry table of two-bit saturating state machine branch predictors. The global history table is indexed by the history of the last twelve branches. A third array is a 4096 entry *chooser* array that selects whether to use local or global branch prediction. The chooser array consists of 4096 two-bit saturating counters that are indexed into by the branch history. Each saturating counter in the chooser array determines whether the current branch should be chosen from the global or local history arrays. When a branch instruction retires (i.e., is completed), the chooser array is updated: If the local and global predictors disagree, the chooser array selects the outcome that was correct.

The 21264 provides both out-of-order issue and out-of-order execution. Four instructions are fetched per clock cycle (limiting the maximum sustained throughput to four instructions per cycle). The fetch unit in stage 0 of Figure 8.14 can retrieve 16 bytes from the two-way set-associative instruction cache (we cover caches in detail in Chapter 9).

The 21264's out-of-order dispatch is aggressive. Instructions in the 20-entry integer queue that can be executed are issued immediately. Instructions that have been in the queue the longest time are then given preference over those that have not waited as long for resources.

The 21264 speculatively allocates a register to each instruction that has a register as a destination operand.[3] Register renaming eliminates WAW and WAR dependencies while ensuring correct read-after-write behavior. The effect of register renaming is to ensure that the maximum possible level of instruction level parallelism can be exploited. In Figure 8.14, the integer register files (in the register read phase) have 80 entries. This means that up to 80 instructions can be in partial states of execution at any time.

---

[2] "Exploring Alpha Power for Technical Computing," *Compaq Technology Brief*, April 2000.

[3] Kessler, R.E., McLellan, E.J., and Webb, D.A., "The Alpha 21264 microprocessor architecture," Compaq Computer Corp., Shrewsbury, MA, *International Conference on Computer Design: VLSI in Computers and Processors,* 1998. ICCD '98. Proceedings.

FIGURE 8.14   The Alpha pipeline

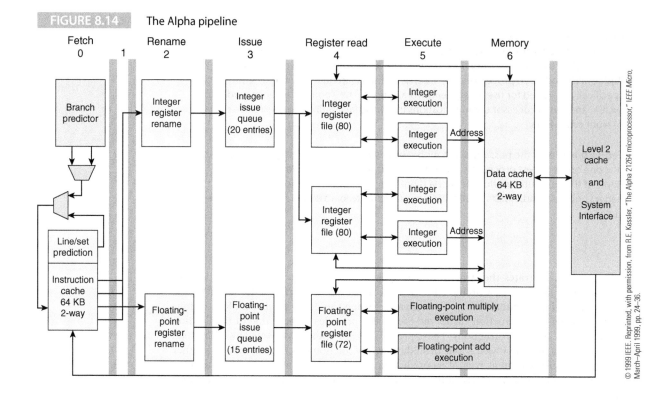

The Integer issue queue (stage 3 of the pipeline in Figure 8.14) is responsible for maintaining a list of instructions in the window waiting for execution. At each cycle, instructions are selected for execution as the resources they require become available. A *scoreboard* uses the renamed register numbers to keep track of available registers. When functional units or data become available, the scoreboard notifies all relevant instructions (i.e., those waiting for data or resources). Instructions are issued as soon as all resources (including bypassed data) are available. Four integer and two floating-point instructions can be issued from the queue every cycle. If more than four instructions are available the oldest four are selected.

The integer execution units are not identical; for example, only one ALU can perform shifts and multiplications, and only one ALU can deal with branches.

The final state in the Alpha pipeline deals with the in-order retire mechanism. Although the instructions are issued and executed out-of-order, they have to be retired in order. Once an instruction begins execution it is able to retire only when all previous instructions have retired. The retiring of an instruction makes it non-speculative[4] and ensures that its effects will be visible to the programmer; that is, its effects on flag bits become valid at the point of retirement.

The Pentium

It is easy to give the impression that superscalar execution is exclusively the province of the pipelined RISC processor. Over the years, pipelining and superscalar technologies have been applied to CISC processors (especially the IA-32 family) with increasing vigor. We now take a brief look at Intel's Pentium P6, launched in late 1995. This was to become a major step on Intel's path to its current dominant position in the computer world.

---

[4]R.E. Kessler, "The Alpha 21264 Microprocessor", *IEEE Micro*, March–April 1999.

FIGURE 8.15   Simplified structure of the P6 execution engine

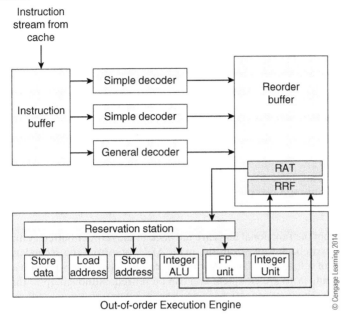

Out-of-order Execution Engine

© Cengage Learning 2014

Figure 8.15 provides a simplified diagram of the P6 processor[5] that Intel launched to extend its product line and to compete with IA-32 compatible processors from AMD and NextGen. The Pentium was Intel's first processor to introduce a superscalar architecture by means of a pair of integer pipelines. The P6 takes a much more radical approach by providing a three-way superscalar structure that can simultaneously dispatch five instructions. The P6 also adopts some of the techniques pioneered by companies such as AMD; for example, several IA-32 instructions are translated into a RISC-like format and executed by a RISC core processor. The P6 implements out-of-order execution and register renaming.

The P6 translates complex, variable-length x86 instructions into simpler, fixed-length micro-operations (called *uops* in Intel literature). Instructions from the instruction buffer are fed to three instruction decoders (Figure 8.15). One decoder deals with all types of x86 instruction and two decoders are designed to handle only simple x86 instructions such as register-to-register operations. At this stage, instructions are dealt with in sequence. Since the decoding mechanism requires a mix of simple and complex instructions to keep the decoders happy, a sequence of register-to-memory instructions in the source code will degrade performance. The choice of how many complex decoders and how many simple decoders came from detailed analysis of the frequency of IA32 instructions seen in desktop applications. By the mid-1990s the advantages of RISC were quite evident, and a lot of compilers (especially optimizing compilers) chose to use simpler instructions unless a more complex instruction was better suited to the given task.

The microoperations (uops) are passed to the 40-entry-deep reorder buffer (ROB). As instructions (i.e., the uops) are executed, they write results and condition codes back to the ROB. Associated with the ROB are two register files, the *real register file* (RRF) and the *register alias table* (RAT). The RAT maps the architectural registers onto the physical registers. When a uop is logged in the reorder buffer, the register alias table is used to determine whether the corresponding source operand(s) should be taken from the real register file or ROB.

[5]Linley Gwennap, "Intel's P6 uses decoupled superscalar design," *Microprocessor Report*, Vol. 9, No. 2, February 16, 1995.

If a source register in the current uop matches a destination register of an earlier instruction that is currently in the reorder buffer, the source register is replaced by a pointer to the corresponding ROB entry. When a source register is replaced by a pointer, the RAT is also updated with the uop's destination register.

Microoperations from the reorder buffer are transferred to another buffer, the *reservation station*, which holds up to 20 uops. The uops have to wait in the reservation station until all of their destination operands are available. Once a uop has all of its resources, it can be dispatched to a functional unit. Figure 8.15 shows that *up to* five uops can be dispatched per machine cycle because there are two integer stages, a load, a store address, and a store data stage. The two calculations may be two integer operations or an integer and a floating-point operation.

Results from the functional units are fed back to the reorder buffer. A machine cycle can execute three writebacks: one from the main ALU, one from the secondary integer ALU, and the result of a load uop. Once destination results have been written back to the ROB, the corresponding uops are complete and can be retired. Although up to three uops can be retired per cycle, the uops are retired in program sequence, so a uop is not retired until all of the uops before it have been retired.

Intel's next major update to the IA-32 architecture was the Pentium 4, which was introduced in November 2000 with an initial clock speed of 1.5 GHz that reached a speed of 2.5 GHz within 18 months and over 3 GHz in two years. The Pentium 4 continued from where the P6 left off. The Pentium 4 architecture was given the name *NetBurst* by Intel, although the term NetBurst refers to a bag of enhancements, such as the Pentium 4's very long pipeline, rather than to any specific technology. Whereas the Pentium 6 has a 10-stage pipeline, the Pentium 4's hyperpipeline has 20 stages. This statement needs qualification. The actual pipelines are even deeper. There is no clear documentation of the exact pipeline depth, since there are so many complications with the trace cache, out-of-order completion, and the ROB, that it is hard to say how deep the pipeline is. It has been suggested that the typical depth for the Pentium 4 was 48 stages.

Fundamentally, the Pentium 6 and the Pentium 4 are similar. The differences are those of scale rather than direction. The Pentium 4 has the same type of in-order front end and out-of-order execution logic as the Pentium 6, but on a much grander scale. The Pentium 4 front end includes a sophisticated form of instruction cache known as an *execution trace cache*. (Chapter 9 covers caches in more detail.)

Conventionally, instructions are fetched from the system memory, placed in the instruction cache and then brought into buffers and decoded. With a sufficiently large instruction buffer, instructions are mostly read, decoded, re-read and re-decoded from the buffer. It makes good sense to perform the decoding only once and then store the decoded instruction. Remember that IA-32 instructions have variable lengths and the decoding process is complex.

The Pentium 4's trace cache replaced the conventional level 1, L1, cache now integrated on-chip in most processors. (Note that modern processors have L1, L2, and even L3 caches on chip. We discuss this in the next chapter.) Instructions are fetched from the trace cache, and new instructions are fetched and decoded only following a miss to the trace cache. The trace cache can hold approximately 12K uops, which corresponds to a conventional 8K to 16K byte instruction cache (with a similar hit ratio).

We have already explained how the P6 decodes IA-32 instructions into uops. The execution trace cache takes the uops generated by the instruction decoder and forms them into sequences called *traces*. A trace line is composed of six uops. The trace cache is organized to allow the target of a branch to be located in the same trace line as the branch instruction — even if the address of the target is a long way from the branch.

Not all IA-32 instructions can be decoded into simple uops. Instructions such as string moves require more extensive support. When a complex instruction is encountered, the trace cache accesses the microcode ROM that issues the stream of uops required to complete the interpretation of the complex instruction. Once an instruction has been executed, control is returned to the trace cache. Note that the microprogram ROM is also used to provide the sequence of uops needed to deal with exceptions, faults, and interrupts.

Figure 8.16 describes the Pentium 4's internal organization. The in-order front end includes an instruction buffer, a branch predictor, an IA-32 instruction decoder, the trace cache, and a microprogram ROM. The P4's NetBurst architecture has a more complex out-of-order execution mechanism than the P6. At any instant, up to 126 instructions (i.e., uops) can be in the buffer, including up to 48 loads and 24 stores. The physical register file has 128 entries that are managed by the register alias table (RAT) to implement register renaming.

FIGURE 8.16   The Pentium 4 structure

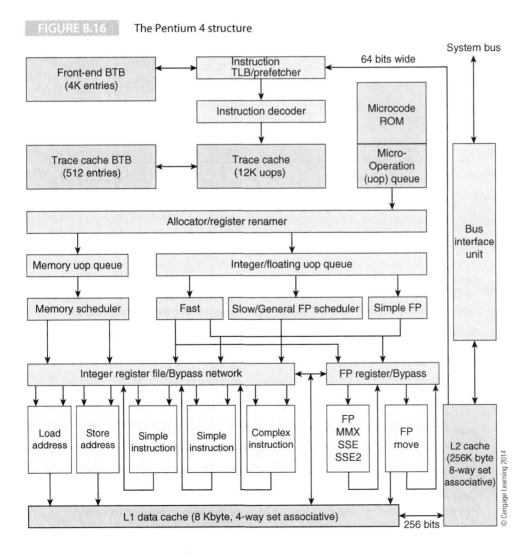

Example of Explicit Register Renaming

Consider the superscalar processor below that has a four-entry instruction window and a four-entry reorder buffer. Up to two instructions can be issued simultaneously to the four processing units. The two instructions must be of *different types,* as there is only one processing unit for load/store, arithmetic, logical, and branch operations. Assume that the following code is currently in the instruction window.

```
1.   ADD  r0,r1,r2      ;arith
2.   ADD  r1,r0,r3      ;arith
3.   AND  r0,r0,#0xFF   ;logical
4.   OR   r1,r2,#0xFF   ;arith
```

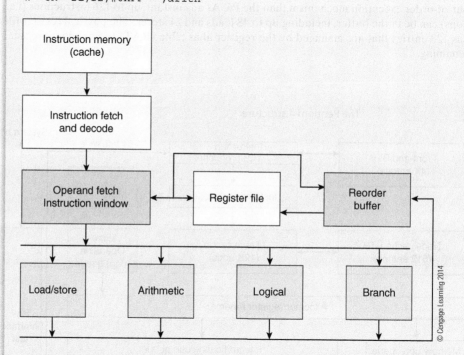

In explicit renaming, the destination operands will be renamed as follows. We use the notation rr to indicate a named register, and renamed registers are allocated sequentially. Source operands are not renamed unless they also appear as a destination operand.

```
1.   ADD  rr0,r1,r2
2.   ADD  rr1,rr0,r3
3.   AND  rr2,rr0,#0xFF
4.   OR   rr3,r2,#0xFF
```

The sequence of actions that take place during execution can be described as follows.

Instruction 1. Register r0 is renamed as rr0, and an entry allocated in the reorder buffer. Operands r1 and r2 are brought from the register file, and the instruction issued to the arithmetic unit.

Instruction 2. Register r1 is renamed as rr1, and an entry allocated in the reorder buffer. Operand r3 is brought from the register file. Operand r0, renamed as rr0, is not available. The instruction conflicts with instruction 1 because there is only one arithmetic unit. Consequently, it has to wait to be issued.

Instruction 3. Register r0 is renamed as rr2, and an entry allocated in the reorder buffer. This instruction stalls because rr0 is currently invalid in the reorder buffer.

Instruction 4. Register r1 is renamed as rr3, and an entry allocated in the reorder buffer. The source operand r2 is fetched from the register file, and the instruction is issued to the logical unit. Note that it is out of order.

Instruction 1 completes in the arithmetic unit, and rr0 is written to the reorder buffer. Register rr0 is retired to r0 in the register file.

Instruction 2 reads rr0 from the reorder buffer and is issued to the adder.

Instruction 3 reads rr0 from the reorder buffer and is issued to the logical unit.

Instruction 4 completes in the logical unit, writes rr3 to the reorder buffer, and retires rr4 to r1 in the register file.

Instruction 2 completes in the arithmetic unit, writes rr2 to the reorder buffer, and retires rr3 to r1 in the register file.

Instruction 3 completes in the logical unit, writes rr3 to the reorder buffer, and retires rr4 to r0 in the register file.

The Pentium—A Brief History

By an accident of history, IBM adopted Intel's 4.77 MHz 8088 processor in its personal computer in 1981. I employ the expression *accident of history* because some, including engineers at IBM, felt that the then new Motorola 68000 (which had a true 32-bit internal architecture) would have been a better choice. It is said that IBM had two significant reasons for choosing the 8088 over the 68000. First, the 68000 was a very recent product, and Motorola had not yet developed a range of peripheral interface chips. Second, IBM and Intel were involved in a cross-licensing agreement with IBM getting the right to manufacture the 8086 family and Intel getting the rights to IBM's bubble memory (bubble memory was a magnetic storage technology that demonstrated much promise but never caught on). The 8088 was a low-cost version of Intel's first 16-bit processor, the 8086. That is, the 8088 had an 8-bit data bus, making it much cheaper to construct a computer system at a time when memory and interfacing components were relatively expensive. In 1984, the IBM AT was launched with a true 16-bit Intel 80286 processor that could address 16 Mbytes of memory.

In 1985, Intel introduced the 80386 that provided a 32-bit architecture capable of addressing 4GB of memory. The '386 became the cornerstone of future Intel architectures, including the '486 and the Pentium family. Other semiconductor manufacturers produced compatible versions of the '386, such as AMD and Cyrix.

The Pentium processor was introduced in 1993 with two variants: a 60 MHz version and a 66 MHz version. The Pentium minimally extended the instruction set architecture of the '486 but radically extended its microarchitecture with the addition of two five-stage integer pipelines, a six-stage floating-point pipeline, and two-issue superscalar capability. Moreover, the Pentium incorporated dynamic branch prediction. In 1997, the Pentium's ISA was extended by incorporating multimedia extensions (i.e., short vector or SIMD operations).

(continued)

In 1995, Intel introduced the Pentium Pro (also known as the P6). The Pentium Pro provided a significant performance enhancement due to its improved microarchitecture that enhanced its superscalar capabilities by permitting out-of-order execution and register renaming. The Pentium Pro considerably extended the pipeline from five to twelve stages. Branch prediction was also improved from an approximately 75% hit rate to 90%. The Pentium Pro did not incorporate Intel's MMX technology due to limitations on chip size and transistor count at that time.

Confusingly, the Pentium P6 was followed by the Pentium II in 1997, which restored the MMX mode. Otherwise, little was changed. Intel's next advance was the Pentium III in 1999, when the MMX instruction set was expanded to become the SSE instruction set that included floating-point operations.

The seventh generation of IA32 processors was the Pentium 4 in 2000. This was to be the end of the line for the Pentium name, because later processors would have names like Intel Core 2 and Core i7. The P4 was the first radically new processor (in terms of its microarchitecture) since the P6. The P4's microarchitecture was called *Netburst*. As in the case of previous generations, the ISA was only gently upgraded—first by SSE2 multimedia extensions and then by SSE3 extensions. The NetBurst microarchitecture increased the pipeline to 20 stages. The P4 used an execution trace cache that cached micro-operations rather than instructions; that is, IA32 op-codes were translated into primitives and the primitives cached. Ultimately, the NetBurst architecture proved to be a dead-end because of its power consumption; it was not possible to go beyond 3.8 GHz. In 2006, Intel replaced the NetBurst microarchitecture with its Core microarchitecture and dropped the name Pentium.

In 2003, Intel launched its *M series* of Pentium processors; these were low-power devices intended for use in mobile applications. Significantly, the M series was derived from the later Pentium III rather than the Pentium 4. The M series included features from the P4 by incorporating its front size bus interface and SSE2 support. The microarchitecture was also improved by incorporating a better instruction decoding and issue mechanism. Intel further developed its M series to create the Core 2 Duo and Core 2 Quad processors in 2006 and 2008, respectively. Although these were desktop rather than mobile processors, they owe more to the P3 microarchitecture than the P4 microarchitecture.

Never a company to adopt a nice consistent naming system, Intel's next generation of IA32 processors was called the Core processors with Core i3, Core i5, and Core i7 launched in 2008. These processors all incorporated Intel's Nehalem microarchitecture. The Nehalam microarchitecture is really a mixed bag of techniques largely aimed at increasing computational efficiency and reducing power consumption. For example, a turbo-boost mode permits parts of the processor to operate at higher than the nominal clock rate during intensive computational activities. Nehalam also restores multithreading technology (which was absent in the M series and its derivatives).

In 2011, Intel brought out its replacement for the Nehalem architecture: Sandy Bridge. Sandy Bridge improves performance across a variety of fronts rather than by implementing a radically new microarchitecture. For example, the memory, graphics, and PCI express controllers that are normally incorporated in the North Bridge interface on the motherboard are now on-chip. The ISA is extended by the addition of the Advanced Vector Extensions (AVX) instruction set with its 256-bit operand support. Intel has made a significant change to its branch prediction mechanism.

An interesting feature of Intel's literature is that it introduces the *tick-tock* development model, which like Moore's law, codifies the way in which the semiconductor industry operates. Intel calls each new shrinking of feature size a *tick*, and each new microarchitecture a *tock*. For example, a new microarchitecture is implemented with a 45 nm process and later is shrunk down to 32 nm.

In the next section, we introduce a new class of superscalar processor, the *very long instruction word* (VLIW) processor, which avoids the complexity of superscalar issues by handing it over to the compiler, which then chooses the instructions to be executed in parallel prior to runtime. That is, the compiler avoids sequences of instructions that have data dependencies. Note that some people don't use the term *superscalar* to describe a VLIW processor; however, if you define superscalar to indicate a processor with greater than one instruction per clock cycle, then the VLIW falls into this category.

8.1.3   VLIW Processors

A VLIW processor allows the programmer or compiler to specify multiple instructions that are to be executed concurrently. The instruction word is long (i.e., *wide*) simply because it is made up of several individual instructions. For example, you could construct a 96-bit VLIW processor that is able to execute three 32-bit MIPs instructions concurrently.

Some use the term *molecule* or *bundle* to describe the VLIW instruction that consists of several operations called *atoms* or *syllables*, respectively. You could say that the VLIW architecture emulates a superscalar processor, because a superscalar processor creates a VLIW architecture by issuing instructions in parallel. This statement raises an interesting issue. A superscalar processor is a matter of organization, whereas a VLIW processor is a matter of architecture. To compare the notions of superscalar with VLIW means comparing organization with architecture. However, there are important differences between these two paradigms. VLIW processors identify parallelism at *compile time*, because the instructions generated by a compiler are suitable for parallel execution and no dependencies have been detected. That is, the compiler has the time and resources to optimally arrange instructions to exploit the most parallelism. A superscalar processor has to dynamically issue multiple instructions and deal with the problems of control and resource dependencies dynamically in the processor's hardware at run time.

> ### VLIW—The Wrong Name
>
> The defining characteristic of a VLIW processor is the bundled instruction that contains a group of instructions that can be executed in parallel. It is the job of the programmer or compiler to ensure that the individual instructions in a bundle can be executed in parallel. The long instruction word is a consequence of instruction bundling and not a cause. You could design a conventional processor with a long instruction word (e.g., because you allowed a 64-bit constant), but it would not be a VLIW processor. Perhaps the VLIW should have been called BWP (bundled word processor).

> ### History of the VLIW Processor
>
> The VLIW processor was introduced by Joseph Fisher following his earlier work on the design of horizontal microprogramming and on trace scheduling. In some ways, horizontal microcode is rather like a very long instruction word, because you can carry out multiple independent operations in parallel. In particular, Fisher felt that the processor and compiler should be co-designed and work together in harmony.
>
> Fisher left Yale University to create Multiflow, Inc. in 1984 and to market a VLIW computer using a 256-bit instruction format with each instruction defining eight 32-bit operations. VLIW processors run into the *terminological problem* that an instruction read from memory is really a bundle of individual instructions, which means that the word *instruction* becomes confusing since it is not clear whether it is referring to an instruction group/bundle or to one of the individual operations within the instruction.
>
> Multiflow had a limited success, but the company ceased trading in 1990. Another contemporary VLIW processor was the Cydrome, designed by Bob Rau, which also had a 256-bit instruction with six 40-bit op-codes. Like the Multiflow, the Cydrome was not a

(continued)

commercial success. However, many of those who worked on early VLIW processors had a very significant effect on the development of the computing industry. Both Fisher and Bob Rau later ended up at Hewlett-Packard, where they continued to work on computer architecture research. Fisher received the IEEE's Eckert-Mauchly award in 2003 for his pioneering work on VLIW architectures.

The concept of the VLIW received a boost when the Transmeta Corporation was founded in 1995 to market a VLIW-based IA-32 compatible architecture with low-power consumption. Transmeta achieved fame (notoriety) because it employed some big names in computing, such as Dave Dizel and Linus Torvalds, and remained silent about its product for a considerable time. Ultimately, their VLIW never achieved the level of performance expected of it and Transmeta went from Silicon Valley to "Silicon Cemetery" in 2009.

Texas instruments employed a VLIW architecture in their C6000 special-purpose digital signal processing chips intended for use in embedded applications such as modems and image processors. The TI VLIW architecture is called VelociTI and executes eight 32-bit operations in a 256-bit instruction.

Later in this chapter we look at Intel's VLIW processor called the Itanium, which has its origins in HP's EPIC project.

Figure 8.17 describes the organization of a VLIW processor. A long instruction has several fields or slots (or atoms or syllables), each containing an individual operation. Typically, VLIW instruction formats are limited by restrictions on the operations that may be assigned to the three slots. For example, the first two slots might be reserved for integer operations, whereas the third slot may hold only a floating-point operation. Such an arrangement would mean that the sequence

```
ADD   r1,r2,r3
ADD   r4,r5,r6
SUBF  r7,r8,r9     ;floating point subtraction
```

could be fitted in one instruction,

```
ADD   r1,r2,r3: ADD   r4,r5,r6: SUBF  r7,r8,r9
```

whereas the sequence

```
ADD   r1,r2,r3
ADDF  r4,r5,r6
SUBF  r7,r8,r9     ;floating point subtraction
```

would require the two instructions

```
ADD   r1,r2,r3
NOP                ;the second integer slot
ADDF  r4,r5,r6
NOP                ;the first integer slot
NOP                ;the second integer slot
SUBF  r7,r8,r9
```

This can be represented as

```
ADD   r1,r2,r3: NOP: ADDF  r4,r5,r6;  first long instruction
NOP:                 NOP: SUBF  r7,r8,r9;  second long instruction
```

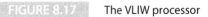   The VLIW processor

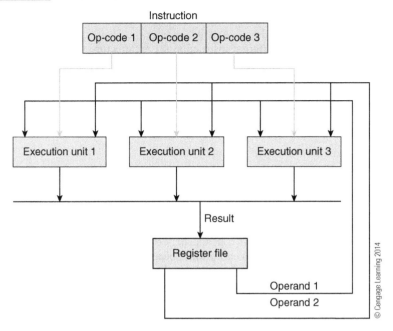

In this example, we have assumed that the compiler could not find anything to put in the NOP slots. This demonstrates an inherent weakness of ILP mechanisms: unless instructions can be found to feed all of a processor's operational units, it cannot achieve high levels of efficiency. As we shall see in Chapter 13, the same problem arises with multiple cores. Indeed, you might argue that the difference between a VLIW architecture and a multi-core processor is the granularity of the tasks given to the execution units. A VLIW execution unit gets an instruction, and a core gets a thread (stream of instructions).

The advantage of the VLIW architecture is that it is relatively easy to design in comparison to an out-of-order superscalar. You do not have to worry about detecting instruction dependencies as, *by definition*, they don't occur because the compiler generates code without dependencies. The disadvantages of VLIW architecture are its susceptibility to long instruction latencies and its poor code density. If an instruction has a long latency (e.g., a load following a cache miss), the processor is held up until the load can be completed. If a VLIW with m operations per bundle has an instruction with an n-cycle latency, up to $m \cdot n - 1$ execution slots are lost. If the compiler detects data dependencies and cannot generate

VLIW and Superscalar Processors

The VLIW processor is not just a variant of the superscalar processor. It's a different machine organization. Superscalars attempt to execute two or more operations in a serial instruction stream concurrently and must go to great lengths to ensure that data dependencies do not modify the program's semantics.

VLIWs move that complexity to the compiler; they don't check data dependencies. Consequently, VLIW processors require optimizing compilers that take account of the processor's microarchitecture and its limitations and operational latencies. Moreover, code portability across different processors becomes impossible because, by its very nature, a compiler for a VLIW processor creates code for a specific machine architecture both at the architectural and the microarchitectural levels.

multiple instructions in the same word, the system degenerates into a conventional non-superscalar processor that executes a lot of NOPs in parallel. VLIW processors are sensitive to the relative frequency of load instructions. Too many loads can lead to the need for a lot of NOPs to pack holes in the code.

Because a VLIW has a different instruction set architecture to a non-VLIW machine, existing code must be recompiled. A superscalar machine doesn't require recompilation, because the binary codes for scalar and non-scalar versions of the same programs are the same. You can sum this up by saying that a processor is called superscalar because of its *organization*, whereas a VLIW is so-called because of its *architecture*. Later in this chapter, we introduce the IA64 architecture that has a VLIW-like architecture.

Interrupts and Superscalar Processing

Before leaving superscalar processors, we have to make a comment on interrupt processing. Interrupt processing in a conventional CISC processor is relatively straightforward. When an interrupt or exception occurs, the only major decision is whether to execute the current instruction to completion or to store sufficient state information to allow the interrupted instruction to be restarted once the interrupt/exception has been serviced. Even this relatively simple arrangement can prove a headache. For example, early Motorola 68K processors could not continue from certain exceptions because too much information had been lost; such exceptions required that the current task be aborted.

The situation with both pipelined and superscalar processors is rather more complex. When an interrupt occurs, we have more than one instruction to consider. In a so-called *precise interrupt*, all instructions that are in the pipeline prior to the interrupt or exception are committed and executed to completion, and instructions following the exception are restarted after the exception has been dealt with. In other words, if an exception occurs between instruction X_i and instruction X_{i+1}, all state information up to and including X_i must be complete (that is, all instructions must be retired). No instruction after X_i must be modified. Consequently, an interrupt/exception is precise if the saved process state is consistent with the sequential architectural model so that the superscalar processor behaves exactly like a non-superscalar processor with respect to exceptions.

Saving Energy by Short-Circuiting Operations

Performance is not the only consideration in computer design. Power consumption is of vital importance in both fixed (desktop) and mobile applications. This box looks at just one way of reducing power consumption by modifying the microarchitecture to remove trivial computations.

Computers spend time performing calculations and expend even more valuable energy while doing it. A corollary of this is that if the calculation isn't necessary then save time and money (energy) by not doing it. Such an approach might seem difficult, because surely, all computations are essential. Otherwise we wouldn't have written the program in the first place.

Consider the following code. The fine details like setting up the loop don't matter. We are adding two arrays together **A = A + B**.

```
Loop
        LDR  r0,[r4,#4]!
        LDR  r1,[r5,#4]!
        ADD  r0,r0,r1
        STR  r0,[r4]
        B    Loop
```

Suppose that the arrays are very large and that array *B* is sparse and contains very few non-zero elements, all of which are 1. We could use predicated execution and carry out the add and store only if the element from *B* is not zero, as follows.

```
Loop
      LDR    r0,[r4,#4]!
      LDRS   r1,[r5,#4]!    ;;read B and set Z-bit
      ADDNE  r0,r0,r1       ;;if B = 0, do nothing
      STRNE  r0,[r4]        ;;if B = 0, do nothing and don't store
      B      Loop
```

We haven't achieved a lot with this code. But we have ensured that the store operation doesn't take place if we would be adding a zero, and at least we've avoided the dangers of a cache miss or a page fault. Islam, Själander, and Stenström demonstrate that we can improve the energy efficiency of computers by eliminating some operations.[*] These are called *trivial operations,* because their execution does not affect the result.

Trivial operations frequently involve the *null value* (zero) or the *identity value* (one), as the following examples demonstrate.

Operation	Example	Value	Result
Addition	$Y = X + Y$	$X = 0$	Y
Subtraction	$Y = Y - X$	$X = 0$	Y
Multiplication	$Y = Y * X$	$X = 1$	Y
Division	$Y = Y/X$	$X = 1$	Y
OR	$Y = Y \text{ OR } X$	$X = 0$	Y
AND	$Y = Y \text{ AND } X$	$X = 1$	Y

© Cengage Learning 2014

Islam, et al. indicate that the frequency of trivial operations in real programs is anything but trivial. They present results using SPEC CPU2000 data ranging from over 20% of the code to under 5%.

If we can both identify trivial operations and eliminate them at the microarchitectural level, energy can be saved by omitting these operations. Of course, the additional energy required to detect the trivial operations must be less than that saved by not executing these operations. Islam, et al. conclude that the potential energy saving is in the region of 9% when executing SPC CPU2000 code.

A simple way of dealing with trivial operations is to detect them at the decode stage of the pipeline before they are issued. Clearly, the earlier the stage at which trivial operations can be detected, the more energy that can be saved. Once an operation is identified as trivial, it is not necessary to issue it, execute it, and (more importantly) perform a write-back.

[*] Islam M.J., Själander M. and Stenström P., "Early detection and bypassing of trivial operations to improve energy efficiency f processors," *Microprocessors and Microsystems 32*, 2008, pp.183–196.

8.2   Binary Translation

The subtitle of this text is *"themes and variations."* The theme of this chapter is high-performance computing and instruction-level parallelism. We're now going to look at a variation called *binary translation*, which offers a means of improving performance by automatically converting one machine's executable code into the code of another machine. We provide a brief overview of the concept and then introduce the Transmeta Crusoe that implemented a form of binary translation and was widely expected to have a significant impact on computer architecture. In the event, Transmeta's Crusoe was to provide little more than a short-lived excursion into on-chip binary translation.

Binary translation is used in two contexts. The first is in relation to machine *portability*, where binary translation allows you to port code from one machine to a different machine (including a different operating system) without re-compiling the code. The other context is in the translation of one machine's code into another machine's code by the processor itself, as part of code execution; that is, binary translation is embedded into the target machine's own hardware or firmware. *Static* binary translation involves translating the source code before it is executed on the target processor, and *dynamic* binary translation involves translating code and optimizing it at runtime.

Growth of interest in binary translation as a means of implementing high-speed processors is due to the complex hardware required by the superscalar processors and the microcode mechanisms used by the Pentium 4.[6] In the late 1990s, Cristina Cifuentes, Mike Van Emmerik, and Norman Ramsey created a framework for general-purpose binary translation called UQBT (University of Queensland Binary Translator) that they used as a resourceable and retargetable binary translator.[7] Although general-purpose binary translation is not the theme of this chapter, it's interesting to look at the work carried out by Cifuentes, et al. They provide an interesting example of the generation of binary code by a compiler for a Solaris machine using a Pentium processor and its binary translation into code that will run on a SPARC-based Solaris computer. The following provides the C code for a recursive program that generates Fibbonaci numbers and the disassembly of the binary output.

```
int fib (int x)
{
        if (x > 1)
                return (fib(x - 1) + fib(x - 2));
        else return (x);
}


int main (void)
{       int number, value;

        printf ("Input number: ");
        scanf ("%d", &number);
        value = fib(number);
        printf("fibonacci(%d) = %d\n", number, value);
        return (0);
}
```

---

[6]Michael Geschwind, et al, "Dynamic and Transparent Binary Translation," *Computer*, March 2000.

[7]Cristina Cifuentes and Mike Van Emmerik, "UQBT: Adaptable Binary Translation at Low Cost," *Computer*, Vol. 33, No. 3, March 2000, pp. 60–66.

The IA-32 code

```
fib()
8048960:    55                      pushl    %ebp
8048961:    8b ec                   movl     %esp,%ebp
8048963:    53                      pushl    %ebx
8048964:    83 7d 08 01             cmpl     $0x1,8(%ebp)
8048968:    7e 2e                   jle      0x2e <8048998>
804896a:    8b 45 08                movl     8(%ebp),%eax
804896d:    48                      decl     %eax
804896e:    50                      pushl    %eax
804896f:    e8 ec ff ff ff          call     0xffffffec <fib>
8048974:    83 c4 04                addl     $0x4,%esp
8048977:    8b d8                   movl     %eax,%ebx
8048979:    8b 45 08                movl     8(%ebp),%eax
804897c:    05 fe ff ff ff          addl     $0xfffffffe,%eax
8048981:    50                      pushl    %eax
8048982:    e8 d9 ff ff ff          call     0xffffffd9 <fib>
8048987:    83 c4 04                addl     $0x4,%esp
804898a:    8b c0                   movl     %eax,%eax
804898c:    8d 14 18                leal     (%eax,%ebx),%edx
804898f:    8b c2                   movl     %edx,%eax
8048991:    eb 0d                   jmp      0xd <80489a0>
8048993:    90                      nop
8048994:    eb 0a                   jmp      0xa <80489a0>
8048996:    90                      nop
8048997:    90                      nop
8048998:    8b 55 08                movl     8(%ebp),%edx
804899b:    8b c2                   movl     %edx,%eax
804899d:    eb 01                   jmp      0x1 <80489a0>
804899f:    90                      nop
80489a0:    8b 5d fc                movl     -4(%ebp),%ebx
80489a3:    c9                      leave
80489a4:    c3                      ret
```

© 2000 IEEE. Reprinted, with permission, from Cristina Cifuentes, Mike Van Emmerik, "UQBT: Adaptable Binary Translation at Low Cost," *Computer*, Vol. 33, No. 3: March 2000, pp. 60–66.

The code includes Pentium-specific prolog and epilogue code fragments required by functions to set up the stack frame (shown in gray together with the code required to pass parameters on the stack). The instructions in blue are redundant jumps and no-operation instructions. After each call, the stack pointer, esp, is restored with addl $0x4,%esp. Note that % indicates a register. The disassembled SPARC code generated by the UQBT package is given by

```
fib()
8050b54:    9d e3 bf 90             save     %sp, -112, %sp
8050b58:    80 a6 20 01             cmp      %i0, 1
8050b5c:    04 80 00 08             ble      0x8050b7c
8050b60:    01 00 00 00             nop
8050b64:    7f ff ff fc             call     fib
8050b68:    90 06 3f ff             add      %i0, -1, %o0
8050b6c:    a0 10 00 08             mov      %o0, %l0
8050b70:    7f ff ff f9             call     fib
8050b74:    90 06 3f fe             add      %i0, -2, %o0
8050b78:    b0 02 00 10             add      %o0, %l0, %i0
8050b7c:    81 c7 e0 08             ret
8050b80:    81 e8 00 00             restore
```

Cifuentes points out that the new version is shorter because the original binary program was poorly organized. Moreover, the Pentium prologue and epilogue code has been replaced by special SPARC save and restore instructions. The function argument is placed in a register rather than being pushed on the stack, and some instructions have been moved to SPARC delay slots. Cifuentes says, "*It's hard to believe that it's the same program, but the semantics of the original program have been preserved, using different instructions to perform the same tasks.*"

There are three levels of binary translation. The lowest and most primitive level is called *emulation*, whereby each binary op-code from the source machine is simply interpreted on the target machine. For example, the binary code 24AC 0012 3344 may be read from the source file and the emulator determines that this is equivalent to add [r1],00123344_{16}. The emulator then carries out this operation on the target machine. Emulation is slow, inefficient, clumsy and crude. Otherwise, it works fine.

The next level of binary translation is called *static translation*. In this case, the source code is translated into the target code and then that is executed. This mechanism is faster than emulation and it permits a measure of optimization because the instructions are not being considered in isolation. A good static translator does not offer a simple instruction-by-instruction translation process whereby instruction x on machine p is just represented by instruction sequence y on machine q. Instead, it reads the source code and *raises the semantic level* (the reverse process to that of a conventional compiler) to create control flow graphs; that is, it analyses what the code actually does. Once this step has been taken, efficient object code for the target processor can be generated.

The highest level of binary translation is *dynamic binary translation*, which may start off using emulation. However, as the process runs, the executed code is monitored in real time and dynamically modified as it is translated and optimized. The dynamic binary translation is therefore an adaptive process.

8.2.1   The Transmeta Crusoe

The best known processor employing dynamic binary translation to execute IA-32 code was the Crusoe processor produced by the Transmeta Corporation in 2000. Crusoe was intended to compete with Intel's processor in the low-power laptop and embedded processor markets. I am including this processor because it represents a significant footnote in the development of microprocessors. Although the path Crusoe followed was not successful, the underlying technology may later be used to accelerate performance.

Crusoe processors are not exclusively hardware based; the dynamic binary translation process relies on embedded software to perform the translation. This approach reduces the transistor count and saves energy. Also, it makes it much easier to upgrade processors as better translation technologies are developed. Transmeta calls the software layer that performs the binary translation "*code morphing*" because it dynamically transforms x68 code into VLIW code.

A VLIW processor sits at the heart of the Crusoe processor and can execute four instructions per clock cycle. The VLIW's instruction set is free of the complex CISC-based legacies of the x68's IA-32 architecture, permitting the core of Transmeta's first generation Crusoe to use approximately 25% of the number of transistors in a corresponding IA-32 processor.

The VLIW processor uses a 128-bit word, called a *molecule*, which is divided into four 32-bit *atoms*. The expressions *molecule* and *atom* are Transmeta's own terms. Later we will see that the Intel IA64 Itanium architecture shares some of the Crusoe's features but uses the term *bundle* rather than molecule to describe the long instruction word. Figure 8.18 illustrates the Crusoe's *six* instruction formats. Two formats are 128-bit bundles, and two are 64-bit bundles. By providing both 128-bit and 64-bit bundles, Transmeta ensures that no bundle will have to introduce two NOPs when the compiler can't find four instructions to execute in parallel.

FIGURE 8.18   The TM500's six instruction formats

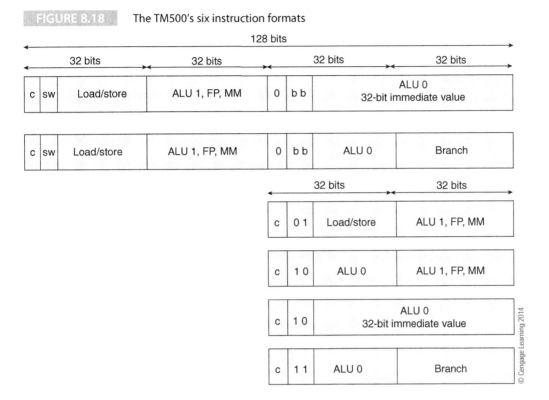

The Crusoe doesn't have unlimited resources. It provides four functional units (an ALU, a floating-point unit, a load/store unit, and a branch unit) and 64 general-purpose registers. These registers synthesize the IA-32's user-visible registers and provide additional temporary registers for the software morphing (binary translation) process. Because the Crusoe VLIW processor does not use superscalar technology (instruction issue, out-of-order execution, and register renaming), it is a much simpler device with a correspondingly lower transistor count and, therefore, a lower power consumption.

By hiding the Crusoe's instruction set architecture behind the code-morphing software layer, the actual details of the VLIW architecture are not exposed to the compiler and changes to the underlying architecture can be made without modifying existing compilers (the code-morphing software must be rewritten of course). Indeed, Transmeta's first two processors, the TM5400 and the TM3120, had different core VLIW architectures that were not compatible with each other. The TM3120 was optimized for embedded Internet applications and the TM5400 for high-performance mobile PCs.[8] Figure 8.19 describes the internal organization of the TM5400.

Although the Transmeta VLIW processor failed, one processor with a VLIW-like architecture has been designed for use in very high-performance systems: Intel's 64-bit Itanium with its IA64 architecture. I have included the Itanium, because it represents a significant departure from Intel's IA32 architecture and includes a lot of interesting technology. Whether the continuing march of Intel's IA32 and its post-Pentium, post-Core processors will also marginalize the Itanium remains to be seen.

---

[8]Alexander Klaiber, *The Technology Behind Crusoe™ Processors: Low-power x86-Compatible Processors Implemented with Code Morphing™ Software*, Transmeta Corporation, January 2000.

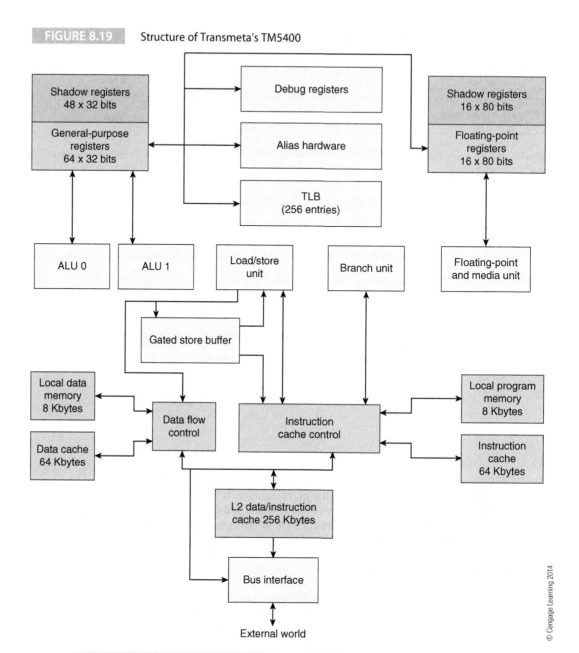

FIGURE 8.19   Structure of Transmeta's TM5400

The Demise of a Processor

I cannot resist including the following comment that appeared on a notice board about the initial over-hyping of the Crusoe:

> Kinda sad how the upshot of all the nifty news technology is the rather boring (though admittedly valuable) goal of low power consumption. The original press conference read sort of like the Cheese Shop Sketch.
> "The Crusoe can emulate any chip at all."
> "Like a PowerPC?"
> "Theoretically."
> "Or an Alpha?"

"Technically."
"Or a Dragonball?"
"Probably."
"So what can it actually emulate?"
"Any chip at all, so long as it's an x86."

Tom Halfhill made the same comment in a more muted fashion when he wrote, "It's probably more than just a coincidence that Crusoe chips have 80-bit-wide floating-point registers, the ability to perform partial-register writes, and native support for the same data types and single-instruction multiple data (SIMD) operations found in Intel's MMX extensions."*

Although the Crusoe enjoyed limited success because of its low power consumption, it had too many limitations to compete in the competitive IA32 market, (for example, poor bus interfacing, poor L2 cache design, no SSE extensions, and poor code translation performance in some applications).

Transmeta designed a second-generation 256-bit VLIW called *Efficeon* that also morphed IA32 code into its native code. Efficeon, shipped in 2004, addressed many of Crusoe's shortcomings and was intended to compete with the Pentium 4. It was extensively reviewed and benchmarked by Van's Hardware Journal on April 4, 2004. The review's conclusion could not be more damning:

> *"…The Transmeta Efficeon is that company's last, best chance for survival. Moreover, it represents the acme of ideologically pure VLIW development. Featuring a laundry list of very real architectural advances beyond the Crusoe, Efficeon raised the hopes of many that Transmeta had finally turned the corner on performance.*
>
> *Unfortunately, the Efficeon is a staggering failure by nearly every measure.*
>
> *Performance is unambiguously lackluster. In fact, Efficeon is only slightly faster than Crusoe. If it weren't for the other Transmeta products and the 366MHz AMD Geode, thrown in for comic relief, the Crusoe would be dead last even when compared to the miniscule VIA C3.*
>
> *For this slight performance nudge beyond its predecessor, Transmeta pays for it with a die size that is over twice as large and at least four times as expensive. And the design will hardly ramp beyond 1GHz, a glacial speed compared with all other modern processors. As the fat cherry on top of this black sundae of disaster, the Efficeon thermally throttles at the slightest provocation. Unlike Intel's P4 throttling apparatus, the Efficeon leverages "LongRun" to actually reduce clock speed and voltage. While this is a superior approach, it still means the same thing to end users: reduced performance. And Efficeon significantly throttles back even under relatively light loads.*
>
> *We suspect that the reason Efficeon throttles so horrendously is because Transmeta greatly desired to reduce its recommended Thermal Design Power so that it could secure fanless 1GHz designs. The problem is that as soon as you start needing that 1GHz that you paid for, the Efficeon is throttling down to 933MHz, 800MHz or even lower."*

Efficeon was a response to the Crusoe's weakness, but it too proved to be a disaster. Transmeta was acquired by Novafora in January 2009, and Novafora collapsed in July 2009.

Van Smith, "Benchmarking Transmeta's efficeon," from http://www.vanshardware.com/reviews/2004/04/040405_efficeon/040405_efficeon.htm.

---

* Tom R. Halfhill, "Transmeta breaks x86 low-power barrier," *Microprocessor Report*, February 14, 2000.

8.3 EPIC Architecture

We now introduce the IA64 or *Itanium* architecture, which is Intel's "big brother" to the IA-32 architecture that encompasses the 80386, 80486 Pentium, and iCore families. *Successor* to the IA-32 it may well be, but *heir* it is not. The introduction of the Itanium processor in 2001 represented one of the most significant developments in microprocessor architecture in a decade. Although the Itanium is referred to as the IA64 architecture in order to contrast it with Intel's IA-32 architecture, there is no relationship between the IA-32 and IA64. A similar situation occurred with Motorola's 68K family. The 68K was used in computers from Sun workstations to the Apple Mac, Amiga, and Atari. When Motorola introduced the PowerPC to replace the 68K, they adopted an entirely new architecture whose origins were in IBM. The origins of the IA64 architecture lie within Hewlett-Packard's *Explicitly Parallel Instruction Computing* (EPIC) program.

As we have seen, superscalar processors dump the problems of reordering instruction execution on the chip itself. The VLIW IA64 transfers that burden to the programmer (i.e., compiler), allowing a simpler chip organization. The EPIC architecture, on which the Itanium is based, employs the *long instruction format* that we introduced earlier. Although the concept of a very long instruction word processor has existed for a considerable time (IBM was working on VLIW architectures as early as 1986),[9] it was the EPIC architecture that helped transform VLIW from a topic at computer architecture conferences into a commercially viable product.

Although the public doesn't associate HP with microprocessors, Hewlett-Packard (HP) has had a long history of developing state-of-the-art processors with innovative features—in particular, the PA-RISC architecture. Computer architects at HP realized that conventional RISC superscalar processors would probably not be able to execute more than about three instructions per clock cycle. They developed mechanisms to allow the compiler to communicate decisions to the hardware and compilers that not only exposed parallelism but also enhanced and exploited it.

The HP Corporation decided to form a strategic alliance with Intel because it was thought that HP would probably not be successful alone. HP could not easily afford the immense cost of the fabrication facilities for a large complex architecture using submicron technologies. Equally, HP was dubious as to whether enough software would be written to make a new architecture viable. The likelihood of an independent manufacturer being able to launch a new processor in a world dominated by Intel's IA-32 seemed remote. By working with Intel, HP brought together Intel's experience of computer architecture, chip design and fabrication, and melded it with HP's flair for systems design and a well-respected brand image.

Itanium is aimed at the high-performance workstation market. The first version of the Itanium in 2001 proved somewhat disappointing. A second version of the Itanium, code-named McKinley, was launched in 2002 to better compete with other manufacturers in the RISC workstation market. The McKinley version of Itanium 2 was constructed with 0.18 micron technology. In 2003, Intel launched a second version of the Itanium 2, the Madison, constructed with 0.13 micron technology and a massive 6 MB on-chip cache. By 2006, a dual-core Itanium was available.

In 2010, Intel announced its Itanium 9300 series that included multi-core technology (two to four cores) plus an enhanced memory interface. In spite of the Itanium's pedigree, it has not yet grabbed a large share of the market for workstations. In 2010, Microsoft announced that it was withdrawing future support for the IA64 architecture. Intel responded by stating that most Itanium-based systems used Unix rather than Windows. However, in February 2010, Intel announced that 80% of the Global 100 corporations had chosen Itanium-based servers for their most mission-critical applications. In 2011, Intel announced an eight core version of the IA64.

---

[9] K. Ebcioglu, "Some design ideas for a VLIW architecture for sequential-natured software," *Parallel Processing*, pp. 3–21
M. Cosnard, M.H. Barton, and M. Vanneschi (Editors), Elsevier Science Publishers B.V. (North-Holland) © IFIP, 1988.

Itanium Pipeline

In this section, we are concentrating on the Itanium's instruction set architecture. In this panel, we provide a brief overview of the structure of the Itanium II's pipeline.

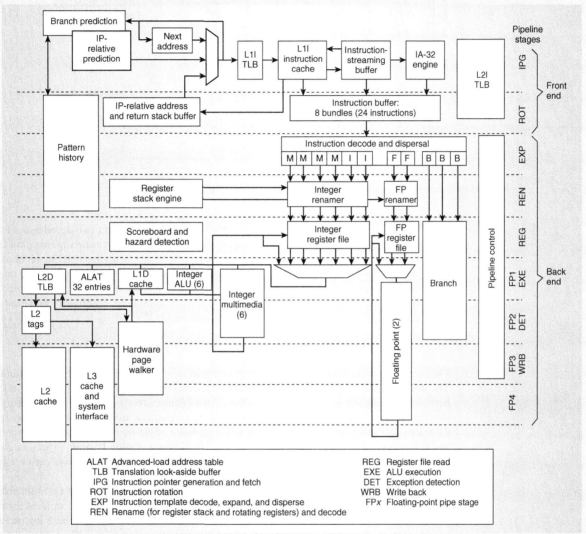

ALAT	Advanced-load address table	REG	Register file read
TLB	Translation look-aside buffer	EXE	ALU execution
IPG	Instruction pointer generation and fetch	DET	Exception detection
ROT	Instruction rotation	WRB	Write back
EXP	Instruction template decode, expand, and disperse	FP*x*	Floating-point pipe stage
REN	Rename (for register stack and rotating registers) and decode		

Reprinted with permission of Intel Corporation. © 2003 IEEE. Reprinted, with permission, from McNairy, C.; Soltis, D., "Itanium 2 processor microarchitecture," IEEE Micro Magazine, Volume: 23, Issue: 2, pp. 44–55.

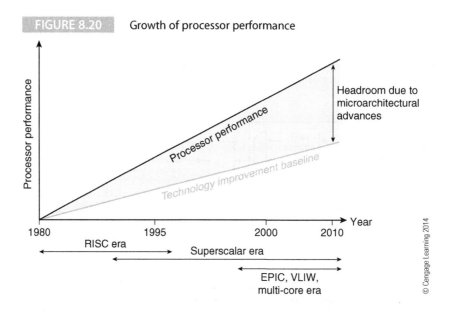

FIGURE 8.20   Growth of processor performance

Figure 8.20 shows the progression of computing performance over two decades as a function of improvements in semiconductor technology and in architecture and organization. The trace at the bottom demonstrates the expected increase in performance if RISC technology alone had become standard and no future changes made other than by improving manufacturing techniques. This diagram illustrates how there has been a steady improvement in performance due to enhancements in the underlying semiconductor technology and jumps in performance due to new microarchitectural enhancements.[10]

8.3.1   Itanium Overview

We are now going to look at a practical implementation of the EPIC architecture, the Itanium. The Itanium is such an immensely sophisticated processor that we can provide only an outline here. Its defining feature is *explicit parallelism* that requires the programmer to organize the instruction flow to ensure that instructions can be carried out in parallel without incurring data dependency hazards. That is, it's the programmer's job to handle instruction sequencing and resource control rather than the processor. The first-generation Itanium processor could handle up to six operations per machine cycle. By 2011, Intel was shipping quad core versions of the Itanium—each of which could handle twelve instructions/cycle.

Let's start by briefly describing the instruction format that is so unusual by the standards of conventional CISC and RISC processors. Figure 8.21 describes the format of an IA64 instruction, or, to be more precise, a bundle of three 41-bit instructions. Each of these instructions has an op-code, three register fields, and a predicate field that is used to implement predicated execution—rather like ARM. As well as the three instructions, there's a 5-bit template that

FIGURE 8.21   IA64 Instruction format

Template	Instruction slot 0				Instruction slot 1				Instruction slot 2			
	Opcode	R3	R2	R1	Predicate	Opcode	R3	R2	R1	Predicate	Opcode	R3	R2	R1	Predicate

0   4 5   45 46   86 87   127

---

[10] "Inside the Intel Itanium 2 Processor." A Hewlett Packard White Paper, July 2002.

tells the processor something about the structure of the three following op-codes. Note that the three 41-bit instructions do not necessarily have to have the format of Figure 8.21.

The Itanium is a superscalar RISC processor with a very large *rotating* register set. I am using *superscalar* here in its fundamental meaning of executing more than one instruction per clock cycle rather than its wider meaning that included the techniques required to handle

Software Pipelining

Software pipelining is analogous to the hardware pipelining used to implement processors, but is traditionally carried out by the programmer rather than by hardware. We shall soon see that the IA64 implements hardware support for software pipelining. The fundamental difference between hardware and software pipelining is one of granularity or scale. In hardware pipelining, the fundamental actions/operations are instruction fetch, decoding, execution, etc. The set of actions add up to make an instruction. In software pipelining, the fundamental actions are machine level instructions, and a set of fundamental actions add up to a trip around a loop in software. That is, hardware pipelining overlaps instruction execution, whereas software pipelining overlaps trips around a loop.

Consider a loop that reads x_i, processes it, and then stores the result in y_i. In conventional programming, you read x_i, process it, store it in y_i, and increment the pointers to arrays X and Y as the following demonstrates.

```
For (i=1, i<64,i++){
    p = x[i] * 4;
    y[i]= p }
```

In software pipelining, the programmer picks up element x_i, processes element x_{i-1}, and stores element x_{i-2} in the *same* iterative loop. This approach means that you don't have to wait for x_i to be loaded while you are operating on element x_{i-1}. Software pipelining requires a significant overhead to implement and is not always cost-effective on some machines. The IA64 provides the hardware support required to make software pipelining a more attractive feature to the programmer. Essentially, the IA64 keeps track of registers x_{i-1}, x_i, x_{i+1}, etc. and automatically renames them from loop to loop. The figure below illustrates the concept of software pipelining. We discuss this in greater detail later.

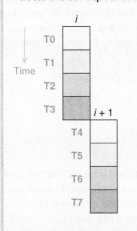

(a) Conventional interration: One trip finished before the next starts.

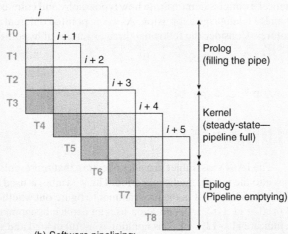

(b) Software pipelining: Next trip begins in the next time slot. Ultimately, each stage of a different trip is being executed in the same time slot.

© Cengage Learning 2014

Register Windowing

Some early RISC processors (e.g., the Berkeley RISC) increased the number of user-accessible registers beyond 32 by employing register windowing. When a procedure is called, that procedure is allocated eight private registers, r16 to r23, that can not be accessed from other procedures. In other words, registers r16 to r23 are private to each procedure and can be used only for local variables.

However, register windowing has not been adopted by other processors because it is regarded as inefficient. Once all the on-chip registers have been allocated, further procedure calls require existing on-chip registers to be dumped in main memory, and that is slow and inefficient.

out-of-order operations. The term *rotating* appears in Intel's literature and indicates that registers can be allocated to functions in an analogous fashion to the Berkeley RISC's windowing. Unusually, the IA64 provides hardware support for *software pipelining*. The panel provides an overview of software pipelining that we discuss later.

Like the ARM, the Itanium employs *predication* (conditional execution) in which each instruction is executed only if certain conditions are met; however, the Itanium's predication mechanism goes well beyond that of the ARM. The Itanium also provides speculative loads in which data is fetched from memory before it is required. Finally, the Itanium has a very long instruction format that combines three individual operations. The IA64 is sufficiently complicated to prevent all but a few assembly language programmers writing code for this device.

We look at the IA64's register set, its instruction format and the way in which three operations are encapsulated in each instruction word, its *speculative* memory access mechanism, and its software pipelining. Before we address these topics, let's say a few words about the IA64's assembly language conventions.

IA64 Assembler Conventions

IA64 assembly language is not significantly dissimilar to other assembly languages in terms of the conventions used, but it does incorporate features that make it a little more modern than traditional assembly languages. I find it surprising that computer languages have never really moved into today's post-ASCII typographical world. For more than three decades, we have had powerful word processors and everyone is familiar with all the fun things they can do with text.[11]Unfortunately, such innovations have not been applied to programming languages, where monospaced Courier still reigns.

Some might argue that it is better to allow the programmer to write $12B4_{16}$ to indicate hexadecimal rather than the clumsy 0x12B4 or $12B4? Typography, which could have been used to make the meaning of assembly language clearer. For example, the following hypothetical examples demonstrate how typography and color could be used to clarify the meaning of an RTL language expression. As we've pointed out, real assemblers don't incorporate these features. Consider the following three examples of hypothetical assembly language notations.

$[R0]^{0:7} \leftarrow [R0]^{0:7} + [M_{[R1]}]$	The 8-bit contents of the memory location pointed at by R1 are added to R0.
$ADD^{.B}$ R0 ← R0,D4	Add the byte in R0 to the byte in D4 and put the result in R0.
$ADD^{.B}$ R0 ← R0,P$^{[R1]}$	Add the byte in R0 to the byte pointed at by R1 and put the result in R0.

The IA64's assembler provides a feature that represents a leap from the programming dark ages into the merely medieval period. The = symbol is used to indicate the destination operand rather than leaving it to the programmer to figure out whether add r1,r2,r3 means r1 = r2 + r3 or r3 = r1 + r2. IA64 notation uses an explicit assignment and we write add **r1** = r2,r3 to indicate r1 = r2 + r3. This notation is useful in load and store operations, where we have

ld4 **r1** = [r2] and st4 **[r3]** = r4   Note that the use of bold font to indicate a destination operand is my convention and not part of the assembly language.

[11] Just so.

IA64 instructions often include one or more *qualifier* fields that are separated from the base mnemonic and from each other by periods. For example, the load instruction ld4 can be written as

```
ld4.s.ntl r1 = r2,[r3]
```

where the s indicates a *speculative* load and the ntl provides the cache mechanism with a hint about the location of the data. Comment fields are indicated by double backslashes //. For example,

```
ld4 r1 = [r2]          //Load r1 with the data pointed at by r2
```

A feature peculiar to the Itanium VLIW is the use of a symbol to indicate the extent of a region of parallelism. The Itanium uses two semicolons to indicate a *stop* between regions of parallelism, and the { and } braces enclose instruction bundles. We will cover stops and bundles later. A simple example of a stop is

```
add r1 = r2,r3         //r1 = r2 + r3
add r4 = r6,r7 ;;      //r4 = r6 + r7 We need a stop because of the RAW hazard in
                           the next instruction
add r5 = r4,r1         //r4 = r6 + r7
```

A fragment of IA64 code looks like

```
{
        cmp.ne   p1 = rx,ry     // Perform the comparison x == y with result in p1
   (p1) br        else          // If true then do the else part
}
{.label then
        add       rp = rz, 1    // z++
        add       rp = rp, rq   // p += q
        br        exit          // jump over else part
}
{
.label else
              add rr = rr, 1 // r++
}
```

IA64 instructions can be *predicated*; that is, they are executed or not executed depending on the contents of a predicate register. A predicate register p0 to p63 is placed in parentheses in front of the instruction that is predicated. For example, the construct

```
(p1) add r1 = r2,r3
```

means "*if the contents of predicate register p1 are true then execute* add r1 = r2,r3." This is analogous to the ARM processor's conditional execution mechanism, except that the IA64 has 64 predicate registers. Because predication is so important, we discuss it in more detail in Section 8.3.7.

8.3.2 The Itanium Register Set

The IA64 has one of the most extensive *user-visible* register sets of any microprocessor. Traditional RISC architectures have 32 registers. The Itanium's register set can be divided into seven distinct groups, as Figure 8.22 demonstrates. General registers Gr_0 to Gr_{127} can be written as r0 to r127 and provide 128 64-bit-wide integer registers that you'd expect in almost any architecture. Registers Fr_0 to Fr_{127} are 82-bit floating-point registers (1 sign-bit, 17-bit exponent, and 64-bit mantissa).

Data from memory is loaded in the lower-order bits of a register (bit 0 is the lowest-order bit and bit 63 is the highest). Loading less than an eight-byte quadword into a register results in the data being *zero extended*. General-purpose register r0 (we will write r0, etc. rather than Gr_0 for the sake of clarity) is read-only and contains the permanent value 0 to provide

FIGURE 8.22   The Itanium's register set

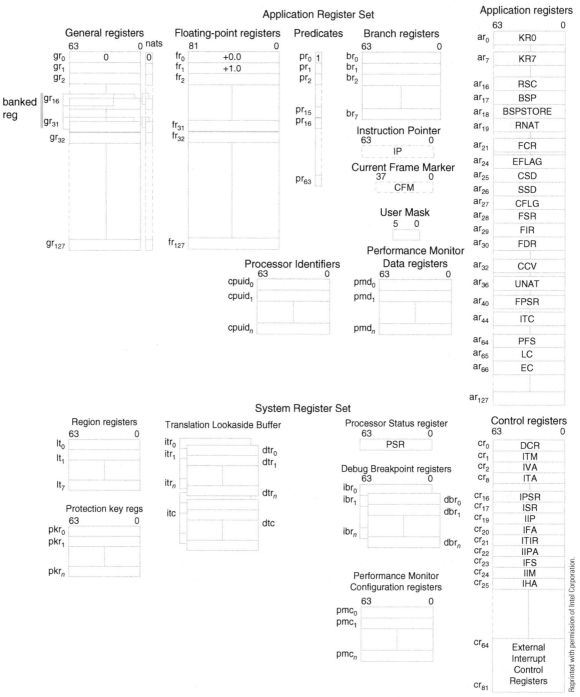

a convenient means of generating the constant zero or *suppressing* an operand in an instruction that requires two operands. For example, a *move* instruction can be implemented by add **r3** = r5,r0, because this is equivalent to

$$[r3] \leftarrow [r5] + [r0]$$
$$[r3] \leftarrow [r5] + 0$$
$$[r3] \leftarrow [r5]$$

Registers r0 to r31 are called *static general registers* and are visible to all processes, that is, they can always be accessed. The remaining 96 registers r32 to r127 can be assigned to a procedure as a *stack frame*. When a stack frame is assigned to a procedure, its registers can be accessed only from that function. We return to stack frames and the allocation of registers to procedures later.

A radical aspect of the IA64's general-purpose register file is its ability to dynamically renumber registers and treat register addresses as variables; that is, the registers themselves form a data structure. This is an echo of the less sophisticated windowing system of the original Berkeley RISC architectures. We return to this *rotating register* mechanism when we deal with the IA64's register allocation mechanisms for procedure calls and its support for software pipelining.

The IA64 implements a register stack. A register frame includes input (in), local (loc), and output (out) registers. A frame is set up by means of the stack frame allocation instruction, alloc. When a new function is called, the stack frame is modified, and the former *output* registers become the new *input* registers.

The Itanium has 64 *predicate* registers, used to make instructions conditionally executed, eight branch registers, 128 application registers, processor identifier registers, and performance monitor registers. The 64-bit instruction pointer (IP) provides a 2^{64}-byte logical address space; that is, 1.8447×10^{19} bytes can be accessed. You can read the IP but you cannot explicitly load it. Because all Itanium instructions are arranged as 256-bit *bundles*, instruction addresses fall on 16-byte boundaries, and the IP is incremented by 16 after each instruction fetch.

The Not a Thing Bit

The 128 64-bit general-purpose registers are in fact *65 bits wide,* because a flag bit called *not a thing* (NaT) is associated with each register. This bit is used to indicate the validity of the register's contents (a bit like the NaN bit in floating-point arithmetic). The NaT bit is used in conjunction with the Itanium's load mechanism to indicate the validity of a *speculatively loaded* operand. The IA64 loads a register from memory and sets the NaT bit if the load was unsuccessful because of a page fault. This mechanism allows program execution to continue even if a load failed. If the data loaded later proves to be unneeded, there is no point halting the computer unnecessarily. If the data is eventually required, the NaT flag validates the data at the point it is actually used.

NaT bits are propagated when a register with a NaT bit set takes part in an operation. For example, an operation on two variables sets the NaT bit in the destination operand if one of the source operands has its NaT flag set. This approach can be exploited. If the end result of a chain of calculations is later abandoned because it was in a not-taken path of a conditional expression, time has not been lost in dealing with the problem that led to the NaT. The NaT mechanism allows deferred or late decisions about the validity of data. Consider the example:

```
if (a == b) then x = y + 5;
            else z = t + 5;
```

Suppose that loading *t* from memory generates a fault, and the NaT flag for *t* is set. The calculation z = t + 5 takes place in parallel, and the value of *z* will also be a NaT because the NaT from the previous load is propagated. This sequence of operations is in the *else* path of the construct. If the test (a == b) is true, the *then* path is executed, and the NaT in the *else* part can be safely ignored.

Predicate and Branch Registers

The Itanium IA64 provides 64 *predicate registers* p0 to p63 that hold a one-bit Boolean variable. Instructions are executed only if the predicate register contains the value true. For example, (p3) mov **r2** = r4 is executed only if predicate register p3 contains 1. Predicate register p0 is hardwired to *true* and forces the execution of an instruction. For example, writing add **r7** = r8,r9 is interpreted as (p0) add **r7** = r8,r9.

The predicate registers are user-selectable and you can write (p2) add **r1** = r2,r3 or (p12) add **r1** = r2,r3 at will.

The Itanium IA64 has eight 64-bit *branch registers* b0 to b7 that supply a destination address to branch instructions and are accessed only by special branch instructions. The ARM can just copy a general register into the program counter r15 by mov **r15**,r4.

When calling a function, the return address is stored in branch register b0 by convention. It must be saved to local registers by the called function if the function needs to call other functions itself (i.e., like the ARM processor's s r14 register). A function is called with

```
(p1) br.call subr
```

which calls a function if predicate register p1 is true.

Other Itanium Registers

Computers have a few special registers, such as a program counter, condition code register, interrupt vector register, and so on. The IA64 has several special-purpose registers plus an array of 128 64-bit applications registers. The special-purpose registers include the 64-bit instruction pointer that is not user accessible. Code is stored in *little endian* byte order and must be aligned on 16-byte boundaries. IA64 data is aligned on a boundary appropriate to its size. For example, bytes are aligned on byte boundaries. An exception is the 10-byte floating-point number that must be aligned on a 16-byte boundary. Big or little endian formats are controlled by the big endian enable bit in the user mask. A 38-bit current frame marker (CFM) controls the allocation of registers to the current function's stack frame. The current frame marker holds the state of the register rotation.

The IA64, like the Pentium, provides a CPU identity register (CPUID) that allows the software to determine the type of processor installed and its operating capabilities. The 128 *application registers* perform global functions. Not all of these registers are currently dedicated to specific functions. Two important special registers are AR65, the loop count (LC) register, and AR66, which is the epilogue count (EC) register. These registers are used in conjunction with a loop-counting instruction to mechanize loops. Although the loop count register is Ar_{65}, by convention, the IA64 assembler uses the notation ar.function to indicate a specific application register, where *function* describes the register's application. For example, we can copy the contents of register r1 into the loop counter with the instruction mov **ar.lc** = r1. A simple example of a loop is

```
        mov   r1,loopCount       //save loop counter in r1
        mov   ar.lc = r1         //initialize the loop counter
Next    ...                      //here be the loop body
        ...                      //YAN (yet another instruction...)
        br.cloop Next            //Loop termination–branch on loop count
                                        exhausted
```

Note how the loop is terminated by br.cloop, which decrements the counter register and automatically branches back until the loop count terminates.

8.3.3   IA64 Instruction Format

The IA64 instruction is 41 bits wide, which is an unusual size in a world where most RISC instructions are 32 bits wide and CISC instructions are a multiple of 8 bits long. Like RISC processors, the IA-32 instruction has a simple encoding with relatively few variations on the basic instruction format. This statement needs qualifying, since there are a lot of variations in the encoding formats of certain classes of instruction. However, many of the instructions do share a common simple format. Figure 8.23 describes the structure of a basic IA64 register-to-register instruction. The most significant 14 bits provide an op-code that defines the operation to be carried out. Three 7-bit operand fields support three-operand, register-to-register instructions with a set of 128 registers. The least-significant 6 bits of the op-code select one of 64 predicate registers that determines whether this instruction is to be executed or not.

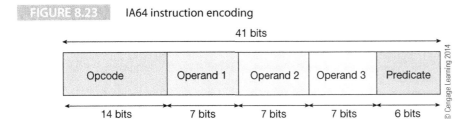

FIGURE 8.23   IA64 instruction encoding

The large 14-bit op-code space isn't used to provide a massive number of distinct instructions but to support an instruction set whose instructions have numerous *options*. For example, branch instructions may take a *completer* suffix that indicates whether a branch is likely to be taken. IA64 literature employs the expression *completer* to describe a suffix that defines the way an instruction is executed.

A big headache for the computer architect and ISA designer is the *literal operand*. You want a large literal, but you don't want to devote too many bits to it—a paradox of mutual exclusivity. The Itanium solves this problem by combining the three 7-bit operand fields to give a $3 \times 7 = 21$-bit literal.

Like the ARM, the IA64 assembler makes use of *pseudo instructions* to provide the programmer with operations that are constructed using other instructions. For example, we have already seen that the mov **r1** = r2 instruction is really the IA64 add **r1** = r2,r0 instruction that adds zero to r2.

The Itanium's op-codes are not always uniquely encoded. Consider what Jim Turley had to say.[12]

> *It's hard not to think that Intel's institutionalized taste for baroque and ungainly, not to mention bizarre instruction set features didn't creep in here somewhere. With so much elegance going for it, IA64 falls down in the evening gown competition. First, IA64 op-codes are not unique - they're reused up to four times. In other words, the same 41-bit pattern decodes into four completely different and unrelated operations depending on whether it's sent to the integer unit, the floating-point unit, the memory unit, or the branch unit. A C++ programmer would call this overloading. An assembly program would call it nuts. You'd think that Itanium's designers would have been satisfied with 2^{41} different op-codes, but no...*

This comment is not fair. It may be true that some instructions share the same encoding. However, if you regard the *template* as a qualifier that is part of the instruction, it's unreasonable to state that op-codes are not unique. The Itanium template is a five-bit field that describes the nature of the op-codes in a 128-bit instruction bundle; that is, the template tells the processor how the three instructions in the bundle are to be interpreted (e.g., what execution units are going to be needed). Consequently, it is not necessary to have different instruction encodings because the template has pre-defined the type or class of each of the three instructions in the bundle. We return to this topic later.

8.3.4   IA64 Instructions and Addressing Modes

We don't intend to cover the IA64's instruction set in any detail; moreover, the basic instructions are conventional. However, we are going to take a quick look at some of the most interesting aspects of the Itanium instruction set. Itanium instructions are interesting because they have reintroduced CISIC into RISC by including instructions that perform quite sophisticated operations that were once the hallmark of classic CISC processors like the 68K (ColdFire) family. Some of the Itanium's shift operations act rather like bit field operations and allow

[12]64-Bit CPUs: What You Need to Know, February 8, 2002, http://www.extremetech.com.

you to manipulate groups of bits. These operations are useful in packing and unpacking data structures and in graphics operations where bits represent individual pixels.

Shift Instructions The format of the IA64 shift instructions is

(r_i)	shift_{type}	$r_j = r_j, r_k$
(r_i)	$\text{shift}_{type}.u$	$r_j = r_j, \text{count}_k$
(r_i)	shrp.u	$r_j = r_j, r_j, \text{pos}_k, \text{len}_l$
(r_i)	extr.u	$r_j = r_j, \text{pos}_k, \text{len}_l$
(r_i)	dep.z	$r_j = r_j, r_j, \text{pos}_k, \text{len}_l$

The mnemonic shift_{type} stands for shl (shift left) or shr (shift right). A simple shift left instruction is defined as shl **r1** = r2, r3, where r2 contains the source to be shifted and r3 contains the number of bits to be shifted. The programmer can use a version of the shift that specifies an integer shift count, shl **r1** = r2, count. However, this is a pseudo instruction built on the merge instruction dep.[13] The completer .u indicates an unsigned right shift (i.e., the default right shift is signed and the sign bit is propagated right).

A variation on the shift is the *shift and add*; for example, the

```
shladd r1 = r2, count, r3
```

instruction shifts the first source operand left by count bits and adds the second source operand to the shifted result before depositing the total in the destination register. The number of places that the first source operand can be shifted is specified by the integer count as 1, 2, 3, or 4. This instruction is rather like the ARM processor's ADD **r1**,r2,r3, LSL #2 and can be used to scale array offsets by 2, 4, 8, or 16 before adding them to a base.

The IA64 mechanizes extended shifts by means of a double shift that operates over two source operands. The *shift right pair* instruction

```
shrp r1 = r2, r3, count
```

concatenates source operands r2 and r3 into a 128-bit value that is shifted right by count bits, where count is in the range 0 to 63. The least-significant 64 bits of the result are deposited in destination register r1.

The extract instruction, extr, extracts a *bit field* from a general-purpose register. The bit field is either zero or it is sign-extended and deposited in the destination register. Extract bit field operations require a means of specifying the bit field; both the 68020 and IA64 do it by specifying a start and a width. The format of signed extr instruction is

```
extr r1 = r2, start, length
```

where r1 is the destination, r2 the source, start a 6-bit integer that defines the beginning of the bit field in r2, and length a 6-bit integer that defines its width. The value of start is in the range 0 to 63, and the value of length is in the range 1 to 64. The unsigned form of the instruction that uses zero-extension when the bit-field is moved is extr.u.

The *merge* or *deposit* instruction, dep, is the inverse of extr and takes a right-justified bit field from the source register which it deposits in an arbitrary location in the destination register. The format of the merge instruction is

```
dep r1 = r2,r3,start,length
```

where r1 is the destination, r2 the source, start a 6-bit integer that defines the beginning of the bit field in r3, and length a 4-bit integer that defines its width. Figure 8.24 demonstrates the action of a dep **r1** = r2,r3,36,16 instruction. This operation can be used to pack data structures. It can also be used to perform graphics operations such as inserting one image into another.

---

[13]The actual instruction used is dep.z r1 = r2, count, 64-count.

FIGURE 8.24   Example of the use of the dep (deposit) instruction to merge bits

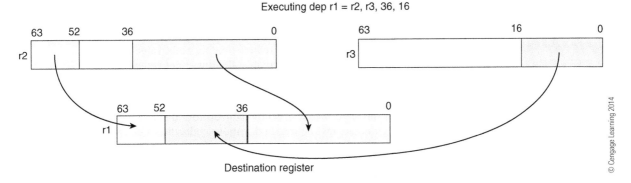

Multiplication and Division Unusually, the IA64 does not provide simple integer multiply and divide instructions (although integer multiplication has been added to the fourth-generation of the Itanium). Indeed, the Itanium provides no integer division instructions at all. A 16-bit parallel multiplication instruction is provided as part of the Itanium's multimedia support that you could use as a conventional multiplication instruction, as Figure 8.25 demonstrates.

The Itanium does provide limited support for integer multiplication with two instructions that perform simultaneous multiplications on two pairs of 16-bit integers to give two 32-bit products. These instructions perform *signed* multiplication.

```
Pmpy2.l r1 = r2,r3      //left form
Pmpy2.r r1 = r2,r3      //right form
```

The difference between these two assembly language instructions lies in the suffix .l (left) or .r (right). The right form of the instruction takes bits 47:32 of the two source operands and creates a 32-bit product in bits 63:32 of the destination, and bits 15:0 of the two source operands and creates a 32-bit product in bits 31:0 of the destination. The left form takes bits 63:48 from both source registers to create bits 63:32 of the destination, and bits 31:16 of the source to generate bits 31:0 of the destination. Figure 8.25 describes the operation of these two instructions graphically. Note that they have a certain similarity to multimedia extensions.

FIGURE 8.25   IA64 parallel multiplication

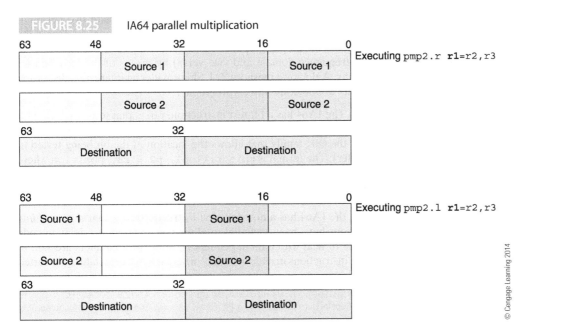

Because of the lack of integer multiply instructions, in order to carry out conventional 64-bit integer multiplication and division, it's necessary to take the source integers, convert them into floating-point values, perform the multiplication or division, and then convert the floating-point results back to integer form. For example, integer multiplication can be carried out by

```
setf.sig  f1 = r1           //create a floating-point value from r1
setf.sig  f2 = r2           //create a floating-point value from r2
xma.l     f3 = f1,f2,f0     //generate the product as a floating-point value
getf.sig  r3 = f3           //convert floating-point to integer in r3
```

The setf.sig and getg.sig instructions convert to and from floating-point and xma.l is a multiply and accumulate instruction. In the above code, the value of floating point register f0 is added to the product or to f1 and f2 without effect because f0 always contains zero.

Multimedia Instructions The IA64 supports short vector multimedia operations, rather like the corresponding IA32 extensions. Parallel add and subtraction operations act on bytes, 16-bit words, and 32-bit words. Both conventional (modulo or wrap-round) and saturating arithmetic mode are supported. However, the IA64 is unusual in that it provides unsigned saturating, signed saturating (twos complement minimum and maximum values), and a mixed mode in which one operand is signed and one is unsigned.

Parallel averaging instructions are provided; for example, pavg takes pairs of elements, adds them together, and right shifts each result by one bit (i.e., divides it by 2 to get the average of the two elements). The instruction, pavgsub, performs a similar operation on the difference between pairs of elements.

A *parallel sum of absolute differences* instruction, psad, takes the sum of the absolute difference of the pairs of operands. Finally, the parallel maximum and parallel minimum instructions return the maximum or minimum of the set of one-byte (or two byte) elements. Curiously, one-byte elements are treated as unsigned values, and two-byte elements as signed values.

The IA64 provides a user-programmable *multiplex* instruction, mux2, that allows you to specify the order of the words in a register. The format of the instruction is

```
mux2  r1 = r2,sequence
```

where *sequence* is an integer specifying the order of the words. The word order from left-to-right is 11, 10, 01, 00 so that reversing the word order would use the sequence 00 01 10 11 or $1B_{16}$. Hence, the operation mux2 r1 = r2, 0x1B would copy the words in r2 into r1 in reverse order. Figure 8.26 demonstrates the effect of this instruction with two permutations. Word and bit permutations have applications in many areas of information processing where high-speed is essential, for example, graphics, signal processing (converting data from the time domain into the frequency domain and vice versa) by means of the Fast Fourier Transform, and cryptography. A doctoral thesis by Z. J. Shi provides an interesting discussion of bit permutation, computer architecture and cryptography.[14]

Miscellaneous Instructions The IA64 has a bit test instruction, tbit, that tests a selected bit in a register. The bit location is defined by a literal which makes dynamic bit testing impossible (unlike, for example, the 68K family that allows the location of the bit being tested to be specified as a data register). The format is tbit.trel p1, p2 = r3, position where trel is either nz (test for not zero) or z (test for zero). The specified bit position in r3 is tested and, if zero, (assuming tbit.z) the value of p1 is set to 1 and p2 is set to 0. Otherwise, p1 is set to 0 and p2 to 1.

Like most processors, the IA64 has a no-operation instruction, nop, that *does nothing*. Traditionally, nop instructions have been included to allow you to provide a delay in code (or a bubble in the pipeline to deal with data dependency). Some programmers are said to have inserted no operation instructions in code at a place where a *bug fix* can later be inserted

[14]Zhijie Jerry Shi, "BIT PERMUTATION INSTRUCTIONS: ARCHITECTURE, IMPLEMENTATION, AND CRYPTOGRAPHIC PROPERTIES" Doctoral dissertation, Dept. of Electrical Engineering, Princeton University, 2004. http://www.engr.uconn.edu/~zshi/publications/shi_thesis.pdf.

FIGURE 8.26   Effect of the MUX2 word multiplexing instruction

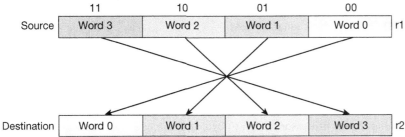

(a) Effect of MUX2 r1 = r2,0x1B is to reverse the word order

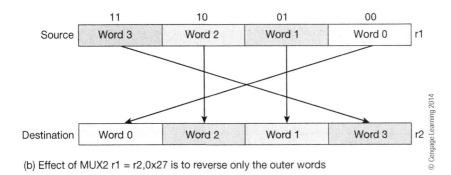

(b) Effect of MUX2 r1 = r2,0x27 is to reverse only the outer words

© Cengage Learning 2014

(a practice that is rarely performed today). In superscalar code, the no operation can be used to enforce strict execution order. Because IA64 instructions are predicated, the nop is predicated. Consider the form

```
(p1) nop
```

If predicate p1 is false, the no-operation is not executed. If the predicate p1 is true, the no-operation is executed. Please do not think too deeply about this—that's the job of the philosopher.

The nop instruction can take a 21 bit immediate operand. This operand is not used by the nop and serves only as a programmer-defined marker in the code stream. Finally, the nop takes a completer .i, .b, .m, or .f. These completers define the execution unit to which the instruction is assigned. For example, nop.i is assigned to the integer unit, whereas nop.b is assigned to the branch unit. So, you decide which execution unit gets to not execute an instruction. I'm not making this up! The nop can be used to pack instructions when three operations cannot be found to put in a bundle.

Addressing Modes
The IA64 has a fairly simple memory accessing mechanism and supports only register indirect addressing. You can also apply an optional post-incrementing. The remarkable range of memory access operations supported by ARM processors is absent in the IA64. Typical IA64 memory accesses are

```
ld8 r1   = [r4]       //load the 8 bytes pointed at by r4 into r1
ld8 r1   = [r4],r2    //load the 8 bytes pointed at by r4 into r1 and update r4 by
                          adding r2
st8 [r2] = r5         //store the 8 bytes in r5 at the location pointed at by r2
st4 [r6] = r9,4       //store the 4 bytes in r9 at the location pointed at by r6 and
                          post increment r6 by 4
```

Note that the load and store operations have a size completer that indicates the number of bytes loaded or stored (1, 2, 4, or 8); for example ld4, st8. The format of the post indexed store is (to me) odd because the increment is stored on the wrong side of the equation; that is, st4 **[r6]**=r9,4 rather than st4 **[r6]**,4=r9.

8.3.5   Instructions, Bundles, and Breaks

We now return to two concepts that we have already briefly introduced, the *bundle* and the *break* that are related to the IA64's VLIW-like architecture. Recall that IA64 bundles are 128 bits long and the processor reads a 16-byte *instruction bundle* containing three op-codes,[15] Figure 8.27. The three op-codes occupy positions slot 1, slot 2, and slot 3. It's not always possible to fill each slot with a useful instruction and sometimes you have to pad slots with no operations, NOPs. Because an individual IA64 instruction (op-code) is 41 bits long, a bundle has $128 - 3 \times 41 = 5$ bits left over to provide extra information about the current bundle.

FIGURE 8.27   Structure of an IA64 instruction bundle

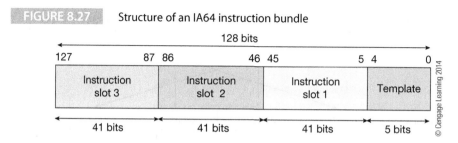

The five low-order bits of an instruction bundle are called its *template*, because they indicate the type of the current bundle and map the instruction slots onto the IA64's processing units. You could say that the template is an instruction-bundle sub-address used to route each op-code to its ultimate destination.

The IA64 is a highly specific form of a superscalar processor with independent execution units that can deal with individual instructions which permits their parallel execution. Table 8.1 demonstrates the relationship between instruction type and execution unit. The IA64's execution units are the I-unit that performs integer operations, the F-unit that performs floating-point operations, the M-unit that performs memory related operations (load and store), and the B-unit that performs branch operations.

Table 8.2 relates the encoding of each bundle's template to the instruction type in terms of processing units. For example, template 0 indicates that the instruction in slot 0 requires the memory unit and that the instructions in both slots 1 and 2 require integer units. Some of the instructions are in blue to indicate the presence of a *stop* or *break* in the instruction sequence that defines the limit of a region of parallel execution.

TABLE 8.1   IA64 Instruction Types and Execution Units

Instruction Type	Description	Execution Unit Type
A	Integer ALU	I-unit or M-unit
I	Non-ALU integer	I-unit
M	Memory	M-unit
F	Floating-point	F-unit
B	Branch	B-unit
L+X	Extended	I-unit/B-unit

[15]Once again, we are forced to use *instruction* to indicate the 256 word read by the Itanium containing three *op-codes*.

TABLE 8.2   IA64 Instruction bundle template encoding

Template	Encoding	Type	Slot 0	Slot 1	Slot 2
0	00000	MII	M	I	I
1	00001	MII	M	I	I
2	00010	MI_I	M	I	I
3	00011	MI_I	M	I	I
4	00100	MLX	M	L	X
5	00101	MLX	M	L	X
6	00110				
7	00111				
8	01000	MII	M	M	I
9	01001	MII	M	M	I
10	01010	M_MI	M	M	I
11	01011	M_MI	M	M	I
12	01100	MFF	M	F	I
13	01101	MFF	M	F	I
14	01110	MMF	M	M	F
15	01111	MMF	M	M	F
16	10000	MIB	M	I	B
17	10001	MIB	M	I	B
18	10010	MBB	M	B	B
19	10011	MBB	M	B	B
20	10100				
21	10101				
22	10110	BBB	B	B	B
23	10111	BBB	B	B	B
24	11000	MMB	M	M	B
25	11001	MMB	M	M	B
26	11010				
27	11011				
28	11100	MFB	M	F	B
29	11101	MFB	M	F	B
30	11110				
31	11111				

IA64 code contains markers called *stops* that indicate the barrier between regions of code that can be executed in parallel. The use of the stop allows the programmer to explicitly deal with read after write (RAW) or write after write (WAW) dependencies. Consider the sequence of operations described in Figure 8.28. There are four groups of instructions where the individual instructions in a group can be executed in parallel without suffering from dependencies. For example, the first group contains instructions 1, 2, and 3 that can be executed at the same time.

A STOP is placed between groups to limit the extent of the parallelism; for example, the instructions in group 1 must be executed before the instructions in group 2. The sequence in which instructions are executed is determined by the compiler, which places the STOPs in the instruction stream. A STOP ensures that a RAW (read after write) hazard does not take place by ensuring the competition of instructions up to the STOP.

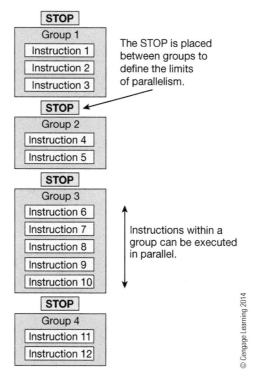

FIGURE 8.28   Regions of possible parallelism in code

The STOP is placed between groups to define the limits of parallelism.

Instructions within a group can be executed in parallel.

© Cengage Learning 2014

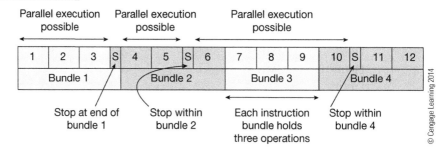

FIGURE 8.29   Instruction bundles and STOPs

Parallel execution possible   Parallel execution possible   Parallel execution possible

Stop at end of bundle 1   Stop within bundle 2   Each instruction bundle holds three operations   Stop within bundle 4

© Cengage Learning 2014

What is the relationship between STOPs in the instruction stream defining regions of parallelism and instruction bundles? None! The concepts of bundles and STOPs are *independent* and a STOP may be placed anywhere within an instruction bundle. Figure 8.29 illustrates this concept with the above instruction sequence. As you can see from Figure 8.29, some bundles contain an internal STOP and others don't.

Before we look at Itanium assembly language, we provide another example that illustrates the nature of bundles, groups, and stops. Figure 8.30 demonstrates a fragment of code consisting of six groups extending over 11 bundles. The groups are defined by the regions of parallelism, the bundles are determined by the instruction format, and the stops are placed between groups to signify the range of possible parallelism. We say *possible* parallelism, because the operations in a group can be executed in parallel but the processor may not have sufficient resources to execute them all at the same time. Multiple bundles of instructions can be dispatched in each clock cycle.

FIGURE 8.30   Instruction bundles and groups

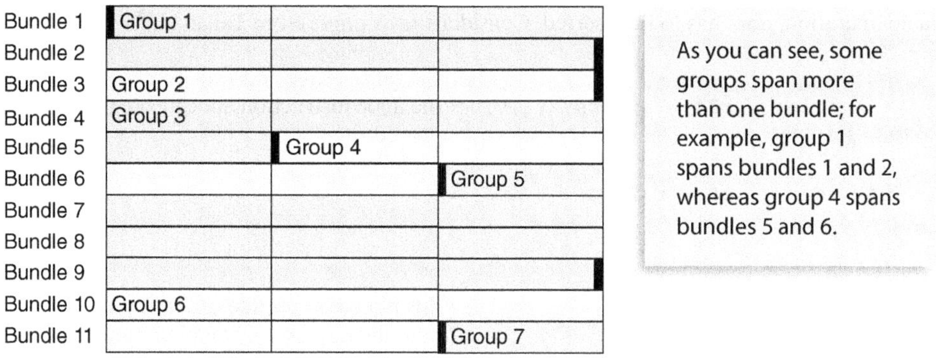

Bundle 1	Group 1		
Bundle 2			
Bundle 3	Group 2		
Bundle 4	Group 3		
Bundle 5		Group 4	
Bundle 6			Group 5
Bundle 7			
Bundle 8			
Bundle 9			
Bundle 10	Group 6		
Bundle 11			Group 7

As you can see, some groups span more than one bundle; for example, group 1 spans bundles 1 and 2, whereas group 4 spans bundles 5 and 6.

© Cengage Learning 2014

IA64 Bundles, STOPs, and Assembly Language Notation

As we have indicated, the compiler decides where STOPs go and how instructions are bundled. The next step is to show how this is indicated in the code. A STOP is indicated in Itanium assembly language by *two* semicolons; for example, consider the following sequence.

```
add r6    = r7,r8
ld4 [r13] = r10    ;;    //load r10 with the word pointed at by r13
sub r10   = r10,r12
```

The subtraction cannot take place in parallel with the add and the st4 (i.e., store) instructions because of the STOP that limits the range of parallelism. You use the STOP to prevent register dependency errors; that is, applying a ;; between two instructions ensures that a dependency violation cannot take place. However, the unnecessary use of STOPs leads to degradation in the processor's performance. Assembler directives have been designed to prevent the IAS (Itanium assembler) from generating unnecessary STOPs when data dependency is apparent rather than real.

The correct placement of STOPs and the generation of bundle templates is not a task easily undertaken by the human programmer. We must stress that the IA64 is a machine whose code is intended to be generated by a compiler. The bundle is indicated by enclosing instructions in { and } braces, for example

```
{ .MII
      ld4 r10 = r[9]
      add r12 = r13,r14
      sub r20 = r15,r14
}
```

Here, we have a memory operation followed by two integer operations. The .MII at the start of the bundle indicates the template type, MII = M-unit, I-unit, I-unit, to be used by this bundle.

A superscalar processor deals with data dependencies by resequencing instruction execution. The IA64 leaves the hard work to the programmer or the compiler. It is up to the programmer or compiler to generate code without dependencies. If you don't, the results can be unpredictable. IA64 literature calls this situation a *dependency violation*.

The IA64 assembler is not a conventional assembler that translates mnemonics into machine code and deals with symbolic operands and references. The IA64 assembler provides support for *explicit parallelism* and prevents dependency violations. The assembler operates either in an explicit mode that requires the programmer to define bundle boundaries or in an automatic mode that leaves the handling of instruction bundling to the assembler.

Because the number of valid templates is limited, it follows that the number of legal combinations of instructions is also limited. If a valid instruction cannot be found to fill a slot, a no operation, nop, has to be inserted. Consider the example from Tal, et al.[16]

```
{  .mii
       alloc    r34 = ar.pfs,2,1,0,0  //the alloc instruction allocates registers to a
                                                procedure
       cmp.lt   p5,p6 = r32,r33 ;;   //stop
   (p6) add     r8 = r32,r0
}

{  .mib
   (p6) add     r8 = r33,r0           //this is a move pseudo operation
       nop.i  0                       //note the nop's .i completer and 0 literal
       br.ret.sptk b0 ;;              //this is a return from subroutine
                                      //the .sptk is a branch hint

}
```

In this example, we have two instruction bundles. The first bundle has a .mii template to indicate M-unit, I-unit, and I-unit. Some instructions can be placed in more than one slot, whereas other instructions are limited to a specific slot; for example, the allocate instruction, alloc, must go in an M slot. In this .mii bundle a stop has to be inserted after the compare instruction because the predicate generated by the compare, (p6), is required by the following addition.

The next bundle has an integer operation, add, and a branch. Bundles with a B unit are .mib, .mbb, .mmb, .mfb, and, .bbb. In this case, a .mib bundle is used. A nop.i no operation instruction is used in the middle slot. The add instruction is assigned to an M slot because M slots can handle both memory and integer operations.

Another example from Tal demonstrates how the sequence

```
ld4 r4  = [r33]
add r8  = 5,r8
mov r2  = r56
add r32 = 5,r4
mov r3  = r33
```

is translated into the bundled sequence

```
{  .mii
       ld4    r4 = [r33]
       add    r8 = 5,r8
       mov    r2 = r56          ;;
}

{  .mmi
       nop.m
       add    r32 = 5,r4
       mov    r3  = r33         ;;
}
```

The first group uses a .mii template because the memory reference and two integer operations map naturally onto this template. The second bundle has two integer instructions. Since all bundles (see Table 8.2 on page 525) apart from the .bbb bundle begin with an M slot, the two possible templates are .mmi and .mii.

---

[16] A. Tal, V. Bassin, S. Gal-On, and E. Demikhovsky. "Assembly Language Programming Tools for the IA64 Architecture," *Intel Technology Journal Q4*, 1999.

8.3.6   Itanium Organization

Figure 8.31 illustrates the Itanium's structure. There are nine functional units:[17] two memory units, two integer units, two-floating-point units, and three branch units. Figure 8.32 from Intel's Itanium Processor Manual demonstrates the execution of a code stream consisting of four three-instruction bundles. Remember that the compiler is responsible for ordering the instruction stream into bundles that can be executed by the Itanium. The Itanium can hold two bundles in its *dispersal window* and issue up to six instructions per cycle using the two M slots, the two I slots, the two F slots, and the three B slots.

FIGURE 8.31   Internal organization of the Itanium

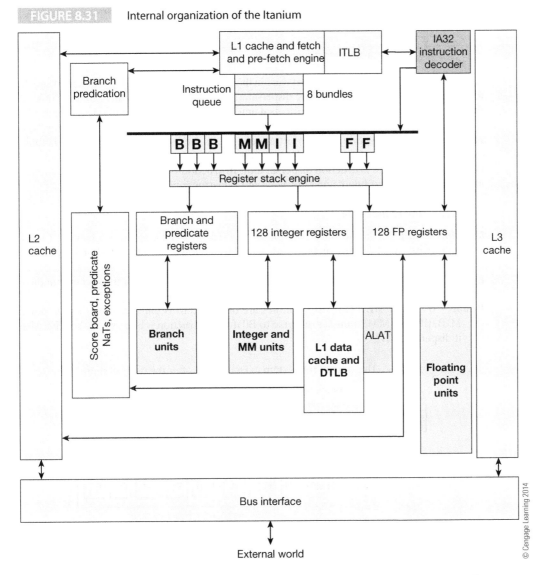

Two bundles with the templates MFI and MIB are currently within the Itanium's dispersal window and are waiting for execution. The following two instruction bundles with templates MII and MIB are in the input stream. As Figure 8.32 demonstrates, the first six instructions are all assigned to functional units.

---

[17]This figure refers to the first-generation Itanium. The Itanium 2 has more functional units and is, therefore, capable of a greater degree of parallelism.

FIGURE 8.32   Relationship between the Itanium's functional units and bundles

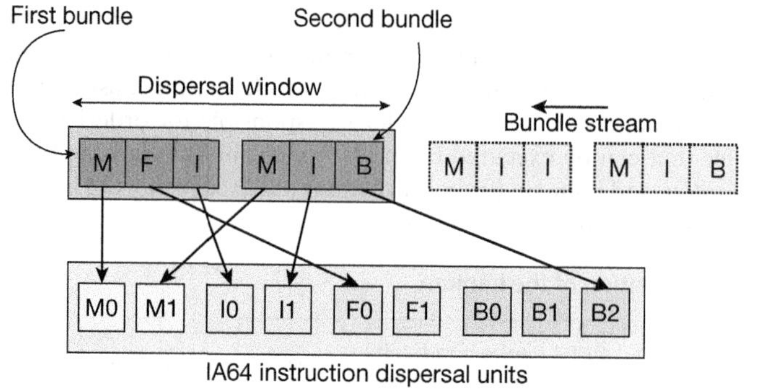

First bundle   Second bundle

Dispersal window

Bundle stream

This figure was taken from Intel's literature. It shows the sequence of instruction flow from right to left.

M F I M I B   M I I M I B

M0 M1 I0 I1 F0 F1 B0 B1 B2

IA64 instruction dispersal units

© Cengage Learning 2014

In Figure 8.33, the two bundles that were in the dispersal window have been executed and the following two bundles brought into the dispersal window. Intel uses the term *rotation* to indicate that a new bundle is brought into the dispersal unit. In this case a double rotation has occurred.

In Figure 8.33, all three instructions in the first bundle, MII, in the dispersal window have been issued to the execution unit. Unfortunately, not all the instructions in the second bundle can be issued because of the lack of integer units. In the next rotation, Figure 8.34, only one new bundle enters the instruction dispersal unit. Instruction issue continues with the I and B instructions of the first bundle and the M, I, and B instructions of the second bundle.

The dispersal (issue) of instructions to functional units is governed by a set of rules that are given in Intel's literature.

M/I Slots—An I slot in the third position of the second bundle is always dispersed to I1. Otherwise, an M or I instruction is dispersed to the lowest numbered M or I unit that is currently free.

F Slots—An F slot in the first bundle is dispersed to F0, and an F slot in the second bundle is dispersed to F1.

B Slots—B slots in MBB and BBB bundles are dispersed to B units in order. The B slot in MIB, BFB, or MMB bundles disperse to B0 if the instruction is brp or nop.b; otherwise, it disperses to B2.

FIGURE 8.33   The state of the system in Figure 8.32 after the next rotation

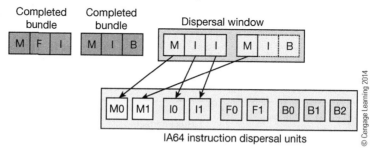

Completed bundle   Completed bundle   Dispersal window

IA64 instruction dispersal units

© Cengage Learning 2014

FIGURE 8.34   The state of the system in Figure 8.33 after the next rotation

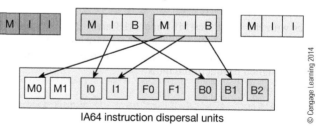

IA64 instruction dispersal units

© Cengage Learning 2014

The McKinley—The Itanium 2

The Itanium 2, released in 2002, was the second EPIC-based IA64 processor produced by the Intel-HP alliance. The architectures of the Itanium 1 and 2 are the same; however, considerable improvements took place in the Itanium's organization. Table 8.3 illustrates some of the enhancements.

One of the disadvantages of the Itanium 1 was the relatively limited number of bundles that could provide full issue (all syllables executed in parallel). Table 8.4 demonstrates the templates that provide full issue on the Itanium 1 and Itanium 2. Light blue represents Itanium 1 and dark blue Itanium 2. As you can see, only 26 combinations are valid for the Itanium 2. For example two bundles with consecutive templates MII, MBB. The Itanium 2 adds a further 50 combinations to this total.

TABLE 8.3   Differences Between the Itanium 1 and 2

Feature	Itanium 1	Itanium 2
Integer/ALU units	4	6
Floating-point units	4	4
Streaming FP units	4	2
Multimedia units	4	6
Load/store units	2	4
L1i instruction cache	16 Kbytes	16 Kbytes
L1d instruction cache	16 Kbytes	16 Kbytes
L2 cache	96 Kbytes	256 Kbytes
L3 cache	2 Mbytes off-chip	1.5 to 6 Mbytes

© Cengage Learning 2014

TABLE 8.4   McKinley Instruction Dispersal Matrix

	MII	MLI	MMI	MFI	MMF	MIB	MBB	BBB	MBB	MFM
MII										
MLI										
MMI										
MFI										
MMF										
MIB										
MBB										
BBB										
MBB										
MFM										

Combination that provides full issue

Combination that provides full issue in McKinley only

Full issue not possible

© Cengage Learning 2014

The Itanium 9300 Tukwila Processor

Intel's next version of the Itanium was somewhat delayed and did not appear until 2010. This was called the Tukwila rather than the Itanium 3. The Tukwila has a transistor count of over two billion and was aimed at the high-performance, high-reliability enterprise server market. Tukwila was a four-core processor with two-way hyperthreading[18] and could execute eight threads of code simultaneously. The 9350 version of this processor had a 24 Mbyte level 3 cache (and a price tag of $3,800 in 2010).

The Itanium Poulson Processor

In 2011, Intel announced a post-Tukwila processor that extended the Itanium IAS further. The Poulson has eight cores (see Chapter 13) and provides 54 MB of on-die memory (cache). The term on-die indicates that the memory is not necessarily on the same silicon chip as the processor but is within the same physical housing as the processor. Poulson employs 32 nm technology and has 3.1 billion transistors in comparison with Tukwila's 2.05 billion transistors and 65 nm technology. Shrinking feature size from 65 nm to 32 nm (the largest atom has a diameter of about 0.5 nm) means that eight cores can be located on a 544 mm^2 chip consuming 170 W, whereas Tukwila's chip has an area of 700 mm^2 and consumes 185 W. On the other hand, Tukwila has to get rid of 0.26 W per mm^2, whereas Poulson has 0.31 W to dissipate per mm^2. This demonstrates that reducing chip size can create other problems. Poulson has also increased the maximum number of instructions that can be issued in a cycle from 6 to 12.

Poulson is the first Intel processor with support *instruction replay technology* that increases system integrity. A buffer in the instruction path holds instructions and is able to re-execute them and allow the processor to recover from system errors. It is the architectural equivalent of the rollback function associated with operating systems that enable you to go back to a known good point after an error. Poulson is also the first Itanium processor to provide integer multiplication.

Is the IA64 a VLIW Processor?

The IA64 uses instruction bundles and parallel instruction units without internal dynamic instruction scheduling, just like typical VLIW processors. However, the IA64 does have several features that distinguish it from VLIW processors. A VLIW processor executes all components of an instruction (all syllables of a bundle) in parallel, whereas the IA64 can use a stop to force non-parallel execution within a bundle.

VLIW processors dedicate specific slots (syllables within the bundle) to particular operations such as integer or floating-point. The IA64 forces no restrictions on the placing of the syllables of a bundle (subject to the availability of a template for the intended combination of syllables). Finally, the VLIW processor requires that a branch be at the end of a bundle, whereas the IA64 permits a branch to be in any slot; indeed, it permits three branch instructions in a bundle.

8.3.7   Predication

We spent much of the previous chapter looking at the havoc caused by conditional branches in pipelined processors. The IA64 uses predicated computing to minimize the number of branches required in a program and, therefore, to minimize the number of penalty cycles caused by mispredicted branches. We now provide two examples to demonstrate the IA64's use of predication before we discuss the fine details. Appending (p1) to an IA64 instruction causes it to be executed if and only if the Boolean value in predicate register p1 is true. Consider the following construct.

```
if (x =! y)  s = 4;
          else s = s + 4;
```

We can express this construct in IA32, 68K, ARM and IA64 forms as in Table 8.5.

---

[18]Hyperthreading offers a means of increasing performance without adding more execution units. User and state registers (like the PC) are replicated and a thread or stream of instructions are executed. When a thread is halted by a load latency (due to a cache miss), another thread can be swapped in using the data in the alternate set of registers while the first thread waits. We discuss hyperthreading later in Chapter 13.

TABLE 8.5

IA32 Code		68K Code		ARM Code		IA64 Code	
	CMP **ax**,bx		CMP D0,D1	CMP r1,r2		cmp.ne p1,p2 = r1,r2 ;;	
	JNZ Then		BNE Then	ADDEQ **r3**,r3,#4		(p1) mov **r3** = 4	
	ADD **S**,4		ADD #4,**S**	MOVNE **r3**,#4		(p2) add **r3** = r3,4	
	JMP Exit		BRA Exit				
Then	MOV **S**,4	Then	MOVE #4,**S**				
Exit		Exit					

The IA64 code is branchless, more compact than the CISC code, and the two predicated instructions can be carried out in parallel in the same cycle. A simple example of the effective uses of predication is provided by Geva and Morris.[19] Consider the code fragment

```
if (x == y) {
    z++;
    p += q;
} else {
    r++;
}
```

This sequence can be translated into the following IA64 code. To keep things simple, we will use register names consisting of "r" plus the variable name. For example, rx is the register that holds *x*.

```
{
    cmp.ne    p1 = rx,ry   // Perform the initial compare x == y
    (p1) br else            // If true then do the else part
}
{.label then
    add rz = rz, 1          // z++
    add rp = rp, rq         // p += q
    br  exit                // jump over else part
}
{
.label else
    add rr = rr, 1          // r++
}
```

This is conventional but inefficient code because it is divided into blocks separated by conditional or unconditional branches. If the conditional test is mispredicted, the cost is high. Why? Because the IA64 is a parallel processor and a lost cycle potentially throws away up to three instructions.

The following version of the algorithm makes better use of prediction by setting *two* predicate registers when the test is performed and then executing three simultaneous operations—two of which are predicated on true and one on false. The second block performs all the computation in a single cycle.

```
{
    cmp.eq    p1,p2 = rx,ry    // Perform the initial compare x == y
} {                            //This block executes in one cycle
    (p1) add rp = rz, 1        // z++
    (p1) add rp = rp, rq       // p += q
    (p2) add rr = rr, 1        // r++
}
```

---

[19]Robert Geva (Intel) and Dale Morris (Hewlett-Packard), "IA64 Architecture Disclosures White Paper." http://cpus.hp.com/technical_references/ia64_arch_wp.pdf.

The `cmp.eq` instruction compares the contents of rx and ry and sets predicate registers p1, p2 to 1,0 if the outcome of the test is true and 0,1 if the outcome is false. The next three instructions are executed if the associated predicate register is true.

Because predicated execution is such a strong feature of the IA64's architecture, we now look at its compare instruction and the application of predicated instructions in greater detail.

Compare Instructions in Detail

Conventional architectures like the IA32 perform compare operations by subtracting one operand from another and then setting the appropriate flag bits of a status register. A later conditional branch can then use the appropriate condition codes. This paradigm is used by RISC processors such as the ARM. For example,

```
CMP     r1,r2              ;compare r1,r2 by [r1] – [r2]
ADDEQ   r4,r5,r6           ;if [r1] – [r2] = 0 then [r4]=[r5]+[r6]
```

A single flag register doesn't work well with superscalar and highly parallel processors like the IA64. When you do a conditional operation, precisely which of the parallel instructions set the flags being tested? The IA64 solves the problem by depositing the Boolean result of a comparison in one of the 64 predicate registers p0 to p63. The format of an Itanium conditional operation is

```
cmp.cond p_i,p_j = r_x,r_y
```

The cond op-code completer specifies the condition being tested for (e.g., equality, greater than, etc), p_i, p_j are two predicate registers, and r_x, r_y are the registers being compared. Consider the instruction where the outcome of the comparison is stored in two predicate registers—one predicate register is set if the test is true and the other predicate register is set if the test is false.

```
cmp.eq p1,p2 = r6,r9
```

If the contents of registers r6 and r9 are equal, then the single-bit Boolean variable in p1 is set to 1 and the single-bit variable in p2 is set to 0. The first predicate register receives the value generated by the test and the second predicate value receives the complement of the value generated by the test. This arrangement allows you to test for a condition and its inverse simultaneously. Consider the example

```
if (x == y) then s = s + 3
            else s = s - 4
```

This operation can be implemented as

```
        cmp.eq p1,p2 = r1,r2 ;;   //r1 = x and r2 = y
(p1) add r4 = r4,3                 //THEN part: if true then add
(p2) sub r4 = r4,4                 //ELSE part: if false then subtract
```

Note how the mutually exclusive then and else parts of the if construct are computed in parallel at the same time. This construct enables you to use *branchless* code. The full format of the compare instruction is

```
(r_i) cmp.crel.ctype p_i,p_j = r_i,r_j
```

where (r_i) is the predicate used to decide whether this compare instruction is to be executed. The completers crel and ctype specify the condition being tested and the *type* of the test. Valid relationships are given in Table 8.6.

If you write an Itanium instruction without specifying a predicate register, the default p0 predicate register is selected and that contains the permanent hardwired value 1, which forces

TABLE 8.6	Itanium Comparison Instructions			
cmp.eq	equal	a == b		
cmp.ne	not equal	a != b		
cmp.lt	less than	a < b	signed	
cmp.le	less than or equal	a ≤ b	signed	
cmp.gt	greater than	a > b	signed	
cmp.ge	greater than or equal	a ≥ b	signed	
cmp.ltu	less than	a < b	unsigned	
cmp.leu	less than or equal	a ≤ b	unsigned	
cmp.gtu	greater than	a > b	unsigned	
cmp.geu	greater than or equal	a ≥ b	unsigned	

© Cengage Learning 2014

execution. You can use predicated registers in chained or compound conditionals by means of the operation

```
cmp.cond.boolean pi = rx,ry
```

where boolean is the logical operation OR, or AND, or the completer unc (*unconditional* which is described later). The Boolean completer modifies the way in which the predicate registers are set or cleared. The OR completer sets the predicate register if the test is true but does not clear the predicate register if the test is false.

Similarly, the AND completer clears the predicate register if the test is false but does not set it if the test is true. Consider the construct

```
if ((a == 0)||(b == 1)||(c =! 2)) {x = 5;}
```

We can write

```
          cmp.ne    p1 = r0,r0;;   //trick code to set p1 = 0
          cmp.eq.or p1 = 0,r1      //do a == 0 (set p1 if true)
          cmp.eq.or p1 = 1,r2      //do b == 1 (set p1 if true)
          cmp.ne.or p1 = 2,r3;;    //do c =! 2 (set p1 if true)
   (p1)   mov r4 = 5               //if P1 true then r4 = 5
```

In this example, we set predicate register p1 to false by means of cmp.ne p1 = r0,r0, which compares r0 with r0 (i.e., hardwired 0 with 0) and gets the value false on a comparison for not equal.

The instruction cmp.eq.or p1 = 0,r1 compares r1 with 0 and then sets the value of p1 if the condition is true. The next two instructions also set p1 if the tested condition is true. Finally, the (p1) mov r4 = 5 instruction is executed if *any* element of the three-part test was true. Let's say this again—after the first operation, all three tests are carried out in the same clock cycle.

Let's look at another example of predication from Geva and Morris. Figure 8.35 illustrates an if-then-else construct with a second nested if-then-else in the first else path. Predication provides efficient code because compare instructions *themselves* may be predicated (as we found with ARM code).

The construct of Figure 8.35 can be written as

```
{
          cmp.gt p1, p2 = ra, rb        // block 1
} {
   (p1)   add    rc = rc,1              // block 2
   (p2)   add    rd = rd, rc            // block 3
   (p2)   cmp.eq.unc p3, p4 = re, rf    // block 3
} {
   (p3)   add    rg = rg, 1             // block 4
   (p4)   add    rh = rh, -1            // block 5
```

FIGURE 8.35   Structure of nested if-then-else constructs

```
if (a > b) {          //block 1
    c++;              //block 2
}
    else {
        d += c;       //block 3
        if (e == f) { //block 3
            g++;      //block 4
        }
            else {
                h--;  //block 5
                }
}
```

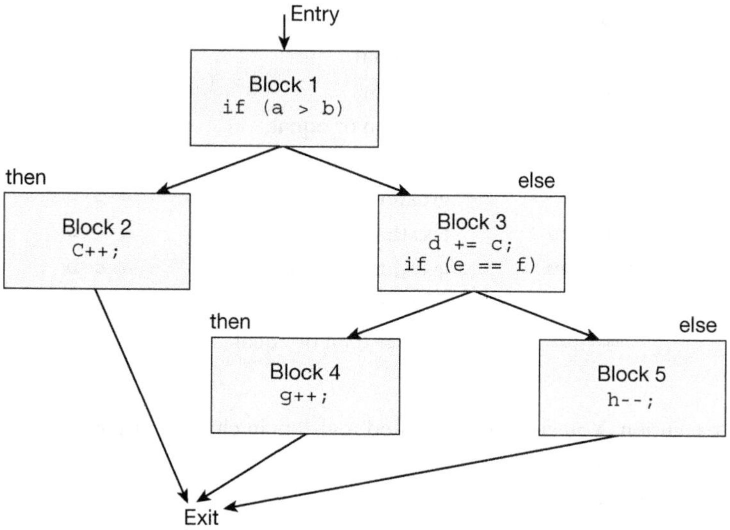

© Cengage Learning 2014

Note how the two components of block 3 are both predicated by p2. The second component of block 3 is the test

    (p2)   cmp.eq.unc **p3**, **p4** = re, rf

which carries out the test if (e == f) and sets predicate registers p3 and p4. However, the additional completer, unc (unconditional) overrides the normal behavior of the comparison. If the qualifying predicate is true, the cmp.eq.unc behaves normally, and p3, p4 are set to 0,1 or 1,0 depending whether the outcome of the test was false or true, respectively. Now, if the qualifying predicate in p2 is *false* and the .unc completer is specified, the two predicates are forced to 0. In this case, if (p2) is zero then both (p3) and (p4) are forced to zero.

The .unc completer can be used to prevent *both* paths of a nested conditional being executed—in this case blocks 4 and 5.

Predicated execution can be used as an effective way of implementing multiway braches as the following example demonstrates.[20]

```
{ .mii
    cmp.eq **p1**,**p2** = r1,r2
    cmp.ne **p3**,**p4** = 4, r5
    cmp.lt **p5**,**p6** = r8,r9
}
{.bbb
(p1) br.cond label1
(p3) br.cond label2
(p5) br.cond label3
}
```

© 2000 IEEE. Reprinted, with permission, from J. Huck et al., "Introducing the IA-64 Architecture," *IEEE Micro*, vol. 20, no. 5, Sept.–Oct. 2000, pp. 12–23.

In this example, three comparisons are performed in the first cycle and a branch made in the second cycle if any of the conditions is satisfied.

Table 8.7 demonstrates how the completers UNC, AND, OR, and ANDOR operate. The *qualifying predicate* column defines the value in the predicate register that predicates the current instruction and the *compare result* columns correspond to the outcome of the conditional test. This table gives the values loaded into the two predicate registers under these circumstances.

---

[20]Huck, et al., "Introducing the IA64 Architecture," *IEEE Micro*, September–October 2000.

TABLE 8.7   The Action of Completers UNC, AND, OR, and ANDOR

| Type | Qualifying Predicate = 1 | | | | Qualifying Predicate = 0 | |
| | Compare Result = True | | Compare Result = False | | | |
	First Target	Second Target	First Target	Second Target	First Target	Second Target
Normal	0	1	1	0		
UNC	0	1	1	0	0	0
AND	0	0				
OR			1	1		
ANDOR			1	0		

© Cengage Learning 2014

Predication—A Further Example

Consider the following code from Snavely, et al.:

```
       cmp.eq p1,p2 = r3,r4
(p1)   ld8  r5  = [r1]
(p2)   ld8  r6  = [r2] ;;
(p1)   add  r5  = r5,1
(p2)   add  r6  = r6,1 ;;
(p1)   st8  [r1],r5
(p2)   st8  [r2],r6
```

The compare instruction is always executed and sets p1, p2 to 0, 1 or 1, 0, depending on whether r3 = r4 (1,0) or r3 =! r4 (0,1). The next six instructions are guarded by either p1 or p2 (instructions guarded by p2 are shown in blue). Note how both the then and else paths of the comparison can be executed in parallel due to the mutual exclusion between p1 and p2. However, note also that the code is inefficient, because stops have to be placed after each of the pairs of actions because of the register dependencies.

N. Snavely, S. Debray, and G. Andrews, "Predicate Analysis and If-Conversion in an Itanium Link-Time Optimizer," *Proceedings of the Workshop on Explicitly Parallel Instruction Set (EPIC) Architectures and Compilation Techniques*, 2002.

Preventing False Data Dependency in Predicated Computing

The Itanium provides a completer to deal with false data dependency in cases where operations are mutually exclusive and, therefore, the computer does not have to worry about the case in which two actions are both executed. As we have already stated, the Itanium assembler (or compiler) analyzes the code and inserts STOPs to prevent data dependency. Sometimes a situation occurs in which the data dependency is *apparent* rather than real because the dependent instructions are mutually exclusive. Consider the example:

```
(p1)  add r1 = r2,r3
(p2)  add r1 = r4,r5
```

The assembler may examine this code and conclude that a write-after-write dependency exists for register r1 and insert a STOP after the first addition in order to ensure that r2 + r3 is written to r1 before r4 + r5 is written to r1.

Exposing the Architecture and Organization

You throw instructions at a superscalar processor and the powerful magic inside executes the instructions. This is not so with advanced VLIW architectures like the IA64 Itanium. The example of false data dependency demonstrated the extent to which to the programmer/compiler is exposed to the architecture and internal organization of the Itanium. Here you have to tell the processor that some instructions are mutually exclusive and that you can execute them in parallel because only one member of the group will determine the outcome. Such techniques increase the processor's power but at the price of requiring a greater knowledge of the architecture/organization.

Suppose that this code is part of an *if then else* statement preceded by cmp.eq **p1,p2** = r10,r11. It is impossible for both p1 and p2 to be set to 1 simultaneously and both additions may take place in parallel (because only one addition results in a write to r1). The Itanium assembler can receive help in the form of a .pred.rel assembler directive that tells it how to treat predicated instructions. Consider the example:

```
.pred.rel "mutex", p1,p2,p3    //tell the assembler that p1, p2, p3 are mutually
                                 exclusive

(p1) add r1 = r2,r3
(p2) add r1 = r4,r5
(p3) add r1 = r6,r7            //only one of these three instructions will be executed
```

Having defined mutually exclusive predicate registers, we may wish to reassign them and we need to tell the assembler that mutual exclusivity is at an end. Suppose that p1 and p3 are no longer mutually exclusive. We use the assembler directive

```
.pred.rel "clear", p1, p3    //clears p1 and p3 relations
```

Another assembler directive is .pred.rel "imply", p1,p2 that informs the compiler that if p1 is true, p2 is also true. If p1 is false, nothing can be determined about the state of p2.

Branch Syntax

A typical branch instruction is B_{cond} target, where B_{cond} represents *branch on a specific condition* and target is the destination of the branch; for example, the ARM branch BEQ Label5. As well as conditional branches, there are unconditional branches, the equivalent of the high-level *goto*.

You could say that the IA64 doesn't actually need a conditional branch instruction because of its predication mechanism. The Itanium doesn't do things the easy way, and the IA64's branch instruction is rather more complex, with the format

```
(P1) BR.bytpe.bwh.ph.dh target
```

The btyte completer defines the type of the branch and the other completers are optional because they aid processor efficiency by providing hints about the outcome of the branch. Typical branch types are

cond   branch on specified condition in predicate register
call   call procedure
ret   return from procedure
cloop   counted loop (if the LC register is not zero, it is decremented and the branch taken)
ia   IA32 context switch (the Itanium branches to a region of IA32 code)

The branch target is either a signed 21-bit literal that is shifted left four times (i.e., multiplied by 16) and used to generate a program counter relative address[21] or the contents of a branch register. This gives a branching range of 16 Mbytes in each direction. Note that the literal offset is multiplied by 16 to ensure that a target address is on a 16-byte boundary (for 128-bit instructions).

The *branch with hint* completer, bwh, is used to indicate the likely outcome of the branch; that is, the compiler is able to pass information to the Itanium's branch prediction logic. Its four possible vales are

spnt   static: assume not taken
sptk   static: assume taken
dpnt   dynamic: assume not taken
dptk   dynamic: assume taken

---

[21]Itanium literature uses the term IP or *instruction pointer* rather than program counter. I use the term 'PC relative' for consistency with other chapters.

In spite of its complexity, this instruction can be used quite simply. For example,

```
(p3) br.cond next    //if [p3] = 1 then goto "next"
```

If you know that the branch is going to be taken almost all the time, you could write it as

```
(p3) br.cond.sptk next    //if [p3] = 1 then goto "next"
```

to indicate that it is to be predicted (statically) as taken. As an example of a branch, consider the following simple loop.

```
        mov     r1,count        //set up r1 as the loop counter
Next                            //body of the loop
        add     r1 = -1,r1;;    //decrement the loop counter
        cmp.eq p0,p1 = r1,0     //check for end of loop
(p1)    br.cond.sptk Next;;     // continue until zero (note the hint)
                                //continue here after the loop
```

Note that the predicate register pair in the compare is p0,p1, where p0 is hardwired to zero. We are not interested in the 'true' condition of the test—only in the 'false' condition when r1 is not zero and we set p1 to 1 and take the branch back to the start of the loop. Note also the use of two stops. There is a stop after the loop count decrement because we are going to test it and we can't do that in parallel, and there is a stop after the branch instruction.

As well as conventional relative addressing, the branch instruction supports register indirect addressing. Other processors, for example, the 68K used an address or data register for this purpose. The Itanium, however, uses one of its 16 special-purpose branch registers br0 to br15; for example,

```
(p2)    br br5    //if [p2] = 1 then go to [br5]
```

8.3.8   Memory Access and Speculation

We have seen that three principal obstacles to high-speed computation are the branch that causes a pipeline flush, data dependency that forces serial execution of code, and the load latency that requires you to wait for memory. The Itanium provides some help with memory access.

Loading data from cache or memory incurs a latency penalty and it makes sense to fetch data before it is needed. The IA64 lets you load data from memory into a register *before* you know that the data is valid,[22] and it lets you load data into a register from within a block of code that is the target of a branch *before* the branch has been executed. These loads are called *advanced* loads and *speculative* loads, respectively.

The full format of the IA64's load instruction is rather complex, although you don't have to specify all options; that is,

```
(qp) ldsz.ldtype.ldhint r_dest = [r_source]
```

The base mnemonic is `ldsz` where sz is the size of the operand in bytes and may be 1, 2, 4, or 8; for example, the mnemonic `ld4` loads a 4-byte value into one of the 128-bit general-purpose registers. Loading is into the least-significant end of the destination register and unused bits are set to zero. The source address is register indirect so that

```
ld2   r1 = [r9]
```

loads register r1 with the 16 bits pointed at by register r9. The *type* and *hint* fields defined by completers `ldtype` and `ldhint` define how the load instruction behaves. We now look at the two completers .s (*speculative*) and .a (*advanced*) in more detail.

---

[22]By valid, we mean that the required data is currently available; for example, it may not be currently cached and an access to the data will generate a cache miss. Even worse, the data may not be in immediate access store and an access will generate a page fault.

Control Speculation

The load instruction is a nuisance because it has a latency that can be fairly short, if the data is in the highest level of the cache, or very long if the data is well down the memory hierarchy. To add insult to injury, many of the loads take place at the beginning of a block of code where they do most damage, because data has to be read before it can be used.

Figure 8.36a illustrates the situation in which a branch is made to a block of code that includes a load instruction. At best, the load is fetched from high-level cache. At worst, the data is not even in main store and a page-fault occurs, forcing the operating system to get the data from disk.

You could move the load instruction up to an earlier point in the program, called *hoisting* the instruction. Hoisting sometimes allows you to avoid some of the load latency, but you pay a horrendous price if you read a location, generate a page fault, and then find that you never really wanted the data because you don't branch to the block that uses it.

Figure 8.36b shows how the Itanium provides *control speculation* by using the ld.s instruction to perform a load before the data is needed and then a chk.s instruction immediately before the data is required.

Consider the effect of an ld2.s r1 = [r9] instruction. This does what you would expect: It loads register r1 with the contents of the memory location pointed at by r9. The notion of speculation means that the load is not yet needed, it is *opportunistic* The load may or may not be carried out at a later point in the program. Because the load is speculative, exceptions that the load might have generated are *deferred*. There's no point in dealing with exceptions for an instruction whose execution may not be necessary. When an exception is deferred, the destination register's *not-a-thing bit* (NaT) is set.

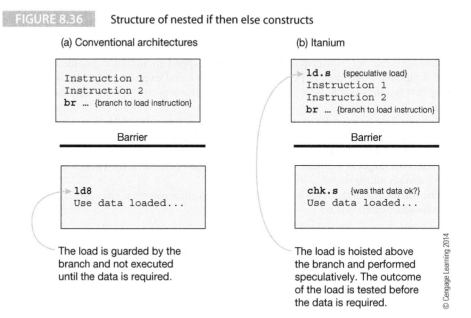

FIGURE 8.36   Structure of nested if then else constructs

(a) Conventional architectures

```
Instruction 1
Instruction 2
br  ... {branch to load instruction}
```

Barrier

```
ld8
Use data loaded...
```

The load is guarded by the branch and not executed until the data is required.

(b) Itanium

```
ld.s    {speculative load}
Instruction 1
Instruction 2
br  ... {branch to load instruction}
```

Barrier

```
chk.s    {was that data ok?}
Use data loaded...
```

The load is hoisted above the branch and performed speculatively. The outcome of the load is tested before the data is required.

© Cengage Learning 2014

At a later point in the program, the load that was speculatively executed may be needed if we branch to the region of code where it is used. We can't just go ahead and use the value that was loaded speculatively. We have to check whether the earlier speculative load was *successful*. We use the *load instruction with a check*, chk.s, to determine whether the load was successful. One parameter required by the check instruction is the address of the code that deals with a failed speculative load. Consider the following fragment of code.

```
        ld8.s    r7 = [r8]     //read the pointer speculatively
                               //more code
        cmp.eq   p1 = r9,r0    //is r9 == 0?
  (p1)  br.xyz                 //this is the branch that guards the load
                 .
                 .
                 .
xyz:    chk.s    r7,Error      //verify the load
xyz1:            .
                 .
                 .
Error:  ld8      r7 = [r8]     //try again
        br       xyz1          //and return
```

We read a pointer if the contents of register r9 are zero. However, `ld8.s r7 = [r8]` reads the pointer speculatively *before* it is needed. If the load is successful, all is well. If the load is unsuccessful due to a page fault, the register's NaT bit is set and processing continues.

If the branch is taken, the speculative load is checked by `chk.s r7,Error`. If the load was successful, the value in r7 is used. If the load was not successful and the NaT bit in r7 is set, a branch is made to `Error` where the code needed to perform the recovery lies. What has this arrangement gained? If a page fault does not occur, the load latency required by the load is hidden because the load took place before it was needed. The only downside is that we've added extra instructions for the error checking and we've tied up register r7 from the time of the speculative load.

The Advanced Load

An *advanced load* also performs a load operation before the operand is needed. This operation is indicated by appending the completer `.a` to a load instruction. Consider the following fragment of code.

```
ld4.a    r7 = [r8]         //read the 32-bit word here
add      r4 = r4,1    ;;   //increment r4
st4      [r9] = r4         //store the new r4
ld4.c    r7 = [r8]    ;;   //test the load – was it ok?
add      r6 = r7,8         //use the data we've read
```

In this example, the `ld4.a r7 = [r8]` instruction performs the initial advanced load. Later, the `ld4.c r7 = [r8]` instruction checks or verifies whether the advanced load was successful. If the advanced load was successful, the operation continues. If the operation failed, the load must now be repeated.

So, how does an advanced load differ from a speculative load? The advanced load deals with the problem of *disambiguation*; that's what it does. Disambiguation means *removing* the ambiguity surrounding two addresses. Let's look at an example from Geva and Morris where a simple test is performed on an element pointed at by *a and a flag updated if the test is true.

```
unsigned char flag;      //a global variable declared in the program
         .
         .
         .
int test (int *a, *b)    //a function with two pointers as parameters
{
    if (*a)              //test the value pointed at by *a
        flag += 1;       //if it's not zero then increment global variable flag
    return (*b - 1);     //decrement the variable pointed at by *b
}
```

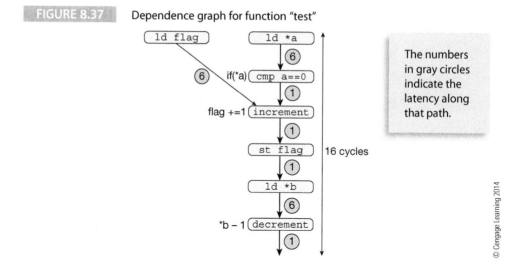

FIGURE 8.37   Dependence graph for function "test"

The numbers in gray circles indicate the latency along that path.

16 cycles

© Cengage Learning 2014

In this fragment of code, three values have to be loaded from memory: flag, *a, and *b. Geva and Morris assume that these variables are cached and a load requires six cycles. Figure 8.37 illustrates how the code might be structured conventionally (i.e., without hoisting). The code takes 16 cycles.

We can load flag and *a at the same time. However, we can't load *b until *after* the flag has been stored. Why? Because *b and flag might be the *same* location. Clearly, if we were to load *b before storing the updated flag, the semantics of the program would be altered.

Now, you and I know that *b and flag are not the same variables and no problem exists. We can perform these two memory accesses in parallel without changing the program's semantics. Unfortunately, the Itanium has not taken a course in C; nor has it read the source code and decided that flag is an unsigned char and *b a pointer to an integer variable. Since the Itanium accesses memory via pointer-based addressing, it cannot know whether or not two pointers are pointing at the same variable. Consequently, it has to be cautious.

Data speculation and the use of the ld.a instruction solve the problem by performing an *advance* load before the data is needed. The source address is automatically saved in the ALAT (advanced load address table) that records the register number, the address accessed, and the size of the data transfer.

When a store operation takes place, the address of the destination operand is passed to the ALAT where an associative match is performed (i.e., every store address is checked against all entries in the table). If this address is in the ALAT, the corresponding entry is removed from the table.

When the variable that was loaded using the ld.s instruction is ready to be used, it is tested with a ld.c instruction where the .c completer indicates *check*. The load check instruction reads the ALAT to look for an entry corresponding to the specified register number. If it finds an entry, all is well and the previously loaded value can be used. If the load check instruction does not find an entry corresponding to the register number, then a collision must have occurred. The contents of the register that was loaded may have been overwritten since they were read by the advance load. In this case, the advance load was unsafe and a new load must be carried out.

Figure 8.38 shows the dependency graph for the same function but using an advanced load. In this case, the execution time has been reduced to 9 cycles.

FIGURE 8.38   Dependence graph for function "test" using data speculation

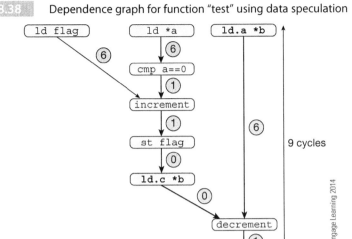

9 cycles

© Cengage Learning 2014

8.3.9   The IA64 and Software Pipelining

The Forth Rail Bridge in Scotland was opened in 1890 and requires continuous painting to preserve it against the elements. They start painting it at one end, and by the time the painters reach the other end, it's time to begin painting it all over again. The same is true of microprocessor development. Innovation is continuous: As soon as one round of innovation is complete, like the Forth Rail Bridge, it's time to start all over again.

Having apparently squeezed every drop of performance out of microcomputers with pipelining and branch predication, the IA64 begins a new round of performance acceleration with *software pipelining*, a technique for speeding up iterative loops.[23]

First, let's look at the simple loop. Consider the vector operation $\mathbf{Y} = \mathbf{X} + 1$ where $y_i = x_i + 1$. Before continuing, we have to make a comment concerning terminology. When dealing with loops, we encounter three concepts: the clock cycle, the instruction bundle, and the *iteration* cycle that consists of one pass round a loop. IA64 literature employs the term *trip* to indicate a cycle of iteration to avoid confusion with clock cycle. We will follow this convention. The vector operation $\mathbf{Y} = \mathbf{X} + 1$ can be represented in C code as

```
for (int i = 0; i < n; i++) {
    y[i] = x[i] + 1;
}
```

This expression can be translated into IA64 code in the following way. We use the Itanium's *loop counter register*, LC, to mechanize this loop. Register r10 is a pointer to vector **X** and register r11 is a pointer to vector **Y**.

```
        mov ar.lc = r1      ;; // set up the loop counter (initial count in r1)
Repeat: ld4 r4    = [r10],4  ;; // load element xᵢ into r4 and update the X pointer
        add r4    = r4,1     ;; // calculate yᵢ = xᵢ + 1
        st4 [r11] = r4,4        // store and update the Y pointer
        br.cloop Repeat      ;; // decrement the LC register and loop until zero
```

---

[23] I am not implying that the notion of software pipelining is new. Software pipelining was used by the 1980s and is also associated with the Cray supercomputers. What is interesting is the way in which the Itanium architecture has been designed to facilitate software pipelining.

This is, of course, inefficient code because most of its instructions cannot be executed in parallel due to the load latency and data dependencies − notice the stops in blue. The time taken to execute the code is

```
Repeat: ld4 r4    = [r10],4      ;; //Cycle 0, load latency 2 cycles
        add r4    = r4,1         ;; //Cycle 2, use stop because of following
                                        dependency on r4
        st4 [r11] = r4,4         //Cycle 3
        br.cloop Repeat          ;; //Cycle 3, parallelism ok here
```

A cycle of iteration takes four clocks because of the load latency and data dependency. The only parallelism is the ability to execute the store and the branch instructions at the same time. Note the STOPs. There is a STOP after the initial load and a stop after the add instruction. We can improve the situation by unrolling the loop and carrying out two operations per iterative cycle, that is,

```
Repeat: ld4 r4    = [r10],4      ;; //Cycle 0 get x_i
        ld4 r5    = [r10],4      ;; //Cycle 1 get x_{i+1}
        add r4    = r4,1         ;; //Cycle 2 y_i = x_i + 1
        add r5    = r5,1         //Cycle 3 y_{i+1} = x_{i+1} + 1
        st4 [r11] = r4,4         ;; //Cycle 3 store y_i
        st4 [r11] = r5,4         //Cycle 4 store y_{i+1}
        br.cloop Repeat          ;; //Cycle 4
```

On each trip round the loop, we perform two vector operations—the second operation is shown in blue for clarity. We have doubled the work per *trip* for the cost of one extra clock cycle. We can go further by employing separate pointers to elements x_i and x_{i+1} as follows

```
        add r12   = 4,r10        //preamble create second X pointer
        add r13   = 4,r11        ;; //preamble create second Y pointer
Repeat: ld4 r4    = [r10],4      //Cycle 0 get x_i using first pointer
        ld4 r5    = [r12],4      ;; //Cycle 0 get x_{i+1} using second pointer
        add r4    = r4,1         //Cycle 2
        add r5    = r5,1         ;; //Cycle 2
        st4 [r11] = r4,4         //Cycle 3 store y_i using first pointer
        st4 [r13] = r5,4         //Cycle 3 store y_{i+1} using second pointer
        br.cloop Repeat          ;; //Cycle 3
```

We have more fully exploited the IA64s parallelism and executed two iterations for the price of one. Loop unrolling can be extended and more operations executed per trip. We can't go much further because a significant source of delay is the load latency. If you load x_i in the current iteration (i.e., trip) and then use x_i, you have to wait for the load to complete. *Software pipelining* hides latencies and is analogous to hardware pipelining in the sense that the current trip involves operations from more than one trip of the iteration.

Suppose we designed *dedicated* hardware to execute the vector operation $\mathbf{Y} = \mathbf{X} + 1$. A pipelined implementation would be storing y_i, incrementing x_{i+1}, and reading x_{i+2} at the same time. We can do exactly the same in software by means of the following loop that is expressed in pseudocode.

```
Repeat: load     x_{i+2}
        add      x_{i+1} = x_{i+1},1
        store    y_i
        br.cloop Repeat
```

We first load element x_{i+2}, which is not required in the *current* cycle of iteration, so we don't have to worry about latency. We then calculate $x_{i+1} = x_{i+1} + 1$. By using the element

x_{i+1} that was read in the *previous* trip, we don't have to wait for the source data before doing the addition. Finally, we store y_i that was created last time round the loop.

Software pipelining is composed of three parts that have to be treated separately: the initial part or *prolog* when the pipe is being filled; a central part or *kernel* when all stages of the pipeline are active; and an *epilog* when no more data is entering the pipeline and the last operation is working its way through. Consider the previous example where the vectors **X** and **Y** each have seven components, x_0 to x_6. Table 8.8 illustrates the effect of software pipelining. Software pipelining can be difficult to describe adequately because the distinction between *iteration cycle* and *pipeline stage* can be confusing. Time slots T_0 to T_8 represent cycles of iteration or trips round the loop. Slots 1, 2, and 3 are the three stages in the software pipeline.

TABLE 8.8   Software Pipelining

Time	Slot 1	Slot 2	Slot 3
T_0	Get x_0		
T_1	Get x_1,	Add 1 to x_0	
T_2	Get x_2,	Add 1 to x_1,	Store y_0
T_3	Get x_3,	Add 1 to x_2,	Store y_1
T_4	Get x_4,	Add 1 to x_3,	Store y_2
T_5	Get x_5,	Add 1 to x_4,	Store y_3
T_6	Get x_6,	Add 1 to x_5,	Store y_4
T_7		Add 1 to x_6,	Store y_5
T_8			Store y_6

© Cengage Learning 2014

Events in Table 8.8 can be separated into *prolog*, *kernel*, and *epilog* phases (Table 8.9). Table 8.9 shows that, on the first trip, element x_0 is fetched from memory. On the second trip round the loop, the second element x_1 is fetched and the previous element x_0 is incremented. The first kernel stage occurs at T_2 when x_2 is being fetched as x_1 is being incremented and y_0 is being retired (stored in memory). We can express the three operations executed during the first kernel trip in time frame T_2 in the *hypothetical* form

```
ld4  r4   = [r10],4     //load x₂
add  r5   = r5,1     ;; //add 1 to x₁
st4  [r11] = r6,4       //store y₀
```

TABLE 8.9   The Prolog, Kernel, and Epilog Phases of Software Pipelining

Time	Slot 1	Slot 2	Slot 3	Phase
T_0	Get x_0			Prolog
T_1	Get x_1,	Add 1 to x_0		Prolog
T_2	Get x_2,	Add 1 to x_1,	Store y_0	Kernel
T_3	Get x_3,	Add 1 to x_2,	Store y_1	Kernel
T_4	Get x_4,	Add 1 to x_3,	Store y_2	Kernel
T_5	Get x_5,	Add 1 to x_4,	Store y_3	Kernel
T_6	Get x_6,	Add 1 to x_5,	Store y_4	Kernel
T_7		Add 1 to x_6,	Store y_5	Epilog
T_8			Store y_6	Epilog

© Cengage Learning 2014

As well as this basic kernel code that sits in the main loop, we need separate code blocks to implement the prolog and the epilog. Moreover, the kernel code is incomplete because we need assignments to "make the new x_i the old x_{i+1}." For example, in the above fragment of hypothetical code the current register r6 contains the value in r5 generated on the previous trip.

The IA64 makes software pipelining a practical proposition by implementing two key features. First, it automates the prolog and epilog processes and second, it deals with the pipelining mechanism whereby the value of a variable generated in one trip becomes the value used in the next trip. If you look at the above code, the current value of r5 is the value of r4 in the previous cycle of iteration (trip), and the new value of r6 is the value of r5 in the previous trip.

The fundamental pipelining operation is implemented by a mechanism that IA64 literature calls *register rotation*. At each stage in the iteration, register i becomes register $i + 1$. Consider the following example of pipelined adding in a software pipelined loop.

```
add r40 = r10,5      //q_{i+1} = x + 5
add r41 = r40,12     //q_i  = q_i + 12
```

On the trip round the loop, register r40 accumulates the contents of r10 + 5. On the next trip round the loop, the second instruction adds 12 to the contents of what was r40. That is, the *current* r41 is the *old* r40. *Register rotation* implies that the registers "*move*" so that register r_{i+1} in the last trip becomes register r_i in the current trip. Register rotation and software pipelining make it difficult for the human programmer to read assembly language code because the names of registers change during the iteration.

The other factor in software pipelining, the prolog and epilog code, is implemented by a combination of register rotation and the IA64's *predication* mechanism. Recall that instructions are predicated so that

```
(p16) add r20 = r30,5
```

indicates that the instruction is executed only if the single Boolean variable in predicate register r16 is true. Suppose a loop has the following four stages (we are not interested in the details of the code here).

```
(p16) code 1    // pipeline stage 1
(p17) code 2    // pipeline stage 2
(p18) code 3    // pipeline stage 3
(p19) code 4    // pipeline stage 4
```

During the prolog, code 1 is executed in the first trip. Code 1 and code 2 are executed in the second trip. Code 1, code 2, and code 3 are executed in the third trip, and so on. By *marching* a pattern of 1s through the predicate registers we can achieve this goal. On trip 1, [p16] = 1 and all other predicates are 0. On trip 2, [p16] = 1 and [p17] = 1, and so on.

The reverse process takes place during the epilog as the pipeline is flushed or emptied; that is p16, p17, p18, p19 contain 0,1,1,1 after the last code 1 has been executed, 0,0,1,1 after the last code 2 has been executed, and so on.

General-purpose, floating-point, and predicate registers can take part in rotation, although not in precisely the same way. Predicate registers p16 to p63 and floating-point registers f32 to f127 may rotate. The number of general-purpose registers that rotate must be programmer defined in the region r32 to r127 in multiples of 8 by the alloc instruction. This instruction controls the mapping between logical and physical registers and is used prior to calling a procedure.

We can now write the code for the operation **Y** = **X** + 1 (omitting setting up and looping).

```
L1
    (p16) ld4 r32     = [r10],4        //load xᵢ and update pointer
    (p17)                              //dummy
    (p18) add r35     = r34,1          //calculate yᵢ = xᵢ + 1
    (p19) st4 [r11]   = r36,4          //store yᵢ and update pointer
          br.ctop  L1                  //Repeat
```

This code demonstrates just how much software pipelining exposes the programmer to the fine details of the IA64's architecture and its instruction latencies. The first instruction,

```
    (p16) ld4 r32 = [r10],4
```

loads register r32 with the 32-bit value pointed at by register r10 and post-increments the pointer by 4. A load instruction has a latency of two cycles, which means that the load will not be completed and the contents of r32 are not available until two further trips round the loop have been made.

The empty line beginning (p17) is grayed out because it is a dummy line for the purpose of this example; it has been included to show the missing cycle created by the load latency. Note that the third line, (p18) add r35 = r34,1, uses register r34 as a source operand because this is register $r32$ after two rotations. Similarly, the instruction (p19) st4 [r11] = r36,4 stores the contents of register $r36$ in memory because register $r35$ is rotated to become r36. The br.ctop L1 instruction decrements the loop counter register and branches back to label L1 until the counter becomes zero.

The IA64 uses the EC counter, that counts down through the epilog sequence once the main loop counter has been exhausted. The complete sequence is therefore

```
    mov   lc = 6                       //set up loop count = cycles – 1
    mov   ec = 4                       //set up ec count = stages + 1
    mov   pr.rot = 1<<16               //set up predicate registers
L1
    (p16) ld4 r32     = [r10],4        //load xᵢ and update pointer
    (p17)                              //dummy
    (p18) add r35     = r34,1          //calculate yᵢ = xᵢ + 1
    (p19) st4 [r11]   = r36,4          //store yᵢ and update pointer
          br.ctop  L1                  //Repeat
```

The initial code sets up the two counters and initializes the predicate registers with the instruction. Table 8.10 walks through the execution of the code. Instructions that are not executed during a trip are grayed out.

TABLE 8.10   Example of Software Pipelining

Cycle	Port and Instruction				Predicate Registers				Counter	
	M	I	M	B	b16	b17	b18	b19	LC	EC
0	ld4	add	st4	br.ctop	1	0	0	0	6	4
1	ld4	add	st4	br.ctop	1	1	0	0	5	4
2	ld4	add	st4	br.ctop	1	1	1	0	4	4
3	ld4	add	st4	br.ctop	1	1	1	1	3	4
4	ld4	add	st4	br.ctop	1	1	1	1	2	4
5	ld4	add	st4	br.ctop	1	1	1	1	1	4
6	ld4	add	st4	br.ctop	1	1	1	1	0	4
7	ld4	add	st4	br.ctop	0	1	1	1	0	3
8	ld4	add	st4	br.ctop	0	0	1	1	0	2
9	ld4	add	st4	br.ctop	0	0	0	1	0	1
					0	0	0	0	0	0

Registers and Function Calls

The IA64 uses register management to implement a windowing function for function calls and returns. General registers r0 to r31 act as global registers (using the Berkeley RISC terminology) and registers r32 to r127 can be managed; that is, they can be renamed under software control. IA64 literature refers to these registers as *stacked* registers.

Unlike the Berkeley RISC that implements a fixed windowing scheme, the IA32 allows the programmer to directly manipulate register allocation to functions via the allocation instruction, alloc, that gives you complete control over the mapping process. You use alloc to specify the number of registers a function requires and the distribution between parameter-passing registers and local registers.

When the procedure call is made, the registers are renamed. The local registers from the previous register stack are hidden and the output registers in the calling procedure renumbered. The first operation carried out in the target procedure is to create its own local registers. This action is done using the alloc instruction to define the input parameter registers, the local registers and any output registers it uses to communicate with procedures that it will call. This instruction cannot be predicated. The format of an alloc instruction is

```
alloc rframe = ar.pfs, input, local, output, rotating
```

The previous procedure state register is copied to general register r_{frame} to preserve the calling procedure's status. The remaining four parameters are integer values that define the size of the stack frame. The parameters define the size of the input parameter space, the size of the local parameter space, the size of the output parameter space, and the number of rotating registers. Note that the size of the frame is determined by input + local + output, and the size of the local region is given by i + 1.

Figure 8.39 demonstrates the operation of the IA64's register renaming mechanism. It is very similar to the original Berkeley MIPS windowing mechanism; the principle difference being the programmer's ability to select the size of the windows at run time.

Like the MIPS, the IA64 is prone to run out of windows after several nested calls have been made. However, the IA64 deals with the problem by means of its so-called *register stack engine* that handles overflow by dumping registers in memory and automatically restoring them.

The IA64 uses a VLIW architecture with explicit parallelism, predicated execution, and speculative loading to improve performance and help reduce the effect of load latencies. In Chapter 13, we look at another way of handling load latency, *multithreading*.

FIGURE 8.39   The IA64 register stack

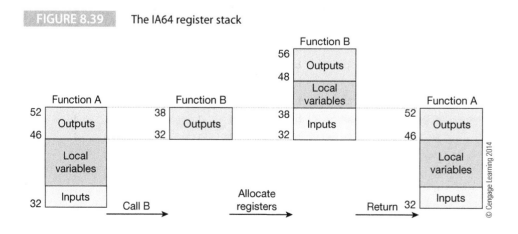

Summary

Once you have speeded up a processor by increasing the clock rate, by implementing pipelining, and by providing complex short vector instructions (e.g., MMX), there's only one place left to go; *instruction level parallelism* (before resorting to simultaneous multithreading and multiple processors on the same chip). In this chapter, we have looked at two forms of instruction level parallelism: the superscalar processor and the VLIW processor of which the IA64 Itanium is an example.

Superscalar processing employs multiple execution units that can execute instructions in parallel. This is a nice idea in theory, but it is complicated by the need to deal with data dependencies between instructions. Superscalar processors have to implement logic that issues batches of instructions and determines which can be executed immediately and which have to wait until the resources they need become free. Equally, instructions have to be completed or retired in the appropriate order if the semantics of the program are to be preserved.

The Itanium processor implements instruction level parallelism by using a long instruction word that holds three individual operations but doesn't provide the hardware to ensure that instructions are executed in the appropriate order. Instead, the compiler is responsible for ensuring that the instructions are organized so that those that can be executed in parallel are bundled into groups. We have looked at the Itanium, because it incorporates some of the interesting features we've seen in other processors such as predicated execution.

The next four chapters deal with computer memory and input/output. In Chapter 13, we return once again to performance and look at multithreading and multi-core processors.

Problems

8.1 Superscalar processing could be added to any existing processor without having to recompile source code. Why?

8.2 The performance of a computer can be expressed as the time taken to execute a task. That time can is given by

Time = cycles/instruction × seconds/cycle
× instructions/task

Many factors affect the design of a processor; for example, technology, ISA, the compiler, pipelining, superscalar technology. Discuss how these factors affect the above equation.

8.3 If a VLIW form of an existing processor were to be produced, why would the source code have to be recompiled?

8.4 Mainstream techniques for accelerating the performance of the computer are the RISC instruction set philosophy, superpipelining, superscalar technology, VLIW technology and code morphing. Briefly outline the features of these processor acceleration techniques.

8.5 In 1991, Nick Tredennick said:

"Superpipelining is a new and special term meaning pipelining. The prefix is attached to increase the probability of funding for research proposals. There is no theoretical basis distinguishing superpipelining from pipelining. Etymology of

the term is probably similar to the derivation of the now-common terms, methodology and functionality as pompous substitutes for method and function. The novelty of the term superpipelining lies in its reliance on a prefix rather than a suffix for the pompous extension of the root word."

What do you think of this comment?

8.6 You are redesigning an existing RISC-style pipelined processor to make it a three-way superscalar processor that permits three instructions to be executed at the same time.
   a. Do you expect the new processor to achieve three instructions per clock?
   b. What level of design complexity do you expect the new processor to have in comparison with the pipelined version?

8.7 Is the IA64 Itanium a VLIW processor? Give your reasons.

8.8 What is *register renaming* and why is it used in superscalar processors?

8.9 Why does register renaming prevent write-after-write hazards (WAW) but not read-after-write (RAW) hazards?

8.10 What is the theoretical upper bound on the CPI metric for a RISC processor without superscalar facilities?

8.11 What is the limitation on the number of registers (i.e., the user-visible register file) in a processor?

8.12 Describe the difference between superscalar and VLIW architectures.

8.13 What is software loop unrolling and why does it help to speedup execution?

8.14 What is *predication* and how can it be used to speed up program execution?

8.15 The superscalar and VLIW processors share fundamentally different relationships with their compilers. Why?

8.16 Consider the following fragment of assembly language that is to be executed on a generic three-way superscalar processor that can accommodate one memory access per clock. Rewrite the code to improve its performance.

```
LDR  r1,[r2]
LDR  r3,[r4]
MUL  r5,r3,#12
ADD  r3,r3,r1
STR  r3,[r4]
ADD  r1,r1,#4
STR  r1,[r2]
ADD  r7,r7,#14
MUL  r7,r1,#5
SUB  r6,r6,#4
```

8.17 Both the ARM processor and the IA64 architecture have predicated instruction execution. In what way is the IA64's predication mechanism superior to the ARM processor's?

8.18 In the context of the IA64, what is the difference between instruction and data speculation?

8.19 Demonstrate how a compiler might translate the following fragment of code into a form that can run on a VLIW processor that specifies three operations per instruction word (i.e., three instructions per bundle in Itanium terminology). You may perform instruction reordering. Assume that the latency for a load is three cycles and the latency for a multiply is two cycles (i.e., the load takes three cycles in total). Use a NOP whenever an instruction slot cannot be filled.

```
MOV  r7,#4
LDR  r1,[r2]
LDR  r3,[r4]
LDR  r5,[r6]
ADD  r1,r1,r3
ADD  r1,r1,r5
DIV  r1,r1,r7
ADD  r2,#8,r2
ADD  r4,#8,r4
ADD  r6,#8,r6
```

Repeat the same exercise, but assume that the VLIW processor can perform only one memory access operation per cycle. Assume that the memory unit is fully pipelined; that is, it has a latency of three

cycles but the next memory operation can begin in the following cycle.

8.20 Consider the following code. Show how it might be compiled into generic assembly language assuming that the compiler unrolls the loop three times. All values are integers. Assume that register indirect with offset addressing is available but pointers have to be explicitly updated.

```
for (int i = 0; i < 30; i++) {
z[i] = 3 * (x[i] + y[i]);
}
```

8.21 The following fragment of code is to be executed on two different superscalar processors. The processors have two integer units and two load/store units (i.e., up to two memory accesses and two integer operations can be executed concurrently. The fetch window is eight instructions. Show how it would be executed, cycle-by-cycle, on the following.

a. Superscalar with in-order issue and in-order execution.

b. Superscalar with out-of-order issue and out-of-order execution.

```
LDR   r3,[r0]      ;get x[i]
ADD   r0,r0,#4     ;update pointer
STR   r3,[r6]      ;store q[i]
ADD   r6,r6,#4     ;update pointer
ADD   r8,r8,r3     ;keep running total
LDR   r4,[r1]      ;get y[i]
ADD   r1,r1,#4     ;update pointer
ADD   r4,r4,r3     ;x[i] + y[i]
ADD   r4,r4,r4     ;2(x[i] + y[i])
STR   r4,[r2]      ;store z[i]
ADD   r2,r2,#4     ;update pointer
SUBS  r5,r5,#1     ;dec loop counter
```

Assume that the latency for each instruction is one cycle, except for a load which is two cycles. Assume that the multiply operation has a latency of two cycles and that the multiplier cannot be reused until the previous instruction has been completed.

8.22 What is the disadvantage of register renaming in a superscalar processor?

8.23 What is *software pipelining*?

8.24 How does the IA64 increase the efficiency of software pipelining?

8.25 Identify the types of register data dependencies in the following piece of code.

```
L1: MUL  r2, r3, r4
L2: ADD  r3, r2, r4
L3: ADD  r4, r2, r4
L4: SUB  r3, r4, r3
```

8.26 A delayed branch can enhance the performance of a simple RISC processor, but it may not increase the performance of a superscalar processor. Why?

8.27 What is the difference between *dispatching* and *scheduling*?

8.28 Are VLIW and EPIC processors different or is EPIC just HP's term for VLIW?

8.29 The Alpha 21x64 family, the Motorola 68K family, and the AMD 29000 family demonstrate that a high-quality architecture or superior performance does not guarantee commercial success. A company's ability to control the market is far more important than a processor's technical merits. Is this statement true?

8.30 What is the rôle of a template in an IA64 Itanium instruction bundle?

8.31 Why does the Itanium have more than one type of NOP (no operation), for example nop.i and nop.b?

8.32 Consider the following generic code.

```
LDR    r1,[r2]
LDR    r3,[r4]
ADD    r6,r1,r3
SUB    r7,r1,r3
MUL    r3,r6,r7
STR    [r2],r3
```

Indicate all the potential data hazards in this code.

8.33 The throughput of a superscalar processor depends on the number of instructions it can issue per clock cycle. Early superscalar processors were able to issue two instructions per clock. Later processors could issue four instructions per clock. Although massive progress has been made in processor parameters such as clock speed or cache size, it is unlikely that we will see 64-way or 128-way superscalar processors. Why?

8.34 IA64 general-purpose registers r0 to r127 have 65 bits, because they include a special NaT (not a thing) bit in addition to the normal 64 data bits. What is the purpose of the NaT bit and how is it used?

8.35 Explain the meaning of the term *memory disambiguation* in the context of memory access.

8.36 What is the effect of the .crel completer when it is suffixed to an IA64 comparison instruction, for example, cmp.crel p1,p2 = r5,r6

8.37 Translate the following into IA64 code making use of predication.

```
for (i = 0; i < 100: i++)
{ if    (X[i] = Y[i])
            Y[i] = Y[i] + 1;
        else X[i] = X[i] - 1;
}
```

8.38 A superscalar processor can be best thought of as a pipelined processor where the pipeline is replicated. For example, a four-way superscalar processor has four parallel pipelines allowing four instructions to be executed in parallel. True or false—and why?

8.39 All other factors being equal, which is easier to design: a superscalar processor or a VLIW processor?

8.40 An existing superscalar processor with 32 registers in the register file is redesigned and the instruction length increased from 32 to 64 bits which allows 1,024 general-purpose registers. Providing a very large register file can have a useful side-effect in a superscalar processor, which may help reduce the processor's complexity. What is that?

8.41 We said (when discussing the potential limitations of VLIW computers) "If a VLIW with m operations per bundle has an instruction with an n-cycle latency, up to $m \cdot n - 1$ execution slots are lost." Demonstrate the truth or otherwise of this statement.

8.42 Why is a stack machine a particularly poor candidate for the application of superscalar principles? Note that a stack machine makes use of push and pull operations, and its data processing operations are applied to the top of the stack.

IV THE SYSTEM

"The whole is greater than the sum of its parts."
Anon

"We build our computer (systems) the way we build our cities: over time, without a plan, on top of ruins."
Ellen Ullman

A computer is composed of four physical elements: CPU, memory, buses, and I/O. Without each of these four components the other three parts would be junk—albeit technologically impressive junk. The first element of a computer is the *CPU* that does all the information processing, although today's high-performance computers have graphics cards with their own powerful processors. If we couldn't build *fast* processors on *tiny* chips, the computer revolution would never have occurred. The second part of a computer is its *memory* that stores large volumes of data. Over the years, a spectrum of quite radically different *memory technologies* has been developed to store data. Some of these technologies store large volumes of data that is slow to retrieve, and some store small quantities of data that can be accessed rapidly. By using clever design techniques, we can construct memory systems that appear to be fast and yet store large volumes of data economically. In Chapter 9, we show how memory technologies can be combined to design memories that appear both fast and large.

The processor and memory elements of a computer are vital, but on their own, they are as good as a head without a body. A practical computer requires two other components. First, it needs *buses* to link its functional parts—just as we need highways to link homes, cities, and businesses. Data has to be transferred into a computer and distributed between its various parts. Some buses are *internal* highways that copy data from the CPU to memory, and some are *external* highways that transfer data from devices such as camcorders or scanners to the computer. Buses have made ubiquitous computing possible.

The final component of a computer is made up of the *peripherals* that input data from, or transmit it to, the outside world. A peripheral can be a mouse, a scanner, a printer, a GPS receiver, or a Dolby surround-sound music system. Because of the range and complexity of peripherals, we cannot go into their design and construction in this text. However, we do examine the techniques used to perform input and output transactions.

In Part IV, we look at two of the elements of the computer system we've rather neglected so far: *the memory and the bus*. The bus is generally combined with Input/ Output (I/O) because there's an overlap between the way in which we move information between two points in a computer and the way in which we move information into and out of a computer. Memory and bus systems are so complex and so sophisticated that in order to do them justice we devote three chapters to these topics (memory gets two chapters because we separate immediate access from serial access memory). We go into more detail than is normal in conventional texts because the computer revolution owes its success as much to memory devices and buses as to sheer computing power. Only those of a certain age can appreciate a pre-plug-and-play world when just connecting a computer to a simple printer was a nightmare.

The first three chapters of Part IV look at a computer's memory subsystem. Chapter 9 begins with the *cache memory* that we have already mentioned several times. Cache memory bridges the gap between the fast processor and the slower main store by holding a copy of frequently used data. We will demonstrate how small cache memories dramatically increase a computer's performance and how cache systems are organized. This chapter also introduces *memory management,* which is concerned with integrating magnetic disk memory into a computer's high-speed semiconductor main memory. Memory management techniques bring data from disk into the computer whenever it is needed, automatically and transparently to the user. We have combined cache memory and memory management because they share the same fundamental principles of operation. Cache memory has been placed before basic memory principles and technology because of the vital contribution cache memory makes to a computer's performance, and because you don't need to know how memory chips work internally in order to understand the way in which a cache memory operates.

Chapters 10 and 11 of the memory section are concerned with the memory systems themselves and explain how we can store information by modifying the electrical or magnetic properties or matter. Chapter 10 looks at high-speed random access memories, and Chapter 11 covers secondary storage, which is largely made up of hard disks and CD/DVD optical stores. These two chapters describe a wide range of technologies in some depth because of the role memory plays in determining a computer's performance and because memory technology is one of the most interesting areas of computer science and engineering.

The final chapter in Part IV, Chapter 12, is devoted to getting information into and out of a computer. We begin by looking at the strategies used to get data into and out of a computer, such as *polling,* where the computer continually tests a status bit until the peripheral is ready, and *direct memory access,* where dedicated logic transfers data directly between a peripheral and memory without the intervention of the CPU. Chapter 12 discusses some of the issues associated with data transfer, such as the *buffering* required to ensure that data doesn't pile up before we can read it and the *closed-loop transfer,* which indicates the receipt of each data element as it arrives.

The second part of the I/O section introduces the bus, which is a topic of enormous size and scope. We look at the principles of the bus and describe *arbitration,* which is the technique that allows multiple processors or peripherals to request access to the bus and for the bus to be granted to just one of these contenders. One of the reasons for the popularity of the PC is the introduction of fast, low-cost serial interfaces, such as USB and FireWire, that not only provide high-bandwidth buses but are also *stackable;* that is, you can wire peripherals together in a chain.

Cache Memory and Virtual Memory

"Any sufficiently advanced technology is indistinguishable from magic."
Arthur C. Clark

*"What's in a name? That which we call a rose
By any other name would smell as sweet."*
Shakespeare, Romeo and Juliet

Memory Hierarchy

We begin with *cache memory,* which makes a computer look as if it has a much faster memory than it really has. Memory should be non-volatile, cheap, fast, small, and consume virtually no power. In reality, each memory technology has its own particular characteristics, some of which are mutually exclusive. For example, you can get fast, expensive memory or slow and cheap memory. Memory systems invariably use several different technologies with each technology making its own contribution to the computer. Taken together, these technologies give the user the appearance of a system with fast, non-volatile, low-cost memory. We can categorize the various memory technologies into distinct layers that form a hierarchy.

Figure 9.1 describes the classic pyramidal memory hierarchy. The memory at the top of the pyramid has the highest speed (lowest access time), and the memory at the bottom has the lowest speed. We use a pyramid because it implies, generally correctly, that the quantity of the memory at the top is smaller than the memory at the bottom.

The on-chip registers provide the fastest memory in a computer and hold the working data required by the processor. It is the speed of register access that makes the load/store register-to-register architecture so popular. Registers are fabricated with the same technology as the CPU, run at the same clock rate, and have no long data paths between them and the rest of the CPU. Moreover, on-chip registers can be directly accessed from the CPU, whereas all external memory is accessed via a pathway that involves memory management, address translation, and complex data buffering and control mechanisms. Consequently, register memory is fast. CPUs have only a handful of registers to store working data and status information. Registers can't hold programs.

Below the registers in the memory hierarchy we have *cache memory,* or more strictly, *level 1 cache memory.* The size of cache memory may be orders of magnitude smaller than the main memory, but the nature of real programs and the distribution of data mean that a typical application might use a small working set of the instructions and data for 95% or more of the time. Once, cache memory was located on the motherboard, but advances in chip technology have made it possible to locate substantial caches on the processor chip.

Figure 9.1 shows that there are *two* cache memories. If the data being accessed is not in the level 1 cache, the next level down in the memory hierarchy is accessed, which is the *level 2 cache.* Not all systems have two levels of cache: some systems have *three* levels of cache memory.

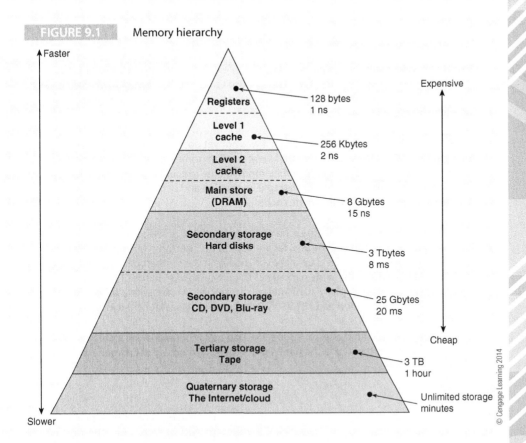

FIGURE 9.1 Memory hierarchy

Faster

Expensive

Registers — 128 bytes
1 ns

Level 1
cache — 256 Kbytes
2 ns

Level 2
cache

Main store
(DRAM) — 8 Gbytes
15 ns

Secondary storage
Hard disks — 3 Tbytes
8 ms

Secondary storage
CD, DVD, Blu-ray — 25 Gbytes
20 ms

Cheap

Tertiary storage
Tape — 3 TB
1 hour

Quaternary storage
The Internet/cloud — Unlimited storage
minutes

Slower

© Cengage Learning 2014

If data isn't in the cache, it must be retrieved from the computer's main memory, which is the next level down in Figure 9.1. Modern PCs and workstations almost invariably use DRAM to implement the main random access memory. Today's PCs might implement somewhere between 1 GB and 48 GB of DRAM.

Because main store is volatile, it's necessary to store programs and data in nonvolatile memory. One of the cheapest storage mechanisms is the *hard disk* that stores data in the form of magnetic patterns on a flat, rotating platter. A hard disk can store more than 4,000 GBs of data, that is, 4 TB. However, the access time of the hard disk is of the order of 5 ms, which is very fast by human standards but 10^6 times slower than main store. Today, the mechanical hard disk is being replaced by the faster and more robust solid state disk (SSD). We discuss both these technologies in Chapter 11.

Because the main store can't normally hold all the programs and data required by the processor, computers implement a memory management system called *virtual memory,* whereby the main memory contains only the data currently being used and the data not in current use remains on a hard disk. When the processor accesses data that's not in the main store, the operating system steps in and swaps in a *page* of typically 4 K to 64 KB of data from the hard disk to the main store. A virtual memory system lets you run programs many times larger than the main store without incurring a significant reduction in performance; that is, a virtual memory system with 512 MB of DRAM and a 100 GB hard disk performs almost as well as a memory with 100 GB of DRAM. Virtual memory also provides a means of protecting data.

The next step down in the memory hierarchy is the *optical store* implemented by the CD, DVD, or Blu-ray. Optical memory stores data in the form of indentations along a spiral track on a plastic disk that are read using a laser that bounces light off the indentations. The optical memory is slower than the hard disk, mainly because the CD or DVD rotates at a tiny fraction of the speed of the platter in a hard disk drive.

History of Cache Memory

The use of cache memory was proposed by Maurice Wilkes and Gordon Scarott in 1965. Cache memory was in use in the 1970s, but it wasn't until the 1980s that the performance of main memory was beginning to significantly lag behind that of on-chip registers, making cache memory so attractive to the designer.

In the early days of microprocessors when it was not yet possible to fabricate large memories on CPU chips, cache memories were small or non-existent. The Motorola 68020 had a 256 byte instruction-only cache that improved the execution time of small loops. Motorola's next offering, the 68030, had split instruction and data caches, both 256 bytes (the 68040 was to have two 4096-byte caches).

In the Intel world, external off-chip cache was used by 80386 processors. The 80468 was the first Intel processor to include an 8 KB on-chip cache. Shortly after, some 80486-based systems introduced second level L2 caches of 265 KB on the motherboard. Intel's Pentium Pro included L2 cache in the same package as the processor (i.e., in the CPU housing but not on the same silicon chip). By 2010 Intel was producing i7 chips with 8 MB of on-chip cache (although Intel's Xeon had 16 MB of L2 cache when it was launched in 2008).

Intel's quad-core Itanium processor 9300 series had a 24 Mbyte L3 cache in 2010. To put this into context, the Itanium cache has 24 times the total memory of first-generation PCs.

Figure 9.1 is inaccurate in at least one sense: optical storage is slower than magnetic storage but it is not generally larger. The CD holds about 650 MB of data, and a Blu-ray disk can store 25 GB. A few years ago these were truly large quantities of data, but progress in hard disk technology has been so dramatic that, in the space of a decade, the capacity of a hard disk has gone from about 100 MB to over 4 TB (an increase of 2^{22} to 4×2^{30}, that is, a 4×10^8-fold increase in storage capacity). Today's hard drives store much more data than optical disks, and optical storage is positioned in Figure 9.1 by speed rather than by size.

Below the optical storage mechanisms in Figure 9.1, we have *magnetic tape* or *cartridge storage*. These storage technologies use magnetic recording techniques in conjunction with a long stream of magnetic tape rather than a rotating platter. Magnetic tapes can hold vast quantities of data but their access time is in the order of minutes or even hours, so they are used only as backup and archival storage mechanisms.

Below tape comes the *Internet* and *cloud storage* that provide remote distributed storage. Data is stored externally, usually on virtual servers belonging to a third party; that is, you buy a storage service and store your information via the Internet and WWW. You don't need to know where the data is. All that matters is that the data is safe from catastrophes that might befall your own installation. Moreover, the data is accessible from anywhere that is connected to the Internet.

Before we delve into memory systems technology in Chapters 9 and 10, we have to examine the way in which memory is *related* to the computer system. If the access time of a register in the CPU is less than 1 ns and the access time of a CD drive is over 200 ms, it is going to take a lot of effort to manage a system with such diverse memory devices—all trying to do the same job.

This chapter looks at the relatively small *cache memory* that is combined with a larger and slower memory to make the combined system look like a large and fast memory. Magic. Then we look at *virtual memory* that manages addresses so you don't have to worry about where data is in memory, allows you to run several programs simultaneously in different regions of memory, and automatically loads data from disk when the computer needs it.

We cover cache and virtual memory systems in this chapter because they perform the *same* task. Both allow a small quantity of fast memory to act as if it were much larger by integrating a large quantity of slower memory into the system. Both technologies are concerned with mapping addresses from the computer onto the actual location of data in a memory. The difference between cache and virtual memory is largely a matter of speed and control mechanisms. Cache memory operates in the nanosecond range and is managed automatically by hardware, whereas virtual memory operates in the millisecond range and is managed by the operating system.

Let's look at how cache and virtual memory fit into a system. Suppose a computer executes the following operations.

```
                     ; time = time + 1
LDR  r1, = time      ; r1 points to "time"
LDR  r2, [r1]        ; read time
ADD  r1,r1,#1        ; increment time
STR  r2, [r1]        ; store time
```

These instructions read the operand `time` from memory. All the CPU has to do is to put the address `time` on the address bus and then read the data. In principle, it couldn't be simpler. Figure 9.2 illustrates the type of system that may execute this code. The address of an operand from the computer (in this case `time`) is a *logical* address. This address goes to the fast cache memory where it *tries* to access the operand. If the data is in the cache, it is accessed from the cache. Otherwise, if the data is in the slower main memory, it will be brought from the main memory and loaded both into the computer and the cache. This transfer is implemented in hardware and is invisible to the programmer and operating system.

FIGURE 9.2   Cache and virtual memory

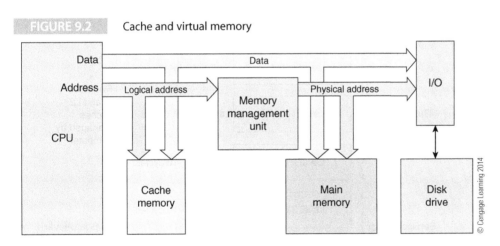

© Cengage Learning 2014

Sometimes, the data isn't even in main memory; it's on the hard disk. When the data is on disk, the virtual memory mechanism copies it from disk to the main memory. It will, of course, also be copied to the cache memory. The data that is transferred from disk to main memory may go almost *anywhere* in the main memory. It is the job of the *memory management unit* (MMU) to translate the *logical* address from the computer (i.e., the place where the computer thinks the data is located) into the *physical* address of the data (i.e., the place where the operating system actually loaded it in memory). Virtual memory management is one of the principal tasks of the operating system and requires close cooperation between the special-purpose hardware of the memory management unit and operating system.

Importance of Cache Memory

It is impossible to over stress the importance of cache memory in today's world. The ever widening gap between the speed of processors and the speed of DRAM has made the use of cache mandatory.

Suppose a high-performance 32-bit processor has a superscalar design and can execute four instructions per clock cycle at 1,000 MHz (i.e., a cycle time of 1 ns). In order to operate at this speed, the processor requires $4 \times 4 = 16$ bytes of data from the cache every 1 ns. If data is not in cache and it has to be fetched from 50 ns DRAM memory over a 32-bit bus, it would take 4×50 ns $= 200$ ns to fetch four instructions. This corresponds to the time it would normally take to execute $4 \times 200 = 800$ instructions.

We are now going to discuss the nature of the memory hierarchy implicit in Figure 9.1 and then demonstrate how cache memory speeds up a computer's operation. Part of this chapter is devoted to the factors influencing the performance of cache memory and how you can improve the speed of a system by optimizing both cache hardware and user software.

The second part of this chapter looks at virtual memory and the way in which logical or virtual addresses from the processor are mapped onto the physical addresses of data in memory.

9.1   Introduction to Cache Memory

The main memory that holds programs and data in von Neumann computers should be as fast as the CPU requires. If the CPU accesses data or an instruction and the memory can't supply it within the current cycle, the memory must return a signal telling the CPU to wait. A CPU waits by inserting *idle* or *wait* states in its machine cycle to prevent it from continuing on to the next operation. During a wait state, the CPU ceases normal operation, and slow memory can seriously degrade its performance. In this section, we show how cache memory offers a way of substantially increasing a processor's performance without incurring an excessive economic penalty. Figure 9.3 reminds us of the *memory hierarchy* and describes delays in terms of clock cycles.

FIGURE 9.3   Memory hierarchy

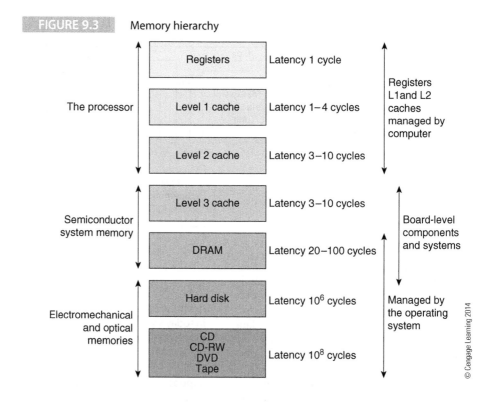

In the early 1990s, it was common to see comments in journals expressing the sentiment, *"There's no point in building or using high-speed CPUs if all they do is wait faster."* One solution to the designer's dilemma is to build faster memory that can keep up with the CPU. This might sound easy to do, since CPU and memory technologies are linked to each other because they both employ the same design and fabrication processes. It's tempting to argue that a CPU with a cycle time of 5 ns requires a memory component with an access time of 5 ns. There are two flaws in this apparently persuasive argument. First, it is a historical fact that the rate of increase in the speed of processors has far outstripped the increase in DRAM speeds over the last two decades. Second, a CPU with a machine cycle time (i.e., the time taken by the processor to execute a read or write access to external memory) of 5 ns might have a clock cycle time of only 2.5 ns, because each CPU machine cycle is executed in two clock cycles. Suppose that the CPU, with its 5 ns machine cycle time, spends a clock cycle performing internal housekeeping tasks. The processor requires data in only one clock cycle measured from the time at which the address is first available to the time at which data from the memory is latched. The access time of the memory needed to keep up with the CPU is therefore 2.5 ns. Consequently, there is a demand for memories with access times well below those of the machine cycle times of the processors they serve.

Although modern technology does indeed produce memory components with access times below 5 ns, the cost of such devices prohibits their use in large memory systems. The large-scale economic production of personal computers and workstations in the mid-1990s demanded the use of tried and tested mainstream memory components (e.g., 64 Mbit DRAMs with an access time of 50 ns).

There's nothing mysterious about cache memory; it's just a very high-speed memory that can be accessed rapidly by the processor. The element of magic comes from the ability of systems to employ a modest amount of high-speed memory (e.g., 256 KB in a system with 512 MB of DRAM) and expect the processor to make over 95% of its accesses to the cache rather than to the DRAM.

Cache memory can be understood in everyday terms by its analogy with a diary, address book, or iPhone used to record telephone numbers. A telephone directory contains hundreds of thousands of telephone numbers, and nobody carries one around with them. People keep a list with a hundred or so telephone numbers. Although the fraction of all possible telephone numbers on this list might be less than 0.0001%, the probability that the next call will be to a number in the address book is high, as you are most likely to call a friend.

Cache memory operates on the same principle as the list in an address book or on an iPhone by locating frequently accessed information in the cache rather than in the much slower main memory. Unlike the personal list, the computer can't know in advance what data is most likely to be needed. Instead of dividing data into important categories the way people do, computer caches operate on a *learning principle*. They learn by experience what data is most frequently used and then transfer it to the cache.

Ideally, as DRAM-based memories become faster and faster, the need for complex and expensive cache systems should disappear. In the 1970s, I stated such a view. Time proved me very wrong.

What's in a Name?

The word *cache* is pronounced *cash* or *cash-ay* and is derived from the French word meaning *hidden*. Cache memory is invisible to the programmer and appears as part of the system's memory space.

Cache memory is also called *look-aside* memory, because it is physically arranged in parallel with main memory and appears *at the side* of main memory. However, the term is more common in the expression *translation look-aside buffer* that describes a specialized cache that holds address translation vectors.

Figure 9.4 shows the general trend in memory and microprocessor performance over the last two decades. Memory has become faster at about 7% compound per annum. Processors have become a lot faster over the same time with their performance increasing at about 55% per annum. Because the gap in processor-memory performance has widened, cache memory systems are more important today than when they were first invented.

The effect of different rates of change in the performance of memory and processors was highlighted by Wulf and McKee[1] who argued that lag in the speed of DRAM compared to processors will eventually prove to be a limit on computer performance. They said that processing time is the sum of the time performing internal operations plus the sum of the time spent accessing external memory, and therefore, the memory access component will eventually dominate. Any further progress in processing power will be pointless. They coined

---

[1] William Wulf and Sally McKee, "Hitting the Memory Wall: Implications of the Obvious," *Computer Architecture News*, 23(1):20–24, March 1995.

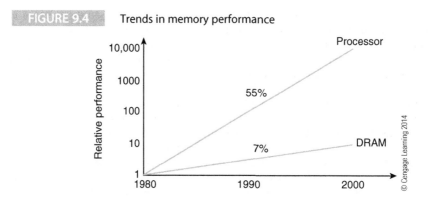

FIGURE 9.4   Trends in memory performance

the expression *"hitting the memory wall"* to suggest that there is a finite limit to progress in conventional microprocessor systems design.

Cache has no *intrinsic* value. Buses distribute data; hard disks store large volumes of data. Caches simply hide memory latency. If memories were faster, we wouldn't need cache.

9.1.1   Structure of Cache Memory

The general structure of a cache memory system is provided in Figure 9.5. A block of cache memory sits on the processor's address and data buses in parallel with the much larger main store. Data in the cache is also maintained in the main store (i.e., DRAM).

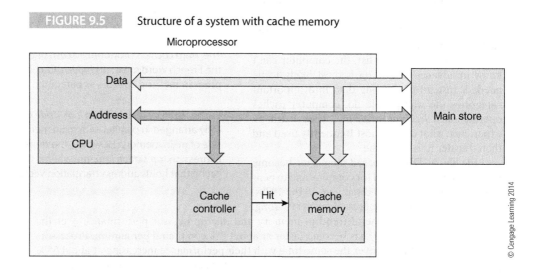

FIGURE 9.5   Structure of a system with cache memory

Principle of Locality of Reference

To return to the analogy with the telephone list, adding a friend's number to the list does not delete their number from the directory. Suppose that the computer accesses a memory location. The probability that any given memory location will be accessed is not constant, since some locations are more likely to be accessed than others. Because of the nature of programs and their data structures, the data required by a processor is often highly clustered throughout memory. For example, the stack may be accessed very regularly, and some functions are called more often than others. This phenomenon is called the *locality of reference* and makes the use of cache memory possible.

Some addresses are said to exhibit *spatial locality* because they are clustered within the same region of memory (e.g., data structures). The programmer and compiler have a considerable degree of control over special locality. Suppose a program has variables P, Q, and R, where P is an integer, R is an array of 8 integers, and Q is another integer. Moreover, suppose that P and R are accessed frequently and Q rarely. If the data is declared in the order P, R, and Q, the two frequently accessed items are next to each other in memory and may be cached together.

Some addresses are said to exhibit *temporal locality* because they are accessed over and over again within a short time span (e.g., the locations accessed within a loop). A loop such as

```
(for i = 0; i < 127; i++){
    P = P + R[i];
}
```

accesses the same variables in a regular pattern over time.

The principle of locality is a guide rather than a law. Some programs display both spatial and temporal locality, while others don't. A program with a very large matrix with data arranged at random might not exhibit spatial locality. A database of mail-order consumers indexed by geographical location may well have a high degree of spatial locality, because some communities may be frequent users of the service. A simple example of both temporal and spatial locality is given by the generation of the inner product

$$\sum a_i b_i$$

When the computer accesses a_0, a_1, a_2, a_3, ... , it is exploiting *spatial* locality, because the consecutive elements will invariably be adjacent in memory. The elements a_i and b_i may be widely separated in memory, but they are *temporally* adjacent, because the values of a_i and b_i are accessed at approximately the same time.

A cache memory uses a *cache controller* to determine whether the operand accessed by the CPU resides in the cache or whether it must be obtained from the main memory. When an address is applied to the cache controller, the controller returns a signal called *hit* that determines whether the cache access was successful or not. *Hit* is asserted if the data is currently in the cache. The logical complement of *hit* is *miss*, and a miss indicates that the data is not in the cache and that the cache must be reloaded from memory.

Modern high-performance systems now have multiple levels of cache called L1 cache, L2 cache, and L3 cache. The L1 or level 1 cache is the smallest and fastest cache. If data isn't in the L1 cache, the L2 cache is searched. If the data isn't there, the L3 cache is searched. Multiple levels of cache are cost-effective because they provide better performance without increasing the size of the fastest cache. We will return to multiple levels of cache.

9.2   Performance of Cache Memory

We need to know how much the addition of cache memory affects a computer's performance before we can decide whether adding cache memory is cost-effective. We begin with a simple model that omits the fine details of a real cache system: details that vary markedly from system to system. In particular, the model assumes that cache entries are all one word wide, whereas practical caches store a line (group of words).

The principal parameter of a cache system is its *hit ratio*, h, that defines the ratio of hits to all accesses and is determined by statistical observations of the system's operation. The effect of locality of reference means that the hit ratio is usually very high, often in the region of 98%. Later in this chapter, we examine some of the factors that affect the value of the hit ratio and discuss ways of keeping h high.

Before calculating the effect of a cache memory on a processor's performance, we need to introduce some terms.

Access time of main store	t_m
Access time of cache memory	t_c
Hit ratio	h
Miss ratio	m
Speedup ratio	S

The *speedup ratio* is defined as the ratio of the memory system's access time without cache to its access time with cache. For N accesses to memory, the total access time of a memory without cache is given by Nt_m.

For N accesses to a memory system with a cache, the total access time is given by $N(ht_c + mt_m)$. The miss ratio, m, is defined as $m = 1 - h$, since if an access is not a hit it must be a miss. Therefore, the speedup ratio for a system with cache is given by

$$S = \frac{N \cdot t_m}{N(h \cdot t_c + (1 - h)t_m)} = \frac{t_m}{h \cdot t_c + (1 - h)t_m}$$

This expression assumes that all operations are memory accesses, which is not true because processors also perform internal operations. We will return to this point later. If we are not interested in the absolute speed of the memory and cache memory, we can introduce a parameter, $k = t_c/t_m$, that defines the ratio of the speed of cache memory to main memory. The speedup ratio in terms of h and k is given by

$$S = \frac{1}{h \cdot k + (1 - h)} = \frac{1}{1 - h(1 - k)}$$

Figure 9.6 plots the speedup ratio S as a function of the hit ratio h when $k = 0.2$. As you might expect, the speedup ratio is 1 when $h = 0$ and all accesses are made to the main memory. When $h = 1$ and all accesses are to the cache, the speedup ratio is $1/k$.

The most important conclusion to be drawn from Figure 9.6 is that the speedup ratio is a sensitive function of the hit ratio. Only when h approaches about 90% does the effect of the cache memory become really significant. This result is consistent with common sense. If h drops below about 90%, accesses to main store take a disproportionate amount of time and the effect of fast accesses to the cache has little effect on system performance.

The equation for the speedup ratio, S, that we've just derived is really the speedup ratio of a *parallel* memory-cache system, because it is calculated by assuming that each memory

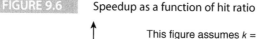

FIGURE 9.6   Speedup as a function of hit ratio

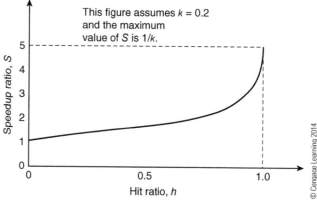

Cache Complexity

Performance calculations involving cache memory are invariably simplified because of all the various factors that affect a real system. The first cache calculations that we introduce here assume that information is obtained from either cache or memory a word at a time. In practice, whenever a data element is not in cache, we have to reload an entire *line* rather than a single entry. We also have to consider the difference between cache read and write operations, which may be treated differently.

Moreover, most high-performance systems have separate data and instruction caches at the highest level to increase the processor CPU bandwidth by allowing simultaneous instruction and data transfers. These caches are written as I-cache and D-cache, respectively.

Finally, we are assuming a simple main memory system with a single access time. In practice, modern high-performance systems with DRAM have rather complex access times because information is often accessed in a burst, and the first access of a burst of accesses may be longer than successive accesses.

cycle begins with a *simultaneous* (i.e., parallel) access to the main memory and the cache; that is, an address goes to both the cache and the main store. If a hit occurs, the access to main store is terminated. If the cache does not respond, data is returned by the main store.

The speedup ratio achieved by real microprocessors is not as optimistic as the previous equations suggest. A real microprocessor operates at a rate determined by its clock speed, the number of clock cycles per memory access, and the number of wait states introduced by the memory. There is little point in speeding up the cache memory beyond that needed to achieve zero wait states. Even if you use a very fast cache, you cannot reduce a memory access to less than that of a bus cycle without wait states.

Consider the following example.

Microprocessor clock cycle time	10 ns
Minimum clock cycles per bus cycle	3
Memory access time	40 ns
Wait states introduced by memory	2 clock cycles
Cache memory access time	10 ns
Wait states introduced by cache	0

This data tells us that a memory access takes (3 clock cycles + 2 wait states) $\times$ 10 ns = 50 ns, and an access to cache takes 3 $\times$ 10 ns = 30 ns. The actual access times of the main memory and the cache don't appear in this calculation. The speedup ratio is given by

$$S = \frac{50}{30h + 50(1 - h)} = \frac{50}{50 - 20h}$$

Assuming an average hit ratio of 95%, the speedup ratio is given by 1.61 (i.e., 161%). This figure offers a modest performance improvement but is less than that provided by calculating a speedup ratio based only on the access time of the cache memory and the main store (i.e., 2.46).

We have omitted the effect of internal operations that don't access memory. Let's look at a Texas Instruments' application note that gives the average cycle time of a microprocessor, taking account of non-memory operations as well as accesses to data in the cache or main memory store.

$$t_{\text{average cycle time}} = F_{\text{int}} \cdot t_{\text{cyc}} + F_{\text{mem}} \left[h \cdot t_{\text{cache}} + (1 - h)(t_{\text{cache}} + t_{\text{wait}}) \right]$$

where

F_{int} = fraction of time the processor spends doing internal operations
F_{mem} = fraction of time processor spends doing memory accesses
t_{cyc} = processor cycle time
t_{wait} = wait-state time caused by cache miss
t_{cache} = cache memory access time
h = hit ratio

If we put some figures into this equation, we get

$$t_{average\ cycle\ time} = 70\% \times 10\ ns + 30\% \times [0.9 \times 5\ ns + 0.1(5\ ns + 50\ ns)]$$
$$= 7\ ns + 3\ ns = 10.0\ ns$$

Even this equation doesn't tell the whole truth, because real systems don't move data between cache and main store a single word at a time. The basic unit of storage in a cache is not the word but the *line* that is composed of about 4 to 64 bytes. When a miss occurs, it results in a *line* of data being transferred from memory to the cache. Hence, there is an additional penalty associated with a miss, that is, the time taken to refill a line.

At this point perhaps, we should make a comment on the memory wall we referred to earlier. Suppose the average time, in cycles, taken by a certain system is

$$Time_{ave} = CPU_{use} \cdot t_{CPU} + Memory_{use}\ [h \cdot t_{cache} + (1 - h)(t_{memory})]$$

If the CPU spends 80% of the time accessing non-memory instructions and the CPU time is 1 cycle, the cache access time is 1 cycle, the memory access time is 10 cycles, and the hit ratio is 0.95, we get

$$Time_{ave} = 0.80 \cdot 1 + 0.20 \cdot 0.95 \cdot 1 + 0.20 \cdot (1 - 0.95) \cdot 10$$
$$= 0.80 + 0.19 + 0.10 = 1.09\ cycles$$

Suppose that, over time, the processor speeds up by a factor of ten, while the cache memory speeds up by a factor of five and the DRAM by a factor of two. The ratio of CPU:Cache:DRAM access times is no longer 1:1:10 but 1:2:50. The average time is now

$$Time_{ave} = 0.80 \cdot 1 + 0.20 \cdot 0.95 \cdot 2 + 0.20 \cdot (1 - 0.95) \cdot 50$$
$$= 0.80 + 0.38 + 0.50 = 1.68\ cycles$$

If we assume that, in the second case, the clock is running at ten times the clock in the first case, the speedup ratio is 10.9/1.68 = 6.488. The clock and CPU are ten times faster, but the throughput has increased by a factor of only six.

We revisit cache memory performance when we include the effect of write misses on the cache and when we look at the effect of misses on cache performance.

Other Ways of Looking at Performance

There are several ways of expressing the performance of cache memories. Some writers express performance in terms of miss rates and penalties. Some include CPU performance in the equations. Sometimes difference in the way in which cache equations are expressed depends on the assumptions made about the system. Below are several cache equations.

1. Memory stall cycles = memory accesses × miss rate × miss penalty
2. t_{CPU} = (CPU execution cycles + memory stall cycles) × t_{cyc}
3. AMAT = average memory access time = hit time + (miss rate × miss penalty)

Example

A computer has separate instruction and data caches. The data cache has a hit rate of 95%, and the instruction cache has a hit rate of 98%. The miss penalty (i.e., additional cycles to access main memory) is 100 cycles. The intrinsic CPI rating of the processor is 1.5 (i.e., not counting memory misses), and 25% of the instructions are loads and stores.

What is the average number of miss cycles?

This is given by misses due to the I-cache plus misses due to the D-cache; that is,

$$\text{Average memory miss cycles} = 0.02 \times 100 + 0.25 \times 0.05 \times 100 = 2.0 + 1.25$$
$$= 3.25 \text{ cycles}$$

Note that I-cache and D-cache miss ratios are 0.02 and 0.05, respectively, and the factor 0.25 in the D-cache contribution is because only 25% of instructions are memory data accesses.

What is the total CPI?

The total CPI is given by the CPU and memory components; that is, $1.5 + 3.25 = 4.75$ CPI.

What would the speedup be if (a) the cache were perfect and (b) the cache did not exist?

(a) With a perfect cache, the CPI would be that of the CPU alone, which is 1.5. The speedup would be (time with stalls)/(time without stalls) = $4.75/1.25 = 3.8$.

(b) If there were no cache, the average time per instruction would by 1.5 (CPU time) + 100 (instruction time) + 0.25 x 100 (data time) = $1.5 + 100 + 25 = 126.5$ CPI. The speedup would be $3.8/126.5 = 0.03$, which represents a 33-fold decrease in speed.

9.3   Cache Organization

We now look at the structure or organization of cache memories. If a cache holds only a tiny fraction of the available memory space, what data goes into it and where do you put it? Cache memory systems are harder to design and to integrate into computer systems than main stores, partially because of the high speed of cache memories and partially because of their complexity. Indeed, until the 1980s cache memories were seen only in minicomputers and mainframes. It wasn't until the late 1980s, with the advent of the 68020/30 and the 80386/486 generation of microprocessors, that cache memories began to appear in personal computers. Today, the ability to put more than a billion transistors on a chip ensures that all high-performance processors include a large cache memory on-chip.

The fundamental problem of cache memory design is how to create a memory that contains a set of data elements that can come from anywhere within a much larger main store. There are many ways of solving this mapping problem, although all practical cache systems use a *set associative* organization. Before we describe this, we look at two possible cache organizations that will help us to understand the operation of the set associative cache.

Example – Cache Loading

We look at DRAM in the next chapter. However, for the purpose of this example, all you need to know is that the first access in a group takes longer than successive accesses. Suppose that the DRAM in main store is clocked at 200 MHz, and it takes six clock cycles to access the first element in a burst. Successive elements in the burst are retrieved at one-clock intervals. The memory bus is 64 bits wide, and the cache memory has a line width of 64 bytes; that is, the cache is loaded with 64-byte lines.

How long does it take to load a line following a cache miss?

1. The memory is 64 bits (8 bytes) wide, and eight reads have to be made to load a cache line of 64 bytes.
2. The clock speed is 200 MHz, corresponding to a cycle time of 5 ns. The DRAM takes six cycles to access the first 64-bit word and then one cycle for each of the remaining seven words.
3. The total time is 6×5 ns $+ 7 \times 5$ ns $= 30$ ns $+ 35$ ns $= 65$ ns.

9.3.1   Fully Associative Mapped Cache

The first question we need to ask when designing any memory system is, *How large or small should the basic unit of data be?* Main memories handle data in units that are equal to the fundamental wordlength of the machine. For example, a 64-bit machine with 64-bit registers uses 64-bit memory. If the computer wishes to read less than a full word, it reads a word and then ignores the bits it doesn't want.

Although the computer can read a word from a cache memory, the word is not the basic unit of storage in a cache. The unit of storage is the *line* that contains several consecutive words. Suppose a cache were organized at the *granularity* of a word. If an instruction were accessed, and it wasn't currently in the cache, it would have to be fetched from the main store. However, the next instruction would probably cause a miss too. We need a bigger unit than the word.

A *cache line* consists of a sequence of consecutive words, allowing several consecutive instructions to be read from the cache without causing a miss. When a miss does occur, the *entire line* containing the word being accessed is transferred from memory to the cache. The optimum line size for any system depends on the total size of the cache, the nature of the code, and the structure of the data.

We would like a cache that places no restrictions on what data it can contain; that is, data in the cache can come from anywhere within the main store. Such a cache uses *associative memory* that can store data anywhere in it because data is accessed by its *value* and not its *address* (location).

Figure 9.7 illustrates the concept of an *associative memory*. Each entry has two values, a *key*

The Cache Jargon Nightmare

Computer science has a reputation for using terminology inconsistently; for example, a typical ROM memory is also a RAM, or a word is 16-bits to a 68K programmer and 32-bits to an ARM programmer.

In the world of cache memory, life gets truly difficult with different terminology used to describe the subfields of an address. For example, some call the basic unit of data stored in a cache a *line*. Others call it a *block*.

A *line* is made up of individual *words*. The location of a word in a line can be described by its *lines address* or by its *block offset*.

In direct mapped cache, the location of a line is expressed by a *set* address and a *line address* (my terminology) or by a *tag* and an *index*. The following diagram illustrates these terminologies.

tag	index	block offset
set	line	word address

© Cengage Learning 2014

FIGURE 9.7   Principle of the associative memory

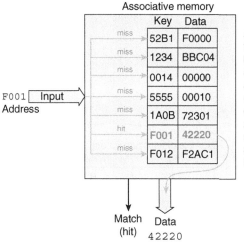

An associative memory is unordered in the sense that data can be stored in any location; that is, there is no concept of address. Each data item is identified by a key and is retrieved by its key. In this example the key is F001 and that retrieves data item 42220.

All keys in an associative memory are matched in parallel at the same time. If a key is found, the match (or hit) line is asserted.

© Cengage Learning 2014

and a data element; for example, the top line contains the key 52B1 and the data F0000. The data is not ordered in the sense that an entry can go anywhere in the memory. The key is the primary means of retrieving the data. An associative memory is accessed by applying a key to the memory's input and *matching* it with all keys in the memory in parallel. If the key is found, the data at that location (i.e., the data *associated* with the key) is retrieved. For example, suppose the computer applies the key F001 to the system in Figure 9.7. This key is applied to all locations in the memory *simultaneously*. Because a match (i.e., a *hit*) takes place with this key, F001, the memory responds by indicating a match and supplying the value 42220 at its data terminals.

An associative cache would look like the scheme in Figure 9.7 where the key would be the current address from the processor and the data would be the data stored at that location. The difference between a conventional memory and an associative memory is that a conventional memory contains a block of sequential memory elements 0, 1, 2, ..., whereas an associative memory contains a set of elements that are not ordered or sequential.

Let's look at some of the details of a fully associative cache memory. Figure 9.8 describes the associative cache that allows any line in the cache to hold data from any line in the main store. In this example, the memory is divided into lines of two words (we've used two words per line for simplicity; a real cache might have lines containing eight or more words).

An associative cache can be of any size, and there's no relationship between the number of lines in the cache and the number of lines in the main memory. Consider a system with 16 MB of main store and 64 KB of associatively mapped cache. If the size of a line is four 32-bit words (i.e., 16 bytes), the main memory is composed of $2^{24}/16 = 1$ M lines and the cache is composed of $2^{16}/16 = 4{,}096$ lines.

An associative cache permits any line in the main store to be loaded into any one of its lines. In this case, line i in the associative cache can be loaded with any one of the 1 M possible lines in the main store. Therefore, line i in the cache requires a *tag* to uniquely label it as being associated with line i from the main store. Since, in this example, there are 1 M lines in the main store, the cache tag must be 20 bits long to specify which of the 2^{20} lines it represents.

When the processor generates an address, the word bits are used to select a word location in both the main memory and the cache (in this case there are two word-select bits). The 20 line address bits $A_{23} - A_{04}$ from the CPU can't be used to select a line in the cache. Why? Because the associative memory can store any of the 1 M memory lines in any one of its 4,096

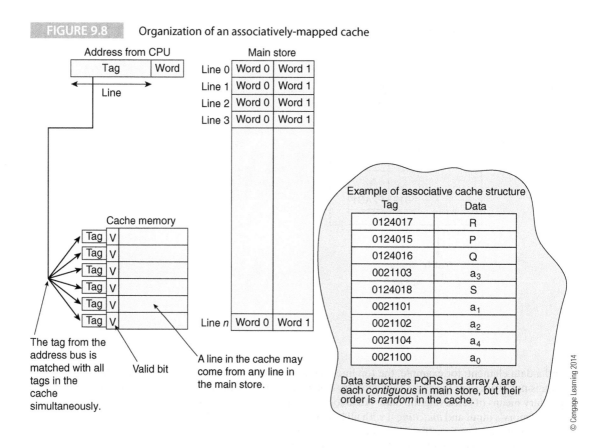

FIGURE 9.8   Organization of an associatively-mapped cache

The tag from the address bus is matched with all tags in the cache simultaneously.

Valid bit

A line in the cache may come from any line in the main store.

Example of associative cache structure

Tag	Data
0124017	R
0124015	P
0124016	Q
0021103	a_3
0124018	S
0021101	a_1
0021102	a_2
0021104	a_4
0021100	a_0

Data structures PQRS and array A are each *contiguous* in main store, but their order is *random* in the cache.

lines, and the cache doesn't know whether the line being accessed is currently in the cache. Even if it knows that the line is in the cache, it doesn't know where it is.

Figure 9.9 demonstrates why we can't find a simple solution to the problem of locating a specific line in the cache. In this example, we've used the same system with a 24-bit address bus accessing 16 MB of DRAM in the main store and a 64K-line associatively mapped cache. Each line in the cache holds 16 bytes corresponding to four 32-bit words. The total number of possible lines that can be addressed by the CPU is $2^{24}/2^4 = 2^{20}$ and a 24-bit address is made up of A_{23} to A_{04} (the line address), A_{03} to A_{02} (the word address), and A_{01} to A_{00} (the byte address). Consequently, a line requires a 20-bit tag to uniquely identify it.

How do we map one of 1 M line addresses from the CPU onto one of 4,096 lines in the cache? Figure 9.9 demonstrates one solution in which a look-up table uses a 2^{20} by 12-bit word random access memory to hold the 4,096 pointers to a line in the cache. This memory is said to be *sparse* because only 4,096 of its 1 M locations contain pointers to lines that are currently in the cache. All the other locations are empty because the corresponding line in main store is not currently in the cache. Of course, in order to implement this scheme, we would need an extra bit in each line of the look-up table to indicate whether that current line was in the cache or not (this would be a hit/miss bit).

The system of Figure 9.9 can't be implemented economically, because the memory required to hold the pointers to the cache can be as large as or larger than the main store itself! Moreover, the look-up table would have to be very fast to avoid increasing the cache access time. In this example, we have a 64 KB cache and a tag look-up table that is 1M words of 12-bits or 1.5 MB. Clearly, such a scheme is not practical. The solution is the associative memory.

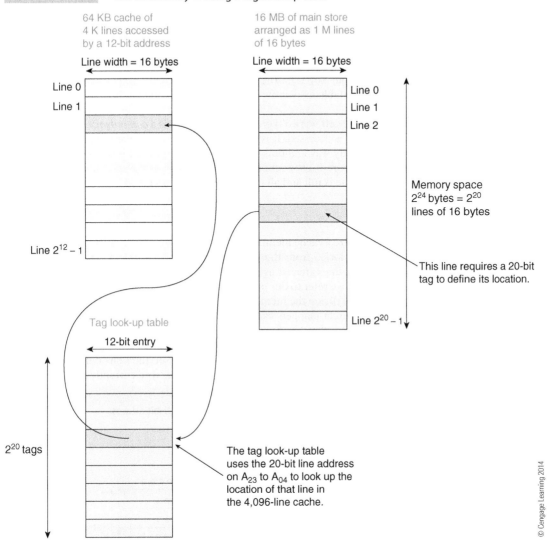

FIGURE 9.9   The infeasibility of using a tag look-up table

64 KB cache of
4 K lines accessed
by a 12-bit address

16 MB of main store
arranged as 1 M lines
of 16 bytes

Line width = 16 bytes

Line width = 16 bytes

Line 0
Line 1

Line 0
Line 1
Line 2

Line $2^{12} - 1$

Memory space
2^{24} bytes = 2^{20}
lines of 16 bytes

This line requires a 20-bit
tag to define its location.

Line $2^{20} - 1$

Tag look-up table

12-bit entry

2^{20} tags

The tag look-up table
uses the 20-bit line address
on A_{23} to A_{04} to look up the
location of that line in
the 4,096-line cache.

Associative Memory

The associative cache is so called because it uses an *associative memory* to hold the tags. An associative memory has an *n*-bit input but not necessarily 2^n unique internal locations. The *n*-bit input to the associative memory is matched with a tag field in each of its locations *simultaneously*. If the input address matches a stored tag, the data associated with that location is output. Otherwise, the associative memory produces a miss output.

Continuing the previous example, an associative cache stores all the 4,096 20-bit tags (one for each line currently cached) in an associative memory. When the CPU performs a memory access, the high-order 20 bits of the address bus are applied to the associative memory's inputs. If it contains the corresponding tag in one of its locations, it responds by providing the appropriate cache line.

Because a line from main store can be located *anywhere* within an associative cache, what happens when the cache is full? Which line must be deleted in order to make room for the new entry? A practical associative cache would mark its entries with when they were last used to let us throw away the oldest entry (or some other parameter could be used to identify the line to be ejected from the cache). Replacement algorithms for cache are similar to those used by virtual memories (see the next section) [For example, least-recently used (LRU), first-in first-out (FIFO), or random]. The least-recently used algorithm is intuitively best because it seeks to remove data that has not been accessed for the longest time (use it or lose it!). However, LRU algorithms are not easy to implement because to do so would require lines to be tagged with their time of last access.

An associative memory is expensive because it requires a large amount of parallel logic in order to match an input key (i.e., the current address) with each stored key simultaneously. Commercially available associative memories are too small to be used to make practical cache systems.

An associative cache suffers two types of miss. The first is a *compulsory miss* which takes place the first time a line is accessed. It is *compulsory* as it has to take place because the line is initially empty. The only way compulsory misses can be reduced is by preloading a cache with data before it is required. The second type of cache miss is the *capacity miss* that occurs when the associative cache is full and all lines are currently occupied.

9.3.2   Direct-Mapped Cache

The easiest way to organize a cache memory employs *direct mapping,* which relies on a simple algorithm to map data block i from the main memory into data block i in the cache. In a direct-mapped cache, the lines are arranged into units called *sets,* where the size of a set is the same size as the cache. If we refer to our previous example, a computer with a 16 MB memory and a 64 KB cache would divide the memory into 16 MB/64 KB = 256 sets.

To illustrate how direct-mapped cache works, we'll create a memory with 32 words accessed by a 5-bit address that has a cache holding eight words. The line size will be two words. From our previous definition, the number of sets is memory size/cache size = 32/8 = 4. A 5-bit address is s_1, s_0, l_1, l_0, w, where the s bits define the set, the l bits define the line, and the w bit defines the word. Figure 9.10 demonstrates how the word currently addressed by the

FIGURE 9.10   Accessing cache and main store with a 5-bit address

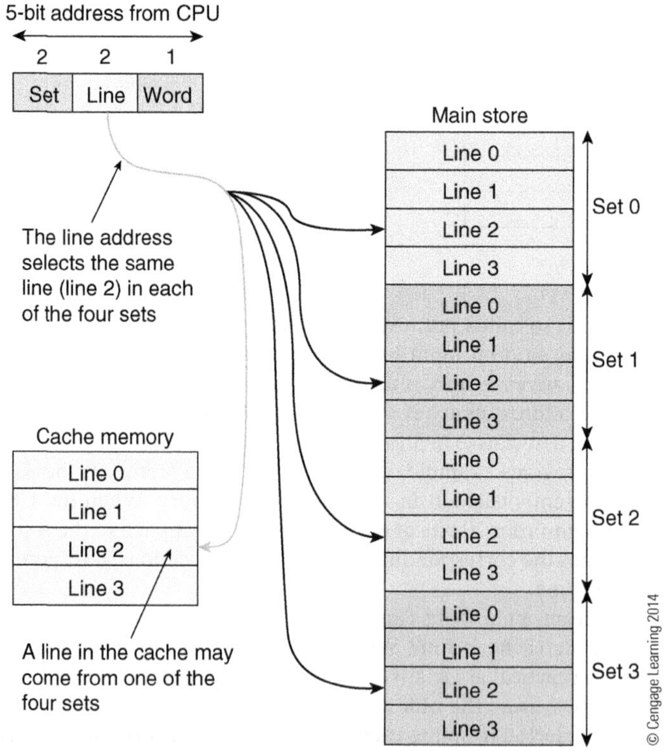

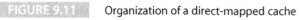

FIGURE 9.11   Organization of a direct-mapped cache

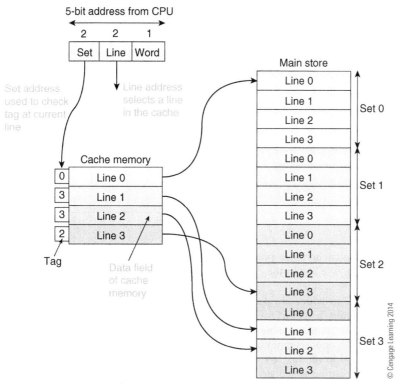

processor is accessed in memory via its set address, its line address, and its word address. For the purpose of this discussion, we need consider only the set and line as it doesn't matter how many words there are in a line.

The arrangement of Figure 9.10 is called a *direct-mapped cache* because there is a direct relationship between the location of a line in cache and the location of the corresponding line in memory. In the example of Figure 9.10, the cache memory has a 2-bit line address and therefore holds $2^2 = 4$ lines. If a direct-mapped cache has a b-bit line address field, the cache must hold 2^b lines of data.

When the processor generates an address, the appropriate line in the cache is accessed. For example, if the processor generates the 5-bit address 01100, line 2 is accessed. Figure 9.10 reveals that there are four possible lines numbered two—a line 2 in set 0, a line 2 in set 1, a line 2 in set 2, and a line 2 in set 3. Suppose that, in this example, the processor accesses line 2 in set 1. The obvious question to ask is, "How does the system know whether the line 2 accessed in the cache is the line 2 from set 1 in the main memory?"

Figure 9.11 demonstrates how the ambiguity between lines is resolved by a direct-mapped cache. Each line in the cache memory has a tag or label that identifies which set that particular line belongs to. When the processor accesses a memory location whose line address is 3, the tag belonging to line 3 in the cache is sent to a comparator. At the same time, the set field from the processor is also sent to the comparator. If they are the same, the line in the cache is the requested line, and a hit occurs. If they are not the same, a miss occurs and the cache must be updated.

When line i is accessed and a miss occurs, the old line i in the cache is either discarded or rewritten back to main memory, depending on how the updating of main memory is organized. We examine this aspect of a cache memory's memory later.

Another way of viewing a direct-mapped cache is provided by Figure 9.12 where the main store is depicted as a matrix of dimension *set* × *line*, in this case 4 lines × 4 sets. Alongside this matrix is the cache memory that has the same number of lines as the main memory. Lines

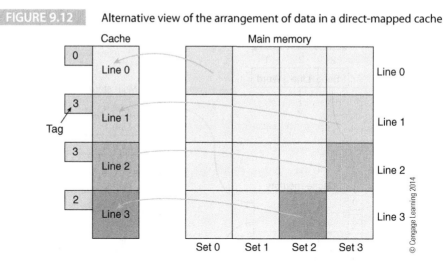

FIGURE 9.12   Alternative view of the arrangement of data in a direct-mapped cache

currently in the cache corresponding to lines in the main store are shaded. This diagram demonstrates how a line in the cache can come from any one of the sets with the same line number in the main store.

Figure 9.13 provides the skeleton structure of a direct-mapped cache memory system. The cache memory is a block of high-speed RAM that holds data. The *cache tag RAM* is a special device that contains a high-speed random access memory and a data comparator. The cache tag RAM's address input is the *line address* from the processor that accesses or *indexes into* the location in the tag RAM containing the tag for this set. The data in the cache tag RAM at this location is sent to the comparator and matched with the set address on the address bus. If the set field from the processor matches the tag of the line being accessed, the cache tag RAM returns a hit signal.

FIGURE 9.13   Structure of a direct-mapped cache

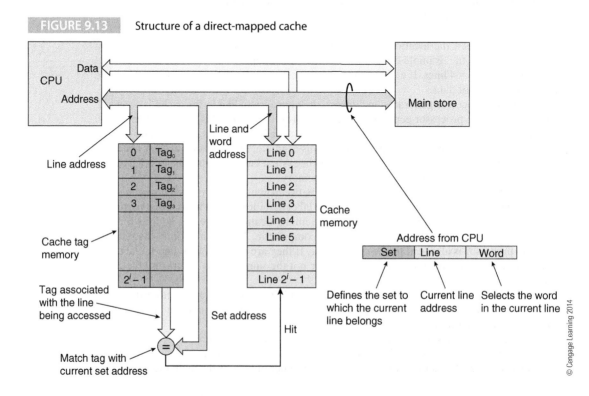

The direct-mapped cache requires no complex line replacement algorithm. If line x in set y is accessed and a miss takes place, line x from set y in the main store is loaded into the frame for line x in the cache memory. No decision concerning which line from the cache is to be rejected has to be made when a new line is to be loaded.

An advantage of direct-mapped cache is its inherent parallelism. Since the cache memory holding the data and the cache tag RAM are independent, they can both be accessed simultaneously. Once the tag field from the address bus has been matched with the tag field from the cache tag RAM and a hit has occurred, the data from the cache will also be valid.

The disadvantage of direct-mapped cache is its sensitivity to the location of the data to be cached. We can relate this to the domestic address book that has, say, half a dozen slots for each letter of the alphabet. If you make six friends whose surname begins with S, you have a problem the next time you meet someone whose name also begins with S. It's annoying because the Q and X slots are entirely empty. Because only one line with the number x may be in the cache at any instant, accessing data from a different set but with the same line number will always flush the current occupant of line x in the cache.

Even when a cache is not full, lines may have to be swapped in and out because two lines with the same number but from different sets are accessed. This situation can lead to a low cache utilization and a high miss ratio. A miss where a line is replaced even though the cache is not full is called a *conflict* miss because there is a conflict between a new line and a currently cached line. We will shortly see that the performance of the direct-mapped cache can be improved quite easily.

Although a direct-mapped cache can have a very poor performance if data is arranged badly, statistical measurements on real programs indicate that the very poor worst-case behavior of direct-mapped caches has little significant impact on their average behavior.

Figure 9.14 illustrates the operation of very simple hypothetical direct-mapped cache in a system with a 16-word main store and an 8-word direct mapped cache. Only accesses to instructions are included to simplify the diagram. This cache can hold lines from one of two sets. We've labeled cache lines 0 to 7 on the left in black. On the right we've put labels 8 to 15 in blue to demonstrate where lines 8 to 15 from memory locations are cached. The line size is equal to the wordlength and we run the following code.

```
        LDR    r1,[r3]      ;Load r1 from memory location pointed at by r3
        LDR    r2,[r4]      ;Load r2 from memory location pointed at by r4
        BL     Adder        ;Call a subroutine
        B      XYZ          ;
Adder   ADD    r1,r2,r1     ;Add r1 to r2
        MOV    pc,lr        ;Return
```

Strange But True – Direct-Mapped Cache Can Beat Associatively-Mapped Cache

Suppose we have a tiny direct-mapped cache with four lines. Imagine a loop in which accesses are PQRST (five accesses). The memory access sequence is PQRSTPQRSTPQRSTPQ...

It is impossible to cache the whole sequence. P,Q,R,S are cached and then T replaces P to leave T,Q,R,S. On the next iteration, P replaces T and we have P,Q,R,S again. Once round the loop results in the hit/miss sequence h,h,h,m,m.

If the same cache were fully associative using a LRU algorithm, then the sequence would be P,Q,R,S. The next element would replace the oldest element in the cache P to give T,Q,R,S. Now, the next element of the second trip would be P. Using the LRU algorithm, P would displace Q, Each new element would result in a miss and the fully associative cache would achieve endless misses.

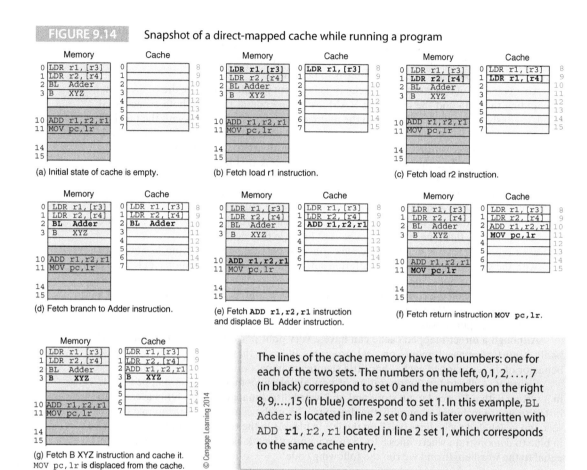

FIGURE 9.14   Snapshot of a direct-mapped cache while running a program

(a) Initial state of cache is empty.

(b) Fetch load r1 instruction.

(c) Fetch load r2 instruction.

(d) Fetch branch to Adder instruction.

(e) Fetch ADD r1,r2,r1 instruction and displace BL Adder instruction.

(f) Fetch return instruction MOV pc,lr.

(g) Fetch B XYZ instruction and cache it. MOV pc,lr is displaced from the cache.

© Cengage Learning 2014

The lines of the cache memory have two numbers: one for each of the two sets. The numbers on the left, 0,1, 2, ..., 7 (in black) correspond to set 0 and the numbers on the right 8, 9,...,15 (in blue) correspond to set 1. In this example, BL Adder is located in line 2 set 0 and is later overwritten with ADD r1,r2,r1 located in line 2 set 1, which corresponds to the same cache entry.

Figure 9.14 shows only instruction fetch cycles. Figure 9.14a shows the initial state of the system. Figures 9.14b to d show the fetching of the first three instructions, each of which is loaded into a consecutive cache location. When the subroutine is called in Figure 9.14d, a branch is made to the instruction at location 10. In this direct-mapped cache, line 10 is the same as line 2. Consequently, in Figure 9.14e, the ADD overwrites the B instruction in line 2 of the cache. This is called a *conflict miss* because it occurs when data can't be loaded into a cache because its target location is already occupied.

In Figure 9.14f, the MOV **pc, lr** instruction in line 11 is loaded into line 3 of the cache. Finally, in Figure 9.14g, the return is made and the B   XYZ instruction in line 3 is loaded into line 3 of the cache, displacing the previous cached value.

Figure 9.14 demonstrates that, even in a trivial system, elements in a direct mapped cache can be easily displaced. If this fragment of code were running in a loop, the repeated displacement of elements in the cache would degrade the performance.

Example—Cache Size

A four-way set-associative cache uses 64-bit words. Each cache line is composed of four words. There are 8192 lines. How big is the cache?

1.   The cache has lines of four 64-bit words, that is, 32 bytes/line.
2.   There are 8192 lines giving $8192 \times 32 = 2^{18}$ bytes per direct-mapped cache (256 KBs).
3.   The associativity is four which means there are four direct-mapped caches in parallel, giving $4 \times 256K = 1$ Mbyte of cache memory.

9.3.3   Set-Associative Cache

The direct-mapped cache we've just described is easy to implement and doesn't require a line-replacement algorithm. However, it, doesn't

FIGURE 9.15   Organization of set-associative cache

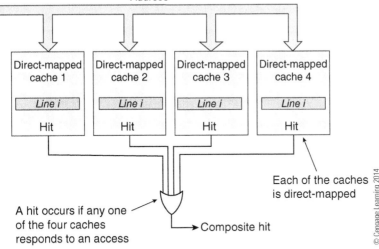

allow two lines with the same number from different sets to be cached at the same time. The fully associative cache places no restriction on where data can be located, but it requires a means of choosing which line to eject once the cache is full. Moreover, any reasonably large associative cache would be too expensive to construct. The *set-associative* cache combines the best features of both these types of cache and is not expensive to construct. Consequently, it is the form of cache found in all computers.

A direct-mapped cache has only one location for each line i. If you operate two direct-mapped caches in *parallel*, line i can go in either cache. If you have n direct-mapped caches operating in parallel, line i can go in one of i locations. That is an n-way set-associative cache.

In an n-way set-associative cache, there are n possible cache locations that a given line can be loaded into. Typically, n is in the range 2 to 8. Figure 9.15 illustrates the structure of a four-way set-associative cache that consists of four direct-mapped caches operated in parallel. In this arrangement, line i can be located in any of the four direct-mapped caches. Consequently, the chance of multiple lines with the same line number leading to a conflict is considerably reduced. This arrangement is *associative* because the address from the processor is fed to each direct-mapped cache in parallel. However, instead of having to perform a simultaneous search of thousands of memory locations, only two to eight direct-mapped caches have to be accessed in parallel. The response from each cache (i.e., a hit) is fed to an OR gate that produces a hit output if any cache indicates a hit.

Figure 9.16 repeats the example of Figure 9.15 with a set-associative cache. Everything is the same except that the direct-mapped cache has only four lines, but there are two caches making the total eight lines as before. This is a two-way set-associative cache, where a line may be cached in the upper (light blue) or lower (dark blue) direct-mapped cache.

Everything is the same until we get to Figure 9.16e when instruction ADD **r1**, r2, r1 at address 10 is mapped onto line 2 (set size 4), which is currently occupied by the BL Adder. The corresponding location in the *second* cache in the associative pair is free, and therefore, the instruction can be cached in location 2 of the lower cache without ejecting line 2 from the upper cache. In Figure 9.16f, the MOV **pc**, lr has a line 3 address and is cached in the upper cache. However, when the B XYZ instruction in line 3 of the main memory is executed, line 3 in the upper cache is taken, and it is placed in line 3 in the lower cache.

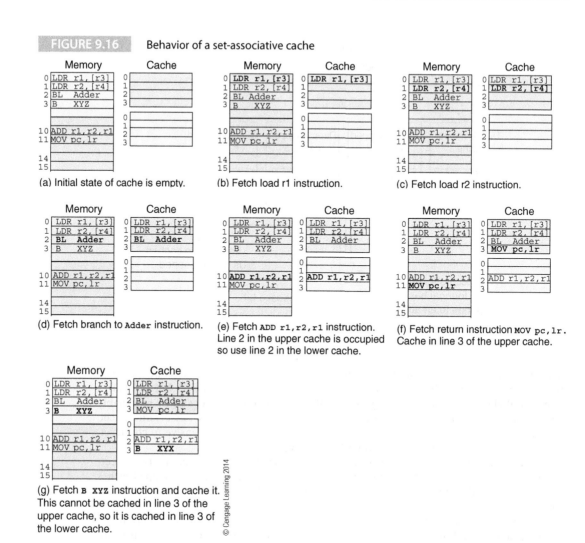

FIGURE 9.16   Behavior of a set-associative cache

(a) Initial state of cache is empty.

(b) Fetch load r1 instruction.

(c) Fetch load r2 instruction.

(d) Fetch branch to `Adder` instruction.

(e) Fetch `ADD r1,r2,r1` instruction. Line 2 in the upper cache is occupied so use line 2 in the lower cache.

(f) Fetch return instruction `MOV pc,lr`. Cache in line 3 of the upper cache.

(g) Fetch `B XYZ` instruction and cache it. This cannot be cached in line 3 of the upper cache, so it is cached in line 3 of the lower cache.

© Cengage Learning 2014

Table 9.1 from IDT Application note AN-07 (*Cache tag RAM chips simplify cache memory design*) demonstrates the effect of cache organization on the miss ratio. The miss ratio has been normalized by dividing it by the miss ratio of a direct-mapped cache in order to demonstrate the results relative to a direct mapped-cache. A four-way set-associative cache is about 30% better than a direct-mapped cache. Increasing the associativity makes little further improvement on the cache's performance.

Figure 9.17 from a Freescale Semiconductor application note demonstrates the relationship between associativity and hit rate for varying cache sizes for a GCC complier. As you can see, the degree of associativity is a significant factor at only very small cache sizes. Once caches reach 256 KB, the effect of associativity becomes insignificant.

TABLE 9.1   Effect of Cache Organization on Miss Ratio

Cache Organization	Normalized Miss Ratio
Direct-mapped	1.0
Two-way set associative	0.78
Four-way set associative	0.70
Eight-way set associative	0.67
Fully associative	0.66

© Cengage Learning 2014

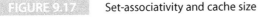

FIGURE 9.17   Set-associativity and cache size

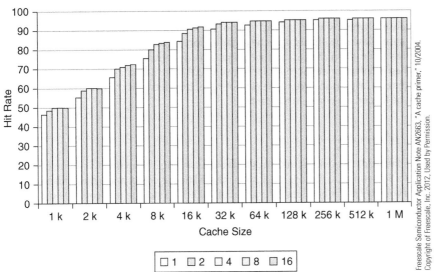

Categories of Miss

When calculating the efficiency of a cache system, we are interested in the hit rate, because it's the level of *hits* that make the cache effective. When designing cache systems or trying to improve cache systems, we are interested in the *miss ratio*— rather than the hit ratio—because there is only one source of hits (the data was in the cache), whereas there are several sources of misses. We can improve cache performance by asking, "Why wasn't the data that resulted in the miss not already in the cache?"

Cache misses are divided into three classes: *compulsory*, *capacity*, and *conflict*. The compulsory miss is so called because it cannot be avoided. A compulsory miss occurs because of the inevitable miss on the first access to a block of data. Some processors provide a means of avoiding the compulsory miss by anticipating an access to data and bringing it into cache before it is required. When the data is eventually accessed, a compulsory miss doesn't take place because the data was already in the cache in advance of its first access. This is a form of *pre-fetching* mechanism.

Another cause of a miss is the *capacity miss*. In this case, a miss takes place because the working set (i.e., the lines that make up the current program) is larger than the cache, and all of the required data cannot reside in the cache. Consider the initial execution of a program. All of the first misses are compulsory misses because the cache is empty, and new data is cached each time the address of a previously uncached line is generated. If the program is sufficiently large, there comes a point when the cache is full and the next access causes a capacity miss. Now the system has to load new data into the cache and eject old data to make room.

(continued)

A third form of cache miss is the *conflict miss*. This is the most wasteful type of miss because it happens when the cache is not yet full, but the new data has to be rejected because of the side effects of cache organization. A conflict miss occurs in an m-way associative cache when all m associative pages already contain a line i, and a new line i is to be cached. Conflict misses account for between 20% and 40% of all misses in direct mapped systems. A fully associative cache cannot suffer from a conflict miss, because an element can be loaded anywhere in the cache.

Cache Pollution

A cache holds frequently used data to avoid retrieving it from the much slower main store. Sometimes, an access results in a miss, a line is ejected, and a new line is reloaded. This line may never be accessed again, but it takes up storage that could be used by more frequently accessed data. We call this effect *cache pollution*.

Cache Considered Important

The photomicrograph from Intel below is an image of the Core i7 die. This is an advanced processor with four individual processors (we cover multiple processors in Chapter 13). The section of the processor at the bottom with the regular pattern is the level 3 cache shared by the four processors. What this photograph demonstrates is that chip manufacturers consider cache as a very important component of a processor. Here, the chip area devoted to cache is greater than that of a single processor.

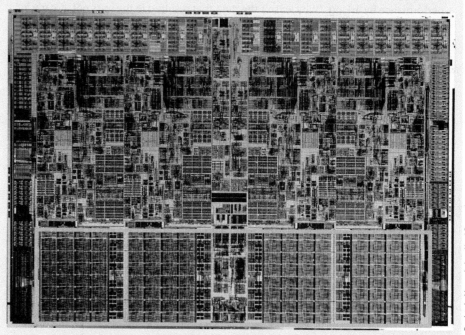

9.3.4   Pseudo-Associative, Victim, Annex and Trace Caches

Because cache memory is so important, considerable research has been carried out into finding ways to improve it, particularly into ways of dealing with anomalous behavior (for example, when an item is cached, ejected, and cached again,…).

A variation on the direct-mapped cache is the *pseudo-associative* cache. This uses a direct-mapped cache but gives conflict misses a second chance by finding *alternative accommodation*. The sidebar discusses the nature of *conflict misses* and other types of misses in greater detail. When a direct-mapped cache returns a conflict miss, the associative cache makes another attempt to store the data using a new address generated from the old address. Typically, the new address is obtained by inverting one or more high-order bits of the current address. Although this is an ingenious way of bypassing the direct-mapped cache's limitation, it does require a second cache access following an initial miss.

A *victim cache* is a small cache that holds items recently expelled from the cache (i.e., the *victims*). The victim cache is accessed in parallel with the main cache and is, ideally, fully associative. Because it is so small, it is possible to construct a fully associative victim cache because the number of entries that are searched simultaneously is very small.

A small victim cache reduces the conflict miss rate of a direct-mapped cache because it can hold an item expelled by another item with the same line number. The victim cache can also be used when the main cache is full and capacity misses are being generated. The victim cache holds data that has been expelled from the main cache, and therefore, it does not waste space, because data is not duplicated both in the main cache, and the victim cache.

A typical example of the application of a victim cache is in nested loops.[2] Consider a loop that calls a procedure where the start of the loop and the procedure are *some distance apart*. The loop begins, and the procedure is called. The cache is full, and lines must be flushed to make room for new instructions to be cached. When a return to the loop is made, the cache is flushed yet again, and so on. A victim cache can hold the flushed instructions and ensure that they are held when the loop and function call sequence is being executed. Jouppi's work on the victim cache demonstrates that it can be remarkably effective even at very small sizes. Some benchmarks demonstrate that 80% of conflict misses are removed when using a victim cache with as few as four entries. When Jouppi looked at the total reduction of misses due to a victim cache, the results varied widely depending on the benchmark. Averaged over the benchmarks, the improvement in the miss rate was about 15% for a four-entry victim cache, with one benchmark producing a large 70% reduction.

A modification of the victim cache uses *selective victim caching* where entries to the victim cache may come from either lines expelled from the cache or incoming lines from memory. When a new line is fetched, a prediction algorithm is used to determine whether it should be loaded in the cache or the victim cache. The purpose of the prediction is to try to avoid polluting the cache with lines that are unlikely to be used. The prediction mechanism requires that lines have state information associated

Stream and Stride Buffers

Another caching mechanism is the *stream buffer,* which exploits the principle of locality beyond the length of a line. When a miss occurs, the stream buffer prefetches data beyond the current line. If the processor later accesses this data, it fetches it from the stream buffer. A stream buffer can reduce compulsory misses by fetching data before it is referenced.

The price paid for stream buffers is the increase in memory bus traffic caused by prefetching data that may never be required.

The *stride buffer* is a related mechanism, except that the stride buffer exploits data patterns in structures such as arrays. In ordered data structures, the next and successive elements are likely to be fetched from a fixed offset from the previous element (e.g., at addresses $X, X + 8, X + 16, X + 24, …$). The stride buffer operates by prefetching the next value at the appropriate offset from the current value.

---

[2] Norman P. Jouppi, "Improving Direct-Mapped Cache Performance by the Addition of a Small Fully-Associative Cache and Prefetch Buffers," WRL Technical Note TN-14 Digital Western Research Laboratory.

with them that record the line's history. Using a two-bit predictor, one bit is used to indicate that the line was never accessed last time it was in the cache, and another bit is used to provide inertia and prevent excessive switching between cache and victim cache.

Another special cache proposed by John and Subramanian[3] is the *annex cache*. Like the victim cache, the annex is a special-purpose cache that sits at the mouth of the level 1 cache. The victim cache sits at the exit, and the annex cache sits at the entrance. Whereas the victim cache gives data flushed from a cache a second chance, the annex cache requires that data that wants to go into the cache has to prove its worthiness.

Annex cache helps to reduce cache pollution by preventing data from entering the cache that is rarely accessed. A cache is operating inefficiently if frequently used data is expelled to make room for an item that is never accessed again. On startup, all entries enter the cache in the normal way. After the initial phase, all data loaded into the cache comes via the annex. A line from the annex cache is swapped into the main cache only if it has been referenced twice after the conflicting line in the main cache was referenced. Data is admitted into the annex cache only if it has demonstrated a right of residence indicated by temporal or spatial locality. Sophisticated cache mechanisms, such as the annex cache, are not easy to implement in practice, because adding layers of complexity requires a lot of control circuitry to be added to the inherently simple set associative cache.

Trace Cache

The *trace cache* is a special-purpose cache devised by Intel. Trace cache first appeared in the IA32 family with the Pentium 4. In fact, the Pentium 4's trace cache (which replaces level 1 cache) is only 8 Kbytes in contrast with the earlier Pentium III's 16 Kbyte level 1 cache.

Trace cache goes a step further than the conventional cache and *unpacks* instructions; that is, it stores *decoded* instructions. The trace cache, therefore, performs not only the role of a cache but can reduce execution time by cutting out parts of the instruction decoding process. The Pentium 4 execution trace cache supplies the pipeline with six micro-operations every two clocks.

The Pentium 4 trace cache has special features built in to it to deal with the IA32's convoluted multilength instruction architecture. You could say that a trace cache is a form of *intelligent cache*. The IA32 and the 68K architectures both support very long instruction formats that have to be decoded into tens or even hundreds of micro-operations. Because it would not take too many of these long instructions to fill the trace cache, Intel has provided a simple solution. Complex op-codes are decoded and their micro-operations stored in ROM. When a long instruction is encountered, the micro-operations are not cached. Instead, a call to the appropriate procedure in the microcode ROM is inserted in the trace cache.

The Pentium 4 trace cache also includes a limited amount of branch prediction logic that operates over the scope of the trace cache itself and is independent of the Pentium's normal front-end branch target address predictor. The trace cache branch predictor takes a decoded branch instruction and gets the decoded micro-operations at its predicted target address (assuming they are in the trace cache).

[3] L. John, A. Subramanian, Annex cache: "A cache assist to implement selective caching." *Microprocessors and Microsystems*, Vol. 23, Issues 8–9, December 1999.

9.4   Considerations in Cache Design

As we have already stated, the design of cache memories is complicated because there are many factors to take into account, some of which depend on the precise nature of the computer system. In this section, we look at some of the factors that affect the design of a cache system.

9.4.1   Physical versus Logical Cache

In a computer system with a memory management unit, the cache memory can be located either between the CPU and the MMU or between the MMU and physical memory. Figure 9.18 describes these two alternatives. If the data at the CPU's data terminals is cached, the data is *logical data*, and the cache is a *logical cache*. However, if the data is cached after address translation has been performed by the MMU, the data is *physical data*, and the cache a *physical cache*. We now describe the implications of logical and physical caches and discuss the trade-off between them.

Memory Management

We cover memory management later in this chapter. Here, we need to provide a brief overview.

Virtual memory is a mechanism that loads data from disk into the main store when it is needed by the processor. It relies on a *memory management unit* (MMU) to translate addresses generated by the CPU into the addresses of the corresponding operands in main store. In computers with memory management, logical addresses of operands are generated by the computer and are translated into the physical addresses used by memory. Because an address from the CPU is translated into the location of the data in memory, it is possible to move data about in memory without changing code as long as the MMU changes the logical to physical mapping.

A physical data cache has a longer access time than a logical cache because the data cannot be accessed until the MMU has performed a logical-to-physical address translation. A logical cache is faster than a physical cache because data can be accessed in the cache without having to wait for an address translation.

Suppose that in a multitasking system a context switch occurs and a new task is executed. When the new task is set up, the operating system loads the appropriate address translation

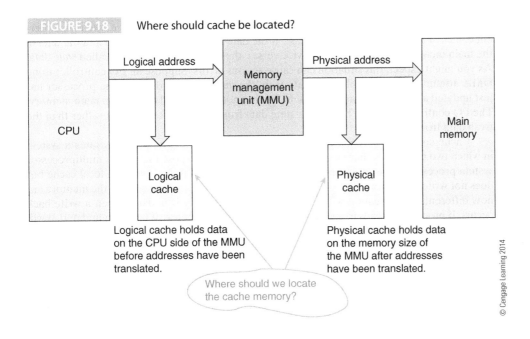

FIGURE 9.18   Where should cache be located?

Logical cache holds data on the CPU side of the MMU before addresses have been translated.

Physical cache holds data on the memory size of the MMU after addresses have been translated.

Where should we locate the cache memory?

© Cengage Learning 2014

table into the MMU. When the logical-to-physical address mapping is modified, the relationship between the cache's data and the corresponding physical data is broken; the data in the cache cannot be used and the logical cache has to be flushed. A physical cache doesn't have to be flushed on such a context switch.

However, the penalty you pay for a physical cache is the additional time required to perform the logical-to-physical address translation before beginning the memory access. In practice, if you make the cache page the same size as a memory page, you can perform a line search to the cache in parallel with the virtual address translation. Microprocessors generally use physical cache in order to reduce the need to flush the cache after a context switch.

9.4.2   Cache Electronics

Although we look at main memory systems in the next chapter, we must make a comment about the circuit design of cache memory here. There are two major classes of semiconductor random access memory: *static* and *dynamic*. Static memory uses conventional digital logic to store a single-bit in a flip-flop—very much as we described in Chapter 2. A static memory is characterized by its low power consumption, high speed, and ability to retain data as long as the power is maintained. Consequently, cache memories are usually constructed with static RAM. Unfortunately, it takes six transistors to store one bit, and therefore, a static memory of a given physical size (i.e., the area of silicon it takes up) is much bigger than a DRAM cell. This means that static memory is more expensive and has a smaller capacity than DRAM, and we can't build very large cache memories.

Dynamic memory (DRAM) stores data as a charge on a capacitor in a single transistor cell. This makes DRAM very cheap and compact, and we can construct very large memories. Unfortunately, DRAM requires a lot more electronics to control it, and the charge on the capacitor leaks away in a few milliseconds. In order to retain data in a DRAM, a memory cell must be periodically read and the data written back to it every 4 ms or so. DRAM is not suited to cache memory construction.

9.4.3   Cache Coherency

Data in the cache also lives in the main memory. When the processor modifies a data element by executing a write cycle, it must modify both the copy in the cache and the copy in the main memory, although not necessarily at the same time. Circumstances may arise when two different copies of the same data element exist. If the data in the cache is modified and the data in the main memory is not modified (or vice versa), the old unchanged data is called *stale* data. As you might expect, this situation can cause serious errors. Suppose an I/O controller using DMA attempts to move a block of data from the main store to disk and the processor has just updated a copy of the data in its cache but has not yet updated the copy in main memory. The I/O controller will then be moving state data from the main store to disk rather than the fresh data from the cache.

Cache coherency is sometimes called *data consistency*. Figure 9.19 illustrates a system in which two processors share a common memory block. Suppose that in this multiprocessor system processor 1 executes a memory write operation and updates its own local cache but does not write to memory. The copy of the data in the cache and the copy in the memory are now different. This situation will continue until the memory is updated when a write back occurs. If processor 2 reads from the same memory location before it has been updated, it will access old or stale data from the memory.

A similar problem occurs when several processors have their own local cache memories. Suppose processor X updates both its own local cache and the common memory. Processor Y may keep a cached version of the same data in its own cache. Processor Y does not know that its cached data is now stale. The term cache coherency implies that the data in the various cache and main memories are all in step (i.e., there is no stale data). Keeping data in the

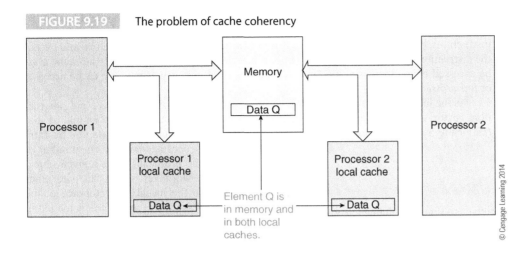

FIGURE 9.19   The problem of cache coherency

cache and main memory in step (i.e., ensuring cache coherency) is one of the principal design considerations of multiprocessor systems.

Some processors ensure cache coherency by means of a technique called *bus snooping*. A processor monitors the bus along which addresses and data flow and detects write accesses to locations in the main memory that it has a copy of in its own cache. When cached data in main memory is modified, the contents of the processor's local cache can be marked as invalid, or its own cache can be updated. We return to this topic in Chapter 13 when we cover multiprocessor systems.

9.4.4   Line Size

The *line* is the basic unit of storage in a cache memory. An important question to ask is *how big should a line be for optimum performance*? A lot of work has been carried out into the relationship between line size and cache performance, sometimes by simulating the operation of a cache in software and sometimes by monitoring the operation of a real cache in a computer system.

The optimum size of a line is determined by several parameters, not least of which is the nature of the program being executed. The bus protocol governing the flow of data between the processor and memory also affects the performance of a cache. A typical computer bus transmits an address to its main memory and then sends or receives a data word over the data bus—each memory access requires an address and a data element. Suppose that the bus can operate in a *burst mode* by sending one address and then a burst of consecutive data values. Clearly, such a bus can handle the transfer of large line sizes better than a bus that transmits a data element at a time. Another factor determining optimum line size is the instruction/ data mix. The optimum line size for code may not necessarily be the same as the optimum line size for data.

> ### Miss Ratio
>
> We have talked about cache memories in terms of *hit ratio*, which is the fraction of accesses that find the data in the cache. Typical hit ratios are in the region of 0.90 to 0.98.
>
> The miss ratio represents the fraction of accesses where the data is not cached and is given by $m = 1 - h$. Clearly, it doesn't really matter whether we use h or m. However, when comparing two high-performance caches, the value of h may be 0.98 or 0.99. The difference between these is only 1%. Now, if you look at the corresponding miss ratios, the value of m is 0.02 and 0.01, respectively, and there is a 100% difference between them. Consequently, computer scientists find it more illuminating to work with m.

Suppose that the line size is very small. CISC microprocessors like the Intel IA32 family have variable-length instructions ranging from two bytes to 10 or more bytes. With very long instructions, it's possible for part of the current instruction to be cached in one line and part of the instruction to be cached in another line. When such an instruction is read, the cache must be accessed twice. Increasing the line size reduces the frequency of multiple cache accesses or *line crossers*.

As the line size is increased, a cache's efficiency rises, because a data object (e.g., instruction, vector, or list) is composed of a group of consecutive bytes and the principle of spatial locality is better exploited. However, as the line size continues to increase, the hit ratio eventually falls, because reducing the number of lines reduces the probability that a given object will be cached. Moreover, a large line size relies very heavily on the locality of reference of the data. When a miss occurs and a line is loaded in the cache, it may not contain frequently accessed data, yet it may displace a line that is often accessed.

Cache and the Programmer

The way in which code is written has a significant influence on cache performance. In a C array, elements are stored in row order. For example x[0,1] is next to x[0,0]. Accessing elements in row order makes good use of spacial locality because one element accessing element x[i,j] causes element x[i,j+1] to be cached if its in the same line. Accessing elements in column order may cause a miss on each access because consecutive elements do not fall in the same line.

Figure 9.20 illustrates the relationship between line size and cache size for data accesses. These classic results are obtained by simulation and they relate to a period when cache capacities were vastly smaller than today. The *miss ratio*, $1 - h$, is plotted in Figure 9.20 rather than the hit ratio. Each trace corresponds to a particular cache size (from 32 bytes to 32 KB). Figure 9.21 provides the corresponding results for an instruction cache. The miss ratio of the data cache first gets better (i.e., decreases) then worsens (i.e., grows) as the line size approaches the size of the cache itself, whereas the miss ratio of an instruction cache increases with line size. This effect demonstrates that locality of reference applies more strongly to instructions than to data. In general, there is an optimum line size for a given cache size; for a 256-byte cache, it is 64 bytes. The larger the cache, the larger the optimum block size.

FIGURE 9.20   Average miss ratio for a data cache as a function of line size (each trace corresponds to a given cache size).

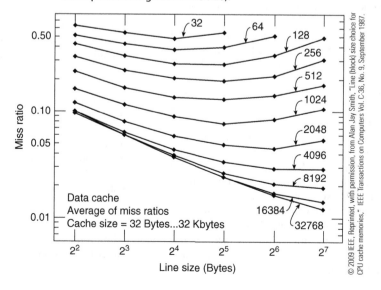

© 2009 IEEE. Reprinted, with permission, from Alan Jay Smith, "Line (block) size choice for CPU cache memories," IEEE Transactions on Computers Vol. C-36, No. 9, September 1987.

Average miss ratio for an instruction cache as a function of line size (each trace corresponds to a given cache size)

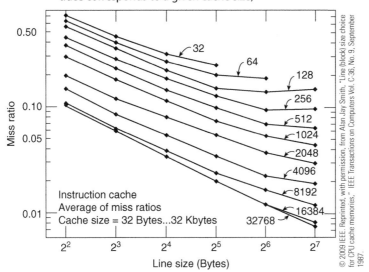

9.4.5   Fetch Policy

Several strategies can be used for updating the cache following a miss (for example, demand fetch, prefetch, selective fetch).[4] The *demand fetch* strategy retrieves a line following a miss and is the simplest option. The *prefetch strategy* anticipates future requirements of the cache (e.g., if line $i + 1$ is not cached, it is fetched when line i is accessed). There are many ways of implementing the prefetch algorithm. The *selective fetch* strategy is used in circumstances when parts of the main memory are non-cacheable. For example, it is important not to cache data that is shared between several processors in a multiprocessor system—if the data were cached and one processor modified the copy in memory, the data in the cache and the data in the memory would no longer be in step. We will return to this topic later in Chapter 13.

If you want your data sooner, perhaps you should go and get it earlier. This maxim can be applied to cache systems by *prefetching* data. Some microprocessor instruction sets include a *prefetch* instruction that generates an operand address but does nothing else. The operand appears on the bus, and the cache system automatically caches the data at that address. The instruction does nothing else and is a dummy operation intended to trigger a prefetch.

If, a few instructions later, you access the data at the prefetch address, the corresponding data is already in the cache. This prefetching operation can be done manually by the programmer or automatically as part of the compiler's optimisation pass. Prefetching is not an exact science. If you delay prefetching, the data will not be in the cache when the CPU needs it. On the other hand, if you prefetch data too early, the cache may throw out the data to make room for a new line before the CPU has had chance to access the cached data. Such an early prefetch that is flushed before the data is used is an example of *cache pollution*.

Prefetching is most closely associated with loops because they are control structures that are repeated and you know what data you are going to need in advance. The simplest mode of prefetching is to include a prefetch address ahead of an array element access. Consider the evaluation of the expression $s = \Sigma a_i$. The corresponding code is

```
for (i = 0; i < N; i++){
    S = a[i] + S;
}
```

---

[4]S. P. VanderWiel and D. J. Lilja, "When caches aren't enough: data prefetching techniques," *IEEE Computer*, July 1997, pp. 23–30.

Following the example of Wiel and Lilja,[5] we will use the construct `fetch (&address)` to indicate a prefetch operation that issues an address. The simplest example of prefetching is to call the address before it is used in the loop; that is

```
for (i = 0; i < N; i++) {
    fetch (&a[i + 1]);                  /* perform the prefetch */
    S = a[i] + S;
}
```

We generate the address of the *next* reference so that the item at location i + 1 has been referenced by the time we go round the loop again. We can improve on this code in two ways. First, the initial element is not prefetched, and second, the loop is inefficient because there is only one active operation per cycle. Consider the following:

```
    fetch (&a[0]);                      /* prefetch the first element */
for (i = 0; i < N; i = i + 4) {
    fetch (&a[i + 4];                   /* perform the prefetch */
    S = a[i]   + S;
    S = a[i+1] + S;
    S = a[i+2] + S;
    S = a[i+3] + S;
}
```

In this case, four operations are carried out per cycle. We need do only one prefetch, because the line of data loaded into the cache on each fetch contains the 16 bytes required to store four consecutive elements.

9.4.6   Multi-Level Cache Memory

In the late 1990s, memory prices tumbled, semiconductor technology let you put very complex systems on a chip, and clock rates reached 500 MHz (a cycle time of only 2 ns). Cache systems increased in size and complexity, and computers began to implement two-level caches with the first-level cache in the CPU itself and the second-level cache on the motherboard. A two-level cache system uses a small amount of very high-speed, the L1, cache and a larger amount of fast memory to hold a second level, L2, cache. In other words, there are two cache memories in series: L1 and L2. The faster, smaller cache, L1, is searched first. If it does not hold the required data, the larger but slower L2 cache is searched. If that also does not have the data, main memory is accessed. The access time of a system with a two-level cache is made up of the access time to the L1 cache plus the access time to the L2 cache plus the access time to main store; that is,

$$t_{ave} = h_1 t_{c1} + (1 - h_1) h_2 t_{c2} + (1 - h_1)(1 - h_2) t_m$$

where h_1 is the hit ratio of the L1 cache and t_{c1} is the access time of the L1 cache. Similarly, h_2 and t_{c2} refer to the L2 cache. We obtain this expression by summing the probabilities:

$$t_{ave} = \text{access time to L1 cache} + \text{access time to L2 cache} + \text{access time to main store}$$

The access time to the L1 cache is $h_1 t_{c1}$. If a miss takes place at the L1 cache, the time taken accessing the L2 cache is $(1 - h_1) h_2 t_{c2}$ if a hit at L2 occurs. If the data is in neither cache, the access time to memory takes $(1 - h_1)(1 - h_2) t_m$. The total access time is, therefore,

$$t_{ave} = h_1 t_{c1} + (1 - h_1) h_2 t_{c2} + (1 - h_1)(1 - h_2) t_m$$

This equation is simplified because it doesn't take account of cache writeback and cache reload strategies. Consider the following example. A computer has an L1 and an L2 cache. An access to the L1 cache incurs no penalty and takes 1 cycle. A hit to the L2 cache takes 4 cycles. If the data is not cached, a main store access, including a cache reload, takes 120 clock

---

[5] S.P. Vander Wiel and D.J. Lilja, "When caches aren't enough: data prefetching techniques," *Computer*, July 1997, pp. 23–30.

FIGURE 9.22   Hit rate as a function of L1 and L2 size

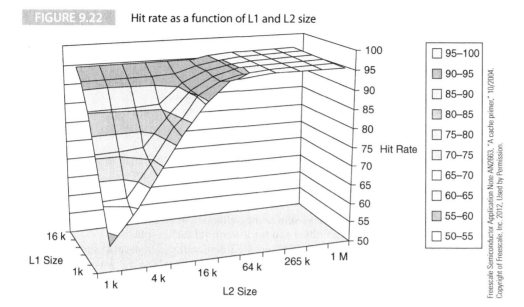

Freescale Semiconductor Application Note AN2663, "A cache primer," 10/2004.
Copyright of Freescale, Inc. 2012, Used by Permission.

cycles. If we assume that the hit rate for the L1 cache is 95% and the subsequent hit rate for the L2 cache is 80%, what is the average access time?

$$t_{\text{ave}} = h_1 t_{c1} + (1 - h_1) h_2 t_{c2} + (1 - h_1)(1 - h_2) t_m.$$
$$t_{\text{ave}} = 0.95 \times 1 + (1 - 0.95) \times 0.80 \times 4 + (1 - 0.95) \times (1 - 0.8) \times 120$$
$$= 0.95 + 0.16 + 1.20 = 2.31 \text{ cycles}.$$

Figure 9.22 from Freescale Semiconductor's application note AN2663 gives the hit ratio as a function of cache size on a three-dimensional graph for both L1 and L2 cache sizes. The peak hit rate is 96%, which is ultimately a function of the particular code being executed (a GCC compiler). The application note concludes that a 16 KB L1 with a 1 KB L2 gives almost the same results as a 1 KB L1 with a 16 KB L2 (although no one would design a system with a larger L1 cache than an L2 cache).

9.4.7   Instruction and Data Caches

Data and instructions are at the heart of the von Neumann concept; that is, they occupy the same memory. Cache designers can choose to create a *unified cache* that holds both instructions and data or to implement separate caches for data and instructions (the *split cache*). It makes good sense to cache data and instructions separately, because they have different properties. An entry in an instruction cache is never modified, except when the line is initially swapped in. Furthermore, you don't have to worry about swapping out instructions that are overwritten in the instruction cache, because the program does not change during the course of its execution. Since the contents of the instruction cache are not modified, it is much easier to implement

Self-Modifying Code

In the very early days of computing when ISAs were primitive, programmers took advantage of the von Neumann machine by writing self-modifying code. As its name suggests, it indicates that you change the code dynamically at runtime.

For example, if your computer didn't have indexing, you could write:

```
100 LOAD  r0,2325
101 ADD   100,#1
```

The first line loads r0 with the contents of memory location 2325. The second line adds 1 to the contents of instruction 100; that is, to the previous instruction itself! If the operand address is at the end of the instruction, it becomes 2326 and you've just used self-modifying code.

Self-modifying code is difficult to read and debug and entirely shunned by most programmers. Moreover, it often cannot work in environments with cache or memory management because you cannot access the code in memory directly.

an instruction cache than a data cache. Split instruction and data caches increase the CPU-memory bandwidth, because an instruction and its data can be read simultaneously. Separate instruction and data caches are essential in pipelined systems if instruction and operand accesses are to take place concurrently. We can summarize the advantages of both split and unified caches as follows.

- I-cache can be optimized to feed the instruction stream.
- D-cache can be optimized for read and write operations.
- D-caches can be optimized (tuned) separately.
- I-caches do not readily support self-modifying code.
- U-cache supports self-modifying code.
- D-caches increase bandwidth by operating concurrently.
- U-caches require faster memory.
- U-caches are more flexible (an I-cache may be full when the D-cache is half empty).

Most of today's processors have split caches, although in systems with multilevel caches, the higher level caches may be unified and the lower level caches split.

AMD's Barcelona architecture in Figure 9.23 demonstrates how cache memories have developed. Barcelona is a multi-core system where each core has both its own 64 KB L1 cache and a 512 KB L2 cache. All four cores share a common 2 MB L3 cache. The L1 cache is made up of two split 32 KB caches: one for data and one for instructions. Traditionally, multilevel caches are arranged so that the lowest level cache successively moves up the ladder following a cache miss (L1 interrogates L2, then L3, then main memory, until the missing data is located). In the Barcelona architecture, the L1 cache is the target of all cache loads and all fetches are placed in L1. The L2 cache holds data evicted from the L1 cache. Because of the tight-coupling between L1 and L2 caches, the latency incurred in transferring data back from L2 to L1 is low. We return to multi-core architectures in Chapter 13.

The L3 cache is shared between the cores. Data is loaded directly from the L3 cache to the L1 cache and does not go through L2. Data that is transferred may either remain in L3 if it is required by more than one processor, or it may be deleted if it is not shared. Like L2, the L3 cache is not fed from memory but from data spilled from L2.

Figure 9.24 illustrates Intel's Nehalem[6] architecture, a contemporary of Barcelona. The L1, L2, and L3 cache sizes are 32K, 256K, and 8M bytes, respectively.

FIGURE 9.23   AMD's Barcelona architecture

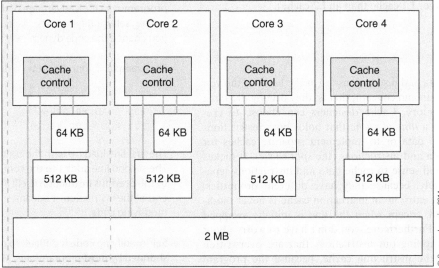

[6] Trent Rolf, "Cache organization and memory management of the Intel Nehalem computer architecture," University of Utah, http://rolfed.com/nehalem/nehalemPaper.pdf.

© 2009 IEEE. Reprinted, with permission, from D. Molka, D. Hackenberg, R. Schone, and M.S. Muller, "Memory Performance and Cache Coherency Effects on an Intel Nehalem Multiprocessor System," in 2009 18th International Conference on Parallel Architectures and Compilation Techniques, September 2009.

FIGURE 9.24   Intel's Nehalem architecture

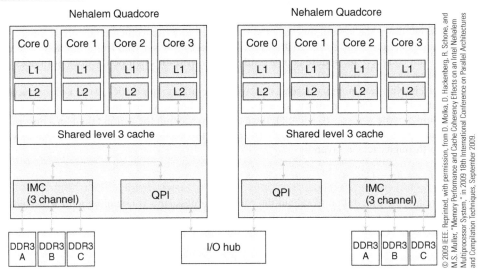

As well as instruction and data cache, some computers implement more specialized caches. For example, the *branch target cache* was introduced when we were covering pipelining. A branch target cache stores information concerning branches, such as branch addresses and instruction op-codes at the target address. Similarly, it is possible to cache subroutine return addresses in a special return address cache in order to reduce the overhead of a return from subroutine when the return address is on the stack.

9.4.8   Writing to Cache

Up to now, we've considered only read accesses to cache (the most frequent form of access). Now we look at the rather more complex write access. When the processor writes to the cache, both the line in the cache and the corresponding line in the memory must be updated, although is not necessary to perform these operations at the same time. However, you must ensure that the copy of a cached data element in the memory is updated before it is next accessed; that is, the copies of a data element in cache and memory must be kept in step.

We have already stated that the average access time of a system with a cache that's accessed in parallel with main store is $t_{ave} = ht_c + (1 - h)t_m$. If data is not in the cache, it must be fetched from memory and loaded both in the cache and the destination register. Assuming that t_l is the time taken to fetch a line from main store to reload the cache on a miss, the effective average access time of the memory system is given by the sum of the cache accesses plus the memory accesses *plus the re-loads due to misses*:

$$t_{ave} = ht_c + (1 - h)t_m + (1 - h)t_l$$

The *new* term in the equation $(1 - h)t_l$ is the additional time required to reload a line in the cache following each miss. This expression can be rewritten as

$$t_{ave} = ht_c + (1 - h)(t_m + t_l)$$

Accessing the element that caused the miss and filling the cache with a line from memory can take place concurrently. The term $(t_l + t_m)$ becomes $\max(t_l || t_m)$ and, because $t_l > t_m$, we can write

$$t_{ave} = ht_c + (1 - h)t_l$$

Let's now consider the effect of write accesses on this equation. When the processor executes a write, data must be written both to cache and to the main store. Updating the main memory at the same time as the cache is loaded is called a *write-through policy*. Such a strategy slows down the system, because the time taken to write to the main store is longer than the time taken to write to the cache. If the next operation is a read from the cache, the main store can complete its update concurrently (i.e., a write-through policy does not necessarily suffer an excessive penalty).

Relatively few memory accesses are write operations. In practice, write accesses account for about 5 to 30% of memory accesses. In what follows, we use the term w to indicate the fraction of write accesses ($0 < w < 1$). If we take into account the action taken on a miss during a read access and on a miss during a write access, the average access time for a system with a write-through cache is given by

$$t_{ave} = ht_c + (1 - h)(1 - w)t_l + (1 - h)wt_m,$$

where t_l is the time taken to reload the cache on a miss (this is assuming a no-write-allocate policy where a value is not cached on a write miss).

The $(1 - h)(1 - w)t_l$ term represents the time taken to reload the cache on a read miss access and the $(1 - h)wt_m$ represents the time taken to write to the memory on a write miss. Since the processor can continue onto another operation while main store is being updated, this $(1 - h)wt_m$ term can often be neglected because the main store has time to store write-through data between two successive write operations. This equation does not include the time taken to load the cache on a write miss, because it is assumed that the computer does not update the cache on a write miss.

The performance of the cache can be improved by employing a *write buffer* to hold the data waiting to be written through to memory. A typical write buffer holds four address/data pairs. Of course, you have to take care to ensure that the data in the write buffer is accessed if the processor executes a read to a location whose data has just been updated in the buffer but not in memory. One solution is to permit the write buffer to complete the memory update before performing a read operation.

An alternative strategy to updating the memory is called *write-back*. In a cache system with a write-back policy, a write operation to the main memory takes place *only when a line in the cache is to be ejected*. That is, the main memory is not updated on each write to the cache. The line is written back to memory only when it is flushed out of the cache by a read miss. We can now write:

$$\begin{aligned} t_{ave} &= ht_c + (1 - h)(1 - w)t_l + (1 - h)(1 - w)t_l \\ &= ht_c + 2(1 - h)(1 - w)t_l \end{aligned}$$

Note the term $(1 - h)(1 - w)t_l$ is repeated because a read miss results in writing back the old line to be swapped out to memory and loading the cache with a new line.

Each line in a cache memory includes flag bits that describe the current line. For example, each line may have a *dirty* bit that indicates whether the line has been modified since it was loaded in the cache. If a line has never been modified, it doesn't need writing back to main store when it is flushed from the cache. The average access time for a cache with such a write-back policy is given by

$$t_{ave} = ht_c + (1 - h)(1 - w)t_l + (1 - h)p_w wt_l$$

where p_w is the probability that a line will have to be written back to main memory.

Figure 9.25 provides a decision tree for a memory system with a cache that uses a write-back strategy. This figure adds up all of the outcomes for a system that updates the cache on a read miss and writes back a line if it has been modified. On a write miss, the line in the cache is written back and the cache loaded with the new line. These parameters give an average access time of

$$t_{ave} = ht_c + (1 - h)(1 - w)(1 - p_w)t_l + (1 - h)(1 - w)p_w \cdot 2t_l + (1 - h)w \cdot 2t_l$$

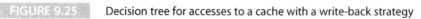

FIGURE 9.25   Decision tree for accesses to a cache with a write-back strategy

We have provided several expressions for the average access time of systems with differ-ent cache arrangements. These are approximate equations and the actual behavior of a real system will depend on its specific arrangements.

Early Restart, Critical World First, and Non-blocking Cache

One way of improving cache performance is to execute an early restart. That is, as soon as the requested word in a line has been fetched from memory, the CPU uses it and restarts execution rather than waiting for the entire line to be filled.

Another way of improving performance, called *critical word first*, is to request the word that caused the miss first, permit the processor to continue, and then fill the line.

The non-blocking cache is another attempt to reduce the misery caused by a miss. Normally, when a miss occurs, the cache is loaded with a replacement line and the pro-cessor stalled. However, it is possible that the next memory access is not to the line being replaced. In that case, the processor can continue. This is possible with the type of split transaction bus that we discuss in Chapter 12.

9.5   Virtual Memory and Memory Management

Memory management is the point at which the operating system and hardware meet, and it is concerned with managing the main store and disk drives. In many ways, memory management is a type of scaled up cache technology.

When computers first appeared, an address generated by the computer corresponded to the location of an operand in physical or *real* memory. Even today, 8-bit microprocessors in controllers do not generally use memory management. The *logical address* generated by high-performance computers in PCs and workstations is not the *physical address* of the operand accessed in memory. Consider the instruction LDR **r2**, [r3] that copies the contents of the memory location pointed at by register r3 into register r2, and assume that register r3 contains 0x00011234. The data might actually be copied from, say, memory location 0x00A43234 in DRAM-based main store. The act of translating 00011234 into 00A43234 is called *memory management*. In this section, we explain why memory management is necessary and how it is achieved.

Virtual memory is a term borrowed from optics, where virtual describes an image that appears to be in a place where it is not (for example, a telescope may make an object appear as if it's just in front of you when it's a long distance away). Virtual memory space is synonymous with *logical address space* and describes the address space that can be accessed by a computer. A computer with 64-bit address and pointer registers has a 2^{64}-byte virtual (logical) address space even though it may be in a system with only 2 GB (2^{31}) of physical main store memory.

Memory management has its origins in the 1950s and 1960s and describes any technique that takes a logical address generated by the CPU and translates it into the actual (i.e., physical) address in memory. Memory management allows the physical address spaces of DRAM and hard disk to be seamlessly merged into the computer's virtual memory space.

9.5.1   Memory Management

Computers using operating systems like Windows or UNIX make extensive use of memory management techniques. Figure 9.26 describes the structure of a system with a memory management unit (MMU). In principle, it's a very simple arrangement: the logical address from the CPU is fed to an MMU that translates it into the physical address of the operand in memory. This translation involves a look-up table that converts logical addresses into physical addresses. Because a very large table indeed would be required to translate each logical address into a physical address, memory space is divided into *pages,* and each address on a

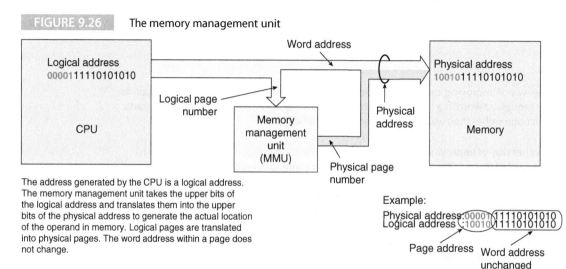

FIGURE 9.26   The memory management unit

Logical address
0000111110101010

CPU

Word address

Logical page number

Memory management unit (MMU)

Physical address

Physical page number

Physical address
1001011110101010

Memory

The address generated by the CPU is a logical address. The memory management unit takes the upper bits of the logical address and translates them into the upper bits of the physical address to generate the actual location of the operand in memory. Logical pages are translated into physical pages. The word address within a page does not change.

Example:
Physical address: 0000111110101010
Logical address: 1001011110101010

Page address   Word address unchanged

logical page is translated into the corresponding address on an a physical page. A page is, typically, 4 KB. For example, if the page size is 4 KB and the processor has a 32-bit address space, the logical address 0xFFFFAC24 might be 0x00002C24.

The size of a processor's logical address space is independent of the addressing mode used to specify an operand. Nor does it depend on whether a program is written in a high-level language, assembly language, or machine code. In a 32-bit system, the instruction LDR **r4**, [r6] lets you address a logical space of 4 GB. No matter what technique is used, the processor cannot specify a logical address outside the 4 GB range 0 to $2^{32} - 1$, simply because the number of bits in its program counter is limited to 32.

Physical address space is the address space spanned by all of the actual address locations in the processor's memory system. This is the memory that is in no sense abstract and costs real dollars and cents to implement. In other words, the system's main memory makes up the physical address space. The size of a computer's logical address space is determined by the number of bits used to specify an address, whereas the quantity of physical address space is frequently limited only by its cost.

We can now see why a microprocessor's logical and physical address spaces may have different sizes. What is much more curious is why a microprocessor might, for example, employ memory management to translate the logical address 0x00001234 into the physical address 0x861234. The fundamental objectives of memory management systems are the following.

1. To control systems in which the amount of physical address space exceeds that of the logical address space (e.g., an 8-bit microprocessor with a 16-bit address bus and a 64 Kbyte logical address space with 2 MB of physical RAM).
2. To control systems in which the logical address space exceeds the physical address space (e.g., a 32-bit microprocessor with a 4 GB logical address space and 64 MB of RAM).
3. Memory protection, which includes schemes that prevent one user from accessing the memory space allocated to another user.
4. Memory sharing, where one program can share the resources of another program (e.g., common data areas or common code).
5. Efficient memory usage in which the best use can be made of the existing physical address space.
6. Freeing programmers from any considerations of where their programs and data are to be located in memory. That is, the programmer can use any address he or she wishes, but the memory management system will map the logical address onto an available physical address.

A real memory management unit may not attempt to achieve all of these goals (the first two are mutually exclusive). The second goal (i.e., dealing with logical address spaces greater than the physical address space) is especially important to designers of 64-bit systems. When memory management is applied to this problem, it is frequently referred to as *virtual memory technology*. Virtual memory is almost synonymous with logical memory.

The problem of the available physical address space being smaller than the processor's logical address space is caused by economics and has always plagued the mainframe industry. In the late 1950s, mainframes were available with large logical address spaces but they were restricted to tiny 2K or so blocks of RAM. A group of computer scientists at Manchester University in the United Kingdom proposed a memory management technique, now known as virtual memory, to deal with this situation. A section of the logical (or virtual) address space is mapped onto the available physical address space, as shown in Figure 9.27. In this example, the 256 KB section of logical address space, in the range 78 0000 to 7B FFFF is mapped onto the physical memory in the range 0 0000 to 3 FFFF. As long as the processor accesses data in the logical address space that is currently mapped onto the existing physical address space, all is well. We have been doing this all the way through this text, because we have assumed that an address from the CPU is passed directly (i.e., unchanged) to the system's address bus.

When the processor generates the logical address of an operand that cannot be mapped on to the available physical address space, we have a problem. The solution to this problem

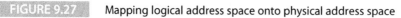

FIGURE 9.27   Mapping logical address space onto physical address space

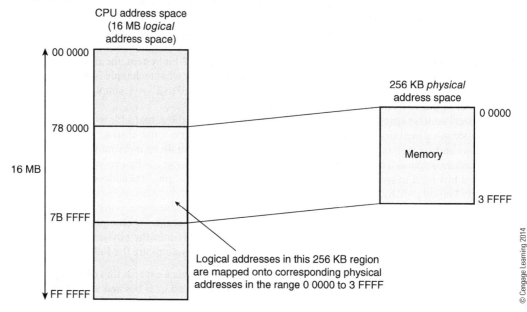

adopted at Manchester University was delightfully simple. Whenever the processor generates a logical address for which there is no corresponding physical address, the operating system stops the current program and deals with the problem. The operating system fetches a block of data containing the desired operand from its disk store, places this block in physical memory (overwriting any old data), and tells the memory management unit that a new relationship exists between logical and physical address space. In other words, the program or data is held on disk and only those parts of the program currently needed are transferred to the physical RAM. The memory management unit keeps track of the relationship between the logical address generated by the processor and that of the data currently in physical memory. This entire process is very complex in its details and requires harmonization of the processor architecture, the memory management unit, and the operating system. People dream of simple virtual memory systems and have nightmares about real ones.

A Tale of Two Processors

First-generation processors (forgetting the earlier 4004) had 16-bit registers and address buses providing a $2^{16} = 64$ KB address space. Although only 64 KB of memory could be addressed, in the late 1970s that represented a lot of memory, and microprocessor applications generally used less physical memory than the 64 KB available.

When Motorola introduced its 68K, the address registers were 32 bits, giving it a 2^{32} byte address space. This address space was said to be linear in the sense that it was continuous and not partitioned in any way. However, in order to reduce manufacturing costs, the 68K had only 24 address pins allowing a $2^{24} = 16$ MB physical address space. A 32-bit address in hexadecimal is represented by $XXYYYYYY_{16}$ where YYYYYY describes the physical address and XX represent 8 "don't care" bits, because they can't be accessed from the address bus. At the time, 16 MB of memory was considered an extraordinarily large physical address space.

Intel beat Motorola to the market place with both their 8-bit 8080 and their 16-bit 8086. Unlike the 68K with its 32-bit address registers, the 8086 had only 16-bit address registers (including the PC), restricting the logical address space to 64 KB. The 8086 was able to access 2^{20} bytes of memory by a technique called *segmentation*. When the 8086 generates a 16-bit address, it is added to a 20-bit value to create a new 20-bit address spanning 1 MB. The 8086 has four *segment registers* allowing the programmer to access four 1 MB segments in physical memory. The segments are code, data, extra, and stack and are pointed at by the CS, DS, ES, and SS segment registers, respectively.

A 20 bit address is obtained by taking the 16-bit address from an address register plus the 16-bit address from a segment register shifted left four places; that is, the address is given by R +16S, where R is a pointer and S a segment register. Although rather cumbersome, Intel's addressing mechanism provided a means of implementing separate code/data/stack address spaces and made relocation easy, because code fragments less than 64 KB don't have to be recompiled if they are moved elsewhere in memory as 16-bit addresses don't change.

9.5.2   Virtual Memory

Virtual memory systems serve four purposes: they support systems with larger logical memory spaces than physical address spaces, they map logical addresses onto physical addresses, they allocate physical memory to tasks running in logical address space, and they make it easier to construct multitasking systems.

It would be foolish to pretend that justice can be done to the topic of virtual memory in a text of this size. Virtual memory systems are designed by teams of designers and programmers and require many hours to produce, because the management of virtual memory is not only complex but is also found almost exclusively in systems with multiuser or multitasking operating systems. Here we provide only a basic overview of virtual memory.

Memory Management and Multitasking

Multitasking systems execute two or more tasks or processes concurrently by periodically switching between tasks. Clearly, multitasking is viable only if several tasks reside in main memory at the same time. If the tasks had to be transferred from hard disk every time they run, the time required to swap in a new task would be prohibitive.

Figure 9.28 demonstrates how logical address space might be mapped onto physical address space in a system with two tasks. Tasks A and B reside in physical memory at the same time, as Figure 9.28 demonstrates. Each task has its own logical memory space (e.g., program and stack) and can access shared resources lying in physical memory space. Programmers are entirely free to choose their own addresses for the various components of their tasks. Consequently, task A and task B in Figure 9.28 can each access the same data structure in physical memory, even though they use different logical addresses. That is, each task is aware of only its own copy of data that it shares with another task.

The memory management unit maps the logical addresses chosen by the programmer onto the physical memory space, and the operating system is responsible for setting up the logical-to-physical address mapping tables. Whenever a new task is created, the operating system is informed of the task's memory requirements. The operating system searches the available physical memory space for free memory blocks and allocates these to the task. You can imagine that, after a time, the physical memory space may become very fragmented with the various physical blocks belonging to each task interwoven in a complex pattern. A good operating system attempts to perform memory allocation efficiently and should not permit large numbers of unused blocks of physical memory. The way in which this fragmentation is dealt with depends on both the type of memory mapping implemented and on the operating system.

FIGURE 9.28   Address mapping in a multitasking environment

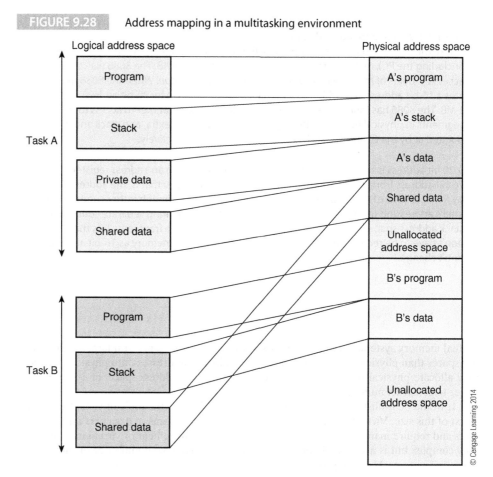

A powerful feature of memory mapping is that each logical memory block can be associated with various permissions. For example, memory can be made read-only, write-only, accessible only by the operating system or by a given task, or shared between groups of tasks. By ensuring that physical memory blocks can be accessed only be pre-defined tasks, you can ensure that one task cannot corrupt another. There are two fundamental ways of implementing memory management. One uses fixed-sized blocks of memory called *pages,* and the other has variable-sized blocks of memory called *segments.*

Address Translation

Memory management provides two distinct services. The first is to map logical addresses onto the available physical memory. The second function occurs when the physical address space *runs out* (i.e., the logical-to-physical address mapping cannot be performed because the data is not available in the random access memory).

Figure 9.29 shows how a page-memory system can be implemented. This example uses a microprocessor with a 24-bit logical address bus and a 512 KB memory system. The 24-bit logical address from the processor is split into a 16-bit displacement that is passed directly to the physical memory plus an 8-bit page address. The page address specifies the page (one of $2^8 = 256$ pages) currently accessed by the processor. The displacement field of the logical address accesses one of 2^{16} locations within a 64 KB page.

The page table contains 256 entries—one for each logical page. For example, in Figure 9.29, the CPU is accessing the 8-bit logical page address 00000111_2. Each entry contains a 3-bit page frame address that provides the three most-significant bits of the physical

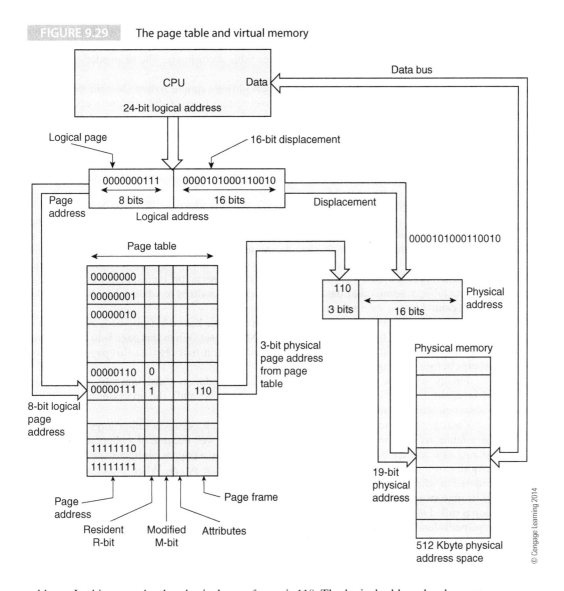

FIGURE 9.29   The page table and virtual memory

address. In this example, the physical page frame is 110. The logical address has been condensed from 8 + 16 bits to 3 + 16 bits and logical address 00000111 0000101000110010 is mapped onto physical address 110 0000101000110010.

Although there are 256 possible entries in the page frame table (one for each logical page), the physical page frame address is only 3 bits, limiting the number of unique physical pages to eight. Consequently, a different physical page frame in random access memory cannot be associated with each of the possible logical page numbers. Each logical page address has a single-bit, R-field labeled resident associated with it. If the R-bit is set, that page frame is currently in physical memory. If the R-bit is clear, the corresponding page frame is not in the physical memory, and the contents of the page frame field are meaningless.

Whenever a logical address is generated and the R-bit associated with the current logical page is clear, an event called a *page fault* occurs. Once a memory access is started that attempts to access a logical address whose page is not in memory because the R-bit was found to be clear, the current instruction must be suspended, because it cannot be completed.

A typical microprocessor has a bus error input pin that is asserted to indicate that a memory access cannot be completed. Whenever this happens, the operating system intervenes to deal with the situation. Although the information the CPU is *attempting* to access is not currently in the random access physical memory, it is located on the disk. The operating system retrieves the page containing the desired memory location from the disk, loads it in to the physical memory, and updates the page table accordingly. The suspended instruction can then be executed.

This procedure is simple isn't it? Well, not entirely simple. When the operating system fetches a new page from the disk, it must overwrite a page of the random access physical memory. Remember that one of the purposes of virtual memory is to permit relatively small physical memories to simulate large memories. If we are going to replace old pages by new ones, we require a strategy to decide which old pages are to go. The classic paging policy is the *least recently used* (LRU) algorithm, where the page that has not been accessed for the longest time is overwritten by the new page (i.e., if you haven't accessed this page recently, you are not likely to access it in the near future).

The LRU algorithm has been found to work well in practice. Unfortunately, the operating system must know when each page is accessed if this algorithm is to work, which somewhat complicates the hardware (each page has to be date-stamped after use). Another problem the operating system has to deal with is the divergence between the data stored in RAM and the data held on disk. If the page fetched from disk contains only program information, it will not be modified in RAM, and therefore, overwriting it causes no problems. If, however, the page is a data table or some other data structure, it may be written to while it is in RAM. In this case, it cannot just be overwritten by the new page.

In Figure 9.29, we can see that each entry in the table has an M (modified) bit. Whenever that page is accessed by a write operation, the M-bit is set. When this page is to be overwritten, the operating system checks the M-bit, and if set, it first rewrites this page to the disk storage before fetching the new page.

Finally, when the new page has been loaded, the address translation table updated, the M-bit cleared, and R-bit set (to indicate that the page is valid), the processor can rerun the instruction that was suspended.

Clearly, the effort involved every time a page fault occurs is rather large. As long as page faults are relatively infrequent, the system works well because of a phenomenon called locality of reference. Most data is clustered so that once a page is brought from the disk, the majority of memory accesses will be found within these pages. When the data is not well ordered or when there are many unrelated tasks, the processor ends up by spending nearly all of its time swapping pages in and out, and the system effectively grinds to a halt. This situation is called *thrashing*. Thrashing refers to any computer activity where accessing a resource repeatedly leads to a catastrophic decrease in performance. However, thrashing largely refers to the situation in which a virtual memory system breaks down because of the need to load and reload pages.

Two-Level Tables

The arrangement of Figure 9.29 is impractical in modern high-performance processors. Suppose a 32-bit computer uses an 8 KB page that is accessed by a 13-bit page offset (the offset is the location within a page). This leaves $32 - 13 = 19$ bits to select one of 2^{19} logical pages. It would be impossible to construct such a large page table in fast RAM (notice that this is the same problem facing the designer of memory cache).

Figure 9.30 describes how it is possible to perform address translation without the need for massive page tables by using *multi level* page tables. The logical (virtual) address from the computer is first divided into a 19 bit page number and a 13-bit page offset. The page number is then divided into a 10-bit and a 9-bit field corresponding to first-level and second-level page tables. These two tables would require $2^{10} = 1,024$ and $2^9 = 512$ entries, respectively.

The diagram in Figure 9.30 is simplified, because a real page table would contain a lot more information about the address translation process than just the pointers. Figure 9.31 illustrates the PowerPC's address translation tables.

The structure of a real page table includes more than pointers to other tables. A page table entry contains a descriptor that points to the next level in the hierarchical address translation table. The final descriptor in the chain points to the actual physical page and contains

FIGURE 9.30   A two-level page table

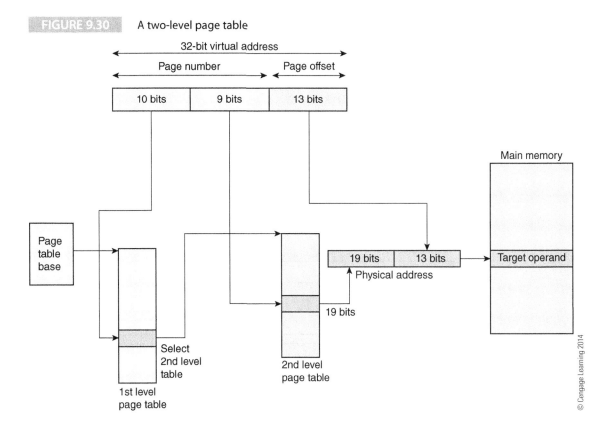

information the MMU requires about that page. In practice, a typical memory management unit may contain table descriptors with the following information.

Descriptor Type—The *descriptor* type tells the MMU whether another level in the table exists.

Write Protect—The *write protect* bit indicates that pages pointed at by this descriptor may not be written to. If W = 1, all subsequent levels in the trees and their associated page descriptors are write protected.

U—The *used* (U) bit is initially cleared to zero by the operating system when the descriptor table is set up. When the descriptor is accessed for the first time, the MMU automatically sets U to 1. The U-bit is used in virtual memory systems when deciding whether to write back a physical page to disk when it is swapped out.

S—When the *supervisor* bit is set, pages pointed at by this descriptor can be accessed only from the supervisor mode (i.e., operating-system level privilege). The supervisor state is the state in which the operating system runs, and it has a higher level of privilege than the user state. For example, I/O such as disk drives can be accessed only from the supervisor state.

Shared Globally—When set to 1, the shared globally (SG) bit indicates that the page descriptor may be shared. That is, if SG = 1, all tasks within the system may access the physical page. SG tells the MMU that only a single descriptor for this page need be held in the page table cache. The *translation look-aside buffer* (TLB) is just a term for a small associative cache that can rapidly perform an address table by searching all entries simultaneously.

Write Access Level—The write access level (WAL) indicates the minimum privilege level allowed for pages located via this descriptor.

Read Access Level—The three read-access-level bits perform the corresponding read function to the WAL bits.

FIGURE 9.31   The PowerPC memory management system

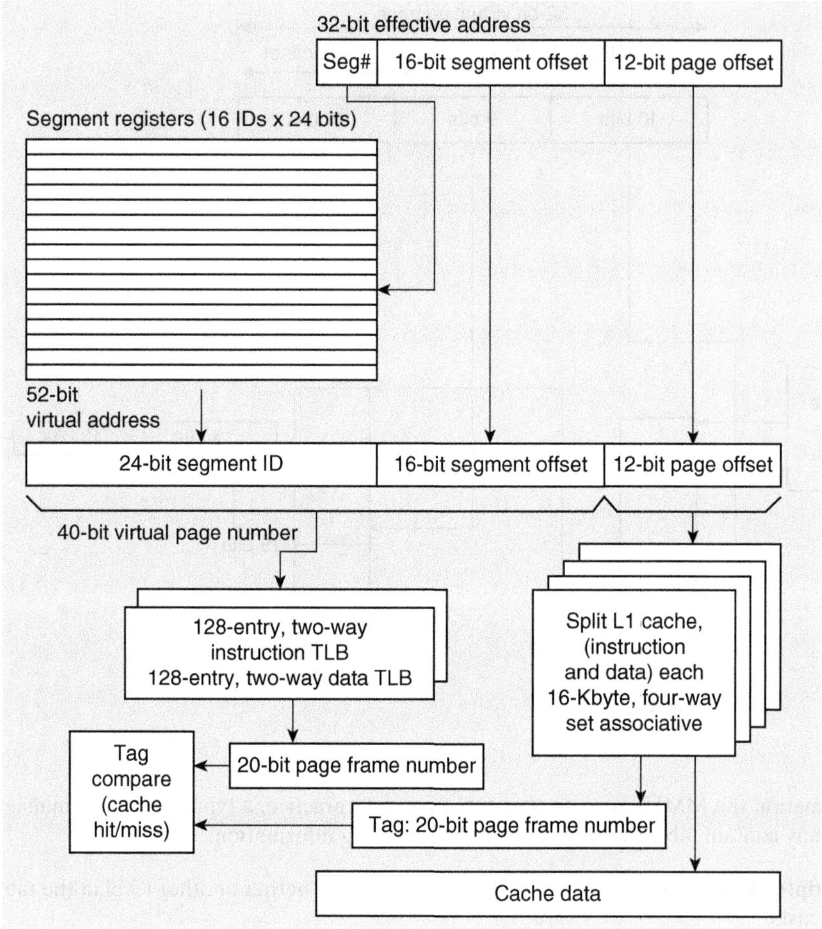

Limit—The limit field provides a lower or upper bound on index values for the next level in the translation table; that is, the limit field restricts the size of the next table down. For example, one of the logical address fields may have 7 bits and therefore will support a table with 128 entries. However, in a real system, you might never have more than, say, 20 page descriptors at this level. By setting the limit to 5, you can restrict the table to 32 entries rather than 128.

Lower/Upper—The lower/upper (L/U) bit determines whether the *limit* field refers to a lower bound or to an upper bound. If L/U = 0, the limit field contains the unsigned upper limit of the index, and all table indices for the next level must be less than or equal to the value contained in the limit field. If L/U = 1, the limit field contains the unsigned lower limit of the index, and all table indices must be greater than or equal to the value in the limit field. In either case, if the actual index is outside the maximum/minimum, a limit violation will occur. The end result of a table walk is the page-descriptor that is going to be used to perform the actual logical-to-physical address translation.

Walking through a multilevel page table (e.g., as shown in Figure 9.30) produces the page descriptor that will be used in the actual logical-to-physical address translation. In addition to the table descriptor bits listed previously, a page descriptor may have the following control bits not found in a table descriptor:

Modified (M) Bit—Indicates whether the corresponding physical page was written to. The M-bit is set to zero when the descriptor is first set up by the operating system, since the MMU may set the M-bit but not clear it. Note that the used bit is set if a *table descriptor* is accessed, while the M-bit is set if the page is accessed.

Lock (L) Bit—Indicates that the corresponding page descriptor should be made exempt from the MMU's page replacement algorithm. When L = 1, the physical page cannot be replaced by the MMU. Thus, we can use the L-bit to keep page descriptors in the address translation cache.

Cache (CI) Inhibit Bit—Indicates whether or not the corresponding page is cacheable. If CI = 1, then this access should not be cached.

Summary

We use memory to store instructions and data in a computer. Because of the nature of technology and manufacturing processes, there's no single device or technology that can yet fulfill all of the needs of a personal computer or workstation. In the simplest terms: If memory is fast, it's expensive, and if it is cheap, it's slow.

When mainframe computers were being developed to support the needs of the government, industry, and the military, it soon occurred to designers that the limitations of real memory systems could be reduced, or rather, *hidden*. It's a truism that you can store data in slow memory if you don't actually need it. In computing terms, this means keeping the data that's being frequently accessed in fast memory and archiving data that's not immediately required in slower memory. This is a simple and elegant idea and has had a vital impact on the performance of computers. On the other hand, it can be fiendishly difficult to implement in practice.

There are two fundamental types of memory in a computer: the random access semiconductor memory (usually DRAM) of the main store and the serially accessed magnetic or optical memory of the secondary store. Both of these memory systems employ similar techniques to overcome their speed limitations: *cache memory* is used to speed up main store and *virtual memory* to speed up secondary store. Cache memory and virtual memory mechanisms are similar in principle but differ considerably in detail and implementation.

The first part of this chapter looked at the cache memory used to speedup main store. A relatively small amount of high-speed cache (e.g., 1 MB in 4 GB of DRAM) can dramatically increase system performance. Because data and instructions are not accessed at random and real data and instructions demonstrate both *temporal* and *spatial* locality, we can expect that more than 90% of the CPU's accesses to memory result in the data/instruction being found in the cache. When it isn't, the data has to be read from memory and a copy loaded into cache.

We have examined several aspects of cache memory, such as how it is organized; that is, how data in physical memory can be mapped onto a location in the much smaller cache. Although direct-mapped cache is the easiest to design, we have demonstrated that it has limitations that can reduce its efficiency and have shown how it can be used to create a set-associative cache memory—the type of cache found in all microprocessor systems.

The second part of this chapter has looked at *virtual memory*. Hard disks are slow—many orders of magnitude slower than the DRAM of main store. Virtual memory systems divide memory into pages of typically 4 KBs. A 4K block of data can be loaded from a disk drive into any 4K page frame in main store; that is, the physical memory space is not continuous like the virtual or logical memory space of the processor. When the computer makes a memory access, the memory management unit translates the logical page address into the physical page address, and the data is read off that page in memory. However, if the logical page being accessed is not in main memory, a page fault is issued by the memory management unit, and the operating system steps in to load the missing page from disk into a page frame in main memory. Of course, if all physical pages are currently in use, the operating system has to sacrifice one of these pages, eject its data, and load the new page.

Problems

9.1 What is cache memory?

9.2 Why do computers use cache memory?

9.3 What is the meaning of the following terms.
 a. Temporal locality
 b. Spatial locality

9.4 From first principles, derive an expression for the speedup ratio of a memory system with cache (assume the hit ratio is h and the ratio of the main storage access time to cache access time is k, where $k < 1$). Assume that the system is an ideal system and that you don't have to worry about the effect of clock cycle times.

9.5 For the following ideal systems, calculate the speedup ratio S. In each case, t_c is the access time of the cache memory, t_m is the access time of the main store, and h is the hit ratio.
 a. $t_m = 70$ ns, $t_c = 7$ ns, $h = 0.9$
 b. $t_m = 60$ ns, $t_c = 3$ ns, $h = 0.9$
 c. $t_m = 60$ ns, $t_c = 3$ ns, $h = 0.8$
 d. $t_m = 60$ ns, $t_c = 3$ ns, $h = 0.97$

9.6 For the following ideal systems, calculate the hit ratio h required to achieve the stated speedup ratio S.
 a. $t_m = 60$ ns, $t_c = 3$ ns, $S = 1.1$
 b. $t_m = 60$ ns, $t_c = 3$ ns, $S = 2.0$
 c. $t_m = 60$ ns, $t_c = 3$ ns, $S = 5.0$
 d. $t_m = 60$ ns, $t_c = 3$ ns, $S = 15.0$

9.7 Real microprocessors operate in time slots of a given duration (i.e., an integer multiple of the clock period). For example, if the clock has a speed of 100 MHz, the clock cycle time is 10 ns, and all operations must take an integer number of 10 ns units. The following data gives the clock cycle time of a processor t_{cyc} and the access time of the main memory t_m (including all overheads such as address decoding). In each case, calculate the *actual time* required to access the memory.
 a. $t_{cyc} = 20$ ns, $t_m = 75$ ns
 b. $t_{cyc} = 10$ ns, $t_m = 75$ ns
 c. $t_{cyc} = 50$ ns, $t_m = 100$ ns
 d. $t_{cyc} = 50$ ns, $t_m = 105$ ns

9.8 For the following systems that use a clocked microprocessor, calculate the maximum speedup ratio you could expect to see as h approaches 100%.
 a. $t_{cyc} = 20$ ns, $t_m = 75$ ns, $t_c = 15$ ns
 b. $t_{cyc} = 20$ ns, $t_m = 75$ ns, $t_c = 25$ ns
 c. $t_{cyc} = 10$ ns, $t_m = 75$ ns, $t_c = 15$ ns

9.9 In practice, a computer spends a fraction of the time performing internal operations as well as memory accesses. Consequently, the effective speedup ratio is reduced, because cache memory has no effect on internal operations. The average time required to execute an instruction can be written as

$$t_{ave} = F_{internal} \cdot t_{cyc} + F_{memory}[ht_c + (1 - h)(t_c + t_d)] \cdot t_{cyc}$$

where

$F_{internal}$ = fraction of time spent doing internal operations (0 to 1)

F_{memory} = fraction of time spent doing memory accesses operations (0 to 1)

t_{cyc} = clock cycle time

t_c = cache access time expressed in clock cycles

t_d = delay penalty paid when not accessing cache memory expressed in clock cycles

For the following systems, calculate the average cycle time
 a. $F_{internal} = 20\%$, $t_{cyc} = 20$ ns, $t_c = 1$, $t_d = 3$, $h = 0.95$
 b. $F_{internal} = 50\%$, $t_{cyc} = 20$ ns, $t_c = 1$, $t_d = 3$, $h = 0.9$

9.10 For the system in Problem 9.9, with the parameters

$$F_{internal} = 40\%, t_{cyc} = 20 \text{ ns}, t_c = 1, t_d = 4$$

calculate the value of the hit ratio required to halve the average instruction time.

9.11 In a direct-mapped cache memory system, what is the meaning of the following terms.
 a. Word
 b. Line
 c. Set

9.12 How is data in main store mapped on to each of the following?
 a. A direct-mapped cache
 b. A fully associative cache
 c. A set-associative cache

9.13 Why is it so difficult to construct an associative cache?

9.14 Why is the set-associative cache so popular?

9.15 With the aid of a diagram, show how a cache-tag RAM is used to implement a direct-mapped cache memory. Discuss the advantages and disadvantages of a direct-mapped cache over an associatively mapped cache.

9.16 What is burst-mode operation (in the context of a cache memory)?

9.17 What is cache coherency?

9.18 In principle, cache memory is a very simple concept. You simply keep a copy of frequently accessed data in high-speed RAM. In practice, few elements of a computer are harder to design than a cache memory system. Discuss the truth, or otherwise, of this statement.

9.19 Discuss the factors that an engineer would take into account when selecting a suitable line size for a cache memory.

9.20 What are *level 1 cache* and *level 2 cache* memories (i.e., L1 and L2 caches)?

9.21 The cache system can be located between the CPU and the MMU (i.e., a logical cache) or between the

MMU and the system random access memory (i.e., a physical cache). What factors determine the optimum location of cache memory?

9.22 Why is it harder to design a data cache than an instruction cache?

9.23 When a CPU writes to the cache, both the item in the cache and the corresponding item in the memory must be updated. If data is not in the cache, it must be fetched from memory and loaded in the cache. If t_1 is the time taken to reload the cache on a miss, show that the effective average access time of the memory system is given by

$$t_{ave} = ht_c + (1 - h)t_m + (1 - h)t_1.$$

9.24 A cache memory may be operated in either a serial or a parallel mode with respect to the main memory. In the serial access mode, the cache is examined for data, and if a miss occurs, the main storage is then accessed. In the parallel access mode, both the cache and the main store are accessed simultaneously. If a hit occurs, the access to the main store is aborted. Assume that the system has a hit ratio h and that the ratio of cache memory access time to main store access time is k ($k < 1$). Derive an expression for the speedup ratio of both a serial access and a parallel access cache.

9.25 If a serial mode cache is used and a 5% penalty in speedup ratio over the corresponding parallel access cache can be tolerated, what value of the hit ratio is necessary to achieve this? Assume that the main store access time is 30 ns and that the cache access time is 3 ns.

9.26 A system has a level 1 cache and a level 2 cache. The hit rate of the level 1 cache is 90%, and the hit rate of the level 2 cache is 80%. An access to level 1 cache requires one cycle, an access to level 2 cache requires four cycles, and an access to main memory requires 50 cycles. What is the average access time?

9.27 A computer has a cache with an access time of 1 cycle and an average hit rate of 95 percent. The miss penalty is 100 cycles.
  a. What is the average cycle time for this computer?
  b. A level 2 cache with a hit rate of 80 percent and a penalty of six cycles is added to this system. What effect does this have on the average cycle time?
  c. A computer with L1 and L2 caches executes 10,000 memory accesses. While the test program is running, 500 misses are recorded to the L1 cache and 300 to the L2 cache. What are the miss rates of the L1 and L2 caches?

9.28 In the context of multilevel caches, what is the difference between a local miss rate and a global miss rate?

9.29. Why is the miss rate often quoted (and used) in preference to the hit rate?

9.30 What is a *victim cache*? How is it used?

9.31 What type of misses does a *victim cache* reduce?

9.32 What are the essential differences between *victim* and *annex* caches?

9.33 A processor with memory management has a 4K page size. It has a 32K cache memory with 16-byte cache lines. In order to speed up memory access, you decide to arrange the cache so that the cache is accessed at the same time a logical-to-physical address translation is taking place. In order for the scheme to work, what level of associativity must be implemented?

9.34 Suppose a unified cache has the following characteristics:

Read/write *penalty*	1 cycle
Miss rate	3%
Load instructions (read data)	20%
Store instructions (write data)	5%
Miss *penalty*	20 cycles

What is the average access time?

9.35 A 64-bit processor has a 8-MB, four-way set-associative cache with 32-byte lines. How is the address arranged in terms of set, line, and offset bits?

9.36 A computer with a separate data cache has a write-back cache memory. Cache line size is 64 bytes. Read access accounts for 80% of memory traffic. The processor, memory, and data buses are all 64-bits wide. Main memory latency (first access) is 20 cycles, and successive accesses take two cycles. The cache hit rate is 96%.
  Calculate the cost of a cache miss.

9.37 There are three causes of cache miss: compulsory, capacity, and conflict. Define the meaning of these terms. Briefly explain what can be done to minimize their effect.

9.38 Why is memory management necessary in a system that uses hard disks?

9.39 What forms of *protection* can memory management provide?

9.40 Memory management has a protection function. Does such a facility exist with cache memory?

9.41 What are the fundamental differences between cache memory (as found in a CPU) and cache memory found in a hard disk drive?

9.42 What are the differences between *write-back* and *write-through* caches, and what are the implications for system performance?

9.43 A computer with a 32-bit address architecture has a memory management system with single-level 4 KB page tables. How much memory space must be devoted to the page tables?

9.44 Consider a fully associative 16-byte cache with four lines of four words. The cache uses a LRU (least recently used) algorithm to deal with line replacement. When the cache is initially empty, lines are added from line 0 onward.

Given the following sequence of hexadecimal addresses, indicate whether a hit or miss takes place. Show the state of the cache at the end of the reads.

00 03 05 08 13 14 11 04 0F 0C 23 00 01 02 04 06 05 07 09 21

9.45 A computer runs an instruction set with the characteristics in the following table.

Class	Instruction Frequency	Cycles
Arithmetic operations	70%	1 cycle
Conditional operations	15%	2 cycles
Load	10%	2 cycles
Store	5%	2 cycles
Hit rate	95%	
Cost of a cache miss (read)	10 cycles	
Write-through time	5 cycles (writes to memory are not buffered)	

What is the average number of cycles per instruction?

9.46 Consider the following code that accesses three values in memory scalar integers x and s, and an integer vector $y[i]$. What is the memory latency in clock cycles for a trip round the loop (after the first iteration)? Assume that the array is not cached and each new access to the array results in a miss.

The system has both L1 and L2 caches. The access time of the L1 cache is two cycles, the access time of the L2 cache is 6 cycles and main memory has an access time of 50 cycles. In this case all memory and cache memory accesses take place in parallel.

```
for (i = 0; i< 100; i++)
{
x = y[i];
s = s + x;
}
```

9.47 Assuming the same systems as problem 9.46, what is the average memory latency for a trip round the loop if prefetching is used and each access to main memory results in the next line being loaded into L2 cache? Assume that preloading the L2 cache incurs no further memory access penalty.

9.48 A 64-bit computer has a 128 KB 8-way associatively mapped cache. The cache has 128 sets and a line is 16 words. How many tag bits does each address require?

9.49 What type of cache is particularly useful in reducing thrashing in the cache due to repetitive swapping in and swapping out?

9.50 A 16-bit CPU has a cache with 32 lines, each of 16 bytes. The CPU accesses a byte at the decimal address 3210. This results in a miss and a line is loaded. Where is the line loaded in the cache?

9.51 Given the following data and assuming a clock rate of 1,000 MHz.

Memory	Hit Time	Miss Rate
L1 cache	1 cycle	2%
L2 cache	8 cycles	5%
DRAM	20 cycles	0.1%
Disk	10 ms	

Calculate the average memory access time. Assume that L2 and DRAM are accessed in parallel with L1.

9.52 Consider a computer with a 256-byte address space and a two-way 32-byte set associative cache. The computer word size is a byte, each cache line contains four bytes, and each cache four lines. If the cache is initially empty and the following sequence of hexadecimal addresses is read, show the corresponding sequence of hits and misses.

48, 0C, 48, 4C, 5C, 3A, 20, 21, 22, 24, 81, 49, 30, 34, 27, 3E, 24, 28, 2C, 40

9.53 People are always looking for more effective cache mechanisms, particularly for ways of reducing the miss penalty (for example, using means of annex caches or victim caches). A student makes the following suggestion. Not all data is the same. Some numbers are used more frequently than others, particularly small numbers. So, why not arrange the cache so that if there are two candidates for eviction, the higher value is ejected first? What do you think?

9.54 A computer has a cache with a hit ratio of 95% and a line size of four 32-bit words. The average processor cache access rate is 100 million/s. Twenty percent of CPU operations are loads/stores with 30% writes and 70% reads). The cache employs a write through mechanism and a line is replaced on a write miss. What is the bandwidth of the 32-bit CPU-memory bus used under these conditions?

9.55 Consider Problem 9.54 except that the cache uses a write-back mode. On average 25% of cache lines are dirty (have been modified).

9.56 A computer has a 256 word memory and a 16 word cache. The cache line size is one word. The following sequence of addresses is read in series:

0,1,2,3,4,5,10,13,16,19,21,4,8,12,30,40,41,42,35,1,3,13

Show how the cache memory would be accessed assuming that all lines are initially invalid. In each cache, mark the access as hit, capacity miss,

compulsory miss, or conflict miss. Do this for each of the following.
a. Fully associative cache
b. Direct-mapped cache
c. Two-way set associative cache

9.57 A computer with a 24-bit address bus has a main memory of size 16 MB and a cache size of 64 KB. The wordlength is two bytes.
a. What is the address format for a direct-mapped cache with a line size of 32 words?
b. What is the address format for a fully associative cache with a line size of 32 words?
c. What is the address format for a four-way set-associative cache with a line size of 16 words?

9.58 A system has a memory access time of 50 ns and a cache access time of 2 ns. The instruction time is 4 ns (not counting memory access) and the average instruction requires 0.25 memory accesses. If the hit rate is 0.90, what is the average instruction time?

9.59 Why is the hit rate of an L2 cache usually lower than that of an L1 cache?

9.60 A computer has a memory access time of 38 ns and does not use cache. Cache with an access time of 10 ns is added. The computer then runs 90% faster. Estimate the hit ratio.

9.61 A system has an L1 cache with a hit rate of 87% (hits takes one cycle). The L2 cache has a hit rate of 90% and a penalty of 10 cycles. The main store has an access time of 200 cycles. What is the average access time?

Main Memory

"Memories are all we really own."
Elias Lieberman

"The life of the dead is placed in the memory of the living."
Cicero

"Every passing hour brings the Solar System forty-three thousand miles closer to Globular Cluster M13 in Hercules—and still there are some misfits who insist that there is no such thing as progress."
Kurt Vonnegut, The Sirens of Titan

"Computers should be made of glass—they're so full of bottlenecks."
Alan Clements March 2001

"The great thing in the world is not so much where we stand, as in what direction we are moving."
Oliver Wendell Holmes

"To aim is not enough; you must hit."
German proverb

"...computer hardware progress is so fast. No other technology since civilization began has seen six orders of magnitude in performance-price gain in 30 years."
Fred Brooks, Jr

10.1   Introduction

Our next step in examining a computer's memory and I/O systems is its *main memory* or *immediate access store*. In many ways, this is the most boring part of a computer, sitting there alongside the processor chip just storing data. Memory doesn't appear to do anything clever, and you never hear computer geeks chatting about it with the same excitement they show when discussing overclocking or multiple processors. Put simply, the main store grabs data and holds it until either the cache requests it or someone turns off the power. However, we are now going to describe some of the remarkable progress made in memory technologies over the years. We examine problems that have arisen and been overcome and introduce new technologies and device-physics that are being developed.

In this chapter, we look at the operating principles of immediate access memory, the difference between *static* and *dynamic* classes of memory, and the difference between *volatile* and *nonvolatile* memory. We also introduce some of the new technologies that are beginning to play a more significant role in memory systems. As well as the characteristics of memory devices, we look at some of the considerations that the memory systems designer has to take into account. Most of those reading this book will probably never design a CPU, but some may well have to design a memory system for an embedded computer or similar device. Digital systems range from computers to cameras to cell phones, and memory systems are often constructed from off-the-shelf components.

Dynamic—The Misnomer

The main store of most computers uses *dynamic memory* (DRAM). The term *dynamic* is rather misleading. The normal English usage of *dynamic* implies a positive, even aggressive, level of performance. However, its use in DRAM means the reverse. Data is stored as an electric charge that leaks away over a few milliseconds. In other words, DRAM loses its data in a few milliseconds. Data can be retained in the DRAM only by continually reading it before it disappears and writing it back. The term *dynamic* indicates this property. (Perhaps the 'D' in DRAM should stand for *drippy*).

In principle, a computer's memory is the easiest component to understand; it's the place where programs and data live. In practice, a memory system is invariably *nonhomogeneous* and is composed of devices from DRAMs to hard disks that are fabricated with a wide range of technologies. Indeed, the spread in the performance of memory devices is wider than any other element in a computer system. A memory component may hold a few words or a hundred gigabytes; it may have a read time as short as 1 ns or as long as several seconds; it may cost a few cents or a thousand dollars.

Because *memory* is such a gigantic topic, encompassing different storage technologies, we have divided the memory section into two chapters. This chapter looks at the *immediate access store* or *primary storage* where programs are located while they are running. The next chapter looks at the *secondary storage* system used to hold programs and data that are not currently being executed. In general, primary storage uses semiconductor technology, and secondary storage uses magnetic or optical technologies—although even secondary stores are now using solid-state disks.

Memory performance has increased rapidly at a rate of about 7% over the last two decades, as Figure 10.1 demonstrates. On the other hand, processor performance has improved at a truly remarkable rate of 60% a year, which has dwarfed memory technology and made the memory system a major bottleneck in modern processors.

In this chapter, we look at the operating principles of both static and dynamic semiconductor memory and demonstrate how they are interfaced to the processor. In particular, we examine the timing diagrams of memory devices and show the sequence of operations that take place during a data transfer. This material will help you understand some of the concepts we later introduce in Chapter 12 when we cover input/output techniques. A section of this chapter is devoted to DRAM, which has complex interfacing requirements and has appeared in several forms over the years.

FIGURE 10.1   Changes in memory technology over two decades

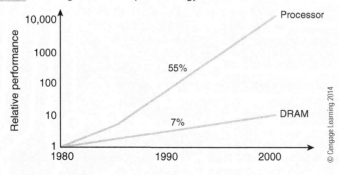

10.1.1   Principles and Parameters of Memory Systems

A memory device does exactly what you would expect it to do; it *remembers* data. You can implement a memory by exploiting *any* property of matter that allows you to make a change and then later detect that change. Most computer books will tell you that the first memory systems were the punched cards of the Hollerith tabulator or the wooden cards of the Jaquard loom. Of course, all systems of writing from Egyptian hieroglyphics to pen-on-paper are also memory systems.

The range of physical properties that have been exploited to store data is quite remarkable. Although we've all heard echoes off the walls of a canyon, few people would think of using the echo as the basis of a memory system. In the late 1940s, columns of mercury in tubes were used to store data as *sound in motion*. At one end of the tube, data in the form of a sequence of ultrasonic pulses was transmitted down the tube traveling at the speed of sound *in mercury* (1450 m/s). When the sound reached the far end of the tube, it was picked up by transducers, amplified, and fed back to the other end of the tube. This was a form of dynamic storage, because the data was always physically in motion. Even today glass delay lines are used in some television signal-processing applications to delay a signal by a fixed amount.

Between the 1950s and 1970s, data was stored as a magnetic field in a tiny bead (or *toroid*) of a magnetic material called a *ferrite core* (hence the term *core memory* that often crops up in computer literature). Today's hard disk drives still use the same magnetic phenomenon to store data. Because of its importance, we look at magnetic recording in detail in the next chapter.

From the mid-1970s on, semiconductor memory provided the standard form of main memory either as semiconductor *static RAM* (SRAM) or as *dynamic RAM* (DRAM). Today, only small embedded systems use static RAM, and PCs employ about 2 to 48 GB of DRAM. If DRAM is the *theme* of this chapter, the *variations* are provided by some of the newer forms of semiconductor memory, such as semiconductor *ferroelectric* memory that stores data as the position of an atom within a crystal and *ovonic* memory that stores data by switching a glass-like material called a *chalcogenide* between an amorphous state and a polycrystalline state. We now introduce a vocabulary that allows us to describe memory systems and technologies.

Random Access and Sequential Access Memory

A fundamental distinction between the various memory technologies used in computers is the way in which data is accessed: *directly* or *sequentially*. Memory that is directly accessed is called *random access memory* (RAM), because you can access any data element at random and the time taken to perform the access is constant and effectively independent of the physical location of the data. Such memories are also called *immediate access memories* (IAS). Of course, these memories aren't really *immediate* access—nothing is immediate; they are just so much faster than other types of memory.

Sequential access memory requires you to access each memory element in turn until you locate the element you are seeking. An example of a sequential access memory is magnetic tape storage, where you have to read the tape until you find the item you want. The acoustic mercury delay line memory that we mentioned earlier is also a sequential access memory. Random access memories are invariably faster than sequential access memories, but they are also much more expensive. Most semiconductor memory, such as DRAM or flash memory, is random access. The shift register is, of course, a serial access memory.

We often speak of the *speed* of memory or say how *fast* or *slow* it is. These terms refer to how long it takes to access data. The key parameter of memory is t_{acc}, which is its access time.

Random and Serial Access

You should be aware that serial access is sometimes used differently by hardware and software people. A memory is serial access if the device reading it has to step through several elements to find the desired data. In the software world, a file on a disk is considered random access if you can access an element without reading other elements in a data structure. However, the underlying storage device, the disk, is serial access because it is rotating and you have to wait for the required data to pass under the read head.

Volatile and Nonvolatile Memory

If you store data in an ideal memory, it stays there until you explicitly modify it. Such a memory is called *nonvolatile*. For example, when you write data to a hard disk, the data remains on the hard disk indefinitely. Some memory technologies retain stored data only as long as they receive electric power—pull the plug and the data is gone. These memories are said to be *volatile* because data evaporates in the absence of power. The main stores of most PCs and workstations are composed of DRAM volatile memory. If the memory were nonvolatile, you wouldn't have to boot the computer (i.e., transfer the operating system from nonvolatile memory on disk to volatile memory within the computer) each time you switch the power on. We will later look at a class of nonvolatile memory, called *flash memory*, that is now a mainstream technology and will introduce emerging forms of nonvolatile memory.

Read/Write and Read-Only Memory

If you can write data into memory or read data from memory and perform both operations in approximately the same amount of time, the memory is said to be *read/write* memory. A computer's main store is, of course, composed of read/write memory. If the memory can easily be read but its contents can't be modified, it is said to be *read-only* memory. Read-only memory is invariably nonvolatile.[1]

There is, of course, no true read-only memory. If there were, it would be impossible to put data into it in the first place. Perfect read-only memory would be in the same position as the perfect solvent that can't be stored in any container without dissolving it. Data is loaded into the *mask-programmed* ROM at the time of its manufacture, because the physical structure of each memory cell determines whether it holds a 1 or a 0. Mask-programmed ROM is cheap in volume but cannot be modified later. It was once used to hold bootstrap loaders, operating systems, and the BIOS. Today, it has been rendered obsolete by flash memory.

Practical read-only memory is better described as *read-mostly memory* that can be modified a limited number of times. Moreover, it requires a more complex and slower write operation than a read operation. Examples of read-mostly memory are EPROM, EEPROM, and flash memory. EPROM stands for *electronically programmable read-only memory*, and EEPROM stands for *erasable and electronically programmable read-only memory*. We will look at some of these technologies in more detail later.

Static and Dynamic Memory

Random access, read-write, and volatile memory can be divided into two subclasses: *static* and *dynamic*. These classes refer to the structure of semiconductor memory cells and their properties. Static memory uses cross-coupled transistors to create an RS flip-flop that stores the data in the state of the flip-flop. Dynamic memory (DRAM) employs a semiconductor technology that stores data as an electrostatic charge in a capacitor. Static memory is faster, more expensive, and less dense (bits per chip) than its dynamic counterpart. Dynamic memory is much cheaper than static memory but is more difficult to use in actual circuits. This statement is less true today than in the past. Nowadays, DRAM control is built into CPUs, motherboard bridge chips, and the DRAMs themselves. In the 1980s, building a DRAM controller into your computer was a daunting task.

Pins

Memory components are interfaced through pins (connections). How many connections does a memory need? We have to worry about this because it determines the complexity of the printed circuit or motherboard.

Consider a static RAM organized as 64K 8-bit words. Such a device requires 16 address lines to access one of 64K words and eight data lines. It requires eight data lines because it is byte-organized. Finally, it requires two power-supply lines and a read/write pin to select a memory read or memory write operation and a CS (chip select) line to enable the chip to take part in a data transfer (because it may be one of hundreds of chips in a memory array).

Consequently, the minimum number of connections is 16 + 8 + 2 + 2 = 28.

---

[1] I was going to say that volatile read-only memory is an oxymoron. However, it occurs to me that there is at least one important application of volatile ROM. Can you think of it?

The data in a dynamic memory cell is lost after a few milliseconds unless it is continually rewritten in an operation called *refreshing*. Dynamic memory has different read and write access times, and typical DRAM is not truly random access because adjacent memory cells are faster to access than cells selected at random.

Because DRAM forms the bulk of most PC and workstation memories, the performance and characteristics of DRAM strongly determine the overall performance of computers. We examine DRAM in greater detail later.

Memory Parameters

The smallest unit of memory is the *memory cell* that stores a single bit. Semiconductor memories are organized as an array of n rows by m columns; that is, they contain $n \times m$ cells. The *width* of the memory, m, is the number of bits per word in the memory. When a read or write operation is executed, all m bits of a word take place in the operation simultaneously. The *length* of the memory, n, is defined as the number of addressable locations (i.e., the memory employs n address lines to select 2^n locations).

The width of the memory component is not necessarily the same as the width of the bus or the width of basic data units in the computer. For example, a computer may have a 64-bit data bus and use 4-bit wide memory components. This arrangement requires $64/4 = 16$ memory components arranged side-by-side to *span* the 64-bit data bus, because each memory component contributes four data bits. If each 4-bit memory device has 4 M uniquely addressable locations, we can say that a memory device has a capacity of 4 bits $\times$ 4 M locations = 16 Mbits = 2^{24} bits = 2 MB, and the total storage capacity of the memory system is 16 chips $\times$ 16 Mbits = 2^{28} bits = 32 MB.

Figure 10.2 illustrates how *width* can be used three times in the same diagram—each time with a different meaning. The CPU has 64-bit registers and is properly called a 64-bit machine; that is, it has a 64-bit architecture. The bus between the CPU and memory is 32 bits wide, and a 64-bit data element is fetched from memory in two consecutive bus transactions. You could say that the 64-bit *architecture* has a 32-bit *organization*. Finally, the memory array is composed of four 8-bit chips. Each of these chips contributes 8 bits of the data in a read cycle; that is, the chips are accessed in parallel.

A memory's principal timing parameters are its *read access time*, its *write access time*, and its *cycle time*. The read access time is the time taken to access a memory location and

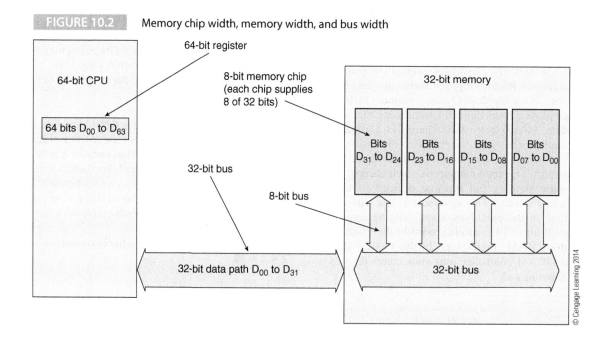

FIGURE 10.2   Memory chip width, memory width, and bus width

© Cengage Learning 2014

to retrieve its contents. The write access time is the time taken to write data into the device. The cycle time is the minimum period that must elapse between two consecutive memory accesses.

Ideally, a memory's read, write, and cycle times should all be the same—this is generally true for semiconductor static RAM. Some memory (e.g., DRAM) has a longer cycle time than a read or write access, because certain internal operations have to take place between consecutive accesses. We have already pointed out that read-mostly devices, such as flash EPROMs, have very much longer write access times than read access times.

A particularly important memory parameter is its *power consumption*. Semiconductor read/write memories require power to operate and to store data. Power consumption is important to those designing battery-operated portable equipment; this is a major concern in today's world of pervasive and personal computing.

The power that's consumed by a memory device (or any other component) ends up in the form of heat and has to be dissipated into the surrounding environment. Power dissipation is of importance because the amount of power being dissipated determines the operating temperature of the memory (or CPU). When the temperature of an object is greater than the ambient temperature, the quantity of the heat dissipated is related to the area of the device dissipating the heat, the efficiency of the heat-dissipating surface, and the difference between the ambient temperature and the device. The power dissipated by a device, $P_{dissipate}$, is given by

$$P_{dissipate} = K \cdot A \cdot (t_{device} - t_{ambient})$$

where K is a constant and A is the area of the dissipating surface.

Since the probability of a component failing is an exponential function of temperature (failure rate doubles for every 15°C rise in temperature), power consumption should be as low as possible. Chips that dissipate a large amount of power, such as processors, are fitted with a *heat sink* to increase their surface area; if the heat dissipating area, A, is increased, the temperature difference ($t_{device} - t_{ambient}$) required to dissipate the heat is reduced.

Table 10.1 summarizes the typical characteristics of the broad groups of memory device we will be looking at in this chapter.

TABLE 10.1	Semiconductor Memory Characteristics		
Characteristic	DRAM	Static RAM	Flash Memory
Static	No	Yes	Yes
Volatile	Yes	Yes	No
Typical size	256 Mbits	64 Mbits	256 Mbits
Organization	4 bits × 64 M	8 bits × 8 M	8 bits × 32 M
Access time	25 ns	2 ns	40 ns
Application	Main store	Cache memory	BIOS, digital film, MP3

© Cengage Learning 2014

10.1.2   Memory Hierarchy

Although we've already introduced *memory hierarchy* in the previous chapter, Figure 10.3 reminds us of this concept because it's the key to understanding the organization of computer memory systems. Here we look at *random access memory*, including both static and dynamic memory. After we've examined *read/write memory*, we introduce the family of *read-mostly* memory that includes flash memory. If low-cost, high-speed random access memory has improved computers by making them much faster and more able to process video information in real time, flash memory has given us the new generation of personal digital systems ranging from hand-held organizers to MP3 players.

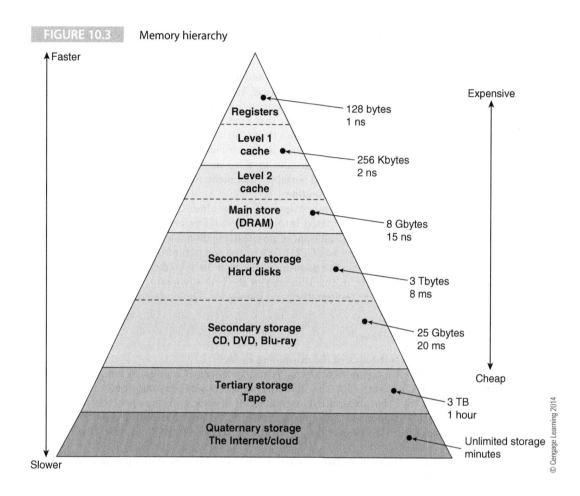

FIGURE 10.3   Memory hierarchy

10.2   Primary Memory

In this section, we are going to look at the immediate access store or primary memory. Without a low-cost, high-speed memory, it would be difficult to implement large operating systems and sophisticated applications, such as Photoshop®. We begin with static memory, SRAM, because it is easier to understand than DRAM and because it was developed before DRAM. Moreover, cache memories are normally implemented with SRAM.

FIGURE 10.4

The principle of static RAM

10.2.1   Static RAM

Figure 10.4 illustrates how static RAM works *conceptually*. Two invertors are connected end-to-end in a ring. The input to gate 1 is A, and its output is $B = \bar{A}$, which is the input to gate 2. The output of gate 2 is A, where $A = \bar{B} = \bar{\bar{A}} = A$. This is also the input to gate 1. The input to gate 1 is fed back to produce the input of gate 1. This is a *self-sustaining* memory element. Whatever state the input to gate 1 is in initially, that state is fed back to maintain itself. This *cross-coupled* circuit is essentially the same as the RS flip-flop we introduced in Chapter 2.

FIGURE 10.5   Operation of the static RAM cell

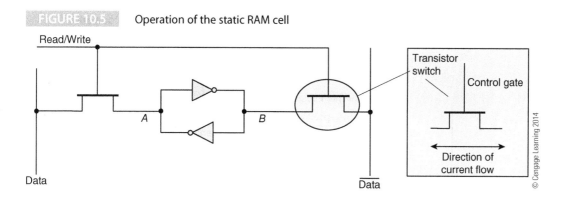

Figure 10.5 illustrates how the two cross-coupled inverters of Figure 10.4 are converted into a practical static memory cell. Two transistors[2] operating as on/off switches are connected to the invertors on the left and right of the circuit and are used to access the memory element (the shaded inset on the right shows that the transistor has three terminals; the signal level on the control gate determines whether the path between the other two terminals is open or closed). When the Read/Write line is negated, these two transistor switches are open and the Data and $\overline{\text{Data}}$ lines are not driven (i.e., controlled) by the invertors.

Data is written into the cell by applying a bit to Data and its complement to $\overline{\text{Data}}$, and then asserting the Read/Write line. Inverter nodes A and B are forced to the desired values and are held in that state when the Read/Write line is negated. A read operation is performed by asserting Read/Write to connect the invertors to the output terminals and reading the level of the signal at the A node.

Because an inverter is implemented by two transistors in series, this static cell requires six transistors to store a bit. A real static memory is arranged as an array of m by m cells. You select an individual cell by addressing its row and its column. Consequently, an n-bit static memory cell contains $6n$ transistors plus the transistors necessary to decode an address into a row and column and to perform the various signal steering and housekeeping operations.

Some static memories use only four transistors per cell. However, the significant point about the n-bit static memory cell is that it requires $4n$ or $6n$ transistors. The dynamic memory cell requires only one transistor per cell, which means that dynamic memories are at least four times the density of static memories. Figure 10.6 gives the circuit diagram of a six-transistor static memory cell. We have included a DRAM cell as an inset for comparison. Static memory is faster than dynamic memory.

Static RAM

Static RAM is, in many ways, less important than it once was. Because of its simpler CPU interface than DRAM, it was once the memory of choice for many (particularly small) computer systems. Moreover, static RAM can retain data in a power-down mode using a small battery, which is a feature that was of great value before flash memory became so widely available. Today, static RAM is of less importance to the designer of large systems because of the economics of DRAM-based memories.

Static RAM is covered here because it is still important (e.g., in small or very low-power systems) and its timing characteristics provide a gentle introduction to the more complicated behavior of the DRAM. Moreover, static RAMs are interfaced to microprocessors rather like some of the memory-mapped peripherals we will discuss in Chapter 12.

However, because static memory can be very fast, it is still used to fabricate cache memories. As early as 1981, experimental static RAMs with access times as low as 0.6 ns were being investigated (these used gallium-arsenide rather than silicon).

---

[2]We don't have the time to go into the operation of transistors here. Just imagine it as a device with three terminals (connections) as described in the inset of Figure 10.5. Two terminals appear to be connected to a switch; that is, a circuit can be made between them or broken between them. The third terminal is a control terminal that changes the state of the switch. If the terminals are A, B, and C (where C is the control terminal), then the transistor behaves as: IF $C == 1$ THEN A is connected to B ELSE A is disconnected from B.

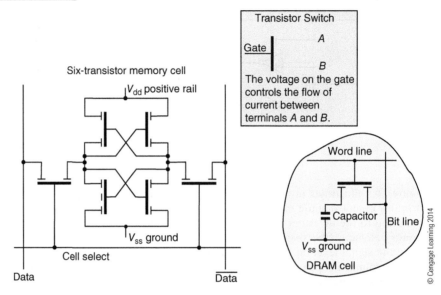

FIGURE 10.6   Circuit of the six-transistor static RAM cell with DRAM comparison

Progress in semiconductor technology is always continuing. In 2001, researchers at Fujitsu Laboratories in Japan announced the design of a single-transistor static memory cell composed of indium-gallium-arsenide (an esoteric semiconducting material in contrast with the more common silicon) operating at $77°K$ or $-196°C$. Such a temperature is far too low for current practical applications. A mechanism called q*uantum tunneling* is used to store data. Other technologies such *nanotubes* are being investigated as possible contenders for future high-density, low-power memories.

A practical semiconductor static RAM chip is composed of an array of individual memory cells. Figure 10.7 illustrates a 16-bit memory array (a real static memory array might contain 2^{24} cells). The 4-bit memory address A_0 to A_3 is divided into a row and a column address. A two-line to four-line decoder decodes the two-bit row address and asserts one of the horizontal lines.

Simultaneously, the column decoder asserts one of the four columns. A one-bit memory cell is located at each row and column intersection and takes part in a read or a write cycle depending on the state of the $\overline{\text{Write}}$ signal. This array illustrates the operation of memory in terms of the logic components we introduced earlier. The actual implementation of a memory array would use the memory cells of Figure 10.6.

Best of Both Worlds

Cyprus Semiconductor has produced a 1 Mbit semiconductor static RAM organized as 128K words of 8 bits or 64K words of 16 bits. The access time is 25 ns. The key feature of this device is the incorporation of a nonvolatile element in each static cell that uses *quantum trap technology* to store data. When the device powers down, data in a static cell is stored in the nonvolatile element and later restored at power-up. Consequently, the device achieves all the advantages of static memory as well as the advantages of non-volatile memory.

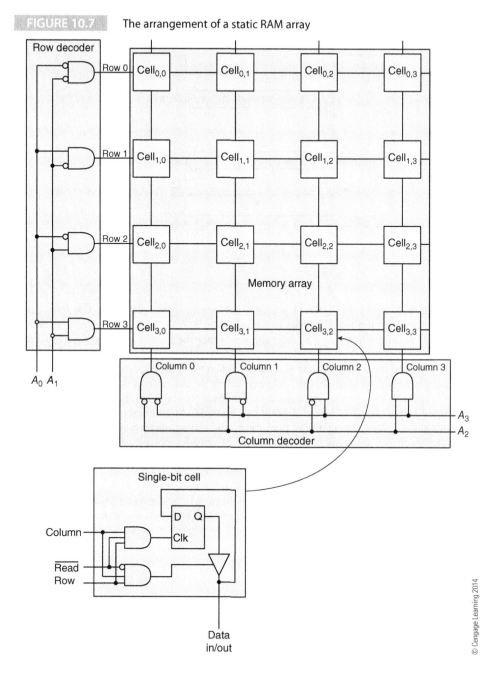

FIGURE 10.7   The arrangement of a static RAM array

The Static RAM Memory System

Let's look at how static memory is used. RAM chips are organized as one-bit wide, nibble wide (4 bits), byte wide (8 bits), or 16-bit wide devices. Figure 10.8 illustrates a 64K word by 8-bit static RAM with a capacity of 512K bits. It has 16 address lines from A_0 to A_{15} that select one of the 2^{16} = 64K memory locations and eight data lines from d_0 to d_7 that transmit eight bits to the processor in a read cycle and receive eight bits from the processor in a write cycle.

A static RAM is accessed by providing an address on its address bus, setting the values of its control inputs to 1s or 0s to define the current operation, and then either applying data to the data bus in a read cycle or reading data off the data bus in a write cycle.

FIGURE 10.8   The static RAM chip

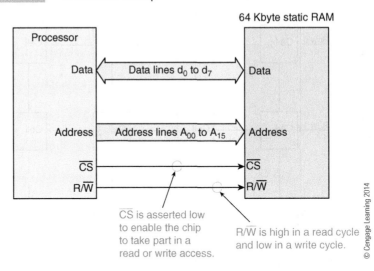

$\overline{CS}$ is asserted low to enable the chip to take part in a read or write access.

R/$\overline{W}$ is high in a read cycle and low in a write cycle.

© Cengage Learning 2014

The operation of a static RAM is controlled by two signals, R/$\overline{W}$ and $\overline{CS}$. R/$\overline{W}$ determines whether the chip is taking part in a read cycle, R/$\overline{W}$ = 1, or a write cycle, R/$\overline{W}$ = 0. The active-low chip select line, $\overline{CS}$, determines whether the memory is to take part in a read or write access, or whether it remains in an idle state. In normal operation, the $\overline{CS}$ line is inactive-high and the signal on R/$\overline{W}$ is ignored. When $\overline{CS}$ is asserted active-low, the memory takes part in a read cycle or a write cycle depending on the state of the R/$\overline{W}$ line. In a read cycle, $\overline{CS}$ controls the memory's tristate buffer outputs (covered in Chapter 2).

Figure 10.9 provides a *timing diagram* of a static RAM and demonstrates the sequence of events taking place during a read operation. The timing diagram specifies the minimum times for which you must apply signals to the memory for correct operation, and it states the maximum period of time that may elapse between the initiation of an action and its conclusion. The lines with arrows indicate cause and effect; for example, when $\overline{CS}$ goes low at C (cause) the data bus drivers are turned on at E (effect).

Assume that initially (left-hand side of the diagram) $\overline{CS}$ is inactive high and that R/$\overline{W}$ is at a 1 state for the entire duration of the operation. R/$\overline{W}$ is not shown in the diagram.

FIGURE 10.9   The static RAM read cycle timing diagram

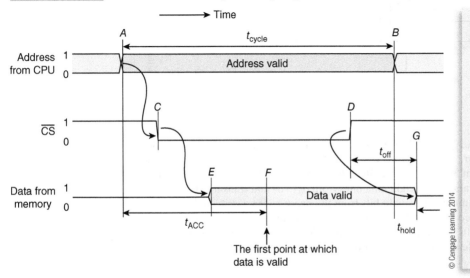

The first point at which data is valid

The parameters define the time to perform an action or the time required by an action. t_{cycle} is the minimum time required to perform a read cycle. t_{ACC} is the access time, which is the longest time you have to wait for data. t_{off} is the time taken for the data bus to float when $\overline{CS}$ goes high. t_{hold} is the time that the data will be valid after the address changes.

© Cengage Learning 2014

At point A in Figure 10.9, the contents of the address bus change state to provide a valid address for the current read operation. The change of state is indicated by the cross-over of the lines on the address, and the blue shading indicates the period for which the current address is valid. Once the address is stable, the chip select input is asserted at point C to begin a memory access. Since R/$\overline{\text{W}}$ is high, the access is a read cycle.

The effect of R/$\overline{\text{W}}$ high and $\overline{\text{CS}}$ low is to cause the memory's data bus drivers to put data onto the memory's data bus. The data bus leaves the floating state, and data appears on the bus at point E. However, at this point, the data is not valid, because the access time of the chip hasn't elapsed.

At point F which is t_{ACC} seconds after the address first *becomes* valid, the data may be read by the host processor. At point D, the processor negates $\overline{\text{CS}}$ to finish the read cycle, and at point G, the data bus once more assumes its high-impedance floating state. The period between the address changing at point B and the data changing at point G is called the *data hold time*. The data hold time is the period for which the data lingers on the bus after the address and/or control signals have changed. Some systems require a positive (non-zero) data hold time in order to ensure that the data is reliably captured.

The Write Cycle

Figure 10.10 illustrates the write cycle timing diagram of a generic static RAM, which is a little more complex than the corresponding read cycle (this statement is true of all memory devices). The details vary from chip to chip, but the basic principles are the same. The data to be stored is applied to the data bus, and then $\overline{\text{CS}}$ and R/$\overline{\text{W}}$ are asserted low to trigger the write cycle.

In Figure 10.10, the address becomes valid at point A and remains valid until the end of the write cycle at point B. We will assume that the $\overline{\text{CS}}$ input goes low at point C. At this time, R/$\overline{\text{W}}$ is still high and the data bus is floating.

At point C, an access has begun, but as far as the RAM is concerned, it is a read cycle because R/$\overline{\text{W}}$ is still high. So, at point G, the RAM starts to drive the data bus with data. Of course, this data is invalid because the access time hasn't yet elapsed.

At point E, the host processor drives R/$\overline{\text{W}}$ low to indicate that the current cycle is a write cycle. The memory now terminates the read cycle it was beginning to initiate and stops driving the data bus at point H. As you can see, the memory has placed spurious data on the data bus between points G and H. The systems designer has to take care that no other device drives the bus at the same time.

FIGURE 10.10   The static RAM write cycle timing diagram

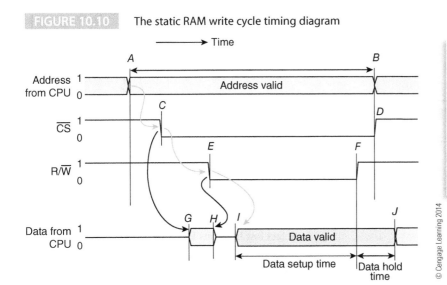

The *data setup time* is the minimum time for which the data must be valid and stable before it is captured.

The hold time is the minimum time for which the data must be held after it has been captured.

At point I, the processor drives the data bus with the data to be stored. The key event in a write cycle occurs at point F when the rising edge of the $R/\overline{W}$ signal is used to capture the data. The data must be valid for its setup time before the rising edge of $R/\overline{W}$ and remain valid for the *data hold time* after the rising edge. Real static RAMs are designed to terminate a write cycle when the *first* of $\overline{CS}$ or $R/\overline{W}$ goes high at the end of a write cycle.

Having described static RAM, it's natural to look at *dynamic* RAM. However, before we do that, this is a convenient point to introduce two issues that affect the design of all memory systems, whether they be static, dynamic, read-write, or read-only memory. These are the control of byte and word accesses and the use of address decoding.

Byte/Word Control

In the 1970s, microprocessors were byte-oriented, and data buses were eight bits wide. Address buses were 16 bits wide, and an address on A_{15} to A_{00} selected one of $2^{16} = 64K$ unique bytes. When 16-bit processors were introduced, the situation became more complex because of the need to access both bytes and 16-bit words (remember that memory is byte-addressed, and you can access an individual byte even though the fundamental wordlength may be two, four, or more bytes).

Because microprocessor designers wanted the best of both worlds with the ability to access individual bytes as well as 16-bit words, they implemented byte-control mechanisms allowing access to the individual bytes of a selected word. A typical mechanism uses the address bus to select a 16-bit word (or a 32-bit or a 64-bit) word and then uses *byte control lines* to select one or more of the bytes at that address. Figure 10.11 demonstrates a possible arrangement in which one of 2^{15} 16-bit words is selected by address lines A_{01} to A_{15}.

Address line A_{00} is not required, because two bytes are always selected by each address; that is, the addresses are $0, 2, 4, 6, 8, \ldots, 2^{15}$. The two *byte select strobes*, BS_0 and BS_1, select either the odd byte of the addressed word or the even byte of the addressed word, or both bytes (Table 10.2).

FIGURE 10.11   Byte and word control

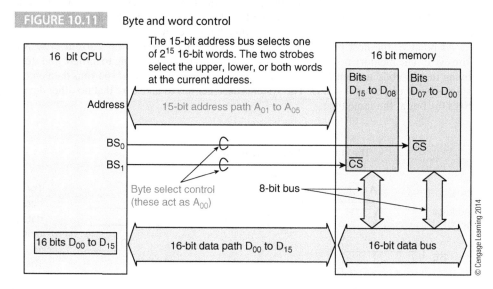

The 15-bit address bus selects one of 2^{15} 16-bit words. The two strobes select the upper, lower, or both words at the current address.

16 bit CPU

16 bit memory

Address — 15-bit address path A_{01} to A_{05}

BS_0

BS_1

Byte select control (these act as A_{00})

8-bit bus

Bits D_{15} to D_{08}

Bits D_{07} to D_{00}

$\overline{CS}$

$\overline{CS}$

16 bits D_{00} to D_{15}

16-bit data path D_{00} to D_{15}

16-bit data bus

© Cengage Learning 2014

TABLE 10.2   Byte Selection in a Memory

BS_1	BS_0	Operation
1	1	No operation
1	0	Select low byte bits D_0 to D_7
0	1	Select high byte bits D_8 to D_{15}
0	0	Select both bytes, bits D_0 to D_{15}

© Cengage Learning 2014

Dealing with Ranges of Memory Addresses

The address inputs of a memory device *span* its *address space* by including every addressable location; for example, if the memory has six address pins, the address range is 000000, 000001, 000010, . . . , 111111.

In order to avoid cumbersome binary arithmetic, hexadecimal arithmetic is used to express addresses; for example, a 6-bit address device is spanned by the range of addresses 00 to 2F.

Suppose you have a computer with a 64 MB address space spanned by address lines A_{00} to A_{25} and you are using 8 MB memory modules spanned by address lines A_{00} to A_{22}. A location within a memory module is selected by address lines A_{00} to A_{22}, and address lines A_{23} to A_{25} from the computer are used to select this module out of eight possible 8 MB blocks of address space (8×8 MB $= 64$ MB). The address spaces that can be occupied by the memory module are

    000 0000   to   03F FFFF
    040 0000   to   07F FFFF
    080 0000   to   0BF FFFF
    0C0 0000   to   0FF FFFF
    100 0000   to   13F FFFF
    140 0000   to   17F FFFF
    180 0000   to   1BF FFFF
    1C0 0000   to   1FF FFFF

If we wish to locate the 8 MB block in the address space 0C0 0000 to 0FF FFFF, the most-significant three bits of the address bus would have to be 011. That is, A_{25}, A_{24}, A_{23} would have to be 0, 1, 1 to select an address in that block. A simple logic element could accomplish this address decoding operation.

Suppose you have a small 8-bit microprocessor-based controller with a 16-bit address bus and two 8 K $\times$ 8 flash memories plus two 4 K $\times$ 8 static RAMs. We will locate the flash memory at high addresses (where the interrupt vector tables are) and the static RAM at low memory. Below is a decoding table for this arrangement. R1, R2, S1, and S2 are the signals that select the four memory components. As you can see, R1 responds to addresses in the range 0000-0FFF, R2 to 1000-1FFF, S1 to C000-DFFF, and S2 to E000-FFFF.

	Address lines																Device selects			
	A_{15}	A_{14}	A_{13}	A_{12}	A_{11}	A_{10}	A_{09}	A_{08}	A_{07}	A_{06}	A_{05}	A_{04}	A_{03}	A_{02}	A_{01}	A_{00}	R1	R2	S1	S2
0000-0FFF	0	0	0	0	x	x	x	x	x	x	x	x	x	x	x	x	1	0	0	0
1000-1FFF	0	0	0	1	x	x	x	x	x	x	x	x	x	x	x	x	0	1	0	0
2000-2FFF	0	0	1	0	x	x	x	x	x	x	x	x	x	x	x	x	0	0	0	0
3000-3FFF	0	0	1	1	x	x	x	x	x	x	x	x	x	x	x	x	0	0	0	0
4000-4FFF	0	1	0	0	x	x	x	x	x	x	x	x	x	x	x	x	0	0	0	0
5000-5FFF	0	1	0	1	x	x	x	x	x	x	x	x	x	x	x	x	0	0	0	0
6000-6FFF	0	1	1	0	x	x	x	x	x	x	x	x	x	x	x	x	0	0	0	0
7000-7FFF	0	1	1	1	x	x	x	x	x	x	x	x	x	x	x	x	0	0	0	0
8000-9FFF	1	0	x	x	x	x	x	x	x	x	x	x	x	x	x	x	0	0	0	0
A000-BFFF	1	0	x	x	x	x	x	x	x	x	x	x	x	x	x	x	0	0	0	0
C000-DFFF	1	1	x	x	x	x	x	x	x	x	x	x	x	x	x	x	0	0	1	0
E000-FFFF	1	1	x	x	x	x	x	x	x	x	x	x	x	x	x	x	0	0	0	1

Address Decoding

In this section, we take a brief look at the art of address decoding, which is the way address components are mapped onto a processor's physical address space. This material uses concepts from Chapter 2 (binary/hexadecimal numbers and simple logic). Because address decoding is more the province of the embedded system designer rather than the computer architect, some students may wish to omit this on a first reading.

In the best of all possible worlds, a computer would have an n-bit address bus spanning 2^n words and a memory component with n address inputs and 2^n locations. In practice, a real system may have several different memory devices that are connected to the processor's buses. Suppose a computer has a processor with a 32-bit address bus spanning 2^{32} bytes, and a memory module (a memory component is a single chip, whereas a *memory module* is a circuit board containing several memory chips) in this computer provides 2^{29} bytes (512 MB) of storage; that is, it would take $2^{32}/2^{29} = 2^3 = 8$ modules to completely fill the processor's memory space. We need a means of mapping the address space of a memory module onto the processor's address space. A circuit called an *address decoder* detects an access to a particular memory module and asserts its $\overline{CS}$ signal whenever the CPU generates an address falling within the module's address space.

Figure 10.12 illustrates why we need address decoding. Assume that a processor has an address space of 4,096 MB spanned by address lines A_{00} to A_{31}; that is, $2^{32} = 4,096$ MB. The system uses three 512 MB memory modules—each spanned by address lines A_{00} to A_{28}.

Figure 10.13 shows how three 512 MB modules can be mapped onto the processor's memory space. Although we have mapped the modules into contiguous regions of memory space, there's no reason why physical memory has to be mapped onto adjacent blocks. Of course, if the memory isn't contiguous, the programmer or operating system has to take care when loading programs and data. Module 1 has been assigned the address space 0000 0000 to 1FFF FFFF, module 2 is assigned 2000 0000 to 3FFF FFFF, and module 3 gets 4000 0000 to 5FFF FFFF. Each of these blocks is, of course, 512 MB and is spanned by A_{00} to A_{28}.

Figure 10.14 shows how we can design the logic required to perform the mapping of memory modules onto a processor's address space. The 29 address lines from each of the three modules are connected to the same address lines from the processor, A_{00} to A_{28}. This arrangement leaves three address lines from the processor, A_{29} to A_{31}, to select the three memory modules; that is, these address lines generate the signals that select (enable) the memory modules.

FIGURE 10.12   The relationship between processor address space and memory space

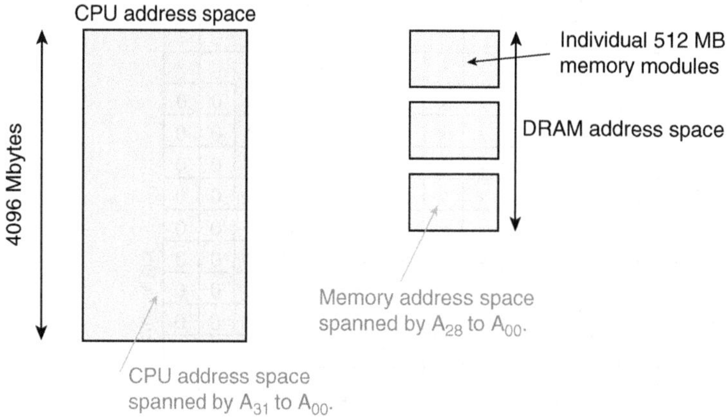

CPU address space

4096 Mbytes

CPU address space spanned by A_{31} to A_{00}.

Individual 512 MB memory modules

DRAM address space

Memory address space spanned by A_{28} to A_{00}.

This diagram shows the processor's 4,096 MB memory space and the memory space of each of the three 512 MB memory modules. Address decoding maps the address space of each of the memory modules onto that of the processor.

© Cengage Learning 2014

FIGURE 10.13   Mapping memory space onto processor space

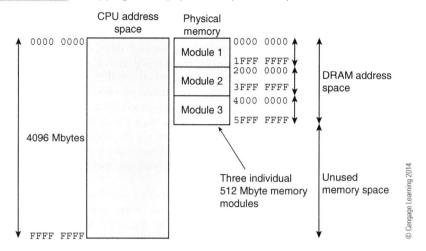

FIGURE 10.14   Implementing an address decoder

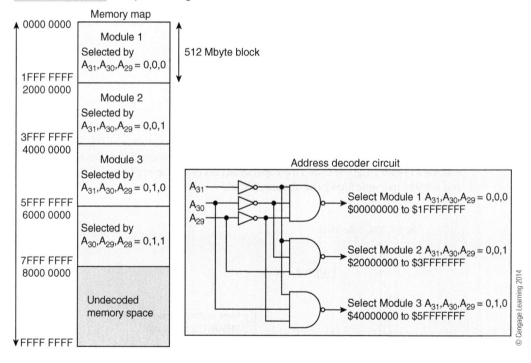

The first block of memory spans the 512 MB region $0000\ 0000_{16}$ to $1FFF\ FFFF_{16}$. This block is selected whenever A_{31}, A_{30}, $A_{29} = 0,0,0$. Similarly, the second block of memory extends from $2000\ 0000_{16}$ to $3FFF\ FFFF_{16}$ and is selected whenever A_{31}, A_{30}, $A_{29} = 0,0,1$.

We do not go further into address decoding here, as that is the province of the systems designer and electronic engineer. Although address decoding can be used to match the memory space of memory components to a CPU, it can do more. For example, you can electronically switch between multiple blocks of memory to ensure that some tasks are physically isolated from each other (a form of memory management).

10.2.2   Interleaved Memory

If you have a memory system with an access time of t_{acc}, there's nothing you can do to reduce that access time. You can reduce the *effective access* time by *interleaving*. If a memory location is accessed at time T_0, the data becomes available t_{acc} seconds later. Another location in the *same* memory can't be accessed until at least t_{acc} has elapsed. But, you can access a *different* module at any time; that is, you access multiple memories in parallel.

Figure 10.15 illustrates interleaving in which two banks of memory are arranged side by side in parallel. If you access bank 1 at T_0, the data becomes available t_{acc} seconds later. If you access bank 2 at time $T_0 + t_{cyc}$ seconds (where t_{cyc} is the cycle time), the second data element is available at $T_0 + t_{cyc} + t_{acc}$ seconds. If t_{cyc} is less than t_{acc}, the second access is completed earlier than it would have been without interleaving. Interleaving is effective only if you can generate the address of an operand in a different bank while an operand in the current bank is being retrieved.

One memory interleaving technique is called *low-order interleaving*, in which the low-order bits of a memory perform memory bank selection. Suppose a processor has a 32-bit address bus and a 64-bit data bus. Address bits A_{03} to A_{31} select a word in memory and address bits A_{00} to A_{02} (via byte selection strobes) select one or more bytes in the current word. By using address bit A_{04} to perform bank selection, all words at odd addresses are fetched from one bank and all words at even addresses are fetched from the other bank. If two address bits were used to perform interleaving, it would be possible to have four banks of memory that could be accessed in parallel. In modern PCs, dual- or triple-channel memory performs essentially the same function as interleaving by allowing the simultaneous access of two or three DRAM memory modules.

Size/Performance of Memory

Although static RAM has been eclipsed by DRAM, static RAMs are still in widespread use. Like DRAM, its production has been centered round a few manufacturers. The figure below using information from a paper by Reneas Electronics presented at ISSCC2009 (3/3) Volume 85 (April 16, 2009) shows the capacity/access time graphs of static RAM, DRAM, and MRAM (magnetic RAM to be discussed later).

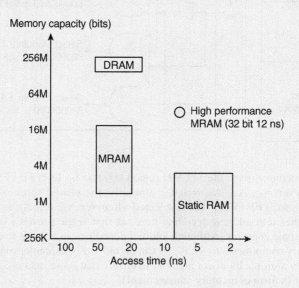

FIGURE 10.15   Interleaving memory

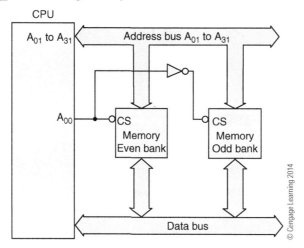

We next examine the operating principles of the DRAM because of its continued evolution through several generations of devices that all rely on the same fundamental operating principle but use different techniques to reduce average access times. We describe basic DRAM and discuss variants such as page-mode DRAM, nibble-mode DRAM, EDO memory, SDRAM, or DDR DRAM.

10.3 DRAM

Most PCs and workstations implement their main memories with DRAM, which has played a key role in the development of high-performance computers, where its operating parameters are continually changing as new variants are developed. An understanding of DRAM is necessary to appreciate where computers are heading in the near future. We begin this section by describing the DRAM's operating principles, look at its timing requirements, introduce the DRAM family, and then describe how they are used in PCs.

The operating principle of *dynamic random access memory* (DRAM) is very simple. Data is stored in the form of an electrostatic charge within a transistor. When a material is uncharged, the number of electrons in the material equals the number of protons; that is, negative electrons cancel out the charge on positive protons and there is no overall charge. When a material is charged, electrons are either added or removed. If there are too few electrons, the charge is positive, and if there are too many electrons, the charge is negative. Information can be stored a bit at a time as a charge or no-charge on a tiny capacitor. All we need is a means of writing the charge into a capacitor, removing it, and detecting its presence or absence.

DRAM History

DRAM is just a little older than the microprocessor itself. Indeed, the world's first commercial DRAM chip was Intel's 1103 1024-bit memory.

The concept of the one-bit dynamic memory cell using a single transistor dates back to R. H. Denning's work at IBM in 1966. The first DRAM was fabricated with positive channel metal oxide semiconductor (PMOS) technology, which is now obsolete. A few years later, 4K-bit DRAMs were built with negative channel metal oxide semiconductor (NMOS) technology. The introduction of complementary metal oxide semiconductor (CMOS) technology was a breakthrough because of its much reduced power consumption. DRAMs still use CMOS technology.

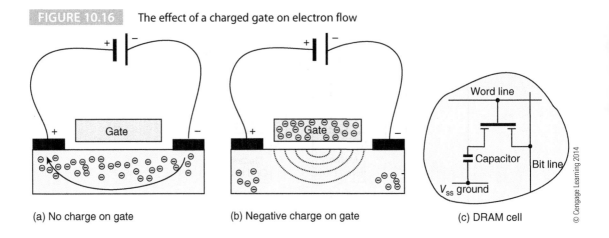

FIGURE 10.16   The effect of a charged gate on electron flow

(a) No charge on gate   (b) Negative charge on gate   (c) DRAM cell

Figure 10.16 illustrates the basic principle of the field effect transistor from which most logic devices are fabricated. A tiny region of *doped* silicon on a chip has two connections to the positive and negative terminal of a battery. The term *doped* means that an impurity has been added to the silicon to provide a supply of electrons that are free to carry a current through the silicon (the electrons in pure silicon are bound to the atoms and can't move through the material). In Figure 10.16a, a current flows between the two terminals.

Above the region of silicon that conducts the electricity is a conductor labeled *gate* in Figure 10.16. We will assume that this conductor is electrically insulated from the silicon channel through which the electrons flow between the terminals in Figure 10.16a.

In Figure 10.16b, a negative charge is applied to the gate. This charge creates an electrostatic field that penetrates the silicon channel. Because the charge from the gate is negative, the electrons in the channel are repelled away from the gate (like charges repel and unlike charges attract). In Figure 10.16b, the charge is so strong that the channel is said to be *pinched off*, and no current can flow through the channel.

By putting or not putting a negative charge on the gate, the current flowing through the channel can be controlled. This ability to cut off a flow of electrons is used by all gates. It is also used by the DRAM and the class of read-mostly memory that includes EPROM, EEPROM, and flash EPROM.

Suppose we apply a negative voltage to the gate of a transistor in order to place a charge on the gate and turn the transistor off. If the voltage is removed, the transistor does not begin to conduct again until the charge is dissipated. If the gate is insulated from the channel, the time taken to dissipate the charge may be several milliseconds.

First-generation DRAMs used a three-transistor memory cell. One transistor switches a charge onto the capacitor and two transistors read the value of the charge. Figure 10.16c illustrates a modern one-transistor DRAM cell that uses only one transistor and a capacitor per bit.

This stored charge eventually leaks away, and any data stored in the cell is lost leaving all cells in the same state. In order to exploit the memory effect of a stored charge, a practical memory has to read the state of the transistor every few milliseconds and then rewrite the charge back into the transistor. This operation is called *refreshing*.

Refreshing the first generation of DRAM memories was a nightmare and required a lot of overhead in the form of complex circuits and timers. Wherever possible, designers used static RAM to avoid the horrors of DRAM systems design. Today, the situation is better because the refresh requirements of DRAMs have been greatly simplified by putting the refresh logic on the DRAM chip itself.

Threat of Radiation

DRAMs suffer from the so-called *alpha-particle problem*. The stored charge is exceedingly small, and an alpha particle (created by the radioactive decay of an atom) passing through a memory cell can cause sufficient ionization to corrupt the stored data. The alpha particle creates a *soft error*, because the cell has lost its stored data but has not been permanently damaged. Alpha-particle contamination occurs largely in encapsulating material.

Semiconductor manufacturers minimize the problem by careful quality control of the material used to encapsulate the chip. However, soft errors due to cosmic radiation can't be prevented. The soft-error rate can't be reduced to zero and the probability of a soft error in a 1 Gbit DRAM memory is of the order of one in 2^{11} per hour (i.e., about the same chance as winning the lottery…).

Systems requiring high reliability use error-correcting memories where each word stores extra bits that can be used to repair a soft error. A typical error-detecting and correcting code (EDC) is the Hamming code.

Figure 10.17 illustrates a DRAM's structure. At first sight, it looks very similar to a static memory chip except for two active-low control inputs, *row address strobe* ($\overline{\text{RAS}}$) and *column address strobe* ($\overline{\text{CAS}}$) instead of a chip select ($\overline{\text{CS}}$). The DRAM interface differs from the static RAM interface in one important aspect: Dynamic memories reduce the number of address pins by using a *multiplexed address bus*. That is, an address is loaded in two steps. For example, a 256 Mbit chip with a 28-bit address has a 14-bit address input and the 28-bit address must be loaded as two consecutive 14-bit values. At the start of a memory access, the chip is provided with a 14-bit *row address* to select the row in which the desired cell is located, and then a second 14-bit address is applied to the same 14 pins to supply a *column address*. A pulse on $\overline{\text{RAS}}$ captures the row address, and a pulse on $\overline{\text{CAS}}$ captures the column address. The cell at the selected row and column location is then accessed.

Reducing the address bus requirement to 14 pins instead of 28 requires two *strobes* (clocks) to latch the address. The row address strobe ($\overline{\text{RAS}}$) latches the 14-bit row address and then the column address strobe ($\overline{\text{CAS}}$) latches the 14-bit column address. Address multiplexing and the control of $\overline{\text{RAS}}$ and $\overline{\text{CAS}}$ strobes are carried out off-chip with logic supplied by the user.

FIGURE 10.17   The inputs and outputs of a DRAM memory element

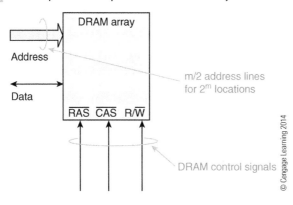

FIGURE 10.18   Structure of a DRAM chip

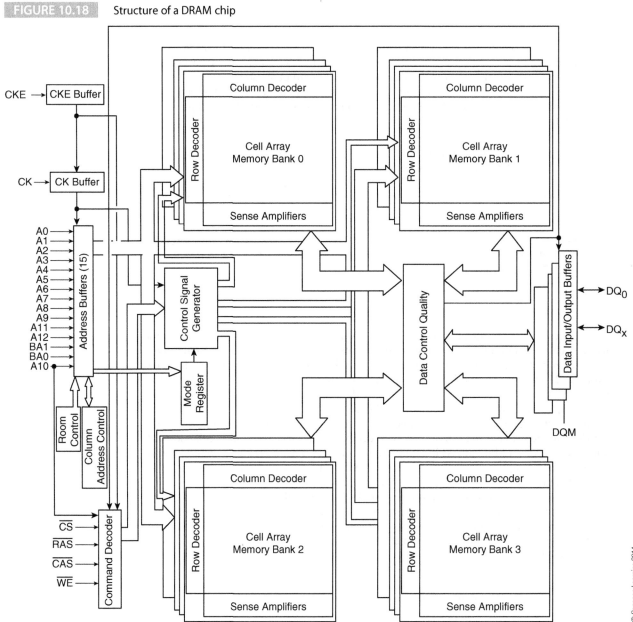

Figure 10.18 illustrates the internal structure of a DRAM memory. The most significant point to note is that there is not a single memory array. Commercial chips consist of typically four memory arrays or banks.

Figure 10.19 illustrates the structure of a 4 MB DRAM system spanned by A_{00} to A_{21} constructed from 32 one-bit by 1 M location DRAM chips. Address lines from the computer select one of 2^{20} 32-bit words using address bits A_{02} to A_{21}. The DRAM memory subsystem uses a multiplexer to select either the row or column address from the computer's address bus. A control unit (see Figure 10.19) generates the DRAM's $\overline{RAS}$, $\overline{CAS}$, and $\overline{W}$ signals from the CPU's control signals.

Having introduced the basic concept of DRAM, the next step is to look at its timing, because that is the key to all the later variants of DRAM.

FIGURE 10.19 Structure of a DRAM system; note that byte selection logic is not shown

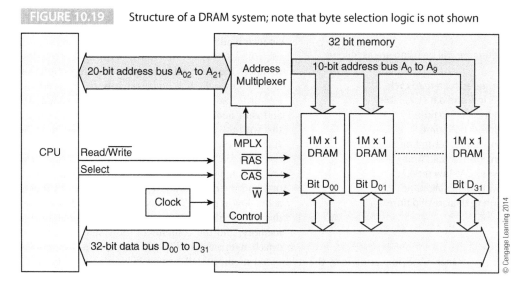

10.3.1   DRAM Timing

We begin with the timing of the classic DRAM of the 1980s from which all modern variants are derived. We have to understand the nature of DRAM timing in order to appreciate how successive generations have been made faster, although the nature of the one-transistor DRAM memory cell has not changed. Figure 10.20 presents a simplified read-cycle timing diagram of a basic DRAM chip. We'll assume a 1 Mbit by 1 organization with 2^{20} locations.

Twenty address lines, A_{00} to A_{19}, from the CPU are fed to a multiplexer to select either A_{00} to A_{09} (the row address) or A_{10} to A_{19} (the column address). The ten outputs of the CPU address multiplexer (MPLX) are connected to DRAM address inputs A_0 to A_9.

A read cycle in Figure 10.20 lasts from A to B and has a minimum duration of t_{RC}, which is the *read cycle time*. Unlike static memory with equal cycle and access times, dynamic memory has a longer cycle time than its access time. A DRAM can't begin a new access as soon as the current one has been completed because it performs an internal operation, known as *pre-charging*, between accesses.

FIGURE 10.20   DRAM read cycle timing

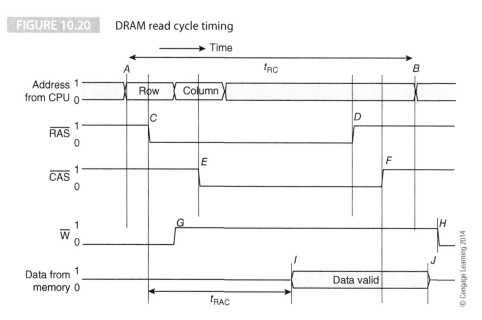

DRAM Parameters

Some of the timing parameters we will be using are as follows:

t_{RC}	maximum time required for a read cycle
t_{RAC}	time between $\overline{RAS}$ low and data available
t_{ASR}	minimum row address setup time
t_{RAH}	minimum row address hold time
t_{ASC}	minimum column address setup time
t_{CAH}	minimum column address hold time
t_{RCD}	minimum $\overline{RAS}$ low to $\overline{CAS}$ low time
t_{CAC}	minimum $\overline{CAS}$ low to data valid time
t_{OFF}	minimum $\overline{CAS}$ high to data invalid time

The first step in a read cycle is to provide the chip with the lower-order bits of the CPU address on its ten address inputs, A_0 to A_9. Then, at point C, the *row address strobe* ($\overline{RAS}$) is brought active-low to latch the row address into the chip's internal latches. Once the row address has been captured, the low-order address from the processor is redundant and isn't needed for the rest of the cycle. Compare this with the static RAM where the address must be stable for the entire read or write cycle.

The ten higher-order address bits from the CPU are then applied to the address inputs of the memory, and the column address strobe ($\overline{CAS}$) is brought active-low at point E to latch the column address. Now the entire 20-bit address has been acquired by the memory, and the contents of the system address bus can change, since the DRAM has captured the address.

Once $\overline{CAS}$ has gone low, the addressed memory cell responds by placing data on its data-output terminal, allowing the CPU to read it. At the end of a read cycle, $\overline{CAS}$ returns inactive-high, and the data bus drivers are turned off, floating the data bus. $\overline{RAS}$ and $\overline{CAS}$ may both go high together or in any order. It doesn't matter which goes high first, as long as all other timing requirements are satisfied. To make our explanation of the DRAM more tractable, we will divide it into its component parts beginning with a discussion of the role of the address pins.

Details of the DRAM's address timing requirements are given in Figure 10.21. The row address must be stable for a minimum of t_{ASR} seconds (i.e., *row address setup time*) *before* the falling edge of the $\overline{RAS}$ strobe. As the minimum value of t_{ASR} is quoted as 0 ns, the row address has a zero setup time and does not have to be valid prior to the falling edge of $\overline{RAS}$. In the worst case, it must be valid coincident with the falling edge of $\overline{RAS}$. After $\overline{RAS}$ has gone low, the row address must remain stable for t_{RAH} seconds, which is the *row address hold time*, before it can change. The hold time restricts the time before which the column address may be multiplexed onto the chip's address pins.

FIGURE 10.21   Address timing of a DRAM

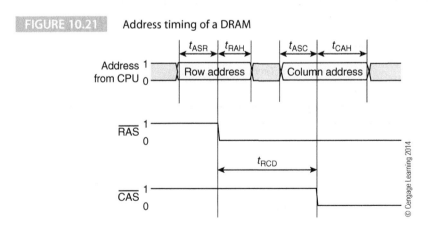

© Cengage Learning 2014

Once the row address hold time has been satisfied and the column address multiplexed onto the memory's address pins, $\overline{CAS}$ may go low. The column address setup time, t_{ASC}, is typically 0 ns minimum; that is, $\overline{CAS}$ may go low at the *same* time that the column address becomes valid. After $\overline{CAS}$ has gone active-low, the column address must be stable for a further t_{CAH} seconds, which is the column address hold time, before it may change. Once t_{CAH} has been satisfied, the address bus plays no further role in the current access.

The row address must be valid for t_{ASR} seconds *before* the falling edge of the row address strobe and remain valid t_{RAH} seconds *after* it. Similarly, the column address must be valid t_{ASC} second before and t_{CAH} seconds after the falling edge of the column address strobe. The minimum time between the falling edge of $\overline{RAS}$ and the falling edge of $\overline{CAS}$ is t_{RCD}, which is made up of the row address hold time, the multiplexer switching time, and the column address setup time.

An important parameter in Figure 10.21 is t_{RCD}, which is the *row-to-column strobe lead time*, derived from other parameters. The minimum value of t_{RCD} is determined by the row address hold time plus the time taken for the address from the multiplexer to settle, plus the column address setup time. The maximum value of t_{RCD} is a *pseudomaximum* that, if exceeded operationally, extends the access time of the memory. The relationship between $t_{RCD}(\max)$, t_{RAC}, and t_{CAC} is $t_{RCD}(\max) = t_{RAC} - t_{CAC}$.

Having latched an address by asserting $\overline{RAS}$ and $\overline{CAS}$ in turn, data appears at the chip's data pin as depicted in Figure 10.22. Only the $\overline{RAS}$, $\overline{CAS}$, and the data signals are included in Figure 10.22 for clarity. We assume that $\overline{W}$ is high for the duration of the read cycle and that the address set up, hold times, and all relevant parameters have been satisfied.

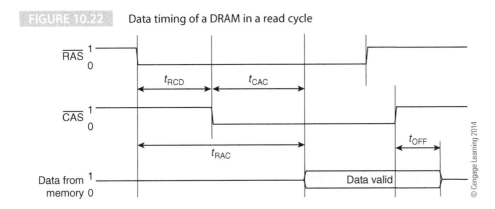

FIGURE 10.22   Data timing of a DRAM in a read cycle

Following the falling edge of $\overline{RAS}$, the data at the data pin is valid no later than t_{RAC}, which is the access time from row address strobe. The row access time is achieved only if other conditions are met, as we shall see. The column address strobe serves two functions: it latches the column address that interrogates the appropriate column of the memory array, and it turns on the data output buffers. After the falling edge of $\overline{CAS}$, data is not available for at least t_{CAC} seconds, which is the access time from $\overline{CAS}$ low.

When $\overline{CAS}$ goes high at the end of a read cycle, the data bus drivers are turned off and the bus floats t_{OFF} seconds later (t_{OFF} = output buffer turn-off delay). $\overline{RAS}$ does not play any part in the ending of a read (or write) cycle. $\overline{RAS}$ may be negated before or after $\overline{CAS}$, as long as its timing requirements are met.

Data becomes valid not more than t_{CAC} seconds after the falling edge of $\overline{CAS}$ and not more than t_{RAC} seconds after the falling edge of $\overline{RAS}$. At the end of a cycle, the data bus buffer is turned off no later than t_{OFF} seconds after the rising edge of the first of $\overline{RAS}$ or $\overline{CAS}$.

The final aspect of the DRAM's read cycle timing diagram we have to consider is the timing requirements of the row and column address strobes described in Figure 10.23. The $\overline{RAS}$ and $\overline{CAS}$ clocks latch addresses and control internal operations and the three state buffers.

A fundamental parameter of Figure 10.23 is t_{RC}, which is the *read cycle time*—the minimum time that must elapse between successive memory cycles. The $\overline{RAS}$ strobe must be asserted for at least t_{RAS} seconds (the row address strobe pulse width) during each read access. This has a typical maximum value of 10,000 ns that is related to the need to refresh the device and creates no problems, as it is many times longer than a microprocessor's read cycle. After $\overline{RAS}$ has been negated, it must remain high for at least t_{RP} seconds, which is the *row address strobe precharge time* (a characteristic of dynamic memories relating to an internal operation

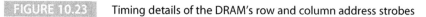

FIGURE 10.23   Timing details of the DRAM's row and column address strobes

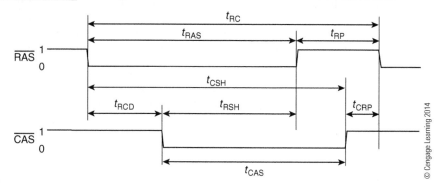

in the chip). The final constraint on the timing of $\overline{\text{RAS}}$ is its hold time with respect to $\overline{\text{CAS}}$, t_{RSH}. $\overline{\text{RAS}}$ must remain low for at least t_{RSH} seconds after $\overline{\text{CAS}}$ has been asserted.

The column address strobe timing requirements are similar to those of the row address strobe. $\overline{\text{CAS}}$ must be asserted for no less than t_{CAS} seconds, it must be negated for at least t_{CRP} seconds before the falling edge of the next $\overline{\text{RAS}}$ clock, and it must be asserted for at least t_{CSH} seconds measured from the falling edge of the current $\overline{\text{RAS}}$ clock.

Write-Cycle Timing

A DRAM's write cycle is rather more complex than the corresponding read cycle, because stringent requirements are placed on both its $\overline{\text{W}}$ and data inputs. Having worked through the read cycle timing diagram, we don't need to go through the same material again. Figure 10.24 gives a simplified DRAM write cycle timing diagram. This is an *early write cycle*, because the $\overline{\text{W}}$ input is asserted *before* $\overline{\text{CAS}}$ goes low. Some DRAMs implement a write cycle in which the $\overline{\text{W}}$ input is asserted *after* $\overline{\text{CAS}}$ goes low. The timing requirements of the $\overline{\text{RAS}}$, $\overline{\text{CAS}}$, and address inputs are identical in both read and write cycles.

The write command, $\overline{\text{W}}$, is latched by the falling edge of the $\overline{\text{CAS}}$ clock and has a setup time of t_{WCS} seconds. Once asserted, $\overline{\text{W}}$ has a minimum down-time of t_{WP} seconds and must not be negated until at least t_{WCH} seconds after the falling edge of $\overline{\text{CAS}}$. $\overline{\text{W}}$ must be asserted at least t_{RWL} (write command to row strobe lead time) before the rising edge of $\overline{\text{RAS}}$ and

FIGURE 10.24   DRAM write cycle timing

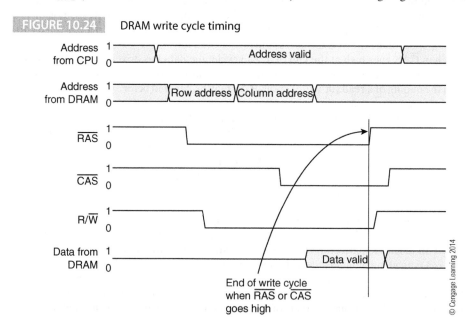

at least t_{CWL} (write command to column strobe lead time) seconds before the rising edge of $\overline{CAS}$. The critical event in a write cycle is the falling edge of $\overline{CAS}$, which latches the $\overline{W}$ and the data input to the DRAM.

Data is written into the memory on the *falling edge* of the $\overline{CAS}$ clock. The requirements for data-input timing are entirely straightforward and involve only three parameters. The data to be written into the memory must be valid for t_{DS} seconds (the data setup time) before the falling edge of $\overline{CAS}$ and maintained for t_{DH} seconds (the data hold time) following the falling edge of $\overline{CAS}$.

10.3.2   Developments in DRAM Technology

We've described basic DRAM. Now we look at the successive generations of improved DRAMs starting with the *page*, *nibble*, and *static column* modes. These variations exploit the way in which the address input is multiplexed between rows and columns and overcome some of the limitations caused by precharging between accesses.

The *page mode* permits a fast access to any column location in a given row, as Figure 10.25 demonstrates. Suppose that a 1M bit DRAM is accessed in a normal read or write cycle. The 10-bit row address is first applied, and $\overline{RAS}$ is then brought low to latch it and to select a row. Next the column address is applied, $\overline{CAS}$ asserted, and a location accessed. The page mode permits successive accesses to the *same row* simply by pulsing $\overline{CAS}$ and latching a new column address on each falling edge of the $\overline{CAS}$ strobe.

The period between successive page-mode, column-access cycles is t_{PC}, which is about half the cycle time. The page mode permits *bursts* of accesses to a single-row address, which compares favorably with the cycle time of basic DRAM. The page mode is particularly useful for transferring bursts of data from consecutive locations (e.g., during block transfers or in raster-scan graphics). Page mode and other *burst modes* are described by the notation 4:2:2:2, which means that the latency for the first access is four cycles and the following data elements are each obtained after a further two cycles. Page-mode DRAM is obsolete today (as are most DRAM modifications prior to DDR2 and DDR3).

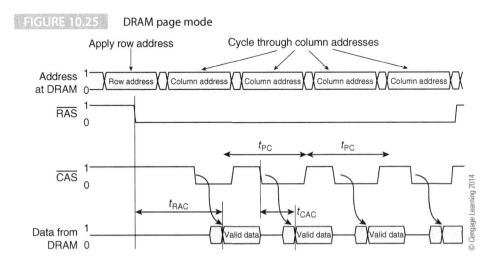

FIGURE 10.25   DRAM page mode

DRAMs with a *nibble mode* share some of the features of a page mode but are even faster. A nibble mode access begins exactly like a normal DRAM access with the capture of the row address followed by the column address. If the $\overline{CAS}$ strobe is *cycled*, up to four successive locations can be read from (or written to), without providing new column addresses.

Figure 10.26 illustrates the timing of the *nibble mode* access. Unlike the page mode, the nibble mode latches just a single column address at the start of the burst. The next three

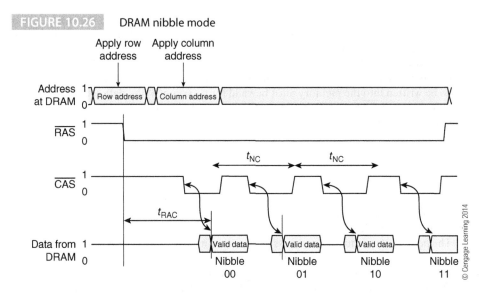

FIGURE 10.26   DRAM nibble mode

accesses (made by cycling $\overline{CAS}$) take place in the sequence 00, 01, 10, 11, 00, 10, etc. The DRAM itself *automatically* generates the sequential addresses internally. For example, if we access location 0x0 1234 and then cycle $\overline{CAS}$, we will access locations 0x0 1234, 0x0 1235, 0x0 1236, and 0x0 1237 (in that order). The first cycle of a nibble mode takes as long as any other read or write cycle. Subsequent cycles can be performed in less than half the normal cycle time.

As system speeds increased, DRAM manufacturers developed other ways of decreasing the access times of DRAMs. *Extended data-out* DRAM (EDO) is a variation on the page-mode DRAM that allows shorter page cycle times than page-mode DRAM. EDO devices don't turn off their output drivers when $\overline{CAS}$ goes high. Data is valid on the falling edge of CAS, so the designer can use the $\overline{CAS}$ signal to strobe data.

SDRAM

Fast-page mode and EDO DRAMs are variations on a theme. The first radical change in DRAM technology occurred around 1997 with the introduction of the *synchronous DRAM* (SDRAM) that has a different interface than previous DRAMs. By 1998, SDRAM had taken about 50% of the world market for DRAM. SDRAM uses a system clock to perform synchronization and incorporates a more complex interface that can receive *encoded commands* from the host processor. For example, SDRAM uses a combination of control signals to encode a command such as read, write, or *precharge*. The SDRAM access time is similar to that of other members of the DRAM family; however, its burst access time is considerably shorter.

Control signals and commands are latched on the rising edge of the clock, which simplifies the design of the system. A control register in the SDRAM defines its operational parameters, such as the burst length (i.e., the number of words accessed per read or write cycle). That is, the SDRAM is a programmable device whose operational parameters can be adjusted to suit specific systems.

SDRAMs have multiple memory banks that operate independently. Figure 10.27 describes a SDRAM read cycle. The command is read on the rising edge of the clock (15 ns period = 66.67 MHz) together with the row address. The column address is read on the next rising edge of the clock. Data appears 60 ns after the start of the read cycle—an access time similar to the DRAM. However, the next three elements in the current data burst appear in successive 15 ns slots.

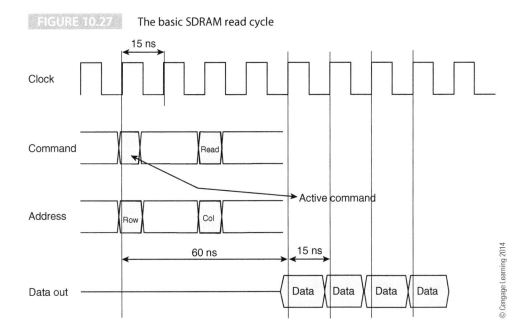

FIGURE 10.27   The basic SDRAM read cycle

RAMBUS RAM

We've put Rambus® RAM in a panel because it represents a development path in memory technology that proved to be a dead end. The specification for RDRAM was created by a company called Rambus working closely with Intel to promote RDRAM technology in high-performance systems. RDRAM is a systems approach in the sense that it requires both a memory technology and a bus technology, hence *Rambus*. The acronym RDRAM stands for Rambus direct RAM, and the memory modules themselves are called RIMMs (for Rambus inline memory modules).

Rambus is narrower than conventional data buses (the first version was 8 bits wide), but it operates at a much higher clock rate. The first-generation Rambus operated at 400 MHz in comparison with the PC's 100 MHz bus. Using the same double-clocking mechanism as the DDR memory, the Rambus has a bandwidth of 2×400 Mhz $\times$ 16 bits/8 = 1.6 Gbytes/s.

The difference between RDRAM and SDRAM lies in the Rambus's protocol and its physical and electrical characteristics (we look at bus protocols in depth in Chapter 12). DRAM multiplexes row and column information. RDRAM divides address lines into two groups and transmits row and column information at the same time.

In a conventional system, each memory module is connected to the memory bus. Rambus uses *transmission line technology*, which is concerned with the way in which fast pulses behave on buses and the way in which pulses are reflected at the ends of buses to degrade the bus's performance. Signals enter at one side of a RIMM module, access the RDRAM, and leave at the other side to travel along the bus via each memory connector. All memory slots on the motherboard must be populated to ensure that Rambus signals can be propagated from end-to-end along the bus. A special pass-through module called a *continuity RIMM* is used to fill unpopulated slots on the motherboard. The electrical design of the Rambus allows it to operate at far higher clock rates than conventional buses.

(continued)

The table below relates bit width, number of channels, and clock rate to the data transfer bandwidth on the RAMBUS.

Name	Width	Channel	Clock Rate	Bandwidth
PC600	16	Single	300 MHz	1200 MB/s
PC700	16	Single	355 MHz	1420 MB/s
PC800	16	Single	400 MHz	1600 MB/s
PC1066	16	Single	533 MHz	2133 MB/s
PC1200	16	Single	600 MHz	2400 MB/s
RIMM 3200	32	Dual	400 MHz	3200 MB/s
RIMM 4200	32	Dual	533 MHz	4200 MB/s
RIMM 4800	32	Dual	600 MHz	4800 MB/s
RIMM 6400	32	Dual	800 MHz	6400 MB/s

© Cengage Learning 2014

Although Rambus was launched with much fanfare, it was never successful following its appearance in 1999. The complexity of its interface increased manufacturing costs, and in practice, it appeared to offer no advantages over other emerging forms of DRAM memory. In the end, it proved to be another (expensive) footnote in the history of computer design.

DDR DRAM

DDR DRAM or *double-data rate SDRAM* is little different to conventional SDRAM in terms of its internal storage mechanism. The difference between DDR DRAM and SDRAM lies in its interface. DDR SDRAM performs a data access on *both the rising and falling edge* of the clock; that is, it delivers data at twice the clock rate. Figure 10.28 illustrates the DDR's read-cycle timing.[3] Once the first access of a burst has been made, data is available at each edge of the clock. The parameter *CAS latency* (CL) defines the time, in clock cycles, between the point at which the column address strobe is asserted and the point at which data becomes valid. This parameter is quoted as one of DDR's principal parameters. Three other parameters associated with DRAM are RL, AL, and BL. (RL is the read latency, AL is the additive latency, and BL is the burst length.)

DRAM Speed Terminology

There is some confusion in the way memory standards for PCs are described. DDR SDRAM is described as PC1600, PC2100, PC2700, etc., which defines the *bandwidth* of the memory. SDRAM standards are written as PC66, PC100, and PC133, which describe the clock speed of the SDRAM. A DDR SDRAM clocked at 100 MHz is called DDR200, and a system with a 64-bit data bus can transmit eight bytes at a time to give a bandwidth of $8 \times 200 = 1,600$ MB/s (hence the designation PC1600).

DDR2 and DDR3 DRAM

All variant technologies such as DDR DRAM have a lifespan. DDR peaks then DDR2 takes over. DDR2 peaks and DDR3 is introduced. DDR4 is waiting in the wings. In the PC world, each variant technology is accompanied by new families of motherboard chipsets to interface them to the host CPU.

DDR2 takes the rising and falling edge clocking of DDR one step further and performs four data transfers per clock cycle. A DDR2 memory module with a 64-bit data bus operating at a clock of 266 MHz is able to transfer data at a peak rate of $64 \times 266 \times 10^6 \times 4/8 = 8,512$ MB/s. The factor 4 in the expression indicates the four data transfers per clock. The *data transfer rate*

---

[3]General DDR SDRAM functionality, Micron Technical Note TN-46-05.

FIGURE 10.28   The DDR DRAM read cycle

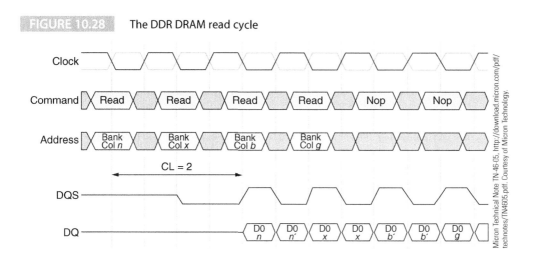

is one eighth this value because a transfer involves eight bits. The peak information transfer rate is 1,066 million transfers/s.

DDR2 and later members of this family are programmable in the sense that they have configuration registers that are loaded with operational parameters by the host system. For example, you can define how the DDR2 SDRAM is to carry out memory refreshes (recall that each cell has to be periodically updated and the stored data rewritten).

DDR3 is a further extension of the double data rate principle in which the data rate is doubled yet again to provide a data rate of eight times that of the clock. As well as increasing the data rate, DDR3 modules can accommodate individual chips with capacities up to eight gigabits. Samsung announced the first four gigabit DDR3 DRAM using 50 nm technology in 2009, barely 40 years after a newly formed company called Intel released the first 1,024-bit DRAM. Who, then, could ever have imagined that the capacity of memory chips would grow by a factor of over four million in their own lifetime? DDR3 memory is internally organized as eight banks rather than DDR2's four banks. Increasing the number of memory banks permits a degree of interleaving. The internal core speed of DDR3 is little different to that of DDR2.

A DDR3-1600 module operating at a clock of 200 MHz and a cycle time of 5 ns executes a peak of 1,600 million transfers/s corresponding to a data rate of 12,800 MB/s. Figure 10.29 demonstrates a DDR2 burst read operation (taken from the *Samsung DDR2 SDRAM Device Operating & Timing Diagram* manual).

An important trend in dynamic memory design is the reduction in operating voltage. DDR2 operated at 1.8 V, first-generation DDDR3 at 1.55 V, and later DDR3 at 1.35 V. A drop of 1.8 to 1.35 V provides a 25% reduction in power.

FIGURE 10.29   The DDR2 DRAM burst read mode

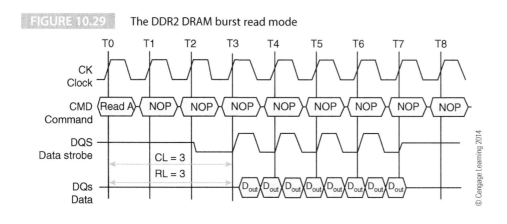

Fly-By Wire

The layout or *topology* of DDR3 memory modules on a motherboard differs from that of DDR2 memory. The change is necessitated by the increase in clock and signaling rate, which means that circuit effects that were of passing interest in earlier computers now have very significant effects.

In a DDR2-based system, signals are fed to all modules in parallel. In DDR3-based systems, signal integrity is improved by providing a fly-by topology; that is, the signals are fed to the first module and then to the second module, and so on. Thus, the signals fly-by each module serially.

One effect of this new topology is to require that the timing parameters of each module be different; that is, the module must be tuned. At startup, the DRAM controller adjusts the parameters of each DRAM module in a process called *read and write leveling*.

Developments in DRAM technology have enabled the PC world to flourish by providing the large, cost-effective, and immediate-access memories required both for sophisticated operating systems running multiple tasks and for multimedia applications processing audio and video. We now look at the read-only memory that has had only a modest impact on the PC world but a much greater impact on pervasive computing (for example, on MP3 players and digital cameras).

DDR4

In 2011, just as DDR3 was gaining in popularity and replacing DDR2 in new designs, DDR4 appeared. Samsung was the first company to make a DDR4 module. The new DDR4 technology is both faster and more power efficient than previous members of the DRAM family. Power savings were made by using new circuit design technologies and dropping the power supply from the 1.55 V of DDR3 to 1.2 V. We've come a long way since the 1103 DRAM that required three supply rails of +5, −5, and +12 V. The module's data transfer rate was 2,133 bits/s.

Having covered mainstream read/write memory devices, we are going to look at read-only memory before returning to emerging (now emerged) technologies that are shaping the future of main memory.

DRAM Generations

The table below illustrates the progress made by successive generations of DRAM. The table gives the DRAM family name, its clock rate (except for basic DRAM and EDO DRAM that is not clocked), and the data rate assuming a 16-bit data bus. Ofcourse, data rates are quadrupled for 64-bit data buses.

DRAM Family	Clock/Command Data Rate MHz	Data Bus Rate MB/s
Fast page-mode DRAM		10 to 33
EDO DRAM		33 to 66
Synchronous SDRAM	66 to 133	66 to 133
DDR SDRAM	100 to 200	200 to 400
DDR2 SDRAM	200 to 400	400 to 800
DDR3 SDRAM	400 to 800	800 to 1,600
DDR4 SDRAM	800 to 1,600	1,600 to 3,200

DRAM Module Form Factors

The following diagrams describe the form factors of three generations of DRAM module. Each has a key (a notch) in a different position to ensure that the correct module is inserted into the correct motherboard.

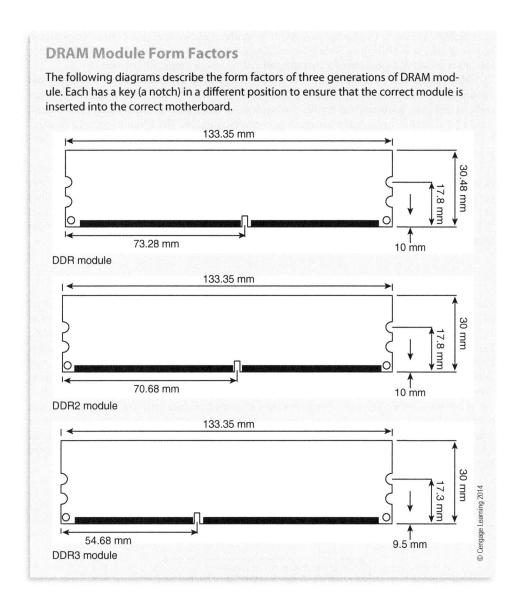

DDR module

DDR2 module

DDR3 module

10.4   The Read-Only Memory Family

Read-only memory is a device whose contents can be accessed but not modified. Today we often use the term *read-only memory* in a wider sense to include *read-mostly memory* that is written to occasionally and read frequently (such as a PC's BIOS that is written to only when it is upgraded by a new firmware revision). We need read-only memory to hold information that must be in a computer before it is switched on. For example, the bootstrap program that loads the operating system or other system parameters. Read-only memory is vital in diskless systems (for example, cell phones, MP3 players, and digital cameras). ROM is needed to hold both programs and user data such as music, videos, and images.

True read-only memory is either programmed during its fabrication or before it is used. *Mask-programmed ROM* is programmed during one of the final stages of its manufacture by using a mask (i.e., stencil) that contains the data to be stored. A mask-programmed ROM has a typical capacity of 128 M bits and is arranged as either 16 M words of 8 bits or 8 M words of 16 bits. Its access time is 100 ns, which is relatively long in comparison to DRAM (50 ns) or static RAM (10 ns). The *fusible-link ROM* is another read-only device. It uses a tiny metallic fuse in each cell. During its initial programming in a special programmer, each link is either left intact or blown by means of a current pulse of sufficient amplitude to melt it.

Mask-programmed ROM is suitable only for large-scale production runs of systems that are not going to require reprogramming, and fusible link ROM is used only in special applications. A low-cost, high-volume application such as an electronic dictionary and phrase book would be a suitable application for mask-programmed ROM. The computer world requires read-only semiconductor memories that can store large amounts of data and yet be reprogrammed with relatively little effort.

In this section, we look at three members of the family of semiconductor read-only devices that can be programmed electrically. These are the EPROM (the first electrically programmable read-only memory), the EEPROM, and the flash EPROM. All three families are very similar—the only real difference is in their characteristics and the mechanisms used to program and erase them. However, in today's world, the flash EPROM is the dominant form of read-only (read-mostly) memory.

10.4.1   The EPROM Family

Erasable and programmable read-only memory (EPROM) stores information that is never, or only very infrequently, modified. The EPROM was invented by Dov Frohman at Intel in 1971 and relies on an electrostatic charge trapped in a transistor cell (a bit like a DRAM cell). Figure 10.30 illustrates an EPROM memory cell consisting of a single field effect transistor. Current flows between the V_{ss} and V_{dd} terminals through a positive channel in the same way as the transistors we've described earlier. By applying a charge to a gate electrode, the current flowing in the channel can be turned on or off. The floating gate is insulated from any conductor by means of a thin layer of silicon dioxide—an almost perfect insulator. By placing or not placing a charge on the floating gate, the transistor can be turned on or off to store a one or a zero in the memory cell. EPROM is very similar to DRAM—the difference is the way in which the charge used to store data is added and removed and the time taken for the charge to leak away, which is about 10 ms for a DRAM and decades for EPROM.

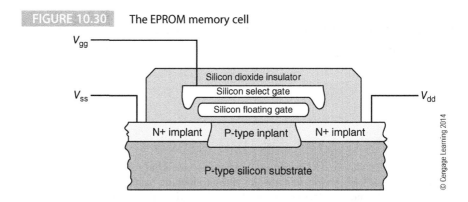

FIGURE 10.30   The EPROM memory cell

How do we place a charge on the *entirely insulated* floating gate? A second gate is located close to the floating gate but insulated from it. By applying a high voltage (i.e., 12 to 25 V) to this second gate, electrons cross the insulator and travel to the floating gate. Although 12 V doesn't seem like a high voltage, the *potential gradient* is large enough to force electrons through an insulator when applied across a very tiny gap.

Once an EPROM has been programmed, its data remains trapped on floating gates for ten or more years. To remove the charge, you have to expose the chip's surface to ultra-violet light by placing it under a UV lamp. EPROMs are mounted behind transparent windows made of quartz, because glass is opaque to UV light. All of the data is erased at once.

First-generation EPROMs had small capacities and were programmed simply by writing the data to each location in turn. Second generation high-capacity EPROMs used *smart programming algorithms* to apply a short programming pulse to each cell and repeat the operation until the data has been correctly written. Some EPROMs are *one-time programmable* (OTP), because they lack a quartz window (which is expensive to make) and are programmed once and for all in the field.

The EPROM provided a means of developing computer firmware in the laboratory. Unfortunately, its slow programming and its tedious erasure mechanism means that the EPROM is unsuited to consumer applications that require even occasional reprogramming. Today, the UV erasable EPROM is largely obsolete, and the few remaining EPROMs are suffering death by eBay.

The EEPROM

The electrically erasable and reprogrammable read-only memory (EEPROM) provides a link between the original EPROM and today's flash memory. We introduce it to show the development of the EPROM. The major difference between the EEPROM and the flash EEPROM is in the way data is erased. The EEPROM was developed at Intel by George Perlegos in 1978.

In an EPROM, the trapped electrons in the insulator are removed by the photons of UV light. In an EEPROM, the insulating layer is so thin that a *quantum mechanical* effect, called *Fowler-Nordheim* tunneling, transports electrons across it when the chip is erased. When a voltage is applied across the insulating layer, electrons on the floating gate are able to tunnel through the layer, even though they don't have enough energy to cross the barrier. The voltage across the insulating layer is approximately 10^7 V/cm. Table 10.3 illustrates the difference between the three programmable devices we describe here.

TABLE 10.3   EPROM Family Differences

Device	EPROM	EEPROM	Flash Memory
Normalized cell size	1.0	1.0 to 1.2	3.0
Programming mechanism	Hot electron injection	Hot electron injection	Tunneling
Erase mechanism	UV light	Tunneling	Tunneling
Erase time	20 minutes	1 s	5 ms
Minimum erase	Entire chip	Entire chip (or sector)	Byte
Write time (per cell)	< 100 μs	< 100 μs	5 ms
Read access time	200 ns	100 ns	35 ns

© Cengage Learning 2014

EEPROMs (or E^2PROMs) are more expensive than flash EEPROMs and have smaller capacities. Like flash memory, they are read-mostly devices with a lifetime of 10,000 erase/write cycles. EEPROMs have access times as low as 35 ns but still have long write cycle times (10 ms).

Flash Memory

The flash EEPROM (today, most people just call it flash memory) can be programmed and erased electrically and provides a convenient means of storing firmware in computers, digital electronic devices, and portable applications. It was invented by Fujio Masuoka at Toshiba in 1980. Figure 10.31 illustrates the structure of a flash memory cell. The thickness of the silicon oxynitride insulating layer (ONO) between the floating gate and the surface of the MOS transistor is about 300 Å in an EPROM but only 100 Å thick in a flash EEPROM ($1\text{Å} = 1 \times 10^{-9}$ m or 1 nm). The human eye can see light that falls in the visible spectrum of 390 to 750 nm, which means that the thickness of a floating gate is about one half of the wavelength of the highest frequency light (blue) that we can see.

When an EPROM is programmed, the charge is transferred to the floating gate by an *avalanche effect* that causes electrons to burst through the oxynitride insulating layer. These electrons are called *hot electrons* because of their high kinetic energy. A flash EEPROM

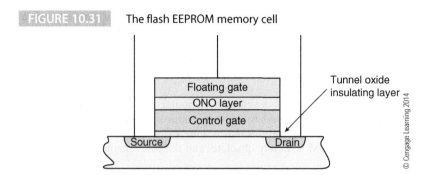

FIGURE 10.31   The flash EEPROM memory cell

is programmed by electrons tunneling through the insulator and is erased in the same way. Table 10.3 compares the writing/erasing mechanisms of the three programmable devices.

You can't erase individual cells in flash EEPROMs. A flash EEPROM is divided into sectors with a capacity of typically 1,024 bytes. Some devices let you erase individual sectors. First-generation flash EEPROMs were guaranteed to perform only 100 erase/write cycles, but devices are now available with lifetimes of at least 10,000 cycles. Actual flash EEPROM lifetimes are probably an order of magnitude greater.

Flash EEPROMs are programmed in the same way as the EPROMs. Their read interface to the system is like that of a static RAM, and they have a conventional active-low, write-enable input. A write cycle is rather like a conventional write to a static RAM except that the duration of the write is very much longer. A flash EEPROM has an on-chip timer and associated control circuits that automatically ensure the appropriate signal delays without the use of external hardware. Some flash EEPROMs can be programmed a byte at a time, whereas others require an entire sector (e.g., 1,024 bytes) to be written in one operation. The erase interface of flash EEPROMs varies from manufacturer to manufacturer. They can be erased in one operation or erased a sector at a time.

EPROMs were typically available as 32 Mbits devices arranged as 4 MB. Modern flash memory is available as 16 Gbits ($2G \times 8$) in its NAND form and as 256 Gbits ($16M \times 16$) in its NOR form. The NAND and NOR forms of flash memory refer to the internal arrangement of cells, and we will return to this aspect of flash memory later.

Flash Technology

The two pillars of modern ubiquitous computing are the USB bus and flash memory. The USB bus (discussed in Chapter 12) allows us to connect a wide range of digital systems together (digital camera to PC, cell phone to iPad, and so on) with an absolute minimum of fuss. Data transfer rate and information exchange protocols are handled automatically and are invisible to the user. Similarly, flash memories have grown from 8 MB devices to flash cards that can hold 512 GB in 2010. Interestingly, you can buy a 256 GB flash drive that combines both USB and flash technology to give you a portable storage system capable of holding 256 GB in the palm of your hand (a lifetime's text and program storage—only images and multimedia require a lot more storage).

The continued progress in flash technology includes the introduction of solid-state disk drives in laptop computers in 2010. Replacing hard disks in laptops and notebooks with flash technology increases performance (data transfer and access time), power consumption, and reliability (there are no vulnerable moving parts). We look at this in detail in the next chapter.

Multi-Level Flash Technology

As we know, both DRAM and the EPROM families store data as a charge on a capacitor that controls conduction through a channel. The only difference between these memory technologies is their structure and operational parameters. In the late 1990s, Intel developed a *multi-level flash cell* (MLC) that extended flash technology by storing a *measured charge* on the control gate. By carefully controlling the stored charge and by reading back the amount of stored charge, MLC is able to write more than one bit into a cell; that is, multi-level technology stores data in an *analog* form. The relationship between the capacitance of a floating gate, the stored charge, and the gate voltage is $V_{FG} = Q_{FG}/C_{TOT}$, where V_{FG} is the floating gate voltage, Q_{FG} is the charge on the gate, and C_{TOT} is the capacitance.

When programming conventional flash memory, you can blast as much charge into a cell as you want, as long as it is sufficient to change the state. Multi-level cells have to be programmed precisely; you can't permit overshoot in a MLC cell because it would convert one stored level into another. A different bit pattern is assigned to each charge (for example, 00, 01, 10, 11). A traditional flash memory or SLC flash senses the gate voltage, compares it to a threshold, and then assigns a 1 or 0, depending on whether the voltage is above or below the threshold. MLC technology compares the gate voltage to several reference levels and assigns a binary pattern to each level. The charge on the floating gate changes by about 1 V for each 10,000 electrons stored.[4] Figure 10.32 shows the distribution of states in SLC and MLC cells.

A cell with four reference levels stores two bits per cell; a cell with eight reference levels stores three bits per cell; and so on. The number of levels that can be stored depends only on the ability to store a precisely measured charge and to accurately compare the voltage level on the gate with a precise reference. MLC reduces the effective cell area and the die size for a given bit density. This ultimately leads to a significantly reduced unit cost per megabyte. Today's MLC memory products are capable of storing two bits per memory cell.

Aging Transistors

Mechanical devices with moving parts, such as motors or moving heads (in disks), wear out over time. Conventional wisdom states that devices like ICs that use transistors never wear out or degrade. This is not true.

In early 2011, many were surprised to learn of a defect in a support chip designed for the second-generation Core i5 and Core i7 processors. The problem affected only the 3 Gbps SATA disk interfaces after a period of operation (estimated as three years). The fault was traced to a transistor whose voltage was too high, causing it to degrade over time.

The Sandy Bridge problem was a stark reminder that, unlike diamonds, transistors are not forever. Transistors degrade for several reasons. Over time, electrons drift out of the conduction channel to get trapped in the dielectric insulator layer and affect the switching threshold. The dielectric layer can breakdown over time due to electrical stress. It is even possible for atoms of the copper or aluminum used to connect the chip to gradually diffuse into the silicon and modify its properties.

Flash memory is even more prone to failure because of the large electrostatic field required to force electrons onto and off the control gate that holds the data.

Current MLC memory is considered as a consumer product rather than an industrial product, because it cannot work over the industrial temperature range and it is not as reliable as SLC flash. The maximum number of write cycles to MLC is typically 10% of the number of cycles to SLC flash.

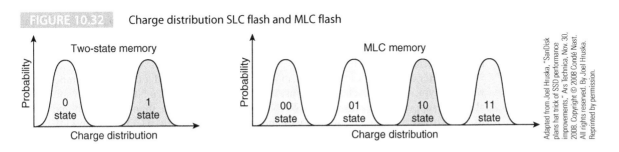

FIGURE 10.32   Charge distribution SLC flash and MLC flash

NAND and NOR Flash

Modern flash memories are described as being *NOR flash* or *NAND flash*. NOR flash was introduced by Intel in 1988 and NAND flash by Toshiba in 1989. The distinction between these two varieties of flash memory lies in the arrangement (i.e., interconnection) of cells. In general, when people speak of flash memory, they are usually referring to NOR flash. Figure 10.33 illustrates the basic difference between NOR and NAND structures.[5] As their

---

[4] Al Fazio and Mark Bauer, "Intel StrataFlash Memory Development and Implementation," *Intel Technology Journal*, Q4'97.

[5] M-Systems White Paper, "Two Technologies Compared: NOR vs. NAND." July 03 91-SR-012004-8L.

FIGURE 10.33   NOR and NAND flash technologies

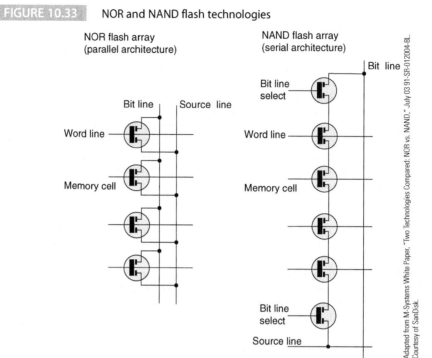

NOR flash array
(parallel architecture)

NAND flash array
(serial architecture)

Adapted from M-Systems White Paper, "Two Technologies Compared: NOR vs. NAND," July 03 91-SR-012004-8L.
Courtesy of SanDisk.

names suggest, the storage arrays look like NOR gates or NAND gates where the cells are wired either in parallel (NOR) or series (NAND).

Profound differences exist between these two technologies in their *operating characteristics* and, therefore, their applications. An advantage of NOR flash is that code can be executed directly from it [called the eXecute in Place (XIP) principle]. Code cannot be executed from a NAND flash and has to be first transferred to static memory or DRAM. NOR flash is used largely to implement relatively small memories, such as 4 MB.

NAND flash can provide higher cell densities but requires a more complicated system interface (hence the inability to run code directly). NOR flash has erase blocks of up to 128 KB that take as long as 5 s to erase, whereas NAND erase blocks might be 32 KB and take only 4 ms to erase. The system interface of a NOR flash is very similar to that of a static RAM, whereas NAND flash has a serial bit-by-bit interface and not all vendors implement the same serial data transfer protocol.

Another advantage of NAND flash is its ability to withstand a far higher number of re-writes of typically 1-M cycles in contrast with the 100K cycles of a NOR flash. In spite of its ability to support more erase cycles, NAND flash is less reliable than NOR flash and has to use an error-correcting code to deal with individual bit errors[6] (like CD ROM and DVD that we discuss in the next chapter).

NAND flash memories contain more data blocks than necessary (i.e., its storage capacity is greater than the nominal value). This redundancy is necessary because some of these blocks may be *bad block*s. The system software continually monitors the operation of the memory and swaps out bad blocks as they are encountered (a similar process can take place in hard disk drives). Figure 10.34 from a Toshiba report provides an illustration of the relative differences of NAND and NOR technologies.[7]

---

[6] The very nature of flash memory storage means that some cells can suffer from 'bit-flipping' when the state of a bit spontaneously changes or the state of a stored bit is incorrectly reported. This problem is more common with NAND flash than NOR flash, which makes NAND flash more suited to multimedia operations where the occasional incorrect bit is of no real significance.

[7] TOSHIBA, "NAND vs. NOR Flash Memory Technology Overview." Toshiba America Electronic Components, Inc., Irvine, CA.

FIGURE 10.34   Comparison of NOR and NAND flash

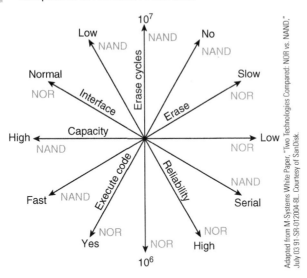

Adapted from M-Systems White Paper, "Two Technologies Compared: NOR vs. NAND," July Q3 91-SR-012004-8L. Courtesy of SanDisk.

Wear Leveling in Flash Memories

Flash memory cells have a finite life because of the wear and tear on the insulating layer that takes place when electrons are injected into it or removed from it. If all cells in a flash memory array were accessed and erased equally, all parts of the array would *age* at the same rate. However, by its very nature, memory is not accessed uniformly. For example, in an MP3 player, some music may remain unheard, while other tracks are frequently played. Consequently, some cells in an array may wear out prematurely, while other cells are still capable of many more erase cycles.

A concept called *flash wear leveling* has arisen to mitigate against the uneven distribution of erase cycles. Static wear leveling moves fixed data to higher-use regions of the flash memory and allocates the less frequently used areas to frequently changing data; that is, an attempt is made to balance the number of accesses to the array. More sophisticated, dynamic leveling algorithms exist to monitor the use of the flash array and then to swap out high-use blocks of data with low-use blocks of data when the need arises.

Compact Flash

Traditionally, semiconductor memory components have been available either in the dual in-line packages that you plug into sockets or solder directly to a PCB or in tiny packages that are designed to be soldered to boards by automatic machinery. Such packages can't be used in consumer applications that require interoperability. In 1994, CompactFlash memory cards were introduced to enable users to plug memory into domestic and personal computer-based systems. The most important of these systems are the digital camera and the MP3 player.

The CompactFlash Association (CFA) was established in October 1995 to promote CompactFlash (CF) technology and work with vendors to standardize CF technology.

The CompactFlash memory is a small, but not ultra-small, unit measuring 43 mm × 36 mm (1.7″ × 1.4″) and is only 3.3 mm (0.13″) thick. It is about half the size of the PC cards that we describe in Chapter 12. Indeed, you can buy very low-cost PC card adaptors for

(continued)

CompactFlash in order to copy pictures from your digital camera to your laptop. CF cards are available in both the Type I format we've described and Type II format that is 5-mm thick.

The advantage of CompactFlash modules is their high reliability, lack of susceptibility to shock, and low cost. Wisely, they were designed to support the dual-voltage standards of 5 and 3.3 V. This makes the CompactFlash modules compatible with both modern and older systems.

The CompactFlash format has also been adapted for other functions in addition to flash memory (for example, microdrives and I/O devices such as Ethernet and Bluetooth wireless adaptors). CF cards that provide extended functions such as I/O are called CF+ cards.

The CompactFlash card interfaces to a host system using a 50-pin connector that's similar to the PC card connector. The CF interface to an external system is not a simple memory interface (i.e., consisting of data, address, and control lines). The CF interface provides a bus interface, which means that intelligence can be built into the cards.

The CF card's intelligent controller manages the interface protocol, data storage and retrieval, error-correcting codes, and diagnostics. The interface uses configuration registers and a software management system. From the point of view of the host system, the CF card looks like a standard ATA (i.e., IDE) disk interface.

Compact flash is relatively large by comparison with some of today's tiny MP3 players and point-and-shoot cameras, so a new generation of flash cards called *secure digital* (SD) were developed in 1999. (SDHC cards are SD cards with a capacity greater than 2 GB). In turn, even smaller, micro-SD cards were created in 2005 for applications such as cell phones. SD cards have simpler serial interfaces than CF cards.

Memory Stick

Sony, one of the leading companies in the design of stylish consumer products, created its own standard for interchangeable memory, the *Memory Stick*, in 1998. This device has all the attributes of a product aimed at the consumer market. It is compact, light, easy to carry and handle, has a simple but reliable electrical interface, and can be made read-only to protect data.

First-generation Memory Sticks launched in 1998 had a capacity up to 64 MB. By 2002, the maximum capacity had increased to 512 MB and the interface upgraded to 20 Mbits/s. Memory Sticks are now available with a capacity of 32 GB and the Memory Stick PRO standard allows for up to 2 TB However, Sony's support for SDHC cards in 2010 must cast a shadow over the Memory Stick's long-term future.

The Memory Stick adopts the PC FAT file management system for compatibility with PC-based systems. However, Sony designed the Memory Stick not only for transporting data between applications, but also for *controlling* the flow of copyrighted data. Companies in the music and film industries have benefited greatly from the Internet and new technologies because they have helped expand markets. Equally, the music industry is unhappy with a medium that provides massive interconnectivity together with digital technology that allows audio and video files to be copied without loss of quality. Sony's *MagicGate* technology has been incorporated into its Memory Stick to prevent unauthorized data storage. MagicGate copyright protection consists of two technologies: *authentication* that determines whether the device legally supports MagicGate and *encryption* that protects the contents. Each Memory Stick has a unique ID number which is employed by MagicGate.

Unlike flash cards with their parallel interface, the Memory Stick has a serial interface between the card and the host. The advantage of a serial interface is its simplicity. The Memory Stick provides only ten interface pins, including pins for future expansion. Such a simple interface makes it much easier to use a slim form factor. This is important for tiny personal appliances and ensures that the interface is mechanically reliable and is not subject to the problems of intermittent contacts caused by frequent insertion and removal.

The Memory Stick consists of a flash EPROM, which is a serial interface controller that manages the storage, and the MagicGate copyright protection mechanism. The controller provides an interface to the flash memory that can be modified to suit new memory technologies as they evolve. The controller performs the serial-to-parallel conversion and vice versa and handles error detection and correction. Three signals implement the serial interface:

SCLK   (the serial clock)
SDIO   (the common bidirectional serial data path into and out of the memory)
BS   (the burst state signal).

The host always initiates communication with the Memory Stick by means of SCLK and BS. The serial data signal SDIO transfers data in 512-byte frames with a CRC error detecting mechanism. The *burst state* signal indicates the start of data transfer. BS is also used to distinguish between RDY/BSY messages and interrupt messages. The burst state signal classifies the data on SDIO bi-directional data line into four types

Mode	BS	Operation
BS0	low	Interrupt—no data transmitted.
BS1	high	Transmission control protocol state—TCP from host copied to memory stick.
BS2	low	Data transfer state for write protocol; transferring data to Memory Stick. Handshake state for read protocol. Waiting for RDY signal.
BS3	high	Handshake state for write protocol; waiting for RDY signal. Data transfer state for read protocol. Reading data from Memory Stick.

Transfer protocol control commands are transferred in state BS1 and define the type of transferred data and the correct protocol for states BS2 and BS3. The signals on SDIO then transfer data and perform handshaking based on the protocol.

The Memory Stick is used in audio–visual devices such as digital cameras, voice recorders, and MP3 players. Since information will represent still or moving images and voice or music, most data transfers will be relatively large. There's little point in optimizing data transmission around very short data exchanges. The minimum data exchange is therefore set at 8K bytes. Data is stored in a PC-compatible format using the *file allocation table* (FAT) mechanism. However, Sony has identified several significant audio–visual file formats and predefined the directory management mechanism for compatibility with these applications.

10.5   New and Emerging Nonvolatile Technologies

In this section, we look at some of the emerging technologies in the world of nonvolatile memory systems. Some technologies are in an early stage and others are already in production. We first look at two *mature* emerging technologies: ferroelectric RAM and Ovonic memory.

FRAM or ferroelectric RAM, a form of semiconductor random access nonvolatile memory, began to emerge during the late 1990s. Although, the prefix *ferro* in the term *ferroelectric* implies something to do with iron or magnetism, ferroelectric is a misnomer and has nothing to do with magnetism. DRAM stores an electric charge by displacing the electrons in a material. Disk and tape systems store a magnetic field by aligning the *spin* of adjacent electrons in the recording medium. FRAM stores data by changing the *polarization* in a material by moving individual atoms within a crystal lattice.

The ferroelectric effect describes the ability of a material to store an electric *polarization* in the absence of an applied electric field. The term is used because it is entirely analogous to the property of magnetic materials to retain a magnetic field in the absence of an applied magnetic field.

In order to understand the ferroelectric effect, you have to appreciate a few simple facts about the properties of matter. Materials can be divided into two classes: conductors and insulators. The difference between these is that electrons are free to move through a conductor but are not free to move in an insulator. If you apply an electric field to a conductor such as copper, the electrons in the conductor move under the influence of the field. If an insulator is placed in an electric field, the electrons are not free to move through the insulator. However, the electric field does have an effect on the molecules that make up the insulator; it disturbs the symmetrical distribution of positive and negative charges in the molecular structure of the insulator. This change in the position of electrons and nuclei under the influence of an electric field is called *polarization*.

Figures 10.35 and 10.36 demonstrate the effect of an electric field on a single atom. In Figure 10.35, the negative electrons orbit the positively charged nucleus, and the average

FIGURE 10.35   The structure an individual atom

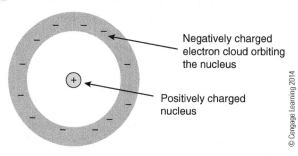

FIGURE 10.36   The effect of an electric field on an individual atom

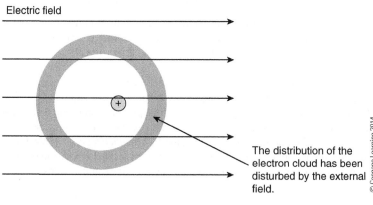

FIGURE 10.37   An atom in an electric field appears polarized

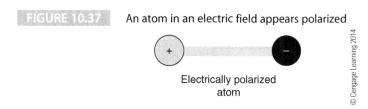

Electrically polarized
atom

© Cengage Learning 2014

distribution of the atom's charge is zero because the positive and negative charges cancel. In Figure 10.36, an electric charge is applied and the symmetry is disturbed. The average position of the *electron cloud* orbiting the nucleus is disturbed by an amount that depends on the strength of the field and the atom gains a net electric charge. Figure 10.37 shows how the atom can be modeled by a *dipole*, which consists of two point charges (this is where the term polarization originates). The dipole is the electric equivalent of the magnet.

Just as magnets align themselves north to south, dipoles align themselves negative to positive, as Figure 10.38 demonstrates. A string of dipoles can be modeled as a single, more powerful dipole, as Figure 10.39 demonstrates. An insulator in which dipoles form under the influence of an electric field is called a dielectric. The strength of an insulator's polarization is indicated by its *dielectric constant,* which is expressed relative to the dielectric constant for a vacuum. The most important effect of a polarized material, from our point of view, is that it can store data.

FIGURE 10.38   The alignment of individual dipoles

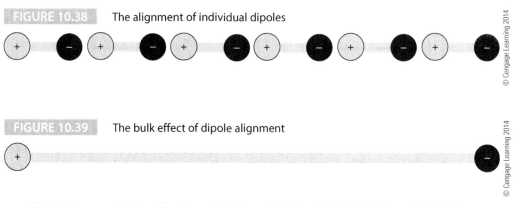

© Cengage Learning 2014

FIGURE 10.39   The bulk effect of dipole alignment

© Cengage Learning 2014

FRAM is a practical application of the ferroelectric effect. A thin ferroelectric film is used as the dielectric in a tiny capacitor to store data. Ferroelectric films electrically polarize in one of two directions, depending on the direction in which the electric field is applied.

The use of ferroelectric technology was investigated at Stanford University in the early 1960s and S.Y. Wu, et al. investigated the use of ferroelectric materials in conjunction with semiconductor technology in 1974. In 1988, Ramtron International Corporation announced the first commercial ferroelectric random access memory (FRAM).

The ferroelectric materials used in FRAM memories belong to the class of crystals called perovskite. Figure 10.40 illustrates the structure of a perovskite crystal that is expressed chemically as ABO_3, where O represents an oxygen atom and atoms A and B determine the specific perovskite. For example, one commonly used ferroelectric material is PZT (lead zirconate titinate), which is a mixture of $PbZrO_3$ and $PbTiO_3$.

Figure 10.41 demonstrates the effect of an electric field on a perovskite crystal. When the field is applied, the atom in the center of the crystal (in this case a zirconium or titanium atom) is moved into one of two stable positions. This is the mechanism by which the ferroelectric perovskite stores data.

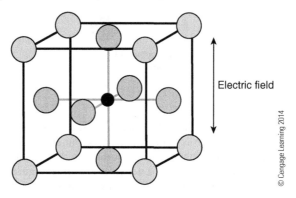

FIGURE 10.40   Structure of the perovskite crystal

Electric field

© Cengage Learning 2014

FIGURE 10.41   Effect of an electric field on a perovskite crystal

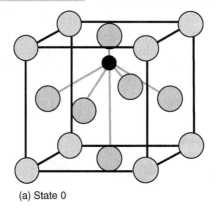

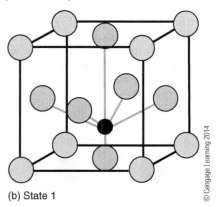

(a) State 0

(b) State 1

© Cengage Learning 2014

10.5.1   Ferroelectric Hysteresis

Consider a system with a slice of ferroelectric material sandwiched between two metal plates, shown in Figure 10.42a. Suppose we apply a voltage across the dielectric and it becomes polarized, as Figure 10.42b demonstrates. Removing the voltage, as in Figure 10.42b, has no effect on the dielectric's polarization. Similarly, applying a small voltage across the metal plates in either direction has no overall effect.

If the voltage across the plates is reversed and its magnitude increased, as in Figure 10.42c, the polarization of the dielectric switches over and the material changes state. The *pulse of current* that flows when the atoms in the ferroelectric material move from one end of the crystal to the other can be detected. We can't detect the state or polarization of one of these devices, but we can detect whether it *changes* state; that is, FRAM memory has a destructive readout.

In Figure 10.42d, the power is once again removed, and the ferroelectric material retains its new charge. We have everything we need for a memory device: a means of forcing it into one of two states and a means of detecting which state it *was* in by applying a voltage across its terminals.

A practical FRAM memory cell is constructed using slightly modified DRAM technology. We have just seen that writing data into an FRAM cell is achieved by applying a charge across the ferroelectric material. In order to read the data stored in the cell, the transistor is turned on, and an electric field applied across the capacitor. If the cell is already polarized in the same direction as the applied field, nothing will happen. If, however, the cell is charged in the reverse direction, a current flows through the switching transistor.

FIGURE 10.42   Ferroelectric storage

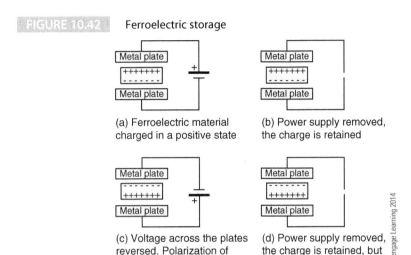

(a) Ferroelectric material charged in a positive state

(b) Power supply removed, the charge is retained

(c) Voltage across the plates reversed. Polarization of material changes

(d) Power supply removed, the charge is retained, but in the opposite direction

© Cengage Learning 2014

The FRAM cell changes state in about 100 ns, which is faster than some conventional semiconductor nonvolatile memories, such as EPROM, but slower than either static RAM or DRAM. The act of reading a FRAM cell is, of course, destructive, because data is sensed by the cell changing state or not changing state. Consequently, a FRAM read cycle must always be followed by a write cycle to write back the data that may or may not have changed.

Table 10.4 compares semiconductor nonvolatile memory technologies and FRAM. Unfortunately, many nonvolatile storage mechanisms, including FRAM, are limited to a finite number of changes of state. The PZT-based ferroelectric material has excellent properties, a high polarization, and is easy to mass-produce. It suffers from a phenomenon called *fatigue,* which limits the number of write cycles. Newer materials, such as SBC (i.e., $SrBi_2Ta_2O_9$), can be polarized over 10^{12} times without fatigue. In order to replace traditional volatile RAM, devices such as FRAM would require a lifetime in the region of 10^{15} cycles.

TABLE 10.4   Comparison between FRAM and Other Memories. FRAM Technology Backgrounder: http://www.fujitsu.com/emea/services/microelectronics/fram/technology/. Data from Fujitsu Semiconductor Limited.

	FRAM	EEPROM	Flash Memory	EPROM	Mask ROM	DRAM	SRAM
Retention	10 years	10 years	10 years	10 years	Unlimited	Volatile	Volatile
Cell structure	1T + 1C	2T	1T	1T	1T	1T + 1C	6T
Read time	180 ns	200 ns	< 120 ns	< 150 ns	< 120 ns	70 ns	70 to 85 ns
Write voltage	2 to 5 V	14 V	9 V	12 V	—	3.3 V	3.3 V
Rewrite method	Overwrite	Erase or write	Combination of write and erase	UV light erase	—	Overwrite	Overwrite
Rewrite cycle	180 ns	10 ns (by byte)	1 s (by sector)	0.5 ms (by byte)	—	70 ns	70 to 85 ns
Data erasure	Unnecessary	Necessary (byte erase)	Necessary (sector erase)	Necessary (UV erase)	—	Unnecessary	Unnecessary
Write cycles	> 10^{12}	10^5	10^5	100		Unlimited	Unlimited
Standby current	20 µA	20 µA	5 µA	100 µA	30 µA	1000 µA	7 µA

1T + 1C: 1 transistor/1 capacitor
    1T: 1 transistor
    2T: 2 transistors
    6T: 6 transistors

FRAM memories are not subject to the high levels of electrical stress that are common in EPROMs, where a charge has to pass through an insulator. EPROMs are prone to early catastrophic failure in the insulating material around the gate that holds the charge. Another advantage of the FRAM is that it is *radiation hard*. That is, it's less affected by high-energy ionizing radiation than some semiconductor devices and is therefore better suited to applications in harsh environments such as satellites. The practical advantage of FRAM from the user's point of view is that it combines the advantages of flash memory with those of DRAM. At the moment, flash memory is still very much a read-mainly technology. Moreover, FRAM can support about seven orders of magnitude more write cycles than flash memory.

By 2008, FRAM had come a long way, and the Ramtron International Corporation was awarded the Electronic Products China Product of the Year Award with its FM22L16 4-Mbit FRAM memory organized as a 256K × 16 nonvolatile RAM. Its access time is 55 ns, and its cycle time is 110 ns. It is electrically compatible with industry standard static RAM. That is, you can simply replace existing static volatile RAM with nonvolatile FRAM without modifying the hardware.

Figures 10.43 and 10.44 give the FRAM's read-and write-cycle timing diagrams. Active-low enable CE1 and active-high CE2 must be asserted in a read or write cycle. A 19-bit

FIGURE 10.43   FRAM read-cycle timing diagram (RAMTRON FM23MLD16 512Kx16)

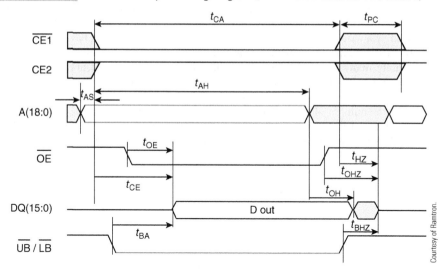

FIGURE 10.44   FRAM write-cycle timing diagram (RAMTRON FM23MLD16 512Kx16)

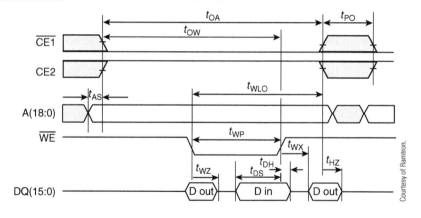

address is fed in on A_{18} to A_{00}, and the joint input/output is given by DQ_{15} to DQ_{00}. The remarkable aspect of these figures is their unremarkableness; that is, they are very similar to any other semiconductor static RAM.

10.5.2   MRAM—Magnetoresistive Random Access Memory

Magnetoresistive Random Access Memory (MRAM) exploits both electrical and magnetic properties of matter. Like some of the other modern technologies its future is uncertain, not least because it is in competition with the other forms of nonvolatile static RAMs. The first commercial MRAM memory was a 4-Mbit device introduced by Freescale in 2006.

MRAM has a similar access time to semiconductor static memory (5 to 40 ns) and a 10 ns write time. Unlike flash memory, MRAM cells can perform an unlimited number of write cycles. Indeed, MRAM has many of the characteristics of a near ideal memory (compared with competing technologies).

Figure 10.45 illustrates the structure of an MRAM memory element which consists of a *magnetic tunnel junction* (MTJ), which is composed of three layers: an oxide layer (MgO) sandwiched between two magnetic layers (CoFeB). One of these magnetic layers is fixed (i.e., the direction of the internal magnetic field does not change), and the other is free to rotate its magnetization as shown by the double arrows in the top layer in Figure 10.44. The oxide barrier layer is very thin, being of the order of 1.2 nm.

FIGURE 10.45   MRAM cell

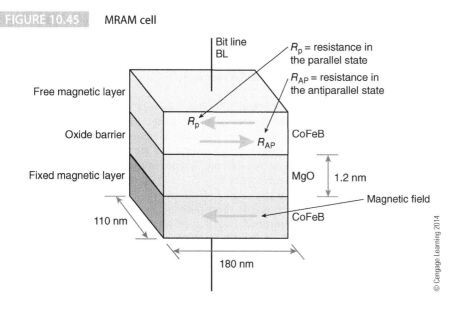

The magnetization of the free layer can be switched to be parallel or anti-parallel to the bottom layer. The magnetization required to perform the switching of the free magnetic layer is generated by passing a current down a write line located close to the MRAM cell. The current in the line generates the field required to magnetize the layer. When the magnetization of both layers is in the same direction, the cell has a low electrical resistance (i.e., along the bit line (BL) in Figure 10.44), and when the magnetization is anti-parallel, it has a high resistance. The resistance of the cell can readily be measured and its binary state determined. Moreover, this memory cell does not suffer the destructive readout of the FRAM cell. This technology is also used in the read heads of some hard disk drives that we discuss in the next chapter. However, by the end of 2011, MRAM technology was beginning to appear in commercial products. With its high-speed and effectively unlimited read/write cycles, MRAM promises to provide a universal memory component that fits all needs (apart from high-speed cache).

10.5.3   Ovonic Memory

Another property of matter that can be used to store information is its *phase*. Ovonic memory uses the phase change of a thin-film material to store data. This phase change is reversible, so the cells can be erased and rewritten. An Ovonic material is either in a structured *crystalline* state or in an *amorphous*, unstructured non-crystalline state. One class of materials that have suitable phase-change properties is the *chalcogenide* glasses, which were first investigated by Ovshinsky at Bell Laboratories in 1968.

If a chalcogenide glass is melted and cooled rapidly, it enters an *amorphous phase*. If it is heated to slightly below its melting temperature at a relatively slow rate, the amorphous material reverts to its initial *crystalline phase*. An amorphous material is characterized by short-range atomic order, high reflectivity, and high electrical resistivity; whereas a polycrystalline material is characterized by a long-range atomic order, low reflectivity, and a low electrical resistance.

An Ovonic cell can be read because there is a forty-fold difference in the electrical resistance of the material in these two states. We will discover in the next chapter that phase-change mechanisms are also used in optical storage systems such as re-writable CDs and DVDs. Because the chalcogenide can be fabricated as a thin film, it can be incorporated in a semiconductor cell and an Ovonic memory constructed just like any other semiconductor memory.

Unlike flash memory, phase-change memory can be *bit organized* in the sense that individual bits can be re-written. Flash memory is block organized, so you have to erase an entire block (sector) in order to make a change. Ovonic memory is static and has no need for a refresh; the data readout is nondestructive. By controlling the amplitude of the current pulse used to write data into a cell, it is possible to select one of several amorphous states (or rather degrees of amorphousness)—each with its own electrical resistance. Consequently, Ovonic memory is a candidate for multi-bit storage cells that will boost the density of bits/chip.

Ovonic cells are capable of more than 10^{13} phase changes without degradation, and it may be possible to employ different degrees of phase change to store multiple values (like MLC flash technology). A further advantage of Ovonic memory is its stability. At normal operating temperatures, it is estimated that the contents of a memory cell will remain stable for over 300 years. Similarly, the projected lifetime in terms of read/write cycles is greater for a phase-change device than a flash memory cell.

Flash memory and DRAM that store data in the form of a charge are susceptible to the effects of ionizing radiation which preclude them for some space and military applications. Phase-change memory devices are intrinsically radiation-hard.

An important question to ask of Ovonic technology is its *scalability*: Can devices be made increasingly smaller? Since a memory cell has a heater that is used to change the state of the Ovonic material, will decreasing the size of cells lead to the heat from one cell corrupting its neighbors? Fortunately, theoretical calculations on the heat flow within the material show that it does scale with size. That is, as the cells become smaller, the heat flow between cells is also less, and therefore, scalability is not an issue.

In June 2009, Samsung Electronics announced that it was jointly developing phase-change memory with Numonyx (Numonyx had been manufacturing a 90 nm 128 Mbit PRAM since 2008, but on a small scale). In September 2009, Samsung announced that it had begun production of a 60 nm 512 Mbit PRAM aimed at the mobile phone market. By October 2009, Numonyx had demonstrated the ability to vertically stack multiple phase-change memory arrays on a single silicon die and was intending to produce a Gbit PRAM using 45 nm technology. Table 10.5 from Numonyx illustrates the relative parameters of phase change memory in comparison with other technologies in 2010.[8]

---

[8] S. Eilert, M. Leinwander and B. Crisenza, "Phase Change Memory (PCM): A new memory technology to enable new memory usage models." Micron Technology Inc, June 2011.

TABLE 10.5   Comparison between PCM (Phase-Change Memories) and Other Storage Mechanisms. © 2009 IEEE. Reprinted, with permission, from S. Eilert, "Phase Change Memory (PCM): A new memory technology to enable new memory usage models," Memory Workshop, 2009.

Attributes	DRAM	PCM	NAND	MLC NAND	HDD
		Comparison of High-Density Memory Technologies			
Nonvolatile	No	Yes	Yes	Yes	Yes
Erase Required	Bit	Bit	Block	Block	Sector
Software	Simple	Simple	Complex	Very complex	Simple
Power	~W/GB	$100 \rightarrow 500$ mW/die	~100 mW/die	~100 mW/die	~10 W
Write Bandwidth	~GB/s	$1 \rightarrow 100+$ MB/s/die	$10 \rightarrow 100$ MB/s/die	~10 MB/s/die	$200 \rightarrow 400$ MB/s
Write Latency	~20 to 50 ns	~1 μs	~100 μs	~800 μs	~10 ms
Write Energy	~0.1 nJ/b	<1 nJ/b	0.1 to 1 nJ/b	<1 nJ/b	>10 nJ/b
Read Latency	50 ns	50 to 100 ns	10 to 25 μs	25 to 50 μs	~10 ms
Read Energy	~0.1 nJ/b	<<1 nJ/b	<<1 nJ/b	<<1 nJ/b	>10 nJ/b
Idle Power	~W/GB	<<0.1 W	<<0.1 W	<<0.1 W	<10 W
Endurance	∞	10^8	$10^5 \rightarrow 10^4$	$10^4 \rightarrow ?$	∞
Data Retention	ms	Not f (cycles)	f (cycles)	f (cycles)	Not f (cycles)

The Memristor

One of the strangest potential memory devices is the *memristor* (memory resistor), which was first hypothecated in 1971. Leon Chua examined the three classic passive elements of electric circuit theory: the resistor, inductor, and capacitor and suggested that there should be a fourth element: the *memory resistor*. At that time, this hypothesis was pure speculation.

Chua stated that a memory resistor would demonstrate the same relationship between magnetic flux and electric charge that a resistor displays between voltage and current. He suggested that a memristor would *remember* the value of the current passing through it even after that current ceased to flow. A memristor would be a two-terminal electrical device with a memory property, making it a possible candidate for data storage.

After remaining a theoretical curiosity for over 30 years, Stanley Williams at Hewlett Packard announced the discovery of a prototype memristor in 2008. The semiconductor titanium dioxide (TiO_2) has a high resistance in its pure state (like silicon). Equally like silicon, the addition of impurity atoms, called dopants, change its electrical properties significantly. However, unlike silicon, the dopants are not stationary in an electric field; they drift in the direction of the current. Consider a thin film of titanium dioxide that is doped on one side only (i.e., it's a sandwich of doped and undoped TiO_2). Applying an electric field across a thin film of titanium dioxide causes the dopants in the doped layer to migrate into the pure TiO_2 layer and thus reduce its electrical resistance. If you then remove the field, the undoped layer will *remember* the duration and intensity of the field.

The first experimental memristor was a three terminal device. Memristor technology has potential in both the digital world as a storage element and the analog world as a component of neural networks. Whether memristors ever manage to overtake FRAM, MRAM, or Ovonic memory remains to be seen. However, in recent years, other forms of memristor have been investigated. For example, there are materials that exhibit similar properties to titanium dioxide, such as certain polymer films. Equally, phenomena other than ion migration have been investigated, such as electron spin. The need for new nonvolatile memory technologies is pressing, because flash technology appears to be reaching the limit of its density.

Strukov, Snider, Steward, Williams, "The missing memristor found," *Nature,* Vol. 453, 1 May 2008, pp. 80–83.

Summary

Memory is the "Cinderella" of the computer world. Many computer users will know the difference between an Intel processor and an AMD processor, but they will probably not appreciate the difference between Rambus DRAM, DDR3 DRAM, and SDRAM, for example. Equally, they may not be aware of the trends in memory systems design and the increasing role played by new forms of nonvolatile memory.

However, without memory, all of the fast CPUs in the world would be useless—except, perhaps, as devices that can calculate the value of π to thousands of decimal places. Today, huge memories are required to hold the seemingly unending quantities of data required for multimedia applications. Moreover we want low-power memories for cell phones and portable applications and nonvolatile memories for MP3 players and digital cameras.

In this chapter, we've looked at the technology and organization of the memory components used to construct immediate access stores. We have seen how semiconductor memory is even more affected by trade-offs in speed, price, and performance than in processor technology. We've discovered that you can have nonvolatile memory that retains its data when you switch the computer off, but you have to pay a large premium in terms of the very long time required to write data into the memory. We've seen that you can have low-cost, high-density DRAM, but you have to put up with a longer access time than static RAM, and you have to use more complex logic to control it. Static RAM is fast and easy to use, but it is more expensive and less dense than DRAM.

In this chapter, we introduced three relatively new memory technologies: FRAM, MRAM, and Ovonic memory. These demonstrate how engineers are constantly seeking properties of matter that can be altered electrically and used to store data.

Problems

10.1 What is the meaning of the following terms (when applied to memory systems technology)?
   a. Random access
   b. Serial access
   c. Dynamic RAM
   d. Static RAM
   e. Read-mostly memory
   f. Access time
   g. Non-volatile memory
   h. Cycle time

10.2 A computer has a memory space of 1 MB.
   a. How many address lines are required to span this address space, assuming it is byte-addressed?
   b. If this computer has a 16-bit data bus and can access bytes and 16-bit words, suggest ways in which the byte/word selection may take place.
   c. This computer has a block of 512 KB of 32-bit-wide memory built using 64-Kbit static RAM chips that are each 4-bits wide. How many RAM chips are required to implement the memory?

10.3 What is the meaning of *memory hierarchy,* and why is this concept of importance to the designer of PCs and similar workstations?

10.4 A memory component spans the address range 0x00400000 to 0x007FFFFF. What is its capacity?

10.5 At the beginning of this chapter, we stated that volatile read-only memory was an oxymoron. However, there is one application where a volatile read-only memory might be very beneficial under certain conditions. Can you think what this application might be?

10.6 A designer can use two different static memory devices to construct a computer with a 16-bit CPU. Both memory chips can hold 2^{22} bits. Suppose it is necessary to construct a computer with a 16-bit data bus and 128 MB of storage. The memory chip is available as 256K words of 16 bits or 4 M words of 1 bit. Which arrangement would provide the most physically compact storage system? *Hint*: We are interested in the number of pins per chip.

10.7 A company designs a computer in 2012. At its launch, the memory is twice as fast as the CPU. It is estimated that each year the CPU will get faster by 20%. Equally, it is estimated that each year the memory will get faster by 10%. After how many years will the CPU have to wait for the memory to provide data?

10.8 Why is all ROM RAM but not all RAM ROM?

10.9 Why is static RAM more suited to cache memory than DRAM?

10.10 What is the typical amount of main store, cache, and hard disk provided by current high-performance, state-of-the-art, cutting-edge personal computers?

10.11 We state that a static memory cell requires at least four transistors and a DRAM cell can use one transistor per bit. Could a memory ever use less than one transistor per bit to store data? Can you imagine a mechanism that could store more than one transistor per bit?

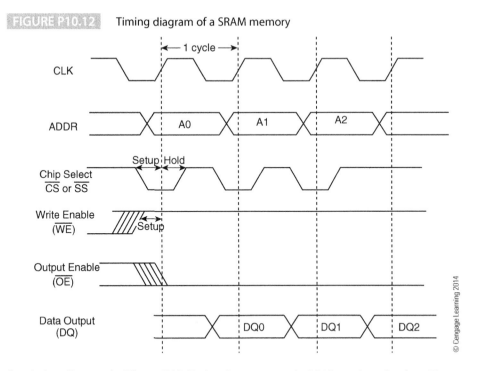

FIGURE P10.12   Timing diagram of a SRAM memory

10.12 Given the timing diagram in Figure P10.12 that is derived from the data sheet of a commercial SRAM, explain its action in words.

10.13 What are the basic differences between static RAM and dynamic RAM?

10.14 Why does a DRAM have fewer pins (I/O connections) than a static RAM of the same size (i.e., capacity in bits)?

10.15 In a conventional DRAM, what is the function of the column address and row address strobes?

10.16 In the context of DRAM timing, what is a pseudomaximum? What DRAM parameter is a pseudomaximim? What are the implications of a pseudomaximum for design engineers?

10.17 What fundamental properties of matter are exploited to implement memory systems? Can you think of any properties that have not yet been exploited and suggest ways in which they may be used to construct memory systems in the future?

10.18 What are the differences between Ovonic and ferroelectric memories?

10.19 A DDR SDRAM module in a PC is clocked at 133 MHz and is interfaced to a 64-bit data bus. What is the designation of the memory system in terms of the PCxx00 standard?

10.20 The speed of DRAM is increasing by about 7% a year, while the performance of processors is increasing by about 60%. What is the consequence of this state of affairs in the short term? What is the consequence in the long term?

10.21 You have a microprocessor chip with a 16-bit data bus. The computer accesses words; that is, it performs only 16-bit reads and writes. You are going to build a minimal computer which requires the processor and two 8-bit memory components. Unfortunately, you have only one 8-bit memory component, which means that you can't build a system. Or can you? Explain how you could construct a system with a 16-bit processor and an 8-bit memory.

10.22 A computer has a 64-bit data word and a 32-bit address. Interleave addressing is used with four banks. The total amount of memory is 1,024 MB. Show how the processor's 32-address bits are partitioned (i.e., divided into four fields).

10.23 A computer with a 24-bit address bus uses a logic circuit whose function is $F = A_{23} \cdot A_{22} \cdot \overline{A_{21}} \cdot A_{20} \cdot \overline{A_{19}}$ to select a memory component. What range of memory addresses are decoded by this circuit?

10.24 A computer with a 64-bit data bus uses the following memory chips. In each case, the chip is specified by locations × data width. For each of these chips, state the minimum number of chips required and the size of the corresponding memory block.
   a. 4 M × 1
   b. 1 M × 4
   c. 256K × 16

10.25 A DDR3 DRAM is specified as 9-9-9-24. What does this mean?

10.26 Suppose you could include cache memory in DRAM chips. How would you organize it, and what would the advantages be? What changes might have to be made to the computer system architecture?

10.27 An embedded microcontroller with a 20-bit address bus implements the following four blocks of memory.

Draw an address decoding table to satisfy the following memory map, and design an address decoder to select each of these devices.

a. RAM1   0 0000–3 FFFF
b. RAM2   4 0000–7 FFFF
c. ROM1   E 0000–E 7FFF
d. ROM2   F 0000–F FFFF

10.28 A CPU with a 24-bit address bus and 16-bit data bus implements the following memory blocks:

1 M byte of ROM using 256K × 8-bit chips
8 M bytes of DRAM using 2M × 4-bit chips

Design an address decoder to implement this arrangement.

10.29 It has been reported that flash memory was reaching the limit of its density. Why do you think that this may be so?

10.30 What is *wear leveling*, and why does it have to be undertaken?

10.31 You have been asked to design the on-board computer for a deep-space vehicle. What special considerations would you have to take into consideration, and how would these affect your design?

10.32 Figure P10.32a gives the timing diagram of a microprocessor during a write cycle. Figure P10.32b gives the timing diagram of a memory-mapped peripheral during a write cycle. Using the timing information in Table P10.32, verify that the interface of Figure P10.32 will function correctly. Note that a memory-mapped peripheral is a peripheral that looks exactly like a block of static RAM as far as the processor is concerned (we cover such peripherals in Chapter 12). In this question, active-low signals are indicated by an asterisk rather than an overbar; for example, W* indicates NOT write.

TABLE P10.32

(a) Processor timing

Parameter	Name	Min	Max
t_1	Address setup	10 ns	
t_2	Address hold	15 ns	
t_3	Address strobe asserted	100 ns	
t_4	Address strobe low to data strobe low	20 ns	
t_5	Address strobe low to R/$\overline{\text{W}}$ low	20 ns	30 ns
t_6	Address strobe high to R/$\overline{\text{W}}$ high	0 ns	10 ns
t_7	Data setup time	5 ns	10 ns
t_8	Data hold time		5 ns

(b) Peripheral timing

Parameter	Name	Min	Max
T_1	Address setup	5 ns	
T_2	Address hold	2 ns	
T_3	Chip select asserted	40 ns	
T_4	$\overline{\text{CS}}$ low to $\overline{\text{WE}}$ (write enable) low	10 ns	
T_5	$\overline{\text{WE}}$ high to $\overline{\text{CS}}$ high	5 ns	
T_6	Data setup time to $\overline{\text{WE}}$ low	20 ns	
T_7	Data hold time from $\overline{\text{WE}}$ high	3 ns	

© Cengage Learning 2014

FIGURE P 10.32

(a) Processor timing diagram

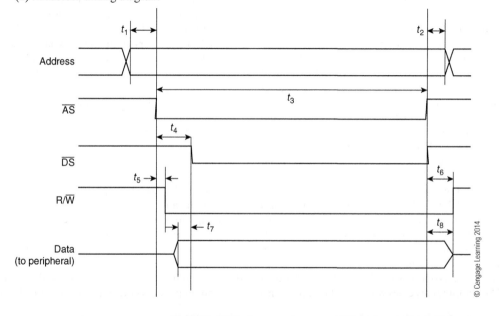

© Cengage Learning 2014

FIGURE P 10.32   *continued*

(b) Peripheral timing diagram

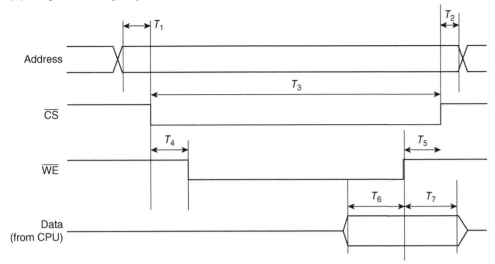

(c) Circuit of the CPU to peripheral interface

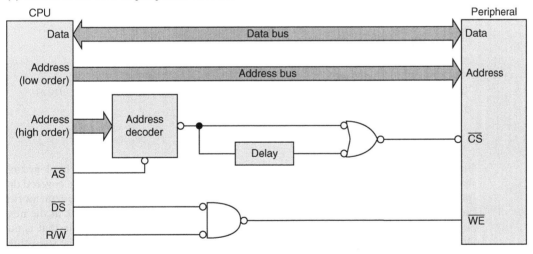

Note:

$\overline{AS}$ = Address strobe (low when address from CPU valid)

$\overline{DS}$ = Data strobe (low when data from CPU valid)

$R/\overline{W}$ = Read/not write (high in read cycle, low in write cycle)

$\overline{CS}$ = Chip select (low when peripheral to take place in a read or write cycle)

$\overline{WE}$ = Write enable (low when peripheral to take place in a write cycle)

Secondary Storage

"The whole is greater than the sum of its parts."
Anon

"We build our computer (systems) the way we build our cities: over time, without a plan, on top of ruins."
Ellen Ullman

"I can see the future now – and it has no moving parts."
Bartoz Kijanka

"Tape will outlive us all."
Title of an HP White Paper

"Do not put your CDs in the washing machine—but if you insist on doing so, please use the cycle for delicate fabrics."
Steve Meretzky

In previous chapters, we've described the rapid developments in processor technology that have given us computers capable of processing video data in real time, and we've covered the progress in semiconductor technology that's provided the large immediate access memories needed to hold the data processed by modern CPUs. Now we're going to look at the next stage in the chain, the *secondary storage units* that hold vast quantities of data that is not currently being processed. These storage devices are the magnetic hard disks, the solid-state disks, and the optical CD/DVD/Blu-ray drives.

When hard disks first appeared in PCs, they had a capacity of 5 MB. Today, a processor may have an 8 MB cache, the three TB[1] hard disk is already a commodity product, and Samsung was the first company to demonstrate a 4 TB hard drive in 2011. Toshiba announced plans to build 5 TB drives in 2013 with an areal density of 1Tbit/in^2. That's an increase in hard disk capacity of 800,000 in just over two decades. On the other hand, the access time of hard disks has not kept pace with growth in capacity. Today's hard disks are barely faster than those of a decade ago, although the advent of the solid-state disk does promise to provide a new generation of fast disks.

---

[1] 1 MB = 2^{20}, 1 GB = 2^{30}, 1 TB = 2^{40}. If a novel is 100,000 words and the average English word is five letters long, a 3 TB hard disk can store about $3 \times 2^{40}/2^{19}$ = 6 million novels.

Because secondary storage is so vital to the development of computers, we devote a chapter to its technology and characteristics. We begin with a discussion of the nature of the magnetic recording process and demonstrate how disc drives operate. In particular, we show how their electromechanical characteristics determine their performance. Now that semiconductor technology has invaded the world of mass storage with the introduction of the *solid-state disk* (SSD), we provide an introduction to the SSD.

The second part of this chapter examines optical storage media comprising CDs, DVDs, and Blu-ray disks that provide low-cost, nonvolatile, and transportable secondary storage. These media are very similar to magnetic disks except that changes in the optical characteristics of matter are exploited rather than the magnetic properties.

11.1   Magnetic Disk Drives

IBM shipped the first disk drive in 1956 as part of the 305 RAMAC system. Its platter was 24 inches in diameter; it had a capacity of 5 MB, was larger than a washing machine, and cost thousands of dollars. By 1983, the first PC disk drive was introduced by Seagate, which also stored 5 MB but cost a mere $1,500 and just managed to fit inside a PC. Today, hard disk drives have capacities of 4,000,000 MB. Some cost less than $50.

The disk drive uses a technology we have understood since the 1940s. Indeed, the magnetic disk is a direct descendent of the phonograph invented by Thomas Edison in 1877. Edison originally stored sound along a track on a cylinder covered by tin foil (later wax). The Edison phonograph stored sound by physically deforming the side of the groove to store sound vibrations, whereas the magnetic disk stores data by magnetizing the surface of a track. The CD/DVD/Blu-ray stores data by changing the optical properties of the surface of a track.

A disk drive uses a flat rotating platter covered with a very thin layer of a material that can be locally magnetized in one of two directions: North–South or South–North. This platter rotates under a write head that magnetizes the surface to create a circular track of 1s and 0s. When the data is retrieved, a read head that's normally co-located with the write head detects the magnetization of the surface and uses it to reconstruct the recorded data. Couldn't be simpler.

In practice, the construction and operation of real disk drives is immensely complex, because the size of the magnetized regions is very small and the disk rotates at a high speed. Indeed, the details of a modern disk drive are truly awesome: The magnetic layer is of the order of 2,000 atoms deep and the read/write head itself flies in a rotating layer of air 0.2 μm above the surface of the platter. On top of the magnetic layer is a lubricating layer of a fluorocarbon that is about one molecule thick.

We begin by examining the principles of magnetic recording before looking at the construction and characteristics of disk drives. We include milestones in the design of disk drives that have pushed capacities from 10 GB to 4 TB in a few years. Figure 11.1 from IBM plots the *areal density* against the year for several disk drives and illustrates the remarkable increase in recording density in the 45 years from 1956 to 2001. The term *areal density* specifies the packing density of bits and is often measured in bits per square inch. Areal density increased from approximately 2×10^{-3} Mbits/in.^{2} to 4×10^4 Mbits/in.^{2}, which is an increase of the order of 10^7 in. about 50 years. By 2010, Toshiba was shipping disks with an areal density of 540 Gbits/in.^{2} (i.e., 54×10^4 Mbits/in.^{2}), and only one year later Toshiba announced a 2.5-in. hard drive for use in laptops with a density of 744 Gbits/in^2.

Disk capacity has gone hand-in-hand with the development of Microsoft's operating systems. Figure 11.2 demonstrates how both disk capacities and operating systems have grown with time. Without large capacity disks, today's operating systems would not be possible. Of course, there are those who would argue that much of today's software is *bloated* and should be smaller.

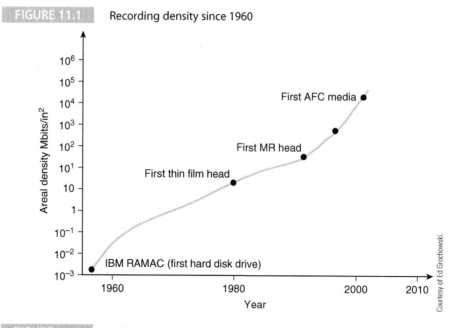

FIGURE 11.1   Recording density since 1960

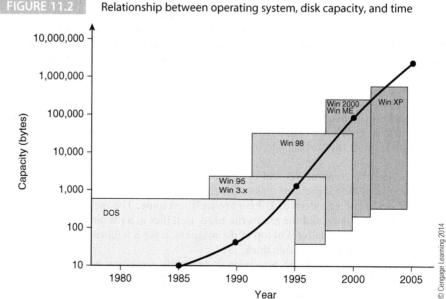

FIGURE 11.2   Relationship between operating system, disk capacity, and time

11.2   Magnetism and Data Storage

Magnetic storage technology is rather *quaint*; it belongs to the early days of computer technology and relies on electromechanical mechanisms with moving parts. Magnetic recording techniques have been used for a long time. For example, the wire sound recorder recorded speech on a reel of steel wire before being replaced by the tape recorder after WW2. However, magnetic storage technology stubbornly refuses to go away in spite of its inherent limitations. Indeed, in December 2000, an article on magnetic storage in the *IEEE Spectrum*[2] appeared with the title "Magnetic Storage: The medium that wouldn't die."

---

[2]R. Comerford, "Magnetic Storage: The medium that wouldn't die." *IEEE Spectrum*, December 2000, pp. 36–39.

The magnetic properties of matter are probably the most obvious means of storing data because magnetism is an excellent binary recording medium: magnetic particles can be magnetized N–S or S–N. When certain substances are magnetized, they remain magnetized until they are magnetized in the opposite sense, which makes magnetic storage mechanisms inherently nonvolatile.

In this section, we introduce the notion of magnetization at the atomic level and show how a bulk material can be magnetized in one of two states by means of a write head. We then demonstrate how this property of matter can be used to construct the disk drive.

The origin of magnetism lies in the atomic structure of matter—in particular, the behavior of electrons in atoms. An electron has two motions: its orbit round the nucleus and its *spin*. The principal cause of magnetization is the spin of electrons. Although the term *spin* implies rotation, it is misleading if you think of an electron spinning like a top. Electrons have two quantized spin values that are called *spin up* and *spin down*.

In a *ferromagnetic* material, the spins of individual atoms couple; that is, there is an interaction between neighboring atoms. When an external magnetic field is applied, ferromagnetic atoms tend to align with the field. When the external magnetic field is removed, a ferromagnetic material can retain some of the magnetization in the direction of the applied field.

The quantum interactions between electrons in a ferromagnetic material have a range that extends beyond the individual atoms. This interaction causes the magnetic moments of atoms within a region called a *domain* to align in parallel. Domains vary in size from 30 nm to 150 μm.

In a bulk ferromagnetic material, the individual domains are aligned at random, as Figure 11.3a illustrates, and there's no overall magnetization. Figure 11.3b demonstrates the effect of applying an external field. Domains that are magnetized in the same direction as the external field remain magnetized in that direction. However, domains that are magnetized in other directions rotate the direction of their magnetization in the direction of the external field.

Because the internal field in the material is the sum of the external field and the field due to the domains, the internal field rapidly increases as more and more domains become oriented in the direction of the external field. Suddenly, the number of domains rotating in the direction of the external field increases dramatically as the internal field builds up like an avalanche. Soon, all domains are magnetized in the same direction, as in Figure 11.3c, and bulk material is said to be magnetized. If the external field is removed, the material remains in the magnetized state, because the fields from the domains are sufficient to keep the domains from realigning themselves.

FIGURE 11.3   Magnetization and the domain

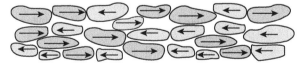

(a) No external field

Magnetization of domains is at random. No overall magnetization of the bulk material.

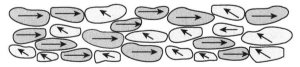

(b) Weak external field applied

Magnetization of some domains rotates.

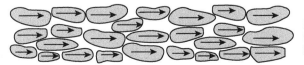

(c) Strong external field applied

Magnetization of domains aligned in the same direction. The bulk material is magnetized.

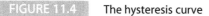

FIGURE 11.4   The hysteresis curve

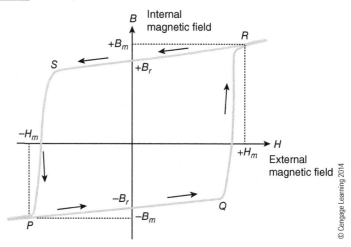

Raising the temperature of a material increases the thermal motion of the atoms. In magnetic materials, the thermal motion eventually becomes so great that the cohesive force of the domains is overcome and the material one again becomes demagnetized and the domains randomly oriented. The temperature at which this occurs is called the *Curie point* and is over 1,000°C for iron.

If we plot the internal field against the external field for a ferromagnetic material, we get Figure 11.4, which is called the *hysteresis curve*.

The horizontal axis H represents the external field. In the absence of an external field (i.e., $H = 0$), the value of the internal field B is either $+B_m$ or $-B_m$; that is, the material is magnetized in one of two states. Suppose the initial state of the material with $H = 0$ is $B = +B_r$, and H is increased along the positive axis. The curve is followed to the point R. When the external field is removed, the system returns to $+B_r$.

Now suppose that we are in the state $+B_r$ and the external field H is increased in the negative direction towards S. If the external field is removed, the material returns to $+B_r$. Now, if the applied field is increased beyond point S, the domains begin to switch direction and the material rapidly changes its magnetization and moves to point P on the curve. If the external field in now removed, the material returns to $-B_r$ and the direction of magnetization remains in this stable state but is inverted.

Figure 11.4 demonstrates that a material in state $+B_r$ can be changed into state $-B_r$ by applying an external field less than $-H_m$. Similarly, a material in state $-B_r$ can be changed to the state $+B_r$ by applying an external field greater than $+H_m$.

A material can be magnetized by applying a sufficiently large positive or negative field. The next step is to explain how this operation is performed in practice and how we can detect the state of a magnetic material.

Another property of a magnetic material is its *coercivity*, which is the strength of the field that must be applied to reduce the remnant flux to zero. Magnetic materials are often divided into two classes: *hard* and *soft*. A hard magnetic material has a high coercivity and can be permanently magnetized. A soft magnetic material has a low coercivity and is not magnetized when an external field is removed. Read and write heads are constructed from magnetically soft materials, whereas the magnetic coating that stores data is constructed from a magnetically hard material.

11.2.1   The Read/Write Head

Figure 11.5 describes the structure of a read/write head used for writing and reading data on magnetic recording media such as hard disks, floppy disks, and tape. Very early recording

FIGURE 11.5   The original form of the read/write head

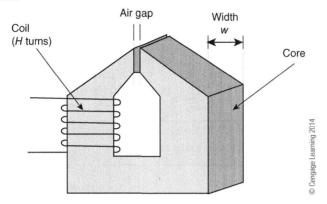

© Cengage Learning 2014

heads consisted of toroids of magnetically soft metallic ferromagnets with a few turns of wire around them.

High-frequency magnetic fields induce *eddy currents* in the write heads that reduce the head's efficiency. Second-generation heads used non-conducting ferromagnetic materials called *ferrites* (a ceramic compound of iron and nickel or iron and zinc).

Constructing read/write heads is difficult because of the complex interacting requirements of the recording system. The air gap must be as small as possible, because that determines the size of the field that leaks from the write head into the magnetic recording medium. If the gap is large, the area of magnetization is also large, and the maximum number of bits that can be stored on the disk is reduced.

It's also necessary to build the heads with a ferromagnetic material that has a very high saturation (i.e., the largest field that can be generated in the material). A strong field is required in order to magnetize the particles on the recording medium. The head material must also have a low remnant magnetization (i.e., the residual field after the write current has been turned off). If the remnant magnetization is too high, previously magnetized bits can be disturbed by the remnant field. If the same head is used for reading as well as writing, it must have a high value of *permeability* (a material's permeability is a measure of its ability to conduct magnetic flux: the higher the permeability, the easier it is to magnetize a material). Read heads should also have a low saturation *magnetostriction*. Magnetostriction describes a phenomenon whereby a change in magnetic field changes the material's physical dimensions, and vice versa. If a read head suffers any form of physical shock, magnetostriction generates a spurious field and hence a spurious current in the coil. On top of all of these magnetic properties, the head must be physically robust and resistant to both wear and corrosion.

The Recording Process

We now look at the way in which data is recorded in a magnetic material. Figure 11.6 describes the recording process. A coil of wire is wound round a ring of metal. When a current is passed through the coil, a magnetic field is created in the coil, and this in turn induces a field in the ring. The ring contains a tiny air gap, and the field has to flow across the gap. In practice, it leaks out into the surrounding world. If the gap is close to a ferromagnetic material, this external field can magnetize it. The only difference between a hard disk drive and a tape (or cassette) recorder is that in one case the ring is placed above a hard platter and in the other a band of tape coated with a magnetic material is moved past the gap in the ring.

Figure 11.7 illustrates the effect of switching the current in the write-head coil and the corresponding magnetization of the surface passing under the write head.

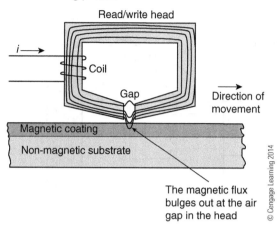

FIGURE 11.6   The recording process

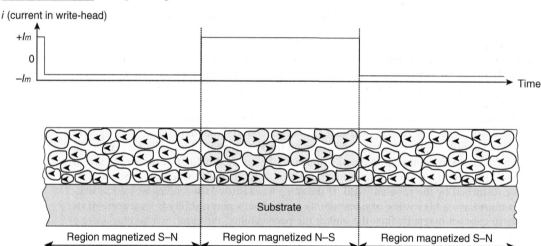

FIGURE 11.7   Magnetizing the surface of a material

i (current in write-head)

Time

Region magnetized S–N   Region magnetized N–S   Region magnetized S–N

11.2.2   Limits to Magnetic Recording Density

There are physical limitations on the ultimate areal density of magnetic recording mechanisms; that is, there is a finite limit to the maximum number of bits that can be stored in a square inch. Fortunately, as the *anticipated* theoretical limit is reached, physicists and engineers seem to find ways of extending this theoretical maximum (echoes of Moore's law).

In the 1990s, scientists believed that a phenomenon called the *superparamagnetic* effect imposed a limit on magnetic recording density. The minimum amount of magnetic material that can be used to store information is the magnetic *grain* (a single-domain particle) from which bulk magnetic materials, such as the surface of a disk, are constructed. When grains reach sizes of the order of 10 nm or so, thermal effects can cause these grains to spontaneously demagnetize at room temperature. This corresponds to a maximum areal density of about 6 Gb/cm^2 or 0.93 Gb/in.^{2}. Fortunately, several ways of avoiding the superparamagnetic effect have been discovered.

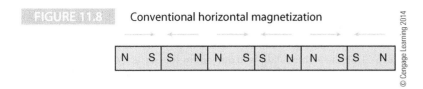

FIGURE 11.8   Conventional horizontal magnetization

© Cengage Learning 2014

Figure 11.8 illustrates surface recording with horizontal magnetization. This figure represents the worst case with the smallest possible regions of magnetization, which are magnetized alternately N–S and S–N. An alternative to horizontal magnetization is the vertical or perpendicular magnetization of Figure 11.9 in which the magnetic domains are magnetized at right angles to the surface of the recording medium. Perpendicular recording reduces the demagnetizing influences of adjacent bits, because they are oriented so they do not oppose each other. They form part of a closed magnetic field.

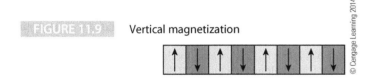

FIGURE 11.9   Vertical magnetization

© Cengage Learning 2014

A different write-head design (i.e., compared to the head used in conventional horizontal magnetization) is required for writing. A "monopole head" is required to write magnetic transitions vertically within the media.

The proximity of the tiny magnetic particles to each other tends to demagnetize adjacent particles. Figure 11.10 illustrates a means of reducing the size of vertical particles without demagnetization developed by Fujitsu. This technology can provide an eight-fold increase in areal density over conventional techniques and permit densities in the region of 50 Gb/cm^2. By about 2,000 esoteric magnetic media with areal densities approaching 700 Gb/cm^2 (100 Gb/in.^{2}) were being *proposed*. By 2011, disks with area densities of over 700 Gb/in.^{2} were being *sold*.

One approach to delaying the inevitable effects of superparamagnetism is to modify the media's magnetic properties by increasing the energy barrier required to reverse the state of a grain. The arrangement in Figure 11.10 uses a layer of magnetically soft material beneath the magnetic recording surface.

Traditional magnetic surfaces have been homogenous; that is, they are composed of a single metal film or a coating of magnetic particles in a binder. Hard disk designers are investigating the use of *patterned media* to store data. By using a bit pattern consisting of islands of magnetic material surrounded by a nonmagnetic matrix the effect of demagnetization between adjacent areas is reduced.

FIGURE 11.10   Vertical magnetization with a magnetic backing

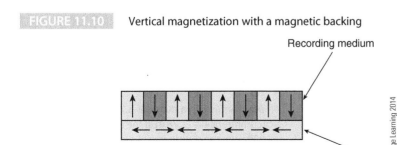

Recording medium

Magnetically soft substrate

© Cengage Learning 2014

11.2.3   Principles of Data Recording on Disk

The next step is to look at how data is stored on a disk and how the recording and playback process works. Figure 11.11 shows how data is arranged on the surface of a disk. The read/write head can move or *step* in towards the center or out towards the periphery. As the disk rotates, the head describes a circle called a *track*. A track is too large a unit of data to be practical, so the track is divided into individual *sectors*. A sector is the smallest unit of data that can be read from or written to the disk. The structure of data on a disk has important implications for the performance of disk drives. Consider, for example, the *granularity* of data; small sectors are inefficient because a large file would take up many sectors, each of which has an overhead. On the other hand, large sectors are inefficient if you wish to store small units of data. For example, if sectors were 8 KB and you were using a lot of 3 KB files, each sector would waste 5 KB. Typical disk drive sectors are 512 bytes.

In order to reduce the physical size of a disk drive and increase its data capacity, disk manufacturers located several platters together on the same spindle, and the read/write heads that read each surface are connected to the same actuator so that all heads step in or out together. Figure 11.12 illustrates a system with three platters. Early disk drives didn't use the top- and bottom-most surfaces to store data, and the arrangement of Figure 11.12 has four surfaces. However, modern disk drives use all surfaces.

The actuator arrangement of Figure 11.12, in which the heads move in or out along a radius, uses a head assembly that travels along a metal track. Such an arrangement is no longer in widespread use because it's complex and slow. A simpler actuator mechanism employs

FIGURE 11.11   Magnetizing the surface of a material

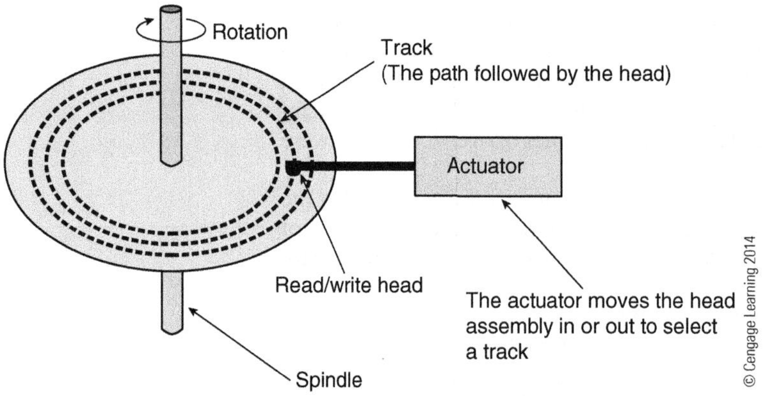

Rotation

Track
(The path followed by the head)

Actuator

Read/write head

The actuator moves the head
assembly in or out to select
a track

Spindle

© Cengage Learning 2014

Platter Size

The standard hard disk in a PC is called a 3.5 in. drive. You would expect the platter to be less than 3.5 in. It isn't. The diameter of a platter is 3.75 in. And, yes, a 3.5 in. drive isn't really 3.5 in. wide; it's four inches wide. The term 3.5 in. refers to its *form factor* and not its physical width.

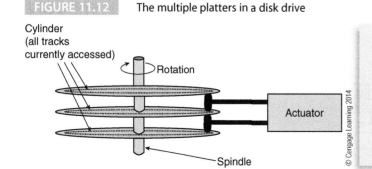

FIGURE 11.12   The multiple platters in a disk drive

Cylinder
(all tracks
currently accessed)

Rotation

Actuator

Spindle

© Cengage Learning 2014

Speed of Disks

How fast does the edge of a disk rotate?

If the diameter of a platter is 3.75 in., its circumference is 11.78 in. If the disk rotates at 7,200 rpm, the outer edge moves 7,069 feet/min or 424K ft/hour. That's 80.3 mph.

FIGURE 11.13   Picture of a disk drive

Courtesy of Seagate Technology.

a pivoted arm that swings across the disk's surface (rather like the tone arm of the old black vinyl phonograph). Figure 11.13 provides a photograph of a Seagate disk drive that uses a similar head-tracking mechanism.

In order to produce a tiny magnetic field on the surface of a disk, it's necessary to have tiny magnetic particles, a tiny head gap, and a tiny gap between the head and the surface of the recording medium. Any technological development has to take place in tandem in all of these areas, as one parameter can't be improved without corresponding improvements in the others.

It's relatively easy to create a very small gap in a write head using modern manufacturing technology. Constructing a platter rotating at over 50 mph under a write-head that's only 10×10^{-9} m above the surface is a bit tricky. Indeed, it would be impossible if it weren't for the *boundary effect*.

When a disk rotates in an atmosphere, the gas at its surface rotates at the speed of the disk, because surface roughness drags the molecules of the gas with it. At some distance above the surface, the gas is not moving. Consequently, there is a velocity gradient between the surface and free space above the surface.

The read/write head forms part of a structure known as a *slider,* which positions the head over the requested track. The slider also connects the heads to the actuator via the *suspension arm*, carries the signals from the head, and provides the rigidity or stiffness required to keep the head in place. The slider is sculpted to have suitable aerodynamic characteristics so that it "flies" above the surface of the disk in the moving boundary later.

The lift generated by an airfoil (i.e., the slider) is proportional to the square of the speed of the air moving over its surface. As the slider moves towards the disk, the air is moving faster, and more lift is generated, which keeps the slider at an almost constant distance above the surface. You could say that the slider is following the surface of the disk on an *air bearing.*

The flying head is able to track the undulating surface of a disk with remarkable precision. The suspension must provide a force on the slider in a direction into the disk to counteract the upward aerodynamic forces of the air bearing that cause the slider to fly over the disk's surface. This force must act precisely in the proper location, or a twisting force will cause one of its corners to be too close to and the other too far from the disk's surface.

Designing an effective air bearing is harder than you think. As the head moves between the inner and outer radius of the disk, there's an approximately two-to-one change in the velocity of the disk's surface, and therefore, the speed of the rotating air. Modern air-bearing designs are able to compensate for this and keep the head about 1.0 μin. above the magnetic surface.

The suspension must allow the slider to gimbal (i.e., rotate) in the pitch and roll directions so that it can stay close to the surface despite undulations of the disk.

Figure 11.14 illustrates the way in which the surface-head gap has been reduced between 1993 and 2004. The improvement is somewhat under an order of magnitude with contemporary heads tracking a disk surface at a height of about 10 nm (i.e., 10^{-8} m), which is an unimaginably small gap.

Should the head fail to follow the surface of a disk and hit the surface at about 50 mph, it damages the magnetic coating and destroys data. Such an event is called a *head crash*, which is the origin of the term now used to indicate any sudden and catastrophic computer system failure. Figure 11.15 gives the classic illustration of just how small the gap is from the read/write head to the surface. On the same scale, we have the gap together with a human hair, a smoke particle, and a fingerprint. Yes, the flying height of the head is less than the height of a fingerprint.

FIGURE 11.14   Head-surface spacing

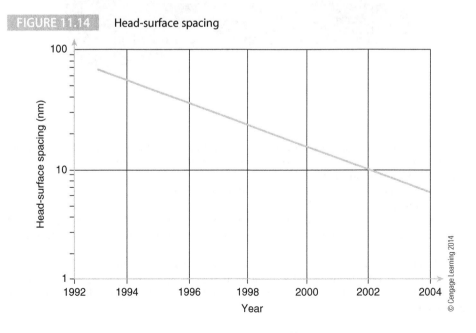

© Cengage Learning 2014

FIGURE 11.15   Illustration of the relative size of the head surface gap

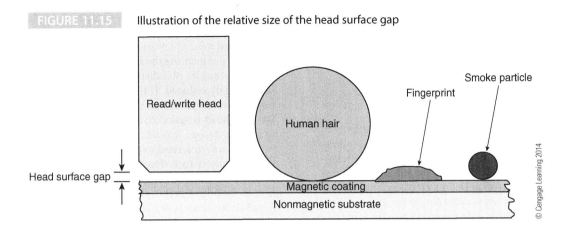

If designing the head suspension mechanism with its minuscule flying height and track-to-track spacing in the hard disk in the computer that sits on your table is a difficult job, spare a thought for those who have to design suspension, slipper, and head mechanisms for use in *laptops* and other portable computers. The head has to maintain its flying height while the disk drive itself in undergoing vibration and acceleration because it's being carried.

What happens to the head when the drive is not in use? Early disk drives used a *landing zone* to park the head during the power-down stage. This region was not, of course, used to store data. Improvements in head and disk surface technology have led to very smooth heads and smooth disk surfaces. The degree of smoothness is so great that the head and disk surface can stick together if they come into contact when the disk is not rotating (a form of *cold welding* that takes place when two atomically smooth surfaces come into contact).

Today, the head is retracted and lifted off the disk, as Figure 11.16 demonstrates. When the power to the motor is removed, it generates a *back EMF* as it slows down. This voltage is used to move the slider to a ramp and lift it clear of the disk's surface.

FIGURE 11.16   The head loading and unloading mechanism

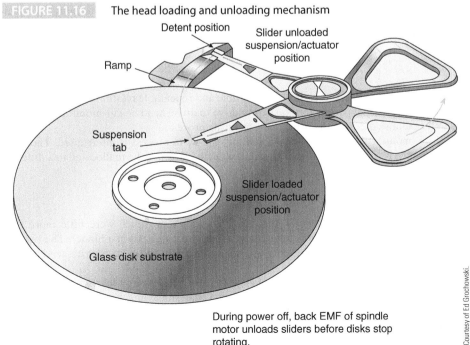

During power off, back EMF of spindle motor unloads sliders before disks stop rotating.

The amount of data that can be stored on a disk is a function of the number of tracks per inch and the number of bits per inch along a track. The product of these two parameters gives the *areal density* in bits per square inch. You can raise areal density only by increasing the number of bits on a track or the number of tracks per inch. As the number of bits per inch increases, the size of individual bits has to be correspondingly reduced. If the bits get smaller, the read signal generated as the disk rotates under the read head is reduced, and it becomes harder to reliably decode the data.

One way for read/write heads to maximize areal density is to reduce the distance from the head to the disk—the head's *fly height*. A reduction in fly height makes the bit pattern's output signal stronger and easier to detect.

Unfortunately, there are several factors that can adversely affect the flying height, such as changes in altitude or temperature, the effects of contamination, external shock, and vibration. Changes in a disk's optimum environmental parameters can degrade the drive's error performance when reading data. Equally, if the write head flies too high over the surface of the disk, the magnetic field may be insufficient to reliably write to the media. This causes a soft error that can be corrected.

Western Digital introduced a mechanism called *Fly Height Monitoring* to detect and provide a warning of potential failure. This monitor uses a write condition detector that detects when the transducer's flying height deviates from its nominal position and suspends any write operation.

When raw data is read as an analog signal from the surface of a disk, magnetic flux transitions produce a pulse in the read head. One parameter that defines a pulse is its *shape*; that is, the ratio between its area and its peak. The shape of a pulse can be used to monitor head height. As a head flies higher, the ratio between pulse height and width increases, creating a measurable and repeatable relationship for each head. Because the individual parameters vary from drive to drive, head height-monitoring circuitry has to be calibrated during the drive manufacturing process.

If the monitor detects an unsafe condition, the write process is interrupted. The write operation may be repeated. If it is still unsuccessful, the data may be reallocated to a different region of the disk.

Thermal Recalibration

Some early disk drives used *step-actuators* that moved the head on a linear carriage in or out by a fixed step at a time. In those days, the tracks were relatively far apart, and track following wasn't a major issue. A rotary actuator that swings the arm across the disk by rotating through a given angle has now replaced the step actuator. Today's track spacing is much denser than a few years ago, and it's harder to turn a spindle through a precise angle than it is to step a head in or out. Head positioning accuracy and track following are now more of a limiting factor than in the past.

It gets worse. As the various components of a disk drive heat up or cool down, they do so at different rates, and the head may wander off track. An operation called *thermal calibration* is periodically carried out by some disk drives to recalibrate and reposition the head. Recalibration may take place when the temperature of the drive changes by a predetermined amount or after a certain interval. This operation can cause a gap of about 500 ms in the flow of data between the disk and host controller. Such a break in the data stream is not normally significant, but it is unacceptable in audiovisual applications when a short pause in sound or movement in a video is disturbing. During the mid-1990s, when audiovisual technology was becoming more popular, some disk manufacturers produced a range of AV disks in which the thermal recalibration process was effectively hidden.

Today, more accurate head positioning mechanisms using feedback have been introduced, and it is no longer necessary to perform thermal recalibration.

Platter Technology

Today's platters are more complex than their predecessors, which were little more than aluminum disks coated with a magnetic material (an oxide of iron) in a binder. The two key parameters that determine data density are the flying height of the read/write head and the size of the magnetic particles.

Some platters are made of low-defect glass, because glass is more thermally stable (it has a lower coefficient of expansion than aluminum), smoother, and harder than aluminum. Figure 11.17 is a photomicrograph from IBM showing the relative surface smoothness of both aluminum and glass surfaces. Glass is more rigid than aluminum for the same weight of material. Improved rigidity reduces noise and vibration at high speeds. The rigidity of glass allows

FIGURE 11.17   Platter surface smoothness

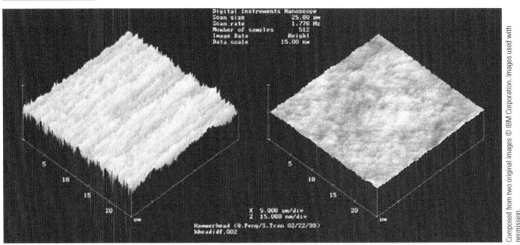

platters to be made thinner and lighter, which reduces the load on spindle motors. Moreover, a lighter platter means faster spin-up times.

The surface is applied by *sputtering* multiple films on the disk. Sputtering is a process whereby the platter is placed in a high vacuum and the coating vaporized to create a thin film on the disk's surface.

Modern platters contain five or more layers (Figure 11.18). The uppermost layer is a lubricating layer that enhances the durability of the head–disk interface. Below the lubricating layer lies a thin protective carbon-based overcoat. The lubricating layer is about 1 nm thick and the overcoat 15 nm thick. The recording surface consists of two layers: the recording layer (often a compound of cobalt and chromium) and a chromium underlayer. Finally, a glass substrate provides the surface that holds these other five layers.

FIGURE 11.18   Cross section through a platter

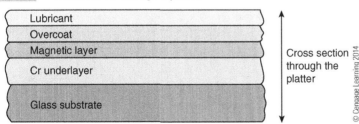

Lubricant
Overcoat
Magnetic layer
Cr underlayer
Glass substrate

Cross section through the platter

© Cengage Learning 2014

The GMR Head—A Giant Step in Read-Head Technology

The conventional read head has an important limitation, *inductance*. In order to detect the changing magnetic flux, the coil round the head requires a lot of turns to pick up a sufficient signal. Increasing the number of turns raises the coil's *inductance*. Inductance is a property of circuits that resists the rate at which a current changes. Increasing the inductance of a read head reduces the rate at which it can read changes in the magnetic flux at the disk's surface. Fortunately, a magnetic property of matter (described later) was discovered that made it possible to do away with the inductive read head. Moreover, removing the read head means that the write head can be optimized for writing.

A magnetic field causes a tiny change in the electrical resistance of certain materials, which is a property called the *magnetoresistive* (MR) effect. Detecting changes in magnetic flux from a disk using the MR effect has advantages over conventional inductive read heads, because the inductance of an MR head is lower and it's possible to read data more rapidly.

IBM pioneered the use of MR heads in 1 GB disk drives in 1991, and by 1994, IBM had demonstrated that an areal density of 3 Gbits/in.² was possible. Unfortunately, the electrical output of an MR head is very low. In the late 1980s, researchers discovered that some materials exhibited massive changes of up to 50% in their resistivity in the presence of a magnetic field. This property is called the *giant magnetoresistive* (GMR) effect and is found in materials consisting of alternating very thin layers of metallic elements. It was soon realized that the GMR effect could be used to build effective read heads, and IBM was one of the first commercial organizations to attempt to exploit the GMR effect.

The advantage of GMR heads is their greater sensitivity to magnetic fields from the disk, making it possible to detect smaller recorded bits. Equally, bits can be read at a higher speed, and the effect of electrical noise is much reduced in comparison with MR heads. By about 1998, IBM was using GMR heads to push areal densities beyond 11.6 Gbits/in.². These heads have a sensor thickness of 0.04 μ, and IBM claims that halving the sensor thickness to 0.02 μ will allow possible densities of 40 Gbit/in.². The advantage of higher recording densities is that disks can be reduced in physical size and power consumption, which in turn increases data transfer rates. With smaller disks for a given capacity combined with lighter read/write heads, the spindle speed can be increased further, and the mechanical delays caused by necessary head movement can be minimized

In a read/write head assembly (Figure 11.19), the read element consists of a GMR sensor between two magnetic shields. These magnetic shields reduce unwanted magnetic fields from the disk so that the head detects only the magnetic field from the recorded data bit under the head. In a *merged head,* a second magnetic shield also functions as one pole of the inductive write head. The advantage of separate read and write elements is that both elements can be individually optimized. A merged head is less expensive to produce and performs better in a drive, because the distance between the read and write elements is less.

FIGURE 11.19   The structure of a GMR read/write head

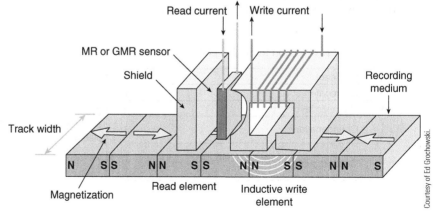

Courtesy of Ed Grochowski.

Pixie Dust

At the beginning of 2001, IBM announced a breakthrough in disk technology that could increase areal densities by a factor of four. IBM used a sandwich with three layers to store data. The top layer is a ferromagnetic material that stores the data. The lower layer is an *antiferromagnetic* layer. Antiferromagnetism occurs when atoms align themselves antiparallel to a magnetic field—the opposite of ferromagnetism. However, antiferromagnetics is a very weak effect. Between these two layers sits *pixie dust*, which is a three-atom thick layer of the element ruthenium. Ruthenium is a rare metal belonging to the same group as platinum, and only about twelve tons are produced annually—so making the layer only three atoms thick will make a little ruthenium go a long way.

This sandwich is called an *antiferromagnetically-coupled* (AFC) media and is capable of areal densities of up to about 100 Gb per in.^{2}. IBM claims that AFC media avoids the high-density data decay. The ultra-thin ruthenium layer forces the adjacent layers to orient themselves magnetically in opposite directions. The opposing magnetic orientations make the entire multilayer structure appear much thinner than it actually is. Thus, small, high-density bits can be written easily on AFC media, but they will retain their magnetization due to the media's overall thickness.

As early as 1990, IBM scientists had discovered that a thin layer of ruthenium atoms created the strongest anti-parallel coupling between adjacent ferromagnetic layers of any nonmagnetic spacer-layer element. The structure was used in the first giant magnetoresistive read element for disk drives in 1997.

Figure 11.20a describes the conventional magnetic surface, and Figure 11.20b illustrates the AFC media with its layer of ruthenium pixie dust.

FIGURE 11.20   Conventional and AFC media

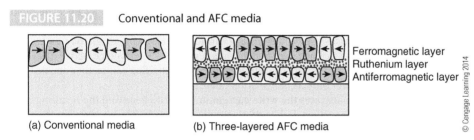

Ferromagnetic layer
Ruthenium layer
Antiferromagnetic layer

© Cengage Learning 2014

(a) Conventional media   (b) Three-layered AFC media

The Optically Assisted Head

One way of increasing bit density uses an *optically assisted head* to improve the positioning of the read/write head by borrowing techniques from the optical read/write CD. Some compounds containing rare earth elements (e.g., gadolinium) have stable magnetic properties at room temperature and a low Curie point. These materials can be magnetized just like any other substance used to coat a disk. If they are heated and then subjected to a magnetic field, the heated region can easily be magnetized in a different direction. The normal magnetization process localizes the region of magnetization by controlling the size of the magnetic field. The magnetization of surfaces constructed from these compounds is controlled by the size of the region heated—it doesn't matter if the magnetic field spills over, because only the heated region of the surface is magnetized.

Figure 11.21 illustrates the principal of the optically assisted write head. A tiny laser beam performs the surface heating. An optical fiber delivers the laser beam to the disk's

FIGURE 11.21   The optically assisted head

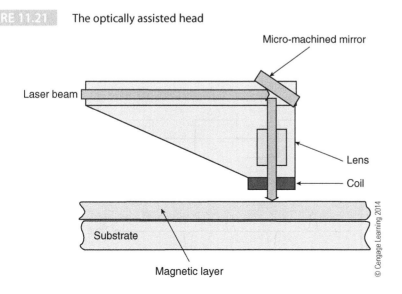

Micro-machined mirror

Laser beam

Lens

Coil

Substrate

Magnetic layer

© Cengage Learning 2014

surface via an optical fiber and a lens assembly. Pulsing the laser beam allows you to selectively heat the surface. A current is passed through the coil above the surface of the disk to magnetize the heated spot in one direction or another.

A micro-machined mirror controlled by a servomechanism directs the laser spot onto the surface of the disk. Because you can deflect the mirror electronically, you can steer the beam on the surface of the disk (i.e., switch from tack to track). This technology promises areal densities of 100 Gbits per square inch.

11.3   Data Organization on Disk

Having described the magnetic recording process and the structure of disk drives, the next step is to look at how data is stored on the surface of the disk. The write head directly writes bits onto the surface of a disk. First generation read heads were identical to the write head, and until the discovery of the MR effect, the same head was often used for reading and writing.

When a read head passes over a magnetized surface, the *changing* magnetic flux induces a current in the coil and a voltage across the coil's terminals. The voltage across the coil's terminal's is proportional to the *rate of change* of the magnetic flux rather than its absolute value; that is, you can detect only a *change* in flux density.

Figure 11.22a and b illustrates the write current in the write head and the resulting magnetization of the recording surface. Below these graph in Figure 11.22c, we have a trace of the voltage induced in the coil when the recorded surface passes under the head. This figure is idealized and the detail in the inset demonstrates how a recorded pulse might look in reality.

You can't store a long string of 1s and 0s on the surface reliably, because only changes in flux lever create a signal in the head. If you record 00000 or 11111, both sequences would produce the same output—nothing. Suppose, for example, you stored the string 000111111111110000. The read head would detect only two flux transitions: the initial 0 to 1 and the final 1 to 0 (these are marked in blue).

A GMR head can detect absolute magnetization, because even a constant field creates detectable low or high resistance in the magnetoresistive element. However, there is no delineation between the 1s and 0s in a long string.

Digital data-recording mechanisms (both magnetic and optical) encode data prior to recording in order to avoid situations in which the recorded information is difficult to read back. In particular, they avoid longs runs or constant magnetization; that is, they ensure that the recoded flux changes state regularly. This restriction is required to extract a data

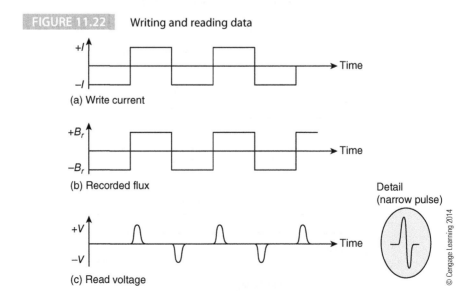

FIGURE 11.22   Writing and reading data

(a) Write current

(b) Recorded flux

Detail
(narrow pulse)

(c) Read voltage

© Cengage Learning 2014

clock from the disk. Most recorded codes are known as *self-clocking,* because the bit pattern recorded on the magnetic surface contains sufficient information to allow hardware to recreate or *regenerate* a clock wave that can be used to sample the incoming data.

The design of codes for recording data is a fine art, because of the many conflicting requirements:

- You want to increase the efficiency of the code by reducing the number of flux reversals required to record each bit (the best you can do is one transition per bit).
- You want to make the recorded symbols for 1 and 0 as unlike each other as possible in order to make it easy to tell the difference between 1s and 0s in the presence of noise and other extraneous signals.
- You want to ensure that there is no significant gap between flux transitions to make the code self-clocking.
- You want to avoid patterns that contain low-frequency components, because the analog circuits that process data from the read head do not handle low frequencies well.

Any recording code is a compromise. Figure 11.23 illustrates an encoding method once used by floppy disk drives called *modified frequency modulation* (MFM). Floppy disk drives operate on the same principles as the hard disk except that the head is in contact with the recording surface and the speed of rotation is very much slower.

Figure 11.23 shows the sequence of bits to be recorded; that is, 010100111. A clock pulse marks the boundary between each of these bits. The first step in the encoding process (line labeled *Data pulses* in Figure 11.23) is to generate a pulse whenever the data bit to be stored is a 1. If these pulses were used to store data directly, a problem would occur whenever the input stream contained two or more consecutive 0s, because there would be no recorded data.

MFM solves the problem of a lack of signal when there is a stream of 0s by recording a 1 at the cell boundary between two consecutive 0s. This rule ensures that a string of 0s still creates flux transitions, yet the inserted pulse is not interpreted as a 1, because it falls between cell boundaries rather than in the middle of a cell. The bottom line of Figure 11.23 is the current in the write head that changes direction on each pulse.

Modified frequency modulation encoding was widely used in hard drives in the 1980s but has been replaced by better codes. *Run length limited* (RLL) codes, are more complicated than MFM, but they reduce the number of flux reversals required to store data. These codes map recorded bit sequences onto the patterns of flux reversals with the restriction that the maximum run of zeros is limited. For example, the 2,7 RLL code limits the sequence of zeros in the codes to between 2 and 7. A more modern RLL code is 3,9 RLL, which limits the run length of 0s to nine.

FIGURE 11.23   MFM recording

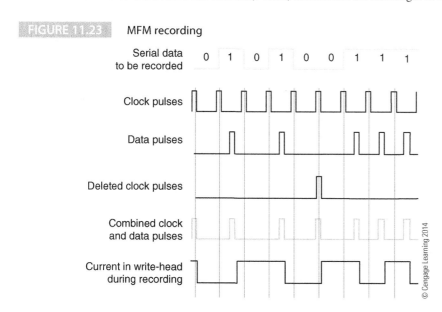

Today's disk drives employ sophisticated means of encoding data prior to its recording and then decoding it after read back. In particular, Hitachi pioneered the use of *partial-response maximum likelihood* (PRML) technology in the 1990s to process data read from the disk. Early recording techniques encoded the data, wrote it, and then read back pulses from the read head using a simple peak detector to read a 1 or a 0. However, when you operate at a high speed, the data read back from the disk due to a flux transition is a complex analog waveform rather than a nice square pulse. Moreover, if you write bits close together, the signals from each flux transition overlap in time and interfere with each other. The mutual interaction between bits is called *intersymbol interference,* and this is a problem that has plagued those designing modems for telephone channels since the 1960s. PRML technology involves reading back the analog signal from the read head and using a knowledge of the channel characteristics (i.e., a knowledge of the waveform generated by a single pulse) to reconstitute the original data. The term *maximum likelihood* indicates that the decoder selects the recorded data pattern that was most likely to generate the received data. This technique is also called Viterbi decoding, and it is highly successful in the presence of noise.

11.3.1   Tracks and Sectors

We have already stated that the path of the disk's surface describes a track. A track is divided into *sectors,* which are the smallest units of data that can be written or read, because there is no way that you can locate a single bit on a track and then modify it. Because the speed at which a disk rotates is not precisely constant (real-world manufacturing doesn't allow that), it is necessary to read a block of data at a time.

The disk drive is a mechanical device, and the write head cannot be located precisely over a sector. There is always some left or right misalignment of the head, and the sector may be written slightly ahead of or behind the old sector that is being overwritten.

When data is read from a sector, the recorded signal contains a tiny fraction of the data written in the previous write operation (and the write operation before that too). The signal picked up from the remnants of these old sectors that are under, but slightly skewed from, the current sector is too weak to affect operation of the disk and is not a problem unless the head is badly aligned.

Suppose you take the analog signal from a read head and construct the corresponding data pattern. If you then use a knowledge of the head's characteristics and reconstitute the recorded signal from the digital data, you will have a perfect copy of the data that was recorded but a copy that is free of any extraneous signal. If you subtract the signal reconstituted from the data from the signal received from the read head, you have the tiny signal that corresponds to the data written to the same sector on an earlier occasion.

Advanced File Formatting

In a move to increase the data density of disk drives, Samsung championed the *Advanced Format* that rejects the standard 512 byte sector in favor of a 4096 byte sector. A large sector reduces the volume of overhead. For example, only one error-correcting code is required per 4096 bytes instead of one per 512 bytes.

The trend to the 4K advanced file format led to all disk drive manufacturers adopting this standard in new products from 2011. However, since not all operating systems can handle 4 KB sectors, it is necessary that the disk be able to emulate 512 byte sectors; that is, they have a 4,096 byte physical sector and a 512 byte logical sector. Both Windows 7 and Mac OS are *4K aware* and are able to fully use the advanced file format.

As we've just stated, sector-skew means that a write operation doesn't totally eradicate the old data. Because the level of the signal from previous writes to the same sector is too weak to corrupt data, there's no practical problem. However, a forensic expert from the FBI may well be able to read the file that you deleted, and he or she may not be willing to believe that the package you ordered from Colombia really was a secret ingredient in mom's meatloaf.

You can buy programs that securely delete data. These deal with partially deleted data by continually writing random data patterns to the same track until the multiple write operations have entirely obliterated any previously written data.

This is not the same as *undeleting* files, where a deleted file is recovered because it has simply been removed from the directory and its contents are still there on the disk.

FIGURE 11.24   The structure of a track

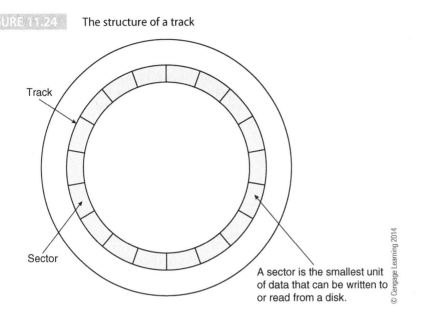

A sector is the smallest unit of data that can be written to or read from a disk.

Figure 11.24 illustrates the structure of a track. A sector should be as large as possible for the purposes of storage efficiently. Since each sector contains housekeeping information, small sectors are inefficient because they waste disk space. A sector should be as small as possible because of storage efficiency. Since a sector is the smallest unit of data that can be written, the unit of granularity of a file is the sector. If you increase a file size, you can increase it only in units of one sector. If the smallest sector is, say, 4 KB, it means that, on average, a file will contain a last sector that is only half full. If a disk has hundreds of thousands of files, the wasted space is significant. The above two statements are, of course, mutually contradictory. The optimum sector size is a compromise.

A sector is a data structure that holds a basic unit of data. When a disk is first used, the sectors are written onto the surface of the disk; that is, a sector is a software structure and not a physical feature of the disk.

Figure 11.25 shows the structure of a track that is written to the disk when the disk is first *formatted* (this is a floppy-disk sector structure that is easy to understand). Until this structure is laid down, the disk cannot be used to record data. The overhead needed to store data has no equivalent in semiconductor memories.

FIGURE 11.25   Example of a sector structure

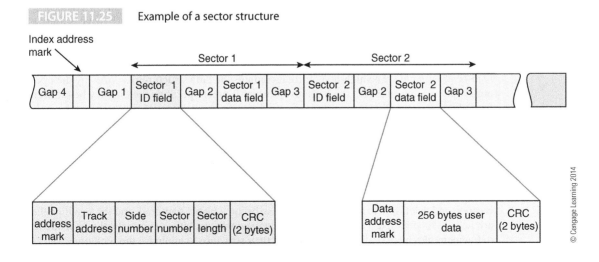

Gaps containing no useful data are required between the various fields of a track in order to give the head time to synchronize itself with the stream of bits coming from the read/write head. Because magnetic media are inherently unreliable, error-detecting codes are required to detect data that has been corrupted. In Figure 11.25, the track contains index, address, and data *marks*. These marks are special *illegal* binary patterns that are recorded on the disk (they are illegal because they break the data encoding or modulation algorithm used to store data). Consequently, the hardware can easily detect these marks and use them to synchronize reading or writing operations.

Zoning

The circumference of a track is $\pi{\cdot}d$, where d is the diameter of the track. In it there are n sectors per track; the length of a sector is $\frac{\pi{\cdot}d}{n}$. The approximate size of a bit is given by $\frac{\pi{\cdot}d}{(m{\cdot}n)}$, where m is the number of bits per sector. Because the value of d varies between the inner and outer tracks, the width of a bit varies correspondingly. If the size of a bit is sufficiently large to be detected on the innermost track, it is too large on the outermost track, and the storage efficiency is compromised.

Disks deal with the problem of different track lengths by *zoning*, whereby adjacent tracks are grouped into zones and each zone has a different number of sectors. Some of today's disks divide the surface into 30 or more zones.

Zoning affects the rate at which data is read from the disk. With fewer sectors along the innermost track, the data transfer rate may be 60% less than when reading a sector at the outermost edge.

Formatting a Disk

The first generation of hard disks and floppy disks required the user to perform a *low-level format*, because they had no track/sector structure written on them. This is not true for today's modern high-performance hard drives, because they are formatted at the factory. A disk drive may have a more complex track structure than the one we described. For example, there may be fewer sectors on tracks closer to the center where the circumference is smaller (zoned recording), or some of the sectors may be mapped out (i.e., invisible) because they contain areas where the recording surface is faulty. Early disk drives appeared before today's disk interfaces, device drivers, and operating systems. You had to create your own data structures, perform formatting, and read/write data at the bit level. Today, a disk drive is more of a black box that receives commands from the host operating system. The user does not have to worry about how data is encoded and stored.

Once a disk has a low-level format that contains the track and sector structures, it can be *high-level formatted*. The high-level format contains the data structures required by the operating system. Consequently, the high-level format for one operating system may be different from the high-level format on a disk using a different operating system. When you format a disk in a PC, you are performing a high-level format.

Figure 11.26 demonstrates how a file is read from a disk. A file is composed of a sequence of sectors. The sectors themselves may be arranged as a *linked list*, or a directory may define the sequence of sectors belonging to a file. Figure 11.26 shows that a lot of time can be wasted when reading a file because of the need to perform a new seek operation whenever the track number changes. When files are first created, they are allocated sequential sectors. After a period of file creation and deletion, the free sectors on a disk become highly scattered, and the resulting files are heavily fragmented. Fortunately, operating systems can either automatically or manually defragment files by periodically reorganizing their structure to minimize the seek time.

FIGURE 11.26   A file may contain sectors distributed across the disk

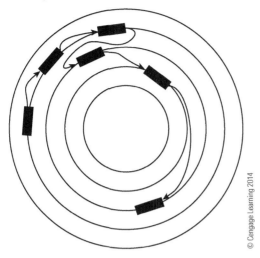

© Cengage Learning 2014

Interleaving

The sectors on a track are numbered $0, 1, 2, \ldots, n - 1$ and are called the *physical* sectors. Suppose several sectors are read by the operating system. If sector x is read first, it is necessary for the disk drive electronics to do considerable processing between the end of reading sector x and beginning to read sector $x + 1$. Today this is not a problem, because the disk rotates painfully slowly in comparison with the high-speed 16-bit processors embedded in disk electronics. The inter-sector gap is sufficient for any housekeeping. This was not always so. Earlier disk drives did less internal processing, and computers were much slower. By the time the processor was ready to read sector $x + 1$, the head had already moved over this sector (or even the sector beyond that), and it was necessary to wait for sector $x + 1$ to come round again. The solution adopted was to map consecutive logical sectors on interleaved physical sectors. For example, a 1:2 interleave factor with a 17-sector disk might be

$$0, 9, 1, 10, 2, 11, 3, 12, 4, 13, 5, 14, 6, 15, 7, 16, 8, \ldots$$

Suppose the disk is currently reading logical sector 4. The next logical sector is 5, but the next physical sector is 13. This arrangement gives the computer time to perform its housework before sector 5 moves under the head. Current disks use a 1:1 interleave factor, and interleaving is no longer of concern to the user. Moreover, caching and buffering makes interleaving redundant, because an entire track can be buffered.

11.3.2   Disk Parameters and Performance

Up to now, we haven't discussed in any detail the operational parameters or the performance of disk drives. Table 11.1 describes the characteristics of a 120 GB disk drive, a 2 TB drive, and a 3 TB drive.

The end user is interested in three aspects of a drive: how much data it can store, how long it takes to access the data, and how the data is moved into the host computer. The *capacity* of a disk is given by surfaces × tracks × sectors × bytes/sector.

The *access time* of a disk is composed of two major components: the time taken to access a given track (the *seek* time) and the time to access a given sector once its track has been reached (the *latency*). The latency is easy to calculate. Assuming that the head has stepped to a given track, the minimum latency is zero (the sector is just arriving under the head). The worst-case latency is the period of one revolution (the head has just missed the sector and has to wait for it to go round). On average, the latency is $1/2t_{rev}$, where t_{rev} is the time for a single revolution of the platter. If a disk rotates at 7,200 rpm, its latency is given by

$$\tfrac{1}{2} \times 1/(7{,}200 \div 60) = 0.00417 \text{ s} = 4.17 \text{ ms}$$

TABLE 11.1   Parameters of the 120GXB and 7K200 Hard Drives

Parameter	120GXB	Deskstar 7K2000	Deskstar 7K3000
Interface	ATA-100 compatible	SATA 3 Gb/s	SATA 6 Gb/s
Capacity	120 GB	2 TB	3 TB
Sector size (bytes)	512	512	512
Recording zones	31	31	
Data heads (physical)	6	10	10
Data disks	3	5	5
Max. areal density (Gbits/sq. inch)	29.7	285	411
Max. recording density (KBPI)	524	1,457	
Track density (TPI)	56,700	195,000	
Data buffer	2 MB	32 MB	64 MB
Rotational speed (rpm)	7,200	7,200	7,200
Latency average (ms)	4.17	4.17	
Media transfer rate (max. Mbits/sec)	592	1,621	1,656
Interface transfer rate (max. MB/sec)	100	300	600
Sustained data rate (MB/sec)	48 to 23 (zones 0–30)	134	
Seek time average (ms)	8.5	8.2	
Seek time track-to-track (ms)	1.2	0.6	
Seek time full-track (ms)	15.0		

The average rotational latency can be reduced only by increasing the speed of the disk. The energy to rotate a disk increases with the square of its speed. Moreover, the stress on a rotating disk also increases with the square of its speed. The maximum speed at which disks rotate is determined by their energy requirements and mechanical characteristics. It is probable that the nature of the materials from which the disk is constructed will impose an ultimate limit on their speed. Disk speeds have changed relatively little over the years. In 2004, the basic low-cost commodity hard disk rotated at 5,400 rpm (replacing the older 3,600 rpm standard), and high-performance disks rotated at 7,200 rpm. A few expensive hard disks were available at 10,000 rpm, while the state-of-the-art was 15,000 rpm. Little changed over the next decade.

What is the average seek time? Suppose the disk has N sectors and it takes t_{step} seconds to step from track-to-track. If the head is parked at the edge of the disk after each seek, the average number of tracks to step over when seeking a given track would be $N/2$, and the average seek time is given by $\frac{1}{2} \times N \times t_{step}$. Figure 11.27a illustrates this situation.

However, suppose that the head were automatically moved to the center of the tracks (track number $N/2$) after each access, as Figure 11.27b shows. When a new seek operation is

FIGURE 11.27   Seek time and the initial head position

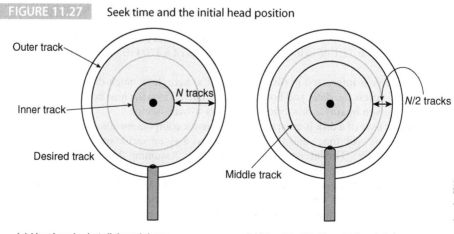

(a) Head parked at disk periphery
Average number of tracks stepped = N/2

(b) Head parked in middle of disk
Average number of tracks stepped = N/4

issued, the head can move either left or right. In this case, the average number of tracks to step is $N/4$, and the average seek time becomes $\frac{1}{4} \times N \times t_{step}$.

In practice, the read/write head remains where it is after an access, and some seeks will be long and others short. Let's calculate the average access time. Suppose that the head is parked over track i, where $i = 0$ to $N - 1$, and has to step to a new track. The number of steps the head must move for each of its possible destinations is given by

Head starting point	0	1	2		$i-1$	i	$i+1$		$N-2$	$N-1$
Tracks stepped	$i-1$	$i-2$	$i-3$		1	0	1		$N-i-1$	$N-i$

The total number of tracks stepped for each possible starting position of the head can be obtained by summing the distance moved for all possible track positions. That is,

$$i - 1 + i - 2 + \ldots + 3 + 2 + 1 + 0 + 1 + 2 + 3 + \ldots + N - i$$

Using the formula for the sum of k terms $1 + 2 + 3 + \ldots + k = \frac{1}{2}(k)(k + 1)$, we get the following expression for the total number of tracks moved for all possible starting positions.

$$\begin{aligned} \text{Number of tracks stepped} &= \frac{1}{2} \times i\,(i + 1) + \frac{1}{2}(N - i)(N - i + 1) \\ &= \frac{1}{2}\,(i^2 + i + N^2 - N_i + N - N_i + i^2 - i) \\ &= \frac{1}{2}\,(2i^2 + N^2 - 2Ni + N) \end{aligned}$$

The average number of tracks stepped per seek operation is this value divided by the number of tracks. That is,

$$\text{Average tracks stepped} = \frac{1}{2}\,(2i^2 + N^2 - 2Ni + N)/N.$$

The last step is to perform the calculation using all possible values of i and take the average. That is,

$$\frac{1}{N}\sum_{0}^{i=N-1}\frac{(2i^2 + N^2 - 2Ni + N)}{2N}$$

You can simplify this expression by arguing that if N is large we can write

$$\frac{1}{2N^2}\sum_{0}^{i=N}(2i^2 + N^2 - 2Ni)$$

Treating i as a continuous variable gives the average number of tracks stepped per seek as $N/3$. In practice, the situation is not as simple as we have implied, because the head doesn't step across the surface of the disk at a constant rate. As the voice-coil actuator rotates and the arm swings across the surface of the disk, the arm accelerates and decelerates rather than moving at a constant velocity. Figure 11.28 illustrates the effect of an arm sweeping across the surface of the disk from the outermost track to the innermost track.

Figure 11.28 demonstrates the movement of an arm with a rotary actuator where it spends part of its time accelerating up to speed, part of its time moving at approximately common speed across the disk's surface, and part of its time decelerating as it approaches its target destination.

It's difficult to give exact times for track-to-track head movements. Wilkes and Ruemmler[3] investigated the modeling of disk drives and demonstrated that a better approximation to a linear head tracking speed is to assume that the head accelerates, coasts at constant speed, and then decelerates. The relative contributions of these three regimes depend on how much the head has to move. For example, a short movement doesn't have a period of coasting because the head never builds up speed.

Wilkes and Ruemmler provide a simple model for the HP C2200A drive operating at 4,000 rpm. The seek time for a short seek of d tracks is given as $3.14 + 0.5597\sqrt{d}$ ms. For a long seek, it's $10.8 + 0.012d$ ms. As you can see, seek time is linear for large numbers of tracks and proportional to the square root of the number of tracks for short movements. These figures are what we would expect: The fundamental relationship between distance and acceleration is distance $= \frac{1}{2} \cdot \text{acceleration} \cdot t^2$.

---

[3] Chris Ruemmler and John Wilkes, "An Introduction to Disk Drive Modeling," *Computer*, March 1994, pp. 17–28.

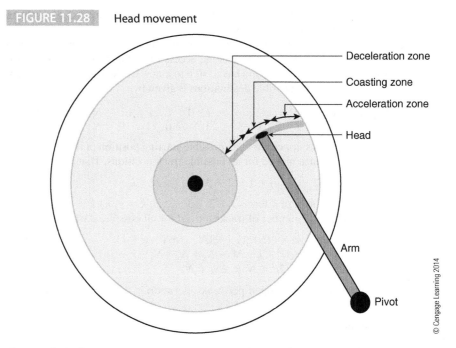

FIGURE 11.28   Head movement

Accessing Sectors

If a disk drive reads physically sequential sectors sequentially, data flows at a data rate given by (sector size)/(time to read a sector). In practice, this upper limit cannot normally be achieved. Files become fragmented as they are created and modified. Moreover, a complex computer activity might require the computer to access many different types of data during the execution of a program: code, subroutine libraries, help files, and so on. Systems with multitasking operations are in an even worse situation. They have to cope with requests from entirely different programs running unrelated data. In the worst case, a disk might have to perform a random seek operation between sector access (something that is unlikely to happen in practice because of data buffering and caching).

Although the sequencing of data from a disk belongs to the realm of operating systems, we provide a short account of the way in which operating systems access data here in order to demonstrate the architecture–hardware–software trade-offs.

Suppose an operating system makes a series of requests to a disk drive for sectors on tracks 50, 150, 32, 16, 125, 8, 130, 50, 60, and 200. Figure 11.29 provides a head movement

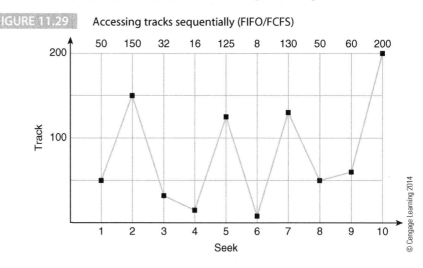

FIGURE 11.29   Accessing tracks sequentially (FIFO/FCFS)

FIGURE 11.30   Accessing nonsequentially (SCAN)

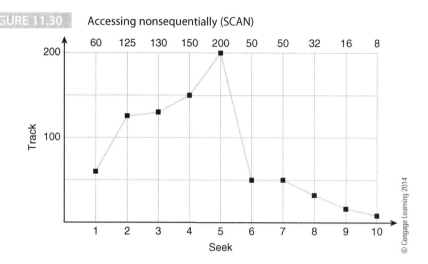

graph of the time plotted against the position of the head if the requested tracks are accessed in the order in which they were received, that is, *first-come, first-served* (FCFS).

Operating systems often provide support for disk drives. For example, Figure 11.30 illustrates the same situation as in Figure 11.29 except that the tracks are buffered and the surface of the disk is first swept in one direction and then back in the other direction. Table 11.2 gives the total number of tracks swept during these two sweeps. As you can see, reordering the stepping sequence greatly reduces the head movement. Typical disk-scheduling algorithms are described next. In each case, the sequence of seeks is given for the series of requests 10, 19, 3, 14, 12, and 9.

FIFO—First-in-first out (also called *first come first served* or FCFS). This algorithm processes requests in the order in which they are received. It is fair to all processes. Sequence = 10, 19, 3, 14, 12, 9. Average seek length = 8.2.

SSTF—Shortest seek time first. The next seek is the one that is closest to the current head position. This algorithm selects the next request as the one requiring the least movement of the head. Because newer requests may be serviced before older requests, it is not fair. Sequence = 10, 12, 14, 19, 9, 3. Average seek time = 5.0.

SCAN—This approach implements the *elevator algorithm* by taking the closest request in the direction of travel. It satisfies all outstanding requests in the current direction of head motion before reversing direction. Sequence = 10, 14, 19, 12, 9, 3. Average seek time = 5.0.

LOOK—This is a modest variation on SCAN where the software looks ahead and changes direction when there are no requests beyond the current sector.

C_SCAN—The circular scan algorithm moves the head in one direction, sweeping across the disk. However, the head then moves back to the other edge of the disk and starts again. This is a unidirectional version of SCAN. Sequence = 10, 14, 19, 3, 9, 12. Average seek time = 6.8.

TABLE 11.2   The Number of Tracks Moved When Accessing Ten Sectors

											Total
Sequential Access											
Track	50	150	32	16	125	8	130	50	60	200	
Step		100	118	16	109	117	122	80	10	140	812
Scan Across											
Track	60	125	130	150	200	50	50	32	16	8	
Step		65	5	20	50	150	0	18	16	8	332

© Cengage Learning 2014

FSCAN—This algorithm is intended to deal with *arm stickiness*. Two request queues are required. Initially, at the start of a scan, all requests are in one queue and the other queues is empty. Once a scan is in progress, all new requests are put in the other queue. This mechanism defers all new requests until the existing ones have been serviced. It is, of course, a fair access mechanism.

Note that the efficiency of these algorithms is compromised by the addition of rotational latency between each sector seek.

The Internal Disk Cache

Disk drives now contain a RAM buffer called a *disk cache* that holds data read from the disk. This *internal cache* is not the same as the disk cache that is set up by the operating system—even though it performs a similar function. A hard disk's internal cache is used to hold data from recent read operations. The cache can even be programmed to prefetch data that might be required in the near future; that is, the disk can read ahead when given a command to read a specific sector. Typically, once a sector read has commenced, sectors are cached until the current buffer is full. Some internal cache memories employ a mechanism called *active segmentation* in which the size of the blocks of data cached is variable. This allows the cache to adapt to the type of accesses being executed.

A cache improves the performance of a drive by reducing the number of head movements and seek operations by fetching data from the cache memory rather than the disk itself. Typical drives have internal caches of 8 MB to 64 MB. Disk caches can reduce the drive's average access time; however, uncorrelated random accesses to a fragmented disk can't make good use of the cache.

It is also possible to cache write accesses to a disk. Caching data during a write saves time, because the disk drive can return a *finished* signal to the host controller as soon as the data has been buffered. The disk then writes the data from the cache to the disk's surface in its own time. This mechanism is known as *write-back* caching. IBM implemented full write caching with the Ultrastar 18LZX and 36ZX disk drives in 1999. The feature may be turned on and off from the server by means of the drive interface.

Unfortunately, write-back caching can lead to nasty surprises. If the power fails between the caching of the data and its writing, the data is lost. Even worse, the host computer has already been informed that the write operation has taken place and is unaware of the problem. This causes file consistency problems that can have serious consequences. Write caching is therefore not widely used. It is an option with some disks used when the system has an uninterruptable power supply.

Transfer Rate

Another parameter of interest to the disk user is the *rate* at which data is transferred to and from the disk. This parameter is easy to calculate. If a disk rotates at R revolutions per minute and has s sectors per track and each sector contains B bits, the capacity of a track is $B \cdot s$ bits. These bits are read (or written) in $60/R$ seconds. Therefore, the data rate is given by

$$\frac{BsR}{60} \text{ bits/s}$$

A typical drive, for example the Deskstar 7K3000, rotates at 7,200 rpm and has a maximum data transfer rate from the disk of 1,656 Mbits/s while reading the media. The disk interface maximum transfer rate in 600 MB/s (i.e., 4,800 Mbits/s).

11.3.3   SMART Technology

CPUs are complex semiconductor devices with approximately 10^7 to 3×10^9 transistors. Commercial pressures force manufacturers to minimize the design, development, and testing cycles in order to put them in production in the shortest possible times. It's not so much a wonder that CPUs work well; it's a wonder they work at all. If a CPU fails, it's a considerable nuisance. You order a new chip, open the case, take the old CPU out, and plug in the new one.

Hard disks are complex electromechanical devices operating at the frontiers of technology. Electromechanical systems with moving parts are far more unreliable than their semiconductor counterparts. The major disk manufacturers developed a technology called SMART that monitors the performance of hard disk drives, can predict the probability of failure, and therefore provide the user with an advance warning of possible failure. The acronym SMART stands for *self-monitoring, analysis, and reporting technology*.[4]

Both IBM and Compaq independently developed their own precursors to SMART before bringing their expertise and technology together. IBM's *predictive-failure analysis* (PFA) used the measured value of head flying height and other parameters to indicate possible failure. Compaq developed IntelliSafe technology with Seagate and Quantum to measure drive parameters and compare them with pre-set thresholds. If the thresholds are exceeded, a status message is sent to the host. The parameters and the threshold values vary from drive to drive, but the way in which the status is sent to the host is consistent across all systems. The implementation of SMART technology, like IntelliSafe, varies from system to system, because the architecture of drives is continually being developed and some drives are in more *mission-critical* applications than others.

Some electronic failures are sudden and cannot be predicted. Mechanical problems are much more predictable. In everyday life, the automobile driver can observe the condition of the car's tires, the slack, stiffness and play in the steering wheel, and oil level. All of these parameters give valuable information about the state of the car. Some of a disk drive's parameters that can be used to indicate possible failure are given in the following panel.

SMART Parameters

SMART technology used to monitor the performance of hard disk drives employs several parameters related to the operation of a drive. Ultimately, these parameters are all about detecting the sluggishness as a drive ages. Some of these parameters are listed here.

Head Flying Height—One of the most critical parameters of a disk drive is the height of the read/write head above the surface. This is very small indeed, of the order of a micro-inch. The head-to-surface height is controlled by the flow of air over the surface of the disk as it rotates. Too low of a gap and the head is in danger of crashing into to the surface of the rotating disk and damaging it. Too high a gap and the signal from the surface is too weak; there is a possibility that signals from multiple bits will be picked up, and the write signal might be insufficient to magnetize adequately the surface. This is an easy parameter to monitor by measuring the strength of the signal in the read head.

Data Throughput—Data throughput indicates the rate at which the drive provides data. Several mechanical problems can contribute to a reduction in the average data throughput.

Spin-up Time—The spin-up time is the time taken for a disk drive to reach its operating speed. If this increases, it could indicate problems with the control electronics or, more probably, wear in the spindle of the drive.

Spin Retry Count—If a drive fails to reach its operational speed within a given period, a new attempt is made. The spin retry count measures the number of attempts made to reach operational speed. This parameter indicates problems with the spindle and motor or possible problems with the power supply.

---

[4] "Get SMART for reliability," Paper TP-67D, July 1999, Seagate Technology, Scotts Valley, CA.

(continued)

Re-Allocated Sector Count—When a read or write verification error occurs (possibly due to a defect in the magnetic surface), that sector is reallocated and the data transferred to a new area. Consequently, bad sectors are automatically mapped onto good sectors (which leads to fragmentation and the slowing of data transfer). The drive keeps a count of the number of remapped (reallocated) sectors, and any increase in this value indicates that there may be problems with the reliability of the drive.

Seek Error Rate—In a seek operation, the head moves to the desired track. If a seek fails, it must be tried again. Seek errors occur due to thermal problems (expansion of the arm and disk), because of friction in the positioning mechanism, or because of damage to the surface of the disk. If the seek error rate increases, there is an increased probability of failure.

Seek Time Performance—The seek time is an indication of the time required to locate a given track. An increase in seek time indicates the probability of a mechanical error.

Drive Recalibration Recount—Drives go through an automatic calibration process to ensure correct track tracking. If this process fails, it is repeated (i.e., recalibration). If recalibration is requested too frequently, it is possible that there is a mechanical fault or a problem with the power supply (or even the read/write head itself).

Seagate's discussion of SMART technology in "Get SMART for reliability," paper TP-67D, also covers enhancements. In particular, the paper states that 40% of the drives returned to Seagate with suspected problems are later found to be fully functional. These drives are returned because of other problems in the host system (for example, software errors or the effects of viruses).

In 1999, Seagate introduced the *drive-self test* (DST), which offers a more proactive means of predicting disk failure. Various tests are embedded in the drive's firmware. For example, the *quick test* takes two minutes and reads the first 1.5 GB, and the *extended test* provides a more thorough investigation by completely scanning the media. Extended SMART technology also logs and saves a record of the most recent errors reported by the drive. This log can also be used as a diagnostic.

Effect of Temperature on Disk Reliability

The reliability of electronic components is highly temperature dependent. The reliability of both the electronics and the mechanics, such as the spindle motor and actuator bearings, degrades as temperature rises. Operating a disk drive at extreme temperatures for long periods dramatically reduces its life expectancy. An interesting paper, *IBM's Drive Temperature Indicator Processor (Drive-TIP) helps ensure high drive reliability,*[5] discusses the relationship between temperature and disk reliability.

Figure 11.31 shows the relationship between temperature and the failure rate of systems. Failure rate is an exponential function of temperature and is generally governed by the Arrhenius equation rate $= Ae(-E/kT)$ where A and E are constants, k is Boltzman's constant, and T is the absolute temperature. In general the failure rate of disk drives increases by about 3% per °C.

High temperatures can induce several failure modes. For example, a thermal tilt of the disk or actuator arms can cause off-track writes that corrupt data on adjacent cylinders. High temperatures can cause outgassing of the lubricants in the spindle motor and voice coil motor, leading to jamming or even a head crash.

---

[5] Gary Herbst, "IBM's Drive Temperature Indicator Processor (Drive-TIP) helps ensure high drive reliability," IBM Storage Systems Division, 5600 Cottle Road, San Jose, CA, 1997.

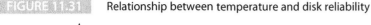

FIGURE 11.31   Relationship between temperature and disk reliability

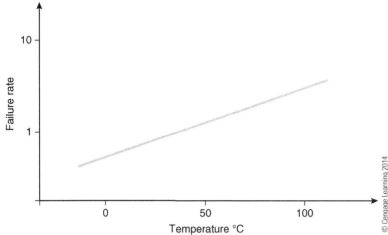

Heat is generated in the computer by power dissipated in the chips and disk drives, and this heat must be removed by fans or natural convection. A failed fan or a clogged vent can cause the temperature to rise and lead to disk failure. IBM has provided temperature monitoring in some of their disk drives to provide advance warning. The temperature in the drive is automatically monitored, and the drive controller is alerted when the drive exceeds its maximum allowable temperature. In practice, two temperatures are monitored. The lower temperature is selected by the user, and the higher temperature is the drive's maximum allowable temperature of the base casting at 65°C.

The temperature is read when the drive is powered on and every 25 minutes thereafter. When the first temperature trip point is exceeded, the sampling period changes from 25 minutes to 15 minutes. Also, an entry is made in the permanent drive error log that includes the temperature and power-on hours (POH) when it occurred. As long as the temperature remains above the first trip point, it will continue to create log entries. If the temperature exceeds the 65°C trip point, the sampling period changes from 15 to 10 minutes.

IBM's drive-temperature monitoring works with the SMART standard developed to monitor and predict device performance and reliability. If the warning is recognized, prompt corrective action can save data before it is too late.

Shake, Rattle, and Roll

All mechanical systems are sensitive to physical disturbances, such as vibration. As you can imagine, it is not easy to keep read/write heads centered over tracks that are spaced at a pitch of 50 nm (2,000 times narrower than a human hair at 100 μ) on hard disks rotating at 7,200 rpm. Disk drives use feedback control to center the head over a track; that is, an error signal due to incorrect positioning of the head is used to move the arm in such a way as to reduce the error.

When a drive suffers from external vibration, the head moves off position and generates a bigger error signal, which moves the head back. Unfortunately, this is a *post hoc* solution; the correction happens *after* the disturbance. An Hitachi white paper, "Rotational Vibration Safeguard," describes a new technique used to minimize the effects of vibration. Two vibration sensors (G-sensors or accelerometers) are located on the disk drive's electronics board to detect movement. The signals from these sensors control the head actuator and move it back into position *before* the head has drifted off track. In actual tests where a disk was shaken to reduce performance to about 30%, the application of the rotation vibration safeguard mechanism increased performance to 90%.

11.4   Secure Memory and RAID Systems

In the 1970s and 1980s, hard disks had considerably smaller capacities than today and were much more expensive in terms of dollars per megabyte. The personal computer revolution led to a very rapid decline in the cost of small and medium-sized hard disks—a decline that wasn't *initially* matched by a corresponding price reduction in the large drives found in professional and commercial computer systems.

In the late 1980s, researchers realized that low-cost disk drives could be used in new ways, and in 1987, Patterson, Gibson, and Katz at UC Berkeley published a paper, "A Case for redundant arrays of inexpensive disks, RAID,"[6] proposing a means of exploiting the inexpensive disk systems found in PCs. The expression *array of inexpensive disks* implies a regular structure built around commodity off-the-shelf disk drives, while *redundant* implies a degree of fault tolerance; that is, the failure of a single drive should not bring down the entire system.

The RAID concept rapidly moved out of the research laboratory, and by the mid-1990s, RAID systems were being advertised in personal computer magazines. Today, most PCs have motherboards with RAID support. Patterson, et al. proposed several ways of organizing clusters of drives at RAID level 0, RAID level 1, etc. The various RAID levels provide different functionalities; some emphasize speed and some reliability.

Probability of Failure

Suppose a drive has a 1 in n probability of failure over a given period of time. Let's call this p, where p might be 0.00001 for 1 in 100,000 hours. If we have m drives in an array, what is the probability of failure?

In order for one drive to fail, $m - 1$ drives must not fail. The chance of a drive not failing is $1 - p$. The probability of two drives not failing is $(1 - p)(1 - p)$ because probabilities are multiplicative. Therefore, the probability of $m - 1$ drives not failing is $(1 - p)^{m-1}$. The probability of exactly one drive failing is $p[1 - (1 - p)^{m-1}]$. However, as it does not matter which of the m drives fails, the probability of one drive failing is $mp(1 - (1 - p)^{m-1})$.

We can extend this to the probability of exactly two failures. This requires two drives to fail and $m - 2$ to not fail. Moreover, as any two out of m drives can fail, the number of possible failures is given by mC_2, which is $m!/(m - 2)!$. We can write the probability of two drives in an array of m failing as: $_mC_2\, p^2(1 - (1 - p)^{m-1})$

In practice, not all levels of RAID are widely used today, and some hybrid systems combining features from two RAID levels have been implemented.

Disk drives are inherently serial storage devices. In a RAID system, multiple disk drives are operated in parallel, and the bits of a single file are divided between the individual drives. RAID arrays can be used to improve the *performance* of disk systems or to improve their *reliability*. By replicating data across several disks, the failure of a single disk can be made invisible to the system. RAID systems are important in applications where data security is vital (e.g., banking).

The reliability of n disk drives is about $1/n$ that of a single drive, because there are n of them to fail. By the same token, a four-engine aircraft is more likely to suffer engine failure than a single-engine aircraft. Since aircraft can continue to fly with three engines, the loss of one engine is not catastrophic. The same goes for RAID systems. If data is distributed between the disks of an array, the failure of one disk doesn't cause the system to crash. This statement is not true of RAID level 0, which is a special case.

To understand RAID, we need to remember two concepts. First, the data recorded on disks is stored in *sectors*. Second, each sector has a *frame check sequence* that can detect one or more errors in a sector. This ability to detect errors means that a RAID array can rapidly respond to the failure of an individual drive unit. The key concept in RAID technology is *striping*. Disk space is divided into units called *stripes* that may be as small as a single sector or as large as several megabytes. These stripes are spread out or *interleaved* across several disks in parallel. The way in which data is allocated to the stripes and the stripes allocated to the individual drives determines the *level* of the RAID system.

---

[6] D. A. Patterson, G. Gibson, and R. H. Katz, "A case for redundant arrays of inexpensive disks (RAID)," *SIGMOD '88 Proceedings of the 1988 ACM SIGMOD International Conference on Management of Data*, Volume 17 Issue 3, June 1988.

FIGURE 11.32   RAID level 0 striping

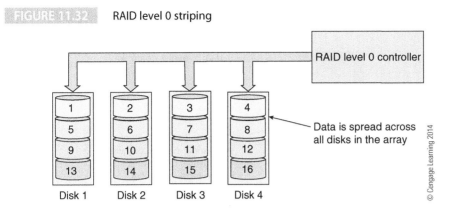

Data is spread across all disks in the array

© Cengage Learning 2014

RAID level 0 is the only level not using redundancy to provide additional security. A RAID level 0 with n drives divides data into n stripes that are applied in parallel to the n drives. Figure 11.32 demonstrates level 0 striping with four drives. The size of each disk drive should be the same, otherwise the effective size of a RAID array defaults to the size of the smallest drive in the array.

The advantage of a RAID 0 array is its high throughput. For n drives, the capacity is n times that of a single drive, and the speed is higher because read operations can take place in parallel. The RAID controller may be implemented in either hardware or software. However, as we said earlier, it is common for RAID controllers to be built into PC motherboards, leaving the user with little to do other than to plug in the disk drives and configure the BIOS.

The capacity of a RAID level 0 array with n drives is simply n times the capacity of one drive; that is, no capacity is lost due to redundancy and the storage efficiency is 100%. However, there is no fault tolerance, and the loss of any drive in the array renders all of the data invalid. Read and write performance is excellent.

Because the loss of one disk brings down the entire system, a RAID level 0 array makes sense only if the data is frequently backed up and provision is made for failure between backups. In that case, RAID level 0 is both efficient in terms of its use of disk space and fast because of its inherent parallel read and write mechanism.

> ## BIOS
>
> A PC's *basic input/output system* (BIOS) resides in flash memory and acts as a bootstrap loader on power up. The function of the BIOS is to load the operating system from the hard drive, flash drive, or optical disk. The BIOS can also be used to define system parameters, such as the CPU clock speed, CPU voltages, and DRAM timing. The traditional BIOS is disappearing and being replaced by the *unified extensible firmware interface* (UEFI). The BIOS to UEFI change is necessary, because the BIOS was designed in an era of 16-bit processors and small drives. UEFI is designed to handle 32-bit and 64-bit processors as well as drives over 3 TB. UEFI found its way into mainsteam motherboards in 2011.

RAID Level 1

Figure 11.33 illustrates a RAID level 1 array, which is called *mirroring* because it replicates copies of stripes on multiple drives (here we are using only two drives in parallel). Data security is increased because you can remove one of the drives without losing data. Not only does a RAID level 1 system increase data security, it improves access time. Suppose a given stripe is accessed. During a read access, the system will read the stripe from the first disk that is able to provide it.

The write time is the longer of the two parallel writes. Fortunately, most accesses are reads rather than writes. Moreover, it is possible to cache writes and allow the disk to do the writing when it is free. A RAID level 1 array is expensive because it duplicates data, but it can provide cost-effective security in an age where a large-capacity, high-speed disk is of the order of $100.

The efficiency of a two-disk RAID level 1 system is 50% because data is simply duplicated. Duplication provides excellent fault tolerance. If a drive fails, the system can continue working normally. All you have to do is to remove the failed drive, install a new one, and then rebuild the lost data. Most RAID level 1 controllers support automatic rebuilding of a failed drive.

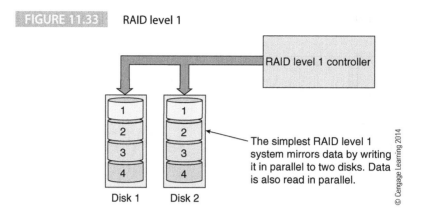

FIGURE 11.33   RAID level 1

RAID level 1 controller

Disk 1   Disk 2

The simplest RAID level 1 system mirrors data by writing it in parallel to two disks. Data is also read in parallel.

© Cengage Learning 2014

There is a hybrid system called RAID level 0+1 or RAID level 0/1 that combines features of levels 0 and 1 by providing both fast data access and protection against drive failure. Figure 11.34 illustrates a system with two sets of three drives. A stripe is written across drives 1, 2, and 3 to provide a RAID level 0 service. However, because drives 1, 2, and 3 are mirrored as 4, 5, and 6, the arrangement provides the security of a level 1 system. This arrangement is the most costly form of RAID.

FIGURE 11.34   RAID level 0/1

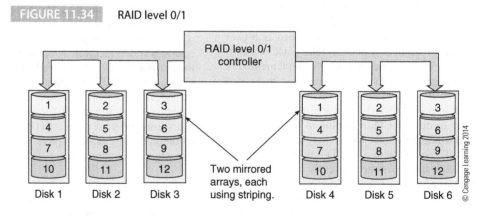

RAID level 0/1 controller

Disk 1   Disk 2   Disk 3

Two mirrored arrays, each using striping.

Disk 4   Disk 5   Disk 6

© Cengage Learning 2014

RAID Level 2 and Level 3

RAID levels 2 to 6 all distribute data in the form of stripes like level 0, but they do not fully replicate the data like level 1. In other words, levels 2 to 6 fall between the extremes of level 0 (no redundancy) and level 1 (redundancy by replication).

RAID levels 2 and 3 employ multiple synchronized disk drives; that is, the spindles are synchronized so that sector i passes under the read/write head of each of the disks in the array at the same time. Level 2 and 3 arrays provide true parallel access in the sense that, typically, a byte is written to each of the disks in parallel. The difference between levels 2 and 3 is that level 2 uses a Hamming code to provide error detection and correction, whereas level 3 provides only a simple parity-bit error detecting code. The parity check data in a RAID level 3 is stored on one disk, whereas the Hamming code of a RAID level 2 may be spread over more than one drive. Figure 11.35 illustrates the concept of RAID level 3, which is also called *bit-interleaved parity*.

Hamming Codes

The Hamming code is the simplest member of the class of error detecting and correcting codes. A Hamming code uses several redundant bits, each of which is a parity bit formed over a subsection of the bits of the word to be protected. If a single error occurs, one or more of these parity bits will no longer be correct, and they can be used to locate the bit in error and to correct it.

Modern error detecting and correcting codes are far more powerful than Hamming codes and can be used to correct multiple errors. On the other hand, Hamming codes are very easy to implement and perform well in an environment where only one disk is likely to fail at any instant.

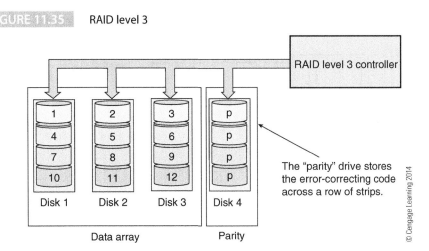

FIGURE 11.35   RAID level 3

The "parity" drive stores the error-correcting code across a row of strips.

Disk 1   Disk 2   Disk 3   Disk 4

Data array   Parity

© Cengage Learning 2014

A single-bit parity code can't normally be used to correct an error. But it can in a RAID level 3 array. Suppose a disk drive fails. The stripes recorded on the failed disk are, therefore, missing. However, the stripes on the parity disk can be used to reconstruct the missing data. Table 11.3 illustrates a level 3 array with four data disks and a parity disk. If disk 3 fails, we have the situation of Table 11.4.

TABLE 11.3   Illustration of a RAID Level 3 Array

Bit 1	Bit 2	Bit 3	Bit 4	Parity Disk
0	1	0	0	1
1	1	0	0	0
0	1	1	1	1
1	0	1	0	0

© Cengage Learning 2014

Because we know the parity bit across each row, we can recalculate the missing data. For example, in line 1, the bits are 0, 1, ?, 0, 1. Since the parity bit is odd, there must be an odd number of 1s in the data bits. Therefore, the missing bit must be 0.

TABLE 11.4   The Effect of Disk 3 Failing in the Array of Table 11.3

Bit 1	Bit 2	Bit 3	Bit 4	Parity Disk
0	1	?	0	1
1	1	?	0	0
0	1	?	1	1
1	0	?	0	0

© Cengage Learning 2014

RAID Level 4 and Level 5

RAID levels 4 and 5 are similar to levels 2 and 3. However, in these cases, the individual disks are not synchronized and operate independently of each other. The stripes are much larger than levels 2 and 3. In level 4, *block interleaved parity*, the parity stripes are stored on a single disk, whereas in level 5, the parity stripes are interleaved and stored on all disks in the array. You can update the parity information in RAID 5 systems more efficiently by changing the corresponding parity bits only when the data changes.

RAID level 5, as shown in Figure 11.36, is a popular configuration, providing striping as well as parity for error recovery. The parity block is distributed among the drives of array, which gives a more balanced access load across the drives. A minimum of three drives is required for a RAID level 5 array.

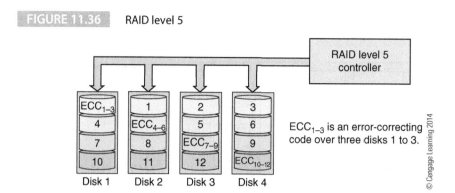

FIGURE 11.36   RAID level 5

ECC$_{1-3}$ is an error-correcting code over three disks 1 to 3.

© Cengage Learning 2014

Because it takes time to maintain parity information, the write performance of RAID level 5 systems is reduced, so they are not the ideal choice if you require the highest level of performance.

Failure of RAID 5—An Example

Permabit Technology provides an interesting example of the cost of RAID failure in their white paper.[7] They start by considering a RAID level 5 system constructed from 500 GB SATA drives with a mean time between failure (MTBF) of 1,000,000 hours and a bit error rate of 1 in 10^{14} or 12.5 TB. This means that a read will fail approximately once for every 12.5 TB of data read.

Suppose a RAID level 5 array has seven data disks and one parity disk. The useful data capacity will be 7×500 GB = 3.5 TB. The eight drives have a MTBF of 125,000 hours (one-eighth of the MTBF of a single drive). In normal operation, uncorrectable reads may occur, and these will be corrected by the RAID mechanism.

Suppose one of the eight drives fails. Because of the properties of RAID 5, the seven remaining disks will continue to operate in a degraded mode until the faulty drive has been replaced and the data reconstructed. The nightmare now begins. In order to perform a reconstruction, the seven working drives containing 3.5 TB of data must be read in their entirety. With a bit error rate of one in 12.5 TB, there is a 28% chance that an uncorrectable error will occur during the reconstruction process and data lost. In other words, with such a large volume of data to reconstruct following the loss of a single drive, the lack of further redundancy means that there is an almost 25% chance that all the data cannot be recovered. Clearly, RAID is not a complete protection against error.

RAID Level 6

RAID level 6 is an extension of RAID level 5 that uses the same arrangement as Figure 11.36 with one change. As well as the error-correcting information stored by a RAID level 5 array, a second parity block is stored on each drive. RAID level 6 has similar properties to RAID level 5. However, the additional parity information makes it possible to correct errors during the rebuilding after a single drive failure.

Dealing with Failure in a RAID Array

When a hard drive in a RAID array fails, it has to be replaced. RAID controllers can be designed to be hot swappable; that is, you can pull out the failed drive and insert a replacement without powering down and rebooting. When you replace a drive, the new drive has to be configured for the array. In RAID terminology, this operation is called *rebuilding*. This rebuilding operation has to take place while normal operation is continuing (that's the whole point of the RAID system).

---

[7] "RAIN-EC: Permabit's Revolutionary New Data Protection for Grid Archive," January 2008.

It's relatively easy to rebuild a drive in a RAID 1 system, because all you have to do is copy data from a working drive to the new mirror drive. Rebuilding data on a RAID level 5 array takes a lot more time, because you have to synthesize all of the data by reading the appropriate stripes from the other disks and then performing an exclusive OR operation with the data. The table below summaries the various RAID levels.

Level	Characteristics
RAID 0	The fastest and most efficient arrangement. No fault-tolerance is provided. Requires a minimum of two drives.
RAID 1	Data is mirrored (duplicated). This is the best choice for performance-critical, fault-tolerant environments. Requires a minimum of two drives.
RAID 2	This mode is not used with today's drives that include embedded ECC mechanisms.
RAID 3	This mode can be used to speed up data transfer and provide fault tolerance by including a drive with error correcting information that can be used to reconstruct lost data. Because this mode requires synchronized-spindle drives, it is rarely used today. Requires at least three disks.
RAID 4	Little used.
RAID 5	This mode combines efficient, fault-tolerant data storage with good performance characteristics. However, performance during drive failure is poor and rebuild time slow due to the time required to construct the parity information. Requires a minimum of three drives.
RAID 0 + 1	Striped sets in a mirrored set with four or more disks. Provides fault tolerance and improved performance.
RAID 1 + 0	Mirrored sets in a striped set. Like RAID 0 + 1 but with better performance.
RAID 5 + 1	A mirrored striped set. Requires at least three disks.

JBOD

Another mnemonic found in the world of storage is *just a bunch of disks* (JBOD). JBOD is not a close relative of RAID; it's more of a distant cousin (as one writer commented). JBOD technology allows you to combine multiple drives into a single volume (the exact opposite of partitioning a disk).

The advantage of JBOD technology is that it allows you to combine disks (say, from an older system that is no longer in use) to create a new larger single disk. Unlike RAID technology, JBOD neither increases performance nor improves reliability.

11.5   Solid-State Disk Drives

We've seen that the electromechanical disk drive is a wonder of technology with storage capabilities of 4 TB in a small 3-½ form factor. However, its days may be limited by the introduction of *solid-state drives* (SSD) that mimic the hard drive electrically. The solid-state hard drive uses semiconductor flash technology to store data and an electrical interface that makes it physically compatible with hard disk interfaces; that is, you just plug an SSD into a hard disk SATA socket. The serial advanced technology attachment (SATA) is a low-cost high-speed serial interface that now interfaces mass storage devices from disk drives to host controllers.

Politics of SSD

Commercial organizations tend to be most enthusiastic about 'green' issues when doing so is profitable. For example, hotels ask you to reuse towels to save water, which is environmentally good and also saves the hotel energy/water.

Similarly, solid-state disk drives are heavily promoted as green by drive manufacturers. Although such a promotion is self-serving, it is true. Because a SSD has no moving parts (in particular, no rotating platter), its energy consumption is intrinsically lower than a hard disk drive. Moreover, the increased reliability of SSDs means that they will need replacing less frequently, saving the energy incurred in manufacturing replacement drives.

The SSD has considerable advantages over the electromechanical disks—the most important of which are higher performance, lower power consumption, lower weight, and greater tolerance to shock. In 2010, SSDs were finding their way into high-end executive laptops (in a 2-½ in. form factor) and specialist high-end applications (in 3-½ in. form factors). By 2012, SSDs were standard in most premium laptops. The limitations of the SSD are two-fold: their considerable cost premium over hard drives and their limited storage capacity.

Solid-state disks are constructed with the type of flash memory technology we discussed in the previous chapter; it's only in recent years that the cost of flash memory has declined to the point at which large (over 128 GB) memories are economically feasible.

Because solid-state disks have no moving parts, they are truly random access devices. There is no rotational latency and no seek time. Consequently, the fragmentation problem associated with hard disks simply goes away. It's not necessary to periodically defragment an SSD when files are scrambled throughout the memory space. As early as 2007, a white paper from IDC[8] indicated the potential savings from an SSD-based notebook.

- IT labor savings for PC deployment   2.4%
- Reliability savings for outsourced repair   1.4%
- Reliability savings for repairs   7.5%
- Reliability savings for user productivity due to hard drive loss   17.2%
- Savings from power enhancements   16.9%
- User productivity savings   54.5%

Price of SSD

Like all new technologies, the SSD entered the world with a massive price premium compared to magnetic hard disks. Since the SSD's introduction, its price has rapidly declined.

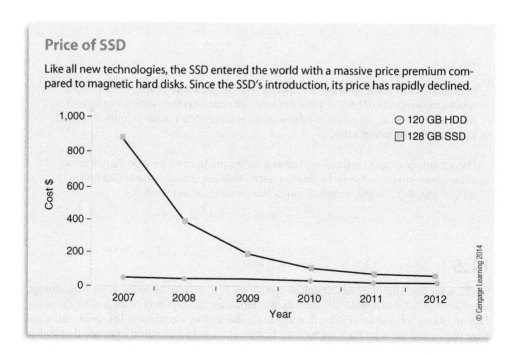

© Cengage Learning 2014

---

[8] J. Janukowicz and D. Reinsel, 'Evaluating the SSD total cost of ownership,' *IDC*, November 2007.

Of course, these figures apply to a particular corporate environment at a particular instant and are the basis of a lot of assumptions. However, the principle is valid. The replacement of hard drives by SSDs is going to bring benefits, and the perception of those benefits will lead to the development of new generations of SSDs.[9]

Special Features of SSDs

The term *features* in this heading is intended to be ironic, as in the notorious expression "It's not a bug, it's a feature." Magnetic disks have effectively equal read, write, and erase times. Of course, they have to, because the surface moves under the read/write head and the time available to read or write is fixed by the mechanics of the system. Solid-state memory uses different read and write mechanisms. For example, to read a flash memory cell, you just inject a current into a silicon channel and then detect whether the stored charge on a floating gate is letting the current through or not. To write data into the cell, you have to inject a charge onto the floating gate. This can be a multiple step process, because some systems inject a charge in discrete units and test whether the data is stored (several writes may be performed before sufficient charge has been stored). Erasing the cell is even more complicated because the charge has to be stripped off the floating gate.

As a consequence of the different read, write, and erase mechanisms, SSDs don't have equal read and write times. In general, the read time of an SSD is considerably faster than the write time. Figure 11.37 demonstrates the performance of several SSDs for 2 MB random-access data transfers for both read and write accesses. The horizontal length defines the data rate in MB/s. Note the considerable variation between different models of SSD.[10]

Figure 11.38 from storagereview.com demonstrates the sequential data transfer performance of several solid state disks. These figures are for both read and write operations using 2MB sequential and random data transfers. Note the variation in performance across several models of SSD. Note also that the write performance can be up to 50% lower than the read performance.

A second feature of SSDs is a limitation on the number of write cycles. When flash memory cells are erased, the dielectric insulator surrounding the gate is damaged, and the cell fails after a certain number of cycles. Moreover, repeated write cycles can lead to *charge trapping*, which leads to cells becoming stuck at zero. When flash memory is used for the BIOS or as storage in digital cameras or MP3 players, flash endurance is not likely to be a problem, because the minimum erase endurance is of the order of 10,000 cycles and 100,000 cycles is typical. However, when flash memory is used as secondary storage, the picture changes.

A technique called *wear leveling*[11] has been devised to help militate against the limitation on the number of write cycles in flash memory by spreading the load across the entire device. A form of memory management is used to map a requested block from the drive controller address onto a physical block address within the memory array. The controller keeps track of how often the physical memory blocks are being used and changes the mapping table to ensure that all physical blocks get a fair share of the load.

Airlines Promote SSDs ... Indirectly

Sometimes the reasons for the growth in technology are either not noticed or not talked about. A former colleague of mine went to work for a multinational organization involved with computing, networks, and data distribution. He asked me what I thought drove much of computer technology. I gave the usual answers. He then told me a major factor was the provision of in-room movies in chain hotels. It takes a lot of bandwidth to allow hundreds of people in a large hotel to watch individual videos on demand.

Similarly, my own purchase of a notebook with a SSD was determined by restricting the weight of my hand luggage on some airlines. Ultra-portable notebook computers often do not provide a DVD drive to save weight/size. However, some high-end notebooks provide a DVD and an SSD drive.

---

[9] SSDs also have negative benefits. If a thief breaks into a conference hall where delegates have left their notebooks while out to lunch, the thief is going to run off with SSD-based notebooks.

[10] Oliver B., "Comparison of 32GB SSD vs 10,000 RPM HDs", http://www.xlr8yourmac.com/.

[11] "Wear-leveling techniques in NAND flash devices", Micron Technical Note TN-29-42.

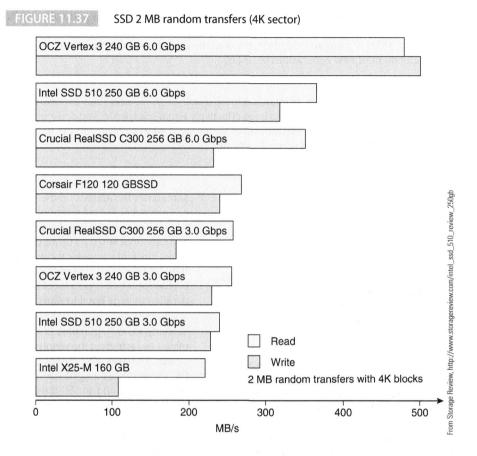

FIGURE 11.37   SSD 2 MB random transfers (4K sector)

Micron's TN-29-42 technical note provides an example of wear leveling. Suppose a system does not use wear leveling, and 50 blocks of data are updated six times an hour. Moreover, suppose that these 50 blocks are allocated a 200-block region of memory. Since each block can use one of four blocks in the device (i.e., 200/50), the endurance is 10,000 cycles $\times$ 4 = 40,000 cycles. At an update rate of six files/hour, the endurance of the memory is approximately 280 days. However, if wear leveling is used and all 4,096 blocks in the memory are available, the endurance is scaled by 4,096/200 to give approximately 15 years.

There are many ways of implementing wear-leveling. For example, the memory space can be partitioned into static and dynamic regions. Static regions are used to store data that is relatively unlikely to change (for example, programs). Wear-leveling is not required to manage static data. Dynamic regions are used to store data that is frequently being changed and are managed by the wear-leveling software.

An information sheet published by Toshiba[12] makes the interesting point that a 128 GB Toshiba ML.C SSD has a five-year estimated storage capacity of 80 TB. This class of disk uses both NAND technology and multi level data storage (i.e., a cell holds more than one bit) and supports fewer write cycles than other technologies. According to Toshiba, this corresponds to over 20 GB/data per day written over the lifetime of the disk, which is an unreasonably large amount for most end-user applications of a mobile computer. If the SSD is larger, the amount that can be written increases correspondingly. Note that a 512 GB SSD based on SLC technology would support the writing of 2 TB daily.

---

[12] "Solid State Drives: Separating myths from facts," Toshiba America Electronic Components, Inc., V1.3, June 2009.

FIGURE 11.38   SSD random reads and writes (the vertical axis is data rate in MB/s)

(a) Sequential transfer rates for 2 MB blocks (read/write)

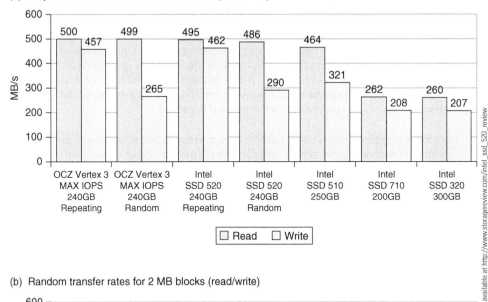

(b) Random transfer rates for 2 MB blocks (read/write)

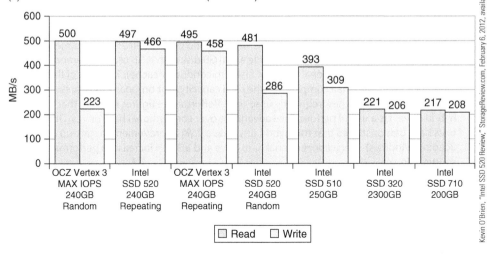

Kevin O'Brien, "Intel SSD 520 Review," StorageReview.com, February 6, 2012, available at http://www.storagereview.com/intel_ssd_520_review

The Intel Optimizer

Conventional hard disks support read operations and write operations. However, SSDs cannot be directly written to (overwriting old data like the HDD), because an erase operation must first be applied to any 4 Kbyte page that is to be written to. As we have pointed out, an SSD requires a relatively long read–modify–write cycle. Moreover, the ATA command set supported by HDDs does not include a block erase command.

A white paper from Intel describes the Intel SSD Optimized software that makes use of a new data set management command that is not part of the ATA8-ACC-2 specification.

(continued)

The optimizer collects information about the file system from the operating system and uses it to ensure that free space on the SSD is used optimally. There is no difference between Windows' view of the filing system and the HDD's view, but there is a difference between the way in which an SSD views free space and the way in which Windows views it. This disparity arises because the SSD distinguishes between free space that has been erased and can be written to and free space that contains unwanted data but which is not ready to be written to. Because the optimizer recovers free space by erasing its data and making blocks ready for use, system performance is improved.

Note that the optimizer does not delete files from the recycle bin, as that would prevent you from recovering accidentally deleted files. Files in the recycle bin are retained until the user performs an *empty recycle bin* command—at which time the optimizer recovers the space occupied by files.

The Hybrid Drive

By 2010, SSDs were appearing in high-end applications but were too expensive for large-scale applications. Seagate attempted to bridge the gap between fast but costly silicon and slow but cheap magnetic disks with a hybrid drive. Their 2½ inch Momentus XT combined both technologies to provide a 500 GB disk with 4 GB of flash memory (plus an unusually large conventional 32 MB of fast semiconductor cache). The 4 GB of flash memory is relatively small compared to the total capacity, but on-board software monitors disk usage and caches frequently used data. Performance figures indicate that the hybrid drive has a useful performance advantage over conventional hard drives. Other tests have demonstrated that this hybrid drive has a 25% improvement in start-up time (loading Windows) over a conventional hard drive and a 300% increase in performance in shutting down.

11.6   Magnetic Tape

Magnetic tape provides a means of archiving large quantities of data. In a world where the production of data is estimated to increase by 60% compounded annually, archival storage is vital. It's impossible to miss tape drives in science fiction films of the 1960s or 1970s. Computer rooms were filled with large cabinets containing spinning reels of wide magnetic tape that whizzed round and then stopped dead before starting up again or even reversing. In those days, the preferred backing medium was magnetic tape because it was relatively cheap and could store large volumes of data.

Data is stored on tape as multiple parallel tracks (typically nine; i.e., eight data bits and a parity bit). Magnetic tape recording technology is virtually identical to disk technology except that there is a single track that is nine bits wide along a long (typically 2,400 feet) flexible magnetic tape. A tape drive required large and powerful motors to spin reels of tape rapidly and to stop them equally rapidly. Magnetic tape was available on 10.5-in. reels until the 1980s when data was stored at, typically, 128 characters per inch. Of course, all tape systems have a long latency, because the tape has to be moved past a read/write head and the desired data may take several seconds or even minutes to locate. Consequently, tape is a purely archival medium.

Tape drives grew smaller and tape cartridges (similar to audio cassettes and VCR tapes) were developed. In order to store more data on the tape, the information was stored along diagonal tracks on the tape by using rotating helical read/write heads. This same writing mechanism was used in domestic VCRs. The quarter-inch cartridge (QIC) set of standards introduced in 1972 provided a great leap forward over the reel-to-reel machines and supported 1.35 GB tapes with 30 tracks at 51K bits/in. linear density and a 120 in/s tape speed. By 2010, the quarter-inch cartridge was largely obsolete. Today, linear *serpentine* recording is widely used to store data on tape. The term serpentine hints at the zigzag nature of the recording with some tracks recorded left to right and some right to left on the tape. Consequently, when the tape reaches its end, it does not have to be rewound, but can simply change direction to continue reading or writing data.

Ultrium Technology

Ultrium technology is the generic name given to the set of linear tape-open (LTO) standards developed jointly by Quantum, HP, and IBM to provide a common standard in a world where users were often forced to buy a proprietary recording technology that forced them to remain with one supplier.

The LTO standards have three important features:

1. They are regularly updated to keep pace with both demand and technological development.
2. They use linear serpentine technology (in contrast with helical recording).
3. They offer the very high storage capabilities required in today's world.

It's a popular myth that tape and cartridge recording technologies are obsolete. The death of tape is much exaggerated. In January 2011, HP reported that the worldwide market for their linear tape-open (LTO) tape drives declined by about 30% between late 2008 and late 2009, but increased by 45% (of the 2008 figure) by the end of 2010. This amounts to a renaissance in the use of magnetic tape. Moreover, tape has a much lower *total cost of ownership* (TCO) than disk drives. A large organization, such as law enforcement or a medical institution, has vast quantities of data to store. The total cost of ownership includes the equipment, media, floor space, maintainace, and energy consumption. The TCO for data over a 12-year period in a large organization might be fifteen times greater for disk-based storage than for tape-based storage.[13]

Towards the end of the 1990s, the LTO (linear tape-open) standard was developed. The word *open* indicates that the standard is not proprietary like the earlier standards that were owned by IBM or HP. The first standard, LTO-1, introduced a 100 GB cartridge. By 2010, the LTO-6 standard had been launched, and that provided a capacity of 1.5 TB and a data speed of 140 MB/s using 896 tracks at a linear density of 15,142 bits/mm. The LTO standards have been scaled up to 12.8 TB/cartridge in version LTO-8, which has not yet been released. LTO-5 1.5 TB cartridges are not cheap. In 2011, the cost of a cartridge was of the order of $50, which is compatible with hard-disk storage.

Manufacturing modern high-density tapes is not easy. The distance between tracks is so small that *tape dimensional stability* becomes an issue. Magnetic tape can change size under the influence of heat and humidity. This means that tracks that were aligned with the parallel heads at the time the data was written may not be aligned when the tape is read at a later date. This has led to the search for highly stable tape substrates. For example, some manufacturers use a polyamide that is a close relative of Kevlar. An additional advantage of this is that, if you are attacked, you can knit a bullet-proof vest from your cartridge.

An interesting option of LTO cartridges is a write once, read many times (WORM) mode that supports cartridges that can be written to only once. The underlying recording medium is the same as normal cartridges, but the physical interface prevents re-writing, and there is a different servo track to verify that data has not been modified. These cartridges are intended for the tamper-proof storage of data that must not be modified for legal purposes.

Magnetic tape continues to be an attractive archival mechanism for data because it is removable and transportable. The optical storage media we introduce in the next section are also removable and transportable but do not yet offer the ability to store data in the multi-terabyte region. Up to the early 1990s, magnetic tape used oxide media and ferrite

---

[13] *Clipper Notes*, report #TCG2010054LL, December 2010.

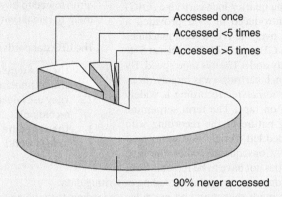

A Paradox, A Paradox, A Most Ingenious Paradox

The following slide taken from an Ultrium presentation is interesting because it makes the point that 90% of network data is never accessed. If this data is stored on disks, it is taking up expensive real estate and consuming power.

Source: Bruce Master, "The Evolving Roles of Tape and Disk," *Ultrium consortium presentation*, 2009.

based read/write heads. Over the past two decades, advances in head technology (the same advances we have seen in hard disk drives), much improved magnetic media, and an increase in track density has seen the areal density of tape systems increase by over three magnitudes. For example, Sony has developed magnetic particle technology to provide high-density recording by adopting 30 nm metal particles that are individually encapsulated in an alloy/ceramic coating to prevent oxidation and gradual data loss.[14] Similarly, it is necessary to provide a very thin magnetic layer in order to reduce demagnetization effects. Current magnetic layers are of the order of 200 nm (a human hair is 60,000 nm in diameter).

11.7   Optical Storage Technology

It's not surprising that the magnetic properties of matter were used to store digital data, because magnetism is an inherently binary property and has been used in various systems for a very long time. The application of light-based recording mechanisms to digital storage is even less surprising. The photographic plate has been used to store analog images for many years. In 1834, Henry Fox Talbot created permanent negative images using paper impregnated with silver chloride. Positive images were made by *contact printing* the negative onto another sheet of photosensitive paper.

Conventional chemical-based photographic techniques could indeed be used to store digital data. Suppose that you had a 2 in. × 2 in. sheet of plastic covered with a photosensitive material. You could write, for example, $2 \times 1200 \times 2 \times 1200$ dots on it using the same technology found in a laser printer, develop it, and read it back using the technology of the document scanner. Such technology might provide a capacity of about 720,000 bytes/sheet. Even improving the technology by a factor of ten would yield only 7.2 MB/sheet.

The optical digital storage medium was an invention waiting to happen and was inevitable. It was just a matter of waiting to see which technology would predominate and which companies would get to the marketplace first. Here we introduce the three optical recording

---

[14] "Sony, metal particle, and A3MP tape: Nanoscale technology for terabyte storage", Sony Electronics Inc., Park Ridge NJ., January 2009.

mechanisms used by digital computers: the CD, DVD, and Blu-ray. We will look at each of these in turn, because they all employ the same underlying technology. The difference between them is largely one of *scale*; as time has progressed, it has become possible to scale down the size of the features on the disk that stores the data.

11.7.1   Digital Audio

Optical storage has been around for longer than many think. Philips developed their LaserDisc in the early 1970s and launched it in the 1980s. LaserDiscs shared some of the attributes of today's *compact discs* (CDs), but the LaserDisc used analog technology. The system was read-only, and few movies ever made it to disc. LaserDisc was a failure because it was not cost-effective and the technology was not sufficiently mature.

Digital audio as we know it owes much to the cooperation between two giants of the consumer electronics industry. During the 1970s, Philips and Sony joined forces to develop optical storage—not least because Philips had considerable experience in optical technology and Sony had expertise in encoding and error correction mechanisms.

An important achievement of Philips and Sony was the standardization of the fundamental recording and playback mechanisms. This was no mean feat because of the complexity of the technology and the myriad of parameters involved. Two fundamental parameters are the size of the CD itself and the *sampling frequency*. The disc's diameter is 12 cm, and the sampling frequency was chosen to be 44.1 kHz. Although this seems to be a strange number, it's the sampling rate required for high-quality audio, and the same frequency is used in television systems. It has been reported that the basic parameters of the CD were juggled to yield a playing time of 74 minutes, which is sufficient to include the von Karajan version of Beethoven's Ninth Symphony on a single CD. On the other hand, the veracity of this story has also been denied by some of those involved with the development of the CD. Anyway, it deserves to be true.

The first really practical, low-cost, high-density storage mechanism based on light was the CD, which was introduced in 1981 as a means of storing high-quality sound *digitally*. Prior to the introduction of the CD, music was either stored on cassettes or on black vinyl long-playing (LP) records. The LP was large, heavy, and held only about 25 minutes of music per side. Moreover, it was an analog medium and highly prone to damage (for example, scratches and warping).

The structure of a CD is similar to a magnetic disk, because information is stored along a track. Unlike the hard disk, a CD's track is continuous and arranged as a continuous spiral, as Figure 11.39 illustrates. The spiral has approximately 20,000 turns, which corresponds to a length of about three miles. The effective track density is 16,000 turns per inch, and the theoretical maximum areal density is 1 Mb/mm^2 or 80 MB per square inch.

FIGURE 11.39   The spiral track

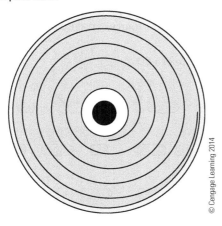

© Cengage Learning 2014

FIGURE 11.40   The structure of a CD

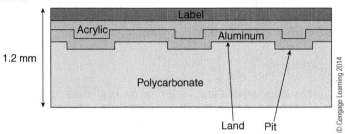

Digital information on a CD is imprinted on the surface of a 1.2-mm thick plastic disk in the form of a pattern that contains indentations, called *pits*, of varying length. The pits are coated with a metallic mirror and a protective layer. The region of the surface that is not indented is called *land*. The disc's central hole is 15 mm in diameter, and data is stored from 25 to 58 mm from the center.

Data is read from the disk's surface by illuminating it with a tiny spot of light and then detecting how much light is reflected back from the surface. A change in reflected light intensity occurs every time the laser spot moves from the pit onto the land and vice versa.

Figure 11.40 illustrates the structure of the surface of a CD. Whoever coined the term *pit* was lexically challenged. A pit rises above the surface, and it's what us normal folk call a *bump*. The term *pit* refers to an indentation when looking down at the data-carrying layer from above the label side of the disk. The laser beam that reads the data sees a *bump*, rather than a pit. Figure 11.40 demonstrates that there are four layers; the disk itself is almost 1.2 mm thick and is made of a transparent polycarbonate plastic. The pits are coated with a thin layer of aluminum followed by a protective acrylic layer with the label printed on top. This label side is much more sensitive to scratches and abrasion than the clear side.

Scratches on the transparent surface of a CD don't damage the data, although the light scattered by scratches can increase the error rate. Because data is stored along tracks, the medium is less sensitive to *radial* scratches than to circumferential scratches. If you polish a disk, the strokes should be radial!

The areal density of a CD is a product of the number of bits stored along a track and the pitch of the tracks. The size of the individual bits is determined by the size of the spot of light projected onto the disk's surface. Figure 11.41 illustrates the structure and dimensions of the pits and land on a CD's surface. A bump is approximately 0.5×10^{-6} m, and the pitch of the tracks is 1.6×10^{-6} m. The height of the bump (pit) is 1.25×10^{-7} m.

11.7.2   Reading Data from a CD

As we've said, a beam of light illuminates the pits and land along a track, and the amount of reflected light is used to read the data. The pits can be made very small by adapting existing semiconductor technology. Unfortunately, it's much harder to create a truly tiny spot of light because of the way in which most light sources produce a divergent beam that spreads out.

FIGURE 11.41   The surface of a CD

In order to create and control the smallest possible spot of light, it is necessary that the light beam be both *coherent* and *monochromatic*. A light source is *monochromatic* if the light waves all have the same frequency—unlike white light that contains frequencies distributed across the visible spectrum. A light source is *coherent* if all of the light waves that make up the beam have the same *phase* as well as the same frequency. The laser is a device that produces a monochromatic coherent light beam, and hence, the laser is used to illuminate the surface of the CD. The laser is a low-cost semiconductor laser rather like a conventional LED (light emitting diode).

The CD recording and playback mechanisms are most ingenious. Figure 11.42 illustrates the energy distribution of a spot of laser light on the surface of a disk. A perfect spot would have hard edges between light and dark. However, because of the wave-like properties of light, the edge of a spot is not sharp; it consists of a series of rings of light that gradually fade.

Figure 11.42 also illustrates the relative dimensions of the spot, the tracks, the pits, and land. When light hits a flat region on the surface of the CD (the land), a large fraction of the light is reflected back. Suppose that the size of the spot is somewhat larger than an individual bump. When light hits a bump on the surface, some light is reflected back from the top of the bump and some from the surface around the bump.

If the light source weren't coherent, it would matter little whether the reflected light came from the surface (i.e., land) or the bump (i.e., pit). However, in a CD, the height of the bump above the land is one quarter the wavelength of the laser light. Suppose that the light hitting the top of a bump travels a total distance x from the light source to the surface and then to the observer. The light that hits the bottom of the bump has to travel a little further distance; that is $\lambda/4$ in each direction. Thus, the total difference traveled by the light reflected off the land is $x + \lambda/2$. Because the spot doesn't cover a pit entirely and the path length between the reflected light from the pit and from the land is half the wavelength of the light, the beams tend to cancel. This is a property of coherent light.

Light is reflected well by the land, but light from a pit is severely attenuated. Consequently, it is easy to detect land and pits and, therefore, data can be read off the disc. Reading data from a CD does not require any physical contact between the surface of the disc and the read head assembly.

Because the CD reader uses a medium that is removable, touched by people's fingers, and generally left lying around collecting dust, it's impossible to prevent the surface being contaminated by particles that are orders of magnitude bigger than the stored bits. There's no way in which a CD drive could achieve the engineering precision of the *magnetic* disk

FIGURE 11.42   Distribution of the beam's energy

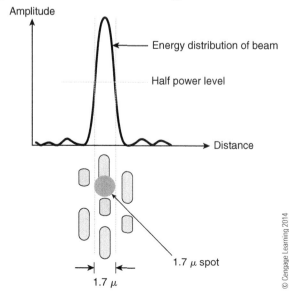

that requires a read-head-to-disk separation of the order of one micron. Fortunately, the problems of spot size are solved by the optical properties of the clear polycarbonate used to manufacture CD ROMs.

Figure 11.43 demonstrates the how the effects of surface contamination on a CD are much reduced. Light from a semiconductor laser passes through an objective lens and is focused onto the data-carrying surface of the disk. The size of the actual spot on the pits and lands is 1.7 μm, whereas the size of the spot on the upper clear surface of the disk is 800 μm. The spot on the clear surface is nearly 500 times larger than the spot on the pits and land. This means that the system is relatively tolerant of surface contamination because slight imperfections are out of focus.

The size of the spot on the pits and land is given by

$$r_{\text{spot}} = 1.22 \frac{\lambda}{NA}$$

where NA is the *numerical aperture* of the objectives lens and λ is the wavelength of the laser light. In a CD ROM, the wavelength of the light is 780 nm, and the numerical aperture is approximately 0.45.

The numerical aperture is given by $NA = n \sin \theta$, where n is the refractive index of the polycarbonate. A material's refractive index is a measure of its ability to bend light rays and is defined as the speed of light in a vacuum divided by the speed of light in the medium. The numerical aperture is a measure of the light-gathering capacity of the lens system and determines its resolving power and depth of field. We can relate this to the camera by pointing out that the numerical aperture of a lens is inversely proportional to its f-stop number.

To a first approximation, the geometry of Figure 11.43 demonstrates that the relationship between disk thickness and surface spot size is $\tan \theta = s/2d$, where d is the thickness of the disk and s is the diameter of the spot. For small values of θ, $\tan \theta = \sin \theta$, and therefore,

$$s = 2d \sin \theta = 2d\, NA/n.$$

where

d = thickness of the disk
NA = numerical aperture of the objective lens
n = reflective index of the disk material

Another important consideration in the design of CD readers is *depth of focus*; that is, the range over which the spot is in focus. The smaller the depth of focus, the harder it is for the laser spot to track the moving surface of the disk. The depth of focus can be expressed as

$$d_{\text{focus}} = \lambda \cdot n/NA^2$$

where λ is the wavelength of the light, n is the refractive index of the material, and NA is the numerical aperture.

The wavelength of lasers used in CD drives is in the region 670 to 690 nm (infrared). Shorter wavelengths allow smaller pits and land and therefore higher bit densities, as we shall see later. We can use Figure 11.43 to estimate the maximum theoretical capacity of a CD ROM. If we assume that a spot can read or write a single bit, the total capacity is given by the area of the disk divided by the area of a single spot. If R_{outer} is the outer radius of the disc and R_{inner} is the inner radius of the disk, the disc's data storage area is $\pi R_{\text{outer}}^2 - \pi R_{\text{inner}}^2$.

The area of a spot is given by πr_{spot}^2. The disc's capacity is therefore

$$\text{Capacity} = \frac{R_{\text{outer}}^2 - R_{\text{inner}}^2}{\left(\dfrac{1.22\lambda}{2NA} \right)^2}$$

If we plug typical values into this equation,[15] we get a theoretical maximum capacity of 2.5 Gb for a CD. In practice, we can't realize this value, partially because of the redundancy introduced by error correcting codes.

---

[15] Lane, P. M. and Van Dommelen, R., "Compact Disc Players in the Laboratory," *IEEE Transactions on Education*, Vol. 44, No. 1, February 2001, pp 47–60.

FIGURE 11.43   CD optics and spot size

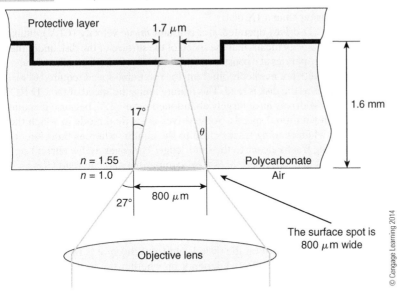

© Cengage Learning 2014

What is a Laser?

A laser is a device that emits coherent light at a single frequency in a narrow, nondiverging beam. Essentially, a laser is a nuclear reactor for light; that is, like atomic fission, it relies on a chain reaction. Instead of using neutrons to split atoms, it uses photons of light to generate other photons.

Passing a current through a semiconductor diode causes the generation of electron and hole pairs (a hole is a region of a crystal lattice where an electron is absent just as a hole in the road is a region where the road isn't). When an electron and hole meet, they combine and annihilate each other. However, when they do combine, they emit energy. By using suitable semiconductor materials such as indium phosphide or gallium arsenide, the energy emitted by an electron-hole is a photon of light.

Here's the clever bit. A photon stimulates the recombination of other electron-hole pairs in close proximity, which then produce a photon of the same phase and frequency (light amplification). The ends of the crystal are partially reflective, which means that the light in the laser bounces from end-to-end, generating further photons.

The acronym laser stands for *light amplification by stimulated emission* and refers to any process/mechanism that generates a chain reaction releasing more and more photons. The power output of a laser in a storage device is typically 5 to 10 mW in read-only devices and up to 250 mW in burners. Laser power goes up to 100 W in surgical lasers and 100 kW in military lasers.

Disk Speed

The speed of an audio CD is governed by the speed at which data is required from the disc. This speed corresponds to a data rate of 150 Kbits/s and is called 1X. At this speed, the surface of the disk moves under the read head at 1.25 m/s. Because computer users want to read data as fast as possible, CD drives have become faster since their introduction. A 4X drive provides data at four times the rate of a standard audio disc. Drives operating at 48X are now commonplace. However, the test results published by organizations that benchmark drives demonstrate that these drives don't provide the sustained data rates you might expect. A 48X disk is not 48 times faster than a 1X disc.

First-generation CD drives operated at a constant *linear* velocity (CLV), unlike the hard disk. A constant liner velocity means that the speed of the surface of the disk under the read head is constant. A hard disk operates at a constant *angular* velocity. Since the radius of the tracks at the center and edge of the disk are markedly different, the rotational speed required to read the track at a constant rate varies as the disk is read. This feature limits the speed of the CD ROM drive.

Modern CD ROM drives have largely abandoned pure CLV, because it is quite difficult to implement at high rotational speeds. Some drives use a dual mode in which their angular velocity is constant when reading tracks close to the center, whereas their linear velocity is constant when reading tracks closer to the outer edge. However, today current optical drives are DVD (or Blu-ray) drives that are backward-compatible and can read CDs.

The Optical Read-Head

We now look at the complete optical system that reads data from an optical disk. Figure 11.44 illustrates the path taken by the light from the laser to the surface of the disc and then back to the photoelectric sensors. Essentially, light from the laser passes through an arrangement of lenses and is focused to a spot on the disk. Light from the spot is reflected back along the same path; the amount of light reflected depends on whether the spot is hitting a pit or land.

FIGURE 11.44   Reading data

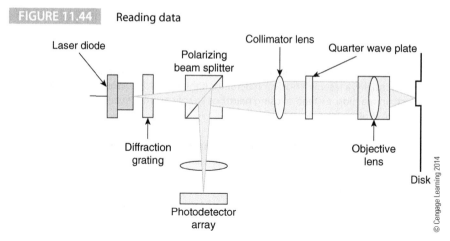

© Cengage Learning 2014

When the light returns from the disc, it hits a beam splitter, and some of it is reflected down to the sensor where it is detected. A photodiode measures the amount of light being reflected from the surface of the disk. If the light from the surface of the disk is coming from a bump, some of the light has traveled a further $\lambda/2$ and is 180° out of phase, resulting in a fall in the signal level at the detector.

Focusing and Tracking

To read data reliably, an optical drive's read head must follow the spiral track accurately. It is easy to move the device optics radially on runners to step in or out along a radius. It is harder to move the beam to the required spot with the necessary precision. This applies to movement in the XY plane when seeking a track and in the Z plane when focusing.

The objective lens in the read head is mounted on gimbals and can move in two planes: left and right for tracking and in and out for focusing. A magnetic field from an electromagnet is used to position the lens to perform the fine tracking and focusing.

The way in which the data is tracked is ingenious (there are several variations). In order to understand the tracking and focusing, you have to understand the beam sensor. Figure 11.45 illustrates the optical sensors—all *six* of them. Light from the laser first passes through a *diffraction grating*, which is a transparent material indented with parallel lines. The effect of a diffraction grating is to split the single beam into a main beam and two side beams.

FIGURE 11.45   The optical sensors used to read data and perform focusing/tracking

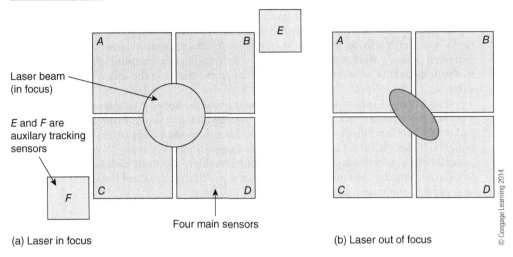

Laser beam (in focus)

E and F are auxilary tracking sensors

Four main sensors

(a) Laser in focus

(b) Laser out of focus

© Cengage Learning 2014

Sensors A, B, C, and D pick up the main spot. Sensors E and F pick up the two side beams. The outputs of the two side beam sensors are subtracted to get $track_{error} = E - F$. If the beam is centered, the tracking error is 0. If $track_{error} > 0$, the beam must be moved left, and if $track_{error} < 0$, the beam must be moved right.

The main beam falls on sensors A, B, C, and D, where the sum of the outputs of these sensors is used to regenerate the data from the disk. Differences in outputs between pairs of sensors are used to focus the spot.

In Figure 11.45, the beams from the diffraction grating pass through a collimating lens that makes the beams parallel, a *quarter wave plate*, and an objective lens that focuses the beams onto the disk. These beams are reflected back along the same path they took and then through the beam splitter to the six photo sensors.

The objective lens is *cylindrical* rather than *spherical*, which introduces *astigmatism* in the focusing; that is, the focal point is different in the vertical and horizontal planes. If the beam is in focus, the spot is circular, and all four central sensors receive equal amounts of energy. If the objective is too close to the surface, the beam is elliptical, as Figure 11.45b demonstrates, and the signal $(A - D)$ is greater than $(B + C)$. The difference can be used to move the lens back from the disk. If, however, the lens is too far away from the surface, the effect of the astigmatism is to rotate the elliptical spot by 90° and make $(B + C)$ greater than $(A + D)$.

Buffer Underrun

Sectors on a magnetic disk can be located relatively rapidly in contrast with the optical drive. Moreover, magnetic disks can be written to very rapidly. Writing data to optical media is more complex. If you make an error during writing, the media can be ruined and must be thrown away. Damaged CDs were so widespread in the late 1990s that the term *coaster* was commonly employed to describe the use of a damaged CD-R[16] as a mat for coffee cups. An optical drive requires a continuous stream of data while it is writing to the CD-R. The tiniest gap in the data stream is fatal. Because data from the host processor is stored in a buffer in the CD drive during writing, the situation in which the buffer empties is called *buffer underrun*.

Such buffer underrun was one of the biggest problems of CD recording, and it happened whenever the CD drive was fast and the data gathering slow or the processor was interrupted during writing. By 2000, CD-R and CD-RW drive manufacturers were competing with each other to offer ways of handling the problem of buffer underrun.

---

[16] A CD is pressed and cannot be modified. A CD-R is a writable CD that can be written once. A CD-RW is a rewritable CD that can be written, erased, and re-written.

One of the first manufacturers to tackle the problem was Sanyo, who employed the term *BURN-Proof*™ (Buffer UndeRuN-Proof) technology to describe their system. Sanyo's mechanism periodically checks the state of the data buffer. If the buffer is emptying too rapidly, the CD-RW's firmware intervenes to deal with the situation. All you have to do is to finish writing the current frame and then wait until the buffer refills. Plextor extended Sanyo's BURN-Proof technology by periodically sampling the recorded data to check its quality, allowing the drive to change the speed of the disk to a more suitable value.

The Yamaha Corporation developed a solution to the buffer problem called SafeBurn™. Yamaha's system both seeks to avoid buffer overrun and to optimize writing speed. SafeBurn technology uses a relatively large 8 MB buffer (at a time when competitors were using 2 MB buffers). If the data level in the buffer does begin to fall dangerously low, writing is suspended exactly as in the case of BURN-Proof technology. The third component of Yamaha's SafeBurn technology is a mechanism that checks the characteristics of the disk being written to in order to optimize recording parameters.

11.7.3   Low-Level Data Encoding

The encoding of data on a disc is subject to several severe constraints—the most important of which concerns the *distribution of energy in the spectrum of the signal* from the optical pickup. In particular, there should be no DC component in the signal; that is, its average value must be 0 (this is a consequence of the nature of electronic circuits that can transmit only time-varying signals and not a constant signal level). Consequently, the low-level data encoding must be designed to control the number of consecutive 0 bits or 1 bits and allow the clock signal to be regenerated from the data signal.

Source data is stored in units of 8-bit bytes. These data bytes are each encoded into 14 bits by means of 8-to-14-bit modulation (EFM). Each 14-bit code can represent $2^{14} = 16,376$ different values, although only $2^8 = 256$ of these values are used. EFM is a form of *run length limited* code that reduces the bandwidth of the signal. The EFM ensures that there are no more than ten and no less than two consecutive 0s in the data stream on the disc. In fact, there are 267 *legal* 14-bit patterns conforming to the rule: the number of consecutive 0s in the data stream must be between three and nine, inclusively. This means that there are $267 - 256 = 19$ bit patterns that are legal from the modulation mechanism but do not describe a valid data byte. Some of these codes can therefore be used as special markers in the data stream.

A logical 1 value in the data is interpreted as a change of state in the signal from the disk (i.e., a transition from land to pit or vice versa) so that the 1s are represented by the starts and ends of recorded pits. Figure 11.46 illustrates the relationship between the data stream and the disk surface.

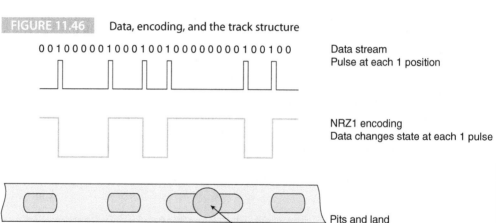

FIGURE 11.46   Data, encoding, and the track structure

001000001000100100000000100100   Data stream
Pulse at each 1 position

NRZ1 encoding
Data changes state at each 1 pulse

Pits and land

Laser spot

Unfortunately, it is possible for the end of one group and the beginning of the next group to be misinterpreted as a valid code word. To stop this happening, a 3-bit delimiter is placed between each group of 14 bits. These 3 bits are called *merging bits*. The 3 bits chosen to form the merging pattern are not always the same. The specific merging pattern is selected to ensure that the average number of 1s and 0s remains approximately the same (this constraint ensures that there is no DC component in the data from the optical head).

The low-level data structure on a CD is rather complex due to the error-correcting code mechanism required to convert optical recording from an essentially unreliable storage medium into a reliable one. Optical media are exposed to the atmosphere and handled by users making them easily contaminated. Moreover, the recording medium is subject to manufacturing defects, and the playback mechanism is subject to vibration and other disturbances. Consequently, it's quite likely that the data read from an optical disk will have a considerably higher error rate than a hard disk. Indeed, the difference in error rate between magnetic and optical storage is several orders of magnitude.

Optical media employ several error detecting and correcting mechanisms. Error correction is a feature that distinguishes audio discs from CD-ROMs holding digital data. A higher level of error correction is applied to data discs. For this reason, a music CD apparently stores more bits than a data CD.

The fundamental error-correcting mechanism used in optical storage is called a Cross Interleaved Reed-Solomon (CIRC) code. Cross interleaving is simply a technique that rearranges the order of bytes as they are recorded to ensure that a long error burst affects bytes from many different data frames by a small amount rather than massively corrupting the bits of one or more adjacent frames. In other words, a long error burst is redistributed as a lot of short error busts. These short bursts can usually be corrected.

We first look at the format of audio discs. The digitized sound information originates in the form of pairs of multiplexed 16-bit values with one sample for each of the left and right channels. This data is divided into bytes to create a 24-byte frame. The CIRC mechanism takes these 24 bytes of data in a frame and generates a 28-byte output. The additional four bytes are the error-correcting information created by the Reed Solomon coder. The 28 data bytes are then scrambled and passed to a second RS encoder to generate a 32-byte frame by adding a further four bytes of error correction. This encoding mechanism is designed to protect against single bit and multiple bit burst errors. A one-byte subcode is added to the frame for reasons that will be explained later.

This information is encoded using eight-to-fourteen modulation and a 24-bit synchronization pattern added to the front. A 3-bit merging pattern is located between eight 8-bit bytes but not within the synchronizing pattern. The total number of bits is

24 synchronizing bits + 3 merging bits	
1 subcode byte	$1 \times (14 + 3)$ bits
24 data bytes	$24 \times (14 + 3)$ bits
8 parity code bytes	$8 \times (14 + 3)$ bits

which adds up to 588 bits. Note how 588 bits have been used to encode $24 \times 8 = 192$ bits.

These frames are grouped into 98-frame blocks containing $98 \times 24 = 2352$ bytes of audio data. The audio CD drive operating at a 1X speed and reads 75 blocks per second, corresponding to a data rate of 2353 bytes $\times$ 8 bits $\times$ 75 blocks/s = 1,411,800 bits/s (i.e., 1.4 Mbits/s).

The number of audio samples per second is 12 samples/frame $\times$ 98 blocks $\times$ 75 blocks/s = 88,200 samples/s (or 44.1K samples per channel per second). This value is, of course, the fundamental data rate around which the audio CD was constructed. The data rate from the CD is given by 98 frames $\times$ 588 bits/frame $\times$ 75 blocks/s = 4,321,800 bits/s (4.32 Mbits/s).

Figure 11.47 illustrates the three levels of structure on an audio CD. In Figure 11.47a, the block is composed of 98 frames. Each frame contains a single-byte subcode that contains information about the current track, etc.

It takes 98 frames to build up the entire subcode. The bits of the subcode represent the eight channels P, Q, R, S, T, U, V, and W. The P subcode is used to specify a separator between music tracks, and the Q subcode contains the disc's table of contents and the track location. The other subcodes are not generally used. The subcode block is assembled from

FIGURE 11.47   Data structure on an audio CS

Frame 1	Frame 2		Frame 98

(a) A block is composed fo 98 frames. Blocks are read at the rate of 75 blocks/s

Synchronization	Subcode	12 bytes data	4 bytes parity	12 bytes data	4 bytes parity

(a) A frame has a synchronization code, subcode byte, 24 data bytes, and eight parity bytes

Synchronization 24 bit code	3 merging bits	14-bit code representing one byte of information	3 merging bits	▪ ▪ ▪ ▪	

(c) Fine structure of the frame: 8-bit values are encoded as 14 bits with three merging bits between groups

© Cengage Learning 2014

the 98 subcode bytes in a block, beginning with two special sync patterns that are 14-bit codes that do not conform to the normal 265 data codes. This means that a subcode block can be uniquely identified. A subcode block also contains its own 16-bit cyclic redundancy code to provide error protection.

Figure 11.47b shows the frame structure with its 24 bytes of audio data, synchronizing header, subcode bytes, and parity check bytes generated by the Reed Solomon encoder. Figure 11.47 is misleading, because the order of the data is scrambled by the CIRC coder.

Figure 11.47c illustrates the bit level structure with the 24-bit unique synchronizing code and successive 14-bit information units, each separated by a 3-bit merging code.

The encoding of CD-ROMs is more complex because an extra level of data correction has been added to further reduce the uncorrectable error rate. Audio signals are more tolerant of errors than digital data. Dropping the occasional bit in a steam of audio samples is of no great significance, because the resulting sound disturbance will be very short lived and probably partially filtered out by the audio processing circuits. Dropping even a single bit in a data signal can have a far more profound effect, particularly if the data represents code.

The structure of a CD-ROM is similar to the audio ROM in order to provide compatibility between the two systems. Consequently, the same frame structure and low-level encoding is used. The 98-frame blocks are now called *sectors* and store 2,048 bytes of user data. This arrangement means that, of the 2,336 audio bytes that can be stored in a block, $2,336 - 2,048 = 288$ bytes are available for the extra layer of error correction.

When an audio CD is read, the data rate is given by 75 sectors/s $\times$ 2,048 bytes/frame = 153,600 bytes/s, which is rounded to 150 KB/s and is, of course, the standard 1X data rate.

CD-ROMs have three data modes. Mode 0 has the same basic structure as the other modes, but all of its data fields are 0s (i.e., the data fields are blank). Mode 1 is used for data CDs and mode 2 for audio CDs. Table 11.5 illustrates mode 1 data structures, and Table 11.6 illustrates mode 2 data structures.

TABLE 11.5   Mode 1 CD-ROM Sector Data Structure

Mode 1	
Sync field	12 bytes
Header field	4
User data field	2,048
Error detection code	4
Reserved	8
Error correction	276

© Cengage Learning 2014

TABLE 11.6	Mode 2 CD-ROM Sector Data Structure	
Mode 2		
Sync field	12 bytes	
Header field	4	
User data field	2,048	
Auxiliary data field	288	

© Cengage Learning 2014

Because mode 1 provides data blocks with $2K = 2^{10}$ bytes, the CD looks like other mass storage devices to the host system. As Table 11.5 demonstrates, a data track is divided into sectors containing 2,352 bytes. Each sector includes a 12-byte synchronizing field and a header that contains the sector's address and its mode. The sector address is unusual, because an address is defined by *time*. It's rather like defining an aircraft's position by the time it takes to reach a certain point at a constant speed from its origin. An address is specified as minutes:seconds:sector. The minutes field is *excess* $A0_{16}$, so that the three-byte code $C0_{16}, 32_{16}, 21_{16}$ indicates a position 32 minutes, 50 seconds, sector 33; that is, the minutes field is greater than the actual value by $A0_{16}$. In a mode 2 frame, all 2,336 bytes are available for user data. This mode can be used for video or audio encoding.

11.7.4 Recordable Disks

The recordable CD is the subject of Part II of the "Orange Book,"[17] and the first CD-R machines were introduced in 1993. Recordable CD drives are not greatly different from the read-only varieties. The principal difference is in the media. A recordable CD uses a twin-layer technique with a reflective layer behind a dye layer. If light hits the dye layer, it's absorbed; if it hits the reflective later, it's reflected back. Data is recorded by using a laser to burn a hole in the dye layer through to the reflective layers. For this reason, people sometimes use the expression *burning a CD* to indicate writing to a blank CD.

Originally, the dye was cyanine-based. Today, more modern dyes have been used to cope better with the degradation of data due to ultraviolet light and to operate at much higher writing speeds. The reflective layer may be gold because of its great stability and freedom from corrosion. A spiral track is imprinted on a CD-R disk during its manufacture with the same width and pitch as a CD.

Re-Writable CDs

To make CD-RW technology compatible with existing CD drives, it was necessary to find a means of creating and deleting areas of differing reflectivity along the track of a disc. The two obvious candidate technologies were *phase-change* and *magneto-optical*. Panasonic and others pioneered magneto-optical storage. However, it is phase-change technology that has been universally adopted for CD-RW devices.

Increasing the power of a tightly focused laser beam locally heats the surface of the data-carrying layer. This layer contains a compound of silver, indium, antimony, and tellurium that can exist in two stable states: crystalline and amorphous. When this material is crystalline, it reflects the laser light better than when it is amorphous.

The CD-RW disc itself is similar to the conventional CD. The substrate is a 1.2-mm polycarbonate disc, and the track (i.e., spiral groove) is molded on the disk with the time information. The recording layer is sandwiched between two *dielectric layers* to control the thermal characteristics of the phase-change layer when it is heated during the writing or erasing process. A reflective layer is provided behind the data and dielectric layers.

---

[17] The Orange Book is an informal name for the set of recordable CD standards defined by Philips and Sony which was published in 1990. There are other optical storage standards in the Red Book (CD audio), Yellow Book (CD ROM), and Blue Book (enhanced CD).

CD Standards

As the CD has developed, new standards have been created to suit its changing characteristics and applications. Traditionally, each new set of standards is called a *book*, and the specific book is denoted by a color. The first standards were the Red Book standards for the Compact Disc Digital Audio system. The Yellow Book standards refer to the CD-ROM used to store digital data, and the Yellow Book standards were extended to include the CD-ROM/XA extension that defines a CD-ROM format containing computer data as well as encoded audio and video information.

The next standard was the Green Book, which defined the *Compact Disc Interactive* (CD-I) format. CD-I provides a platform for mass consumer-interactive multimedia applications.

The Orange Book provides a standard for write-once compact discs. The European Computer Manufacturer's Association has created the ECMA 168 standard that incorporates the Orange Book. This standard governs write-once CDs and CD-ROMs, including multi-session recordings, and extends the file naming convention to cover Unix style names. Finally, the Blue Book covers standards for the LaserDisc, which is used to store video and sound data but is now largely obsolete.

The Orange Book CD-R standard divides the CD into a *System Use Area* (SUA), which defines the format and type of data on the disc and an information area. The SUA is divided into a *power calibration area* and a *program memory area*. The power calibration area provides a testing ground for the laser. This area allows the drive to calibrate the laser power for each recordable disc that is inserted. A disc is calibrated by setting a bit to 1.

The information area that holds the user data begins with a lead-in that contains a table of contents in the Q subchannel. Synchronizing takes place in the lead-in area. After the data comes the lead-out area, containing data and defining the end of the CD.

The *High Sierra* file standard provides an interface between the host operating system and a CD ROM. This standard has been adapted as ISO 9660 by the International Standards Organization and is intended for read-only media. It omits mechanisms required by random access read/write media.

The High Sierra file system is hierarchical, and each CD has a root directory. Because the CD is an intrinsically slow medium and a hierarchical structure may require a long search from directory-to-directory, High Sierra uses a separate path table on each volume describing the file hierarchy in a form that can be cached. File naming conventions are more restrictive than Windows, and filenames are limited to eight characters, a dot, and a three-character extension. Moreover, file names can't include special characters other than underscores.

The ISO 9660 standard was devised for CD-ROMs and is not ideal for recordable and rerecordable media. A new standard, ISO 13346, has been designed to cover recordable media. This standard is also called the *Universal Disk Format* (UDF).

The most important feature of ISO 13346 is its support for *packet writing*. The ISO 9660 logical file system has to know which files are going to be written at the start of a writing session. This information is needed to create a pointer to the physical location of files on the disc. ISO 13346 uses packet writing to allow files to be added to a CD-R or CD-RW disc at any time. At the end of each packet-writing session, a *virtual allocation table* is added to the disc to describe the physical locations of each file. Each newly created virtual allocation table includes data from the previous table.

The laser in a CD-RW drive operates at three powers. During reading, it provides the beam that detects the edges of the pits and operates in its lowest power mode. During writing, the laser operates in its highest power mode when it heats the recording layer sufficiently to create an amorphous state. The write power level heats the surface locally to about 600°C, and the compound melts. When it cools rapidly, the liquid freezes and shrinks to create a pit.

When the CD-RW drive is erasing data, the laser operates in a lower power mode than its write mode, and the laser heats the surface sufficiently to turn the data layer into its crystalline state. The phase-changing material is heated to about 200°C, where it crystallizes and the atoms take up an ordered state. There is a limit to the number of write and erase cycles that the material can undergo and still continue to provide two optically distinguishable states. CD-RW is still a read-mostly medium rather than a true read/write medium.

Magneto-Optical Storage

An alternative to the phase-change technology is *magneto-optical* recording. Magneto-optical (MO) systems are not fully compatible with conventional CD drives, and the rise of the low-cost CD-RW drive has led to a decline in MO technology.

Recall that the Curie temperature defines the point at which a magnetic material loses its magnetization. Some substances have a Curie point of 200°C, which means that they can be demagnetized by heating with a laser. Figure 11.48 illustrates the principle of an MO system. The data-carrying surface of the disc is a ferromagnetic material with a low Curie point. In normal operation, the domains on the surface are magnetized perpendicular to the disc's surface. If the surface is heated by a laser, it is demagnetized. However, because there is an electromagnet under the disc, applying a field will magnetize the surface when it cools.

An MO disc can be read because there is an interaction between magnetism and optics. When a polarized light passes through a magnetic material, a change in the polarization of the light takes place. This phenomenon is called the *Kerr effect*.

FIGURE 11.48   Magneto-optical recording

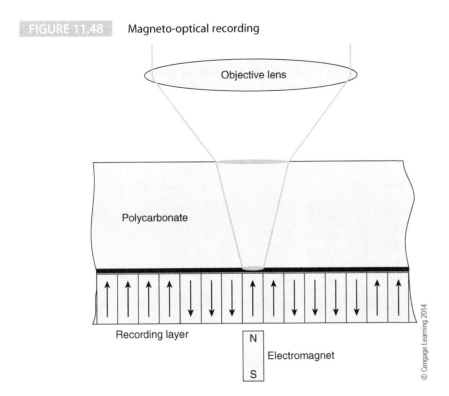

© Cengage Learning 2014

A similar optical system to that of a CD can be used to read MO discs. When light is reflected from the disk, its plane of polarization is rotated by about 0.5° if the surface is magnetized in one direction and by 0.5° in the other direction if the magnetization is reversed. Consequently, there is a difference of 1° in the polarization of the light from the two magnetic states. The optical system and sensors are able to detect this difference.

11.7.5   The DVD

The DVD is a development of the compact disc that means *Digital video disc* or *Digital versatile disc*; no one seems to know which of these is correct. What matters is that the CD was optimized to store about 75 minutes of high-quality audio, whereas the DVD can store a movie. Unlike the CD-ROM, the DVD is available in several capacities, depending on whether there are one or more data-carrying layers. DVD technology was developed in the early 1990s by a group of companies including Toshiba, Time Warner, Sony, and Philips.

Some of the leading players in the development of DVD technology had close links with Hollywood, which strongly influenced the emerging standard. In particular, the DVD was designed to provide 133 minutes of encoded video information (sufficient to cover most mainstream movies). The DVD provides both high-quality sound and audio and includes up to three separate audio channels, allowing the same DVD to be used with audiences of different nationalities. The sound channels include Dolby encoding and can be used as the basis for the *home theater* technology that began to emerge in about 1991.

DVD technology is virtually the same as CD technology. In fact, you could say that the DVD is the CD constructed with technology that has advanced and matured over ten years. The pits on DVD discs are packed more tightly, and the minimum pit size is 0.4 μm rather than 0.8 μm used on a CD. The laser light wavelength is reduced from 780 to 640 nm. Similarly, the track spacing is reduced from 1.6 to 0.74 μm. Figure 11.49 illustrates the structure of CD and DVD tracks.

The low-level modulation used by a DVD is eight-to-sixteen encoding—instead of the CD's eight-to-fourteen modulation. The DVD format is actually more efficient, because by selecting 256 out of 65,536 patterns, it's possible to avoid a DC component in the recorded data and the three merging bits are no longer required. Consequently, only 16 bits/byte are required in contrast with the CD's 17 bits/byte.

When a DVD is first made, it is 0.6 mm thick or half the thinness of a CD. A second disk that is also 0.6 mm thick is then bonded to it to produce a final DVD consisting of two 0.6 mm

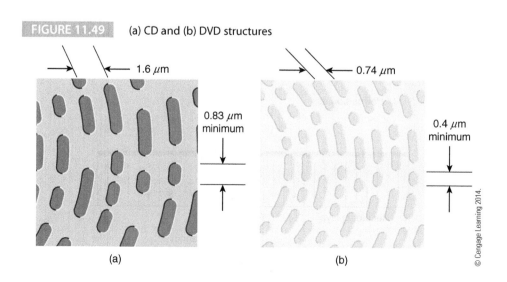

FIGURE 11.49   (a) CD and (b) DVD structures

1.6 μm

0.83 μm minimum

0.74 μm

0.4 μm minimum

(a)

(b)

FIGURE 11.50   The multilayer, multiple sided DVD structures

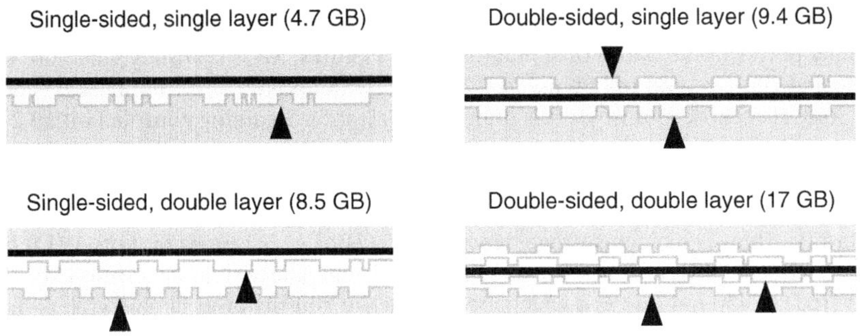

Single-sided, single layer (4.7 GB)

Double-sided, single layer (9.4 GB)

Single-sided, double layer (8.5 GB)

Double-sided, double layer (17 GB)

Based on Figures 7, 9, 10, 11 from Fred A. Byers, "Care and Handling of CDs and DVDs: A Guide for Librarians and Archivists," NIST Special Publication 500–252, 2003.

layers. One layer is the DVD proper, and the other is a blank or dummy layer. Moreover, it is possible to have two different data layers in the same disc. DVD supports the following four basic formats illustrated by Figure 11.50.

- DVD-5 uses a single-sided, single-layer disc.   Total capacity = 4.7 GB
- DVD-9 uses a single-sided, dual-layer disc.   Total capacity = 8.5 GB
- DVD-10 uses a double-sided, single-layer disc.   Total capacity = 9.4 GB
- DVD-18 uses a double-sided, dual-layer disc.   Total capacity = 17.0 GB

Recordable DVDs

Recordable DVDs followed DVDs as inevitably as recordable CDs followed CD ROMs. However, the technology didn't settle down as rapidly as CDs, where the CD, CD-R, and CD-RW followed each other with a reasonable degree of backward compatibility. Within a few years, the DVD reader was followed by the DVD-RAM, DVD-RW, DVD+RW, and DVD-R.

DVD-R is a write-once medium with a capacity of 4.7 or 9.4 GB that can be used in most compatible DVD drives. It first appeared in 1997 with a lower-capacity 3.95 GB version. In 1998, the first rewritable device appeared called the DVD-RAM, which relied on both phase-change and magneto-optical techniques to write data to the disc. First-generation devices had a 2 GB capacity, but that rose to 4.7 GB by 1999. This system was not compatible with other DVD formats.

Pioneer developed a DVD-RW system with the object of creating a format compatible with DVD and DVD-R. This format, which first appeared in 1999, uses phase-change technology for reading, writing, and erasing information. A 650 nm laser heats a phase-change alloy to change it from crystalline and amorphous states, exactly as in CD-RW systems. Another rewritable DVD appeared in 2002 called the DVD+RW format. This claimed to be more compatible than its DVD-RW competitor.

11.7.6 Blu-ray

Just as the DVD replaced the CD, Blu-ray technology is replacing the DVD. Blu-ray was driven by the introduction of high-definition television (HDTV), which required more storage capacity than the DVD could provide. Without a new storage medium, high-definition home cinema would have been impossible (other than via off-air broadcasting). Alas, two different solutions to the problem of increasing optical media capacity were proposed: HD DVD and Blu-ray. Both systems were launched, each backed by media giants. Blu-ray was championed by Sony, Panasonic, Philips, LG, Pioneer, and Apple. HD DVD was championed by Toshiba, Hitachi, Microsoft, and NEC.

These two incompatible formats provided an echo of the struggle between the VHS and Betamax standards for VCR tapes decades earlier. Two formats force the consumer to make a choice and stores to stock films in each. Ultimately, the major studios had the greatest bargaining power in the battle of standards. Sony Pictures, MGM, Disney, and 20th Century Fox selected Blu-ray, and only Universal Studios (with about 9% of the market) chose HD DVD. Sony also chose Blu-ray for its popular PlayStation 3 gaming console (with 3.2 million consoles in the USA alone) to increase the demand for Blue-ray products. Another nail in the coffin of HD DVD was Wal-Mart's promotion of Blue-ray. So, Blu-ray prevailed, and the world was spared a long-lasting format battle.

Blu-ray achieves its high storage density of 25 GB (i.e., 5.3 times that of DVD) by using a higher-frequency laser with a wavelength of 405 nm. The visible spectrum extends from 620 nm (red) to 450 nm (violet), which means that the Blu-ray laser is blue/violet—hence its name. The DVD laser is 650 nm (red), and the CD laser at 780 nm falls in the infrared spectrum. Blue-ray disk are physically different from CDs and DVDs because the Blu-ray data layer is below a front cover layer of only 0.1 mm. The numerical aperture of lenses used to focus the beam in Blu-ray systems is higher (0.85) than for DVDs (0.6) and CDs (0.45). Recall that a high numerical aperture allows a smaller spot size.

Like DVD, Blu-ray supports dual layers to provide a capacity of 50 GB. Figure 11.51 provides images of the three generations of recording media demonstrating their relative densities.

FIGURE 11.51   Scanning electron microscope image of three generations of optical disks.

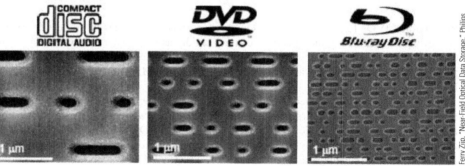

Ferry Zijp, "Near-Field Optical Data Storage," Philips Electronics, 2007. Courtesy of Philips.

A significant difference between Blu-ray and DVD is the structure of the disk and position of the recording layer. In a DVD, the data-carrying layer is sandwiched in the middle of the disk (in a CD it's at the bottom). In Blu-ray media, the data layer is on the top of the 1.1-mm thick substrate and is protected by a 0.1-mm thick cover. The advantage of locating the data layer near the top of the disk increases the disk's tolerance to *tilt*. Not all disks will be exactly flat (i.e., normal to the laser beam), so there will be a degree of tilt. Blu-ray disk are able to tolerate a tilt of up to 0.64°.

DVDs are manufactured by taking two disks, putting the data layer on one disk, and then sandwiching them together. A Blu-ray disk is easier to manufacture, because the data layer is put on the top followed by a protective layer. This protective layer is harder than the polycarbonate substrate of CDs and DVDs and is intended to protect against *light abrasions*. Modern Blu-ray discs may have complex coating structures. For example, Verbatim's disks have a hard coat layer, a cover layer, and a protection layer above the data layer. Under the data layer there is another protective layer, a reflective layer, and then the polycarbonate substrate.

Summary

In this chapter, we've looked at secondary storage systems. These hold very large quantities of data at remarkably low costs. Unfortunately, the access time of secondary storage systems is of the order of 10^6 times greater than semiconductor memory, which means that computers have to use effective memory management systems to hide the latencies of secondary storage devices. The capacity of some secondary storage has increased at a truly phenomenal rate, reaching about 60% per year in the early 2000s. Although the speed of secondary storage systems has improved over the years, access times have fallen very slowly, and we are unlikely to see any real progress unless there is a radical change in the technology.

The principal secondary storage technologies are magnetic and optical. Each of these technologies occupy a niche in the computer world. Magnetic technology provides better access time, greater capacity, and better read/write performance than optical storage. These characteristics make magnetic storage ideal for the disk drives that store programs and data while a computer is running. Optical storage systems provide low-cost transportable storage with the CD (about 600 MB), the DVD (about 4.7 GB), and Blu-ray (25 GB).

We have looked at the use of hard disks to create low-cost, large storage systems called RAID, that can have both greater reliability and better performance than individual disks alone. We have also introduced solid-state disk technology, which is replacing the rotating disk drive but brings with it its own performance limitations.

Magnetic storage is a mature technology. Progress is being made largely in the capacity of drives and in their interfaces. Optical storage is still evolving with Blue-ray technology replacing DVDs and CDs. However, optical technology is not primarily driven by the PC world but by Hollywood and the major studios, because it is primarily used to distribute sound and vision media.

Problems

11.1 What limits the areal density of a hard disk drive?

11.2 This chapter refers to an article on magnetic storage in the December 2000 edition of *IEEE Spectrum* with the title "Magnetic Storage: The medium that wouldn't die." The title implies that magnetic storage should have become obsolete. Explain why many may have believed this, and why magnetic storage has continued to thrive.

11.3 Explain the following terms as they are applied to magnetic storage technology.
   a. Domain
   b. Permeability
   c. Magnetostriction
   d. Remnant field
   e. Ferromagnetic
   f. Curie temperature
   g. Hard
   h. Soft

11.4 What are the major obstacles faced by the designer of hard drives?

11.5 Making hard drives for laptops and portable computers imposes a different set of design criteria on the engineer. What are the special considerations that the designer of mobile hard disks has to contend with?

11.6 Hard disk drives once used combined inductive read/write heads. Today, high-performance drives have GMR read heads.
   a. What is a GMR read-head?
   b. What are the advantages of separate read and write heads?

11.7 Why is it difficult to fully erase data from a magnetic medium such as tape or disk?

11.8 Why does data have to be encoded before it can be written to a disk?

11.9 The following data sequence is to be MFM encoded for recording on a diskette. Draw the resulting waveform that would be presented to the write head.

0101001100001111

11.10 What are the criteria that determine the optimum size of a sector? Are these criteria permanent or do they change with time? What other developments in computing affect the answer to this question?

11.11 A hard disk drive has a track density of 524,000 tpi and an area density of 29.7 Gbits/in.^{2}. What is the density of bits along the track?

11.12 A hypothetical disk drive had the following parameters. The seek time for a short seek is given as $1 + 0.2\sqrt{n}$ ms, and the seek time for a long seek is $3 + 0.003n$ ms. Assume that short seeks are less than 200 tracks.

Suppose the drive accesses the following track numbers sequentially. What is the approximate seek time required (neglecting rotational latency)?

1234, 1235, 1237, 1200, 2456, 2470, 1240, 1242, 100, 2120

11.13 Suppose that access to the tracks in Problem 11.12 was rearranged with the seek beginning with the lowest track number and continuing in ascending order to the highest track number. What would the seek time be then?

11.14 A hard disk has the following parameters.
- 6 surfaces
- 20K tracks per surface
- 256 sectors per track
- 512 bytes per sector
- Rotational speed 7,200 rpm

a. What is the capacity of the drive?
b. What is the rotational latency?
c. Once a sector has been accessed, what is the rate at which data is read?

11.15 How can disk drives be made faster (i.e., what options are available to the manufacturer)?

11.16 A manufacturer wants to create a disk with a rotational latency of 1 ms. What rotational speed is necessary to achieve this value?

11.17 Describe the following disk head scheduling algorithms: FIFO, SSTF, SCAN, and C-SCAN.

11.18 A hypothetical disk drive has the following properties.

Rotational speed	7,200 rpm	
Sectors/track	256	
Track-to-track		
seek times	2 ms	Tracks stepped < 10
	1 ms	Tracks stepped < 20
	0.4 ms	Tracks stepped < 40

What is the approximate time required to access the following sectors, assuming that a new seek is not required if successive sectors are on the same track and the difference between sector numbers is greater than 1? The following data is presented as (t, s) where t is the track number and s is the sector number.

(12, 25), (12, 27), (12, 39), (12, 40), (13, 90), (11, 108), (26, 25), (26, 24), (200, 19)

11.19 What is SMART technology, and how can it help the corporate hard disk user?

11.20 Suppose that a disk drive has 5,000 cylinders numbered 0 to 4,999. The drive is currently serving a request at cylinder 143, and the previous request was at cylinder 125. The queue of pending requests in FIFO order is

86, 1470, 913, 1774, 948, 1509, 1022, 1750, 130

Starting from the current head position, what is the total distance (in cylinders) that the disk arm moves to satisfy all the pending requests for each of the following disk-scheduling algorithms?
a. FCFS
b. SSTF
c. SCAN

11.21 The most popular (i.e., commercially successful) RAID levels are 1 and 5. Why?

11.22 If the probability of failure of a single drive is 0.000001 per 1000 hours, what is the probability of failure of exactly one drive in an array of four?
a. One drive in an array of four
b. Two drives in an array of four

11.23 When calculating the reliability of systems with multiple disks such as RAID arrays, it is conceptually obvious that, if one drive has a probability of failure of one in a thousand, the probability of two drives failing is one in a million because independent probabilities multiply. Why is this statement probably false?

11.24 Why is a CD so relatively insensitive to the effect of contamination by dust particles that are far larger than the bits stored on the disk?

11.25 Why is a laser required to read data from an optical disk?

11.26 Why is it necessary to use a code such as 8-to-14 to encode data bits before recording them on a CD?

11.27 What is the size of a spot on the recording surface of a CD if the wavelength of the laser light is 780 nm and the numerical aperture is 0.45?

11.28 What is the difference between constant linear and constant angular velocity? Why is one used in magnetic recording and the other in optical recording?

11.29 A disk rotates at 7,200 rpm, and the head is over a track 2 in. from the center. What are the corresponding angular and linear velocities?

11.30 How are writable CDs implemented?

11.31 If magnetic tape (pre-1980s) was ½ in. wide and stored data in nine tracks at 128 characters/in., what was the areal density in bits per square inch? What is the areal density ratio between this and the platters of modern disk drives?

11.32 In 1996, NEC announced a 64-MB flash memory chip with a 98 mm^2 die size. What is the corresponding areal density in bits/in.^{2}?

11.33   By 2002, a 125-mm^2, 1-Gb flash memory had been announced. What is the corresponding areal density in bits/in.^{2}?

11.34   In September 2010, Toshiba launched the first embedded flash memory modules (i.e., for incorporation on circuit boards rather than as plug-in modules). These devices had a capacity of 128 GB in a $17 \times 22 \times 1.4$ mm FBGA package. What is the corresponding areal density in bits/in.^{2}?

11.35   Suppose the embedded flash memory modules of Problem 11.34 were to be incorporated in standard 3.5-in. form factor disk-drive housing. How much data could be stored, assuming that half the volume would be taken up by cooling heat sinks and interface circuits?

11.36   What is the minimum spot size of a DVD laser if its wavelength is 650 nm and the lens has a numerical aperture of 0.6?

11.37   What is the minimum spot size of a Blu-ray laser if its wavelength is 405 nm and the lens has a numerical aperture of 0.85?

11.38   Calculate the theoretical maximum capacity of a single-layer Blu-ray disk, assuming that a single laser spot can burn a 1 or 0 and that the wavelength of the laser is 405 nm, the numerical aperture of the lens 0.85, and the inner and outer tracks have radii of 23 mm and 59 mm, respectively.

11.39   Why are there no RAID systems using optical memory?

11.40   What are the relative advantages and disadvantages of SSDs and HDDs?

11.41   What is wear-leveling? Why is it needed in SSDs?

11.42   A hard disk is made of a material with a linear coefficient of expansion of 1.5×10^{-6}/°C. The track-to-track spacing is 1,000 nm. Assuming that the head is perfectly aligned over a track at a radius of 2 in. and that the error rate increases if the head strays 20% off track, how much temperature rise can we afford? What do you conclude from your answer?

Input/Output

"I could have been a contender."
Marlon Brando, *On the Waterfront*

"Never put off until run time what you can do at compile time."
David Gries

"Anyone can build a fast CPU. The trick is to build a fast system."
Seymour Cray

"Though this be madness, yet there is method in it."
Shakespeare, Hamlet

"Falsus in uno, falsus in omnibus."
Roman legal principle

"Nothing is certain but death, taxes – and now, apparently, the continued success of USB."
Cheryl Coupé

Up to now, we've introduced the architecture of a computer, discussed how performance can be measured and enhanced, and looked at memory systems. What we haven't thought about is how data gets into and out of the computer and how data is moved about within the computer. Before computers can process and store information, they have to read it from the world outside. Similarly, computers have to be able to transfer information to external devices. We now examine how data is moved into and out of a computer and how this data is distributed between the computer's functional parts by means of a *family of buses*.

A computer isn't just a single monolithic chunk of silicon; it's made up of interconnected subsystems, such as memory arrays, disk drives, video display systems, and one or more processors. Each of these subsystems carries out specific tasks and has to communicate with the other subsystems in the computer. We are interested in how both intra-system and extra-system communications take place. Let's start by asking whether I/O is part of computer architecture or computer organization. I/O largely belongs to *computer organization* because it is concerned with the means by which data is transferred from one place to another. Some architecture texts concentrate on the fine details of a processor's ISA or its internal organization and trip lightly over buses and I/O. Here we delve a little more deeply into I/O, because its contribution to system performance is every bit as important as the CPU. Moreover, a study of I/O includes important topics that are essential to an understanding of digital systems (for example, buffering, handshaking, and protocols).

The first part of this chapter provides an overview of I/O operations and the way in which data is moved from one point to another. The second part examines several I/O systems from the PCI bus to the USB interface. Section 12.1 looks at how computers handle I/O transactions. We employ the term *transaction* because I/O involves a dialogue between the processor and the input or output device.

The principal theme of this chapter is the *bus,* which moves data within a computer and between a computer and its external peripherals. There is little point in making CPUs and memory faster if the bus that links them can't supply the CPU with data at the required rate. The growth of the modern personal computer owes as much to bus technology as to processor development. People like performance, but they buy *functionality*. It's the fact that you can connect a computer to the Internet, to your digital camera, to multiple printers, scanners, your MP3 player, and your iPad that makes the personal computer so desirable. If you couldn't easily connect it to all of these external systems by one or more buses, the computer would remain forever a word processor, database engine, or games machine.

Today we speak of bus *architectures* and *system infrastructures* because the bus is no longer a simple collection of signal paths that link two or more parts of a computer together. Over the years, buses have evolved to provide greater functionality than before. For example, buses have to cope with multiple CPUs in a multiprocessing environment. Furthermore, a modern high-performance computer system doesn't have a single bus; it has a hierarchy of buses—each of which is optimized for a specific purpose. Some buses provide very high-speed data transfers between a CPU and its random access memory, whereas other buses send data at more leisurely rates over long paths to printers and modems.

12.1 Fundamental Principles of I/O

Let's look at the underlying principles and vocabulary concerned with I/O transactions. Figure 12.1 describes a generic system with a CPU, I/O controllers, and peripherals and a system bus that links the CPU to memory and peripherals. The word *peripheral* appears twice in Figure 12.1; it is used both to describe an external device such as a printer or a mouse connected to a computer and to describe the controller that provides an appropriate interface between the external peripheral and the CPU.

Figure 12.2 provides an alternative view of the computer system hierarchy. The processor (CPU) and memory lie at the heart of the system. The *peripheral interfaces,* connecting the processor and its memory to peripherals, are shown in two boxes: one includes *internal*

FIGURE 12.1   The processor, bus, and I/O system

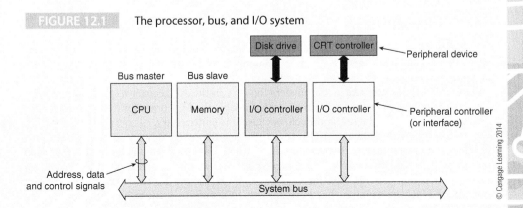

© Cengage Learning 2014

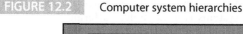

FIGURE 12.2   Computer system hierarchies

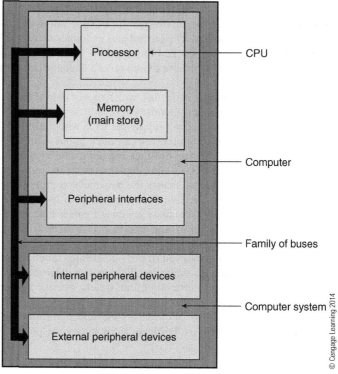

peripherals, such as disk drives, and the other includes *external* peripherals, such as modems, printers, and scanners. There is no fundamental difference between internal and external peripherals.

An I/O controller can best be thought of as a *protocol converter* because it has to conform both to the needs of the computer's bus protocol and to the needs of the external peripheral's protocol. It may convert between different data formats (for example by changing voltage levels, encoding signals, converting parallel data to serial data, and so on). I/O controllers can rival the CPU in their complexity.

Some computers employ special machine-level instructions and control signals to handle the I/O transaction. For example, a hypothetical microprocessor might send a byte in register r3 to a disk drive called *channel 5* by means of the instruction:

```
OUTPUT r3,Chan5    ;send r3 to a disk via channel 5
```

A limited set of I/O instructions is incorporated in the instruction set of Intel's processors, allowing members of the 8080 and 8086 families to transfer data between a register and an 8-bit port. For example, IN **AL**, 4CH transfers the byte at port address $4C_{16}$ to register AL, and the instruction OUT **4EH**, AL transfers the byte in the AL register to port address $4E_{16.}$ The processor generates the control signals that enable the I/O controller to detect an access. Few microprocessors take this approach, because it is unnecessary.

Processors with dedicated I/O instructions use the address bus to define a data port to which data is sent in a write cycle and from which data is obtained in a read cycle. Such processors are said to have both *memory space* and *I/O space*. Dedicated I/O architectures, such as that provided by the 8080, require specific hardware and consume instruction space (each instruction that performs an I/O operation could be doing something else).

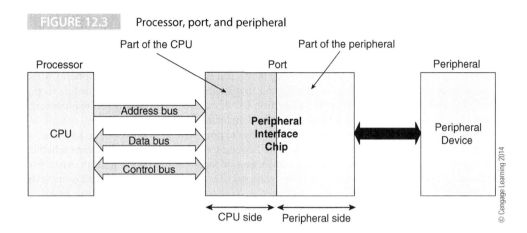

FIGURE 12.3   Processor, port, and peripheral

Memory-Mapped Peripherals

Microprocessors don't require I/O facilities because there's no fundamental difference between an I/O transaction and a memory access. Outputting a word to a peripheral is exactly the same as storing a word in memory, and getting a word from a peripheral is exactly the same as reading a word from memory. Treating I/O transactions as memory accesses is called *memory-mapped* I/O. This doesn't mean that we can forget about I/O because it's just like accessing memory, since the properties of random access memory are radically different from the properties of typical I/O systems. When implementing I/O structures, we have to take into account the characteristics of the I/O devices themselves. For example, when writing a file to a disk drive, you might have to send a new byte of data every few microseconds. Figure 12.3 shows what a typical memory-mapped I/O port (*peripheral interface chip*) looks like to the processor.

To the host CPU, this peripheral appears as the sequence of consecutive memory locations described by Figure 12.4. The left-hand side of the peripheral interface chip shaded gray in Figure 12.3 looks exactly like a memory element as far as the CPU is concerned. The other half of the peripheral interface chip, shown in blue, is the *peripheral side* that performs the specific operations required by the interface. For example, a disk controller interface might seek a particular sector, or a serial interface chip might convert a byte into a sequence of pulses that can be transmitted over a single wire. The peripheral interface chip is connected to the peripheral proper (e.g., disk drive, mouse, keyboard, or printer).

FIGURE 12.4   Memory-mapped registers

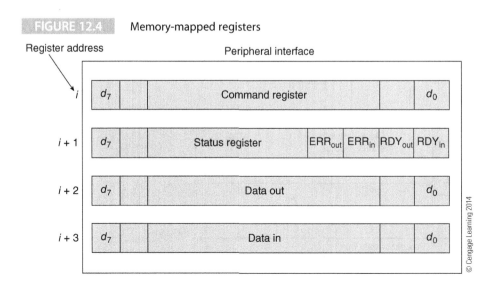

The memory-mapped port of Figure 12.4 has four consecutive registers at addresses $i, i + 1$, $i + 2$, and $i + 3$. We have assumed that the peripheral is an 8-bit device and that its consecutive locations are each separated by one byte. In a system with a 32-bit data bus, the addresses of the registers would be $i, i + 4, i + 8$, and $i + 12$. The first location at address i contains a command register that defines the operating mode and characteristics of the peripheral. Most memory-mapped I/O ports can be configured to operate in several modes, according to the specific application. By providing multifunction I/O devices, the semiconductor manufacturer makes one chip that caters to a large segment of the market.

The second location at address $i + 1$ contains the port's status, which is set up by the associated peripheral. This status information can be read by the processor to determine whether the port is ready to take part in a data transaction or whether an error condition exists. For example, a printer connected to a memory-mapped I/O port might set an error bit to indicate

The Self-Clearing Flag

The diagram below illustrates the structure of a possible self-clearing mechanism. The Q_1 output of flip-flop FF1 is set when the peripheral asserts its D_1 input in response to an event.

Assume that a peripheral has generated data and a peripheral strobe has clocked FF1 to latch the state of the data. At this point, Q_1 contains the data bit (i.e., status flag) available for reading by the host processor.

The processor reads the state of FF1 by generating the appropriate address of the memory-mapped port. A valid address enables the tristate gate and puts the output Q_1 onto one of the data bus lines for reading by the processor. The read signal also sets flip-flop FF2 when it is next clocked by a negative edge.

The Q_2 output of FF2 is connected to the D_3 input of FF3 so that Q_3 is asserted on the next rising edge of the clock. Because the output of FF3 is connected to the reset (CLR) input of flip-flop FF1, the status flag is automatically cleared when Q_3 is asserted active-high. When the read signal is negated, flip flops FF2 and FF3 return to their normal inactive-low states, and the CLR signal is removed from FF1.

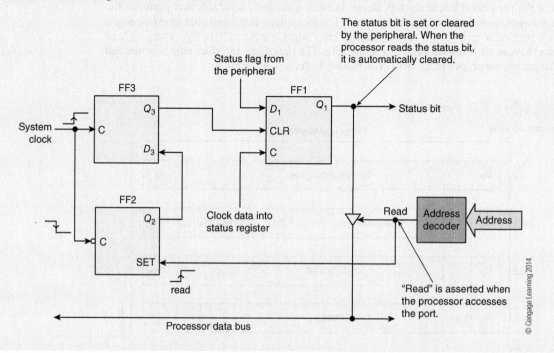

The figure below gives the timing diagram of this circuit. Self-clearing circuits have to be used carefully, because an unintended read to a flag clears it. This situation arises as a side effect under certain circumstances. For example, some processors perform a dummy access to location $l + 1$ when you access location l.

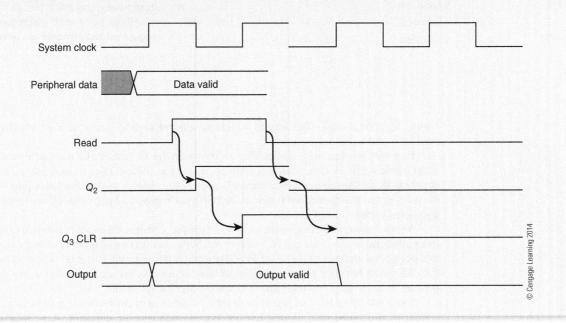

that it is out of paper. In this example, we've created generic status bits,[1] such as ERR_{out}, ERR_{in}, RDY_{out}, and RDY_{in}.

The locations at addresses at $i + 2$ and $i + 3$ are the addresses used to send data to or receive data from the peripheral, respectively. It is through these locations that the processor actually communicates with the I/O port.

Some peripherals have *self-clearing* status flags that are automatically reset. Suppose a flag, such as data ready, device busy, or data error, is set by an I/O *event*. Although you can manually clear it once you've read it, a better technique is to use a flag that is cleared when it has been read. The previous panel demonstrates the logic of the self-clearing flag.

12.1.1   Peripheral Register Addressing Mechanisms

Not all peripheral interface devices adopt the addressing model of Figure 12.4 with uniquely addressable registers. Observe that the command and data-to-peripheral registers are write-only, and the status and data-from-peripheral registers are read-only. Consequently, a single address line can distinguish between two pairs of registers (i.e., the command and status pair, and the data-in and data-out pair). In this case, the processor's read and write signals distinguish between the read-only and write-only registers.

Table 12.1 demonstrates this register-addressing scheme. The peripheral provides four internal registers, but the processor sees only two unique locations, N and $N + 4$. In this case, the CPU's $R/\overline{W}$ output is used to select one of two pairs of registers. When $R/\overline{W} = 0$, the write-only registers are selected, and when $R/\overline{W} = 1$, the read-only registers are selected.

---

[1]As in the case of a condition code register, a status bit is often referred to as a flag.

TABLE 12.1   Register Selection Using the CPU's R/W Output

Register Address	Function	CPU Address	R/W
i	Status	N	1
$i + 1$	Data out	$N + 4$	1
$i + 2$	Control	N	0
$i + 3$	Data in	$N + 4$	0

Some computers have explicit read and write signals. Other computers have a R/$\overline{W}$ (Read/NotWrite) signal that is high during a read and low during a write.

© Cengage Learning 2014

Figure 12.5 emphasizes the way in which peripheral register space can be divided into read-only and write-only regions.

You may wonder why semiconductor designers try to reduce the number of uniquely addressable registers. The reason is simple—it takes n address lines to uniquely address 2^n registers. Reducing the number of uniquely addressed registers means that fewer pins or connections to the peripheral are required. A chip's packaging and pin connections represent a significant component of its cost.

Another means of reducing the number of uniquely addressed memory locations assigned to a peripheral involves the use of a *pointer bit*. Suppose that two or more read/write registers in a peripheral share a *common address*. By using a bit in another register as a *pointer*, you can distinguish between two registers at the *same* address by associating one of them with the pointer bit set to zero and the other with the pointer bit set to one.

Some interfaces have such a large number of internal registers that it may be too expensive to provide sufficient address lines to access each register uniquely. A peripheral can employ just two addressable registers to control all of its internal accesses: a *pointer register* and a single data register. The programmer accesses the peripheral's internal registers by loading the pointer register with the offset of the required register and then reading from or writing to the single data register.

This technique requires only one address line to distinguish between internal registers but suffers a penalty in the form of a reduced access rate. Internal pointer-based addressing is cost-effective for peripherals, such as display controllers, whose configuration registers are infrequently accessed once they have been initialized.

A variation on the pointer-based addressing mode involves the use of an *automatically incrementing* internal pointer. Figure 12.6 illustrates a register file that is addressed using the output of a counter. After the peripheral interface has been reset, the internal pointer (i.e., counter) is loaded with zero. Each successive access to the interface increments the pointer and therefore selects the next register in sequence. Peripherals with auto-incrementing pointers are useful when the registers will always be accessed in sequence. For example, In a display controller, the resolution of the display, the horizontal and vertical pixel counts, and the frame rate, etc. are always loaded into a register set in strict order.

FIGURE 12.5   CPU space and register space

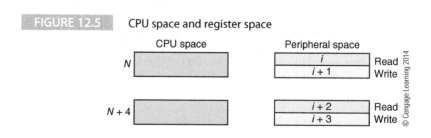

© Cengage Learning 2014

FIGURE 12.6   Accessing multiple registers via an auto-incrementing pointer

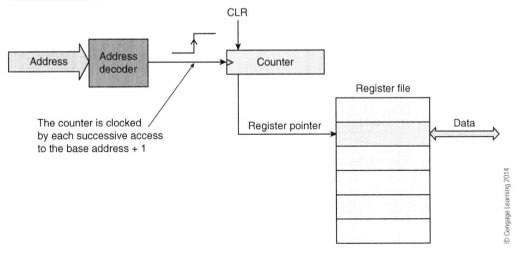

12.1.2   Peripheral Access and Bus Width

The next aspect of memory-mapped peripherals we need to examine is the relationship between the peripheral's data bus width and the host computer's data bus width. Many peripherals have 8-bit wide buses and are interfaced to computers with 32 bits.

Life is easy when 8-bit peripherals are connected to 8-bit data buses with 8-bit processors or when 16-bit peripherals are connected to 16-bit buses with 16-bit processors. Things get more complicated when low-cost, 8-bit peripherals are interfaced to 16- or 32-bit buses. Two problems can arise when you interface an 8-bit peripheral to a 16-bit bus: *endianism* and the mapping of 8-bit registers onto a processor's 16-bit address space. Consider the arrangement in Figure 12.7 where an 8-bit peripheral is interfaced to a 16-bit bus. The peripheral is

FIGURE 12.7   Eight-bit peripherals with a 16-bit bus

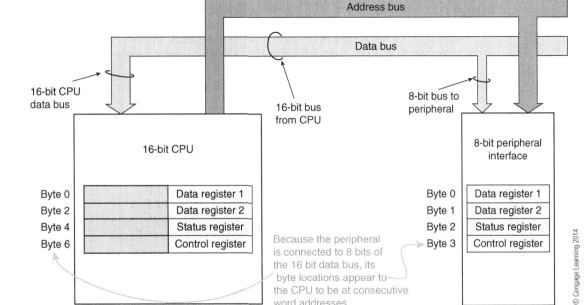

FIGURE 12.8   Mapping 8-bit peripherals onto a 16-bit bus

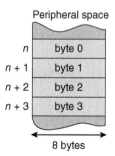

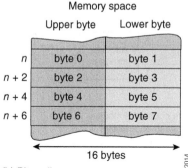

(a) Byte-addressed peripheral. Four locations occupy consecutive bytes.

(b) Bit-endian memory space. Memory locations are 16 bits wide and occupy two bytes. The peripheral's space is mapped onto consecutive odd bytes.

connected to half the bus's data lines. If the processor supports true 8-bit bus transactions, all is well, and the registers can be accessed at their byte addresses (at byte offsets 0, 1, 2, and 3).

If the processor supports only 16-bit bus operations when a 16-bit value is written to memory, all 16 bits are put on the data bus. When the processor performs a byte access, it still carries out a word access but informs the processor interface or memory that only 8 bits are to be transferred. A separate control or address signal is required to specify whether the byte being accessed is the *upper* or *lower* byte at the current address.

In this case, the peripheral is hardwired to one half of the data bus and can respond only to either odd or even byte addresses. In a big-endian environment, the peripheral would be wired to data lines [0:7] and accessed at the odd address, whereas in a little-endian environment, the peripheral would be wired to data lines [0:7] and accessed at even addresses. The peripheral's four addresses would appear to the computer at byte offsets of 0, 2, 4, and 6.

Some processors have implemented dedicated instructions to facilitate data transfer operations to byte-wide peripherals. For example, the 32-bit 68K processor has a MOVEP, meaning *move peripheral*, instruction that copies 16- or 32-bit values to or from an 8-bit memory-mapped peripheral. Figure 12.8 shows a peripheral with four internal registers together with a section of the CPU's address map, where the peripheral's data space is mapped onto *successive odd* addresses in this big-endian processor's memory space.

Figure 12.9 shows a peripheral with four 8-bit registers that are memory-mapped at address 0x08 0001. The four 8-bit registers appear to the programmer as locations

FIGURE 12.9   Example of a byte-wide, memory-mapped peripheral

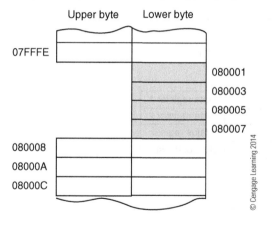

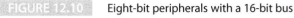

FIGURE 12.10   Eight-bit peripherals with a 16-bit bus

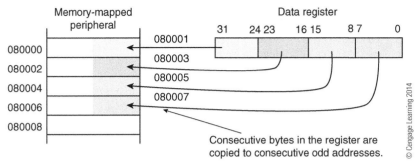

Consecutive bytes in the register are
copied to consecutive odd addresses.

© Cengage Learning 2014

0x08 0001, 0x08 0003, 0x08 0005, and 0x08 0007. Locations 0x08 0000, 0x08 0002, 0x08 0004, and 0x08 0006 do not exist and cannot be accessed. MOVEP automatically moves a 16- or 32-bit value between a data register and a byte-wide, memory-mapped peripheral. The contents of the register are moved to consecutive even (or odd) byte addresses. For example, MOVEP.L D2, (A0) copies the four bytes in register D2 to addresses [A0] + 0, [A0] + 2, [A0] + 4, and [A0] + 6, where A0 is an address or pointer register.

Figure 12.10 demonstrates how a MOVEP.L D0, (A0) instruction copies four bytes in D0 to successive odd addresses in memory, starting at location $08\ 0001_{16}$. The suffix .L in 68K code indicates a 32-bit operation, and .B indicates a byte operation. The most-significant byte in the data register is transferred to the lowest address. Without the MOVEP instruction, it would take the following code to move four bytes to a memory-mapped peripheral.

```
MOVE.L   #Peri,A0        ;A0 points to the memory-mapped peripheral
MOVE.B   D0,(6,A0)       ;Move least-significant byte of D0 to the peripheral
ROR.L    #8,D0           ;Rotate D0 to get the next 8 bits
MOVE.B   D0,(4,A0)       ;Move the next byte, bits 8 to 15, to the peripheral
ROR.L    #8,D0           ;and so on ...
MOVE.B   D0,(2,A0)
ROR.L    #8,D0
MOVE.B   D0,(0,A0)
ROR.L    #8,D0           After four rotations D0 is back to its old value
```

Each time a byte is transferred, it is copied to an address two bytes away from the previous address. The ROR.L #8,D0 instruction following each data movement rotates the 32-bit value in D0 8 bits to move the next byte into the least-significant position. By using the MOVEP instruction, the entire code can be reduced to

```
MOVE.L   #Peri,A0        ;A0 points to the memory-mapped peripheral
MOVEP.L  D0,(A0)         ;Move the longword in D0 to the peripheral
```

Notice how compact this code is. The MOVEP instruction is a nice instruction that performs a useful task. But it's not a *necessary* instruction.

Preserving Order in I/O Operations

RISC architectures provide only memory load-and-store operations and do not implement instructions that facilitate I/O operations. However, there are circumstances where RISC organization and memory-mapped I/O clash. Recall that some memory-mapped peripherals have configuration and self-resetting status registers or auto-incrementing pointers. It's important to access such peripherals in the appropriate programmer-defined sequence. Because superscalar RISC processors take an opportunistic approach to memory access, data can be stored in memory out of order. Such out-of-order memory accessing doesn't cause problems with data storage and retrieval, but it can disrupt memory-mapped I/O.

The PowerPC implements an *enforce in-order execution of I/O*, eieio, instruction that has no parameters but ensures that all memory accesses previously initiated are completed. Consider this example where two loads are followed by an addition.

```
lwz   r5,1000(r0)      ;load r5 from memory[1000]
lwz   r6,1040(r0)      ;load r6 from memory[1040]
add   r7,r5,r6         ;r7 = r5 + r6
```

When these instructions are executed, the processor may swap the order in which r5 and r6 are loaded from memory. As long as the first two loads are executed before the add instruction, the outcome is not dependent on the order of the loads. Suppose that addresses 1000 and 1040 are memory-mapped locations. If, for example, the peripheral is designed so that a read access to address 1,000 updates a register at 1,040, the sequence of the two load instructions becomes all-important, and reversing their order may lead to an incorrect result.

Consider the following example where we have to update a peripheral. Because the register is accessed via a pointer, we write the register address to the peripheral's pointer register before writing data to the register being pointed at. In this example, we want to load peripheral register number 35 with the value 99. The PowerPC code is

```
addi  r5,r0,35         ;r5 = 35 (note r0 is always 0 in the PowerPC)
addi  r6,r0,99         ;r6 = 99
stw   r5,1234(r0)      ;store 35 at memory location 1234 (the pointer)
stw   r6,5678(r0)      ;store 99 at memory location 5678
```

The two writes must be executed in the correct order. To ensure this, the PowerPC has three synchronization instructions: eieio, sync, and isync. The isync forces instructions or memory transactions to complete before continuing; that is, instructions prior to isync are executed and fetched instructions are discarded. Then, a new fetch begins. The instruction eieio forces all posted writes to complete prior to any subsequent writes. The sync instruction forces all previous reads and writes to complete on the bus before executing any instructions after it. We can ensure that the previous code runs in the correct order by inserting an eieio between the writes.

```
addi  r5,r0,35         ;r5 = 35
addi  r6,r0,99         ;r6 = 99
stw   r5,1234(r0)      ;M[1234] = 35; we're changing register 35
eieio                  ;Make sure r5 is written before proceeding
stw   r6,5678(r0)      ;M[5678] = 99; new register value is 99
```

Side Effects

Before leaving memory-mapped I/O, we must introduce the notion of instruction *side effects*. An instruction should do exactly what it is supposed to do and nothing more. Consider the 68K's CLR <ea> that clears the operand at the specified effective address; that is, [ea] ← 0. Simple, isn't it? However, the clear operation is internally implemented as

```
[Temp]  ← [ea]         ;dummy read (a side effect of the CLR)
[ea]    ← 0            ;the actual clear operation
```

The first operation is a *dummy read* followed by a memory clear. Recall that some peripherals use read and write accesses to distinguish between pairs of registers. This CLR would perform a dummy read of one register and then write zero into the other register. Is this a problem? The correct register would be cleared, and the other register at the same address would be read. Although a read operation to memory is harmless, reading a *status* register with *self-clearing* flags would clear them. Here we have an instruction that looks innocent enough but could lead to inexplicable side effects in a memory-mapped system.

12.2   Data Transfer

Three concepts are vital to an understanding of data transfer: *open-* and *closed-loop* transfers and *data buffering*. In an open-loop transfer, information is sent on its way and its correct reception is assumed. In a closed-loop transfer, the receiver actively acknowledges that the data has arrived. Data buffering is concerned with handling disparities between the rate at which data is transmitted and the rate at which it is *consumed* by the receiver.

12.2.1   Open-Loop Data Transfers

The simplest method of transmitting data is to put the data on a bus and assert a signal or *data strobe* to indicate that it is available. Figure 12.11 illustrates an *open-loop* transmission between a peripheral interface component and an external peripheral (e.g., a printer). The processor moves data to the peripheral interface with its address and data buses, and the peripheral interface puts the data on the bus.

The peripheral interface asserts a *data available* strobe ($\overline{\text{DAV}}$) to indicate to the peripheral that the data at its input terminal is valid. The peripheral reads the data, and the peripheral interface negates its $\overline{\text{DAV}}$ strobe to complete the transfer.

FIGURE 12.11   Open-loop data transfer

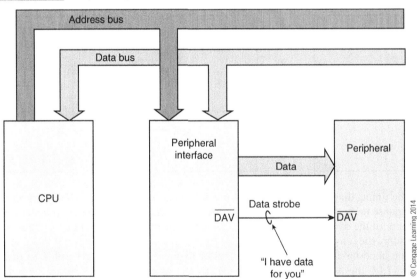

Figure 12.12 provides a timing diagram for this information exchange, which is called is *open loop* because there is no feedback to acknowledge that the data has indeed been received. If the peripheral is off line, busy, or just very slow, the data may not be read during the time for which it is available (i.e., $\overline{\text{DAV}}$ asserted). Open-loop data transfers are also called *synchronous* transfers, because the device receiving the data must be synchronized with the device sending the data.

FIGURE 12.12   Timing diagram of an open-loop data transfer

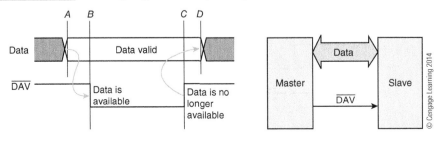

12.2.2   Closed-Loop Data Transfers

In a *closed-loop* transfer, the device receiving the data returns an acknowledgment to the sender to close the loop. Figure 12.13 is identical to Figure 12.11, except that there is a feedback path labeled *data transfer acknowledge* ($\overline{\text{ACK}}$). Figure 12.14 provides the corresponding timing diagram. The peripheral interface makes the data available and asserts $\overline{\text{DAV}}$ at *B* to indicate that the data is valid just as in an open-loop data transfer. The peripheral receiving the data sees $\overline{\text{DAV}}$ asserted and reads the data. In turn, the peripheral asserts $\overline{\text{ACK}}$ to inform the interface that the data has been accepted. The interface de-asserts (negates) $\overline{\text{DAV}}$ to complete the data exchange. This sequence of events is known as *handshaking*. Handshaking supports slow peripherals, because the transfer is held up until the peripheral indicates its readiness by asserting $\overline{\text{ACK}}$.

FIGURE 12.13   A closed-loop data transfer

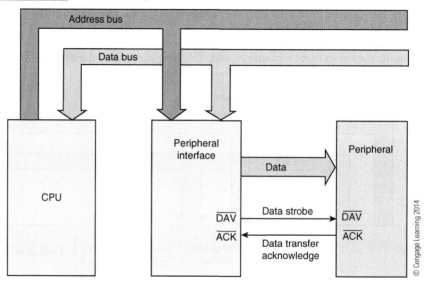

© Cengage Learning 2014

The timing diagram in Figure 12.14 is called a *handshake* because the assertion of $\overline{\text{ACK}}$ is a response to the assertion of $\overline{\text{DAV}}$. The advantage of a closed-loop data transfer is that the originator of the data knows that it has been accepted, and data cannot be lost because it was not read by the remote peripheral.

The handshaking closed-loop protocol of Figure 12.14 can be taken a step further. In Figure 12.14, the assertion of $\overline{\text{DAV}}$ is met by the assertion of $\overline{\text{ACK}}$ from the peripheral. At this point, it is assumed that the data has been received and the data exchange ends. Figure 12.15 shows a *fully interlocked handshake* in which the sequence of events is more tightly defined and each event triggers the next event in sequence.

FIGURE 12.14   Timing diagram of a closed-loop data transfer

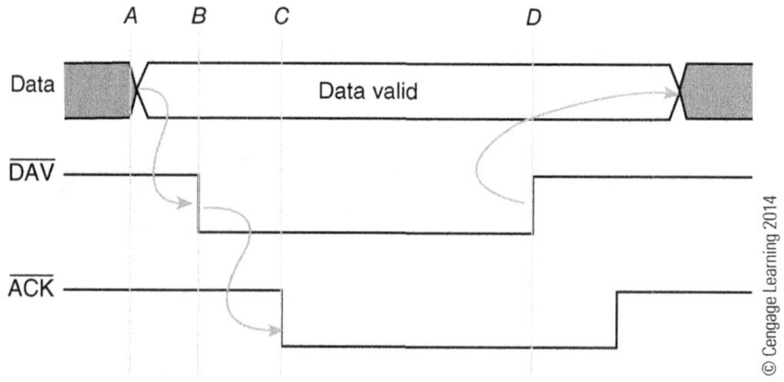

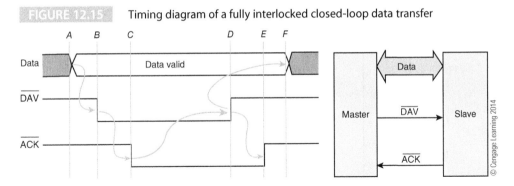

FIGURE 12.15   Timing diagram of a fully interlocked closed-loop data transfer

At point B in Figure 12.15, $\overline{\text{DAV}}$ is asserted to indicate valid data, and at point C, $\overline{\text{ACK}}$ is asserted to indicate its receipt. The sequence continues with the negation of $\overline{\text{DAV}}$ at point D. The peripheral interface can negate $\overline{\text{DAV}}$ because the peripheral's assertion of $\overline{\text{ACK}}$ indicated that $\overline{\text{DAV}}$ had been recognized. The negation of $\overline{\text{DAV}}$ indicates to the peripheral that its acknowledgement has been detected. Consequently, the peripheral negates $\overline{\text{ACK}}$ at E. The peripheral interface also removes the data at point F after negating $\overline{\text{DAV}}$. Point F may come before point E, because the removal of the data is a response to the negation of $\overline{\text{DAV}}$ rather than to the negation of $\overline{\text{ACK}}$.

In any data transfer involving handshaking, a problem arises when the transmitter asserts data available but data acknowledge isn't asserted by the receiver in turn (e.g., because the equipment is faulty). When the transmitter wishes to send data, it starts a timer concurrently with the assertion of $\overline{\text{DAV}}$. If $\overline{\text{ACK}}$ isn't asserted by the receiver after a given time has passed, the operation is aborted. The period of time between the start of an action and the declaration of a failure state is called a *timeout*. When a timeout occurs, an interrupt is generated, forcing the computer to take action.

12.2.3   Buffering Data

When data is transmitted over a bus, you either have to use it immediately, while it is valid, or capture it in a memory device. Figure 12.16 illustrates three input circuits that might be used in a peripheral. The circuit of Figure 12.16a reads the *instantaneous* values on data inputs I_0 to I_3. That is, the current data values are read and it is necessary for the transmitter to maintain the data values while they are being used. Such a simple input mechanism is rarely implemented, because it is difficult to guarantee that the data will be stable for the time it is needed.

Figure 12.16b illustrates *single-buffered* or *latched* input, where the inputs are connected to D flip-flops. When the data is to be read, the flip-flops are latched and the input captured. In this arrangement, the only requirement is that the input data be valid t_{setup} seconds before it is latched and then held for t_{hold} seconds following the clock (these are the fundamental parameters of a latch).

Single-buffered input captures data and holds it *until the next time the latches are clocked*. What happens if the rate at which data is arriving is close to the rate at which data is being read from the input latches? Suppose that new input data arrives every t_{input} seconds and the peripheral reads the data every t_{cycle} seconds. If t_{cycle} is less than t_{input}, then all is well. Suppose that the peripheral can't read data for a short time (for example, a disk drive may have to perform a new seek or a thermal recalibration operation). In this case, the incoming data is not read before the next sample arrives, and it is lost. Single-buffered input cannot be used reliably in situations where the rate of arrival can exceed the rate of removal.

Figure 12.16c provides a solution to the problem where new data arrives before the previous value has been read. Initially, incoming data is latched exactly as in Figure 12.16b. However, the data in the input latches is then copied to a second set of latches—the output latches—where it is buffered for a second time. This arrangement means that the input side of the buffer can be capturing data while the output side is waiting for the old data to be read. Of course, this arrangement fails if three data elements arrive in quick succession.

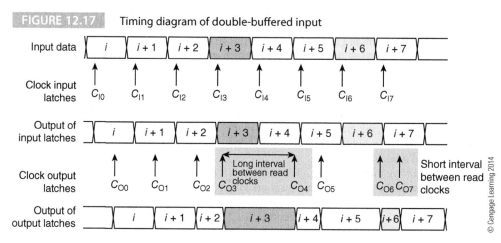

FIGURE 12.16   Buffering data

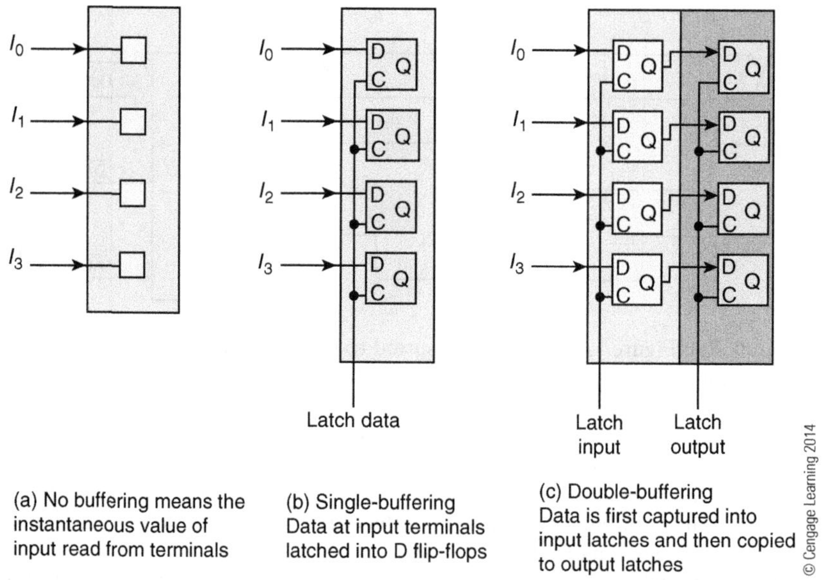

Latch data

Latch input   Latch output

(a) No buffering means the instantaneous value of input read from terminals

(b) Single-buffering Data at input terminals latched into D flip-flops

(c) Double-buffering Data is first captured into input latches and then copied to output latches

© Cengage Learning 2014

Figure 12.17 gives the timing diagram of a double-buffered input system. The input arrives at fixed time intervals. Input samples are clocked into the input latches at regular intervals by clock C_{Ii}, where i is the clock pulse number.

The data is latched into the output latches by clock C_{Oi}. In this case, the data is read at irregular intervals. Two examples are highlighted. Input sample $i + 3$ is latched early into the output latches, but is not read until very late. Similarly, input sample $i + 5$ is read late and $i + 6$ read early. As you can see, the interval between clocks C_{O3} and C_{O4} is greater than the period between two successive inputs, yet no data is lost because of the double buffering.

FIGURE 12.17   Timing diagram of double-buffered input

| Input data | i | $i+1$ | $i+2$ | $i+3$ | $i+4$ | $i+5$ | $i+6$ | $i+7$ |

Clock input latches C_{I0} C_{I1} C_{I2} C_{I3} C_{I4} C_{I5} C_{I6} C_{I7}

| Output of input latches | i | $i+1$ | $i+2$ | $i+3$ | $i+4$ | $i+5$ | $i+6$ | $i+7$ |

Clock output latches C_{O0} C_{O1} C_{O2} C_{O3} Long interval between read clocks C_{O4} C_{O5} C_{O6} C_{O7} Short interval between read clocks

| Output of output latches | i | $i+1$ | $i+2$ | $i+3$ | $i+4$ | $i+5$ | $i+6$ | $i+7$ |

© Cengage Learning 2014

The FIFO

A more general solution to data buffering is provided by the *first-in-first-out* (FIFO) memory. The FIFO is an *n*-level buffered memory that can be built into any microcomputer or peripheral's core. Data is written into a FIFO queue one value at a time and read out in the same order. Once the data has been read, it cannot be accessed again. A FIFO can be empty, partially filled, or full; they usually have output flags to indicate fully empty or partially full.

The simplest FIFO structure is a register with an input port that receives the data and an output port that provides data. The data source provides the FIFO input and a strobe. Similarly, the reader provides a strobe when it wants data from the FIFO. Figure 12.18 describes a FIFO

FIGURE 12.18   The FIFO

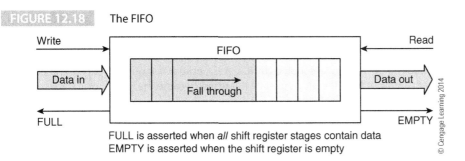

FULL is asserted when *all* shift register stages contain data
EMPTY is asserted when the shift register is empty

© Cengage Learning 2014

with two control outputs: FULL indicates that no more data can be accepted, and EMPTY indicates that no more data can be read. You can think of the register-based FIFO as a self-shifting shift register. When data arrives at the input terminals, it ripples down the shift register until it arrives at the next free location.

Figure 12.19 demonstrates a 10-stage FIFO. Initially, the FIFO contains data values 3, 9, and 7, where 7 is the oldest data element in the FIFO and 3 is the newest. In Figure 12.19b, the value 8 is written into the FIFO, and in Figure 12.19c, the value 6 is written. Each new value ripples through from the input and joins the back of the queue. In Figure 12.19d, a read takes place, and the value 7 is removed from the FIFO. At this stage, all data elements move one place right. The sequence continues with another write and a read operation. The following panel describes a FIFO that is available as a single chip.

FIGURE 12.19   Example of data movement in a FIFO

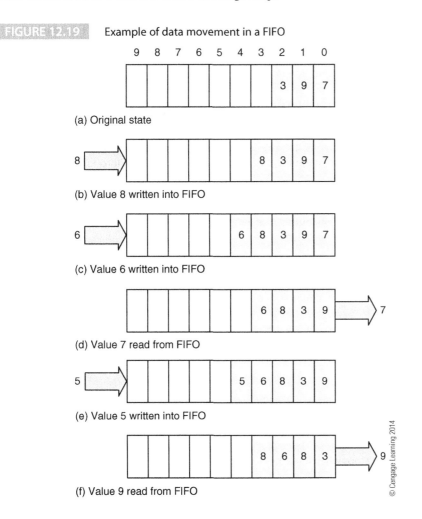

(a) Original state

(b) Value 8 written into FIFO

(c) Value 6 written into FIFO

(d) Value 7 read from FIFO

(e) Value 5 written into FIFO

(f) Value 9 read from FIFO

© Cengage Learning 2014

A FIFO

The figure below illustrates a SN74SS225 FIFO that uses a shift register. Here, we've shown only four 1-bit wide stages. The upper part shows four RS flip-flops made from cross-coupled NAND gates clocked via inputs C1, C2, C3, and C4. When one of these stages is clocked, the input from the previous state is latched. Because the inputs to all stages are X and its complement, the RS flip-flops behave like D flip-flops.

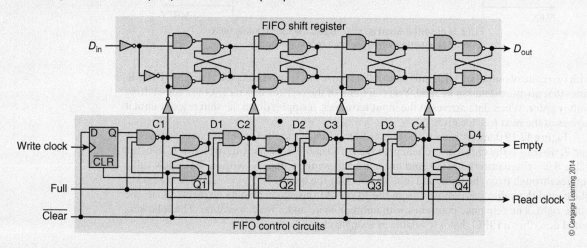

This shift register relies on an effect normally considered harmful. When a D or RS flip-flop is clocked, the data at its input ripples through to its output, and if the data changes while the flip-flop is being clocked, the data will ripple through or fall through the flip-flop. That's why master-slave and edge-triggered flip-flops were devised. However, a FIFO makes good use of the level-triggered flip-flop's ripple-through effect.

Suppose that an input is applied to the shift register composed of the four flip-flops and all clocks C1 to C4 are negated. Nothing happens. If clock input C1 is asserted, the data input is latched into the first stage. If clock C2 is asserted, the input captured by stage 1 is also captured by stage 2. In order to turn this shift register into a FIFO, all we have to do is to control the clock inputs of the four stages. If the input position is 1 and the next free location at the output end of the queue is i, all we have to do is clock stage 1, 2, …, i to cause the data to ripple through the FIFO.

The lower part of the circuit is the control section of the FIFO that has five terminals: a Clear input that initializes the system by putting the FIFO into an empty state, a Write clock input that enters data into the FIFO, a Read clock input that removes the oldest data item in the FIFO, an Empty output that indicates there is no data to read, and a Full output that indicates all stages in the FIFO are currently occupied.

The control section itself is arranged as a shift register and its outputs are the clock signals used by the FIFO. If the Q outputs of the first i stages of the control shift register are set to 1, the first i clocks are high, and input data ripples through the first i stages of the FIFO. This provides a marker that determines where the next data element is to go in the FIFO.

The key to the control shift register is its feedback mechanism. The output of each stage is fed back to the stage on the left. Consider the output stage when the FIFO is not empty. The state of D4 is 1 (i.e., not empty) and clock C4 is 1 (i.e., the output stage of the FIFO is not being clocked and data is held in the output stage). The Q4 output is the complement of D4 and is currently 0.

When the read clock is asserted low, the Q4 output will be forced high. However, because Q4 is now high, the output of the triple-input NAND gate to the final control state will go high, because one of the gate's three inputs in now high. The output of this NAND gate is fed back to the previous, *and symmetric*, control stage to behave as a write clock for that stage. Consequently, a read clock applied to the right-hand side of the control shift register ripples right-to-left through the control circuit. However, the ripple effect stops when it reaches a stage that is occupied, and its output is high.

The disadvantage of the register-based FIFO is its hardware complexity, inflexibility, and its variable-length fall-though time. When the n-stage FIFO is empty, it is going to take $n \cdot t_{stage}$ seconds for the data to ripple from the input to output stages, where t_{stage} is the ripple-through delay of a single stage. A register-based FIFO provides a cost-effective solution when the depth is small. For deeper FIFOs, alternative structures are necessary.

Three parameters define the FIFO: its width, depth, and speed. The minimum throughput time, t_{min}, is determined by the access time to read data, t_A, and the fall-through time of the FIFO, t_F. Therefore, $t_{min} = t_F + t_A$. In terms of frequency, $f_{max} = 1/(t_F + t_A)$.

FIGURE 12.20   Logical arrangement of a memory-based FIFO

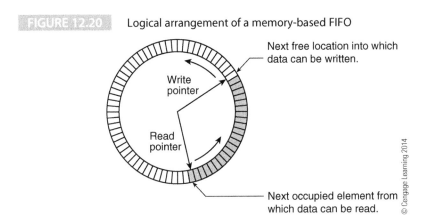

Next free location into which data can be written.

Write pointer

Read pointer

Next occupied element from which data can be read.

© Cengage Learning 2014

The FIFO is usually built around a random access memory element that is arranged as a circular buffer (Figure 12.20). A read pointer and a write pointer keep track of the data in RAM. Figure 12.21 illustrates the structure of a dual-port RAM FIFO. The advantage of the RAM-based FIFO over register-based FIFOs is that the fall-through time of a RAM-based FIFO is constant and independent of its length. This factor can be important in the case of very long FIFOs with several thousand stages.

FIGURE 12.21   Structure of a memory-based FIFO

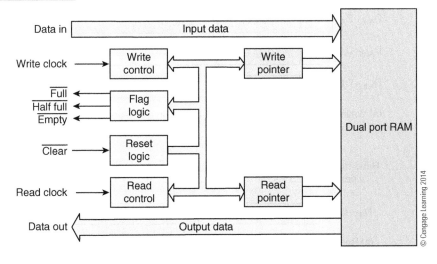

FIGURE 12.22   Using a FIFO to link systems with different bus widths and endianisms

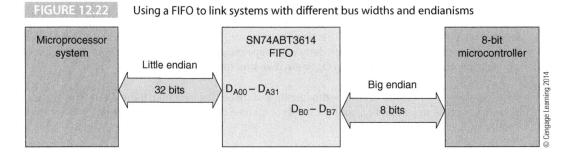

Semiconductor manufacturers make two types of RAM-based FIFO: *asynchronous* and *synchronous*. When its shift-in and shift-out signals are derived from independent sources, the FIFO is operating asynchronously. When the shift-in and shift-out signals are derived from a common clock, there is a precise, known relationship between them, and the FIFO is operating in the synchronous mode. The synchronous FIFO is preferred to the asynchronous FIFO.

FIFOs provide more than the buffering of I/O transactions. Figure 12.22 demonstrates the use of Texas Instruments FIFO in a system with a 32-bit computer using little-endian I/O and an 8-bit port using big-endian I/O. This FIFO is user-configurable and can be set up to perform *bus matching*; that is, its input and output buses may have different widths. Here, its port A interface is 32-bits wide and its port B interface is 8-bits wide. Moreover, you can program it to perform the byte swapping required when data is copied from a little-endian to a big-endian system. Figure 12.23 gives the timing diagram for the case when two 32-bit words are read into the FIFO and eight 8-bit byes are read from it.

We have now covered the vocabulary and basic strategies for moving data into and out of a computer. The final step in this introduction is to provide an overview of I/O performance.

FIGURE 12.23   Timing diagram for Figure 12.22

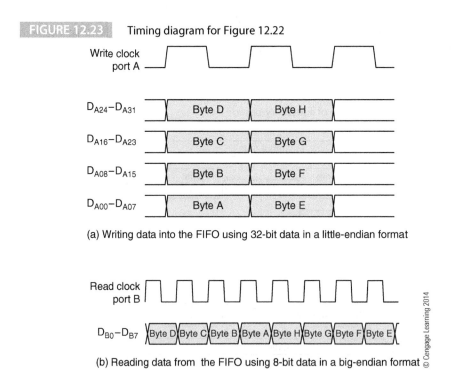

12.3 I/O Strategy

A computer implements I/O transactions in one of three ways. It can perform an individual I/O transaction at the point the operation is needed using *programmed I/O*. It can execute another task until a peripheral signals its readiness to take part in an I/O transaction using *interrupt-driven I/O*. It can ask special-purpose hardware to perform the I/O transaction by using *direct memory access* (DMA) hardware. Computer systems may employ a mixture of these strategies. We first look at simple, programmed I/O mechanisms whereby data is moved into or out of the system by executing the appropriate input or output instruction. The main part of this section describes the interrupt mechanism used by a peripheral to request a data transfer. Interrupt-driven I/O is more efficient than programmed I/O because it takes place only when the peripheral is ready. The third I/O strategy, DMA, is the most complex, because it requires a controller subsystem to take over control of the buses and initiate a data transfer between the peripheral and memory. The processor may be halted during a DMA operation or it may interleave its own operations with DMA cycles, depending on the actual configuration. For example, while the CPU is fetching instructions and data from its own internal cache, the DMA controller can be transferring data directly between main memory and a disk drive.

12.3.1   Programmed I/O

A typical memory-mapped peripheral has a flag bit that is set by the peripheral when it is ready to take part in a data transfer. In programmed I/O, the computer interrogates the peripheral's status register and proceeds when the peripheral is ready. We can express this operation in pseudocode as

```
REPEAT
     Read peripheral status
UNTIL ready
Transfer data to/from peripheral
```

The operation "REPEAT Read peripheral status UNTIL ready" constitutes a *polling loop*, because the peripheral's status is continually tested until it is ready to take part in the I/O transaction. In the following example, status bit RDY is set if the peripheral has data. If we take the I/O model of Figure 12.4 and translate the pseudocode into generic assembly language form to perform an input operation, we get

```
      ADR    r1,i0         ;Register r1 points to the peripheral
      MOV    r2,#Command   ;Define the peripheral operating mode
      STR    [r1],r2       ;Set up peripheral by loading the command
Rpt1  LDR    r3,[r1,#2]    ;Read input status word into r3
      AND    r3,r3,#1      ;Mask status to RDY_IN bit
      BEQ    Rpt1          ;Repeat until device ready
      LDR    r3,[r1,#4]    ;Read the data into r3
```

The register offsets in the peripheral are 2 and 4, rather than 1 and 2, because the processor is a byte-addressed device and we have assumed that the registers of the I/O port are 16-bits wide. In practice, the code would be more complex because most status registers include error bits. The previous code might include error checks as follows.

```
      ADR    r1,i0         ;Register r1 points to the peripheral
      MOV    r2,#Command   ;Define the peripheral operating mode
      STR    [r1],r2       ;Configure the peripheral
Rpt1  LDR    r3,[r1,#2]    ;Read input status word into r3
      MOV    r4,r3         ;Take a copy of the status word in r4
      AND    r4,r4,#Error  ;Mask status to global error bit
      BEQ    BigFault      ;Deal with the error
```

```
MOV    r4,r3                ;Restore the status for the next test
AND    r4,r4,#RDY           ;Mask status to RDY_IN bit
BNE    Rpt1                 ;Repeat until device ready
MOV    r4,r3                ;Restore the status for the next test
AND    r4,r4,#OK            ;Look for an operational error
BEQ    TinyError            ;Deal with it
LDR    r3,[r1,#4]           ;Read the data.
```

The input routine tests a status bit `Error` that is set if the port is not working correctly. Next, the ready status bit is checked. If the ready bit is set to indicate new input, another error bit, `OK`, is tested to see whether the current input is in error.

If we assume that status bits are self-clearing, we have to load the status word into a register and then copy it when we need it. Each time we test a bit of the status word, we reload it from r3.

Program-driven polled I/O isn't widely used because it's inefficient. Suppose that the average instruction takes t_{inst} seconds and an I/O transition takes place every $T_{I/O}$ s. During the I/O transaction, the processor is in a polling loop and the processor could have carried out $T_{I/O}/t_{inst}$ useful instructions. For example, if the processor carries out 10,000,000 operations/s with $t_{inst} = 100$ ns and the I/O device is a keyboard operating at $1/T_{I/O} = 10$ characters/s, the value $T_{I/O}/t_{inst}$ is 1,000,000. That is, for every input operation, the processor carries out one million instructions that perform no useful work. In the next section, we look at an I/O strategy that avoids polling by relying on a peripheral to interrupt the processor when it's ready.

12.3.2   Interrupt-driven I/O

A more efficient I/O strategy uses an *interrupt handling* mechanism to deal with I/O transactions when they occur. That is, the processor carries out another task until a peripheral requests attention. When the peripheral is ready, it *interrupts* the processor, carries out the transaction, and then returns the processor to its pre-interrupt state. Figure 12.24 describes a system using interrupt-driven I/O. The two peripheral interface components are each capable of requesting the processor's attention. Most peripherals have an *active-low* interrupt request ($\overline{IRQ}$) output that runs from peripheral to peripheral and is connected to the processor's $\overline{IRQ}$ input. *Active-low* means that a low voltage indicates the interrupt request state. The reason that the electrically low state is used as the active state is entirely because of the behavior of transistors; that is, it is an engineering consideration that dates back to the era of the open-collector circuit that could only pull a line down to zero.

Whenever a peripheral wants to take part in an I/O transaction, it asserts its $\overline{IRQ}$ output and drives the $\overline{IRQ}$ input to the CPU active low. The CPU detects that $\overline{IRQ}$ has been asserted and responds to the interrupt request if it has not been *masked*. Most processors have an *interrupt mask register* that allows you to turn off interrupts if the CPU is performing an important operation. Interrupts may be masked when the processor is performing a critical task. For example, a system using real-time monitoring of fast events would not defer to a keyboard input interrupt (even a fast typist is glacially slow compared to a computer's internal operation). Similarly, recovery from a system failure such as a loss of power will be given priority.

The way in which a processor responds to an interrupt is device-dependent. The two peripherals in Figure 12.24 are wired to the common $\overline{IRQ}$ line, and the CPU can't determine which device interrupted. The CPU identifies the interrupting device by *polling* each peripheral's status register until the interrupter has been located. Interrupt polling provides *interrupt prioritization*, because important devices whose interrupt requests must be answered rapidly are polled first.

In Figure 12.24, each memory-mapped peripheral has an *interrupt vector register* (IVR) that tells the processor how to find the appropriate interrupt handler. Typically, the IVR supplies a pointer to a table of interrupt vectors.

FIGURE 12.24   Basic structure of interrupt-driven I/O

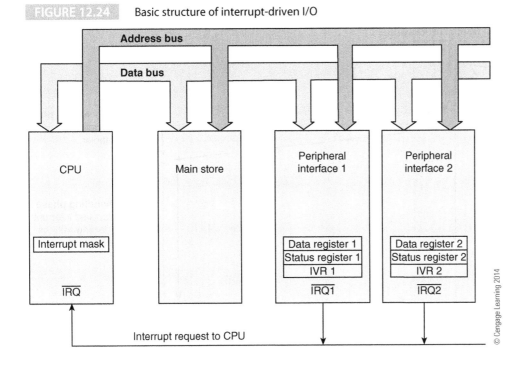

Interrupt Processing

An interrupt is an *asynchronous* event because the processor can't know when a peripheral, such as a keyboard, is going to generate an interrupt. When an interrupt occurs, the computer first decides whether to service it or whether to ignore it. The time between the CPU receiving an interrupt request and the time at which it responds is called the *interrupt latency*. When the computer responds to the interrupt, it carries out the following sequence of actions.

- It completes the current instruction. Instructions are *indivisible* and must be executed to completion.
- The contents of the program counter are saved to allow the program to continue from the point at which it was interrupted. CISC processors save the program counter on the stack so that interrupts can, themselves, be interrupted without losing their return addresses. Most RISC processors save the PC in a link register.
- The *state* of the processor must also be saved. A processor's state is defined by the flag bits of the condition code, plus other status information. Suppose an instruction sets the Z-bit and the next instruction tests the Z-bit. Clearly, if an interrupt takes place between the setting and testing of the Z-bit, the interrupt mechanism must ensure that the state of the Z-bit is unmodified.
- A jump is then made to the location of the interrupt handling routine, which is executed like any other program. After this routine has been executed, a return from interrupt is made, the program counter restored, and the system status word returned to its preinterrupt value.

Figure 12.25 illustrates how a typical CISC processor responds to an interrupt request. The action *Stack PSR* indicates that the processor status register (PSR) is pushed on the stack. The interrupt is transparent to the interrupted program, and the processor must be returned to the state it was in immediately before the interrupt took place. We now briefly define some of the key concepts used in any discussion of interrupts and exceptions.

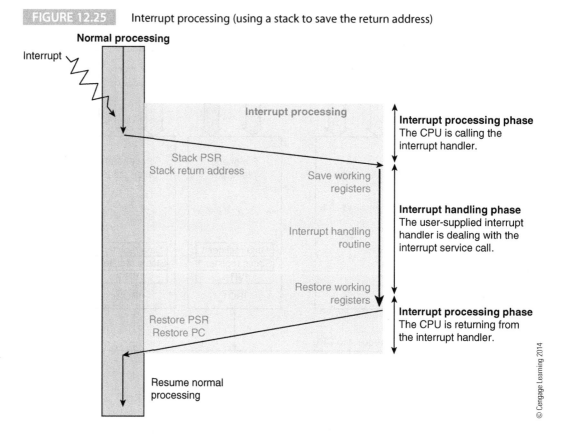

FIGURE 12.25   Interrupt processing (using a stack to save the return address)

Nonmaskable Interrupts

An interrupt *request* is so-called because it implies that it may be denied or *deferred*. Sometimes it may be necessary for the computer to respond to an interrupt no matter what it is doing. Some microprocessors have a *nonmaskable* interrupt request (NMI) that can't be deferred and must always be serviced. A nonmaskable interrupt is reserved for events such as a loss of power. In this case, a low-voltage detector generates a nonmaskable interrupt as soon as the power begins to decay. The NMI handler routine forces the processor to deal with the interrupt and to perform an orderly shutdown of the system before the power drops below a critical level and the computer fails completely.

An environment in which more than one device may issue an interrupt request requires a mechanism to distinguish between an important interrupt and a less important one. For example, if a disk-drive controller generates an interrupt because it has some data ready to be read by the processor, the interrupt must be serviced before the data is lost and replaced by new data from the disk drive. On the other hand, an interrupt generated by a keyboard interface probably has from 200 ms to several seconds before it must be serviced. Therefore, a request for attention from a keyboard can be forced to wait if interrupts from devices requiring immediate servicing are pending.

Prioritized Interrupts

Microprocessors often support prioritized interrupts (i.e., the chip has more than one interrupt request input). Each interrupt has a predefined priority, and a new interrupt with a priority lower than or equal to the current one cannot interrupt the processor until the current interrupt has been dealt with. Equally, an interrupt with a higher priority can interrupt the current interrupt. We will look at the IA32's interrupt handling mechanisms later.

Interrupts and Exceptions

In this chapter, we are interested in interrupts, which are requests for attention from peripherals. Interrupts are most frequently associated with I/O operations. However, interrupts can be generated by other hardware events, such as the timeout of a timer (e.g., a periodic interrupt that can be used to switch tasks in a multitasking system).

Interrupts can be used to deal with unusual events, such as failure of a subsystem or loss of power.

Such hardware interrupts are part of a larger class of events called *exceptions*. The interrupt, reset, and page-fault are all exceptions that originate in hardware, although the reset is unusual because there is no such thing as a *return from reset*. As well as hardware exceptions, there are software exceptions—these have their origin in the processor's software. Software exceptions behave exactly like hardware exceptions, where the only difference is that the origin is in the software and the address of the appropriate exception handler is automatically supplied (unlike hardware interrupts, where either the device supplies a vector or the processor hunts for the address in a polling loop).

There are two types of software exception: processor initiated and programmer initiated. Processor initiated exceptions are invariably due to an error. For example, these exceptions can be an attempt to divide by zero, the attempted execution of an illegal instruction (non-valid op-code), a privilege violation (a user attempting an operation reserved for the operating system), or a misaligned access (accessing a 32-bit word at an odd byte address).

User-generated software exceptions include system calls and emulator calls. In a system call, a special instruction called a supervisor call or trap is inserted in the code to call a function that is to be performed by the operating system, such as a console I/O transaction or a disk read/write. The difference between a trap and a subroutine is that a trap does not require an explicit target address like a subroutine, because the trap's target is built into the hardware of the processor. Moreover, traps are portable and can be used in any system with the same operating system.

The emulator trap provides a means of executing instructions that haven't yet been designed (or are not built into the actual chip running the code). For example, a low-cost processor may lack a hardware floating-point unit. Emulator traps can be used to replace floating-point instructions. When the processor encounters an emulated instruction, the operating system is called, and the instruction is emulated in software. This mechanism can be used to ensure software compatibility across a family of processors.

Nested Interrupts

Interrupts and other processor exceptions have all of the characteristics of a subroutine, where the return address is stacked at the beginning of the call and then restored once the subroutine has been executed to completion. The interrupt is a subroutine call with an automatic target address supplied in hardware or software and a mechanism that preserves the state of the condition code as well as the program counter.

Just as subroutines can be nested, so can interrupts. Figure 12.26 demonstrates how a level 1 interrupt is called and processed and a return is made to normal processing. A level 1 interrupt occurs a second time. However, in this case, a level 2 interrupt takes place before the level 1 interrupt handler has completed its task. In this case, the level 1 interrupt handler is interrupted, and the level 2 interrupt processed. Once the level 2 interrupt has been dealt with, a return is made to the level 1 interrupt handler, and this interrupted interrupt is completed.

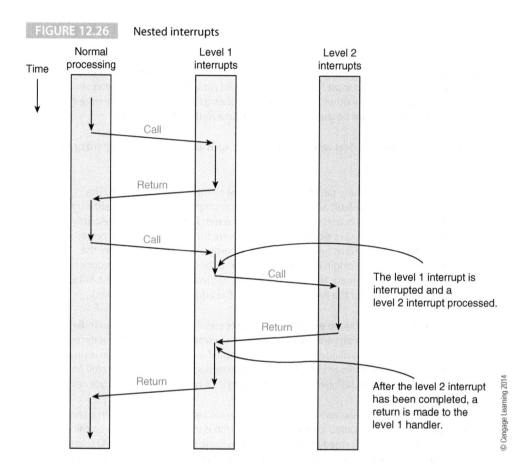

FIGURE 12.26   Nested interrupts

Consider the following example of interrupt prioritization and nested interrupts. A system has six levels of interrupt prioritization and an interrupt may occur every 10 μs. We will assume that all interrupts take 15 μs to complete. Table 12.2 gives the sequence of interrupts in terms of their priority, and Figure 12.27 demonstrates how they are executed (in this rather hypothetical example).

TABLE 12.2	Example of Prioritized Interrupts
Time Slot	Interrupt Level
0	None
10 μs	4
20 μs	5
30 μs	None
40 μs	3
50 μs	2
60 μs	4
70 μs	5
80 μs	None
90 μs	1
100 μs	None
110 μs	None

FIGURE 12.27   Example of interrupt prioritization

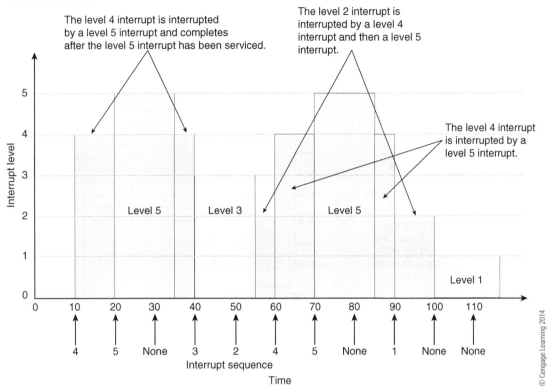

Vectored Interrupts

When a processor with a single interrupt request line detects a request for service, it doesn't know which device made the request and can't begin to execute the appropriate interrupt handler until it has identified the source of the interrupt. A *vectored interrupt* solves the problem of identifying the source by forcing the requesting device to identify itself to the processor. Without vectored interrupts, the processor must examine each of the peripherals' interrupt status bits.

When the processor detects an interrupt request, it *broadcasts* an *interrupt acknowledge* to all potential interrupters. Each possible interrupter detects the acknowledge from the CPU, and the interrupting device returns a *vector* that is used by the CPU to invoke the appropriate interrupt handler. Figure 12.28 demonstrates how the 68K family implements *prioritized, vectored* interrupts. There are seven levels of interrupt requests, and level i is serviced in preference to level j if $i > j$. The scheme permits *nested* interrupts so that an interrupt at level i can be interrupted by a new interrupt at level j only if $j > i$.

The $\overline{IRQ}$ output from each peripheral is wired to one of the seven levels of interrupt request $\overline{IRQ1}$ to $\overline{IRQ7}$. The level at which a peripheral generates an interrupt is user-defined. A priority encoder converts interrupt request inputs into a 3-bit code on the CPU's interrupt priority level pins $\overline{IPL0}$ to $\overline{IPL2}$. The CPU compares the level of the requested input with the interrupt mask in its status register, i_0 to i_2. If the interrupt mask is below the level of the interrupt, the interrupt is processed. The CPU acknowledges the interrupt by setting function control outputs FC2, FC1, FC0 to 1, 1, 1 and by putting the level of the interrupt being acknowledged on address bus lines A_3, A_2, A_1. External logic decodes the function code, and these three address lines to generate seven levels of interrupt acknowledge, $\overline{IACK1}$ to $\overline{IACK7}$.

Suppose peripherals at interrupt levels 2 and 6 simultaneously request service. $\overline{IRQ2}$ and $\overline{IRQ6}$ are asserted at the same time. The priority encoder logic selects $\overline{IRQ6}$ and sends the code 6 to the CPU. If the interrupt mask is set to 5 or below, an interrupt request is signaled on FC2, FC1, FC0, and the value 6 is placed on the three lower-order address lines so that $\overline{IACK6}$ is asserted.

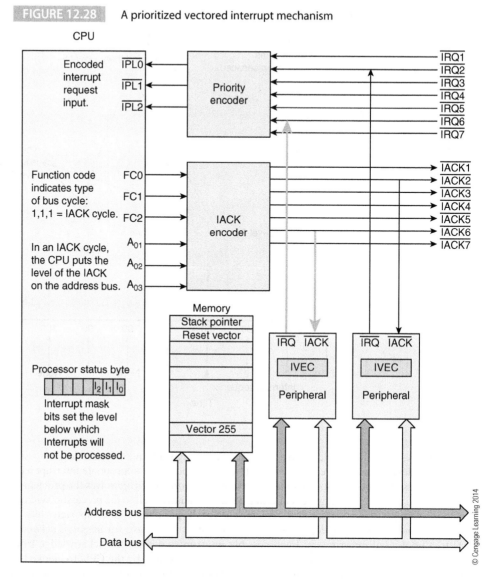

FIGURE 12.28   A prioritized vectored interrupt mechanism

The peripheral requesting service at level 6 detects that $\overline{\text{IACK6}}$ has been asserted and responds to the interrupt acknowledge by putting the contents of its 8-bit interrupt vector register on the data bus. For example, the peripheral might supply the value 64_{10} to the CPU.

The CPU reads this *interrupt vector number* from the interrupting peripheral and multiplies it by 4 to create a 32-bit address in the range from 0000_{16} to 00FF_{16}. The CPU then uses this 32-bit value to index into a table of pointers to interrupt handlers in memory; in this example, the interrupt number 64_{10} would be multiplied by 4 to get 0100_{16}. The contents of this location represent the first location of the interrupt handler for this peripheral and are read from memory and loaded into the program counter.

Interrupt Timing

Interrupt processing requires an overhead: namely, the time to recognize and respond to the interrupt (*interrupt latency*), the time taken to call the appropriate interrupt handler, and the time taken to return from the interrupt handler to the interrupted program.

In a high-performance workstation environment, the function of the interrupt can be performed by other mechanisms such as dedicated I/O systems with their own processors or message-passing mechanisms on buses. However, a processor's interrupt-handling structure is often of vital importance in embedded applications where a processor may be monitoring many real-time systems. We have already seen that the ARM family implements a special *fast interrupt request* (FIQ) mode that provides the interrupt handler with a new set of registers, avoids the time lost saving working registers on the stack at the start of interrupt processing, and then restores them prior to a return from interrupt.

If the time taken to begin interrupt processing is t_{int} and the time taken to return from an interrupt is t_{ret}, the total time taken to perform an I/O transaction is $t_{int} + t_{I/O} + t_{ret}$. The I/O efficiency is $t_{I/O}/(t_{int} + t_{I/O} + t_{ret})$. The true situation may be more pessimistic because interrupt handling is carried out by the operating system. Consequently, the time taken to process an interrupt might be much longer than the time taken by the CPU itself to process the interrupt because of the operating system overhead.

Interrupts and the ARM

The ARM supports the seven forms of exception described below. There are three hardware-initiated exceptions (interrupts). A reset occurs at power up or after an unrecoverable crash. An interrupt request occurs in response to the IRQ input asserted, and a fast interrupt request occurs in response to FIQ asserted. The other exceptions are an access to invalid memory, a breakpoint, a software interrupt, and an invalid op-code (i.e., trying to execute data). Each of these interrupts is associated with a specific priority, so that if two exceptions occur simultaneously, the one with the highest priority wins.

1. **Reset**—priority 1 supervisor mode, vector 0x00.
2. **Invalid Address Access**—priority 2 abort mode, vector 0x10.
3. **FIQ**—priority 3, FIQ mode, vector 0x1C.
4. **IRQ**—priority 4, IRQ mode, vector 0x18.
5. **Breakpoint**—priority 5, Abort mode, vector 0x0C.
6. **SWI**—priority 6, supervisor mode, vector 0x08.
7. **Illegal Instruction**—priority 6, undefined mode, vector 0x04.

The ARM stores the entry instruction (which is usually a branch or jump to the actual exception handler) of the appropriate exception handler at the stated address in memory. For example, the reset address (initial pointer into the bootstrap loader) is at 0x00000000.

Recall that the ARM responds to exceptions by executing a mode change, in which the status word and the return address in r14 and r13 are saved. New shadow, r14, and r13 registers are then created for the current exception. These new registers are called banked registers in ARM terminology. The fast interrupt request banks registers r8 to r12 as well as r13 and r14.

The processor status register (PSR) has two interrupt control bits: the IRQ bit enables and disables interrupt requests and the FIQ bit enables and disables *fast* interrupt requests. When ARM is operating in its user mode, the only way it can enter the privileged mode is via an SWI instruction or an exception.

Consider this interrupt request. An IRQ is detected by a low-to-high electrical transition on the IRQ pin. The interrupt is processed if the interrupt request bit in the CSR is enabled, and the IRQ mask bit is disabled to prevent further interrupts. The current value of the processor status register is saved and new R13 and r14 registers switched in. The appropriate interrupt handler is then executed, and a return from interrupt made.

Nested interrupts are possible in the interrupt handler if you save the current values of r13 and r14, the status register, and re-enable the IRQ bit.

Interrupt Handling in the PC Environment

This panel provides an overview of interrupt handling in PCs. Early PCs, employing 8086 16-bit microprocessors, supported a less sophisticated interrupt-handling system than the 68K family. The 8086 provided a single interrupt request input (INTR). When INTR is asserted, the processor finishes executing the current instruction and outputs two interrupt acknowledge pulses on its INTA pin if its interrupt enable flag is set.

The 8086's primitive interrupt processing mechanism was enhanced by an external chip, the 8259A programmable *peripheral interrupt controller* (PIC). The PIC operates in conjunction with the CPU and the PC system to provide a prioritized and vectored interrupt-handling system, as illustrated below.

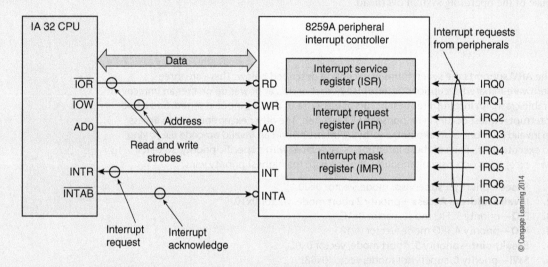

Interrupt request lines IRQ0 to IRQ7 are connected to the PIC's 8-bit interrupt request register (IRR). Asserting interrupt request line IR$_i$ sets bit IRR$_i$. The PIC is programmable, allowing the user to mask any of the interrupt request bits. The PIC can implement a *round-robin* rotating priority in which all interrupt levels get a fair access to the processor, because the interrupt currently being serviced goes to the end of the line in the next round of arbitration.

When an interrupt request is pending, the INT output (connected to the CPU's INTR input) is asserted to request attention. When the PIC receives the first INTA acknowledge pulse from the CPU, it clears the highest priority bit in the IRR register and sets the corresponding bit in its interrupt in-service register (ISR). The highest priority request is removed from the queue in the interrupt request register and passed to the interrupt service register. The remaining interrupts are pending, and it is possible to receive another interrupt at the same level of priority while the current interrupt is being processed. The second interrupt acknowledge pulse from the CPU forces the PIC to place an 8-bit interrupt number on the data bus, which the CPU uses to call the appropriate interrupt handler.

Bit ISR$_i$ has to be reset to indicate that the current interrupt has been serviced. You can select two ways of clearing ISR$_i$. In the *automatic end of interrupt mode*, the ISR bit is self-clearing and is reset at the end of the second INTA pulse. In the manual mode, the CPU must clear the IRS bit at the end of the interrupt handling routine. Note that the CPU can choose to clear the highest in-service bit of those set or a specific in-service bit. If nested interrupts are implemented, the highest level in-service bit should be cleared, because this corresponds to the interrupt that was most recently acknowledged and serviced.

The PIC and CPU combination fully supports nested interrupts. If an interrupt request whose priority is lower than the interrupt being serviced occurs, it is ignored and remains pending. If however, an interrupt with a higher priority is received, this will be processed before the current interrupt request has been serviced. An interrupt can't be interrupted until the INTA sequence has been completed.

The 8259A is cascadable, which means that each of the PIC's eight interrupt request inputs can be connected to the INT output of another PIC; that is, you could connect eight PICs to one PIC to allow 8 x 8 = 256 prioritized interrupt levels. The 8259A PIC is a legacy device that is used in conjunction with older PC technology, such as the now obsolete ISA bus (the PIC controls interrupt requests from devices on the ISA bus). Today's PC's use sophisticated bus controller subsystems that deal with interrupt processing.

12.3.3   Direct Memory Access

The most sophisticated means of dealing with IO uses *direct memory access* (DMA) in which data is transferred between a peripheral and memory without the active intervention of a processor. In effect, a dedicated processor performs the I/O transaction by taking control of the system buses and using them to move data directly between a peripheral and the memory. DMA offers a very efficient means of data transfer, because the DMA logic is dedicated to I/O processing and a large quantity of data can be transferred in a burst (for example, 128 bytes of input).

Figure 12.29 describes a system that uses DMA to transfer data to disks. A *DMA controller* (DMAC) controls access to the data bus. Suppose that a block of data is to be copied from the disk to the processor's memory. The DMA controller must first be loaded with the destination of the data in memory and the number of bytes to be transferred; that is, you have to program the DMA controller before it can be triggered. In practice, the DMA controller is configured by the operating system.

FIGURE 12.29   Overview of DMA logic

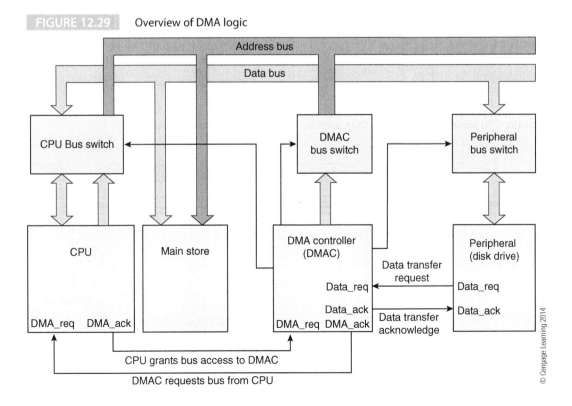

© Cengage Learning 2014

Figure 12.29 shows the address and data paths between the CPU and memory. Three bus *switches* control access to the data bus by the CPU, memory, and DMA controller. A bus switch (usually tristate gates) can be turned on or off to enable or disable the information path between the bus and the device interfaced to the bus switch. Normally, the CPU bus switch is closed, and the DMAC and peripheral bus switches are open. The CPU transfers data between memory and itself by putting an address on the address bus and reading or writing data. Figure 12.30a illustrates the situation in which the CPU is controlling the buses, and Figure 12.30b demonstrates how the DMA controller takes control of the data bus to perform the data transfer itself.

There is no real distinction between I/O by DMA and a *multiprocessor* system. A DMA controller is little more than a processor with a fixed instruction set that copies blocks of data between peripherals and memory. You could say that a DMA controller is a *string processor* with two instructions: MOVE string, **Memory** and MOVE Memory, **string**. Having described the strategies for data transfer, the next step is to examine some of the factors involved in point-to-point data transmission; that is, the terminology and protocols.

FIGURE 12.30   Data flow in normal and DMA cycles

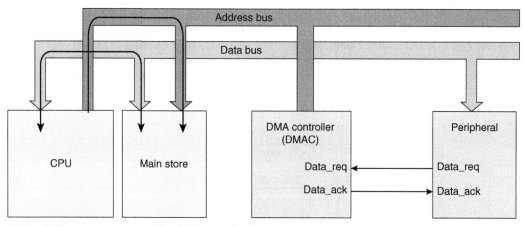

(a) The CPU accesses memory. The DMA controller is not connected to the address bus

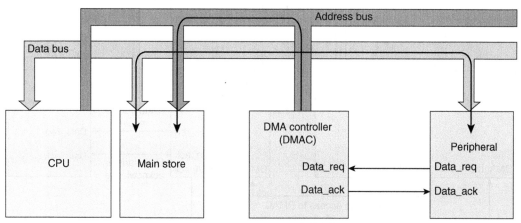

(b) The CPU is disconnected from the address and data buses. The DMA controller supplies an address and the peripheral drives the data bus

12.4   Performance of I/O Systems

We now look at the impact of I/O on computer performance. Since a computer is either performing calculations or accessing peripherals, we can say that the total time required to execute a job is expressed as

$$\text{Total time} = \text{CPU time} + \text{I/O time}$$

This expression is simplistic, because I/O operations can often be overlapped with computational operations. For example, a multimedia system may be reading data from a video camera, processing the data previously read, and storing the data that was last processed on disk.

Some computer applications such as weather simulation are called *CPU intensive*, because they require much CPU time and relatively little I/O time. Other applications are I/O intensive (for example, downloading a movie from the Internet).

Even the fastest of disk drives are orders of magnitude slower than CPUs. Moreover, the performance of processors is increasing faster than the performance of I/O subsystems. For example, the access times of hard disks has increased minimally over the years, whereas processor speeds might increase by 10 to 40% in a year.

Computer designers are interested in improving I/O performance for several reasons. The I/O bottleneck is becoming more and more significant as multimedia applications are developed. Ironically, people can tolerate a moment's delay in loading a database, spreadsheet, or document, but even a short pause of 1/50 s in playing music or showing a video is noticeable.

Consider the following hypothetical example. Suppose that, currently, a multimedia application is 70% computation intensive and takes up 60% of a computer's I/O capacity. Assume that this application will demand 20% more resources (both CPU and I/O) each year as it is updated and new versions are released. So the computer's processing performance increases by 40% per year, and the computer's I/O performance increases by 4%. How will the situation develop over the next five years? Table 12.3 demonstrates how the situation changes with time.

TABLE 12.3   Relative Growth of Computer and I/O Power and Demands

Year	Application CPU	Application I/O	Computer CPU	Computer I/O	Headroom CPU	Headroom I/O
0	70	30	100	50	30	20
1	84	36	140	52	56	16
2	100.8	43.2	196	54.08	95.2	10.88
3	120.96	51.84	274.4	56.24	153.5	4.4
4	145.15	62.21	384.16	58.49	239.01	none
5	174.18	74.65	537.82	60.83	363.64	none

© Cengage Learning 2014

In the second year, the demand for performance is 84 units CPU and 36 units I/O. During this time, the computer develops and the CPU can deliver 140 units and its I/O 52 units. As you can see, the computer performance outstrips the demand for computation and all is well. However, the increasing demand of I/O performance causes the computer to gradually lag behind the requirements until the computer can no longer run this application.

Sheer, sustained data-transfer speeds are not the only issue in the performance of I/O systems. More than anywhere else in computer systems, *latency* becomes a major factor in the performance of I/O systems. Consider the interrupt, which has a set-up time (call and stack return address), a processing time, and a return time (unstack the return address and restore system status).

Suppose a computer has an interrupt set-up and return time of L(i.e., the interrupt overhead) and executes instructions at the average rate of p instructions/s. If an interrupt handler has n instructions, the efficiency of the interrupt system is

$$E = \frac{\dfrac{100n}{p}}{L + \dfrac{n}{p}} = \frac{100n}{pL + n}$$

If the value of L is large or the value of n small, the efficiency of the interrupt mechanism is low. In a system with few interrupts, this is not a major concern. However, in a system with frequent interrupts, such as from an embedded real-time controller, interrupt efficiency may be the dominating factor in systems design.

Suppose that a computer is processing audio data that's being received from a disk drive in 4096-byte blocks. Data is read from the disk at an average rate of 20 Mbits/s, and each data block requires an interrupt processing overhead time of 100 μs. How efficient is the data transfer? The time taken to read a block is $4096 \times 8 \times 0.05 \; \mu s = 1638.4 \; \mu s$. Each block requires a 100 μs overhead, corresponding to an efficiency of $100 \times 1638.4/(100 + 1638.4) = 94.2\%$.

Now that we have covered the basis of input/output, the next step is to examine how data is moved between both internal subsystems (memory and CPU) and between the computer and external peripherals.

12.5   The Bus

We now discuss the bus, or rather *the family of buses*, that are essential to the operation and performance of all modern computers. The term *bus* is a contraction of the Latin word *omnibus*, which means *for all*. A bus behaves like a highway in that it is used by multiple devices. In a computer, all of the devices that wish to communicate with each other use a bus. We begin our discussion with the high-speed internal bus found inside the computer before introducing the family of serial buses that connect the computer to external devices such as printers.

Figure 12.31 illustrates the organization of a computer with three buses. The *system* bus of is made up of the address, data, and control paths from the CPU. Memory and memory-mapped I/O devices are connected to this bus. Such a bus has to be able to operate at the speed of the fastest device connected to it—normally the main store. The system bus demonstrates that a *one size fits all approach* does not apply to computer design because it would be hopelessly cost-ineffective to interface low-cost, low-speed peripherals connected to a high-speed bus. Later we shall see how low-cost buses arose because of the need to connect low-cost peripherals.

In this chapter, we look at the organization or *topologies* of buses; that is, we describe how they connect functional units together and how different buses are themselves interconnected. We also discuss the role of the signals that are required to implement a meaningful dialog between the various systems using the bus and look at the protocols and timing of these signals.

In systems with more than one CPU (or at least more than one device that can *initiate data transfer actions* like a CPU), the bus has to decide which of the devices that want to access the bus should be granted access to it. This mechanism is called *arbitration* and is a key feature of modern system buses. We look at the principles governing arbitration and the way in which buses implement arbitration.

A device that can take control of the system bus is called a *bus master*, and a device that can only respond to a transaction initiated by a remote bus master is called a *bus slave*. In Figure 12.31, the CPU is a bus master, and the memory system is a bus slave. One of the I/O ports has been labeled *bus master* because it can control the bus (e.g., for DMA data transfers), whereas the other peripheral is labeled *bus slave* because it can respond only to read or write accesses. The connection between the disk drive and its controller is also labeled *bus* because it represents a specialized and highly dedicated example of the bus. The CPU is

FIGURE 12.31   The bus

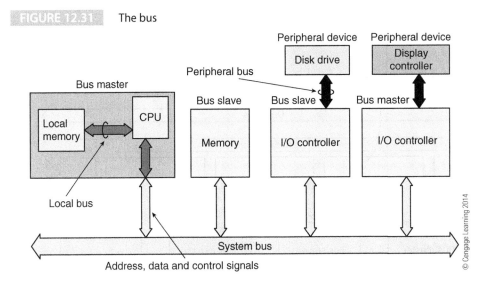

shown in a box with memory and a *local bus*. Some computer systems are arranged so that the CPU is directly connected to a memory system on the same card via a local bus. This arrangement means that the CPU does not have to access the system bus for all of its data.

We have pointed out that one of the reasons for the rise of the personal computer is the ease with which it can be interfaced to systems as diverse as the high-speed hard disk, the video camera, the keyboard, or the mouse. Consequently, we cover some of the peripheral buses that are specifically designed to connect a processor to external subsystems. These buses include the SCSI bus, FireWire, and the Ethernet.

We briefly mention one of the unsung heroes of the computer bus: *plug-and-play*. In the 1970s and 1980s, connecting a new peripheral to a computer was a veritable nightmare. You had to worry about the details of the physical connection, the hardware configuration of the computer, and the hardware configuration of the peripheral. You had to assign address space to the peripheral, interrupt request numbers, DMA channels, handshake mechanisms, and so on. Today we largely take for granted technologies such as plug-and-play, where the peripheral and computer automatically negotiate with each other and assign resources to the peripheral that don't conflict with any other peripheral's resources.

12.5.1   Bus Structures and Topologies

A simple bus structure is illustrated by the *CPU plus memory plus local bus* in Figure 12.32. The address bus, data bus, and control bus are often lumped together, and the whole structure is called *the bus*. Only one device at a time can put data on the data bus. A simple structure like this is found in many microcontrollers and first-generation microcomputers. Data is transferred between CPU and memory or peripherals such as input/output devices and disk systems. The CPU may be the permanent bus *master,* and only the CPU can put data on the bus or invite memory/peripherals to supply data via the bus. Note that a bus master may not necessarily use the data bus itself; it may take control of the bus on behalf of some other agent.

Figure 12.32 demonstrates the subdivision of a bus into units, an address bus that specifies a location in memory that is to be read from or written to, a data bus that copies data from one point in the system to another, and a control bus. The term *control bus* is used ambiguously. Sometimes the control bus refers to the signals that control the flow of information during a read or write cycle. Sometimes the control bus refers to an entirely separate sub-bus that performs special functions, such as interrupt control or arbitration.

High-performance computer systems, such as the PC, have a more complex bus structure than that of Figure 12.31. Figure 12.33 illustrates a bus structure that employs two buses

FIGURE 12.32   The structure of a general-purpose bus

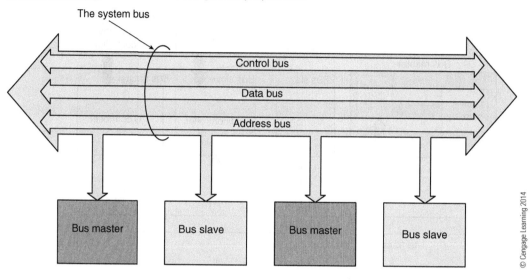

FIGURE 12.33   The structure of a general-purpose bus

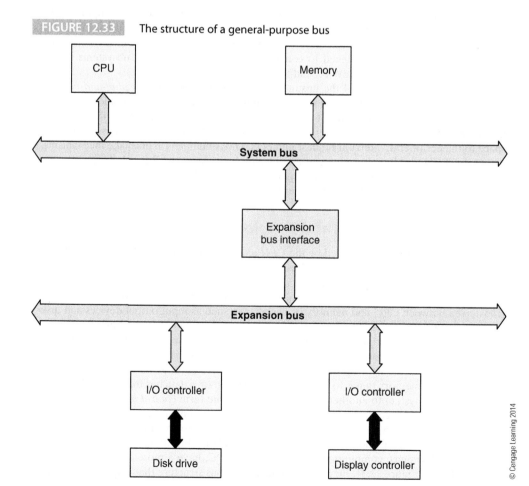

linked by an *expansion interface*. Each of these separate bus systems may have entirely different levels of functionality: one might be optimized for high-speed processor-to-memory transactions and the other to support a large range of plug-in peripherals. The PC's expansion bus was originally intended to allow low-cost peripherals to be plugged into the PC to create an open system and provide flexibility.

12.5.2   The Structure of a Bus

When two entities engage in an exchange of information via a bus, one of them initiates and controls the exchange of data. The system that controls the bus is called the *bus master,* and the system that is accessed by the bus master is known as the *bus slave.* At any instant, there can be only *one* active bus master, because today's buses cannot support multiple simultaneous transactions. A bus can support multiple bus slaves simultaneously if the bus master *broadcasts* information to several slaves. This description is simplified. Modern buses like the PCI bus increase throughput by allowing *split transactions* in which one bus master accesses the bus and then another bus master uses the bus before the first bus master has completed its transaction. Consequently, two or more bus masters can be active during a data transfer, and their data transfers overlap. No two bus masters can, of course, drive the bus at the same instant.

We have already seen that a bus can be viewed as a number of sub-buses: an *address bus* that specifies the source or destination of the data to be transferred, a *data bus* that transfers the data, and a *control bus* that is responsible for controlling the flow of data on the bus. The control bus is made up of a number of individual components. For example, the control bus controls the transfer of data, is used by potential bus masters to gain control of the address and data buses, and is used to implement interrupt handling mechanisms. We now look at each of these sub-buses in greater detail

PC Bus History

The original PC expansion bus provided an 8-bit data highway operating at what would now be considered the derisory clock speed of 4.77 MHz. The later AT version of the PC introduced a 16-bit expansion bus called the *ISA bus* using a 8.33 MHz clock. This was not fast enough for professional applications, so yet another expansion bus—the *extended industry standard architecture* (EISA) bus—was introduced. EISA supported 32-bit data paths and had a bandwidth of 33 MB/s at the same clock rate as the ISA bus. EISA was not a great success, because EISA systems and peripherals were expensive. Consequently, the bus was found principally in file servers.

IBM created a proprietary bus architecture called the IBM Microchannel that died because it was proprietary (i.e., manufacturers had to pay a royalty to IBM) and, therefore, were expensive to use. A more popular low-cost, 32-bit bus called the VESA was introduced to support high-performance peripherals. Very shortly afterwards, another competitor, Intel's 64-bit Peripheral Component Interconnect (PCI) bus was introduced, and the PCI bus won the race for standardization.

The PCI bus is a *mezzanine* bus falling between the processor's own bus and the peripheral expansion bus. You can connect a secondary peripheral expansion bus to the PCI bus. The driving force behind the PCI bus was the need to perform very high-speed data transfers to graphics controllers. Unlike IBM's disastrous Microchannel architecture, Intel put the PCI bus in the public domain and peripheral manufacturers could develop PCI systems without paying royalties or being involved in patent litigation.

(continued)

In 2004, Intel introduced the PCIe (PCI express) bus to interface high-speed peripherals. This operated at 250 MB/s. By 2007, the PCIe bus was in its third revision and could support up to 8 Gigatransfers/s (although the third revision of the bus did not become available until 2011). The PCIe bus became the preferred interface for graphics cards from 2009 onward.

As well as the PCIe bus, there are other buses that are found in professional, high-performance computer systems. These buses are called *backplane buses* because they run the length of a computer system and all modules (circuit cards) are plugged into them. Unlike the PC bus that is part of the motherboard along with the CPU, these other buses have little or no active circuitry, and the CPU is plugged into the bus just like any other memory or peripheral. Typical professional buses in the backplane category are the VMEbus, the Multibus, the NuBus, and the Futurebus+.

The backplane bus can connect together multiple processors, and you can even regard it as a type of local area network for CPUs. The modules or cards connected to such a bus often contain their own local buses connected to local memory and peripherals.

The Data Bus

Three of the parameters that specify a data bus are its *width*, its *speed*, and its *latency*. The width of a data bus is the number of bits it can transfer in a *bus cycle*. In the mid-1970s, typical data buses were eight bits wide, whereas today 64-bit buses are commonplace. Some internal buses within chips are 128 or more bits wide. Note the distinction between *clock* cycle and *bus* cycle—a clock cycle is the shortest event that takes place in a computer, whereas a bus cycle is a complete transaction on the bus that may take several clock cycles to complete.

A bus's *bandwidth* indicates its throughput and is expressed as bytes/s. Clearly, the wider the bus, the greater the throughput. For example, if a 16-bit bus can transfer data at 100 MB/s, doubling the width of the bus to 32 bits, it doubles its throughput to 200 MB/s. The intrinsic speed of a bus is governed largely by its *physics* (i.e., its construction) and the nature of the devices connected to the bus (see the panel *Bouncing on Buses on page 758*). Some logic devices are capable of higher speed operation than others.

A bus's *latency* is the time taken to set up a data transfer. Latency may be low in a system where there's a permanent bus master but much longer in a system where the device wishing to transfer data has to wait for the arbitration mechanism to grant it access to the bus.

Bus Speed

Suppose device A at one end of a bus transmits data to device B at the other end. Let's go through the sequence of events that take place when device A initiates the data transfer at time $t = 0$ (see Figure 12.34). Initially, A drives data onto the data bus at time t_d, where t_d is the delay between device A initiating the transfer and the data appearing on the bus. Data on the bus propagates along the bus at about 70% of the speed of light or about 1 ft/ns. The actual speed of a signal on a bus is governed by the electrical nature of the bus (its dimensions and the physical properties of the substance surrounding the conductors). This is a fundamental limitation of matter, and whatever technological changes take place in the future, signals cannot be encouraged to go faster than light. The only way to reduce propagation delay is to reduce the length of the bus.

When the data reaches B, it must be *captured*. Latches and memory elements are specified by their *setup* and *hold* times. The data setup time, t_s, is the time for which the data must be available at the input to system B for it to be recognized. The data hold, t_h,

FIGURE 12.34   Transmission timing

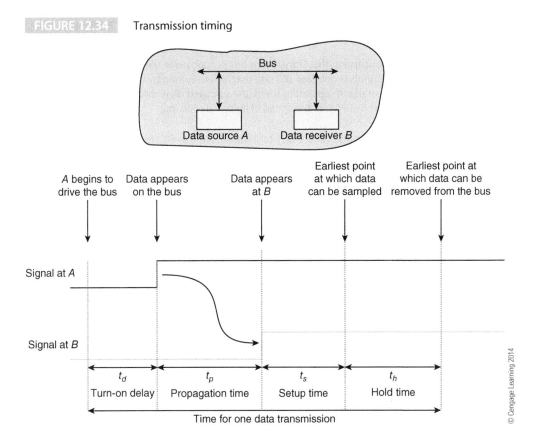

© Cengage Learning 2014

time is the time for which the data must remain stable at system B's input after it has been captured.

The time taken for a data transfer, t_T, is therefore $t_T = t_d + t_p + t_s + t_h$. Inserting typical values for these parameters yields $4 + 1.5 + 2 + 0 = 7.5$ ns, corresponding to a data transfer rate of $1/7.5$ ns $= 10^9/7.5 = 133.3$ MHz. A 32-bit-wide bus can transfer data at a maximum rate of 533.2 MB/s. In practice, a data transfer requires time to initiate it, called the latency, t_L. Taking latency into account gives a maximum data transfer rate of $1/(t_T + t_L)$.

Higher data rates can be achieved with *pipelining* by transmitting the next data element before system B has completed reading the previous element. Figure 12.35 demonstrates the application of pipelining to the previous example. Data must be stable at the input to system B for at least $t_s + t_h$ seconds; then a new element may replace the previous element. Pipelining allows an ultimate data rate of $1/(t_s + t_h)$.

FIGURE 12.35   Pipelining a data transfer

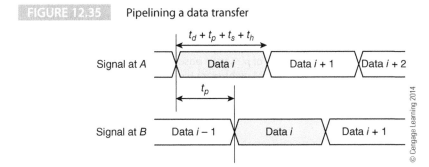

© Cengage Learning 2014

Bouncing on Buses—Signal Transmission

It's a common misconception that signals on a bus propagate at the speed of light. In all physical media, signals propagate at a rate that is determined by the geometry of the conductors and the material separating them (the so-called dielectric). Signals on typical buses propagate at about 70% of the speed of light *in a vacuum*.

What happens when a pulse (0 to 1 or 1 to 0 transition) travels along a bus and reaches the end? If you switch a light on in a room, the light comes on. Now, suppose you first remove the bulb… and then switch the light on. What happens? Nothing appears to happen, but consider the situation at the switch. At the instant you throw the switch, the electricity does not *know* that you have removed the bulb. A pulse of current, *I*, flows down the wires until it reaches the open circuit left by the removal of the bulb. At this point, the current cannot flow further, because it has nowhere to go. We also know that in the steady state the current in the wire cannot continue to flow. So, when the current pulse reaches the open circuit, a pulse of –*I* must flow back along the line to cancel the incoming or incident current. The current flowing back from the open circuit is called a *reflection*. Now, if the height (voltage) of the incoming pulse is *V* in order to drive a pulse of –*I* back from the bulb, a voltage of 2 V must appear at the bulb.

Exactly the same happens on computer buses. If you apply a pulse to the bus by driving it from a 0 to a 1 or vice versa, a pulse will flow in both directions along the bus and be reflected at its ends. This pulse may be reflected many times up and down the bus (being reflected at the ends or at other points where there is an impedance change between the bus and another bus or stub). In practice, the reflections die down in a few nanoseconds, and in everyday life, we are unaware of this phenomenon. However, in computer buses, where a few nanoseconds can be longer than the duration of a clock pulse, reflections can play havoc and cause incorrect operation because they can cause 1s to be interpreted as 0s and vice versa.

The effect of reflections can be reduced by careful design of the bus, its physical layout, and the use of terminating devices that absorb reflections. If a load (resistor) is connected across the end of a bus and the impedance of the load is the same as that of the characteristic impedance of the bus, reflections do not occur. The characteristic impedance of a bus is typically in the range 50 to 200 Ω. The coefficient of reflection Γ (the fraction of a pulse reflected) is given by the following equation, where Z_L is the termination impedance and Z_S is the characteristic impedance.

$$\Gamma = \frac{Z_L - Z_S}{Z_L + Z_S}$$

When the first transatlantic cable was laid in 1858, the length of the bus was over three thousand miles and the rate at which signaling was possible was far less than expected (the first cable broke down after 400 messages, and it was not until 1886 that a new working cable was in use). The problem of investigating the unexpectedly slow rate of single transmission was handed over to Professor William Thomson, who formulated the basis of transmission line theory and the behavior of pulses on buses. You could say that this was the beginning of electronics. Thomson was knighted for his work and later became Lord Kelvin.

FIGURE 12.36   Multiplexing address and data

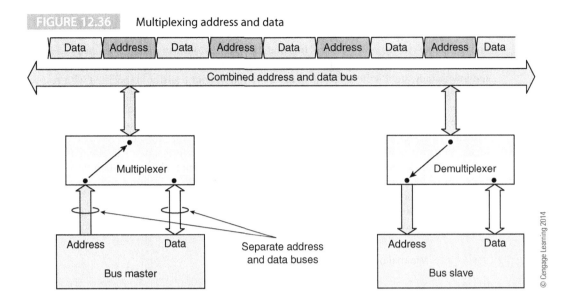

The Address Bus

When a CPU accesses memory, the bus master controlling the data transfer must provide the *address* of the source or destination of the data. Some computer systems have an explicit address bus that operates in parallel with the data bus. For example, when the processor writes data to memory, a 32-bit address is transmitted to the memory system on the address bus at the same time the data is transmitted on the data bus. Some systems combine the address and data buses together into a single *multiplexed* address/data bus that carries both addresses and data (albeit alternately). Such a bus is said to be *time-division* multiplexed because time is divided into address slots and data slots.

Figure 12.36 describes the multiplexed address/data bus that is cheaper to implement than conventional non-multiplexed buses, because it requires fewer signal paths and the connectors and sockets require fewer pins. Multiplexing addresses and data onto the same lines requires a *multiplexer* (high-speed electronic switch) at one end of the transmission path and a *demultiplexer* at the other end. Multiplexed buses can be slower than non-multiplexed buses and are often used when cost is more important than speed. This is especially true when the multiplexing and demultiplexing is built into the processors and interface components themselves.

The efficiency of both non-multiplexed and multiplexed address buses can be improved by operating in a *burst mode* in which a sequence of data elements is transmitted to consecutive memory addresses. Burst-mode operation is used to support cache memory systems. When a line in a cache is to be loaded from memory, the address of the first word is transmitted to the memory. The memory responds by providing the word at the specified address, followed by the word at the next address in sequence, and so on. These sequential addresses can be generated at the memory.

Figure 12.37 illustrates the concept of burst mode addressing where an address is transmitted for location i and data for locations $i, i + 1, i + 2,$ and $i + 3$ are transmitted without a further address.

FIGURE 12.37   Burst mode and data

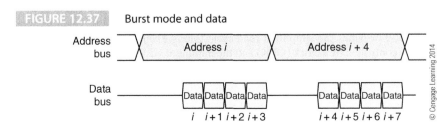

The Control Bus

The control bus regulates the flow of information on the data bus. Figure 12.38 describes the simplest 2-line *synchronous*[2] control bus that uses a *data-direction* signal and a *data validation signal*. The data direction signal is often called R/$\overline{\text{W}}$ and is high to indicate a read operation and low to indicate a write operation (the direction of data transfer is specified with respect to the bus master that originated the data transfer). During a read cycle, data flows from the bus slave to the bus master, and during a write cycle data flows from the master to the slave.

Some systems have separate read and write strobes rather than a composite R/$\overline{\text{W}}$ signal. Individual $\overline{\text{READ}}$ and $\overline{\text{WRITE}}$ signals have the advantage that they can indicate *three* bus states: an active read state, an active write state, and a bus free state ($\overline{\text{READ}}$ and $\overline{\text{WRITE}}$ both negated). A composite R/$\overline{\text{W}}$ signal introduces an ambiguity, because when R/$\overline{\text{W}}$ = 0, the bus is always executing a write operation, whereas when R/$\overline{\text{W}}$ = 1, either a read operation is being executed or the bus is free.

The active-low *data valid signal* ($\overline{\text{DAV}}$) in Figure 12.38 is asserted by the bus master to indicate that a data transfer is taking place.

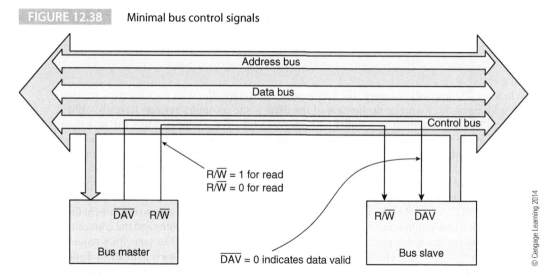

FIGURE 12.38   Minimal bus control signals

© Cengage Learning 2014

The synchronous open-ended transfer mechanism of Figure 12.38 is fast. The bus master does not know whether the slave has correctly responded to its access. The slave might have not been able to comply with the request for data. Consequently, the bus master must operate at the speed of the *slowest* slave taking part in data transfers. The *asynchronous* bus that we met when discussing handshaking solves both of these problems.

An asynchronous data transfer requires *two* data-flow control signals: a data available strobe to indicate that data is ready and a data acknowledge strobe to indicate that the data has been read. Recall that a bus transaction sometimes can't be completed (for example, the master might address a non-existent memory location), and the data available strobe never receives a response. When this happens, the master must terminate the transaction itself after a suitable time-out period. Typically, when the master asserts data available, a timer is started. If the timer reaches a predetermined value, the master is forced to abandon the access. The processor executes a special exception called a *bus error* that calls the operating system.

Let's look at an example of an asynchronous data transfer—a processor memory read cycle. Figure 12.39 provides the simplified read cycle timing diagram of a 68020 processor. The processor is controlled by a clock (CLK) and the minimum bus cycle takes six clock states labeled S0 to S5.

---

[2] The bus is synchronous in the sense that a master clock is required to synchronize signals. For example, the bus master assumes that the slave has responded *n* clock cycles after data available has been asserted.

FIGURE 12.39   Example of an asynchronous data transfer

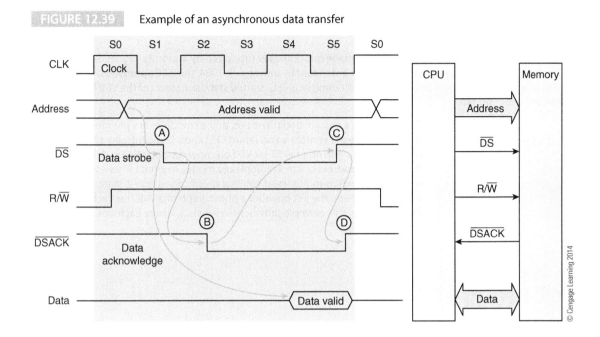

Once the processor has placed an address on the address bus in state S0, the CPU asserts a data strobe ($\overline{\text{DS}}$) at point *A* in state S1 to trigger the data transfer. The data strobe is detected by the bus slave (i.e., the memory) and used to initiate the read cycle if R/$\overline{\text{W}}$ is high.

The bus slave asserts $\overline{\text{DSACK}}$ (data strobe acknowledge) in response to $\overline{\text{DS}}$ at *B* during bus state S2 to indicate that the bus cycle may continue. If $\overline{\text{DSACK}}$ isn't asserted before bus state S3, the processor inserts *wait states* until $\overline{\text{DSACK}}$ is asserted. That is, the clock states will be S0, S1, S2, W, W, W, W, S3, S4, and S5 (assuming that the bus slave requires an extra four states to complete the access).

At point *C* in bus state S5, the processor negates the data strobe, and at point *D,* the bus slave negates $\overline{\text{DSACK}}$ to complete the asynchronous handshake procedure.

In the next section, we are going to look at one of the most important aspects of modern computer buses—*arbitration,* which is the mechanism used to decide which bus master can take control of the bus when two or bus masters compete for control.

12.6   Arbitrating for the Bus

In a system with several potential bus masters connected to a common bus, a mechanism is needed to deal with simultaneous bus requests. The process by which multiple requests are recognized and priority given to one of them is called *arbitration*.

There are two approaches to dealing with multiple requests for a bus—*localized* arbitration and *distributed* arbitration. In localized arbitration, an arbitration circuit receives requests from the contending bus masters and then decides which of them is to be given control of the bus. In a system with distributed arbitration, each of the masters takes part in the arbitration process and the system lacks a specific arbiter—each master monitors the other masters and decides whether to continue competing for the bus or whether to give up and wait until later. We describe how three buses implement these techniques—the VMEbus found in high-performance professional systems (e.g., industrial control), the NuBus that was used in the Apple Macintosh and several professional systems, and the PCI bus.

The Backplane Bus

The VMEbus is an asynchronous, non-multiplexed backplane bus originally supported by Motorola, Signetics, Mostek, Philips, and Thomson-EFCIS and rapidly was adopted as a standard by industry. In 1984, the IEEE approved the VMEbus as the IEEE 1014 bus. The International Electrotechnical Commission (IEC) started standardization of the VMEbus in 1982 and called it the IEC 821 bus.

http://en.wikipedia.org/wiki/File:VMEbus.jpg. Courtesy of Sergio Ballestrero.

The VMEbus is a passive backplane bus, and all modules are plugged into it; that is, it doesn't contain an on-board CPU and a bus controller chipset like the PCI bus. The original 32-bit VMEbus now supports 64-bit data paths (by using the address bus in a multiplexed mode). In what follows, we are referring to the non-multiplexed VMEbus specification unless stated otherwise. The figure on the left provides a photograph of a VMEbus backplane. As you can see, this example provides twelve slots, where each slot uses two connectors.

The following figure shows a modern high-performance motherboard found in some PCs. As you can see, a large fraction of the computer's hardware is located on the motherboard. When PCs first appeared, the motherboard simply linked together functional modules; if you wanted audio input/output you had to buy a sound card and plug it in. It was even necessary to buy an interface for a simple serial data link. This board, using Intel's Z77 chipset, is intended for high-performance systems. It contains a complete audio interface, high-speed Ethernet connection, and even provides all the logic necessary to support HDMI. There are PCI and PCI express expansion slots, USB 3.0 interfaces, and slots capable of supporting 32GB of fast DDR3 DRAM. This board provides eight serial SATA interfaces to disk (SSD, magnetic, and optical) with four capable of operating at 6 GB/s.

Alan Clements

12.6.1   Localized Arbitration and the VMEbus

In this section, we are interested in how a device goes about requesting the bus and how the bus grants this request. We will use the VMEbus that supports several types of functional module as an example. We are interested in the *bus master* that controls the bus, the *bus requester* that requests the bus, and the *arbiter* that grants the bus to a master. A bus requester is employed by a bus master when it wants to access the VMEbus. A VMEbus is usually housed in a box with a number of slots into which modules can be plugged (rather like the slots used by the PCI bus).

The VMEbus's arbitration sub-bus is composed of the fourteen lines described in Figure 12.40. Four lines, $\overline{BR0}$ to $\overline{BR3}$ (bus request 0 to bus request 3), are used by a bus requester to indicate that the bus master associated with the requester wants the bus. Four bus grant lines are used by the arbiter to grant control of the bus to the requester. Two other lines, bus clear ($\overline{BCLR}$) and bus busy ($\overline{BBSY}$), control the arbitration process.

The VMEbus arbiter is located in a *special* position on a VMEbus called slot 1. This first location in a VMEbus *must* be occupied by the arbiter. All bus request lines run the length of the VMEbus, and any master can place a request on one of these lines. The particular level of the request made by a bus requester is user-determined; that is, the user decides which of the four bus request lines are to be connected to a module's request output.

The arbiter reads the bus request inputs from all of the slots along the bus, decides which request is to be serviced, and then informs other modules of its decision via its bus grant outputs. The way in which the arbiter decides which of the competing bus masters gets the bus is dealt with after we have described how the arbitration bus operates.

The VMEbus supports four levels of arbitration. We will soon see that each of these four levels can be further subdivided. The bus request lines run the length of the VMEbus and terminate at the arbiter in slot 1. Figure 12.41 demonstrates the relationship between these lines, the *arbiter,* and bus *requester* modules.

When one or more bus requesters wish to access the VMEbus on behalf of their associated bus masters, they assert the bus request lines to which they have been assigned. For example, the card in slot 3 might assert bus request line $\overline{BR1}$, and the card in slot 5 might assert bus request line $\overline{BR3}$. The arbiter in slot 1 examines all of the incoming bus requests and decides whether one of them is to succeed.

FIGURE 12.40   VMEbus arbitration signals

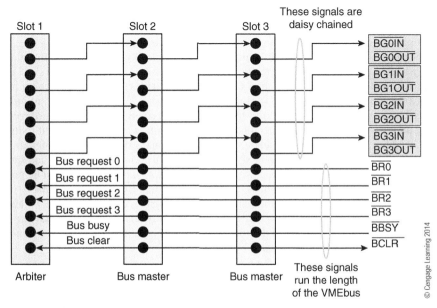

© Cengage Learning 2014

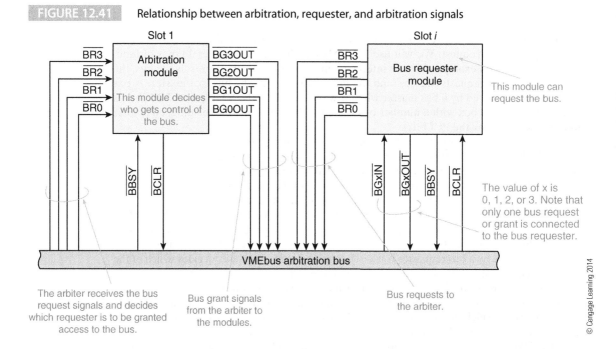

FIGURE 12.41   Relationship between arbitration, requester, and arbitration signals

If a request on, say, $\overline{\text{BR2}}$ is successful, the arbiter sends a *bus grant message* on its level 2 bus grant output, $\overline{\text{BG2OUT}}$. In what follows, we will write $\overline{\text{BGxIN}}$, $\overline{\text{BGxOUT}}$, and $\overline{\text{BRx}}$ to avoid referring to specific levels. By convention in VMEbus literature, the x stands for levels 0 to 3.

The $\overline{\text{BGx}}$ lines do not run along the entire length of the bus. Instead the VMEbus employs a chain of lines called *bus grant out* and *bus grant in*. In Figure 12.40, the $\overline{\text{BGxIN}}$ and the $\overline{\text{BGxOUT}}$ lines are broken and run only from slot-to-slot rather than from end-to-end. A $\overline{\text{BGxOUT}}$ line from a left-hand module is passed out on its right as a $\overline{\text{BGxIN}}$ line. Therefore, the $\overline{\text{BGxOUT}}$ of one module is connected to the $\overline{\text{BGxIN}}$ of its right-hand neighbor. The arrangement of the $\overline{\text{BGxIN}}$ and $\overline{\text{BGxOUT}}$ lines in Figure 12.40 is called *daisy-chaining*. A continuous bus line transmits a signal in both directions to all devices connected to it. The daisy-chained line is *unidirectional*, transmitting a signal from one specific end to the other. Each module connected to (i.e., receiving from and transmitting to) a daisy-chained line may either pass a signal on down the line or inject a signal of its own onto the line.

Figure 12.42 shows a system in which the requester in slot *j* requests the bus at level 1 when no other device is requesting the bus. When $\overline{\text{BR1}}$ is asserted, the arbiter in slot 1 detects it and asserts $\overline{\text{BG1OUT}}$, which passes down the bus until it reaches slot *j*. The arbiter in slot 1 sends a bus grant input to the card in slot 2. The card in slot 2 takes this bus grant *input* and passes it on as a bus grant *output* to the card in slot 3, and so on. In this way, a card receives a bus grant input from its left-hand neighbor and passes it on as a bus grant output to its right-hand neighbor. However, a card might choose to terminate the daisy-chain signal-passing sequence and not transmit a bus grant signal to its right-hand neighbor, as we shall soon see. If a slot is empty because no card is plugged into it, *bus jumpers* (i.e., links) must be provided to route the appropriate $\overline{\text{BGxIN}}$ signals to the corresponding $\overline{\text{BGxOUT}}$ terminals.

A requester module makes a bid for control of the system data transfer bus by asserting one of the bus request lines, $\overline{\text{BR0}}$ to $\overline{\text{BR3}}$. Only one line is asserted, and the actual line is chosen by assigning a given priority to the requester. This priority may be assigned by on-board, user-selectable jumpers or dynamically by software.

The arbiter in slot 1, on receiving a request for the bus, asserts one of its $\overline{\text{BGxOUT}}$ lines, and a bus grant signal then propagates down the daisy-chain (the level of the grant corresponds to the level of the request that won the current round of arbitration). Each $\overline{\text{BGxOUT}}$ arrives at the $\overline{\text{BGxIN}}$ of the next module. If that module does not require access to the bus,

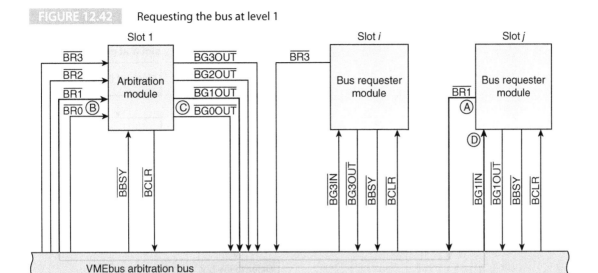

FIGURE 12.42   Requesting the bus at level 1

The requester in slot *j* is requesting the bus at level 1.
No other request is pending and the arbiter is granting a level 1 request. The BG1OUT line from the arbiter is daisychained. We have shown a direct connection, in blue, to demonstrate the message path.

Ⓐ   The requester in slot *j* requests the bus at level 1

Ⓑ   The bus request arrives at the arbiter in slot 1

Ⓒ   The arbiter decides that the requester at level 1 is to get the bus and asserts bus grant at level 1

Ⓓ   The requester in slot *j* receives the bus grant and can now take control of the bus

it passes on the request on its $\overline{\text{BGxOUT}}$ line. If, however, the module does wish to request the bus, it does not assert its $\overline{\text{BGxOUT}}$ signal and takes over control of the bus itself. Daisy chaining provides automatic prioritization, because bus requesters further down the line do not receive a bus grant—this is called *geographic prioritization*.

Let's walk through a VMEbus arbitration sequence and introduce some of the other signals that take part in the process. Figure 12.43 provides a *protocol flowchart* for the VMEbus arbitration procedure that describes the sequence of events that take place during an arbitration procedure.

Initially, in Figure 12.43, a bus master in slot *M* at a priority less than *i* is in control of the bus. This current bus master asserts the bus busy signal ($\overline{\text{BBSY}}$), which runs the length of the bus. As long as any master is asserting $\overline{\text{BBSY}}$ no other master may attempt to gain control of the VMEbus. An active bus master in a VMEbus *cannot* be forced off the bus.

Suppose a bus requester in slot *N* requests the bus at a priority level *i*. This request is at a higher level than the current master. The arbiter detects the new higher level and asserts its $\overline{\text{BCLR}}$ (bus clear) output. The bus clear line from the arbiter informs the current master in slot *M* that another master with a higher priority now wishes to access the bus. The current master does not have to relinquish the bus within a prescribed time limit. Typically, it will release the bus at the first convenient instant by negating $\overline{\text{BBSY}}$. Note that the VMEbus provides both geographic prioritization determined by a slot's location in the daisy-chain and an *optional* prioritization by bus request level as we shall soon see.

$\overline{\text{BCLR}}$ is driven only by arbiters that permanently assign *fixed* priorities to the bus request lines. Other arbitration mechanisms, such as the round robin arbitration scheme to be described later, have no fixed priority, and the arbiter does not make use of the bus clear line.

When the arbiter detects that the current master has released the bus, the arbiter asserts $\overline{\text{BGiOUT}}$ to indicate to the requester at level *i* that it has gained control of the bus. The arbiter knows only the *level* of the request and not which slot it came from. The bus grant message ripples along the bus, entering each module as $\overline{\text{BGiIN}}$ and leaving as $\overline{\text{BGiOUT}}$. When this

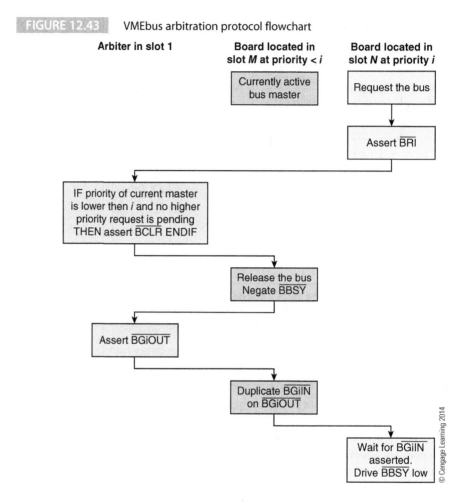

FIGURE 12.43   VMEbus arbitration protocol flowchart

Arbiter in slot 1

Board located in slot *M* at priority < *i*

Board located in slot *N* at priority *i*

Currently active bus master

Request the bus

Assert $\overline{\text{BRI}}$

IF priority of current master is lower then *i* and no higher priority request is pending THEN assert $\overline{\text{BCLR}}$ ENDIF

Release the bus Negate $\overline{\text{BBSY}}$

Assert $\overline{\text{BGiOUT}}$

Duplicate $\overline{\text{BGiIN}}$ on $\overline{\text{BGiOUT}}$

Wait for $\overline{\text{BGiIN}}$ asserted. Drive $\overline{\text{BBSY}}$ low

© Cengage Learning 2014

message reaches the requester in slot *N* that made the request at level *i*, the message is not passed on.

Instead, the requester asserts $\overline{\text{BBSY}}$ to show that it now has control of the bus. What would have happened if a requester also at level *i* but located nearer to the arbiter than slot *N* had also requested the bus at approximately the same time? The answer is that the requester closer to the arbiter would have received the bus grant first and have taken control of the bus.

Releasing the Bus

The requester may implement one of two options for releasing the bus: *release when done* (RWD) and *release on request* (ROR). Option RWD requires the requester to release the bus as soon as the on-board master stops indicating bus busy; that is, the master remains in control of the bus until its task has been completed, which can lead to undue bus hogging. The ROR option is more suitable in systems where it is unreasonable to grant unlimited bus access to a master. The ROR requester monitors the four bus request lines. If it sees that another requester has requested service, it releases its $\overline{\text{BBSY}}$ output and defers to the other request. The ROR option also reduces the number of arbitrations requested by a master, as the bus is frequently cleared voluntarily.

The Arbitration Process

Let's look at an example of the arbitration sequence. Figure 12.44 demonstrates what happens when two requesters at different levels of priority request the bus. Initially, both requesters *A* and *B* assert their bus request outputs simultaneously. Requester *A* asserts $\overline{\text{BR1}}$ and *B* asserts

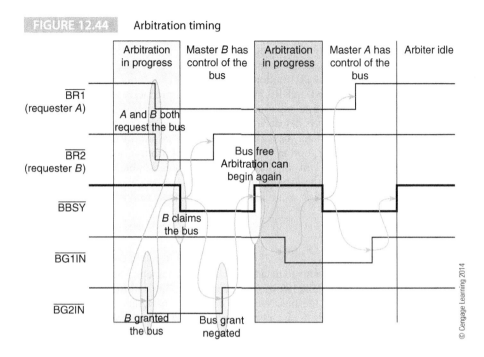

FIGURE 12.44 Arbitration timing

BR2. Assuming that the arbiter detects both $\overline{\text{BR1}}$ and $\overline{\text{BR2}}$ low, the arbiter will assert $\overline{\text{BG2IN}}$ on slot 1, because $\overline{\text{BR2}}$ has a higher priority than $\overline{\text{BR1}}$. When the bus grant signal has propagated down the daisy-chain to requester B, requester B will respond to $\overline{\text{BG2IN}}$ by asserting $\overline{\text{BBSY}}$. Requester B then releases $\overline{\text{BR2}}$ and informs its own master that the VMEbus is now available.

After detecting that $\overline{\text{BBSY}}$ has been asserted, the arbiter negates $\overline{\text{BG2IN}}$. At this point, both $\overline{\text{BR2}}$ and $\overline{\text{BG2IN}}$ are inactive-high, because $\overline{\text{BBSY}}$ and the bus grants are interlocked, as shown in Figure 12.44. The arbiter is not permitted to negate a bus grant until it detects $\overline{\text{BBSY}}$ low. When master B completes its data transfer, requester B releases $\overline{\text{BBSY}}$. The negation of $\overline{\text{BBSY}}$ is conditional on $\overline{\text{BG2IN}}$ remaining high and at least 30 ns having elapsed since the release of $\overline{\text{BR2}}$. The 30 ns delay ensures that the arbiter will not interpret the old active-low value of $\overline{\text{BR2}}$ as another request. Requester B will wait until the 30 ns interval has elapsed and will then release $\overline{\text{BBSY}}$.

The arbiter detects the release of $\overline{\text{BBSY}}$ and arbitrates bus requests once more. $\overline{\text{BR1}}$ is still active-low and is the only bus request line asserted. The arbiter grants access to Requester A by asserting $\overline{\text{BG1IN}}$. Requester A responds by asserting $\overline{\text{BBSY}}$. When master A has completed its data transfer, requester A releases $\overline{\text{BBSY}}$, provided $\overline{\text{BG1IN}}$ has been received and 30 ns have elapsed since the release of $\overline{\text{BR1}}$. Since no bus request lines remain asserted when requester A releases $\overline{\text{BBSY}}$, the arbiter remains idle.

VMEbus Arbitration Algorithms

Three strategies that the arbiter in slot 1 can use for dealing with the prioritization of the bus request lines are given below. Note that these are suggested options—the user may employ any other strategy.

1. **Option RRS (round robin select)** The RRS option assigns priority to the masters on a rotating basis. Each of the four levels of bus request has a turn at being the highest level. The four levels of bus request, $\overline{\text{BR0}}$ to $\overline{\text{BR3}}$, are treated cyclically with $\overline{\text{BR3}}$ following $\overline{\text{BR0}}$; that is, the sequence of successive highest levels of priority is $\overline{\text{BR0}}$, $\overline{\text{BR3}}$, $\overline{\text{BR2}}$, $\overline{\text{BR1}}$, $\overline{\text{BR0}}$, $\overline{\text{BR3}}$, $\overline{\text{BR2}}$, . . .: At any instant, one of the four levels is made the highest level so that a requester at that level may gain control of the bus. If a requester at the current highest level does not wish to use the bus, the next level downwards is made the new highest level and so on. For example, if the current highest level is $\overline{\text{BR2}}$, in the next cycle the highest level will be $\overline{\text{BR1}}$.

Suppose a requester is granted control of the bus. After the bus has been released, the next level downwards is made the new highest level and the cycle continues. All levels of bus request become the highest priority in turn, and no level is ever left out. Round robin select is a *fair* method of arbitration, as all the masters are granted equal access to the bus.

2. **Option PRI (prioritized)** The PRI option assigns a level of priority to each of the bus request lines from $\overline{BR3}$ (highest) to $\overline{BR0}$. Whenever a master requests access to the bus, the arbiter deals with the request by comparing the new level of priority with the current level. A higher priority request always defeats a lower priority request. Option PRI is not a fair strategy, as a low-level request may never be serviced if higher priority devices are *greedy*.

3. **Single level (SGL)** The SGL option provides a minimal arbitration facility using bus request line $\overline{BR3}$ only. The priority of individual modules is determined by *daisy-chaining*, so that the module next to the arbiter module in slot 1 of the VMEbus rack has the highest priority. As the position of a module moves further away from the arbiter, its priority reduces.

12.6.2   Distributed Arbitration

Not all buses use a centralized arbiter to decide which of the competing bus masters is to get control of the bus. A mechanism called *distributed arbitration* allows arbitration to take place simultaneously at all slots along the bus. We now describe a backplane bus that supports distributed arbitration—the NuBus—which is a general-purpose synchronous backplane bus with multiplexed address and data lines that is also known as ANSI/IEEE STD 1186-1988. It was conceived at MIT in 1970 and later supported by Western Digital and Texas Instruments (1983). Apple implemented a subset of NuBus in their Macintosh II.

Address space spanned by the NuBus is partitioned with each card in a NuBus system being allocated a unique slice of the total address space; that is, the NuBus implements *geographic addressing*. A 32-bit NuBus address is expressed in hexadecimal as $YXXXXXXX_{16}$, where Y defines one of 15 slots. The 256 MB address space for which $Y = 1111_2$ extending from $F0000000_{16}$ to $FFFFFFFF_{16}$ is reserved and is called *slot space*. This slot space is subdivided into sixteen blocks of 16 MB, and one of these blocks is allocated to each slot. A slot, therefore, has a 16 MB block of unique address space associated with it. It follows that the NuBus cannot support more than 16 slots.

Each slot along the backplane has its own unique identification code (ID) that is hardwired into the backplane. The ID of each slot is fixed and is determined by the backplane connector itself and not by the card. For example, if you plug a card into slot eleven, active-low backplane lines $\overline{ID3}$ to $\overline{ID0}$ are hardwired at the connector to provide the value 0100 (i.e., the inverse of $1011_2 = 11_{10}$). Slot identification lines $\overline{ID3}$ to $\overline{ID0}$ don't run along the backplane and are simply connected to ground or to the positive power supply voltage at each connector to provide the appropriate slot number.

NuBus Arbitration

Like the VMEbus, NuBus supports multiprocessing and multiple bus masters. We are going to explain how the NuBus implements distributed arbitration by exploiting the properties of an open-collector gate. NuBus has four arbitration lines, $\overline{ARB0}$ to $\overline{ARB3}$, and one prioritization algorithm—when a group of modules compete for the bus, the module with the highest ID number wins. A would-be master begins arbitration by asserting its bus request line, $\overline{RQST}$.

The key to NuBus arbitration is each module's *unique slot* number that ranges from 0_{16} to F_{16}. When a card in a slot arbitrates for the bus, the card places its slot number on the bus, and as if by magic, any other requester with a *lower* slot number stops arbitrating for the bus. Equally, if a slot with a higher number wants the bus, the requesting slot stops requesting the bus; that is, if a card arbitrates for the bus and then finds that a card with a higher priority is also arbitrating for the bus, *it backs off*.

In order to appreciate how distributed arbitration works, you have to understand the *open-collector* gate. Up to now, we've described two types of gates: the conventional gate

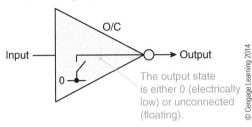

FIGURE 12.45   The open-collector gate

Input ⟶

O/C

Output

The output state is either 0 (electrically low) or unconnected (floating).

© Cengage Learning 2014

whose output is always actively driven to 0 or 1 and the tristate gate whose output is 0, 1, or floating. Historically, the open-collector gate precedes the tristate gate and is used to allow more than one device to drive the same bus.

Figure 12.45 illustrates an inverter with an open-collector output. The gate's output can be actively forced to a low voltage. When the input of the gate is 1, the internal transistor switch is closed, and its output is forced low just like a normal inverter.

When the input is 0, the transistor switch is open, and the output of the open-collector gate is left *floating* because it is internally disconnected from the high- or low-level power rails. That is, the open-collector gate has an active-low output state and a *floating state* and can pull a bus down into a low state, but it can't pull the bus up into a high state.

Let's contrast the behavior of the tri-state and open-collector gates. You can connect several tri-state gates to a single bus line. However, only one of these gates can be enabled at any instant because only one may drive the bus to a low or high level. You can also connect multiple open-collector outputs to the same bus. If all the open-collector gates are in the floating state, the output of the bus will also be floating (it will actually be high because a resistor is used to pull the bus up into a high state when it is floating). If one or more of the gates go into a 0 state, the bus is forced into a 0 state. This arrangement is sometimes called *wired-OR logic*, because the bus is low if any of the outputs are low. The wired OR term applies to *negative logic*, because the output is low if any input is low; in positive logic terms, a wired OR gate is an AND gate.

Figure 12.46 illustrates the key circuit used in a distributed arbiter that has an input X and an output Y. The circuit is also connected to one of the arbitration control lines on the bus. In what follows, we are interested in the relationship between the circuit and the state of the bus. Let's begin by looking at the circuit's basic logic transfer function between its input and output. If you examine the logic of this circuit, the output is given by

$$Y = \overline{P} \cdot \overline{Q}$$

Since $P = X$ and $Q = \overline{X}$, it follows that

$$Y = \overline{P} \cdot \overline{Q} = \overline{X} \cdot \overline{\overline{X}} = X \cdot \overline{X} = 0$$

FIGURE 12.46   Distributed arbitration mechanism

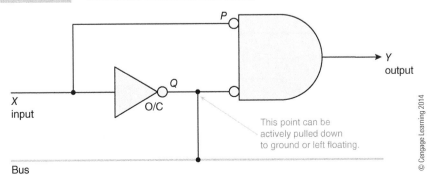

P

Q

O/C

Y output

X input

This point can be actively pulled down to ground or left floating.

Bus

© Cengage Learning 2014

At first sight, it seems that this circuit does nothing, because its output is always 0 and is independent of its input. However, because the output of the open-collector inverter is connected to the bus, this circuit will drive the bus to 0 if $X = 1$ (i.e., $Q = 0$). If the X input is 0, the output of the open-collector gate will be floating, and the bus will not be driven. By default, the bus will be pulled up into a high state by a resistor.

Here's the clever part. Suppose the X input is 0 and that the level on the bus is low because another device is driving it low. In this case, the *output* of the open-collector inverter will also be forced low by the bus. Now, both inputs to the AND gate will be 0, and the Y output will be 1. That is, the Y output is 0 unless the input X is 1 and the bus is being driven low. We have a mechanism that can actively drive the bus low or detect when another device is driving the bus low when we are attempting to drive it high. This mechanism forms the basis of distributed arbitration.

Figure 12.47 shows how the distributed arbiter of Figure 12.46 operates by considering all possible input conditions together with the state of bus line. Remember that the bus can be floating (not driven) or actively pulled down to a low level. When it is floating, a resistor weakly pulls the bus up to a high level.

Figures 12.47a and b assume that the bus is floating and is not being driven. In this case, the output of the circuit is always 0 and is independent of its input. In a real system, this situation does not exist, because the bus will always be actively pulled down to an electrically low level or weakly pulled up to an electrically high level by a resistor.

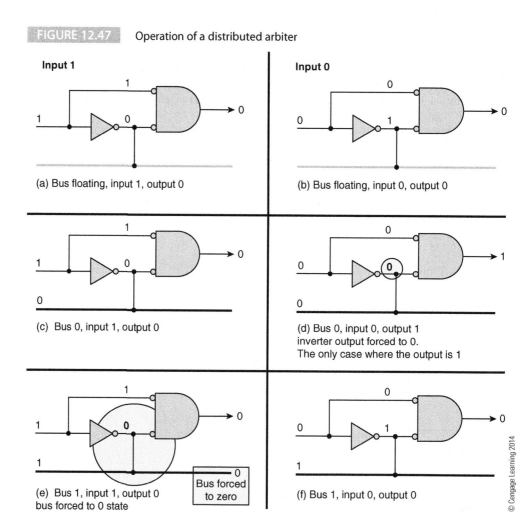

FIGURE 12.47   Operation of a distributed arbiter

In Figures 12.47c and d, we consider the situation in which the bus is being actively driven to 0 by some other device connected to the bus. In Figure 12.47c, the bus is actively being driven low, but the state of the open-collector is also low, so there is no conflict between the output of the open-collector inverter and the bus. The output of the circuit is 0.

In Figure 12.47d, the input is 0, and the output of the open-collector gate is floating. However, the state of the bus is low, and the output of the inverter is pulled down to an electrically low state. The output of the circuit is now 1. The output is telling the system that another device is driving the bus low in contradiction to the input.

In Figures 12.47e and f, the bus is in a high state because no other device is driving it low. Figure 12.47e is the interesting case. Here the input is 1 and the output of the open-collector gate is electrically low. This drives the bus to a low state. In this case, the circuit is driving the bus. The output of the circuit is 0. In Figure 12.47e, the input is 0 and the output of the inverter is floating, so there is no conflict with the state of the bus.

As you can see, there are two special cases. In one, the bus is active-low and the output of the inverter is high, which results in the inverter's output being pulled down. In the other case, the bus is high and the output of the inverter is active-low, which results in the bus being forced low.

Table 12.4 summarizes the action of this circuit. The input to this circuit represents the condition *I want the bus* or *I don't want the bus*. If the bus is not being driven low, this circuit will drive the bus low itself if its input is 1. This circuit produces a 0 output unless its input is 1 and the bus is being actively driven low by some other device.

Figure 12.48 illustrates the details of a NuBus arbiter that uses the principle of the circuit in Figure 12.46. A potential master that wants to use the bus places its arbitration level on the 4-bit arbitration bus, $\overline{ID3}$ to $\overline{ID0}$. Note that we have drawn only three of the four stages of the arbiter.

TABLE 12.4   Summarizing the bus states of Figure 12.47

Situation	Bus Condition	Result
I want the bus.	Bus free (high level).	Output is 1. Get the bus and drive it low.
I want the bus.	Bus busy (low level).	Output is 1. I do not want the bus.
I do not want the bus.	Don't care.	Output is 0.

© Cengage Learning 2014

FIGURE 12.48   Distributed arbitration

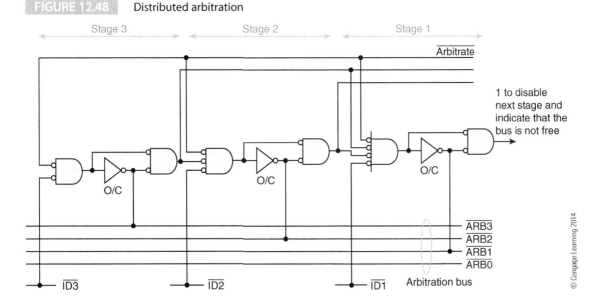

© Cengage Learning 2014

Since NuBus uses negative logic, the arbitration number is inverted so that the highest level of priority is 0000 and the least is 1111. The key to NuBus arbitration is simple—if a competing master sees a higher level on the bus than its own level, it ceases to compete for the bus. Each requester simultaneously drives the arbitration bus and observes the signal on the bus. If it detects the presence of a requester at a higher level, it backs off.

In Figure 12.48, lines $\overline{ID0}$ to $\overline{ID3}$ define the slot location, and therefore, the priority level of the master and lines $\overline{ARB0}$ to $\overline{ARB3}$ are the arbitration lines running the length of the bus. The signal labeled *Arbitrate* permits the master to arbitrate for the bus, and the output $\overline{GRANT}$ is asserted if the master wins the arbitration. NuBus uses nonsequential arbitration logic, and arbitration takes place in parallel with normal bus activity.

Suppose three masters numbered 0100 (4), 0101 (5), and 0010 (2) simultaneously put the codes 1011, 1010, and 1101, respectively, onto the arbitration bus. As the arbitration lines are open-collector, any output at a 0 level will pull the bus down to 0. In this example, the bus will be forced into the state 1000. The master at level 2 that puts the code 1101 on the arbitration bus will detect that $\overline{ARB2}$ is being pulled down and leave the arbitrating process. The arbitration bus will now have the value 1010. The master with the code 1011 will detect that $\overline{ARB0}$ is being pulled down and will leave the arbitration process. The value on the arbitration bus is now 1010, and the master with that value has gained control.

Since NuBus implements a prioritized arbitration system, a high-priority slot can monopolize the bus and stop a low-priority slot from ever gaining control. Such bus hogging is eliminated by *deferral*. Once a slot has gained bus mastership and then relinquished it, that slot will not attempt to reestablish bus mastership until all pending bus requests have been dealt with. The NuBus does have special mechanisms that permit a bus master to maintain its bus mastership. A special NuBus control cycle, called an *attention cycle,* can be executed to request continuing bus ownership.

We have just looked at the backplane bus and described examples of both *distributed* and *localized* arbitration protocols. In fact, there are so many variables in bus design that almost every engineer dreams of creating the ultimate bus, just as all authors want to write "The Great American Novel". We are now going to look at the more sophisticated PC bus, the PCI bus, and its derivative, the PCI express.

 ## 12.7   The PCI and PCIe Buses

Before we look at the PCI bus proper, we need to put it into context. The original PC bus used a 62-pin connector, supported a 20-bit address bus, and ran at 4.772 MHz. This bus was really an extension of the CPU's own interface. Table 12.5 describes the pin functions of the original PC bus. We have included this here to demonstrate how simple the bus was and how it was constructed in a purely *ad hoc* fashion with little thought of future expansion or enhanced functionality.

12.7.1   The PCI Bus

The *peripheral component interconnect local bus* (PCI bus) represents a radical change to the PC's systems architecture. Intel designed this bus for use in Pentium-based systems towards the end of 1993. The PCI bus is not only much faster than previous buses; it greatly extends the functionality of the PC architecture. Indeed, the PCI bus is central to the PC's expandability and flexibility. The PCI allows users to plug cards into the computer system to increase functionality by adding modems, SCSI interfaces, video processors, sound cards, and so on. More specifically, the PCI bus lets these cards communicate with the CPU via a bus interface circuit known as a *North bridge.* Bus interface circuits have come to be known collectively as a *chipset.* All PCs with PCI buses require such a *chipset.*

TABLE 12.5	The First-Generation PC Bus Signals
Pin group	**Function**
CLK	4.772 MHz clock.
OSC	14.318 MHz clock.
RESET	The reset line resets devices connected to the bus.
SA0 to SA19	20-bit address bus spanning 1 Mbyte address space.
D0 to D7	8-bit data bus.
AEN	When active-high address enable indicates a DMA or refresh operation.
ALE	The address latch enable indicates the presence of a valid bus cycle.
$\overline{\text{SMEMR}}$	When active-low indicates a valid memory read cycle.
$\overline{\text{SMEMW}}$	When active-high indicates a valid memory write cycle.
$\overline{\text{IOR}}$	When active-low indicates a valid I/O port read cycle.
$\overline{\text{IOW}}$	When active-high indicates a valid I/O port write cycle.
$\overline{\text{IOCHRDY}}$	Used by a bus slave to extend a bus cycle.
$\overline{\text{0WS}}$	The zero wait state line is used to indicate no wait states.
$\overline{\text{IOCJCHK}}$	Parity status used to indicate memory errors.
DRQ1 to DRQ3	DMA request lines used by peripherals.
$\overline{\text{DACK1}}$ to $\overline{\text{DACK3}}$	DMA grant lines from the CPU to peripherals.
T/C	Terminal count indicates the end of a DMA operation.
$\overline{\text{REF}}$	The refresh signal indicates to DRAM that a refresh cycle should be executed.
IRQ2 to IRQ7	Interrupt request lines.

© Cengage Learning 2014

The PCI is called a *local bus* to contrast it with the address, data, and control signals from the CPU itself. Connecting systems directly to the CPU provides the fastest data transfer rates, and a bus connected directly to a CPU is called a *front side bus*.

The PCI bus supports *plug-and-play* capabilities in which PCI plug-in cards are automatically configured at power up and resources such as interrupt requests are assigned to plug-and-play cards transparently to the user. The original PCI bus operated at 33 MHz and supported a 32-bit and 64-bit data bus. PCI bus Version 2.1 supports a 66 MHz clock.

The PCI bus is connected to the PC system by means of a single-chip *PCI bridge* and to other buses via a second bridge. This arrangement means that a PC with a PCI bus can still support the older ISA bus. Today, the ISA bus is obsolete and the PCI bus has been replaced by the PCI express bus.

Figure 12.49 illustrates the relationship between the PCI bus, the bridge, processor, memory, and peripherals. The processor is directly connected to a bridge circuit that allows the processor to access peripherals via the PCI bus. The PCI system consists of the PCI local bus itself, any cards plugged into the bus, and central resources that control the PCI bus. These central resources perform, for example, arbitration between the cards plugged into the bus. The PCI bus specification itself makes it clear that North bridge chipset is a mandatory part of a PC. Historically, the north bridge also handled the AGP video interface. Moreover, the partition of a system into CPU, North bridge, and South bridge chipsets was imposed by manufacturing limitations. Today, more and more interface functions that were once carried out by external bridges are being provided by the CPU.

FIGURE 12.49   Relationship between PCI bus, bridge, and processor

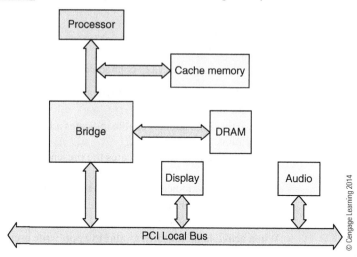

The Family of PCI Connectors

There are several versions of the PCI connector depending on the width of the slot (32-bit or 64-bit) and the voltage level (5 V circuits or more modern 3.3 V circuits). Of course, plugging a PCI card into the wrong type of slot could damage the card or the motherboard. In order to ensure that cards are correctly placed (the right way round and the right voltage version), each PCI socket on the motherboard has one or more *keys* or bridges across the connector. A card cannot be plugged into that socket unless there is a corresponding slot cut-out of the card as the figure below demonstrates.

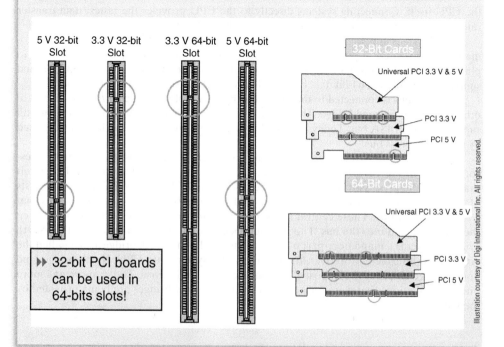

FIGURE 12.50 North and South bridges and the PCI bus

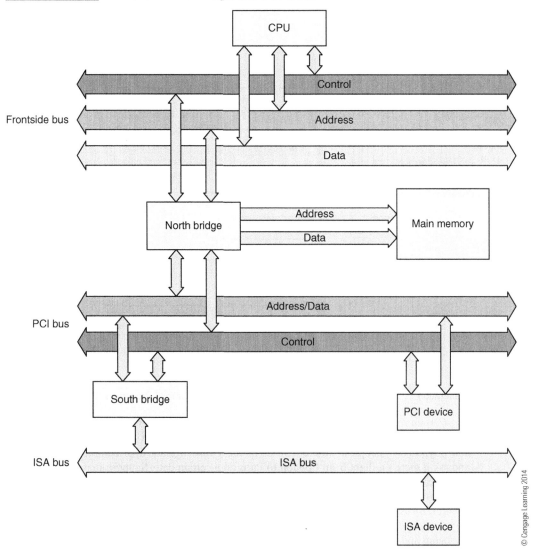

Figure 12.50 shows a system diagram of a PC with a PCI local bus and an ISA bus.[3] A second bridge, commonly called the *South bridge*, links the PCI and ISA buses. Figure 12.51a illustrates the relationship between the Pentium 4, its Intel chipset, and the PCI bus. Figure 12.51b illustrates the more modern Intel Core i7 Processor interface.

Each card that plugs into the PCI bus may contain a bus master and is able to take control of the PCI bus in order to control the flow of data. Any card or *agent* with a bus master function must provide two signals to the PCI bus: REQ (request the bus) and GNT (receive a grant from the bus). The central resource (i.e., North bridge) is responsible for dealing with PCI bus arbitration and handling the REQ/GNT pairs from each of the agents. When a device on one of the cards wants to use the bus, it asserts its REQ output. The REQ signals from all PCI cards are fed to an arbiter (part of the logic of the chipset) that decides which card is to get the bus. The arbiter asserts the GNT input to the card that won the arbitration process, allowing that card to access the bus.

---

[3]For a time, the PCI bus had a link to the older ISA bus to allow people to use their old ISA cards in their new PC. The ISA bus is now obsolete.

FIGURE 12.51 **The IA32 chipset.** (a) Reproduced with permission from Computer Desktop Encyclopedia © 1981–2012. The Computer Language Co. Inc.,(www.computerlanguage.com) (b) Reprinted with permission of Intel Corporation.

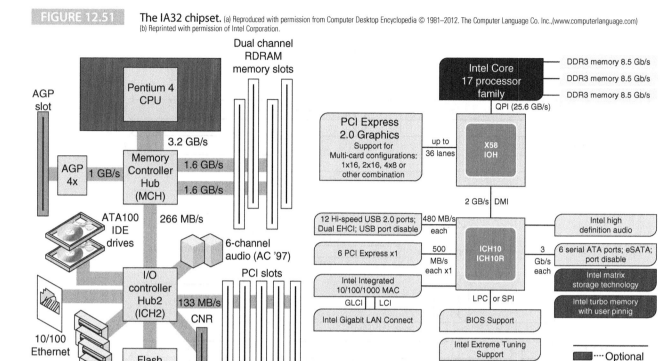

(a) Pentium 4 generation   (b) Core i7 Processor

Certain computer operations must be *indivisible* or *atomic*; that is, the operation cannot be interrupted. This restriction is required by multiprocessor systems to implement a *locked transaction*. Suppose two people in different places phone a travel agent at approximately the same time and ask if there's a flight from London to Paris on a certain date. Suppose that only one seat is available. The first person asks and is told there is a seat. The second person asks and gets the same answer. Both the first and second person then book the same seat with disastrous consequences. By using a *locked transaction* the person who asks the question first prevents anyone else from accessing the data until they have decided whether to book the seat. Locking a bus is exactly analogous and can be used to prevent two processors trying to control the same device.

Figure 12.52 shows how PCI bus arbitration is implemented. The REQ and GNT signals are connected to an arbiter that forms part of the north bridge. This arbiter reads the bus requests on the REQ0 to REQ3 inputs and returns a grant message on the GNT0 to GNT3 line corresponding to the arbitration winner. When a PCI agent arbitrates for the bus, the arbiter asserts the BPRI signal to inform the host processor that a PCI agent (i.e., a priority agent) requires the host bus.

Data Transactions on the PCI Bus

We now look at the way in which the PCI bus transfers data on its multiplexed address and data lines. The PCI bus compensates for the address/data bus bottleneck in several ways. First, it can operate in a burst mode in which a single address is transmitted, and then the address/data bus is used to transmit a sequence of consecutive data values. Second, the PCI bus supports *split transactions*; that is, one device can use the bus and another device can access the PCI bus before the first transaction has been completed. Split transactions mean that the bus is used more efficiently. Finally, devices connected to the PCI bus can be buffered, which allows data to be transmitted before it is needed.

PCI bus literature has its own terminology (some of which is shared by SCSI systems). A device that acts as a bus master is called an *initiator* and a device that responds to a bus master is called a *target*. Some of the key signals of the PCI bus are given in Table 12.6.

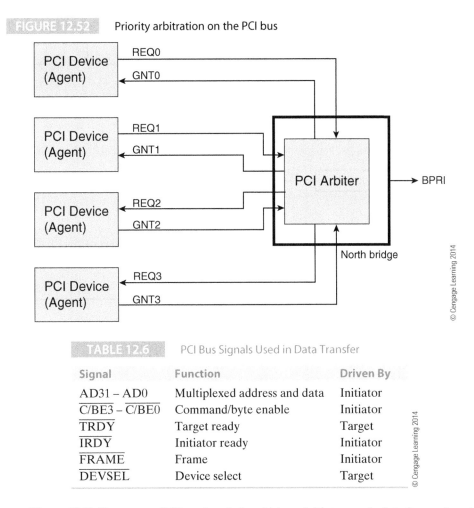

FIGURE 12.52   Priority arbitration on the PCI bus

© Cengage Learning 2014

TABLE 12.6   PCI Bus Signals Used in Data Transfer

Signal	Function	Driven By
AD31 – AD0	Multiplexed address and data	Initiator
$\overline{C/BE3}$ – $\overline{C/BE0}$	Command/byte enable	Initiator
$\overline{TRDY}$	Target ready	Target
$\overline{IRDY}$	Initiator ready	Initiator
$\overline{FRAME}$	Frame	Initiator
$\overline{DEVSEL}$	Device select	Target

© Cengage Learning 2014

Figure 12.53 illustrates a PCI read cycle in which an initiator reads data from a target on the PCI bus. PCI operations are divided into phases. This transaction begins with of an *address phase* in which the initiator addresses the target by putting an address on the A/D bus. At the same time, the initiator puts a 4-bit code on the command/byte enable bus and asserts $\overline{FRAME}$ to indicate a valid address and transaction code. The PCI bus doesn't need explicit read or write signals because the dual-function command/byte enable lines are used (in the command mode) to indicate a read or a write access. $\overline{FRAME}$ remains asserted until the initiator is ready to complete the final data phase (there may be more than one data phase in a burst transaction). The data phase is completed when the target samples $\overline{IRDY}$ asserted, $\overline{FRAME}$ negated, and the target has also asserted $\overline{TRDY}$. All potential targets have to latch the address on the next rising edge of the clock because the address will be removed from the multiplexed address/data bus.

Target devices decode the address and the command on the *command/byte enable* bus. The address/data bus is now turned round and is available to the target. When the target detects an access, it claims the transaction by asserting the device select line, $\overline{DEVSEL}$. The PCI protocol specifies that a target must claim the bus within a predetermined time, otherwise the transaction is terminated. The initiator stops driving the command/byte enable bus with the current command value and puts out the appropriate byte enable signals for the duration of the transaction. These byte-enable signals permit one to four bytes to take part in the current transaction. The initiator also asserts its target ready signal, $\overline{TRDY}$, to indicate that the data on AD31 to AD0 is valid. In this example, a single data transaction takes place (i.e., the current transaction is not a burst-mode operation). Figure 12.54 illustrates a PCI read cycle in which the address phase is followed by three data phases.

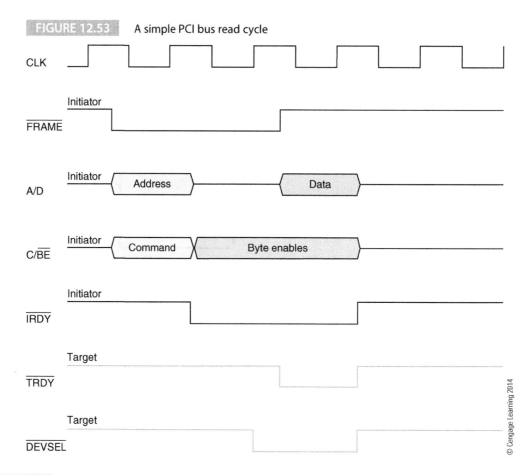

FIGURE 12.53   A simple PCI bus read cycle

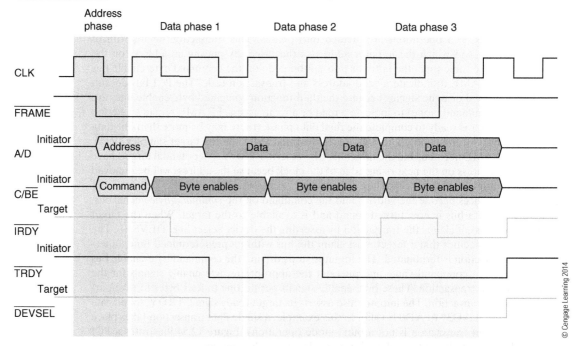

FIGURE 12.54   A PCI bus burst read cycle with three data phases

Multiprocessing and the PCI Bus

We will look at multiprocessors in Chapter 13. However, this panel introduces at the PCI bus's support for multiprocessing (note that the PCI bus was designed before multi-core processors became generally available). Putting two or more microprocessors on the motherboard itself, as opposed to a plug-in card, solves the problem of heat dissipation and simplifies access to the fast front side bus.

Adding processors to a motherboard creates two problems. How is mutual arbitration achieved between the individual processors, and what about the needs of rest of the system? Remember that the rest of the system includes devices and subsystems that also require access to the system bus for the purposes of DMA, etc. The diagram below illustrates a PC-based multiprocessor system with two processors connected to the processor-memory bus and a bridge between the I/O bus and processor bus.

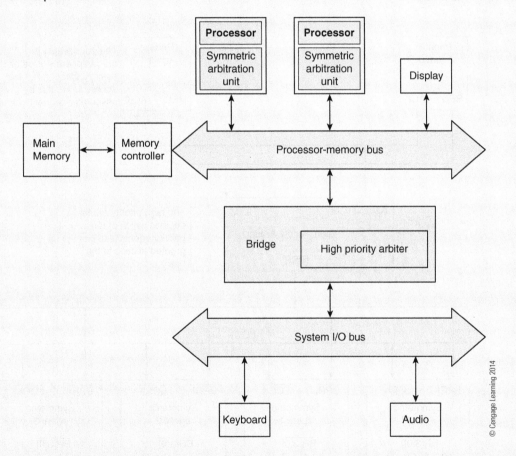

The nature of the multiprocessor system with two to four processors arranged side-by-side suggests that the processors should be treated equally, and therefore, a fair arbitration algorithm should be implemented (that is, *symmetric multiprocessing*, SMP). Each processor has a personal identifier that is used in the arbitration process. When a processor wants to access the bus, it indicates its number to all the other processors. Each processor maintains a *next-in-line* counter that indicates which processor has the right to go next. The winning processor gets the bus, and all processors update their next-in-line counters to ensure that the new winner is now at the bottom of the list.

Instead of giving each processor a binary code to indicate its identity, Intel adopts the NuBus solution and hardwires an identification code on the processor's pins. Intel puts the next-in-line counter on the chip along with the arbitration mechanism, making it easy to design a motherboard and accommodate one or more processors.

(continued)

The rest of the PC system still needs to access the bus from time to time. Intel treats the four multiprocessors as a *single entity* and gives it a single *I want the bus* priority request input called BPRI. When a peripheral wants to access the bus, it makes a request to the appropriate arbiter that generates a BPRI signal to request the bus. This BPRI signal is fed to the four multiprocessors in parallel and they give up the bus. They do collectively what a single processor did on its own.

Whenever one of these processors wants the bus, it arbitrates with the other processors and gets it (subject to the fairness constraints of the SMP arbitration mechanism). When a priority agent wants the bus, that agent gets it from the cluster of processors exactly as if they were dealing with a single processor. The diagram below illustrates the relationship between symmetric multiprocessing and priority arbitration, in which a device other than a processor requests access to the bus.

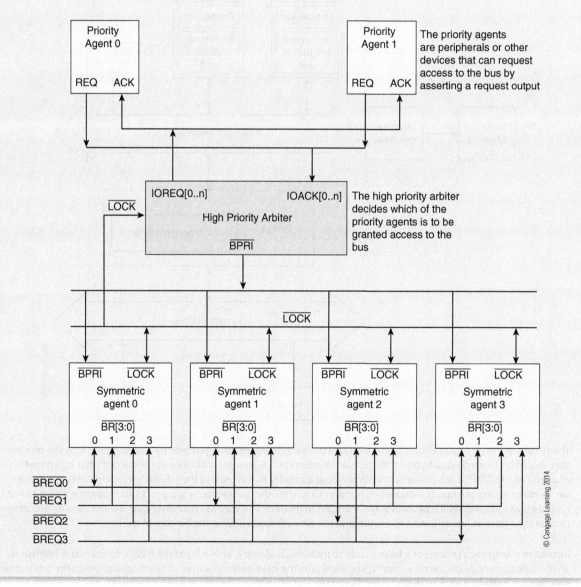

Although multi-core processors have become commonplace today, the PCI bus was used in conjunction with Pentium processors to implement multiprocessor systems. The panel provides further details of Pentium based multiprocessor systems.

12.7.2   The PCI Express Bus

The PCI Express bus was developed in a world of high technology that demanded ever increasing performance and ever decreasing cost. Its design goals were to cost less than the existing PCI bus, use off-the-shelf technology (boards, connectors, and circuits), support mobile, desktop and server markets, and be compatible with existing PCI-based systems. So, no pressure there. Version 1.0a of the PCI Express was introduced in 2003, and the third major revision, PCI Express 3.0, was released in November 2010. The greatest difference between PCI and PCIe is that the PCI express uses *serial transmission* to transfer data from point to point.

Figure 12.55 demonstrates the difference between the PCI bus and PCI Express protocols. The PCI bus protocol has echoes of the ISO standard for the Open Systems Interconnection (OSI) model, which attempts to divide any communications system into seven abstract layers, where each layer performs a certain function for the layer above it. The advantage of a layered protocol is that any layer can be replaced with new technology without having an impact on the layers above or below it. For example, the physical layer could be changed from copper tracks on a PCB to a fiber optic link with no changes to the data link or transaction layers.

FIGURE 12.55   PCI and PCI Express abstract architectures

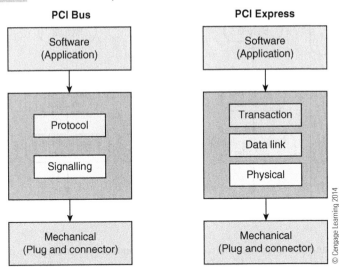

The lowest level of the PCI Express protocol is the *physical layer*, which is responsible for transferring the bits from point-to-point. Instead of the PCI's conventional parallel data/address bus, the PCI Express uses a serial bus where data is transmitted bit-by-bit along a single line or along a pair of lines using differential encoding. A particularly interesting aspect of the PCI Express is that two serial data paths are provided—one for each data direction—that is, a PCI Express card can both read and write data to the bus simultaneously and support full-duplex operation. The two signal paths are collectively called a *lane*, and it is possible to implement multiple lanes. Performance scales linearly with lane numbers, so you can have a ×1 bus, a ×2, bus, a ×4 bus, a ×8 bus, and so on. A single lane supports a peak data rate of 250 MB/s in each direction. If you have a 16-lane system using duplex transmission, the total effective data rate is 8 GB/s. Figure 12.56 illustrates the concept of lanes.

Conventionally, information at the electrical level in digital systems is specified with respect to the ground or chassis; that is, a signal at greater than 3.0 V is interpreted as high, and a signal at less than 0.3 V is interpreted as low. PCI Express uses two signal paths to

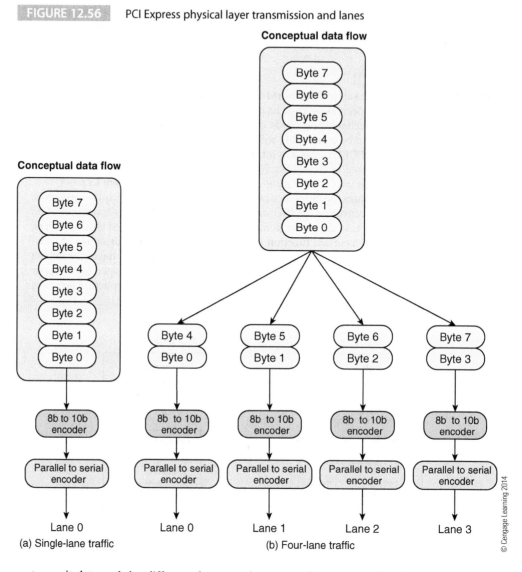

FIGURE 12.56   PCI Express physical layer transmission and lanes

Conceptual data flow

Byte 7
Byte 6
Byte 5
Byte 4
Byte 3
Byte 2
Byte 1
Byte 0

Conceptual data flow

Byte 7
Byte 6
Byte 5
Byte 4
Byte 3
Byte 2
Byte 1
Byte 0

| Byte 4 | Byte 5 | Byte 6 | Byte 7 |
| Byte 0 | Byte 1 | Byte 2 | Byte 3 |

8b to 10b encoder

Parallel to serial encoder

Lane 0

(a) Single-lane traffic

8b to 10b encoder | 8b to 10b encoder | 8b to 10b encoder | 8b to 10b encoder

Parallel to serial encoder | Parallel to serial encoder | Parallel to serial encoder | Parallel to serial encoder

Lane 0 | Lane 1 | Lane 2 | Lane 3

(b) Four-lane traffic

© Cengage Learning 2014

transmit data and the *difference* between the two conductors contains the information; for example, the signals may be[4] $+V$, $-V$ or $-V$, $+V$. The advantage of differential transmission is that it is more immune to interference (noise and other signals induced by capacitive or inductive coupling). This form of signaling is called *low voltage differential signaling* (LVDS). If both conductors of a pair pick up interference, it does not affect the information, which is determined by the difference between the two conductors. The encoding of the bit stream across the serial link ensures that a clock signal is embedded in the data stream and the data stream can be used to recover a clock signal; this means that designers do not have to worry about the distribution of clock signals and delays between data and clocks caused by different path lengths in the signals (an important factor when signaling at 2.5×10^9 bits/s).

The bit encoding is called 8b/10b because each 8-bit byte is transmitted at 10 bits in order to equalize the number of 1s and 0s transmitted and to ensure that a clock signal can be recovered from the data signal. Further details of 8b/10b encoding are provided in the panel (see page 785).

---

[4] An LVDS data path uses current switching. A current of either $+3.5$ mA or -3.5 mA is injected into the bus at the transmitter. At the far end of the bus, there is a 100 Ω termination resistor. Ohm's law, $E=IR$, indicates that the steady-state voltage between the two conductors is 350 mV.

Serial versus Parallel Transmission

Information can be transmitted serially bit-by-bit or in parallel. A 64-bit bus can transmit 64 bits of data between two points at the same time. The prototype serial data transmission system was the electric telegraph that was devised to send messages round the world using Morse code. Later, the serial telegraph gave way to the serial telephone connection. Computers used serial data links largely to move information between the processor and slow peripherals such as the printer.

As time passed, computer buses were developed, and more and more sophisticated high-speed parallel buses were launched. It was axiomatic that the parallel transmission of m bits was m times faster than the serial transmission of m bits. And yet . . . we have Serial ATA technology for hard disk drives, USB and Firewire buses, and the PCI express bus—all of which use serial transmission.

Of course, parallel transmission is much, much more expensive than serial transmission because of the design of buses and the cost of connectors. However, there are other, more subtle factors involved in data transmission. First, bus lines act as antennas both radiating information when data is transmitted along the bus and receiving information (interference) from neighbouring bus lines. At high speeds, the problem of electromagnetic interference becomes severe—even more so on a parallel bus with multiple lines. A serial bus requires only two lines, which makes it easier to shield the bus from interference.

High speed parallel buses suffer another problem—data skew. When 32 bits are applied to 32 data paths and the data transmitted to 32 receivers somewhere along a bus, the 32 bits do not reach their destinations at exactly the same time because of varying delays in bus drivers, signal propagation paths, and receivers. The range of times at which the signals are received is called data skew. In the past, data skew was not a significant problem because signalling (clocking) speeds were low. Today, with signalling speeds greater than 1GHz, data skew is a major problem. Sophisticated design techniques and deskewing circuits are necessary. Serial data cannot suffer this problem and data must be received in the order it was transmitted. Of course, it is possible to suffer skew between data and the clock (a clock is required to break the data stream into individual bits). However, many serial data transmission systems encode the serial data in such a way that a clock signal can be derived from the data itself.

Parallel data systems are inherently half-duplex, which means that data can be transmitted from one point to another in one direction at a time. Of course, you can turn the channel around, but that takes time for the transmitter and receiver to reverse roles. Although a serial channel is also half-duplex, it is easy to provide two serial channels, one in each direction, to achieve full-duplex simultaneous two way transmission.

The PCI express bus is interesting in the sense that it provides the best of both worlds. It provides a pair of serial channels to give a full-duplex capability at a very high signalling speed. However, it also permits multiple lanes (each composed of two serial channels) to provide an effectively parallel bus.

Finally, parallel buses require complex control structures for data transmission protocols and auxiliary functions such as bus arbitration and interrupt handling. A serial bus is forced to adopt a one-size-fits-all strategy, and all functions are transmitted as serial messages without the need for additional control lines.

PCIe Data Link Layer

Before we go into detail about the data link layer, it's necessary to describe the format of data transmitted over the PCIe bus (or over other serial buses with layered protocols, including the Internet itself). Data transmitted across systems that support layered protocols looks a bit like a Russian doll with multiple layers of encapsulations. At one end of a link, the application takes a packet of data and wraps it between delimiters. Then, the application layer hands the package to another layer (e.g., the data link layer) and that layer in turn wraps up the data with its own terminators. The data link layer passes the data to the physical layer and that too adds beginning and end flags. When the data is transmitted, it now has three flags at each end of the message—each flag added by a layer. Figure 12.57 illustrates the concept of encapsulation using a system with three protocol levels or layers. Each protocol layer adds a header and a tail to the information passed from the layer below. Equally, each layer strips the header and tail off before passing the message to the next level.

FIGURE 12.57   Encapsulating a message

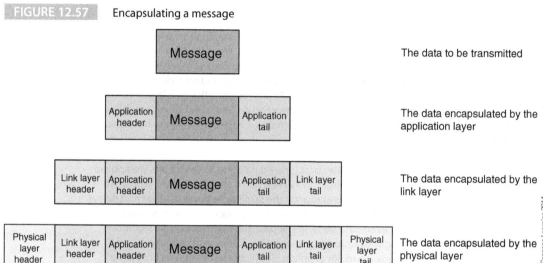

Figure 12.58 illustrates the PCIe bus message structure where the elements of a message are shown in blue and the protocol layers in grey. The highest level is the *transaction layer* that consists of a header and the actual message itself. The header defines the nature of the data message and includes information such as the address of the data—we will look at the header in more detail later. The transaction layer's tail is an error-detecting code (ECRC). The transaction layer provides four address spaces: memory, I/O, configuration, and *message space*. The term *message space* indicates the set of messages that perform control functions; that is, a bus operation is specified as an encoded command (much like a CPU op-code) rather than by means of a dedicated control line. The message space is used to support interrupts and resets (and any other forms of hardware management). By using message space, the physical control lines of buses like the VMEbus or NuBus become unnecessary.

FIGURE 12.58   The serial data structure on the PCIe bus

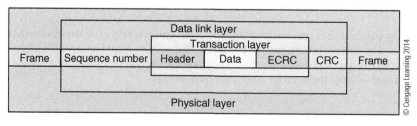

8b/10b Encoding

8b/10b encoding is a means of transmitting serial data using 10 bits to carry 8 bits of information and is a relatively modern data encoding/decoding mechanism (Widmer and Franaszek at IBM in 1983). The additional two bits (or 25% redundancy) per byte improves the performance of the transmission mechanism. The ten-bit code is constrained to contain five 1s and five 0s, or four 1s and six 0s, or six 1s and four 0s. This ensures that there are no long series of only 1s and 0s. Moreover, a mechanism called *running disparity* is used to ensure that there is an equal number of 1s and 0s on average; this is necessary to ensure that there is no dc component in the signal.

The 8b/10b coding algorithm operates by taking an 8-bit byte, splitting the 8 bits into a 3-bit group and a 5-bit group. These two groups are encoded separately with the 5-bit group becoming a 6-bit code word and the 3-bit group becoming a 4-bit code word. These two code words are merged into a 10-bit code by rearranging the order of the bits. Note that in addition to the 256 valid code words generated by this algorithm, there are a further 12 code words reserved for control functions.

The data link layer adds a header consisting of a sequence number to the frame from the transaction later, and appends a CRC (cyclic redundancy check) that is responsible for ensuring error detection at the data link layer level; that is, if any of the information on the frame at the data link layer is corrupted, the error will be detected and a retransmission requested. The data link layer transmits a package only when it knows that there is a buffer available to receive it—this avoids the retransmission of lost packages (a common feature of earlier serial data link protocols such as HDLC).

The specification of data exchanges on the PCIe bus is moderately complicated. Figure 12.59 gives the general structure of a packet header that consists of 12 or 16 bytes. As you can see, there are eleven fields (of which two are reserved for future use). This structure means that all the hardware overhead associated with conventional buses becomes redundant (arbitration, interrupt, handshaking, etc.) at the price of increased latency and reduced efficiency due to the data overhead.

FIGURE 12.59   PCIe Packet header structure

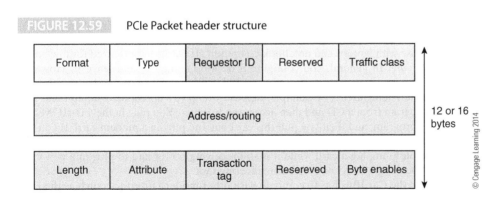

12 or 16 bytes

© Cengage Learning 2014

12.7.3   CardBus, the PC Card, and ExpressCard

Just as the desktop computer has evolved, so has the portable computer. The heavy *luggable* computer of the 1990s led to the notebook and then the netbook. The netbook itself has been challenged by the non-computer computer, the tablet, and iPad-class computers. This train of evolution has required corresponding developments in memory technology for bulk storage, processor technology for performance and energy efficiency, display technology for low power and high resolution, and interface technology. In this section, we look at how interfaces have developed to allow small external devices to interface to a computer's system bus. Although these interfaces can be used in desktop computers, they are primarily the province of the notebook. Post-netbook computers do not, in general, fall into this group. First, they are just too small to support the type of interfaces we describe. Second, in my opinion, there has been a retrograde step towards proprietary interfaces (for example, Apple's iPad interface). Finally, the use of WiFi allows interconnectivity without wires. In what follows, we look at interfaces to PCI-based computers and then to PCI express based computers.

The introduction of the CardBus and the *PC Card* that slots into this bus greatly enhanced the functionality of the laptop computer. In short, the CardBus is an extension bus intended for laptop computers that allows you to plug tiny, credit-card size modules into a laptop. These modules provide a massive amount of functionality, from miniscule hard drives to wireless LAN adaptors to satellite-based global positioning systems. Of course, almost as fast as external devices can be designed and manufactured, they are moved from the outside of a computer to its inside. For example, all laptops and netbooks now include internal WiFi, and even 3G wireless connections are now being included as standard in some laptops. GPS will be included in order to compete with GPS in cell phones.

PCMCIA is a registered trademark of the non-profit Personal Computer Memory Card International Association, which is a trade association that promotes PC Card technology, creates technical standards, and provides information.[5] Some literature refers to the PC Card as a PCMCIA card, but this use is technically incorrect because PCMCIA is a trade association and not a card. Typical functions provided by PC Cards are

- A/D conversion and data acquisition
- CD-ROM interface
- Cellular phone interface
- Ethernet LAN adapter
- GPS (Global Positioning System) card
- Infrared wireless LAN adapter
- Joystick interface
- Memory including Flash, SRAM, and hard drives
- Modem and Ethernet combination cards
- Modem and ISDN cards
- Radio LAN adapter
- SCSI adapters

The CardBus is a 32-bit bus that was standardized by the PCMCIA in 1996 and is largely intended for use with laptops and notebook computers. It has been optimized for low power and *hot swapping*. In general, you configure a desktop computer and leave it. However, because a laptop's power consumption is so critical, you do not wish to burden it with functions that are not being used. For example, suppose you have a notebook computer and wish to install a program from a CD and then access the Internet. You plug in the CD-ROM card and load the program, and then you pull this card out and plug in a modem card. If you had to power-down and reboot between each card change, you'd be waiting forever. Hot swapping allows you to unplug a card while it is still receiving power and to plug in a new card. Electrically speaking, this is not a trivial engineering task.

Although the CardBus has to compete with other modern interfaces such as the USB bus and FireWire, it does have a significant advantage. The CardBus connector is both an interface and a *device bay*. The PC Card slot can neatly hide away a complex plug-in function without trailing wires or the external power supply required by all too many peripherals. Some manufacturers have adopted this approach to USB and have incorporated some systems within an enlarged USB plug (such as security devices and flash memory systems).

---

[5] PCMCIA ceased to exist in 2009, because the ExpressCard has replaced the PC Card from about 2007 onward.

PC Card History

Like almost every other aspect of the PC, the CardBus has evolved. Some early laptop computers implemented PCMCIA cards based on the ISA bus operating at 8.33 MB/s. In 1985, the Japan Electronic Industry Development Association (JEIDA) began the process of standardizing PC Card Technology and released four specifications by 1990. The Personal Computer Memory Card International Association (PCMCIA) was founded in 1989 by a consortium of companies to standardize memory cards to promote trade.

In mid-1990, PCMCIA standard 1.0 was released. This defined the card's 68-pin electrical interface and the Type I and Type II PC Card form factors. The card form factor and its 68 pin connectors was originally defined by JEIDA in 1985. The original standard was designed for interfaces to *memory*. By the end of September 1991, the PCMCIA standard was updated to version 2.0, which extended the 68-pin interface to provide I/O functionality. Standard 2.0 added support for dual-voltage memory cards. By the end of 1991, version 2.01, which added the Type III card and specified an enhanced software interface, was released.

The next upgrade was in mid-1993 when the API software interface and the Card Information Structure (CIS) were upgraded. In 1995, the PCMCIA card disappeared and the PC Card standard was introduced. This standard required a Card Information Structure on every PC Card. In 1997, PCMCIA 5.0 with a 32-bit 33MHz bus was implemented in laptops (these allowed 32-bit transfers at a rate of 132 MB/s).

The CardBus card configuration software is called *Card and Socket Services* and provides plug-and-play functionality via the automatic allocation of system resources. The configuration software also implements the hot-insertion that lets you change cards without powering down and rebooting. Socket Services can be added to a computer as a component of a device driver, or it can be built into the BIOS. The Card Services software layer is an application programming interface that sits above the socket services. This API allows you to include Internet applications within PC cards.

CardBus Cards

All CardBus PC Cards are attached to their host system through a 68-pin connector. The cards are enclosed by a metal shroud to enhance their signal integrity. Older PC Cards had an 8- or 16-bit interface operating at ISA bus speeds. CardBus provides a 32-bit multiplexed address and data path operating at up to 33 MHz and providing a peak bandwidth of 132 MB/s (in contrast with a 64-bit PCI bus that provides a peak 528 MB/s at a 66 MHz clock).

The CardBus data transmission protocol is virtually the same as the PCI bus. Although earlier PCMCIA devices could act only as bus slaves, the CardBus supports bus masters, which removes any limitation on their functionality. Because the CardBus and PCI bus share the same protocol, it is relatively easy to construct an interface between them.

Although CardBus and PCI interfaces and protocols are very similar, some differences are necessary in order for the CardBus to support older PCMCIA cards. For example, the electrical interface specified by CardBus does not support the high-speed signals (i.e., signals with edges rising and falling at greater than 1 V/ns). Because power consumption is related to clock rate, the CardBus provides software control of the clock frequency. Dual-voltage operation at 5 or 3.3 V is implemented by sensing the card's needs when it is first attached and then adjusting the voltage. The CardBus standard makes provision for further reductions in operating voltage in future revisions.

ExpressCard Cards

PC Cards were rather large for the new generation of highly mobile, lightweight laptops appearing around 2005. In 2003, a new family of plug-in cards operating at 2.5 Gbit/s called ExpressCards was developed. ExpressCard 2.0 was introduced in 2009 with a maximum transfer speed of 500 MB/s. Figure 12.60 shows the two ExpressCard formats alongside a CardBus card. Both cards have the same connector interface: one version is 34mm wide and the other 54mm wide. It is interesting that the ExpressCard supports two interfaces: a PCI Express bus interface and a USB 2.0 serial bus interface. The ExpressCard standard specifies that the host system should be able to supply 1,000 mA at 3.3 V, an auxiliary 3.3 V supply at 250 mA, and a 500 mA at 1.5 V. As in the case of USB bus (see later), this current is available to power ExpressCard modules.

The next part of this chapter looks at external buses that connect a computer to systems that may be outside the case such as printers and even other computers. We begin with the SCSI bus that is used to connect high-performance peripherals to a computer.

FIGURE 12.60   ExpressCard formats

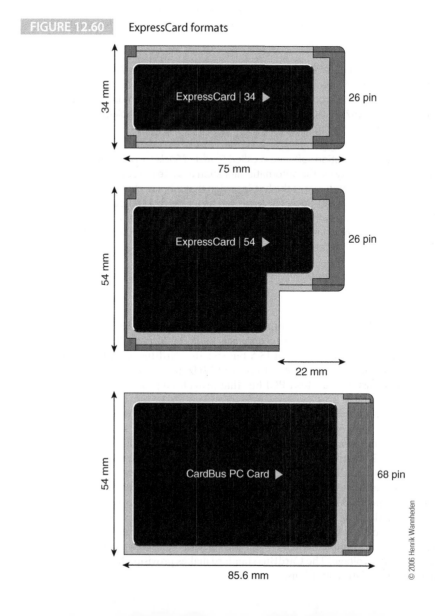

12.8   The SCSI and SAS Interfaces

One of the earliest external buses designed to link a computer and peripherals is the SCSI bus that has proved remarkably enduring in a world of change. At one time, it was the preferred bus in professional and high-end systems. Today, it is in decline in the face of very low-cost high-performance buses such as USB and FireWire. We include it here because of its role in computer development and the influence it has had on the design of other buses. We also introduce the modern Serial Attached SCSI (SAS) that has adopted low-cost serial technology. SAS-based products first became available in 2005.

The *Small Computer System Interface* (SCSI) is an 8-bit parallel bus dating back to 1979, when the disk manufacturer Shugart was looking for a universal interface for its family of hard disks. The SCSI bus is a parallel data bus that incorporates an information exchange protocol optimized for the bus's intended use, the linking of disk drives and other storage systems to a host computer. Figure 12.61 illustrates the concept of the SCSI bus, which was originally called the SASI bus (Shugart Associates Systems Interface). In 1981, Shugart and NCR worked with ANSI to standardize the SCSI bus, which became X3.131-1986 in 1986.

FIGURE 12.61   The SCSI bus

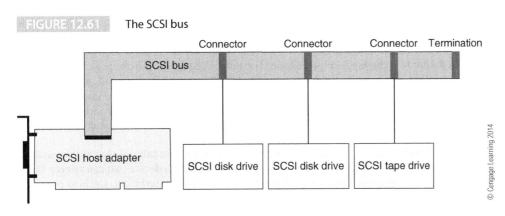

The original SCSI-1 bus operated at 5 MHz permitting up to seven peripherals to be connected together. A family of SCSI buses with a common architecture and different levels of performance (Table 12.7) has been developed. The specification was revised in 1991 providing a *fast* SCSI-2 bus at 10 MHz and a *wide* bus with a 16- data path. Ultra SCSI or SCSI 3 was the next step with a clock rate of 20 MHz. All SCSI systems support asynchronous data transfers, but SCSI 2 also supports faster synchronous data transfers. USB 3.0 bus provides a theoretical limit of 4.8 Gbps or 600 MB/s.

TABLE 12.7   Versions of the SCSI Bus

Version	Width	Data Rate MHz	Throughput MB/s
SCSI-1	8	5	5
Fast SCSI	8	10	10
Fast Wide SCSI	16	10	20
Ultra SCSI	8	20	20
Wide Ultra SCSI	16	20	40
Ultra-2 SCSI	8	40	40
Wide Ultra-2 SCSI	16	40	80
Ultra-3 SCSI	16	80	160
Ultra 320 SCSI	16	160	320
Ultra 640	16	320	640

As well as 8-bit and 16-bit versions of the SCSI bus, there are three electrical interfaces. SCSI buses support a conventional *single-ended* electrical interface in which the level of a data signal is transmitted with respect to a ground reference level of 0 V. The long leads to external SCSI disks act as antennas and pick up extraneous signals leading to errors and incorrect operation. A better mechanism than transmitting data as signals referenced to ground is to use *differential signaling* in which data is transmitted as the voltage across a *pair* of wires (like the PCI express). Differential signaling requires twice the number of connections but is more immune to interference because interference increases the signal in both leads of a pair, without affecting the difference in the signal levels between adjacent conductors. This ability to reject interference that appears in both wires is called *common-mode rejection*. Differential-mode transmission permits transmission paths up to 25 m in contrast with the maximum path length of 3 m imposed on conventional *single-ended* SCSI systems. Because the SCSI is a bus, it requires termination units at both its ends to prevent reflections (see the panel at the beginning of this section).

Another electrical interface is a low-voltage differential (LVD) SCSI that also uses differential signaling but at a lower voltage. Low-voltage signaling technology improves the performance of the receiver, reduces the power consumption, and allows the interface to be integrated with the SCSI chips. Some SCSI systems use the *fiber channel*, which is a serial interface using fiber optics. Although optical systems require data converters and parallel to serial (and serial to parallel) converters, they are highly immune to interference.

Given its origin, it is not surprising that the SCSI bus has become a popular disk interface in medium- to high-speed professional systems. A SCSI disk drive includes an internal SCSI interface and requires no more than a connector and a length of ribbon cable to link it to a SCSI bus. The same statement is, of course, also true of the once popular but now obsolete IDE interface found in PCs. SCSI bus interface cards for PCs are widely available at very low cost, although high-performance fast and wide SCSI PCI cards can be quite expensive.

The SCSI bus has its own terminology, and SCSI literature talks of the *initiator* and the *target* (terms also used in PCI bus terminology). An initiator is a device that can select a target on the SCSI bus and send commands to it. The initiator is the interface to the host processor (i.e., the SCSI bus controller), and the target is a peripheral device using the SCSI bus (e.g., a disk drive or a printer). The SCSI bus supports up to eight initiators and targets. A system may have more than one initiator, but only one may be active at a time—most systems have a single initiator. Each device connected to the SCSI bus has an address or SCSI ID. SCSI devices are allocated eight *logical unit numbers* (LUNs) that allow a single SCSI device to be partitioned into one to eight individual logical devices.

The SCSI bus operates on a *message-passing* principle, where the initiator tells the target what action it wants to be performed. The reason for implementing this mode is bound up with the SCSI's role as a processor-disk bus. It takes a relatively long time for the target to respond to a command from the initiator (for example, seek a given track on the disk). Moreover, the command transmitted by the initiator can be quite long and consist of ten or more bytes. A feature of the SCSI bus is that it can be *released* and used by another device while the current target is busy carrying out its allotted task.

SCSI Signals

The SCSI 1 bus consists of 18 data lines, data-flow control lines, and control lines described in Table 12.8. The bus operates as a *state machine*—at any instant the bus is in one of a fixed number of states. Figure 12.62 provides a simplified state diagram for the SCSI bus and Table 12.9 describes the states.

In the *bus free* state, the bus is idle and no device is using it. This phase is indicated by the negation of both the $\overline{\text{SEL}}$ and $\overline{\text{BSY}}$ signals. During the *arbitration* phase, one or more devices attempt to take control of the bus. In the *selection or reselection* phase, an initiator asks a target to carry out a given task. In the *information transfer* phase, data is exchanged between the target and the initiator.

TABLE 12.8   The SCSI 1 Bus Signals

Pin	Function
DB(0) to DB(7)	The 8-bit data bus
DB(P)	A data parity bit
$\overline{\text{I/O}}$	A data direction signal
$\overline{\text{REQ}}$	A request signal used by the target to request a transfer cycle
$\overline{\text{ACK}}$	An acknowledge signal used by the initiator to complete a data transfer
$\overline{\text{C/D}}$	A control/data message asserted to indicate the transfer of control or data
$\overline{\text{MSG}}$	A signal that indicates the information being transmitted is a message
$\overline{\text{BSY}}$	A busy signal that is asserted to indicate the bus is not free
$\overline{\text{SEL}}$	A signal that is asserted during a bus selection operation
$\overline{\text{ATN}}$	An attention signal asserted by an initiator that wishes to access a target
$\overline{\text{RST}}$	A reset signal that clears the bus and resets all initiators

© Cengage Learning 2014

FIGURE 12.62   State diagram for the SCSI bus

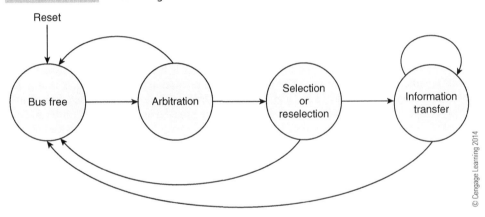

© Cengage Learning 2014

TABLE 12.9   SCSI Bus States

State	Description
Bus free	In the bus free state no devices are controlling the bus or transferring data. This state is indicated by the negation of the $\overline{\text{SEL}}$ and $\overline{\text{BSY}}$ lines.
Arbitration	When a device wishes to take control of the bus and become an initiator, it enters the arbitration state by asserting $\overline{\text{BSY}}$ and putting its ID on the data bus. If no device with a higher priority claims the bus, it goes ahead and claims the bus by asserting $\overline{\text{SEL}}$.
Selection	In the selection state the initiator selects (i.e., specifies) a target device and issues commands asking it carry out a specific operation. Selection is done by issuing the logical OR of the target device and the initiator on the bus.
Reselection	Because a target is able to give up the bus during a long operation, a reselection state is needed during which the target reclaims the bus.
Command	The command phase is used by the target device to request a command for the initiator. The target sets the $\overline{\text{C/D}}$ signal low to indicate a command and sets $\overline{\text{I/O}}$ high to indicate an output operation.
Data	During the data phase, data is transmitted between the initiator and target.
Message	The interface is controlled by messages sent between the initiator and target.
Status	The target returns a code to the initiator, indicating an operation's status.

© Cengage Learning 2014

SCSI Bus Transactions

We now examine a typical SCSI bus transaction. Suppose the host processor wishes to read a file from a hard disk. The processor first accesses the SCSI initiator, which is, of course, the interface between the CPU and the SCSI bus. The initiator will almost certainly be a single integrated circuit and appear to the host processor as a memory-mapped peripheral.

The initiator examines the $\overline{\text{BSY}}$ line to determine whether the bus is currently free. If it is free, the initiator arbitrates for the bus. Each of the SCSI bus's eight data lines is associated with one of eight levels of arbitration (one for each possible SCSI bus user). The SCSI bus supports *seven* peripherals—the SCSI bus initiator has its own address which accounts for the eighth device.

All a requesting device need do is to assert the data line it was assigned during installation. The selected data line is asserted active-low by means of an open-collector output, exactly as in the case of NuBus arbitration. If the device taking part in the arbitration process finds that one or more higher-order data lines are being driven low by other devices, it leaves the arbitration process and defers to the device with the higher priority.

Once the initiator has gained control of the bus, it enters the *selection* state that allows it to address or select another device on the bus. The initiator asserts the select line, $\overline{\text{SEL}}$, together with the data bit corresponding to the identity of the device it is selecting and the data line corresponding to its own identity. For example, if the initiator is device number 7 and it wishes to select device number 3, the initiator asserts $\overline{\text{SEL}}$ and sets DB(7) and DB(3) low. The initiator can now release the $\overline{\text{BSY}}$ line.

The selected device (i.e., the target) detects that it is being accessed and asserts $\overline{\text{BSY}}$ itself to take control of the bus. The initiator then releases the two data lines that it had been asserting, and negates $\overline{\text{SEL}}$. Note that the selection and *reselection* phases are similar. Reselection takes place when a target was previously selected and the operation terminated by a bus free phase before the target's operation had been completed. Once a device has been selected or reselected, the bus enters its information transfer phase. This phase is really one of several operations: command, message out, message in, status, data out, and data in.

Data transfer can now take place over the bus between the initiator and the target, using the $\overline{\text{REQ}}$ and $\overline{\text{ACK}}$ handshake signals to control the flow of data. $\overline{\text{REQ}}$ requests a data transfer and $\overline{\text{ACK}}$ acknowledges it, just like the two-line asynchronous data transfer protocol we described earlier. The two control signals $\overline{\text{I/O}}$ and $\overline{\text{C/D}}$ determine the direction of data flow and whether the information on DB(0) to DB(7) is device data or a command. Initially, the target asserts $\overline{\text{C/D}}$ to request a command from the initiator and $\overline{\text{REQ}}$ to trigger a data transfer (at this stage the target knows it's been addressed but doesn't know why). The initiator places a byte on the data bus and asserts $\overline{\text{ACK}}$, which is a handshake to $\overline{\text{REQ}}$. The target may then request more command bytes, depending on how it interprets the first command byte.

When a long command is issued to a target, the initiator can release the bus to other traffic and then perform a reselection phase later. This leaves the SCSI bus free while the target is busy carrying out its task.

The SCSI bus proved very popular because it provided high-performance and was well-suited to medium- to high-speed peripherals such as hard disk drives, optical (CD) drives, printers, and optical scanners at a time before USB, FireWire, and WiFi were widely available.

SCSI Messages and Commands

SCSI bus messages between initiators and targets vary in length between one and three or more bytes. A typical message sent by a target to an initiator is *command complete* indicating that a command has been finished and a status byte has been returned to the initiator. Similarly, a *disconnect* message from the target tells the initiator that the connection is going to be suspended and will have to be remade later.

The SCSI bus also supports commands that act on the devices themselves (e.g. the disk drives and printers). Commands are transmitted from an initiator in the form of a frame of sequential bytes. A typical SCSI command format contains the following information.

Operation Code—A one-byte command that defines the operation. The operation code also includes a 3-bit group code that indicates the type of the command.

Logical Unit Number—A one-byte value, LUN identifies the device attached to the target. This feature permits multiple functional units to be connected to one target.

Logical Block Address—An optional two- to four-byte value that provides the starting address for the data transfer.

Transfer Length—An optional one-byte parameter that defines the number of logical blocks to be transferred in a read or a write operation.

Parameter List Length—This provides the length of the list of parameters used to configure the operation on a LUN.

Allocation Length—An optional parameter that specifies the maximum number of bytes an initiator has allocated for returned data.

Control—A one-byte control that determines how the SCSI bus is going to behave. For example, it can be used to indicate that the bus should not be freed after the current command because the initiator is going to issue another command.

The actual commands in the SCSI bus specification relate to the operation of disk drives, printers, CD-ROMs, and several other peripherals. However, some universal commands are provided that act on all peripherals. For example, the *inquiry* command asks the target to supply its own parameters and those of its associated peripherals.

The SCSI bus is interesting because it blurs the distinction between the bus and its interface and the devices connected to the bus. Moreover, it includes a command subset that is highly specific to the devices that are expected to use this bus.

Serial Attached SCSI (SAS)

SCSI was a *white elephant*—venerable and revered but not particularly useful by the 1990s when USB and FireWire appeared. However, two decades of development had gone into extending SCSI and SCSI-based systems, and some did not want to through away a proven technology simply because it was slow, bulky, and expensive.

Serial attached SCSI (SAS) is an attempt to retain the best of SCSI while moving closer to the world of USB and PCIexpress. Serial attached SCSI throws away SCSI's antiquated physical layer and replaces it with a low-cost, high-performance serial interface. This serial interface offers a 12 Gbits/s throughput which is faster than the maximum read/write rates of hard disks. When introduced in 2004, SAS supported 3 Gb/s which was increased to 6 Gb/s in 2007 and 12 Gb/s in 2010. The topology of SAS is point-to-point, unlike SCSI which is a multipoint bus.

The physical layer of SAS uses differential signaling and cables up to 10 m (33 ft) are supported. The bulky termination networks required by parallel SCSI cables are no longer necessary. SAS defines two low-level layers, physical and PHY, that divide the traditional physical-layer-level functions (plus some traditional link-layer level functions) into two layers. The physical layer level is concerned only with connectors and voltage levels, where the PHY layer is concerned with data encoding, link initialization, speed negotiation, and reset sequences. SAS uses 8b/10b encoding.

An important aspect of SAS is that cables and connectors are physically compatible with the SATA interface now used by all modern hard disk drives. Consequently, the same low-cost connectors can be used in both conventional PCs and SAS-based systems. Moreover, SAS supports a SATA Tunneling Protocol (STP) that enables conventional hard drives to be connected to SAS architectures.

(continued)

Each SAS device has a unique address which is assigned at the time of its manufacturer; that is, each device has its own address when it is purchased. Addresses are assigned by an international organization called the Name Address Authority (NAA), which overseas SAS address allocation.

SAS is an excellent example of the convergence of technologies (serial transmission and SCSI networks) and a good example of how the computer industry overcomes bottlenecks (e.g., the inherent slowness of parallel SCSI).

12.9   Serial Interface Buses

Once upon a time, a serial bus swapped speed for simplicity. Parallel buses require multiple data paths and correspondingly complex plug-socket arrangements and cables, whereas a serial bus carries data a *bit at a time* using two signal paths: one for the data and a ground return. A serial data link using a fiber optic cable requires a single data path. In the 1970s when RS232C serial connections were slow, you had to use a parallel data bus if you wanted speed. Early PCs had a parallel port with eight data lines using a DB-25 connector that you could use to interface to printers using 25-way ribbon cable. Some printers still have such basic parallel interfaces, although they are becoming increasingly obsolete. Modern serial data links are simple, fast, and effective.

The great advantage of the serial data bus is its ease of use. The buses we describe next[6] have had a tremendous impact on computing because they provide low-cost, high-perfor-

RS232C – The Nightmare

Probably most computer users today have no idea what RS232C is, or was. In 1969, the Electronic Industries Association created a standard for the interface between two different devices: DTE (data terminal equipment) and DCE (data communications equipment). These two entities were the modem that transmitted data over the telephone line at a rate of about 9,600 bits/s and the computer. RS232 was developed long before personal computing.

RS232C interfaces transmitted data serially a bit at a time using asynchronous serial transmission (the data did not use a synchronizing clock and each character was transmitted as 8 to 11 pulses representing ASCII-encoded characters). The data connecters were DB-25 25-pin plugs and sockets (25 wires being used to transmit two channels of data, one on each direction at 9,600 bits/s). As well as the two signal paths, several control signals such as a *ring indicator* were included.

First-generation microprocessor systems used existing RS232C standard connections to link other devices to computers such as printers. The situation was a nightmare because RS232C was being employed for purposes unintended by those who designed it, with the result that interfacing often involved manually setting switches in the computer or even having to solder wires between pins. The notion of *plug and play* did not exist and you had to get a driver for each device you connected to a serial link. Often, simply getting a printer to work was a real achievement. Today, you take a printer and a computer and connect them together with a USB cable. End of story.

---

[6]In fact, we have already described a serial bus, the PCI Express when we were discussing parallel backplane buses because it builds a virtual parallel bus on top of a physical serial data path.

mance solutions to linking computers with peripherals and even with other computers. We begin with an introduction to the serial bus that was to have a profound effect on computer communications: the Ethernet.

12.9.1   The Ethernet

We introduce serial buses by briefly describing the *Ethernet*, developed to support local area networks at 10 Mbits/s. The Ethernet dates back to 1978 and now has the IEEE standard number 802.3. Today, it is the standard for low-cost local area networks operating at 100 Mb/s or 1 Gb/s.

In an Ethernet, all devices are connected to a single cable and no special control lines are required. A device, or node, can transmit serial data onto the common bus that is connected to all other devices. The Ethernet transmission cable is now available in four versions: a thick coaxial cable, a thin coaxial cable, a very low-cost *twisted pair* of unshielded conductors, and fiber optics. Table 12.10 defines the nomenclature for these connections. The 10 refers to the speed of the link, the *Base* refers to *baseband* transmission (as opposed to *modulated* carrier systems), and the 5/2/T/F refer to the media type. The nomenclature 1000BaseF refers to Ethernet media operating at 1 Gb/s using fiber optics.

TABLE 12.10   Ethernet Physical Media Naming Conventions

Name	Media Type	Maximum Segment Length	Maximum nodes/segment
10Base5	Thick coaxial	500 meters	100
10Base2	RG58 (thin coaxial)	185 meters	30
10BaseT	UTP (twisted pair)	100 meters	1024
10BaseF	Fiber optic	2,000 meters	1024

© Cengage Learning 2014

The data is transmitted in the form of *packets* or *frames*, as seen in Figure 12.63. An Ethernet packet consists of seven fields starting with an 8-byte (64-bit) *preamble* that synchronizes the clock at the receiver with the transmitted bit stream. The first seven bytes have the bit pattern 10101010. The last byte of the preamble is the *start of frame delimiter* with the special pattern 10101011 that indicates the start of a frame.

The 48-bit destination and source address fields indicate where the packet is from and where it is going. The length field defines the size of the packet, which is between 46 and 1500 bytes long. However, since the minimum data field must contain at least 46 bytes, data fields shorter than 46 bytes are padded to bring the size up to 46 bytes. Finally, a 32-bit frame check sequence provides a powerful error-detecting code.

Unlike a parallel bus, the Ethernet is just a message exchange mechanism (i.e., there are no control signals). One node on the Ethernet simply posts a frame to another node on the Ethernet. How the data field in a package is interpreted is not part of the Ethernet specification. The physical layer of the Ethernet uses a *baseband* cable with phase encoded data transmitted at 100 Mb/s or 1 Gb/s. The term baseband means that the digital data is transmitted directly without the need for modems.

No two nodes can access the Ethernet simultaneously without their messages interfering destructively with each other. When two messages do overlap in time, a *collision* occurs and

FIGURE 12.63   Ethernet frame

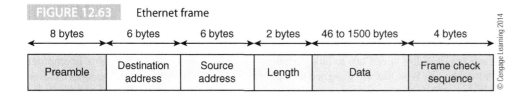

© Cengage Learning 2014

both messages are lost. Any node wishing to communicate with another node just goes ahead and transmits its message. If another node is transmitting at the same time, or joins in before the message is finished, the message is lost. The loss of a message is detected by an equally crude technique: If the sender does not receive an acknowledgment within a defined time-out period, it assumes that its message has been corrupted in transmission.

Without any control over when a node may transmit, there's nothing to stop two or more nodes transmitting simultaneously. The simplest form of contention control would be to let the transmitters retransmit their messages. Such a scheme cannot work, as the competing nodes would keep retransmitting the messages which would keep getting scrambled.

A better strategy on detecting a collision is to *back-off* or wait a random time before trying to retransmit the frame. Now it is less likely that the competing nodes would reschedule the transmissions for precisely the same time. Networks operating under this form of contention control are well suited to bursty traffic. That is, the arrangement works as long as the average traffic is very low (much less than the maximum capacity of the bus). If the amount of traffic rises, there comes a point where collisions generate repeat messages that generate further collisions and further repeats, and the system eventually collapses.

The Ethernet's contention control mechanism allows the node to listen to the bus before trying to send its frame. If a node is already sending a message, other nodes do not attempt to transmit. In Ethernet terminology, this is called *deference*. A collision occurs only if two nodes attempt to transmit at nearly the same instant. Once a node has started transmitting and its signal has propagated throughout the network, no other node can interrupt. For almost all systems this danger zone, the propagation time of a message from one end of the network to the other, is very small.

A further modification of this arrangement is to allow transmitters to listen to the bus *while* they are transmitting. Suppose a transmitter starts transmitting and, at the same time, another transmitter does likewise. After a very short time, both transmitters become aware that the bus is in use and abort their messages. In this way, the effect of a collision is reduced, because the transmitters stop as soon as they detect the collision. Once a station has started transmitting, it acquires the channel, and after a delay equal to the end-to-end round trip propagation time of the network, a successful transmission without collision is guaranteed.

The contention mechanism adopted by the Ethernet is called *carrier sense multiple access with collision detect* (CSMA/CD). When a station realizes that its packet is being corrupted by another packet, it reinforces the collision by transmitting a jam packet. If it stopped transmitting immediately, the other transmitter might not detect the collision. The collision would be detected indirectly much later by the error-detecting code that forms part of the transmitted frame. This process is inefficient and wastes time. Sending a short jam packet makes the collision visible to all listeners. After the jam has been sent, another attempt is made after a random delay. If repeated attempts fail, the random delay is increased as the sender tries to adapt to a busy channel. The minimum data packet size of 46 bytes is dictated by the need to detect collisions.

Unlike other communications systems, the Ethernet relies on the statistical nature of messages. It is possible, but improbable, that a node on the Ethernet might never get to transmit a message because it was always interrupted.

Ethernet is still used to link computers to cable modems and to other computers in a local area network. Today, both printers and memory (called network attached storage) can be connected to Ethernet-based networks. In recent years, WiFi has replaced Ethernet-based networks in the home because WiFi avoids the need for unsightly cables. However, Ethernet has continued to develop and most PCs now support a Gigabit Ethernet. The standard for a 10 Gb/s Ethernet was published in 2002 that supports both copper connections and optical fiber connections. In 2010, the IEEE published standards for 40 and 100 Gb/s versions of Ethernet.

Two serial buses are strongly associated with the personal computer: the USB bus and the FireWire bus. Both of these buses perform similar functions. The USB was originally designed as a low-cost, low-speed bus to link peripherals such as scanners and printers to a computer, while the FireWire bus was designed to support much faster peripherals such as external disks and camcorders. The development of the faster USB 2.0 bus in 2000 blurred the distinction between these two serial buses. By 2010, systems with USB 3.0 interfaces operating at up to 4,800 Mb/s were beginning to appear. We now describe the FireWire bus.

12.9.2   FireWire 1394 Serial Bus

The FireWire bus (now called the IEEE 1394 bus) is an example of a bus that began life as a company project and later became an industry standard. Apple developed FireWire in 1986 as a replacement for the then parallel SCSI bus for use in the professional audio and video world. FireWire is an Apple trademark—the same bus is called iLink by Sony.

Several factors have led to the design of the FireWire bus. The first factor is *cost*. Over the past four decades, microcomputers have been embedded in almost all domestic products—from TVs to washing machines. This is particularly true in the audio-visual area. If such devices are to be interconnected, any link must be relatively cheap—the consumer doesn't expect to pay a fortune for an add-on. The second factor is *size*. Electronic devices are getting smaller and smaller. When computers were large, the space taken by a bulky connector on the back was of little importance. Small devices require correspondingly small connectors (e.g., a hand-held video camera, a computer games console, MP3 player, or cell phone).

The third factor is *speed*. Year by year the rate at which computers are clocked and the speed at which data is transferred between digital devices has increased relentlessly.[7] In 1975, a 600 baud[8] modem was regarded as fast—by 1995, the 28.8 Kb/s modem was routinely used to interface computers to the Internet, and by 2000, the 512 Kb/s cable modem could be found in many households. Today many computer users have cable modems operating at 20 M or more bits per second. The fourth factor is *reliability*. Systems often fail due to faulty connectors, largely caused by repeated insertion and removal of plugs or the continuous flexing of cables. A reliable connector should have as few signal paths as possible.

In the early 1990s, a serial bus, initially called the P1394 High Performance Serial Bus, was proposed (the "P" indicates a provisional standard). A serial bus has many advantages over a parallel bus like the SCSI bus. In particular, a serial bus has only two conductors (one if a fiber optic path is used), which reduces the cost of cabling and the cost and size of connectors. The P1394 bus was designed to take advantage of the best available technology and can support several different *physical layers*; that is, the P1394 bus is not tied to one type of physical implementation. Some the most important features of the serial bus are as follows.

- Automatic assignment of node (i.e., the device connected to the bus) addresses—there is no need for address switches or other means of assigning addresses to nodes.
- Variable-speed data transmission—from over 24 Mbit/sec for TTL backplanes to 400 Mb/s for the cable medium. The IEEE 1394b specification doubled the number of bits per packet to increase the bus rate to 800 Mb/s.
- The cable medium allows up to 16 physical connections or *cable hops*, each of up to 4.5 meters.
- A fair bus access mechanism that guarantees all nodes equal access.
- Consistent with IEEE Std 1212–1991, IEEE Standard Control and Status Register (CSR) Architecture for Microcomputer Buses (ANSI).
- The 1394 Serial Bus limits the number of nodes on any bus to 63. However, up to 2^{16} nodes are supported by means of multiple buses linked via bus bridges.

---

[7]Of course, this statement must be qualified by stating that the maximum clock rate of processors has ceased to grow rapidly after about 2010. However, performance rates are continuing to be boosted by improvements in microarchitecture, specialized instruction sets, and the use of multicore processors.

[8]Data rates are normally expressed in bits per second. The term Baud comes from the days of telegraphy and is a measure of the switching or signalling rate (i.e., the number of symbols transmitted per second). If data is transmitted as two-level binary signals, data rate and Baud rate are the same. However, if, say, 8-level signals were being transmitted, the data rate would be three times the Baud rate, because each element could carry three bits.

The specification of the 1394 serial bus borrows features from the ISO model for OSI (open systems interconnection). In particular, the serial bus standard employs layers of abstraction, each of which carries out a specific task. The three layers of the 1394 interface are essentially identical to those of the PCI express bus.

Physical Layer—This level defines the way in which information is transmitted on the bus. It is concerned with the electrical properties of the medium and the signaling techniques. The physical layer takes *logical symbols* from the link layer above it, transports them across the interface, and delivers them to the link layer at another node.

Link Layer—This level describes the way in which packets of data are transmitted on the bus. The link layer takes data from the physical layer and deals with data framing, addressing, and error checking. The link layer passes the data to the transaction layer above it. In serial bus terminology, a complete link layer operation: arbitration, packet transmission, and acknowledgment is called a *subaction*.

Transaction Layer—This level is concerned with the end-to-end protocol used by nodes on the bus to communicate with each other. The transaction layer at one node receives data from the application and passes it to the link layer. At another node, the corresponding transaction layer takes data from the link layer and passes it to the application.

Figure 12.64 describes the 1394 serial bus's layered protocol. Each layer provides a specific service. Because each layer communicates with the layer above or below it in a tightly specified manner, it is possible to replace any layer by a system that performs the same function. That is, the 1394 serial bus is technology independent. For example, if a faster serial path is developed, the physical layer can be implemented with the new technology. The link and transaction layers that make use of the service provided by the physical layer are not affected.

FIGURE 12.64   FireWire's layered protocol

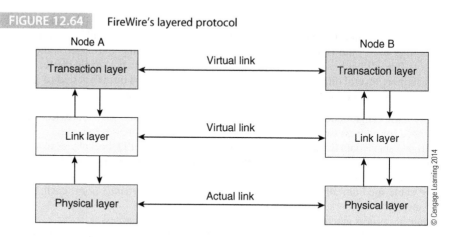

The general structure of the 1394 serial bus is described by Figure 12.65. The 1394 standard defines *two* serial bus environments. One is a *backplane environment* that exists within a processor system, and the other is a *cable environment* that links modules outside systems. The backplane environment has a bus topology and operates at either 25.576 or 49.152 Mb/s (depending on the technology). The cable environment supports rates of 98 to 400 Mbit/s (1394a) or 800 Mb/s (1394b). These two environments have different physical-layer-level topologies, arbitration mechanisms, and transmission rates. In 2007, S1600 and S3200 versions of FireWire were announced offering bit rates of 16 and 3.2 Gb/s, respectively.

As Figure 12.65 demonstrates, individual nodes on the serial bus may be located in different backplane environments. It is not immediately clear from Figure 12.65 that the

FIGURE 12.65   Structure of a FireWire system

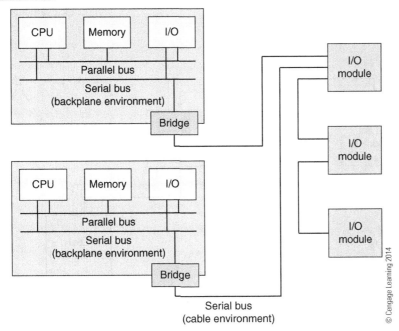

© Cengage Learning 2014

Serial Bus Topology

Data transmission systems are characterized by their topology that describes the way in which the individual nodes are related. The Ethernet is a bus because all nodes are connected to it and information is sent from one node to all other nodes on the bus—there is no routing mechanism to determine how information propagates on the bus. Another system is the ring in which all nodes are connected to each other and information flows from node to node. The following figure describes the bus and ring topologies.

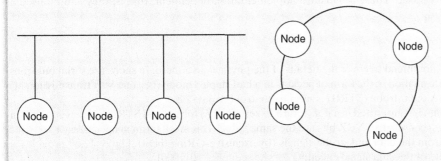

(a) In a bus all nodes are connected to a common data path

(b) In a ring each node is connected only to a right hand and a left hand neighbor

© Cengage Learning 2014

Another topology is the star topology that has a single central node through which all traffic passes. All other nodes are connected to this central node.

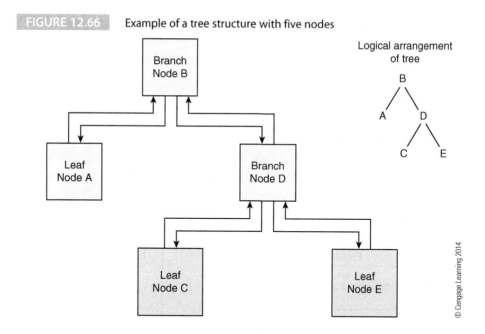

FIGURE 12.66   Example of a tree structure with five nodes

topology of the serial bus is really a *tree*. Figure 12.66 illustrates a simple tree structure with five nodes. Each node is either a *branch*, directly connected to more than one neighbor, or it is a *leaf* with only a single neighbor. Many applications of the serial bus *daisy-chain* the nodes together in a special case of a tree structure.

Serial Bus Addressing

Systems capable of supporting multiple nodes must distinguish between individual nodes. The 1394 serial bus provides 64-bit addressing, where the upper 16 bits of each address represent the ID (identification) of a node, allowing up to 64K nodes. The serial bus divides this ID into two subfields: the higher-order 10 bits specify a bus ID and the lower-order 6 bits specify a physical ID allowing 64 devices per cable. The remaining 48 bytes of an address is divided between register space, ROM ID space, initial unit space, and initial memory space. These 48 bits are implementation dependent; that is, they can be used as required.

The Physical Layer

We do not intend to cover the details of the physical layer here. In short, the serial bus transmits information in the form of packets in a half-duplex mode (i.e., one-way transmission at a time). A data strobe (STRB) is used to control the data flow.

Data is transmitted in a non-return to zero (NRZ) format and STRB changes state whenever two consecutive NRZ bits are the same value. This mechanism makes it easy to derive a clock from the data and strobe signals (by exclusive-ORing them). Figure 12.67 provides an example of the data signal encoding for the sequence 10110001.

The 1394 Serial Bus uses two data channels (see the IEEE 1394 panel) called TPA (twisted pair A) and TPB (twisted pair B) to provide a transmission path in each direction. The data path and strobe paths are crossed over with the strobe for TPA on TPB and vice versa. This arrangement is unusual; the twisted pair carries a data signal and a strobe signal belonging to the *other* channel.

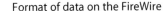

Format of data on the FireWire

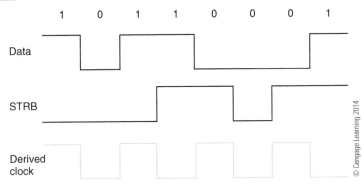

Hot Swapping and Hot Insertion

Some connectors cannot be removed or inserted when a computer is up and running (for example, the legacy PS/2 keyboard interface). Part of the problem lies in the hardware. For example, if a connector is inserted, the pins will not be inserted at the same time (it might seem instantaneous to a human inserting a card into a socket, but there may be many microseconds between the insertion of pins). That could cause unreliable operation. Similarly, if the software polls the device only at startup, removing a device or plugging in another will go unrecognized.

One of the unsung heroes of modern computing that most take for granted is *hot swapping* or *hot plugging*. Once the only way of plugging a new device into a computer, or any other electrical system, was to power down, plug in, and then to switch on again. Equipment sometimes had stickers written in unfriendly block capitals darkly warning you of the dire consequences of attempting to plug something in while the power was on. If you took a still smoldering computer to the repair shop, they'd look at you, shake their heads and say, "You plugged something in with the power on, didn't you?"

Things have changed. Hot plugging means you can plug in a USB device or pull the plug out without having to worry about the consequences. However, it is not a trivial matter because the requirements of both electrical and logical connections have to be dealt with. The connector and plug arrangement has to be designed so that signals are connected in the correct order. For example, the ground connection is usually made first. Some pins may be longer than others to ensure that first contact is made with, say, ground. This is partially to ensure that static charges are removed and partially to provide a reference voltage for the remaining connections. It is often important that signals be connected in the correct order and delay circuits are used to ensure that critical connections are not made until the rest of the circuit is up and running. In particular, tri-state bus drivers have to be turned off until they are correctly activated when they drive the bus.

Hot swapping can also be applied to hard disk drives. This is of great importance in highly reliable RAID memory systems where a faulty disk drive can be pulled out and a new drive added without having to perform a power down and make the system unavailable.

IEEE 1394 Hardware

The 1394 bus uses a six-element cable. There are two twisted pairs of data conductors to provide two independent data channels as well as two power supply lines. A seventh conductor surrounds the inner six conductors to shield the entire cable electrically.

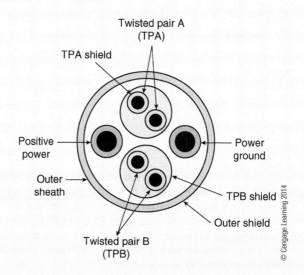

There are two types of 1349 bus connector: four pin and six pin. The six-pin connector provides full functionality, whereas the four-pin connector lacks the two power supply leads. The power leads can supply between 8 and 40 V at up to 1.5 A (40 V is relatively high in a world where most devices operate on 3.3 or 5 V). The figure below describes these connectors, which are somewhat more expensive than USB connectors.

IEEE 1394 PLUG PINOUTS		
Signal	6-pin Plug Pin	4-pin Plug Pin
Power +	1	–
Power Gnd	2	–
TPB –	3	1
TPB +	4	2
TPA –	5	3
TPA +	6	4

Arbitration

The serial bus implements several forms of arbitration (arbitration on the backplane serial bus is treated differently to arbitration on the cable bus). Here we describe *fair arbitration*, which occurs on the cable bus. Arbitration is *geographic* because the node closest to the root on a cable will always win.

The fair arbitration protocol is based on the concept of a *fairness interval* that consists of one or more periods of bus activity separated by short idle periods called *subaction gaps* followed by a longer idle period known as an *arbitration reset gap*. At the end of each subaction gap, bus arbitration determines the next node to transmit an asynchronous packet. Figure 12.68 illustrates this concept.

FIGURE 12.68   FireWire's arbitration protocol

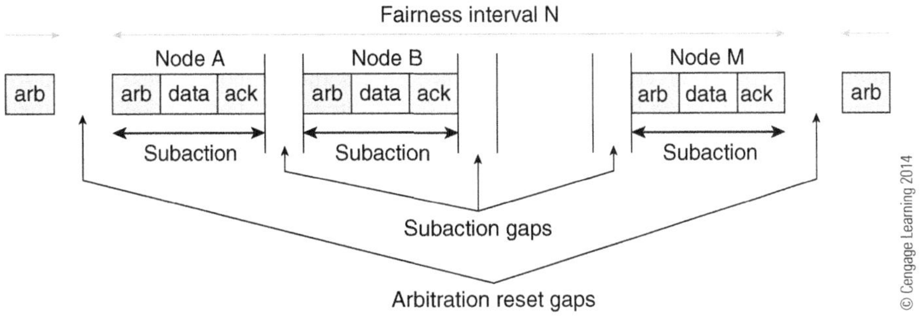

Fairness interval N

| arb | | Node A | | | | Node B | | | | Node M | | | arb |

Subaction — Subaction — Subaction

Subaction gaps

Arbitration reset gaps

When using fair arbitration, an active node can initiate sending an asynchronous packet exactly once in each fairness interval. An active node can arbitrate only if its arb_enable signal is set. The arb_enable signal is set to one by an arbitration reset gap and is cleared when the node wins the arbitration. This disables further arbitration requests for the remainder of the fairness interval. A fairness interval ends when arbitration by the final fair node is successful; this generates an arb_reset_gap since all nodes now have their arb_enable signals reset and cannot drive the bus. The arb_reset_gap re-enables arbitration on all cards and starts the next fairness interval. This process is illustrated in Figure 12.69.

The backplane environment also supports a form of urgent arbitration that permits a node to get more than its fair share of the bus time. However, the node demanding urgent arbitration is not permitted to hog the bus.

FIGURE 12.69   Fair arbitration

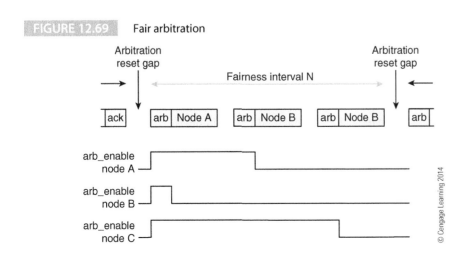

Initialization

Nodes on the serial bus don't require switches to select their addresses, because addressing is dealt with automatically and dynamically. Whenever a node joins the bus, a bus reset signal forces all nodes into a state that clears all topology information. Initially, the only information known to a node is whether it is a branch, a leaf or is isolated (unconnected).

After initialization, the serial bus topology is translated into a tree, where one node is designated a root and all of the physical connections have a direction associated with them pointing towards the root node. The direction is set by labeling each connected port as a *parent* (connected to a node closer to the root) or *child* port (connected to a node further from the root). Any loop in the topology is detected by a configuration time-out in the tree-ID process. The node that has all of its connected ports identified as children becomes the root.

The next step is to give each node an opportunity to select a unique physical ID and identify itself. Sending the self ID is done by transmitting one to four very short packets at the base rate onto the cable including the physical ID and some management information. The physical ID is simply the count of the number of times a node passes through the state of receiving self ID information before having its own opportunity to do so (i.e., the first node sending self-ID packet(s) chooses zero as its physical ID, the second chooses one, and so on). The identification process begins at the root node of the tree. At each node down the tree, the lowest numbered port is tested first. Note that a node is not required to decode the self ID packet, it merely has to count the number of self identify sequences since the bus reset.

The management information included in the self ID packet includes codes for the gap timer settings, the power needed to turn on the attached link layer, the state of the various ports (unconnected, connected to child, and connected to parent), and data rate limitations.

The self ID process uses a deterministic selection process where the root node passes control of the media to the node attached to its lowest numbered connected port and waits for that node to send an ident_done signal, indicating that it and all of its children have identified themselves. The root then passes control to its next highest port and waits for that node to finish. When the nodes attached to all the root's ports are finished, the root itself does a self identify. The child nodes use the same process in a recursive manner.

The Link Layer

The link layer uses the services of the physical layer below it to transmit information between the nodes in the form of packets. Data can be transmitted *asynchronously* by sending

Isochronous Serial Data Transmission

There's a saying that it's better to travel in hope than to arrive. Isochronous communication puts this principle into practice and ensures bandwidth and real-time communication at the expense of accuracy; that is, an isochronous message protocol guarantees that a source will have regular access to the channel but the normal error checking and retransmission mechanism (used by other protocols) will be abandoned. The function of isochronous transmission is to ensure continuity of transmission rather than accuracy in cases such as video conferencing, where the loss of a few frames of video is of no great consequence.

A typical way of implementing isochronous transmission is to assign regular time slots to a device in order to guarantee that the device receives a guaranteed fraction of the available bandwidth. If a reserved slot is not used, its bandwidth is lost. If the source has more data than can be placed in its reserved slot, that data is lost. Both the 1394 serial bus and USB support isochronous communication protocols.

The fundamental difference between asynchronous and isochronous communication is that asynchronous protocols guarantee delivery and reliability (via retransmission of lost data) whereas isochronous protocols guarantee only timing and do not attempt to recover lost data.

a packet to a specific address and receiving an acknowledgment. Data can also be transmitted *isochronously* by sending a sequence of fixed-length packets with simplified addressing and no acknowledgment.

The packets used by the serial bus include a header with source and destination addresses, packet type, and an error-detecting code (i.e., CRC).

In the next section, we introduce the *universal serial bus*, USB. Initially, USB was a low-cost version of FireWire offering a lower data rate and reduced performance at a far lower cost. FireWire was for video cameras and USB for keyboards and mice. However, the gradual improvement in USB standards has all but dealt FireWire a fatal blow.

12.9.3   USB

I would argue that the two factors contributing most to the success of the personal digital revolution that encompasses desktop computers, laptops, notebooks, MP3 players, both still and digital cameras, as well as cellular phones are *flash memory* and the *universal serial bus* (USB). Flash memory provides robust, high-density, low size, non-volatile storage at low prices, and USB technology allows you to connect almost any modern digital device to a computer—or even to connect two digital devices together without a host computer (e.g., a camera and a printer). Indeed, the USB is the single most successful digital interface ever, with over one billion USB devices sold by 2009. In this section, we provide an overview of the USB.

The universal serial bus was devised a consortium of companies and later established as a standard interface. Essentially, a USB bus uses low cost connectors and cabling to connect a computer to a range of peripherals from the mouse/keyboard/printer/scanner to memory devices such as external hard-drives and flash memory devices (so called pen drives). In many ways, USB is an alternative to FireWire.

USB is *host controlled*; that is, unlike other buses such as PCI, VMEbus, or NuBus, it does not support a multimaster arrangement. There can be only one USB master (or host) per bus. Figure 12.70 describes the *tiered star* topology of a USB system.

> ### USB History
>
> USB was developed by a consortium of Compaq, DEC, IBM, Microsoft, NEC, and Nortel in 1994. The first USB specification, 1.0, was introduced in 1996 and supported a data rate of 12 Mb/s. USB 1.1 was released in 1998 to deal with problems related to hubs. USB 1.1 was widely adopted.
>
> In 2000, USB 2.0 emerged to provide a maximum data rate of 480 Mb/s. USB had jumped into FireWire territory and it became the *de facto* standard for most PC interfaces to printers, external drives, keyboards, mice, and so on.
>
> It wasn't until 2009 that version 3.0 USB saw the light of day with an operating speed of 300 MB/s (i.e., 2,400 Mb/s), displacing FireWire from its eight-end niche. USB 3.0 is a giant leap forward over version 2.0 and requires a new cable format and technology.

At the top of the hierarchy in Figure 12.70 sits the host, which communicates with the computer and controls the USB bus. The host is connected to a hub, which is a device that distributes the USB bus to lower levels in the hierarchy. A hub may be connected directly to a peripheral (labeled *function* in Figure 12.70), to several peripherals, or to another hub. Each hub may be connected to a lower-level hub as Figure 12.70 demonstrates. A hub may be a stand-alone device (i.e., a USB port expander that has one input and several outputs) or it may be built into a keyboard, display unit, or even an external disk system.

USB – The First Two Generations

USB began life in the mid-1990s with versions 1.0 and 1.1 providing speeds of 12 Mbps (full speed mode) and 1.5 Mbps (low speed mode). Version 2.0 was launched in 2000 and provided 480 MB/s (high speed) as well as the two 1.1 mode speeds. Version 2.0 of the USB bus is backward compatible with earlier versions. Because the performance of version 2.0 is so much greater than version 1.1, we do not discuss earlier versions here.

FIGURE 12.70   The USB bus topology

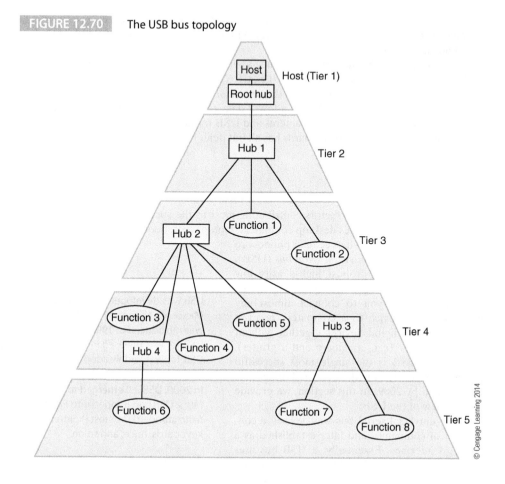

Electrical Characteristics

USB uses a simple multi-core connector with dedicated connectors. However, since the introduction of USB, peripherals have become smaller and smaller and USB cables have been forced to follow this trend with the result that there are now four basic sizes. Figure 12.71 illustrates both USB plugs and sockets. The computer end of the link uses type A sockets and type B plugs, which are fairly substantial. Type B plugs and sockets are used at the other end of the link (i.e., at a hub or a peripheral such as a printer). Mini-B plugs and sockets were developed for digital cameras, cell phones, and portable disk drives. Figure 12.72 provides a photograph of the USB plugs, and Figure 12.73 illustrates the structure of a USB system.

USB cables use four conductors. Data is transmitted differentially between a twisted pair of wires labeled D+ and D− in Figure 12.74. Recall that differential mode transmission increases reliability be rejecting common mode interference (a voltage induced in both wires does not affect the potential difference between the wires). The twisted pair is enclosed

FIGURE 12.71   USB connectors

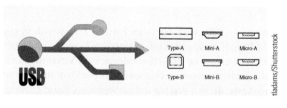

FIGURE 12.72   Photograph of USB connectors

Alan Clements

FIGURE 12.73   USB mechanical environment

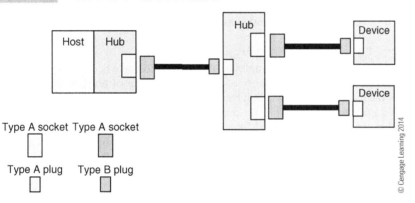

© Cengage Learning 2014

FIGURE 12.74   USB cable

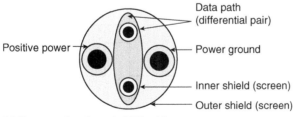

(a) Cross section through USB cable

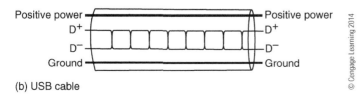

(b) USB cable

© Cengage Learning 2014

More Power

The USB's ability to deliver power was intended to support hubs and peripherals like the mouse and keyboard. In practice, many manufacturers have taken advantage of this facility, For example, the USB's built in power supply has been used to charge cell phones and MP3 players.

A new power mode was added to the USB specification called *battery charging*. In this mode, a host can supply up to 1.5A when communicating at 12 Mbps or 0.9A when communicating at 480 Mbps. Furthermore, by 2010, many of the world's cell phone manufacturers had provided micro USB ports to charge their phones.

in a metal shield to further reduce the dangers of picking up stray signals.[9] The specified maximum length of the cable is 5 meters. Of course, you can use a hub at the end of a 5m cable to increase the length of the USB path.

The USB cable includes two separate power cables that carry current to the remote devices (e.g., a mouse or keyboard).[10] USB peripherals may be *bus-powered* or *self-powered*. Today, many portable hard drives can be connected to a USB socket and run entirely off the USB bus. In a USB system with multiple hubs, it's necessary to manage the distribution of power. The USB power wires are at 0 and +5 V and can supply a maximum current of 500 mA. If the hub is bus-powered, it can provide a total of 500 mA to *all* downstream ports. If it's self-powered, it can provide a maximum of 500 mA to *each* downstream port. A device can't take more than 100 mA from the USB until after it has been identified by the host in a process called enumeration.

USB connectors are said to be *hot-pluggable*; that is, you can plug in a peripheral or unplug it without powering the host system down.

Physical Layer Data Transmission

At the electrical level, the USB employs NRZI data encoding where a logical 1 is represented by no change of level and a 0 is represented by a change of level. Sending a string of 0s requires the greatest bandwidth, and transmitting a sequence of 1s results in a constant level with no signal transitions. Figure 12.75 illustrates a NRZI sequence (this encoding format is very old in comparison with the 8b10b encoding used by the PCIe bus and is relatively little used because it is non-self-clocking and has a dc bias). The two signal levels on a USB bus are referred to as J and K.

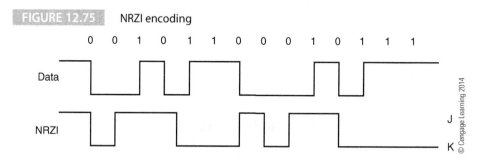

FIGURE 12.75   NRZI encoding

© Cengage Learning 2014

Because a long string of 1s would mean that there was no signal transition on the bus at all, a technique called *bit stuffing* is used to ensure a maximum interval between transitions. A zero bit is automatically inserted after every six consecutive ones prior to NRZI encoding. At the receiver, a transition is guaranteed at least every seven bit periods, which is sufficient to maintain transmitter-receiver synchronization. Since every group of six consecutive 1s is followed by a 0, the receiver simply deletes the 0 after six 1s.[11]

---

[9]If you are using a USB connection at its 1.5 Mbps low speed mode, you can employ simple unshielded and untwisted conductors up to a length of 3 m.

[10]FireWire has two versions—one that includes a power supply and one that does not.

[11]Bit stuffing is best known for its use in HDLC transmission, where a string of six 1s is used as a token or flag. To prevent a token appearing in the data stream, a 0 is inserted after five 1s. At the receiver, flags are removed (as they are the only sequence of six 1s) and then the 0 following any five 1s is deleted from the data stream as it must be a stuffed 0.

Logical Layer

Communication on a USB bus takes place between the host and a peripheral device. Logically, a communications channel exists between the host and device, and in USB-speak, this channel is called a *pipe*. The USB host can support 32 active pipes at any instant (16 up-stream and 16-down-stream pipes). The pipe is terminated at a peripheral by an *endpoint*. Although the physical structure of a USB system is a *tiered star* (i.e., there is a central node with other nodes connected toit, and any node can be the center of a local node), it is *logically* a star network (see Figure 12.76). All nodes have to communicate with each other via the single central node.

Less Speed

The USB has to share its bandwidth among several devices and provide bandwidth for protocol overheads. For example, although high-speed USB 2.0 has a theoretical maximum data rate of 57.3 MB/s (480 Mbp/s), the actual rate is in the region 20 to 25 MB/s.

When you first plug a USB device into a hub, the host begins an identification process (called *enumeration*) where the new device is reset, an address assigned to the device, and its information read. The host controller operates in a master-slave mode and polls each of the devices on the USB network for data. The USB is a polled bus and lacks an interrupt request mechanism.

Information is transmitted as packets, which may be token packets, data packets, or handshake packets. A USB transaction begins when the host transmits (broadcasts) a *token*[12] packet that defines the type and direction of the current transaction, the device address of the recipient, and the device's endpoint number. The actual data transfer then takes place according to the direction specified in the token. After the transaction, the recipient of the data sends a handshake package to indicate the receipt of the data.

Endpoints

A term associated with USB is *endpoint*, which is related to USB's logical topology, the star. Each USB device is logically connected to the host, and you can regard each device as having a set of *pipes* (data paths or channels) that run between itself and the host. Each of these data paths is terminated by an *endpoint* at the device. More specifically, an endpoint is a buffer that holds data that's been received from the host or which is waiting to be sent to the host.

There are 16 possible endpoints per device numbered 0 to15. Each data endpoint is unidirectional and either sends data to or receives it from the host. An IN endpoint sends data to the host and an OUT endpoint receives it. A control endpoint is bidirectional and is used to configure the device. Each device must assign endpoint 0 as a control endpoint. Note that there are actually 30 unidirectional data paths, because endpoints 1 to 15 can be configured as 1 IN, 1 OUT, 2 IN, 2 OUT, ...

USB supports four types of data transfer: control transfer, bulk data transfer, isochronous transfer, and interrupt data transfer. The control transfer performs operations such as the configuration of devices and the reading of status information. The bulk transfer mode is intended for the delivery of large quantities of data and employs three phases: token packet, data packet (up to 512 bytes), and a handshake packet. Bulk transfer is reliable in the sense that the data is protected by error-detecting codes and arrangements exist for dealing with lost data.

The isochronous transfer mode is a *best effort* mode that uses packets with a maximum size of 1024 bytes. There is no handshake phase and delivery is not guaranteed (i.e., a *best effort* delivery service). This mode supports a maximum bandwidth of 24 MB/s and is intended for applications such as streaming audio or video.

The interrupt transfer mode is used by devices that transfer data periodically such as a mouse or keyboard. This mode has a *bounded latency* (i.e., worst-case value) and involves a three-phase token/data/handshake sequence.

---

[12]The term 'token' is borrowed from the world of serial data links with a ring topology. To avoid contention, only one node has a packet called a token. If that nodes wishes to go ahead, it does so. If it does not wish to use the ring, it passes the token to the next node down the line. This concept comes from the single-line working of steam trains. A stretch of single line would have only one token – a ring and an engineer would take it to ensure that no other train could use that section of the track.

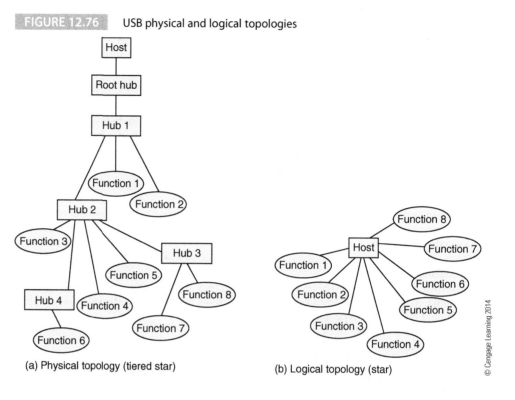

FIGURE 12.76   USB physical and logical topologies

(a) Physical topology (tiered star)

(b) Logical topology (star)

© Cengage Learning 2014

On-the-Go USB

The USB 2.0 doesn't support *peer-to-peer networking*; it covers only host to peripheral communications—the host is usually a computer. Today, the range of devices that require external communications is quite astounding varying from fax machines and copiers, to cell phones, to MP3 players, to audio-visual systems. It is not always convenient to implement a host-based system (for example, if you want to directly print an image from your cell phone).

The On-The-Go USB mode is a peer-to-peer (point-to-point) protocol which allows two USB devices to communicate without a host computer and provides direct device-to-device communications. On-the-go (OTG) operation was adopted in December 2001 when USB 2.0 revision 1a appeared. This extension to USB provided for communications between several OTG devices or between an OTG device and a conventional USB host. The revision also introduced a new range of USB plugs and sockets called Micro-A and Micro-B. The revised standard introduces the dual-role device (i.e., OTG) that can act as either a host or a peripheral. Moreover, a dual-role device must be able to supply a limited current to the bus of at least 8 mA (remember that most ORG devices are battery-powered).

OTG devices, when acting as a host, may support only a *targeted peripheral list*; that is, the OTG device may operate only in conjunction with certain specified peripherals – it is not intended to be a general-purpose host with full USB capabilities. A significant element of OTG technology is *the Host Negotiation Protocol* (HNP) that allows the transfer of control between two OTG devices.

USB 3.0

The most radical change to the universal bus took place in 2010 with the introduction of USB 3.0, which provides a ten-fold increase in performance and uses less power. Even more remarkably, USB 3.0 is physically compatible with USB 2. USB 3.0 is interesting because it is not really a development or extension of USB 2.0, but a *replacement bus* that coexists with USB 2.0; that is, a USB 3.0 bus incorporates a USB 2.0 bus as well. Figure 12.77 illustrates the structure of the USB 3.0 cable. The two data carrying conductors of USB 2.0 are maintained as well as the two power conductors. Two new differential pairs of conductors (SSRX and SSTX) have been added carry the new USB 3.0 data in a full duplex bidirectional mode. The additional functionality of USB 3.0 is called the *SuperSpeed* bus and provides a maximum speed of 4.8 Gb/s. To put this into context, it takes USB 2.0 13.9 minutes to transfer an HD movie, whereas USB 3.0 can perform the transfer in only 70 s. In short, USB 3.0 is an impressive feat of engineering that take a great leap ahead in terms of functionality and performance, while maintaining backward compatibility with a vast existing market of USB 2.0 users.

FIGURE 12.77   USB 3.0 Cable

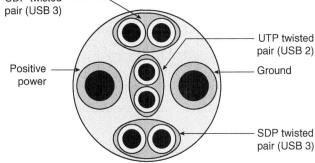

(a) Cross section through USB 3.0 cable

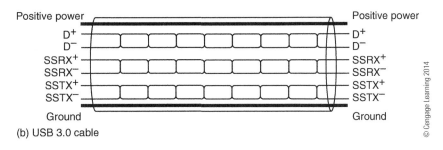

© Cengage Learning 2014

(b) USB 3.0 cable

USB 3.0 is compatible with USB 2.0 in the sense that you can plug a USB 2.0 plug into a USB 3.0 socket on the host computer and the link will behave exactly like a conventional USB 2.0 bus; that is, USB 3.0 defaults to USB 2. The key to this mechanism lies in the plug and socket arrangement. A USB 2.0 plug has a tiny board with four connectors on one side that mate with the socket. The USB 3.0 arrangement has two rows of connectors—one exactly matching the USB 2.0 standard and one providing the four new USB 3.0 signals plus a ground pin. In other words, USB 3.0 offers a true dual-bus mechanism—one identical to USB 2.0 and a new SuperSpeed bus.

Figure 12.78 illustrates the logical structure of the USB 3.0 bus in terms of protocol layers (like the ISO seven layer model for open systems interconnection). The SuperSpeed bus that adds all the new functionality to USB 3.0 is similar to the PCIexpress bus and uses 8b/10b encoding at the physical layer level. Data is *scrambled* on SuperBus. This is not a security mechanism but a means of converting data into a sequence that appears random in order to improve the electrical properties of the data link.

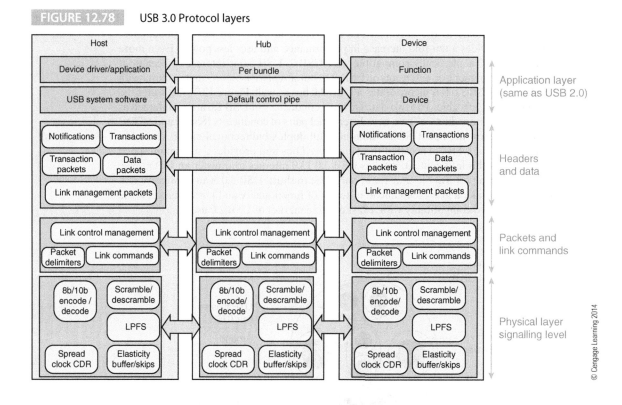

FIGURE 12.78   USB 3.0 Protocol layers

Figure 12.78 demonstrates end-to-end communication between a host and a device that goes through a hub. USB 2.0 uses hubs to expand the network. You can buy hub expanders with one USB input that is connected to the host and several USB outputs, or you can get, for example, video displays that have a USB port to the computer and two or three expansion USB ports for a mouse, etc. The USB 2.0 hub simply passes all messages on its input port to all output ports. The USB 3.0 SuperSpeed hub plays a more active role and routes packets only between the host and device taking part in the communication. Hubs are also able to *store and forward* SuperSpeed packets. The important point that Figure 12.78 makes is that USB 3.0 has a sophisticated tiered set of communications protocols. Consequently, any layer can be changed or improved without affecting the system; that is, there is room for USB 3.0 to grow.

Summary

In this chapter, we have looked at the subject of input/output. All too often, this topic is neglected in favor of the more glamorous processor and instruction set architecture. However, without the high level of connectivity of today's computers, we would not have seen such a growth in personal and ubiquitous computing. Modern computing is built on three pillars: low-cost, low-power, and high-performance processors; flash memory; and interconnectivity, including USB and WiFi.

We began this chapter with an introduction to the core concepts of input/output activities, buffering data, and handshaking. We also discussed the three strategies by which information is moved into and out of a computer: program-driven polling, interrupt-driven I/O, and high-speed direct-memory access that transfers data directly between peripherals and memory.

The core of this chapter is devoted to the bus, which is the vital high-speed highway between a processor and its memory and peripherals or between a processor and other processors. Like any other computer component, buses have had a long history of development and improvement. We have looked at several issues in bus design, in particular, the way in which various devices contend for or arbitrate for access to the bus.

The final part of this chapter looked at the serial bus, which is the modern wonder that lets you plug almost anything into a computer with having to worry about either the electrical interface or the way in which data is transferred.

Problems

12.1 A peripheral interface chip may have several internal registers. Explain the techniques available to reduce the number of address lines required to access these registers. Can you think of any techniques (not discussed in this text) that could be used to allow n address lines to select more than 2^n registers?

12.2 A computer with interrupt-driven I/O has an interrupt response time of 4 μs (i.e., it takes 4 μs to invoke the interrupt handler and begin executing the target code). Similarly, it takes 2 μs to return from the interrupt and begin executing the interrupted program.

   a. If this computer executes 10 instructions per microsecond and the interrupt handler is 10 instructions long, how efficient is interrupt handling?

   b. If interrupt handling is approximately 80% efficient, how large should the interrupt handler be (in terms of instructions)?

12.3 A computer spends 80% of its time executing user code and 20% of its time performing input/output operations. If you regard I/O as an unnecessary overhead, the efficiency of the computer is 80/(80 + 20) = 80%.

   Suppose that CPU performance is increasing at a compound rate of 40% per year and that the performance of I/O systems is increasing at a compound rate of 10% per year. In another two years, a new computer can be bought with both faster processing power and I/O performance. How efficient would the new computer be in two years' time?

12.4 An I/O system using DMA has a latency time of 10 ms between the peripheral requesting a data transfer and the transfer beginning. This system transfers data at the rate of one byte per μs. What is the peak efficiency of this arrangement for the following conditions?

   a. Data is transmitted in 1 KB blocks

   b. Data is transmitted in 10 KB blocks

12.5 This question asks you about the way in which the performance of I/O systems has changed and requires you to carry out your own research. Plot the maximum speed per data transfer as a function of time for at least two I/O buses of your own choosing.

12.6 If Ethernet addresses were uniformly distributed among all the people on the planet, would there be enough address space for each human? If there is more than one address per person, how many addresses would be assigned to each individual?

12.7 An interrupt-based I/O system uses polled interrupt handling where the processor has to interrogate each peripheral's interrupt status flag after an interrupt request. Suppose that an interrupt request takes 2 μs, an interrupt return takes 1.5 μs, and it takes 0.5 μs to poll a potential interrupter. The interrupt handling routine takes 5 μs. If an interrupt must be responded to within 50 μs, what is the maximum number of peripherals that can be supported?

12.8 Suppose you had a computer system with four interrupting devices, P1 to P4, where P1 has the highest priority and P4 the lowest. Design a logic unit with four interrupt request input, P1 to P4, and four interrupt request outputs, I1 to I4. The logic should covert any combination of inputs on P1 to P4 into a single request on I1 to I4 that reflects the highest priority input. For example, if P1 and P3 are asserted, output I1 is asserted and I2 to I4 are not asserted. Assume positive logic (i.e., the request level is 1).

12.9 Many years ago, I read that a computer had been designed for military and aerospace applications. An interesting feature was that it lacked any form of interrupt handling mechanism. Why do you think the designers took that decision?

12.10 A memory-mapped input device has two registers: status and input. The status register is at address PeriAddress and the data register at the next word address. The input mechanism is to poll bit 2 in the status register and then read the data when the status bit is asserted. When a data element has been read, it is stored in a simple table using pointer-based addressing. The loop is exited once the value 0 has been input. Write a simple polling loop to input data into the table.

12.11 What are the relative merits of open-loop and closed-loop data transfers? What factors would you take into account if you had to choose one over the other for a specific application?

12.12 A high-speed USB 2.0 port uses 30% of the available bandwidth. The bus is used to read a flash card with 700 images—each consisting of 25 MB. If the card is not the limiting factor, how long does it take to read the data?

12.13 In what ways is the USB bus better than the RS232C bus?

12.14 A microprocessor has a *prioritized vectored* interrupt handling mechanism. In this context, what are the meanings of the two words in italics?

12.15 Can the interrupt handling system make a computer more vulnerable to the dangers of malware (e.g., a virus)? If so, how and why?

12.16 Suppose you were designing a special-purpose I/O chip to relieve a high-performance workstation of the burden of I/O. What special instructions would you add to the computer's instruction set? What additional hardware interface would you include—if any?

12.17 Why is double-buffered input sometimes required?

12.18 Under what circumstances is a FIFO necessary in an I/O environment?

12.19 Figure P12.19 gives the timing diagram of the IEEE 488 bus's data transfer. This bus was first designed for use in an intelligent automated instrumentation (laboratory) environment. Multiple asynchronous devices can be connected to the IEEE 488 bus. Not all the devices operate at the same rate. A patented three-wire handshake is used to permit communication between one device and multiple receivers (or *listeners* in 488 terminology). The control signals use negative logic (the 0 state is the high voltage state) driven by open-collector gates (a device can drive a line to its low voltage state). The three control signals that control the sequencing of data transfer

are NRFD (not ready for data, which indicates that a device cannot currently accept data), DAV (data available, which indicates that a talker has data ready), and NDAC (not data accepted, which indicates that a listener has accepted data). Remember that the use of open-collector bus drivers implements a wired-OR circuit in which the line is pulled to a low level if any driver is pulling it low. Explain in plain language how the three-wire handshake works.

12.20 The clocked circuit in Figure P12.20 is an arbiter. Suppose two devices, 1 and 2, can request a resource such as a bus at any instant. The problem is to ensure that only one requester is granted the resource— even if both requests are made simultaneously. The circuit of device 2 has two request inputs plus a clock and two grant outputs. Explain how the circuit is able to perform its function reliably. Note that the circuit uses edge-triggered flip-flops and that it is immune to metastablilty.

12.21 An application is 80% computation intensive and takes up 60% of a computer's capacity (both CPU and I/O). Assume that this application will demand 20% more resources each year as it is updated; the computer's processing power increases by 20% per year; and the computer's I/O performance increases by 5%. How will the situation develop over the next five years in terms of processing power and I/O usage?

12.22 Why did IBM's Microchannel architecture fail commercially?

12.23 What is a *split-transaction* in the context of I/O, and when may it be necessary?

12.24 You have a system that transmits data between two points in packets of 1024 bytes. The transmitted bit rate is 10 Mbps. Before a packet is transmitted, a handshake must take place with the recipient, and this imposes a 1ms delay. You decide to redesign the system, and you are able to either reduce the latency by 20% or increase the bit rate by 10%. Which is the better option?

FIGURE P12.19

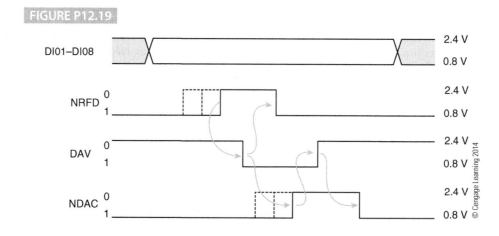

FIGURE P12.20

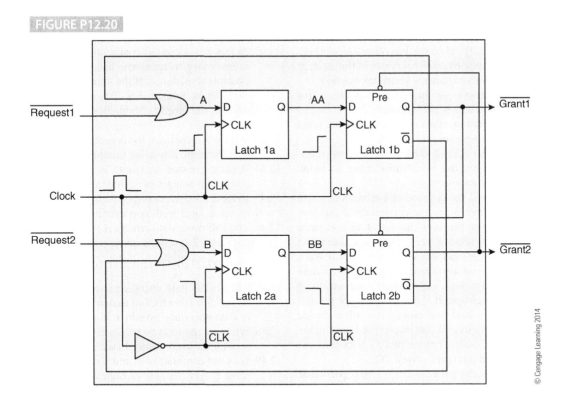

12.25 A bus is 80 cm long and signals travel down it at 70% of the speed of light in a vacuum. If it takes 2 ns for a bus driver to turn on and a receiver has a 1 ns setup time and a 2 ns hold time, how long does it take to perform a single end-to-end transaction?

12.26 You are in a laboratory at the top of a mountain 100 miles away from a base station. There are two links with the base station: a microwave link that transmits data at the speed of light and a wire link that transmits data at two thirds the speed of light. You decide to transmit a common data stream using both media for security. Because of the difference in transmission speeds, the base station receives the microwave data earlier than the cable data. Suppose you decide to perform some error processing on the microwave data before comparing with the cable data. How many operations can you perform if each operation takes 2 ns?

12.27 A memory-mapped input device in an ARM environment returns a 32-bit, four-byte value in the order ABCD. The processor needs the byte in the order DCBA. Write a fragment of ARM code to perform the transformation.

12.28 What are the relative merits of parallel buses with non-multiplexed addresses?

12.29 A fully populated NuBus has 16 possible interrupting devices. Device numbers 12, 1, 7, and 9 all interrupt at the same time. Show the sequence of operations that take place on the bus as each device requests and then gives up or retains the bus during the distributed arbitration until device 12 wins.

12.30 What is the difference between distributed arbitration and centralized arbitration?

12.31 A bus has a characteristic impedance of 75 Ω. It is terminated by a resistance of 200 Ω. What is the reflection coefficient at the termination?

12.32 A microprocessor is connected to one end of a transmission line. The transmission line has a characteristic impedance of 100 Ω. The transmission line is driven by the microprocessor at the near end, which has an impedance of 50 Ω. The line is terminated by 300 Ω at the far end.
   a. If a step voltage of 5 V is applied by the microprocessor, what is the magnitude of the pulse that initially propagates along the bus?
   b. What is the coefficient of reflection at each end of the bus?
   c. What is the voltage level at the far end of the bus after time t_{pd}, where t_{pd} is the end-to-end propagation delay of the bus? That is, calculate the voltage at the far end of the bus as the initial pulse reaches the end.

12.33 Explain how the VMEbus implements arbitration.

12.34 The VMEbus specifies features that are part of the VMEbus standard and features that are not part of the standard (i.e., the choice of bus release algorithm). What are the strengths and weaknesses of this strategy (i.e., mandatory requirements versus user options)?

12.35 What is an atomic or indivisible operation (or instruction), and why is it necessary?

12.36 Why is the machine status saved during an interrupt call and not during a subroutine call?

12.37 Explain why the RS232 asynchronous serial interface was probably the worst misfortune to befall early personal computers.

12.38 This chapter has not covered the Bluetooth wireless interface. Investigate the properties of this bus. Indicate where it fits into the world of computer interfaces and its relative merits with respect to wired buses, such as USB, and wireless busses, such as WiFi.

12.39 A computer has an average interrupt response time of p, takes q to process an interrupt, and takes s to return from interrupt. If a polled interrupt mechanism is implemented and takes x to poll a device and y to service a device that interrupted, how many devices can be polled before polling becomes less efficient than an interrupt-driven I/O?

12.40 Why do superscalar processors present a particular problem when implementing memory-mapped I/O. How can that problem be overcome?

12.41 Memory-mapped I/O may fail to operate in a system with cache memory. Why? What is the remedy?

12.42 A bus has a characteristic impedance of 100 Ω. It is necessary to terminate the bus with the highest possible impedance. If the maximum reflected pulse that can be tolerated is 30% of the incident pulse, what is the largest termination impedance that can be used?

12.43 Why is it necessary to terminate a SCSI bus? Note that the same is true for other buses.

12.44 What is a nested interrupt? What are the advantages and disadvantages of nested interrupts?

12.45 Does the Ethernet physical layer protocol represent an open- or closed-loop operation?

12.46 The following data stream is received on a USB 2.0 data link. What is the actual source data stream?

0001011111101111110111111011101111101111110111

12.47 The 8b/10b data encoding used by the PCI express bus is 20% less efficient than using some other forms of data encoding. So why is it used?

12.48 What are the relative merits of USB and FireWire as serial links in computer-based systems?

12.49 Why not combine USB and FireWire lines in a single cable to get the best of both worlds? Do you think that this is a good idea?

Example of Bus Reflections

As we only mention reflections on buses briefly earlier in this chapter, this example is provided to demonstrate the effect of reflections on a bus. For the bus below, calculate the voltage at the near and far ends of the transmission line (i.e., bus) from the time at which a step function is applied (i.e., $t = 0$) to the near end to four units of delay later (one unit of delay, t, is the end-to-end propagation time of the bus).

Note $Z_s = R_g$. The source end of the bus has a reflection coefficient of

$(Z_s - Z_O)/(Z_s + Z_O) = (33.3 - 100)/(33.3 + 100) = -(2/3)/(4/3) = -0.5.$

The far end of the bus has a reflection coefficient of

$(Z_l - Z_O)/(Z_l + Z_O) = (300 - 100)/(300 + 100) = 200/400 = +0.5.$

The initial pulse travelling down the bus is reduced by the potential divider formed by the source resistance, R_S, and the impedance of the bus, Z_O. If the pulse at the generator is +5 V, the pulse on the bus is 5 × 100/(33.3 + 100) = 3.75 V. We can now consider the behavior of the pulse as it is reflected.

The incident pulse of 3.75 V reaches the far end and is reflected. The reflected voltage is 3.75 × 0.5 = 1.875 V. Therefore, the total voltage at the far end of the bus is 3.75 incident voltage) + 1.875 (reflected voltage) = 5.625 V. This is greater than the pulse used to drive the bus. In the limiting case of an open circuit, the voltage at the end doubles.

The reflected pulse (i.e., 1.875 V) returns to the source end and is reflected again. The reflected component from the source end is 1.875 × −0.5 = −0.9375 V. The total voltage at the source end is now 3.75 (original signal) + 1.875 (incoming signal) − 0.9375 (outgoing signal) = 4.6875 V.

At each successive reflection, the amplitude of the pulse is reduced by 50% (i.e., by −0.5 at the source end and by +0.5 at the destination end. The time for the signal to travel the length of the bus is t_{pd}. The successive pulses travelling on the bus are

Time	Pulse	V_{near}	V_{far}
$t = 0$	+3.7500	3.7500	
$t = t_{pd}$	+1.8750		5.625
$t = 2t_{pd}$	−0.9375	4.6875	
$t = 3t_{pd}$	−0.4688		4.2188
$t = 4t_{pd}$	+0.2344	4.4531	

V

PROCESSOR-LEVEL PARALLELISM

This is the final section of this text, where we look at a new direction in microcomputer architecture: the world of the multi-core processor. We could have covered all the material in this part in previous chapters. However, the change from single-core to multi-core processing is so important and so significant that it merits a section of its own.

We have already stated that power is an important consideration in computer design—both at the level of portable equipment and at the level of high-performance computers. Microprocessors with multiple processing units currently offer a better power: performance ratio than conventional single-core processors. Moreover, multi-core processors cannot be considered to be simply faster single-core processors, because they are dependent on system topology (the relationship between the multiple processors) and have their own special software requirements (the need to share a task between individual cores). Finally, multi-core processing is driven by two specific applications: games consoles (the need for low-cost high-performance computing) and video processing (as required for high-speed display systems).

Processor-Level Parallelism

"Many hands make light work."
Proverb

"Too many cooks spoil the broth."
Proverb

"If it were done when 'tis done, then 'twere well it were done quickly."
Shakespeare, Macbeth

"Let all things be done decently and in order."
Corinthians 14:40

"This shift toward increasing parallelism is not a triumphant stride forward based on breakthroughs in novel software and architectures for parallelism; instead, this plunge into parallelism is actually a retreat from even greater challenges that thwart efficient silicon implementation of traditional uniprocessor architectures."
The Berkeley View, December 2006

"The Old Man left us tens of thousands of mainframes, hundreds of thousands of minis, horrific millions of PCs; Great Johnnie left us thousands of computer science departments!"
Herb Grosch comparing John von Neumann the computer scientist with John Watson Senior, Head of IBM (1993)

"Lasciate ogne speranza, voi ch'entrate."
Dante, Inferno, Divine Comedy (Abandon hope all ye who enter here)

"If you were plowing a field, which would you rather use? Two strong oxen or 1024 chickens?"
Seymour Cray

In this final chapter, we introduce parallel processing—again. We've already encountered the notion of parallel processing, because it's one of the core concepts of computer science. The term *parallel* is synonymous with *concurrent* and *simultaneous,* and it describes several related concepts. This chapter begins with a review of the nature of parallelism in a computer before we look at why we need to implement parallel processing. We also provide a short history of parallelism, because it is not a new topic. Indeed, parallel computing has been around for 50 years. What has changed is that it has moved from an esoteric topic involving computers costing millions of dollars into a feature of many modern games consoles costing hundreds of dollars. We also spend a little time discussing the relationship between power and clock frequency, because rising clock rates have led to increasing power levels, forcing manufacturers to adopt multiple processors running at lower clock rates.

Following the introduction, we introduce three key topics: the taxonomy of multiprocessing, the topology of multiprocessor networks, and their memory systems. Memory is an important consideration, because data has to be shared between processors and that is an important factor—if not *the* key consideration—in multiprocessor design.

We begin our examination of parallel processing with hardware multithreading, which is a halfway house between single core computing and multi-core computing. Multithreading replicates some of the features of a processor but not all of them. You could say that multithreading was the inverse of the superscalar processor, because the superscalar processor replicates the execution units, whereas a multithreaded computer replicates the registers and program counter, allowing several tasks to be run using shared execution units. After multithreading, we introduce the rapidly expanding world of multi-core processors. We also look into the new world of multiprocessors that are now incorporated in graphics cards: the GPUs.

The final section of this chapter addresses the principles of parallel programming and discusses some of its problems and limitations. We now briefly review the concept of parallelism.

Themes and Variations in Reverse

This text is subtitled *Themes and Variations* because we cover a topic or theme and then look at some of the variations. For example, we use the ARM as the principal means of introducing a computer architecture, but we also describe the IA64 VLIW architecture. This chapter is a little different. You could say that we have turned the world upside down by introducing the variations before the theme.

In some ways multiprocessing is a topic apart. Multiprocessing originally belonged to the world of supercomputers. It was frequently taught as a graduate-level advanced topic. Papers written on multiprocessor interconnections, or topologies, were amazing to behold, with immensely complex structures, such as hypercubes or Banyan networks. Today, the ability to fabricate multiple processors, along with their interconnections and caches on a single chip, has revolutionized the world of multiprocessing. After a long history of variations, the theme has arrived.

Because of the history of multiprocessing, this chapter does not begin with the multi-core processor. We take the historic approach, first looking at classical interconnection networks and then discussing the issues in memory technology. Finally we introduce the theme—the multi-core processor.

Pseudo-Parallel—Multitasking systems execute programs in parallel. For example, an operating system like Windows lets you surf the Internet while playing music from an mp3 file and copying data from one hard disk to another. I've called this *pseudo-parallel* because it is not parallel processing in a literal sense. Multitasking is sequential processing where tasks are switched so rapidly that they appear to the human observer to be continuous. In practice, multitasking may involve a degree of parallel processing in the sense that the operation of multiple disk drives may overlap.

Pipelining—In pipelined systems, several instructions are in the process of being executed at the same time. The execution of individual instructions is overlapped. Because there is one execution unit, only one instruction can be completed in a clock cycle. By overlapping the execution of instructions, better use is made of all the processor's hardware, allowing a significant performance enhancement. You could say that pipelining offers the smallest granularity of parallelism where the overlap is in the individual sub-components of an instruction (fetch, decode, execute, etc.).

Superscalar—In a superscalar processor, true parallel processing becomes possible because there are multiple execution units, and several instructions can be executed in the same clock cycle. The parallelism of superscalar processing is invisible to the user, because it takes place within the CPU itself. Superscalar processing is visible only as an improvement in the performance of a processor over non-superscalar versions. Note that compiler technology can be used to ensure that the processor can find more instructions to execute in parallel.

Computing's Core Problem

It may seem verging on madness to say that computers haven't changed a lot in the last 70 years. Especially when you consider that no human artifact has ever made such progress going from tens of tons to grams, from clock speeds of 100 kHz to 5 GHz, from prices of millions of dollars to cents. But… computers still perform the same basic operations of moving data from A to B and simple operations like addition and subtraction. Much of the progress has taken place because of changes in manufacturing technology. Enhancements in organization have arisen to fix problems rather than to change the nature of computing. For example, memory management and cache memory systems may take up a lot of a processor's real estate, but they don't add an iota to computational ability. It's the same with pipelining, branch prediction, and out-of-order execution. These are all band-aids to keep the CPU going. Perhaps the only real architectural advancement has been in multimedia instruction sets.

With billions of transistors on a chip, clock rates limited by heat dissipation, and very large cache memories, there are few places for the computer designer to go. Multi-core processing provides a new way of using up all those transistors. There are no new radically different instruction sets or data types. Just more of the same … and now at the same time.

Hyperthreading—Hyperthreading is rather like multitasking: It offers a form of pseudo-parallelism. In a hyperthreaded computer, there is one execution unit, but other architectural resources are replicated, such as the register set and program counter. At any instant, one set of registers is in use together with the execution unit. When the current thread of execution stalls because a resource is unavailable (e.g., a load from main store is taking place and the data will not be available for another 100 clock cycles), another register set and program counter can be switched in, and execution continues along a new thread (using the single execution unit). Hyperthreading hides the latency of memory loads.

Parallel Processing—True parallel processing uses multiple execution units in parallel to improve performance. The forms and structures of parallel processing are legion and go beyond the scope of this text. Parallel processing is used to create supercomputers by operating hundreds or thousands of CPUs in parallel.

Multi-core CPUs—Multi-core processors are single chips that incorporate multiple CPUs. Advances in manufacturing technologies have allowed multiple processors to be fabricated on a single chip together with their interconnection networks, memory interface, and individual caches. Multi-core processors have been created because they offer a more cost-effective means of increasing computational throughput than by increasing the complexity and clock rates of single processors.

Dimensions of Parallel Processing

Before we continue, it's necessary to set the scene by introducing some of the elements of the parallel processor design space. In a parallel processor, a group of processors work together to solve a task. The first consideration is the relationship between processors. They could be connected together in a row, a ring, or a star. The number of ways of connecting m processors together is large for modest values of m. The relationship between the processors is called the system's *topology*.

Processors can communicate with each other in two ways: via a shared bus or via a message-passing network. A bus provides a very high-speed link (high bandwidth) but it is expensive and complicated. Bus-based multiprocessors share memory, which means that one processor can access another processor's memory directly. It is difficult to scale bus-based multiprocessors to very large numbers of processors. Message-passing multiprocessors use a communications medium like an Ethernet link to pass messages to each other. Clearly, the degree of coupling of a bus-based system is far higher than that of a message-passing system.

The memory systems of multiprocessors fall into two general categories. In a uniform memory access (UMA) system, memory is equally accessible by all systems. In a non-uniform memory access (NUMA) system, the memory is not homogeneous and not all memory is equally accessible to all systems. In general, NUMA structures characterize large multiprocessor systems using message passing, whereas UMA systems characterize smaller, tightly coupled, bus-based systems.

Granularity refers to the size of the individual tasks being performed in parallel and is an important aspect of multiprocessor systems. Coarse-grained threading refers to the situation where the tasks are relatively large (for example, monitoring the mouse or performing a disk transfer). Coarse-grained threads often correspond to processes in operating systems. Such coarse-grained tasks are easy to set up (by the programmer) and control. However, it is difficult to scale coarse-grained multithreading because you can't always find a sufficiently large number of threads to exploit the hardware fully. Fine-grained threading involves much shorter threads of code and is invisible at the applications level.

Because this is a text on computer architecture, we will concentrate on UMA, bus-based systems, particularly those using the new generation of multi-core chips that are really entire multiprocessors on a chip.

A Brief History of Parallel Computing

Until the microprocessor became ubiquitous, parallel processors (often called *supercomputers*) were few and far between because of their complexity, size, expense, and need for special-purpose software. The first parallel mainframe computers appeared in the 1960s from then mainstream companies, like Burroughs and Control Data Corporation (CDC). Some systems contained a single CPU and separate processing units, and others had independent CPUs.

The 1970s were dominated by two giants of the computer world: Seymour Cray and Gene Amdahl. Cray left CDC to found Cray Research and develop a series of supercomputers.

Parallel processors and supercomputers flourished in the 1980s, and in 1987 the Gordon Bell Prize for parallel performance was awarded. The Gordon Bell Prizes are a set of awards given by the ACM and IEEE at the Supercomputing Conference for outstanding achievement in the supercomputer world. Gordon Bell was one of the great pioneers of computing who worked for the Digital Equipment Corporation between 1960 and 1969, where he developed the PDP machines and later the VAX.

IBM Stretch Computer

The IBM 730 Stretch computer was IBM's first transistorized supercomputer and was delivered to Los Alamos in 1961 for use in hydrodynamic modeling. When designed, Stretch was intended to be 100 times more powerful than the IBM 704, but in practice, that performance was never achieved. Techniques that were revolutionary in the 1950s were incorporated in Stretch (for example, pre-decoding, data pre-fetch, and a limited form of out-of-order execution by pre-executing some instructions). Stretch led the way in terms of performance enhancement and the future development of parallelism.

The 1980s saw the introduction of the transputer, developed by the British company Inmos. *Transputer* is a contraction of transistor and computer and was the first microprocessor to be specifically designed for use in parallel systems. The name transputer implied a component that could be used in a system as one of many transputers to create a much larger system (just like the transistor). Transputers had simple serial interfaces, making it easy to connect them together. Transputers were designed to use Occam, a concurrent programming language built on *communicating sequential processes* and closely associated with Tony Hoare, a British computer scientist who based communicating sequential processes on formal methods.

The transputer, while initially promising so much, was not a commercial success; it was too expensive to compete with the low-cost microprocessors used in PCs and did not have a significant performance advantage over the then high-performance RISCs that were appearing. Although its interface and special-purpose programming language were innovative, the world did not beat a path to Inmos's door, and the company vanished.

To some extent, the 1990s saw a relative decline in parallel processing and several companies specializing in supercomputers went out of business. Of course, supercomputers were still being produced and were becoming ever more powerful tools of government, the scientific establishment, and the military. Today's astronomers use supercomputers to help them study the Big Bang and the origin of matter, and the military studies ways of making a bigger bang in order to destroy matter. Meteorologists use supercomputers to study why matter is so wet. The rate at which the performance of conventional microprocessor-based systems increased seemed to lower interest in multiprocessor systems.

Rise and Fall of the Coprocessor

The first generations of 8- and 16-bit processors were rather anemic by today's standards. Because of the limit on the number of transistors on a chip, there was barely enough room for a basic processor, and certainly no space for many of the facilities we take for granted today, such as floating-point arithmetic, memory management, and cache memory.

Without hardware floating-point units, processors had to perform all floating-point operations in software. Both Intel and Motorola introduced external floating-point coprocessors to enhance performance. Essentially, a plug-in, floating-point processor took over the execution of certain instructions. When the host processor was executing code, it handed over control to the coprocessor if it detected a floating-point instruction. This mechanism offered a limited form of parallel processing. The coprocessor provided a solution to those who needed greater performance and were willing to pay for it.

As manufacturing technology improved over the years, the functions provided by coprocessors were often incorporated within the CPU. The ARM explicitly supports coprocessors by providing both an electrical interface and a set of instructions that move data to and from a coprocessor and pass execution control to the coprocessor.

One of the most famous early parallel computers was the Illiac IV, invariably called the *ill-fated Illiac IV* or the *notorious Illiac IV*. Work on the Illiac IV began in 1965, and the system was expected to cost $8 M and operate at 100 MFLOP/s. It was eventually delivered to NASA Ames in 1972 but was not operational until 1976, by which time it had cost $31 M yet ran at only 15 MFLOP/s. It took eleven years to complete. Just imagine how many generations of a processor a company like Intel produces in that time. The Illiac IV was an early SIMD (i.e., *single instruction multiple data*) machine, composed of 64 64-bit processors. In practice, only 16 processors were implemented, which partially accounts for the difference in performance between the Illiac as built and the original concept. The advantage of a SIMD architecture is that only a single control unit is required. The processing is performed by ALUs that are less complicated than control units.[1] However, SIMD architectures are often effective only when handling applications involving structured data, such as arrays.[2] It is said that Illiac IV's failure dampened interest in SIMD machines for two decades.

Supercomputers and parallel processing had little impact on the first generation of microprocessors because the microprocessors were suitable only for undemanding applications such as controllers in washing machines, etc. Limited multiprocessing appeared in PCs that used two to four processors on the same motherboard to create high-performance workstations. Before we look at parallel processing in any more detail, we have to examine why it has suddenly become so important in today's world.

Figure 13.1 provides a general illustration of where parallelism can be implemented in a computer system. The bottom layer—the bit level—indicates parallelism in the functional units of the processors. For example, the ALU may be designed to perform fast addition or fast floating-point operations by using parallel circuits. This level is entirely invisible to the user or operating system. Above the bit level is the microarchitecture level, which includes instruction-level parallelism and superscalar techniques. Again, this level is invisible to the programmer.

The next level up is the *thread level,* where tasks are decomposed into relatively short streams of instructions. This is the level supported by hyperthreading and is visible to the user/operating system because threads have to be allocated. Above the thread level, there is the task level (although the boundary between these levels is not hard). At the task level, the work can be partitioned and executed on the individual processors of a multiprocessor system. Finally, at the top, we have the application level, where entire jobs are run on the computers of a multiprocessor system. The difference between the application level and the task level is the *granularity* of the code executed on the individual processors.

---

[1]This is a bit of an overstatement. Of course you can build simple control units and complex arithmetic units. However, the point is that the control units of machines of the class of Illiac IV were more complex than the multiple arithmetic units that they controlled in a SIMD array.

[2]The SIMD made its comeback in the era of multimedia processing.

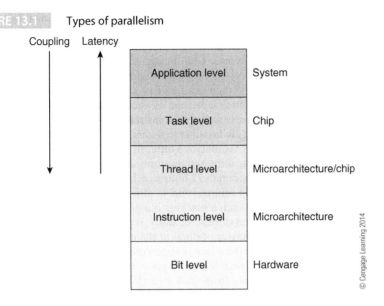

FIGURE 13.1   Types of parallelism

Coupling   Latency

Application level	System
Task level	Chip
Thread level	Microarchitecture/chip
Instruction level	Microarchitecture
Bit level	Hardware

© Cengage Learning 2014

13.1   Why Parallel Processing?

No sane person would ever use a parallel processor if they could avoid it. To implement one you have to replicate processors and provide a high-speed interconnection network to link processors to each other and to systems such as memory and I/O. Moreover, the number of ways in which you can link processors, their *topology*, is vast and the optimum topology for one class of problem is not the same as for another (unlike single processors that have one topology that fits all problems). You have to decide how to *partition* programs so that some code can run on all the processors at the same time, otherwise you will have a situation in which some or all of the parallel processors are doing nothing but consuming electrical power. And the memory... it's a nightmare. Each processor can have its individual or local cache memory. Suppose processor A creates data P and writes it to its own cache A. If processor B wants to use data P and reads it from memory, it may get the old or *stale P* from its own cache B. Somehow, we have to ensure that data remains consistent or coherent across all processors so that when A updates P in its cache, all other cached copies of P are either marked as invalid or updated.

We resort to parallel processing because it's more cost-effective to use two or more processors in parallel rather than design a new processor that is twice as fast. A few years ago processors were operating at just below 4 GHz and were dissipating a large amount of power. Some PCs were using water cooling systems to keep the chip from

Why is Parallel Processing So Difficult to Get Right?

Making fast hardware is not particularly difficult. Getting high performance out of parallel hardware is. If you accelerate the performance of serial hardware, the system runs faster unless you run into bottlenecks with, for example, buses or memory. If you build parallel hardware, you get absolutely no increase in performance unless you can partition a task between individual processors. Doing that is difficult. The way in which tasks are partitioned between processors depends on the nature of the task, the way in which messages are passed between processors, and the synchronization of the tasks.

Not all real-world problems are equally suited to execution on parallel processors. There is a class of programs that are called *embarrassingly parallel* because they can be easily partitioned into subtasks, and the subtasks are assigned to individual processors. Moreover, the subtasks require little or no communication with other subtasks. Such *embarrassingly parallel* computations are common in image processing, cryptography, and weather prediction. Some problems are difficult to parallelize (for example, iterative methods where the next calculation depends on the result of the previous calculation).

melting. Since power goes up with the square of clock frequency (all other factors being equal), it is unlikely that we will see commodity 8 GHz processors in the near future. Equally, no one would argue that slight modifications to a processor's instruction set or to its micro-architecture will double its performance. Chip manufacturers have been forced to follow Moore's law in the only feasible way—put more processors on a die. Single-core processors gave way to dual-core processors and then quad-core processors in the late 2000s.

Today, parallel processing has three faces, and we should distinguish between them. One face is the multi-core processor that is now the mainstream processor in high-performance personal computers; that is, it is largely invisible to the user and adds little to the end-cost of the machine. For most users, its only visibility is a little sticker on their new PC or laptop. The second face of multiprocessing is the ultra-high-performance computer, generating informa-tion that is very time-sensitive. An accurate 72-hour weather forecast is invaluable to farmers and airlines, where knowledge of the future is vital to efficient utilization of resources. A few tens of millions of dollars spent on such a machine is a good investment if you have a large fleet of aircraft that need to fly round storms. Similarly, if you are a drug manufacturer, using supercomputers to simulate the behavior of drugs at the molecular level can help you beat a competitor to the patent office. The third face of supercomputing is all about high-speed data processing, processing on an unimaginable scale. Typical examples are in large-scale nuclear physics (for example, processing data from the Large Hadron Collider, which is a synchro-tron in a 17 mile circumference tunnel under the Franco-Swiss border). The Large Hadron Collider produces a total of 15 petabytes of data/year. Another example of data intensive supercomputing comes from the oil industry. A sensor array of tens of thousands of *thumpers* (hydraulic devices that hit the ocean seabed to send out an impulse that is reflected at discon-tinuities in the underlying rock) generates impulses every 10 seconds that must be digitized, recorded, and analyzed. This corresponds to data rates of several terabytes/day.

As an order-of-magnitude example of the need for parallel computing, consider simulat-ing the weather of North America for a 48 hour period. Assume that the granularity of the simulation is 1 hour (i.e., the model has to be updated every hour), the granularity of points in the model is 0.1 km, and each point requires 100 operations. North America has an area of 20 million km^2, and for the purpose of the simulation, the atmosphere extends 20 km from the ground. The volume of the space is $2.0 \times 10^7 \times 20 = 4.0 \times 10^8\ km^3$. The points are 0.1 km apart, corresponding to 10^3 per cubic km. The total number of points is $4.0 \times 10^8 \times 10^3$ and the number of calculations is $4.0 \times 10^{11} \times 100 \times 48 = 1.92 \times 10^{15}$ for the two-day period. A computer operating at 1 GFLOPS could solve the problem in 23 days (i.e., three weeks too late). A computer at 1 TFLOPS would take 30 minutes, and a computer at 1 PFLOPS would take two seconds. If we require more accurate simulation, doubling the spacial resolu-tion of the simulation points from 0.1 km apart to 50 m apart would increase the number of calculations by a factor of 8. A PC with a single processor chip (albeit a multi-core processor) has a rating of the order of 100 GFLOPS. In mid-2011, a Japanese computer used 68,544 8-core SPARC64 processors (i.e., 548,352 CPUs) to achieve a speed of over 8 PFLOPS. By the end of 2011, IBM was building a supercomputer for the Lawrence Livermore National Laboratory with a performance of 107 PFLOPS using 8.4 million cores.

Clearly, the demand for high-performance computing is unlimited at both ends of the spectrum. The consumer wants endless improvements in multimedia and games performance, and the industrial/academic world simply changes the goalposts as each new target is reached. We are now going to look at another reason parallel processing has become necessary: the *power wall*.

13.1.1   Power—The Final Frontier

The main reason that the microprocessor industry has adopted multi-core processors so enthu-siastically is the need to keep power consumption within limits. Let's begin with a question. Why are oxygen and electrical power similar? Because they are both necessary evils. Oxygen is a corrosive and destructive substance that is very harmful to human tissue. Unfortunately, we can't live without it. Similarly, we need power to operate electronic equipment even

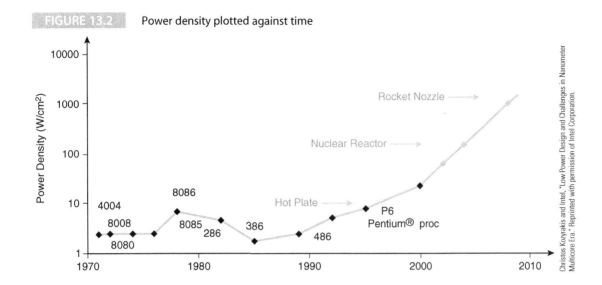

FIGURE 13.2   Power density plotted against time

Christos Kozyrakis and Intel, "Low Power Design and Challenges in Nanometer Multicore Era." Reprinted with permission of Intel Corporation.

though it is expensive to produce, difficult to store and transport, and always ends up as heat. The power dissipated by a digital circuit can be expressed as

$$P_{total} = P_{switching} + P_{resistive} + P_{leakage} = V_{dd}^2 \cdot F \cdot C \cdot k + V_{dd} \cdot I_{resistive} + V_{dd} \cdot L_{leakage}$$

The term in blue in the equation is related to the clock rate and one of the reasons for adopting multi-core technology is to increase performance without increasing clock frequency. Conversely, multi-core processors have more computing power per Watt than single-core processors. Let's look at power in a little more detail, because it is becoming increasingly important in the design of computers.

Figure 13.2 from Intel plots *power density* produced by processors over the years. As you can see, the power dissipated is increasing exponentially. This situation cannot continue, because we cannot remove heat from chips fast enough. Already, some high-performance PCs have water-cooling systems to keep their processors from burning out. The three terms in the above equation represent the power generated by switching in clocked circuits, the power dissipated in resistive elements, and the power lost due to leakage currents in transistors. The symbol V_{dd} represents the supply voltage used by the circuit. Over the years, the voltage required to operate electronic devices has reduced, leading to lower power consumption; on the other hand, the number of active devices has grown much faster. In the 1970s, microprocessors operated with a 5 V power supply (plus other biasing voltages). Today, microprocessors operate at much lower voltage levels and modern processors have core voltages as low as 0.85 V. It is not possible to reduce V_{dd} to a vanishingly small value because the minimum voltage required to operate a transistor is related to the atomic properties of semiconductors. It is probable that the lower limit to the power supply of conventional technology is about 0.6 V.

Just another Brick in the Wall

Modern computing is all about overcoming limitations or *walls* as some call them. We have already covered the memory wall, which is a potential barrier to progress caused by the increasing disparity between the access time of main store and the cycle time of processors.

The trend to multi-core processors has been driven by the frequency wall and the related heat wall. The frequency wall describes the limitation on the increase in clock speeds due

(continued)

to the quadratic relationship between clock frequency and power consumption. The heat wall describes the limitation due to increasing heat dissipation because of higher switching speeds and the ever-increasing number of transistors on a chip.

There's the Amdahl wall that is imposed by the limitation on the extent to which parallelism can be applied to any given problem. There's the atomic wall that manifests itself in three ways: fabrication (transistors are created using photolithography to lay down structures on silicon and there is a limit to how small we can build things), quantum mechanics (as devices get smaller, quantum mechanical effects begin to predominate, for example, the ability of electrons to tunnel through barriers), and statistics (a current is made up of the movement of electrons—as devices get smaller current flows become more governed by the laws of statistics than by conventional current models).

Finally, there's c wall, which is the speed of light. As clock speeds become faster, the time for signals to propagate through circuits becomes a limiting factor.

Other walls may be hiding over the horizon (for example, a bank wall). Transistors have been getting cheaper and cheaper over the years. However, the total number of transistors per chip has been increasing exponentially according to Moore's Law, and the cost of building a fabrication plant has also been rising dramatically. A fabrication facility in Taiwan constructed in 2012 was estimated to cost \$9.3 billion. The increasing cost of fabrication plants could lead to a reduced number of manufacturers, periodic chip shortages, and vulnerability to natural disasters (e.g., the Japanese earthquake of 2011).

One of the components of the total power dissipation is due to *switching* when transistor capacitances are charged and discharged on each clock transition; this can be expressed $V_{dd}^2 \cdot F \cdot C \cdot k$, where F is the clock rate and C is the effective capacitance of the circuit. Dissipation due to switching can be reduced by lowering the voltage and by minimizing internal capacitance. Processor clock rates have been increasing year on year. By 2008, clock rates of over 2 GHz were common. Intel's i7-860 (2011) ran at 3.46 GHz, and some users have overclocked their processors at 4.2 GHz.

By about 2005, high-performance microprocessors were reaching the practical limits of dissipation. Processors in PCs required large heat sinks and fans to remove the heat from the chip; indeed, the failure of a fan could destroy a processor in seconds. Some high-performance PCs resorted to water cooling to keep the chip within its operational limits. There were even cases of people being burned by the dissipation from laptops.

Clock rates could be pushed higher by the use of esoteric materials (highly conductive substrates such as diamond or sapphire). Such solutions are incompatible with mass markets and commodity devices.

One of the factors contributing to the continued success of Moore's law (as applied to improving performance) has been ever increasing clock rates. If computers were to continue to get faster, some means of compensating for the limit on clock rate was necessary. Fortunately, Moore's law, as it relates to the increasing number of components on a chip, has not yet run out of steam. Greater numbers of transistors can be put on chips because of advances in manufacturing technology. Designers could use all those extra transistors to create radically new architectures. Or they could make more *of the same*. They chose the latter approach and implemented multiple processors on the same chip. Processing power could be increased without raising the clock rate. All you had to do was to figure out how to use the new multiple processor devices. The next section reintroduces an old friend, Amdahl's law.

Overclocking—Fool's Gold?

When you buy a microprocessor, you specify a particular component part, and that part is designed to operate at a stated clock frequency. The cost of a microprocessor increases with operating frequency (you pay more for faster). However, today's microprocessor at a high clock rate may cost less than yesterday's microprocessor at a lower clock rate.

If you buy a 3 GHz microprocessor, it will operate at over 3 GHz. If it didn't, some microprocessors would fail at 3 GHz. There has to be a margin. Consequently, some probe the margins by increasing the clock frequency of their microprocessor. Indeed, it's so common that you can now increase clock frequency just by changing BIOS parameters. Operating at a frequency beyond a component's design limit is called *overclocking,* and today, the term applies to a bag of techniques used to tune PCs (e.g., changing the core voltage or the clocking rate of DRAMs). Overclocking is about getting improved performance without paying more. It's the ultimate free lunch.

Overclocking has its dangers. First, heat dissipation increases with frequency and operating at a higher than intended frequency could damage the silicon real estate. Second, when the clock frequency goes up, "Mr. Guarantee" flies out the window. A manufacturer's guarantee is not valid if the equipment or component is operated outside its stated operational parameters. Although some may take the risk to squeeze the last drop of performance out of a CPU-mother board combination for use in a games environment, few would apply overclocking to the CPU in an aircraft's navigation system.

13.2   Performance Revisited

We have already looked at performance in Chapter 6. Here we revisit performance because it is at the heart of the multiprocessor revolution. Paradoxically, Amdahl's law has been the driving power behind the move to multiprocessor systems, but it has also poured cold water on the very idea that multiprocessing can ever provide the type of performance increase, year after year, that we have seen over the last 40 years.

Amdahl's law is simple. If the fraction of a task that is executed serially is f_s, the fraction that can be executed in parallel is $(1 - f_s)$. If we have n processors that can be applied to the parallel portion, the total time is $f_s + (1 - f_s)/n$. A glance at this formula tells us that increasing n, which is the number of processors, reduces only the part that can be parallelized. In the limit, we are left with f_s, and that isn't reducible. If f_s is 50%, using an infinite number of processors will only halve the execution time; not a great return on investment.

An Internet search of the literature on Amdahl's law and computer performance reveals that this law is one of the most contentious areas of computer science. As expressed, it is, of course, uncontroversial. It's the underlying and implicit assumptions that are the source of controversy. Let's begin with $t_{total} = t_{serial} + t_{parallel}$. I'm all in favor of small equations, but how many problems can be decomposed into two components in this way? Consider the following.

- Complicated computational problems can be tackled in many different ways; no two programmers would necessarily come up with the same way of solving a problem.
- Amdahl's equation assumes that the parallel part of the problem demonstrates perfect speedup (i.e., it neglects the cost of parallelization in terms of overheads and intertask communication).
- Amdahl's law assumes that infinite parallelization is possible. In practice, the parallel portion of a problem may *saturate* and reach the stage at which further parallelization is not possible.

- It is assumed that the serial and parallel code is executed at the same rate. This may not necessarily be so. It may be possible to accelerate the serial part of a problem by the use of optimized hardware.
- It is assumed that the serial portion of a program and the parallel portion cannot be overlapped. Consider a simple example. A task has a serial part that takes 10 s and a parallelizable part that consists of 20 operations each taking 5 s. Assume that 4 s of the parallel processes can overlap the serial process. What is the speedup if we use 20 processors?

The non-speedup time is $10 + 20 \times 5 = 110$ s.
The basic Amdahl equation gives us 10 (the serial part) $+ 20 \times 5/20 = 15$ s.
However, as we can overlap 4 s of the parallel processing with the serial processing, we get $10 + 5 - 4 = 11$ s.

As well as these objections, we pointed out in Chapter 6 that some believe that Gustafson's law has replaced Amdahl's law. Gustafson argued that Amdahl's law does not take account of scaling. Suppose you have a problem whose execution time can be expressed as $f_s + (1 - f_s)/n$. Now, suppose we scale the problem up and the new time is $f'_s + (1 - f'_s)/n$, where f'_s refers to the fraction of time the new problem spends executing serial code. The essence of Gustafson's argument is that $f'_s < f_s$; that is, scaling up reduces the fraction of serial time and hence makes it possible to use more processors to reduce the total time. This is an appealing argument. For example, if you were simulating the flow of liquid in a pipe and scaled up the problem, the volume of liquid would increase as the square of the radius, but the surface area of the pipe would increase linearly with the radius of the pipe. Since the more complex calculations occur at the pipe-liquid boundary, the fraction of complex calculations would decrease with the size of the problem. The same is true if we increase the number of sampling points.

Unlike taxis that disappear as soon as it begins to rain, Amdahl's law is often there when you need it. Speeding up programs is not an abstract concept. I really don't care whether parallel processing can speed up the response of my computer from 0.5 to 0.01 s when I'm searching the Internet. However, I do care about increasing the response time when I am using Photoshop to apply complex image processing operations to large high-resolution images. Such images require large amounts of processing to be applied to arrays of pixels. The data structures are regular and many of the operations can be applied in parallel; that is, many of the real computational problems that we need to speed up are well suited to parallelization.

Students taking a course in programming inevitably encounter the concept of *big O notation*, which describes the relationship between the time taken to solve a problem and the size of the problem. You can say that big O notation expresses the worst-case results for problems involving time (or even space, such as the memory requirement of a task as it grows). For example, $O(1)$ indicates constant time; that is, a process takes the same time irrespective of the size of the problem or data. The notation $O(N)$ describes an algorithm whose performance is linearly related to the size of the problem; for example adding integers

Amdahl versus Gustafson

We introduced Amdahl's law when we discussed performance in Chapter 6. The speedup of a parallel system is dependent on two factors: the fraction of the code that can be executed in parallel and the number of processors. The speedup is $1/(f_s + (1 - f_s)/n)$, where f_s is the serial fraction and n is the number of processors.

This tells us that there is little point in throwing processors at a problem, because the serial part will dominate and additional processors will have no effect. Gustafson pointed out that Amdahl's law is pessimistic because the problem size also scales and undermines Amdahl's assumptions. Consider the following argument.

Suppose an artist was designing clouds for graphics in 1990. He or she might have to design a cloud and a suitable silver lining for a screen that was 1024×768 pixels. Using parallel processing could increase processing speed, but only up to the limit set by Amdahl. However, if we move forward to today, a screen might have a resolution of 1920×1200 pixels. This represents a 300% scaling of the problem. However, if you consider the cloud, the area of the cloud has increased with respect to its silver lining; that is, the problem has scaled and it has become more advantageous to use a larger number of processors.

Professors seem to be divided over whether Gustafson's argument is valid. I would suggest that Gustafson's observations apply to the very systems that most need speeding up, such as image and multimedia processing.

$0 + 1 + 2 + 3 + \cdots + n - 1$ demonstrates $O(N)$, because the time taken scales with n. Generally, problems of $O(N)$ are welcome.

Problems that are described as $O(N^2)$ have a square-law relationship with size; that is, if you double the data set, you quadruple the computing time. A good example would be the addition of matrices. If you add together two $n \times n$ element arrays, there are n^2 additions. However, if you multiply two $n \times n$ arrays, the number of multiplications is proportional to n^3 and the problem belongs to the class $O(n^3)$.

Some problems have a complexity expressed by $O(2^N)$; that is, each additional element doubles the size of the problem, which means that it is often impracticable to compute solutions to large problems in real time. One class of problems has a complexity of $O(\log N)$; that is, the time is proportional to the logarithm of n. This implies that doubling the size of the problem adds only a unit increment to the time. For example, a data set of 1,000 elements may take 4 s whereas a dataset of 2,000 elements may take 5 s. Obviously, $O(\log N)$ problems scale well.

Just as algorithmic problems can be described by big O notation, so can tasks running on parallel processors. A task is said to be linear if the time scales inversely with the number of processors; that is, you double the number of processors and you halve the time. However, it is also possible for networked computers to demonstrate a superlinear performance with respect to the number of processors; that is, performance increases faster than the number of processors. In principle, superlinearity is impossible. In practice, it can be achieved due to other factors than processing power. In a single processor system with a large data set, the data may not fit into the available cache, and the processing time is extended due to cache misses. However, in a parallel processor, each processor may be able to cache all the data it needs for its slice of the problem with the result that there are very few misses and the speedup results from both the increased computer power and the reduction in cache misses. Figure 13.3 illustrates the scalability of parallel processors.

 **FIGURE 13.3** Scalability of parallel processors

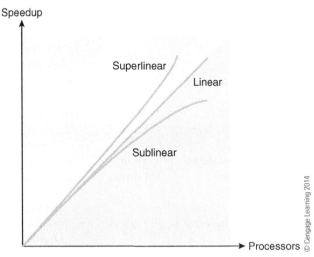

Performance Measurement

How do we measure multiprocessor performance? We could use any technique that measures single processor performance. Just because there are a lot of processors on the chip in our computer shouldn't make us change our benchmarking methodology. Or should it? You don't use a high-performance parallel processor primarily to check your email. We have the iPad for that. You use high-performance computers to solve computationally intensive tasks. In other words, benchmarks should ideally approximate to real world tasks in order to make a realistic comparison of machines.

SPEC has devised the MPI2007 benchmark for testing high-performance multiprocessor systems. In particular, MPI2007 tests the type and number of processors and the interprocessor communications mechanism—the memory architecture, the compilers, and the shared file system. This benchmark requires the use of the *Message-Passing Interface* (MPI) environment. Table 13.1 lists the individual benchmarks of the MPI2007 suite. As you can see, they are all representative of very large-scale computing, such as weather forecasting. The benchmarks were chosen to be compute bound with little I/O activity. MPI2007 exploits the process-level parallelism of MPI; it does not use thread-level parallelism.

MPI2007 uses the same methodology as SPEC2006; the performance measurements are references to a standard machine, and the overall result is the geometric mean of the individual normalized results. Two different reference machines are used: one for *medium* data sets and the other for *large* data sets. The medium data set reference is an 8-node cluster of Celestica dual single-core AMD Opteron processors at 2.2 GHz. The large data set reference uses a 64-node cluster of Intel-based systems, each using two quad-core Xenon processors at 3.2 GHz.

TABLE 13.1   *MPI2007* Parallel Processing Benchmarks

Benchmark	Suite	Language	Application Domain
104.milc	Medium	C	Physics: Quantum Chromodynamics (QCD)
107.leslie3d	Medium	Fortran	Computational Fluid Dynamics (CFD)
113.GemsFDTD	Medium	Fortran	Computational Electromagnetics (CEM)
115.fds4	Medium	C/Fortran	Computational Fluid Dynamics (CFD)
121.pop2	Medium, large	C/Fortran	Ocean Modeling
122.tachyon	Medium, large	C	Graphics: Parallel Ray Tracing
125.RAxML	Large	C	DNA Matching
126.lammps	Medium, large	C++	Molecular Dynamics Simulation
127.wrf2	Medium	C/Fortran	Weather Prediction
128.GAPgeofem	Medium, large	C/Fortran	Heat Transfer using Finite Element Methods (FEM)
129.tera_tf	Medium, large	Fortran	3D Eulerian Hydrodynamics
130.socorro	Medium	C/Fortran	Molecular Dynamics using Density-Functional Theory (DFT)
132.zeusmp2	Medium, large	C/Fortran	Physics: Computational Fluid Dynamics (CFD)
137.lu	Medium, large	Fortran	Computational Fluid Dynamics (CFD)
142.dmilc	Large	C	Physics: Quantum Chromodynamics (QCD)
142.dleslie	Large	Fortran	Computational Fluid Dynamics (CFD)
145.lGemsFDTD	Large	Fortran	Computational Electromagnetics (CEM)
147.l2wrf2	Large	C/Fortran	Weather Prediction

A Dissenting View

Views on parallel processing vary widely. Some see it as the solution to the decline in the rate of growth of processor performance (i.e., the end of Moore's law). Some see it as a temporary Band-Aid to improve performance until something new turns up in computer hardware. Here is a comment from a paper by Patrick Madden, "Parallel Computing: The Elephant in the Room." If you look at the progress made by contemporary parallel processors, you might find these comments overly pessimistic.

"While the computer industry seems to be pinning its hopes for sustained growth on massively parallel computing, these efforts will fail for a majority of application areas.

Computers are simply the embodiment of Turing's mathematical concept. Based on this framework, there is no fundamental difference between a parallel computer and one that is serial. A serial computer can easily emulate an arbitrarily large number of parallel computers, with no more than a constant factor disadvantage. Just as changing the type of paper one might write on does not alter the foundations of mathematics, neither does switching from serial to parallel computing.

Software simply implements algorithms on the available hardware. Complexity theory shows clearly that how you solve a problem is far more important than what you solve the problem with. Different algorithms have different levels of innate parallelism, but few scale linearly with the number of processors. Just as changing the font or notation one might write with does not alter the foundations of mathematics, neither does switching from one program-ming language to another.

The algorithmic constraints apply to all software, not just portions that implement "classic" functions such as sorting and searching. No matter if it is simply coordinating the processing, linking subroutines together, handling user input, or providing output, there are serial con-straints. Amdahl's observation, which has been backed up by literally decades of experimental evidence, is that the portion of work that is serial is non-trivial.

In all but a handful of applications, the upper bound on performance gains through parallel-ism is depressingly low.

These constraints, and their implications, are based on mathematics. Apart from shifting to an alternate universe where mathematics does not apply, there's little that can be done."

13.3   Flynn's Taxonomy and Multiprocessor Topologies

Let's take a step back and introduce some of the fundamental concepts of multiprocessing. The notion of using two or more computers in parallel to improve throughput is not new and has existed for over four decades. For a long time, parallel computing was an esoteric topic, largely taught at postgraduate level. When computers were expensive mainframes, only large organizations could employ parallel processing—typically for scientific processing (weather simulation) or engineering (aerospace and military). The introduction of relatively low-cost microprocessors made the design and construction of parallel processors more feasible and generated new interest in parallel processing.

Parallel processing includes two concepts: the *topology* or *interconnection* of the multiple processors and the distribution of tasks between individual processors. In 1972, Michael

J. Flynn wrote a classic paper that was to provide a standard taxonomy for the description of parallel systems.[3] Flynn's classification was based on two criteria: *instruction execution* and *data flow*. This classification provides only a very general and basic division between four generic computing structures and does not cover the topology of multiprocessors. Flynn's classifications are given here.

SISD   The *single instruction single data* processor describes the conventional uniprocessor that executes instructions sequentially. Today's highly pipelined superscalar processors are SISD machines only at the abstract level because of the inherent parallelism in their structure.

SIMD   The *single instruction multiple data* processor describes a system whose instructions act on multiple data elements in parallel. Such machines were called *array processors*. Modern microprocessors can be regarded as SIMD machines when they implement multimedia or short-vector operations because they are performing a single operation on multiple streams of data (the multiple streams being the individual sub-components of a word in a register). SIMD machines have a single processor that determines instruction flow (instruction fetch and branch control) and multiple arithmetic or processing units that carry out data operations.

MISD   *Multiple instruction single data stream* processors represent the rather improbable combination of multiple instruction streams and single data streams. (Do several processors argue about which gets to process the single data element?) An example of a MISD processor might be a highly secure fault-tolerant processor employing multiple processors to operate on a single stream of data and then comparing the outputs of each processor in order to ensure reliable operation (such a system is employed in fly-by-wire aircraft).

MIMD   *Multiple instruction multiple data stream* processors define systems in which two or more independent or autonomous processors operate on different streams of data. Distributed system and multicore processors are classified as MIMD machines.

We have seen how SIMD technology can be used to enhance computer performance with short vector operations, but SIMD does not provide a solution to more general processing problems. That requires MIMD technologies that employ multiple processing units. Before we look at multi-core processors, we provide a little background on multiprocessing topology. Figure 13.4 describes computer structures from the point of view of Flynn's classification. In this chapter, we are concerned with MIMD-class multiprocessor systems.

FIGURE 13.4   Computer classifications—the Flynn view

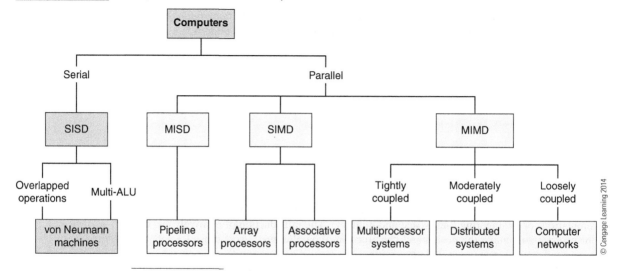

[3]M. Flynn, "Some Computer Organizations and their Effectiveness," *IEEE Transactions on Computers*, Vol. C-21, No. 9, September 1972.

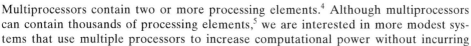

13.4   Multiprocessor Topologies

Multiprocessors contain two or more processing elements.[4] Although multiprocessors can contain thousands of processing elements,[5] we are interested in more modest systems that use multiple processors to increase computational power without incurring the power dissipation penalties of high clock rates. The two types of multiprocessor of interest to us are board-level multiprocessors and chip-level multiprocessors. Board-level multiprocessors use several separate chips, typically up to four, on the motherboard. Chip-level multiprocessors are now called multi-core processors and employ several processors on the same chip.

Let's look at some classic multiprocessor topologies. Figure 13.5 describes two extremes: the unconstrained or ad-hoc topology and the fully-connected topology. In the former, processors are connected if they need to communicate with each other and unconnected otherwise. In a fully connected topology, each processor has a direct connection to every other processor. As we saw in the chapter on computer buses, both these topologies are generally undesirable.

> ## Multiprocessor Coupling
>
> Several parameters describe multiprocessors – Flynn's classification, their topology, and the degree of coupling. The term *coupling* describes the degree of interconnection between the individual processors. The two extremes are loosely connected processors and tightly connected processors. The most loosely connected processors are those that are connected via, for example, the Internet. Data is exchanged relatively slowly (relative to the processor's clock speed) and the granularity of data is high (entire files). In a tightly coupled multiprocessor, data bandwidth is high and latency low because the processors share the same memory. Multi-core chips are very tightly coupled multiprocessors.

Figure 13.6 describes crossbar switching (the type of routing once used in manual telephone exchanges). The notation used is P_{ri} indicating the ith processor in the row and P_{cj} indicating the jth processor in the column. The notation $S_{p,q}$ indicates a switch between row p and column q.

Figure 13.7 illustrates a single butterfly switch and a network made up of four butterfly switches. A butterfly switch can either connect the two horizontal pairs of processors left-to-right or switch processors top-left to bottom-right and vice versa. Structures using butterfly switches can be iterated to create fast, effective, and scalable switching mechanisms.

FIGURE 13.5   Unconstrained and fully connected topologies

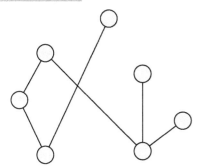

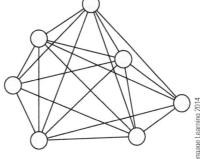

© Cengage Learning 2014

(a) The unconstrained (ad hoc) topology   (b) The fully connected topology

---

[4]Not if you are a lawyer. I was once involved in a patent litigation case where one company was suing another for violating a multiprocessor patent in a single-processor computer. When I expressed surprise, a lawyer told me, "If a 747 shuts down three engines, it does not become a single-engine aircraft."

[5]The Cray XT5 Jaguar (2009) had 224,000 processing elements and was capable of operating at 1.7 petaFLOPS (1.7×10^{15} floating-point operations/s).

Vector Computers

One of the first classes of supercomputer was the *vector computer* of the 1970s and 1980s that has influenced the design of modern computers (for example, the growth of SIMD instruction extensions and the Itanium IA64's software pipelining). A vector computer is a SIMD machine with a single instruction acting on data in parallel and hardware support for operations like $Z = s \cdot X + Y$, where $z_i = s \cdot x_i + y_i$ for $i = 0$ to $m - 1$. Typical vector machines are heavily pipelined. Of course, vector machines are intended for mathematical operations.

Vector machines have several advantages over conventional scalar processors. A single vector operation can replace a lot of scalar operations. Data hazards are reduced because of the sequential nature of the data. Moreover, caching can be exploited to the full. However, pipelined vector computers can suffer from bubbles when two vector operations are performed consecutively. For example, if you execute $Z = X \cdot Y$ followed by $P = Q + R$ the vector unit must complete the $Z = X \cdot Y$ pipeline first and you have to wait until $z_{m-1} = x_{m-1} y_{m-1}$ has passed through the vector unit.

The Cray-1 designed by Seymour Cray (Seymour Cray and Gene Amdahl were the two giants of supercomputing in the 1970s) was probably the most famous vector supercomputer. The Cray-1 was a 64-bit machine with eight scalar registers and eight 64-bit 64-element vector registers plus eight 24-bit address registers. The address registers are used as memory reference registers (pointers or index registers), although they can serve other purposes, such as counters in loops and shifts, as well as I/O operations.

Vector instructions have a 7-bit op-code and three 3-bit register select fields allowing the selection of three operands. Each operand specifies one of eight registers. In that sense, the Cray-1 is rather like ARM or MIPS. Instructions themselves are either 16-bit (operation and three registers) or 32-bit (a 16-bit instruction plus a 16-bit literal).

The eight scalar registers are conventional and act as the source or destination of operands for scalar logical and arithmetic operations. They may hold either fixed or floating-point values.

The eight vector registers are of interest to us in this chapter because they provide the parallel processing facility by handling eight 64-component vectors. As well as the vector registers themselves, CRAY-1 has a vector mask register (VM) that controls the way in which the vectors are accessed. CRAY-1 was able to achieve its high-level of performance (it was, for a time, the world's fastest computer) partly because of its physical structure and semiconductor technology and partly because of its vector unit; that is, they are able to perform multiple operations in parallel using pipelined functional units. For example, ADDV.D **V3**,V1,V2 performs a vector addition.

Vector computers are fast but rather specialized. As we have seen, some of the concepts of vector computing have been incorporated into conventional processor instruction sets as small vector operations. Cray Research merged with Silicon Graphics in 1996 and continues to supply supercomputers. Some of Cray's recent computers have abandoned vector processing in favor of multi-core processing. For example, the Cray XE7 uses AMD's 12-core Opteron processors with up to 2,304 cores/cabinet and a peak performance of 20.2 teraflops per cabinet. The original Cray-1 supercomputer delivered 160 Mflops, whereas a single Intel 6-core i7 delivered 109 Gflops in 2010.

FIGURE 13.6   Crossbar switching

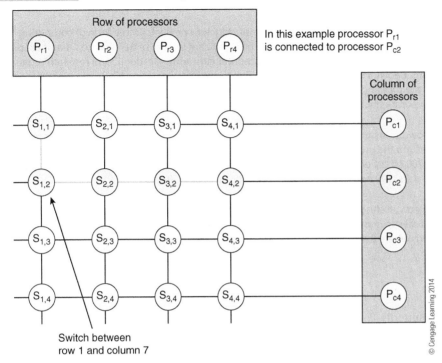

Row of processors

P_{r1} P_{r2} P_{r3} P_{r4}

In this example processor P_{r1} is connected to processor P_{c2}

Column of processors

P_{c1} P_{c2} P_{c3} P_{c4}

$S_{1,1}$ $S_{2,1}$ $S_{3,1}$ $S_{4,1}$
$S_{1,2}$ $S_{2,2}$ $S_{3,2}$ $S_{4,2}$
$S_{1,3}$ $S_{2,3}$ $S_{3,3}$ $S_{4,3}$
$S_{1,4}$ $S_{2,4}$ $S_{3,4}$ $S_{4,4}$

Switch between row 1 and column 7

© Cengage Learning 2014

FIGURE 13.7   The butterfly switch

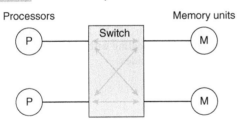

Processors   Memory units

P   Switch   M

P   M

(a) Simple butterfly switch

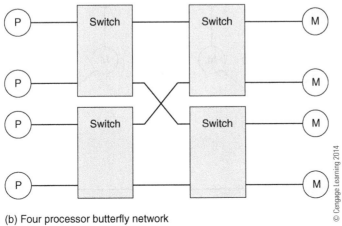

P   Switch   Switch   M

P   M

P   Switch   Switch   M

P   M

(b) Four processor butterfly network

© Cengage Learning 2014

Beowulf Clusters

In the mid-1990s, Becker and Sterling introduced a new affordable form of computing that was called the Beowulf project. The significant feature of Beowulf was that it used low-cost components to construct computer clusters for academic and research organizations. Beowulf was a COTS-based solution to parallel processing (COTS means *commodity off the shelf* and you often hear it when organizations (e.g., military, NASA) are stressing how cost-effective they are being).

Beowulf clusters used Intel-based PCs, low-cost Ethernets, and Ethernet switches to perform the processor-to-processor linking. By 1997, clusters of 16 P6 Pentiums running at 200 MHz were being used. The same year, NASA researchers combined two Beowulf clusters (199 nodes) to achieve a throughput of 10.9 Gflop/s.

A Beowulf cluster uses a single node as a server; all the other nodes are used to create a virtual supercomputer. In this way, the Beowulf cluster differs from networks of workstations where the individual workstations retain their normal functionality.

The Butterfly Parallel Processor was developed at BBN Labs in Cambridge, MA to link 64 tightly-coupled shared memory multiprocessors. It was intended for computer vision applications.

Figure 13.8 demonstrates the cluster, where groups of processors are linked by a switch to a common bus. The cluster can be used to provide highly reliable computing by allowing the processors in a cluster to check each other's operation.

Figure 13.9 illustrates the Banyan tree that is constructed with arrays of butterfly switches. Banyan networks support multicast operations in which one processor can be simultaneously connected to several other processors. Finally, Figure 13.10 describes the binary tree architecture. In this example, processor P_{0100} is communicating with processor P_{0110} by routing the connection back two nodes and then forward two nodes.

FIGURE 13.8   The cluster

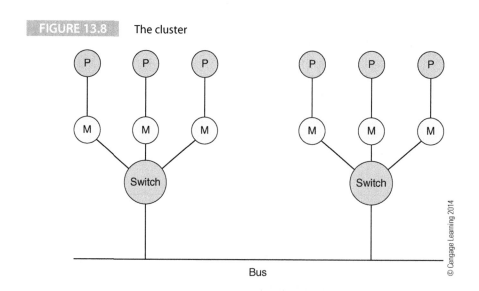

Bus

© Cengage Learning 2014

FIGURE 13.9   The Banyan tree network

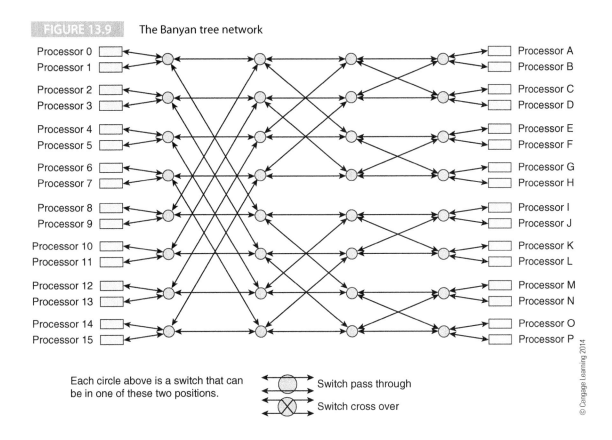

Each circle above is a switch that can be in one of these two positions.

Switch pass through

Switch cross over

FIGURE 13.10   Binary tree

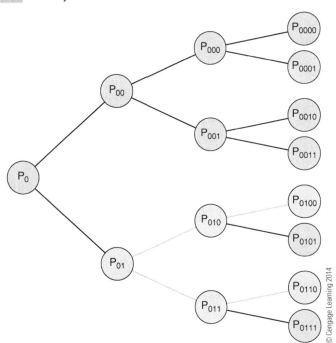

FIGURE 13.11   The 2D mesh

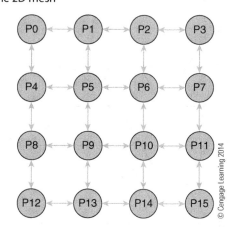

Figure 13.11 illustrates the two-dimensional mesh, where processing units are arranged on a regular grid. This is an important topology, because it is used by the networks on a chip (described later) that fabricate arrays of processors on a single chip. Figure 13.12 shows an extension of the 2D array called the torus, where processors along the edges are connected to processors at the opposite edge to improve connectivity.

A glance at a text on multiprocessing will reveal a wealth of alternative multiprocessor technologies—each with its own unique properties, such as cost, switching complexity, resistance to failure (some topologies rely on a single node, as in Figure 13.10, while others are robust and permit multiple paths between processors), and fitness for purpose. The last factor is an indication of how well the software maps onto the hardware. A multiprocessor architecture intended for matrix algebra might not be optimum for fast Fourier transforms (the basis of much signal processing). In this chapter, we are going to concentrate on the relatively simple, tightly-coupled, SMT systems provided by multi-core processors. Before that we look at two topics: the structure of memory in multiprocessor systems and the special problems posed by cache memory.

FIGURE 13.12   The torus

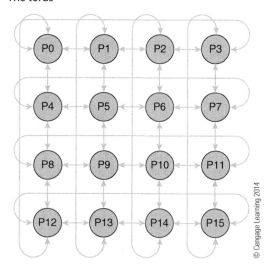

Network Parameters: Degree and Diameter

Two parameters used to characterize networks are *degree* and *diameter*. The degree of a node in a network is the number of other nodes to which it is directly connected and is a measure of the network's connectivity. A node in a 4 × 4 mesh can have a connectivity of 2 (corner), 3 (edge), or 4 (inside), whereas the degree of all nodes in a torus mesh is 4.

The network diameter is the maximum distance between two nodes in a network. For example, the diameter of a star is 1, whereas the diameter of a 4 × 4 mesh is 6, because there is no direct route between corners.

Figure 13.13 illustrates the fat tree topology, so-called because it employs multiple buses. The use of a fat tree avoids blocking when pairs of processors communicate with each other and block communication paths. By doubling the number of communications paths at each switching unit, it becomes possible to allow any processor to communicate with any other processor. The fat tree also supports *all-to-all* multicasting in which any processor can broadcast to all other processors simultaneously. The structure of Figure 13.13 is a *binary* fat tree and is just one member of a class of processors characterized by multiple bus structures.

FIGURE 13.13   The fat tree

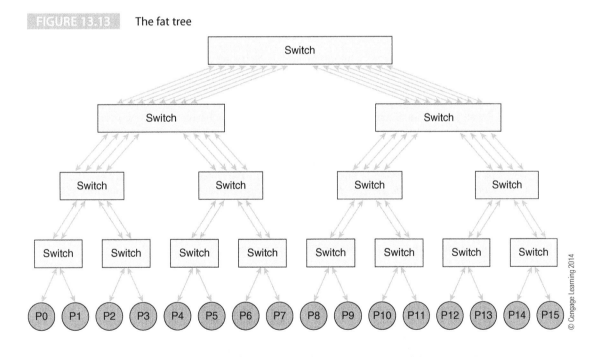

© Cengage Learning 2014

Network Switching

Network switching refers to the way in which data flows through nodes. Small networks can be permanently connected; that is each node is always connected to the bus, like the backplane in a PC (see Chapter 12 on I/O). Larger networks use switching mechanisms to route data between nodes. In a circuit-switched network, a connection is made between nodes for the exchange of data. The link is generally a full parallel bus that allows address and data to be transferred.

(continued)

In a packet-switched network, data is usually transmitted over a serial data link, a bit at a time, in the form of packets of data containing routing information (very much like the Ethernet). In circuit-switching, a data link is owned by a node for the duration of a transfer, whereas packet-switched networks can use the same link to carry information between different pairs of nodes. This is also called a *store and forward* routing because a node buffers an incoming message and then sends it on to the next node along its route.

Wormhole routing is a more efficient form of packet switching that has a lower latency and requires smaller buffers at the nodes through which a message is routed. A packet has to be read, buffered, and decoded, which both adds a delay and requires local storage. In wormhole routing, the data stream is encoded so that its leading bits are able to perform routing without having to fully buffer the entire message. A packet is divided into a number of units called *flits* (flow control digits). The header flit controls the routing. When a node receives a header flit, the node is immediately able to direct the flit onto the node taking it to its destination. The following flits follow the header along the path forged by the header. Because flits are not packages with full routing information, a sequence of flits cannot be interrupted by other messages.

13.5   Memory in Multiprocessor Systems

A computer has a memory hierarchy usually including cache and main store. A multiprocessor system begs the question, "*Who gets what memory and where?*" In this section, we look at the ways in which multiprocessor memory systems may be categorized. The simplest possible memory arrangement would be to have a single global memory unit that could be accessed by all processors. Such a simple scheme is unworkable because only one processor at a time would be able to access memory, leaving all other processors with nothing to do. Each processor must be able to access memory.

Parallel processors can be divided into two groups according to the way in which memory is accessed. One group is known as *uniform memory access* (UMA) and the other is *nonuniform memory access* (NUMA). Systems with UMA structures treat memory symmetrically and all processors have equal access to memory, whereas systems with NUMA memory treat memory unsymmetrically. In a NUMA system, the access time of some memory may be much longer than other memory. NUMA systems typically involve distributed clusters of processors connected by buses or local area networks.

13.5.1   NUMA Architectures

NUMA-based parallel networks may or may not use cache memory that is accessible by multiple processors. For example, in a weakly-coupled cluster of processors, an individual processor may have its own memory and its local cache. This local memory and cache may not be directly accessible by other processors, because information is transferred between processors via messages rather than as actual memory accesses. Some NUMA-based systems are called cache-coherent NUMA (CC-NUMA), where processors maintain their own cache memories that are accessible by other processors. CC-NUMAs have relatively low latencies and high bandwidths at the expense of complicated methods of ensuring cache coherency. If a processor updates its own local cache, the corresponding data in other caches must either be updated too or declared invalid.

Figure 13.14 illustrates a generic CC-NUMA structure. As you can see, this consists of three clusters of computers interlinked by a network. Note that some processors have L2 caches and some do not. NUMA systems are characterized by their *ad-hoc* approach to networking. For example, it takes longer for $Processor_1$ to access $Processor_6$'s memory than to access $Processor_2$'s memory.

FIGURE 13.14   A NUMA structure

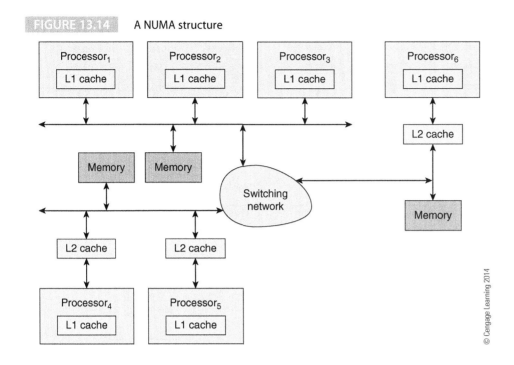

© Cengage Learning 2014

Although it is tempting to dismiss NUMA architectures in favor of UMA architectures (e.g., multi-core processors), the NUMA architecture is not going to go away in the near future. UMA architectures are not sufficiently scalable. If all processors have equal access to memory, then the data path to memory must be symmetric for all processors. That simply does not scale well as the number of processors increases because of the expansion in the number of possible data paths between processors.

13.5.2   Cache Coherency in Multiprocessor Systems

Cache memory is an example of a necessary evil; a little cache speeds the performance of a computer system dramatically, but at a high price in terms of additional system complexity. Nowhere is this truer than in multiprocessor systems. Figure 13.15 illustrates the cache coherency problem. This is a hypothetical system—not least because it does not operate correctly.

Figure 13.15 shows a system with three processors, each with its own cache, and a common memory. There are five states in the figure. Initially, in state (a), there is a variable Q in memory with the value 7. In state (b), Processor 1 reads Q and caches it locally. In state (c), Processor 3 also reads Q and caches it. In state (d), Processor 3 writes 12 to Q in its local cache and to memory. In state (e), Processor 2 reads Q and caches it. Finally, in state (f), Processor 2 reads Q and increments it by 2, caches it, and updates main memory. At this point, we have three different cached values of Q, because each processor's cache has not taken account of the activity of the other processors.

Coherency Requirements

In order to ensure coherency between memories in a multiprocessor system, three criteria must be satisfied:

- Correct program order must be preserved. If Processor x reads data p it must read the most recent version of p written by Processor x. That is, the processor should behave as a single processor.
- All write operations must be visible to all processors. If Processor x writes data p and then Processor y reads data p, Processor y must read the value written by Processor x.
- Causality must be preserved in the sense that all processors must see a series of writes to $p, q, r, \ldots$ in the same order. For example, suppose Processor x sets $p = 1$. Processor y reads p and, if 1, sets it to 2. If Processor z then reads p, it should see 2 and not the old value 1 written by x.

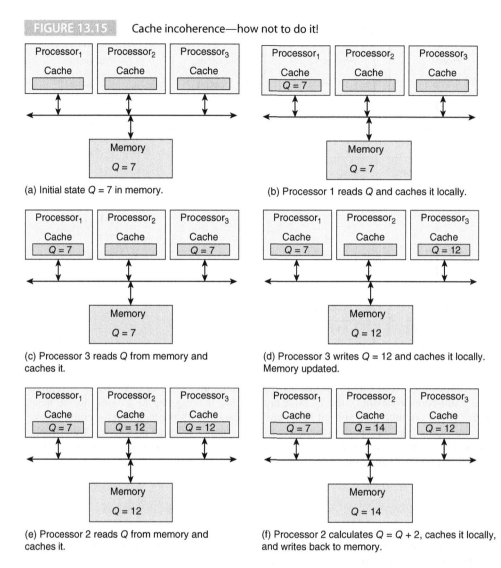

FIGURE 13.15   Cache incoherence—how not to do it!

(a) Initial state $Q = 7$ in memory.

(b) Processor 1 reads Q and caches it locally.

(c) Processor 3 reads Q from memory and caches it.

(d) Processor 3 writes $Q = 12$ and caches it locally. Memory updated.

(e) Processor 2 reads Q from memory and caches it.

(f) Processor 2 calculates $Q = Q + 2$, caches it locally, and writes back to memory.

The MESI Protocol

There are many ways of dealing with cache coherency and several mechanisms have been proposed. One protocol that has become very common in today's multiprocessor systems is the MESI protocol that was introduced into Intel's Pentium processors. The MESI protocol is found in UMA multiprocessor systems that are bus-based because it relies on a mechanism called *snooping*. When a processor performs a write, the write is broadcast on the bus, and all cache memories observe the write. A consequence of snooping is that coherency mechanisms that rely on it are not scalable beyond about 128 processors because of the extra bus traffic that snooping generates.

MESI is just one of a bunch of cache coherency protocols with suitably obscure acronyms such as MSI, MESI, and MOESI. These protocols take their acronyms from the states in which a cache memory can exist. MSI is the basic protocol, MESI is the protocol used by Intel, and MOESI is used by AMD (However, Intel began to use the MESIF protocol, which is an extended version of MESI with the Nehalem architecture). Let's look at MSI first. The acronym MSI stands for *modified, shared, invalid* and refers to the three possible states that a line in a cache may take; that is, each cache line is required to store its current state. A cache line in the I (invalid) state does not contain valid data (i.e., the only valid data is in the memory). A cache line that is in the M (modified) state represents a single valid copy of the data; all other copies must be invalidated. Moreover, a line in the M state is *dirty* and must eventually be written back to memory. A cache line in the S (shared) state indicates that unmodified copies of the line may exist in other caches. The M state indicates that the line is currently owned by the cache's local processor and the copy in memory is stale, whereas the S state indicates that the line is an up-to-date version of the copy in memory.

Figure 13.16 gives a state diagram for the MSI protocol. There is a state machine for each line in each cache in the system. Transitions between states take place as a result of a read or write by a local processor or as the result of a bus snooping operation. For example, if a line is in an S state (shared), and its local processor writes to it, it enters the M state (modified). All other copies of this line now must be invalidated.

FIGURE 13.16   MSI state diagram

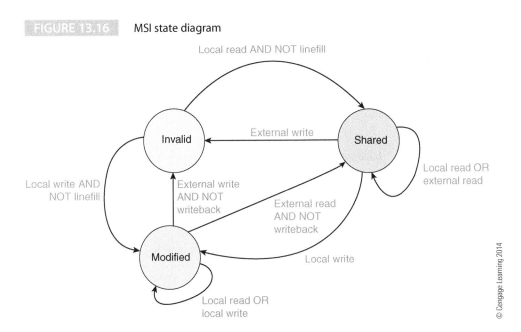

© Cengage Learning 2014

The MESI protocol is also known as the *Illinois* protocol because it was developed at the University of Illinois at Urbana-Champaign. It was adopted by Intel's Pentium processors to support multiprocessing and uses a write-back cache memory. MESI takes the MSI model and adds a new state E (exclusive). The E state denotes a cache line that is present only in the current cache but is clean and equal to the value in memory. The difference between the exclusive and modified states is that the exclusive state indicates that the data is clean. Of course, a write to a line in the E state from a local processor will force it into the modified state. Figure 13.17 gives the state diagram for the MESI protocol as seen from the local CPU bus; that is, these are the state transitions caused by the local processor. Figure 13.18 gives the state diagram for the same cache line caused by other processors and snooping activity on the bus. Some texts present both sets of information on a single figure.[6]

Here is a brief overview of the MESI protocol. At power-on, all cache lines are marked invalid. Suppose a processor performs a write. The corresponding cache line is updated and marked as exclusive because that is currently the only copy of the data. The memory is updated. If this block is written to, it enters the modified state because there is no copy elsewhere and the version in memory is now stale.

---

[6]Juan Gómez-Luna, Ezequiel Herruzo, and José Ignacio Benavides, "MESI Cache Coherence Simulator for Teaching Purposes," *CLEI Electronic Journal*, Vol. 12, No. 1, Paper 5, April 2009.

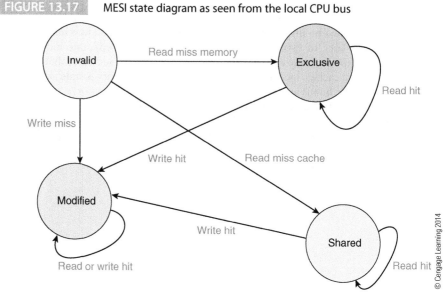

FIGURE 13.17   MESI state diagram as seen from the local CPU bus

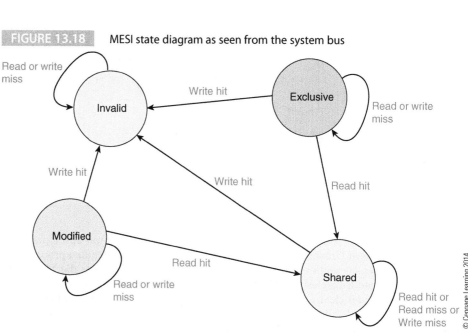

FIGURE 13.18   MESI state diagram as seen from the system bus

If a block is in the exclusive state and suffers a read miss, it is updated from memory, enters the shared state and now exists in more than one cache. If the processor writes to a line that is in the modified state and a cache miss occurs, the modified line in the cache line must be written back to memory. The new data is written into the current cache line and that line marked as exclusive (E state) because it is the only copy of the data. The advantage of the E state is that the amount of bus traffic is reduced in comparison with the MSI protocol. Table 13.2 presents another view of the protocol as a table.

The MOESI protocol extends MESI by adding a further state O (owner), which has the effect of reducing bus traffic for invalid to modified transitions. In the owner state, there is a master copy of the data in the cache (i.e., it is shared) but you can modify the master copy without generating a bus message. This allows the processor owning the data to supply it

TABLE 13.2	Summary of MESI State Changes*	
Bus Transaction	**Operation by Local Processor**	
Read hit	No change	
Read miss	$I \rightarrow S$, or $I \rightarrow E$	
Write hit	$S \rightarrow M, E \rightarrow M$	
Write miss	$I \rightarrow M$	© Cengage Learning 2014

*Note that a change from the M state always results in memory being updated

directly to a second processor making a request. The second processor no longer has to wait for the owner to write back the requested data to memory first. The owning processor can supply data directly to the requesting processor and therefore improve performance. This protocol is implemented by AMD platforms.

False Sharing

Cache memories in multiprocessors can suffer from what has been described as an insidious problem called *false sharing*. As you might guess, false sharing concerns processors that appear to share data across different memory systems when the sharing is not real. The consequence of false sharing is a severe reduction in performance caused by attempts to manage a problem that does not exist.

False sharing is a result of the granularity of cache lines. Suppose that a given cache line contains more than one data object and these objects (instruction or data structures) belong to different threads of execution. Suppose a thread accesses a cache line and modifies object *A* in that cache line. When a cache line is modified, it is invalidated. However, if another thread has object *B* in the same cache line, it finds that the line has been invalidated even though its object has not been accessed. The second thread is paying a penalty for sharing part of a cache line with other threads.

Consider the C declaration int a,b; that declares integers a and b. These will be assigned to contiguous memory by most compilers, and they will probably be stored in the same cache line. If a processor writes to a, its cache line will be invalidated. When b is accessed, its cache line is invalid even though b has not previously been updated. You can avoid this effect by forcing a and b to be in different lines by adding padding; that is, int a, pad[65],b. We have separated a and b by padding with an array (which is not accessed). Now you can access variable a without having to worry about making variable b's cache entry invalid.

13.6   Multithreading

We've looked at two types of instruction-level parallelism: *superscalar*, where the processor grabs a batch of instructions and works out how they can best be rearranged to execute as many at once as possible, and *VLIW*, where the processor executes groups of instructions in parallel that are chosen by the compiler. Where next?

Two significant bottlenecks in computing are the branch penalty and load latency. We've already seen that you can considerably reduce the branch penalty by predicting the outcome of a branch and fetching instructions from the target address or by using predicated execution. Load latency is harder to deal with, because you can't avoid reading the data you process. If you are lucky and the data is in the cache, the load latency may be as little as one or two cycles. If the data is not cached, the load latency may be tens of cycles if it's in main store, and millions of cycles if it's on disk. As time passes, the penalty imposed by load latencies is getting relatively worse as processors are getting faster at a greater rate than memory.

You can't eliminate latency, but you can sometimes *hide* it. Instead of waiting for data to be loaded, the processor can exploit the processor's computing capacity by doing something else. This approach to latency hiding is called *multithreading*.

The notion of multithreading has been part of operating systems technology for a long time; tasks are divided into streams of instructions called threads and the processor switches between threads, making it look as if the processor is executing all the threads at the same time. Multithreading hides the latency of cache misses and allows you to run several streams or threads at once.

The difference between multithreading and multitasking is one of scale. Multitasking involves longer streams of instructions than multithreading and task switching occurs less frequently than with multithreading. A task switch may take place on a page fault, whereas a thread switch may take place on a cache miss.

The overhead involved in switching tasks (Called a *context switch* in the world of multitasking) is considerable because it is necessary to execute all pending instructions—save the registers and flush the cache.

Thread switching is efficient because dedicated hardware keeps multiple copies of registers to eliminate the need to save and restore registers on context switches. Multithreading can be coarse- or fine-grained. In Figure 13.19a, threads are switched after a group of instructions have been executed, and the pipeline is drained each time a switch takes place. Figure 13.19b demonstrates fine-grained multithreading where switching takes place at the end of each cycle.

FIGURE 13.19   Coarse-grained and fine-grained multithreading

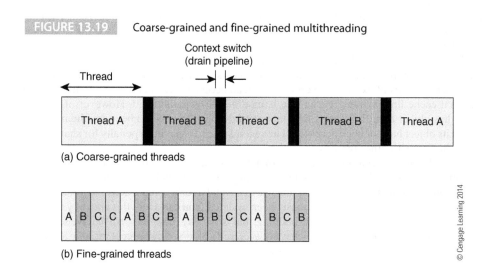

(a) Coarse-grained threads

(b) Fine-grained threads

© Cengage Learning 2014

Thread level parallelism is applied to multiprocessor systems with two or more processors that allow several different threads to be executed in parallel. Such a scheme is inefficient if sufficient streams can't be found to keep all the processors busy. Note the analogy with superscalar processing: a superscalar processor is inefficient if instructions can't be found to keep all units busy, whereas a multiprocessor system is inefficient if sufficient threads can't be found.

Multithreading can be applied to processors that incorporate instruction level parallelism. Consider the *idealized* situations of Figures 13.20a and 13.20b. In Figure 13.20a, multithreading is performed with each thread occupying all three execution stages in the same clock cycle. Figure 13.20b represents *fine-grained simultaneous multithreading* in which each execution unit is available to any thread. At any instant, instructions from three different threads may be executing on three execution units.

Figure 13.20 is misleading because it represents the optimal situation in which execution units are always being utilized. Real processors suffer from hazards and situations in which all execution units cannot be used in the same cycle because insufficient instructions can be found to execute in parallel. Figure 13.21 illustrates these limitations where a white square indicates that the corresponding unit is not executing an operation in that cycle. For example, in cycle 2, execution unit 3 is idle, and in cycle 6, all units are idle.

FIGURE 13.20   Multithreading on a processor with three execution units

A	B	C	C	A	B	C	B	A	B	B	C	C	A	B	C	B
A	B	C	C	A	B	C	B	A	B	B	C	C	A	B	C	B
A	B	C	C	A	B	C	B	A	B	B	C	C	A	B	C	B

(a) Fine-grained threads parallel execution

A	B	C	C	A	B	C	B	A	B	B	C	C	A	B	C	B
C	A	B	C	B	A	B	C	B	C	C	B	A	B	C	A	B
B	C	C	A	B	C	B	B	C	C	A	B	A	B	C	B	A

(b) Fine-grained simultaneous multithreading

The letters, A, B, C represent tasks being carried out. The rows represent the individual processors (execution units). The columns represent the time slots (cycles); that is, the diagram is read from left to right.

© Cengage Learning 2014

FIGURE 13.21   Latencies in multithreading

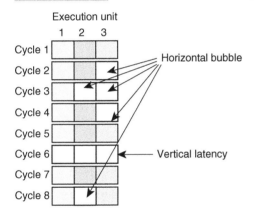

This is essentially the same figure as 13.20b, except that it has been rotated 90 degrees for simplicity and some execution units (in white) are idle due to a stall.

© Cengage Learning 2014

A *horizontal bubble* occurs when not all execution units can be utilized. For example, a system with two integer units and a floating-point unit will not use one slot whenever a floating-point instruction is not part of the instruction stream. Clock cycle 6 is labeled *vertical latency* because an instruction in cycle 6 stalls all units for one clock.

The aim of fine-grained simultaneous multithreading is to keep as many units of the processor occupied executing useful code for as much of the time as possible.

A necessary condition for efficient simultaneous multitasking (SMT) is the absence of any significant overhead associated with thread switching. Clearly, if each thread is using a different processor context (i.e., PC, status and registers) it is necessary for each thread to have its own resources. That is, SMT requires the duplication of registers and the tagging of instructions with the thread number.

Figure 13.22, from Eggers, et al., illustrates the differences between a conventional superscalar processor, a fine-grained multithreaded processor, and a simultaneous multithreaded processor.[7] As you can see, the superscalar processor is very inefficient because of latencies and data dependencies. The fine-grained superscalar improves the situation by interleaving threads and avoiding some of the latencies in a superscalar processor. Finally, the SMT processor is able to fill more of the execution units by finding instructions from any available thread.

---

[7]Susan J. Eggers, et al., "Simultaneous multithreading," *IEEE Micro*, September/October 1997, pp. 12–18.

FIGURE 13.22   Summary of instruction and thread level parallelisms

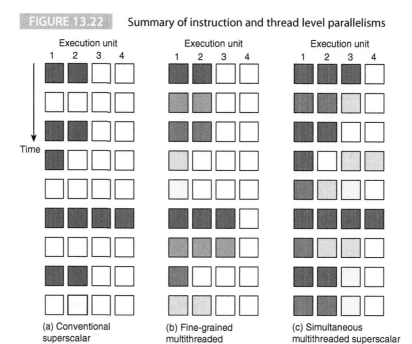

(a) Conventional
superscalar

(b) Fine-grained
multithreaded
superscalar

(c) Simultaneous
multithreaded superscalar
processor

© 1997 IEEE. Reprinted, with permission, from Susan J. Eggers et al. "Simultaneous multithreading: a platform for next-generation processors." *IEEE Micro,* September/October 1997, pp. 12–18.

Throughput and Latency

Multithreading affects both the latency and throughput; it increases both of them. Consider the following example of two threads A and B, each of which accesses memory.

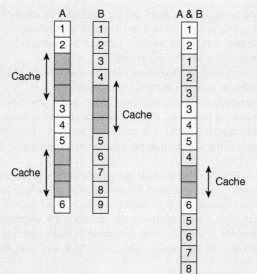

Task A running as a single thread has a latency of 12 cycles and an efficiency of 6/12 (i.e., 6 cycles in 12). Task B has a latency of 12 cycles and an efficiency of 9/12.

Suppose we use multithreading and switch threads on a cache access. When A and B are run in a multi-threaded mode, the total time for the two tasks is 17 units rather than 12+12 = 24.

Task A has a latency of 12 (no change), and Task B has a latency of 17 (an increase). However, the efficiency is now 15/17.

© Cengage Learning 2014

Intel has incorporated SMT in some members of its IA-32 family.[8] Intel uses the expression *Hyper-threading* to indicate SMT.

Intel's first implementation of Hyper-threading technology became available on the Xeon, an IA432-compatible processor intended for high-end workstations. The Xeon has two logical processors per physical processor; that is, it has the resources to support two threads.

The overhead in providing two logical processors is minimal, about 5% of the chip area. It is necessary only to reproduce the architectural state consisting of registers, including the general-purpose registers, the control registers, the advanced programmable interrupt controller (APIC) registers, and some machine state registers. The logical processors share nearly all other resources on the physical processor, such as caches, execution units, branch predictors, control logic, and buses. According to Intel, the incorporation of hyper-threading technology increases the Xeon's performance by 30%.

Intel Hyperthreading

Intel's hyperthreading first appeared in its Xeon processor in 2002 and was followed by the HT versions of the Pentium 4. Hyperthreading disappeared in the Core architectures that followed the P4 (because these were based on an earlier Pentium Pro architecture). Hyperthreading resurfaced in the Core i7 in 2008.

13.7   Multi-core Processors

Having set the scene, we now introduce the multi-core processor. Over four decades, the number of transistors on a chip has increased in accordance with Moore's law. Clock frequencies have also increased dramatically over the same period. Figure 13.23 provides a reminder of the improvement in performance of processors over the years and demonstrates how the era of instruction level parallelism is giving way to the domain of the multiple core. Of course, previous enhancements can still be incorporated in multi-core processors. Once again, we, have to state that the multiple-core processor is a response to the growing physical limitations of semiconductors caused by increasing clock frequencies.

FIGURE 13.23   Processor performance

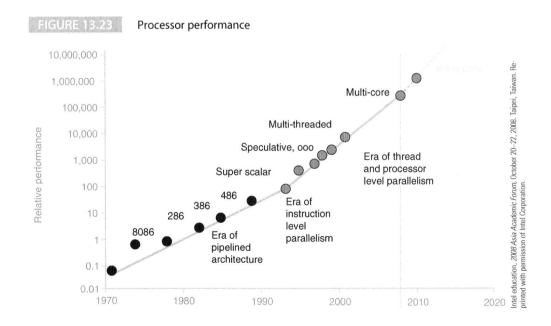

Intel education, *2008 Asia Academic Forum*, October 20–22, 2008, Taipei, Taiwan. Reprinted with permission of Intel Corporation.

---

[8]Deborah T. Marr, et al., "Hyper-threading technology architecture and microarchitecture," *Intel Technology Journal*, Q1, 2002.

What is a Core?

The term *core* in multi-core computing is rather misleading, because its meaning is not precisely defined. At one extreme, a core could be the same as a microprocessor and a multi-core chip could have two or more identical processors on the same chip. Such an arrangement would imply that the CPU, cache memory, memory management system, and even the bus interface were duplicated for each chip. At the other extreme, a core could have just a processing unit, while a multi-core chip could have a CPU plus several processing units (without associated program control and sequencing). Finally, a multi-core chip could have several CPUs that share certain common facilities, such as memory and the external interface.

Homogenous and Heterogeneous Processors

Multi-core processors can be divided into two categories: homogeneous or heterogeneous. The distinction is rather like that between uniform and nonuniform memory access. A homogeneous processor employs a set of cores where every core is an identical copy of all the other cores. A heterogeneous multi-core processor includes cores that differ in ISA, chip area, performance, or power dissipation. Early multi-core processors were generally homogeneous with one core being duplicated. With the passage of time, the benevolent Moore's law has enabled sufficient numbers of active devices to be fabricated on a chip to develop systems where different types of core can be optimized for different functions. This is particularly true of the advanced processors found in commodity applications such as games consoles.

13.7.1   Homogeneous Multiprocessors

Like so many innovations in the microprocessor world, the multi-core CPU gradually emerged into the light rather than suddenly bursting on the scene. It was yet another inevitable step. Right from the very early days of 16-bit microprocessors, the multiprocessor was beginning to appear in an *asymmetric* form via the co-processor. Co-processors are asymmetric in the sense that they provide additional processing power but in a different form (floating-point, graphics, or physics engines). The Pentium was designed to support symmetric multiprocessing and incorporated hardware mechanisms to allow dual-processor motherboards. Moreover, the number of transistors on a chip increased from about three million in 1993 to two billion in 2011. The next step was obvious: put two processors in the same physical housing. IBM introduced a high-end, dual-core processor that put two 64-bit POWER processors on its POWER4 chip in 2001. Intel introduced its mass-marketed dual-core processor in 2005 (a Pentium 840 Extreme Edition). At almost exactly the same time, AMD brought out its Opteron 800 Series.[9] Intel's first dual-core processor was really two processors in the same package that were not well integrated. Intel rapidly brought out the *Core 2 Duo* to replace the *Dual Core Pentium*. The Core 2 Duo was specifically designed for dual-processor systems.

Figure 13.24 summarizes the growth of the multiprocessor chip and its relationship with cache that we described in Chapter 9. Figure 13.24a describes the three blocks of a conventional processor. The CPU state incorporates the registers (including the PC and status register) that define the current state of the processor and the thread (line of code) it is executing. In Figure 13.24b, we have added a second set of CPU state registers to allow two threads to be executed concurrently by switching between state registers whenever one thread is stalled by a data load.

Figure 13.24c describes a multiprocessor system where two processors share the same environment (motherboard). The dotted blue line indicates that they are in the same system but not on the same chip. Figure 13.24d describes a multi-core processor which is the same as (c) but the processors are now on the same chip. Figure 13.24e is an extension of (d) in which the cache is shared between processors. This indicates an increasing degree of coupling between individual processors on the same chip. In Figure 13.24f, we have added hyperthreading to allow each of the individual processors to execute multiple threads. Figure 13.24g increases the number of cores to four and shows that we can choose how to distribute the individual cache levels. In this case, each processor has a private level 1 and 2 cache but shares the much larger L3 cache.

---

[9]The near simultaneous introduction of Intel and AMD multi-core processors demonstrates two points. The first is the inevitability of multi-core processing and the second is, given that Intel and AMD were operating in the same markets with similar levels of technology and in similar cultural environments, it is not surprising that both organizations made similar innovations at roughly the same time.

FIGURE 13.24   Progressive development of the multi-core processor

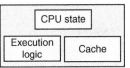

(a) Basic single-core
processor

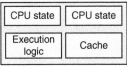

(b) Single-core processor
with hyperthreading

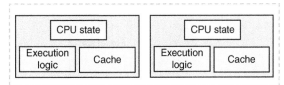

(c) Multiprocessor—two or more separate processors

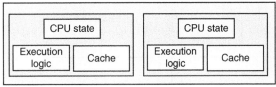

(d) Multi-core processor with two or more processors housed
in the same package

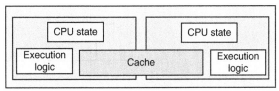

(e) Multi-core processor with cache shared by both
processors

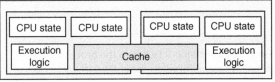

(f) Multi-core processor with each processor's
cache shared by both processors and each processor
implementing hyperprocessing.

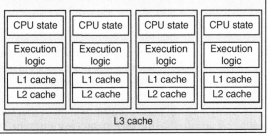

(g) Multi-core quad processor with each processor having its
own L1 and L2 cache and a shared L3 cache

© Cengage Learning 2014

SMP Background

In 2001, Intel introduced the Pentium III Xeon intended for use in servers. This processor was intended to be used in two-processor or four-processor *glueless* multiprocessors. The term *glueless* refers to the logic and auxiliary systems necessary to link processors with each other or with memory and implies that the interface logic is on-chip and that the designer does not have to spend a lot of time creating a special multiprocessor system. In 2002, Intel introduced the Pentium 4 family and the corresponding Xeon family targeted at desktop PCs and high-end workstations and servers, respectively.

The Pentium III Xeon was intended for use in symmetric multiprocessing, (SMP) systems that have a simple structure based on a common bus shared between processors. The term *symmetric* implies that all processors are equal and logically interchangeable. The following figure illustrates the concept of SMP. Some multiprocessor systems have a topology that reflects the application; that is, the system is optimized for an application such as signal processing. An SMP architecture treats all processors equally, and it is the operating system that has the task of partitioning the computation between the individual processors.

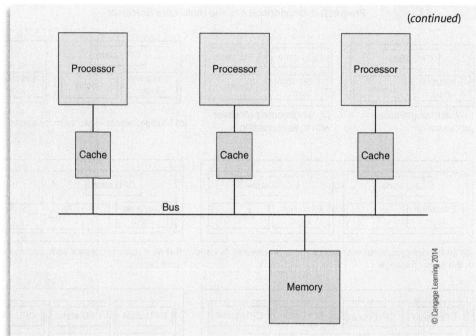

(continued)

© Cengage Learning 2014

The weakness of SMP is that the bus becomes a bottleneck between the processors and between the processors and memory. SMP is successful when the individual processors make good use of their internal data/instruction caches and limit the number of accesses to main memory. Programs can be written to take advantage of SMP structures. However, a consequence of this is that such a program will have a degraded performance when run on a uniprocessor system. This is a problem in the personal computing arena where someone playing a game, optimized for an SMP architecture, on a uniprocessor, is having to pay a performance penalty.

Intel Nehalem Multi-Core Processor

Intel's Nehalem architecture was introduced in November 2008 with the Core i7 processor and was to become Intel's flagship PC processor for several years. Figure 13.25 describes a multiprocessor system using two Nehalem chips. Each chip has four cores where a core has a CPU, L1, and L2 cache, and all four cores share a common L3 cache. A *quick path interconnect* (QPI) provides an external interface allowing clusters of chips to be connected together.[10] The QPI interface can transfer data on both the rising and the falling edge of the clock (at 3.2 GHz) to achieve a bandwidth of 25.6 GB/s.

Technically, QuickPath is not a bus (remember that *bus* is a contraction of the Latin *omnibus* which means *for all*), because it supports only point-to-point communication. However, QuickPath has a multilayered architecture that has been optimized for very high-speed data exchanges. For example, the physical layer uses twenty differential signal paths plus a dedicated clock path in each direction.

AMD Multi-Core Processors

AMD is Intel's principal competitor and has long worked to maximize its share of the market. Part of that strategy has been a move from single-core to multi-core processors. AMD's multiprocessing strategy depends strongly on HyperTransport technology that uses high-speed serial data links to exchange information in a similar way to PCIexpress but with a lower overhead. Table 13.3 from AMD illustrates the rapid growth in multi-core technology. This table gives the size of the manufacturing process (feature size), the number of cores, the L2 and L3

---

[10]"An Introduction to the Intel® QuickPath Interconnect," Intel Whitepaper, January 2009.

FIGURE 13.25   Intel eight core multiprocessor using two Nehalem chips

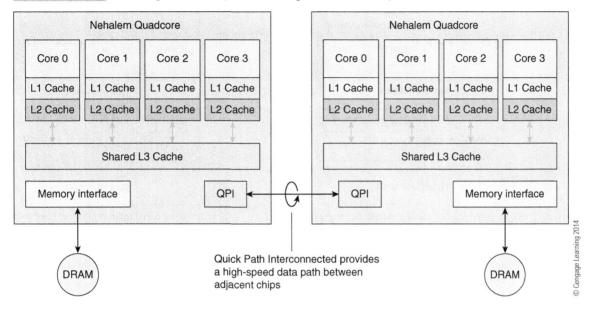

TABLE 13.3   Development of AMD's Multi-Core Processors

Year	2003	2005	2007	2008	2009	2010	2011
Name	Opteron	Opteron	Barcelona	Shanghai	Istanbul	Magny-Cours	Interlagos
Feature size	90 nm	90 nm	65 nm	45 nm	45 nm	45 nm	32 nm
Cores	1	2	4	4	6	12	16
L2 Cache	1 MB	1MB	512 KB	512 KB	512 KB	512 KB	2 MB
L3 Cache	–	–	2 MB	6 MB	6 MB	12 MB	16 MB
HTT	3×1.6 GT/s	3×1.6 GT/s	3×2.0 GT/s	3×4.0 GT/s	3×4.8 GT/s	4×6.4 GT/s	4×6.4 GT/s
Memory	$2 \times$ DDR 300	$2 \times$ DDR 400	$2 \times$ DDR2 677	$2 \times$ DDR2 800	$2 \times$ DDR2 1066	$4 \times$ DDR3 1333	$4 \times$ DDR3 1866

cache sizes, the number of HyperTransport Technology channels between adjacent chips and the link bandwidth, and the memory interface (number of DRAM modules, DRAM type, and DRAM speed). As you can see, progress has taken place on all fronts over the years.

Figure 13.26 demonstrates a system using two Athlon 64 FX dual processors linked by HyperTransport buses. Figure 13.27 depicts a more sophisticated example, where eight dual-core processors can be combined with HyperTransport technology to create a NUMA multi-processor network with crosslinked processors.

FIGURE 13.26   AMD's use of HyperTransport technology in multiprocessors

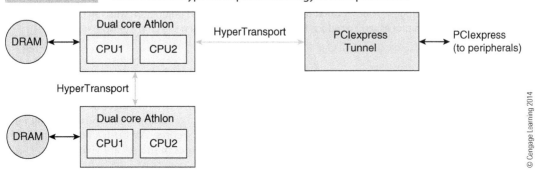

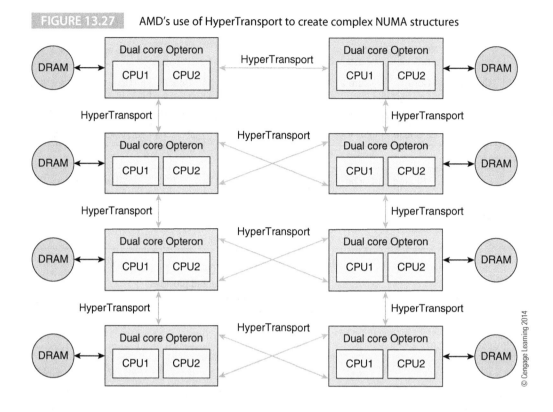

FIGURE 13.27   AMD's use of HyperTransport to create complex NUMA structures

CPU Throttling

For a long time, computers ran at the highest speed possible. That was axiomatic—no sane person would buy a processor and then deliberately run it at a slower speed than necessary. However, in an age where power is limited by batteries or the equipment is in danger of becoming too hot, reducing the clock rate from the maximum becomes essential.

CPU throttling or dynamic frequency scaling is a technique used to lower power consumption by reducing the frequency of the clock. Throttling can be used to reduce the clock rate of processors that are idle or to reduce CPU clock rates when long-latency memory accesses are taking place. Another power-saving technique is to power down areas of the processor not currently in use.

Intel calls its power reduction system *SpeedStep* technology and AMD uses the term *Cool'n'Quiet* in desktop environments and *PowerNow!* in mobile environments. As its name suggests, Cool'n'Quiet technology also works in conjunction with fan control in desktop processors to reduce fan speed and noise when the chip's temperature falls.

ARM Cortex A9 Multi Core

ARM produce a range of processors based on the ARM's core architecture but with various power:performance:cost tradeoffs. The Cortex A9 is a general-purpose processor intended for mobile applications and is available with 1, 2, 3, or 4 cores. A9 Cortex processors can be configured for specific applications. Some of the attributes of the A9 are given here.

- Superscalar pipelined processor core.
- NEON media processing engine – ARM's own SIMD instruction set with 32 64-bit-wide registers that can be used as 16 128-bit registers.
- Floating-point unit.
- Thumb-2 technology – ARM's compressed code mechanism.
- Jazelle technology – this is ARM's on-chip support for Java and enables the direct execution of Java's native bytecode. This allows ARM processors to have three states—each executing a different native instruction set architecture.

Figure 13.28 describes the structure of the Cortex A9. Note that ARM is an IP company, and this processor can be configured in several different ways depending on the wishes of the user. Other special-purpose coprocessors, such as cryptographic accelerators, can be incorporated. Bus snooping control is incorporated on chip, but there is no L2 cache.

FIGURE 13.28   Structure of ARM's Cortex A9 multi-core processor

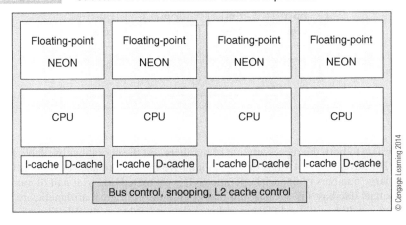

IBM Power7

IBM championed RISC architecture from the very beginning. Credit for RISC architectures is given to John Cock at IBM for his pioneering work on the 801 project (a RISC architecture that was never implemented) in 1974. IBM's first commercially successful RISC processor was the POWER architecture that later formed the basis for the PowerPC used by Apple. IBM's first computers using this architecture appeared in 1990 (the RS/6000 series). The POWER architecture itself continued to be developed for IBMs high-performance workstations, with POWER7 being introduced in 2010.

Figure 13.29 describes the organization of POWER7 which has eight cores using 1.2 billion transistors. The individual cores can be turned off to save energy. Similarly, the clock frequencies of the cores can also be modified to *reallocate* energy. The L2 caches are each 256 KB, and the shared L3 cache is 32 MB. The cores themselves support out-of-order execution and provide two fixed-point ALUs, two floating-point units, and a *decimal floating-point* unit. A decimal FPU is rather unusual today and first appeared in the POWER6 in 2007. In this context, the word

The War is on All Fronts

Where is the battle to improve computer performance taking place? That answer is that it's on all fronts and the POWER7 demonstrates the truth of this. POWER7 is a development of previous versions of this architecture, and like its contemporaries, it takes an aggressive approach to multiprocessing.

However, the inclusion of a dedicated decimal floating-point unit to accelerate large numbers of financial transactions in commercial databases demonstrates how progress takes place in a number of directions simultaneously. Each new generation of processors incorporates lessons learned from previous generations, lessons from the state-of-the-art globally, and innovations peculiar to the current generation.

This multifaceted progress is one of the driving forces behind the modified versions of Moore's law stating that computer performance is increasing year by year.

FIGURE 13.29   Organization of the POWER7

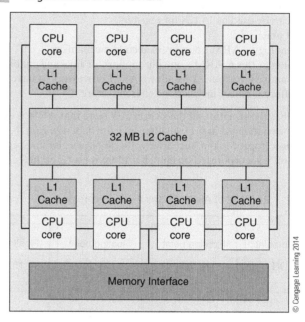

decimal refers to both the representation of numbers and the operations available to be performed. The advantage of DFP is that it can improve the performance of systems that carry out very large numbers of financial operations. It is estimated that over half of numeric data in commercial databases is decimal and that operations in decimal arithmetic are normally carried out in software 100 to 1,000 times slower than in hardware. POWER7 conforms to the IEEE 754R standard and defines 32-, 64-, and 128-bit formats. The 128-bit format has a 34-bit precision and a 16-bit exponent.

The GPU

What do you get if you really take multi-core processing to heart and put very large numbers of processors on a single chip—not two or four or eight, but hundreds? You get a GPU or *graphics processing unit.*[11] Personal computers typically have video cards that deal with the control of display units, although Intel's Sandy Bridge version of its iCore processors does include on-chip hardware video-processing facilities that allow the design of low-cost PCs. A video processor card can include a remarkable amount of computational power because video processing requires the application of immensely large numbers of operations for video data.

The graphics processing unit, the GeForce256, was invented by NVIDIA in 1999. NVIDIA coined the term to describe a single-chip processor that included integrated transform, lighting, triangle manipulation, clipping, and rendering engines. All of these operations are applied to the digital representation of images, but within a very few years, people realized that GPU chips could be used for many common video applications. The Ge256 has a 256-bit graphics core and a 128-bit memory interface.

In recent years, GPUs have added a range of new functions, including video decoding and video post-processing. Another highly-specialized processor relying on multiple cores for high-performance processing is the *physics engine*. A physics engine does for physical

---

[11]It is rather interesting that, in the popular imagination, people make a lot of fuss over Intel or AMD's multiprocessors with four or eight cores when graphics manufacturers like NVIDIA are churning out chips with 512 cores largely unnoticed.

systems what the GPU does for images; indeed, some GPUs can be programmed as physics engines. For a long time, physicists have been concerned with dynamics such as the motion of the atmosphere (weather forecasting) or fluid flow (turbine blade design). Today, games players demand high-speed real-time dynamics. If an object explodes, its pieces should all obey the laws of Newtonian dynamics. Other physics operations are collision detection (for example, when a ball hits a wall) and deformation physics, which deals with the way in which an object behaves under crushing forces (for example, when Tom hits Jerry's head with a brick).

The goal of GPU technology is *Extreme High Definition Gaming* (XHD). In the 1980s, crude, chunky games were the state of the art. Today, the effort of both software and hardware designers is focused on high resolution (not least because of the tremendous drop in the cost of high resolution displays). "Extreme" high-resolution gaming refers to widescreen displays with resolutions of over 2,560 × 1,600 pixels, which is seven times greater than the 1,080 HD used by Blu-ray.

Initially, GPUs were not easy to apply to general computing problems, because they had been constructed specifically for graphics processing and you accessed them via *application programming interfaces* (APIs). An API is essentially a system call from software to some resource which may be a library function or the gateway to hardware acceleration, like a GPU. By 2006, NVIDA's extended APIs were available to GPUs that allowed programming in C. Now, users could access the power of the GPU directly.

In 2008, NVIDIA introduced the GT2000 series that increased the number of *streaming processor cores* to 240. Figure 13.30 describes the organization of the cores in a later 512-core version of the GPU. Each core can execute a new integer or floating-point instruction at each clock. The cores are organized as 16 groups, called *streaming multiprocessors* (SMs), of 32 cores. A part of the GPU called the *GigaThread global scheduler* is responsible for allocating blocks of threads.

The heart of a GPU card is its *graphics pipeline,* which is responsible for taking a description of an image from the computer and presenting a raster-scan (i.e., line-by-line or bitmap) image to the video display. A photograph is just a bitmapped image that cannot be separated into parts, it is simply stored and displayed pixel by pixel after suitable scaling. A graphics image of the type that you see in computer games and animations is a complex set of objects and lighting effects; that is, a graphics image exists as a description of a scene that the GPU has to draw. Consider an image with a person standing in front of a house. The person is a collection of drawing objects overlaid on another collection of drawing objects. If the person moves, even though a million pixels may have been shifted on the final images, the computer needs to tell the graphics card only that an object has moved. It is the job of the graphics pipeline to create the new image dynamically.

The source image from the computer is presented to the graphics card as triangles, which are the fundamental component of a video image. All complex shapes are composed of triangles. The higher the resolution, the greater the number of triangles. The triangles themselves are transmitted as vertices. The GPU has to process the vertices and then

Pipelining—Time and Space

The GPU pipeline differs from the type of pipelining we have associated with CPUs when discussing instruction level parallelism. When a CPU performs software pipelining, it sequentially executes the code for each stage of the pipeline. The pipeline stages are distributed in time.

The GPU pipeline is distributed in space; that is, the individual processors are divided between the different functions to be carried out by the graphics pipeline.

Early GPUs implemented different stages of the graphics pipeline with different processors in order to achieve the best performance. As time has passed, the complexity of operations in all pipeline stages has increased, and there has been a movement towards a so-called unified shader architecture in which all stages share the same programmable core.

CUDA been a contender

Compute Unified Device Architecture (CUDA) is the software interface to GPUs developed by NVIDIA. Essentially, it is a user interface to the graphics pipeline that can be used by the software developer to create code that runs on a GPU card. More importantly, CUDA gives the designer access to the GPU's highly pipelined architecture.

FIGURE 13.30 NVIDA GT200 series architecture (this is not an empty figure—it consists of 512 CPUs)

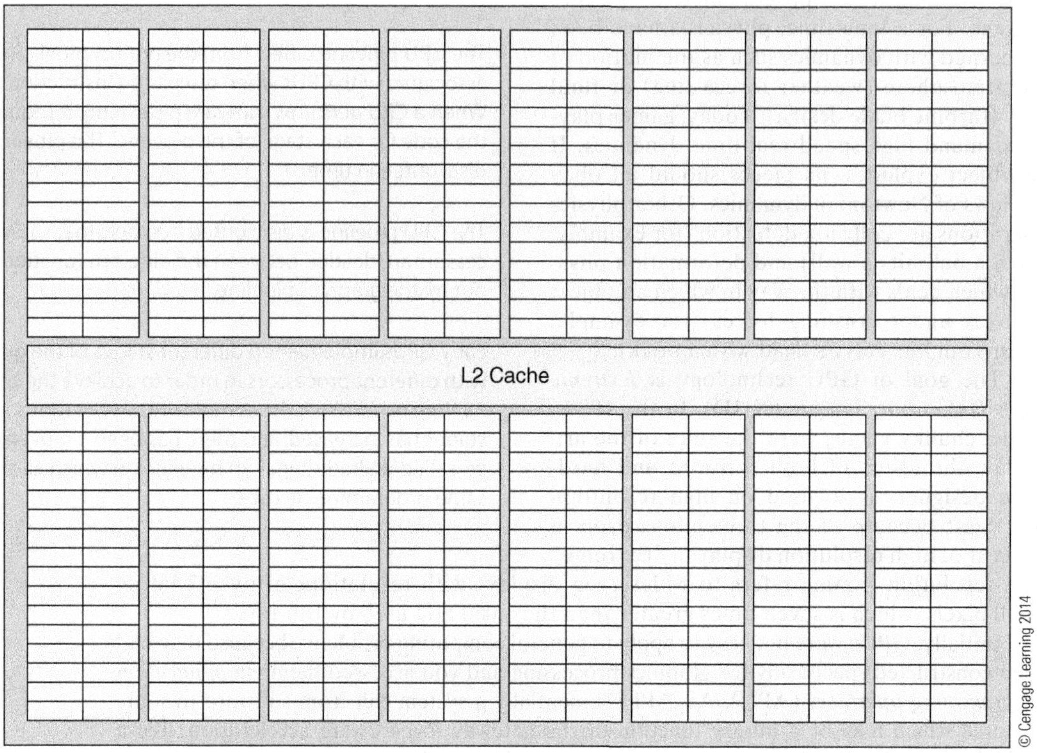

L2 Cache

© Cengage Learning 2014

operates as a pipeline with each stage performing successive transformations on the image. Figure 13.31 illustrates the concept of a graphics pipeline—the details vary from system to system. Another operation taking place in the pipeline is the generation of lighting effects. It is necessary to work out how each point behaves under the influence of a complex lighting system with light from many sources. The rasterization stage takes the images as described by the triangles and converts them into images that can be stored in memory and displayed.

The GPU is highly optimized for its role as a video processing machine, which can lead to remarkable performance enhancements over a CPU. For example, a GPU may achieve a throughput of over 100 times that of a CPU. GPUs achieve their high throughput by

FIGURE 13.31 The GPU pipeline

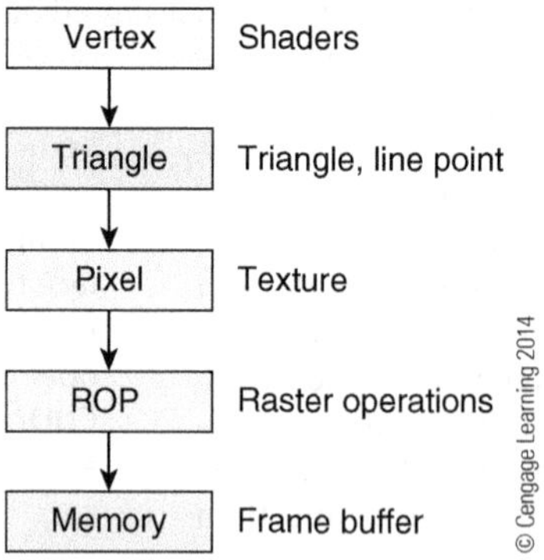

Vertex	Shaders
Triangle	Triangle, line point
Pixel	Texture
ROP	Raster operations
Memory	Frame buffer

© Cengage Learning 2014

dedicating chip area to processing and by not incorporating the massive cache and control logic overhead associated with CPUs.

Not everyone is quite as enthusiastic about GPUs. A paper by G.C. Caragea, et al. appeared on Usenix.org that presented the results of an investigation into the performance of general-purpose multi-core processors versus CPUs when executing *irregular workloads*.[12] Although this paper refers to a specific multi-core processor, XMT, its general conclusion is that GPUs show high performance increases on regular workloads but much poorer performance on irregular workloads (i.e., those demonstrating highly irregular patterns of memory access).

> ## Games Consoles Considered Harmful
>
> Low-cost games consoles, like the Sony PlayStation 3 that use the Cell processor or Microsoft's Xbox 360 which also has a multi-core processor supporting SIMD extensions, are creating headaches in some quarters. *The Economist** reports that sophisticated computer simulation games and high-performance software are being run on low-cost consumer consoles. With such technology being widely available, wealthy countries that have long enjoyed a technological monopoly on high-performance computing are realizing that it is now available to all.
>
> ---
>
> * "Military technology used to filter down to consumers. Now it's going the other way," *The Economist*, Dec. 10, 2009.

13.7.2   Heterogeneous Multiprocessors

In 2007, Kumar, Tullsen, and Jouppi in an article in *Computer*[13] stated that

> "*Heterogeneous chip multiprocessors present unique opportunities for improving system throughput, reducing processor power, and mitigating Amdahl's law. On-chip heterogeneity allows the processor to better match execution resources to each application's needs and to address a wider spectrum of system loads—from low to high thread parallelism—with high efficiency.*"

Heterogeneous multiprocessing permits the use of multi-ISA architectures with individual software being targeted on the most appropriate ISA on the chip. Here, we look at an example of the heterogeneous multiprocessor.

The Cell Architecture

The Cell Broadband Engine Architecture (Cell) is a microprocessor architecture developed by a consortium of Sony, Toshiba, and IBM (STI). The project started in 2001 with a four-year budget of $400 million. The consortium's first commercial product was used in Sony's PlayStation 3 game console released in 2006.

Cell was developed to overcome the three walls of computer architecture: the power wall (limitations due to increasing power dissipation), the frequency wall (the problems of deep pipelines at high frequencies), and the memory wall (the problem of DRAM memory latency at high speeds). The Cell processor used in the PlayStation 3 has 234 million transistors, which is comparable with the Itanium 2 developed at approximately the same time.

Cell takes an unusual approach to multiprocessing by employing different architectures on the same chip[14] that is, the Cell is said to be a *heterogeneous processor*. The organization of the Cell is given in Figure 13.32. The *main* processor is called a *Power Processing Element* that is compatible with IBM's Power 970 architecture. This has a split L1 cache with 32 KB data, 32 KB instruction, and a 512 KB level 2 cache. This processor also incorporates the AltiVec (now called VMX) instruction set extensions to provide SIMD multimedia support. The SIMD processor has 32 128-bit vector registers.

---

[12]G. Caragea, F. Keceli, A. Tzannes, and U. Vishkin, "General-purpose vs. GPU: Comparison of many-cores on irregular workloads," *Proc. USENIX Workshop on Hot Topics in Parallelsim*, 2010.

[13]Rakesh Kumar, Dean M Tullsen and Norman P. Jouppi, "Heterogeneous chip multiprocessors," *Computer*, November 2005, pp. 32–38, Vol. 38, No. 11.

[14]Michael Gschwind, Peter Hofstee, Brian Flachs, Marty Hopkins, Yukio Watanabe, and Takeshi Yamazaki (IBM), "A novel SIMD architecture for the Cell heterogeneous chip-multiprocessor," *Hot Chips* 17, 2005.

FIGURE 13.32   Cell architecture

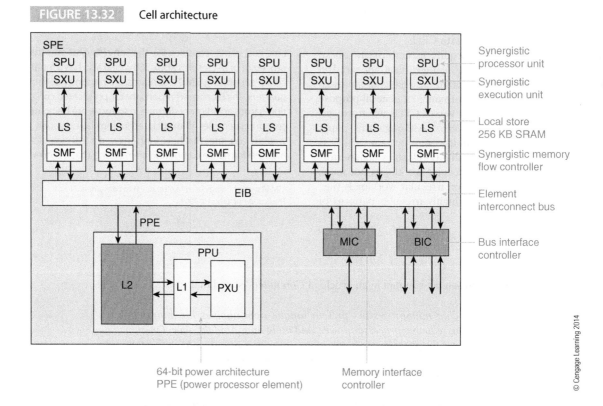

As well as the central power processing element, the Cell has eight *synergistic processor elements* (SPEs) arranged as a network with a high-bandwidth internal element interconnect bus (EIB). This bus is also connected to an interface to external memory and an I/O controller. The data bus itself consists of four 16-byte wide data rings that support multiple simultaneous transfers per ring. The peak bandwidth is 96 bytes per clock cycle. The EIB can sustain a data rate of 200 GB/s.

The eight SPEs are special purpose RISC-based SIMD processors optimized for *data-rich operations*. The SPEs each have 128 registers in a 128-bit register file and 256 KB of local store. SPEs are designed to process applications level data and do not run the operating system; that's the function of the power processor element core (PPE). Most instructions operate on 128-bit operands divided into four 32-bit words. This fits in well with the Cell's role as a high-performance multimedia processor and its role in PlayStation3. Anyone interested in the history of computer technology will find it quite remarkable that a low-cost commodity item like the PlayStation contains a state-of-the art network on a chip processor with a computational throughput unimaginable a few years ago.

The PPE core provides the 64-bit Power architecture together with the L1 and L2 caches, an instruction control unit, a load and store unit, a fixed point integer unit, a floating-point unit, a branch unit, and a memory management unit.

13.7.3   Networks on a Chip

In 2011, the Wikipedia entry for *Network on a Chip* (NoC) called it an *emerging paradigm* that links multiple processor cores on a chip with multiple point-to-point routes. In other words, the NoC represents the next stage in multi-core processors with more processing units than the first generation of multi-core processors and with more complicated inter-processor communication methods. There is an overlap between NoC and SoC; the latter refers to a *system on a chip* and implies a single chip solution to an engineering problem that may include multiple CPUs, memory, I/O, and special-purpose circuits that might be necessary in cell phones or GPS receivers. A SoC is normally intended for a specific purpose, whereas a NoC is a general-purpose processing device.

In 2007, Intel announced its NoC processor, the *Teraflops Research Chip,* that was to be the first general-purpose microprocessor to break the Teraflops barrier.[15] This *terascale* processor was intended to lead Intel's research into the area of multi-core processing well beyond the modest number of cores in the i7 and to achieve a lower power/core than existing processors. Teraflops uses an 80-core *tiled architecture* with a two-dimensional mesh of cores—each implemented with 100 million transistors using a 65 nm process technology. Teraflops is a research vehicle and a platform for processor evaluation rather than a commercial processor. The cores or tiles are arranged as an 8×10 array.

Each Teraflops tile has two floating-point units, 3 KB of instruction memory arranged as 256 96-bit instructions, and 2 KB of data memory sufficient to hold 512 single-precision numbers. The register file has 32 entries with six read ports and four write ports. The floating-point units, FPMAC0 and FPMAC1 are two independent, fully pipelined, single-precision, floating-point, multiply accumulate devices that can provide a peak performance of 20 GFLOPS. Clearly, Teraflops is not intended for conventional programming. It is optimized for scientific computing and graphics. The underlying ISA is VLIW with 96-bit instructions. Each long instruction word contains eight operations (i.e., an instruction bundle has eight instructions). The instruction set is tiny. Table 13.4 describes the Teraflops instruction set. A Teraflops tile also includes a router to enable messages to be passed to and

Tiles

Cores in multi-core processors are often referred to as *tiles.* In general, a tile includes a CPU, memory, and a communications interface that allows communications between tiles. The communications technology usually uses packet switches or wormhole routing rather than a conventional switched bus.

The term tile is used because cores in a multi-core processor are frequently set out like the tiles on a roof (that is, a rectangular array of tiles often forming a mesh network).

The term tiling has a second use in high-performance computing that refers to software, rather than hardware. Tiling also refers to the transformation of a loop construct to include smaller blocks of data that can be cached and hence reduce the miss rate. Such a tile is a block of memory that fits within the cache.

TABLE 13.4   Teraflops Instruction Set

LOAD, STORE	Move a pair of 32-bit floats between the register file and data memory.
LOAD0, STORE0, OFFSET	Move a pair of 32-bit floats between the register file and data memory at address plus OFFSET.
BRNE, INDEX	The native loop capability. INDEX sets a register for loop count and BRNE branches while the index register is greater than zero.
JUMP	Jump to the specified program counter address.
SENDI[H|A|D|T]	Send instruction header, address, data, and tail to a core.
SENDD[H|A|D|T]	Send data header, address, data, and tail to a core.
WFD	Stall while waiting for data to arrive from any tile.
MULT	Multiply operands.
ACCUM	Accumulate with previous result.
STAL	Stall program counter (PC) while waiting for a new PC.
NAP	Put FPUs to sleep.
WAKE	Wake FPUs from sleep.

© Cengage Learning 2014

---

[15]Tim Mattson, Rob van der Wijngaart, Michael Frumkin, "Programming Intel's 80 core terascale processor," *Proceedings of the 2008 ACM/IEEE Conference on Supercomputing*, SC08, Austin Texas, November 2008.

FIGURE 13.33   Structure of the Teraflops tile

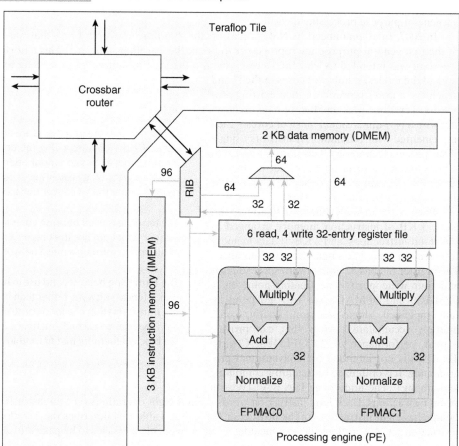

© Cengage Learning 2014

from other tiles. Figure 13.33 gives a diagram of the structure of a Teraflops tile. The communications unit is a five-port packet-switched router that links a tile to its four neighbors and to the clock. The data timing is called *mesosynchronous*, which means that the data is insensitive to phase errors because the timing is derived locally from the data signal. Local clocks are synchronized to a master clock, and buffering is used to ensure that the data and the locally generated clock are in synchronism. If the data stream were treated as a synchronous signal (as might happen in systems with much lower clock rates), it would be difficult to ensure that the clock phase relationship would be the same at all 80 tiles.

An advantage of Terascale architectures, like Teraflops, is their partitionablility[16] and, as a consequence, reliability and fault-tolerance. The tiled topology combined with the ability to route messages through the mesh network means that tasks can be partitioned and the tiles allocated particular subfunctions. However, tiling provides fault tolerance and the potential for graceful degradation. If one or more tiles fail, they can be switched out and tasks assigned to good tiles.

Figure 13.34 provides an example of fault-tolerant routing. At reset and initialization time, all of the tiles in the array can be individually tested. Errors may be present in individual tiles or in the links between tiles. Figure 13.34 illustrates the situation in which an array has four dead tiles and a faulty link. Note that in Figure 13.34 several nodes have

[16]"Azimi, et al., Integration Challenges and Tradeoffs for Tera-scale Architectures." *Intel Technology Journal*, Vol. 11, Issue 03, August 2007.

FIGURE 13.34   Example of fault tolerant routing

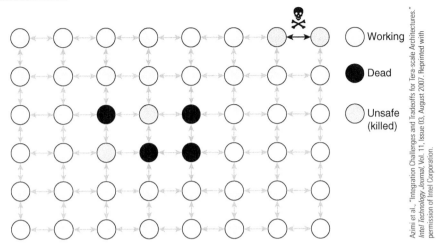

Azimi et al., "Integration Challenges and Tradeoffs for Tera-scale Architectures." *Intel Technology Journal*, Vol. 11, Issue 03, August 2007. Reprinted with permission of Intel Corporation.

been marked as unsafe. These are removed from the array. In the center, they are removed to make the region containing the dead tiles a rectangle in order to simplify routing algorithms. The two unsafe tiles at the top right-hand corner have been removed because of the faulty link.

An interesting feature is the ability to issue instructions that send floating-point units to sleep or wake them up. This facility can be used to save power and reduce heat dissipation. It can be used dynamically to move active floating-point units round the chip to ensure that heat dissipation is averaged across the die. Power management is remarkably aggressive with the ability to put floating-point units into 90% power down mode and to reduce the power consumption of most of the other elements of a tile by factors ranging from 10 to 72%.

A limitation of the instruction set is the provision of a single-level loop construct that uses the INDEX, OFFSET, and BRNE instructions. However, Teraflops was designed as a research machine.

13.8   Parallel Programming

Technology gives us processing power in the form of low-cost processors or even multiple processors on a single chip, and technology provides the high-speed interconnection systems required to create networks of processors. However, we are left with the fundamental problem of taking a program and breaking it up into separate pieces that can be processed by multiple processors and then somehow putting the pieces back together to create the final result. Essentially, we want to take *m* processors and make them look like a single processor that is *m* times faster than the individual processors. Moreover, do we do this manually (i.e., the programmer must partition the problem) or do we do it automatically (i.e., leave it up to the software or operating system)?

Figure 13.35 demonstrates parallel processing. Suppose you have a list of integers and wish to find the maximum. In a serial system, we could use the following code:

```
int temp = 0;
for (int i = 0; i< 28; i++){
if(x[i] > temp){temp = x[i];}
}
```

FIGURE 13.35   Example of parallel processing

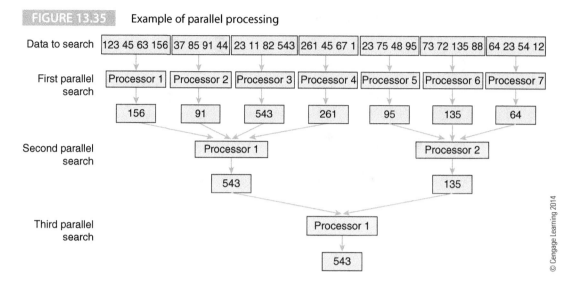

We can readily convert this to ARM code and express as

```
        MOV    r0,#0           ;set up initial maximum value
        MOV    r1,#28          ;set up the counter
        ADR    r2,Table        ;r2 points to the table of values to search
Loop    LDR    r3,[r2],#4!     ;get element and increment pointer
        CMP    r3,r0           ;test new element
        MOVGT  r0,r3           ;if greater then keep
        SUBS   r1,#1           ;decrement counter
        BNE    Loop            ;and continue if not all done
```

As you can see, the loop has five instructions and takes, say, seven cycles, allowing two stalls for the load. The total time taken is proportional to the number of elements to search.

Suppose we parallelize the problem by grouping the integers, giving each group to a processor to find the maximum in a group, and then repeating this operation until the maximum has been located. Figure 13.35 illustrates this concept. The numbers are divided between seven processors, and each processor searches a subgroup of four integers to find the largest. Then the seven integers in the first search are divided between two processors (in a group of 4 and 3) to give two results in the second round. Finally, a single processor selects one of the winners of the second round.

In this example, each search is of four numbers (rounds 1 and 2) and two numbers (round 3). How does this compare with the original problem? Each search is three cycles (startup) + 7n, where n is the number of integers. Using one processor, n is 28 and the total time is 3 + 7 × 28 = 199 cycles.

Using parallel processing and omitting all overheads due to managing the parallelism, the time is 3 + 4 × 7 (first stage) + 3 + 4 × 7 (second stage) + 3 + 2 × 7 (third stage) = 31 + 31 + 17 = 79 cycles. We have more than halved the time to get a speedup ratio of 2.5, but at a cost of seven processors.

Suppose we scale the problem to 1,000,000 numbers to search and we have ten processors to do the searches. In the first stage, we would search ten groups of 100,000 integers in parallel. In the second stage, we would search ten groups of one. The number of cycles without any parallelism would be 3 + 1,000,000 × 7 = 7,000,003. In the parallelized version, we would have

$$3 + 100,000 \times 7 + 3 + 10 \times 7 = 700,003 + 73 = 700,076$$

In this case, the speedup is 7,000,003/700,076 = 9.9989. Note how scaling the problem has provided us with a speedup of almost 10 instead of the 2.5 when we had only 28 integers to search.

Clearly, the performance increase in parallel programming is determined by the nature and the size of the problem.

Can all problems be parallelized? We have just demonstrated that searching can be parallelized to provide a near optimum speedup if the size of the problem is sufficiently large. However, not all algorithms can be transformed into a parallel version. Consider, for example, the Fibonacci series which is defined by the relationship $F(i + 2) = F(i + 1) + F(i)$; that is, the ith term in the series is given by the sum of the previous two terms. The first two members of the Fibonacci series are 0 and 1 by definition. The Fibonacci series is 0, 1, 2, 3, 5, 8, 13, 21, 34, 55, 89, ... Because each new element is the sum of the preceding two elements, it is difficult to parallelize this algorithm.

13.8.1   Parallel Processing and Programming

We've looked at some issues in the structure and organization of *parallel processing* and the next step is to introduce the notion of *parallel programming*. Of course, this is a topic that is worthy of an entire course itself. Here we can only introduce this topic and describe how it relates to parallel structures. Figure 13.36 illustrates one way of looking at the hierarchy of parallelism in computers. The top two levels do not generally concern the programmer because they are handled automatically by the hardware. However, even if they are invisible to the programmer, their side effects are not. For example, pipelining introduces bubbles due to mispredicted jumps and ILP can be restricted by data dependencies. Consequently, compilers try to reduce the number of jumps, and programmers try to use algorithms that best use cache structures by exploiting spatial locality.

FIGURE 13.36   Parallel programming hierarchy

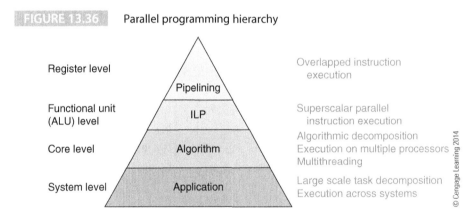

The third level down in Figure 13.36 is the core level (that is, the multi-core processor and the NoC). At this level, cores are tightly coupled and an algorithm can be finely divided between the processors. Communications bandwidths are high and latencies are low. At the bottom level, we have system level parallelism that describes entire systems in separate housings (e.g., clustered systems like Beowulf). Here, the communications bandwidth may be low and the latency high, and parallelism is effective only if the task can be divided into large chunks and each requires extensive computing. *Grid computing*, a close relative of cluster computing, also lives at this level. In grid computing, a large number of systems are interconnected and each works on part of a task with very little intercomputer interaction. Grid computing is sometimes used as a scavenging mechanism to exploit slack time on desktop computers (e.g., overnight when a computer is largely idle), for example, using the SETI project.[17] SETI is NASA's *Search for Extra-Terrestrial Life* project that analyzes radio signals from the world's largest radio telescope in Arecibo, Puerto Rico. This radio telescope generates massive amounts of data that can be processed in an attempt to detect very weak artificial

---

[17]David P. Anderson, "SETI@home: An Experiment in Public-Resource Computing," *Communications of the ACM*, Vol. 45 No. 11, November 2002, pp. 56–61.

signals created by a technologically advanced civilization amongst the random signals generated by natural interactions between matter and energy (background radio noise, pulsar emissions, quasar emissions, cosmic microwave background emissions, etc.).

The student of computer architecture is interested in the algorithm level of Figure 13.36, because it is at that point that the hardware and software interact and ultimately will determine the feasibility of large arrays of cores. If efficient parallel programming cannot be implemented, the future of multi-core processing becomes less secure—apart from applications like video display processing where very high levels have parallelism and have been demonstrated by the GPU manufacturers.

A typical domain where parallel processing appears to be useful is linear algebra (that is, operations on vectors and arrays). Consider the multiplication of two square $n \times n$ matrices $C = AB$. The multiplication of these arrays can be expressed as

```
for (i = 0; i < n; i++)
    for (j= 0; j < n; j++) {
        C[i,j] = 0;
        for (k = 0; k < n; k++)
            C[i,j] = C[i,j] + A[i,k]*B[k,j]
}
```

Figure 13.37 illustrates the matrix multiplication of two 4×4 matrixes and gives the calculation of element $c_{1,2}$. The inner product used to calculate $c_{1,2}$ would be used to calculate the 16 values of elements $c_{0,0}$ to $c_{3,3}$. We have seen in Chapter 5 that transformations on vector images are carried out by means of linear algebra involving 4×4 matrices. In a multi-core system, the total number of transformations could be split between the individual processors. In systems with very large arrays where dimensions may be of the order of 50,000, the formation of inner products could be assigned to individual cores.

FIGURE 13.37   Matrix multiplication

$$
\begin{bmatrix} c_{0,0} & c_{0,1} & c_{0,2} & c_{0,3} \\ c_{1,0} & c_{1,1} & c_{1,2} & c_{1,3} \\ c_{2,0} & c_{2,1} & c_{2,2} & c_{2,3} \\ c_{3,0} & c_{3,1} & c_{3,2} & c_{3,3} \end{bmatrix}
=
\begin{bmatrix} a_{0,0} & a_{0,1} & a_{0,2} & a_{0,3} \\ a_{1,0} & a_{1,1} & a_{1,2} & a_{1,3} \\ a_{2,0} & a_{2,1} & a_{2,2} & a_{2,3} \\ a_{3,0} & a_{3,1} & a_{3,2} & a_{3,3} \end{bmatrix}
\times
\begin{bmatrix} b_{0,0} & b_{0,1} & b_{0,2} & b_{0,3} \\ b_{1,0} & b_{1,1} & b_{1,2} & b_{1,3} \\ b_{2,0} & b_{2,1} & b_{2,2} & b_{2,3} \\ b_{3,0} & b_{3,1} & b_{3,2} & b_{3,3} \end{bmatrix}
$$

$$c_{1,2} = a_{1,0} \cdot b_{0,2} + a_{1,1} \cdot b_{1,2} + a_{1,2} \cdot b_{2,2} + a_{1,3} \cdot b_{3,2}$$

© Cengage Learning 2014

OpenMP

Parallel languages have been designed to simplify (that is, for the programmer) the generation of parallel code and the assignment of processors. There are several parallel programming languages, two of which are particularly important: OpenMP and MPI. These are not new languages in the sense that they have their own data structures and constructs; they are extensions to existing languages that allow the programmer to exploit parallelism. Open multiprocessing (OpenMP) is an *application programming interface* (API) that can be used with C, C++, and FORTRAN on Windows, Max OS, Unix, and Linus platforms. The first OpenMP specification for FORTRAN, OpenMP 1.0, was published in 1997, and OpenMP 2.0 for both FORTAN and C/C++ was published in 2005. Versions 3.0 and 3.1 were released in 2008 and 2011, respectively. The FORTRAN and C/C++ versions of OpenMP are not identical, and there are small differences in functionality between the two languages.

Let's look at matrix multiplication (Figure 13.37) again. OpenMP uses the statement `#pragma omp parallel for` to indicate that the *for loop* should be implemented in parallel as the following code demonstrates.

```
for (i = 0; i < n; i++)
    #pragma omp parallel for
    for (j= 0; j < n; j++) {
        C[i,j] = 0;
        for (k = 0; k< n; k++)
            C[i,j] = C[i,j] + A[i,k]*B[k,j]
}
```

OpenMP offers a means of parallelizing programs by providing *pragmas* that can be applied to existing code. A pragma has the format #pragma omp <name>, where <name> specifies the actual pragma; in this case it is for. Code is executed normally as a *master thread* until a for pragma is encountered. When the for parameter is encountered, the region enclosed by the pragma is parallelized, and a team of *worker threads* generated to perform the parallel computation. The programmer does not have to worry about assigning individual tasks to processors.

Note that parallelizing is not as straightforward as you might think. Suppose you are generating factorials and decide to use the recursive relationship $i! = i*(i - 1)!$ and you write the following.

```
#pragma omp parallel for
for (i=2; i < 100; i++)
{
    factorial[i] = factorial[i-1];
}
```

This code will fail, because you have two instances of factorial and they could end up being run on different processors leading to a *race condition*. You have to ensure that all data is private to the parallel region. For example, if you use a temporary variable temp, it should be declared within the parallel region. That is,

Incorrect

```
int temp;
.
#pragma omp parallel for
for (i=2; i < 100; i++)
{
    factorial[i] = factorial[i-1];
}
```

Correct

```
.
#pragma omp parallel for
int temp;
for (i=2; i < 100; i++)
{
    factorial[i] = factorial[i-1];
}
```

Figure 13.38 illustrates the way in which parallelization takes place in OpenMP. At the top of the *diagram* the single master thread runs initially. When a parallelization command like #pragma omp parallel is encountered, a number of *worker threads* are automatically

FIGURE 13.38   Threads in OpenMP

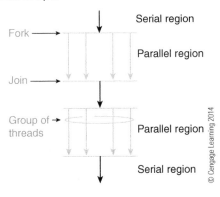

© Cengage Learning 2014

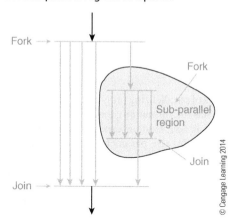

FIGURE 13.39   Embedded parallel regions in OpenMP

created. The serial part of the program forks to the parallel worker threads until they rejoin and serial processing continues. The region of parallelism is defined by the #pragma command, and is terminated by the end of the block.

OpenMP also supports parallelism within parallelism (*nested* or *sub-parallelism*), as Figure 13.39 demonstrates, where one thread in a parallel region is itself parallelized. The number of threads assigned to a parallel region is automatic. However, the programmer can control the number of threads used. Some of the functions that control thread usage are

`int omp_get_numthreads (void)`	Return the number of threads that are in the team executing the current parallel region.
`void omp_set_num_threads` `  (int num_threads)`	Set the number of threads to be used in the next parallel region.
`int omp_get_max_threads (void)`	Returns the maximum value that can be returned by a `omp_get_num_threads` call.
`omp_get_thread_num (void)`	Returns the thread number of the thread within the team making this call. The master thread is always number 0.
`omp_get_thread_limit (void)`	Returns the maximum number of threads available to a program.
`Int omp_get_num_procs (void)`	Returns the number of processors available to the program.

These functions give the programmer a degree of control over how threads are used and assigned in an OpenMP program.

13.8.2   Message Passing Interface

Another parallel programming system is *message passing interface* (MPI), which began life in 1991 at a meeting of researchers in parallel processing. MPI 1.0 was released for public use in 1994. MPI 3.0 was released in 2003.

MPI is a set of functions that can be incorporated in any programming language. For example, it is currently used with C/C++ and FORTRAN. When a task is executed using MPI, it is replicated on each processor with each task having its own local (private) address space. Each processor works on a subset of the problem, exchanging data with other processors when required. Although both OpenMP and MPI enable you to write parallel code,

FIGURE 13.40   Message passing in MPI

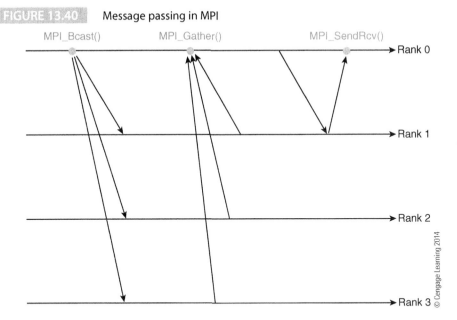

OpenMP works on SMP systems, whereas you can use MPI on distributed memory systems like clusters. Moreover, OpenMP has a finer grain (i.e., the granularity is smaller) than MPI.

Figure 13.40 describes message passing in MPI, where each task is called a *rank*. Messages can be broadcast to all ranks, received from them (gathered), or sent and received between two ranks.

Although MPI has over 100 built in functions, programs can be constructed using the following basic calls.

`MPI_Init`	Initializes MPI.
`MPI_Finalize`	Terminates MPI.
`MPI_Comm_size`	Determines the number of processes.
`MPI_Comm_rank`	Determines the label of calling process.
`MPI_Send`	Sends a message.
`MPI_Recv`	Receives a message

These allow you to set up an MPI program, determine the number of processors, send messages between tasks (the messages are used to send data between ranks), and terminate the MPI program.

13.8.3   Partitioned Global Address Space

A relatively recent innovation in parallel programming is the Partitioned Global Address Space (PGAS) programming model. In 2010, Saraswat, et al.[18] stated

"the multi-core discontinuity is forcing programmers to deal with a variety of concurrent architectures, such as clusters of SMPs/multi-cores, heterogeneous accelerators such as the Cell and GPUs, and integrated high core-count architectures such as Blue Gene".

This is not the first time that architecture has advanced ahead of software to leave a gap between hardware and software capabilities.

[18]V. Saraswat, G. Almasi, G. Bikshandi, C. Cascaval, D. Cunningham, D. Grove, S. Kodali, I. Peshansky and O. Tardieu, "The Asynchronous Partitioned Global Address Space Model," *The First Workshop on Advances in Message Passing*, Toronto, Canada, June 2010.

Sarawat, et al. are arguing for the creation of a programming model that can address the issues raised by multi-core programming and heterogeneous programming. They point out that it took the computer industry a decade to accept object oriented programming. It's just not the same with hardware where things change faster. Sometimes it seems that if Intel designs a chip on Monday, it's in every new computer on Tuesday. On Wednesday, AMD announces their version, and on Friday, it's being dumped on eBay as people prepare for Intel's next version. Of course, in reality, a company like Intel involves the developer community in chip design long before their next generation of processors is actually marketed. However, it is not uncommon for a manufacturer to have a team working on the next generation of processors when the current generation is being developed, because they can rely on Moore's law to ensure that the necessary process technology will be available when their processor goes into production.

The partitioned global address space model provides a unified address space and an environment that is suited to both SMP tightly coupled architectures and to more loosely coupled cluster architectures. Several special-purpose languages have been developed to support the PGSA memory model. The most significant language is *unified parallel C* (UPC). This language is an extension of ISO standard C. UPC supports pointers to both private address space and to shared address space. Any standard C program is also a valid UPC program, except that you can run N copies as N threads in parallel. UPC also supports explicit sharing. For example, shared int x[4][THREADS]; indicates that array x will be shared.

As well as a C-like programming language, a parallel programing language based on Java has been developed for a PGAS environment called X10.

13.8.4   Synchronization

Multiprocessor systems with shared memories have to ensure that the threads running on various processors and sharing variables in memory do not change the semantics of a program; that is, the result of executing a program should be the same on a multiprocessor as on a single processor. Always. This is, of course, the same problem encountered with superscalar machines using out-of-order execution. Consider the following example of the problems of synchronization where two threads run concurrently.

Thread T1	Thread T2
$X \leftarrow X + 1$	$X \leftarrow X + 2$
LDR **r0**,[r1]	LDR **r0**,[r1]
ADD **r0**,r0,#1	ADD **r0**,r0,#2
STR r0,**[r1]**	STR r0,**[r1]**

Suppose that the semantic intent is $X \leftarrow X + 1$ followed by $X \leftarrow X + 2$, so that X becomes $X + 3$. However, if thread switching is not carefully synchronized between threads, we can end up with X incremented by 1, 2, or 3. For example, if LDR **r0**,[r1] is executed in thread 1 and then thread 2 executed, X will be assigned $X + 2$ in thread 2 but the return to thread 1 will use the previous value of X held in r0 and then increment X to $X + 1$.

The problems of synchronization have been around for a long time (for example, in multitasking operating systems and distributed data bases). A solution to the synchronization problem involves *mutual exclusion*; that is, the ownership of the variable is assigned to a processor or thread until the variable has been used and ownership can be relinquished. When a variable is assigned to a process, it is said to be *locked*.

Locking a variable is not easy. Suppose a variable or a block of data has a flag F that indicates whether it is free. A processor can read F and test whether it is free. For example,

```
LDRS   r0,[F]        ;test variable F with a load and set condition codes
STREQ  r1,[F]        ;if zero then claim F by loading 1 into it (assumed in r1)
```

Although this code looks appealing, it is flawed. Most of the time it will work. However, suppose processor A executes the first instruction, and *F* is free (0). The *Z* bits will be set in processor A indicating that the variable is free. Now, suppose processor B tests flag *F* at almost exactly the same time. Processor B will read *F* and also set its Z-bit. Then, both processors will set *F*, each assuming that it is the sole owner. One way of dealing with this problem is to introduce the *atomic* or *indivisible instruction*. Such an instruction performs a test and then changes a variable in a single operation that cannot be interrupted by another processor.

The Freescale 68K has an indivisible test and set instruction, TAS. You can implement a locked flag with TAS F; the processor reads variable *F* and sets the *Z*- and *N*-bits accordingly. The access is locked and no other processor can access *F* between performing the test and setting *F*. The 68K achieves this by using a special bus cycle—a *read-modify-write* cycle that does not release the bus until both the read, internal test, and write cycles have been completed.

The ARM has an atomic swap instruction that exchanges a word between memory and a register. For example, the instruction SWP **r0**,r1, [r2] moves the contents of the memory location pointed at by r2 into r0 and moves the contents of r1 to the memory location pointed at by r2. In other words, it reads the current value of *F* and stores a new value of *F* in one operation. The SWPB operates in the same way but with a byte operand. A typical use is

```
ADR  r1,F              ;load the flag address
MOV  r0,#1             ;get ready to set the flag
SWPB r0,r0,[r1]        ;swap r0 and flag F
CMP  r0,#0             ;test the result
```

The Spinlock

A *spinlock* provides a means of ensuring that threads are correctly synchronized. The term *spin* implies a loop that is continually executed until the requested resource (memory) becomes available. Spinlocks can be implemented by means of the indivisible operations we have just described. When a process requests memory that is being used by another process, the blocked process is said to spin until the resource becomes available. Spinlocks can have variable granularity. For example, they can be applied to entire subsystems such as I/O resources or they can be applied to individual memory locations.

OpenMP provides a critical construct to ensure that a region of code is executed only by one thread at a time. In C/C++ the pragma #pragma omp critical is placed before the structured block to be protected. If one thread is in the process of executing a critical region and another thread attempts to execute the same code in parallel, the new thread will be blocked until the first thread exits the critical region. Consider the following example.

```
#include <omp.h>
main()
{
int x;
x = 0;
#pragma omp parallel shared(x)
    {
    #pragma omp critical
    x = x + 1;
    }   /* end of parallel section */
}
```

Summary

In the last chapter, we have looked at the next and inevitable step in computer development: the multi-core processor encapsulating several CPUs in the same package. Historically, parallel processing has been around for a long time, but it has been the province of the supercomputer designer. Parallel processors have often been the playthings of government agencies, oil companies, NASA, weather forecasters, and the military. Parallel processing in all its forms is here to stay, not least because it is contributing to two ends of the computing spectrum: parallel processing is enhancing mass-produced commodity computers, and it is being used to create ever more powerful super computers to solve problems at the frontiers of science and engineering.

The increasing number of transistors per chip has made the construction of several processors in a single package a possibility. The increasing heat dissipation due to the square-law relationship between clock frequency and energy has made multi-core processing inevitable. We looked first at the background to parallel processing, the topologies of parallel processors, and introduced some of the ways in which cache memories can be organized in multiprocessor systems. We have also discussed the use of multiprocessing in graphics processing where thousands of cores are integrated on a chip. We looked at the way in which some semiconductor manufacturers have approached the design and organization of their multi-core processors.

The final section very briefly looked at the software aspect of parallel programming. Extensions to existing languages have been designed to make parallel programming as easy as possible. However, we still have a long way to go if we are to fully exploit the parallelism made available by the hardware.

Problems

13.1 Parallel processing—why here, why now? For a long time, parallel processing was an esoteric graduate-level topic. Suddenly, it became mainstream. Explain why this has happened.

13.2 If the Earth's atmosphere is considered to extend from the surface to 20 km, the earth's radius is 6,200 km, and the atmosphere is partitioned into cells spaced 0.5 km apart, how many floating-point calculations will be required to run one iteration of a global weather simulation, assuming that each cell requires 200 floating-point operations? If it requires 20 cycles (iterations) of simulation to calculate the weather over the next 24 hours and a forecast is needed in 30 minutes, what must the FLOPS rating of the computing system be?

13.3 Does parallel processing in the form of multi-core processors represent the future … or the past?

13.4 What is a multi-core processor?

13.5 Is Amdahl's law a friend or foe of parallel processing?

13.6 A multiprocessor computer can execute a program using multiple processors. If the problem can be divided into parts that can run on the following number of processors, what is the speedup?

Number of Processors	40	20	20	2
Percent of program serial	20	2	10	40

13.7 For the MESI protocol, fill in the next state for each of the entries in the following table. Note that some entries are not possible and can be left empty.

Current state of line *i* in cache of processor *j*	Read of line *i* by processor *j*		Write to line *i* by processor *j*	
	Line *i* is valid in other caches	Line *i* is not valid in other caches	Line *i* is valid in other caches	Line *i* is not valid in other caches
M				
E				
S				
I				

13.8 A write-back of data occurs from only one state in the MESI protocol. Which state is that?

13.9 What is the advantage of the E (exclusive) state in the 4-state MESI protocol over the 3-state MSI protocol?

13.10 Is an *n*-core multiprocessor *n* times as fast as the equivalent single core microprocessor? Give your reasoning.

13.11 What, if any, is the difference between a multi-core processor and a co-processor?

13.12 What is *false sharing* in a multiprocessor system, and how does it affect performance?

13.13 What is the difference between a multi-core processor and a GPU?

13.14 A 2D square mesh computer is fabricated on a chip with 16 processors. Messages pass between node processors via the mesh. What is the longest routing between any two processors?

13.15 In what ways do console computers like Xbox and PlayStation represent the cutting edge of modern computing?

13.16 A program is ported from a single-core processor to a dual core processor and open MPI used to parallelize it. It runs more slowly on the dual-core processor than on the single-core processor. Why?

13.17 The following six topics are associated with OpenMP or MPI. Indicate which is which.
- Message passing between processes
- Shared memory
- Distributed memory
- Does not run on clusters.
- Multiple threads in single process

Bibliography

Chapter 1

History

W. Aspray, "The Intel 4004 microprocessor: What constituted invention?" *IEEE Annals of the History of Computing*, Vol. 19, No. 3, 1997

F. Faggin, "The Birth of the Microprocessor", *Byte*, March, 1992

F. Fagin, M. E. Hoff, S. Mazor, and M. Shima, "The History of the 4004," *IEEE Micro*, Vol. 16, No. 6, December 1996

M. Garetz, "Evolution of the Microprocessor," *Byte*, September, 1985

H. H. Goldstine and A. Goldstine, "The Electric Numerical Integrator and Computer (ENIAC)," *IEEE Annals of the History of Computing*, Vol. 18, No. 1, 1996

J. P. Hayes, *Computer Organization and Design*, McGraw-Hill, 1998.

K. Katz, "The Present State of Historical Content in Computer Science Texts: A Concern," *SIGCE Bulletin*, Vol. 27, No. 4, December 1995

S. H. Lavington, "Manchester Computer Architectures, 1940–1975," *IEEE Annals of the History of Computing*, Vol. 15, No. 1, 1993

Derek de Solla Price, "A History of Calculating Machines," *IEEE Micro*, Vol. 4, No. 1, February 1984

B. Randell, "The Origins of Computer Programming," *IEEE Annals of the History of Computing*, Vol. 16, No. 4, 1994, pp 6–13

A. Tympas, "From Digital to Analog and Back: The Ideology of Intelligent Machines in the History of the. Electrical Analyzer, 1870s–1960s," *IEEE Annals of the History of Computing*, Vol. 18, No. 4, 1996

M. V. Wilkes, "Slave memories and dynamic storage allocation," *IEEE Transactions on Electronic Computers*, Vol. 14, No. 2, April 1965

M. R. Williams, "The Origin, Uses and Fate of the EDVAC," *IEEE Annals of the History of Computing*, Vol. 15, No. 1, 1993

Computer Architecture Education

E. Brunvand, "Games as Motivation in Computer Design Courses: I/O is the Key," *SIGCSE'11*, Dallas, Texas, March 2011

O. Mutlu, "Modernizing the Computer Architecture Curriculum at Carnegie Mellon: A Multi-Core-Systems Centered Approach," Carnegie Mellon University, www.ece.cmu.edu/~omutlu/pub/mutlu_multi_core_teaching.pdf

S. Sohoni, D. Fritz, and W. Mulia, "Transforming a Microprocessors Course through the Progressive Learning Platform," *Proceedings of the 2011 Midwest Section Conference of the American Society for Engineering Education*, 2011

A.Tew, B. Dorn, W. D. Leahey, Jr., and M, Guzdak, "Context as Support for Learning Computer Organization," *ACM Journal on Educational Resources in Computing*, Vol. 8, No. 3, October 2008

J. Marpaung, L. Johnson, and W. Flanery, "Work-in-Progress: Enhancing Students' Interest, Motivation and Academic Abilities using Video Games," *Proceedings of the 2011 Midwest Section Conference of the American Society for Engineering Education*, 2011

Chapter 2

Binary Arithmetic and Digital Logic

C. Abzug, "Representation of Numbers and Performance of Arithmetic in Digital Computers," 2008, https://users.cs.jmu.edu/abzugcx/public/Discrete-Structures-II/Representation-of-Numbers-and-Performance-of-Arithmetic-in-Digital-Computers-by-Charles-Abzug.pdf

M. Arora, *The Art of Hardware Architecture: Design Methods and Techniques for Digital Circuits*, Springer, 2011

S. Carlough, A. Collura, S. Nueller, and M. Kroener, "The IBM zEnterprise-196 Decimal Floating-Pointer Accelerator," *20th IEEE Symposium on Computer Arithmetic*, 2011

A. Clements, *Principles of Computer Hardware*, 4th Edition, Oxford University Press, 2006

M. D. Ercegovac and T. Lang, *Digital Arithmetic*, Morgan Kaufmann, 2004

D. Goldberg, "What Every Computer Scientist Should Know About Floating-Point Arithmetic," *ACM Computing Surveys*, Vol. 23, No. 1, March 1991

D. Harris and S. Darris, *Digital Design and Computer Architecture: From Gates to Processors*, Morgan Kauffman, 2007

G. L. Herman, C. Zilles, and M. C. Loui, "Flip-Flops in Students' Conceptions of State," *IEEE Transactions on Education*, Vol. 55, No. 1, February 2012

K. Hwang, *Computer Arithmetic: Principles, Architecture and Design*, New York: John Wiley & Sons, 1979

M. M. Mano and M. D. Ciletti, *Digital Design*, 4th Edition, Pearson Education, 2007

B. Parhami, *Computer Arithmetic: Algorithms and Hardware Designs*, 2nd Edition, New York: Oxford University Press, 2010

J. F. Wakerly, *Digital Design*, 4th Edition, Pearson Education, 2006

Chapter 3

S. P. Dandamudi, *Introduction to Assembly Language Programming: For Pentium and RISC Processors*, 2nd Edition, Springer, 2005

C. Hamacher, Z. Vranesic, S. Zaky, and N. Manjikian, *Computer Organization and Embedded Systems*, 6th Edition, McGraw-Hill Higher Education, 2011

E. Larson and M. O. Kim, "A Simple but Realistic Assembly Language for a Course in Computer Organization," *Frontiers in Education Conference*, Saratoga Springs, NY, 2008

L. Null and J. Lobur, *The Essentials of Computer Organization and Architecture*, 3rd Edition, Jones & Bartlett Publishers, 2010

D. Page, *A Practical Introduction to Computer Architecture*, Springer, 2009

W. Stallings, *Computer Organization and Architecture*, 9th Edition, Prentice Hall, 2012

A. S. Tanenbaum and T. Austin, *Structured Computer Organization*, 6th Edition, Pearson Education, 2012

ARM Microprocessor

Application Note 04: Programmer's Model for Big-Endian ARM, ARM document number ARM DAI 0004C, 1994

A. Clements, "ARMs for the Poor: Selecting a Processor for Teaching Computer Architecture." *Frontiers in Education Conference*, Washington, D.C., October 2010

F. Franchetti, S. Kral, J. Lorenz, and C. W. Ueberhuber, "Efficient Utilization of SIMD Extensions," *IEEE Proceeding Special Issue on Program Generation, Optimization, and Platform Adaptation*, Vol. 93, No. 2, February 2005

S. B. Furber, *ARM Systems-on-chip Architecture*, Addison Wesley, 2000

S. B. Furber, J. D. Garside, D.A. Gilbert, "AMULET3: a high-performance self-timed ARM microprocessor", Proc. International Conference on Computer Design: VLSI in Computers and Processors, Austin, Texas, 1998

S. B. Furber, J. D. Garside, and D. A. Gilbert, *The ARM Cortex-A9 Processors*, ARM White Paper, September, 2009

L. Gwennap, "AltiVec vectorizes PowerPC," *Microprocessor Report*, Vol. 12, No. 6, May 1998.

W. Hohl, *ARM Assembly Language,* CRC Press, 2009

J. Rokov, *ARM Architecture and Multimedia Applications*, www.fer.unizg.hr/_download/repository/Kvalifikacijski-Rokov.pdf

N. Sloss, D. Symes, and C. Wright, *ARM System Developer's Guide*, Morgan Kaufmann, 2004

V. S., Vinnakota, *ARM Programming and Optimisation Techniques*, http://www.idt.mdh.se/kurser/cdt214/Programming_examples_for_ARM.pdf

J. Yiu and A. Frame, *32-Bit Microcontroller Code Size Analysis*, www.**arm**.com/files/pdf/**ARM_ Microcontroller**_Code_Size_(full).pdf

Chapter 4

Variations

A. Clements, *Microprocessor Systems Design*, CL-Engineering, 1997

L. Goudge and S. Segars, "Thumb: Reducing the Cost of 32-bit RISC Performance in Portable and Consumer Applications," *Compcon '96*, Santa Clara, CA, 1996

T. R. Halfhill, "MicroMIPS Crams Code," *Microprocessor Report*, November 16, 2009

M. Hampton and M. Zhang, "Cool Code Compression for Hot RISC," groups.csail.mit.edu/cag/6.893-f2000/project/hampton_final.pdf

K. D. Kissell, *MIPS16: High-density MIPS for the Embedded Market*, 1997

A. Krishnaswamy and R. Gupta, "Efficient Use of Invisible Registers in Thumb Code," *MICRO-38, Proc. 38th Annual IEEE/ACM International Symposium on Microarchitecture*, November 2005

C. Lefurgy and T. Mudge, "Code Compression for DSP," Compiler and Architecture Support for Embedded Computing Systems (CASES 98), Washington, D.C., December 1998

MC68020 MC68E020 Microprocessor's User's Manual, Freescale Semiconductor, Inc.

microMIPS™ Instruction Set Architecture, MIPS Technologies, October 2009 http://www.mips.com/auth/MD00690-2B-microMIPS-APP-01.00.pdf

MIPS32® 1074® CPU Family Software User's Manual, Document MD00749, MIPS Technologies, Inc., June 2011

R. Phelan, *Improving ARM Code Density and Performance*, ARM, 2003

Programmer's Reference Manual, M68000PM/AD REV.1, Freescale Semiconductor, Inc.

D. Sweetman, *See MIPS Run*, 2nd Edition, Morgan Kaufmann, 2006

V. M. Weaver and S. A. McKee, "Code Density Concerns for New Architectures," *ICCD Conference*, 2009

X. Xu and S. Jones, "Code Compression for the Embedded ARM/THUMB Processor," *IEEE International Workshop on Intelligent Data Acquisition and Advanced Computing Systems: Technology and Applications*, Lviv, Ukraine, September 8–10, 2003

Chapter 5

Multimedia Extensions

J. Corbal, M. Valero, and R. Espasa, "Exploiting a New Level of DLP in Multimedia Applications", IEEE/ACM Symposium on Microarchitecture, November 1999

M. Hassaballah, S. Omran, and Y. B. Mahdy, "A Review of SIMD Multimedia Extensions and their Usage in Scientific and Engineering Applications," *The Computer Journal*, Vol. 51, No. 6, pp. 630–649, November 2008

Intel Processor Identification and the CPUID instruction, Intel, Application note AP–485

Intel® SSE4 Programming Reference, Intel D91561-003, July 2007

S. Larin, "H1119Introduction to AltiVecTen easy ways to Vectorize your code," Freescale Semiconductor, Smart Developer Forum, Dallas 2004,

R. B. Lee, "Accelerating multimedia with enhanced microprocessors," *IEEE Micro*, Vol. 15, No. 2, April 1995

R. B. Lee, A. M. Fiskiran, Z. Shi, and X. Yang, "Refining instruction set architecture for High-Performance multimedia processing in constrained environments," *Proceedings of the 13th International Conference on Application-Specific Systems, Architectures and Processors*, July 2002

R. B. Lee, "Subword Parallelism with MAX-2," *IEEE Micro*, Vol. 16, No. 4, August 1996

M. Mittral, A. Peleg, and U. Weiser, "MMX™ Technology Architecture Overview," *Intel Technology Journal*, Q3, 1997

H. Nguyen and L. K. John, "Exploiting SIMD Parallelism in DSP and Multimedia Algorithms Using the AltiVec™ Technology," *International Conference on Supercomputing*, 1999

S. Oberman, G. Favor, F. Weber, "AMD 3DNow! technology: architecture and implementations," *IEEE Micro*, Vol. 19, No. 2, March–April 1999

A. Peleg, S. Wilkie, and U. I Weiser, "Intel MMX for Multimedia PCs", *Communications of the ACM*, Vol. 40, No. 1, January 1997

S. K. Raman, V. Pentkovski, and J. Keshava, "Implementing Streaming SIMD Extensions and the Pentium III Processor,"*IEEE Micro*, Vol. 20, No. 4, July–August 2000

A. Shahbahrami, B. Juurlink, and S. Vassiliadis, "A Comparison Between Processor Architectures for Multimedia Applications," *Proc. 15th Annual Workshop on Circuits, Systems and Signal Processing (ProRISC)*, 2004

A. Shahbahrami, B. Juurlink, and S. Vassiliadis, "Efficient Vectorization of the FIR Filter,"*Proceedings of the 16th Annual Workshop on Circuits, Systems and Signal Processing, ProRisc 2005, Veldhoven*, The Netherlands, November 2005

N. Slingerland and A. J. Smith, "Performance analysis of instruction set architecture extensions for multimedia," *Proceedings of the 3rd Workshop on Media and Stream Processors*, December 2001

N. T. Slingerland and A. J. Smith, "Multimedia extensions for general purpose microprocessors: a survey," *Microprocessors and Microsystems*, Vol. 29, No. 1, February 2005

D. Talla, L. K. John, and D. Burger, "Bottlenecks in multimedia processing with SIMD style extensions and architectural enhancements," *IEEE Transactions on Computers*, August 2003

S. Thankkar and T. Huff, "The Internet streaming SIMD extensions," *Intel Technology Journal*, 1999

Tremblay, M. O'Connor, J. M. Narayanan, and V. Liang He, "VIS speeds new media processing," *IEEE Micro*, Vol. 16, No. 4, 1996

Chapter 6

Computer Performance

V. Agarwal, M. S. Hrishikesh, S. W. Keckler, and D. Burger, "Clock rate versus IPC: The End of the Road for Conventional Microarchitectures," *27th Annual International Symposium on Computer Architecture*, Vancouver, Canada, June 2000

D. Citron, A. Hurani, and A. Gnadrey, "The Harmonic or Geometric Mean: Does it really Matter?" *ACM SIGARCH Computer Architecture News*, Vol. 24, No. 4, September 2006

P. G., Emma, "Understanding some simple processor-performance limits," *IBM J. Res. Development*, Vol. 41, No. 3, 1997

The impact of proper PC maintenance on computer performance, White paper, Auslogics Software Oty Ltd., Australia

B. Jacob and T. Mudge, *Notes on Calculating Computer Performance*, University of Michigan Tech Report CSE-TR-231-95, 1995

K. E. Knight, "Evolving Computer Performance 1963–1967," *Datamation*, January 1968

J. R. Mashey, "War of the Benchmrk Means: Time for a Truce," *ACM Sigarch Computer Architecture News*, Vol. 32, No. 4, September 2006

Standard Performance Evaluation Corporation, http://www.SPEC.org

Chapter 7

Microprogramming

M. Cutler, and R. Eckert, "A Microprogrammed Computer Simulator," *IEEE Transactions on Education*, Vol. 30, No. 3, 1987

J. L. Donaldson, R. M. Salter, J. Kramer-Miller, S. Egorov, and A. Singhal, "Illustrating CPU Design Concepts with DLSim 3," *Frontiers in Education Conference*, San Antonio, Texas, October 2009

D. Jackson, "Evolution of Processor microcode," *IEEE Transactions on Evolutionary Computation*, Vol. 9, No. 1, February 2005

Ryo-Il Kang and Katsufusa Shono, "Two-bit microcomputer for educational use," *Microprocessors and Microsystems*, Vol. 15, No. 6, July/August 1991

A. S. Tanenbaum, *Structured Computer Organization*, 3rd Edition, Prentice-Hall, 1990

J. S. Warford and R. Okelberry, "Pep8CPU: A Programmable Simulator for a Central Processing Unit," *38th ACM Technical Symposium on Computer Science Education*, Covington, Kentucky, March 2007

Computer Organization

J-L Baer, *Microprocessor Architecture: From Simple Pipelines to Chip Multiprocessors*, Cambridge University Press, 2009

M. Bekerman, A. Mendelson, "A performance analysis of Pentium processor systems," *IEEE Micro*, Vol. 15, No. 5, October 1995

D. W. Clark and W. D. Strecker, "Comments on 'the case for the reduced instruction set computer' by Patterson and Ditzel," *Computer Architecture News*, Vol. 8, No. 6, 1980

H. Corporaal, *Microprocessor Architectures from VLIW to TTA*, Wiley, 1998

H. G. Cragon, *Memory Systems and Pipelined Processors*, Jones and Bartlett, 1996

S. P. Dandamudi, *Guide to RISC Processors for Programmers and Engineers*, Springer, 2005

J. Eliott, "IBM Mainframes – 45 Years of Evolution," IBM System Z, 2009-03-20

P. G. Emma and E. S. Davidson, "Characterization of branch and data dependencies in programs for evaluating pipeline performance," *IEEE Trans. Computing*, Vol. 36, No.7, 1987

A. Hartstein and T. R. Puzak ."The optimal pipeline depth for a microprocessor," *Proceedings of the 29th Annual International Symposium on Computer Architecture (ISCA)*, 2002

M. G. H. Katevenis, *Reduced Instruction Set Architectures for VLSI*, The MIT Press, 1985

V. M. Milutinovic, *High Level Language Computer Architecture*, Computer Science Press, 1989

D. A. Patterson and D, R. Ditzel, "The case for the reduced instruction set computer," *Computer Architecture News*, Vol. 8, No. 6, 1980.

D. A. Patterson and J. L. Hennesey, *Computer Organization and Design*, Revised 4th Edition, Morgan Kauffmann, 2011

D. Sima, T. Fountain, and P. Kacsuk, *Advanced Computer Architectures: A Design Space Approach*, Addison-Wesley, 1997

V. R. Wadhankar, G. H.Raisoni, and V. Tehre, "A FPGA Implementation of a RISC Processor for Computer Architecture," *National Conference on Innovative Paradigms in Engineering & Technology (NCIPET-2012)*, IJCA 24

Branch Prediction

V. Agarwal, M. S., Hrishikesh, S. W. Keckler, and D. Burger, "Clock rate versus IPC: The end of the road for conventional microarchitectures," *Proceedings of the 27th Annual International Symposium on Computer Architecture (ISCA)*, June 2000

Am29000 User's Manual, Advanced Micro Devices, Sunnyvale CA, 1989

B. Calder and D. Grunwald, "Fast & accurate fetch and branch predication," *Proceedings of the 21st Annual International Symposium on Computer Architecture (ISCA)*, 1994

I. C. K. Chen and T. N. Mudge, "The bi-mode branch predictor," *The 30th Annual IEEE-ACM International Symposium on Microarchitecture*, December 1997.

J. A. DeRosa and H. M. Levy, "An evaluation of branch architectures," *Proceedings of the 14th Annual*

International Symposium on Computer Architecture (ISCA), 1987

D. R. Ditzel and H. R. McLellan, "Branch Folding in the CRISP Microprocessor: Reducing the Branch Delay to Zero," *Proceedings 14th Annual Symposium on Computer Architecture (ISCA)*, 1987

M. Katevenis, *Reduced Instruction Set Computer Architectures for VLSI*, MIT Press, 1985

K. M. Kavi. "Branch folding for conditional branches," *IEEE CS Technical Committee on Computer Architecture (TCCA) Newsletter*, December 1997

J. K. F. Lee and A. J. Smith, "Branch Prediction Strategies and Branch Target Buffer Design" *Computer*, Vol. 21, No. 7, January 1984

D. J. Lilja, "Reducing the Branch Penalty in Pipelined Processors," *IEEE Computer*, Vol. 21, No. 7, July 1988

S. McFarling, *Combining Branch Predictors*, Digital Tech Report: WRL-TN-36:

A. Mital and B. Fagin, "The Performance of Counter- and Correlation-Based Schemes for Branch Target Buffers," *Transactions on Computers*, Vol. 44, No. 12, December 1995

T. Mudge, I. Chen, and J. Coffey, *Limits to Branch Prediction*, CSE-TR-282-96, The University of Michigan, January 1996

S. T. Pan, K. So, and J. T. Rahmeh, "Improving the accuracy of dynamic branch prediction using branch correlation", ASPLOS-V, 1992.

D. A. Patterson, C. H. Sequin, "A VLSI RISC," *Computer,* Vol. 15, No. 9, 1982

J. Pierce and T. Mudge, "Wrong-Path Instruction Prefetching," *MICRO 29, Proceedings of the 29th Annual International Symposium on Microarchitecture*, 1996

C. H. Sequin and D. A. Patterson, "Design and Implementation of RISC I," Technical Report CSD-82-106, UC Berkeley

E. Sprangle, R. S. Chappell, and M. Alsup, "The agree predictor: a mechanism for reducing negative branch history interference," *Proceeding of the 24th Annual International Symposium on Computer Architecture (ISCA)*, 1997

T. Yeh and Y. Patt, "Alternative implementations of two-level adaptive branch prediction," *Proceedings of the 19th Annual International Symposium on Computer Architecture (ISCA)*, 1992

T. Yeh and Y. Patt, "Two—level adaptive training branch-prediction," *MICRO 24, Proceedings of the 24th Symposium on Computer Architecture*, New York, 1991

Chapter 8

Superscalar Processors

64-Bit CPUs: What You Need to Know, February 8, 2002 http://www.extremetech.com

D. Alpert and D. Avnon, "Architecture of the Pentium Microprocessor," *IEEE Micro*, Vol. 13, No. 3, June 1993

Alpha 21264 microprocessor hardware reference manual, COMPAQ, July 1999

S. Eyerman, L. Eeckout, J. E. Smith, and Tejas Karkhanis, "A Mechanistic Performance Model for Superscalar Out-of-Order Processors," *ACM Transactions on Computer Systems (TOCS)*, Vol. 27, No. 2, May 2009

Exploring Alpha Power for Technical Computing, Compaq Technology Brief, April 2000

M. J. Flynn and W. Luk, *Computer System Design: System-on-Chip*, Wiley, 2011

L. Gwennap, "Digital 21264 Sets New Standard," *Microprocessor Report*, Vol, 10, No. 14, October 26, 1996

L. Gwennap, "Intel's P6 uses decoupled superscalar design," *Microprocessor Report*, Vol. 9, No. 2, February 16, 1995

S. Hu, I. Kim, M. H. Lipasti, and J. E. Smith, "An Approach for Implementing Efficient Superscalar CISC Processors," *12th Int. Symposium on High Performance Computer Architecture*, February 2006

M. J. Islam, M. Själander, and P. Stenström, "Early detection and bypassing of trivial operations to improve energy efficiency of processors," *Microprocessors and Microsystems*, Vol. 32, 2008

W. M. Johnson, *Superscalar Processor Design*, Technical Report No. CSL-89-383, Computer Systems Laboratory, Stanford University, Stanford, CA, June 1989

R. E. Kessler, "The Alpha 21264 Microprocessor," *IEEE Micro*, Vol. 19, No. 2, March–April 1999

R. E. Kessler, E. J. McLellan, and D. A. Webb, "The Alpha 21264 Microprocessor Architecture," *Proceedings of the International Conference on Computer Design: VLSI in Computers and Processors*, 1998

E. McLellan, "The Alpha AXP Architecture and 21064 Processor," *IEEE Micro*, June 1993

A. Mendelson and N. Suri, "Designing High-Performance & Reliable Superscalar Architectures—The Out of Order Reliable Superscalar (O3RS) Approach," *Dependable Systems and Networks*, 2000

S. Palacharla, N. P. Jouppi, and J. E. Smith, "Quantifying the complexity of superscalar processors," Technical Report, CS-TR-96-1328, University of Wisconsin Technical Report, 1996.

J. P. Shen and M. H. Lipasti, *Modern Processor Design: Fundamentals of Superscalar Processors*, McGraw-Hill Series in Electrical and Computer Engineering, 2004

D. Sima, T.J. Fountain, and P. Kacsuk, *Advanced Computer Architectures*, Harlow, UK: Addison Wesley, 1997

D. Sima, "The Design Space of Register Renaming Techniques," *IEEE Micro*, Vol. 20, No. 5, September–October 2000

D. Sima, "Superscalar Instruction Issue," *IEEE Micro*, Vol. 17, No. 5, August 2002

R. U. Tena, *Out-of-order retirement of instructions in superscalar, multithreaded and multicore processors*, PhD Thesis, Universidad Politecnica de Valencia, Spain, 2010

P. Zhou, S. Onder, and S. Carr, "Fast Branch Misprediction Recovery in Out-of-order Superscalar Processors," *ICS '05, 19th ACM International Conference on Supercomputing*, Boston, MA, June 2022

Binary Translation

V. Bala, E. Duesterwald, and S. Banerjia, *Transparent Dynamic Optimization*, HPL-1999-77, HP Laboratories Cambridge, MA, 1999

C. Cifuentes and M. Van Emmerik, "UQBT: Adaptable Binary Translation at Low Cost," *Computer*, Vol. 33, No. 3, March 2000

M. Geschwind, E. R. Altman, S. Sathaye, P. Ledak, and D. Appenzeller, "Dynamic and Transparent Binary Translation," *Computer*, Vol. 33, No. 3, March 2000

T. R. Halfhill, "Transmeta breaks x86 low-power barrier," *Microprocessor Report*, February 14, 2000

A. Klaiber, *The Technology Behind Crusoe™ Processors: Low-power x86-Compatible Processors Implemented with Code Morphing™ Software*, Transmeta Corporation, January 2000

VLIW Processors

S. Aditya, B. Ramakrishna Rau, and R. Johnson, *Automatic Design of VLIW and EPIC Instruction Formats*, HP Laboratories, Palo Alto, CA, 2000

K. Ebcioglu, "Some design ideas for a VLIW architecture for sequential-natured software," *Proceedings IFIP WG 10.3 Working Conference on Parallel Processing*, 1988

J. A. Fisher, "Global code generation for instruction-level parallelism: Trace Scheduling," *Technichal Report HPL-93-43*, Hewlett-Packard Laboratories, June 1993

J. A. Fisher, P. Faraboschi, and C. Young, *Embedded Computing: A VLIW Approach to Architecture, Compilers and Tools*, Morgan Kaufmann, 2005

L. W. Fook, *VLIW Microprocessor Hardware Design: On ASIC and FPGA*, McGraw-Hill, 2008

M. S. Schlansker and B. Ramakrishna Rau, *EPIC: An Architecture for Instruction-Level Parallel Processors*, HP Laboratories, Palo Alto, 2000

P. Song, "Demystifying EPIC and IA-64," *Microprocessor Report*, January 26, 1998

Itanium Architecture

64-Bit CPUs: What You Need to Know, February 8, 2002, http://www.extremetech.com

R. Geva and D. Morris, *IA64 Architecture Disclosures*, White Paper http://cpus.hp.com/technical_references/ia64_arch_wp.pdf

J. Huck, D. Morris, J. Ross, A. Knies, H. Mulder, and R. Zahir, "Introducing the IA64 Architecture," *IEEE Micro*, Vol. 20, No. 5, September–October 2000

Inside the Intel Itanium 2 Processor, Hewlett Packard Technical White Paper, July, 2002

Intel® Itanium® Architecture Software Developer's Manual Volume 1: Application Architecture, Intel, May 2010

Intel® Itanium® Architecture Software Developer's Manual Volume 3: Intel Itanium Instruction Set Reference, Intel, May 2010

C. McNairy and D. Soltis, "Itanium 2 Processor Microarchitecture," *IEEE Micro*, Vol. 23, No. 2, March–April 2003

H. Sharangpani and K. Arora, "Itanium Processor Microarchitecture," *IEEE Micro*, Vol. 20, No. 5, September–October 2000

Z. J. Shi, *Bit Permutation Instructions: Architecture, Implementation and Cryptographic Properties*, Doctoral dissertation, Dept. of Electrical Engineering, Princeton University, 2004. http://www.engr.uconn.edu/~zshi/publications/shi_thesis.pdf

N. Snavely, S. Debray, and G. Andrews, Predicate Analysis and If-Conversion in an Itanium Link-Time Optimizer, *Proceedings of the Workshop on Explicitly Parallel Instruction Set (EPIC) Architectures and Compilation Techniques*, 2002

A. Tal, V. Bassin, S. Gal-On, and E. Demikhovsky, "Assembly Language Programming Tools for the IA64 Architecture," *Intel Technology Journal*, Q4, 1999

Chapter 9

Cache Memory

J. R. Goodman, "Using cache memory to reduce processor traffic," *Proceedings of the 10th Annual International Symposium on Computer Architecture (ISCA)*, 1983

M. D. Hill, *Aspects of Cache Memory and Instruction Buffer Performance*, Technical Report CSD-87-381, UC Berkeley, 1987

M. D. Hill, "A Case for Direct-Mapped Caches, *Computer*, Vol. 21, No. 12, December 1988,

L. K. John and A. Subramanian, "Annex cache: a cache assist to implement selective caching," *Microprocessors and Microsystems*, Vol. 23, Issues 8–9, December 1999

L. K. John and A. Subramanian, "Design and performance evaluation of a cache assist to implement selective caching," *ICCD '97, IEEE International Conference on Computer Design*, 1997

N. Jouppi, "Improving Direct-Mapped Cache Performance by the Addition of a Small Fully Associative Cache and Prefetch Buffers," WRL Technical Note TN-14, Digital, Palo Alto, March 1990

F. Sebeck, *Instruction Cache Memory Issues in Real-time Systems*, Technology Licentiate Thesis, Department of Computer Science and Engineering, Mälardalen University, Västerås, Sweden, September 2002

A. J. Smith, "Cache Memories," *Computing Surveys*, Vol. 14, No. 3, September 1982

Virtual Memory

P. J. Denning, *Before Memory was Virtual*, http://cs.gmu.edu/cne/pjd/PUBS/bvm.pdf

P. J. Denning, "Virtual Memory," *ACM Computer Surveys*, Vol. 2, No. 3, September 1970

K. Elphinstone, S. Russell, and G. Heiser, *Issues in Implementing Virtual Memory*, University of New South Wales Report, UNSW-CSE-TR-9411, September 1994

B. Fuhrt and M. Milenkovic, "A survey of microprocessor architectures for memory management," *Computer*, Vol. 20, No. 3, March 1987

B. Jacob and T. Mudge, "Virtual Memory in Contemporary Microprocessors," *IEEE Micro*, Vol. 18, No. 4, July–August 1998

M. Milenkovic, "Microprocessor Management Units," *IEEE Micro*, Vol. 10, No. 2, March 1990

D. Roberts, J. Chang, P. Ranganathan, and T. N. Mudge, *Is Storage Hierarchy Dead? Co-located Compute-Storage NVRAM-based Architectures for Data-Centric Workloads*, HP Laboratories, HPL-2010-119, 2010

Chapter 10

Computer Memory

Jon Burnett, *DDR3 Design Considerations for PCB Applications*, Application note AN111, Freescale Inc., July 2009

Calculating Memory System Power for DDR3, Micron Technical Note TN-41-01

Challenges and Solutions for Future Main Memory, Rambus White Paper, May 26, 2009 http://www.rambus.com/us/downloads/document_abstracts/products/future_main_memory_whitepaper.html

Design Guide for Two DDR3-1066 UDIMM Systems, Micron Technical Note TN41-08, Micron Technology Inc., 2009

A. Fazio and M. Bauer, "Intel StrataFlash Memory Development and Implementation," *Intel Technology Journal*, Q4, 1997

General DDR SDRAM functionality, Technical Note TN-46-05, Micron Technology Inc., 2001

B. Howard, R. Bacchus, E. L. Pope, and Br. Graham, *DDR3 For Dummies 2nd HP Special Edition*, Wiley, http://files.hypervisor.fr/doc/DDR3forDUMMIESv2.pdf

B. Jacob, S. Ng, and D. Wang, *Memory Systems: Cache, DRAM, Disk*, Morgan Kaufmann, 2007

K. Kilbuck, "Main Memory Technology Direction," Micron Technology, Inc., 2007

R. Mahajan, "Memory Design Considerations when Migrating to DDR3: Interfaces from DDR2," *DesignCon*, 2007

M. Muneeb, I. Akram, and A. Nazir, *Non-Volatile Random Access Memory Technologies (MRAM, FeRAM, PRAM)* http://www.imit.kth.se/info/SSD/KMF/2B1750/2B1750_06_RAMs.pdf

P. Murray and F. Al-Hawari, "Challenges in implementing DDR3 memory interface on PCB systems: a methodology for interfacing DDR3 SDRAM DIMM to an FPGA," *DesignCon 2008*, February 2008.

Nanometer Soft Errors, what lies beneath? www.tayden.com/publications/**Nanometer**%20**Soft**%20**Errors**.pdf

G. Prasad, *DDR3 migration to DDR4 - DIMM Thermal Sensor and SPD changes*, NXP Semiconductors, October 2011 http://sites.amd.com/us/Documents/TFE2011_012NXP.pdf

SDRAM Memory Systems:Architecture Overview and Design Verification, Tektronix, 2009

D. B. Stukov, G. S. Snider, D. R. Steward, and R. S. Williams "The missing memristor found," *Nature*, Vol. 453, May 2008

Two Technologies Compared: NOR vs. NAND, M-Systems White Paper, July 03 91-SR-012004-8L

Understanding Soft and Firm Errors in Semiconductor Devices: Questions and Answers, Actel Corporation, Sunnyvale, CA, 2002

Non-Volatile Memory

J. Brewer and M. Gill, "Nonvolatile Memory Technologies with Emphasis on Flash: A Comprehensive Guide to Understanding and Using Flash Memory Devices," *IEEE Press Series on Microelectronic Systems*, 2008

B. Engel, "Technology, Manufacturing and Markets of. Magnetoresistive Random Access Memory (MRAM)," Everspin Technologies, Inc., Spintronics workshop, Kyoto, Japan, 2011 http://www.csis.tohoku.ac.jp/files/2011_SpintronicsWorkshoponVLSI_Japan_Engel.pdf

H. Li and Y. Chen, "An Overview of Non-Volatile Memory Technology and the Implication for Tools and Architectures," *Design, Automation & Test in Europe Conference & Exhibition*, 2009

NAND Flash 101: An introduction to NAND Flash and How to Design It In to Your Next Product, Micron Technical Note, TN-29-19, 2006

NAND vs. NOR Flash Memory Technology Overview, Toshiba, Inc.

Phase Change Memory (PCM): A new memory technology to enable new memory usage models, Numonyz White Paper

H. Pozidis, N. Papandreou, A. Sebastian, A. Pantazi, T. Mittelholzer, G. F. Close, and E. Eleftheriou, "Enabling Technologies for Multilevel Phase-Change Memory," *European Symposium on Phase Change and Ovonic Science*, Zurich, 2011

E. Ou and P. Leong, "Emerging Non-volatile Memory Technologies for Reconfigurable Architectures," *IEEE 54th International Midwest Symposium on Circuits and Systems (MWSCAS)*, 2011

S. R. Ovshinsky, "The Basis for Electronic Mechanisms in Ovonic Phase Change Memories," *European Symposium on Phase Change and Ovonic Science*, Zurich, 2011

T. Raja and S. Mourad, "Digital Logic Implementation in Memristor-Based Crossbars - A Tutorial," *DELTA '10 Proceedings of the 2010 Fifth IEEE International Symposium on Electronic Design, Test & Applications*, January 2010

The World is Flash, Denali White Paper, April 26, 2010, www.denali.com

P. Zhou, B. Zhao, J. Yang, and Y. Zhang, "A durable and energy efficient main memory using phase change memory technology," *International Symposium on Computer Architecture (ISCA)*, 2009.

Chapter 11

Secondary storage

80 mm (1,46 Gbytes per side) and 120 mm (4,70 Gbytes per side) DVD Re-recordable Disk (DVD-RW), Standard ECMA-338, December 2002 www.ecma.ch

J. Best, *The Femto Slider in Hitachi Hard Disk Drives*, Hitachi White Paper, 2007, www.hitachiGST.com

W. A. Burkhard and J. Menon, "Disk array storage system reliability," *The Twenty-Third International Symposium on Fault-Tolerant Computing*, FTCS-23, 1993

R. K. Chellappa and S. Shivendu, "Economics of Technology Standards: Implications for Offline Movie Piracy in a Global Context," *Proceedings of the 36th Hawaii International Conference on System Sciences*, 2003

P. M. Chen and E. K. Lee, "Striping in a RAID Level 5 Disk Array," *Proceedings of the 1995 ACM SIGMETRICS Conference on Measurement and Modeling of Computer Systems*, 1995

R. Comerford, "Magnetic Storage: The medium that wouldn't die," *IEEE Spectrum*, December 2000

Deliver a superior digital archive solution for the media and entertainment industry, Business White Paper, Hewlett-Packard Development Company, 2011

S,.C. Esener, et al., WTEC Panel Report on The Future of Data Storage Technologies, International Technology Research Institute, World Technology (WTEC) Division, June, 1999

M. E. Fitzpatrick, *4K Sector Disk Drives: Transitioning to the Future with Advanced Format Technologies*, Toshiba America Information Systems, Inc., 2011 or at http://www.toshibastorage.com

Flash Memory Guide, Kingston Technology, http://media.kingston.com/pdfs/FlashMemGuide.pdf

R. Freitas, "Storage Class Memory: Technology, Systems and Applications," *Hot Chips Conference*, Stanford University, August 2010

Get SMART for reliability, Paper TP-67D, Seagate Technology, Scotts Valley, CA, July 1999

E. Grochowski and R. D. Halem, "Technological impact of magnetic hard disk drives on storage systems," *IBM Systems Journal*, Vol. 42, No. 2, 2003

Hard Disk Drive Specification Hitachi Ultrastar 7K3000, Hitachi Global Storage Technologies 2011

G. Herbst, *IBM's Drive Temperature Indicator Processor (Drive-TIP) Helps Ensure High Drive Reliability*, IBM Storage Systems Division, San Jose, CA, 1997

High Density Data Storage - Principle, Technology, and Materials, World Scientific Publishing Co. Pte. Ltd 1999 http://www.worldscibooks.com/materialsci/7014.html

IBM System x Server Disk Drive Technology, IBM Red Paper, 2011

In Search of the Long-Term Archiving Solution— Tape Delivers Significant TCO Advantage over Disk, Clipper Notes, report #TCG2010054LL, December 2010

K. A. S. Immink, *Codes for Mass Data Storage Systems*, 2nd Edition, Shannon Foundation Publishers, 2004

F. Marceteau, O. Rouchon, J. Raber, and G. Tsouloupas, *Media and Technology Appraisal for Long Term Preservation*, Partnership for Advanced Computing in Europe, 2011. www.prace-ri.eu

S. Mishra and P. Mohapatra, "Performance Study of RAID-5 Disk Arrays with Data and Parity Cache," *Proceedings of the 25th International Conference on Parallel Processing, Vol. I, Architecture*, 1996

R. New, *The Future of Magnetic Recording Technology*, Hitachi Global Storage Technologies, April 2008, www.asia.stanford.edu/events/spring08/slides402S/0410-Dasher.pdf

C. Ruemmler and J. Wilkes, "An Introduction to Disk Drive Modeling", *Computer*, March 1994

Sony, metal particle and A3MP tape: Nanoscale technology for terabyte storage, Sony Electronics Inc., Park Ridge, NJ, January 2009.

Tang, Y. Lee, *Magnetic Memory: Fundamentals and Technology*, Cambridge University Press, 2010

R. Wood, Y, Hau, and M. Schultz, *Perpendicular Magnetic Recording Technology*, Hitachi White Paper, www.hitachiGST.com

Z. B. Zvonimir and H. V. Randall, "Storage Technologies," *Proceedings of the IEEE*, Vol. 96, No. 11, November 2008

Optical storage

80 mm (1,46 Gbytes per side) and 120 mm (4,70 Gbytes per side) DVD Re-recordable Disk (DVD-RW), Standard ECMA-338, December 2002, www.ecma.ch

Blu-ray Disc™ Format: 1.C Physical Format Specifications for BD-ROM, 6th Edition, White Paper, Blu-ray Disc Association, October 2010

DVD+ReWritable: How it works, Philips Disk Systems, 1999, http://www.dvdplusrw.org/resources/docs/howitworks.pdf

Gauch, S., "Impacts and dynamics of competing standards of recordable DVD media," *4th Conference on Standardization and Innovation in Information Technology*, September 2005

M. Hall and H. Takemoto, "Blue Laser Recording for High-Density Archival Storage," THIC meeting at the Raytheon ITS Auditorium, October 2004

K. A. S. Immink, "Shannon, Beethoven, and the Compact Disc," *IEEE Information Theory Society Newsletter*, December 2007

W. Kazuo, "Next-Generation Optical Disc Technologies," *Toshiba Review*, Vol. 66, No. 8, 2011

P. M. Lane and R.Van Dommelen, "Compact Disc Players in the Laboratory," *IEEE Transactions on Education*, Vol. 44, No. 1, February 2001

T. D. Milster, *Optical Data Storage*, Optical Sciences Center, The University of Arizona, Tucson, AZ, http://www.optics.arizona.edu/milster/380A%20Lab/Lab%203%20http://www.optics.arizona.edu/milster/380A%20Lab/Lab%203%20-%20CDROM/ODS%20Encyclopedia%20Article%20-%20Draft%20Version.pdf

T. D. Milster, "Status and Trends of Optical Data Storage Technology," THIC Meeting at the Center for Atmospheric Research, August 2007

B. H. Schechtman and S. Dror , "A Roadmap for Optical Data Storage Applications," *Optics and Photonics News*, Vol. 18, No. 32, 2007

K. A. Schouhamer Immink, "The CD Story," *Journal of the AES*, Vol. 46, 1998

W. Straw, "In Memoriam—The Music CD and Its Ends," *Design and Culture*, Vol. 1, No. 1, 2009

Solid State Drives

N. Agrawal, V. Prabhakaran, and T. Wobber, "Design Tradeoffs for SSD Performance," *Proceedings of the USENIX Technical Conference*, June 2008.

L. M. Grupp, J. D. Davis, and S. Swanson, "The Bleak Future of NAND Flash Memory," *Proceedings of the 10th USENIX Conference on File and Storage Technologies*, 2012

L. M. Grupp, A. M. Caulfield, J. Coburn, and S. Swanson, "Characterizing Flash Memory: Anomalies, Observations, and Applications," *MICRO'09*, New York, NY, December 2009

W. Hutsell, J. Bowen, and N. Ekker, *Flash Solid-State Disk Reliability*, Texas Memory Systems, November 2008, http://www.ramsan.com/files/f000252.pdf

J. Janukowicz and D. Reinsel, *Evaluating the SSD Total Cost of Ownership*, IDC, November 2007

J. Janukowicz and D. Reinsel, *MLC Sold State Drives: Accelerating the Adoption of SSDs*, MLC White Paper, September 2008

S. Kumar and R. Vijayaraghavan, *Solid State Drive (SSD) FAQ*, Dell, October 2011, http://www.dell.com/downloads/global/products/pvaul/en/solid-state-drive-FAQ.pdf

S. Lee, B. Moon, C. Park, J. Kim, and S. Kim "A case for flash memory ssd in enterprise database applications," *Proceedings of the 2008 ACM SIGMOD International Conference on Management of Data*, 2008

B. Oliver, *Comparison of 32GB SSD vs 10,000 RPM HDs*, http://www.xlr8yourmac.com

OLTP performance comparison: solid state drives vs. hard disk drives, Principles Technologies Test Report, 2009, High Endurance Technology in the Intel® Solid-State Drive 710 Series, Intel Technology Brief, 2011

Solid State Drives Separating Myths from Facts, Toshiba America Electronic Components, Inc., December 2008, http://www.toshiba.com/taec/news/media_resources/docs/SSDmyths.pdf

The Top 20 Things to Know About SSD, Seagate Technology Paper, www.seagate.com/docs/pdf/ssd_faq.pdf

Wear Leveling Techniques in NAND Flash Devices, Micron Technical Note TN-29-42

RAID

P. M. Chenand D. A. Patterson, "Maximizing performance in a striped disk array," Proceedings of the 17th Annual International Symposium on Computer Architecture (ISCA), 1990

R. H. Katz, "RAID: A Personal Recollection of How Storage Became a System," *IEEE Annals of the History of Computing*, Vol. 32, N. 4, October–December 2010

D. A. Patterson, P. Chen, G. Gibson, and R. H. Katz, "Introduction to redundant arrays of inexpensive disks (RAID)," IEEE Computer Society COMPCON, February 1989

M. Staimer, *Ignore the Impending RAID Catastrophe At Your Own Risk*, White Paper, Dragon Slayer Consulting, 2011

L. Xiao, T. Yu-An, and S. Zhizhuo, "Semi-RAID: A Reliable Energy-Aware RAID Data Layout for Sequential Data Access," *27th Symposium on Mass Storage Systems and Technologies (MSST)*, Denver, CO, 2011

Chapter 12

Input-output (Basics)

A. F. Harvey, *DMA Fundamentals on Various PC Platforms*, National Instruments, Application Note 011, April, 1991

FIFO Architecture, Functions and Applications, Texas Instruments Application Report SCAA042A, March 1999

R. Finger, *Using TI FIFOs to Interface High-Speed Data Converters with TI TMS320 DSPs*, Texas Instruments Application Report SDMA003, June 2001

Implementing FIFO Buffers in FLEX 10K Devices, Altera Application Note AN 66, Altera Corporation, January 1996

Implementing DMA on ARM SMP Systems, Application Note 228, Document number: ARM DAI 0228 August 2009

T. Jackson, *Advanced Bus-Matching/Byte-Swapping Features for Internetworking FIFO Applications*, Texas Instruments Application Report SCAA014A, March 1966

J. B. Johnson, "Application of an asynchronous FIFO in a DRAM data path," Thesis, College of Graduate Studies, University of Idaho, December 2002

K. Kittrell, *FIFO Solutions for Increasing Clock Rates and Data Widths First-In, First-Out Technology*, Texas Instruments, Report SZZA001A, 1996

B. T. Tan, "Generalized protocol for parallel-port handshaking," *Microprocessors and Microsystems*, Vol. 13, No. 9, November 1989

S. M. Taylor, "Data-driven handshake circuit synthesis," Thesis, School of Computer Science, University of Manchester, 2007

Using TI FIFOs to Interface High-speed Data Converters with TI TMS320 DSPs, Texas Instruments Application Report SDMA003, June 2001

Buses

D. Abbott, *PCI Bus Demystified*, LHH Technology Publishing

J. Ajanovic, *PCI Express (PCIe*) 3.0 Accelerator Features*, Intel Corp., 2008, http://www.intel.com/content/dam/doc/white-paper/pci-express3-accelerator-white-paper.pdf

D. Anderson, J. Dykes, and E. Riedel, "More than an interface - SCSI vs. ATA," *Proceedings of the 2nd Annual Conference on File and Storage Technology (FAST)*, March 2003

J. Brewer and J. Sekel, *PCI Express Technology*, Dell White paper, February 2004

A.M. Caulfield, J. Coburn, T. I. Mollov, A. De, A. Akel, R. K. Gupta, A. Snavely, and S. Swanson, "Understanding the Impact of Emerging Non-Volatile Memories on High-Performance, IO-Intensive Computing," *SC10*, New Orleans, November 2010

M. Davidsaver, *Understanding VME Bus*, https://pubweb.bnl.gov/~mdavidsaver/understanding-vme.pdf

R. Dominguez and T. Colligan, *SCSI vs. ATA: Interface Comparison*, Dell Technology Brief, December 1999

FireWire Design Guide, 1394 Trade Association, FWDG 1.0OTA 03/02/2010, 2010

FireWire™ Reference Tutorial (An Informal Guide), 1394 Trade Association, Mukilteo, WA, January 2010

A. Goldhammer and J. Ayer Jr., *Understanding Performance of PCI Express Systems*, Xilinx white paper WP350 (v1.1), September 2008

An Introduction to the Differential SCSI Interface, Texas Instruments Application Note 904, 1998

The GPIB (IEEE-488) Bus, www.optics.arizona.edu/opti680/uLab%20Week%2011.pdf

PCI Express - An Overview of the PCI Express Standard, National Instruments, 2009

PCI Local Bus Specification, Revision 2.3, PCI Special Interest Group, March 2002

NuBus Specification, Texas Instruments document TI-2242825-0001, Irvine, CA, 1983

C. E. Stevens, *USB Attached SCSI Protocol (UASP)*, USB implementer's forum, www.usb.org

B. G. Taylor, "Interfacing to NuBus," *Eurobus Conference*, Munich, May 1989

ULTRA2 SCSI, White Paper, LinFinity Microelectronics, Garden Grove, May 1996

Universal Serial Bus Specification, Compaq, Hewlett-Packard, Intel, Microsoft, NEC, and Philips, September 2000

Universal Serial Bus 3.0 Specification, Hewlett-Packard Company, Intel, Microsoft, NEC, ST-Ericsson, and Texas Instruments, June 2011

VMEbus Specification Manual, http://agata.pd.infn.it/LLP_Carrier/optoisolatore/Tundra/Doc/VME%20bus%20specifications.pdf

VME Technology FAQ, VITA, 2011, www.vita.com/home/Learn/vmefaq/vmefaq.html

M. Wolf, *Computers as Components: Principles of Embedded Computing System Design*, Morgan Kaufmann, 2012

Xilinx PCI Tutorial, Xilinx Inc., 2000

Chapter 13
Multicore Processors

D. P. Anderson, "SETI@home: An Experiment in Public-Resource Computing," *Communications of the ACM*, Vol. 45 No. 11, November 2002

S. Akhter and J. Roberts, *Multi-Core Programming: Increasing Performance through Software Multithreading*, Intel Press, 2006

Azimi et al., "Integration Challenges and Tradeoffs for Tera-scale Architectures," *Intel Technology Journal*, Vol. 11, Issue 03, August 2007

The Benefits of Multiple CPU Cores in Mobile Devices, NVIDIA White paper, 2010

G. Blake, R. G. Dreslinski, and T. Mudge, "A Survey of Multicore Processors," *IEEE Signal Processing Magazine*, Vol. 26, No. 6, November 2009.

G. C. Caragea, F. Keceli, A. Tzannes, and U. Vishkin, "General-Purpose vs. GPU: Comparison of Many-Cores on Irregular Workloads," *Proceedings of the 2nd Workshop on Hot Topics in Parallelism (HotPar)*, 2010

B. Chapman, G. Jost, and R. van der Pas, *Using OpenMP Portable Shared Memory Parallel Programming*, The MIT Press, Cambridge, MA, 2008

J. Coke, H. Baliga, N. Cooray, E. Gamsaragan, P. Smith, K. Yoon, J. Abel, and A. Valles, "Improvements in the Intel® Core™2 Penryn Processor Family and Microarchitecture," *Intel Technology Journal*, Vol. 12, No. 3, October 2008.

N. Copty, *OpenMP Specification 3.0*, Sun Microsystems, Santa Clara, CA, March 6, 2008

S. J. Eggers, J. S. Emer, H. M. Leby, J. L. Lo, R. L. Stamm, and D. M. Tullsen, "Simultaneous multithreading," *IEEE Micro*, Vol. 17, No. 5, September/October 1997

J. Gómez-Luna, E. Herruzo, and J. I. Benavides, "MESI Cache Coherence Simulator for Teaching Purposes," *CLEI Electronic Journal*, Vol. 12, No. 1, Paper 5, April 2009

M. Gschwind, P. Hofstee, B. Flachs, M. Hopkins, Y. Watanabe, and T. Yamazaki, "A novel SIMD architecture for the Cell heterogeneous chip-multiprocessor," *Hot Chips 17, Hot Chips Conference,* Stanford University, Palo Alto, CA, August 2005

J. L. Gustafson, "Reevaluating Amdahl's Lawl," *Communications of the ACM*, Vol. 31, No. 5, 1988

G. Hager and G. Wellein, *Introduction to High Performance Computing for Scientists and Engineers*, Chapman & Hall CRC Computational Science, 2011

M. Hill, M. R. Marty, "Amdahl's Law in the Multicore Era," *Computer*, Vol. 41, No. 7, Jul, 2008

An Introduction to the Intel® QuickPath Interconnect, Intel Whitepaper, January 2009

S. Peyton Jones and S. Singh, *A Tutorial on Parallel and Concurrent Programming in Haskell*, Lecture Notes from Advanced Functional Programming Summer School 2008

Microsoft Research Cambridge, http://research.microsoft.com/en-us/um/people/simonpj/papers/parallel/afp08-notes.pdf

J. A. Kahle, M. N. Day, H. P. Hofstee, C. R. Johns, T. R. Maeurer, and D. Shippy, "Introduction to the Cell multiprocessor," *IBM J. Res & Dev.*, Vol. 49. No. 4/5, July/September 2005

S. W. Keckler (Editor), K. Olukotun, and H. P. Hofstee, *Multicore Processors and Systems*, Springer, 2009

D. B. Kirk and Wen-mei W. Hwu, *Programming Massively Parallel Processors: A Hands-on Approach*, Morgan Kaufmann, 2010

R. Kumar, D. M. Tullsen, and N. P. Jouppi, "Heterogeneous chip multiprocessors," *Computer*, Vol. 38, No. 11, November 2005

J. Leverich, H. Arakida, A. Solomatnikov, A. Firoozshahian, M. Horowitz, and C. Kozyrakis, "Comparing Memory Systems for Chip Multiprocessors," *International Symposium on Computer Architecture (ISCA*, San Diego, June 2007

D. A. Mallon, et al., *Performance Evaluation of MPI, UPC and OpenMP on Multicore Architectures*, Springer Verlag, Berlin, 2009, http://www.des.udc.es/~gltaboada/papers/mallon_pvmmpi09.pdf

D. T. Marr, F. Binns, D. L. Hill, G. Hinton, D. A. Koufaty, M. Upton, and J. A. Miller, "Hyper-threading technology architecture and microarchitecture," *Intel Technology Journal*, Q1, 2002

T. G. Mattson, R. Van der Wijngaart, and M. Frumkin, "Programming the Intel 80-core network-on-a-chip Terascale Processor," SC 2008: *High Performance Computing, Networking, Storage and Analysis (SC) Conference*, 2008

T. Mudge, "Power: A First-Class Architectural Design Constraint," *Computer*, Vol. 34, No. 4, April 2001

P. Pacheco, *An Introduction to Parallel Programming*, Morgan Kaufmann, 2011

D. Z. Pan, *Low Power Design and Challenges in Nanometer Multicore Era,* http://www.ieeevic.org/docs/slides/Victoria_DavidPan.pdf

T. Rauber and G., Rünger, *Parallel Programming: for Multicore and Cluster Systems*, Springer Press, March 2010

J. Sanders and E. Kandrot, *CUDA by Example: An Introduction to General-Purpose GPU Programming*, Addison Wesley, 2011

V. Saraswat, G. Almasi, G. Bikshandi, C. Cascaval, D. Cunningham, D. Grove, S. Kodali, I. Peshansky, and O. Tardieu, "The Asynchronous Partitioned Global Address Space Model," The First Workshop on Advances in Message Passing, Toronto, Canada, June 2010

B. Schauer, "Multicore Processors–A Necessity," ProQuest Discovery Guides, 2008, www.csa.com/discoveryguides/**multicore**/review.pdf

X. Sun and Y. Chen, *Reevaluating Amdahl's Law in the Multicore Era*, Illinois Institute of Technology, Technical Report IIT/CS-2008-12, 2008

A. Tumeo, *Massively Parallel Programming with CUDA* www.ogf.org/OGF25/materials/1605/CUDA_Programming.pdf

M. Winn, *Is the free lunch really over? Scalability in Many-core Systems: Part 1 in a series*, Whitepaper, Intel

S. Zhou, D. Duffy, T. Clune, M. Suarez, S. Williams, and M. Halem, "The impact of IBM Cell technology on the programming paradigm in the context of computer systems for climate and weather models," *Concurrency and Computation: Practice and Experience*, Wiley, 2009

Index